The Geology of
ORE DEPOSITS

John M. Guilbert
University of Arizona

Charles F. Park, Jr.

W. H. Freeman and Company / New York

Front cover: A panoramic view of alteration zoning at the Bajo La Alumbrera porphyry copper-gold deposit in northwestern Argentina. The central reddish hills are potassically altered quartz latite porphyry, the white to buff ring is quartz-sericite-pyrite of the phyllic alteration zone, and the greenish rocks extending to the inner edge of the surrounding hills and in the foreground are propylitically altered andesite. The nature of alteration is discussed in Chapter 5, and porphyry deposits are described in Chapter 11.

Back cover: A synthesis of subsurface zoning relationships in supergene-enriched porphyry copper deposits like Bajo La Alumbrera, also showing how perceptions of geologic relationships developed in this text can be projected downward from surface outcrops and their manifestations. This sketch is based on one in Chapter 17.

Photographs not otherwise credited were taken by the authors.

Library of Congress Cataloging in Publication Data

Guilbert, John M.
The geology of ore deposits.

Includes bibliographies.
1. Ore-deposits. 2. Geology. I. Park, Charles Frederick, 1903– . II. Title.
QE390.G85 1985 553.4 85-10099
ISBN 0-7167-1456-6

Printed in the United States of America

1 2 3 4 5 6 7 8 9 0 HL 4 3 2 1 0 8 9 8 7 6

Contents

Preface

This book presents to the senior level university or college student the principles and data fundamental to understanding the genesis and localization of ore deposits and of the minerals associated with them. It is assumed that the student has at least upper division command of chemistry and physics, as well as mineralogy, petrology, and structural geology. Mineralogy is especially important; the student should be familiar with common minerals, their compositions, and significant interpretative aspects. The text stresses throughout that minerals and their compositions are the memories of formative processes. They are one of the means by which the earth scientist can deduce both what has gone before and where an observer stands in the anatomy of an orebody.

It is an express purpose of this text to develop the concept that no two ore deposits are exactly alike. None of us will find another Butte, Montana, solely by memorizing the descriptive characteristics of that district. But by perceiving the chemical and geologic characteristics of the Butte ores, and by understanding their relationship to other porphyry and Cordilleran Vein orebodies, one is equipped to prospect successfully for "porphyries" and related deposits wherever in the world they occur. Another purpose of the book is to develop and combine the abilities of the explorationist—the mine finder—and of the scholar of ore-forming processes—the researcher. Together they can significantly improve our ability to exploit the nonrenewable resources of the planet Earth effectively and benignly. The book emphasizes that the days of "easy" discovery of outcropping ores are *not* over—tracts of land remain on every continent to be effectively examined as the world's geologic perceptions and needs change. The productive explorationist of the twenty-first century will certainly be the one who understands and can most effectively utilize geochemical, geophysical, and geological knowledge of the Earth's crust.

Modern civilization's dependence upon an increasing volume and diversity of minerals makes the search for new ore deposits ever more difficult. If future geologists are to meet this challenge, they must obtain a broad and thorough training and education; they must develop a blend of ingenuity, imagination, and optimism that will enable them to face the many unsolved problems. This book presents in an orderly fashion the ideas and principles that enable the beginning economic geologist to proceed with strength and confidence revitalized by curiosity and tempered by a realization of the limits of our knowledge.

The physical and chemical characteristics of ore deposits are described and correlated with environments and conditions of deposition. We have selected examples and illustrations to emphasize structural, chemical, and temporal controls and to encourage "three-dimensional" and expanded thinking. Many of the ideas discussed are still controversial, but it is also true that many fundamental questions have been convincingly resolved in the 1970s through the illumination of plate tectonic–lithotectonic relationships. The text espouses the newly reinforced concept that ore deposits in general are better interpreted as extensions of the environments responsible for their enclosing rocks than as exotic freaks of nature generated by some arcane processes from a "deep, unknown source." Improved knowledge of mineralogy, petrology, and lithotectonics—the formation of plutonic, volcanic, sedimentary, and metamorphic rocks—is thus the road of progress in economic geology.

It is only fair to state that this text stresses *geologic* aspects of ore deposits and exploration, not the *techniques* of exploration. The student interested in understanding both the geology of ore deposits and methods of finding them should also have at hand W. C. Peters' text *Exploration and Mining Geology* (John Wiley and Sons, New York, 1978, 696 pp.) and the book *Techniques in Mineral Exploration* by J. R. Reedman (Applied Science Publishers, Great Britain, 1979, 533 pp.) The two journals most frequently cited in this book are *Economic Geology* and *Mineralium Deposita,* full sets of which should be at the serious student's hand.

The Geology of Ore Deposits is based on the third edition of *Ore Deposits,* by C. F. Park, Jr., and R. A. MacDiarmid. Like its predecessor, it emphasizes geology for the beginning student, and the development of "exploration thinking." The overall structure of the geologic descriptions has been modified from the increasingly blurred temperature-pressure categories of Lindgren's classification to descriptive groupings recognized by researchers and explorers alike. The hypothermal class of Lindgren, for example, has essentially vanished; deposits once included in it are now seen to be volcanogenic, sedimentary, or simply of lower mesothermal temperature characteristics. "Porphyry copper" and "porphyry moly" have come to be far more significant species names than "mesothermal."

Descriptions of districts have been almost completely revised, expanded, and updated. As before, the term "hydrothermal" means simply "hot water," and does not imply any particular origin for either the heat or the water. The parts of the book dealing with fluid inclusions and isotopes have been rewritten and examples given; progress in both of these fields has been very rapid. The application of computer techniques and geochemistry to solve problems of economic geology continues to grow. Advances are mentioned here, and more explanation given where reasonable. Future editions will doubtlessly include more such progress as it occurs.

To keep the length of the book suitable for a one-term or one-quarter lecture course, we have had to leave out discussion of many facets of the science. Nevertheless, many sections of the book have been enlarged, and such subjects as plate tectonics and volcanic processes have been expanded.

As with the previous editions, the many helpful comments received from readers are greatly appreciated and have been invaluable in preparing this edition. Special thanks are extended to our colleagues and students at Stanford University and the University of Arizona for their many helpful discussions and comments. Photographs credited to personnel of the United States Geological Survey (USGS) were obtained through the USGS Photographic Library, Federal Center, Denver, Colorado.

John M. Guilbert
Tucson, Arizona

Charles F. Park, Jr.
Stanford, California

My participation in *The Geology of Ore Deposits* stems from early curiosity—as a teenage boy—in the prospects, mines, rocks, structures, and alteration zones of the Patagonia Mountains of southern Arizona, to which I express my gratitude. It was catalyzed by gratifying association with students and colleagues at the University of Arizona. I write it in fond memory of the late Virgil I. Mann of the University of North Carolina and in continuing admiration of Eugene N. Cameron, Professor Emeritus of the University of Wisconsin, to whom my part of its science and teaching content is dedicated. As the product of many months of preoccupation, mess, and papers on the dining room table, I dedicate it as an effort to my wife Mary Clymer Guilbert and our children Anne, Linda, and Paul for their patience, inspiration, and forbearance.

JMG

Introduction

Besides, what we are pleased to call the riches of a mine are riches relative to a distinction which nature does not recognize. The spars and veinstones which are thrown out in the rubbish of our mines may be as precious in the eyes of nature, as conducive to the great object of her economy, and are certainly as characteristic of mineral veins as the ores of silver or gold to which we attach so great a value.

John Playfair, 1802

Ores are rocks or minerals that can be mined, processed, and delivered to the marketplace or to technology at a profit. Ores are generally subdivided into the categories of nonmetallic, energy, and water. In many quarters, *ore* refers only to metals or metal-bearing minerals, but many nonmetallic minerals, such as sulfur and fluorite, are included in modern common usage of the term. Building stone and industrial materials, such as abrasives, clays, refractory materials, lightweight aggregates, and salts, are also ores, but they are classified separately as *nonmetallics, industrial minerals and rocks*, or *IM and R's* (Figure I-1). *Ore minerals* are considered to be naturally occurring compounds valued for their metal content, so further processing after mining—generally including concentration (extractive metallurgy), smelting, and refining—is implied. *Industrial min-*

Figure I-1. A rock mass that is both an *ore* and an *orebody*, since it can be mined from this quarry for use as a building stone at a profit. Sheeting joints parallel to the surface coupled with two nearly vertical joint sets visible in the photo permit easy separation of useful-sized blocks from the Crotch Island granite quarry, Hancock County, Maine. (*Photo by* E. S. Bastin.)

erals, such as salt, garnet, or asbestos, may need some upgrading with respect to associated worthless materials, but they are all used for their own specific physical or chemical properties rather than for anything they contain. They are not "broken down" to be useful. The term *economic mineral* is applied to both ore minerals and industrial minerals.

Not all minerals containing a given element need be classified as ore minerals. For example, most iron silicates, such as biotite and fayalite, are not mined for their contained iron and therefore are not ore minerals, while hematite and magnetite are. However, any mineral of a precious metal is probably an ore mineral. Depending on geologic circumstances, an ore may be a rock containing veinlets, disseminations, or small amounts of useful minerals, or it may be massive, essentially solid metal sulfide or oxide. Although both metallic and nonmetallic minerals are widely distributed in the rocks of the Earth's crust, only under exceptional circumstances are they concentrated in *orebodies* in amounts sufficient to permit economic recovery and in a form that permits that recovery. Most ore minerals are associated with valueless material called *gangue*, and many ores grade laterally or downward into *protore*—mineralized rock that is too lean in ore minerals to yield a profit. As Playfair stated so well in 1802, the economic value of an ore mineral does not set it apart genetically from the worthless pyrite, sericite, calcite, or other gangue mineral or rock with which it is associated and which is unavoidably mined with it (Figure I-2). The study

Figure I-2. Vast Bingham Canyon open-pit mine near Salt Lake City, Utah. Each lower bench is 16 meters high; the upper ones are 25 meters. The dust cloud is from an explosive charge set to shatter the ore, which is taken out in railway gondola cars. The deeper rock is ore because it contains about 2 wt % disseminated sulfides of copper, molybdenum, and iron, with some gold and silver. But that means that about 99% of the rock, 99% of the volume of the hole, is gangue and is discarded into waste piles elsewhere. (*Courtesy of* Kennecott Minerals Company.)

of ore deposits thus becomes a specialized part of the broader field of petrology-petrography. Continued recognition of the fact that ore minerals and gangue minerals are normally "part and parcel" of the rocks that contain them has given rise to a new perception of the subdiscipline called *economic petrology*, which brings the tools of the petrologist—thin-section petrography, polished surfaces for electron microprobe and optical mineragraphic study, physical geochemistry, and mineralogy—more and more to bear on economic geology's problems. It is thus a premise of this book that mineral concentrations—ore deposits where profitability of extraction is real—are generally normal, interpretable extensions of the host rocks that contain them.

▸ MINERAL RESOURCE PROBLEMS

Many demographic and economic influences combine to dictate the prices of metals, and hence the level of activity of exploration, mine development, and mining itself. As this chapter was written, in July 1983, industrial activity was near historically low levels, and the doldrums in metal-consuming sectors such as automobile manufacture and commercial-residential construction had dragged metal prices to all-time inflation-adjusted lows. Three years earlier, however, base- and precious-metal prices were soaring at the highest relative levels in decades. Clearly, metal prices are in general cyclic. Most metals are produced by so many diverse countries and cultures that cartels—and thus price management by the producers—have succeeded only rarely and briefly. Although prices are currently low on world markets, and although use patterns are changing, prices will doubtlessly be pushed up in the 1980s by increasing needs and consumption in underdeveloped nations, by overall global population growth, and by the historic human tendency to translate success and economic well-being into material consumption. The United States finds itself having to compete more assiduously each year with other countries for raw materials—bauxite and chromite, for example—that used to be routed to American industrial centers, but which now go to Japan and eastern and western Europe.

Ultimately and progressively, the world will have to adapt to shortages brought about by the consumption of its nonrenewable natural resources. Economics, politics, human needs and priorities, and geologic characteristics will all determine the geography and volume of mineral production, as they do now. However, many American readers will be surprised to learn that although the United States markets prodigious amounts of copper from a scattering of huge open-pit mines in its western states (Figure I-2), it is a net *importer* of copper. Copper can be produced more cheaply and shipped to eastern ports more economically from Chile, for example, than from southern Arizona. Although the authors intend a global emphasis in this text, some figures pertaining to the United States demand in global perspective are given in Tables I-1 to I-4. Americans should note that the United States is economically, that is, realistically, self-sufficient in only three of the thirty-five commodities listed—a scant 10 percent—lithium, phosphate, and molybdenum. In the 'principal world producers' column of Table I-1, Canada is listed 20 times, the United States 17, the U.S.S.R. 10, South Africa 9, Mexico 7, and Brazil 5 times. The United States is so productive because of its endowment, because it has been the principal seat of twentieth century technologic and economic growth, and because of the international security that a strong domestic reserves position provides. But the United States dependence on foreign sources of supply of many crucial commodities is obvious and problematic, and the commercial-technologic necessity of maintaining friendly trade relations with many nations is clear. More universally, the ultimate need for global-scale harmony, co-operation, and international homogeneity of purpose must be the goal of all nations, governments, and peoples.

A concept frequently referred to in the pages that follow is the distinction between *orebody* and *ore deposit, mineral deposit, ore-mineral occurrence*, or just *ore occurrence*. The first denotes the distribution in a specific volume of "the naturally occurring material from which a mineral or minerals of economic value can be extracted at a reasonable profit" (AGI *Glossary of Geology*, 1980). It thus depends on geography, energy costs, type and degree of dilution, tenor (the grade or amount of a commodity actually present), depth in the crust, and many other variables. *Mineral deposit* carries no necessary profitability implications and usually denotes subeconomic or incompletely evaluated occurrences of ore minerals. *Mineral occurrences* or *ore-mineral occurrences* are uneconomic but still anomalous concentrations of minerals that are common ore minerals elsewhere. A distinction is also made between reserves and resources. *Reserves* include orebodies "in production," ores known by drilling or other specific measurement to exist, or ores reliably inferred to exist in a specific place (Figure I-3). *Resources* include reserves and all other potentially viable mineral deposits that are either unknown or uneconomic at present but that can still reasonably be expected to exist. Note in Figure I-3 that actual mining activity involves only the stippled area. The cross-ruled portion of "undis-

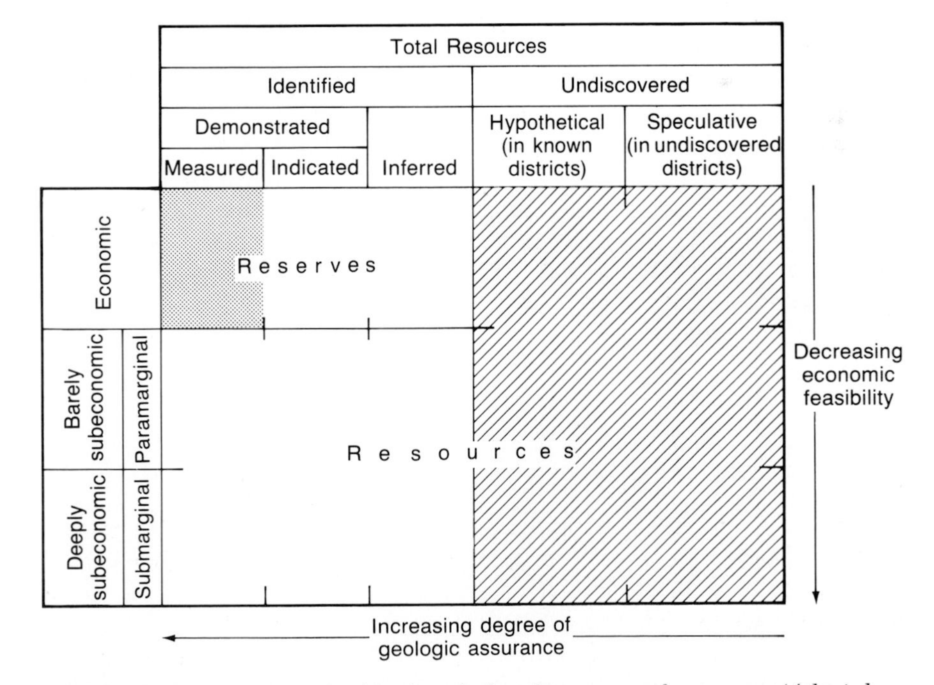

Figure I-3. Geologic-economic classification of mineral reserves and resources. (*Adapted from U.S.G.S. sources.*)

TABLE I-1. U.S. Consumption and Production and World Production, Reserves, and Sources of 35 Mineral Commodities in Order of Increasing U.S. Domestic Self-Sufficiency.

	Newly Mined Material, 1982				Ratio of Reserves to Cumulative Demand; 1976–2000		
Commodity	U.S. Consumption, in Metric Tons	U.S. Consumption, Million $	U.S. Production, %	Approximate World Consumption, 1982, in Metric Tons	U.S. Alone	Entire World	Principal World Producers, 1982 (Most Important First)
Chromium	1,900	15.7	0	NA	<0.1	9.5	U.S.S.R., South Africa, Finland, Zimbabwe
Manganese	240,000	40	0	11,000,000	0	4.6	U.S.S.R., South Africa, Gabon, Brazil
Cobalt	4,500	~100	0	11,400	0	1.2	Zaire, Zambia, Japan, Canada
Tantalum	245	225	0	631	0	1.2	Malaysia, Thailand, Canada
Niobium	1,600	105	0	NA	0	>10.0	Canada, Nigeria, Brazil
Thorium	13	60	0	NA	NA	NA	Australia, Brazil, India, Malaysia
Tin	46,000	660	0.1	160,000	0.1	1.5	Malaysia, Bolivia, South Africa
Platinum group	80e	750e	0.2	200	0	3	South Africa, U.S.S.R., Zimbabwe
Antimony	10,000e	22	0	54,000	0.1	1.8	P.R. China, Bolivia, South Africa
Aluminum	4,943,000	8,284	1	10,657,000	>0.1	5.8	Jamaica, Australia
Asbestos	60,000	30	2	3,900,000	0.2	0.5	U.S.S.R., Canada, South Africa, Zimbabwe, P.R. China
Fluorite	NA	NA	2	4,300,000	0.1	0.4	Mexico, U.S.S.R., Mongolia, P.R. China
Nickel	109,000	768	3	447,000	<0.1	2.2	Canada, New Caledonia, U.S.S.R., Norway
Gold	NA	NA	3	1,230	0.6	1.1	South Africa, U.S.S.R., Canada

Bismuth	862	2	8	2,000e	NA	NA	Mexico, Peru, Canada
Silver	4,000e	6e	25e	14,400	0.4	0.5	Mexico, Peru, Canada
Tungsten	5,400	60	28	22,200	0.4	1.4	Canada, Bolivia, U.S.
Potassium	4,100,000	405	37	31,000,000	0.9	>10.0	U.S.S.R., Canada, Israel, U.S.
Cadmium	3,700	8	40e	9,000e	1.0	1.1	Canada, Mexico, U.S.
Beryllium	50	19	40e	NA	3.1	>10.0	P.R. China, Brazil, U.S.
Zinc	735,000	624	45	3,913,000	0.6	0.8	Canada, Mexico, western Europe
Selenium	500	4	45	1,225	1.9	3.5	Canada, Japan, U.S.
Mercury	1,900	21	46	6,500	0.3	0.9	U.S.S.R., Spain, Italy, U.S.
Titanium	45,000	600	~50	NA	1.4	4.1	Australia, Canada, U.S.
Lead	1,034,000	527	51	3,696,000	1.0	1.1	U.S., Canada, Mexico
Barite	3,400,000	122	53	6,800,000	1.3	1.3	U.S., P.R. China, Peru, India
Copper	2,041,000	3,284	56	6,930,000	1.5	1.6	Chile, U.S., Canada, Zambia, Zaire
Iron	NA	NA	NA	645,000,000	1.5	5.1	U.S., Canada, Australia, Brazil
Sulfur	10,400,000	1,428	82	30,000,000	0.5	0.9	U.S., Canada, France, Mexico
Lanthanides	25,000e	NA	90	NA	9.3	8.0	U.S., Australia, Malaysia, P.R. China
Uranium	8,300	375	125	42,500	NA	NA	Canada, South Africa, Australia
Vanadium	3,900	36	150	28,000	0.02	7.8	South Africa, U.S., Finland, P.R. China
Phosphate rock	31,000,000	1,079	173	140,000,000	3.6	6.3	U.S., Morocco, U.S.S.R., P.R. China
Lithium	1,930	89	200e	5,000	2.8	NA	U.S., Brazil, Namibia, Zimbabwe
Molybdenum	16,000	148	230	60,800	2.9	2.1	U.S., Chile, Canada

e = estimated; NA = not available.

Sources: U.S. Bureau of Mines via National Research Council, 1981, and *Engineering and Mining Journal*, March 1983.

Table I-2. Geologic occurrence and sources of 33 mineral commodities by broad genetic class.

Commodity	Magmatic			Hydrothermal (Chapters 11 and 12)	Volcanogenic (Chapter 13)
	Mafic (Chapter 9)	Intermediate-Felsic (Chapter 11)	Pegmatitic (Chapter 11)		
Chromium	●●●●				
Manganese				●	
Cobalt				●	
Tantalum	●		●●		
Niobium			●●●		
Thorium	●				
Tin		●			
Platinum group	●●●●				
Antimony				●○	●●
Aluminum					
Asbestos	●●			●●	
Fluorine				●●●○	
Nickel	●●				
Gold				●	
Bismuth				●●●●	
Silver		○		○●	●
Tungsten		●●		○●	
Potassium					
Cadmium				○●	
Beryllium			●○○	●	
Zinc				○	●●
Selenium				○●●	
Mercury				○●	●●
Titanium	○○●				
Lead				○	●
Barite				●	○
Copper		○○			○
Iron					○○●
Sulfur				○	
Lanthanides	○○○				
Uranium					
Vanadium	○●				
Phosphate rock	●				
Lithium			○○○		
Molybdenum		○○○○			

Each circle represents 25% of annual world production. Open circles = U.S. sources; solid circles = non-U.S. production (see right-hand column of Table I-1). Elements are arranged from top to bottom in order of increasing U.S. self-sufficiency. "Secretion" includes Mississippi Valley type; "Intermediate-Felsic" includes porphyries.

Source: In part from National Research Council, 1981.

Table I-2. *(continued)*

Commodity	Chemical Sedimentation (Chapter 15)	Mechanical Sedimentation (Chapter 16)	Weathering (Chapter 17)	Secretion (Chapters 19 and 20)	Major Additional Potential Sources
Chromium					Laterites
Manganese	● ● ●				Seafloor nodules
Cobalt	● ● ●				Seafloor nodules
Tantalum		●			
Niobium		●			
Thorium		● ● ●			
Tin		● ● ●			
Platinum group					
Antimony					
Aluminum			● ● ● ●		Aluminum-rich sedimentary and igneous rocks
Asbestos					
Fluorine					Phosphate by-product
Nickel			● ○		Seafloor nodules
Gold		● ●		●	
Bismuth					
Silver					
Tungsten					
Potassium	● ● ● ○				Nonmarine brines
Cadmium				○ ●	
Beryllium					Greisens, skarns
Zinc				○	Zinc-silicate laterite
Selenium				○	Coal by-product
Mercury					
Titanium		●			By-product of copper-moly-porphyries
Lead				○ ○	
Barite	○ ●				Alkalic igneous complexes
Copper	●				Seafloor nodules
Iron	●				
Sulfur	○ ○ ○				Petroleum, coal, gypsum
Lanthanides		○			Phosphate by-product
Uranium		●		● ○ ○	Granites
Vanadium				○ ○	Vanadium-rich shales, oil by-product, layered mafic intrusions
Phosphate rock	○ ○ ○				
Lithium	○				Lithium clays, brines
Molybdenum					

TABLE I-3. U.S. Dependence on Foreign Sources for Some of Its Minerals in 1980.

Less than half imported from foreign sources, so more than 50% self–sufficient	
Copper	Tellurium
Iron	Stone
Titanium (ilmenite)	Cement
Lead	Salt
Silicon	Gypsum
Magnesium	Barite
Molybdenum	Rare earths (lanthanides)
Vanadium	Pumice
Antimony	
One-half to three-fourths imported from foreign sources	
Zinc	Nickel
Gold	Cadmium
Silver	Selenium
Tungsten	Potassium
Three-fourths to 90% imported from foreign sources	
Aluminum	Bismuth
Platinum	Fluorine
Tin	Asbestos
Tantalum	Mercury
More than 90% imported from foreign sources	
Manganese	Niobium
Cobalt	Strontium
Chromium	Sheet mica
Titanium (rutile)	

Source: Adapted from U.S.G.S. sources.

covered resources" should be proportionally much larger, and is in large part what this book is all about.

▸ THE ROLE OF ECONOMIC GEOLOGY

Although the understanding of ore-deposit genesis and occurrence has led to tremendous exploration success in the 1970s in porphyry copper, Climax-type molybdenum, volcanogenic copper-zinc-lead, and unconformity-type uranium deposits, to name but a few, projected consumption rates of metals indicate a secure world reserves position extending into the twenty-first century only for iron. In spite of cyclic supply-and-demand–price vicissitudes, it is clear that corporate and government exploration activity must and will continue and that geologists must be prepared to be increasingly capable, professional, and diligent in their search.

The accelerating growth of the world's population, combined with an improving standard of living throughout the world, is greatly increasing demands for mineral products of all types. These demands will certainly

TABLE I-4. General Outlook for World Reserves and Resources through 2000 A.D. Within Each Group, Commodities Are Listed in Order of Approximate Importance as Determined by Dollar Value of Production.

Group 1—Reserves adequate to fill needs well beyond the year 2000.	
Coal	Phosphorus
Construction stone	Silicon
Sand and gravel	Molybdenum
Nitrogen	Uranium
Chlorine	Gypsum
Hydrogen	Bromine
Titanium (except rutile)	Boron
	Argon
Soda	Diatomite
Calcium	Barite
Clays	Lightweight aggregates
Potash	Helium
Magnesium	Peat
Oxygen	Lithium
Group 2—Identified, but currently subeconomic resources adequate to fill needs beyond the year 2000.	
Aluminum	Vanadium
Asbestos	Zirconium (as zircon)
Nickel	Thorium
Chromium	Titanium (as rutile)
Manganese	Rare earths (lanthanides)
Group 3—Estimated undiscovered resources adequate to fill projected needs beyond the year 2000 and in quantities significantly greater than those of group 2. Research efforts for these commodities should concentrate on geologic theory and exploration methods aimed at discovering new resources.	
Iron	Platinum
Copper	Tungsten
Zinc	Beryllium
Gold	Cobalt
Lead	Cadmium
Sulfur	Bismuth
Silver	Selenium
Fluorine	Niobium
Group 4—Identified subeconomic and undiscovered resources, together probably not enough to fill needs until the end of the century. Research on possible new exploration targets, new types of deposits, and substitutes is necessary.	
Tin	Mercury
Antimony	Tantalum

Source: Adapted from U.S.G.S. sources.

continue to grow. At the same time, the search for ore is becoming more complex; more and more, ore is being sought under cover and at greater and greater depths. In order to obtain sufficient supplies in the future, new geologic, geochemical, and geophysical exploration ideas and techniques must be devised to supplement the old. Recovery, recycling, and mining techniques need to be improved so that large bodies of near-surface minerals that are not now economic can be developed with due regard for ecological and environmental constraints. For these reasons, the successful economic geologist must develop "exploration thinking" requiring imagination, ingenuity, and a degree of optimism, as well as a thorough knowledge of structural geology, stratigraphy, petrology, and mineralogy, and of how fluids migrate underground. Economic geologists should also be familiar with fundamental techniques in geophysics and geochemistry, as these fields are becoming increasingly useful in the search for buried deposits. More importantly, they need to be increasingly able to interpret the field and laboratory significance of observable geologic relationships, and to bring geologic, geochemical, geophysical, mathematical, and computer skills to bear on them.

Under what conditions and as a result of what processes are ores formed? What factors lead to the concentration of certain elements in one environment and not in another? What causes the localization of ore? Probably the best way to answer these questions, and hence the best way to search for new ore bodies, is to study the structure and genesis of known mineral concentrations and then explore geologically favorable analogous areas. One of the precepts of this text is that although no two deposits are exactly alike, they share enough unifying characteristics that they can be grouped into exploration-genetic sets which have lithotectonic-geologic characteristics, and can therefore be successfully hunted and found.

Owing to the complex nature of the Earth, and because many of the processes involved in ore deposition cannot be observed, the study of ore deposits and ore genesis is not an exact science. We work primarily with Earth-scale reaction products, the final result of the interaction of complex geochemical systems of ages, even eons, past. Interpretation of the processes that produced the reaction products that resulted in ore-mineral concentrations is commensurately difficult. Although the study of ore deposits is basically a field science, this does not mean that precise laboratory and experimental data are not desirable or useful; it does mean that laboratory data should be viewed critically and used with wisdom and care. Laboratory data unsupported by field evidence frequently lead to erroneous conclusions, and the final test of all theories and hypotheses in geology is their applicability in the field. But certainly our ability to loop between field observation, geochemical-geophysical laboratory analysis, and process evaluation—all supported by computation and data base management—is a growing and necessary part of the role of the economic geologist searching for, or studying the genesis of, ore deposits.

ONE

The Development of Theories of Ore Deposition

Early humans were acquainted with ore occurrences and extracted useful materials from many of them. They used cinnabar and soft red hematite as pigments long before smelting was devised. Gem stones and native copper, gold, and silver were highly valued as ornaments and as materials for making simple implements. Needless to say, essentially nothing was understood of the origin of minerals in ancient times.

In the early Greek and Roman civilizations, hypotheses about ore genesis were postulated by philosophers, none of whom had firsthand knowledge of ores in the ground. Until the eighteenth century, many purely imaginative or ecclesiastically motivated theories were advanced to explain the presence of minerals and metals in the Earth's crust. Some philosophers envisioned an animated Earth that breathed ores or gave off metallic exhalations as a regular function of "Earth metabolism." Others considered the ores themselves to be alive, growing from seeds—dispersed specks—to massive ores, or ripening from base metals like copper or zinc into the precious metals gold and silver. A popular concept among such "naturalists" was that ores existed in the form of a subterranean "golden tree" whose twigs and branches were various kinds of metals and whose trunk and roots led down to the center of the Earth. The belief generally held by alchemists in the sixteenth and seventeenth centuries was that ore deposits were generated by celestial powers such as the sun's rays or astrologic, planetary influences. These fanciful ideas were maintained not only by alchemists, but also by some of the world's most renowned thinkers, including Aristotle, Pliny, Avicenna, Albertus Magnus, Thomas Aquinas, Robert Boyle, Roger Bacon, and Georges de Buffon. The early hypotheses are of interest to us here principally because their shortcomings emphasize that the study of ore-forming processes must be based upon field observations and petrologic-geologic fact.

In view of the religious beliefs and superstitions that dominated early civilizations, it is not at all surprising that mysterious forces were once called upon to detect and explain ore deposits. Although the ancient Egyptians and Greeks used divining rods for predicting future events, it was not until the fifteenth century that forked twigs were said to locate mineral deposits. The application of the divining rod to prospecting is credited to the Germans of the Harz mining region, who in turn "enlightened" the English (Gardner, 1957). As theories of ore genesis improved, however, so did prospecting techniques, with the result that geophysical techniques (Telford et al., 1976), geochemical methods (Rose, Hawkes, and Webb, 1979), and combinations of the two (Hood, 1979) have totally replaced the hazel stick. The latest concepts and techniques in geochemical and electrical, gravity, magnetic, gamma ray, and seismic geophysical exploration methods are described on pages 775–870 of the *75th Anniversary Volume of Economic Geology* (1981). Even today, however, the ancient practice of witching, or "dousing," for subterranean water or ore deposits is employed by unsophisticated prospectors.

A solid base for modern theories of ore deposits was formulated in the sixteenth century by Georg Bauer, who is generally known by his latinized name Georgius Agricola. Agricola's observations about ore deposits were remarkable for his time, and for this reason he is considered the first real economic geologist. He lived in the Erzgebirge (Ore Mountains) region of Saxony (now southern East Germany, Figure 1-1) and gained intimate first-hand knowledge of the mines in that district. He wrote many treatises on geology and mining, the most significant of which was his comprehensive and pioneering work, *De Re Metallica* (1556). In spite of the fact that most of Agricola's works were in Latin and therefore accessible only to scholars, his writings gave a significant stimulus to the science of ore deposits, which even today bears the stamp of his influence. Agricola's principal contribution was his attempt to classify ores, for without classification no great scientific progress can be possible. His classification was based upon genesis—whether the deposits were alluvial or in situ—and upon form. The classification of in situ deposits included, for example, fissure veins, bedded or horizontal deposits, impregnations, stringers, seams, and stockworks; a vein or seam might be straight, curved, inclined, or vertical. According to Hoover and Hoover, translators of *De Re Metallica,* Agricola was responsible for two fundamental principles in addition to his classification of ore

Figure 1-1. Outline map of central Europe, locating some of the important areas and sites of early progress in the development of economic geology referred to in the text.

deposits. These principles were (1) that "ore channels" are principally secondary features, younger than the country rocks, and (2) that ores have been deposited from solutions circulating in these channels. Agricola's work thus marked the transition from speculation to observation. Moreover, this great geologist argued against the sophistry of the use of the forked hazel twig in prospecting at a time when this technique was first coming into use.

Little advance was made in the study of ore genesis from the time of Agricola until the middle of the eighteenth century. Nevertheless, a few noteworthy works were published and considerable field knowledge was accumulated. Nicolas Steno, the latinized name of Niels Stensen, a Dane who worked in Italy, stands out in the seventeenth century as a scientist of great perception. Responsible for many contributions to general geology, Steno (1669) argued that ores are a product of condensation from vapors ascending through fissures, a remarkably advanced concept that has been retained in some aspects of modern theory.

Before and during this period, people throughout the world looked upon mining as dangerous and, what was worse, degrading work. For centuries, mining was work for convicts and slaves. The ancient Greeks are believed to have maintained a working force of 20,000 slaves in the silver mines near Laurium. Once taken underground, slaves rarely saw the light of day again. At night they were chained to the walls, and by "day" they were beasts of burden; the annual death rate was said to have been 25 percent of the labor force. Since most mining was done in remote parts of the world and communications were primitive or nonexistent, working in the mines was equivalent to being banished from civilization. It was no wonder that for centuries mining was not regarded as worthy of respectable attention.

By the 1700s, however, many trained scientists were puzzling over the problems of ore genesis. Most of the early progress was made in Germany, in the Erzgebirge mining district, a lead-copper-silver vein province between modern East Germany and Czechoslovakia (Figure 1-1), which is still productive. Henkel (1725, 1727) and Zimmermann (1746) acknowledged the importance of hydrothermal solutions, or vapors of deep-seated origin, which they correctly reasoned should contain dissolved rock materials; they even recognized the products of ore deposition by metasomatism, or replacement. In 1749 von Oppel distinguished between veins and bedded deposits, the former being crosscutting features of relatively late, or young, open-fissure origin, the latter being conformably interbedded with stratified sediments and formed with them.

It must have been difficult in those days for a student of ore deposits to differentiate between fact and fancy. Not only had the scientific method not been established, but some of the leading scientists were guilty of proposing theories that did not acknowledge the field data. The result was a mixture of factual contributions and imaginative ideas, and it is only in retrospect that we are able to distinguish between the two and give credit where it is due. A good example of the sometimes observant, sometimes fanciful scientist is Delius (1770, 1773), who first recognized the near-surface alteration of ores by atmospheric agents. He observed the development of weathering-related minerals in a surficial blanket zone as well as a zone of enrichment of copper and silver beneath the altered layer. But he reasoned erroneously that development of these alteration products required heat from the sun, not the percolation of rain waters.

A few years later, Charpentier (1778, 1799), a professor at the Mining Academy at Freiberg in what is now southern East Germany, published two small but well-written books. His writings were based upon astute observations made during many years of study in the Freiberg silver-lead mines. He believed that the metals and minerals in the veins of these mines were the result of alteration reactions between the "country rocks" hosting the veins and water that had passed through them. For supporting evidence he showed how some of the veins graded into wall rock. He likened this process to the silicification of wood and, like Henkel and Zimmermann, correctly proposed that one material had been altered to another. Gerhard (1781), another German scholar, considered veins to be open fissures filled by minerals leached from adjacent country rock. From Charpentier's and Gerhard's writings has come the theory of lateral secretion, which asserts that the metal contents of ore deposits are leached from the adjacent wall rock by circulating waters. The theory of lateral secretion was apparently neglected for almost 100 years, and not until Sandberger (1882) published his *Untersuchungen über Erzgänge* (investigations concerning mineral veins) did the theory become generally appreciated and widely accepted. Sandberger unwisely attempted to show that most mineral deposits were formed according to this theory, but his publication nonetheless influenced contemporary thought profoundly. Actually, the theory has much to commend it; some ore deposits have been shown by careful workers to have formed in this manner (Chapter 19).

During the latter part of the eighteenth century, two outstanding men with diametrically opposed views dominated geological thought: James Hutton, a Scot, and Abraham Gottlob Werner, a German. Their ideas had a far-reaching effect on the science of ore deposits. Hutton (1788, 1795), a "plutonist," thought that both igneous rocks and ore deposits were derived from molten magmas at depth and were transported in the liquid state to their present positions. He was impressed by the similarity between metallic ores and the products observed in smelters. Arguing against the idea of ore deposition from upward-circulating heated water known as hydrothermal solutions, he suggested that ore materials solidified from the molten state, ore magmas having been injected into fissures and cracks of tectonic origin. Massive, mixed sulfides were assumed to have had an origin similar to that of an igneous rock, as evidenced by mutually interpenetrating mineral grains. Hutton went too far with his magmatic theory of ore genesis; he even insisted that metals could *not* have been precipitated from aqueous solutions.

The suggestion that ores are direct magmatic products or are formed as products of differentiation was also advocated by Joseph Brunner (1801), a mine official of Bavaria (now West Germany). Scipione Breislak (1811), an Italian geologist, also called upon the process of magmatic segregation to explain how ore minerals became concentrated in definite layers in igneous rocks. The theory remained essentially unchanged until Joseph Fournet revived it at about the middle of the nineteenth century. His ideas were later revived with little modification by Spurr (1923), but the concept that specialized sulfide melts other than those that are occasional parts of normal fractional crystallization are significant in ore-deposit genesis is now generally discredited.

In contrast to Hutton, Abraham Werner was a "neptunist." He argued that basalts, sandstones, limestones, and ore deposits alike were formed as sediments in a primeval ocean. In 1791 Werner published a résumé of his ideas in the *Neue Theorie von der Entstehung der Gänge* (new theory of the formation of veins). He pictured veins originating as cracks formed on the floor of an ocean by slumping or earthquakes; the cracks were subsequently filled with chemical precipitates, as evidenced—mistakenly, we know now—by the symmetrical banding observed in veins of the Erzgebirge. Werner's personal charm and gift of oratory, as well as his comprehension of mineralogy and local stratigraphy, gained him a devoted following and a reputation as an outstanding teacher and logician. Both directly and indirectly he had great influence on the entire science of geology.

Students and followers of Hutton and Werner amplified the theories of these men. Hutton lacked the ability to present his ideas lucidly, and had it not been for John Playfair (1802), who popularized Hutton's contributions shortly after he died, this great pioneer's influence would not have been so widespread. Charles Anderson (1809) effectively translated Werner's works. Heated arguments and discussions between advocates of Hutton's plutonism and Werner's neptunist theories contributed importantly to the evolution of geology as a science. Observant geologists were quick to discredit both principles as loose generalities. It was readily shown that lavas are not sedimentary formations. Similarly, it was readily shown that minerals, including those that contain metals, are soluble and may be transported in and deposited from aqueous media. Even today, however, this argument between "plutonists" and "neptunists" or, as they are now known by some, "magmatists" and "syngeneticists," has not been resolved in every aspect. No modern geologist could embrace either theory to the exclusion of the other, and the degree to which sedimentation or igneous-spawned hydrothermalism has been responsible for the banded sulfide ores of Rammelsberg, West Germany (Chapter 13) or the Zambian copperbelt (Chapter 15), for instance, is still widely discussed.

Many European mines were operated during the early part of the nineteenth century, but comparatively few advances were made in theories of ore transportation and deposition—geology, especially economic geology, had to await the development of rigor in its sister science chemistry. Mining methods were improved by technical advances and by the development of explosives and such mechanical devices as the hoist (Figure 1-2). One of the greatest steps forward was the invention in 1798 of the Cornish water pump, which for the first time permitted extensive mining to be carried on below the water table. Widely published detailed geological observations from European and Scandinavian universities and scientific societies provided the basic field data for many theories and conclusions. One result was the demonstration of spatial relationships between certain types of ore deposits and intrusive rocks. Recognition of this association established the foundation of modern theories of the origin of hydrothermal ore component transportation and deposition.

It was during this period that geologists began to distinguish between ore deposits of igneous affiliation and those of sedimentary origin. Field

Figure 1-2. An 1805 print showing the level of technology during the early industrial revolution. Explosives were introduced to European mines in 1627 in silver mines near Schemnitz, at what is now Stiavnica. Almost 200 years later, as shown, miners were still lowered by a rope hoist, worked in bare feet, carried torches for light, and used primitive tools. The accident rate, if the "open blasting" shown was normal, must have been high. See Figure 1-1 for location.

data accumulated during the 1800s showed clearly that many large ore deposits are localized in areas of structural complexity and igneous activity that we now call *orogenic belts*, a fact that was then and is still used extensively in exploration. It began to be appreciated that certain types of ore are found so commonly in and around igneous masses of particular composition, and so seldom elsewhere, that it was natural, reasonable, and correct to associate the presence of these ores with definite types of igneous activity. For example, nickel ores are associated with norites and peridotites; disseminated copper deposits are found commonly in monzonite or quartz monzonite stocks; and tin was noted to be associated with siliceous plutonic rocks such as granite and pegmatite. These close associations strengthened the deduction that a genetic relationship exists between igneous rocks and many ores. The arrangement of minerals in zones around igneous centers likewise suggested that ore-bearing fluids spread from channels that tapped deep-seated sources, presumably magma chambers. As the fluids rose through the rocks, they deposited their minerals in favorable structural and stratigraphic environments. Evidence relating ores to volcanic activity has also been obtained from studies of fumaroles and hot springs, where many constituents of ore deposits have been recognized either in the hot fluids themselves or in the wall rocks flanking the channelways.

In the middle of the nineteenth century, the French scholar-scientist Elie de Beaumont made several contributions that are still generally accepted. In addition to his ideas concerning general geology and structural geology, de Beaumont thought that hydrothermal solutions played a part in the formation of ore deposits. Although this suggestion had been made earlier, de Beaumont emphasized and developed it and was first to organize the whole concept of hydrothermal ore-deposit formation around and related to igneous centers. He also recognized and described magmatic segregations and replacement ores formed in the contact metamorphic zones of intrusive igneous rocks. It was during the period of de Beaumont's life that chemical principles began to influence geologists' ideas about ore genesis, thanks to such people as K. G. Bishop and Bernhard von Cotta.

In the latter half of the nineteenth century, many eminent scientists contributed to the theories of ore transportation and deposition. Among the best known were von Cotta, Sandberger, and Stelzner in Germany; Daubrée, and later de Launay in France; Posepny in Bohemia, now Czechoslovakia; J. A. Phillips in England; J. H. L. Vogt in Norway; and S. F. Emmons in the United States. The science was coming of age. The accumulation of field data precluded overgeneralizing; geologists understood clearly that no single theory would explain the genesis of all ore deposits.

Early in the present century, Louis de Launay, J. H. L. Vogt, Waldemar Lindgren, R. Beck, C. R. Van Hise, F. L. Ransome, W. H. Emmons, L. C. Graton, R. H. Sales, J. F. Kemp, W. H. Weed, B. S. Butler, and other scientists formulated many of our modern theories. Some of these, with others including R. A. F. Penrose, created the Society of Economic Geologists in 1905 to facilitate the study and discussion of current problems. As a result of accumulating careful field observations throughout the world, as well as an expanding demand for minerals, economic geology was becoming an important branch of science. It became obvious that structural geology played a vital role in ore localization in the guidance of ore-bearing fluids into areas favorable for the deposition of ore minerals. The process of replacement in and adjacent to vein fissures was widely recognized, and knowledge of the more simple chemical changes was gradually extended. The whole body of physical chemistry as it is now applied to petrology and economic geology is generally dated from J. H. L. Vogt's papers in the late 1920s. Also in the first decades of the twentieth century, genetic classifications of ore deposits were greatly improved. Nearly everyone who worked with ore deposits recognized the need for a systematic arrangement of data that would clarify the differences and similarities among deposits. Some of the most prominent turn-of-the-century geologists in the world gathered from time to time to discuss this problem; their efforts laid the groundwork for present classifications. One symposium in Chicago in 1893 was attended by Waldemar Lindgren (Figure 1-3), whose later endeavors produced a popular genetic classification in four editions of his book *Mineral Deposits* (1907, 1913, 1922, 1933). Lindgren classified deposits according to whether they were products of mechanical or chemical concentration and, if chemical, whether they were deposited from surface waters, from magmas, or

Figure 1-3. Waldemar Lindgren (1860–1939), one of the deans of global economic geology. Born in Sweden and educated in Freiberg, his 31 years with the United States Geological Survey (1884–1915; Chief, 1911–1912) and 21 at the Massachusetts Institute of Technology (1912–1933) afforded extensive travel, field observation and study, and scholarly synthesis. His influence through publications and lectures was profound. His ore deposit classification and concepts held sway for decades and are still widely applicable.

within bodies of rock. The major controversy involved the classification of hydrothermal veins, which fit into Lindgren's group of "chemical deposits introduced into bodies of rock by igneous processes." Within this group Lindgren included both pyrometasomatic (igneous metamorphic) and hydrothermal deposits. He stated that *pyrometasomatic deposits* formed as high-temperature replacement bodies near the border zones of igneous intrusives. *Hydrothermal deposits*—ores formed by hot, aqueous solutions—were further subdivided by Lindgren according to temperature and depth of formation. Those formed at great depth and at high temperature (300 to 500°C) were named *hypothermal*, those formed at intermediate depths and temperatures (150 to 300°C) *mesothermal*, and those formed at shallow depths and relatively low temperatures (50 to 150°C) *epithermal*. By 1933 Lindgren had increased the mesothermal-epithermal temperature boundary from 150 to 200°C, a change reflecting advances in geochemistry. Only two changes in Lindgren's hydrothermal classification have been widely accepted. In 1933 Graton recognized a distinct group of deposits with features that indicated deposition at shallow depths from "nearly spent" solutions; these he named *telethermal*. Buddington (1935) introduced the term *xenothermal* to describe deposits formed at high temperatures and shallow depths. The tendency among American geologists at present is to retain parts of Lindgren's classification but to deny specific genetic implications to the term *hydrothermal*. Hydrothermal simply means hot water, and hydrothermal deposits are not necessarily related to igneous processes.

Previous editions of this book have used each of these categories as chapter headings, but at least three major events in the 1960s and 1970s have required partial relinquishment of Lindgren's classification. First, the explosion of geothermometric and geobarometric data, principally through fluid inclusion research to be described in Chapter 7, have blurred the pressure-temperature (P-T) distinctions between the groupings. Second, the advent of plate tectonics and the vastly improved understanding of the lithotectonics of rocks and their contained ore occurrences have taught us that many ore deposits are not exotic emplacements *into* rocks, but cogenetic parts *of* them, so that many of the disputes of the first half of the twentieth century between endemic and exotic sources have been resolved in favor of the endemic. For example, the Witwatersrand gold-uranium deposits once thought by many non-African geologists to be hypothermal-hydrothermal and thus exotic are now almost universally accepted as mildly metamorphosed alluvial placer sediments (Chapter 16); and the Zambian Copperbelt's hydrothermalist proponents have all but vanished in the face of chemical sediment evidences (Chapter 15). Third, an amazingly improved perception of the role of volcanism, especially submarine volcanic manifestations (Chapter 13), has wrested scores of deposits from Lindgren's hypothermal and mesothermal categories into generally lower temperature ones. Several of Lindgren's terms are still widely and aptly used, such as epithermal, but exploration classifications have evolved into the occurrence categories used in this book, such as *porphyry coppers* and *volcanogenic massive sulfides*. Lindgren's classification served the science well, and he is probably still the most widely quoted and influential economic geologist in the history of the discipline.

One of the toughest problems besetting early economic geologists was how the exceedingly low solubilities of the components of ore minerals in pure water could be reconciled with the fact that hydrothermal fluids do indeed appear to have been responsible for many ore occurrences. Landmark advances in the 1960s and 1970s in the study of fluid inclusions by Edwin Roedder of the U.S. Geological Survey (U.S.G.S.) and by Bernard Poty and Alain Weisbrod of the Centre de Recherches Pétrographiques et Géochimiques in France, have demonstrated that most hydrothermal fluids are in fact *brines* rather than pure water, and theoretical studies by H. Barnes, P. Barton, and H. Helgeson have shown that metals are transported as complex chlorine- or sulfur-based ions. Finally, the thrust of the moment is appropriately into areas of thermodynamics, computer simulation of ore-forming processes of all types, and computer management of exploration- and occurrence-evaluation data bases. Certainly the decade of the 1980s will see dramatic progress in each of these three areas, especially where field data and geologic relationships are part of continuous loops.

Bibliography

Agricola, G., 1556. *De Re Metallica.* Engl. transl. by H. C. Hoover and L. H. Hoover, New York: Dover, 1950.

Anderson, C., 1809. *New Theory of the Formation of Veins by Abraham Gottlob Werner*. Edinburgh.

Breislak, S., 1811. *Introduzione alla Geologia*. Milan: Stamperia Reale.

Brunner, J., 1801. *Neue Hypothese von Entstehung der Gänge*. Leipzig.

Buddington, A. F., 1935. High-temperature mineral associations at shallow to moderate depths. *Econ. Geol.* 30:205–222.

Charpentier, J. F. W., 1778. *Mineralogische Geographie der Chursächsischen Lande*. Leipzig.

——, 1799. *Beobachtung über die Lagerstätte der Erze, hauptsächlich aus den sächsischen Gebirgen*. Leipzig.

Delius, C. T., 1770. *Abhandlung von dem Ursprunge der Gebirge un der darinnen befindlichen Erzadern*. Leipzig.

——, 1773. *Anleitungen zu der Bergbaukunst nach ihrer Theorie und Ausübung*. Vienna.

Gardner, M., 1957. *Fads and Fallacies in the Name of Science*. New York: Dover.

Gerhard, C. A., 1781. *Versuch einer Geschichte des Mineral-Reichs*. Berlin.

Graton, L. C., 1933. The depth-zones in ore deposition. *Econ. Geol.* 28:513–555.

Henkel, J. F., 1725. *Pyritologia oder Kieshistorie*. Leipzig.

——, 1727. *Mediorum Chymicorum Non Ultimum Conjunctionis Primum Appropriatio*, etc. Dresden and Leipzig.

Hood, P. J., Ed., 1979. Geophysics and geochemistry in the search for metallic ores. Geol. Surv. Can. Econ. Geol. Rept. 31, 811 pp.

Hutton, J., 1788. Theory of the earth. *Roy. Soc. Edinburgh Trans.* 1:209–304.

——, 1795. *Theory of the Earth*. Edinburgh.

Lindgren, W., 1907. The relation of ore deposition to physical conditions. *Econ. Geol.* 2:105–127.

——, 1913. *Mineral Deposits*. New York: McGraw-Hill, pp. 178–188.

——, 1922. A suggestion for the terminology of certain mineral deposits. *Econ. Geol.* 17:292–294.

——, 1933. *Mineral Deposits*, 4th ed. New York: McGraw-Hill, 930 pp.

Oppel, F. W. von, 1749. *Anleitungen zur Markscheidekunst nach ihren Anfangsgründen und Ausübung kürzlich entworfen*. Dresden.

Playfair, J., 1802. *Illustrations of the Huttonian Theory of the Earth*. Edinburgh.

Rose, A. W., H. E. Hawkes, and J. S. Webb, 1979. *Geochemistry in Mineral Exploration*, 2nd ed. New York: Academic Press, 657 pp.

Sandberger, F., 1882. *Untersuchungen über Erzgänge*. Wiesbaden.

Sims, P. K., Ed., 1981. *75th Anniversary Volume of Economic Geology*. New Haven, Conn: Soc. Econ. Geologists, Yale Univ.

Spurr, J. E., 1923. *The Ore Magmas*. New York: McGraw-Hill, 915 pp.

Steno, N. S., 1669. *De Solido intra Solidum Naturaliter Contento*. Florence.

Telford, W. M., L. P. Geldart, R. E. Sheriff, and D. A. Keys, 1976. *Applied Geophysics*. London-New York: Cambridge Univ. Press, 860 pp.

Werner, A. G., 1791. *Neue Theorie von der Entstehung der Gänge, mit Anwendung auf den Bergbau besonders den freibergischen*. Freiberg.

Zimmermann, C. F., 1746. *Unteradrische Beschreibung des meissnischen Erzgebirges*. Dresden and Leipzig: Obersächsische Bergakademie.

TWO

The Ore-Bearing Fluids

Four parts of the question "How do ore deposits form?" can be isolated. They concern (1) the source and character of ore-bearing fluids, (2) the sources of the ore constituents and how they are placed in fluids or in solution, (3) the migration of those ore-bearing fluids, and (4) the manner of deposition of the ore minerals from them. We will deal with the first concern in this chapter, with the second and third in Chapter 3, and with the fourth in Chapter 4.

To understand why ore deposits form where they do, it is first necessary to understand the nature of ore-component transporting media, which in almost all cases are liquids or gases. Whether the ores are magmatic, more or less distantly related to magmas, associated with metamorphic processes, or related only to groundwaters and sedimentary processes, they are all intimately associated with the movement of fluids. Although hydrothermal fluids, even if they are deeply derived, may be studied near the surface of the earth, they are likely to have been contaminated with groundwaters and to have changed their characteristics through extensive reaction with their wall rocks along their upward passage. Little is known of the deeper ore-bearing fluids; their nature must be inferred from observations of thermal springs, volcanic gases, and other emanations that may be the end products of ore-forming processes, from studies of the ores themselves and the accompanying gangue, from laboratory studies of appropriate synthetic systems, and from isotopic and geochemical studies.

For the purpose of closer investigation, the ore-bearing fluids can be divided into six categories: (1) silicate-dominated magmas or derived oxide, carbonate, or sulfide-rich magmatic liquids, (2) water-dominated hydrothermal fluids that separate from magmas, (3) meteoric waters, those that have passed through the atmosphere, (4) seawater, (5) connate waters trapped in pore spaces in sediments, and (6) fluids associated with metamorphic processes. Any of the aqueous fluids may be hot or cold and be deep-seated or occur near the surface. If heated and liquid, each aqueous fluid would be considered a *hydrothermal solution* because this term refers to any hot, watery fluid, without regard to origin. If the fluid is a gas, it is called *pneumatolytic*. Above the critical point discussed later, the properties of hydrothermal liquids and gases merge—for these *supercritical fluids*, the distinction between gases and liquids is meaningless. Supercritical fluids or high-pressure gases are therefore commonly regarded as simply hydrothermal. We will consider each of the six ore-bearing fluids in turn.

▸ MAGMA AND MAGMATIC FLUIDS

Magma is "naturally occurring mobile rock material, generated within the Earth and capable of intrusion and extrusion . . ." (AGI *Glossary of Geology*, 1980). Defined less formally, magma is a rock melt or a high-temperature molten mush of liquid and crystals. Solidification of magma produces igneous rocks, the great variety of which, in addition to their diversity of field relations, suggests that the processes by which magmas are generated, transported, and solidified are highly complex. Most magmas are not homogeneous in composition; parts may be rich in ferromagnesian constituents, others in silica, sodium or potassium compounds, volatiles, reactive xenoliths, or other substances. Furthermore, the composition of a magma is thought to be constantly changing due to chemical reactions. Magmas are not static; they are not closed systems in which we should expect general equilibrium. They are probably subject to continuous convective overturn and mixing, especially at the higher temperature, mafic end (Carmichael, Turner, and Verhoogen, 1974). Intermediate to felsic melts may convect in layer-cake-like cells and thereby be vertically stratified. As a magma cools, it crystallizes and separates into fractions by complicated processes of fractional crystallization, or igneous differentiation. Metallic elements can be concentrated by rock-forming mechanisms in various portions of the resulting igneous assemblages, and specific minerals such as chromite may be so abundant locally that the resulting igneous rock itself constitutes ore. Figure 2-1 is a schematic of early portions of fractional crystallization and oxide-sulfide concentration. During differentiation, the more mafic parts of the magma are enriched in chromium, nickel, platinum, and, in certain circumstances, phosphorus and other elements (Chapter 9). In contrast, concentrations of tin, zirconium, thorium, and many other elements are

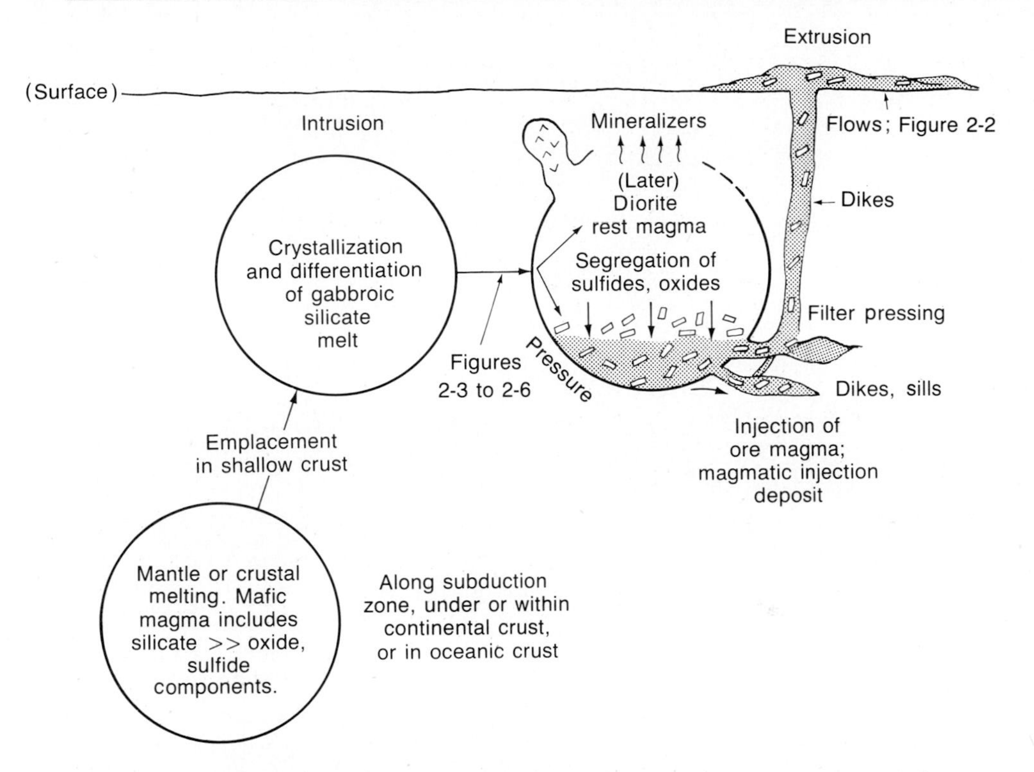

Figure 2-1. Schematic of a sequence of early magmatic events leading to ore magmas and their emplacement. The diagram generally extends to perhaps diorite stages of differentiation in the magma portion at the right. See also Figures 2-3 to 2-7. (*Adapted from diagrams by* A. J. Naldrett.)

found in silicic units (Chapter 11). Titanium and iron concentrations can be generated throughout the range of composition and are therefore found in a variety of igneous rock types. Studies started by Chappell and White (1974) have described petrologic differences within suites of intrusive granitoid rocks previously thought to be generally similar, and lithotectonic-geochemical classes of rocks are being discerned. The granitoids include I, S, A, and M types, where the letters designate source materials and tectonic origins and stand for (conventional) *i*gneous, *s*ediment- or continental-crust-derived, *a*lkalic or *a*tectonic, and *m*antle-derived, respectively. Other types may be identified, and each may be associated with specific ore-deposit-type profiles and tectonic settings. Workers in Europe, China, Japan, and the United States are considering these differences and their significance to the types and occurrences of their ore minerals. For example, Cu-Mo-Zn-Pb-Ag-Au occur preferentially with I-type granites, Sn-W-Be-U-Li with S-type granite. More will be said of this later.

If a partly crystallized magma is subjected to differential external stresses, the fluid fraction can be squeezed away from the crystalline mush.

This process, known as *filter pressing* (Daly, 1933), helps explain the origin of certain ore deposits. At the gabbro stage, for instance, a crystal mush of plagioclase and augite can contain interstitial magma that is ferromagnesian-titanium-oxide-dominated. If that liquid is squeezed out into the surrounding rocks, the process is known as *magmatic injection*. If the solidified liquid is rich enough in a valuable component, the product is known as a *magmatic injection deposit*. The iron deposits of Pea Ridge, Missouri (Emery, 1968), and the titanomagnetite deposits of Sanford Lake (Gross, 1968) appear to be magmatic injection deposits. The fugitive ore-magma melt may remain intrusive, may be extruded (Figure 2-2), or may be so shallow or sill-like in occurrence that the terms *intrusion* and *extrusion* approach meaninglessness. Deposits of apatite-rich magnetite at Kiruna, Sweden, described in Chapter 11 may be in this indeterminate category.

Oxide- or sulfide-dominated magmas or magmatic fractions that solidify directly as ore are called *ore magmas*. Since ore magmas are true melts, and not aqueous solutions, they behave as molten rock would. Figure 2-2 shows a most spectacular and compelling example of a deposit formed from an ore magma in northern Chile, a shallow intrusion and "lava flow" of almost pure magnetite-hematite with minor amounts of apatite (Park, 1961; Ruiz-Fuller, 1965; Rogers, 1968). The idea of ore magmas is well estab-

Figure 2-2. Vertical section in the Laco Sur magnetite-hematite deposit, Chile. The upper surface of the flow is about 1 meter above the hammer, so it is about 2.5 to 3 meters thick here. It consists almost entirely of magnetite, hematite, and minor apatite, is laterally extensive over square kilometers, and in every aspect appears to represent an iron oxide "ore magma" extruded as a flow. (*From* Park, 1961.)

lished, and there is no theoretical or practical reason why transition-metal-rich parts of magmas should not exist. The only current controversies involve the question of the extent of deposits formed by ore magmas. In short, how important are ore magmas, and do they include sulfide melts and silica melts as well as oxide melts?

The process of crystallization, including differentiation and crystal settling, gradually increases the concentration of the more volatile and fugitive constituents in the rest magma if the substances have no means of escape. If a gabbroic melt occupies a magma chamber near the Earth's surface, it will inevitably begin to cool by conduction and convection (Figure 2-1). The first crystals to precipitate—really the result of polymerization of a gradually increasingly viscous melt—are normally olivine and orthopyroxene soon joined by plagioclase. If these crystals settle to the floor or are otherwise sequestered, the rest melt becomes more felsic—its composition point on a phase diagram moves away from a "mafic minerals" or "gabbro composition" apex. As F. G. Smith did some years ago (1963), we might consider a ternary system composed of gabbro silicates, an oxide, and a sulfide component (Figures 2-3 to 2-6). The starting liquid contains only a few tenths of a weight percent of sulfur and is generally strongly reducing at low fugacities (chemical potentials) of oxygen. As the earliest gabbro silicates crystallize—normally enstatite and labradorite—the composition of this liquid shifts away from the gabbro silicate apex and can be described by a straight line so long as the compositions and the ratio of the crystallizing phases are the same (Figure 2-3). Notice that the implication of movement of the composition point is that sulfur and oxygen both build up in the melt. If the melt can contain sulfur and oxygen *in solution* over broad composition ranges, then one liquid will persist. But what actually happens, because mutual solubility of silicate, sulfide, and oxide liquids is restricted, is that as the silicate melt becomes increasingly sulfidic, the magma composition point approaches a saturation surface at and beyond which two or more liquids exist. The coexistence of two such liquids and their separation from one another (Figures 2-1 and 2-4) is called *liquid immiscibility*. Note that there is a large "window"—the shaded area of Figure 2-4—defined by starting compositions of low sulfur and oxygen, which will evolve differentiated silicate magma in a one-liquid field, with sulfur and oxygen dissolved until crystallization of sulfides or oxides occurs along phase boundaries late in crystallization history.

However, we can also see that a separate liquid of either oxide or sulfide character *can* evolve relatively early if the melt starts at relatively high oxygen or sulfur, respectively. Presumably either would form—as the two-liquid boundary is touched—as an emulsion of myriad droplets of one liquid dispersed through the other. The compositions of the two liquids are described by tie lines within the two-liquid fields (Figure 2-5). Elements called *chalcophile* (or *sulfophile* or *thiophile*) would partition, that is, "selectively concentrate," into the sulfide droplets (*b* and *d*, Figure 2-5). Those

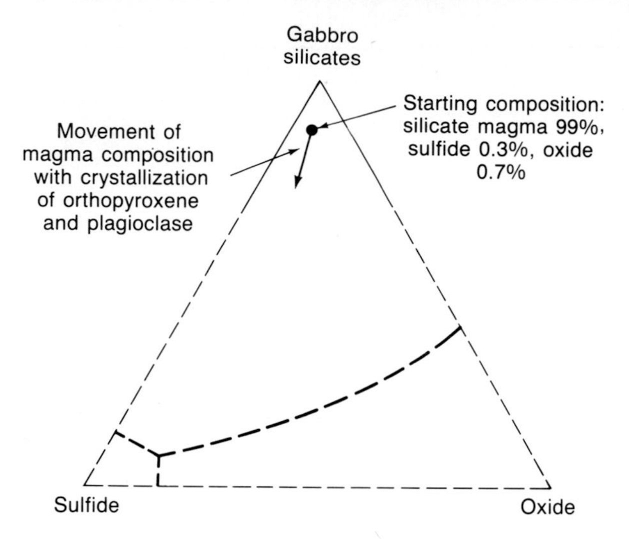

Figure 2-3. Schematic illustration of probable early behavior of a cooling gabbroic liquid. Crystallization and settling out of gabbro silicates moves the composition of the liquid as shown by the arrow. This diagram, as well as those of Figures 2-4 to 2-6, is schematic only and not to scale. Liquid and liquidus paths imply progressively falling temperatures from about 1200 to 800°C.

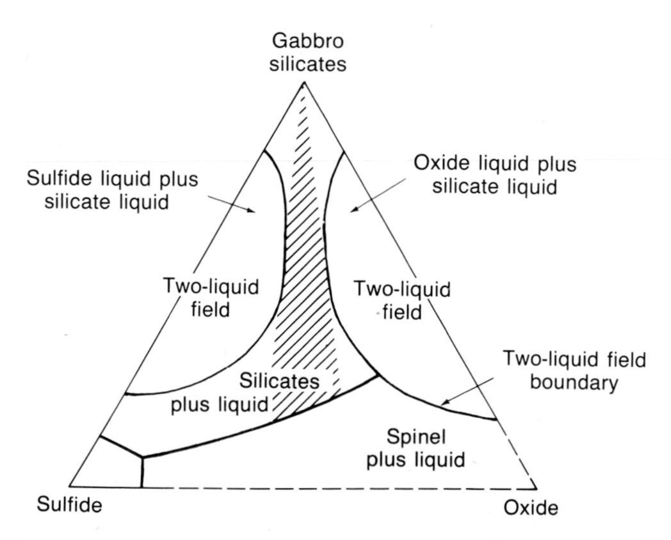

Figure 2-4. The same system as in Figure 2-3, but showing two-liquid fields. The addition of CO_2, for example, may generate three-or-more-liquid fields. The shaded area is described in Figure 2-6.

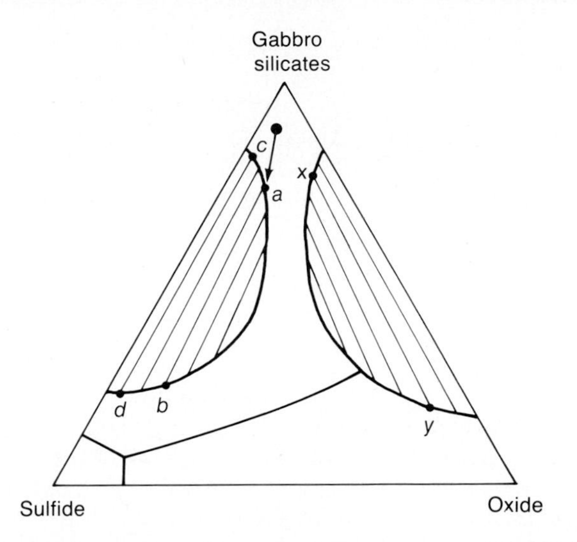

Figure 2-5. Experimentally determined tie lines within the two-liquid fields describe the compositions of each liquid. For example, at the moment of first separation, the silicate liquid composition is *a*, the sulfide liquid *b*. Had the melt contained more sulfur to start with, the pair *c*-*d* might have formed first.

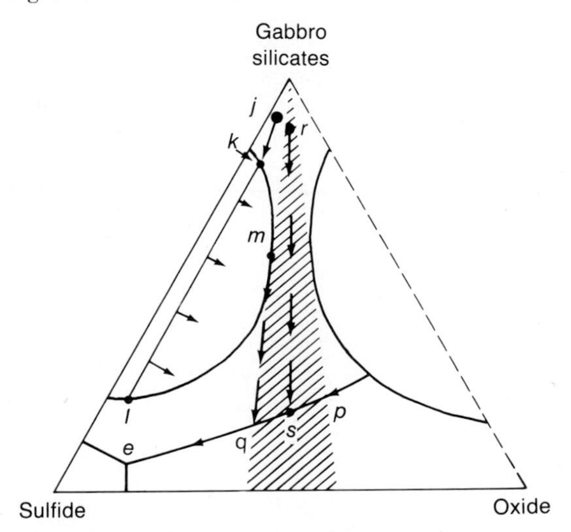

Figure 2-6. Two crystallization paths, one of any starting composition in the shaded field that keeps the melts mutually soluble and involves only solid phases in equilibrium with one liquid and vapor, the second involving two liquids. In the former, silicates crystallized, for example, from *r* to *s* are joined at *p*-*q* on the magnetite-silicate cotectic line by a spinel, with crystallization proceeding to the eutectic *e* where silicates and oxides are joined by minor late sulfides. From point *j* a separation of silicates would move the liquid to point *k*, where a second liquid of composition *l* would form. The two liquids would evolve with further cooling (and by removal of silicates and sulfur-rich liquid) toward the oxide apex. At *m* the system would progress as just described, toward *q* and finally to *e*. But in this latter case, a sulfide melt of composition *l* toward *m*, probably carrying copper and nickel, would be produced and would be transiently available for settling, injection, or both as an ore magma.

elements called *siderophile* (or *oxyphile*) would partition into oxide liquid droplets (y), and *lithophile* (or *silicophile*) elements would remain in the silicate host liquid, (a, c, and x). Thus sulfide liquids greatly enriched in iron, copper, nickel, zinc, and platinum family elements are created or oxide liquids enriched in chromium, iron, titanium, vanadium, and phosphorus appear, while the silicate liquids retain iron, magnesium, aluminum, calcium, sodium, potassium, and other main-line Bowen's reaction-series rock-forming phases. It should be clear from Figures 2-4 and 2-5 that the same normally evolving melt could produce either a sulfide or an oxide second liquid, but not normally both. Also, it should be pointed out first that we are not sure what proportion of melts do actually generate two or more liquids, and second that there are many complicating factors such as the "alkali effect"—increased sodium levels fostering late Fe-PO_4 oxide liquid immiscibility (Chapter 12)—that can only be addressed in petrology texts. Figure 2-6 shows how a system might evolve. We do indeed see tangible evidence of the formation of both oxide and sulfide liquids in different settings, as described in Chapters 9 to 13.

▸ HYDROTHERMAL FLUIDS

The processes just described are generally restricted to mafic magmas. As crystallization proceeds, normally through the "window" p–q in Figure 2-6, the lighter, more alkalic, and more hydrous volatile fractions, along with the compounds that crystallize at lower temperatures later in the crystallization history, accumulate near the top of the magma chamber (Figure 2-1). As these volatiles travel upward ultimately to contribute to the development of an ore-mineral occurrence—and exceptionally to an actual orebody—they carry with them the dregs of the differentiation process of the magma. They are called *magmatic waters*, or *juvenile*, in the sense of new, fresh, and uncontaminated, never having appeared previously at the surface. White (1957b) points out that magmatic waters may actually be largely recycled through remelting of sedimentary and volcanic rocks, and he distinguishes between *plutonic* and *volcanic* water according to the depth of separation of water from magma. Such waters containing volatiles and dissolved minerals of low freezing point are the "mother liquors" of pegmatites and the hydrothermal or gaseous fluids of magmatic affiliation. Collectively called *mineralizers*, they include the more mobile elements present in small but ubiquitous amounts in all magmas and probably all rocks, elements such as copper, lead, zinc, silver, gold, and others; the so-called *LIL*, or *l*arge-*i*on *l*ithophile, elements such as Li, Be, B, Rb, and Cs; and significant quantities of alkalies, alkali earths, and volatiles, especially Na, K, Ca, Cl, and CO_2. They play an all-important role in the late hydrothermal transportation of metals. In general, they are elements of low atomic weight and small ionic radius, though there are some notable exceptions. These elements decrease the viscosity of magmas, lower the freezing points of minerals and assemblages, and make possible the development of minerals and complexes that would not form in a dry melt. In short, they serve as *fluxes* of the melt and vehicles of aqueous transport after separation. The role the mobile elements play in ore fluids can best be inferred from the study of ores and of altered rocks associated with them, from the igneous rocks themselves, and from observations in areas of volcanic and hydrothermal activity. As White (1957b) points out, the compositions of magmatic waters are believed to be determined by (1) magma type and crystallization history, (2) temperature and pressure relations during and after separation from the magma, (3) the nature of other waters that might mix with them as they move, and (4) reactions with wall rocks.

Water is a principal mobile constituent in all magmas, increases in amount with increasing differentiation, and plays a leading part in the transportation of many ore components. Estimates of the amounts of water in magmas range from 1 to 15%. These estimates have been reached by considering volcanic and metamorphic phenomena, by analyzing the water content of volcanic glasses, and by laboratory study of synthetic melts. We know that hydrothermal manifestations are essentially unknown around mafic plutons; they appear to have been water-undersaturated and presumably abstracted water from their surroundings rather than expelled water into them.

The generation of residual aqueous fluids, also called *mineralizers* for obvious reasons, is increasingly apparent and important with progressive differentiation (Figure 2-7). In addition to water, other elements and ions include sulfur, chlorine, fluorine, boron, phosphorus, carbon dioxide, and arsenic. Micas, clay minerals, zeolites, and amphiboles contain small amounts of chemically bonded water; tourmaline and axinite contain boron; scapolites contain chlorine; and many other common minerals in altered rocks around plutons, such as fluorite, apatite, and topaz, furnish evidence of the presence of a wide range of volatile constituents. Many ore and gangue minerals have trapped liquids and gases (the fluid inclusions already mentioned and to be treated in Chapter 7), many of which are apparently primary in origin and preserve the mineralizing-altering fluids for observation.

In a series of experiments of particular interest to economic geologists, Goranson (1931) showed that the concentration of the volatile fractions increases as the differentiation of granitic magmas proceeds. He concluded that these fractions, rich in water, have definite solubility limits under specific conditions of temperature and pressure, beyond which they will constitute a separate phase of the magma. Smith (1948) continued the same type of experimentation by slowly cooling an artificial granitic magma that contained an initial 2% of water. At one point on the cooling curve, the magma separated into two immiscible liquids, one of which was mostly water. Recent statements of overall magmatic water solubilities are by Krauskopf (1967), Burnham (1979), and Burnham and Ohmoto (1980). A key

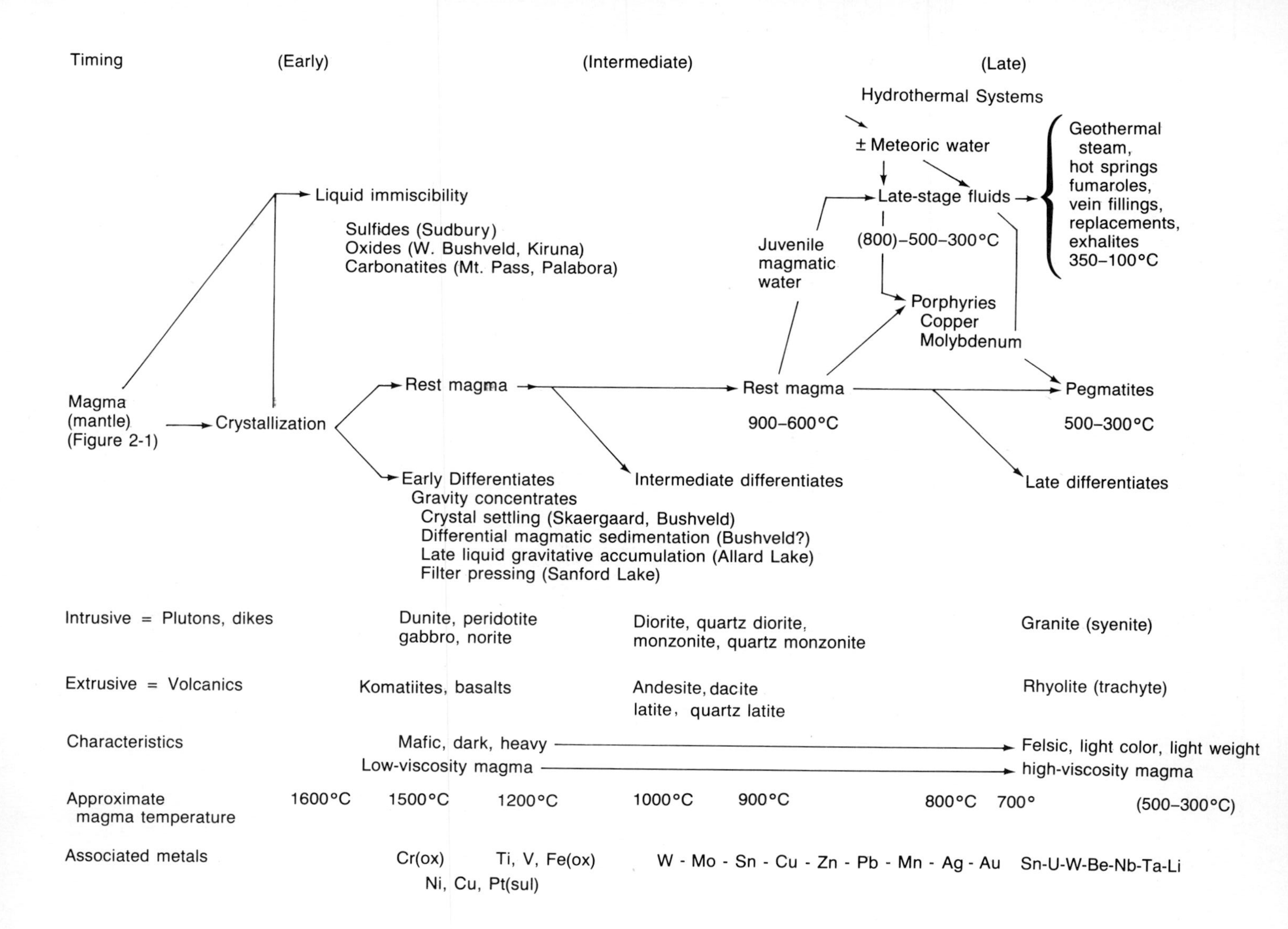

Timing
(Early)
(Intermediate)
(Late)
Hydrothermal Systems
± Meteoric water
Late-stage fluids
Geothermal steam, hot springs fumaroles, vein fillings, replacements, exhalites 350–100°C
Liquid immiscibility
Sulfides (Sudbury)
Oxides (W. Bushveld, Kiruna)
Carbonatites (Mt. Pass, Palabora)
Juvenile magmatic water
(800)–500–300°C
Porphyries
Copper
Molybdenum
Magma (mantle) (Figure 2-1)
Crystallization
Rest magma
Rest magma
900–600°C
Pegmatites
500–300°C
Early Differentiates
Gravity concentrates
Crystal settling (Skaergaard, Bushveld)
Differential magmatic sedimentation (Bushveld?)
Late liquid gravitative accumulation (Allard Lake)
Filter pressing (Sanford Lake)
Intermediate differentiates
Late differentiates
Intrusive = Plutons, dikes
Dunite, peridotite gabbro, norite
Diorite, quartz diorite, monzonite, quartz monzonite
Granite (syenite)
Extrusive = Volcanics
Komatiites, basalts
Andesite, dacite latite, quartz latite
Rhyolite (trachyte)
Characteristics
Mafic, dark, heavy
Felsic, light color, light weight
Low-viscosity magma
high-viscosity magma
Approximate magma temperature
1600°C
1500°C
1200°C
1000°C
900°C
800°C
700°
(500–300°C)
Associated metals
Cr(ox)
Ti, V, Fe(ox)
Ni, Cu, Pt(sul)
W - Mo - Sn - Cu - Zn - Pb - Mn - Ag - Au
Sn-U-W-Be-Nb-Ta-Li

observation echoing through the several figures of this chapter is that late in the history of fractional crystallization of the "typical magma," a water-rich fluid is exsolved and expelled. The amount of water is a function of many parameters—original water content, influx-egress balance, porosity-permeability relationships in the confining walls, net pressure and local pressure differentials in the melt or its chamber walls, and temperature-crystallization dynamics, to name only a few. For example, melts are thought to crystallize from the outside inward. If external forces crack the wall rocks and crystalline shell such that previously confined liquids can escape, breccia columns and mineralized fault veins may result. If such rupture does not occur, internal pressures may build until the water pressure is greater than lithostatic load pressure and an explosion-brecciation event occurs. Or a confined water "bubble" may form by the exsolution of a water phase that cannot at first escape (Norton and Cathles, 1973).

However it occurs, it is most important to perceive how an aqueous fluid might normally separate. When the cap is popped off a beer bottle, the original one-phase solution suddenly—and throughout the volume affected by the pressure drop—becomes two-phase. Just as in the description of sulfide liquid immiscibility earlier, myriad microbubbles of gas form. Early in their history there are so many minute bubbles that most of the liquid is in real chemical contact with gas; as many volatile phases as may exist partition into the gas phase, in which they are thermodynamically more stable than they are in the liquid. Now envision a magma having its "cap lifted"—a water phase, probably a supercritical aqueous one—would separate as an emulsion of droplets, each of which would attract hydrophilic elements and ions down pressure, temperature, and chemical potential gradients. Those elements will be of several overlapping sets, three of which are (1) those that for various geochemical reasons do not fit into the feldspars and mafic silicates and are therefore progressively concentrated into late melts, the LIL elements such as beryllium, lithium, and rubidium, many ore mineral components and volatiles, and halogens; (2) elements that are physicochemically more soluble in water or a gas phase than in silicate melt, including chlorine, sulfur, carbon monoxide and dioxide, and phosphate, borate, tungstate, molybdate, and uranyl complex ions; and (3) elements which can form stable high-temperature complex molecules or ions in water, such as Na^+, K^+, and Ca^{+2} with Cl^-, Cu^+, Cu^{+2}, and Zn^{+2} with Cl and various other anions, and many others that will be treated in Chapter 11. And of course dissociation of water at high temperature produces H^+ and $(OH)^-$ ions in equal numbers. Helgeson (1964), in a pioneering thermochemical study of the system PbS-NaCl-HCl-H_2O at elevated temperatures, showed that the complexes present or those that might form in these

Figure 2-7. (*facing page*) Representation of igneous rock types, differentiation, and hydrothermal fluid involvement associated with fractional crystallization. Probably no one melt passes through all these stages, although the overall flow sheet is realistic.

solutions even with a small number of components include a formidable number:

HCl	H_2O	H_2S
$NaCl$	$NaOH$	HS^-
$PbCl^+$	$PbOH^+$	$NaHS$
$PbCl_2^-$	$Pb(OH)_2$	NaS^-
$PbCl_3^=$	$Pb(OH)_3^-$	Na_2S
$PbCl_4^=$	$Pb(OH)_4^=$	$PbHS^+$
		$PbHS_2^-$

Also, various polysulfide and thiosulfate complexes as combinations of the simple ionic species Pb^{+2}, Na^+, H^+, $(OH)^-$, $S^=$, and Cl^- may form in water. Possible molecular and ionic forms of sulfur, excluding metal complexes, present an even more complicated profile depending in part on the valence of the sulfur present (Table 2-1).

TABLE 2-1.

Valence of Sulfur	Gaseous Molecules	Aqueous Molecules	Aqueous Ions	
+6	Trioxide, SO_3	Sulfuric acid, H_2SO_4	Bisulfate, HSO_4^-	Sulfate, $SO_4^=$
+4	Dioxide, SO_2	Sulfurous acid, H_2SO_3	Bisulfite, HSO_3^-	Sulfite, $SO_3^=$
+2	Monoxide, SO		Bithiosulfate, $HS_2O_3^-$	Thiosulfate, $S_2O_3^=$
0	Sulfur, S, S_2, S_4, S_6, S_8			
$-2/x$		Hydrogen polysulfide, H_2S_x	Polysulfide, HS_x^-	Sulfide, $S_x^=$
−2	Hydrogen sulfide, H_2S	Hydrogen sulfide, H_2S	Bisulfide, HS^-	Sulfide, $S^=$

Source: From Barnes and Czamanske, 1967.

Holland (1972), Roedder (1979), and many others have shown that metal-halogen complex ions are almost certainly the principal "carriers" of most base and precious metals, including copper, zinc, lead, tin, iron, mercury, silver, gold, and platinum in aqueous solutions. Holland (1972) described the separation of an aqueous fluid phase from an intermediate magma, the partitioning into it of volatiles and gases, chloride ions, alkalies, and elements complexed by them, and its collection and separation from the melt as an ore-component transporting medium.

Understanding the nature of the fluid is important to understanding its chemical potency and how it can move through faults, joints, fissures, microfractures, and even the micropermeability of miarolitic cavities, grain boundaries, and cleavages in and between minerals in rocks. The perception of how such solutions—both solutes and solvent—move through rocks is difficult; the subject is discussed more fully in Chapter 3.

What is the chemical nature of a hydrothermal fluid? We have already considered part of that question—a magmatogene fluid is unlikely to be anything like pure water; it contains many of the ions just listed, and certainly scores more involving iron, copper, zinc, and other elements, and can range from trivially to strongly "salty." Figure 2-8 shows that water can dissolve up to 57 wt% NaCl at 500°C, and salinities deduced from the contents of fluid samples microencapsulated in imperfections in vein minerals—the fluid inclusions described later—also range from 0 to 60 or 70% NaCl equivalent, with many in the 30 to 50% range. The fluid also contains variable amounts of the "mineralizer" elements, the complex and simple ions and molecules listed earlier. The pH of the fluid varies with temperature and other variables as the dissociation constant K_w describing the reaction

$$H_2O = H^+ + OH^- \tag{2-1}$$

$$K_w = \frac{a_{H^+} \cdot a_{OH^-}}{a_{H_2O}} = a_{H^+} \cdot a_{OH^-} \tag{2-2}$$

varies from 10^{-14} (neutral pH = 7) at standard conditions to as high as $10^{-11.2}$ (neutral pH = 5.6) at 220°C.

The dissociation constant increases slightly with increase in pressure, causing the pH of a neutral solution to drop to slightly less than 7 (Owen and Brinkley, 1941). Conversely, the addition of a dissolved salt brings about a slight decrease in the dissociation constant. The effects of pressure and dissolved solids are minor, amounting to a few tenths of a pH unit at most. From what has already been said, the real pH of a hypogene hydrothermal fluid is next to impossible to know with certainty because of the many potential association-dissociation reactions involving H^+, but the fluids appear in general to be from neutral to slightly acid. According to Helgeson, the pH of hydrothermal solutions does not deviate markedly from neutrality; hydrothermal solutions that contain high concentrations of chloride probably become weakly acidic as the temperature is decreased from the supercritical region for a given bulk composition (Helgeson, 1964, pp. 83–85). The best field evidence and laboratory observations seem to indicate that ore-bearing fluids are nearly neutral and that strong acids and bases are exceptional, though they do exist under special circumstances. Weissberg (1969) has shown that the waters of several thermal spring areas in New Zealand are close to neutral. These waters transport minute amounts of gold, silver, arsenic, antimony, mercury, and thallium, which are concentrated in the muds and sinters. Trace amounts of lead, zinc, and copper have been found in cores obtained from deeper drill holes there.

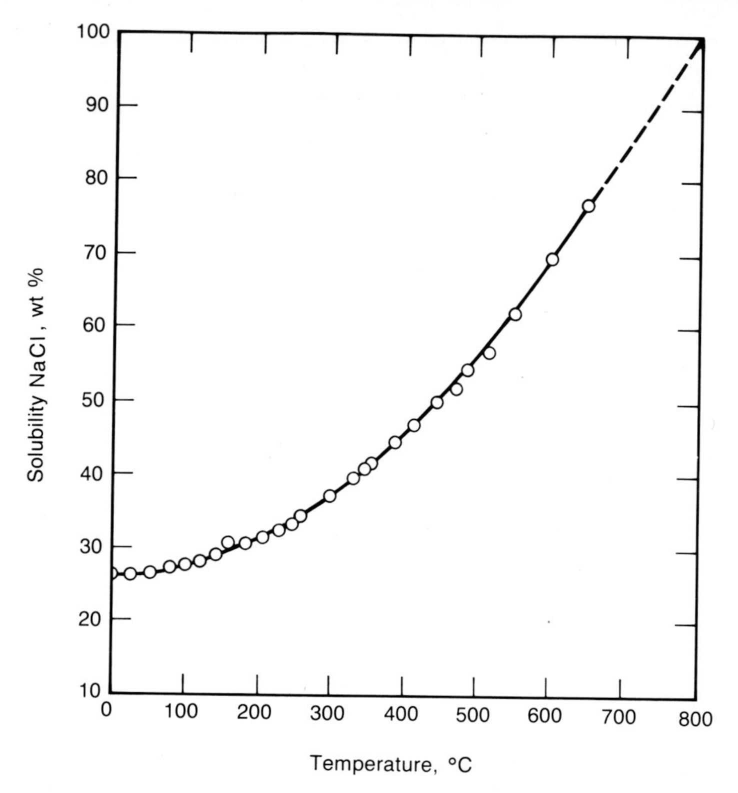

Figure 2-8. Solubility of NaCl in water along the boiling curve of Figure 2-9, the boundary vapor-liquid NaCl. Note that quite high concentrations of NaCl in hydrothermal fluids can be accommodated, for example, 56% NaCl at 500°C. (*After* Holland, 1967.)

Most crustal rocks are composed of either silicate or carbonate minerals, both of which react with acidic solutions to consume H^+ and raise pH, so any acidic solutions buffered by passing through silicate or carbonate rocks will eventually become neutral or alkaline. Thus it is logical to assume that ore-bearing fluids are nearly neutral or basic, at least at pressures close to atmospheric; otherwise they would react immediately with the wall rocks. Such is the case with ordinary groundwaters, many of which are slightly acidic at the surface and become neutral or alkaline at depth. Ore deposition may in part be due to just such a chemical change. Uprising ore-bearing fluids may be acidic at first, with deposition of metals that are soluble in acid solution occurring when the solutions become neutralized. Hydrogen ion activity is exceedingly important in the construction of activity-activity diagrams to describe the evolution of fluids and alteration mechanisms (Chapter 5).

Hydrothermal fluid density is important because it affects viscosity, and hence flow dynamics, and controls the solubility of ore components.

There are a couple of ways to represent the densities of hydrothermal fluids. Helgeson (1964) showed plots of density values plotted over a pressure-temperature (P-T) phase diagram of the system H_2O connected by isodensity lines called *isochores* (Figure 2-9). The critical point given is for pure water. He also showed that lines of equal dissociation constants for KCl (shown), HCl, and presumably NaCl nearly parallel the isochores and that these chlorides are more *associated* at higher P-T conditions. A second way to show density relationships is graphically (Figure 2-10), in which the transitional nature of density above and "around" the critical point as well as the density contrast across the boiling curve are apparent. It appears from an inspection of Figures 2-9 and 2-10 that most natural hydrothermal fluids would be expected to be "supercritical." Physicochemical studies of the water system show that increased percentages of nonvolatile solutes—such as NaCl, KCl, $FeCl_2$, and $CuCl_2$—move the critical point to *higher* P-T values, almost along the projection of the boiling curve, while the presence of volatile substances—such as CO_2, HCl, and CH_4—move it back to *lower* P-T values. With observed salinities ranging up to 60 or 70 wt% NaCl, the critical point is almost certainly generally higher than 374°C at 220 bars (0.22 kbar), thereby extending the possible two-phase boundary boiling curve higher into near-surface geologic conditions.

Two facts of supreme importance to the transportation and deposition of ore minerals can now be presented. The first is that, in general, the solubilities of volatile and nonvolatile solutes in supercritical water vary with density (Krauskopf, 1967); a general decrease in pressure from top to bottom in Figures 2-9 and 2-10—a path from high to low P-T conditions that involves decreased fluid density, especially if the boiling curve from water to vapor is crossed—will generally result in diminished solubility and hence precipitation of solutes. And second, as shown in Figure 2-9, chloride complexes *dissociate* approximately as hydrothermal fluid density decreases, so that alkali- and base-metal-chloride complex ions are "disengaged" at lower P-T; base metals can be taken into solution and stably transported at high P-T and freed for recombination with sulfur to form stable crystalline sulfides at lower P-T. These two effects and their ramifications combine to explain many aspects of hydrothermal transport and deposition, because the described change in P-T environments—again, from high to low overall P-T—is what an aliquot of fluid traveling up a fault-vein fissure would "experience." Finally, boiling is probably important in some hydrothermal systems because of the expectable drastic decrease of solute solubility with the drop in density when a fluid boils. It might thus be expected that some ore deposits—or at least "mineralization"—would develop in a boiling zone where high P-T fluids approach the surface with pressure falling more rapidly than temperature. Such zones are described in the "working" chapters (9 to 20); one is depicted in Figure 7-10.

In a discussion of hydrothermal areas being developed as sources of geothermal energy, Ellis (1970) noted that drilling permits detailed obser-

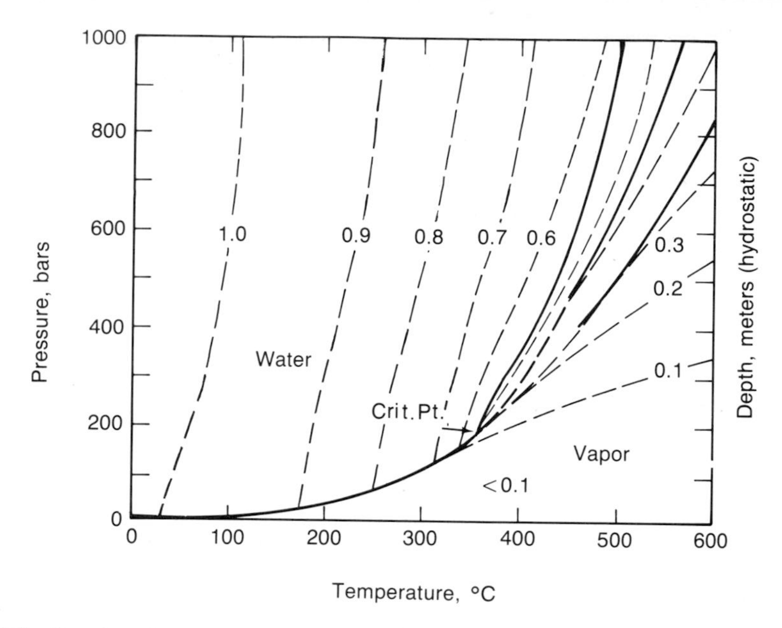

Figure 2-9. Density of water as a function of pressure and temperature (dashed lines). The heavy solid line is the boiling curve, with vapor (steam) below and water above, terminating at the critical point. Light solid lines show dissociation constants for KCl in the supercritical region. (*From* Helgeson, 1964.)

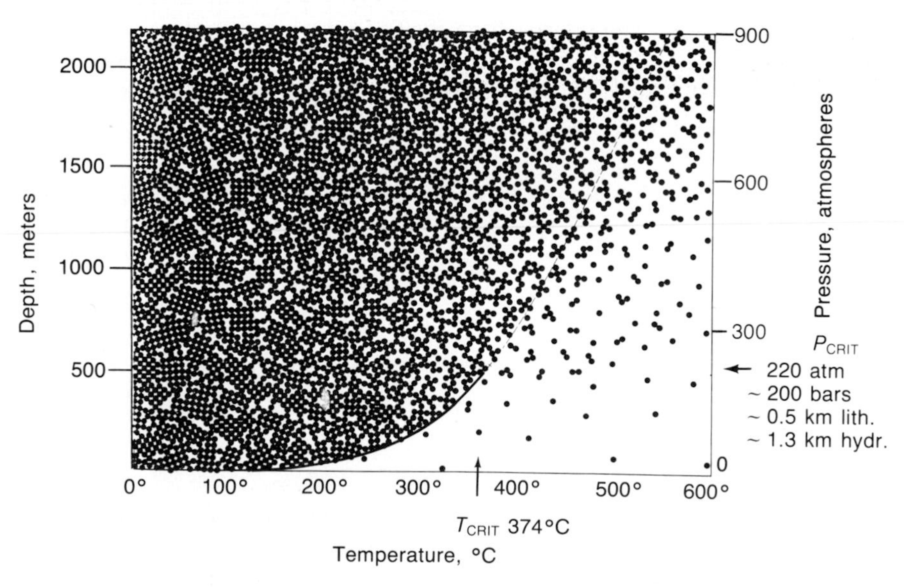

Figure 2-10. Graphic representation of density and "polymerization" of water molecules as functions of pressure and temperature. Note the gradational changes across any series of points to the right or above Tcrit, and the two-phase aspect of the boiling curve. (*From* Smith, 1963.)

vations of deep-water temperatures and pressures. He found that concentrations of metals such as lead, zinc, copper, manganese, and iron in high-temperature waters are related directly to salinity. Naturally hot brines, such as those of the Salton Sea area in California, contain unusually high concentrations of heavy metals and many produce metal-rich scales in drill pipes. In contrast, most volcanic-area thermal waters are very dilute salt solutions at 200 to 300°C with very low concentrations of base metals, silver, and gold. Nevertheless, precipitates from these waters are known to contain a low percentage of antimony, rare ore-grade concentrations of silver and gold, and higher amounts of arsenic, mercury, and thallium. This association is common in near-surface deposits accompanying Tertiary volcanic activity in the western United States and in New Zealand.

For many years, economic geologists were unable to reconcile ore deposit magnitudes with the very large amounts of watery fluid that had to be called upon to transport extremely insoluble metallic sulfides. Calculations of the amounts of water or steam required to transport the metals in individual ore deposits indicated that their conveyance as simple dissolved solids would require unrealistically large volumes of fluids measured in Mississippi River equivalents. The solubility of mercuric sulfide is an extreme example of this fact. The true solubility of HgS at high temperatures is unknown, but between room temperature and 200°C it varies from only about 10^{-15} to 10^{-23} moles per liter, depending upon the pH of the solution (Czamanske, 1959). Solubility increases with increased temperature and with increased acidity, that is, the higher figure given, 10^{-15} moles per liter, refers to a solution at 200°C and a pH of 4 to 5. Since most cinnabar deposits form below 200°C, this temperature should be reasonable for solubility calculations. If HgS were carried in true solution, at 200°C and pH 4, one million times the annual volume of water flowing from the Hudson River would be required to deposit a single ton of cinnabar, provided that all of the dissolved mineral and sulfur could be removed from solution at the site of deposition. At a still lower pH or at a higher temperature, or both, the amount of HgS carried in solution would increase by many orders of magnitude, but would never reach realistic proportions. The mercury must, of course, travel in some form other than a simple solution of HgS in pure water. Although transportation of mercury as a volatile chloride, a soluble chloride or sulfide complex, or a metal vapor can explain the concentrations of HgS found in nature, these mechanisms do not explain occurrences of all metals. Weissberg (1969) believes that the amount of water flow in New Zealand thermal areas is sufficient to account for the formation of many rich gold-silver deposits. The springs have existed for many thousands of years, and even though they carry only minute amounts of metals, concentration by deposition over such prolonged periods could be sufficient to explain economic deposits.

The formation of complexes—species consisting of combinations of ions—increases the solubility of some metals by many orders of magnitude (Barton, 1959; Barnes 1967, 1979). Simple solubilities are based on the number of single ions or common radicals that go into solution in a given amount of water. However, in the presence of other atoms with which an ion can coordinate, the amount of metal that can enter the solution may well surpass its simple solubility limit, that is, metals as parts of complex ions may be many times more soluble than metals as simple ions. Thus mercury may remain in solution as $HgS_2^{=}$, or perhaps as $HgS_2(H_2O)_n^{=}$, $HHgS_2^{-}$, and so forth. In any one of these forms, mercury would be in sufficient concentration to account for its transportation in geologically reasonable volumes of ore-forming solutions. Many metals combine as comparatively stable complexes or complex ions in the form of sulfides, polysulfides, hydrosulfides, halides (especially chlorides), carbonates, hydroxides, oxides, sulfates, and others. The nature of mineral species deposited will depend upon the temperature, pressure, and ratios of ions present at the locus of deposition rather than necessarily on the nature of the complex by which the metal was transported. Hence there is no need for a special complex for each mineral formed; a simple sulfide may be deposited from a solution containing exceedingly complex multimetal complexes. Studies of complex ions have been encouraging, and may explain a number of problems concerning ore-bearing fluids, but such complexes cannot be considered the panacea for all enigmas of ore genesis; at best, they can provide only some of the answers. Nevertheless, the presence of metals as complex ions in solution is especially attractive in light of the general sequence of mineral deposition seen in most sulfide ores, a sequence that is in strong accord with the relative stabilities of complex ions but essentially the reverse of that predicted from simple solubilities (see Chapter 6).

Barnard and Christopher (1966) synthesized chalcopyrite, galena, and sphalerite in the laboratory using chloride complexes. Because chloride complexes are quite soluble and stable, ore components can migrate as metallic chlorides and, as Helgeson (1964) points out, the evaluation of different lines of evidence indicates that chloride complexes may be the most important factor in the transport and deposition of ore-forming metals. He showed that the overall relative stabilities of the chloride complexes of the metals in hydrothermal fluids are in the sequence

$$Cu^{+2} \quad Zn^{+2} \quad Pb^{+2} \quad Ag^{+} \quad Hg^{+2}$$

from high to low P-T, which agrees with the commonly observed sequence of deposition of the ore minerals.

Some geochemists emphasize the possible role of complexes other than relatively simple ones of chlorine and sulfur. Tyurin (1963) considered the thiosulfate complexes $M(S_2O_3)_m^{-n}$ to be the most probable form of transportation of metals. He says that the solubility of these complexes is very high; that both sulfide-sulfur and metals would be transported simultaneously; that the complexes are stable between pH 5 and pH 10, and are most stable in neutral or weakly alkaline solution; that changes in pH and Eh may cause breakdown of the complexes; and that the thiosulfate complexes become unstable in a series that corresponds with the zonal distribution and the temporal succession of deposition (Chapter 6, Table 6-1) established by Emmons (1936). Shcherbina (1964) argued against the presence of thiosulfate ions because thiosulfate compounds are unknown among minerals and the $S_2O_3^{=}$ anion has not been detected either in the water of hot springs or in fluid inclusions. Extensive research is now being carried out on the possible significance of organometallic complexes. It is more and more apparent that hydrocarbon molecules have been involved in some types of mineralization. In Chapters 3 to 7 we return repeatedly to mechanisms of transportation and deposition.

▸ METEORIC WATERS

Water of whatever origin that has passed through and equilibrated with the atmosphere is called *meteoric water*. It is essential to *supergene* processes, those that stem from and involve waters of surficial origin. Water from rain or snow, mist, dew, frost, or wind-borne microdroplets of fresh- or salt-water percolate downward and can react with the lithosphere. They are important because of their involvement with supergene processes, but also because stable isotopic studies in the last 15 years, principally by Hugh Taylor and his students at the California Institute of Technology, have revealed that genetically significant amounts of meteoric waters can be downdrawn by convective circulation and incorporated into ore-forming systems (White, 1957a). An "end-member situation" would be one in which a water-saturated stock—one that would neither give nor receive groundwater—would effect mineralization alteration by acting solely as a heat source causing meteoric waters to convect around it.

Meteoric water has several "tags," or labels. In equilibrating with the atmosphere, water dissolves air and thus nitrogen, oxygen, carbon dioxide, and traces of the rare gases. Carbon dioxide forms several ions and compounds with water, the resulting solutions containing weakly ionized carbonic acid, bicarbonate ion $(HCO_3)^-$, and hydrogen ion (H^+). It commonly contains a few parts per million of NaCl within a few hundred kilometers of coastlines, the result of wave-air mixing in surf, "sea air" mists, and onshore breezes. Near active volcanos, traces of magmatic gases such as SO_2 and HF may be detected, but they are formed only there and seldom importantly. Measured pH and Eh of meteoric water—rain—shows a narrow range from 5 to 5.5 pH units and mildly oxidizing Eh at +0.4 to +0.5 mV away from industrial centers and the "acid rain" of air pollution origin. Geochemically a most important label is the stable isotope tag referred to earlier. Juvenile waters have been found in scores of isotopic analyses to occupy a small area in a plot of the ratios of the stable isotopes of oxygen

($O^{18/16}$), and hydrogen (H^{+2}/H^{+}), or deuterium/hydrogen (D/H). This area is called the *magmatic water box* (Figure 2-11); it is discussed at length in Chapter 7. It has also been determined that all modern ocean waters have a nearly identical, constant oxygen-hydrogen stable isotope ratio which is plotted on Figure 2-11 as SMOW. The word rhymes with hoe and stands for *s*tandard *m*ean *o*cean *w*ater. As is described in Chapter 7, both the evaporation of seawater and the cooling of moisture in air favor the entry and mobility of the lighter isotopes, so O^{16} and H^{1} are increasingly preferentially transferred inland, over mountain ranges, and from the equator toward the poles. Regional values of the stable isotopes therefore characterize meteoric waters, and their presence, the degree to which they have been mixed with juvenile fluids in ore-forming processes, and their movement paths can be determined by careful mass spectrometric measurement. The recognition of this isotopic "tag" has not revolutionized *concepts* of the genesis of ores since downward circulation of marine or atmospheric waters

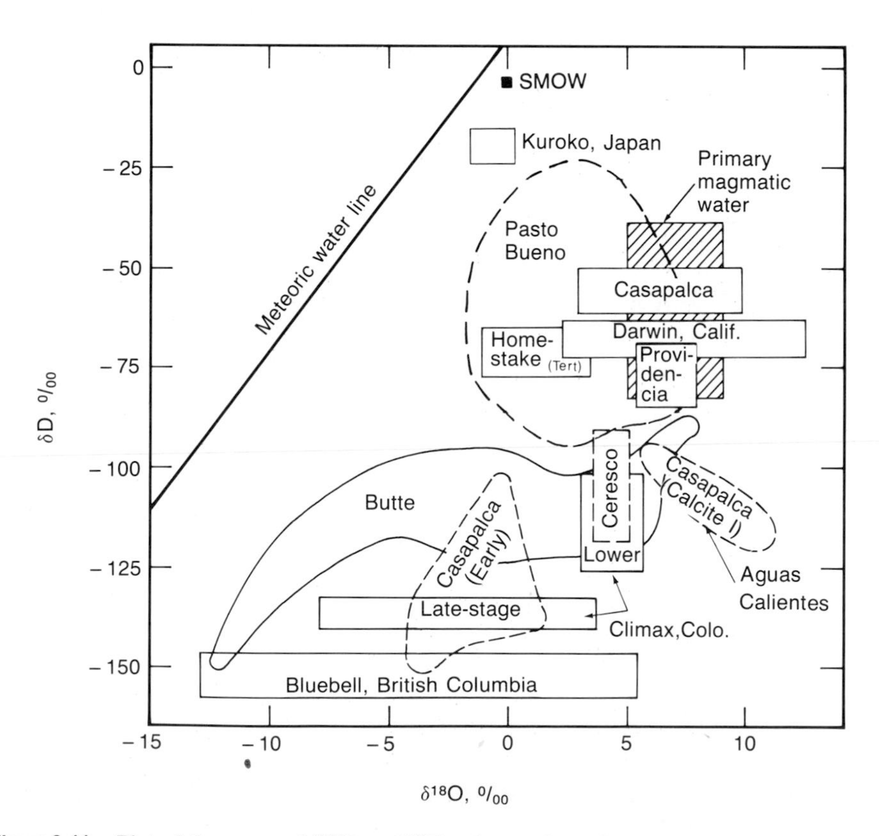

Figure 2-11. Plot of the range of $^{18/16}O$ and $^{2/1}H$ values—also called *del eighteen* and *delta D*—of waters in fluid inclusions from various porphyry, Cordilleran Vein, epithermal, and massive sulfide deposits. The SMOW point and the "primary magmatic water" box are discussed in the text here and in Chapter 7. (*From* Taylor, 1979.)

has been speculated upon since the days of Delius and Werner (Chapter 1); but it has permitted us to move from speculation to demonstration, a dramatic advance.

As meteoric fluids—only trivially modified by atmospheric components, except isotopically—sink into the rind of the Earth, they quickly assume the temperature of the enclosing rocks; water temperature in general increases with depth of penetration. Eh will be reduced and pH slightly increased by reaction with feldspars to produce clay minerals. Although exceptions exist, the content of dissolved mineral substances generally increases as water temperature increases. Descending waters gradually assume chemical and thermal equilibrium with the enclosing rocks. Meteoric waters thus probably contain various amounts of the dominant crustal elements, such as sodium, calcium, magnesium, and the sulfate and carbonate radicals, but not the more fugitive elements such as boron and fluorine that are characteristic of juvenile or volcanic waters (Clarke, 1924). They percolate downward, and may either enter a moving water table or remain essentially sequestered and stagnant for long periods of geologic time. It is safe to say that most water in rocks in the upper kilometer or so of continental environments is of meteoric origin.

▸ SEAWATER

The characteristics of *seawater* as an ore-forming fluid are best described in the contexts of evaporites, phosphorites, submarine exhalites, "manganese nodules," and oceanic crust deposits in later chapters. Seawater assumes both a relatively passive role as a medium of dispersion of dissolved ions, molecules, and suspended particles and an active one as a solvent of ions from rocks beneath the seafloor through which it is convectively drawn. It also participates in direct marine chemical precipitation. The composition of seawater is given in Table 15-1 for comparison with compositions of other aqueous fluids given in Tables 2-2, 4-1, and 20-2, and the involvement of seawater with ore-deposit systematics is discussed in Chapters 10, 13, 14, and especially 15.

▸ CONNATE WATERS

Water trapped in sediments at the time they were deposited is known as *connate water*, a term first applied to brines flowing into the deeper levels of the Michigan copper mines (Lane, 1908). Connate waters are actually fossil waters. White (1957b) recommended that the definition include the fact that connate water "has been out of contact with the atmosphere for at least an appreciable part of a geologic period"; we would want to say "atmosphere and hydrosphere." Connate waters are widely observed in oil field exploration and production as the salty edgewaters of many oil fields, and a great deal of study has been devoted to them by petroleum geologists. Most connate waters are rich in sodium and chloride, but they also contain considerable amounts of calcium, magnesium, and bicarbonate, and many contain strontium, barium, and nitrogen compounds (White 1957b, 1968) and can thus be chemically distinguished. Their identity as brines rich in NaCl should be emphasized (Table 20-2). They also typically contain considerable amounts of light hydrocarbons. They would be difficult to identify isotopically if they had reacted with their host rock minerals, which is expectable; otherwise their stable isotope ratios would be near that of SMOW. As basins are downwarped or sedimentary sections are metamorphosed, the contained connate waters may become heated and activated. When activated, they may become strong solvents of the metals, since they contain large amounts of chlorine. They thus are one source of hydrothermal fluids, and are emphasized in their apparent role in the genesis of Mississippi Valley type deposits in Chapter 20.

▸ METAMORPHIC FLUIDS

Under favorable circumstances, connate and meteoric waters enclosed in rocks buried below the surface of the Earth may be set in motion and made chemically reactive by heat and pressure accompanying magmatic intrusion or regional metamorphism (Shand, 1943). These are the so-called *metamorphic waters* that some geologists believe are active ore carriers. If a depositional basin is gradually heated and compressed by burial and then by regional deformation, its connate waters will be expelled first. These brines—certainly with some ions including the base and precious metals exchanged into them from adsorption sites in sedimentary minerals—would travel along permeable zones. Then as recrystallization progressed, clay minerals—kaolin and montmorillonite family minerals—would give up loosely bound adsorbed or interlayer water, again logically with some base metals. Finally as recrystallization from hydrous to anhydrous phases advanced, a final fluid would be evolved, again probably with some metals. Each of these egresses would be a one-time event; there could be virtually no "pumping" or recycling. Each would be finite in the amount of water generated. Many geologists have concluded that the efficacy of metamorphically derived solutions as ore-forming fluids is inconsequential, it being ordinarily and widely believed that in regional metamorphism metallic and volatile constituents are dispersed rather than concentrated (Taupitz, 1954). In support of this contention, Eskola showed that palingenetic granites—granites formed by the reconstitution of other rocks—are notably free of metallic concentrations. In fact, he proposed that they are distinguished from magmatic granites by their general lack of ore deposits (Eskola, 1932).

Other geologists argue that volatile and mobile constituents are activated during metamorphism and forced from the rock to migrate toward cooler and, in general, less deformed regions. These volatile and mobile elements include most of the metallic constituents, but water constitutes most of the fluid. Trace-element studies of minerals from different metamorphic facies indicate that certain metals are released selectively during regional metamorphism (DeVore, 1955). Accompanying tectonic processes may provide avenues along which the metals and metamorphic fluids travel, producing a hydrothermal system. Water and its dissolved contents are believed to move down the metamorphic gradient, in advance of either regional metamorphism or an intruding magma. Since connate waters and groundwaters commonly contain large amounts of soluble salts, they would become solvents and would tend to remove metals from the rocks once heated and set in motion. Supporting this view is the fact that inclusions in ore minerals generally contain chlorides, indicating that chlorine was present in many of the ore-bearing fluids. These activated and ore-bearing waters might also have combined with fluids released from a magma.

Modern views of the role of metamorphism in mineral deposit formation are reviewed—and generally deemphasized—in Chapter 18. As is emphasized there, high-temperature metamorphism can only rarely be convincingly related to ore formation, but low-temperature metamorphic effects such as dewatering and local mass transfer in response to thermal and pressure gradients in some lateral secretion deposits may have formed ore deposits. They are treated in Chapters 19 and 20.

▸ THERMAL SPRINGS

Subaerial thermal springs, solfataras, and fumaroles have been studied intensively by many geologists, and the store of information applicable to the problems of ore solutions is growing. Among the most comprehensive studies of these phenomena are those of White, Hem, and Waring (1963) and White (1957a, 1957b, 1968), who applied geochemical principles and techniques to the study of Steamboat Springs, Nevada; Salton Sea, California; other areas of thermal activity in the western United States; Providencia, Zacatecas, Mexico; and the metal-containing sediments in the Red Sea. On the basis of stable isotope analyses, whereby the waters issuing from various thermal springs were compared to surface waters of the surrounding terrane, White decided that the contribution of magmatic water to these springs was commonly insignificant—probably no more than 5%—although he notes that none of the districts studied is completely understood. He drew five main conclusions (White, 1968):

1. Most ore deposits are formed by complex processes rather than by simple end-member ones.

2. The ore-bearing fluids of base-metal deposits are Na-Ca-Cl brines.
3. Brines of similar major element composition may form (a) magmatically, (b) as connate waters, (c) by solution of salts formed by the evaporation of any dilute saline water, or (d) by osmosis of salt into dilute meteoric water.
4. In at least three and perhaps all five of the districts studied, the ratio of total dissolved metals to dissolved sulfides in the ore fluids was very high, and the metals were transported as chloride complexes in the presence of only small amounts of sulfide.
5. The density of ore-bearing brines is normally higher than that of near-surface waters.

Groundwater in regions of recent volcanic activity is commonly characterized by a relative abundance of sodium chloride, and apparently grades into acidic sulfate-chloride water. Other waters in volcanic areas contain sodium bicarbonate, sulfuric acid and sulfate ion, calcium bicarbonate, mixed sodium and calcium bicarbonate, and borate radicals. White considers the sodium-chloride waters to be the most closely related to magmatic emanations; the other types of thermal waters are secondary products resulting from reactions with wall rocks, changes in the physical environment, and groundwater mixing. White's analyses of sodium chloride waters are given in Table 2-2. According to White, the volcanic sodium-chloride brines are distinguished from connate and ocean waters by their relatively high content of lithium, fluorine, silica, boron, sulfur, and carbon dioxide in the former, compared to high contents of calcium and magnesium in the latter. Nonvolatile compounds are variably soluble in steam at high pressure; as indicated earlier, high-density steam has solvent properties similar to those of liquid water. In volcanic sodium-chloride waters, a high ratio of lithium to sodium and potassium indicates differentiation in a magma and suggests that the alkalies were transported as alkali halides dissolved in liquid or dense vapor. White believes that even juvenile fluids can be greatly diluted by deeply circulating meteoric water, with mixing at depths of as much as 5 to 6 km. Where circulation of meteoric water is only shallow, halide-bearing vapors may reach low-pressure regions, expand, and precipitate nonvolatile substances; such mechanisms would remove much of the sodium chloride from the vapor system and produce one of the modified calcium-bicarbonate waters or acidic sulfate waters mentioned above (Figure 2-12).

The Valley of Ten Thousand Smokes, an area of fumarolic activity that became active in 1912 where the Aleutian Islands connect to Alaska, provides further information on the nature of hydrothermal fluids (Zies, 1929). The fumaroles formed at the surface of a thick accumulation of hot, rhyolitic pyroclastics which flowed into a U-shaped valley and traveled many kilometers downslope from the vents. Fumaroles along most of the length of

TABLE 2-2. Hot Springs Water Compositions, Temperatures, and Acidities

	Steamboat Springs, Nevada	Morgan Springs, Tehama County, California	Norris Basin, Yellowstone National Park, Wyoming	Upper Basin, Yellowstone National Park, Wyoming	Wairakei, New Zealand
Temperature, °C	89.2	95.4	84	94.5	>100
pH	7.9	7.83	7.45	8.69	8.6
SiO_2	293	233	529	321	386
Fe	—	—	—	Trace	—
Al	—	—	—	0	—
Ca	5.0	79	5.8	4	26
Mg	0.8	0.8	0.2	Trace	<0.1
Sr	1	10	—	—	—
Na	653	1398	439	453	1130
K	71	196	74	17	146
Li	7.6	9.2	8.4	—	12.2
NH_4	<1	<1	0.1	0	0.9
As	2.7	2.2	3.1	—	—
Sb	0.4	0.0	0.1	—	—
CO_3	0	0	0	66	—
HCO_3^-	305	52	27	466	35
SO_4	100	79	38	15	35
Cl	865	2437	744	307	1927
F	1.8	1.5	4.9	21.5	6.2
Br	0.2	0.8	0.1	—	—
I	0.1	<0.1	<0.1	—	—
B	49	88	11.5	3.7	26
S_2O_3	—	—	—	2	—
H_2S	4.7	0.7	0	0	1.1
CO_2	—	—	—	—	11
Totals	2360	4578	1885	1676	3742

Analyses of sodium chloride hot spring waters (in parts per million) from selected thermal springs. High fluorine, lithium, and boron—among others—label these fumarolic waters with respect to connate oil-field brine, for example (Table 20-2). But they are tremendously dilute with respect to the "captive" ore-forming fluids of Table 4-1. Oxygen-hydrogen isotope characteristics of some of these brines are given in Figure 7-14.

Source: From White (1967).

the valley could only have resulted from meteoric water being heated by, circulating into, and interacting with the felsic flow debris (Figure 12-1). Although this type of igneous hydrothermal system is expectably different from fluids related to magma crystallizing at depth, it does provide useful data. Since it is assumed that most of the fumarolic waters were originally meteoric, elements that are not common to rain water must have been derived from the cooling mass of pyroclastic debris. The exhalations of the fumaroles were more than 99% steam, at temperatures up to 650°C. Enough

(a)

(b)

Figure 2-12. Dramatic proof that large amounts of fluids can flow through permeable rocks is offered by the Seismic Geyser in Yellowstone National Park, Wyoming. (*a*) Its steaming pool during a quiet interval between eruptions. Individual beds of sinter are visible; each bed was originally continuous but was fragmented when the geyser increased in vigor. The vent flares upward, probably as a consequence of increasing volume of an erupting, expanding two-phase mixture as pressure decreases upward. (*b*) The same vent in explosive, boiling, bursting eruption. (*Photos by* D. E. White.)

HCl, HF, and H_2S were dissolved in the steam to make the exhalations acidic at the surface. Different mineral assemblages formed around vents as the temperatures dropped, the early phase being characterized by magnetite, the later phase by galena and sphalerite. The incrustations contained iron, lead, zinc, molybdenum, copper, arsenic, antimony, tin, silver, nickel, cobalt, thallium, and bismuth; these metals were combined with sulfur, oxygen, fluorine, chlorine, selenium, and tellurium. Zies suggested that each of these elements can form a volatile compound and that each may have been transported in the gaseous state; such transport has since been generally discounted.

Recent work by Hewett and his associates (Hewett et al., 1960, 1963) indicates that many hot springs are at least partly volcanic in origin. They deposit manganese oxides and minute but detectable amounts of tungsten, as well as other elements such as boron, strontium, and fluorine. Stable isotope studies by Taylor and his associates continue to require that deep circulation of meteoric waters into convective overturn systems are important hydrologic components of both thermal springs and ore deposits.

New insights into thermal spring geochemistry have been provided by the observation on deep seafloors of uprushing "black smoker" hot-water plumes (Corliss et al., 1979; Spiess et al., 1980; Solomon, 1980). These turbid, roiling plumes and their implications are discussed in context in Chapter 13. Their high temperatures of over 350°C, their high volume and velocity, the silica- and sulfide-rich chimneys that some of them have built up to heights of 7 to 10 meters, and the adjacent aprons of pyrrhotite, pyrite, sphalerite, and chalcopyrite, with or without silica and barite, are certain evidence of the flow of mineralizing fluids through rocks. Little is yet known of their geochemical details, but they appear to represent a submarine analog of the fumaroles just considered, wherein seawater, rather than meteoric water, is carried down, laterally inward toward, and finally up and away from a shallowly emplaced thermal source, here a midoceanic rise. Close study of these active submarine thermal springs will doubtlessly contribute to studies of metal leaching, transport mechanisms, fluid competencies and physical states, and other facets of hydrothermal activity.

▸ MINE WATERS

The only comprehensive geologic studies of mine waters published in recent years have been those done on obviously meteoric waters, and little information has been obtained by them which concerns the nature of the ore-bearing fluids (Sato and Mooney, 1960). Although pore moisture remains, measurable groundwater flow in near-surface rocks gradually decreases at depth—many deep mines are dry in their lower levels. Most present mine waters have no relation to the fluids that deposited the ore, although exceptions to this generalization may be found in areas of recent volcanism.

For example, hot water in the lower levels of the Comstock Lode, Nevada, may have been in part of volcanic origin.

But we have perceived recently that heat flows are such that near-surface thermal anomalies can last only hundreds of thousands of years, certainly not tens of millions, and that the temptation of asserting that hot waters in modern mines owe their heat to mineralization phenomena can be justified only where that mineralization is demonstrably geologically young, say less than 0.5 million years. Most hot waters encountered in mines result from exothermic oxidation reactions or younger, unrelated volcanism–hydrothermalism. The study of mine waters is not pursued assiduously, except in evaluating such oxidation phenomena.

Bibliography

American Geological Institute, R. L. Bates and J. Jackson, eds., 1980. *Dictionary of Geological Terms*. Garden City, NY: Anchor Press, 472 pp.

Barnard, W. M., and P. A. Christopher, 1966. Hydrothermal synthesis of chalcopyrite. *Econ. Geol.* 61:897–902.

Barnes, H. L., 1967. *Geochemistry of Hydrothermal Ore Deposits*. New York: Holt, Rinehart and Winston, 670 pp.

——, 1979. *Geochemistry of Hydrothermal Ore Deposits*, 2nd ed. New York: Wiley, 798 pp.

——, and G. K. Czamanske, 1967. Solubilities and transport of ore minerals, pp. 335–381 in H. L. Barnes, Ed., *Geochemistry of Hydrothermal Ore Deposits*. New York: Holt, Rinehart and Winston, 670 pp.

Barton, P. B., Jr., 1959. The chemical environment of ore deposition and the problem of low-temperature ore transport, pp. 279–300 in P. H. Abelson, Ed., *Researches in Geochemistry*. New York: Wiley, 511 pp.

Burnham, C. W., 1979. Magmas and hydrothermal fluids, pp. 71–136 in H. L. Barnes, Ed., *Geochemistry of Hydrothermal Ore Deposits*, 2nd ed. New York: Wiley, 670 pp.

——, and H. Ohmoto, 1980. Late stage processes of felsic magmatism, pp. 1–11 in S. Ishihara and S. Takenouchi, Eds., *Granite Magmatism and Related Mineralization*. Soc. Min. Geol. Jap., Spec. Issue 8, 247 pp.

Carmichael, I. S. E., F. J. Turner, and J. Verhoogen, 1974. *Igneous Petrology*. New York: McGraw-Hill, 739 pp.

Chappell, B. W., and A. J. R. White, 1974. Two contrasting granite types. *Pacif. Geol.* 8:173–174.

Clarke, F. W., 1924. The data of geochemistry. USGS Bull. 779, pp. 63–121.

Corliss, J. B., et al., 1979. Submarine thermal springs on the Galapagos rift. *Science* 203:1073–1083.

Czamanske, G. K., 1959. Sulfide solubility in aqueous solutions. *Econ. Geol.* 54:57–63.

Daly, R. A., 1933. *Igneous Rocks and the Depths of the Earth*. New York: McGraw-Hill.

DeVore, G. W., 1955. The role of adsorption in the fractionation and distribution of elements. *J. Geol.* 63:159–190.

Ellis, A. J., 1970. Present-day hydrothermal systems and mineral deposition, in *Mining and Petroleum Geology, Proc. 9th Commonw. Min. Metall. Congr.*, vol. 2. Melbourne: *Austral. Inst. Min. Met.*

Emery, J. A., 1968. Geology of the Pea Ridge iron ore body, pp. 359–369 in J. D. Ridge, Ed., *Ore Deposits of the United States 1933/1967*, Graton-Sales vols. New York: AIME, 1880 pp.

Emmons, W. H., 1936. Hypogene zoning in metalliferous lodes, 16th Int. Geol. Congr. Rep., pt. 1, pp. 417–432.

Eskola, P., 1932. On the origin of granitic magmas. *Min. Petrogr. Mitt.* 42:478.

Goranson, R. W., 1931. The solubility of water in granite magmas. *Amer. J. Sci.* 222:481–502.

Gross, S. P., 1968. Titaniferous ores of the Lake Sanford district, N.Y., pp. 140–154 in J. D. Ridge, Ed., *Ore Deposits of the United States 1933/1967*, Graton-Sales vols. New York: AIME, 1880 pp.

Helgeson, H. C., 1964. *Complexing and Hydrothermal Ore Deposition*. New York: Macmillan, 128 pp.

Hewett, D. F., and M. Fleischer, 1960. Deposits of the manganese oxides. *Econ. Geol.* 55:1–55.

——, M. Fleischer, and N. Conklin, 1963. Deposits of the manganese oxides. *Econ. Geol.*, Suppl. 58:1–51.

Holland, H. D., 1967. Gangue minerals in hydrothermal ore deposits, pp. 382–436 in H. L. Barnes, Ed., *Geochemistry of Hydrothermal Ore Deposits*. New York: Holt, Rinehart and Winston, 670 pp.

——, 1972. Granites, solutions, and base metal deposits. *Econ. Geol.* 67:281–301.

Krauskopf, K. B., 1967. *Introduction to Geochemistry*. New York: McGraw-Hill, 721 pp.

Lane, A. C., 1908. Mine waters and their field assay. *Geol. Soc. Amer. Bull.* 19:501–512.

Norton, D. L., and L. M. Cathles, 1973. Breccia pipes, products of exsolved vapor from magmas. *Econ. Geol.* 68:540–546.

Owen, B. B., and S. R. Brinkley, Jr., 1941. Calculation of the effect of pressure upon equilibria in pure water and in salt solutions. *Chem. Rev.* 29:461–474.

Park, C. F., Jr., 1961. A magnetite "flow" in northern Chile. *Econ. Geol.* 56:431–436.

Roedder, E., 1979. Fluid inclusions as samples of ore fluids, pp. 684–737 in H. L. Barnes, Ed., *Geochemistry of Hydrothermal Ore Deposits*, 2nd ed. New York: Wiley, 798 pp.

Rogers, D. P., 1968. The extrusive iron oxide deposits, "El Laco," Chile. *Econ. Geol.* 63:700.

Ruiz-Fuller, C., 1965. Geología y Yacimientos Metaliferos de Chile. Inst. de Investig. Geol., Santiago, Chile, pp. 245–247.

Sato, M., and H. M. Mooney, 1960. The electrochemical mechanism of sulfide self potentials. *Geophysics* 25:226–249.

Shand, S. J., 1943. *Eruptive Rocks: Their Genesis, Composition, and Classification.* New York: Wiley.

Shcherbina, V. V., 1964. Problems of the existence of the thiosulfates in hydrothermal solutions (in Russian). *Geol. Rud. Mestorozhd. Acad. Sci. U.S.S.R.* 6(3):110–112. Reviewed by E. A. Alexandrov, *Econ. Geol.* 60:645.

Smith, F. G., 1948. Transport and deposition of the non-sulphide vein minerals, III: phase relations at the pegmatitic stage. *Econ. Geol.* 43:535–546.

——, 1963. *Physical Geochemistry.* Reading, MA: Addison-Wesley, 624 pp.

Solomon, M., 1980. Hot-water plumes on the ocean floor: clues to submarine ore formation. *J. Geol. Soc. Austr.* 27:89–90.

Spiess, F. N., et al., 1980. Hot springs and geophysical experiments on the East Pacific Rise. *Science* 207:1421–1433.

Taupitz, K.-C., 1954. Über Sedimentation, Diagenese, Metamorphose, Magmatismus und die Entstehung der Erzlagerstätten. *Chemie der Erde* 17:104–164.

Taylor, H. P., Jr., 1979. Oxygen and hydrogen isotope relationships in hydrothermal mineral deposits, pp. 236–277 in H. L. Barnes, Ed., *Geochemistry of Hydrothermal Ore Deposits*, 2nd ed. New York: Wiley, 798 pp.

Tyurin, N. G., 1963. The problem of the composition of hydrothermal solutions (in Russian). *Geol. Rud. Mestorozhd. Acad. Sci. U.S.S.R.* 5(4):24–42.

Weissberg, B. G., 1969. Gold-silver ore grade precipitates from New Zealand thermal waters. *Econ. Geol.* 64:95–108.

White, D. E., 1957a. Thermal waters of volcanic origin. *Geol. Soc. Amer. Bull.* 68:1637–1657.

——, 1957b. Magmatic, connate, and metamorphic waters. *Geol. Soc. Amer. Bull.* 68:1659–1682.

——, 1967. Mercury and base-metal deposits with associated thermal and mineral waters, in H. L. Barnes, Ed., *Geochemistry of Hydrothermal Ore Deposits.* New York: Holt, Rinehart and Winston, 670 pp.

——, 1968. Environments of generation of some base metal ore deposits. *Econ. Geol.* 63:301–335.

——, J. D. Hem, and G. A. Waring, 1963. Chemical composition of subsurface waters. USGS Prof. Pap. 440-F, 67 pp.

Zies, E. G., 1929. The Valley of Ten Thousand Smokes. Nat. Geog. Soc., Contrib. Tech. Pap., Katmai ser. 1, No. 4.

THREE

Movement of the Ore-Bearing Fluids

Any level of understanding of the processes that form ore deposits must include an understanding of how various fluids actually move through rocks. Ore-forming fluids do indeed migrate at all levels of the Earth's crust, and large amounts of such fluids, which can be magmas, aqueous liquids, or gases, may be involved. The movement of fluids underground is as significant in ore genesis as it is in the concentration of oil and gas or in the emplacement of dikes and other intrusive igneous masses. Knowledge of the paths traveled by the ore solution and the mode of ore emplacement is fundamental to the understanding of the genesis of an ore deposit.

It should be appreciated that all subsurface fluid movement occurs according to physical or physicochemical laws, and reasonable explanations of how a particular orebody formed must be framed in those terms. The days of near legerdemain and mere fantasy in explaining how ores are transported are past, as is recourse to "underground rivers," "subterranean lakes," and the like. In general, aqueous fluids obey the precepts of geohydrology in that their migration is governed by avenues of net permeability to the fluid that are in turn dictated by its viscosity and density, the abundance of interconnected pores, fractures or fault planes, pressure gradients, and time. Time is a variable only recently appreciated—fluids can move in rocks of exceedingly low permeability if sufficient time at gradient is allowed. And of course magmatic fluids—melts—must obey the rules of igneous petrology.

▶ MIGRATION OF MAGMA

The manner in which ordinary magmas move through rock has been the subject of much discussion in petrology. Although the ultimate reasons for their movement may be unclear, they do move, and they generally move upward in the Earth's crust toward areas of lower pressure and temperature.

Magma is in general buoyant—it is liquid, and hence is less dense than a solid phase of the same composition would be at the same pressure. Also, it normally contains dissolved gases, especially water and especially in the upper parts of magma chambers. Magma may be injected into overlying rocks or force its way between rock layers by actually breaking rocks apart; in fact it is difficult to envisage any other mechanism of emplacement for many sills and dikes. Newall and Rast (1970) edited a book entitled *Mechanisms of Igneous Intrusion.* Its conclusions, modified by more recent studies, indicate that melting of asthenospheric, upper mantle, or lithospheric material is initiated by the rise of hot liquids or gases from deeper in the mantle, by heat generated by radioactivity or introduced from below by conduction, or by the introduction of low-melting-point materials and fluxes into high-temperature environments along subduction zones. Heavier mafic-gabbroic magmas rise because of inner pressures generated by the increase in volume from solid to liquid; lighter felsic magmas also rise in response to their buoyancy with respect to their wall rocks. Other magmas are believed to move by a sort of migration of heat and highly mobile fluids whereby rocks around the upper part of the chamber are slowly melted to become part of the magma. Magmas may rise nonstop to the near-surface or surface, where they freeze intrusively or debouch and freeze extrusively, or their chambers may occupy several levels which become the sites of differentiation, with a potential for occasional remixing. The emplacement of magmas at high levels is structurally governed, and is greatly influenced by the release of volatiles.

In short, we know that magmas are guided by "major structures," by which is meant regional-scale deep-reaching faults or district-scale faults at least tens of kilometers long. The geologic maps of many mining districts show fault planes or zones extending outward from a central intrusion which has obliterated that fault, and a number—such as at Bagdad, Arizona—show an intrusion at the intersection of two major faults. Billingsley and Locke (1935), Mayo (1958), Schmitt (1966), and others have demonstrated this control. Magmas may, of course, also make their own paths by assimilation—the engulfing, melting, and homogenization of blocks of country rocks—or by means such as gas fluxion, or gas "coring"—the collapse of wall rocks into gasfilled pockets—and by tectonic squeezing, as though from a buried toothpaste tube.

Each of these modes of movement of magma is of interest to the ore-deposit geologist. As will be seen, the world-class copper-nickel deposits at

Sudbury, Ontario, appear to involve injection of a sulfide melt from an intermediate but lower level of ultramafic differentiation into a higher level, almost supracrustal lopolithic gabbroic magma chamber. And many so-called porphyry copper intrusive systems of intermediate composition can be shown to have followed regional faults.

It should be noted that although many batholith-scale intrusive bodies lie along major global-scale sutures or regional lineaments, they more generally show assimilation of country rock on a large scale. Smaller stocks are likely to be diapiric in form, like an inverted tear drop, to display demonstrable structural guidance into their final sites, and to have been emplaced by buoyancy interacting with tectonics. As noted, the role of aqueous fluids in lubricating, wedging open, "hydrofracting," and even explosively widening, reaming, or coring out planes or zones of weakness at shallow levels above an upward-advancing magma may be profound. In individual plutons, the combined effects of forceful intrusion, stoping, and assimilation are evident (Compton, 1955, 1960).

Some magmatic fluids become "ore" just by freezing to form, for example, labradorite gabbros or pink or gray granites for building, decorative, or memorial stones (Figure I-1). Others generate water-rich fluids—mineralizers—which are considered in the next paragraphs. But important modes of movement of an ore-bearing fluid are those of filter pressing, late-liquid gravitative accumulation, or magmatic injection. These concepts were broached in Chapter 2, so they need little further discussion here. *Late-liquid gravitative accumulation* refers to the settling out, the sinking of globules of a heavy liquid—usually a sulfide or oxide liquid—which form by immiscibility within and from a parent liquid, usually a silicate melt (Figures 2-1 and 2-3 to 2-6). "Late" refers to the fact that the liquid appears to separate after the melt is in place and has begun differentiating, not that it is "late" in the sense of Bowen's reaction series. *Filter pressing* may or may not actually happen; many petrologists argue that it does. The idea is that a volume of partially crystallized magma should consist of a mush of laths of plagioclase, olivine, or orthorhombic pyroxene with interstitial rest liquid that may be oxide or sulfide rich. If tectonic squeezing injects that two-phase mush into an orifice, a "log jam" of tabular crystals may develop so that the interstitial liquid—then possibly an ore magma—can pass through like wine from a wine press and then freeze to form nearly pure oxide or sulfide ores. The scale seems wrong. How a mesh of centimeter-sized crystals might "hang up" in a meter-sized sill or dike root is not clear, but the pressing of a liquid laterally from a settled mass of crystals and liquid seems quite reasonable. *Magmatic injection* thus involves the relocation by differential tectonic pressure or by buoyancy of a differentiated, settled, or filter-pressed magmatic liquid, resulting in a highly compositionally specialized igneous rock, typically of magnetite-apatite composition, as a dike or sill. Clearly, magmas cannot diffuse through rocks and they cannot travel far through narrow fissures without losing too much heat to remain mobile.

There are thus several compositional-thermal-structural limits on dikes and sills filled by rock melts or ore magmas.

▸ THE ORIGINS OF POROSITY AND PERMEABILITY

Porosity in rocks refers to the ratio of pore volume to total volume of a given rock or soil, whether those pores are interconnected or not. *Permeability* describes the capacity of a rock, sediment, or soil to transmit a fluid, and therefore is a measure of the relative ease of fluid flow through a rock across a pressure gradient. A porous rock thus may or may not be permeable, but all significantly permeable rocks must involve apertures of some sort. And of course it is ultimately permeability that is required for the movement of aqueous fluids (Figure 3-1).

The permeability of a rock or formation is the result of different contributing effects. These effects can be divided into two categories (Table 3-1): Those that are *primary*, or intrinsic to the rock, are part of what the rock is; and those that are *secondary*, or induced, are expressive of the rock's or formation's geologic history. They also can be described as subcapillary, capillary, and supercapillary, as shown in the table. A study of Table 3-1 reveals that in supercapillary openings, water can be expected to flow bodily and therefore to transport whatever solute occurs in it. But if we consider the nature of solute transport, it should be apparent that the relative importance of the diffusion of ions to their bodily transport in a

Figure 3-1. Piece of mineralized limestone from the Eagle Mountains, California. The dark spots are specks to raisin-size masses of magnetite that have plugged original pores and holes in the rock. The holes when open were a measure of the porosity of the rock. The fact that they are now filled by an introduced substance—Fe_3O_4—shows that the rock was also permeable, permitting fluid flow—or at least diffusion of iron and oxygen—from pore to pore and along microfractures. Replacement of the limestone by magnetite also occurred, as discussed later.

moving fluid may well increase as the fluid flow rate decreases. Stated differently, mass transport of a solute can be considered to be the sum of two independent vectors, the "forward" motion of the solvent and the diffusion of ions down chemical gradients within it. So as the fluid flow rate or volume diminishes with finer and finer apertures governing permeability, the relative importance of diffusion may well increase.

TABLE 3-1. Geologic Aspects of Primary and Secondary Porosity and Permeability

	Intrinsic = Primary	Induced = Secondary
Supercapillary, >1 mm Unrestricted flow, turbulent or laminar	*Porosity:* coarse-grained rocks (conglomerates) *Primary sedimentary structures:* cross bedding, sediment arches, breccias, compaction features, fossils *Primary attitudes:* dips, drapes, channel structures *Primary igneous structures:* amygdaloids, flow tops, scoria, cooling cracks, miarolytic cavities *Rock formation contacts:* bedding planes, unconformities	*Faults:* normal, reverse, thrust, wrench, braided, conjugate, etc. *Folds:* axial plane cleavage, slaty cleavage, bedding plane slip, crest-trough rupture *Breccias:* pipes, sheets, dikes; tectonic collapse, explosion, stockworks, shatter breccias *Metamorphic features:* rock slip cleavage, slaty cleavage, boudins, augen
Capillary, 0.01–1 mm Restricted flow, "viscous" fluid via drag on walls Diffusion, osmosis	*Porosity:* medium-grained rocks (sandstone, limestone) *Primary sedimentary structures:* cross bedding planes, algal structures, compaction features, channels *Primary attitudes:* dips, drapes *Primary igneous structures:* cooling cracks, deuteric fractures, vugs, miarolytic cavities *Rock formation contacts:* bedding planes, contacts	*Minifaults* (as above): flanking faults, joints, sympathetic structures, shears *Cleavages:* rock cleavage, bedding plane slip *Stockworks:* fracture nets, shatter breccias, thermal contraction fractures *Chemical effects:* porosity developed by alteration removal of mass, e.g., dolomitization
Subcapillary, <0.01 mm Highly restricted flow Diffusion >> flow Adsorbed water, overlapping force fields	*Porosity:* fine-grained rocks (shale, siltstone, aplite, hornfels) *Intragrain discontinuities:* cleavage, parting, twin planes, zone boundaries *Microfractures:* cooling cracks, deuteric fractures	*Microrupture:* expansion-contraction, thermal effects *Induced mineral cleavage* *Chemical effects:* porosity developed by alteration removal of mass, e.g., hydrolysis, dolomitization

Contributive components of porosity and permeability. "Intrinsic" aspects are created when the rock forms; "induced" components reflect what has happened to it. Not expressed is the obvious contribution to total permeability of increasing the abundance per unit volume of rock of the openings described in the table.

▸ MIGRATION OF HYDROTHERMAL FLUIDS AT DEPTH

Ordinarily, permeability and porosity both decrease with depth in the crust because the pressure of overlying rocks tends to close any openings. The lower limit of freely circulating groundwaters varies considerably and may lie anywhere from a few meters to several hundred meters below the surface, depending upon the nature of the rock. Many deep mines extend below this limit and are dry and dusty in their lower levels; in fact, in some it is necessary to pipe water down for drilling. Because of the lack of obvious permeability, some geologists doubt the ability of large quantities of ascending waters or aqueous ore-bearing fluids to penetrate dense, compact rocks at depth for significant distances. Nevertheless, the preponderance of evidence indicates that ore-bearing solutions in large amounts do move through apparently "tight" rocks at depth. The answer to the conundrum is time. As calculations have shown, almost no rock is really impermeable if flow gradients persist for hundreds of thousands of years. A breccia column with a permeability of 10 millidarcies (defined later) is as permeable as coffee grounds in a percolator. A fracture zone with a permeability of 5 millidarcies can transmit liters per year, still a geologically high flow rate. Permeabilities of 10^{-1} to 10^{-3} millidarcies are normal for moderately deep sedimentary rocks in which fluid flow is marginal at best. Fresh igneous rocks range from approximately 10^{-3} to 10^{-5} millidarcies. Some representative values are given in Tables 3-2 and 3-3 and in Figure 3-2. Not until permeabilities of 10^{-14} millidarcies are recorded—at, say, 10-km depths or in exceptionally "tight" lithologies—can mass transport be ruled out over hundreds of centuries. But as Norton (1982) and others have shown, permeabilities of most normal near-surface rocks—those within 2 to 3 km of the surface—can transmit fluids copiously enough to figure into ore deposit

TABLE 3-2. Some Measured or Estimated Permeabilities of Representative Rocks (in Millidarcies)

	Norton and Knapp (1977)	Davis (1969)
Gabbro		1.4×10^{-5}
Quartz diorite (fractured)	30 to 10^4	
Granite	10^{-3} to 10^{-6}	10^{-9}
Gneiss schist	5×10^{-3}	2×10^{-6}
Gneiss schist	10^{-4}	
Schist	1.5×10^3	10^{-1}
Limestone	2×10^{-4}	10^{-3}
Dolomite	10^{-1}	10^{-3}
Volcaniclastics	1 to 10^4	
Welded tuff	$<2 \times 10^{-5}$	3.3×10^{-4}
Bedded tuff	10^{-1} to 10	10^{-2} to 10^{-3}
Beach sand		10^1 to 10^2
Navajo sandstone		1 to 10^4

TABLE 3-3. Generalized Intrinsic Permeability Values

Permeability		Classification	Material
k(cm^2)	md*		
10^{-3}	10^{8}		
10^{-4}	10^{7}		Gravel
10^{-5}	10^{6}	Pervious	
10^{-6}	10^{5}		
10^{-7}	10^{4}		Sand
10^{-8}	10^{3}		
10^{-9}	10^{2}	Semipervious	
10^{-10}	10^{1}		Sandstone
10^{-11}	10^{0}		
10^{-12}	10^{-1}		
10^{-13}	10^{-2}	Impervious	Limestone
10^{-14}	10^{-3}		
10^{-15}	10^{-4}		
10^{-16}	10^{-5}		Granite

*md = millidarcies.

Source: From Turcotte and Schubert (1982).

formation. A point to be emphasized here is that deep fluid flow is more a function of primary or intrinsic permeability than of induced or secondary permeability (Hubbert, 1940; Muscat, 1937). Capillary and supercapillary openings cannot remain open at depths where rocks behave plastically and the tendency is for newly opened induced openings to be immediately squeezed shut. Figure 3-2 shows water flow through rocks of various permeabilities at different temperatures.

Diffusion is probably a major mechanism of mass transport in deep environments of restricted mechanical fluid flow. *Diffusion* is a spontaneous movement of molecular or ionic particles down concentration gradients that causes one substance to become uniformly intermingled with another. It may take place in the solid, liquid, or gas phase. Water moving through the pores of a rock is therefore not regarded as diffusion, whereas the spread or movement of copper ions through that water is. Experimental and geochemical studies offer little evidence in support of the diffusion theory for the movement of matter at kilometer scales, but it is an appealing and simple answer to problems of local ore transport and wall-rock alteration.

Diffusion of ions or molecules through a liquid phase—for example, through saturated, porous rock—is not difficult to envisage. It is an especially efficient mechanism of transport in replacement—where one mineral dissolves and is progressively supplanted by another precipitated phase—because it permits the movement of ions both toward and away from the replacement front (Figure 3-3). The ions diffuse toward regions of lesser concentration, that is, they move down their own concentration gradients in an attempt to attain chemical homogeneity. As a result, in a growing alteration halo the replacing ions—chiefly H^+—migrate toward the country

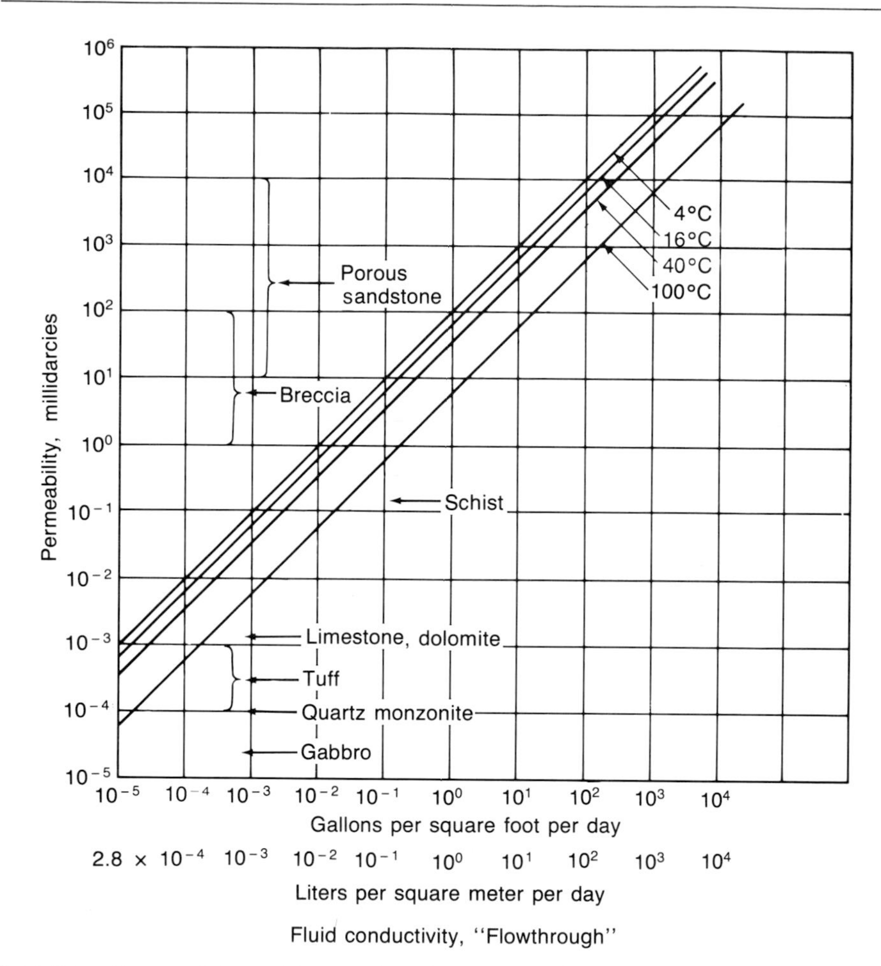

Figure 3-2. Representation of the quantity of pure water that would flow through rocks of various permeabilities at temperatures to 100°C at 1 bar pressure and pressure differential. High temperatures would increase flowthrough; higher pressure, which would permit temperatures over 100°C without boiling, and the presence of most solutes would decrease it. (*Adapted from* Davis, 1969.)

rock, while the replaced ions move away from it and into the fracture for removal from the system. The fluid need not move at all, but may simply be the medium of diffusive transport. For example, as metal ions are deposited at a replacement front, their concentration is automatically lowered, and more ions move in to take their places. Thus the replacement front grows steadily at the expense of the host until the supply of replacing ions is reduced (Holser, 1947).

Dry diffusion of matter through rock and crystalline solids is less likely to be an effective mechanism of transport because rates of diffusion through these media are considerably slower than are rates through liquids. The diffusion of ions through a solid is controlled largely by imperfections

Figure 3-3. Photomicrograph of an example of replacement of galena (PbS; white) by cerussite ($PbCO_3$; dark), a lesson in permeability and diffusion. The rectilinear grid is cubic galena cleavage. Carbonate-bearing fluids impregnated the cleavages, and carbonate ions diffused outward into the body of the galena crystal away from those cleavages. Carbonate radicals $(CO_3)^{=}$ replaced sulfide ions $(S)^{=}$, which backdiffused into the fluid and left the area. Lead ions were only locally redistributed during the anion exchange. Plane reflected light. Juab County, Utah. 80×. (*Photo by* A. S. Radtke.)

in crystal structure. Indeed, a perfect crystal, maintaining ideal order under all circumstances, does not permit ionic diffusion except at prohibitively slow rates (Barrer, 1951), or it permits only very small ions to diffuse. Imperfections may result from foreign ions existing interstitially in normal structures, thus distorting the structure, or they may result from vacancies in a lattice. Diffusion may take place from one structural imperfection to another or, in the case of vacant lattice positions, by the "migration of holes," such that an adjacent ion moves into a vacancy, leaving a hole behind. Thus solid-state diffusion may involve an advancing ion or a retreating vacancy working its way through the crystal structure. The rate of diffusion will, of course, be strongly dependent upon the radius of the particle moving through the crystal, its electrostatics, and its thermal agitation. It will be a function of temperature; near a mineral's melting point its crystal structure may become disordered and expanded. Within 100 or 200°C of a crystal's melting point, diffusion through the crystal may be greatly increased (Holser, 1947; Barrer, 1951).

Diffusion can be measured for several fluids and ions in various media. The rate of diffusion is generally proportional to the concentration gradient, that is, the change in concentration with distance, and to a diffusion coefficient that is a constant for each host material. The concentration gradient depends partly upon the solubility of the substance that diffuses, and the rate of diffusion is accordingly a function of both the fluid or ion undergoing diffusion and the medium through which diffusion takes place (Barrer, 1951). Replacement reactions maintain two-way concentration gradients, with the host material diffusing away from the replacement front and the replacing mineral components diffusing toward it. Deposition of an ore mineral will necessarily reduce the concentration of its components in solution at that point. Indeed, any other mechanism of transfer to and from a replacement front is difficult to imagine.

Field evidence in support of diffusion is found in the trace-element halos in wall rocks near veins. According to diffusion theory, the metal content of wall rocks should increase logarithmically toward the ore deposit (Hawkes and Webb, 1962). Studies of wall-rock aureoles demonstrate this logarithmic pattern in several areas. For example, Morris (1952) found a logarithmic dispersion of copper, lead, and zinc with lesser amounts outward from base-metal veins in dolomite and quartz monzonite of the Tintic district, Utah. Dispersion through dolomite is confined to 3 meters or less from ore, but trace amounts of the heavy metals have moved several times this distance through quartz monzonite. Trace-element aureoles are irregular where fractures permitted fluid flow to modify the diffusion pattern. Diminishing abundance of pyrite outward from veinlets as at Cerro de Pasco, Peru (Figure 3-4), are almost certainly controlled by outward diffusion of iron and sulfur.

The process of solid-state diffusion is well illustrated by the phenomenon of exsolution, or unmixing, in ore minerals. For example, the photomicrograph of Figure 3-5 shows tiny blebs of chalcopyrite scattered through a specimen of sphalerite. How did they get there? It turns out that the tetragonal chalcopyrite structure $CuFeS_2$ is similar to the cubic ZnS structure, which could also be written $ZnZnS_2$. At high temperature, where the crystallochemical properties of copper, iron, and zinc merge, considerable copper and iron proxy for zinc in what will become sphalerite at low temperature. Upon cooling after precipitation, copper, iron, and zinc become more chemically specific as their thermal vibration energies diminish; $CuFeS_2$ regions in the $ZnZnS_2$ matrix form by local diffusion of copper, iron, and zinc. Myriad small, essentially equidistant blebs form, as shown in the figure, with crystallographic planes in the host mineral favored. If the specimen is reheated to a temperature of 400°C, the chalcopyrite blebs disappear, and the sphalerite is essentially homogeneous again. In other words, at elevated temperatures bond strength in a crystal is reduced, permitting the entrance of similar materials; near 400°C chalcopyrite can diffuse, disperse, and homogenize through the expanded structure of sphalerite. Upon slow cooling, the bonds tighten again, the chalcopyrite components are rearranged in accordance with the atomic structure of the sphalerite, and

Figure 3-4. Halos of pyrite replacing wall rock along veinlets which cut a formation called the *Rumilliana Tuff*. The decrease of pyrite outward from the actual fissures reflects progressively greater distances of diffusional transport. Iron and sulfur diffused outward, and silica, Na_2O, K_2O, and others diffused toward the veinlet at the same time. From Cerro de Pasco, Peru. Compare with Figure 3-3 at a much finer scale. The bar scale is in centimeters. (*Photo by* W. K. Bilodeau.)

chalcopyrite will form mineral grains along the boundaries of sphalerite where it is exsolved in great excess (Edwards, 1954).

In discussing the efficacy of diffusion in ore genesis, Edwards (1952) emphasized the small scale of migration in exsolution. Even though ore minerals provide the most favorable structures for solid diffusion—they are far more favorable than complexly multibonded silicate structures—and even though sulfides form under conditions likely to promote maximal diffusion—moderate to high temperatures in the presence of an aqueous phase—the net linear movement of a given ion during exsolution rarely exceeds a few millimeters and is generally measurable in micrometers only. Volume is normally conserved; a reduction in volume of the residual host mineral is compensated for by the volume of the exsolved phase, so that although there is separation, there is no net migration away from the place of deposition.

The migration of fluids through rocks and minerals is greatly facilitated if the rock is stressed. Stress is the normal condition during orogeny,

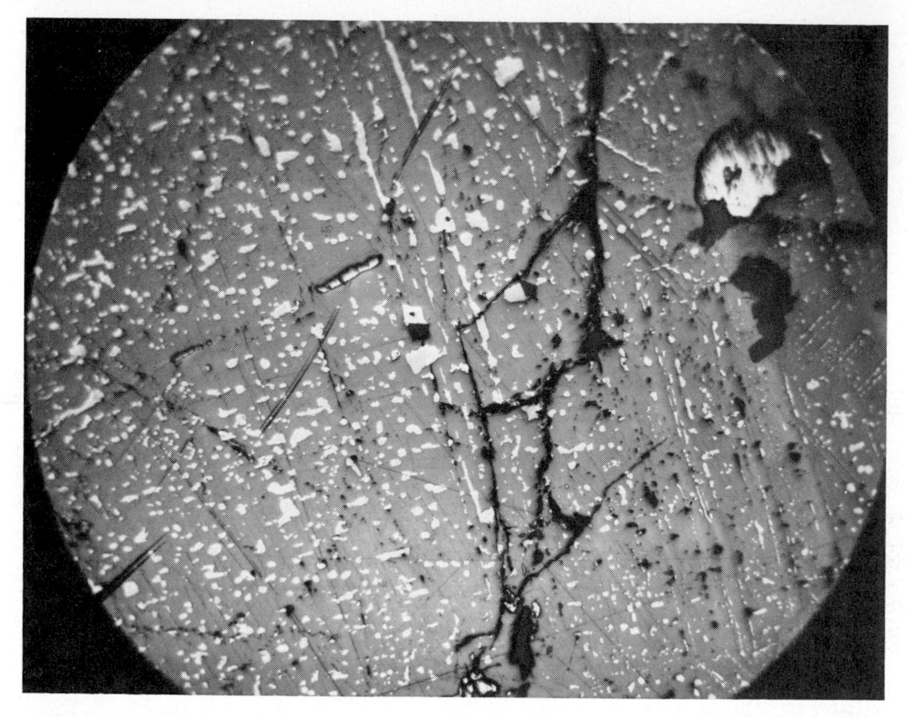

Figure 3-5. Chalcopyrite blebs (white) in sphalerite (medium grey) from Darwin, California. The exsolution texture was developed by the diffusional separation of chalcopyrite from the sphalerite lattice during slow cooling. Note that where the blebs are coarser there are fewer of them. See text. 80×.

when many ores are emplaced. Under a triaxial stress system, fluids tend to migrate along the planes of tension, that is, normal to the axis of minimum principal stress. Fractures, cleavage planes, and crystal boundaries also provide channelways. Edwards (1954) showed that if quicksilver is spread on unstrained brass, it forms only a thin coating, but when the brass is bent, the mercury works into it along microfractures and the brass eventually breaks. So it is with rocks; the more flexure and rupture, the greater the permeability. Although the importance of solid-state diffusion to ore transport remains doubtful, its contribution to local redistribution of ore-mineral components must be acknowledged as a reasonable possibility under favorable circumstances.

More importantly, long-distance migration of ore-bearing fluids is undoubtedly controlled by the relatively open channelways provided by fault and joint systems. But locally, especially near ore deposits, where free circulation may have been impeded, diffusion of metal ions through liquid and solid media certainly contributes to the movement of materials and to the final configuration of ore-deposit components.

▸ MIGRATION OF HYDROTHERMAL FLUIDS AT SHALLOW DEPTH

Induced permeability can be perceived either as "ground preparation"—something that happens to a rock or rock formation causing it to break or otherwise develop openings into which fluids can enter—or as part of the ore-forming process itself. There is increasing evidence (Knapp and Knight, 1977; Knapp and Norton, 1981) that aqueous fluids given off by a shallowly intruded stock can both wedge open and "lubricate" the advancing propagation of fractures around it, and that a stock can create its own aureole of permeability rather than being dependent only upon externally applied shear couples to produce permeability.

That fluids under pressure are able to fracture and work their way through rocks has long been an attractive idea. This method of introduction might be likened to the intrusion of dikes, though it is perhaps more complex because ore fluids are less viscous and more mobile than magmas. Field geologists have suggested that hydraulic pressures in ore-bearing fluids can keep fissures open, allowing the fluids to circulate and permitting time for reaction and deposition (Graton, 1906; Spurr, 1906, 1923; Wandke, 1930). Recently mathematical analyses and model studies have been applied to the problem, with the conclusion that fluid pressures underground can reach significant magnitudes. These studies support the contention that aqueous fluids are able to propagate fractures in rocks and pass through them to areas of lower pressure (Hubbert and Willis, 1957; Hubbert and Rubey, 1959; Norton, 1982). Water also serves as a lubricant along the fractures.

Both the apparent impermeability of many rocks, such as fresh granitic plutons and fine-grained shaly sediments, and results of experimental studies indicate that superimposed permeability due to faults and other secondary structures is more significant to ore transportation and deposition than is intrinsic permeability of the rock (Rove, 1947). For example, no one who has had an opportunity to examine a large thermal spring area, such as the ones at Yellowstone National Park, Wyoming, or at Big Geysers near Healdsburg and Geyserville, California (Figure 3-6), could fail to be impressed by the movement of large amounts of hot fluids through what might seem to be relatively impermeable rocks. At Big Geysers, massive dense graywackes and argillites are thoroughly altered over an area of more than 12 km^2. The altered rock is warm at the surface or within inches of the surface, and even shallow drill holes release superheated steam under very high pressures. In fact, steam in the area is now being used to generate electric power. The thermal zone at Big Geysers follows a large fault. Although permeability at depth along this fault may localize paths of steam migration, the hot fluids in the exposed area have spread for considerable distances away from the fault and into the rocks. The conclusion is ines-

Figure 3-6. Big Geysers area near Healdsburg on the coast of California, 100 km northwest of San Francisco. The country rock was originally Franciscan graywacke and argillite; it has been thoroughly altered along the fault planes that guide fluid flow to carbonates, chalcedony, and clay minerals. The field is eloquent proof that liquids and gases can and do flow through rocks. (*Courtesy of* D. A. McMillan, Jr., Thermal Power Company.)

capable that ore-bearing fluids are able to move through even unfractured rocks by working around individual grain boundaries or by other means.

The relative impermeability of massive carbonate rocks in short time frames has been demonstrated effectively in the Balmat-Edwards district of New York State. In this mine area, a passage way was driven below and within 2 meters of an abandoned flooded shaft. In spite of the pressure exerted by a head of nearly 30 meters in the shaft, the drift remained dry. Similarly, drill holes with 200 meters of hydrostatic head (20 kg/cm^2, or about 17 atm) were plugged where they intersected mine openings, and no water passed through the limestone around the plugs (Brown, 1948). From this evidence Brown concluded that limestones are essentially impermeable to cold watery solutions, even where these solutions are under considerable pressure. Earlier he had suggested that ore now found in limestones had migrated through hot rocks in a gaseous state (Brown, 1941, 1947). He attributed only a minor role to near-surface water, emphasizing that deep mines are dry in their lower levels, that rocks at depth are highly impervious, and that watery fluids would have great difficulty in traversing such material. We shall see in later chapters that although the intrinsic permeability of limestone may indeed be low, even bordering on 10^{-4} md (10^{-15} k(cm^2)) millidarcies, it is brittle at low P-T and readily receives induced permeability. Krauskopf (1957), Taylor (1967, 1979), Norton (1982), and others have shown that fluids at supercritical temperatures and pressures can transport metallic ions and operate through minute openings. Under these conditions the physical states of liquids and gases are essentially the same. Klinkenberg (1941) determined from experiments that flow rates of gases are slightly higher than those of liquids. The flow rate of a gas through rock is not inversely proportional to viscosity, as is the case with a liquid. Klinkenberg's permeability determinations, in which he used gas, are based upon the premise of "slip": where gas is flowing along a solid wall, the layer of gas next to the surface is in motion with respect to that surface, that is, it "slips," or drags only slightly. The volume of gas flowing through a medium is greater than if no slip occurred, that is, if drag were significant. The less dense the gas, the greater the slippage; conversely, the more confining pressure is exerted upon a gas, or the denser the gas, the more nearly it approaches the behavior of a liquid. In the deeper parts of the Earth's crust, gases probably behave essentially the same as liquids.

Ohle (1951) was able to separate the permeability values of the rocks in the east Tennessee zinc district into three groups: dolomite at "low," limestone at "moderate," and "recrystalline" or dolomitic marble, an alteration product of the limestone near mineralized bodies, at "high." Before mineralization, some limestone beds were recrystallized and dolomitized, resulting in a significant increase in permeability. The original dolomite and the recrystalline are virtually the same in composition, but the replacement orebodies are concentrated in the recrystalline. Ohle suggests that the localization of ore was controlled by permeability, and that the ability of the limestone to recrystallize to the more permeable recrystalline made it more favorable for the replacement than the original dolomite. Therefore the replacement orebodies were practically restricted to zones of recrystalline within the limestone. Since his calculations from tests showed that under favorable geologic conditions large volumes of aqueous solutions could pass through the dolomitized marble, Ohle concluded that the amount of dilute solution that could permeate these rocks was sufficient to account for the ore deposits found in them.

As Ohle points out, it is customary to emphasize the relative inefficiency of intergranular flow, compared to flow through open channelways. Consequently most geologists underestimate the volume of fluid that can pass through apparently solid rock, especially sediments. Once the permeability of a rock type has been measured, it is possible to calculate with reasonable accuracy the quantity of a given fluid that will move through that rock under given conditions. Ohle's conclusion that the quantity may be adequate to produce large orebodies seems to contradict the conclusions of those who emphasize the impermeable nature of carbonate rocks, but again time and individual circumstances are probably exceedingly important.

Dolomitized limestones are generally more permeable than undolomitized ones. At the Eagle mine near Gilman, Colorado, Wehrenberg and

Silverman (1965) determined that dolomitization in the immediate area of the mine increased the permeability of the Leadville Formation by a factor of between 10^3 and 10^6. Permeability becomes progressively higher as the ore zones are approached. The shape and extent of the orebodies at Gilman, especially the layerlike "manto" deposits, were influenced not only by dolomitization but also by solution channeling.

The behavior of fluids near the surface of the Earth, within the range of drill holes and mine openings, has been studied carefully by hydrologists, petroleum geologists, and recently by geothermal energy groups. Nevertheless our knowledge of the movement of fluids at shallow depths is incomplete (Scheidegger, 1960; Davis, 1969; Norton and Knapp, 1977). The factors to be considered in the near-surface movement of fluids include the character of the fluid, especially its viscosity and density; the nature of the medium being traversed, especially its porosity and permeability; the hydraulic load, or liquid pressure; and the temperature and temperature gradients across the system. The study can be extremely complex, depending upon the amount of material in solution, the presence or absence of gases, the character and heterogeneity of the rocks invaded, the nature of geologic structures, and the temperature and pressure of the fluids. Many problems encountered in the study of the migration of near-surface fluids are similar to those dealt with in the study of fluids at depth, such as the relative roles of gases and liquids and the importance of diffusion.

The simplest and most fundamental law describing the movement of fluids underground is Darcy's law; another is Richards' equation. Darcy's law relates the amount of fluid passing through a porous permeable medium to the velocity of fluid movement, the permeability of the medium, and the hydraulic gradient of the system (Darcy, 1856). Although many modifications covering special conditions have since been stated, Darcy's original statement has proven to be a reasonably correct approximation. The law may be expressed in physical terms as

$$q_s = k \frac{\rho}{\mu} g_s - \frac{1}{\rho} \frac{\delta P}{\delta s}$$

where

q_s = volume of flow in direction s, in a unit of time
k = net permeability, dependent upon pore geometry and induced features
ρ, μ = fluid density and viscosity
g_s = component of gravity in direction s
$\frac{\delta P}{\delta s}$ = rate of pressure increase in direction s

Richards (1931) introduced an equation that includes capillary forces in partially saturated porous media and which has come to bear his name.

It too was a forerunner to modern analyses of fluid flow, analyses that include more and more geophysical and geochemical terms as computer techniques develop. These formulas, or modifications of them, are used extensively in studies of gas, oil, and water in determining their flow through homogeneous permeable media, such as oil field producing horizons. Increased temperature normally decreases viscosity and lower salinity decreases density, both changes being in the direction of higher flow rates. These variables and others can be incorporated into computer programs that describe normally convective flow in explorationally and scientifically significant systems. For example, Darcy's and Richards' laws are helping to develop an understanding of how fluids percolating in the seafloor below and around midoceanic rises and around hot spots and plutons in continental settings contribute to ore formation.

▸ GROUND PREPARATION

In deposits in which ores formed later than the host rock, favorable premetallization changes can commonly be recognized. Such changes may make the country rock more receptive or more reactive to the ore-bearing solutions. Accordingly, the process is known as *ground preparation.* Many ore deposits formed at the same time as the host rocks are part of the containing rock and therefore do not require ground preparation.

Ground preparation may take place in several ways. Any process that increases permeability, causes a favorable chemical change, or induces brittleness in the rocks may localize precipitation from ore-bearing fluids. The type of ground preparation depends upon the nature of the country rock and the type of preparing agent—heat, fluids, tectonics, or a combination of the three. Silicification, dolomitization, and recrystallization are common examples of ground preparation. Even the removal of a constituent to produce porosity, as magnesium in dedolomitization, may be a form of ground preparation. Ground preparation at some places is clearly an early stage of mineralization. Epigenetic deposits are commonly, though not necessarily, emplaced during a tectonic event; many are late-phase products of associated igneous activity. If so, ground preparation may be caused by early fluids from the same source as later ones. For example, contact metamorphism and ore emplacement may occur sequentially in a limestone section around a cooling pluton. The metamorphism tends to make rocks brittle; subsequent movement may shatter the aureole and produce a clean, permeable breccia that can serve as a porous and permeable receptacle for ore deposition.

Much ground preparation is chemical. Perhaps the most common reaction is the addition and rearrangment of silica in the form of either SiO_2 or silicates. One of the common forms of silica is jasperoid, a cryptocrystalline commonly gray ferruginous silica that is transported by hydrothermal fluids and replaces country rock (Lovering, 1962). Jasperoid abounds at Ely, Nevada, and Tintic and Ophir, Utah; at Metaline Falls, Washington; and in many other mining districts. It forms conspicuous and resistant masses, knobs, and hills along shear zones or faults. It is common in limestone and dolomite, and in some districts the ore is limited to the jasperoid rock, particularly where it has been fractured and shattered (Figure 3-7). Not all jasperoid hosts have been brecciated; the silicified rock may just have been chemically favorable for ore deposition. Gunning (1948) noted that the ore minerals of the Privateer mine, Vancouver Island, Canada, filled fissures in limey and argillaceous country rocks favored by preore silicification. The hardness of the silicified fissure walls permitted clean refaulting. These rocks were relatively inert to the ore-forming fluids. They were not softened by hydrothermal action and therefore allowed essentially continuous fissuring. Those fissures were unobstructed by rock powder, pastelike "gouge," or altered wall rock, so ore minerals grew inward from clean fissure walls. But the veins narrow strikingly where they cut quartz diorite, because the diorite was pulverized during faulting and later altered so that both the gouge and the alteration products reduced permeability.

Permeability and brittleness have been increased in some districts by simple recrystallization of the country rock. Such recrystallization is com-

Figure 3-7. Veins and veinlets of jasperoid cutting limestone. The limestone and its contained jasperoid, both embrittled by the silicification, have been shattered to form the breccia at the upper left and the network of veinlets in the center. Both types of ground preparation gave access of mineralizing—altering fluids to the rock. Note the chloritic alteration halo (Chapter 5) around the central vein. Egan Range, White Pine County, Nevada. (*Photo by* T. G. Lovering.)

mon near intrusive masses. A pluton may also have "preheated" its country rock, permitting the passage of hydrothermal solutions and ore components that might have precipitated in cooler rocks. But the greatest single component of ground preparation is in the realm of structural geology. It may involve overall expansion of a volume of crustal rocks, but it need not. It does involve fracturing—cleaving, jointing, faulting, and fissuring—and commonly folding, brecciation, and dilation. It generally results in structural guidance of solution flow, and commonly results in structural control of the resultant orebody. Often, to understand the structural geology of a prospect or ore district is to understand its economic geology.

▸ STRUCTURAL CONTROL

Detailed studies of structure are essential in exploration, and they unquestionably have led to more discoveries of ore than any other approach. The fact is true because the movement of fluids underground is controlled by permeability, which in turn is a function both of the original character of the rock and of the superimposed structure. Permeability and structure are widely understood and used by petroleum geologists. Oil and gas are commonly channeled to the crest of a dome or to an area of overlap or faulting, causing them to be trapped or ponded against an impermeable layer. The laws that govern the migration of oil, gas, and water apply to all fluids underground, including the ore-bearing hydrothermal fluids. Faults or other permeable features, either primary or superimposed, tap sources of mineral-bearing fluids, allowing them to migrate until they cool and precipitate their mineral content or react with and replace receptive country rocks.

Serious efforts have been made to classify ore deposits according to their structural control. In general, these efforts have not produced an acceptable classification, largely because any structural or sedimentary feature that permits the passage of ore-bearing fluids may contain areas favorable for the precipitation of ore, and many districts contain ores localized by several different styles of structural control. One ore deposit may thus be localized in more than one type of structure. Nevertheless, ore concentrations in many districts follow patterns of deposition that are associated predominantly with one type of structure.

▸ PRIMARY, OR INTRINSIC, PERMEABILITY

In this section we refer again to Table 3-1, which summarizes much of the following detailed discussion. First we describe briefly some districts where primary permeability has guided ore localization, and then we look at some examples of induced effects. Primary structures and textures of rocks can control the distribution of fluids and hence the localization of ores. Because any textural or structural feature that influences porosity and permeability may guide fluids, the variety of primary controls is practically unlimited. A few of the most obvious are (1) clastic or autobrecciated limestone or dolomite, especially where covered by impermeable cap rocks, (2) reef structures, (3) well-sorted conglomerates that permit easy circulation of ore-bearing fluids, (4) broken and scoriaceous tops of lava flows, and (5) permeable sandstones such as channel sands and beach deposits. We will look now at an example of each.

Ohle and Brown (1954) described sedimentary arch structures of depositional origin that controlled ore trends in the southeastern Missouri lead district. The arch structures—called *calcarenite ridges* and described in Chapter 20—apparently were calcite sandbars built by currents that formed ridges of limey sediments parallel to their direction of movement in upper Cambrian seas. Much of the sediment making up the ridges was broken shell material, and the resulting calcarenites were permeable rocks compared to the lime muds in adjacent troughs. Ascending aqueous solutions—probably connate waters from adjacent basins (Figure 20-2)—were directed along these permeable arches and were constrained from direct upward movement by an overlying bed of impermeable limestone. In this same district, other orebodies are ranged around buried knobs and ridges of Precambrian granite against which Cambrian and younger sediments onlap and feather out. The sediments dip away from the granite highs in all directions. Ascending solutions migrated along the basal La Motte sandstone that pinches out against the Precambrian granites. Upon reaching the updip termination of the sandstone, the ore-bearing fluids entered the overlying carbonate sediments, reacted with sulfur encountered there, and deposited their ore minerals.

Brecciation produced by slumping during deposition and compaction of the sediments was yet another permeability control of ore deposition in southeast Missouri. Differential compaction over the clastic shell ridges led to primary oversteepening of slopes and to consequent slumping into the basin regions. These submarine landslides produced huge sedimentary "slump breccia" zones that were favorable for the subsequent ingress of ore-bearing solutions. Individual orebodies in the slides range up to more than 3 km in length and contain several million tons of ore (Snyder and Odell, 1958). Superimposed fracture zones have strongly influenced the distribution of ores in southeastern Missouri, but much new ore has been discovered by mapping and following primary sedimentary features such as the clastics around buried ridges.

The Proterozoic native copper ores of the Keweenaw Peninsula in northern Michigan furnish excellent examples of hydrothermal deposits that have been channeled into their present positions through permeable conglomerates and the broken, vesicular, frothy flowtops of lava flows (Butler and Burbank, 1929). Concentrations of native copper are found in conglomerate beds between the flows and in pores in the fragmental, vesicular surface layers of individual flows. These beds were favorable for ore deposition apparently only because of their extremely high permeabilities; a reductant may also have been present.

Permeable strata formed by channel deposits of conglomerates and sands interbedded with siltstones have been mineralized in the Gas Hills uranium district of central Wyoming. The ore deposits are restricted to the coarse-grained facies and are especially concentrated along the valleys of a buried land surface. Apparently the ore-bearing solutions moved laterally along permeable zones until they were dammed by facies changes or by impermeable shales across the basal unconformity (Zeller, 1957).

Coarse permeable primary zones result from the most obvious textural-structural attributes of rocks, but as Table 3-1 indicates, the primary-intrinsic contribution to permeability extends down to subcapillary effects at the scale of intergrain boundaries and cleavage planes.

▸ SECONDARY, OR SUPERIMPOSED, PERMEABILITY

In most epigenetic ore deposits, the path of circulation followed by ore-bearing fluids has been greatly influenced by structures superimposed on the rocks. Faults and folds are probably the most common secondary structures, although breccia zones, pipes, and other features are locally of great significance. Some basic types of each are shown in Figures 3-8 and 3-9.

Faults are found nearly everywhere, and many ore deposits are directly related to them. Because fault surfaces are typically not smooth, movement along a fault can produce breccia and gouge where the two blocks grind together. This fine-grained gouge frequently plugs the fault surface and hinders the circulation of fluids, either along or across a fault. On the other hand, coarse, clean breccia containing a minimum of "fines" may increase permeability, especially in brittle rocks fractured under light loads (Lovering, 1942). Accordingly, simple minor faults may be better hosts for ore solutions than complex large-displacement faults which are more likely to develop gouge. As a general rule, then, tight gouge-filled fractures are less favorable for ore deposition than are more open fractures. As is usual in geology, there are exceptions to this generalization. For example, the Santa Rosa mine in the Huantajaya district of northern Chile is said to have contained silver ore in clay alongside what seems to be a more permeable fissure filled with quartz. But the generalization fits nicely at Butte, Montana, where the two most important sets of veins are called "East-West-age" and "Northwest-age." The "East-Westers" are tension fault veins with little gouge and abundant open-space-filling textures (Chapter 4); the "Northwesters" are gougy, "tight," and show textures that indicate that ore minerals replaced the gouge. Overall, though, the "Northwesters" are

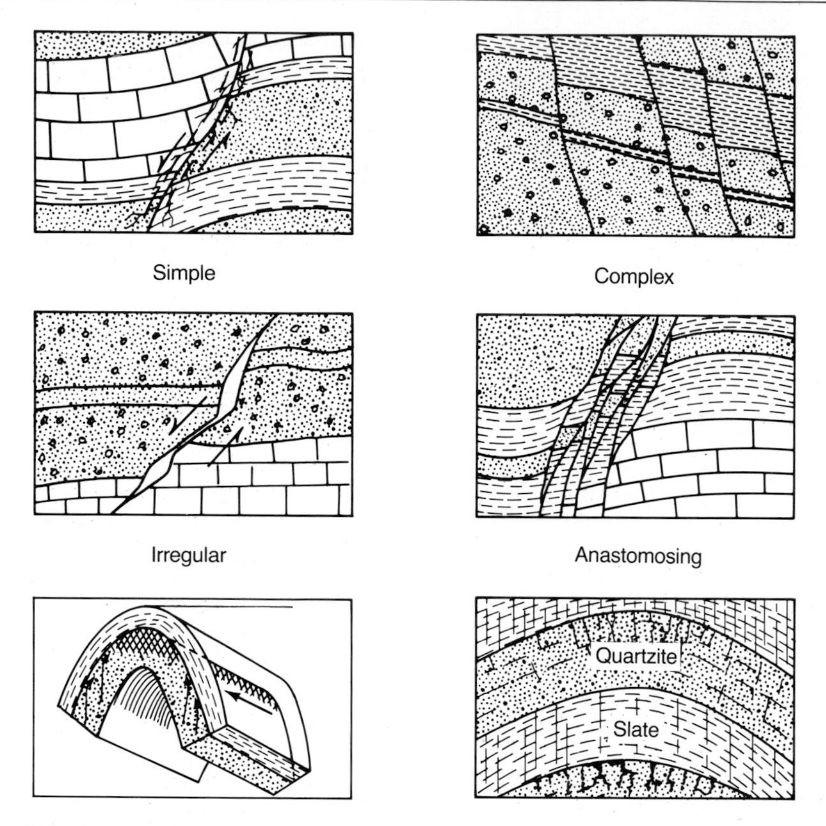

Figure 3-8. Some basic fold and fault relationships that guide the movement of ore-forming aqueous fluids. The fault diagrams show simple, complex, irregular, and anastomosing types. Minor flanking joints and faults in each case broaden the zone of obviously increased permeability. Breakage in a brittle bed contained within others that deform plastically in a fold creates a permeable zone.

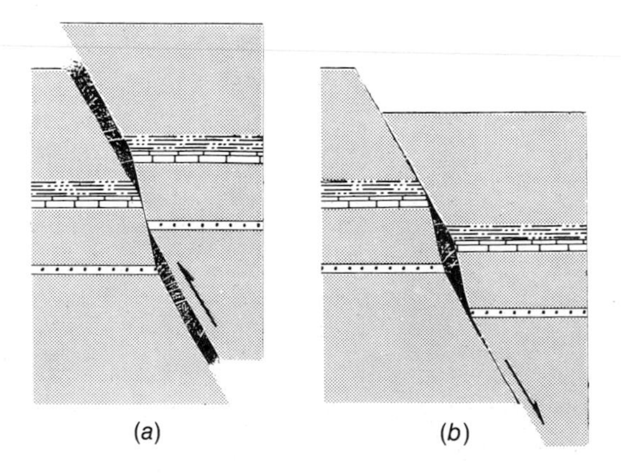

Figure 3-9. Openings (dark) caused by (*a*) reverse and (*b*) normal fault movement along a curved or irregular fault surface. When the openings are mineralized, the vein is said to pinch and swell at thinnings and thickenings, respectively.

narrower, less persistent in mineralization thickness, and lower in overall ore-mineral content, presumably because fluid flow in them was partially restricted.

Veins are tabular bodies of ore and gangue minerals that are long in two dimensions and short in the third, formed along cracks or fissure zones; fault planes are especially favorable loci. Either they result from the simple filling of open fissures or they involve replacement masses along a permeable fracture. Veins are classified as *simple* (Figure 3-10) if they represent mineralization of a single, simple fault fissure, *complex* (Figure 3-11) if they are made up of multiple laminae or layerings along the same fracture or are multiple fillings of the same reopened structure, *irregular* if they are of variable thickness, and *anastomosing* or branching (Figure 3-12) if they are braided or in mutually interlacing patterns. Vein patterns are further described by qualifying terms, such as *conjugate*, two sets of veins that have the same strike but that dip at 90° to each other. A simple vein outcrop is shown in Figure 3-13.

Where many small veinlets are distributed along a tabular zone, the deposit is called a *lead* (pronounced *leed*), a *lode*, or a *fissure zone*, although in modern usage lode refers to vein deposits in general. An important structural locus is the *stockwork*, a three-dimensional zone so laced with closely spaced irregular veinlets as to be pervasively fractured and, commonly, mineralized. An excellent stockwork is shown in the upper part of Figure 3-16, in zone II of Figure 3-18, and in Figure 13-4*d*, *e*, and *f*.

Minor movement along curved fault surfaces causes pinching and swelling, locally separating the *footwall* block from the *hanging-wall* block (see Figure 3-9). Logically, "footwall-hanging wall," "footwall side–hanging-wall side," or simply *foot* and *hanging* refer to pairs of fault blocks or just vein-separated blocks that are underfoot and overhead, respectively. The "roof" of a mine working is called the *back*. Richer portions within veins, pipes, or other ore forms are called *shoots* or *ore shoots;* they are commonly the prize, and may be mined out while subeconomic portions of the same structure are left behind. Ore-bearing fluids are channeled through the more open parts of fissures, the *swells*, and are deflected around the tighter zones, the *pinches*. As a result, *rolls* or changes in attitude of either the strike or the dip of a vein commonly mark the beginning or end of an ore shoot. Such irregularities along veins have long been recognized as highly significant in mineral exploration.

Pipes or *chimneys*, as the names imply, are normally steeply plunging subcylindrical bodies that are relatively short in two dimensions and long in the third. The two names are used interchangeably, though efforts have been made to restrict one or the other to rod-shaped or cylindrical deposits having a specifically defined plunge. For example, many geologists restrict the term *pipe* to steeply dipping rod-shaped bodies, and use the word *manto*, Spanish for "mantle" or "cloak," for flat-lying rod-shaped bodies. However, manto is more correctly used to describe flat, bedded, and sheetlike deposits

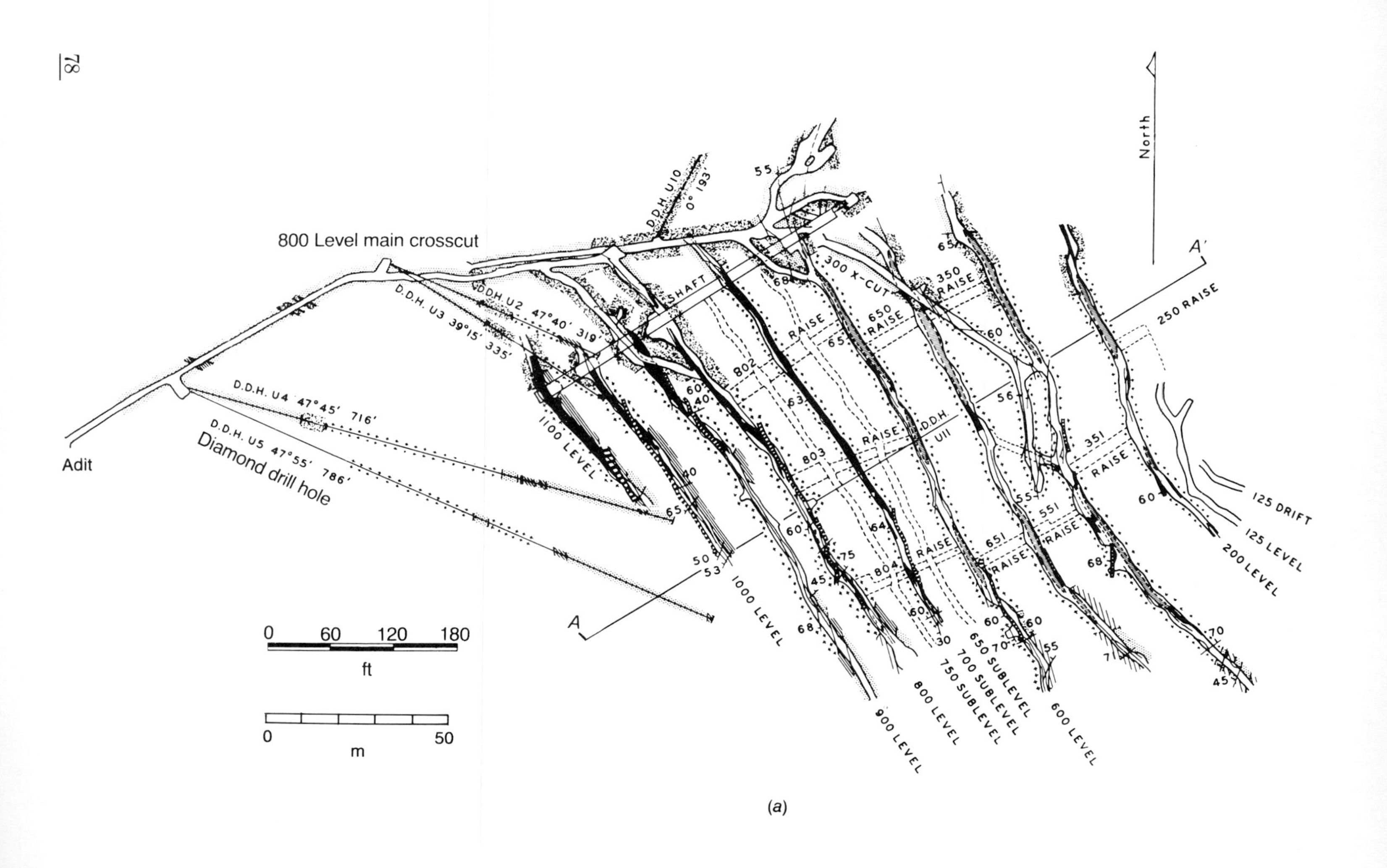
North
800 Level main crosscut
Adit
Diamond drill hole
D.D.H. U10 0° 193'
D.D.H. U2 47°40' 319'
D.D.H. U3 39°15' 335'
D.D.H. U4 47°45' 716'
D.D.H. U5 47°55' 786'
D.D.H. U11
SHAFT
300 X-CUT
350 RAISE
650 RAISE
250 RAISE
351 RAISE
551 RAISE
651 RAISE
802 RAISE
803 RAISE
804 RAISE
125 DRIFT
125 LEVEL
200 LEVEL
600 LEVEL
650 SUBLEVEL
700 SUBLEVEL
750 SUBLEVEL
800 LEVEL
900 LEVEL
1000 LEVEL
1100 LEVEL
A
A'
0
60
120
180
ft
0
50
m

(a)

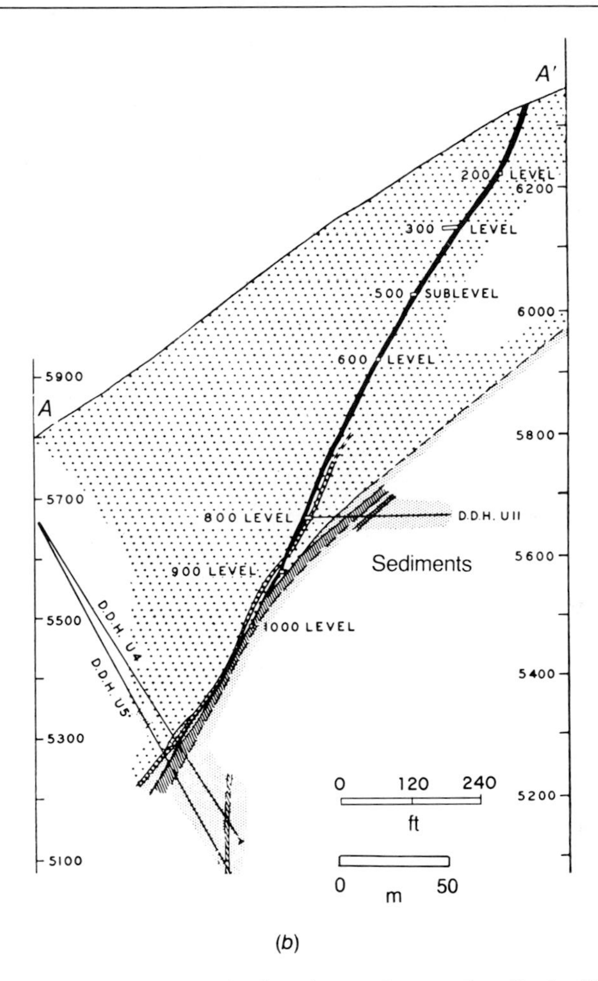

Figure 3-10. (*a*) Mine map showing a simple vein, one fissure mineralized with one quartz-wolframite-pyrite assemblage, at the Red Rose tungsten mine in Canada. The plan view is constructed as though the hanging-wall side were transparent, showing the location and form of the vein at 100-ft levels as they would appear projected vertically, say, to the 1200-ft level. The geometric-structural simplicity is best seen in the cross section (*b*). (*From* Brown, 1957.)

(Prescott, 1915). Pipes are known to change in attitude from nearly vertical to nearly horizontal, so it is probably best to use the term without regard to plunge or dip. Pipes that contain broken rock are known as *breccia pipes*.

Many large, valuable orebodies have been mined from pipes. The origin of pipes is one of the most fascinating problems in the field of ore geology, and its study one of the most rewarding. Ore pipes are found in many environments and result from various combinations of processes. Although the genesis of many pipes is unknown, that of others is readily interpreted. Pipes are commonly formed at the intersections of tabular features, such as faults, fissures, dikes, bedding, lava flows, or joints. Where

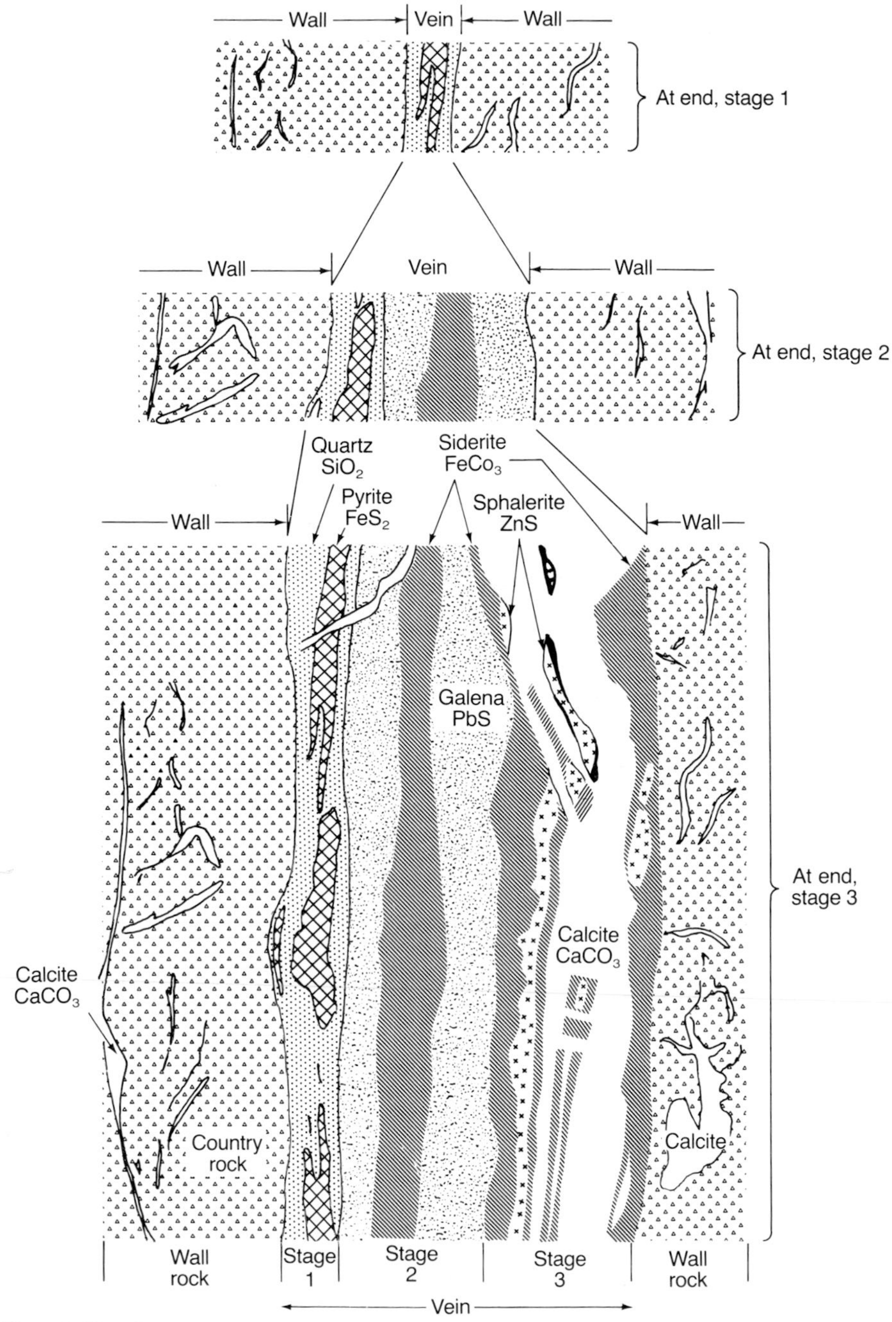

Figure 3-11. Complex vein, 20 cm wide, in the mines of Príbram, Czechoslovakia. The same fault surface shows successive symmetrical fillings of stage 1 (quartz, then centerline pyrite); refaulting and stage 2 (galena and centerline siderite); refaulting and stage 3 (siderite, then sphalerite), some rebreaking and separation, and finally late calcite. (*From* Kutina, 1955.)

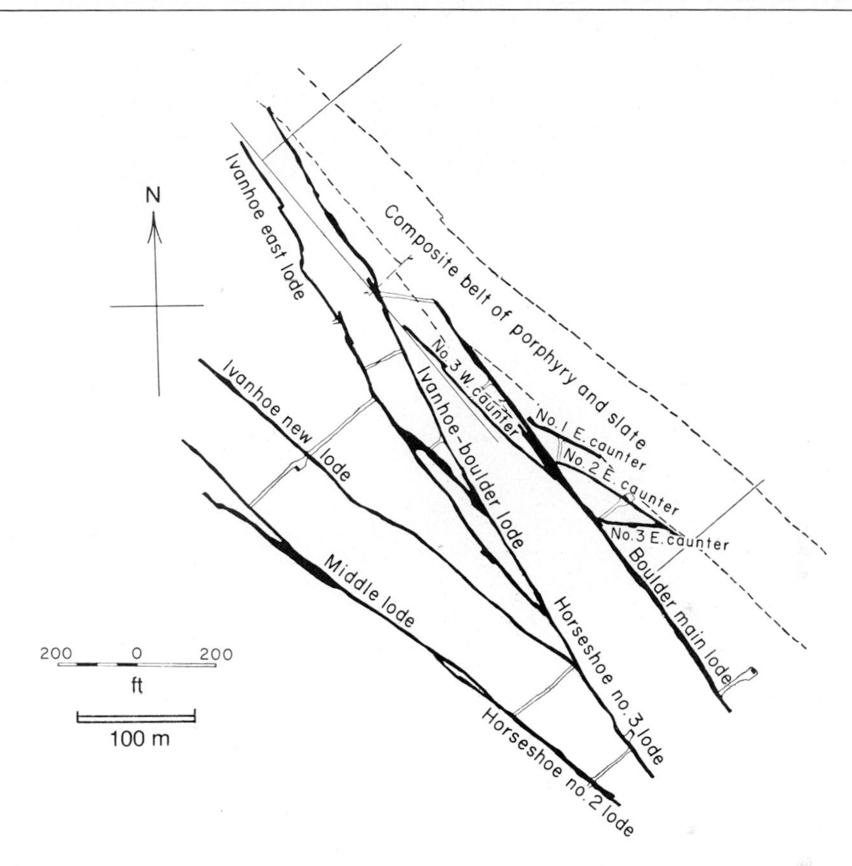

Figure 3-12. Example of a branching, or anastomosing, vein system in a small part of the Kalgoorlie gold district in western Australia. If a vein splits and then rejoins farther along strike, as in the center of the diagram, the separation is termed *cymoid* and the closure a *cymoid loop*. (*From* Finucane, 1948.)

the tabular features are faults, brecciation is likely to be most extensive if the fractures intersect at small angles (Figure 3-14). The term *tectonic breccia* refers both to blocky brecciated material along fault zones and to rocks shattered by large-scale tectonic forces, such as shattering of rocks along thrust zones or in the limbs of folds. The brecciated country rock either predates ore deposition or was formed during the mineralization processes. Pipes are also formed at the crests of folds, especially where the rocks are highly fractured and where permeability has been increased by strata sliding over each other, leaving areas of reduced pressure at the crests; against igneous contacts, either where solutions were channeled along irregularities or where differential movements caused brecciation; in the throats of old volcanoes and in diatremes; along solution channels through carbonate rocks and in "shoestring" sands; along rolls or in changes in dip or strike of a vein; and within small cupolas or other igneous bodies (Ransome, 1911; Butler, 1913; Wagner, 1927; Emmons, 1938; Kuhn, 1941; Blan-

Figure 3-13. Simple siliceous fault vein outcropping in the Swisshelm Mountains of southeast Arizona. It occupies a steep normal fault in Paleozoic limestones and is continuous for several kilometers along strike. (*Photo by* W. K. Bilodeau.)

chard, 1947; Bryner, 1961; Perry, 1961). MacDiarmid (1960) showed how breccia pipes can form along irregular fault surfaces as "pinch and swell" structures, with blocks fragmenting into the swells (Figure 3-15).

Locke (1926) suggested that "mineralization stoping" can form ore pipes. This process involves the ascent of ore-bearing fluids through the crust of the Earth by stoping. Undermined or partly dissolved blocks break loose from the roof and settle into solution caverns, forming a pipe of collapsed breccia. Ore and gangue minerals may then be deposited in the brecciated mass. Some breccia pipes in a caldera at Cripple Creek, Colorado, appear to have formed as a result of stoping by ascending fluids. And many mineralized breccia pipes—and conduits for ore-bearing fluids—are associated with calc-alkaline plutons, which in turn are associated with porphyry base-metal deposits (Chapter 11). Breccia pipes in these settings form by explosive release of volatiles which boil off the magma as it approaches the surface at high temperature but diminishing pressure; by forcible intrusion of an igneous rock, usually a porphyritic one, which entrains fragments of wall rock; by collapse of roof rocks into cavities left by temporary magma withdrawal; by mineralization-alteration stoping; and possibly by collapse of roof rocks into huge vapor bubbles produced by late-stage exsolution of water from a melt. Any one of these breccia columns

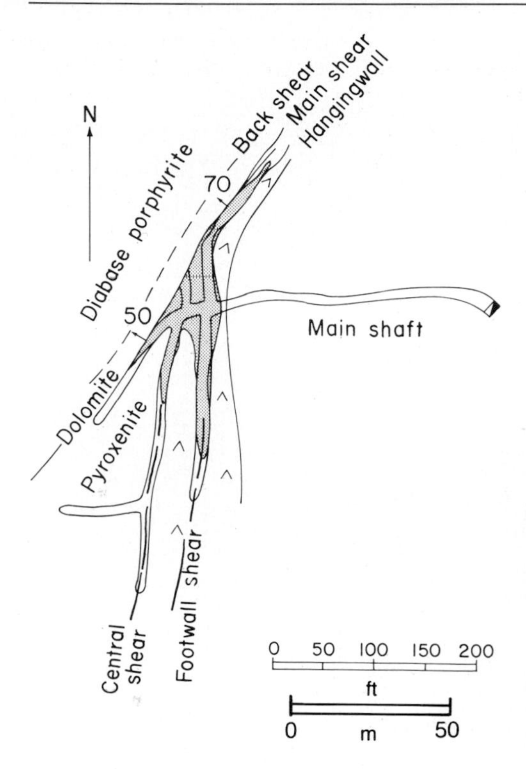

Figure 3-14. Two shear zones, one northeasterly—the Main Shear—and another north-south pair—the Central and Footwall shears—meet at a flat angle and produce a much more extensive pipelike body of breccia ore (stippled) than would exist if they met at 90°. The orebody is at the Magnet mine, Tasmania, Australia. (*From* Edwards, 1960.)

can then guide ore-forming solutions. Figure 3-16 illustrates Sillitoe and Sawkins' (1971) interpretive cross section of a typical Chilean breccia pipe, illustrating their concept of the vertical changes that might be encountered within a pipe. Figure 3-17 shows some of the resulting ore material. Johnston and Lowell (1961) indicate that highly permeable shatter breccia zones—zones in which host rocks are fragmented but the fragments are not rotated (Figure 3-18)—may be developed around breccia pipes, especially explosive ones.

Diatremes, or high-velocity volcanic explosion vents, form where gases expand at explosive rates, causing the gases and rock to rush violently upward. The process is probably similar to the two-phase water-steam action in the throats of active geysers, and it may start where ascending magmas suddenly encounter a porous water-saturated clastic sediment (Williams, 1936). Numerous well-defined diatreme breccia pipes in northeastern Arizona have been described; they are funnel-shaped, with diameters that range in thickness from 1 km near the top to 200 meters at depth (Hack, 1942). Since the diatremes form in areas of igneous activity and represent highly permeable avenues for the escape of hydrothermal fluids, they stand a reasonable chance of becoming loci of pipe-shaped ore deposits.

A fairly common feature in some mineral districts is the presence of pebble dikes, narrow dikelike bodies that contain pebbles—milled and rounded

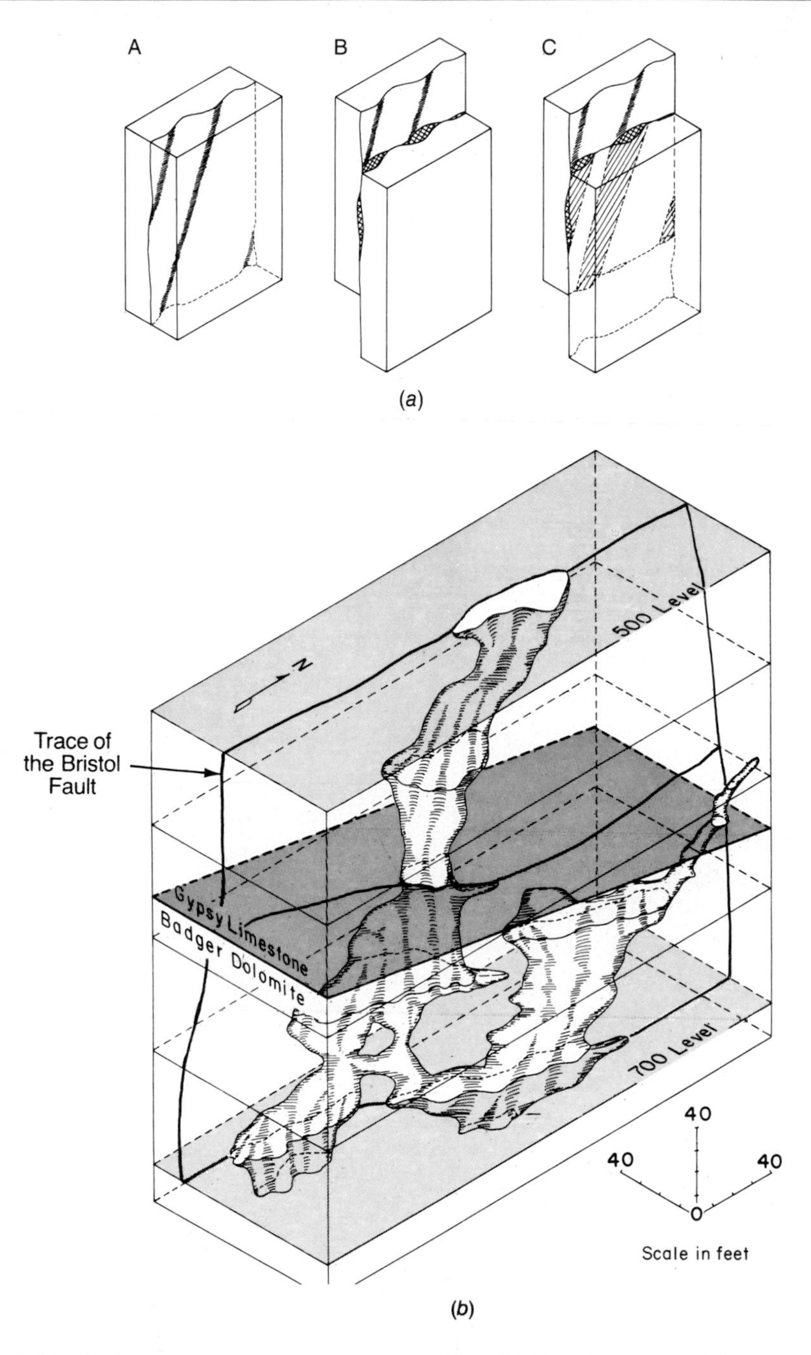

Figure 3-15. (*a*) Formation of breccia pipes by faulting. Block *A* shows an undulating, irregular fault surface before displacement. *B* shows pinches and swells (cross-hatched) after vertical displacement. *C* shows how pipes would form by collapse into those swells. Compare block *C* with the orebody form at the Bristol mine, Nevada (*b*). (*After* MacDiarmid, 1959.)

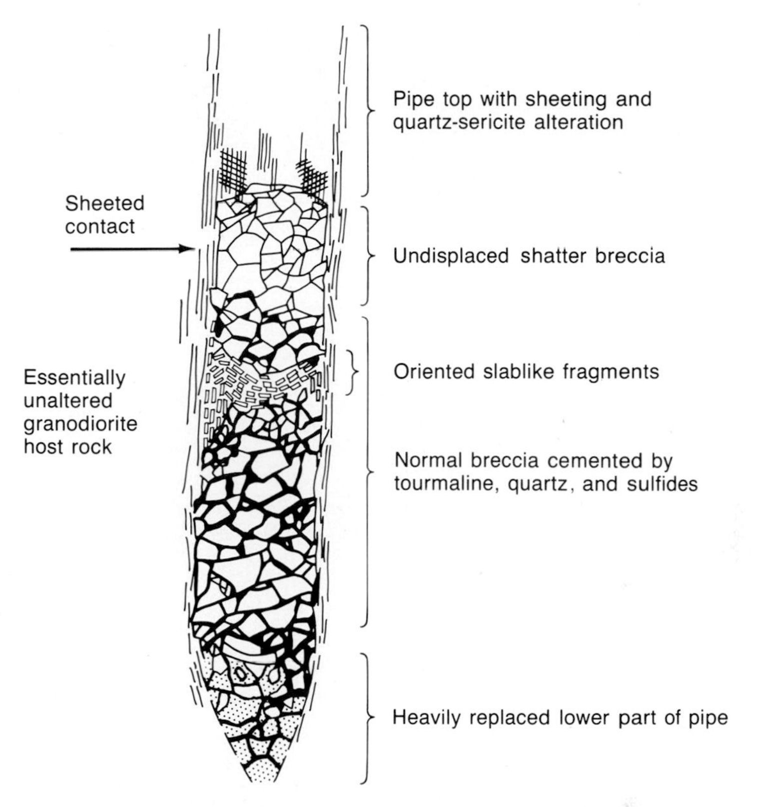

Figure 3-16. Typical breccia pipe associated with porphyry-type copper mineralization in Chile. Notice that several modifications of fragmentation can occur in one pipe. This ore formed in large part by solution and collapse. The upper surface was stoping its way toward the surface when the system shut down. High ore grades are found where sulfides fill the once void spaces between blocks. (*From* Sillitoe and Sawkins, 1971.)

fragments—brought up at the upward "leading edges" of dikes below. Many dikes of this type have been found in Tintic, Utah; Ouray, Colorado; and Silver Bell and Bisbee, Arizona (Bryant, 1974). Geijer (1971) discussed the origin of the puzzling dikeform "ball ores" of Kiruna, Sweden, and concluded that they have the same origin as pebble dikes.

When a sequence of beds is folded, the beds tend to slide over one another and to undergo compression and plastic flow on the flanks and dilation along the crests, which become areas of low pressure. Fractures develop parallel to the axial planes of the folds; they are especially prominent in competent rocks along the crests of anticlines, but they are also found along the troughs of synclines (Figures 3-8 and 3-19 to 3-21). The result of folding and the concomitant fracturing is an increase in permeability in the thickened beds at the crests of folds (Figures 3-20 and 3-21). Mineralizing fluids moving along these permeable crests fill the openings and replace favorable rocks nearby. The resultant orebodies are archlike

Figure 3-17. Slab of mineralized breccia from the Andina mine, Chile. The sericitized quartz latite porphyry fragments are infilled by tourmaline (black) and a mixture of pyrite and chalcopyrite (medium gray). It could have come from the lower middle of the pipe of Figure 3-16. Notice that early veinlets in the fragments do not continue into the breccia matrix, indicating a prolonged alteration-mineralization history. The late fluids from which the breccia filling precipitated must have contained at least Fe, Mg, Cu, Al-Si, B, S, and H_2O. The bar scale is in centimeters. (*Photo by* W. K. Bilodeau.)

or pipe-shaped and, where situated above one another, are called *saddle reefs*. Excellent examples of saddle reefs are the well-known Bendigo gold field, Victoria, Australia, and the gold measures of Nova Scotia (Figures 3-19 and 3-21).

Many iron and gold deposits in the Piedmont region of the southeastern United States are rod-shaped bodies, the long axes of which range in attitude from nearly horizontal to nearly vertical. Some of the rods are as much as 30 meters in cross section and have been mined down the dip of their long axes for more than 300 meters. Most, however, are small, especially in the gold-bearing areas. They may be no more than a few centimeters thick, although at Dahlonega, Georgia, they were so close together that the entire rock was mined at the surface. These orebodies are replace-

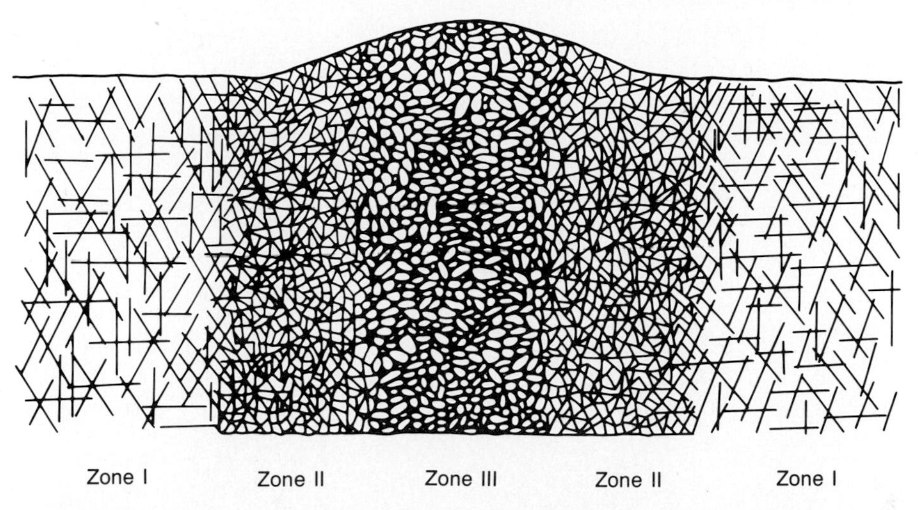

Figure 3-18. Idealized breccia pipe associated with porphyry copper mineralization at Copper Basin, near Prescott, Arizona. Zone I is jointed and fractured wall rock with low permeability and only mild alteration and mineralization. Zone II is a "crackle" or "shatter" breccia, a zone of high fracture density and high permeability, but without rotation of fragments. Zone III is true breccia, with rotated, "milled," rounded fragments. Zone II and III can be transitional or in sharp contact. Where the pipe has formed by explosive activity or by strong solution, the contact will be sharp. Cementing materials (black) are quartz and sulfides. Zones II and III combined range from 20 to 200 meters in diameter. Zone II is also an excellent representation of a stockwork. (*From* Johnston and Lowell, 1961.)

ment saddle reefs formed at or near the crests of small tight folds in schist (Figure 3-22). Much money and effort have been spent to find vertical or downdip continuations of the individual rods on the erroneous assumption that these structures are tabular.

Ore is also recovered from *ladder veins* (Figure 3-23) which, as their name implies, are veins arranged in steplike or ladderlike form. They are ordinarily confined to dikes or competent strata that lie within shales or other incompetent rocks (Grout, 1923). Movement within the shales is taken up by flowage or cumulative displacements along foliation planes, whereas competent rock tends to fracture. Ladder veins have also been found where mineralizing solutions have invaded shrinkage joints in dikes, sills, and lava flows. The igneous dike or sill cools as it contracts, and cracks develop normal to the planar direction. The Morning Star gold mine at Wood's Point, Victoria, Australia, is a classic example of a ladder vein deposit (Threadgold, 1958). Gold-quartz veins fill a conjugate set of reverse faults developed across a 100-meter-thick dike (Figure 3-23).

Carbonate rocks below shales are favorable places to find ore, especially if faulting has taken place along the contact. Where a sequence of

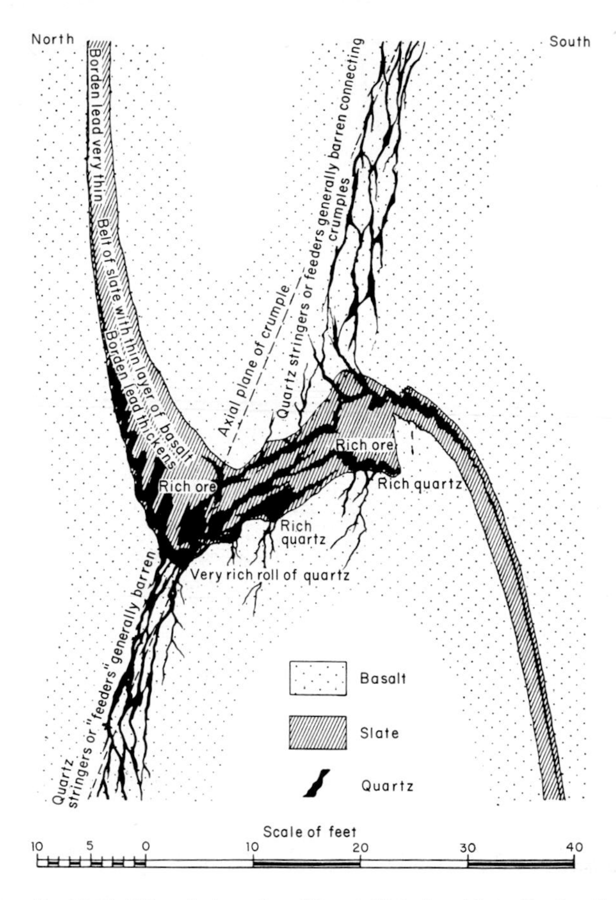

Figure 3-19. Detail of fold, West Lake mine, Mount Uniacke, Nova Scotia, Canada. Ore-bearing fluids were channeled along shear zones that broke along and in the axial plane of the fold, with rich mineralization developed in the abundant openings in the brittle slate along the anticlinal and synclinal axes. (*From* Malcolm, 1912.)

shales and carbonate rocks has been faulted or folded, bedding slip is commonly taken up in the more plastic fissile shales. The movement and drag are normally partly reflected in the limestones or dolomites, especially near the shale contact, increasing the permeability and permitting easier access of solutions.

Yet another type of permeability is accomplished by selective dissolution of limestones by groundwaters or by early hydrothermal fluids. That dissolution can result in enlarged joint and bedding planes (Figure 3-24*a*), in collapse breccias formed where large caverns are opened up (Figure 3-24*b*), and in pipelike zones, such as the remarkable high domal collapse breccias of the east Tennessee zinc district (Figures 3-25 and 3-26), an outstanding example of induced permeability.

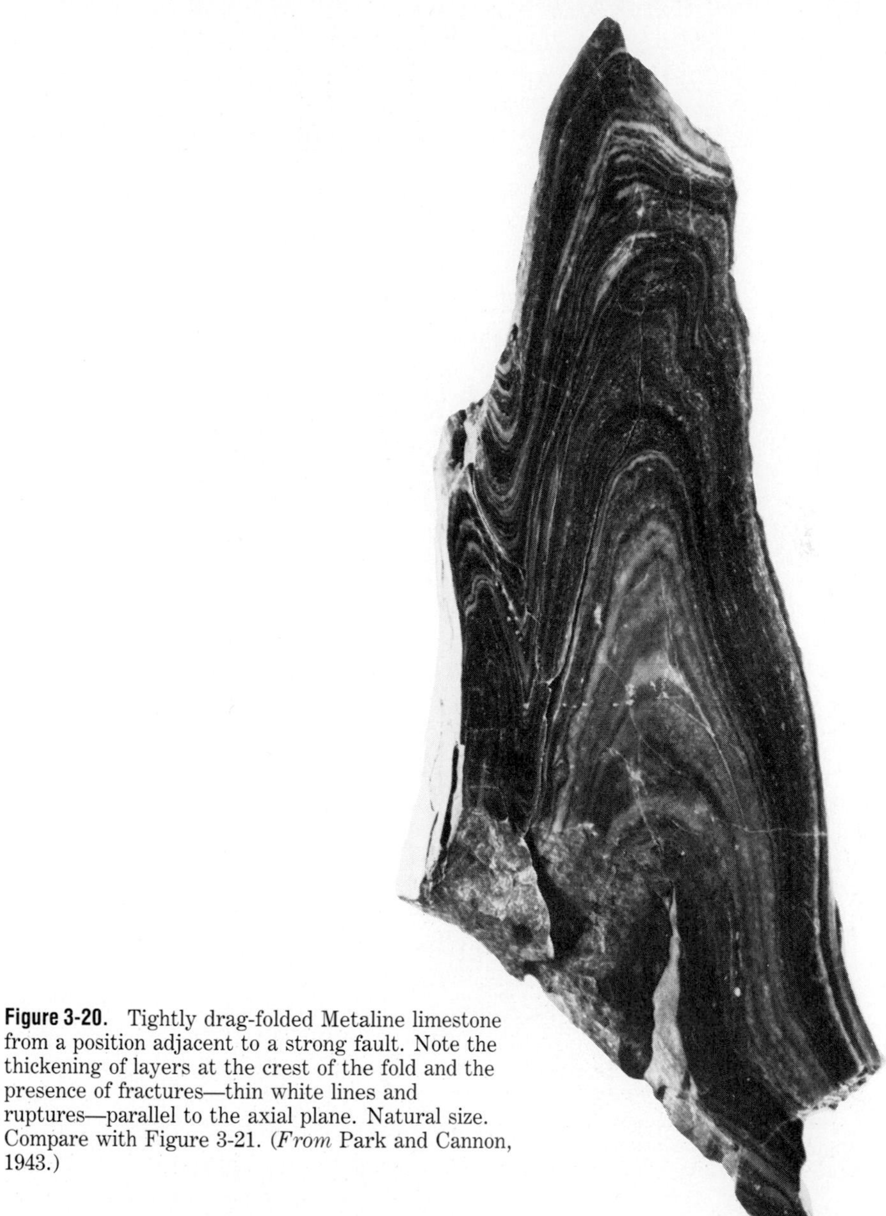

Figure 3-20. Tightly drag-folded Metaline limestone from a position adjacent to a strong fault. Note the thickening of layers at the crest of the fold and the presence of fractures—thin white lines and ruptures—parallel to the axial plane. Natural size. Compare with Figure 3-21. (*From* Park and Cannon, 1943.)

▸ HYDROTHERMAL FLOW MECHANISMS

We have considered diffusion of ore-forming components more or less as though it were unique to deep fluid movement, and structural control as though it might be reserved only to shallow circulation. In truth, of course,

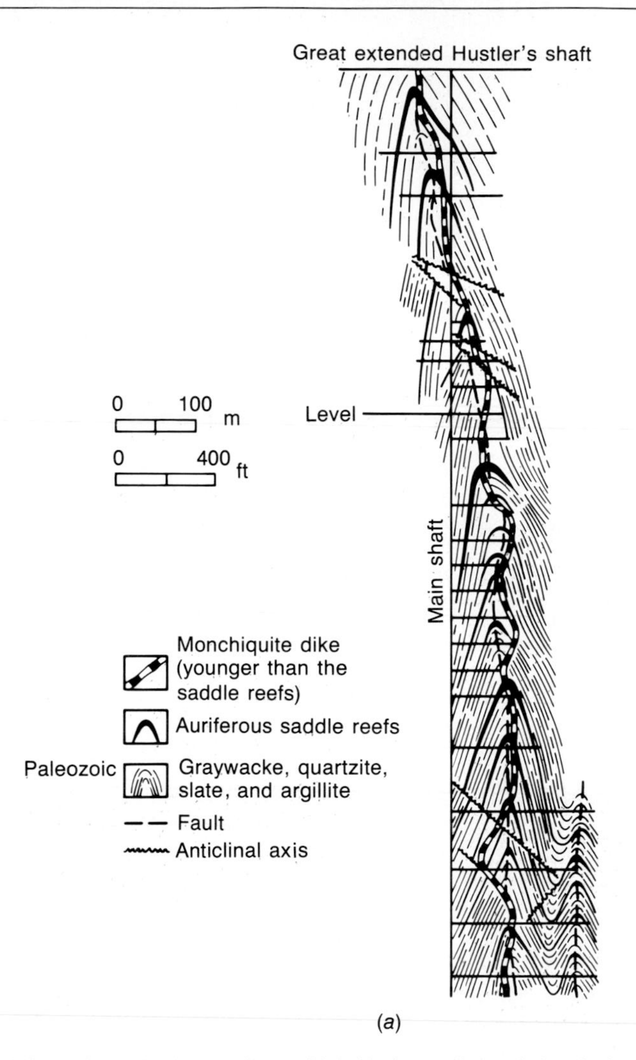

Figure 3-21. Saddle reefs (*a*) in the Bendigo gold field, Australia, and (*b*) (*facing page*) in the Dufferin mine, Salmon River gold district, Nova Scotia. Compare the 50 by 150 meter macrocosm at the Dufferin with the microcosm of Figure 3-20. The saddle reefs refer to the fact that the most pronounced dilation occurred at the crests of folds, so mineralization occurs as though in a stack of saddles. Shearing and attenuation also provided permeability along the limbs of the fold. (*From* Malcolm, 1912 (*a*) and Herman, 1914 (*b*).)

both mechanisms go on in all ore-bearing fluids, and we speak primarily of dominance, of relative importance in the two settings. As Figure 11-4 shows, in the several kilometers of vertical height at the San Manuel-Kalamazoo porphyry copper orebody in Arizona, the style of structural opening shifts from microscopic hairline fractures and disseminations of sulfides at depth

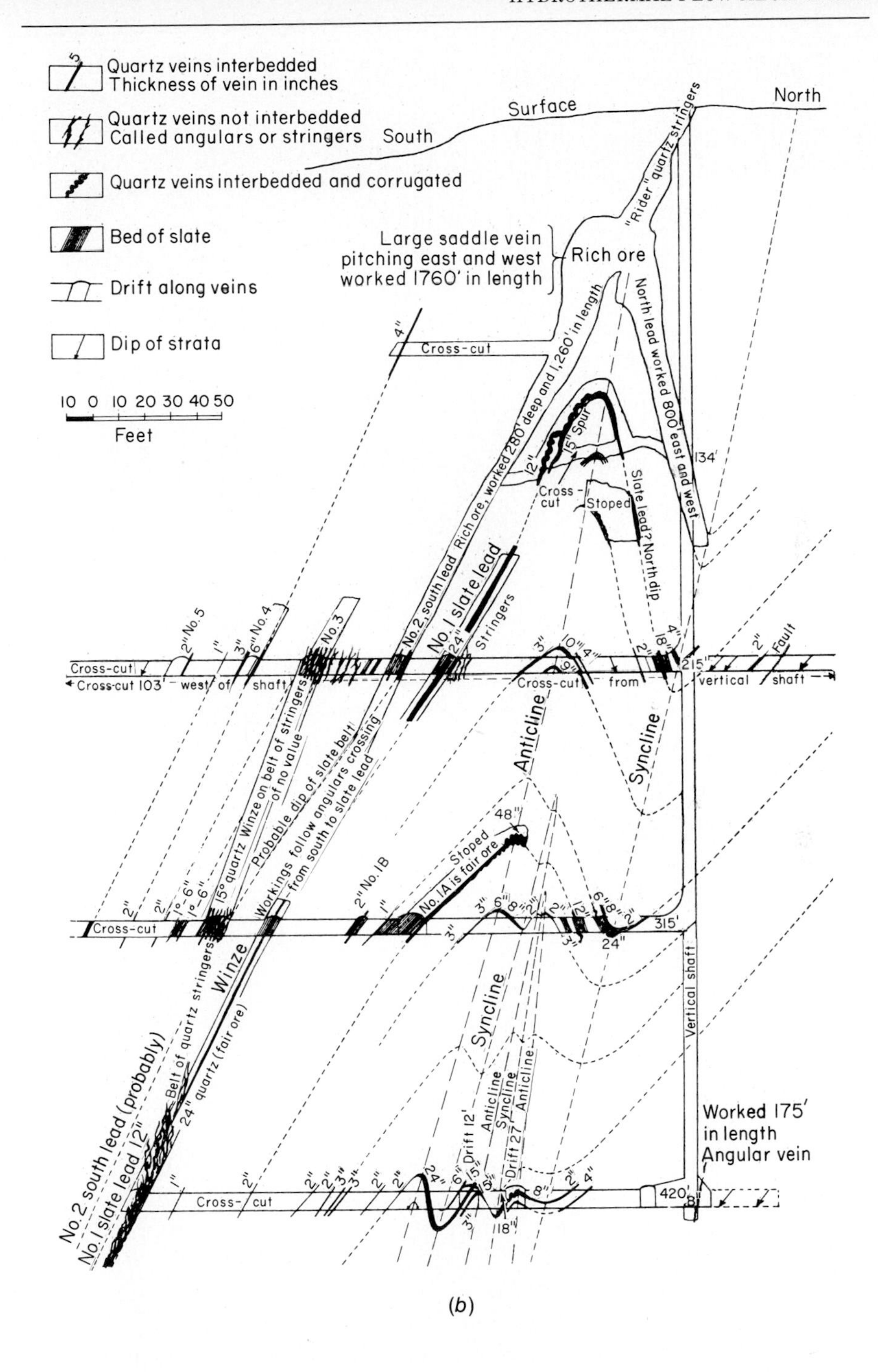
Quartz veins interbedded
Thickness of vein in inches
Quartz veins not interbedded
Called angulars or stringers
Quartz veins interbedded and corrugated
Bed of slate
Drift along veins
Dip of strata
10 0 10 20 30 40 50
Feet
Surface
North
South
"Rider" quartz stringers
Large saddle vein
pitching east and west
worked 1760' in length
Rich ore
North lead worked 800' east and west
Cross-cut
No. 2, south lead. Rich ore, worked 280' deep and 1,260' in length
15" Spur
134'
Cross-cut
Stoped
Slate lead? North dip
No. 1 slate lead
Stringers
No. 5
No. 4
No. 3
Fault
215'
Cross-cut 103' west of shaft
Cross-cut from vertical shaft
Anticline
Syncline
15° quartz Winze on belt of stringers of no value
Probable dip of slate belt
Workings follow angulars crossing from south to slate lead
Stoped No. 1A is fair ore
48"
No. 1B
315'
Winze
Belt of quartz stringers
24" quartz (fair ore)
No. 2 south lead (probably)
No. 1 slate lead 12"
Vertical shaft
Syncline
Anticline
Drift 12'
Drift 27'
Worked 175'
in length
Angular vein
420'
18"

(*b*)

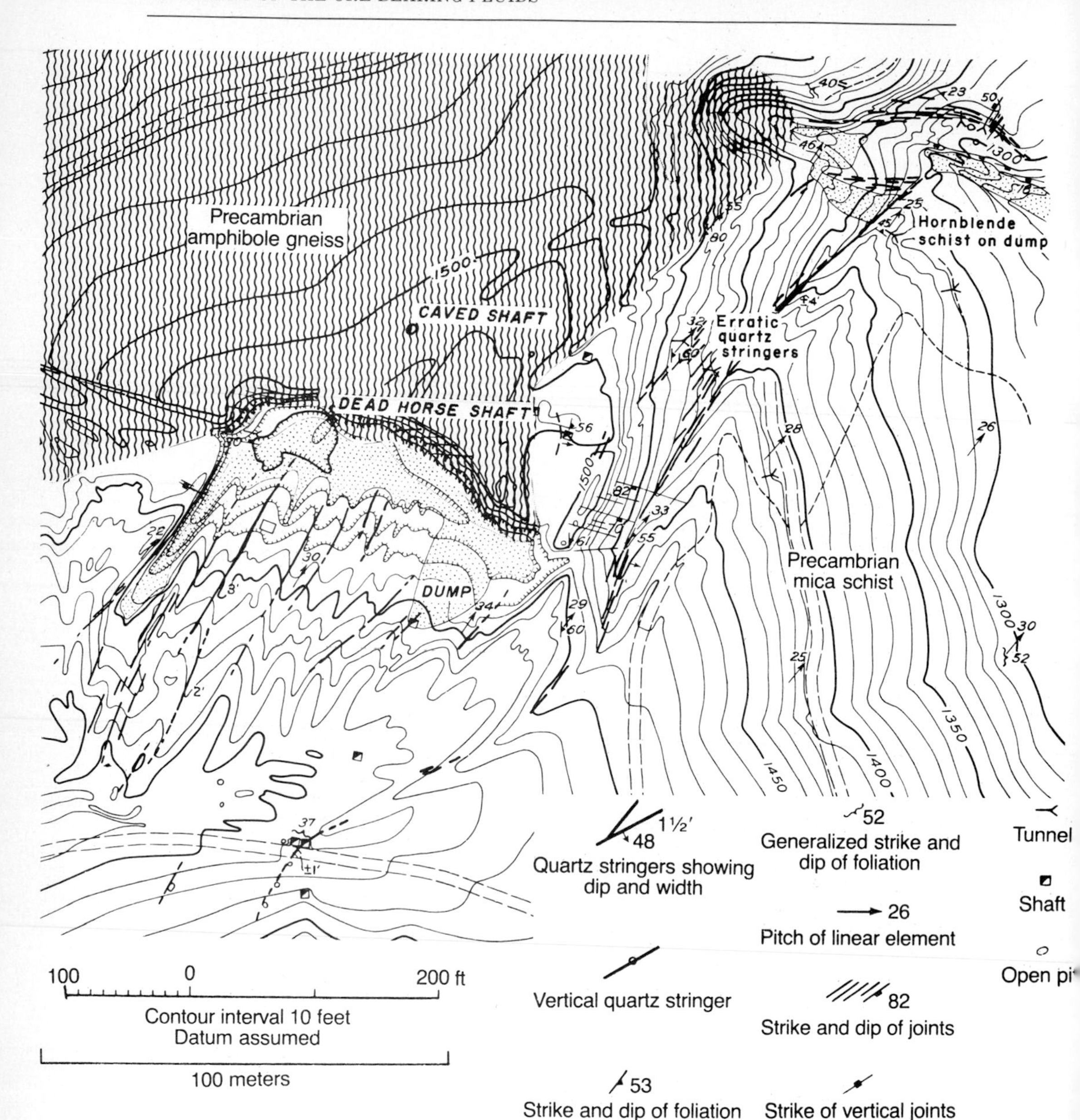

Figure 3-22. The heavy northeasterly lines represent quartz stringers—once with ore-grade gold—formed along the crests of small folds in schist, near Dahlonega, Georgia, one of the first mining districts developed in the United States. Contour lines mimic structure, and the coincidence of veins with fold axes is clear.

to open-fissure veins near the surface. Certainly diffusion was more important at depth, as was mechanical flow in the upper reaches. Again it should be stressed that both processes operate in *all* environments of transport.

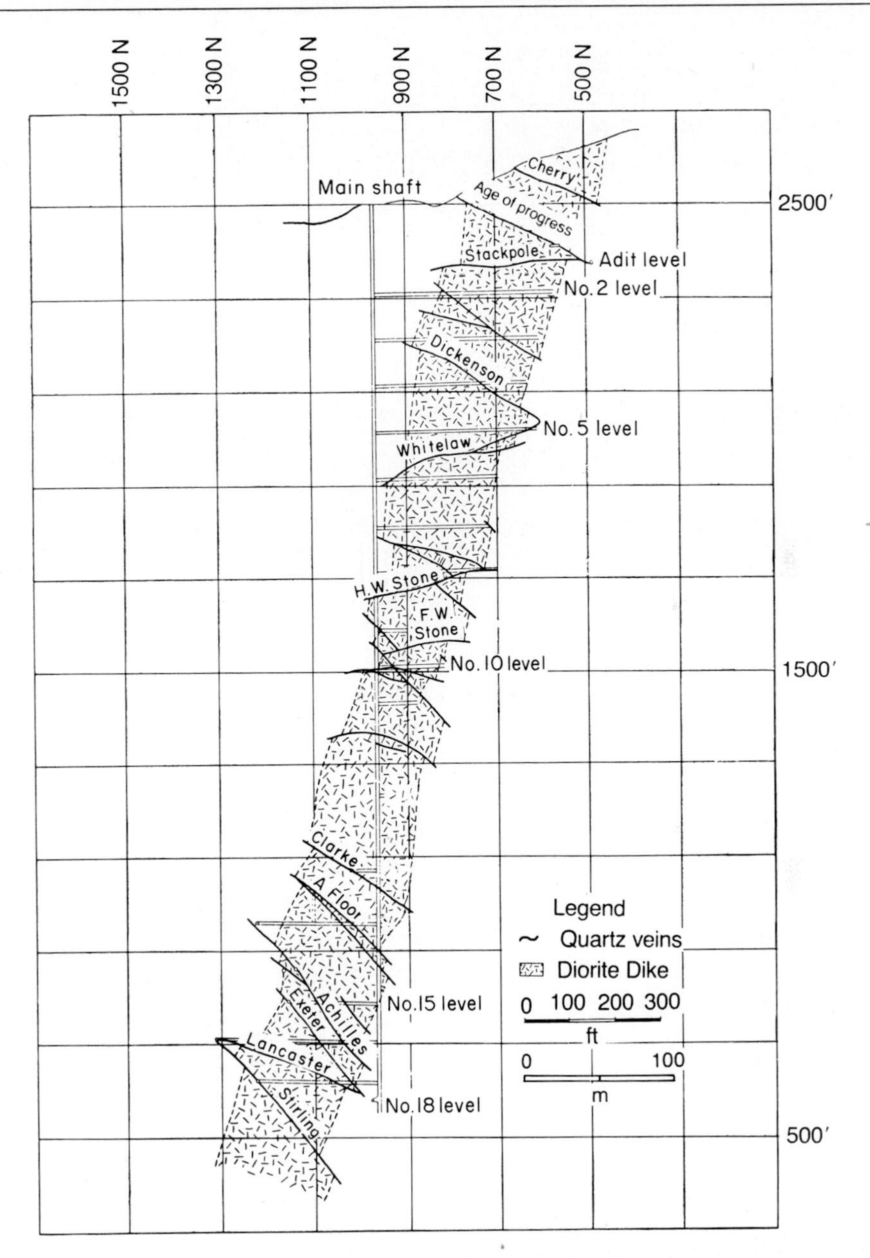

Figure 3-23. North-south cross section through the main shaft, Morning Star dike, Victoria, Australia. Quartz-gold veins that ruptured across a brittle dike formed "ladder veins," and did not continue into the sedimentary walls which failed plastically. (*From* Clappison, 1953.)

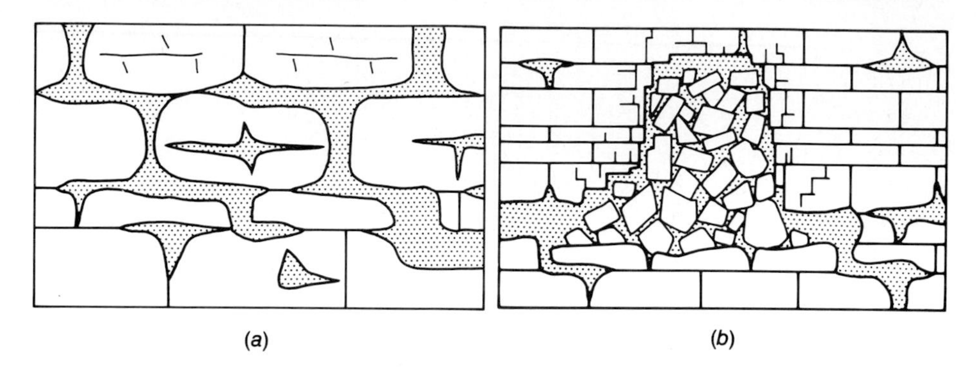

Figure 3-24. Permeability increase (*a*) by solution enlargement of bedding planes and joint planes, and (*b*) by the development of collapse breccias into caverns.

What forces actually drive hydrothermal fluids through their channelways? Barnes (1979) developed a list of six considerations.

1. Upward flow is established when hot to warm fluids dissolved in melts are released as the magma cools (Figure 2-7).

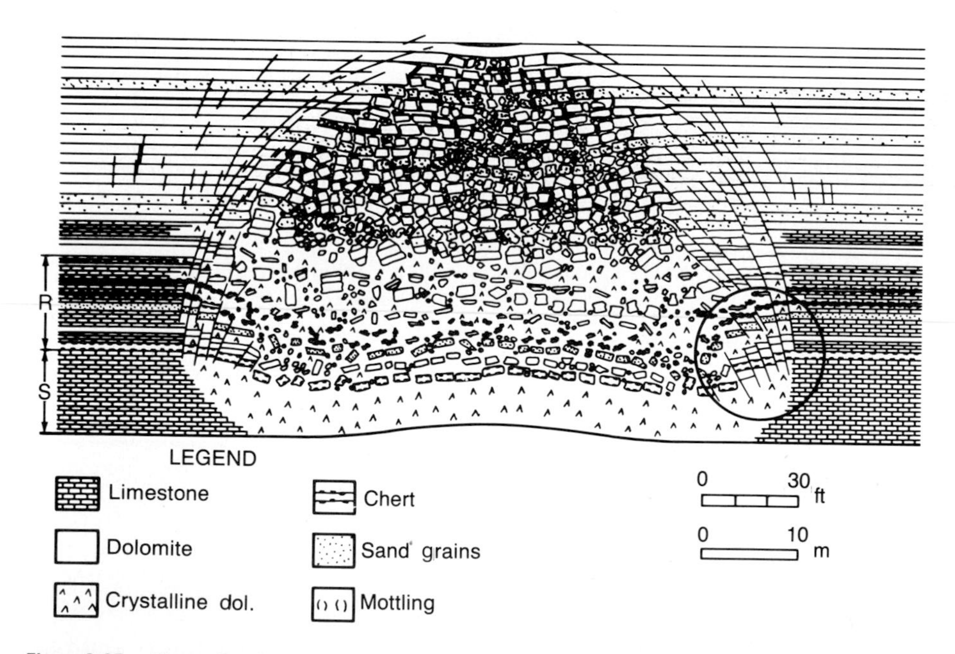

Figure 3-25. Example of permeability greatly enhanced by progressive solution and collapse in carbonate rocks in the east Tennessee zinc district. Note that individual beds can still be traced through the zone of dissolution and collapse. The beds have been dismembered, have settled, and are supported by younger dolomite and sphalerite. The circled area is shown in Figure 3-26. (*From* McCormick et al., 1971.)

Figure 3-26. Slab of mineralized "collapse breccia" from the east Tennessee zinc district. Limestone (medium gray) has been partly dissolved at the left side, leaving plates of undissolved material to settle and stretch-fracture. The open space was filled by sphalerite (dark) as dissolution proceeded. The specimen is representative of a part of the larger scale process of Figure 3-25; the circled area in Figure 3-25 and this specimen are approximately equivalent, although the scales are clearly different. The broken edge at the right is 8 cm long. (*Photo by* W. K. Bilodeau.)

2. Convection of groundwaters can be started when density differences are induced in them by local intrusions. Hot low-density water will rise alongside and over a pluton, and cooler water on the perimeter sinks and sweeps inward to replace that driven upward. Convection and convective mass transfer are probably important in porphyry systems (Chapter 11), epithermal deposits (Chapter 12), and massive sulfide ores (Chapter 13).

3. Solutions migrate when the lithostatic pressure of a compacting sedimentary sequence reduces initial porosity and extrudes connate fluids outward or upward. Mississippi Valley deposits (Chapter 20) form at basin margins where sediments pinch out against shelves; water squeezed from the basin migrates updip, carrying ore-forming constituents along with it (Figure 20-2).

4. "Osmotic pumping" is thought to occur where a semipermeable membrane such as a shale unit separates two solutions of different salinity.

Water will flow through the membrane to equilibrate the system, thereby dewatering the rocks left behind. This force, probably trivial, may contribute to Mississippi Valley type deposition.

5. Flow patterns will develop in bodies of water or in permeable rocks where dense saline fluids sink and displace less dense fluids upward. This mechanism certainly contributes to movements of exhaled volcanic emanations on the seafloor (Chapter 13), to evaporite formation (Chapter 15), and to oil-gas-water arrays in oil fields.
6. Flow occurs when there is a difference in hydrostatic head—hydraulic pressure—between the source and the outlet of an aquifer. Roll-front and some unconformity-type uranium ore deposits (Chapter 20; see Figure 20-14) form in response to this style of fluid flow.

To these six should be added a seventh, namely:

7. Flow may be created by the metamorphic recrystallization of hydrous phases, an extension of item 3, when hydrous minerals are heated by uncommonly steep geothermal gradients and thus high near-surface temperatures (Figure 19-11).

It is, of course, reasonable for more than one of these flow mechanisms to contribute to the overall paleohydrologic system of a given ore deposit. But understanding paleohydrology is vital to exploration and ore genesis understanding, because to perceive the manner and directions of flow of these fluids is to comprehend where the ore minerals may be found.

▸ EXAMPLES OF STRUCTURAL CONTROL

Structurally controlled ore deposits abound. Rather than describe a group of such deposits superficially, we shall discuss three spectacular examples in some detail: the Trepça mine in Yugoslavia, the Tsumeb mine in South-West Africa (Namibia), and the Rambler vein in the Slocan district of Canada.

Trepça Mine, Yugoslavia

The Trepça, or Stari Trg (Old Market Place), lead-zinc deposit is in southern Yugoslavia, slightly over 200 km south of Belgrade (Figure 3-27). The mine area has been studied by Forgan (1948), Schumacher (1950, 1954), and other geologists (Brammall, 1930; Christie, 1950).

Geologic formations consist of the Stari Trg Series, possibly of Ordovician-Silurian age, which forms a complex of schist, phyllite, quartzite, and marbleized limestone. Within this series is one fairly thick bed of pure

Figure 3-27. Location map of the Trepça (pronounced *Trep-cha*) mine in Yugoslavia.

recrystallized limestone called Mazic Limestone, the principal mineralized stratum in the district. The Stari Trg Series is unconformably overlain and partly obscured by Miocene (?) andesite flows and tuffs and is intruded by associated igneous rocks ranging from monzonite porphyry to hornblende andesite. The older rocks of the region are sharply folded. An anticlinal structure, along which the mine is developed, plunges about 40° northwest (Forgan, 1948). A nearly circular pipe containing dacite and breccia (Figure 3-28) transects the Stari Trg Schist–Mazic Limestone contact along the crest of the anticline. This pipe is a volcanic plug. At the surface and in the upper parts of the mine, the pipe contains a core of dacite surrounded by a shell of breccia. The dacite decreases with depth and is absent in the lower levels of the mine. The breccia in the upper levels is composed of subangular fragments of the country rock and rounded fragments of dacite.

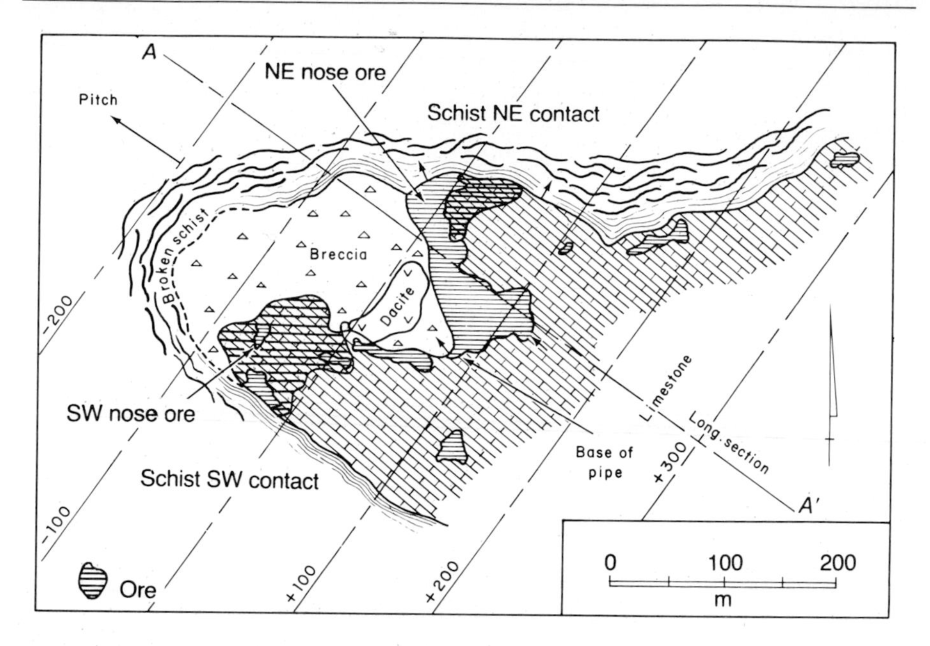

Figure 3-28. Map of the 610-meter level of the Trepça mine, Yugoslavia. The principal ore control was the enhanced permeability of the intruded volcanic breccia along the crest of the northwest-plunging fold structure. The line *A–A′* is Figure 3-29. (*From* Forgan, 1948.)

In the deeper levels even the groundmass of the breccia is composed of crushed sedimentary rocks.

Ore distribution was controlled both by the pipelike structure, which clearly channeled fluid flow, and by the character of the country rock, part of which was reactive to the ore-bearing solutions. The ores are concentrated in the Mazic Limestone around the breccia pipe and along the limestone-schist contact; outside the pipe, the ore gradually thins. The intrusion and the breccia pipe deformed the limestone-schist contact so that the upper contact of the limestone against schist and breccia is M-shaped in plan, the pipe representing the upper reentrant of the letter (Figure 3-28). The two top points of the M were the most favorable loci for ore, though mineralization does spread laterally in the limestone just below. Within the points of the M, the limestone is completely replaced by ore, from the pipe across to the schist. The orebodies themselves are pipe-shaped and plunge northwest, following the breccia pipe (Figure 3-29).

All significant deformation took place before the ore was deposited, as shown by relationships between the orebodies and the breccia pipe and by mineralization in drag folds along the limestone-schist contact. The dacite intruded along the fold crest line, and the jacket of breccia was apparently formed by late explosive activity, and thus is a true volcanic breccia. Ore deposition appears to have followed consolidation of the pipe, as a late phase of the same igneous activity; mineralization took place in the structures

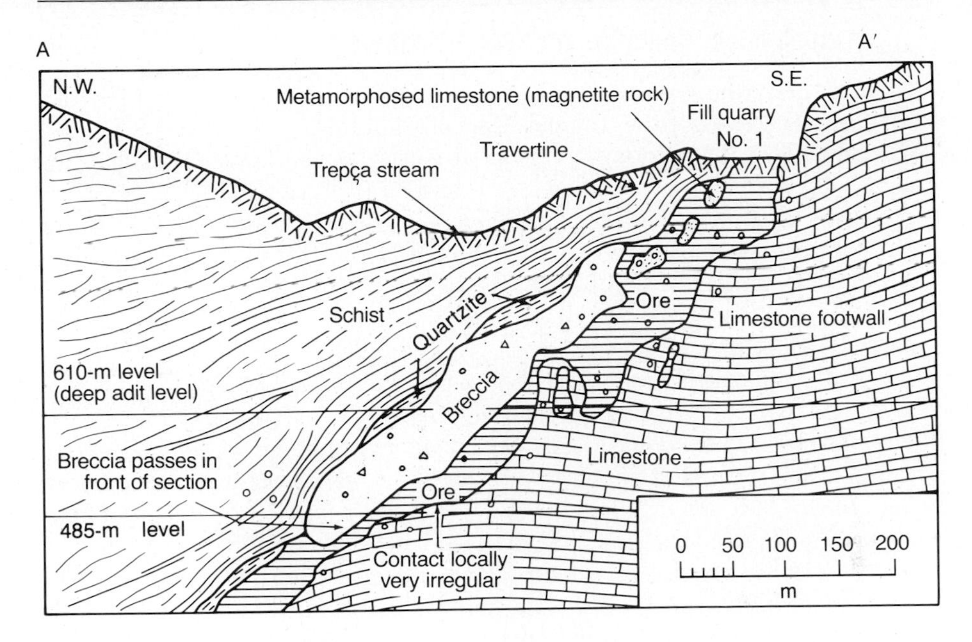

Figure 3-29. Cross section of the Trepça mine along the line *A–A′* in Figure 3-28. The 610-meter level of Figure 3-28 is also shown here. The figures taken together show the overall pipelike shape of the breccia and its ores.

defined by the breccia and the schist-limestone contact (Forgan, 1948). The limestone reacted with the ore solutions and was replaced in preference to the schist, breccia, and dacite. The ore is a massive coarse-grained sulfide mixture of silver-bearing galena, sphalerite, pyrite, and pyrrhotite, with minor amounts of arsenopyrite, jamesonite ($Pb_4FeSb_6S_{14}$), boulangerite ($Pb_5Sb_4S_{11}$), bournonite ($PbCuSbS_3$), and chalcopyrite.

A contact metamorphic skarn type of mineralization is found in parts of the mine and appears to increase with depth. The textures are coarse in the upper levels but finer grained below. Along the flanks of the structure, however, the ore at depth is not greatly different from the ore in the upper levels. These changes in mineralization, following a pattern symmetrical to the breccia pipe, are significant in the history of ore deposition. The mineralizing solutions evidently ascended along the breccia pipe, gradually changing in composition with time, with height in the pipe, and with distance away from the pipe. The deep near-pipe minerals, pyrrhotite and quartz, were deposited early, before the ascending solutions reached the upper levels; the later solutions produced the Zn-Pb-Sb-Mn assemblage sphalerite, jamesonite, and rhodochrosite. The different mineral assemblages reflect changing conditions, from the availability of metal ions in time to a near-surface drop in temperature and pressure, an increase in pH, and a relative increase in the partial pressure of sulfur upward and through time.

Tsumeb Mine, South-West Africa (Namibia)

Another classic example of an ore pipe is the Tsumeb deposit, described by the staff of the Tsumeb Corporation (1961), Söhnge (1963), and Pelletier (1964). The Tsumeb mine is in northern South-West Africa (Namibia), 400 km north of Windhoek, the capital (Figure 3-30).

Precambrian sediments of the Otavi highland overlie older Precambrian metamorphic and igneous basement rocks in the Tsumeb area. Mineralization is confined to a dolomite sequence known as the Tsumeb Stage of the Otavi Formation. This Upper Precambrian dolomite is a fine-grained compact rock containing shale lenses, chert nodules, stromatolites, limestone beds, and oolitic zones. The sediments were folded into open synclines and anticlines and subsequently intruded by dikes, sills, and pipes of a quartz-feldspar rock known locally as "pseudoaplite." The pseudoaplite was emplaced before the ore-bearing fluids ascended along the pipe (Schneiderhöhn, 1929, 1931, 1941; Söhnge, 1952, 1963). The mine is developed along a pseudoaplite pipe that cuts the north limb of a syncline. Petrographically the pseudoaplite consists of rounded and crushed grains of quartz, microcline, and sodic plagioclase. No thermal metamorphism is associated with the pseudoaplite, but it is definitely intrusive (Söhnge, 1952, 1963).

The ore pipe, although it is basically an elliptical structure, pinches and swells irregularly and continues from the surface to a depth of more than 1.4 km, or 1400 meters (Figures 3-31 to 3-34). Along its constrictions it consists of a sparse network of thin veins, but it widens abruptly into plan sections of over 35 by 35 meters. Its widest section, on the 2390-ft level, measures 100 by 200 meters (Söhnge, 1952).

The pipe was formed by fracturing and brecciation preceding intrusion of the pseudoaplite. Brecciation was restricted to the deeper levels, especially where the pipe cuts the axis of a fold. Reverse faulting along the bedding, which dipped about 45° to the south, caused drag folding, local thrusts across bedding, and brecciation. The pipe is nearly vertical above 700 ft and below 2000 ft elevation (230 to 666 meters), but in the intervening section (*Z–Z′* in Figure 3-32) it follows a zone of closely spaced bedding-

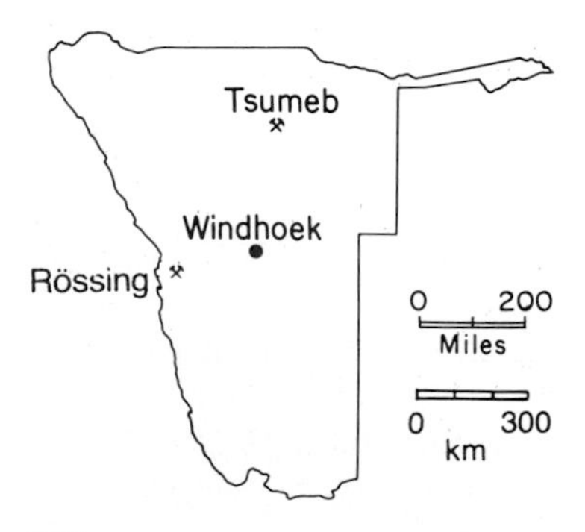

Figure 3-30. Map of South-West Africa (Namibia) showing the location of the Tsumeb mine. Another important Namibian mineral asset is the Rössing uranium mine near the coast west of Windhoek (Chapter 17), one of the largest in the world.

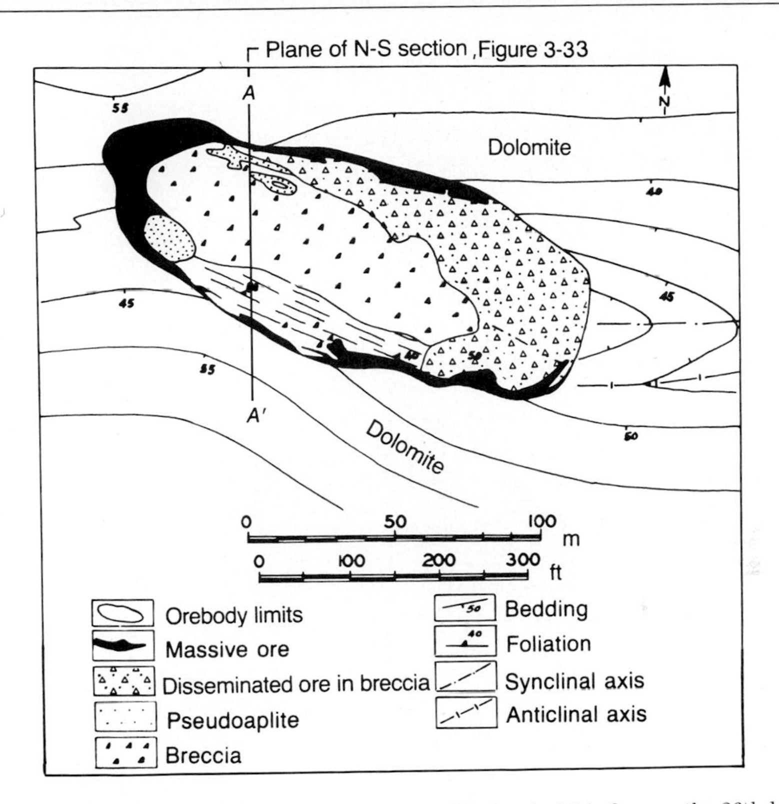

Figure 3-31. Plan view of the Tsumeb breccia pipe. The level of this figure—the 26th level—is shown in Figure 3-32. The uncommon Pb-Zn-Cu-Ge-As-S ores were deposited from fluids presumably moving up the pipe. The wall-rock dolomite may have provided some chemical contrast, but the mechanism of deposition was more physical than chemical. (*From* Tsumeb Corporation Staff, 1961.)

plane faults, thus forming a relatively constricted pipe that plunges about 50° south. The sausage-shaped intrusive pseudoaplite rock was injected after the folding, faulting, and brecciation. It is centrally located in the upper portion of the pipe, but branches and lenses irregularly through the breccia (Tsumeb Corporation Staff, 1961).

The origin of the pseudoaplite and its associated ore minerals has been a long-standing enigma, hence the prefix "pseudo." Söhnge (1963) points out that it is neither a simple igneous nor a simple clastic dike—it seems to be a blend of the two. One possibility is that quartzofeldspathic sediments at depth—perhaps arkosic—were mobilized by deep-seated igneous activity, and the escape of volatiles was made along a previously formed fracture zone. Clastic grains from lower formations would thus be partly dissolved and lifted by *fluidization* from their original positions. A second hypothesis attributes the pseudoaplite and the ores to the intrusion of a highly volatile fraction of a magma that was cogenetic with nearby carbonatites that carry clastic fragments of quartzite derived from below. In either case—and most

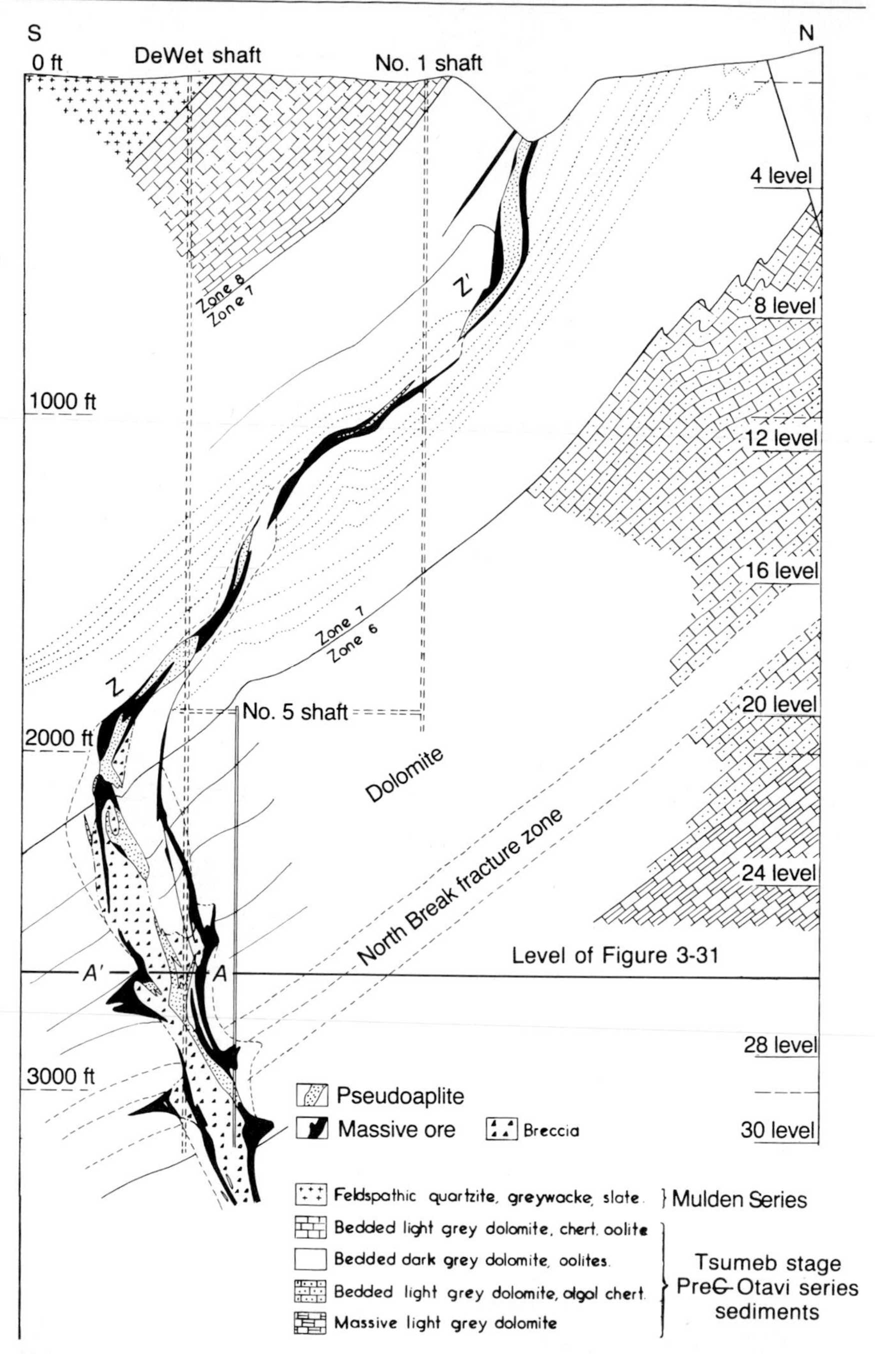

Figure 3-32. Cross section of the Tsumeb breccia pipe. Its role as a "pipeline" to the source of the pseudoaplite magma chamber and its related ore-forming fluids is clear. See also Figure 3-33. A plan view of the pipe at A–A' is given in Figure 3-31. The internal Z–Z' is discussed in the text; the pipe crosscuts bedding except between Z and Z', where it follows bedding plane faults. (*From* Tsumeb Corporation Staff, 1961.)

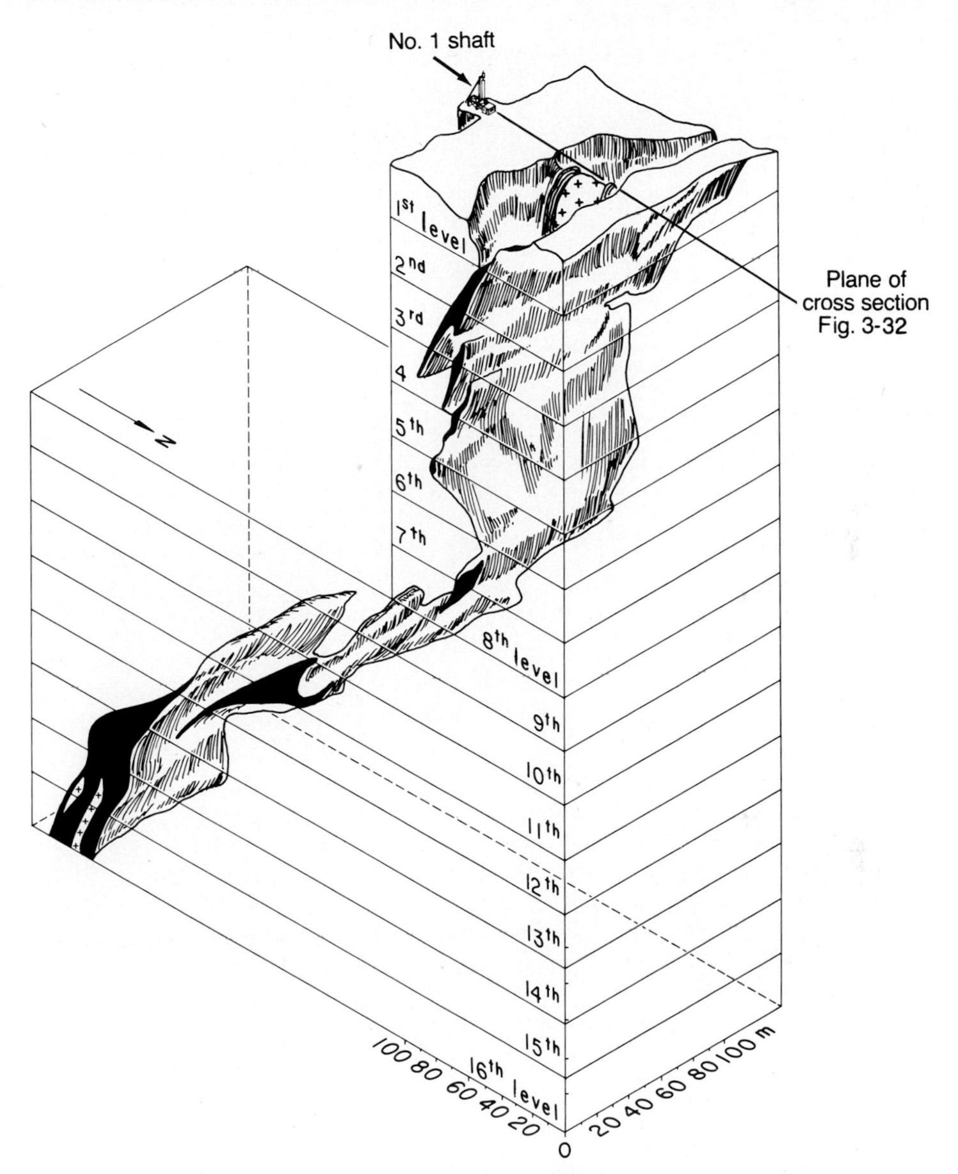

Figure 3-33. Isometric perspective drawing of the Tsumeb breccia pipe down to about half the depth shown in Figure 3-32. The pit and no. 1 shaft are shown in Figure 3-32; the pseudoaplite shows in both figures. (*From* Schneiderhöhn, 1929.)

important to our purposes here—the pipe was a demonstrable guide of particulate material, fluids, gases, and perhaps magma up to the surface from a chamber more than 1 km below.

Metallization follows the periphery of the pipe, sometimes forming rich pods and veinlike masses as well as large tonnages of disseminated low-

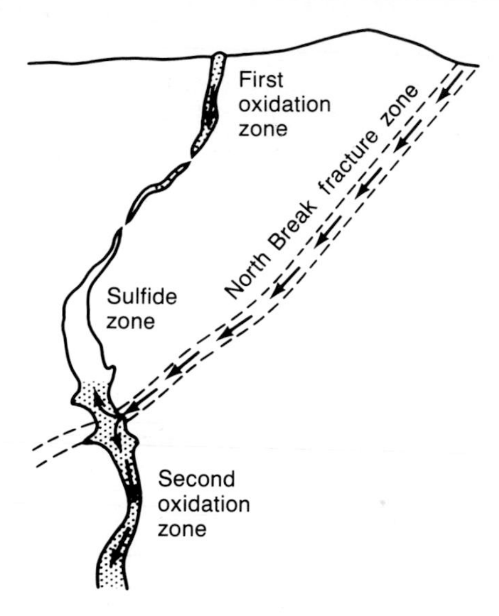

Figure 3-34. Oxidation pattern of the Tsumeb ore pipe. Oxidizing groundwater followed the paths indicated by the arrows, oxidizing sulfides both near the surface and at depth where the North Break fracture zone intersected the pipe. Compare with Figure 3-32. (*After* Wilson, 1977.)

grade ore. The main ore pipe is a horseshoe-shaped complex of veins with massive ore and intervening low-grade disseminations (Figure 3-31). High-grade mantos develop where the pipe intersects breccia zones along bedding slippages and project as much as 100 meters both north and south of the pipe, the "barbs" projecting along bedding in Figure 3-32. The pipe is widest where it crosses the bedding and narrowest where it parallels the bedding (Z–Z' in Figure 3-32).

Individual masses or "shoots" of ore vary in their ratios of lead, copper, and zinc, but in general, lead is most abundant. Massive sulfide bodies contain up to 60% metal. One large sulfide mass contained 26.7% Pb, 12.4% Zn, and 3.6% Cu; high-grade copper ores have produced as much as 23% Cu. The disseminated ores form stockworks, or networks of fissures, in dolomite and breccia and interstitial specks and veinlets in the pseudoaplite. All rock containing over 3.5% metals is considered ore by the Tsumeb Corporation. The mine produces lead, zinc, and copper, with accessory germanium, silver, and cadmium from unusual ore containing galena, sphalerite, tennantite, chalcocite, bornite, digenite, enargite, chalcopyrite, germanite, reniérite, molybdenite, wurtzite, luzonite, greenockite, and stromeyerite. Numerous products of oxidation are also found in the mine, including many rare mineral species, especially germanates and arsenates known only at Tsumeb (Sclar and Geier, 1957; Wilson, 1977).

Although minor amounts of sulfides were present at one time even at the surface, the Tsumeb ores have been oxidized to great depths. Supergene minerals predominate above 400 meters. Between 400 and 830 meters oxidation products diminish strikingly, but even the lowest parts of the pipe show appreciable oxidation. This unusually deepseated oxidation zone results from the presence of permeable horizons in the Tsumeb dolomites, especially the North Break fracture zone (Figures 3-32 and 3-34) where groundwaters migrated along bedding planes and brecciated strata and attacked the sulfides even more effectively than did near-surface waters. Supergene enrichment increased copper values along upper portions of the pipe, although zinc was leached and removed therefrom.

Rambler Mine, British Columbia

The Rambler vein in the Slocan district of southeastern British Columbia (Figure 3-35) is an excellent example of structural control of ore deposition along a fault. The many ore deposits in the Slocan district are nearly all structurally controlled (Cairnes, 1934, 1935, 1948; Ambrose, 1957). Ore-bearing structures include fissures that slice through competent rocks, openings at directional changes along faults (Figures 3-8 and 3-9), brecciation against dike contacts, and brecciation at the intersections of fissures (Figure 3-14). For the sake of simplicity, only the Rambler vein, a combination of the first two types, is described.

The Rambler orebodies (gray in Figure 3-36) occur in a complex vein emplaced along a normal oblique-slip fault with 10 to 30 meters of displacement. Argillaceous sedimentary rocks of the Slocan Series and granodiorite

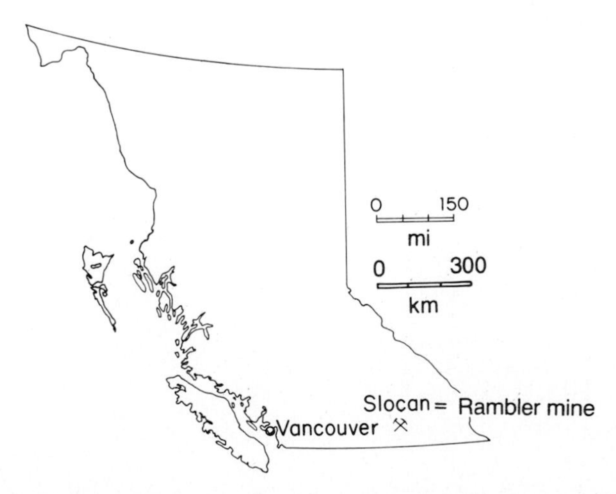

Figure 3-35. Map of British Columbia showing the location of the Slocan district and its Rambler mine.

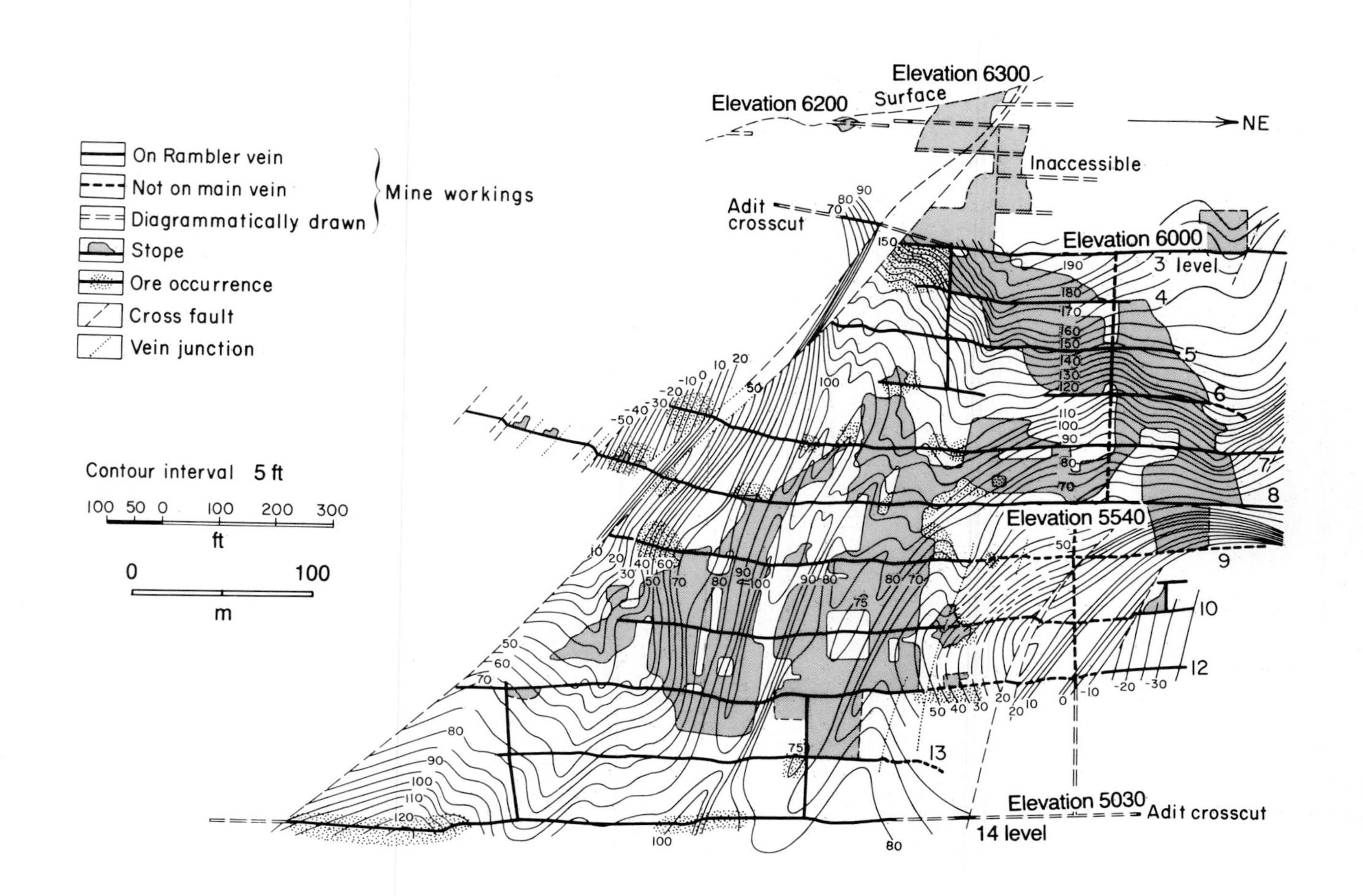

On Rambler vein
Not on main vein
Diagrammatically drawn
Mine workings
Stope
Ore occurrence
Cross fault
Vein junction
Contour interval 5 ft
100 50 0 100 200 300
ft
0 100
m
Elevation 6300
Elevation 6200
Surface
NE
Inaccessible
Adit crosscut
Elevation 6000
3 level
4
5
6
7
8
Elevation 5540
9
10
12
13
Elevation 5030
14 level
Adit crosscut

porphyry sills and dikes are the only rock units near the Rambler mine. The Slocan Series is thought to be Triassic; the granodiorite is Late Cretaceous (Cairnes, 1934).

The Rambler fault cuts both the sediments and the intrusives, and it varies in strike and dip. Its normal strike is northeasterly, with a moderate southeast dip. Where the vein veers eastward it also steepens, forming gaplike openings that thicken at depth.

The vein is thickest along these zones of change in attitude, and narrowest along tight segments where the fault is not deflected. The Rambler orebodies are composite veins of nearly pure sulfides, small sulfide veinlets, and disseminated sulfides in altered wall rock within the fault zone and adjacent to it. The veins range from a few centimeters to as much as 7 meters wide in the major ore shoots (Figure 3-37). Roscoe (1951) studied the details of ore distribution along the Rambler fault and showed that the fault plane was refracted at points of contact between sediment and granodiorite. Wherever the attitude of a granodiorite dike–Slocan Series contact was oriented across the fault plane, the fault was refracted eastward, and an opening was developed in the fault. Mineralization took place along such openings. The orebodies were restricted to dike swarms and were not found where the fault passes through massive granodiorite or thick uniform sediments; moreover, the fault was essentially straight and "tight," except in the dike swarm. By constructing special maps of the vein, Roscoe demonstrated that as total fault offset increased, the central portions of the ore shoots progressively widened. Furthermore the mineralization was contemporaneous with faulting and progressed laterally from the central part of dilatant zones. Recurrent small displacements along the Rambler fault were accompanied by deposition of pyrite, sphalerite, and galena in successive stages. Pyrite was the first mineral to be deposited. As structural offset and dilation along the ore shoot continued, sphalerite began to be precipitated in the central zone, replacing the pyrite, with pyrite then being deposited in the fringe areas. Subsequent movements allowed galena and silver minerals to take over the central portion while both pyrite and sphalerite spread outward. The structural and paragenetic sequences are clear; the ore shoots grew in size and developed progressive zones of mineralization as faulting created an expanding planar zone of permeability. As a result, the ore shoots are disk-shaped bodies of complex veins located where the Rambler Fault is favorably deflected across contacts between compe-

Figure 3-36. (*facing page*) Nearly vertical long-section view of an ore shoot along the Rambler fault vein in the Slocan district, British Columbia. The gray areas are mined-out portions of the vein—note their coincidence with the most dilated vein sections in Figure 3-37. They are also zones of greatest fault aperture where refraction of the fault across granodiorite porphyry dike-sediment contacts changed its strike and dip. See also Figure 3-10 for an example of fault control of a vein system. (*From* Roscoe, 1951.)

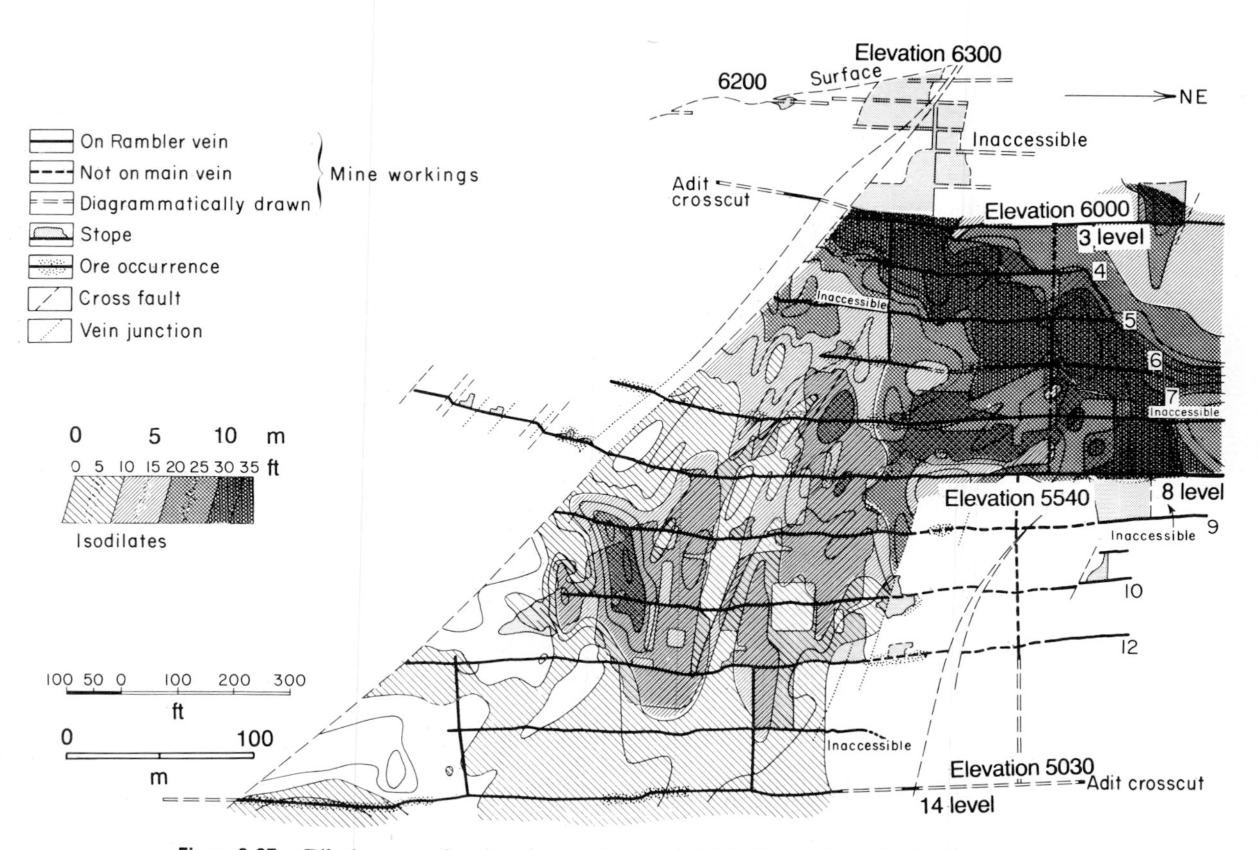

Figure 3-37. Dilation map showing the aperture or total fault opening (dilation) in the plane of the Rambler fault vein. The coincidence of greatest structural aperture and ore deposition, and hence of structural guidance of ore-bearing fluids, is apparent by comparing this diagram with Figure 3-36. (*From* Roscoe, 1951.)

tent and incompetent rocks. The outside, tapered edge of each disk is predominantly pyrite; the central portion is sphalerite and galena. Where a dilatant area opened rapidly early in the faulting and then widened only slightly later, the shoot is predominantly pyrite with a little sphalerite in the core. Conversely, shoots that dilated late in the faulting contain a high ratio of galena to sphalerite and pyrite. Thus the configuration and mineral composition of Rambler orebodies depended strictly upon structural events and not upon wall-rock chemistry. A sequence and mineralogy in many ways similar to the complex veins at Príbram diagrammed in Figure 3-11 resulted.

The Rambler mine has been productive since 1895, though it was shut down between 1926 and 1947. Roscoe states that the average metal content of the ore mined has been about 25 oz of silver per ton (780 ppm), with 7.5% Pb and 5.5% Zn. Argentiferous tetrahedrite and pyrargyrite accompany the galena in significant amounts, and recoverable quantities of silver are found in some of the sphalerite and pyrite; thus silver is a major product of the mine. The waste minerals include quartz and calcite.

Other deposits that show excellent structural control by the development of permeable zones by faulting include the Red Rose tungsten occurrence (Figure 3-10) and the Casapalca, Peru, system (Figures 11-43 to 11-45). Many excellent examples of structural control of ores are given in Bateman (1950).

▸ SOURCES OF ORE-DEPOSIT COMPONENTS

Ever since the first two geologists started sharing their ideas, it seems, there has been controversy over both where the materials of ore deposits ultimately came from and by what agencies they were concentrated where we find them now. Some sources and processes and their results are obvious enough, such as the deposition of halite in evaporite sediments and the transfer and emplacement of mass by either intrusion or extrusion of igneous rocks and their contained orebodies. However, it appears generally true that the more complex an ore-forming process might be, the more alternatives there are for the derivation of both metals and nonmetals. Consider that in a particular vein deposit district, for example, the components that we see in the ore minerals may have been abstracted by a circulating fluid from the immediately adjacent wall rocks themselves or from a larger volume of wall rocks generally lateral to and below the veins (Figure 3-38). They may have been placed in hydrothermal solution by partitioning from a deep differentiating magma in the manner described in Chapter 2, or they may have partitioned into an aqueous phase in the even deeper zone of melting and traveled upward near to, but not necessarily within, the rising melt. Krauskopf (1967) called these two options "cognate" and "collinear," respectively. In either of those cases, the ore deposit component may have entered the melt from source volumes that were deep-seated continental crust or upper mantle materials. Chemical heterogeneity

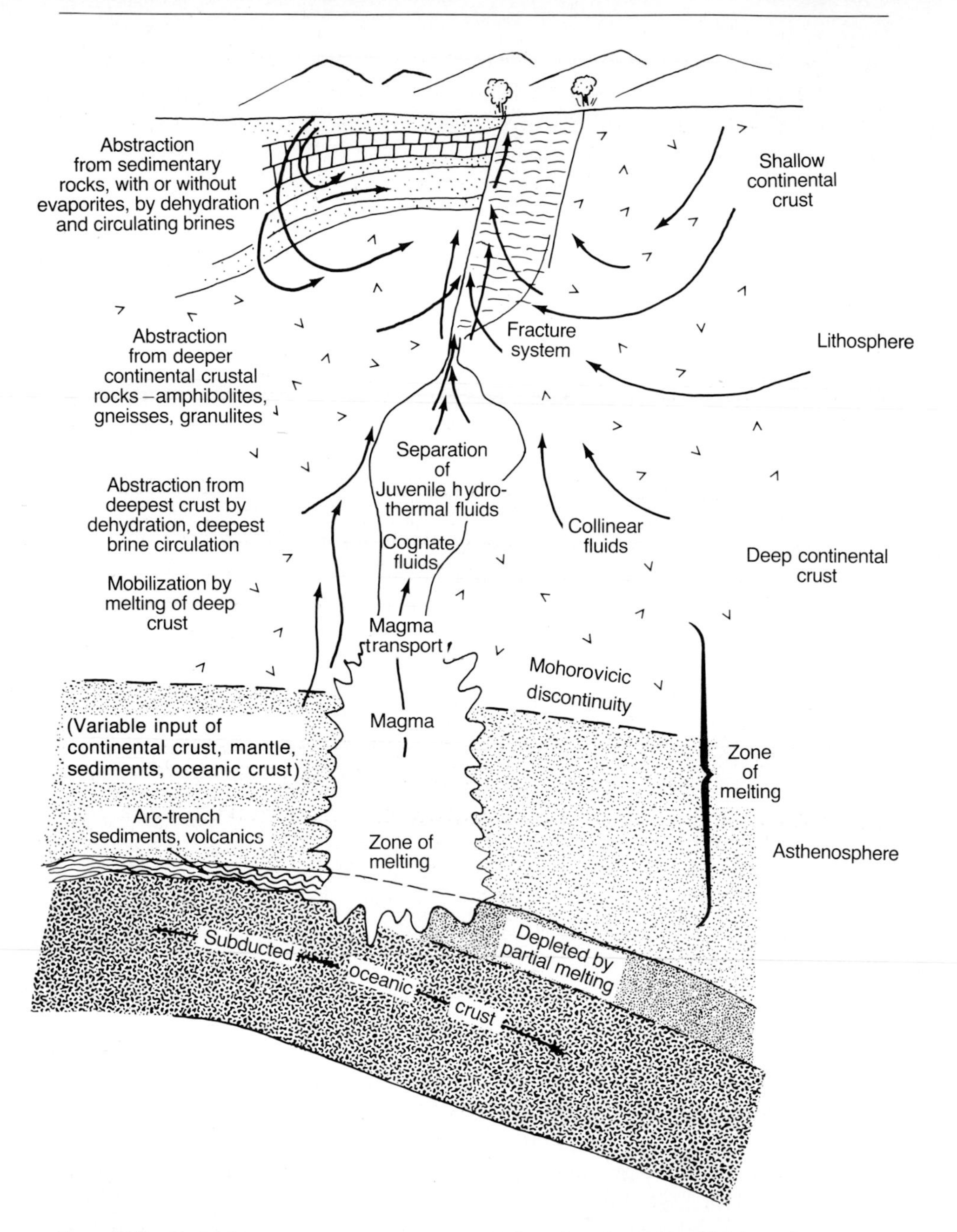

Figure 3-38. Possible sources of hydrothermal ore-deposit components. Not to scale.

of the rocks of the lower continental crust and the upper mantle—specifically zones of higher than average metal or sulfur contents—may play a role in shallower ore-formation processes. They are considered in greater detail in Chapters 8 and 21. If subduction is involved, a downgoing slab might insert oceanic crust and a wedge of arc-trench sediments into the melt zone. Finally, thermal and pressure gradients accompanying deep-seated igneous or tectonic activity may have caused circulation of deep connate and metamorphic waters not actually involved with melting but nonetheless capable of transporting ore-forming constituents.

The question of where the ore components originate may seem at first academic, and to a large extent it is. However, the questions ultimately to be answered will involve exploration-resource availability as well. If all that is required to generate a porphyry copper deposit, a massive volcanogenic sulfide deposit, or a quartz-ankerite-gold vein is a heat source emplaced into wall rocks of appropriate composition and permeability, with other appropriate variables, then the focus of exploration must clearly shift from matters concerning solely igneous activity to those of the broader environment. Some geologists, as far back as C. R. Van Hise in 1900, would have drawn Figure 3-38 with no involvement of igneous activity required. Others are inclined to deemphasize juvenile fluids, and still others discount the importance of collinear and laterally externally derived sources. Most geologists consider the magmatic environment to be an important if not predominant source of mineralizing fluids and ore components in igneous-rock related systems, but the pendulum has been swinging in the opposite sense in the late 1970s. Both Krauskopf (1967) and Skinner (1979) thoughtfully consider these problems in the two Barnes volumes on hydrothermal deposits. They note that a question exists as to the ultimate or immediate source of the metals even for deposits that have obvious igneous affiliations. Many plutons are without associated ore deposits, even where adjacent and seemingly identical stocks are abundantly mineralized. The barren Last Chance stock at Bingham Canyon, Utah, for example, is cut by the comagmatic and geologically similar—but strongly metallized—Bingham Canyon stock (Figures 11-26 and 11-27). Another seemingly simple but still unanswered question that may be fundamental to regional exploration is whether or not we should expect igneous rocks associated with ore deposits to be enriched, normal, or deficient in the metals involved. If orebodies are spawned by special magmas enriched in the ore metals, we should expect a slight concentration of these ions in the rock minerals; that is, the associated pluton should give anomalously high assays of the metals. If deuteric alteration has leached the metals from the igneous rock, or if favorable magmatic differentiation has selectively concentrated metals in the latest fluid fraction, we might expect the rock to be anomalously low in these components (Ingerson, 1954). Further geochemical studies may ultimately resolve this problem. At present the evidence is conflicting and can support either hypothesis—indeed, both mechanisms are probably operative.

Rarely is compelling information available on both the rocks and the ore constituents which permits identification of source. As we see in Chapter 7, we can use lead isotopes to study ore minerals that contain appreciable lead, sulfur isotopes to tag sources of those anions, initial strontium isotope ratios to define source volumes of stocks and batholiths, and oxygen and hydrogen isotopes, in proper circumstances, to evaluate water origins. For example, the fact that ratios of radiogenic and nonradiogenic lead isotopes in major ore deposits commonly fit a simple one-stage growth curve (Figure 7-21) indicates that ore deposit leads were introduced into the crust from time to time as "batches" from a homogeneous source, presumably the upper mantle (Richards, 1971). But determining from what source copper, molybdenum, silver, gold, and other metals are derived is still difficult. Mass transfer determinations and calculations—how much metal may have been added to or removed from a given rock volume—become increasingly difficult as the size and the geologic complexity of the volume increase and the absolute composition contrast between "affected" and "unaffected" wall rocks diminishes. Arguments like those involving abstraction of a few parts per million or a few hundredths of a percent of an element from a large volume of source rock may never be reliably answered. Most process descriptions are not compelling—they state how given redistributions *can* occur, not how they *must* occur. We are still struggling, then, to identify the ultimate origins of most ore system components. Progress is being made in the computer simulation of ore-forming processes, in determining the constraints of metallogenic considerations to source problems, and in understanding quantitative and qualitative aspects of ore-formation processes, but the end is not yet in sight for most ore-component source problems. Stable and radiogenic isotope studies offer perhaps the most powerful near-term resolution of some of these problems.

Part of the source problem relates to the degree of concentration of an ore component that is necessary to place it within technologic and economic reach, in other words, to make it minable. That degree is clearly a function of both an element's natural abundance and its value to society. The *clarke* has been defined—in honor of the geochemist F. W. Clarke—as a number defining the average percentage of an element in the crust of the Earth (Table 3-4). A *clarke of concentration* is then the factor of concentration of a given element in a particular rock or ore volume compared with its crustal abundance. The number of clarkes of concentration required to constitute an ore of a particular element becomes a useful number because it includes abundance, location, and economic factors. Table 3-4 lists the average clarkes of concentration of many elements for the prices of January 1982; recalculation to the prices given using January 1984 or later is simple. Table 3-5 lists the abundances of elements in various igneous rock types. It will be useful to refer back to Tables 3-4 and 3-5 as other aspects of ore-forming processes and results unfold in later chapters.

TABLE 3-4. Clarkes, Clarkes of Concentration, and Prices of Some Important Elements

	Abundance in Earth's Crust,			Approximate Minimum Weight-Percent to Qualify as Ore, 1982	Clarkes of Concentration, 1982	Prices, $/lb			
Element	wt %	= Clarke	ppm or g/ton			January 1978	January 1980	January 1982	January 1984
O	46.60	46.6	466,000						
Si	27.72	27.7	277,200						
Al	8.13	8.1	81,300	22	2.7	0.53	0.66	0.76	0.81
Fe	5.00	5.0	50,000	30	6	0.04	0.05	0.06	0.04
Ca	3.63	3.6	36,300						
Na	2.83	2.8	28,300						
K	2.59	2.6	25,900	20	7.7	0.05	0.06	0.07	0.05
Mg	2.09	2.1	20,900			0.99	1.09	1.34	1.43
Ti	0.44	0.4	4,400	30	68	2.98	3.98	7.65	5.55
	99.03*	98.9	990,300						
Mn	0.095		950	40	420	0.58	0.62	0.70	0.70
Zn	0.0070		70	4	570	0.30	0.37	0.43	0.49
Ni	0.0075		75	1	130	2.10	3.25	3.25	3.20
Cu	0.0055		55	0.5	90	0.64	1.21	0.81	0.69
Mo	0.00015		1.5	0.4	2700	4.31	7.50	8.50	3.90
Ag	0.000007		0.07	2 oz (62 ppm)	990	4.90/oz	38.25/oz	8.43/oz	8.18/oz
Au	0.0000004		0.004	0.06 oz (2 ppm)	500	173/oz	675/oz	409/oz	320/oz
Cr	0.010		100	30	3000	2.63	3.50	3.75	3.75
Sn	0.0002		2	1.5	7500	5.92	8.39	7.93	6.24
U	0.00018		1.8			43.00	41.00	23.00	20.50
Hg	0.00008		0.08	0.1	1000	2.05	5.13	5.43	4.01
S	0.026		260			0.03	0.05	0.07	0.06
Pb	0.0013		13	4	3000	0.33	0.50	0.31	0.25

*This total shows that the remaining elements constitute only 0.97 wt %. Sixty % of the 100-odd elements occur at levels of less than 10 ppm, or 0.001%.

Source: Abundance data from Mason and Moore, *Principles of Geochemistry* (Wiley, 1982); prices from *Engineering and Mining Journal*.

TABLE 3-5. Abundances of Common Ore Metals in Igneous Rocks (In Parts Per Million or Grams Per Metric Ton)

Element	Ultramafic Rocks	Mafic Rocks	Intermediate Rocks	Granitic Rocks	Syenites
Li	0.5	16	22	40	28
Be	0.2	1	2	5	1
Ti	300	12,000	6000	1600	3500
V	40	225	95	42	30
Cr	1800	180	36	15	2
Mn	1600	1750	800	500	850
Co	175	47	14	3	1
Ni	2000	145	35	7	4
Cu	15	90	33	15	5
Zn	40	120	66	50	130
As	0.8	2	2.2	1.5	1.4
Se	0.05	0.05	0.05	0.05	0.05
Zr	37	120	200	185	500
Nb	10	20	20	20	35
Mo	0.25	1.45	1.0	1.2	0.6
Ag	0.06	0.1	0.06	0.04	0.0*X*
Cd	0.1	0.2	0.1	0.12	0.13
Sn	0.5	1.5	1.5	3	*X*
Sb	0.1	0.6	0.2	0.2	0.*X*
Ce	0.*X*	20	80	96	160
Ta	0.5	0.8	2.5	4	2
W	0.5	0.8	1.2	1.9	1.3
Au	0.006	0.004	0.004	0.004	0.00*X*
Hg	0.02	0.09	0.08	0.08	0.0*X*
Tl	0.04	0.2	0.6	1.8	1.4
Pb	0.6	7	15	20	12
Bi	0.001	0.007	0.01	0.01	—
Th	0.005	3.5	7.8	17.5	13
U	0.002	0.75	2.4	3.3	3.0
S	200	300	250	350	300

100 ppm = 0.01 wt %.

Source: Modified from Krauskopf (1969) and other sources.

Bibliography

Ambrose, J. W., 1957. Violamac mine, Slocan district, B.C., pp. 88–95 in *Structural Geology of Canadian Ore Deposits*, vol. 2. Montreal: Can. Inst. Min. Metall., 524 pp.

Barnes, H. L., 1979. *Geochemistry of Hydrothermal Ore Deposits*, 2nd ed. New York: Wiley, 798 pp.

Barrer, R. M., 1951. *Diffusion in and through Solids*. London: Cambridge Univ. Press.

Bateman, A. M., 1950 *Economic Mineral Deposits*. New York: Wiley, 916 pp.

Billingsley, P., and A. Locke, 1935. Tectonic position of ore districts in the Rocky Mountain region. *AIME Trans.* 115:59–68.

Blanchard, R., 1947. Some pipe deposits of eastern Australia. *Econ. Geol.* 42:265–304.

Brammall, A., 1930. The Stantrg lead-zinc mine, Yugoslavia. *Min. Mag.* 42:9–15.

Brown, A. S., 1957. Red Rose tungsten mine, pp. 17–20 in *Structural Geology of Canadian Ore Deposits*, vol. 2. Montreal: Can. Inst. Min. Metall., 524 pp.

Brown, J. S., 1941. Factors of composition and porosity in lead-zinc replacements of metamorphosed limestone. *AIME Trans.* 144:250–263.

——, 1947. Porosity and ore deposition at Edwards and Balmat, New York. *Geol. Soc. Amer. Bull.* 58:505–545.

——, 1948. *Ore Genesis*. New Jersey: Hopewell.

Bryant, D. G., 1974. Intrusive breccias, fluidization and ore magmas, in *Mining Yearbook*. Colo. Min. Assoc., pp. 54–58.

Bryner, L., 1961. Breccia and pebble columns associated with epigenetic ore deposits. *Econ. Geol.* 56:488–508.

Butler, B. S., 1913. Geology and ore deposits of the San Francisco and adjacent districts, Utah. USGS, Prof. Pap. 80.

——, and W. S. Burbank, 1929. The copper deposits of Michigan. USGS, Prof. Pap. 144.

Cairnes, C. E., 1934. Slocan mining camp, British Columbia. Can. Geol. Surv. Mem. 173.

——, 1935. Descriptions of properties, Slocan mining camp, British Columbia. Can. Geol. Surv. Mem. 184.

——, 1948. Slocan mining camp, pp. 200–205, in *Structural Geology of Canadian Ore Deposits*. Montreal: Can. Inst. Min. Metall., 948 pp.

Christie, J. J., 1950. Inside Yugoslavia-II, The Trepça mine's great possibilities. *Eng. Min. J.* (June) 77–79.

Clappison, R. J. S., 1953. The Morning Star mine, Wood's Point, in A. B. Edwards, Ed., *Geology of Australian Ore Deposits*. Melbourne: Australasian Inst. Min. Metall.

Compton, R. R., 1955. Trondhjemite batholith near Bidwell Bar, California. *Geol. Soc. Amer. Bull.* 66:9–44.

——, 1960. Contact metamorphism in the Santa Rosa Range, Nevada. *Geol. Soc. Amer. Bull.* 71:1383–1416.

Darcy, H., 1856. *Les Fontaines Publiques de la Ville de Dijon*. Paris: Victor Dalmont.

Davis, S. N., 1969. Porosity and permeability of natural materials, pp. 54–89 in R. M. DeWeist, Ed., *Flow Through Porous Media*. New York: Academic Press, 530 pp.

Edwards, A. B., 1952. The ore minerals and their textures. *Roy. Soc. N.S.W.J. and Proc.* 85:26–45.

——, 1954. *Textures of the Ore Minerals and Their Significance.* Melbourne: Australasian Inst. Min. Metall.

——, 1960. Contrasting textures in the silver-lead-zinc ore of the Magnet mine, Tasmania. *Neues Jahrb. Miner. Petrogr. Abh.* 94:298–318.

Emmons, W. H., 1938. Diatremes and certain ore-bearing pipes. AIME, Tech. Pub. 891.

Finucane, K. J., 1948. Ore distribution and lode structures in the Kalgoorlie goldfield. *Australasian Inst. Min. Metall. Proc.* 148–149:111–129.

Forgan, C. B., 1948. Ore deposits at the Stantrg lead-zinc mine. 18th Int. Geol. Congr. Rep., pt. 7, pp. 162–176.

Geijer, P., 1971. Sulfidic "ball ores" and the pebble dikes. *Sver. Geol. Unders. Arsb., ser. C Avh. Uppsatser* 65(8):1–29.

Graton, L. C., 1906. Reconnaissance of some gold and tin deposits of the southern Appalachians. USGS Bull. 293.

Grout, F. F., 1923. Occurrence of ladder veins in Minnesota. *Econ. Geol.* 18:494–505.

Gunning, H. C., 1948. Privateer mine, pp. 86–87 in *Structural Geology of Canadian Ore Deposits*, vol. 1, Montreal: Can. Inst. Min. Metall., 948 pp.

Hack, J. T., 1942. Sedimentation and volcanism in the Hopi Buttes, Arizona. *Geol. Soc. Amer. Bull.* 53:335–372.

Hawkes, H. E., and J. S. Webb, 1962. *Geochemistry in Mineral Exploration.* New York: Harper and Row.

Herman, H., 1914. Economic geology and mineral resources of Victoria. Victoria (Australia) Geol. Surv. Bull. 34, 36 pp.

Holser, W. T., 1947. Metasomatic processes. *Econ Geol.* 42:384–395.

Hubbert, M. K., 1940. The theory of ground water motion. *J. Geol.* 48:785–944.

——, and W. W. Rubey, 1959. Role of fluid pressure in mechanics of overthrust faulting, part I. *Geol. Soc. Amer. Bull.* 70:115–166.

——, and D. G. Willis, 1957. Mechanics of hydraulic fracturing. *J. Petrol. Tech.* 9(6):158–168.

Ingerson, E., 1954. Nature of the ore-forming fluids at various stages—a suggested approach. *Econ. Geol.* 49:727–733.

Johnston, W. P., and J. D. Lowell, 1961. Geology and origin of mineralized breccia pipes in Copper Basin, Arizona. *Econ. Geol.* 56:916–940.

Klinkenberg, L. J., 1941. The permeability of porous media to liquids and gases, in *Drilling and Production Practice.* Amer. Petrol. Inst.

Knapp, R. B., and J. E. Knight, 1977. Differential thermal expansion of pore fluids: fracture propagation and micro earthquake production in hot pluton environments. *J. Geophys. Res.* 82:2515–2522.

——, and D. Norton, 1981. Preliminary numerical analysis of processes related to magma crystallization and stress evolution in cooling pluton environments. *Amer. J. Sci.* 281:35–68.

Krauskopf, K. B., 1957. The heavy metal content of magmatic vapors at 600°C. *Econ. Geol.* 52:786–807.

——, 1967. Source rocks for metal-bearing fluids, pp. 1–33 in H. L. Barnes, Ed., *Geochemistry of Hydrothermal Ore Deposits.* New York: Holt, Rinehart and Winston, 670 pp.

Kuhn, T. H., 1941. Pipe deposits of the Copper Creek area, Arizona. *Econ. Geol.* 36:512–538.

Kutina, J., 1955. Genetische Diskussion der Makrotexturen bei der geochemischen Untersuchung des Adalbert Hauptganges in Pribram. *Chemie der Erde* 17:241–323.

Locke, A., 1926. The formation of certain ore bodies by mineralization stoping. *Econ. Geol.* 21:431–453.

Lovering, T. G., 1962. The origin of jasperoid in limestone. *Econ. Geol.* 57:861–889.

Lovering, T. S., 1942. Physical factors in the localization of ore, in W. H. Newhouse, Ed., *Ore Deposits as Related to Structural Features.* Princeton, N.J.: Princeton Univ. Press.

MacDiarmid, R. A., 1959. Geology and ore deposits of the Bristol silver mine, Pioche, Nevada. PhD. dissert., Stanford Univ., CA.

——, 1960. Controls on ore deposition at the Bristol silver mine, Pioche, Nevada. *Geol. Soc. Amer. Bull.* 71:1921.

Malcolm, W., 1912. Gold fields of Nova Scotia. Can. Geol. Surv. Mem. 20-E.

Mayo, E. B., 1958. Lineament tectonics and some ore districts in the Southwest. *Min. Eng.* 10:1169–1175.

McCormick, J. E., L. L. Evans, R. A. Palmer, and F. D. Rasnick, 1971. Environment of the zinc deposits of the Mascot-Jefferson City district, Tennessee. *Econ. Geol.* 66:757–762.

Morris, H. T., 1952. Primary dispersion patterns of heavy metals in carbonate and quartz monzonite wall rocks, part II, in H. T. Morris and T. S. Lovering, Eds., Supergene and hydrothermal dispersion of heavy metals in wall rocks near ore bodies, Tintic district, Utah. *Econ. Geol.* 47:698–716.

Muscat, M., 1937. *The Flow of Homogeneous Fluids through Porous Media.* New York: McGraw-Hill.

Newall, G., and N. Rast, 1970. *Mechanisms of Igneous Intrusion.* Great Britain: Sell House Press, 380 pp.

Norton, D. L., 1982. In S. R. Titley, Ed., *Advances in Geology of the Porphyry Copper Deposits, Southwestern North America.* Tucson: Univ. Ariz. Press.

——, and R. B. Knapp, 1977. Transport phenomena in hydrothermal systems: The nature of porosity. *Amer. J. Sci.* 277:913–936.

Ohle, E. L., 1951. The influence of permeability on ore distribution in limestone and dolomite. *Econ. Geol.* 46:667–706, 871–908.

——, and J. S. Brown, Eds., 1954. Geologic problems in the southeast Missouri lead district. *Geol. Soc. Amer. Bull.* 65:201–222.

Park, C. F., Jr., and R. S. Cannon, Jr., 1943. Geology and ore deposits of the Metaline quadrangle, Washington, USGS, Prof. Pap. 202.

Pelletier, R. A., 1964. *Mineral Resources of South-Central Africa.* London: Oxford Univ. Press, pp. 126–129.

Perry, V. D., 1961. The significance of mineralized breccia pipes. *Mining Eng.* 13:367–376.

Prescott, B., 1915. The main mineral zone of the Santa Eulalia district, Chihuahua. *AIME Bull.* 98:155–198.

Ransome, F. L., 1911. Geology and ore deposits of the Breckenridge district, Colorado. USGS, Prof. Pap. 75, pp. 144–147.

Richards, J. R., 1971. Major lead orebodies–mantle origin? *Econ. Geol.* 66:425–434.

Richards, L. A., 1931. Capillary conduction of liquids through porous mediums. *Physics* 1:318–333.

Roscoe, S. M., 1951. Dilation maps, their application to vein-type ore deposits. PhD Dissert., Stanford Univ., CA.

Rove, O. N., 1947. Some physical characteristics of certain favorable and unfavorable ore horizons. *Econ. Geol.* 42:57–77, 161–193.

Scheidegger, A. E., 1960. *The Physics of Flow Through Porous Media.* New York: Macmillan.

Sclar, C. B., and B. H. Geier, 1957. The paragenetic relationships of germanite and reniérite from Tsumeb, South-West Africa. *Econ. Geol.* 52:612–631.

Schmitt, H. H., 1966. The porphyry copper deposits in their regional setting, pp. 17–34 in S. R. Titley and C. Hicks, Eds., *Geology of the Porphyry Copper Deposits, Southwestern North America,* Tucson, AZ: Univ. Arizona Press, 287 pp.

Schneiderhöhn, H., 1929. Das Otavi-Bergland und seine Erzlagerstätten. *Z. Prakt. Geol.* 37:85–116.

——, 1931. *Mineralische Bodenschätze im südlichen Afrika.* Berlin.

——, 1941. *Lehrbuch der Erzlagerstättenkunde.* Jena: Gustav Fischer, pp. 459–471.

Schumacher, F., 1950. *Die Lagerstätte der Trepça und ihre Umgebung.* Belgrade.

——, 1954. The ore deposits of Yugoslavia and the development of its mining industry. *Econ. Geol.* 49:451–492.

Sillitoe, R. H., and F. J. Sawkins, 1971. Geologic, mineralogic, and fluid inclusion studies relating to the origin of copper-bearing tourmaline breccia pipes, Chile. *Econ. Geol.* 66:1028–1041.

Skinner, B. J., 1979. The many origins of hydrothermal mineral deposits, pp. 1–21 in H. L. Barnes, Ed., *Geochemistry of Hydrothermal Ore Deposits*, 2nd ed. New York: Wiley, 798 pp.

Snyder, F. G., and J. W. Odell, 1958. Sedimentary breccias in the southeast Missouri lead district. *Geol. Soc. Amer. Bull.* 69:899–925.

Söhnge, P. G., 1952. The Tsumeb story: pipe-like, massive-sulphide ore body appears to fill volcanic pipe in depth. *Min. World* 14(6):22–24.

——, 1963. Genetic problems of pipe deposits in South Africa. *Geol. Soc. So. Afr. Trans.* 66:xix–xxii.

Spurr, J. E., 1906. Ore deposits of the Silver Peak quadrangle, Nevada. USGS, Prof. Pap. 55, pp. 112–128.

——, 1923. *The Ore Magmas.* New York: McGraw-Hill, 915 pp.

Taylor, H. P., Jr., 1967. Oxygen isotope studies of hydrothermal mineral deposits, pp. 109–142 in H. L. Barnes, Ed., *Geochemistry of Hydrothermal Ore Deposits.* New York: Holt, Rinehart and Winston, 670 pp.

——, 1969. Oxygen isotope studies of anorthosites with special reference to the origin of bodies in the Adirondack Mountains, New York, pp. 111–134 in Y. W. Isachsen, Ed., *Origin of Anorthosites.* NY State Mus. Sci., Serv. Mem. 18.

——, 1979. Oxygen and hydrogen isotope relationships in hydrothermal mineral deposits, pp. 236–277 in, H. L. Barnes, Ed., *Geochemistry of Hydrothermal Ore Deposits*, 2nd ed. New York: Wiley, 798 pp.

Threadgold, I. M., 1958. Mineralization at the Morning Star gold mine, Wood's Point, Victoria. *Australasian Inst. Min. Metall. Proc.* 185:1–27.

Tsumeb Corporation Staff, 1961. Geology, mining methods, and metallurgical practice at Tsumeb. S. Afr. Inst. Min. Metall., *7th Commonw. Min. Metall. Congr. Trans.* 1:159–179.

Turcotte, D. L., and G. Schubert, 1982. *Geodynamics: Application of Continuum Physics to Geological Problems.* New York: Wiley, 450 p.

Wagner, P. A., 1927. The pipe form of ore deposits. *Econ. Geol.* 22:740–741.

Wandke, A., 1930. Ore deposition in open fissures formed by solution pressure. AIME Tech. Pub. 342. Also 1931. *AIME Trans.* 96:291–304.

Wehrenberg, J. P., and A. Silverman, 1965. Studies of base metal diffusion in experimental and natural systems. *Econ. Geol.* 60:317–350.

Williams, H., 1936. Pliocene volcanoes of the Navajo-Hopi country. *Geol. Soc. Amer. Bull.* 47:111–171.

Wilson, W. E., Ed., 1977. Tsumeb! The world's greatest mineral locality. *Mineral. Rec.* 8(3):128 pp.

Zeller, H. D., 1957. The Gas Hills uranium district and some probable controls for ore deposition, in *Guidebook to Southwest Wind River Basin*, Wyoming Geol. Assoc., 12th Ann. Field Conf.

FOUR

Deposition of the Ores

The careful field observer who has a thorough understanding of why an ore deposit is localized in a given depositional environment has a definite advantage in the search for other deposits in similar environments. Whether an ore deposit is a product of igneous, sedimentary, metamorphic, or weathering processes, its localization is generally the result of many factors. One factor, or series of factors, may account for the delivery of an ore constituent to a particular place, another for its deposition, and still others for the size and scale of the deposit.

Some ores are deposited through physical effects such as gravity; an early formed chromite crystal may settle to the floor of a magma chamber, or a flake of gold may settle to the bottom of a layer of agitated sediments. Other ores are deposited because of chemical changes, such as a change in pH, which result from reactions between ore-bearing solutions and the host rocks. A drop in temperature, pressure, or velocity of the transporting medium, or the admixture of a second solution, may also bring about chemical reactions that cause ore deposition. Localized deposition may also depend on the chemistry of the host rocks. Although the reasons for deposition and the causes of localization are often the same, we can only rarely pinpoint what they are for a given deposit. We have just explored some of the ways in which ore-bearing fluids reach sites of potential deposition. In this chapter we focus on mechanisms of deposition, with emphasis on physical and chemical controls.

▸ DEPOSITION OF MAGMATIC SEGREGATION DEPOSITS

Magmatic segregation deposits form as a direct result of igneous differentiation. The term fundamentally refers only to some form of separation of the components of a melt. We have already considered (Chapter 2) something of the nature of magmatic liquids, how two or three immiscible liquids can form, and how those liquids might accumulate to crystallize in place, be *filter pressed,* or be injected or extruded as *ore magmas.* We really did not consider the mixed systems of solid and liquid magmatic phases, and how some of those liquid phases can cool and ultimately freeze. We deal in detail with such systems in Chapter 9, where layered mafic intrusives such as the Bushveld Igneous Complex of South Africa and the Stillwater Complex of Montana are treated, along with carbonatites, kimberlites, and anorthosites and their related ore deposits. Chapter 10 treats ores related to the formation of oceanic crust, and Chapter 11 those related to intermediate to felsic intrusive rocks. The field of *magmatic segregation deposits* in its broadest sense, then, will be well covered, and only a few basics need be introduced here.

Deposition of magmatic ores—ore minerals or rocks formed from parent liquids that are more generally silicate than oxide or sulfide melts—can occur in five ways: by the settling and accumulation of crystallizing minerals, by direct crystallization on magma chamber walls or floors, by the separation of magmatic liquids and their solidification, by the consolidation of an igneous rock with an entrained accessory economic mineral, or by crystallization of a melt in toto.

The first named has generally been called *magmatic sedimentation* or *differential magmatic sedimentation* if the magma is in convective motion so that specific gravity and grain size and shape variations lead to winnowing and selective deposition. It depends solely on the fact that denser solid-phase crystals will sink to the bottom of a magma chamber if the viscosity of the magma itself is sufficiently low. Layering is possible in static melts, as different phases crystallize consecutively according to the precepts of petrology and phase-rule chemistry, and conspicuous layering is almost inevitable where differential magmatic sedimentation has occurred.

Good examples of compositional layering of differentiates are found in the mafic igneous rocks of the Bushveld Complex of South Africa and the Stillwater Complex of Montana. Layering is especially conspicuous in mafic or intermediate igneous rocks containing chromite, magnetite, or ilmenite and may result from any of several processes during differentiation (Wager and Brown, 1967). One widely acknowledged process is *crystal settling.* As the temperature of a magma begins to fall, crystals begin to form especially near the cooler "ceiling" of the magma chamber. The composition of these crystals depends, of course, upon the composition of the melt with which the crystals are in equilibrium. In general, the ferromagnesian minerals, including the spinel family of chromite and magnetite, crystallize first. Since these early minerals are more dense than the residual magma, they may settle and collect near the base of the magma chamber. As crystallization proceeds, crystals of progressively different composition develop. If they sediment and accumulate, crudely layered *cumulate* igneous rocks may result (Bowen, 1928). Olivine alone produces *dunite.* Olivine and orthopyroxene together constitute *peridotite* (90% or more olivine), *harzburgite,* and *pyroxenite* (90% or more enstatite). Orthopyroxene and calcic plagioclase combine to form *norite. Gabbro* is formed when the orthopyroxene enstatite is followed by the clinopyroxene augite to be precipitated with andesine plagioclase. *Anorthosite* is nearly pure igneous calcic plagioclase. The normal order of crystallization was identified by N. L. Bowen decades ago and is shown in Figure 4-1. Rocks formed by crystal settling are exclusively at the high temperature mafic side of the series because only mafic melts are of sufficiently low viscosity to permit reasonable settling rates of the solid phases. Thus the cumulate rocks involve deposits of chromite (Chapters 9 and 10) because chromite as the mineral [(Mg, Fe) Cr_2O_4] crystallizes only at high-temperature and from mafic melts. If one can conceive of octahedral crystals of chromite, granules of olivine, and tablets of orthopyroxene crystallizing from the upper portions of a magma and raining to its floor—all at white heat—the mechanisms of cumulate rock formation are clear. If crystallization proceeds at normal cotectic mineral volume ratios, as prescribed by the system anorthite-forsterite-diopside familiar to petrologists, chromitiferous peridotites and harzburgites will form early, and successive layers of progressively less mafic–more felsic minerals will accumulate on top of them. Indeed, the peridotites and harzburgites are the source rocks of chromite in the Alpine peridotites discussed in Chapter 10. Layered bodies kilometers thick can accumulate by this process of crystal settling combined with fractional crystallization.

A subset of magmatic sedimentation, or crystal settling, is *differential magmatic sedimentation.* Wager and Deer (1939) developed ideas on the process to account for the finely and multiply layered mafic rocks of the Skaergaard Complex, Greenland, and Cameron and Emerson (1959) applied them to the fine and laterally continuous layered mafic units of the Bushveld Complex. Cameron and Emerson noted many morphologic, physical, kinematic, and compositional variations that suggested to them that 1-to-5-mm euhedrons of chromite and 1-cm beads of olivine, plates of orthopyroxene, and later anhedral plagioclase crystallized in constant proportion along a cotectic, but that variable rates of ponderous convective flow of the magma—centrifugally along the roof, downward at the wall, and inward along a sloping floor—dictated the proportions and ratios of phases that could come to rest on the chamber floor. When flow rates were highest, only the dense high-specific-gravity oxides could settle and accumulate on the floor, and laterally extensive but thin layers of *chromitite*—rocks of 90 to 100% chromite such as the Steelpoort Main Seam (Chapter 9)—formed. At slower flow rates, settling of crystals formed at the cotectic crystallization ratio

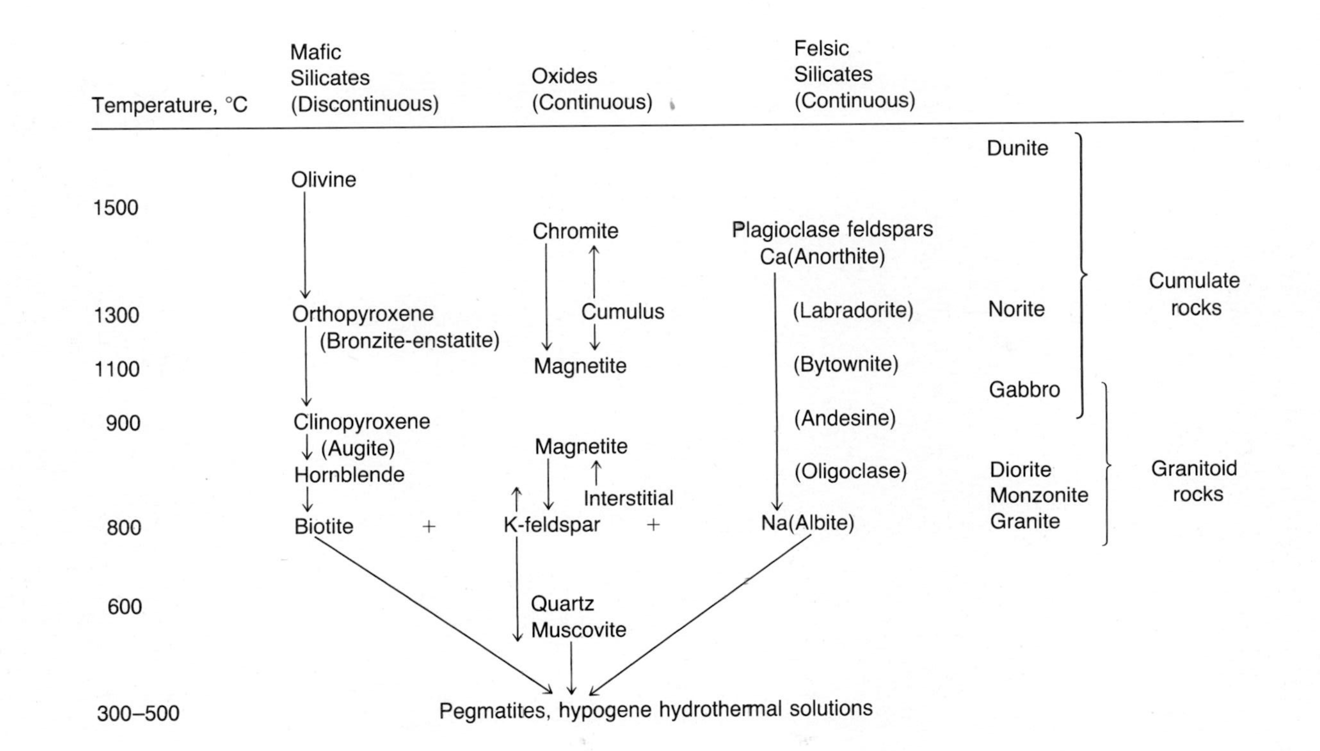

Figure 4-1. Modification of Bowen's reaction series describing the mineral and genetic interrelationships of plutonic rocks. This figure should be read with Figure 2-7 at hand; they overlap completely.

produced a meshwork that resulted in pyroxenites and norites. Differential magmatic sedimentation is not universally acceptable, is currently under review, and will doubtlessly be modified, but many workers are convinced of its applicability.

The second mechanism is that of direct crystallization of igneous minerals on the floor or walls of magma chambers (Jackson, 1961; Irvine, 1977; McBirney and Noyes, 1979). Jackson (1961) explained layering at the Stillwater Complex not by differential magmatic sedimentation but by suggesting that the interplay of the melting point gradient (3° per kilometer) in the melt, the adiabatic gradient (0.3° per kilometer), and heat loss from roof and floor dictated that crystallization and settling would occur at and along the floor, not at and along the ceiling. He thus proposed that layering—and concentration of chromium in chromite and platinoid elements as sulfides—resulted from rhythmic variations in the quantitative aspects of bottom crystallization. Irvine (1977) also appealed to crystallization at or near the magma chamber floor, with layering the result of episodic crystallization of mineral species. He suggested that the episodic nucleation and growth of minerals was controlled by the mixing of basic and silicic magmas.

The separation of magmatic liquids and their ultimate solidification and subsolidus unmixing have already been described. In the context of the deposition of the ores, the concentration of sulfophile elements in a sulfide liquid followed by settling out and "ponding" or injection of that liquid fixes the metals Fe, Cu, Ni, Zn, Pt-Pd-Rh-Re-Os-Ir, and Ag-Au in a district sense. As temperature falls, the elements redistribute themselves locally by the process of exsolution described in the last chapter. An (M_{ss}) monosulfide solid solution formed by freezing of the liquid at high temperatures near 1000°C unmixes at progressively lower temperatures, starting at about 600°C, in the general order pyrrhotite-pentlandite-chalcopyrite-valleriite until the Cu-Ni-Fe-S ions are in their final disposition (Craig and Kullerud, 1968). Oxide phases in layered mafic intrusion layerings readjust, too, probably by means of resorption, overgrowth, and the formation of "chain structures," subhedral linkages of minute interstitial grains (Figure 4-2).

The fourth mode involves entrained or late-crystallizing accessory minerals. The concentration of economic accessory minerals in a massive igneous rock may be great or small, depending upon the kind of mineral and the degree of differentiation. Moreover, economics may determine whether or not an igneous differentiate constitutes ore according to the value of contained accessory minerals such as monazite, cassiterite, or ilmenite. For example, ilmenite is a common accessory mineral in gabbros in the Lake Sanford, New York, district. The area was originally prospected for iron, which is much more abundant, but the titanium impurity made smelting costs prohibitive. Now that titanium is a valuable commodity, the ores, for example in the Tahawas pit at Lake Sanford, are worked mainly for the ilmenite, and iron is a by-product (Killinger, 1942). Monazite

Figure 4-2. Opaque grains of chromite (black) against the white-to-gray twinning of labradorite plagioclase in a specimen from the Bushveld Igneous Complex. The hook-shaped grains of chromite suggest resorption, the elongate clusters at upper center and upper right, mantling and overgrowth to form chain structures of chromite. 20×, crossed nicols. See also Figures 9-2 and 9-3. (*Photo by* C. M. Taylor.)

($CePO_4$), cassiterite (SnO_2), and even diamond (C) are examples of accessory minerals that elevate the economic values of igneous rocks. More felsic rocks are less likely to be layered.

And finally, some igneous rocks are "ores" in themselves, such as the gabbro "black granite" or granite building or *dimension stones* (Figure I-1), and are used as crushed rock for road ballast and as material for concrete aggregate. Magmatic products of this type are dikes and minor intrusives of ordinary igneous rocks, and their magmas migrate and solidify normally. In fact, they are merely igneous rocks of special economic value.

▸ DEPOSITION OF CARBONATITES

The unusual rocks known as *carbonatites* (Chapter 9) are commonly designated as magmatic segregation deposits (Tuttle and Gittins, 1966). A carbonatite is defined as a carbonate-rich igneous rock genetically related to mantle degassing or to alkalic rock-forming processes and alkalic magma evolution (Pecora, 1956). They result from the intrusion, cooling, and crystallization of a true Ca-Fe-Mg carbonate melt, probably with some hydrothermal activity as well. They are typically cylindrically zoned kilometer-scale pipelike bodies shallowly emplaced, normally from great depths in the upper mantle. Carbonatites can be associated with either intrusive or extrusive igneous systems. Many carbonatite deposits have been described in the literature, and some have been of great economic value. One of the most interesting deposits is the Oldoinyo Lengai volcano in northern Tanzania, where several modern lava flows are composed predominantly of sodium carbonate (King, 1965; Dawson, 1966). Other carbonatites, especially those of Africa, as at Palabora in the Transvaal of South Africa, contain valuable copper deposits. Valuable accessory minerals recovered at several places include niobium, rare earth minerals, fluorocarbonates, and phosphates (Deans, 1966).

▸ DEPOSITION FROM HYDROTHERMAL FLUIDS

If one could understand in detail how and why—and therefore where—solutes precipitate from hot aqueous solutions, and if one understood from where ore-forming components come and how fluids migrate through crustal rocks (Chapter 3), then most of the problems of exploration geology and economic geology as a science would be solved. As it is, we have made tremendous progress in all of these areas—and more—in the last two decades. Many research volumes have been written on the subject, most notably two different editions of *Geochemistry of Hydrothermal Ore Deposits* edited by H. L. Barnes and published in 1967 and 1979. We have already dealt with how aqueous ore-bearing fluids are generated, why they flow, and how they are guided or channeled. In this chapter we briefly consider what they are, and then how the separation of the fluid (solvent) and its ore minerals (solute) occurs. The latter involves both mechanical and chemical influences, influences that can be interpreted more or less successfully from the textures of the resultant ores.

There are many ways to consider what hydrothermal fluids are chemically and physically, and how to show them graphically. They are well described isotopically (Figure 2-11) and in terms of the transition from Bowen's reaction-series-type phenomena (Figure 4-1) to the alteration silicates (Figure 5-3). But what makes them *ore-forming fluids*? Here we want to turn to ion species in solution, to complex ions, to competing anions and cations, and to concentrations, or activities. First, analyses of a few samples of captive hydrothermal solutions are presented in Table 4-1. Compare these with the data of Table 2–2 (hot springs and fumaroles) and of Table 20-2 (connate waters), and those ore-forming hydrothermal fluids prove expectably to be orders of magnitude less dilute, that is, far more saline and enriched in sodium, potassium, rubidium, cesium, calcium, strontium, barium, magnesium, sulfur, chlorine, carbon dioxide, and the metals. Burnham (1967) pointed out that the first hydrothermal fluids to separate from a melt of intermediate composition will be greatly enriched in CO_2 and in Cl^-. The partitioning concentration ratio of Cl^- is nearly 10:1 in favor of the aqueous phase, so 0.3 wt % Cl in the melt would produce a watery fluid with 0.6 molal (2.1 wt %) NaCl. This amount of chlorine (Table 4-1) would encourage the partitioning of significant amounts of those ions that form associated complexes and complex ions with chlorine at moderate temperatures, namely, H^+, Fe^{+2}, Cu^+, Zn^{+2}, Pb^{+2}, and others, and a potential ore-forming fluid is created. What are the solubilities or the concentrations of transported metals in a potential ore-forming fluid? As indicated earlier, the solubilities

TABLE 4-1. Analyses of Actual or Potential Ore-Forming Solutions

	Modern Solutions[a]		Ancient Solutions[a]		
Element, Molecule, or Ion	Salton Sea[b] Geothermal Brine	Cheleken[c] Geothermal Brine	Fluid[d] Inclusion, Fluorite	Fluid[e] Inclusion, Sphalerite	Fluid[f] Inclusion, Quartz
Cl	155,000	157,000	87,000	46,500	295,000
Na	50,400	76,140	40,400	19,700	152,000
Ca	28,000	19,708	8,600	7,500	4,400
K	17,500	490	3,500	3,700	67,000
Sr	400	636	—	—	—
Ba	235	—	—	—	—
Li	215	7.9	—	—	—
Rb	135	1.0	—	—	—
Cs	14	0	—	—	—
Mg	54	3,080	5,600	570	—
B	390	—	<100	185	—
Br	120	526.5	—	—	—
I	18	31.7	—	—	—
F	15	—	—	—	—
NH_4	409	—	—	—	—
HCO_3^-	>150	31.9	—	—	—
H_2S	16	0	—	—	—
$SO_4^=$	5	309	1,200	1,600	11,000
Fe	2,290	14.0	—	—	8,000
Mn	1,400	46.5	450	690	—
Zn	540	3.0	10,900	1,330	—
Pb	102	9.2	—	—	—
Cu	8	1.4	9,100	140	—
As	12	0.03	—	—	—
Ag	1	—	—	—	—

Analyses of natural brines believed to be actual or potential ore-forming solutions (concentrations in parts per million). See also Tables 2-1 (hot springs and fumaroles) and 20-2 (connate waters).

[a]Dashes indicate that values were not reported. They should not be read as zero.
[b]Salton Sea geothermal field, California. Brine temperature 300–320°C, density 1.21 g/cm^3, pH (25°C) 5.2.
[c]Well in Cheleken Peninsula, U.S.S.R. Temperature approximately 80°C, pH (25°C) 5.5.
[d]Fluid inclusion from fluorite, Cave-in-Rock district, Illinois.
[e]Average composition of inclusions in late-stage sphalerite from OH vein, Creede, Colorado.
[f]Average fluid inclusions in quartz from core zone, porphyry copper orebody, at Bingham Canyon, Utah.

Source: From Barnes and Czamanske (1967) and Skinner and Barton (1973). See originals for individual data credits.

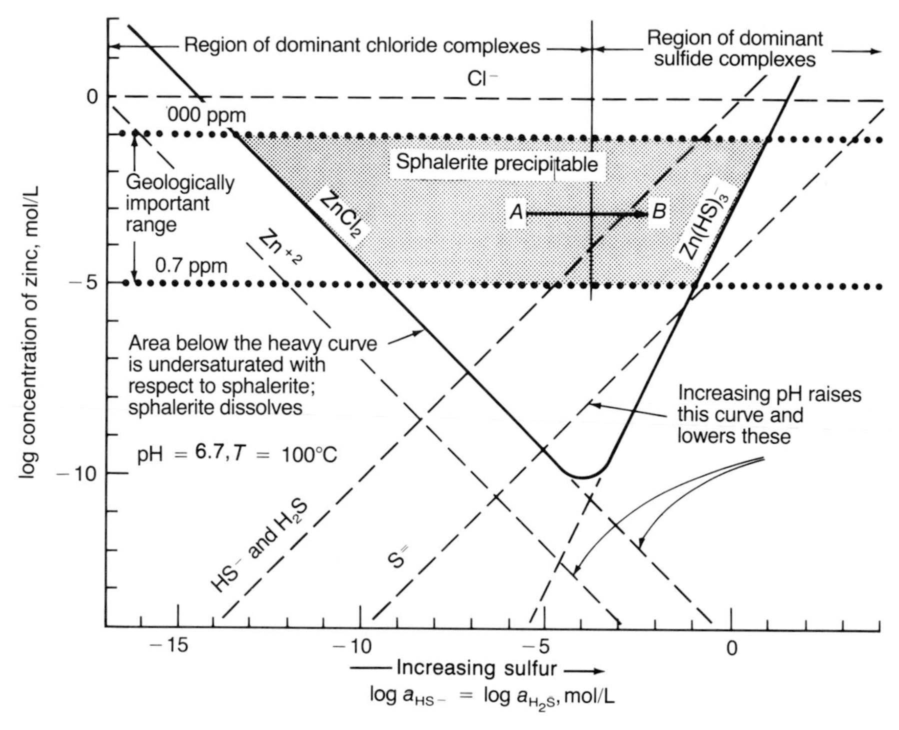

Figure 4-3. Solubility range important to effective zinc transport as $ZnCl_2$ (chloride) to the left of the vertical line and the bisulfide complex $Zn(HS)_3^-$ to the right. The range of 100 to 5000 ppm Zn is probably achieved regularly in hydrothermal fluids. Precipitation from such solutions is expectable. If a fluid at *A* were to increase in bisulfide ion content from 10^{-6} to 10^{-2} at *B*, $ZnCl_2$ would be destabilized, and either sphalerite or $Zn(HS)_3^-$ could form. (*After* Skinner and Barton, 1973.)

of metals in either pure water or water containing enough reduced sulfur species to form sulfide deposits are exceedingly low, and in sulfide or sulfur-enriched fluids, the common ion effect inhibits direct solubility of sulfides. But if chlorine is introduced, the situation changes radically (Figure 4-3). Solutions containing 100 to 500 ppm Zn, as little as 10 ppm Cu, and several hundred parts per million Fe are feasible ore-forming fluids. The most important complexes in terms of metal transport at elevated temperatures appear to be $ZnCl_2$, $PbCl_2$, $PbCl_4^=$, CuCl, $CdCl_2$, $FeCl^+$, Au_2Cl_6, and the bisulfides $HgS{\cdot}(HS)_2^=$ and $Ag(HS)_2^-$, but much detailed chemical work must be done to establish correct stoichiometries of possible complexes. It has been shown (Barnes and Czamanske, 1967) that consistencies in the spatial array and timing of deposition that we call the *zoning* and *paragenetic sequence* of sulfides in orebodies around centrally located stocks (Chapter 6) is such that the transport of metals as vapors—true gases—as colloids, or as simple ions or molecules is unimportant. In general, metal concentrations of 10 ppm or more are all that are necessary to deliver enough metal to permit ore-mineral deposition. Indeed Anderson (1979) accepts 1 to 2 ppm Pb, Zn, and S as sufficient.

The overall nature of the fluid can be discussed in terms of Eh-pH, of $f(O_2)$-$f(S_2)$, of ion activity ratios, or in many other contexts. In any of these "spaces," a favorable zone for ore deposition can be depicted. Two are given here, one relating the fluid to pH and $f(O_2)$ (Figure 4-4) and one to $f(S_2)$-$f(O_2)$ (Figure 4-5), both "averaged" at 250°C and 0.1 molar total sulfur species. In both diagrams, the shaded areas designate regions in which the solutions would be in equilibrium with ore minerals, but capable of precipitating them upon cooling or other change. We will become familiar with these ore minerals in Chapters 5 through 20. The diagrams describe the nature of the transporting fluid *and* the environment of deposition. Minerals deposited at various $f(O_2)$-$f(S_2)$ ratios are best shown in Figure 4-5.

The title of this chapter, however, refers to deposition. What are the causes of deposition? They prove to be physical, or mechanical, and chemical. We can appreciate the difference between the two by considering one fluid moving upward along an irregular but *chemically inert* conduit, and another one traversing a conduit walled by normally reactive, *interactive* wall rocks with their own contained fluids and mobile soluble components. The former may precipitate its load because of boiling, cooling, the Bernoulli

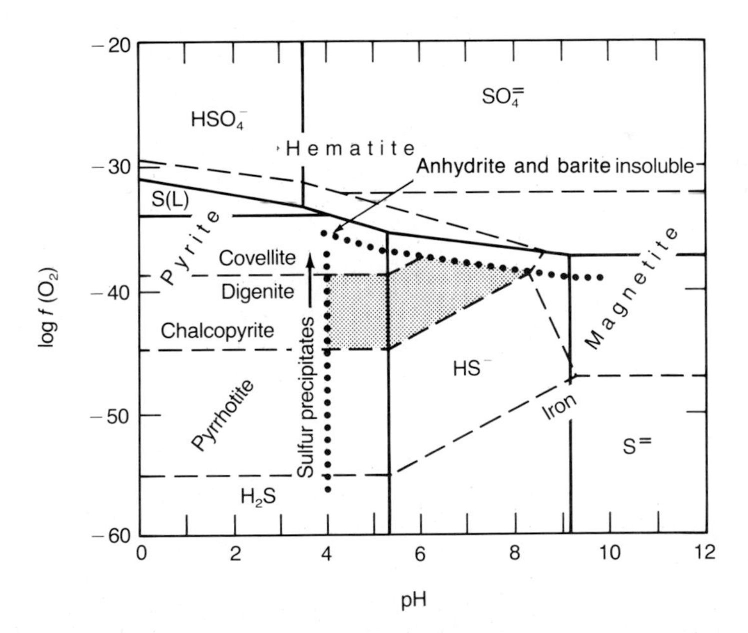

Figure 4-4. Stability fields for the Fe-S-O system and other selected minerals and aqueous species at 250°C based on data at 25°C which were adjusted to 250°C by the van't Hoff equation. Dashed lines represent boundaries between stability fields of minerals and heavy solid lines between aqueous species. The shaded area defines the chemical properties of fluids in equilibrium with pyrite and chalcopyrite, and anhydrite and barite along the dotted line. Total sulfur 0.1 molal. (*After* Barnes and Czamanske, 1967.)

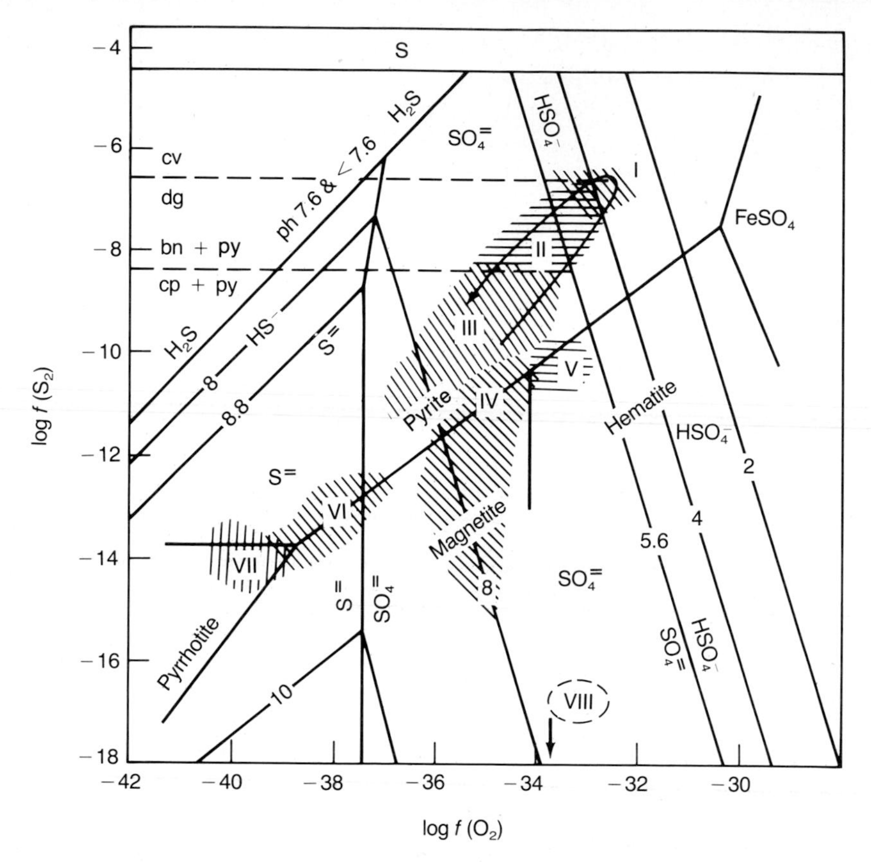

Figure 4-5. Variations in the fugacities of sulfur and oxygen in hydrothermal fluids as they control the mineralogy of phases deposited from them. The hairpin arrow describes the evolution of fluids in time and space at Butte, Montana, and several other districts. The same diagram is related to hydrothermal alteration in Chapter 5. See the expanded caption that is Table 5-1 on page 182. Temperature is 250°C; total sulfur 0.1 molal. Numbered lines are pH contours. (*After* Meyer and Hemley, 1967.)

effect of diminished pressure and solubility where fluid flow velocity is increased by "throttling" in a fissure, and changes in pressure. The precipitated phases will simply coat and successively encrust the walls of the conduit, like layers of boiler scale in a pipe. These deposits are called *open-space fillings*, and they are texturally distinctive, as is shown below. The latter fluid, however, will react with its walls, fault gouge, and minerals deposited earlier, exchanging a profusion of anions, cations, complex ions, gases, liquids, and H^+, $(OH)^-$, and H_2O. It is also subject to all of the modifications of the former, and to dilution, cooling, and oxidation by groundwater. Its precipitated phases are commonly "entangled" with the wall rocks and merge with them, normally with extensive replacement, or chemical substitution. Where we see replacement textures, chemical interaction of the old and new is implicit. So to the extent that the two are separable—and, of course, there are overlaps—"replacement" signifies *chemical* controls and "open-space filling" requires *mechanical* controls. In the next section we examine the observational geologic criteria for the two.

Barnes (1979) treats four principal causes of deposition from hydrothermal fluids, none of which operates alone. They are (1) temperature change, normally cooling, (2) pressure change, normally pressure drop, (3) chemical change, by reaction between a solution and the wall rocks traversed, and (4) chemical changes due to solution mixing.

Toulmin and Clark (1967) discussed purely thermal aspects of deposition. First, temperature change is a balance. Aqueous fluids rising up a fissure will tend to be cooled by mixing with groundwater, by expansion, and by heat loss to cooler wall rocks. They tend to be heated by hotter gases or fluids that may overtake and bubble through them from below, by the generally exothermic precipitation of vein minerals, and probably trivially by the exothermic reactions of wall-rock alteration (Chapter 5). Fluids can be heated if they rise into hotter rocks, such as young volcanic piles. Heating or cooling can cause precipitation in at least four ways: (1) by affecting the solubilities of sulfide, oxide, carbonate, and sulfate minerals, (2) by affecting association and dissociation of aqueous metal-bearing complex ions and thus their stabilities in solution, (3) by influencing ion-pairing hydrolysis constants of ions such as Cl^- and $(HS)^-$ and thus their availability for involvement in metal complexes, and (4) by affecting the fugacity of oxygen, oxidation potential (Eh), and thus the relative ion populations of $SO_4^=$, $(HS)^-$, and H_2S. Quite clearly, the balance referred to above is generally in favor of cooling; the tendencies toward heating involve either special situations or minimal effects, while the tendencies to cool are normal and profound. Also, cooling of a hot metalliferous brine leads to increased *dissociation* of chloride complexes [Equation (4-1)] and increased *reduction* of $SO_4^=$ to $(HS)^-$ and H_2S [Equations (4-2) and (4-3)]. As simple metal ions are released into solution, reduced sulfur becomes available, and solid-phase sulfides of the metals (Me in the equations below) are ready to be deposited. The presence of reduced sulfur is enough to capture metals from complexes whether or not they are becoming unstable for other reasons, such as dilution or falling temperature.

$$MeCl_2(aq) \rightleftarrows Me^{+2}(aq) + 2Cl^- \qquad (4\text{-}1)$$

$$(HSO_4)^- + H^+ \rightleftarrows H_2S(aq) + 2O_2 \qquad (4\text{-}2)$$

$$(HSO_4)^- + H^+ + 4H_2 \rightleftarrows H_2S(aq) + 4H_2O \qquad (4\text{-}3)$$

And the solubility products of the sulfide minerals are so low that sulfide precipitation occurs by

$$Me^{+2} + H_2S \rightleftarrows MeS + 2H^+ \qquad (4\text{-}4)$$

As Barnes (1979) points out, a temperature drop of 20°C or more over a short distance and for a long time is probably needed to cause appreciable precipitation. It is difficult to maintain thermal gradients of that magnitude for long unless mixing of hot upwelling fluids and cool near-surface waters occurs, either on land or on the seafloor, or a throttling-type pressure drop occurs, as discussed below. In any event, sharp drops in temperature are probably the single most important causes of precipitation of both ore and waste minerals in fissures.

A second major effect is that of pressure change. We saw earlier that solute solubilities in supercritical liquids generally increase with pressure. Even if boiling or mixing does not occur, pressure drop takes place as a supercritical fluid moves up a fissure. But it will be a gradual effect, not one likely to produce an economically valuable concentration in a short vertical interval. Two pressure-related effects that are important are boiling and throttling. Boiling is normally thought of as occurring when a fluid is heated past its boiling point, following a horizontal path across the two-phase curve of Figure 2-10. But boiling can occur isothermally or even during cooling if pressure is reduced over an already hot liquid, an essentially vertical downward path in Figure 2-10. This process is called *retrograde boiling*, and doubtless is a common effect in near-surface hydrothermal systems (White, Muffler, and Truesdell, 1971; Sondergeld and Turcotte, 1979). Boiling has two important ramifications: (1) volatile components, generally acids such as HF and HCl, are lost in the escape of the vapor phase, leaving the liquid more alkaline and less capable of metal transport; and (2) the remaining liquid phase becomes concentrated in nonvolatile solutes by loss of water. Both effects lead to the precipitation of dissolved ore components. *Throttling* may occur across an obstruction or restriction in a vein or where the nature of permeability changes the pressure regime from lithostatic to hydrostatic—say, where a network of microfractures at moderate depth merges to constitute a single fault open to the surface, perhaps halfway up in Figure 11-4*c*, for example. An ore-bearing fluid issuing upward into a brecciated zone (Figure 3-14) may deposit its load because of expansion and a resultant loss in pressure. If the fluid does not boil, it may be simply depressurized at constant heat content, and both temperature and solute solubilities drop markedly. Throttling is also called *adiabatic decompression*, or depressurization without heat loss. Pressure effects thus work with temperature effects; lowering either or both tends to cause precipitation of transported components.

The third major effect is that of chemical interaction with wall rocks. These reactions are covered extensively in Chapter 5, but we note several related effects here. Probably most significantly, the alteration of most wall-rock-forming silicates involves hydrolysis and hydration, the consumption of H^+ and H_2O, respectively, and it generates alkali and alkali-earth ions for back-diffusion into the altering fluid. All of these effects tend to cause precipitation in an adjacent vein. Loss of H^+ increases pH in the weakly acidic fluid, destabilizing metal-chloride complexes and permitting sulfide precipitation. Loss of water increases solute concentrations, and increased Na^+ and Ca^{+2} do the same. Addition of Ca^{+2} can cause sulfide precipitation by reducing the activity of Cl^- through ion pairing, thus destabilizing chloride metal complexes. Another powerful H^+ ion "sink" is the solution of carbonates that may occur as essential or accessory minerals in the wall rocks. Reaction of the fluids with sulfur-rich wall rocks—sulfate-rich evaporite sections or sulfide-sulfur-rich and graphite-rich black shales, for example—can have profound effects upon the fluid, such as altering its oxidation state or destabilizing metal complexes by the addition of reduced sulfur.

Fourth and finally, solution mixing can be important in ways other than the cooling effect described, either by adding anions or components such as H_2S or $(CO_3)^=$ that cause precipitation or by causing dilution. We have spoken of solute concentration as a cause of precipitation by the attainment of solubility products of solutes. For some components, dilution can cause precipitation by causing the dissociation of some complexes and leading to the recombination of metals and ions.

All of the above mechanisms can cause deposition of a potential ore assemblage. Understanding the principles can aid substantially in ore deposit evaluation—one can seek out a "boiling zone" in a vein by looking for appropriate fluid inclusions in quartz or sphalerite (Chapter 7), for example. And certainly, further study of ore occurrence can sharpen our perception of the precipitation mechanism perhaps to the point of valid prediction of where precipitation can be expected to have occurred in specific hydrothermal systems.

▸ DEPOSITIONAL TEXTURES

Textures vary among ore deposits, depending upon the nature of the mineralizing fluids, the physical and chemical characteristics of the host rocks, and the mode of emplacement. Textural interpretation can assist greatly in determining the time relationships of successive mineral assemblages in a rock, the overall environment of formation, and the manner of deposition. In epigenetic deposits, textures can help define the sequence and nature of events as in the complex vein shown in Figure 3-11, and they permit the distinction of mechanical and chemical controls as mentioned earlier in this chapter. In syngenetic deposits—such as sedimentary, volcanic-hosted, and mafic igneous ones—textures may reveal accumulation rates and styles, cooling histories, and the nature of precipitation. Resorption of early formed crystals may produce peculiar textures in magmatic segregation ores (Figures 4-2, 9-3, and 9-4), and diagenetic changes may modify the primary textures of sedimentary deposits. The textures of hydrothermal ores tend to be diverse.

Replacement

As defined by Lindgren (1933), replacement, or metasomatism, is "... the process of practically simultaneous capillary solution and deposition by which a new mineral of partly or wholly differing chemical composition may grow in the body of an old mineral or mineral aggregate." Replacement ordinarily implies little or no change in the volume of the replaced rock, although in some rocks considerable shrinkage or expansion takes place. The process of metasomatism is of great significance in the emplacement of epigenetic ore deposits; many ores are deposited almost entirely in this manner, and nearly all ores show some evidence of replacement of preexisting materials. The process is especially characteristic of deposits formed at high temperatures and high pressures where open spaces are scarce and where communication with the surface is impeded.

The efficacy of replacement is often astounding. The intimate preservation of plant cell or growth-ring morphology in petrified wood is well known, as is the fact that wood—a fibrous substance composed of carbon, hydrogen, oxygen, nitrogen, and minor elements—can be replaced by silica, even though there is no apparent similarity between the two substances. The replacement of one mineral or rock by another may be striking and clear; fossils, sedimentary textures, and folded structures are commonly preserved in faithful molecule-by-molecule detail. A compilation of the minerals that can replace one another indicates that there is practically no limit to the direction of metasomatism. As a bold generalization, it might be stated that given the proper conditions, any mineral can replace any other mineral, though natural processes usually result in favored, commonly recurring reactions. The important factor seems to be the chemical difference between the mineral or rock being replaced and the replacement medium, be it liquid, gas, or a wave of diffusing ions. Hence merely because quartz is stable at the Earth's surface, we cannot conclude that quartz will resist metasomatism. In fact, quartz and the silicates very commonly undergo replacement. In contrast, a fluid that reacts with and replaces limestone may be inert to quartz, or vice versa. As a result, selective replacement may be of the most detailed character. Bastin and his colleagues (1931) proposed the following general rules for replacement: (1) sulfides, arsenides, tellurides, and sulfosalts can replace any rock, gangue, or ore mineral, (2) gangue minerals replace rock and other gangue minerals, but do not commonly replace sulfides, arsenides, tellurides, and sulfosalts, (3) oxides replace all rock and gangue minerals but are rarely replaced by gangue minerals, and (4) oxides rarely replace sulfides, arsenides, tellurides, and sulfosalts. These statements reflect the "normal course" of ore deposition, whereby fluids carrying ore components typically invade and replace rocks dominated by silicates. Statements (2) and (4) are correct because the gangue and oxide minerals referred to are part of the ore deposition event and are commingled with it rather than replacing it.

Although replacement has been generally recognized and described for many years, the means by which the actual transfer of materials takes place has been the subject of debate. One fundamental question is how the tremendous volumes of the "old" replaced materials are removed by the same solution that deposits the "new" minerals. Presumably the dissolved materials are transported back from a replacement front through the aqueous ore solutions by diffusion, as in wall rock alteration. In other words, different ions may be diffusing down chemical gradients in quite different directions. The fact that replacement generally appears to take place volume for volume (without increasing porosity) raises another major problem: How can we write chemical equations representing electrically and molecularly balanced reactions with equal volumes of solid materials on each side? Ridge (1949, 1961) attempted to illustrate replacement with equations that balanced molecularly, volumetrically, and electrically. He found that he could write equations acknowledging near-surface chemistry and geology but could develop only rough approximations for deep-seated reactions.

In fact, the "budget" approach to metasomatism requires that atomic structure, space, and valence all be conserved, but there is no real evidence that all of those requirements must be met. Metasomatism can be molecule for molecule, but it may also involve one structure growing while another decays, with components being added and subtracted diffusionally from outside as required by the relative rates of destruction and construction. In other words, there may be no molecular-level *structural* inheritance at all. Barnes (1979) describes the replacement of limestone by sphalerite in a brine medium as involving etching and dissolution of calcite, especially along grain boundaries, merely with the concomitant precipitation of sphalerite and iron sulfides in the space thus created. The reactions were probably

$$CaCO_3 + 2H^+ \rightarrow Ca^{+2} + H_2CO_3(aq) \tag{4-5}$$

The space provided by removal of Ca^{+2} and H_2CO_3 was filled by

$$ZnCl_2(aq) + H_2S(aq) \rightarrow \underset{\text{sphalerite}}{ZnS} + 2H^+ + 2Cl^- \tag{4-6}$$

and

$$FeCl_2(aq) + 2H_2S(aq) + \tfrac{1}{2}O_2 \rightarrow \underset{\text{pyrite}}{FeS_2} + 2H^+ + 2Cl^- + H_2O \tag{4-7}$$

with the H^+ generated in Equations (4-6) and (4-7) aiding in the dissolution expressed by Equation (4-5). Ridge (1961) showed the same reactions summarized as

$$CaCO_3 + Zn^{+2} + S^{=} \rightarrow ZnS + Ca^{+2} + (CO_3)^{=} \qquad (4\text{-}8)$$

and was able to balance it to within 0.01 vol % as

$$\begin{aligned} &20CaCO_3 + 31Zn^{+2} + 31S^{=} \rightarrow 31ZnS + 20Ca^{+2} + 20(CO_3)^{=} \\ &\text{volume} = 1220.6 \text{ cubic angstroms} \rightarrow 1220.47 \text{ cubic angstroms} \end{aligned} \qquad (4\text{-}9)$$

Such careful treatment shows that volume for volume space can be conserved.

Limestone replacement at low temperature and pressure was studied by Garrels and Dreyer (1952). They were able to produce replacement textures similar to natural ones under controlled laboratory conditions. Many variables were examined, and it was concluded that the major control of replacement was the pH of the mineralizing fluid, which in turn controlled the solubility of the carbonate host rock. Garrels and Dreyer suggested that dissolving limestone changes ore-bearing solutions in the direction of ore-mineral precipitation. Accordingly, they consider the solubility of the host to be the key factor in metasomatism. Ames (1961) proposed that the principal factor is the solubility of the replacement product relative to the solubility of the host rock in the same solution, rather than merely the solubility of the host. In any case, the importance of simultaneous dissolving action and precipitation from a single solution is well established.

Garrels and Dreyer also found that numerous small, closely spaced openings and a slightly higher secondary permeability create ideal conditions for replacement. They concluded that the mineralizing solutions are carried along these zones of secondary permeability, but that the solutions move from the channels to the replacement front mainly by diffusion rather than by mechanical flow. It seems clear that replacement is far more important as a mechanism at higher temperatures, and that high pressure inhibits replacement by closure of permeability.

Interpretations of mineral textures to give unique and positive information are difficult. In spite of the great amount of work done on mineral textures and structures, the causes of many textural relationships are poorly understood. Although many textures are clear and unequivocal enough for positive statements, others may have several interpretations. For example, textures such as the one shown as Figure 3-5 have been attributed to the process of exsolution (Schwartz, 1931), replacement (Loughlin and Koschmann, 1942), eutectic crystallization, or, if the two minerals are chalcopyrite and bornite, the removal of part of the iron from chalcopyrite by hot water or steam (Park, 1931). Brett (1964) found that many exsolution textures were deceptive because they look as if they were formed by replacement.

He was able synthetically to produce replacement-type textures by exsolution, and vice versa. Barton and Skinner (1967) showed that textural equilibration times for many sulfide pairs were so short that homogenization of minerals—or at least modification of primary depositional textures in most sulfide ore deposits—is inescapable. However, ores in most types of ore deposits *do* show repetitive, characteristic, and reasonably interpretable textures. Most microscopists and mappers place reliance upon a broad spectrum of textural interrelationships, all the while recognizing the need for cautious interpretation of some of them.

Excellent treatises on the textures of ore minerals are available and should be studied by serious students of ore deposits (Van der Veen, 1925; Schneiderhöhn and Ramdohr, 1931; Schouten, 1934; Bastin 1950; Edwards, 1952, 1954; Ramdohr, 1955, 1980; Craig and Vaughan, 1981; and Picot and Johan, 1982). Textures may be studied megascopically in mine openings or in the field, in place or in hand specimens in the laboratory; they can be evaluated mesoscopically in rough, sawed, or polished pieces with binocular microscope or hand lens; or they can be examined microscopically with finely polished specimens in reflected light.

As established earlier, ores may be deposited by either open-space filling or replacement. Ores deposited in openings probably followed dominantly textural or structural controls; those that replaced preexisting rocks were probably guided structurally and then controlled chemically. Accordingly it is of fundamental importance for the understanding, evaluation, and development of an ore deposit to ascertain whether the minerals originated by replacement or by open-space filling. It should be recalled that a deposit formed by either mechanism alone would be an exception—open-space filling is likely to be accompanied by some replacement, and vice versa. These two major types of textures—replacement and open-space filling—are discussed separately below, but the student must be aware of their interaction.

Replacement Textures

A great deal has been written about textural criteria for recognizing replacement. Pseudomorphs and relict textures are considered diagnostic, but most others are only suggestive and may be formed in other ways. Except for a few diagnostic criteria, it is unwise to base conclusions upon single indications; confidence in any determination is directly proportional to the number of criteria available. A list of 19 of the more reliable criteria is given below. Although the scale of the accompanying illustrations is microscopic, most relationships can also be used megascopically in the field (Bastin et al., 1931; Schouten, 1934).

1. *Pseudomorphs* (Figure 4-6). If the form of a preexisting mineral is preserved, especially if the internal structure is also discernible, a replacement origin is indisputable. The *preservation of original structures*

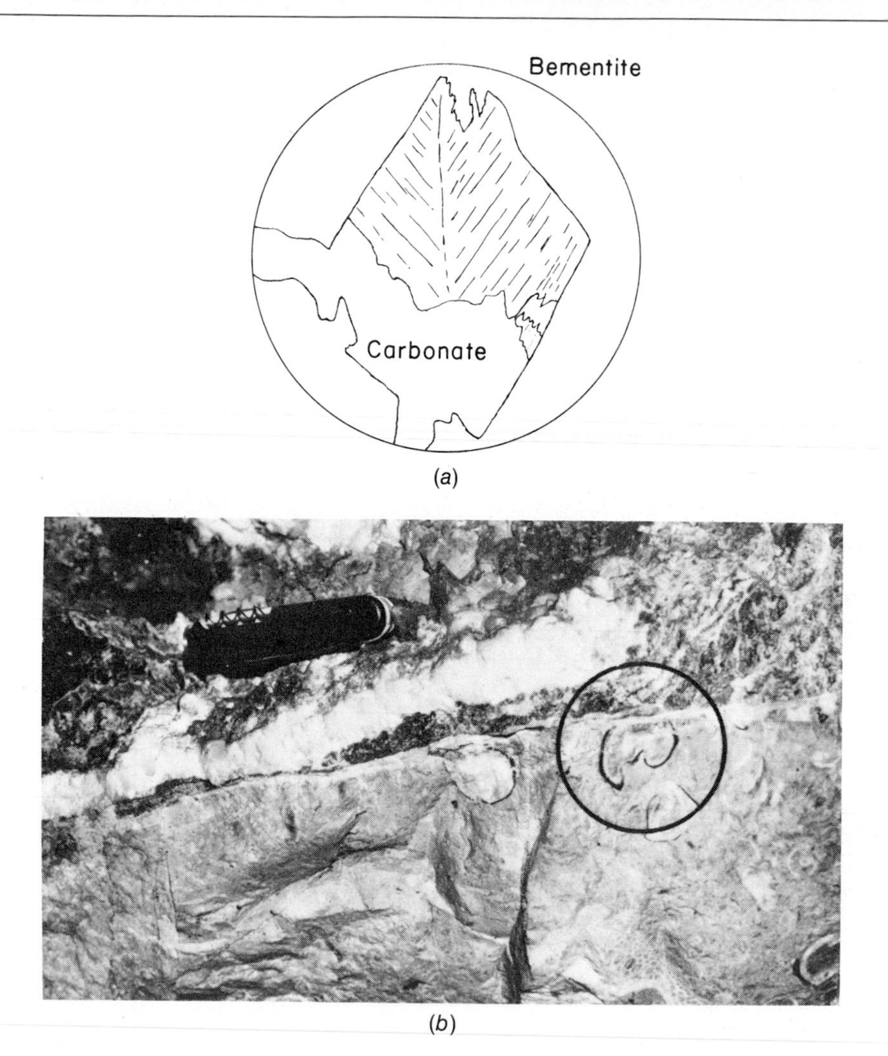

Figure 4-6. (*a*) Pseudomorphism. Bementite (crosshatched) has replaced most of a rhomb of preexisting carbonate. Bementite, a manganese silicate, is monoclinic and could not form a rhombohedron by itself. Olympic Peninsula, Washington. 60×. (*b*) Fluorite has replaced a fossil (circled) in Doctores Formation limestone at the Las Cuevas fluorite deposit, Mexico. The original form of the fossil is obliterated where replaced by fluorite. Replacement is irrefutable since no CaF_2 shells are known. (*From* Ruiz et al., 1980.)

and textures in sedimentary, igneous, or metamorphic rocks, as well as organic remains, may be pseudomorphic. For example, preore fold structures or oolites now pseudomorphed by sulfides are conclusive evidence of replacement.

2. *Widening of a fracture filling to an irregular mass where a fracture crosses a chemically reactive mineral grain or rock* (Figure 4-7). If a

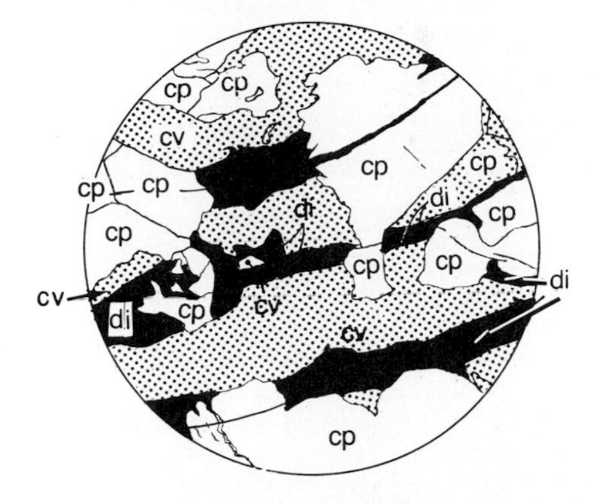

Figure 4-7. Widening of a fracture filling to an irregular mass where the fracture crosses chemically reactive mineral grains or rock layers. Here digenite in the fracture that cuts covellite (cv) and chalcopyrite (cp) has expanded into the covellite and left the chalcopyrite almost unreplaced. Replacement of the covellite is indicated. Cananea, Mexico. 40×. See also Figure 4-30.

veinlet widens across only one variety of mineral, it suggests that this mineral was receptive to replacement and that replacement occurred. Mineralization spreading laterally into a single limestone unit in a sequence of varied sedimentary layers is a large-scale example.

3. *Irregular or vermicular intergrowths at wide places along fractures or at grain boundaries not related to crystallographic directions* (Figure 4-8). The vermicular intergrowths represent a leading edge or front of replacement, not completed. Replacement is not the only mechanism by which vermicular intergrowths are formed; they also develop during crystal growth in a eutectic mixture and by exsolution during the slow cooling of some solid solutions, and they are typically related to crystallographic directions. Only the nonoriented irregular intergrowths pictured can be considered criteria of replacement.
4. *Islands of unreplaced host mineral or wall rock* (Figure 4-9). Isolated, seemingly suspended relict bits of one material in another, especially if

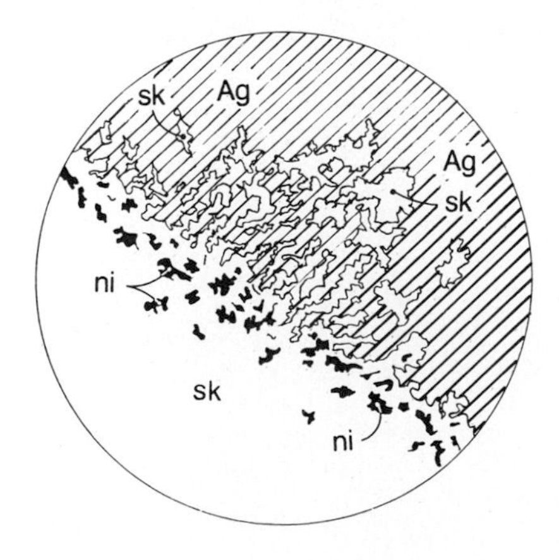

Figure 4-8. Vermicular irregular intergrowths at wide places along fractures or at grain boundaries; not related to crystallographic directions. Skutterudite (sk) and niccolite (ni) have replaced native silver (Ag). The vermicular diffuse intergrowths of skutterudite extending irregularly into the silver are evidence of replacement. The niccolite is probably an intermediate reaction product. The entire field once was presumably all silver. Cobalt, Ontario, Canada. 30×. (*After an unpublished photo by* D. E. Eberlein.)

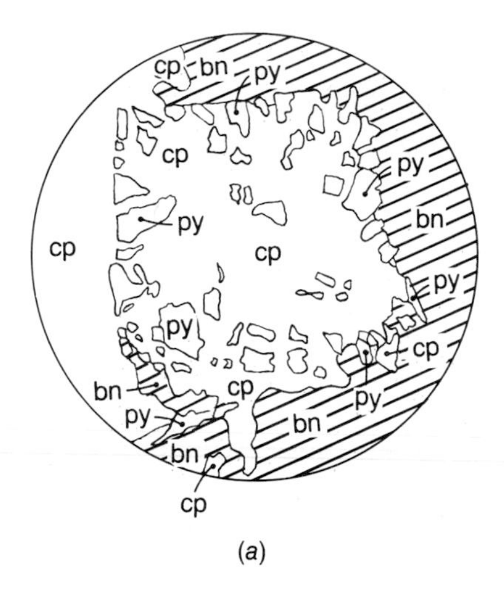

(*a*)

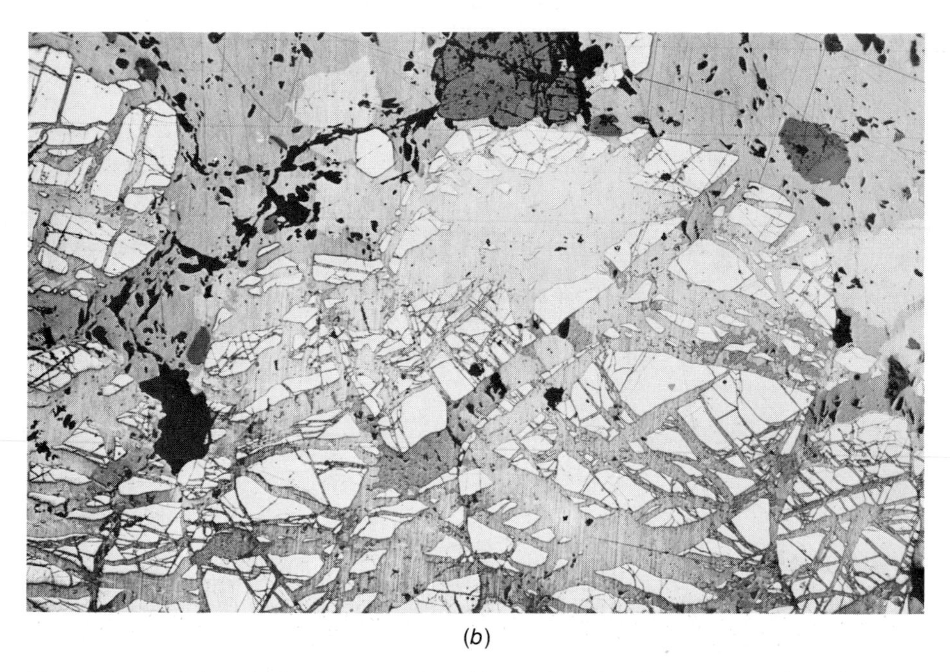

(*b*)

Figure 4-9. (*a*) Oriented islands of one mineral in another. In this texture, pyrite relict fragments form an *atoll structure*, a common texture in Cu-Fe-S vein ores. An early pyrite cube, the outline of which is preserved, has been almost completely replaced by bornite. Chalcopyrite is an intermediate reaction product. Bisbee, Arizona. 40×. (*b*) Photomicrographic example of oriented relict islands of a replaced mineral. Clusters of pyrite relicts (white) in chalcopyrite (light gray, within pyrite) and bornite (medium gray, external to pyrite) clearly reflect earlier grain forms. Dark gray is quartz; black is holes in polished surface. Butte, Montana. (*Photo by* Geological Research Laboratory, Anaconda Company.)

an earlier form or common orientation is perceptible, require a replacement relationship, one that did not carry to completion.

5. *Concave surfaces into the host, the "cusp and caries" texture* (Figure 4-10). The diffusion of ions at the replacement front goes on at different rates, so that some parts of the front form concave reentrants as if the replacing mineral had bitten into the host. The *cusp and caries* designation is a dental analogy—caries are cavities in teeth, clearly

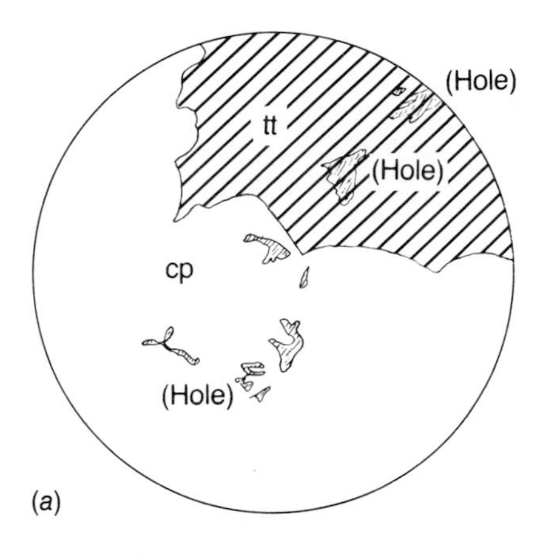

Figure 4-10. (*a*) Concave surfaces of one mineral into another, the *cusp and carie* texture. Chalcopyrite (cp) has replaced tetrahedrite (tt). By analogy with dental effects, the "bites" are caries, the points between them cusps. Coeur d'Alene, Idaho. 45×. (*b*) Excellent example of cusp and carie textures. Hessite (h) ($AgTe_2$) is replacing altaite (a) (PbTe). Ben Butler mine, Red Cliff, Colorado.

(*b*)

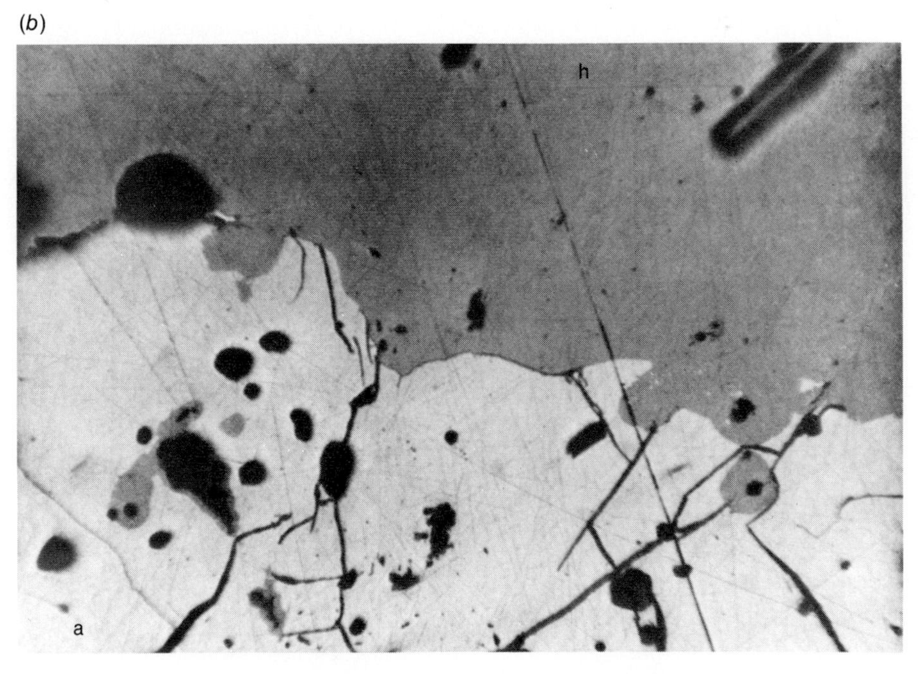

later than the enamel; the cusps are relict protuberances between cavities. The texture is a familiar one to microscopists.

6. *Nonmatching walls or borders of a fracture* (Figure 4-11). If replacement works outward from a central fissure, the opposite fronts of replacement will almost never match in detail and may differ radically.

7. *Rims of one mineral penetrating another along its crystallographic directions* (Figure 4-12). Replacement may work outward from any small fissures or from grain margins and boundaries, but has advanced preferentially along cleavages. For example, galena may be replaced by covellite or cerussite by penetrating it preferentially along directions that are obviously parallel to the cleavages (see also Figure 3-3).

8. *Oriented unsupported fragments* (Figure 4-13). If a piece of one mineral is completely surrounded by another mineral and still maintains its

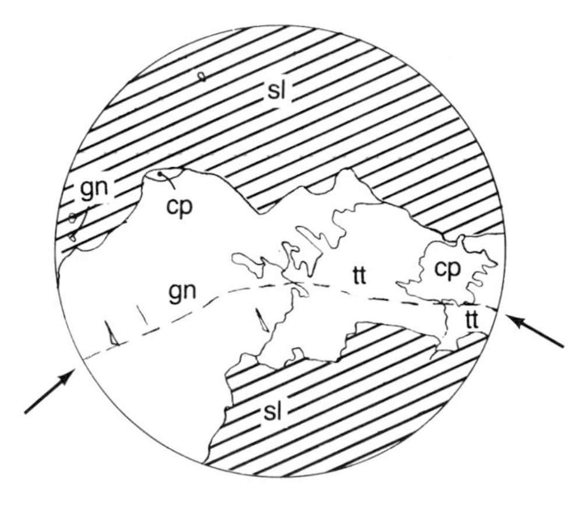

Figure 4-11. Nonmatching walls or borders of a fracture. The original fracture extended between the arrows. Introduced galena (gn), tetrahedrite (tt), and chalcopyrite (cp) have replaced sphalerite (sl). Note the caries of those minerals—the metasomes—into the older host mineral sphalerite. Arrows show the veinlet centerline. Cananea, Mexico. 30×. See also Figure 4-10.

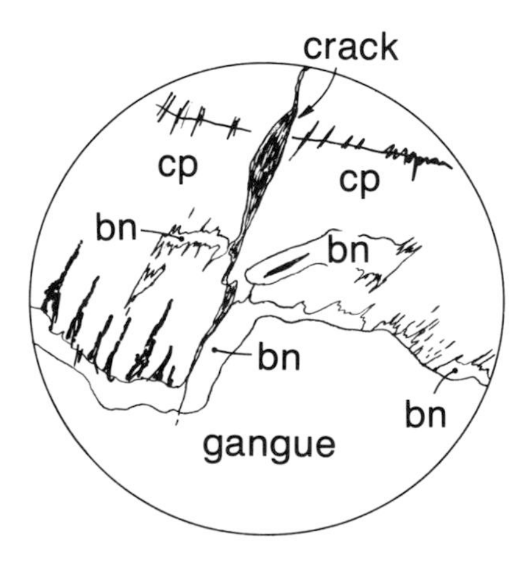

Figure 4-12. Rims of one mineral penetrating another along crystallographic directions. Here the metasome bornite (bn) protrudes from fairly continuous rims along preferred directions in the chalcopyrite (cp) host. Replacement along the fracture echoes the preference. The replacement proceeded from the volume now filled by gangue. Cananea, Mexico. 40×.

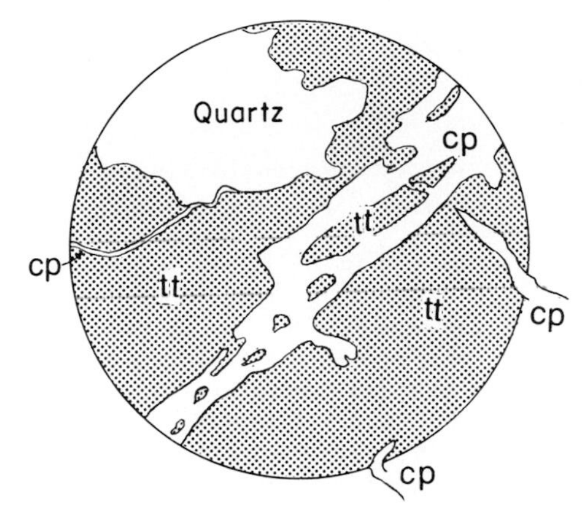

Figure 4-13. Oriented unsupported fragments. If cleavage, anisotropism, or other optical characteristics, trains of inclusions, or any other criteria establish that the "island" was once continuous with the "mainland," then replacement—here tetrahedrite (tt) by chalcopyrite (cp)—is indicated. Coeur d'Alene, Idaho. 45×. See also Figure 4-9.

orientation with respect to more of the same material on the outside, the texture is practically diagnostic of replacement at any scale. The fragment may be any size, and the orientation may be established by crystallographic directions, cleavage, bedding, or foliation.

9. *Selective association* (Figure 4-14). Since replacement is a chemical process, specific selective associations of pairs or combinations of minerals can be expected. If the chemistry of a system changes, it will more likely reflect a changed ratio of related constituents than a complete switch to different ones. Chalcopyrite is more likely to replace bornite by virtue of a changed Cu/Fe ratio or a different $f(S_2)$ than it is to replace quartz. Thus consanguinity of assemblages is part of the diagnosis of replacement.

10. *A younger mineral that transects older structures* (Figure 4-15). The presence of a crystal—a metablast, for example—that interrupts bedding or foliation requires that the structure is older. If the crystal had

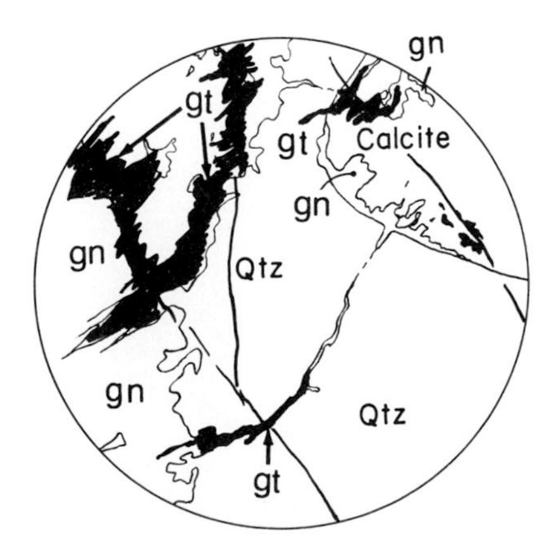

Figure 4-14. Selective association. Galena (gn) (PbS) rather than calcite or quartz is invaded by gratonite (gt) ($Pb_9As_4S_{15}$) when the activity of arsenic in the fluids increased. Adapted from a specimen from Darwin, California. 45×.

Figure 4-15. Metablasts of pyrite that cut foliation in schist from the Mother Lode district, California. The pyrite cubes grew in the rock but do not bend or displace the foliation. They must, then, have replaced the material that occupied their space. About 0.9×.

grown by any process other than replacement, it would have displaced or deformed the structure.

11. *A young phase deposited in obvious relation to microfractures, cleavage planes, or grain boundaries* (Figure 4-16). Ore-forming fluids can be introduced along small fractures. If grains of a young phase grow athwart such fractures and protrude into the walls, replacement is indicated.

12. *Disparity in size of one mineral in another.* Large crystals in a fine-grained groundmass, or vice versa, suggest that the two grew independently or by different processes that might include replacement.

13. *A mineral deposited along what was clearly an advancing alteration-reaction front.* If deposition took place by open-space filling, ore minerals should stop abruptly at the wall of a fissure. Conversely, replacement is indicated by the gradual enlargement and merging of metasomes along a replacement front. Such a zone should also be evident

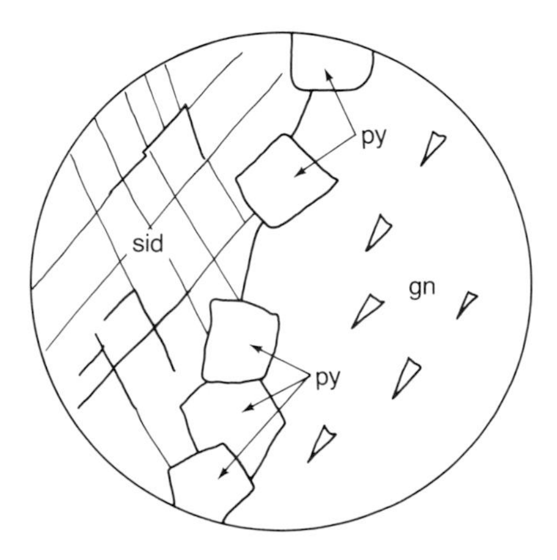

Figure 4-16. Young phase deposited in obvious relation to microfractures, cleavage planes, or grain boundaries. Pyrite (py) lies along an obvious contact between siderite (sid) and galena (gn) showing triangular etch pits. The pyrite must have replaced both galena and siderite. Adapted from a Príbram, Czechoslovakia, specimen. 150×.

from the progressive increase in size of metacrysts and from the gradation of replacement from the wall rock into the vein. Figure 3-4 is an excellent example.

14. *The presence of a depositional sequence in which minerals become progressively richer in one constituent* (Figure 4-17). For example, if polybasite ($Ag_{16}Sb_2S_{11}$) has been acted upon by more silver-rich solutions, the replacement material should show a progressive enrichment in silver. A crystal or fragment of polybasite may grade first to acanthite (Ag_2S) and then eventually to native silver, reflecting the gradual displacement of antimony and sulfur. An intermediate-stage specimen might show relict cores of polybasite rimmed in sequence by acanthite and native silver.

15. *Doubly terminated crystals* (Figure 4-18). If a crystal grows within an open cavity, it is normally attached to a wall and develops crystal faces only at the free end. This restriction does not affect crystals growing by replacement, so that doubly terminated crystals may indicate a replacement origin. This criterion is limited because doubly terminated crystals may also develop within magmas and, exceptionally, by unusual attachment in open spaces.

16. *Gradational boundaries.* Replacement processes may produce either abrupt or gradational contacts between the host rock and the orebody. Since open-cavity filling usually forms abrupt contacts, at least on a microscopic scale, a gradational boundary commonly indicates advancing replacement. (See also Figure 3-4.)

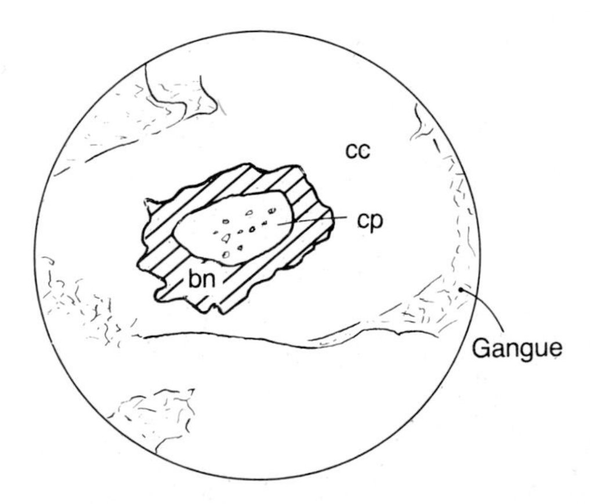

Figure 4-17. Depositional sequence in which minerals become progressively richer in one constituent. The intermediate phase commonly is a reaction rim. Hypogene chalcocite always has a selvage of bornite between itself and the chalcopyrite that it replaces. The sequence of $CuFeS_2$–Cu_5FeS_4–Cu_2S, in which the copper-to-iron ratio is 1–5–∞ and the metal-to-sulfur ratio 1–1.5–2. Adapted from a Tsumeb, South-West Africa, specimen. 45×. See also Figure 4-8.

Figure 4-18. Doubly terminated quartz crystal from the Stibnite Vein, Wolf Creek, Fairbanks district, Alaska, containing stibnite crystals. The quartz is said to be a replacement of gangue minerals in the vein, and probably overgrew and enclosed the stibnite granules. 18×. (*Collected by* P. O. Sandvik; *photo by* W. J. Crook.)

17. *Residual resistant minerals.* Some minerals are refractory to most mineralizing solutions and may be preserved even after the surrounding minerals have been replaced. For example, zircon or corundum found in the same proportions in the sulfides of an orebody as in a nearby schist would support the argument that the ore had replaced some of the schist. The resistant minerals are special types of islands or unreplaced fragments of host rock (criterion 4).

18. *No offset of an intersected linear fracture* (Figure 4-19). When a vein is opened laterally, any intersecting feature is also translated. But if a linear or planar feature projects straight across a vein filling, there can have been no dilation, and filling must be the result of replacement.

19. *No offset along the intersection of fractures* (Figure 4-20). Movement along a fissure offsets any planar or linear feature that it intersects obliquely, but does not commonly offset an intersecting fracture formed

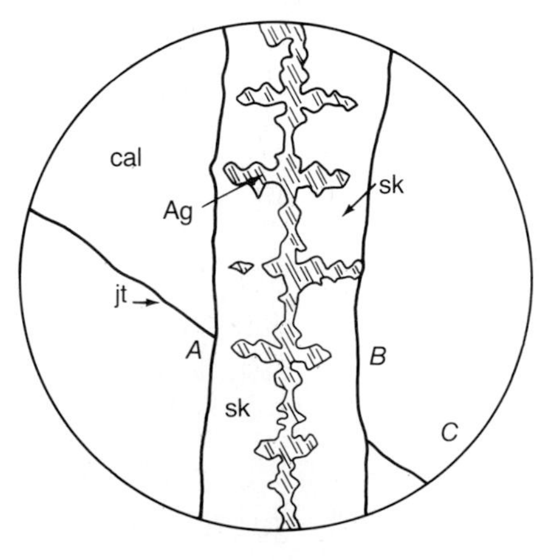

Figure 4-19. No offset of an intersected linear feature. A veinlet of native silver (Ag) flanked by skutterudite (sk) cuts across a straight joint (jt) in calcite (cal). Were the silver and skutterudite open-space fillings in a dilatant opening, the joint would have been spread from *A* to *B* and continue from *B* to *C*. Replacement, or fortuitous movement, is required. Cobalt, Ontario. 30×. See also Figure 4-26.

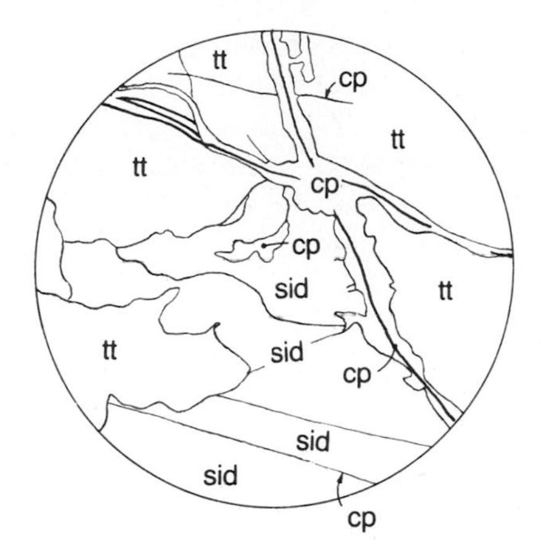

Figure 4-20. Intersecting fractures showing no offset at the intersection. The fractures were probably part of a cognate joint set and show enlargement by replacement rather than by dilational opening. Tetrahedrite (tt), chalcopyrite (cp), and siderite (sid). Coeur d'Alene, Idaho. 30×.

at the same time. Both fractures may be enlarged by replacement of their walls and tend to cross one another without any change in course, but only a coincidence of displacement would permit a rematch of intersecting veins that had been spread apart and open-space-filled.

Exsolution

A "near neighbor" to replacement is the process of *exsolution*. In both phenomena, migration of one or more components, commonly involving diffusion, results in the supplanting of a preexisting phase by a younger one. Exsolution textures and their evaluation are normally only seen by the mineragrapher (the ore microscopist). It is true that exsolution is not strictly speaking a depositional feature, and perhaps should not be included in this chapter. But there are enough situations in which exsolution textures and distributions can be mistaken for those of replacement that their description is justified.

Reexamine Figure 3-5. Were the process of exsolution to be carried further, the miniscule blebs of chalcopyrite would migrate to cleavages and crystallographic planes and collect there, leaving impoverished zones behind. That process can be seen to have begun especially on the right side of Figure 3-5. Taken even further, the chalcopyrite—now essentially all on planes—might resemble a phenomenon of chalcopyrite from outside invading and replacing the sphalerite. Taken to completion, the lowest free-energy state is for the exsolved phase to form separate grains occupying texturally equivalent positions in a mosaic with the original host phase. For many years, veinlets and grains of pentlandite [$(Fe,Ni)_9S_8$] in pyrrhotite ($Fe_{1-x}S$) in the Sudbury Complex (Chapter 9) were perplexing. Some considered them to be of replacement origin, with pentlandite invading from "outside" the pyrrhotite. Others thought them to be the product of well-developed exsolution. A careful textural study of scores of polished pieces of ore sulfides has verified the latter explanation. It matters because it means that pentlandite, a principal contributor of nickel to the ores, can be expected virtually wherever pyrrhotite is found, and pyrrhotite occurs extensively along the floor of the complex. Were the pentlandite not integral to the pyrrhotite, its distribution would almost certainly be more erratically subject to unknown external controls. In general, replacement veinlets widen at intersections, and exsolution veinlets become narrower. In the former case, replacement attacks corners formed by two veinlets from two sides, replacement is more complete, and the zone widens. In the latter, the host contributes the exsolved phase into the veinlet. Close to a corner the exsolved ions are migrating both ways, and the intersection becomes starved. Indeed, some microscopic exsolution patterns involve exsolved crosses the arms of which narrow to a pinpoint at the intersection.

Excellent lists of criteria for distinguishing exsolution and replacement are available in Edwards (1965), Ramdohr (1980), and Craig and Vaughan (1981), but need not be reviewed here.

Open-Space Filling

Open-space filling is common in shallow zones where brittle rocks yield by breaking rather than by plastic flow. The openings in these zones tend to remain open because of low pressure transmitted through surrounding rock. At shallow depths, ore-bearing fluids have relatively free circulation, and their open connection with the surface permits deposition to be brought about by abrupt pressure and temperature changes—including retrograde boiling—as opposed to prolonged contact with surrounding rocks which undergo slow chemical changes in deeper environments. Although ores deposited in vugs and open cavities are generally readily distinguishable from replacement ores, criteria associated with open-space deposition must nevertheless be used with caution because they are sometimes inconclusive (Kutina and Sedlackova, 1961). As stated earlier, it should also be stressed that many ore types show evidences of both replacement and open-space filling and thus an overlap of the two processes. A list of common criteria by which open-space filling may be recognized is given below.

Open-Space Filling Textures

1. *Many vugs and cavities* (Figure 4-21). If veins, breccias, and other partly-filled openings contain many vugs and cavities that can be interpreted as spaces left by incomplete filling of larger open spaces, then open-space filling is indicated. The inward growth of ore and gangue minerals in an open fissure stops when opposite walls meet. Since growth is not uniform along a fissure, the progressively impeded circulation of ore-bearing solutions normally leaves unfilled pockets. These crystal-lined centerline openings are called *vugs*.

2. *Fine-grained minerals on the walls of a cavity with coarser minerals*

Figure 4-21. Quartz-crystal-lined vugs left by incomplete vein filling. Locality unknown, but typical of scores of epithermal veins around the world. 0.9×. See also Figures 4-22 and 4-23.

in the center (Figure 4-22). The first crystals that form along the sides of an open vein are usually fine-grained, probably because of heat loss to the wall rocks and consequent rapid crystallization. Conversely, crystals that form in the centers of such veins are inclined to be coarse and probably form from more dilute, cooler solutions late in the vein history. Whatever the reasons, the phenomenon is common enough to be distinctive of open-space filling.

3. *Crustification* (Figure 4-23). As ore-bearing solutions change in composition, or as environmental factors change, different minerals are deposited along the walls of a vein or cavity. Early formed crystals become encrusted with later minerals. The presence of small euhedral dolomite crystals on top of large euhedral fluorite crystals is a common example. Some veins have a banded appearance owing to crustification.

4. *Comb structure* (Figure 4-23). Along the junction of crystals that have grown from opposite walls of a fissure, there is generally an interdigitated vuggy zone due to the merging of euhedral prismatic crystals. Because this jagged zone of juncture resembles the outline of a rooster's comb, it is known as comb structure.

5. *Symmetrical banding* (Figure 4-24). Crystals deposited along a cavity or fissure ordinarily grow symmetrically toward the center, in which case the orientations, morphologies, and compositions of crystals on opposite sides of a vein are mirror images. As the ore fluid or environment changes, minerals deposited will differ in composition, forming crustification in a bilaterally symmetrical pattern inward from vein walls. Such symmetrical banding may involve mineralogical and color changes, and can produce a banded vein diagnostic of open-space filling.

6. *Matching walls.* If an open fissure has been filled without replace-

Figure 4-22. Fine-grained quartz grading into coarser crystals in a vug. Locality unknown. See also Figure 4-23.

ment, the outlines of opposite walls should match, that is, if the vein material were removed, the unreacted wall rocks on opposite sides should fit together like mated pieces of a jigsaw puzzle. The fit across many veins is readily apparent.

7. *Cockade structure* (Figure 4-25). Mineralization within the open spaces of a breccia, or any other fragmental rock, commonly produces a special pattern of symmetrical banding or crustification known as *cockade structure*. Each opening is a receptacle for sequential deposition. The overall rock pattern is one of host-rock fragments coated with layers of inward-radiating crystals, with triangular patterns predominating.

8. *Offset oblique structures* (Figure 4-26). If a planar or linear feature is cut obliquely by an open fissure, it is offset at right angles to the vein walls because it was spread apart as the fissure opened. Replacement along a fracture would cause no offset along such a preexisting structure, it would project straight across. (See also Figure 4-19.)

9. *Colloform structures*. Colloform banding, composed of fine onionskin-like successively deposited layers (Figures 4-27 and 4-28 and next section), can only form in open space.

(a)

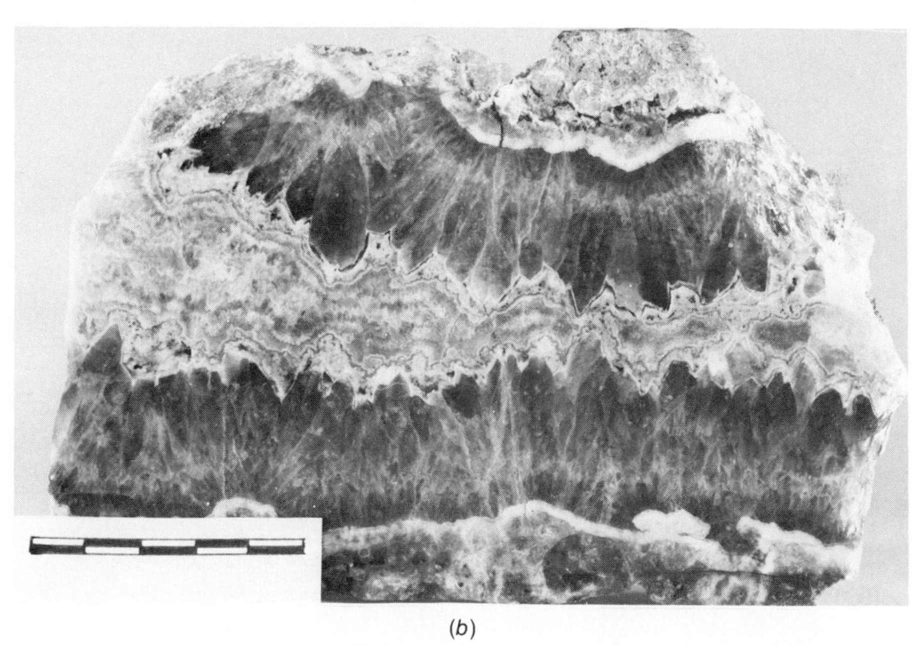

(b)

Figure 4-23. Crustification, comb structure, and centerline vugs. (*a*) Veinlet of aragonite through limestone and shale. Locality unknown. (*b*) Vein from latite porphyry (top and bottom) through amethyst (medium gray) to centerline inward-facing chalcedonic quartz. Commonwealth mine, Pearce, Arizona. The bar scale is in centimeters. (*Photo of b by* W. K. Bilodeau.)

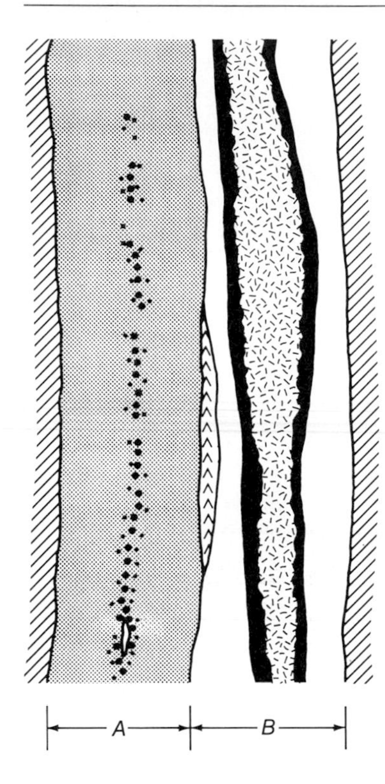

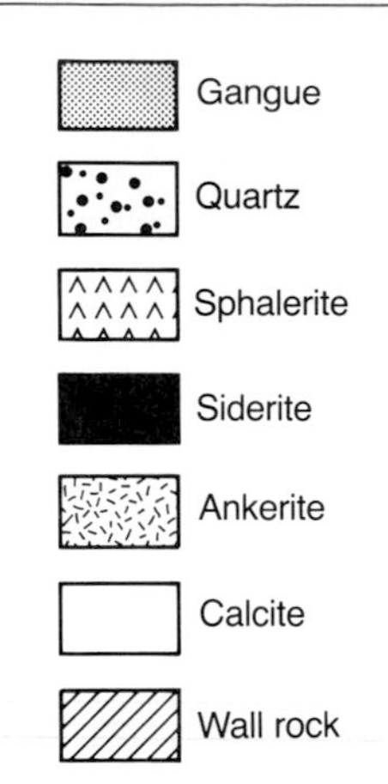

Figure 4-24. Sketch of part of the reopened, symmetrically banded complex Matkobozska Vein, Príbram, Czechoslovakia. An early opening (*A*) was filled with "ore"—an intergrowth of chalcedony and siderite with Ag-Pb-Zn sulfosalts—then quartz. A second opening (*B*) was filled with minor sphalerite, then calcite, siderite, and finally ankerite. See also Figure 3-11. (*From* Kutina, 1957.)

Figure 4-25. Cockade structure of quartz infilling the cavities in a breccia. Locality unknown, but a typical epithermal structure.

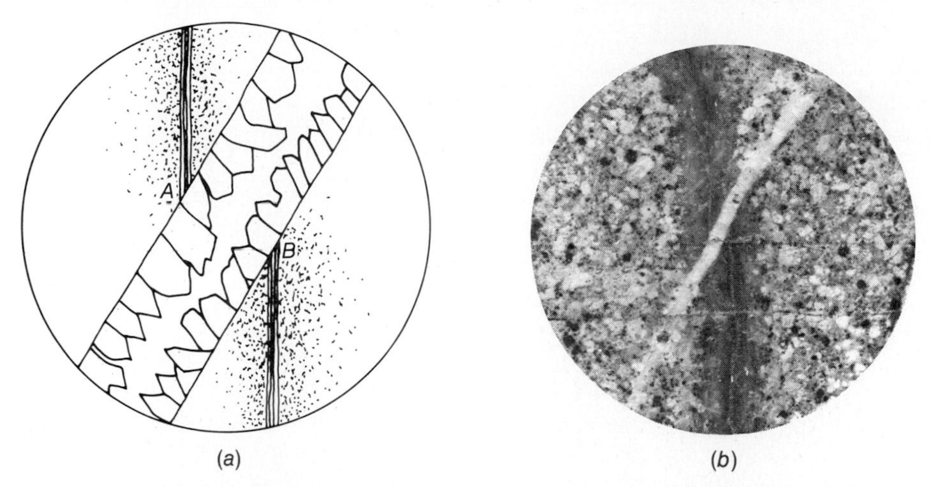

Figure 4-26. (*a*) Offset of an oblique structure. *A* and *B* were adjacent before the vein opened; the two segments no longer lie in a single plane, and the intervening material is open-space filling. Lateral wall movement can of course complicate relationships. Compare with Figure 4-19. (*b*) Specimen from Silver Bell, Arizona, in which the white quartz vein is 1 cm wide. (*Photo by* W. K. Bilodeau.)

Colloidal-Colloform Textures

Amorphous minerals, such as opal, neotocite, "copper pitch," "wood tin," and some garnierite, used to be thought to have been deposited from colloidal suspensions. Moreover, it was believed that many cryptocrystalline minerals—chalcedony, some manganese oxides, pyrite, marcasite, and pitchblende, and the oxidation products of copper, lead, and zinc, such as malachite, azurite, chrysocolla, anglesite, cerussite, and smithsonite—were carried and deposited as colloids that aggregated and crystallized shortly after deposition. Some geologists think that gold, especially in shallow low-temperature deposits, travels short distances in the colloidal state. Most geochemists now consider the existence of colloids in hydrothermal systems highly unlikely.

A colloidal system consists of two phases, of which one, the dispersed phase, is diffused in the other, the dispersion medium. Colloidal particles range in size between ions in true solution and particles in coarse suspension, the general limits being defined at 10^{-7} and 10^{-3} cm. The colloidal material may be solid, liquid, or gas, and may be dispersed in a medium of any of these same states. In the study of ore transport, however, we are concerned essentially with solids dispersed in liquid media. A colloidal system consisting of solid particles dispersed in a liquid is called a *sol*. A given kind of colloidal particle may adsorb cations and behave as a positively charged body, or it may adsorb anions and become negatively charged. Since the particles of a sol all have the same charge, they repel each other and prevent coagulation. If an electrolyte is added to the sol, the colloidal particles become neutralized and flocculate, or deposit. The result can be a banded or finely layered, typically botryoidal or reniform aggregate of the type that characterizes African malachite, for example, and is common in some sphalerite from low-temperature deposits (Figure 4-27). However, we saw in Chapters 2 and 3 that most hydrothermal fluids are indeed brines, or at least saline enough to be the electrolyte that is anathema to colloids and colloidal transport. We now perceive that at most temperatures and in most hydrothermal fluids, at least, colloidal transport and deposition cannot be important. At low temperature and in nonsaline groundwater, colloids can and do persist.

As a result of his study of colloform sphalerite specimens from many of the world's major geological districts, Roedder (1968) decided that most if not all of these ores grew directly as minute druses of continuously euhedral crystals that projected into an ore fluid that was a true solution, not a colloidal dispersion. He found that all of the textural features believed to be diagnostic of colloidal deposition were ambiguous, inapplicable, and therefore invalid as criteria for the samples he studied, and perhaps of most other colloform mineral samples as well. However, Roedder did not explain the fact that such colloform minerals as the manganese oxides, garnierite, and opal at times do not reveal any structural pattern even upon X-ray study; they are truly noncrystalline.

In a thoughtful discussion of Roedder's paper, Haranczyk (1969) concluded from his extensive studies in the Silesian-Cracovian lead and zinc deposits of Poland that colloform textures might form from both true solutions and sols. He proposed the term *hemicolloids* for solutions that are intermediate between true and colloidal ones. The cryptocrystalline, white, claylike form of sphalerite known as *bruncite* (Figures 20-9 and 20-10) was almost certainly deposited as a colloid. Haranczyk later (1971) clearly showed that some of the second-generation minerals in the Silesian-Cracovian ore deposits of Poland could be of colloidal origin. He states that most of the second-generation sulfides show hemicolloidal textures. This conclusion clashes

Figure 4-27. Colloform texture. Globular sphalerite with concentric banding, considered to be of colloidal origin. Orzel Bialy mine, Katowice, Poland. Natural size. Also see Figures 20-9 and 20-10. (*From* Kutina, 1957.)

with the modern concept that these deposits were formed from warm electrolyte brines; clearly, we need more experimentation and morphological data.

The terms *colloidal texture* and *colloform texture* should be used carefully. The former should be used only where deposition from a true colloidal system is demonstrable; colloform suggests a texture that resembles, and could be, colloidal in origin, but cannot be determined to be so. A truly colloidal origin can be postulated if the following are observed: (1) infinitesimal grain sizes in banded textures as in some malachite (Figure 4-28) and most wad, (2) shrinkage or dehydration cracks, as in opal and some limonites, (3) diffusion bands, or Liesegang rings (Figure 4-29), (4) chaotic, amorphous noncrystalline structure of component compounds, (5) the presence of unpredicted, possibly originally adsorbed ions, such as barium in psilomelane, and (6) continuous spherulitic textures demonstrably controlled by surface tension. Interrupted bands normally deny the role of

Figure 4-28. Colloform banding in malachite from Kolwezi, Zaire. Some of the bands can be traced through the whole specimen, but some lobes interfere with others, perhaps as stalagmite-stalactite masses were filled in. Deposition can have been either truly colloidal or as banded microcrystalline accumulations. The bar scale is in centimeters. (*Photo by* W. K. Bilodeau.)

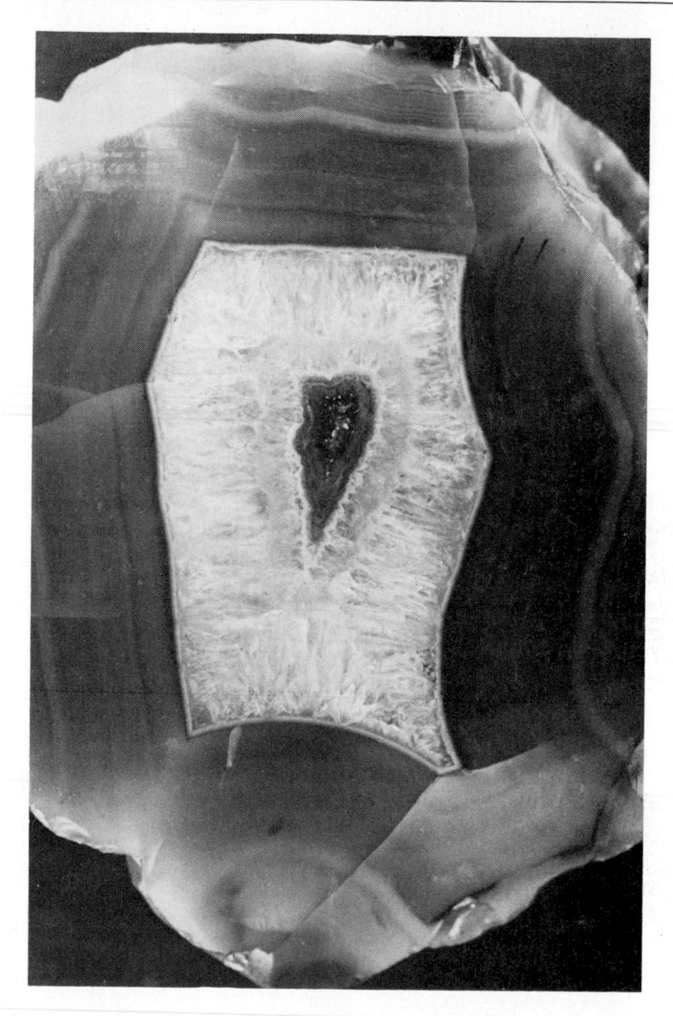

Figure 4-29. Probable diffusion banding in an agate, quite possibly developed while the silica was a colloidal "silica gel." Note also the cockade pattern of the later quartz filling in the central vug. Locality unknown. 1.7×. (*Photo by* W. J. Crook.)

colloids (Figures 20-9 and 20-10). Clearly, all colloform textures are open-space fillings; they may or may not have involved colloids.

Even though the part played by colloids in ore deposition may be trivial, it seems likely that under near-surface conditions and in dilute watery solutions at comparatively low temperatures the role may be appreciable. Many deposits have been described for which a colloidal origin is probable or in which colloids have played a supporting role.

An unusual deposit of colloform magnetite was found in an igneous metamorphic environment at the northern end of Vancouver Island, Canada (Stevenson and Jeffery, 1964). The colloform magnetite possessed the textural features usually ascribed to colloidal deposition, and the deposits appeared to have been formed by a type of gel metasomatism of the limestone. The iron, carried in acid solution, replaced limestone and appears to have been precipitated colloidally during an intermediate state of aggregation between ionic solution and the final precipitate.

Hosking (1964) pointed out the presence of opaline gels that are apparently forming in the deposits of Cornwall, England (Chapter 6). He described the tin lodes of St. Agnes where almost perfect concentrically zoned spherules of "wood tin"—ultrafine-grained colloform cassiterite— are suspended in quartz. These spherules are covered at places with acicular crystals of cassiterite whose long axes are normal to the spherule surfaces. He considers these spherules to have developed as the result of centripetal migration of tin-bearing fluids through a silica gel.

▸ EXAMPLES OF CHEMICAL CONTROL

The last chapter closed with some examples of ore deposits that were emplaced according to mechanical controls with predominantly open-space filling textures. This chapter has introduced many chemical considerations, so it will end with several examples of chemically controlled ore deposition.

A Cobalt, Ontario, Specimen

As we have seen, a favorable structural environment, and even the presence in a structure of ore-bearing fluids, does not necessarily mean that ore will be deposited. Ore-bearing fluids can react continuously with the wall rocks and, like the material they traverse, constantly change in composition. Moreover, factors such as reduction in temperature and pressure may either bring about chemical reactions or decrease solubilities and thus contribute to the deposition of ore minerals. Geochemists cannot always explain why in many places certain beds are mineralized and other apparently identical ones above and below are barren. These barren beds must have compositions and physical characteristics that appear to be identical with the mineralized strata but are not. Even though the causes of chemical controls are not always clear, their existence can often be demonstrated. For example, Figure 4-30 shows a polished slab of diabase cut by a veinlet which contains calcite and native silver. Silver, the dark mineral in the veinlet, is restricted to a large feldspar crystal, but calcite fills the veinlet where it passes through diabase. Note that the width of the veinlet is almost constant across the specimen. The silver must have been deposited as a result of chemical reaction between the fluid and the feldspar crystal, rather than as a result of physical differences between the feldspar and the diabase. Although the fracture served as a channel for introducing the fluid into the environment, the silver was deposited only along that part of the vein that was chemically receptive to it.

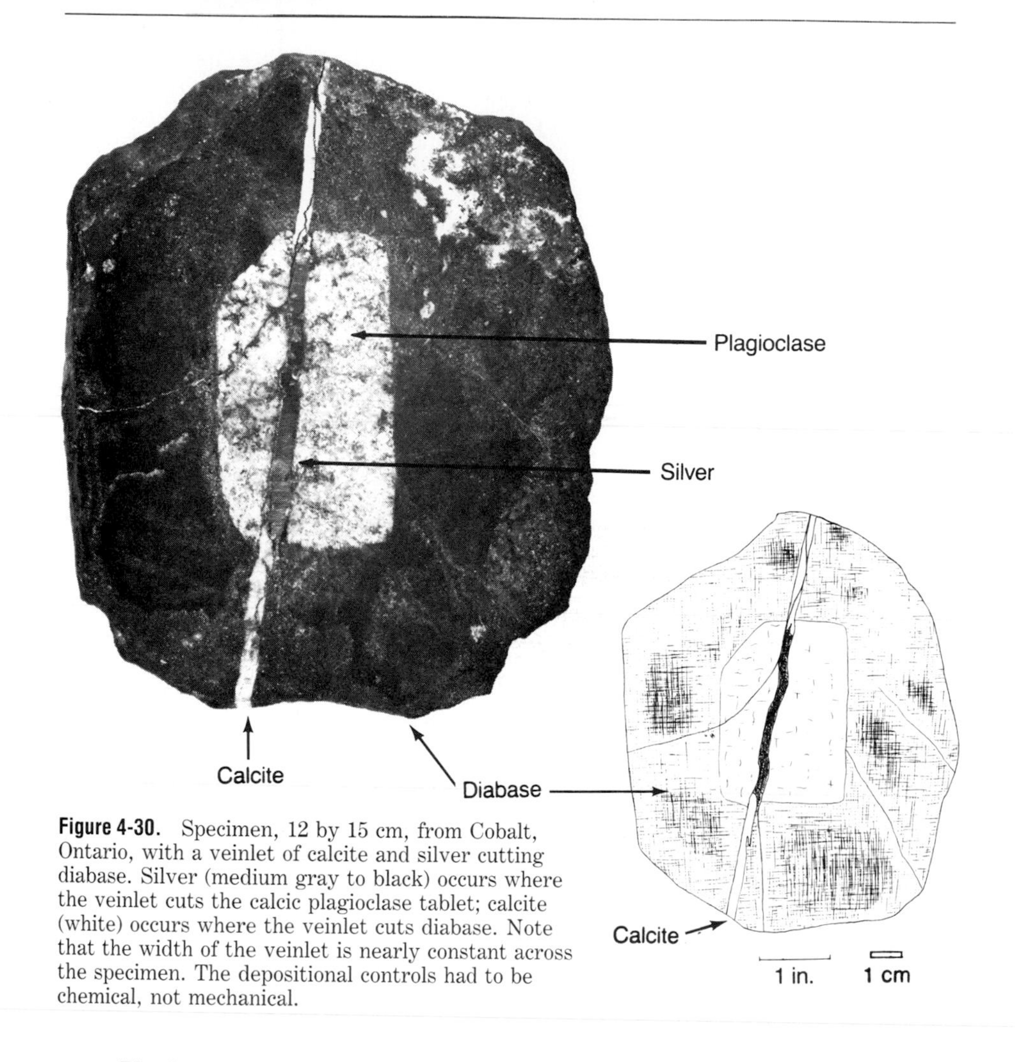

Figure 4-30. Specimen, 12 by 15 cm, from Cobalt, Ontario, with a veinlet of calcite and silver cutting diabase. Silver (medium gray to black) occurs where the veinlet cuts the calcic plagioclase tablet; calcite (white) occurs where the veinlet cuts diabase. Note that the width of the veinlet is nearly constant across the specimen. The depositional controls had to be chemical, not mechanical.

Pioche, Nevada, Lead-Zinc Limestone Replacement Deposits

The limestone replacement deposits of lead-zinc in the Pioche district, Lincoln County, Nevada (Figure 4-31), offer a good example of a chemically controlled zone of deposition. Since operations began in 1869 (Westgate and Knopf, 1932), the district had produced about $100 million worth of ore by the Great Depression and has produced sporadically since World War II. The principal ore deposits are in the Pioche Hills, a northwesterly striking range; related ore has been recovered from small mines scattered along the length of the Highland-Bristol Range to the west. Lower and Middle Cambrian rocks make up most of the mountain ranges, though Tertiary lava flows and tuffs cover large areas on all sides of the district. The Cambrian sequence grades from quartz sandstones and quartzites at the bottom to carbonates at the top. A quartz monzonite stock of probable Tertiary age

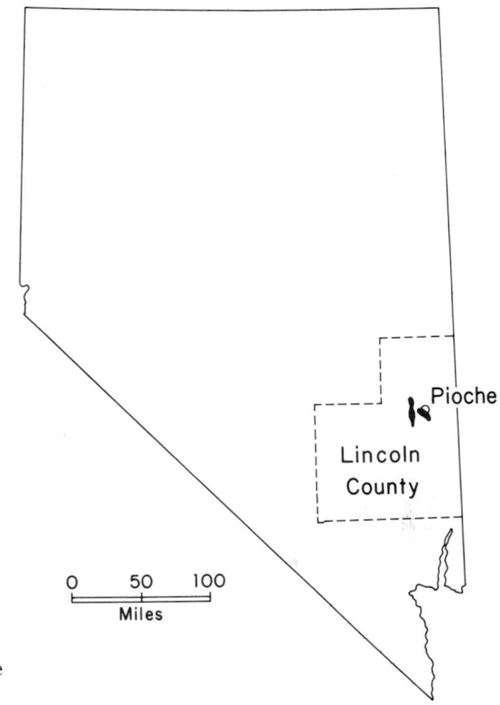

Figure 4-31. Map of Nevada, showing the location of the Pioche district.

intruded the Highland-Bristol Range along its west-central edge, and numerous granite porphyry and diabase dikes are found near the mining areas.

The mountain ranges are typical basin and range structures and are thoroughly broken by normal faulting (Figure 4-32). Ore-bearing solutions appear to have ascended along these faults. In fact, most of the high-grade ore mined in the early days came from structurally controlled fault veins in Lower Cambrian quartzites. Deposition was by replacement, where the mineralizing fluids passed through the quartzites and into the overlying carbonate rocks, generally in the first limestone intersected. Nearly all production since 1924 has come from deposits in these lower limestone beds.

The principal ore-bearing beds are in the Pioche Shale, of Lower to Middle Cambrian age. This formation contains several thin limestone beds that have undergone selective replacement by ore minerals. Carbonate formations overlying the Pioche Shale were also mineralized, but the limestone beds within the shale were the first carbonates met by ascending solutions and were the most abundantly mineralized.

The response of reactive limestone to ore-bearing solutions is strikingly shown by the concentration of ores within the Pioche Shale. The lowermost bed of limestone, known locally as the Combined Metals (CM) Limestone, has provided more than 90% of the replacement ores in the Pioche Hills and is also an important ore producer elsewhere. This limestone stratum is as much as 20 meters thick; although the lower half contains

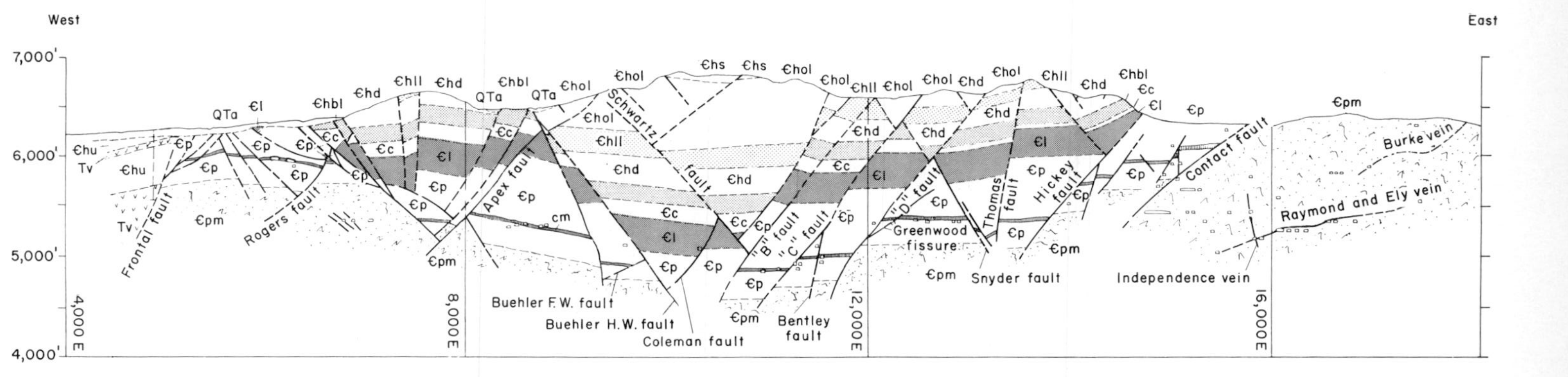

Figure 4-32. East-west cross section through the Pioche Hills, Nevada. Remarkably, the relatively thin CM Limestone unit just above the basement (black in the cross section) was replaced extensively and selectively by sphalerite-galena. Ꞓpm = Prospect Mountain Quartzite; Ꞓp = Pioche Shale, which includes the CM Limestone member (cm) Ꞓl = Lyndon Limestone; Ꞓc = Chisholm Shale; Ꞓhbl, Ꞓhd, Ꞓhll, Ꞓhol, and Ꞓhs are oldest to youngest members of Highland Peak Limestone (Ꞓhu) mapped in this section; Tv = Tertiary volcanic rocks; QTa = Tertiary and Quaternary clastics. (*After* Park, Gemmill, and Tschanz, 1958.) A similarly selective replacement of a favorable bed occurs at the Magma mine (Figure 11-39), where the O'Carroll Bed, a reactive layer in a nonreactive sedimentary sequence, is widely replaced.

most of the ore, in places the whole bed is replaced by sulfides. Scores of replacement orebodies are localized along fractures that acted as channels for the orebearing solutions. Elongate, "bedded" orebodies, which stretch laterally from steep faults up to 150 meters in the plane of the CM horizon, persist along single fissures as much as 3 km along strike, although the CM bed and hence the ore zone is offset at many places along this distance by preore crossfaults (Young, 1948).

The bedded orebodies, or mantos, consist of an intimate mixture of pyrite, argentiferous galena, and sphalerite-calcite gangue. Near the edges of the orebodies manganosiderite is abundant, suggesting that manganese traveled farther from the source faults before being deposited. Similarly manganese concentrations have been mined from some of the zones above the Pioche Shale. Many of the replacement textures described earlier are found. James and Knight (1979) described the ores as averaging about 8% Pb, 18% Zn, and 275 ppm (9 oz/ton) Ag in the discovery area.

At the Bristol mine, 22 km northwest of Pioche, most of the orebodies are in limestones and dolomites above the Pioche Shale. Nevertheless, drilling has shown that the CM Limestone is mineralized with lead and zinc at depth, emphasizing that the bed is exceptionally receptive to ore deposition wherever it is cut by the ore-fluid-conduit fissures. In fact, the limestone beds in the Pioche Shale contain replacement deposits of lead-zinc in other Nevada mining districts, such as the Groom district, nearly 150 km southwest of Pioche (Humphrey, 1945). The fluorite-tungsten-beryllium deposits in the Snake Range, about 100 km to the north, are also in the same lithologic and stratigraphic position (Whitebread and Lee, 1961). Selective replacement of specific lithologies like the CM is not uncommon—the O'Carroll Bed in the Superior district, Arizona (Figure 11–39), is similarly regionally receptive to mineralization.

Where the ore-bearing solutions were confined to quartzite, they deposited their metals in veins. Where these fluids ascended beyond the quartzite without depositing all their dissolved metals, they encountered beds of limestone in the overlying Pioche Shale and younger formations. Since the carbonates were more permeable than the shales, the solutions were able to move laterally. Even more important, they were reactive to the rock and deposited their metals by reactions like Equations (4-5), (4-6), (4-7), and (4-8). If localization of the ore had been a direct function of the permeability of the limestone, mineralization would be concentrated along the upper half of the CM bed, rather than along its lower half. However, the ore-bearing fluids ascended along structures and formed the orebodies where these structures intersected the limestone beds. Thus any mineralized fissures found in rocks above the Pioche Shale are strong evidence of the possibility of replacement deposits along the same structures where they intersect the CM Limestone at depth. James and Knight (1979) suggest that the fluids responsible for mineralization may have been metal-rich brines expelled from laterally adjacent Paleozoic basin sediments themselves, and that the ores are of Mississippi Valley deposit affinity (Chapter 20). Earlier workers have assumed a more conventional Cordilleran Vein type relationship with an igneous hydrothermal source.

Matsuo, Japan, Sulfur Deposits

Solfataric activity may account for the amazing quantity of native sulfur in the Matsuo deposit in central Honshu, Japan. A bleached gypsiferous tuff (Figure 4-33) within a large crater is estimated to contain at least 160 million tons of ore with an average content of about 35% S. An estimate of 50 million tons of native sulfur is therefore reasonable. In addition, large amounts of sulfur are contained in the tremendous quantities of gypsum and in finely divided pyrite, and small amounts of arsenic, bismuth, and antimony sulfides are widely distributed in the ore. The ratio of pyrite to native sulfur is approximately 1:2. If the original tuff at Matsuo had been rich in iron, large bodies of massive pyrite might have formed instead of the native sulfur (Fujita, 1954).

The oldest rock in the Matsuo region, the Upper Miocene Kitanomatagawa dacite, is overlain by a Plio-Pleistocene sequence of lavas and sediments that includes the Hachimantai Formation, the principal host rock of the ores (Figure 4-34). The sulfur and iron sulfide deposits are best developed in a sequence of bleached and altered augite-hypersthene andesites (Takeuchi, Takahashi, and Abe, 1966; Takeuchi and Abe, 1976).

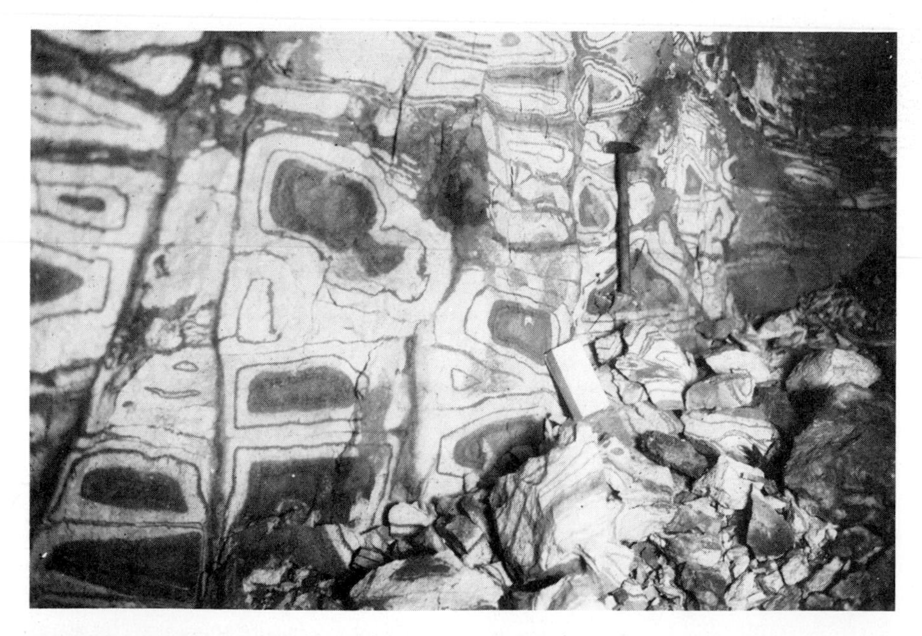

Figure 4-33. Gypsiferous tuff of the Hachimantai Formation, Matsuo mine, Japan. The dark areas are fresh tuff, the light ones being altered and sulfurized. See also Figure 6-3. (*Courtesy of* Matsuo Company.)

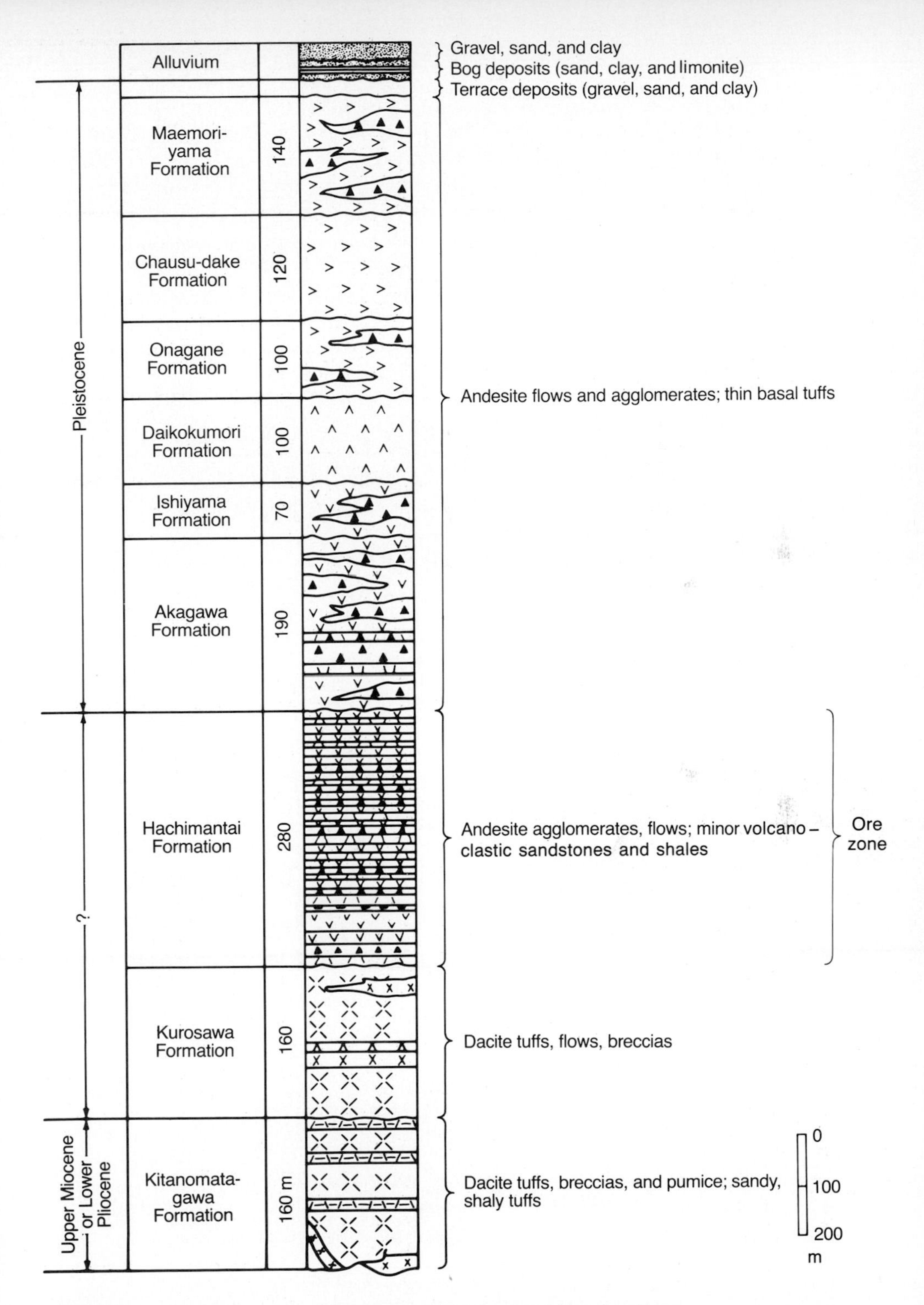

Figure 4-34. Geological column of the Matsuo district, central Honshu, Japan.

The ascending sulfurous fluids were considered by Takeuchi and his associates to have resembled those of present-day solfataras. The fluids have produced ten recognizable alteration zones characterized by (1) silicification, (2) iron sulfides, (3) iron sulfide-sulfur ore, (4) sulfur-iron sulfide ore, (5) sulfur ore, (6) kaolin-alunite-opal, (7) kaolin, (8) montmorillonite, (9) saponite, and (10) quartz. The solutions that caused this alteration were strongly acid and contained H_2O, H_2SO_4, H_2S, H_2SO_3, CO_2, Cl^-, O_2, sulfates, and other metallic and nonmetallic compounds. Where these ascending acidic fluids mixed with groundwaters, they were oxidized to sulfuric acid solutions. When magnetite in the andesites or ferrous iron in the pyroxenes was encountered, the reaction was

$$\underset{\text{magnetite}}{FeO{\cdot}Fe_2O_3} + H_2SO_4 \rightleftarrows FeSO_4 + \underset{\text{hematite}}{Fe_2O_3} + H_2O \qquad (4\text{-}10)$$

and then ferric iron reacted with the fluid to produce native sulfur,

$$\underset{\text{hematite}}{Fe_2O_3} + 2H_2SO_4 + H_2S \rightleftarrows 2FeSO_4 + 3H_2O + \underset{\substack{\text{native}\\\text{sulfur}}}{S^0} \qquad (4\text{-}11)$$

and pyrite,

$$FeSO_4 + H_2S + S \rightleftarrows H_2SO_4 + \underset{\text{pyrite}}{FeS_2} \qquad (4\text{-}12)$$

Both sulfur and pyrite were generated by the reaction of andesite with sulfur species in uprising fluids. It was the replacement of iron minerals, with related reactions, that fixed the sulfur and pyrite.

When iron was absent and aluminum, as in kaolin, was present, the reaction appears to have been

$$\underset{\text{kaolinite}}{Al_2Si_2O_5{\cdot}(OH)_4} + 3H_2SO_4 \rightleftarrows Al_2(SO_4)_3 + \underset{\text{opal}}{2SiO_2{\cdot}H_2O} + 3H_2O$$

producing alunite-jarosite-like phases. Concurrently the oxidation of H_2S by oxygen and sulfuric acid in solution is thought to have formed native sulfur by

$$2H_2S + O_2 \rightarrow 2H_2O + \underset{\substack{\text{native}\\\text{sulfur}}}{2S^0} \qquad (4\text{-}13)$$

and

$$3H_2S + H_2SO_4 \rightarrow 4H_2O + \underset{\substack{\text{native}\\\text{sulfur}}}{4S^0} \qquad (4\text{-}14)$$

Mineral associations in this district are shown graphically in Figure 6-3.

Although chemical interaction between rocks and fluids is not so obvious at other Japanese localities, it is notable that deposition of native sulfur by volcanic-solfataric activity was directly observed at another locality, the Siretoko-Iosan volcano in Hokkaido (Watanabe, 1940). Molten sulfur has been ejected intermittently from a small geyser steam vent. The sulfur flowed down a valley for nearly 2 km, reaching thicknesses of more than 3 meters and widths of 20 to 30 meters (Figure 4–35). The associated water is strongly acidic, owing to the presence of sulfuric acid. Most of the water is meteoric, however, and probably leached the sulfur from previously deposited fumarolic crusts in the underlying volcanic agglomerate. The original volcanic fluids appear to have supplied sulfur in the form of H_2S, meaning that the eruptions of native sulfur were second-cycle products after remobilization. A seasonal and hourly periodicity, reflecting recharge of water, supports this hypothesis. Again, oxidation of reduced sulfur species, especially in H_2S, to S^0 presumably produced the native sulfur. The mode of deposition involves replacement only in part, but is definitely chemically controlled.

Figure 4-35. Valley filled with native sulfur from the Siretoko-Iosan volcano, Hokkaido, Japan. The sulfur crusts, white in the photo, are 30 meters wide, 3 or more meters thick, and are periodically renewed by fresh fumarolic activity. (*From* Watanabe, 1940.)

Bibliography

Ames, L. L., Jr., 1961. The metasomatic replacement of limestones by alkaline, fluoride-bearing solutions. *Econ. Geol.* 56:730–739.

Anderson, G. H., 1979. Experimental data on the solubility of sphalerite and galena in brines. Ann Meet. AIME Prog., p. 50.

Barnes, H. L., 1962. Mechanisms of mineral zoning. *Econ Geol.* 57:30–37.

——, Ed., 1967. *Geochemistry of Hydrothermal Ore Deposits.* New York: Holt, Rinehart and Winston, 670 pp.

——, Ed., 1979. *Geochemistry of Hydrothermal Ore Deposits,* 2nd ed. New York: Wiley, 798 pp.

—— and G. K. Czamanske, 1967. Solubilities and transport of ore minerals, pp. 334–379 in H. L. Barnes, Ed., *Geochemistry of Hydrothermal Ore Deposits.* New York: Holt, Rinehart and Winston, 670 pp.

Barton, P. B., Jr., and B. J. Skinner, 1967. Sulfide mineral stabilities, pp. 238–333 in H. L. Barnes, Ed., *Geochemistry of Hydrothermal Ore Deposits.* New York: Holt, Rinehart and Winston, 670 pp.

Bastin, E. S., 1950. Interpretation of ore textures. Geol. Soc. Amer. Mem. 45.

——, L. C. Graton, W. Lindgren, W. H. Newhouse, G. M. Schwartz, and M. N. Short, 1931. Criteria of age relations of minerals, with special reference to polished sections of ores. *Econ. Geol.* 26:561–610.

Bowen, N. L., 1928. *The Evolution of the Igneous Rocks.* Princeton, N.J.: Princeton Univ. Press; reprinted 1956. New York: Dover.

Brett, R., 1964. Experimental data from the system Cu-Fe-S and their bearing on exsolution in ores. *Econ. Geol.* 59:1241–1269.

Burnham, C. W., 1967. Hydrothermal fluids at the magmatic stage, pp. 34–76 in H. L. Barnes, Ed., *Geochemistry of Hydrothermal Ore Deposits.* New York: Holt, Rinehart and Winston, 670 pp.

Cameron, E. N., and M. E. Emerson, 1959. The origin of certain chromite deposits of the eastern part of the Bushveld Complex. *Econ. Geol.* 54:1151–1213.

Craig, J. R., and G. Kullerud, 1968. Phase relations and mineral assemblages in the copper-lead-sulphur system. *Am Mineral.* 53:145–161.

——, and D. J. Vaughan, 1981. *Ore Microscopy and Ore Petrography.* New York: Wiley, 406 pp.

Dawson, J. B., 1962. Sodium carbonate lavas from Oldoinyo Lengai, Tanganyika. *Nature* 195:1075–1076.

——, 1966. Oldoinyo Lengai—An active volcano with sodium carbonate carbonatite lava flows, pp. 155–168 in O. F. Tuttle and J. Gittins, Eds., *Carbonatites.* New York: Wiley, 591 pp.

Deans, T., 1966. Economic mineralogy of African carbonatites, pp. 385–413 in O. F. Tuttle and J. Gittins, Eds., *Carbonatites.* New York: Wiley, 591 pp.

Edwards, A. B., 1952. The ore minerals and their textures. *Roy. Soc. N.S.W.J. and Proc.* 85:26–45.

——, 1954. *Textures of the Ore Minerals and Their Significance.* Melbourne: Australas. Inst. Min. Metall.

——, 1965. *Textures of the Ore Minerals and Their Significance,* 2nd ed. Melbourne: Australas. Inst. Min. Metall., 242 pp.

Fujita, Y., 1954. Three-way geophysical method points up huge pyrite deposit. *Eng. Min. J.,* Dec., pp. 84–88.

Garrels, R. M., and R. M. Dreyer, 1952. Mechanism of limestone replacement at low temperatures and pressures. *Geol. Soc. Amer. Bull.* 63:325–379.

Haranczyk, C., 1969. Noncolloidal origin of colloform textures. *Econ. Geol.* 64:466–468.

——, 1971. Colloidal transport phenomena of zinc sulfide (brunckite) observed in the Olkusz mine in Poland, in Y. Takeuchi, Ed., *Geochemistry and Crystallography of Sulphide Minerals in Hydrothermal Deposits,* Joint Symp. Vol. IMA-IAGOD Mtgs., 1970. Tokyo: Soc. Min. Geol. Japan, Spec. Issue 2, 500 pp.

Hosking, K. F. G., 1964. Permo-carboniferous and later primary mineralisation of Cornwall and south-west Devon, in K. F. G. Hosking, and G. J. Shrimpton, Eds., *Present Views of Some Aspects of the Geology of Cornwall and Devon.* Penzance: Roy. Geol. Soc. Cornwall.

Humphrey, F. L., 1945. Geology of the Groom district, Lincoln County, Nevada. Univ. Nev. Bull. 39(5).

Irvine, T. N., 1977. Origin of chromitite layers in the Muskox Intrusion and other stratiform intrusions; a new interpretation. *Geology* 5:273–277.

Jackson, E. D., 1961. Primary textures and mineral associations in the ultramafic zone of the Stillwater Complex, Montana. USGS Prof. Pap. 358, 106 pp.

James, J. A., 1949. Geologic relationships of the ore deposits in the Fredericktown area, Missouri. Mo. Geol. Surv. Water Res. Rept. Invest. 8.

James, L. P., and L. H. Knight, 1979. Stratabound lead-zinc-silver ores of the Pioche district, unusual Mississippi Valley deposits, pp. 389–395 in G. W. Newman and H. D. Goode, Eds., *Basin and Range Symp. and Great Basin Field Conf.* Salt Lake City, UT: Rocky Mt. Assoc. Geol.

Killinger, P. E., 1942. Report on the titanium mine at Tahawus, New York. *Rocks and Miner.* 17:409.

King, B. C., 1965. Petrogenesis of the alkaline igneous rock suites of the volcanic and intrusive centres of Uganda. *J. Petrol.* 6:67–100.

Kutina, J., 1957. A contribution to the classification of zoning in ore veins. *Univ. Carolina Geol.* 3(3):197–225.

—— and J. Sedlackova, 1961. The role of replacement in the origin of some cockade textures. *Econ. Geol.* 56:149–176.

Lindgren, W., 1933. *Mineral Deposits.* New York: McGraw-Hill, 930 pp.

Loughlin, G. F., and A. H. Koschmann, 1942. Geology and ore deposits of the Magdalena mining district, New Mexico. USGS Prof. Pap. 200.

McBirney, A. R., and R. M. Noyes, 1979. Crystallization and layering of the Skaergaard Intrusion. *J. Petrol.* 20: 487–554.

Meyer, C., and J. J. Hemley, 1967. Wall rock alteration, pp. 166–235 in H. L. Barnes, Ed., *Geochemistry of Hydrothermal Ore Deposits.* New York: Holt, Rinehart and Winston, 670 pp.

Park, C. F., Jr., 1931. Hydrothermal experiments with copper compounds. *Econ. Geol.* 26:857–883.

——, P. Gemmill, and C. M. Tschanze, 1958. Geologic map and sections of the Pioche Hills, Lincoln County, Nevada. USGS Mineral Invest. Field Stud. Map MF 136.

Pecora, W. T., 1956. Carbonatites—A review. *Geol. Soc. Amer. Bull.* 67:1537–1555.

Picot, P., and Z. Johan, 1982. *Atlas of Ore Minerals.* Amsterdam: Elsevier, 459 pp.

Ramdohr, P., 1955. *Die Erzmineralien und ihre Verwachsungen,* Berlin: Akad. Verlag.

——, 1980. *The Ore Minerals and Their Intergrowths,* 2nd ed., Intern. Ser. in Earth Sci. 35. London: Pergamon Press, 1205 pp.

Ridge, J. D., 1949. Replacement and the equating of volume and weight. *J. Geol.* 57:522–550.

——, 1961. Gain and loss of material in a series of replacements. Spec. Pap. 68, Geol. Soc. Amer. Abstr., pp. 252–253.

Roedder, E., 1968. The noncolloidal origin of "colloform" textures in sphalerite ores. *Econ. Geol.* 63:451–471.

Ruiz, J., S. E. Kesler, L. M. Jones, and J. F. Sutter, 1980. Geology and geochemistry of the Las Cuevas fluorite deposit, San Luis Potosi, Mexico. *Econ. Geol.* 75:1200–1209.

Schneiderhöhn, H., and P. Ramdohr, 1931. *Lehrbuch der Erzmikroskopie.* Berlin: Borntraeger.

Schouten, C., 1934. Structures and textures of synthetic replacements in "open space." *Econ. Geol.* 29:611–658.

Schwartz, G. M., 1931. Textures due to unmixing of solid solutions. *Econ. Geol.* 26:739–763.

Skinner, B. J., and P. B. Barton, Jr., 1973. Genesis of mineral deposits. *Ann. Rev. Earth Planet. Sci.* 1:183–211.

Sondergeld, C. H., and D. L. Turcotte, 1979. A laboratory study of mineral deposition in a boiling environment. *Econ. Geol.* 74:109–115.

Stevenson, J. S., and W. G. Jeffrey, 1964. Colloform magnetite in a contact metasomatic iron deposit, Vancouver Island, British Columbia. *Econ. Geol.* 59:1298–1305.

Takeuchi, T., and H. Abe, 1970. Regularities of hydrothermal alteration of sulphur and iron-sulphide deposits in Japan, pp. 381–383 in Z. Pouba and M. Stemprok, Eds., *Problems of Hydrothermal Ore Deposition*, IUGS Ser. A, no. 2. Stuttgart, 393 pp.

——, I. Takahashi, and H. Abe, 1966. Wall-rock alteration and genesis of sulphur and iron-sulphide deposits in Japan. *Tohoku Imper. Univ. Sci. Rept.*, Ser. 3(geol.), 9(3):381–483.

Toulmin, P., III, and S. P. Clark, Jr., 1967. Thermal aspects of ore formation, pp. 437–464 in H. L. Barnes, Ed., *Geochemistry of Hydrothermal Ore Deposits*. New York: Holt, Rinehart and Winston, 670 pp.

Tuttle, O. F., and J. Gittins, 1966. *Carbonatites*. New York: Wiley, 591 pp.

Veen, R. W. Van der, 1925. *Mineragraphy and Ore Deposition*. The Hague: G. Naeff, 167 pp.

Wager, L. R., and G. M. Brown, 1967. *Layered Igneous Rocks*, San Francisco: Freeman, 588 pp.

—— and W. A. Deer, 1939 (re-issued in 1962). Geological investigations in East Greenland. *Medd. om Grönland* 105(4): 1–352.

Watanabe, T., 1940. Eruptions of molten sulphur from the Siretoko-Iôsan volcano, Hokkaidô, Japan. *Jap. J. Geol. Geog.* 17:289–310.

Westgate, L. G., and A. Knopf, 1932. Geology and ore deposits of the Pioche district, Nevada. USGS Prof. Pap. 171.

White, D. E., L. J. P. Muffler, and A. H. Truesdell, 1971. Vapor-dominated hydrothermal systems compared with hot-water systems. *Econ. Geol.* 66:75–97.

Whitebread, D. H., and D. E. Lee, 1961. Geology of the Mount Wheeler mine area, White Pine County, Nevada. USGS Prof. Pap. 424-C, pp. 120–122.

Young, E. B., 1948. The Pioche district. 18th Int. Geol. Congr. Rept., pt. 7, pp. 98–106.

FIVE

Wall-Rock Alteration and Gangue

The *country rocks* that enclose ore deposits of hydrothermal origin almost always show reaction effects that result from the tendency of hot circulating fluids to equilibrate with rocks flanking the conduits through which they move. Both the fluids and the rocks adjust; the conduits become lined with what amounts to a layer of insulation between fresh external rock and the heated, commonly slightly acid circulating solutions. That layer of "insulation" is called *wall-rock alteration* (Figures 5-1 and 5-2), and the volume that it occupies is the *alteration zone, halo, envelope,* or *selvage.* Alteration effects may be visible for as little as a couple of grain widths beyond a vein wall, or they may extend for kilometers around a network of veins. Normally the envelope around a vein of 1 or 2 meters in width is on the order of 10 to 20 meters, but many exceptions are noted. Some alteration is colorful and texturally and structurally flamboyant, such as that at the Matsuo mine (Figure 4-33). Elsewhere it may be totally inconspicuous, such as in the dolomitization of limestone around Mississippi Valley deposits (Chapter 19).

A fascinating aspect of alteration is its zoning. From the scale of a hairline veinlet with a 2-mm envelope through a 2-meter vein with a 10-meter halo to a district 5 km across, alteration patterns are inclined to be progressively, predictably, repetitively arrayed, a regularity of occurrence called *zoning*. The zoning generally represents a chemical and mineralogical transition or buffer zone from fresh rock outside to the vein environment inside. More is said about zoning below and in Chapter 6. In general, alteration analysis is an exceedingly useful means for the explorationist and the scientist who seek to understand various aspects of ore-deposit occurrence and genesis. Alteration is discernible, its mineralogy and temporal evidence are detectable and usable, and its interpretation and application are fruitful.

Figure 5-1. Thin-section photographs showing what the physical results of wall-rock alteration can be. Fresh quartz latite porphyry (*a*) is converted to a texturally reminiscent assemblage of phyllic alteration (*b*). The fresh rock (*a*) contains clear-cut tablets of zoned and polysynthetically twinned plagioclase (lower left and lower right), Carlsbad-twinned K-feldspar (right center), single-crystal phenocrysts of biotite (upper right), and quartz "eyes" (black, right center). They are all petrographically crisp in an aplitic groundmass (see also Figure 11-5). The phyllically altered equivalent (*b*) is much less crisp. Feldspars and biotite have all been replaced by fine-grained quartz-sericite, the finely stippled areas. Twinning and zoning have been obliterated. Rutile persists as fine white irregular subparallel slashes in sericitized biotite (lower center) and quartz is unaffected. The textural similarities are retained but subdued. The alteration formula (see text) is S-10-10; all susceptible minerals are completely and pervasively phyllically altered. Both specimens from the Kalamazoo porphyry copper deposit, Arizona. 5×, crossed nicols.

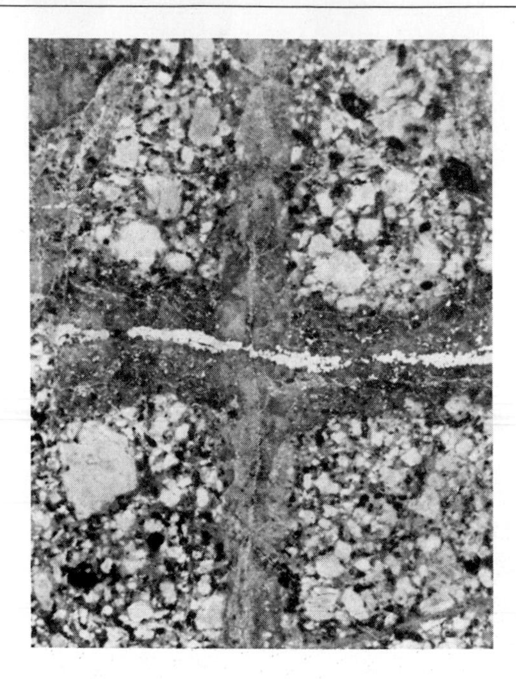

Figure 5-2. Polished slab of porphyry copper ore from Ajo, Arizona. The white phenocrysts are plagioclase partly altered to montmorillonite. The darker, smaller grains are chiefly unaltered biotite. The groundmass consists of fine-grained K-feldspar and quartz. The veinlet cutting the specimen from top to bottom near the center contains quartz and K-feldspar with muscovite (sericite) and chalcopyrite, but little or no biotite; the assemblage is a variety of potassic alteration. Isolated grains of chalcopyrite are also scattered through the groundmass and smaller veinlets. The veinlet running across the specimen consists of pyrite (bright in reflected light) and quartz less than 1 mm wide. Enclosing it is a sericitic envelope about 2 cm wide in which all feldspars have been sericitized and biotite converted to sericite and pyrite. The pyrite clearly cuts the earlier quartz-K-feldspar-chalcopyrite veinlet; this assemblage is called *phyllic*. The K-feldspar of the earlier veinlet is also sericitized within the projected borders of the sericitic envelope, though this is not entirely obvious to superficial examination in the field. (*From* Meyer and Hemley, 1967.)

▸ RELATIONSHIP BETWEEN ALTERATION, GANGUE, AND MINERALIZATION

It is a key point here that alteration mechanisms normally are completely continuous with and a part of the hydrothermal ore-forming process. *Alteration* is defined as any change in the mineralogic composition of a rock brought about by physical or chemical means, especially by the action of hydrothermal fluids. Alteration minerals can be considered as much the result of mineralization processes as are the ore minerals themselves. Indeed, some ores can be considered merely commercially valuable products of wall-rock alteration. Barite, fluorite, and even quartz may be waste products—gangue—at one mine, but the principal products sought at another. And certainly as much is to be learned from studying the solution, transportation, and deposition of the gangue minerals in an ore occurrence as can be learned from the ore minerals. Indeed, most economic petrologists hyphenate the term *alteration-mineralization* to imply the degree to which the two phenomena are inseparable.

The nature of the alteration products depends upon (1) the character of the wall rock, (2) the character of the invading fluid, which defines such factors as Eh, pH, vapor pressure of various volatile species, anion-cation composition, and degree of hydrolysis, and (3) the temperatures and pressures at which the reactions take place. Wall-rock alteration has been recognized for many years as a valuable tool in exploration. The alteration halos around many deposits are more widespread and much easier to locate than the ore bodies within them. As is seen later, the porphyry copper orebody at Morenci, Arizona, is only 2 km^2 in outcrop, but the readily mappable alteration zones cover an area of 6 by 12 km, or 72 km^2! Clearly, gangue minerals themselves may be useful in prospecting. They generally extend beyond the ore shoots and, as a result, may be used to distinguish mineralized veins from barren structures. Similarly, minor changes in the distribution of gangue minerals may indicate which direction along a vein is likely to lead to ore.

Alteration effects may reach old or present surfaces, or they may be "blind" and found only in underground workings or in drill holes. Equipment for rapid and precise determination of alteration mineralogy and significance has become available in the 1970s, and a great deal of work has been done in attempting to determine the exact relationships between various types of ore deposits and the alteration minerals that underlie or surround them. Main problems involve such questions as: What types of alteration products are associated with particular types of ores? How do deep alteration assemblages differ from near-surface ones? What changes are to be expected in the country rocks at various distances from the orebodies? Specifically, where in the alteration halos are the ores most likely to be found? What directional aspects can we discern? What can we learn about mass transfer, the geochemistry of the systems, and how ores actually form?

Several variations on the theme of alteration can be distinguished, and although they are similar in many ways, attempts should be made to separate them. Alteration may result from (1) diagenesis in sediments, (2) regional processes, such as metamorphism, (3) postmagmatic or postvolcanic processes associated with cooling, and (4) direct mineralization processes. For example, *propylitization* is the result of low-pressure-temperature peripheral alteration around many orebodies. The propylite assemblage consists of epidote, chlorite, Mg-Fe-Ca carbonates, and sometimes albite-orthoclase, all involved in partial replacement of wall-rock minerals. The same assemblage can be produced by what the petrologist calls *saussuritization* of mafic to intermediate volcanic rocks—regional metamorphism of greenschist facies. It can also be produced during the cooling history of a basalt-andesite-dacite pile. The three varieties of propylite can be distinguished readily by the serious student of alteration, but the point is clear—hydrothermal alteration can be confused with other phenomena, and vice versa. Geochemistry can be extremely useful; for example, whenever bona fide hydrothermal propylitization exists, anomalous values of 100 to 400 ppm Cu or Zn are common.

▸ TEMPERATURE, PRESSURE, AND COMPOSITION GRADIENTS

As indicated in Figure 2-7, hydrothermal fluids are not generally expelled from magmatic systems until the intermediate to felsic stages. Although mafic stocks and plutons as well as more felsic ones can drive convective groundwater circulation, hydrous alteration effects around mafic bodies are minimal, as are resulting economic occurrences. For example, only recently (Taylor and Forester, 1979; Norton and Taylor, 1979) have alteration effects around the gabbroic Skaergaard Intrusion of Greenland been identified and examined, but no associated ore deposits have been perceived. It is also true that syngenetic deposits in general have no alteration effects; the ore phases were deposited with the host rocks, and no gradients necessarily existed. Chemical and mechanical sedimentary deposits, the volcanogenic ores themselves (but not their pipelike access conduits), most black-shale-hosted ores, and magmatic segregation deposits are generally free of alteration effects. So we deal primarily with epigenetic hydrothermal ore deposits associated with intermediate to felsic volcanism and plutonism when we concern ourselves with hydrothermal alteration. Another important sector, however, has to do with seafloor phenomena. Along modern midoceanic ridges and transform faults and in ancient ocean basins which built toward greenschist terranes (Figures 13-1 to 13-3), geothermal anomalies have driven enormous convective circulations, drawing seawater down into seafloor rocks, pulling it through cubic kilometers of basalts which thereby undergo chloritization and serpentinization, and spewing it back into the sea as more or less metal-, chlorine-, and sulfur-charged "exhaled" (or "exhalative") fluids. Typically, an alteration zone—a pipe or a linear zone—is also formed where the fluids rise to the point of issue. The nature of that alteration is dealt with below.

In any alteration system, temperature, pressure, and composition gradients probably exist between the circulating fluids and the walls. Which is most important? Heat conduction and heat transfer theory dictate that steep thermal gradients are geologically transient and that gradients of more than a few degrees Celsius across a typical alteration envelope would quickly be eliminated by the balance of heat loss and heat gain cited in Chapter 4. Temperatures in and near an active fissure can be expected to be higher than those in the walls, but probably not by much and not for long. Davis (1974) made 1100 fluid-inclusion–filling-temperature measurements at the huge San Manuel-Kalamazoo, Arizona, porphyry copper deposit and found that thermal gradients were almost inconsequential at either the meter or the kilometer scale through the emplacement history of that orebody. We can conclude that steep thermal gradients in general are not "prime movers" in alteration, although total heat flux—whether the system is hot or just warm—certainly is. Gentle thermal gradients—a few degrees per meter—over prolonged periods of 10^3 to 10^5 years probably do contribute to alteration results.

Pressure gradients are probably even less important. Across a given vein envelope at a given level, neither lithostatic nor hydrostatic pressure variation can be called upon to influence mineral zoning. Most ore-forming systems do not extend over more than 1 or 2 km of vertical height at most. Cordilleran Veins, pegmatites, and some porphyry base-metal systems may do so, but no satisfactory assignment of alteration variation has been assigned to the kilobar or two of pressure variation (1 kbar $\cong$ 4 km of crustal rocks) that they underwent from top to bottom. Certainly pressure will affect boiling points, degassing of volatiles, and even precipitation (Chapter 4); and long-term shifts from lithostatic to hydrostatic pressures with retrograde boiling will affect zoning; but pressure gradients at the orebody scale have not been translated to wall-rock alteration effects.

Compositional gradients are clearly another matter. If we consider a fissure opened in granodiorite or quartz monzonite porphyry that is instantaneously filled with hydrothermal fluids, what will happen? We have discussed the ore-forming fluid in terms of alkalies, chlorine, metals, and other components (Table 4-1 and Figures 4-3 and 4-4), but how does fluid composition relate to silicate rock compositions? Reinspection of Figures 2-7 and 4-1 (Bowen's reaction series) indicates that fractional crystallization can be considered to evolve toward and be continuous into what we have called the *hypogene hydrothermal fluid.* Burnham and Ohmoto (1980) provide a diagram (Figure 5-3) that relates that fluid to granodioritic rocks in the context of the activity ratio of potassium (free K^+ ions plus a reservoir of undissociated KCl that varies with temperature) and hydrogen ion (H^+ + HCl). The bottom part of the diagram was first experimentally derived by Hemley and Jones (1964) and refined by Meyer and Hemley (1967) and others. They heated samples of powdered granodiorite in gold-foil capsules with potash brines of known composition and found that many of the common alteration minerals—K-feldspar, the micas biotite, muscovite, and pyrophyllite, the kaolinite-dickite clays, and such minerals as the sulfate alunite and the aluminosilicate andalusite—were produced at specific compositions and ion ratios. They found that reversible reactions can be

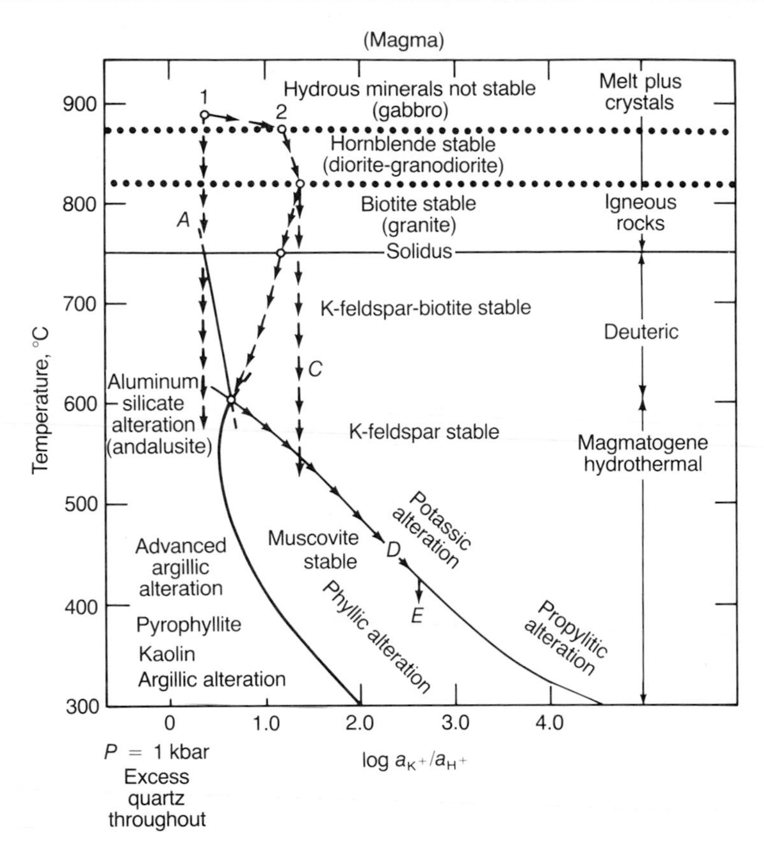

Figure 5-3. Aqueous fluid connection between igneous rocks (above), the deuteric late-stage orthomagmatic rocks and minerals (right center), and some common alteration assemblages (lower left), all as functions of temperature and the ratio of the chemical potentials of K^+ and H^+ at 1 kbar pressure—the pressure of approximately 3-to-4-km depth. A similar diagram using sodium would relate albite, the rare sodium mica paragonite and Na-montmorillonite, and the other rocks and minerals shown. Points 1 and 2 at the top represent aqueous chloride solutions in equilibrium with granodioritic magmas. 1 is low K, and the fluid never precipitates K-feldspar (path *A*). 1 can evolve toward 2 by any means that raises the K^+/H^+ ratio, such as reaction or hydrolysis. From 2, several paths are possible, depending on what happens to the fluid. They all pass through the potassic field to the potassic-phyllic boundary *D* and thence into the phyllic field *E*. (*After* Burnham and Ohmoto, 1980).

written to describe the system, and that there are chemically unifying characteristics. So we can consider that a fluid introduced into the fissure mentioned above will react in characteristic ways with a specific wall rock. The response will be most drastic, or most profound, where the wall rocks are farthest from "equilibrium on contact." It is apparent, then—as is borne out by field and laboratory studies—that chemical gradients are most important, that temperature gradients are probably normally of secondary importance, and that pressure gradients are only rarely significant.

▸ REACTIONS BETWEEN WALL ROCKS AND FLUIDS

If it is accepted that chemical and thermal gradients are the principal determinants of the type and extent of wall-rock alteration, and that hydrothermal briny fluids are at least fundamentally similar, then the particular altered wall-rock mineral assemblages—the reaction products—must be largely governed by (1) the types of alteration reactions that actually proceed, (2) wall-rock composition, and (3) the pressure and temperature of the environment. That the statment is correct is a generalization, of course, but not a bad one—wall-rock components buffer the fluids, the fluid keeps attempting to extend its influence, and layers, or zones, of alteration products expand slowly outward from the fissure as reaction times lengthen.

Reactions that are important to alteration are of many types: (1) hydrolysis, (2) hydration-dehydration, (3) alkali or alkali-earth metasomatism, (4) decarbonation, (5) silication, (6) silicification, (7) oxidation-reduction, and finally, a number of addition-removal interactions such as carbonatization, desulfidation, sulfidation, and fluoridation. *Hydrolysis* refers specifically to the involvement of H^+. It is profoundly important because hydrolysis of wall-rock minerals serves both to convert anhydrous silicates like the feldspars to hydrolyzed silicates like the mica and clay minerals and to buffer the pH of the fluid, in turn affecting solubility and association-dissociation relationships in it. Reactions of the type

$$\underset{\text{K-feldspar}}{3KAlSi_3O_8} + 2H^+(aq) \rightleftarrows \underset{\substack{\text{muscovite}\\ \text{(sericite)}}}{KAl_3Si_3O_{10}(OH)_2} + \underset{\substack{\text{quartz}\\ \text{(crystalline}\\ \text{or aqueous)}}}{6SiO_2} + 2K^+(aq) \qquad (5\text{-}1)$$

$$\underset{\text{albite}}{3NaAlSi_3O_8} + 2H^+(aq) \rightleftarrows \underset{\text{paragonite}}{NaAl_3Si_3O_{10}(OH)_2} + 6SiO_2 + 2Na^+(aq) \qquad (5\text{-}2)$$

are those along the boundary in Figure 5-3 between K-feldspar (or Na-feldspar) and muscovite (or paragonite) in equilibrium with fluid D. It is crucial to realize that the consumption of H^+ to alter feldspar to muscovite affects the pH of the fluid *and* the degree of dissociation of hydrogen-containing complexes including molecular HCl. That in turn—by the common ion effect—affects the degree of association of NaCl, KCl, and metal-chlorine complexes, and the solubility of metals in the solution. So hydrolysis of feldspars is bound symbiotically—or "symchemically"—to the transportation-deposition characteristics of the fluid. Thus we see—through this "domino effect"—how alteration and mineralization are so interdependent. Reaction of metals with $(HS)^-$ to produce sulfides releases H^+ such that mineralization may actually *cause* alteration. Further, hydrolysis controls the progressive "alkali stripping" of muscovite to kaolin (to the left at the base of Figure 5-3) by

$$\underset{\text{sericite}}{2KAl_3Si_3O_{10}(OH)_2} + 2H^+ + 3H_2O \rightleftarrows \underset{\text{kaolinite}}{3Al_2Si_2O_5(OH)_4} + 2K^+ \qquad (5\text{-}3)$$

So it can be seen that the stability of feldspars, micas, and clays—and the transfer from rock-silicate into solution of K^+, Na^+, Ca^{2+}, Mg^{2+}, and other ions—is controlled in large part by hydrolysis. These reactions and the minerals involved are obviously an important key to the interaction and mutual buffering of fluids and alumino-silicate wall rocks. And the potassium, sodium, and calcium analogs all function at the same time in the same system.

Hydration refers to the removal of molecular water from a fluid into a mineral, and *dehydration* to its addition to the fluid. Equation (5-3) just given involves both hydrolysis and hydration. A similar reaction important in seafloor alteration is

$$\underset{\text{olivine}}{2Mg_2SiO_4} + 2H_2O + 2H^+ \rightleftarrows \underset{\text{serpentine}}{Mg_3Si_2O_5(OH)_4} + Mg^{+2} \tag{5-4}$$

A more direct hydration seen in the low-temperature alteration of hematite is

$$\underset{\text{hematite}}{Fe_2O_3} + 3H_2O \rightleftarrows \underset{\text{goethite}}{2Fe(OH)_3} \tag{5-5}$$

Dehydration normally occurs as pressure or temperature increases around alteration assemblages, as shown in Figure 17-1. An excellent example is

$$\underset{\text{kaolinite}}{Al_2Si_2O_5(OH)_4} + \underset{\text{quartz}}{2SiO_2} \rightleftarrows \underset{\text{pyrophyllite}}{Al_2Si_4O_{10}(OH)_2} + H_2O \tag{5-6}$$

Alkali and *alkali-earth metasomatism* are also significant. Magnesium metasomatism, for example, may produce dolomite from limestone by

$$\underset{\text{calcite}}{2CaCO_3} + Mg^{+2}(aq) \rightleftarrows \underset{\text{dolomite}}{CaMg(CO_3)_2} + Ca^{+2}(aq) \tag{5-7}$$

or chlorite from K-feldspar in a rhyolite or arkose by

$$\underset{\text{K-feldspar}}{KAlSi_3O_8} + 6.5\ Mg^{+2} + 10\ H_2O \rightleftarrows \underset{\text{chlorite}}{Mg_{6.5}(Si_3Al)O_{10}(OH)_8} + K^+ + 12\ H^+ \tag{5-8}$$

Alkali metasomatism can either cause readjustment of feldspar compositions by

$$\underset{\text{K-feldspar}}{KAlSi_3O_8} + Na^+ \rightleftarrows \underset{\text{albite}}{NaAlSi_3O_8} + K^+ \tag{5-9}$$

or result in an overall increase in K-feldspar with or without sericite, albite with or without paragonite, or a halo of high Rb^+/K^+ ratios, for example.

Decarbonation reactions are at the center of skarn formation, whereby silicates and oxides are produced by the removal of CO_2 from limestone-dolomite sections and the recombination of components. They can be represented by

$$\underset{\text{dolomite}}{CaMg(CO_3)_2} + \underset{\text{quartz}}{2SiO_2} \rightleftarrows \underset{\text{diopside}}{(CaMg)Si_2O_6} + 2CO_2(g) \qquad (5\text{-}10)$$

or more simply by

$$\underset{\text{(magnesite)}}{MgCO_3} \rightleftarrows \underset{\text{(periclase)}}{MgO} + CO_2(g) \qquad (5\text{-}11)$$

Silicification and *silication* are also common. The words are similar in form, but quite distinctive in meaning. Silicification refers to the addition of silica as quartz or one of its polymorphs such as chalcedony, opal, or jasper, while silication refers to the process of conversion to, or replacement by, silicate minerals. Silicification is the addition or production of a quartz-family mineral, as in the silicification of limestones

$$\underset{\text{calcite}}{2CaCO_3(c)} + SiO_2(aq) + 4H^+ \rightleftarrows 2Ca^{+2}(aq) + 2CO_2 + \underset{\text{quartz}}{SiO_2(c)} + 2H_2O \qquad (5\text{-}12)$$

or as in the simple precipitation of silica in any pore space

$$SiO_2(aq) \rightleftarrows SiO_2(c) \qquad (5\text{-}13)$$

or as in the breakdown of the orthopyroxene bronzite

$$\underset{\text{bronzite}}{MgSiO_3} + CO_2(aq) \rightleftarrows \underset{\text{quartz}}{SiO_2} + \underset{\text{magnesite}}{MgCO_3} \qquad (5\text{-}14)$$

Silication occurs in contact metamorphism by such reactions as

$$\underset{\text{calcite}}{CaCO_3} + \underset{\text{silica}}{SiO_2} \rightleftarrows \underset{\text{wollastonite}}{CaSiO_3} + CO_2(g) \qquad (5\text{-}15)$$

and as Equation (5-10), which describes silication to produce diopside from dolomite and quartz. Desilication-silicification occurs in many skarns according to

$$\underset{\text{garnet}}{Ca_3Fe_2(SiO_4)_3} \rightleftarrows \underset{\text{calcite}}{3CaCO_3} + \underset{\text{hematite}}{Fe_2O_3} + \underset{\text{quartz}}{3SiO_2} \qquad (5\text{-}16)$$

Oxidation-reduction reactions most importantly affect ferrous-ferric iron and sulfur mineralogies and complexes, but they also impinge upon systems containing manganese, vanadium, uranium, and other redox pairs.

Two representative reactions are

$$\underset{\text{magnetite}}{4Fe_3O_4} + O_2 \rightleftarrows \underset{\text{hematite}}{6Fe_2O_3} \tag{5-17}$$

and

$$\underset{\text{ferrous iron biotite = annite}}{2KFe_3AlSi_3O_{10}(OH)_2} + O_2 \rightleftarrows \underset{\text{K-feldspar}}{2KAlSi_3O_8} + \underset{\text{magnetite}}{2Fe_3O_4} + 2H_2O \tag{5-18}$$

Sulfur ranges in valence from S^{+6} in sulfate radicals through zero in native sulfur (S^0) to S^{-2} in bisulfide ion (HS^-), hydrogen sulfide (H_2S), and many sulfide minerals. Traces of native sulfur are found where sulfide minerals are oxidized to sulfates in the weathered portions of ore deposits (Chapter 17).

The final reaction type involves *sulfidation, fluoridation,* and a host of additions and subtractions that will become apparent as Chapters 9 to 20 are contemplated. Just two representative reactions are

$$\underset{\text{ferrous iron biotite = annite}}{2KFe_3AlSi_3O_{10}(OH)_2} + 6S_2 \rightleftarrows 2KAlSi_3O_8 + \underset{\text{pyrite}}{6FeS_2} + 2H_2O + 3O_2(g) \tag{5-19}$$

and

$$\underset{\text{hydroxy sericite}}{\text{OH-mica}} + HF \rightleftarrows \underset{\text{fluorosericite}}{\text{F-mica}} + H_2O \tag{5-20}$$

Having dwelled upon a number of representative reactions, it is well to reemphasize that the most significant overall is the first-named—hydrolysis.

▸ ALTERATION ASSEMBLAGES

One way to describe the alteration process is to go through the reactions that are Equations (5-1) to (5-20) and their analogs and ramifications; another is to consider the reaction products. Schwartz (1947, 1959), Creasey (1959), Meyer and Hemley (1967), Titley and Hicks (1966), Lowell and Guilbert (1970), Guilbert and Lowell (1974), Rose and Burt (1979), and Beane and Titley (1982) have all made contributions in this area. The principal alteration assemblages and associations found in aluminosilicate rocks are summarized in Figure 5-4, mostly from Meyer and Hemley (1967). The terms have already been used (Figure 5-3), and are returned to repeatedly in later paragraphs in this chapter and in later ones. The term *assemblage* implies mutual equilibrium growth of the mineral phases; *association* merely indicates that they occur in contact. Table 5-1 relates some of these assem-

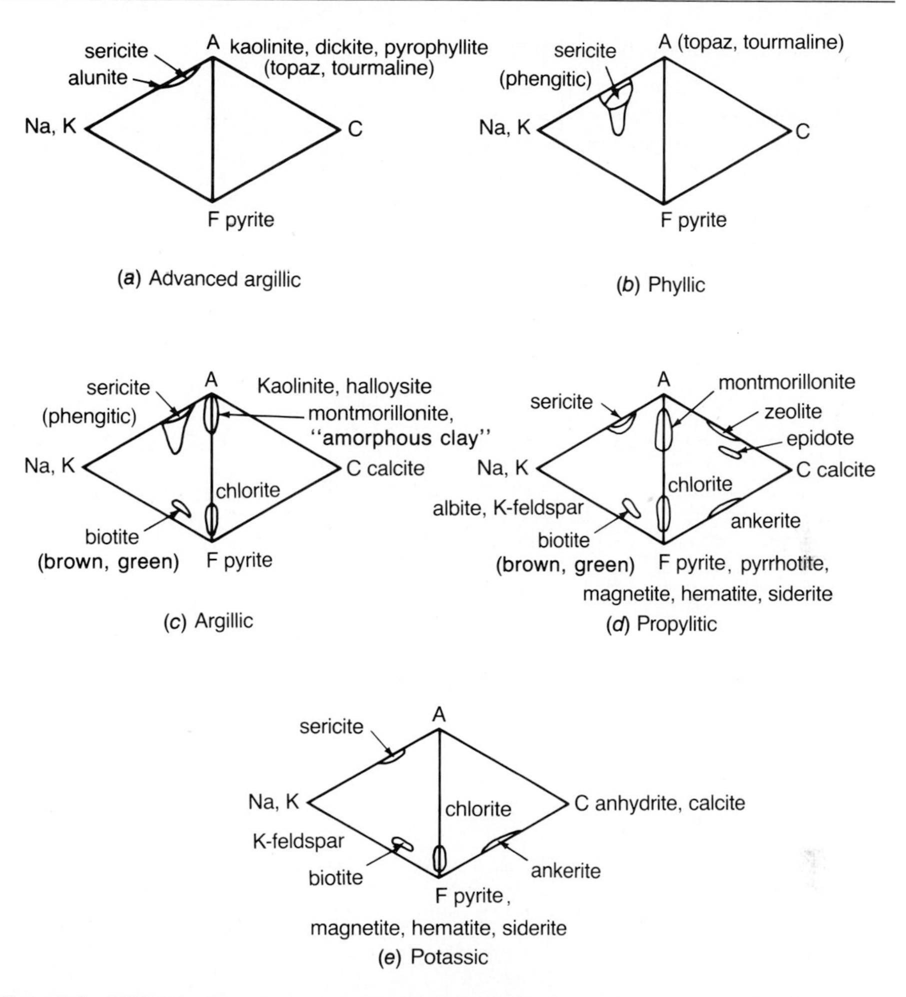

Figure 5-4. Mineral associations in major types of wall-rock alteration in aluminosilicate rocks expressed on ternary AKF and ACF diagrams in which A stands for Al_2O_3, K for sodium and potassium, F for iron and magnesium, and C for calcium. The minerals named on each diagram are the common phases in each type of alteration. Except for the zeolites, and to a lesser extent kaolinite, they have relatively wide pressure-temperature ranges of formation. High-temperature equivalents of the aluminous argillic assemblages would be andalusite-bearing assemblages (see Figure 5-3). See also Table 5-1. Points are stoichiometries; ellipses show extent of known solid solution. (*After* Meyer and Hemley, 1967.)

blages to the sulfide-oxide assemblages depicted in Figure 4-5, and should be considered carefully with that figure at hand.

Potassic or *biotite-orthoclase* alteration, also known less meaningfully as *K-silicate*, involves the presence of introduced or recrystallized K-feldspar in a rock with or without biotite and sericite, commonly with traces of any of the characterizing calcium-salt accessory minerals anhydrite ($CaSO_4$), apatite [$Ca_3(PO_4)_3$], fluorite (CaF_2), calcite or sideromagnesio calcite [$(CaFeMg)CO_3$], and scheelite ($CaWO_4$), chalcopyrite, molybdenite, pyrite, magnetite, or hematite. It requires the addition of some potash, as is implied in the name. K-silicate as a term is less demanding, and could refer to a muscovitic or K-feldspathic assemblage formed without alkali metasomatic addition of potassium. Replacement of hornblende or chlorite by biotite and plagioclase by K-feldspar also require K-fixation and are consid-

TABLE 5-1. Common Ore and Alteration Mineral Assemblages and Representative Ore-Deposit Districts

	Ore Assemblage	Alteration Assemblages and Minerals	Examples
I.	Covellite, digenite, chalcocite, pyrite, enargite	Pervasive phyllic (quartz-sericite-pyrite) or advanced argillic (pyrophyllite-kaolinite-alunite-topaz)	Butte, Montana, Central Zone; Superior, Arizona
II.	Chalcocite, bornite, pyrite; chalcopyrite, bornite, pyrite	Phyllic near veins → argillic (kaolinite-chlorite) → propylitic (chlorite-epidote-calcite) → fresh rock outward	Butte, Montana, Intermediate Zone; Superior, Arizona; some porphyry coppers
III.	Chalcopyrite, pyrite, tennantite	Phyllic ± K-feldspar (*D* on Figure 5-3) → argillic → propylitic → fresh	Butte, Montana, deep chalcopyrite zone; porphyry copper deposits
IV.	Chalcopyrite, magnetite, molybdenite, (hematite), (pyrite); bornite, magnetite, (molybdenite)	Potassic (K-feldspar-biotite-anhydrite-ankerite ± sericite) → propylitic → fresh	Butte, Montana, premain stage; some porphyry copper core zones
V.	Chalcopyrite, (pyrite); hematite	Potassic; red-feldspar propylitic → fresh	Chilean coast range copper deposits; Cobalt, Ontario
VI.	Pyrite, chalcopyrite, magnetite	Chlorite → sericite-calcite-dolomite → albite	Noranda, Chibougamau, Quebec; Kidd Creek, Ontario; Jerome, Arizona
VII.	Pyrrhotite, pyrite, chalcopyrite	Chlorite → sericite-calcite-dolomite	Gold-quartz veins of eastern Canada
VIII.	Native copper, chalcocite, (hematite)	Chlorite, zeolites, albite-K-feldspar, carbonates	Michigan copper; andesite coppers

Expanded caption of Figure 4-5 on which the Roman numerals designate environments of mineralization and relate alteration products to mineralization types and classic localities. Parentheses under Ore Assemblage mean minor or optional. The alteration assemblages are identified on Figures 5-3 and 5-4, and will be encountered repeatedly in Chapters 9 to 20.

Source: Adapted from Meyer and Hemley (1967).

ered *potassic*. Potash metasomatism at high K^+/H^+ (Figure 5-3) is indicated.

Propylitic alteration involves alteration-generation of epidote, chlorite, and carbonates typically replacing plagioclase (epidote, chlorite, calcite) and hornblende-biotite (chlorite, epidote, montmorillonite). H_2O, H^+, CO_2, and S are normally demonstrably added. Some alteration K-feldspar or albite may be added, but alkali-alkali earth metasomatism or leaching is unimportant.

Phyllic or *sericitic* alteration is so named because of the dominance of the phyllosilicate *sericite*, a name reserved for fine-grained white mica (muscovite, hydromica, phengite) which results from alteration and metamorphism. All primary rock-forming silicates such as feldspars, micas, and mafic minerals are converted to sericite plus quartz [Equations (5-1) and (5-2)]. K^+/H^+ is in the sericite field of Figure 5-3, with strong leaching of alkalies except for potassium. Normal accessories include minor to major pyrite, chlorite if magnesium is present, and traces of rutile (TiO_2) or leucoxene ($TiO_2.nH_2O$) from the titanium in biotite, sphene, and other accessories A *biotite phyllic* equivalent is dominated by biotite or biotite-chlorite without K-feldspar addition, and is found in mafic andesitic rocks around many "porphyries" such as at El Salvador, Chile (Chapter 11). If potassium is actually added, this assemblage is best called *potassic*.

Argillic assemblages bespeak the dominance of kaolinite after plagioclase, montmorillonite after amphiboles and plagioclase, and amorphous allophane after both. K-feldspar is metastable and not affected, but alkali and alkali-earth leaching and removal are substantial, except for potassium. Flecks of sericite may be produced, and argillic commonly merges on its inner side with phyllic to form phyllic-argillic transitional assemblages. As Figure 5-3 indicates, argillic alteration forms at relatively low temperatures and low K^+/H^+ ratios, as the $Al_2Si_2O_5(OH)_4$ composition suggests.

Advanced argillic represents low K^+/H^+ and Na^+/H^+ through both low activities of the alkalies and strongly acid, high H^+ fluids. Strong leaching of all alkalies occurs. At higher temperatures above about 300°C, pyrophyllite or pyrophyllite-andalusite occur; at lower temperatures, kaolinite or dickite prevail. Quartz is abundant, and alunite, topaz, zunyite, tourmaline, and other hydro-chloro-fluoro-boro-aluminosilicates also occur. The distribution of advanced argillization is less regular than other alteration types, but it is most commonly found higher in mineralized systems, as at El Salvador, Chile, where K^+ activities are lowest and H^+ are likely to be highest. Enargite and tennantite—both Cu-As-S minerals—are normal associates (Knight, 1977).

Greisen is similar to advanced argillic or phyllic, but implies even more sericite or muscovite and no pyrophyllite. Quartz, muscovite, and topaz dominate, with tourmaline, fluorite, rutile, cassiterite, wolframite, and magnetite as common accessory minerals.

Skarn is an association of calcium-bearing usually iron-rich silicates, including amphiboles, pyroxenes, garnets, epidote-zoisite, and pyroxenoids that have replaced limestone or dolomite, normally with the introduction of large amounts of silicon, aluminum, iron, and magnesium. The term *skarn* originated in Sweden, where it was used to describe high-iron magnetite-hematite-rich assemblages associated with Archean rocks; *tactite* referred to more general calcsilicate assemblages, but the distinction is seldom made now.

It should be stressed here that the alteration assemblage terms defined above should only be used to describe the exact, specific mineral associations described and depicted in Figure 5-4. *Any* departures should be noted by hyphenated expansions, as in biotite-phyllic, or by substituting mineral descriptors such as chlorite-sericite-pyrite. The terminology of alteration, as other terminologies, must be followed rigorously to be generally expressive. Again, it is an assumption of this book that the presence of chlorite, for example, in a particular assemblage is not haphazard, but reflects specific, interpretable, useful pressure-temperature-composition information, in this case higher Mg^{+2} and lower K^{+} activities than normal, a body of information to be developed in Chapters 9 to 20.

▸ PRESENTATION OF ALTERATION DATA

Rose and Burt (1979) showed five ways to represent alteration data in the K_2O-Na_2O-Al_2O_3-H_2O system graphically. They include a ternary plot with apices of Al_2O_3, albite, and K-feldspar, with tie lines isolating areas of stable assemblage; a plot of Na_2O/Al_2O_3 against K_2O/Al_2O_3 which again isolates stable assemblage areas when tie lines are drawn; two similar plots of activities (a_{Na_2O} versus a_{K_2O}) and chemical potential (μ_{Na_2O} versus μ_{K_2O}), with mineral stability field boundaries; and one of the so-called *activity-activity diagrams* (see Figure 5-5) with log a_{Na^+}/a_{H^+} plotted against log a_{K^+}/a_{H^+}. They could have used an Eh-pH plot, or shown yet another diagram, the representation of the lower two-thirds of Figure 5-3, which is the widely used Hemley and Jones (1964) plot of experimental data in the context of potassium-to-hydrogen-ion ratio against temperature. Of all of these, the Hemley–Jones type of diagram has probably been the most used in the 1970s as an interpretative tool, and it has helped to show how chemistry relates hand-sample or petrographic mineralogy to alteration-mineralization. But the activity-activity diagram, although drawn at only one temperature (isothermal), is the more powerful because relationships calculated from it can so easily be represented and compared to field or petrographic observation. First consider that one can use any of many ion pairs as axes—K^+/H^+, Na^+/H^+, or any of Mg^{+2}, Ca^{+2}, Fe^{+2}, Fe^{+3}, Rb^+, Al^{+3}, and so on, all similarly normalized to H^+. Next recall that internally consistent thermodynamic data are being generated for almost all of the silicate and oxide alteration minerals, although the situation is still precarious at elevated temperatures and for complex solid solutions in the micas, for example, at all temperatures. Then think that a long and complicated reaction like Equation (5-1) can be reduced to a simple ion ratio expression, recalling that the activities of end member solids can be taken as 1, and that

$$K = \frac{\text{activities of reaction products}}{\text{activities of reactants}}$$

where K is the equilibrium constant.

Then for Equation (5-1),

$$K = \frac{a_{\text{musc}} \cdot a^{6}_{\text{SiO}_2} \cdot a^{2}_{\text{K}^+}}{a_{\text{K-feld}} \cdot a^{2}_{\text{H}^+}} \quad (5\text{-}21)$$

If silica is considered crystalline quartz, then

$$K = \frac{a^{2}_{\text{K}^+}}{a^{2}_{\text{H}^+}} \quad (5\text{-}22)$$

$$\text{or } \sqrt{K} = \frac{a_{\text{K}^+}}{a_{\text{H}^+}}$$

and great chemical complexity has been reduced to manageable form. This ratio—and those like it in all of the related systems mentioned above—can be evaluated either experimentally or thermodynamically. Data and computer software are now such that one can program almost any set of ion pairs at any low to moderate pressure and temperature, calculate reaction products and correlative solution compositions, and print an activity-activity diagram output on an *X-Y* printer. Even many aspects of solid solutions can be incorporated. But the leverage of the activity-activity diagram is that it describes the chemistry of a fluid as it evolves in equilibrium with alteration assemblages. Figure 5-5 shows the reaction path of liquid *C-D-E* on Figure 5-3. It shows the direction of change not just in K^+/H^+, but in Na^+, K^+, and H^+; and it can easily be related to other systems. Activity-activity diagrams are especially valuable in complicated contact metamorphic or skarn assemblages and multicomponent silicate systems in general. Silicate systems can be related to Cu-Fe-S systems and the link between alteration and mineralization examined (Figure 5-6). And much of their utility is that solid solution and assemblage data can be modified in the programming and data base as relationships are checked or defined in hand samples or in thin-section petrography, and the ties between field and laboratory data management become firmer and more rewarding to the understanding of the processes and results of alteration-mineralization.

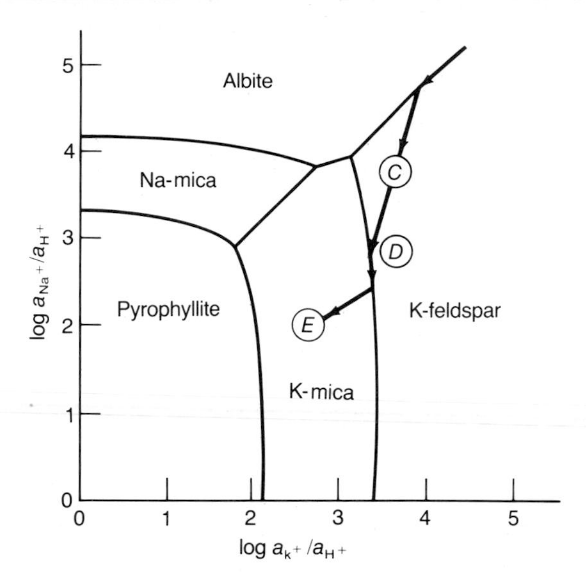

Figure 5-5. Activity-activity diagram showing relationships at 400°C and 1-kbar pressure in the system K_2O-Na_2O-Al_2O_3-SiO_2-H_2O as functions of the activity ratios K^+/H^+ and Na^+/H^+ of a fluid in equilibrium with silicates. The path *C-D-E* shows a progression in isothermal "activity space" equivalent to that in the T-X diagram of Figure 5-3. See also Figure 5-6. (*After* Rose and Burt, 1979.)

▸ QUANTIFICATION OF ALTERATION DESCRIPTION

Alteration data collected in the field, core house, and petrographic laboratory can be integrated and used in a number of ways. Most geologists use terms such as "strongly," "moderately," or "weakly" altered, but the information content of those words is so low with respect to what is actually being observed as to discourage their use. Little interperson use can be made, since one person's "weakly" is another's "moderately." Titley (1982) improved the situation by using the terms "selectively pervasive alteration" to describe the alteration response of a single mineral—say, biotite—dispersed through a large volume of rock, "pervasive alteration" to denote alteration of essentially all of the minerals and textures in a wall-rock mass, and "vein-veinlet" to mark that alteration which is relatable to specific fractures. Other sidenotes provide further specificity in mapping or description. Brimhall (1977) described means of gathering quantitative petrographic alteration data that are suitable for use with computers. Guilbert et al. (1985) presented a quantitative descriptive technique wherein the *intensive* parameters pressure, temperature, and composition are seen, as in metamorphic petrology, as establishing mineralogic characteristics, and thus the alteration mineralogy in a rock. These assemblages are initialed as K (potassic), P (propylitic), S (phyllic), A (argillic), AA (advanced argillic), G (greisen), Q (silicification), SK (skarn), and so on. *Extensive* parameters—

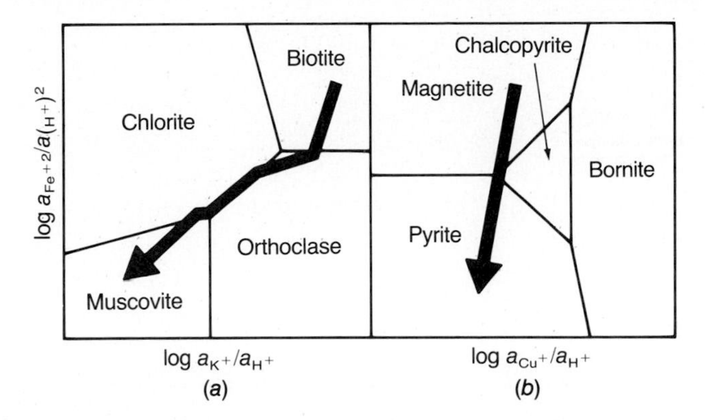

Figure 5-6. Activity-activity diagrams relating alteration and mineralization in porphyry copper deposits, and showing calculated schematic stability relations among (*a*) silicate minerals and (*b*) sulfides and oxides of copper and iron at hydrothermal temperatures and chemical potentials of O_2 and H_2S in the presence of an aqueous solution. Arrows indicate how assemblages change from early to late. (*After* Beane and Titley, 1982.)

heat content, reaction progress, and time—define the degree to which the susceptible minerals in a rock are in fact altered, as estimated by volume percentage on a scale of 1 to 10. A completely sericite-pyrite-quartz-altered rock might then be S-10, a partially affected one only S-4. *Pervasiveness* information is given, also on a scale of 1 to 10, as the degree to which alteration effects permeate the entire rock—1 means "alteration minerals confined to veinlets," and 10 means "pervasive, no perceptible individual veinlet control." In general, both numbers increase as the exploration propitiousness of alteration increases—P-2-4 means "20% of the susceptible minerals propylitized in narrow envelopes flanking veinlets," P-8-10 means that a rock is "80% altered with completely pervasive distribution," and the rock is explorationally more interesting. Of course, departures from standard assemblages can be used, as Ser9-Chl1 for a 90% sericite and 10% chlorite assemblage. Sequence can be denoted by successive notation; for example, the alteration of Figure 5-2 can be described as K-4-8; S-10-2, the former describing the top to bottom veinlet and disseminations, the latter the completely sericitized cross veinlets. The principal advantages of such notation are that one geologist can meaningfully compare and exchange alteration data with another; the diamond drill core shed or field occurrence need never be revisited for evaluation of notes like "moderately altered." Furthermore, comparable data can be gathered from surface or underground, from drill core or thin sections, and most importantly, the information can be computerized for geostatistical comparison with geochemical, geophysical, structural, or other quantitative geologic data, and can be used to help define trends, establish vectors of variation, and refine targets.

Any hydrothermal solution passing through rock pores or along a fissure may alter the country rocks, whether or not an adjacent vein contains ore minerals. Thus not all alteration zones are guides to ore deposits, and mineralization in any region that has undergone a complex history of hydrothermal activity may be complicated by crosscutting and sequential zones of alteration associated with both metalliferous and barren phases, or just barren ones. Most commonly, however, the chemistry of each hydrothermal solution is reflected in the details of wall-rock alteration, slight but perceptible differences that may indicate the proximity of ore-bearing shoots.

▸ DISTRIBUTIONS OF ALTERATION ASSEMBLAGES

Alteration Associated with Magmatic Deposits

Wall-rock alteration around magmatic segregation and magmatic injection deposits is generally inconspicuous and nondiagnostic. Contacts between ore and wall rocks at most magmatic deposit areas are sharply defined, especially in injection-type deposits in which the magma was differentiated at depth prior to emplacement in cold wall rocks. Wall-rock alteration around injection deposits is likely to be very thin or absent; only rarely has the injection been accompanied or followed by hydrothermal activity. Minor hydration is described around the "offset" deposits of the Sudbury Complex in Ontario (Chapter 9), but alteration effects around most magmatic deposits are more aptly described as "regional metamorphism" than wall-rock alteration.

A special case is an assemblage called *fenite*, the result of a process called *fenitization*, which occurs around most carbonatite intrusions (Chapter 9). A narrow zone of blue sodium silicates such as arfvedsonite, barkevikite, and glaucophane, along with phosphates and iron-titanium oxides, is common. These narrow halos—seldom more than a few meters wide—represent sodium metasomatism, hydrolysis, and oxidation, with increased sodium, hydrous silicates, and hematite. They suggest that hydrothermal activity may be important in generating some ores in carbonatites.

Alteration Associated with Porphyry Base-Metal Deposits

One of the advances of the 1970s has been the recognition and elucidation of alteration zoning, geochemistry, isotopes, timing, kinetics, and geothermometry-geobarometry around the porphyry base-metal deposits. Not all of those subjects are to be introduced here, but the alteration characteristics just described can be seen to apply. Broadly, the *spatial* arrangement of alteration zones that is the fabric of these deposits is one of domal or coaxial arrangement outward and upward from a deep core of potassic alteration, a shell of phyllic merging outward to phyllic-argillic or argillic, and an outer zone of propylitic material. The zoning at the Ivaal prospect in Papua New Guinea (Figure 5-7, generalized from Asami and

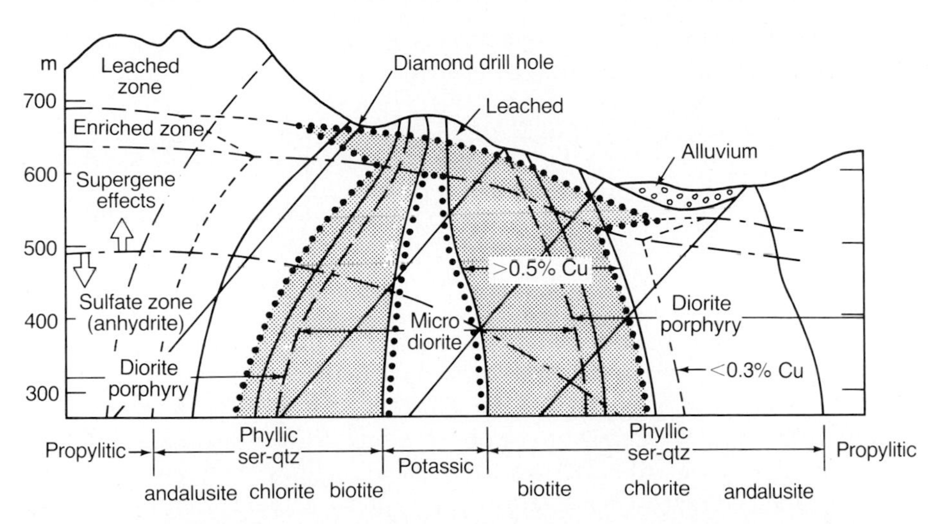

Figure 5-7. Schematic cross section of alteration-mineralization at the Ivaal prospect in the Frieda River district, Papua New Guinea. If andalusite—the high-temperature equivalent—is taken for kaolin, the zoning is geometrically typical of porphyries. Biotite and chlorite in the phyllic zone, normally quartz-sericite-pyrite, are the result of high iron-magnesium inheritance from the dioritic wall rocks. The potassic zone would extend to the limit of biotitization if potash is actually added. The high copper zone is stippled. The sketch is 1 km across, 500 meters high. The data were largely acquired from drill core from the holes indicated. (*After* Asami and Britten, 1980.)

Britten, 1980), is representative, and several diagrams in Chapter 11 (Figures 11-2, 11-4, 11-19 and 11-26) should be consulted. The word *spatial* was emphasized earlier because although the array described is what is commonly found in outcrop, research during the 1970s has shown that the phyllic alteration in many districts is generally later and has "overprinted" an earlier lateral continuum of potassic outward to propylitic. *Early alteration* is thus characterized more as alkali metasomatism with minor hydrolysis (again see Figure 5-3), with profound hydrolysis-hydration developed later as part of the same sequence, amounting as it does to a retrograde collapse of thermal gradients and invasion of meteoric waters (Figure 11-9).

Alteration patterns are similar from one porphyry copper to another, but variations dependent upon structural evolution, host-rock compositions, fluid nature and mixing models, and many other effects produce a rich variation in detail that is interpretable to the professional geologist. Porphyry molybdenum, Climax-type molybdenum, and porphyry tin systematics are dealt with in Chapter 11.

If carbonate sections occur near the porphyry stock or dike swarm that is invariably present with porphyry base-metal deposits, skarn, as described in the next section, is generally developed in them. In terranes such as the North and South American Cordillera, where stocks normally encountered Paleozoic carbonate sections, a map showing porphyry base-metal deposits and skarn deposits is almost identical.

Alteration Associated with Skarn Deposits

Wall-rock alteration products associated with skarn deposits around intrusive masses are extensive in many places and contain conspicuous, diagnostic minerals—grossularite and andradite garnet, wollastonite, epidote, and pyroxenes such as hedenbergite, salite, and diopside, amphiboles, principally tremolite and actinolite, ilvaite, idocrase, minerals of the humite group, serpentine, spinels, scapolite, and many others have been listed. Both silication and silicification are common. The character of the alteration products depends in large part upon the invaded materials and to a lesser extent upon solution composition. For example, the introduction of siliceous fluids into limestone ordinarily results in such calcium-rich minerals as wollastonite and idocrase, whereas the invasion of the same fluid into dolomites forms magnesium-rich minerals, such as serpentine and diopside. The following equations illustrate possible reactions:

$$\underset{\text{limestone}}{CaCO_3} + SiO_2(aq) \rightleftarrows \underset{\text{wollastonite}}{CaSiO_3} + CO_2(g) \qquad (5\text{-}23)$$

$$\underset{\text{dolomite}}{CaMg(CO_3)_2} + 2SiO_2(aq) \rightleftarrows \underset{\text{diopside}}{(CaMg)Si_2O_6} + 2CO_2(g) \qquad (5\text{-}24)$$

Shales in igneous metamorphic zones tend to develop the characteristic fine-grained sugary textures of hornfels. Locally, knots of chlorite, andalusite, garnet, or cordierite form, and epidote may be abundant. Volcanic rocks undergo comparable alteration. For example, the abundant andesites and andesitic tuffs of central and northern Chile contain conspicuous epidote near intrusive masses. Carbonate rocks near intrusions are often thoroughly metasomatized to skarn that consists of various silicate and oxide minerals (Figures 11-20 to 11-28). In some skarn deposits, limestones and dolomites are merely recrystallized to marbles that contain no new constituents. During recrystallization the carbonate rocks tend to incorporate, expel, or coarsen impurities, such as graphite, so that the recrystallized rocks are generally whiter than their unaltered counterparts.

Alteration effects around contacts between igneous and sedimentary rocks range in width from less than 1 cm to several kilometers. In general, the more extensive the metamorphic-metasomatic zone in a mineralized district, the more favorable the area is for ore because heat and hydrothermal fluids have obviously penetrated the rocks. Contact zones that are simply recrystallized and show no other metamorphic effects are said to be "dry," and they seldom contain ore deposits.

Silication is abundant around skarn deposits. Silica may be introduced by hydrothermal fluids or may represent silt, sand, or silica in clays or coarser detrital silicates already in the wall rock. Shales, which are relatively less permeable and thus not as receptive to ore-bearing fluids, may be rendered hard and brittle by silication, and more permeable upon subsequent fracturing. Silication of sedimentary sections is thus one form of ground preparation, whereby soft, impermeable, and unfavorable rocks are made more competent and more receptive to the ore fluids and to the deposition of ores.

As indicated, skarn formation is an integral part of alteration-mineralization around felsic stocks, and is treated extensively in Chapter 11. Figures 11-20 to 11-28 should be consulted. As has been emphasized before, the geologist who knows mineralogy and ore deposits can "read" most successfully the circumstances of stock–wall-rock hybridization in terms of ore deposits.

Alteration Associated with Cordilleran Vein Deposits

Cordilleran Vein deposits refers to fault-fissure-controlled, lead-zinc-copper-mineralized vein districts, such as Casapalca, Peru; Superior, Arizona; the Mayflower mine in the Park City district, Utah; and many others thought to be related to nearby intrusive activity. The Butte district, Montana, would once have been a classic representative, but its remarkable vein ores are now known to be "porphyry"-related; it is included as a porphyry copper deposit in most recent compilations, a "fate" that may await many Cordilleran Vein ores. Butte's vein deposit alteration will be used as an example here in part because so much is known about it, but also because it is a classic example of the ore type. Here, and in other Cordilleran veins, we focus on alteration envelopes that flank fault veins, and are therefore like multilayer sandwiches that are symmetrical about the central plane. Where veins intersect, broad X-shaped zones can develop; and where parallel veins are close together, the outer limits of one alteration envelope may encroach on another with progressive elimination of fresh rock and peripheral zones between them by overlap—whole cubic kilometers of quartz monzonite were so sericitized at Butte.

Sericite is the most persistent and abundant alteration mineral flanking Cordilleran Vein deposits; carbonate minerals are also common, especially calcite and dolomite. Siderite, rhodochrosite, and ankerite are conspicuous in places. Chlorite, most characteristic of epithermal deposits, may also form around these veins; it commonly develops outside the sericite envelopes away from the ore. Silica is generally present, and is in many places abundantly added. Zoning in the direction away from the veins, of the type sketched in Figure 5-8, is typical. Relationships worked out and published by Sales and Meyer (1948, 1950) and Meyer et al. (1968) are given in Figure 5-9; a photograph of an envelope to accompany Figure 5-9 is Figure 5-10. Notice that the sericitic zone is identical to the phyllic zone in porphyries, and that the argillic zone merges outward with propylitic. As the chemical variation part of the figure shows, all boundaries are grada-

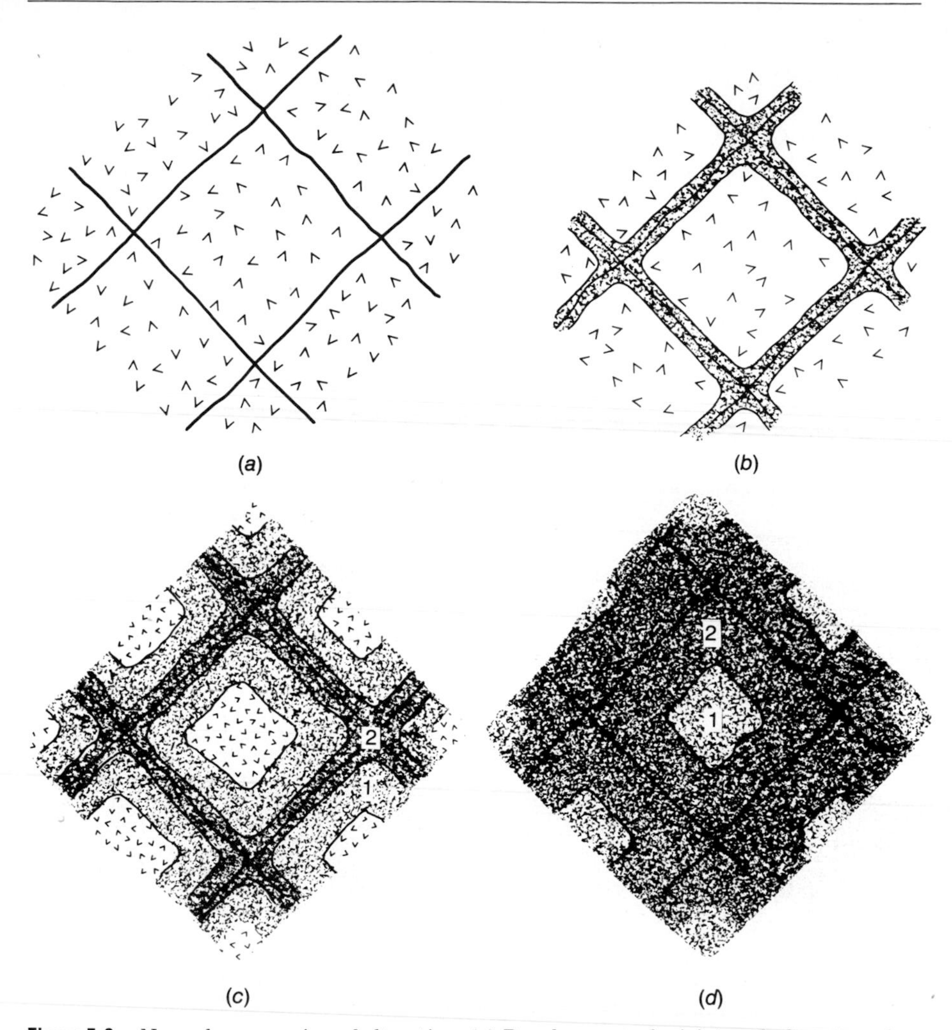

Figure 5-8. Normal progression of alteration. (*a*) Development of a joint or fault pattern in the fresh host rock. (*b*) Earliest alteration to assemblage 1; matrix blocks, the unaltered cores, established. (*c*) Assemblage 1 progresses farther into the host, reducing the volume of the matrix block, and assemblage 2 encroaches upon it from the veinward sides. (*d*) Matrix blocks are now eliminated—no fresh rock remains. Assemblage 2 will continue to expand and encroach until the wall rock is totally converted. The scale can be considered to range from millimeters to tens of meters. If the openings are considered to be faults, and the scale is large, alteration would actually proceed through myriad subparallel joints and fractures. See Figures 4-34 and 5-9.

Figure 5-9. (*facing page*) Diagrammatic summary of alteration on one side of a Main-Stage vein at Butte, Montana. These envelopes occur on both sides of veins and are repeated around every Main-Stage vein (East-West = Anaconda age and Northwest = Blue age) in the 10-km east-west, 6-km north-south, 2-km "high" district. Equations (5-1) to (5-3) show quartz produced by alteration. It appears as fine-grained intergrowths with sericite in the sericitic (phyllic) zone. Notice the strong base leaching of CaO and MgO, the almost total leaching of Na_2O, but the near constancy of K_2O and Al_2O_3. The scale could range from centimeters to tens of meters of width. See also Figure 5-10. (*From* Sales and Meyer, 1948.)

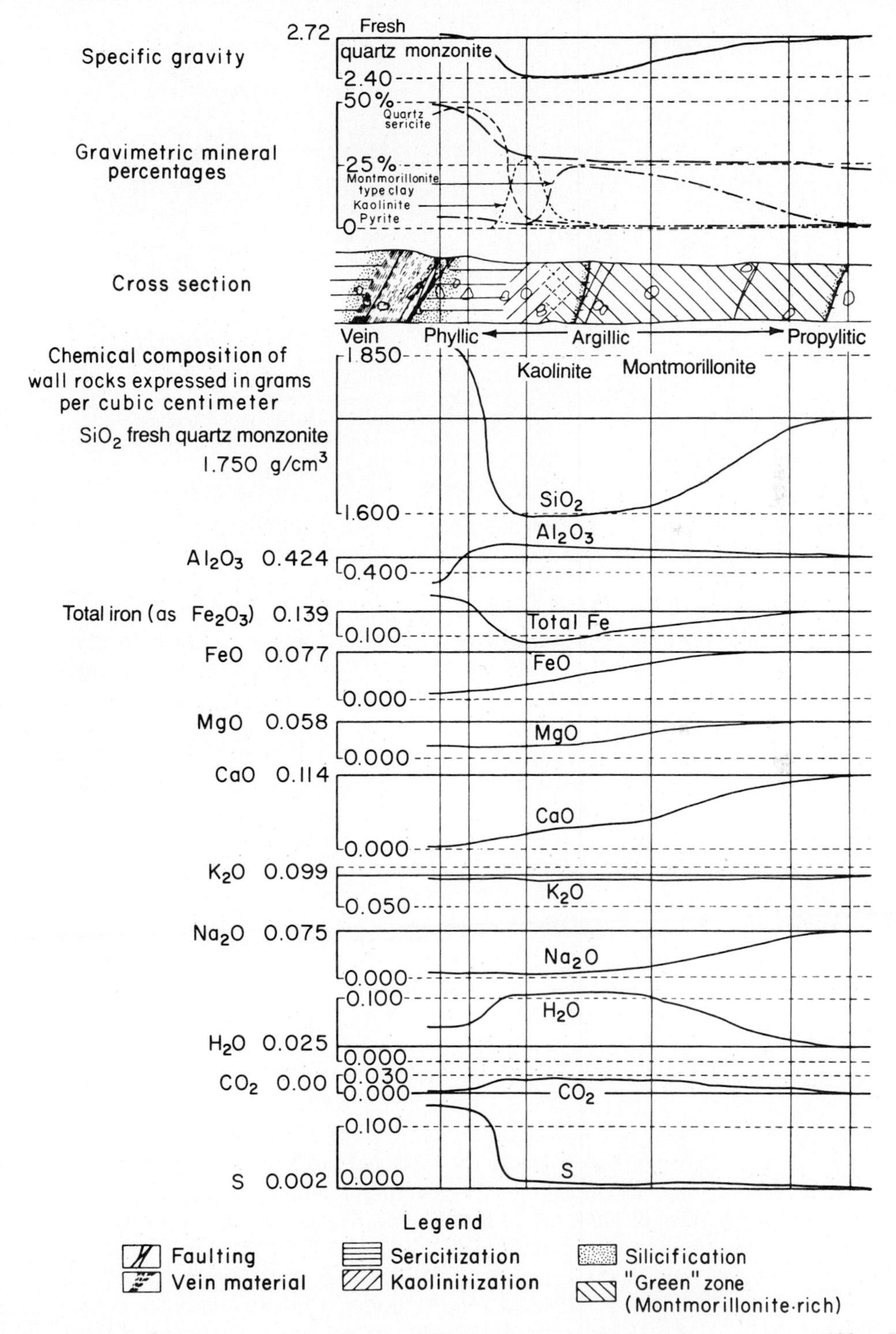

Specific gravity
2.72
Fresh
quartz monzonite
2.40
Gravimetric mineral percentages
50%
Quartz sericite
25%
Montmorillonite type clay
Kaolinite
Pyrite
0
Cross section
Vein
Phyllic
Argillic
Propylitic
Chemical composition of wall rocks expressed in grams per cubic centimeter
1.850
Kaolinite
Montmorillonite
SiO2 fresh quartz monzonite 1.750 g/cm3
SiO2
1.600
Al2O3
Al2O3 0.424
0.400
Total iron (as Fe2O3) 0.139
Total Fe
0.100
FeO 0.077
FeO
0.000
MgO 0.058
MgO
0.000
CaO 0.114
CaO
0.000
K2O 0.099
K2O
0.050
Na2O 0.075
Na2O
0.000
0.100
H2O
H2O 0.025
0.000
CO2 0.00
0.030
0.000
CO2
0.100
S 0.002
0.000
S
Legend
Faulting
Vein material
Sericitization
Kaolinitization
Silicification
"Green" zone (Montmorillonite-rich)

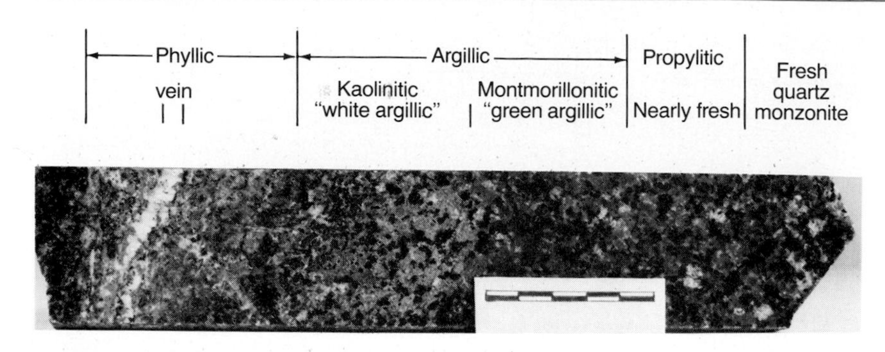

Figure 5-10. Polished piece of hydrothermally altered Butte quartz monzonite, a hand-sample scale representation of the schematics of Figures 5-8 and 5-9. The vein contains quartz (white) and pyrite (medium gray) flanked by quartz-sericite-pyrite alteration. That phyllic assemblage is flanked outward by argillic and propylitic assemblages until fresh rock is reached. Note preservation of granitoid texture even through the alteration-induced mineralogical changes. The bar scale is in centimeters; the piece of diamond drill core is 24 cm long. (*Photo by* W. K. Bilodeau.)

tional, the sharpest being that where K-feldspar is hydrolyzed at the argillic-phyllic boundary. K_2O released there effects sericitization of all other nonpotassic phases such as kaolinite [Equation (5-3) reversed].

Studies of the Butte, Montana, veins demonstrated that the alteration products there were an integral part of the mineralization—the alteration minerals were deposited at essentially the same time and by the same fluids as the ore. Successive zones of sericitized, argillized, and propylitized quartz monzonite lie adjacent to every Main-Stage ore-bearing fissure, regardless of its size, attitude, ore mineralogy, or relative age. Except where overlap of the alteration effects between closely adjacent fissures has eliminated the fresh rock between them, the two major assemblages of alteration always occupy the same relative positions: sericite is adjacent to the vein, and clay minerals are between the sericitized rock and fresh quartz monzonite. Nowhere does the sericite zone grade directly into fresh wall rock. The sequence of alteration is even more striking in detail. The argillic zone is invariably made up of a kaolinitic phase close to the vein and a montmorillonite-rich phase close to the unaltered wall rock. Moreover, the relative proportions of chemical components change systematically from the vein outward, presenting a graphic picture of the hydrothermal alteration process. As shown in Figure 5-9, the sodium, calcium, and silica contents diminish rapidly from the margin toward the vein, and the H_2O content increases; a similar but less striking decrease is also shown by iron and magnesium. These changes reflect the fact that outermost plagioclase was readily hydrolyzed, releasing silica, calcium, and sodium, and that the ferromagnesian minerals were altered to chlorite. Continued alteration leached iron, calcium, and magnesium from the montmorillonite to form kaolinite. Further hydrolysis finally broke down the orthoclase, supplying potassium for the production of sericite. Sulfur content shows a marked increase near the vein; iron liberated by the early destruction of ferromagnesian minerals combined with hydrothermally delivered sulfur to form pyrite, and both iron and sulfur were added. Silica released from feldspars was deposited as quartz near the vein. Thus there was a two-way transfer of ions by diffusion, some ions moving into the wall rock and others moving toward the vein. As long as active fluid circulation continued along the vein, each zone migrated away from the fissures, that is, it grew at its outer edge and simultaneously was encroached upon by the next inner zone at its veinward edge (Figure 5-8).

Such envelopes are found along many Cordilleran veins, and predictable variations on the theme occur where they cut other rock types. Although the Mother Lode district of California may not be a true Cordilleran Vein, the alteration products developed (Knopf, 1929, pp. 41–45) serve as an example. The minerals include ankerite, the dominant alteration mineral in schist, argillite, and serpentinite wall rocks, and sericite, which is probably second in importance except in serpentine, where the chromian mica mariposite is more common. The chief characteristic of the alteration zone is the addition of large amounts of carbon dioxide, the deposited quantity of which apparently depended upon its availability in the fluid and the abundance of iron, magnesium, and calcium in the original rock. Other added elements are potassium, sulfur, and arsenic represented by sericite, pyrite, and arsenopyrite. Large volumes of silica, more than all the silica in the veins, have been removed from silicates, so that it is unnecessary to ascribe this compound to an external source. In places the alteration extends 3 meters on either side of the edges of the veins; it is a valuable guide to the associated vein.

Another common alteration associated with limestone replacement ores is called *jasperoid* (Lovering, 1972). Jasperoid is fine-grained silica commonly in or near veins, typically grayish to tan to red-colored from dispersed hematite and pyrite (Figure 3-6). It may contain enough trace elements near ore occurrences to be a useful exploration guide, but its mere presence is explorationally significant. Preore dolomitization of limestone—followed by the generation of granular "sandy dolomite" volumes that actually flow into mine workings—is also common around dolomite-replacement silver-lead-zinc deposits at Gilman and Leadville, Colorado.

Lovering (1949) studied the lead-zinc-silver replacement deposits at Tintic, Utah. He described five stages of hydrothermal alteration, which he considered sequential, but not necessarily continuous. The stages are (1) an early barren stage, during which limestones were dolomitized and volcanics were chloritized, (2) a midbarren stage, characterized by argillization, (3) a late barren stage closely associated with the ore shoots and characterized by jasperoid, barite, pyrite, and minor chlorite in the carbonate, quartzite, and shale sediments, with allophane, quartz, barite, pyrite, calcite, and minor iron-rich chlorite in the volcanics, (4) an early productive stage closely associated with orebodies but consisting of only an inconspicuous zone of sericite-hydromica representing the introduction of potassium with minor quartz and pyrite, and (5) the productive stage, during which the ore was deposited with quartz, sericite, and pyrite. According to Lovering's interpretation, each of these stages represented a period of hydrothermal activity separated from the others by a time interval, and the close association of ore with more than one of these stages reflected a common source of solutions. Recent studies tend to deny the separation in time of these events and to emphasize their continuous nature.

Alteration Associated with Epithermal Deposits

As is seen in Chapter 12, epithermal precious metal "bonanza" ores such as those of the Erzgebirge, East Germany-Czechoslovakia, Mexico, and the western United States are characterized by enormously high and extensive induced permeability along fault and fracture zones. Alteration is commensurately broad, and complicated by the fact that a retrograde boiling zone can be identified in many districts. In general, a feldspar-stable zone occurs at depth. In some districts it extends almost to the surface, and it is shown in the "model diagram" of Figure 5-10 as an adularia-albite assemblage, *adularia* being low-temperature K-feldspar found in veins. As in the model, alteration then generally flares upward and outward toward the surface, as structural openings proliferate. There, wall-rock buffering becomes important. Chlorite is one of the most abundant minerals in the wall rocks around epithermal deposits in andesites, and in deposits such as the Comstock Lode, Nevada, the enclosing rocks are extensively and pervasively chloritized to a propylitic assemblage. Sericite is more abundant in latite and rhyolite wall rocks, but even here a thin propylite zone may occur at the margins. In general, the more felsic the host rocks, the higher the overall sericite-to-chlorite ratio. Other alteration minerals are quartz, alunite, zeolites, chalcedony, opal, calcite, and other carbonates along the veins. Argillization may be especially widespread in the wall rocks; among the numerous layer silicate minerals, nacrite, dickite, kaolinite, pyrophyllite, illite, and montmorillonite may be present. Alteration products in shallow deposits are ordinarily fine-grained, and many are difficult to separate for identification. Probably the most diagnostic minerals are the clays, pyrophyllite, sericite, and chlorite in the altered rocks, and quartz, locally amethystine, opaline, or chalcedonic in the veins. Notice that in Figure 5-11 one can discern equivalents to the potassic, phyllic, argillic, and near-surface advanced argillic assemblages described earlier.

Figure 5-11. (*facing page*) Alteration effects related to mineralization in a "typical" precious-metal epithermal deposit. Vein contents are listed at the right, with a depth and approximate temperature scale at the left. (*After* Buchanan, 1981.)

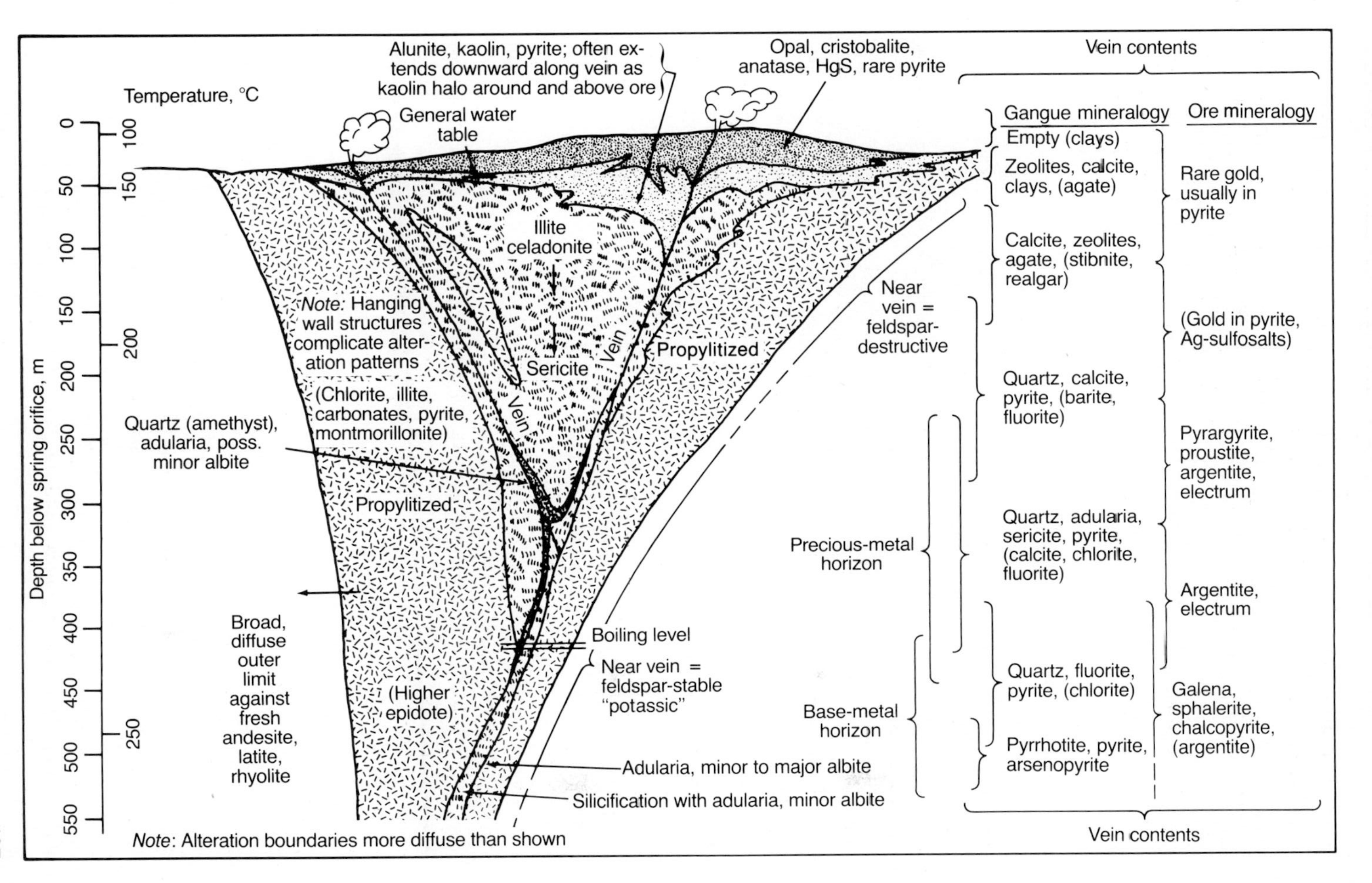

Temperature, °C
100
150
200
250
Depth below spring orifice, m
0
50
100
150
200
250
300
350
400
450
500
550
Alunite, kaolin, pyrite; often extends downward along vein as kaolin halo around and above ore
General water table
Opal, cristobalite, anatase, HgS, rare pyrite
Illite celadonite
Sericite
Vein
Vein
Propylitized
Note: Hanging wall structures complicate alteration patterns
(Chlorite, illite, carbonates, pyrite, montmorillonite)
Quartz (amethyst), adularia, poss. minor albite
Propylitized
Broad, diffuse outer limit against fresh andesite, latite, rhyolite
(Higher epidote)
Boiling level
Near vein = feldspar-stable "potassic"
Near vein = feldspar-destructive
Precious-metal horizon
Base-metal horizon
Adularia, minor to major albite
Silicification with adularia, minor albite
Note: Alteration boundaries more diffuse than shown
Vein contents
Gangue mineralogy
Ore mineralogy
Empty (clays)
Zeolites, calcite, clays, (agate)
Calcite, zeolites, agate, (stibnite, realgar)
Quartz, calcite, pyrite, (barite, fluorite)
Quartz, adularia, sericite, pyrite, (calcite, chlorite, fluorite)
Quartz, fluorite, pyrite, (chlorite)
Pyrrhotite, pyrite, arsenopyrite
Rare gold, usually in pyrite
(Gold in pyrite, Ag-sulfosalts)
Pyrargyrite, proustite, argentite, electrum
Argentite, electrum
Galena, sphalerite, chalcopyrite, (argentite)
Vein contents

Alteration Associated with Pegmatites

Alteration zones around complex granitic zoned pegmatites (Chapter 11) can be thin and sharply defined or wide and gradational, the former being more normal. Some pegmatites contain rare minerals, and the contact zones around these igneous masses may contain traces of the lanthanides, boron, fluorine, beryllium, lithium, and other LIL (Large-Ion Lithophile), silicophile pegmatitic constituents. So there may exist a varied and unique mineralogy of alteration products in pegmatite alteration envelopes. Among the sparse alteration minerals found near pegmatites are beryl, topaz, and fluorite; the most widespread are disseminated feldspars, micas, garnets, and tourmaline (Jahns, 1955). Many pegmatitic fluids are rich in potassium, which may be introduced into the wall rocks in conspicuous amounts; conversion of wall-rock amphiboles to biotite is common.

Jahns (1946) recorded fine-grained muscovite, plagioclase, potash feldspars, and quartz in the poorly delimited, narrow outer zones of pegmatites in the Petaca district of New Mexico. Permeable wall rocks that permitted entry of the pegmatitic fluids were most conspicuously altered; a slabby micaceous quartzite formation showed characteristic alteration for 1 to several meters from the pegmatite, with "tongues" extending along bedding or joint planes. Around some pegmatite bodies, the zone between true pegmatite and unaltered quartzite consists of a hybrid rock such that the contact between igneous rock and wall rock is gradational and almost imperceptible. In the Avon district, Idaho, host-rock mica schists and gneisses were altered to a light-colored aggregate of plagioclase, quartz, muscovite, garnet, and schorlite, which were formed by the recrystallization of original constituents and impregnation and metasomatism by pegmatitic materials. Relict bedding and schistosity are recognized in most places. In general, the alteration zone is less than a third of a meter wide, but at the Muscovite mine it extends about 7 meters (Stoll, 1950).

The Etta spodumene mine near Keystone, South Dakota, is noteworthy in having a pronounced contact zone around a pegmatite. The alteration halo is sharply defined and consists of friable, fine-grained, sugary rock produced by recrystallization and metasomatism of a fine-grained mica schist. The biotite and muscovite of the schist were altered to a granoblastic and poikilitic assemblage of plagioclase, orthoclase, and microcline. Apatite and tourmaline also were formed in the contact aureole, which has an average width of 2 to 3 meters and a maximum width of 5 meters (Schwartz and Leonard, 1927).

Pegmatites commonly occur within parent igneous masses, where the similarity of physical and chemical conditions between the walls and the new pegmatite does not lead to the development of conspicuous alteration zones. In such cases the contact between pegmatite and host rock is likely to be a sharply gradational change in texture rather than in mineralogy or rock type. Conversely, some pegmatite contacts may represent a gradual and simple change in mineralogy, with or without an increase in grain size over that of the host pluton. For example, in the Kaniksu batholith of northeastern Washington, large irregular or oval bodies of muscovite-quartz-feldspar rock outcrop in the normal biotitic intrusive mass. Although these bleached products are usually of the same grain size as the main intrusive, the grain becomes coarser and grades into characteristically pegmatitic materials locally along fractures and in the centers of the bodies. The most notable feature of these pegmatites is the bleaching and removal of iron minerals from their borders. Biotite is changed to muscovite, and much of the rock shows well-developed myrmekitic intergrowths (Park and Cannon, 1943).

In general, alteration zones around pegmatites are thin and of little or no value in prospecting. Even the rare conspicuous halos extend only a few meters from the pegmatite. However, at a few places, for example, in the South-West African (Namibian) pegmatites, studies of the minor elements in barren outer zones can indicate the presence of valuable bodies a few meters away (Hornung, 1962). Canadian and Russian geologists have also studied alteration-geochemical wall rock effects around pegmatites (Beus et al., 1968; Ovchinnikov, 1976; Goad and Cerny, 1981).

Alteration Associated with Volcanogenic Massive Sulfide Deposits

Alteration associated with these deposits (Chapter 13) is unusual because of its "one-sidedness" or unilateral symmetry. The deposits form by the passage of hot metalliferous brines up through pipelike conduits in submarine volcanic sections, from which they issue into marine basin waters (Figures 13-4 to 13-6). Changes in temperature, pressure, dilution by seawater, and other effects lead to the precipitation of the least soluble components near the vent and to dispersion in the seawater medium of nearly colloidal or more soluble components. The least soluble compounds are typically sulfides, so concentrations of pyrite, chalcopyrite, and sphalerite, some with galena and others with pyrrhotite-pentlandite, stibnite, cinnabar, or other minerals, are deposited near—sometimes on top of—the tops of the conduits. A cross section of this array is given as Figure 13-5, the broader scenario in Figures 13-2, 13-3, and 13-6. Alteration associated with these ores is thus restricted to their footwall or underside, except in rare circumstances where hot spring flow continues to penetrate the postore layer of rocks next deposited above the sulfides. The alteration zones are generally circular to subcircular in plan, and normally involve coaxial, cylindrical zones. Since these ores are associated with volcanism and volcanic piles, the host rocks penetrated by conduits are commonly basaltic to andesitic, occasionally latitic to rhyolitic. The former rocks prevail at Kidd Creek, Ontario, and parts of the Noranda district, Quebec; the latter are found at Jerome, Arizona, and at the Millenbach mine at Noranda. Since the hot circulating brines are probably largely convectively driven seawater that has percolated through deeper, more mafic oceanic crust, it is no surprise that Mg^{+2} is more abundant in them than K^{+} or Na^{+}. Magnesium metasomatism is characteristic, and pervasive chloritization, even to the total resorption of original quartz phenocrysts, is not rare. The resultant rock around the conduits is commonly so massive, fine-grained, and dark that it is called *black chlorite*; for a few tens of meters below the ore it may contain stockworklike networks of chalcopyrite and pyrite. The chlorite core of the pipe may yield laterally to propylitelike assemblages in the more mafic rocks and to sericite-rich hybrid transition zones in latite-rhyolite hosts. Larson (1976), Larson and Webber (1977), and Douglas (1982) have described sophisticated geochemical techniques for screening whole-rock analytical data to permit perception of mineralogically inconspicuous zones by the use of magnesium, potassium, sodium, or iron ratios, variously coupled and combined. The most pronounced chemical effects are the 10-to-20-fold increases in magnesium, copper, and zinc, and the megascopic increase in chlorite-sericite. Magnesium-iron variation in chlorites may be petrographically useful.

Subtypes of the volcanogenic groups such as the Kuroko ores of Japan's Miocene section and the "black shale hosted" sedimentary-affiliated deposits may contain more or less gypsum, barite, quartz, and carbonates. Regional or contact metamorphism of chlorite pipes produces uncommon assemblages dominated by Mg-Al silicates, such as cordierite and anthophyllite, still in geometric-geologic position with respect to the ore horizon.

Alteration Associated with Mississippi Valley Deposits

The alteration products associated with sphalerite-galena deposits of the Mississippi Valley lead-zinc type are extensive but quite inconspicuous. They are thought to have been formed by warm connate brines that have migrated from deeply buried basins upward and laterally toward their margins (Figure 20-2). Chemical and thermal gradients between these oil-field-type brines and the rocks they traverse are gentle, and alteration typically consists only of broad, laterally extensive volumes in which original limestone units are dolomitized. The alteration is not discernible to the casual visitor. Sparse veinlets, bedded zones, or bleached fractures contain minor calcite, dolomite, marcasite, and cryptocrystalline silica, but these are seldom helpful in exploration. Distinctive dark gray to black jasperiod is present in several districts.

Alteration Associated with Western States and Roll Front Deposits

These ores (Chapter 20) are formed by the percolation of groundwaters down subsurface paleodrainages and aquifers. The waters are sufficiently oxidizing to oxidize U^{+4} to U^{+6}, in which form it is more soluble; but they reprecipitate uraninite where they are reduced by encountering and flowing through carbonaceous, pyritic, or vegetable trash zones in permeable strata. Their alteration effect is thus one of hydrolysis and oxidation on the upside of the hydraulic gradient; limonites, hematite, and minor sulfates are found there. The oxidized upstream channels are normally reddish to ocherous, the still reduced downstream ones grayish to greenish. The distribution of redox anomalies, such as boundaries within aquifers and interfaces along paleochannels (Figures 20-14 to 20-16), are therefore useful exploration guides.

▸ GANGUE

Gangue comprises all of the noneconomic, unwanted minerals in an ore. It is defined as the valueless rock or mineral matter in an ore, material that cannot be avoided in mining and is later separated and discarded. Clearly, a gangue mineral can be a sulfide, a silicate, a carbonate, or a fluoride, to name but a few, and a mineral considered gangue in one deposit may be regarded as an ore mineral in another. For example, the fluorite present in small proportions in the gold ores at Cripple Creek, Colorado, is gangue, whereas vein fillings of nearly pure fluorite in Illinois and Mexico constitute the district ores (Figure 5-12). Also, a gangue mineral in a particular district may become an ore mineral as a result of technological advances or market

Figure 5-12. Veinlets of light-colored fluorite, 1 or 2 cm wide, cutting Mississippian-age limestone (dark) from the Cave-in-Rock district, Hardin County, Illinois. The veinlets are largely open-space fillings as the crystal faces toward centerlines suggest, but the stipples of white in limestone areas probably bespeak replacement. Most importantly, fluorite, normally a gangue mineral, is here the principal ore mineral. Actual size. (*Photo by* G. H. Bain.)

demands. Rutile is one of those, as is the uranium content of the South African gold ores.

The most common gangue minerals are silica, silicates, and carbonates, with subordinate oxides, fluorides, and sulfates. Studies of gangue minerals are of widespread scientific and economic interest, because the composition, paragenetic sequence, and occurrence characteristics of ore minerals and gangue minerals alike must be considered in any hypothesis regarding the nature of ore-bearing fluids and the history of ore deposition in any single deposit or deposit group. Summaries that focus on gangue minerals in metallic deposits are chapters by Holland (1967) and Holland and Malinin (1979) in Barnes' two *Geochemistry of Hydrothermal Ore Deposits* volumes.

Gangue minerals in sulfide deposits have come under increased scrutiny in recent years because modern economic geologists appreciate the interdependence of ore and gangue minerals better and have perceived that ores and the rocks they occur in are genetically related if not identical. Also, fluid inclusion studies (Chapter 7) are generally easier to perform on the transparent gangue minerals, or on such ore minerals as translucent sphalerite, because most ore minerals are opaque and cannot be used. The temperatures of ore deposition and the composition of the ore-forming fluids are estimated by examining the gangue, recognizing of course the paragenesis of the ore and gangue. Knowledge of a common origin of the ore and gangue minerals can be especially important for prospecting; if the gangue was produced in the same way as the ore, its presence may be used as a guide to ore-bearing structures or potentially favorable environments. Mining engineers, metallurgists, and geologists must be aware that the physical and textural relationships among ore and gangue minerals can affect the milling program, perhaps even dictating the economic threshold between ore and waste. Economic viability can be largely a function of the relative densities, magnetic properties, wetting abilities, and intimacy of mixture between the valuable and waste fractions of the ore.

Certain gangue minerals can be indicators of specific ore characteristics, an association that may be universal or only local. For example, "smoky" quartz may indicate the presence of uranium because that color effect can be due to radioactive bombardment. Similarly, a strong association between dark purple fluorite and economic concentrations of uranium minerals has been observed. An association has also been noted between fluorite and beryllium and between fluorescent varieties of calcite and certain ores. Some calcite fluoresces salmon-pink or red under ultraviolet excitation, and it has been suggested that trace amounts of manganese and possibly a second activator such as lead are responsible for this phenomenon. Whatever causes the fluorescence, a striking relationship between ore distribution and fluorescent calcite has been noted at Pioche, Nevada; Naica, Chihuahua, Mexico; and elsewhere. Calcite is abundant in those districts, but only that in and near mineralized volumes fluoresces (Stone, 1959; MacDiarmid, 1959). A recently developed technique called *cathodoluminescence* uses visible-light fluorescence generated by electron bombardment as a distinctive property. The variable electron-induced fluorescence of quartz, carbonates, and some sulfates and oxides permits, for example, mapping an individual vein-filling lamination or set of laminations (Figure 3-11) of calcite through mines, even through whole districts (Ebers and Kopp, 1979). The cathodoluminescent effect is attributed to minor amounts of manganese or lead in calcite, an effect offset by the presence of minor iron.

Trace amounts of metals can be detected in gangue minerals by spectrographic analyses or various chemical assay procedures. Some nonmetallic minerals that form in the presence of a metal-bearing fluid can be expected to include those metals in trace amounts; the trace metals can, in turn, act as guides to ore. Some metals, such as gold, many even reach ore grades as "impurities" within seemingly barren gangue minerals. For example, gold-bearing pyrite would undoubtedly have been discarded at many mines if the gold content had not been discovered by laboratory analysis.

Specific associations between ore and gangue minerals are well known. Quartz is a common gangue mineral in vein deposits; it is especially prominent with gold. The massive white quartz, or *bull quartz*, of many gold deposits characteristically forms conspicuous ridges across mining districts (Figure 3-13). Because of the common association between bull quartz and gold, early prospectors made a practice of testing all quartz veins for the presence of gold. The wisdom of such a practice becomes obvious when one realizes that a ton of gold ore containing one troy ounce of gold (31.1 grams) is high grade. A ton of such bull quartz is about a third of a cubic meter or 12 ft^3, a block about 2.3 ft or 70 cm on a side. It would contain about 1.6 cm^3—one-tenth of a cubic inch—of gold dispersed throughout the entire mass of quartz, a volume ratio of about one part gold to 172,000 parts quartz. What might appear to be "gangue" may in fact prove to be ore.

Another common ore-gangue association is magnetite with apatite. Iron deposits with this combination have been described from all over the world, and include magmatic segregations as well as hydrothermal vein and replacement ores. The veins in any district may range from pure apatite to pure magnetite. In central Chile, many small mines have been opened on strong apatite veins which grade in depth into apatite-bearing magnetite. Well-known but less common associations are fluorite with lead and zinc deposits; barite with lead, silver, and copper ores; tourmaline and topaz with cassiterite veins; and arsenopyrite with tin, tungsten, and gold.

Barite and fluorite are gangue minerals in many hydrothermal deposits. They form under similar conditions and are frequently associated in a single deposit. Wherever either barite or fluorite is concentrated, the semantic-economic switch from gangue status to ore status can occur. Barite is found in veins, as massive replacement ores, and as seafloor precipitates. Fluorite is mined from fissure veins and replacement deposits in many types of rock, limestone being the most common. The Kentucky-Illinois area, straddling the border of the two states in a northeasterly belt over 100 km long, was once the world's largest source of fluorite; most of the ores filled open fissures along faults that pinch and swell, causing the finely banded orebodies (Figure 5-11) to vary in thickness from 0 to 20 meters within short distances. Also, substantial portions of the Kentucky-Illinois fluorite were deposited as replacements of Mississippian limestone (Weller et al., 1952; Gillson, 1960). The principal gangue mineral with both barite and fluorite ores is normally calcite. Another interesting deposit of fluorite-in-limestone is at Las Cuevas mine in San Luis Potosi, Mexico, where a large body of cavernous fluorite has replaced limestone along a contact with intrusive rhyolite. The main orebody does not crop out, but was discovered by drilling, and is another example of one district's ore mineral being another's gangue. Even quartz is sometimes mined as an ore. Silicon is an increasingly important commodity in electronics and is produced by electrothermic deoxidation-reduction to the metal. Small deposits of pure quartz are normally mined for their silica content, in preference to the dispersed gangue in metalliferous veins. Other quartz deposits are of economic value because they contain large quartz crystals of optical quality; Brazil is the world's most noteworthy producer of optical quartz. And tonnages of quartz, preferably with traces of copper or precious metals, are used as a flux and iron remover in the smelting of copper ore concentrates. At high temperatures, silica reacts with iron to produce an iron-silica melt that floats to the top of the copper matte, where it is removed as slag.

Some use can be made of almost any mineral if it can be found pure enough and in large enough volume. For example, pyrite as a gangue mineral is an economic and ecologic headache in many districts. It is abundant and heavy, and it is unavoidably mined and hoisted or trucked with the ore minerals; once separated and spread on tailings piles, it oxidizes, generates acids, and is both unsightly and ecologically damaging if left untreated. But in some localities it can be recovered and upon roasting yields sulfur and iron suitable for reinforcing rods, sponge iron, and the like; such a process is carried out using gangue pyrrhotite at Sudbury, Ontario, and Ducktown, Tennessee. And at the Leviathan mine in California, and at scores of other places, massive pyrite has been mined specifically for its iron and sulfur content, the latter commonly being recovered for the manufacture of sulfuric acid.

Many commodities such as garnet, wollastonite, pyrophyllite, and even sericite—clearly gangue alteration minerals in most occurrences—are considered industrial minerals where they are found in sufficient purity and tonnage. Obviously the distinction between ore minerals and gangue minerals is purely and simply an economic one.

▸ Summary

If wall rocks are unstable in the presence of hydrothermal ore-forming solutions, both will undergo physical and chemical changes to reach equilibrium. The resulting wall-rock alteration may be subtle, such as the incipient hydration of selected ferromagnesian minerals; or it may be clear-cut, as in the silicification of limestone. The alteration may range from simple recrystallization to the addition, removal, or rearrangement of chemical components. It may take place in advance of emplacement of the ore minerals or during any part of hydrothermal activity.

The nature of wall-rock alteration by hydrothermal processes offers strong evidence in support of the view that many hydrothermal solutions are neutral or slightly acidic at higher temperatures. Helgeson (1964) states that hydrothermal alteration of aluminosilicate rocks is essentially a process of trading H^+ ions for other cations in the rock. The fundamental controls include any processes that affect the activities of reacting constituents in solution. Hemley and Jones (1964) list as the most significant controls (1) the compositional nature of the wall rock, (2) changes in the pressure-temperature state of the aqueous phase, such as by boiling, with possible fractionation of volatile components, (3) mixing of hypogene solutions with supergene solutions or groundwaters, and (4) oxidation of H_2S to strong sulfur-species acids.

Wall-rock alteration may bring about recrystallization, changes in permeability, and changes in color. Carbonate rocks are characteristically recrystallized along the borders of a vein or near an igneous contact. Color changes include bleaching, darkening, and the development of aureoles of various colors. Pastel colors of micas and clays are especially prominent around some ore deposits and may form conspicuous leads to ore. Clay minerals are generally white or light shades of green, brown, and gray, so argillization may produce a noticeable bleaching effect; even a black basalt may be altered to a white or light-green body of clays and other hydrous minerals. Similarly, the formation of chlorite or epidote produces a green color. Silicification, carbonatization, argillization, and hydration are typical of the processes that take place in alteration zones—and they may all operate simultaneously. Although generalizations are hazardous because the possibilities are too diverse, certain reactions in specific environments can be expected. For example, water is usually added to the alteration zone, except where the rocks are completely replaced by silica, and carbon dioxide is generally removed from carbonate host rocks. Furthermore, certain minerals can be routinely and repetitively expected in alteration assemblages such as potassic, propylitic, phyllic, argillic, advanced argillic, and skarn.

Conditions of temperature and composition usually differ at various distances from a fissure or conduit so that different types of alteration can be produced simultaneously in adjacent volumes (Hemley, 1959; Burnham,

1962). In the outer fringes of the alteration zone the feldspars and ferromagnesian minerals may be only partially hydrated. In intermediate reaches they may be argillized, and against the vein sericitized and silicified. Such zoning is conspicuous in some deposits and is an excellent guide to ore. But any hydrothermal solution passing through rock pores or along fissures is likely to alter the country rocks, whether or not the vein ultimately contains ore minerals. Thus not all alteration zones are guides to ore deposits, and mineralization-alteration in any region has to be worked out carefully. The following chapters will build upon these observations.

Bibliography

Asami, N., and R. M. Britten, 1980. The porphyry copper deposits at the Frieda River prospect, Papua New Guinea, pp. 117–140 in S. Ishihara and S. Takenouchi, Eds., *Granitic Magmatism and Related Mineralization.* Soc. Min. Geol. Jap., Spec. Issue 8, 247 pp.

Beane, R. E., and S. R. Titley, 1982. Porphyry copper deposits. Part II, Hydrothermal alteration and mineralization. *Econ. Geol. 75th Anniv. Vol.*, pp. 235–269.

Beus, A. A., et al., 1968. *Geochemical Exploration for Endogenic Deposits of Rare Elements.* Moscow: Nedra, 263 pp.; Engl. transl., Geol. Surv. Can. Library, Ottawa.

Brimhall, G. H., Jr., 1977. Early fracture-controlled disseminated mineralization at Butte, Montana. *Econ. Geol.* 72:37–59.

Buchanan, L. J., 1981. Precious metal deposits associated with volcanic environments in the southwest, pp. 237–262 in W. R. Dickinson and W. D. Payne, Eds., *Relations of Tectonics to Ore Deposits in the Southern Cordillera.* Tucson: Ariz. Geol. Soc. Dig., XIV, 288 pp.

Burnham, C. W., 1962. Facies and types of hydrothermal alteration. *Econ. Geol.* 57:768–784.

—— and H. Ohmoto, 1980. Late stage processes of felsic magmatism, pp 1–11 in S. Ishihara and S. Takenouchi, Eds., *Granitic Magmatism and Related Mineralization.* Soc. Min. Geol. Jap., Spec. Issue 8, 247 pp.

Creasey, S. C., 1959. Some phase relations in hydrothermally altered rock of porphyry copper deposits. *Econ. Geol.* 54:351–373.

Davis, J. D., 1974. Geothermometry, geochemistry, and alteration at the San Manuel porphyry copper orebody, San Manuel, Arizona. Unpub. Ph.D. dissert., Univ. Ariz., Tucson, 269 pp.

Douglas, D. A., 1982. Lithogeochemistry as a guide to volcanogenic massive sulfide deposits. Unpub. M.S. thesis, Univ. Ariz., Tucson, 83 pp.

Ebers, M. D., and O. C. Kopp, 1979. Cathodoluminescent microstratigraphy in gangue dolomite, the Mascot-Jefferson City district, Tennessee. *Econ. Geol.* 74:908–918.

Gillson, J. L., Ed., 1960. *Industrial Minerals and Rocks,* 3rd ed. New York: AIME.

Goad, B. E., and P. Cerny, 1981. Peraluminous pegmatitic granites and their pegmatite aureoles in the Winnipeg River district, southeastern Manitoba. *Can. Mineral.* 19:177–194.

Guilbert, J. M., A. E. Allison, A.H. Stults, and B. J. Suchomel, 1986. Quantitative alteration mapping, an analytical technique in mineral deposit evaluation. *Econ. Geol.* in prep.

—— and J. D. Lowell, 1974. Variations in zoning patterns in porphyry ore deposits. *Can. Inst. Min. Metall. Bull.* 66:1–11.

Helgeson, H. C., 1964. *Complexing and Hydrothermal Ore Deposition.* New York: Macmillan, 85 pp.

Hemley, J. J., 1959. Some mineralogical equilibria in the system K_2O-Al_2O_3-SiO_2-H_2O. *Amer. J. Sci.* 257:241–270.

—— and W. R. Jones, 1964. Chemical aspects of hydrothermal alteration with emphasis on hydrogen metasomatism. *Econ. Geol.* 59:538–569.

Holland, H. D., 1967. Gangue minerals in hydrothermal deposits, pp. 382–436 in H. L. Barnes, Ed., *Geochemistry of Hydrothermal Ore Deposits.* New York: Holt, Rinehart and Winston, 670 pp.

—— and S. D. Malinin, 1979. The solubility and occurrence of non-ore minerals, pp. 461–508 in H. L. Barnes, Ed., *Geochemistry of Hydrothermal Ore Deposits,* 2nd ed. New York: Wiley, 798 pp.

Hornung, G., 1962. Wall rock composition as a guide to pegmatite mineralization. *Econ. Geol.* 57:1127–1130.

Jahns, R. H., 1946. Mica deposits of the Petaca district, Rio Arriba County, New Mexico. N. Mex. Bur. Mines Min. Resources Bull. 25.

——, 1955. The study of pegmatites. *Econ. Geol. 50th Ann. Vol.,* pp. 1025–1130.

Knight, J. E., 1977. A thermochemical study of alunite, enargite, luzonite, and tennantite deposits. *Econ. Geol.* 72:1321–1336.

Knopf, A., 1929. The Mother Lode system of California. USGS Prof. Pap. 157.

Larson, L., and G. R. Webber, 1977. Chemical and petrographic variations in rhyolitic zones of the Noranda area, Quebec. *Can. Inst. Min. Metall. Bull.,* Aug., 70:80–93.

Larson, P. B., 1976. The metamorphosed alteration zone associated with the Bruce Precambrian volcanogenic massive sulfide deposit, Yavapai County, Arizona. Unpub. M.S. thesis, Univ. Ariz., 99 pp.

Lovering, T. G., 1972. Jasperoid in the U.S.—Its characteristics, origins, and economic significance. USGS Prof. Pap. 710.

Lovering, T. S., 1949. Rock alteration as a guide to ore—East Tintic district, Utah. Econ. Geol. Mono. 1.

Lowell, J. D., and J. M. Guilbert, 1970. Lateral and vertical alteration-mineralization zoning in porphyry ore deposits. *Econ. Geol.* 65:373–408.

MacDiarmid, R. A., 1959. Geology and ore deposits of the Bristol silver mine, Pioche, Nevada, Unpub. Ph.D. dissert. Stanford Univ., CA.

Meyer, C., and J. J. Hemley, 1967. Wall rock alteration, pp. 166–235 in H. L. Barnes, Ed., *Geochemistry of Hydrothermal Ore Deposits.* New York: Holt, Rinehart and Winston, 670 pp.

——, E. P. Shea, C. C. Goddard, Jr., and Staff, 1968. Ore deposits at Butte, Montana, pp. 1373–1416 in J. D. Ridge, Ed., *Ore Deposits of the United States, 1933/67*, Graton-Sales Vols. New York: AIME., 1880 pp.

Norton, D., and H. P. Taylor, Jr., 1979. Quantitative simulation of the hydrothermal systems of crystallizing magmas on the basis of transport theory and oxygen isotope data: an analysis of the Skaergaard intrusion. *J. Petrol.* 20:421–486.

Ovchinnikov, L. N., 1976. *Lithogeochemical Methods for Prospecting Rare-Element Pegmatites.* Moscow: Acad. Sci. USSR, 79 pp.

Park, C. F., Jr., and R. S. Cannon, Jr., 1943. Geology and ore deposits of the Metaline quadrangle, Washington. USGS Prof. Pap. 202.

Rose, A. W., and D. M. Burt, 1979. Hydrothermal alteration, pp. 173–235 in H. L. Barnes, Ed., *Geochemistry of Hydrothermal Ore Deposits*, 2nd ed. New York: Wiley, 798 pp.

Sales, R. H., and C. Meyer, 1948. Wall-rock alteration at Butte, Montana. *AIME Trans.* 178:9–35; also AIME Tech Pub. 2400, 26 pp.

—— and ——, 1950. Interpretation of wall-rock alteration at Butte, Montana. *Quart. Colo. School of Mines* 45:261–273.

Schwartz, G. M., 1947. Hydrothermal alteration in the "porphyry copper" deposits. *Econ. Geol.* 42:319–352.

——, 1959. Hydrothermal alteration. *Econ. Geol.* 54:161–183.

—— and R. J. Leonard, 1927. Contact action of pegmatite on schist. *Geol. Soc. Amer. Bull.* 38:655–664.

Stoll, W. C., 1950. Mica and beryl pegmatites in Idaho and Montana. USGS Prof. Pap. 229.

Stone, J. G., 1959. Ore genesis in the Naica district, Chihuahua, Mexico, *Econ. Geol.* 54:1002–1034.

Taylor, H. P., Jr., and R. W. Forester, 1979. An oxygen and hydrogen isotope study of the Skaergaard intrusion and its country rocks: a description of a 55-m.y. old fossil hydrothermal system. *J. Petrol.* 20:355–419.

Titley, S. R., Ed., 1982. *Advances in Geology of the Porphyry Copper Deposits, Southwestern North America.* Tucson: Univ. Ariz. Press, 560 pp.

—— and C. Hicks, Eds., 1966. *Geology of the Porphyry Copper Deposits, Southwestern North America*. Tucson: Univ. Ariz. Press, 287 pp.

Weller, J. M., R. M. Grogan, and F. E. Tippie, 1952. Geology of the Fluorspar deposits of Illinois. Ill. Geol. Surv. Bull. 76.

SIX

Paragenesis, Paragenetic Sequence, and Zoning

One of the useful explorational and scientific aspects of ore deposits is their zoning. It should be clear to readers of the preceding chapters that ore deposits are neither homogeneous from bottom to top and inside to outside nor chaotic mixtures of minerals and textures. The study of the systematics of internal ore-deposit variation is the study of *zoning*, defined as "the spatial distribution patterns of major or trace elements, mineral species, mineral assemblages, or textures in ore deposits." It is apparent from Table 21-1 that ore-deposit zoning can extend to regional scale. It can also be reduced in scale to the consideration of a single ore shoot, barren zone, vein, or veinlet. Zoning may be obvious, as with dramatic changes in mineral species and assemblages, or it may take the form of minor chemical changes that are invisible to the eye. The former is called *assemblage zoning*, *mineral zoning*, or *phase zoning*; the latter is called *cryptic zoning* or *chemical zoning*. *Textural zoning* can occur, as from coarse vein fillings internally to finer grain sizes externally. Overall, zoning deals with arrays of ore deposit components and characteristics in three dimensions.

Temporal changes in a volume of rock undergoing mineralization can result in different apparent timing of mineral assemblages from place to place in an orebody or a district. The word *paragenesis*, from the Greek meaning "born beside," is used to describe any assemblage of ore minerals, with or without gangue, formed at the same time and normally in equilibrium. The chronological order of mineral deposition—the sequence of assemblages—is known as the *paragenetic sequence* of a deposit; and, again, variation in the spatial distribution of paragenesis is known as *zoning*. Paragenesis is determined by mineral studies that commonly focus upon microscopic textural features, but that can and should include megascopic features visible in outcrop or core, such as crosscutting, offsetting vein types with different mineralization or alteration complements. Zoning patterns are manifested by changes along both vertical and horizontal traverses of mineralized areas.

As we have seen, an ore-bearing fluid gradually changes as it migrates from its source. Typically it reacts with wall rocks, and its chemical composition, Eh, pH, sulfur fugacity, and other properties change as it travels into regions of lower pressure and loses heat to cooler country rocks. As these physical and chemical changes take place, the solubility products of the ore and gangue minerals are reached, and they can be sequentially deposited. They thus can leave a detailed record in time and space of the evolutionary trends in an ore-forming solution.

Zoning was defined and contrasted with paragenesis in 1965 by a committee of the International Association on Genesis of Ore Deposits (Kutina, Park, and Smirnov, 1965). Such action became necessary because different meanings were attached to these terms by geologists of different countries. The committee decided that:

> *Zoning in ore deposits is any regular pattern in the distribution of minerals or elements* in space; *it may be shown in a single orebody, in a mineral district, or in a large region. Although zoning is related to the spatial distribution of elements and minerals, both time and space must be considered in the study of zonal phenomena. The term paragenesis, as used in the United States, is the distribution* in time, *or the sequence of, minerals or elements. Paragenesis, as widely used in Europe, is an association of minerals having a common origin, for example, a tin-tungsten association or paragenesis.*

Since 1965 most American mineragraphers and economic geologists have adopted the clearer European meaning of the terms zoning, paragenesis, and paragenetic sequence, and that use is recommended here.

A thorough megascopic and microscopic study of carefully chosen surface, drill core, and underground samples, in conjunction with detailed field mapping, allows the geologist to develop an understanding of the ore and wall rocks, including changes along the three space coordinates and those that took place during the time interval of emplacement. Details in the character of mineralization—correlated with structure, the character of the wall rocks and their alteration products, the chemical reactions involved, and such factors as permeability and porosity of the host rocks—lead to a better understanding of the processes of ore genesis and hence to sound exploration for new ore deposits.

▶ PARAGENESIS

We have seen that ore-forming fluids of whatever origin are chemically complex. Rarely indeed are nearly pure water or the physical chemist's ideal solutions involved, the normal case involving silicate, CO_2-rich, or aqueous fluids with scores of complex ions. It is reasonable to conclude that similar elements will behave similarly, and that when, for example, chloride-based complex ions of the transition metals are dissociating, several of these metals may be available for recombination into solid-phase complexes with sulfur, arsenic, or whatever. In short, mutual precipitation of two or more sulfide or sulfosalt phases is common in a vein, breccia column, or solution cavity. Most ore formation occurs in environments of falling temperature and slightly acid pH which tend to supersaturate the aqueous phase with silica, so quartz is a normal accompaniment of sulfides. And of course, alkalies can recombine with silica and alumina to form many of the gangue minerals, such as K-feldspar and sericite. Holland (1967) and Holland and Malinin (1979) have provided diagrams that indicate the fields of solubility and saturation of sulfate and carbonate accessories such as anhydrite, barite, calcite, and dolomite. It is not surprising that "standard assemblages" of ore and gangue minerals characteristic of particular settings and circumstances are normal, as was suggested in the captions and discussions of Figures 4-4 and 4-5, the $f(S_2)$-$f(O_2)$ and $f(O_2)$-pH representations of normal assemblages in ore deposits at 250°C. These assemblages are also called parageneses.

McKinstry (1963) spent years studying the paragenesis of sulfide deposits by mineragraphic and binocular microscope examination of thousands of polished surfaces of specimens from scores of districts. He noted what pairs, trios, and quartets of minerals occur in textural equilibrium assemblages and thereby established sets of common parageneses, mostly in the Cu-Fe-As-S system (Figure 6-1). Gustafson (1963), working under McKinstry, also noted paragenetic sequence in many districts, and Gustafson—having established the compatibility of pairs and sets of the sulfur and arsenic minerals of copper and iron—was also able to show the theoretical paragenetic sequence through thermodynamically constructed three-dimensional $f(S_2)$-$f(As_2)$-a_{Cu} diagrams (Figure 6-2). Figure 6-3 is another style of paragenesis diagram. Through such observational laboratory synthesis, and now computational studies, scores of "allowable" and "unallowable" parageneses have been established. Many of them will become apparent in Chapters 9 to 20.

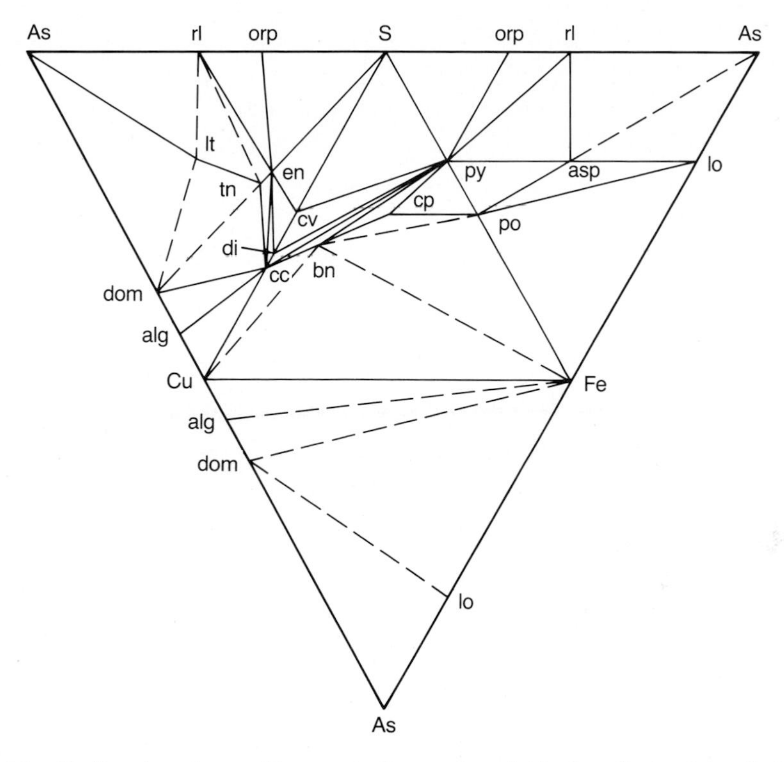

Figure 6-1. Stable mineral assemblages actually seen to exist in deposits involving Cu-Fe-As-S components at temperatures 250 to 350°C. The solid lines enclose stable assemblages, or parageneses: digenite-chalcocite-pyrite is a real paragenesis, but covellite-bornite-pyrite does not occur at hydrothermal temperatures. To form a tetrahedron, photocopy the figure, cut it out, and fold it this side out along the joins Cu-S, Fe-S, and Cu-Fe to bring the As apices together. (*From* McKinstry, 1963.)

▸ PARAGENETIC SEQUENCE

Although broader aspects can usually be established from core and outcrop, studies of mineral sequences are practically restricted to the microscope. Careful analyses of microtextures and microstructures are used to decide the order of mineral deposition, as a review of Figures 4-6 to 4-20 will suggest. Electron microprobe studies are used increasingly to reveal cryptic variations that can be helpful in paragenetic sequence evaluation, and field relationships between mineralized veins that cut across one another are also valuable clues to mineral paragenesis and sequence. In the study of a mine or mineralized district, the geologist records the relative age of each distinguishable mineral pair or paragenesis. It may be found for example, that chalcopyrite always formed earlier than sphalerite, and that sphalerite formed either earlier or at the same time as galena, but that some galena

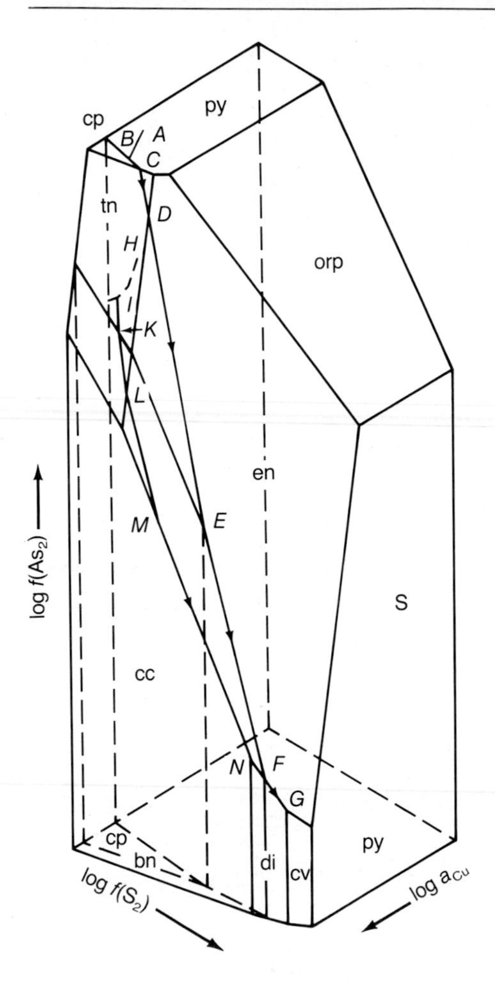

Figure 6-2. Part of the Cu-Fe-As-S system as evaluated by thermochemistry. The arrows mark preferred normal paragenetic evolution in Cordilleran Vein deposits. Successive parageneses are py (from *A* toward *B*), py-cp (from *B* toward *C*), py-cp-tn (from *C* toward *D*), py-cp-tn-en (at *D*), py-cp-en (from *D* toward *E*), and so forth. Enargite-chalcopyrite parageneses are common; tennantite-chalcocite is prohibited. These results conform well with observed assemblages reported in Figure 6-1. (*From* Gustafson, 1963.)

definitely formed later than sphalerite. The geologist may find, further, that pyrite precipitated throughout. Such results can be plotted to show the relative sequence of mineral deposition, and the graph would be a paragenetic sequence diagram similar to the type given in Figure 6-4.

The study of polished sections throughout the world has established a general sequence of mineral deposition in ore deposits. This sequence is based upon mineral stability observations like McKinstry's and upon scores of laboratory studies, and is fairly constant for most deposits of a given type, but varies according to the depth and temperature of formation and the style of origin, be it magmatic, hydrothermal, exhalative, sedimentary, meteoric, or metamorphic. Since the character of an ore-bearing fluid changes gradually as it moves, different minerals will be formed along different parts of a channel. While one mineral is being deposited under certain conditions at one place or level, other minerals are forming elsewhere under different conditions. Thus the deposition of one mineral commonly overlaps

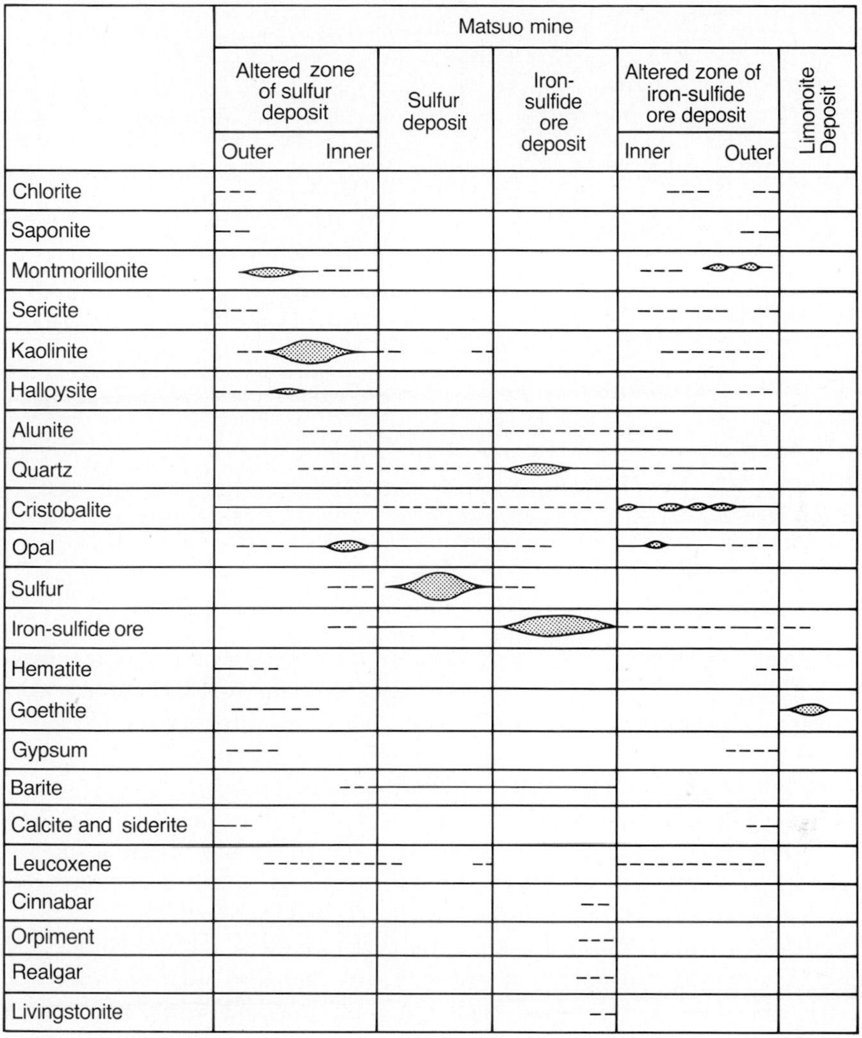

	Matsuo mine						
	Altered zone of sulfur deposit		Sulfur deposit	Iron-sulfide ore deposit	Altered zone of iron-sulfide ore deposit		Limonoite Deposit
	Outer	Inner			Inner	Outer	
Chlorite							
Saponite							
Montmorillonite							
Sericite							
Kaolinite							
Halloysite							
Alunite							
Quartz							
Cristobalite							
Opal							
Sulfur							
Iron-sulfide ore							
Hematite							
Goethite							
Gypsum							
Barite							
Calcite and siderite							
Leucoxene							
Cinnabar							
Orpiment							
Realgar							
Livingstonite							

Figure 6-3. Associations of alteration and ore minerals in the Matsuo mine, Japan. Takeuchi and Abe (1970) discussed hydrothermal alteration associated with sulfur mineralization at the Matsuo deposits described at the end of Chapter 4. They published this table showing ore minerals and alteration products that occur together, presumably as stable parageneses. Each column describes a different part of the deposit and shows what minerals occur there and in what relative amounts. Two columns also show spatial variations from inside to outside.

the deposition of others in both space and time. Slight changes in temperature, pressure, or chemistry of the transporting fluids may also alter the normal course of deposition and cause reversals or interruptions in the process. Furthermore, the paragenesis of each stage in a multiple pulsation system has a chronologic sequence of its own, hence the order of mineral

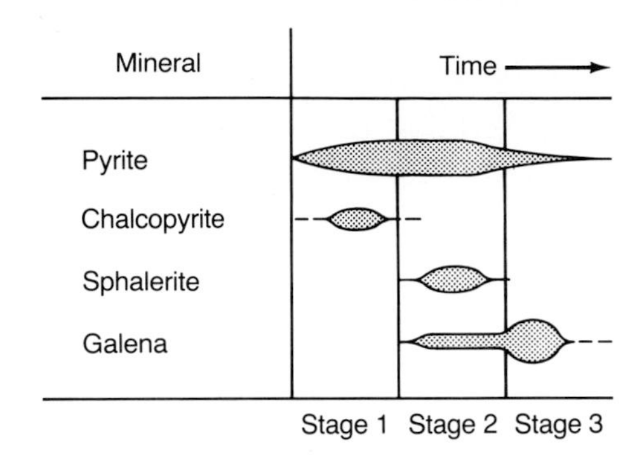

Figure 6-4. Sample diagram of paragenesis and paragenetic sequence, showing the formation of chalcopyrite before sphalerite and galena, with an overlap in the deposition of sphalerite and galena. Pyrite was deposited with all of the ore minerals. Relative time is indicated from left to right, and relative amounts of the phases precipitated are indicated by line width. Stages can be broken out if desirable. Here stage 1 = py-cp, stage 2 = py-sl-gn, and stage 3 = py-gn. Clearly, a fluid which precipitated first copper and iron, then iron-zinc-lead, then iron and lead is indicated, and insights into the solution chemistry, source evolution, and so forth, can be derived. See also Figure 6-8.

deposition in any two pulses may not agree exactly. For example, during mineralization, fissures are commonly repeatedly opened by tectonic activity, allowing discontinuous surges of ore-bearing fluids to enter. Since the chemistry of the fluids may change through time, the mineralogy resulting from each surge may be different. Determining the order of deposition is therefore seldom simple, and the true paragenetic sequence of a mine or group of mines can rarely be ascertained from the study of a few samples; it requires examination of many thin sections and polished sections from samples taken at widely scattered locations within a deposit.

Stated a bit differently, it is important to note, in determining the chemical-depositional history of a deposit, the fact that two or more effects can transpire at once. Suppose that at a given moment, early assemblages a_E, b_E, and c_E are being deposited in the central, intermediate, and outer volumes of an incipient deposit. At a somewhat later time, the system might have cooled, and assemblages a_L, b_L, and c_L are being deposited. At any given *place* in the vein, the assemblages would show overall cooling; they would be a_E-a_L, b_E-b_L, and c_E-c_L. And in the plane of the vein, a progression at any given instant of time would be apparent. Now suppose that a_L and b_E are the same quartz-pyrite-tennantite-chalcopyrite assemblage, one formed later than the other. One hundred million years later it might take a great deal of work to understand that the deposition of an assemblage in one part of a vein was not contemporaneous with the deposition of the same assemblage elsewhere in the same vein.

If it is true that hydrothermal fluids evolve in a more or less normal fashion, that pressure and temperature controls are important, and that solution mixing and dilution are normal in both deep and shallow environments, then a normal progression of minerals, parageneses, and paragenetic sequences should exist. As classifications of ore deposits were constructed in earlier decades to help shed light on ore genesis, several attempts were made to construct an "ideal succession," a sequence from the deepest, highest intensity sites of ore deposition to the outermost, mildest environments. Such "ideal successions" are no more valid in describing a specific ore deposit than is Bowen's reaction series (Figure 4-1) in describing a single igneous rock occurrence. Both represent the total spread, and probably no one vein—or magma—goes through the entire spectrum. But any given deposit—or rock—can be appreciated better by seeing where it fits in an overall flowsheet of parageneses. Lindgren (1933) and Edwards (1947, 1952) both presented such mineral sequences—they were given as Tables 6-1 and 6-2 in earlier editions of this text. But they are now seen to be so damaged by confusion between magmatic and hydrothermal phenomena, by misinterpretations of high- and low-temperature minerals, by misidentity of exsolution-replacement phenomena, and by mistakes inherited from misclassification as to be misleading. "Emmons' reconstructed vein" has fared little better; a new version, extensively revised to include recent progress, is given here as Table 6-1. It is consistent that iron and tin oxides and iron-manganese tungstates are deposited early; sulfides and arsenides of iron and molybdenum sulfide are generally contemporaneous with or slightly later than the oxides; zinc, lead, silver, and combined copper-iron sulfides are intermediate parageneses and are mixed with, or slightly older than, the copper, lead, and silver sulfosalts; native metals and tellurides are typically late; and antimony and mercury sulfides are the latest and farthest removed.

Clearly, such a diagram is as apt to a discussion of paragenesis as it is to one of paragenetic sequence. The schematic of Table 6-1—the reconstructed vein—is one of *paragenesis* in a time-lapse-photo sense. As these words are written, these assemblages could be being precipitated at increasing depths below the solfataras at Mount St. Helens. Over the next 10,000 years the zones (if they exist) will almost certainly expand or shrink, and paragenetic sequences will be evident.

Figure 6-5 shows another style of paragenetic sequence diagram in a skarn zone. It illustrates how parageneses relate to zoning in metasomatic rocks at the margins of a stock. This figure will be recalled in Chapter 11.

▸ ZONING

Changes in molecular or ionic species and in their activities in an evolving fluid produce changes in ore and gangue mineralogy along courses of deposition. Such changes are described as zoning, and are found in sedimentary deposits as well as in magmatic, hydrothermal, and metamorphic ores. In

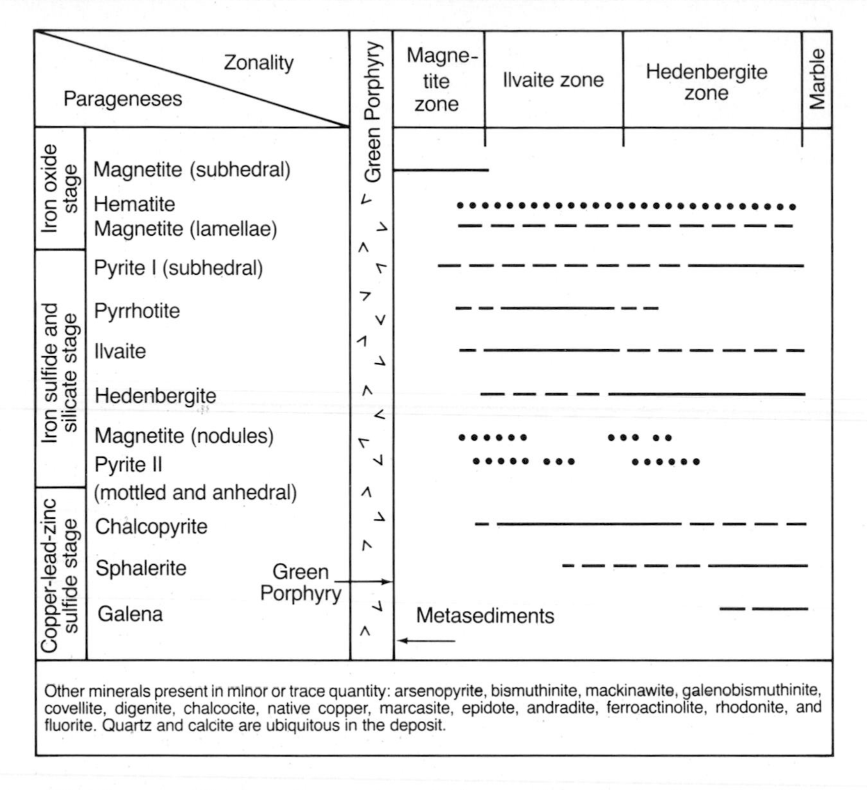

Figure 6-5. Summary diagram of space-time relations as shown by the characteristic minerals of the skarn-sulfide deposit, Valle del Temperino, Italy. Solid, dashed, and dotted lines indicate approximate quantity in decreasing order, another way of representing relative quantities. The passage of time is registered from top to bottom, and distance from the Green Porphyry contact from left to right. It is both a paragenesis and a paragenetic sequence diagram. Skarns are discussed in Chapter 11. (*From* Corsini et al., 1980.)

the idealized schematic of an upward and outward flowing fluid, changes in chemistry, temperature, and pressure along fissures result in the deposition of different minerals in nested, concentric zones at increasing distances from the source. Syngenetic deposits may be zoned parallel to a shore line or along a stream channel. Any detection of a zonal pattern—epigenetic or syngenetic—is important to economic geology because it helps predict changes in the style and grade of mineralization as a deposit is developed and mined. Zones of abstraction and leaching that are parts of lateral secretion, or solution-remobilization (Chapter 19), should be reflected in cryptic or phase zoning.

The theory of zoning was first well stated as a generalization by Spurr (1907), although many workers had previously recognized the phenomenon (de la Beche, 1839; Henwood, 1843; de Launay, 1900; Collins, 1902; Waller, 1904). At the present time, zoning is widely appreciated and is accepted as

TABLE 6-1. Mineralogy of an Idealized Composite Vein from Surface to Depth

1. *Barren:* Chalcedony, quartz, barite, fluorite, carbonates. Some veins carry small amounts of mercury, antimony, or arsenic minerals. 'Fringe' mineralization.
2. *Mercury:* Cinnabar, commonly with chalcedony, quartz, barite, fluorite, or carbonates.
3. *Antimony:* Stibnite, commonly with quartz, locally passing downward into galena, with lead-antimony sulfosalts. Some carry gold-silver.
4. *Gold-silver:* Bonanza gold and gold-silver deposits. Acanthite with arsenic and antimony minerals common. Abundant quartz, chalcedony, amethyst. Tellurides and selenides in places. Relatively small amounts of galena, adularia, alunite with calcite, rhodochrosite, and other carbonates. Epithermal textures; silicification. Some potassic, some sericitic, some propylitic alteration.
5. *Barren interval:* Most nearly consistent barren zone; represents bottom of many Teritiary precious-metal veins. Quartz, carbonates, and small amounts of pyrite, chalcopyrite, sphalerite, and galena.
6. *Silver-manganese:* Quartz gangue with calcite-dolomite-siderite; acanthite-rhodochrosite; silver-arsenic-antimony sulfosalts; minor rhodonite. Low pyrite. Sericitic-argillic-propylitic alteration.
7. *Lead:* Quartz gangue, minor carbonates. Galena generally with silver; sphalerite usually present, increasing with depth; some chalcopyrite, rhodochrosite-rhodonite. Low pyrite.
8. *Zinc:* Quartz, with, in some deposits, Ca-Fe-Mn carbonate gangue. Sphalerite, with galena, chalcopyrite, tennantite-tetrahedrite. Chalcopyrite-pyrite increase downward. *Lithocap:* Some systems at level 6, 7, or 8 are quartz-enargite-tennanite, minor chalcopyrite. Advanced argillic alteration, quartz-pyrophyllite-topaz-alunite-kaolinite. Low pyrite.
9. *Copper:* Tennantite-tetrahedrite-chalcopyrite, commonly argentiferous. Quartz-pyrite gangue. Sericitic-argillic-propylitic alteration.
10. *Copper:* Chalcopyrite, quartz gangue, most with pyrite, some with pyrrhotite. Tennantite-bornite-chalcocite-enargite-downward, commonly with quartz-sericite-pyrite alteration. Generally carry precious metals, especially silver. Huebnerite common. Trace tin.
11. *Molybdenum-tungsten:* Quartz-huebnerite or scheelite-pyrite veins with trace molybdenite-chalcopyrite, or quartz-molybdenite as white quartz-smoky molybdenite veinlets. Some with cassiterite. No alteration or potassic.
12. *Barren or low grade:* Quartz, K-feldspar, biotite, chalcopyrite, molybdenite, trace pyrite, carbonates, anhydrite. Potassic alteration.

Revised "Emmons' reconstructed vein," an idealized composite vein system from the surface (1) to deep-seated conditions (12). No one vein is known to contain all of these associations, but most of them show several intervals in the order given. A typical wall rock would be a related I-type quartz monzonite. Emmons' original reconstruction combined what we now know to be I-type and S-type metallogenic assemblages.

Source: Extensively revised from Emmons (1936).

a tool and as a working hypothesis. Yet the causes of zoning are still being debated, and it is difficult to explain why certain deposits are zoned and others are not (Kutina, 1963; Pouba and Stemprok, 1970).

Zoning in ore deposits is conveniently divided into the following three intergradational classes, which are based upon size and are only in part related in origin.

1. *Regional zoning.* Zoning on a very large scale, as exemplified by the Southern Piedmont region of the southeastern United States and by

ore deposits associated with the Sierra Nevada batholith (Park, 1955) and the Cordilleran orogenic belt in general. This level of order is also called *metallogenic zoning.*

2. *District zoning.* Zoning shown by closely grouped mines or occurrences, a category that includes the well-known mining districts of Butte, Montana (Sales and Meyer, 1949; Meyer et al., 1968; Brimhall, 1979), Cornwall, England (Davison, 1927; Lilley, 1932; Jackson, 1979), and Bingham, Utah (Peacock, 1948; Moore and Nash, 1974; John, 1978).
3. *Orebody zoning.* Changes in the character of mineralization within a single orebody or a single ore shoot (Riley, 1936). Many massive sulfide deposits of volcanogenic origin, such as the Kuroko bodies in the volcanic rocks of Japan, as well as many single fault vein orebodies within zoned districts are in this category. Zoning on the scale of individual veins and veinlets is included here.

Because of the great disparity in size between mineralized regions and individual orebodies, confusion often arises in the discussion of zoning. This confusion led Sampson (1936) to suggest that the term *zoning* be limited to districts. But restricting the term in this way further confuses the usage. The addition of the adjectives *regional, district,* and *orebody* to indicate the relative scale is preferable. This dimensional classification of zoning is not universally accepted by European geologists, some of whom feel that ore deposits usually result from discontinuous surges or pulsations of mineralizing solutions, resulting in a complicated overlapping of zones. They suggest that zoning should be classified according to whether the deposits were developed by a single pulsation or by many, which they describe as *monoascendant* or *polyascendant* fluids, and whether they are in normal or reversed sequence (Kutina, 1957). The broader use is retained here. For example, we may consider zoning in skarns at Carr Fork, Bingham Canyon, as orebody zoning because it occurs over tens of meters (Figure 11-28); we can consider district-scale overall zoning of the Bingham Canyon district at a scale of kilometers (Figures 11-26 and 11-27); and we can consider Bingham Canyon as it compares to other porphyry copper-molybdenum deposits in Nevada, Utah, and Colorado in the context of regional zoning (Table 21-1 and Figure 6-7). Descriptions of examples of zoning at each scale will close this chapter.

Regional Zoning

Regional zoning is best thought of in terms of metallogeny. For many decades, economic geologists have been aware that many types of ore deposits seem to show broad-scale patterned consistencies that have defied explanation until the 1970s, and which are still not completely understood—for example, from west to east across the Cordilleran mountain ranges.

Burnham (1959) found broad, consistent zones of trace element populations in pyrite and sphalerite collected in scores of western American deposits, and bands or belts of relative enrichment in tellurium, beryllium, tin, and many other elements are known to exist (Schilling, 1957; Petrascheck, 1968; Noble, 1970). Many explanations have been proposed, from ore magmas at mantle depths (Spurr, 1923) to nearly horizontal layers of enrichment in the mantle or lower crust that were occasionally tapped by various sorts of orogenic disturbance. During the 1970s, however, a mechanism of appropriate scale was perceived. "Mountain building," "orogeny," and the relationship of different types of ore deposits to belts of different structural and igneous activity had long been appreciated, but plate tectonics appears to have knit it all together with dynamic, explainable geologic processes. Chemical variations in rocks and ores inland from continental borders have long been perceived; now they can be discussed in terms of subduction, zonal melting, oceanic crust, asthenospheric, and continental crust contributions, and so on. Reasons for systematic regional Cu-Mo ratio variations in porphyry coppers and why Climax-type molybdenum deposits and carbonatites occur where they do now make more sense. Certainly not all the answers are in; many questions remain unanswered, and many "answers" must be challenged. Two diagrams, however, serve to show some of the types of studies that are being pressed. Figure 6-6, by Damon, Shafiqullah, and Clark (1981), shows their depiction of regional-scale metallization belts in Mexico as determined from analyses, production records, isotopic geochronological determinations, and geochemical reports; and Figure 6-7 (Westra and Keith, 1981) is the result of analysis of thousands of bits of whole-rock geochemical data, isotopic dating, and mineralization profiles across California, Arizona, New Mexico, and Texas. Keith (1983) holds that regional zoning is genetically related to igneous geochemistry and thereby to subduction zone angle, subduction rates, and other plate tectonic manifestations. Others argue that the impact of near-surface lithologies has not been sufficiently evaluated, but Keith's evidence keeps mounting. Without doubt, the 1980s will be a period of vigorous evaluation of the degree to which plate tectonic, chemical, and structural mechanisms affect ore deposits, how that information can be used scientifically and explorationally, and how concepts and facts of regional zoning can be integrated. Metallogeny and regional-scale ore occurrence will be discussed further as a summary of this book in Chapter 21.

District Zoning

Field observation shows that groups of minerals have characteristically been formed in more or less constant sequence in hydrothermal ore deposits. By gradually piecing together field data on relative zoning in individual regions, districts, and orebodies, geologists have been able to construct theoretical vein systems against which any single deposit may be

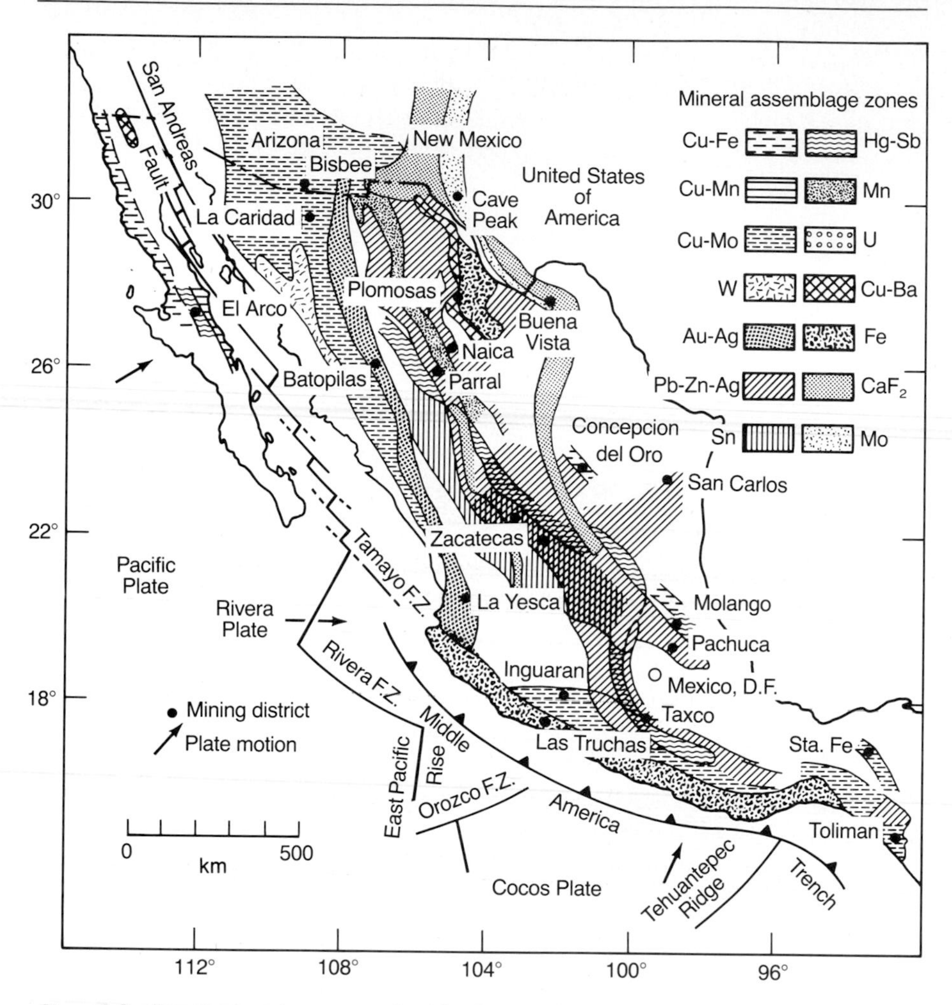

Figure 6-6. Regional zoning as perceived by Damon, Shafiqullah, and Clark (1981). The metal-rich belts parallel the subducting margin and include many prospects and operating mines.

compared. Early observers noted that tin minerals characteristically lie deeper and perhaps "closer to the source" than copper minerals, and that copper minerals, in turn, are inside the silver zone. Expansion of these observations led to the relatively detailed model vein system in Table 6-1. No single field example includes all of the mineral groups in the model, but the vein system is of value in the study and understanding of zoning.

As would be expected, there are many discrepancies between actual vein systems and the theoretical system. Irregularities and reversals result from abnormally rapid deposition, called *dumping*, and the overlapping of zones, to name but two factors. Discrepancies and poorly defined zoning can be caused by the overlapping of deposition from two or more mineral-

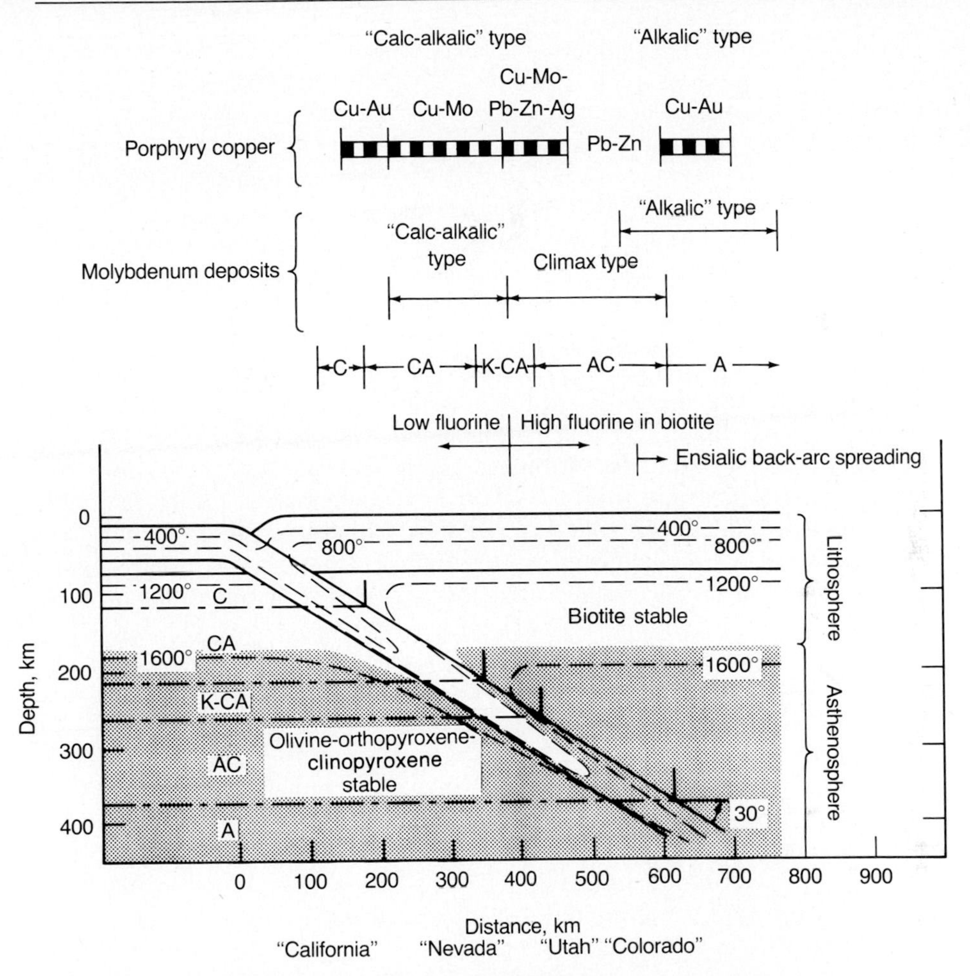

Figure 6-7. Supposed interaction between subduction, igneous rock chemistry, and ore deposits assembled for a section approximately from California eastward through Nevada and Utah into Colorado. See text. It helps explain regional-scale zoning and metallogeny (Chapter 21). (*From* Westra and Keith, 1981.)

izing centers, by the retreat or advance of mineralizing centers during one period of deposition, by repeated periods of mineralization in a single area, by variations in the mixing of juvenile fluids with circulating groundwaters, and by other influences. But studies in the 1970s and 1980s have shown that there is a characteristic suite of minerals associated with intermediate to felsic hydrothermal activity that appears to fit the generalizations of Table 6-1.

Where the minerals of one zone overlap those of another, the deposit or district is said to be *telescoped*. Near the surface, hydrothermal fluids may be subjected to steep temperature and pressure gradients, causing rapid deposition of ore minerals and a shortening, or telescoping, of the ore zone with a profusion of mineral species. At depth, temperature and pressure gradients are less steep. Under these conditions, deposition takes place slowly, the separation of minerals is well defined, mineralogy is inclined to obey the phase rule and be simpler, and zonal boundaries are gradational. Telescoping is therefore restricted largely to deposits formed under shallow conditions, where changes in temperature and pressure are rapid (Borchert, 1951). Some such deposits used to be called *xenothermal*, but the type area—the Bolivian tin-silver deposits—has been shown satisfactorily to be multiply and complexly mineralized, not telescoped (Schneider-Scherbina, 1962).

Where conditions change gradually, as in deep-seated high-temperature high-pressure deposits, zoning is generally inconspicuous and is expressed, for example, by variations in the fineness of gold (Pryor, 1923) or in the amounts of minor constituents. Many districts and mines show no recognizable zoning; ore from the lower levels is apparently identical with ore from the upper levels. Such deposits are called *persistent*. The Mother Lode gold district of California is an example of a persistent district that shows almost no zoning, either at the district scale or in individual orebodies.

The practical application of district zoning is illustrated in the Comstock Lode district of Nevada, where the distribution of mercury suggests horizontal zoning outward from a silver-gold central area to a peripheral mercury area. If similar zoning exists vertically, several surface mercury anomalies may prove to be underlain by silver-gold mineralization (Cornwall et al., 1967), which also indicates that a similar type of exploration might be useful at Tonopah, Nevada. The application of district zoning ideas is proving productive in many areas. The silver ores near the town of Tombstone, Arizona, are being related through analysis of geochemical-geological zoning patterns to a broader system, perhaps 15 km across, involving Cu-Mo-Pb-Zn-Ag and a subsurface porphyry base-metal system by Newell (1975). Many deposit areas have been expanded or simply better understood through zoning analysis. Morenci, Arizona, is an example of the former (Langton, 1973), Ray, Arizona, of the latter (Phillips, Gambell, and Fountain, 1974). Figure 6-8 shows how elaborately zoning can be worked up in a porphyry copper deposit; the significance of the variations is dealt with in Chapter 11.

It is tempting to speculate upon the geochemical reasons for zoning, but we still know too little. The sequence of deposition should be fixed by the relative solubility products of the sulfides and the association-dissociation constants of the metal chloride and bisulfide complexes. We simply have insufficient knowledge to deal quantitatively with these relationships at elevated temperatures. The solubility constants, activities, and concentrations of the metals probably vary locally over small ranges, and detailed data to allow discussion of their *relative* behavior in an aliquot of fluid must be much finer than the present tenfold uncertainties. Barnes (1979) points out that "quantitative understanding of the causes of zoning may require direct measurements of relative solubilities simultaneously from (different parts of) one system at pertinent P-T-X conditions," a formidable task.

We know from Table 6-1 and the geologic content of the chapters to follow that the mobility of the ore metals in the transporting fluid follows a sequence from most mobile (farthest transported) to least mobile (earliest precipitated) Hg-Ag-Mn^{+2}-Pb-(ZnCuFe)-W-Mn^{+4}-Sn. The relative stabilities of covalently bonded complexes of ore metals were thermodynamically approximated by Barnes (1962) with remarkable agreement as

$$\underset{-227}{\mathrm{Hg}} - \underset{-156}{\mathrm{Cd}} - \underset{-154}{\mathrm{Pb}} - \underset{-134}{\mathrm{Cu}} - \underset{-132}{\mathrm{Zn}} - \underset{-126}{\mathrm{Sn}} - \underset{-83}{\mathrm{Ni}} - \underset{-82}{\mathrm{Fe}} - \underset{-81}{\mathrm{Co}} - \underset{-78}{\mathrm{Mn}}$$

the numbers being free-energy values of pertinent solubility reactions in kilocalories. Little other relevant geochemical data can be cited.

Orebody Zoning

Many examples of orebody-scale zoning are confronted in both mineral exploration and mine geology. This level of zoning refers to variations of paragenesis, and hence of mineral species, and to variations in element abundances, which can certainly include the metal or metals sought. Orebody zoning, then, gets to the matter of "grade" and "tenor" and to the difference between ore shoots and subeconomic or barren zones. Ore shoots can usually be discerned by mapping, either by observation of the distribution of ore minerals themselves or by mapping more abundant or obvious minerals that vary in some proportion to the ore metal or metals. Most mines employ geologists whose job is, in part, to map new exposures of mineralization in drill core, underground, or on open-pit benches. Frequently minerals or mineral associations can be found that "telegraph" improvement in assay values or vein widths laterally or below. In some Cordilleran Vein deposits, for example, veinlets carrying quartz-pyrite-bornite among a group of quartz-pyrite-chalcopyrite veinlets are likely to widen a few meters along strike or downdip into a major ore-bearing structure. Most ore-waste management in mining, and diamond drilling evaluation in exploration, is handled by assaying. Samplers regularly chip channel samples across vein headings, or sample churn drill or blast hole cuttings; these samples are assayed for the metals sought, and mine planning proceeds accordingly. Geologists log diamond drill core shortly after it is pulled from a hole, split it lengthwise, and send a split half to the lab for assaying.

Orebody zoning is almost nonexistent in persistent deposits, those formed in environments of apparently gentle chemical and thermal gradients, and is most extreme in epithermal deposits and in those that appear

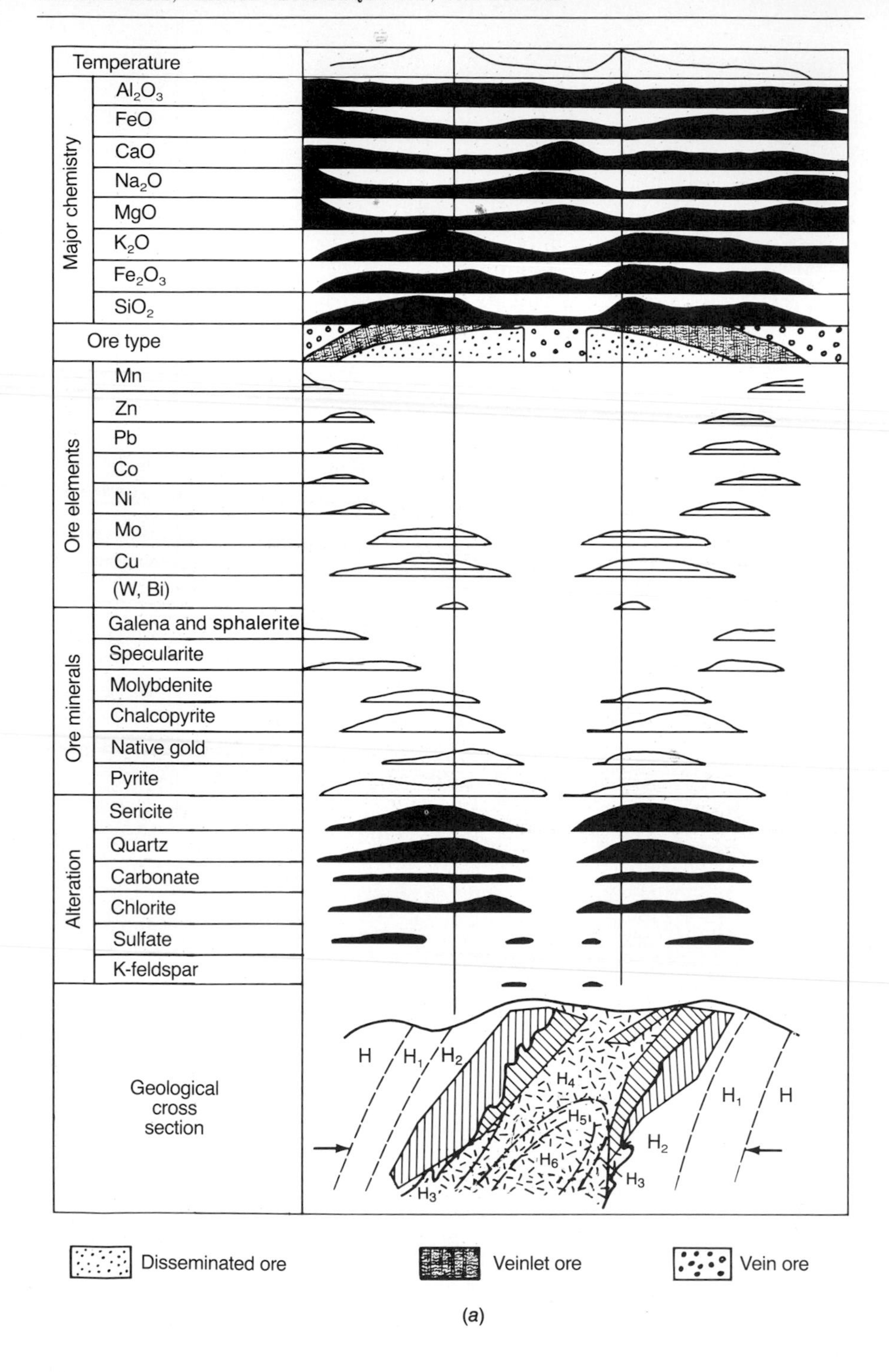
Temperature
Major chemistry
Al2O3
FeO
CaO
Na2O
MgO
K2O
Fe2O3
SiO2
Ore type
Ore elements
Mn
Zn
Pb
Co
Ni
Mo
Cu
(W, Bi)
Ore minerals
Galena and sphalerite
Specularite
Molybdenite
Chalcopyrite
Native gold
Pyrite
Alteration
Sericite
Quartz
Carbonate
Chlorite
Sulfate
K-feldspar
Geological cross section
H
H1
H2
H3
H4
H5
H6
Disseminated ore
Veinlet ore
Vein ore

(*a*)

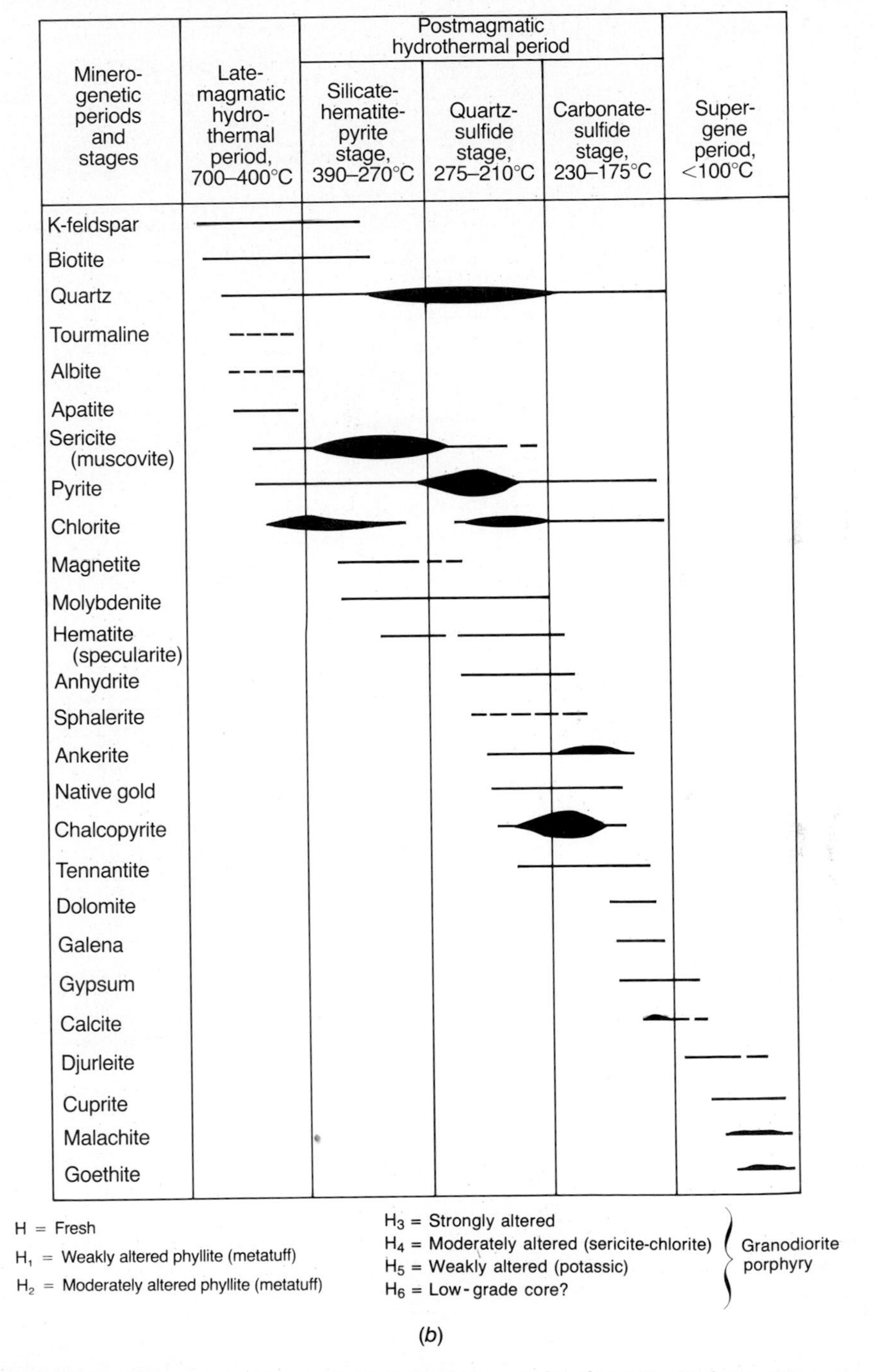

Figure 6-8. (*a*) Expression of the zoning, temperature, major elements, ore type, ore elements, ore minerals, alteration, and geology at a porphyry copper deposit in China. The lateral position of the lenses shows a nesting of shells, or rings, in plan view. The heights of the blips are proportional to quantity. (*b*) Uncommonly clear paragenetic sequence diagram for the same deposit, the Dexing prospect in the People's Republic of China. (*From* Yan and Hu, 1980.)

to have involved structures that were open to the surface and to invasion by meteoric water. At least two interactive trends are noted in many vein mines. One relates to geoisotherms that seem to have been nearly horizontal, or domal, and were established around an intrusive stock or pluton underlying the district—the veins at Casapalca, Peru (Figure 11-12), and Superior, Arizona (Figure 11-37), are of this nature. The other is in the plane of a given vein and produces vertical or near-vertical panels of higher and lower grade and changing mineralogy. The panels may have been caused by preferential fluid flow or gas streaming along more permeable portions of the vein. Perhaps open veins set up near-vertical panels of more vigorous convective and boiling upflow between other panels of cooler condensate and groundwater downflow, rather like the dynamics of a two-dimensional percolator coffee pot.

Other forms of orebody zoning are found in syngenetic deposits. Cryptic zoning is common in layered mafic intrusive bodies, with ratios of Cr_2O_3, Al_2O_3, MgO, and FeO in chromite and MgO and FeO in olivine and orthopyroxene varying invisibly across layers of chromite-rich rock called *chromitite* (Figures 9-7 to 9-9). Ore mineralogy varies strikingly in some chemical-sedimentary base-metal deposits. Copper-to-iron ratios and sulfur availability vary as parts of original sedimentary environments (Figures 15-2 to 15-4) so that the final orebodies show chalcocite-bornite-chalcopyrite-pyrite lateral zoning, as in Zambia. Banded iron formations are also commonly zoned (Figures 13-20, 17-21, and 13-12) from hematite close to the shoreline to progressively deeper water magnetite, siderite, and finally pyrite according to James (1954). Massive sulfide bodies formed in volcanic rocks by submarine hot springs which vent during lapses between extrusive flows are typically copper-rich at the base and zinc-rich at the top (Figure 13-5). If galena and barite occur, they too are likely to increase upward. And finally, orebody zoning is perhaps most striking and integral to the orebody itself in pegmatites (Figures 11-49 and 11-50).

Clearly, orebody zoning is an aspect that can be used in the evaluation of ore-forming processes as well as in exploration practice and strategy.

▸ AN EXAMPLE OF REGIONAL ZONING—THE SOUTHERN PIEDMONT, UNITED STATES, GOLD DEPOSITS

An example of regional zoning is furnished by the ore deposits of the southern Appalachian-Piedmont province from Washington, D.C., southward to the Coastal Plain of Alabama (Figure 6-9; Pardee and Park, 1948). The metalliferous deposits there are zoned around the central core of the Piedmont province where both Precambrian and Paleozoic intrusives are abundant. The central part of the region contains deposits of pyrite and gold. East of this central gold-pyrite zone, Mesozoic Coastal Plain sediments cover the evidence, and regional zoning is best shown to the west. Closely associated with the gold ores, but slightly to the west, are deposits of pyrrhotite and pyrite, as at the Gossan Lead, Virginia. Farther westward are pyrrhotite-chalcopyrite ores of the type found at Ducktown, Tennessee; west of the copper zone are the lead-zinc deposits at Austinville, Virginia, and in the Mascot-Jefferson City district, Tennessee. The outermost zone is defined by barite, which has been mined from several areas west of the lead-zinc ores. Figure 6-9 shows the zonal arrangement of these deposits in relation to the intrusive masses.

It was not known until recently whether the ores were genetically associated with the plutonic rocks or whether both the ores and the intrusives are products of metamorphism or plutonism deep within the Appalachian geosyncline. Regional geologic studies in the last decades have greatly clarified some relationships. The zoning is, of course, real. It has been shown that both Precambrian and early Paleozoic orogeny have produced and influenced ore deposition. Taconic and Acadian subduction zones emplaced porphyry base-metal systems and possible epithermal gold deposits (Worthington and Kiff, 1970) in the Carolinas, shown as "gold" and "copper" areas in Figure 6-9, and exhalite systems produced the Gossan Lead and Ducktown deposits. The region is thus one of great geologic complexity, a complexity that produced many types of at least partially tectonically interrelated deposits. Further studies may reveal a large-scale regional order of the kind implied in Figure 6-7 to have prevailed perhaps several times in the Piedmont region. The district and regional zoning is clear, and regional-scale plate-tectonic-related reasons for it are being developed. Such studies of regional zoning have considerable economic value. In the Southern Piedmont, for example, generally favorable sites for the discovery of lead-zinc deposits lie in the area between southwestern Virginia and southwestern Tennessee, as well as north and south of the known limits of mineralization. If there are any mineral deposits in the eastern part of this region, they are buried under Coastal Plain sediments. As geophysical methods are developed and refined, additional ores may be found beneath these layered rocks.

Several companies have maintained exploration offices in Charlottesville, Virginia; Knoxville, Tennessee; and Atlanta, Georgia, to participate in the combined leverage of improved regional geologic understanding and ore-deposit involvement.

▸ AN EXAMPLE OF DISTRICT ZONING— THE CORNWALL, ENGLAND, TIN DEPOSITS

The tin and copper district that runs through Cornwall and Devonshire, England, has long been recognized as one of the classic examples of district zoning. The mines have been ably and amply described by many geologists since the early comprehensive report of de la Beche (1839). Yet in spite of

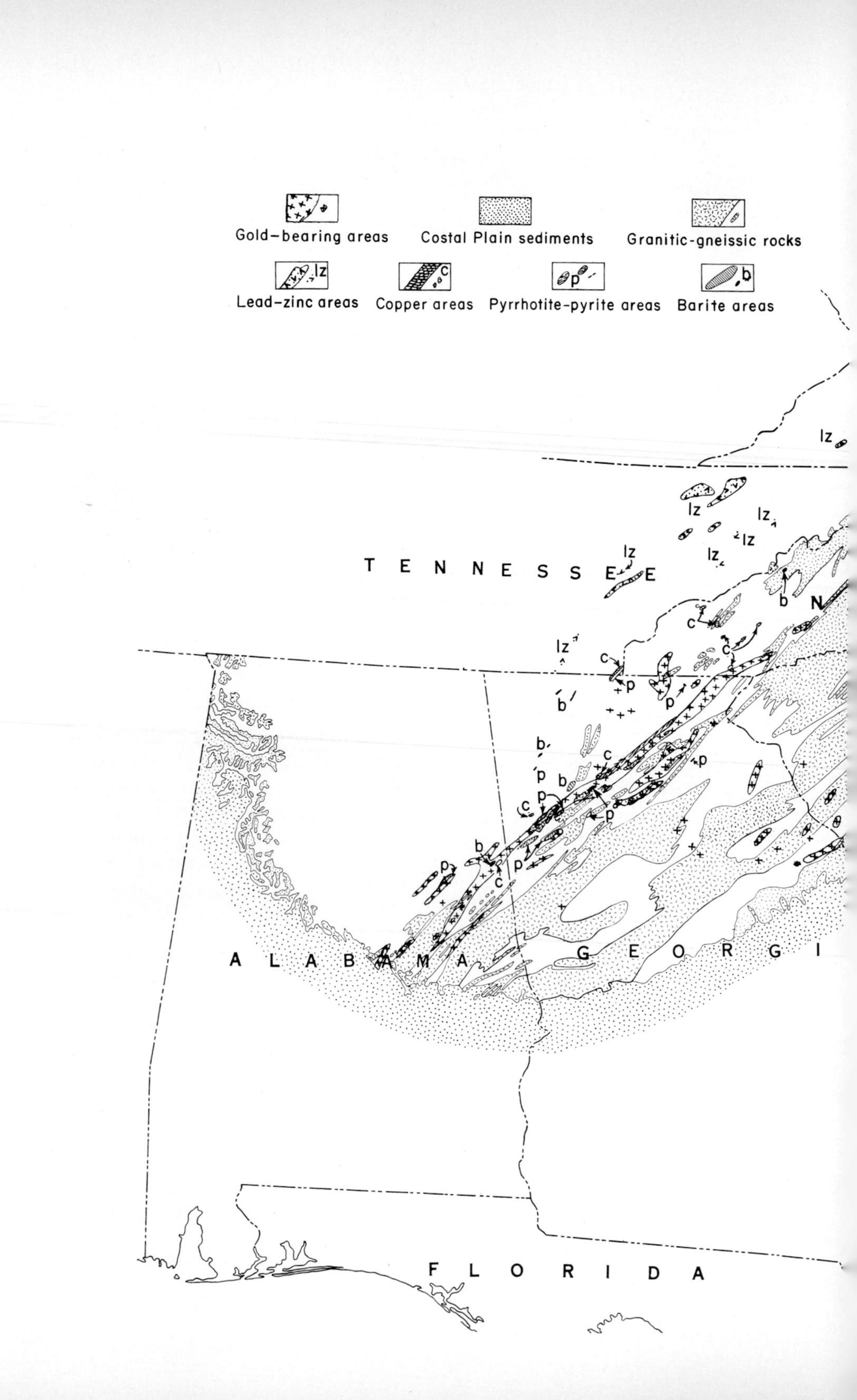
Gold-bearing areas
Costal Plain sediments
Granitic-gneissic rocks
Lead-zinc areas
Copper areas
Pyrrhotite-pyrite areas
Barite areas
TENNESSEE
ALABAMA
GEORGI
FLORIDA

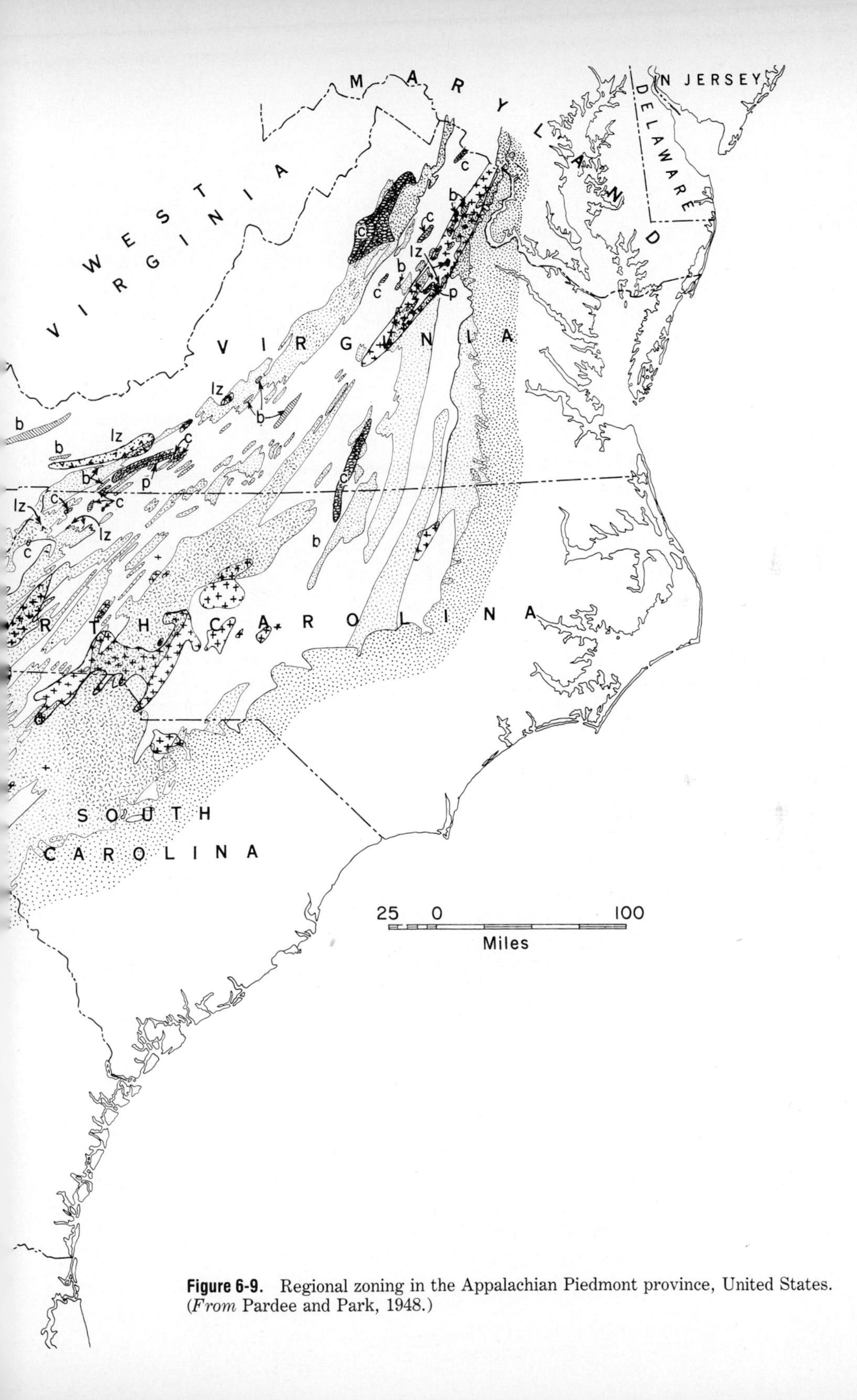

Figure 6-9. Regional zoning in the Appalachian Piedmont province, United States. (*From* Pardee and Park, 1948.)

the voluminous descriptive literature that pertains to the district, surprisingly little had been published concerning the methods of transportation and precipitation of the ores until Jackson's 1979 summary. The generally accepted hypothesis is that zoning resulted from deposition that was largely controlled by declining temperatures and pressures away from a series of intrusive granite stocks. Hosking (1964) and Jackson (1979) state that even though the deposition of a given mineral may have been determined mainly by temperature, it was also dependent upon time relationships, geochemical evolution, and the permeability and nature of the fracture systems through which the fluids moved. The zoning is being restudied and rethought in the 1980s. The patterns described here will not change, but explanations and interpretations almost certainly will. The district has produced well over 2 million tons of tin, 1.3 million tons of copper, 0.35 million tons of lead, 0.25 million tons of arsenic, 0.1 million tons of zinc, 1500 tons of tungsten, and minor amounts of Au, Ag, Sb, U, Ni, Co, Fe, Mn, Bi, Mo, F, and Ba (Jackson, 1979). The region is also well known for its valuable deposits of china clay, a kaolin alteration product that is associated with the regional mineralization. The famous brands of English china—Spode, Lenox, and Minton—owe their reputation for quality to Cornwall clays.

Hosking suggested that two fault zones—running northeast essentially parallel to the long axis of the land mass of southwest England—outlined a broad synclinal area that includes the Cornwall-Devonshire district. Between the fault zones, Paleozoic slates, sandstones, grits, and limestones have been pervasively faulted and folded (Davison, 1921, 1926, 1927; Dewey, 1925, 1935; Hosking and Shrimpton, 1964). Following deformation, a general uplift took place, and the belt between the fault zones was invaded by a two-mica silicic S-type granitic magma of Permian age. Erosion has since revealed five stocklike leucogranite masses and many smaller satellite intrusive granite bodies. It has long been suspected that the five major intrusive outcrops (Figure 6-10) connect at depth. Bott and his coworkers (1958) proved the connection by gravity and other geophysical methods. They showed that the upper surface of the underlying batholith is irregular, that the present outcrops are cupolalike apices of the underlying mass, and that the batholith ascended vertically at the south and then intruded northward, somewhat like a sill. Many pegmatite dikes and small intrusive bodies of quartz porphyry and other felsic rock types were emplaced.

MacAlister (1908) said that the intrusions and their accompanying mineralization developed in three continuous stages: (1) intrusion of granitic magma with accompanying thermal metamorphism of wall rocks near contacts, (2) intrusion of quartz porphyry dikes along shear zones in the metasediments, and (3) deposition of the ores in both sedimentary and igneous host rocks. With various refinements, this statement is valid today. Jackson (1979) subdivided that last event into three mineralization stages: (1) early quartz-wolframite-cassiterite-arsenopyrite-K-feldspar-tourmaline veins, (2) veins and flanking replacements of quartz-cassiterite and sulfides, and (3) late sulfide veins.

Most of the ores have been recovered from complex vein lodes and fissures. These veins cut both the granite and the adjoining slate. As a rule, they dip into the granite, but not always (Figure 6-11); many of the stronger ones trend nearly east-west, parallel to the "grain" of the country rock. In addition to the wide veins in the metasediments, swarms of older greisen-bordered veins occupy the upper hundred meters of the apices of the granite bodies. These veins are ordinary fault zones some of which have been opened repeatedly, and along which the mineralizing fluids appear to have ascended. As the success of many mines has shown, the best ores are commonly found where the vein dips are steepest (Garnett, 1966).

The ores appear in many ways to be genetically associated with the granitic intrusions. Some veins in the slates at substantial distances from the main granite masses contain lenses of granite that intruded along the later mineralized fissures. The presence in veins of topaz and brown-black tourmaline, which are also accessory minerals in the granite, indicates an association between the ores and the igneous rocks. Jones (1931) stated that the granite masses are also to some extent stanniferous, and that as far as could be determined the cassiterite was genetically related to the granites. Cassiterite appears to be as much an indicator of the common parentage of the different igneous outcrop areas as any other mineral common to them. They are all two-mica S-type granites; that they contain tin is a normal association of metal with this rock type.

Hydrothermal alteration products are prominently developed. Five types have been recognized, the first four of which normally occur within the veins and narrowly flanking them: tourmalinization, greisenization, chloritization, silicification, and kaolinization. Tourmaline is ordinarily the brown-black variety called schorl, but in many veins the finer needles are blue. A tourmaline-quartz rock is common as a gangue and alteration assemblage in and beside early cassiterite-bearing veins. In slates that are richer in alumina and lime, axinite and garnet form in place of the tourmaline. Greisenization is restricted to vein envelopes in the granites that are altered to quartz and sericite with minor amounts of topaz, tourmaline, and other minerals. Later chlorite is abundant with some of the ores in both granite and sediments, where it sometimes forms a tough chlorite-quartz rock. Kaolin is broadly developed in feldspars in the granitic rocks that are so thoroughly argillized in places that they form some of the world's largest and purest bodies of kaolinite. Silicification accompanied all other types of alteration; some of the quartz was produced as a by-product of the breakdown of wall-rock minerals, and some was introduced (Hosking, 1951).

Zoning of minerals in the veins and in their alteration products has been recognized at Cornwall for more than a century and a half, and has frequently been used as a tool in exploration. Changes in mineralization are recognized both along vein strike and with depth. According to Hosking (1964), the distribution of major, minor, and trace elements lends strong support to the view that vein minerals were deposited at progressively greater distances from the sources during the major period of ore formation in and around the Permian granites. According to Dines (1933, 1956), any

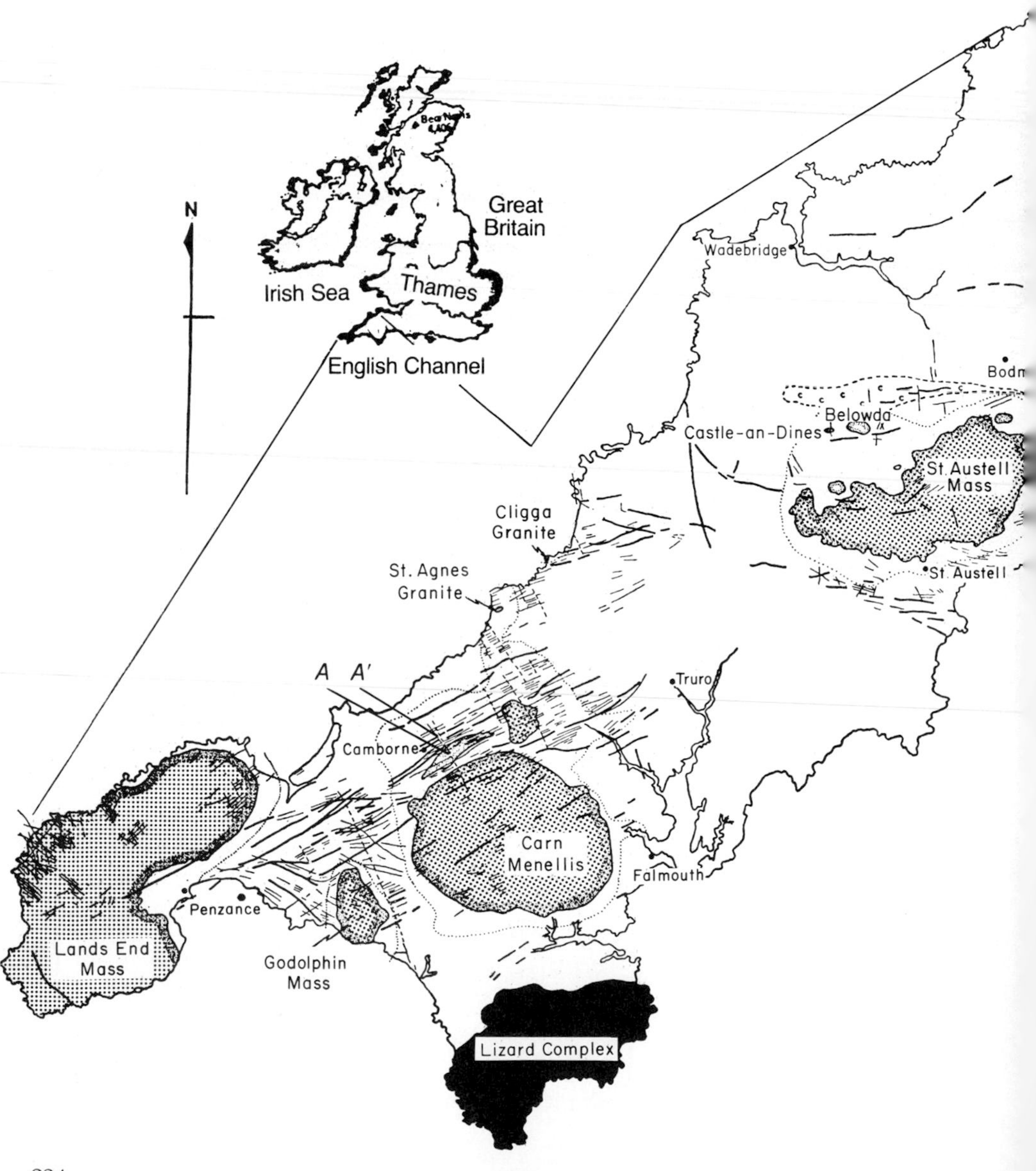

near-surface zone is more extensive than a deeper one because fracturing is more pervasively developed toward the surface.

Four mineralization zones were recognized by Davison (1926). Iron-manganese ores containing small amounts of antimony were deposited farthest from the granitic rocks. Next toward the intrusives are the lead-zinc-silver

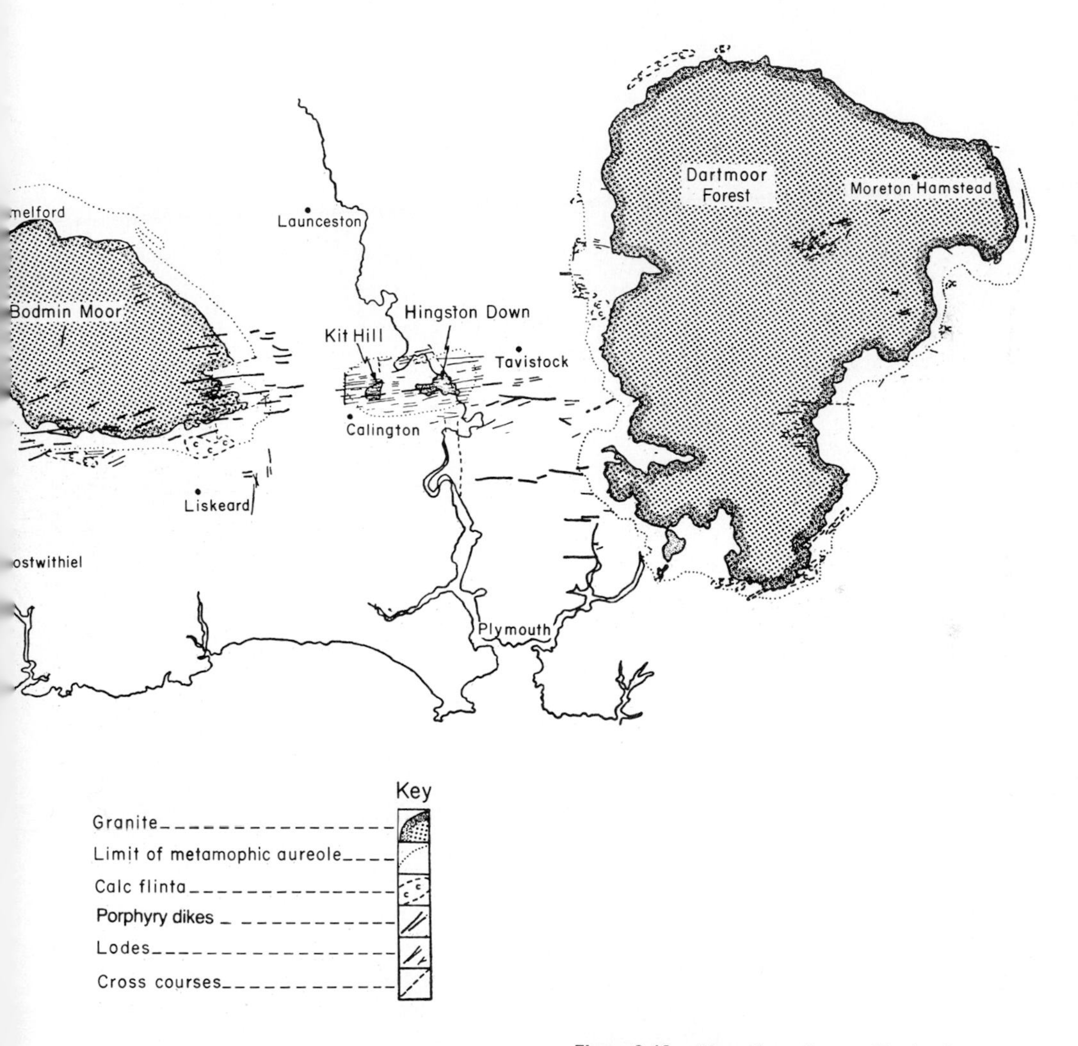

Figure 6-10. Map of southwest England, showing the distribution of the granite outcrops (darkened), metamorphic aureoles (dotted lines), lodes, and dikes. A–A' is the plane of the cross section of the Dolcoath and Roskear mines (now South Crofty) of Figure 6-11. (*From* Hosking, 1964.)

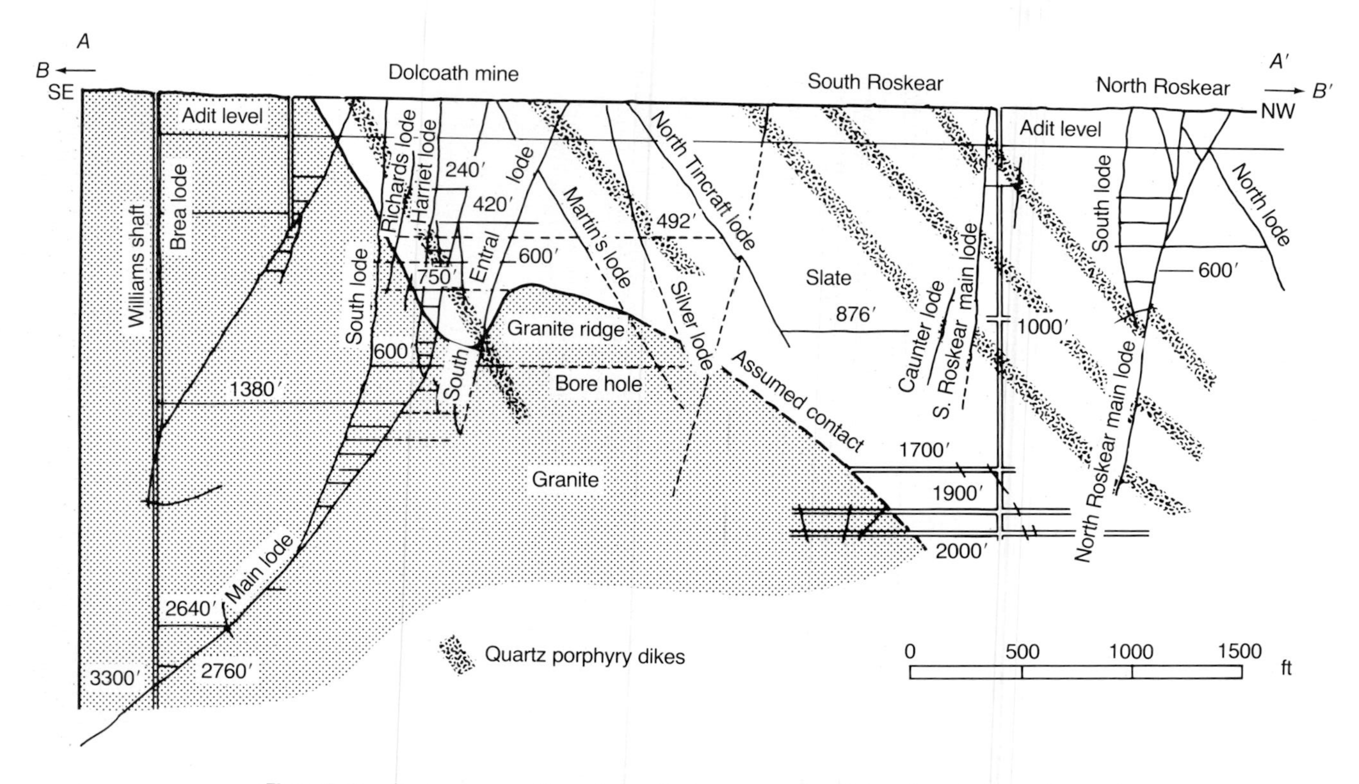

Figure 6-11. Section across the veins in the tin zone of the Dolcoath and the Roskear mines area. The section is southeast-northwest and the veins are seen mostly to dip into the granite, with some dipping away from it. The Main Lode has been the most productive single vein in the province. A–A' is located on Figure 6-10; B–B' refers to Figures 6-12 and 6-13. (*From* Hosking, 1964.)

ores, followed inward by the copper-arsenic-tungsten minerals. Closest to the intrusives—in fact, extending well into them—are the tin ores. As may be seen in Figures 6-12 and 6-13 and in Table 6-2, some of the zones overlap, so that copper and tin, for example, are commonly mined together. In other mines copper and tin are separated by as much as 200 to 300 meters of barren vein material. Before the separation of minerals into zones was understood, the presence of the interval of barren veins led to the abandonment of many mines that produced copper in their upper levels. These mines were reopened later; the barren zones were penetrated, and tin ore was encountered at depth (Table 6-2).

Field data accumulated in recent years indicate that the mineral zones are not precisely parallel to the granite contacts. Deep mining has shown

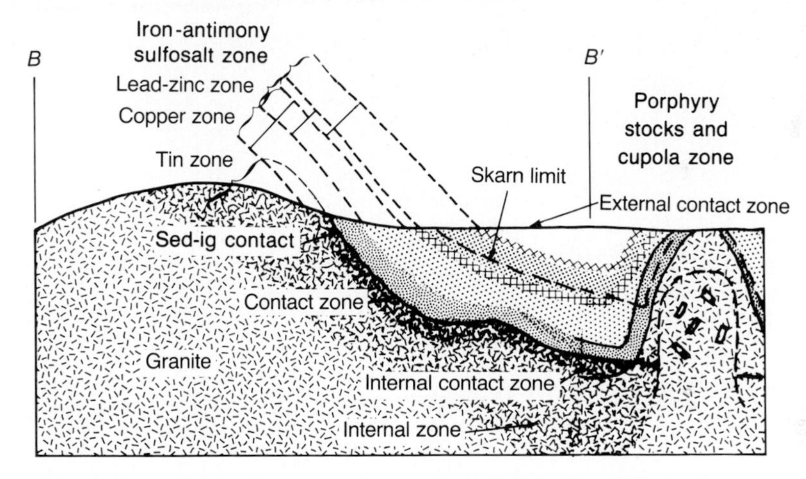

Figure 6-12. Sketch of Cornish mineral zones. The same zones are shown in Figure 6-13. The ore depositional environment zone designations are from Jackson (1979). (*From* Davison, 1926.)

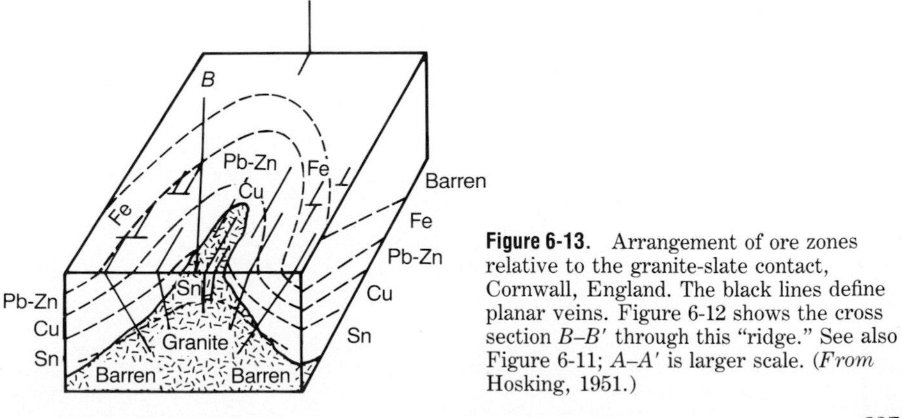

Figure 6-13. Arrangement of ore zones relative to the granite-slate contact, Cornwall, England. The black lines define planar veins. Figure 6-12 shows the cross section *B–B'* through this "ridge." See also Figure 6-11; *A–A'* is larger scale. (*From* Hosking, 1951.)

TABLE 6-2. Paragenesis of the Tin-Tungsten-Copper Deposits of the Cornwall District, England

Gangue	Zone	Ore Minerals	
	Latest minerals 7	Barren (pyrite)	
Iron-antimony sulfosalt zone; Calcite	6	Hematite Stibnite, jamesonite Tetrahedrite, bournonite, pyrargyrite? Siderite Pyrite (marcasite)	Medium- to low-temperature veins, generally at right angles to granite ridges
Chalcedony; Dolomite	5b	Argentite, galena, sphalerite, pyrite	
Lead-zinc zone; Quartz; Barite	5a	Pitchblende, niccolite, smaltite Cobaltite (Native bismuth: bismuthinite?)	
Fluorite	4	Chalcopyrite Sphalerite Wolframite (scheelite) Arsenopyrite Pyrite	High-temperature veins, generally parallel to granite ridges and dikes
Copper zone; Chlorite; Hematite	3	Chalcopyrite (stannite) Wolframite (scheelite) Arsenopyrite Cassiterite (wood tin)	
Tourmaline	2	Wolframite (scheelite) Arsenopyrite (molybdenite?) Cassiterite	
Tin zone	1	Cassiterite, specularite	
	Greisen-bordered veins	Arsenopyrite, stannite Wolframite, cassiterite Molybdenite	Veins often in granite cusps
K-feldspar; Mica	Pegmatites *Earliest minerals*	Arsenopyrite, wolframite Cassiterite Molybdenite	

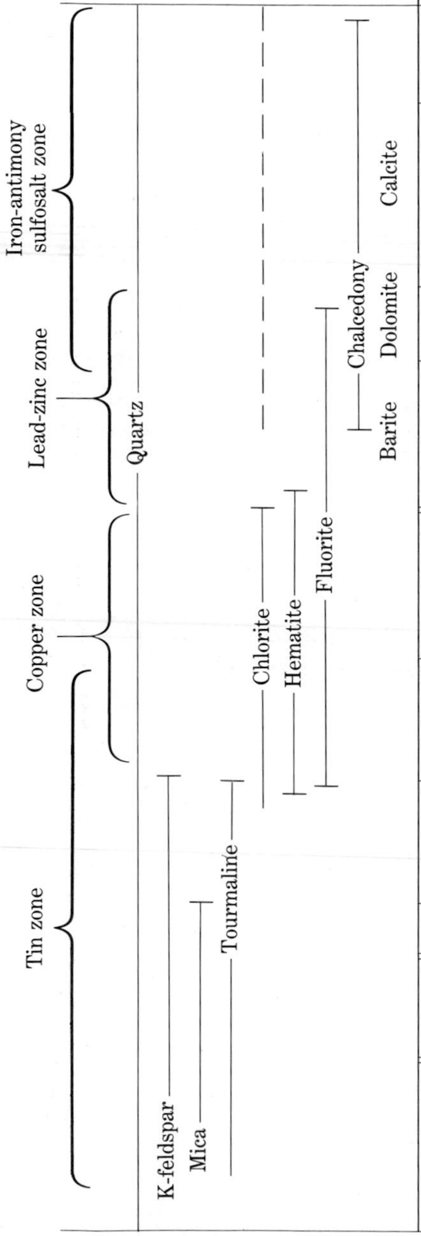

Generalized mineral paragenesis of the mineral deposits of southwest England. Compare these entries with the other Cornwall diagrams and with Table 6-1. The zonal designations refer to Figures 6-12 and 6-13.

Source: Adapted from Park and MacDiarmid (1975)

Economically Important Elements	Composition of Wolframite and Sphalerite	Trace-Element Data	Wall-Rock Alteration
Fe Sb			
Ag, Pb, Zn, U, Ni, Co, Bi		Bi and Sn decrease in galena.	
Sn Cu W As	Fe increases in wolframite and decreases in sphalerite	In, Mn, and Sn decrease and Ge and Ga increase in sphalerite	Greisenized granite Quartz-sericite slate hornfels Tourmalinized granite Tourmaline slate hornfels Chloritized granite Chloritized slate Hematitized granite Hematitized slate Kaolinized granite? Silicified granite Silicified slate Skarn-type rocks derived from greenstones and calc sediments

that the ore zones are related to the granite contacts but are flatter, that is, less steeply inclined (Figure 6-12). Consequently the tin zone is confined to the granite only at the peaks of the stocks; on the flanks it may lie entirely within the metasediments (Dines, 1933; Hosking, 1951). Presumably, the zones represent paleoisogeothermal surfaces, which lay between the ground surface and the inclined granite-slate contact. As we review the four major zones, from inner to outer, in the next paragraphs, check with Table 6-2 and with Figures 6-11, 6-12, and 6-13.

1. *Tin zone.* Tin belongs to the deepest of the recognized zones. It is generally found from depths of about 1300 meters within the granite to a short distance above the granite in the metamorphosed slate. The principal ore mineral is cassiterite, rarely accompanied by minor amounts of other minerals found in the copper-arsenic-tungsten zone. Throughout the district there is a well-defined tendency for the tin content to increase in all lodes at depth, and many deep veins contain cassiterite as the only ore mineral. The character of the gangue also changes. Chlorite, so common in the intermediate zones, is rare at depth; in its place is a fine-grained, dense, blue quartz crowded with tiny black tourmaline crystals. In many mines the blue quartz is brecciated and has been recemented by coarse-grained cassiterite and tourmaline.

2. *Copper zone.* Deposition of the copper-arsenic-tungsten ores took place during and after the formation of the deeper tin minerals. The copper-arsenic-tungsten zone is separable into an upper chalcopyrite-rich region and a lower arsenopyrite-wolframite region. Most of the copper-rich lodes extend from the borders of the granites into the metamorphic aureoles, but not beyond them. Gangue minerals are quartz, chlorite, tourmaline, fluorite, pyrite, muscovite, and jasper. The chlorite is more abundant in the upper zones, and tourmaline is more abundant at depth; small amounts of cassiterite are recognized in the deepest parts of the copper-arsenic-tungsten lodes.

3. *Lead-zinc zone.* The lead-silver parageneses were formed after the tin and copper-bearing lodes and are slightly older than the iron-manganese lodes. Most lead-zinc-silver sulfides and sulfosalts are beyond the outer contacts of the skarn zones, but locally extend into them. Their minerals include argentiferous galena, pyrargyrite, jamesonite, sphalerite, stibnite, bournonite, tetrahedrite, acanthite, pyrite, quartz, dolomite, fluorite, chalcedony, and barite, with minor amounts of uranium, nickel, bismuth, and cobalt as trace elements.

4. *Iron-sulfosalt zone* (Table 6-2; Figures 6-12 and 6-13). The near-surface iron-manganese ores are mostly oxidation products such as limonite, hematite, and manganese oxides developed from the hypogene carbonates siderite and rhodochrosite. A small amount of rhodonite has also been identified. Paragenetic sequence studies indicate that these

minerals were the last of the ores to form. Sulfosalts of Cu-Sb-Ag-Fe-Pb-S are also common. This outer zone is beyond the limits of metamorphism caused by igneous activity and is of only minor economic value.

The Cornwall district appears to be a remarkably consistent example of district-scale zoning, in terms of both spatial relationships (zoning) and temporal relationships (paragenesis, paragenetic sequence), as Table 6-2 indicates. The extent to which Table 6-2 is the S-type-affinity equivalent of the I-type affinity in Table 6-1 is not yet clear, because the metallogenic distinctions have not been established. Cornwall zoning is similar to that described from several European, Chinese, and Australian occurrences. Also, it may well prove true that the zoning and paragenetic relationships are even more complicated than they appear.

▸ AN EXAMPLE OF DISTRICT ZONING— THE TONOPAH, NEVADA, SILVER DISTRICT

The Tonopah district in Nevada (Figure 6-14) is a good example of how subtle the effects of real zoning can be. The Tonopah mines are closely spaced and do not encompass as large an area as the Cornwall district, but variations among them clearly define a district zoning pattern.

Silver was discovered at Tonopah at the beginning of the twentieth century, late enough so that the U.S. Geological Survey was able to witness the development of the district and record the geologic and mineralogic details. Numerous reports were published by some of the foremost geologists in the United States (Spurr, 1905, 1915; Burgess, 1909; Locke, 1912; Bastin and Laney, 1918; Nolan, 1935). From these and later studies, geologists have developed a three-dimensional picture of the mineralization.

The ores at Tonopah were mined principally for silver, though gold is also recovered. Faults provided avenues for mineralizing fluids, and the deposits are mainly replacement veins along these fissures. The host rock is a complex sequence of altered and faulted Tertiary volcanics ranging in composition from rhyolite to trachyte and andesite, including lava flows, dikes, and pyroclastics. The most common ore minerals are acanthite, electrum, polybasite, and pyrargyrite in a gangue of quartz, barite, and a pinkish rhodochrosite or dolomite. Two principal types of alteration have been defined in the district: a central zone of quartz-sericite-K-feldspar and a surrounding envelope of chlorite-carbonate propylite alteration, both of which are superimposed upon early district-wide albitization (Nolan, 1935).

As a result of detailed field studies, Nolan (1935) determined that silver-to-gold ratios vary systematically across the district, as does alteration. Nolan contoured the upper and lower limits of mineralization over the district and found that the productive zone describes a domed shell

(Figure 6-14) that persists without interruption across formation contacts and is only locally modified across faults. Furthermore, higher temperature alteration, represented by quartz-sericite-K-feldspar, coincides with the central part of this domed shell, and chlorite-carbonate propylitic alteration describes a peripheral zone. A plot of silver-to-gold ratios over the contour and alteration map (Figure 6-14) shows relatively low ratios near the top of the dome compared to higher ratios around the outside; silver appears to have traveled farther than gold. The symmetrical pattern of Nolan's compilation, along with the zoning of alteration products and the distribution of silver-to-gold ratios, led him to suggest that the lower and upper surfaces of the productive zone represent approximate isothermal lines and temperature limits of ore deposition, apparently reflecting a deeper hydrothermal source. This hypothesis is supported by the fact that deviations from the symmetrical dome correspond to major fault zones, along which the mineralizing fluids could ascend most readily. Recent work with oxygen isotopes on rocks and ore-bearing materials of the Tonopah district indicate that hydrothermal alteration and mineralization were both produced by heated groundwaters that have undergone very little or no ^{18}O isotope enrichment by the addition of magmatic waters (Taylor, 1974).

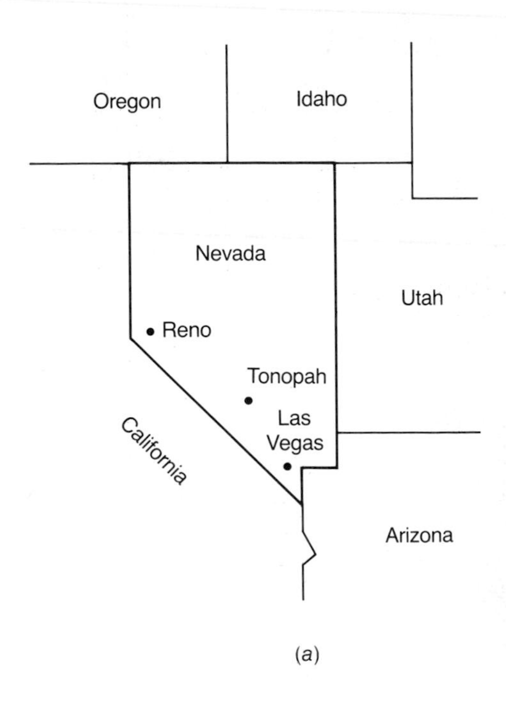

Figure 6-14. (*a*) Location map, (*b*) (*facing page*) contour map, and (*c*) (*facing page*) cross section along *A–A′* of the productive zone at Tonopah, Nevada. Solid lines are contours (in feet) on the upper surface of the productive zone; broken lines represent the lower surface. Crosshatching represents the quartz-sericite-adularia alteration zone; elsewhere the dominant alteration is propylitic chlorite-carbonate. The larger numbers are ratio of silver to gold recorded for all mines that have produced over 100,000 oz (3500 kg) of silver. The smooth progression of silver to gold—with higher peripheral silver values—is an excellent example of district zoning. (*From* Nolan, 1935.)

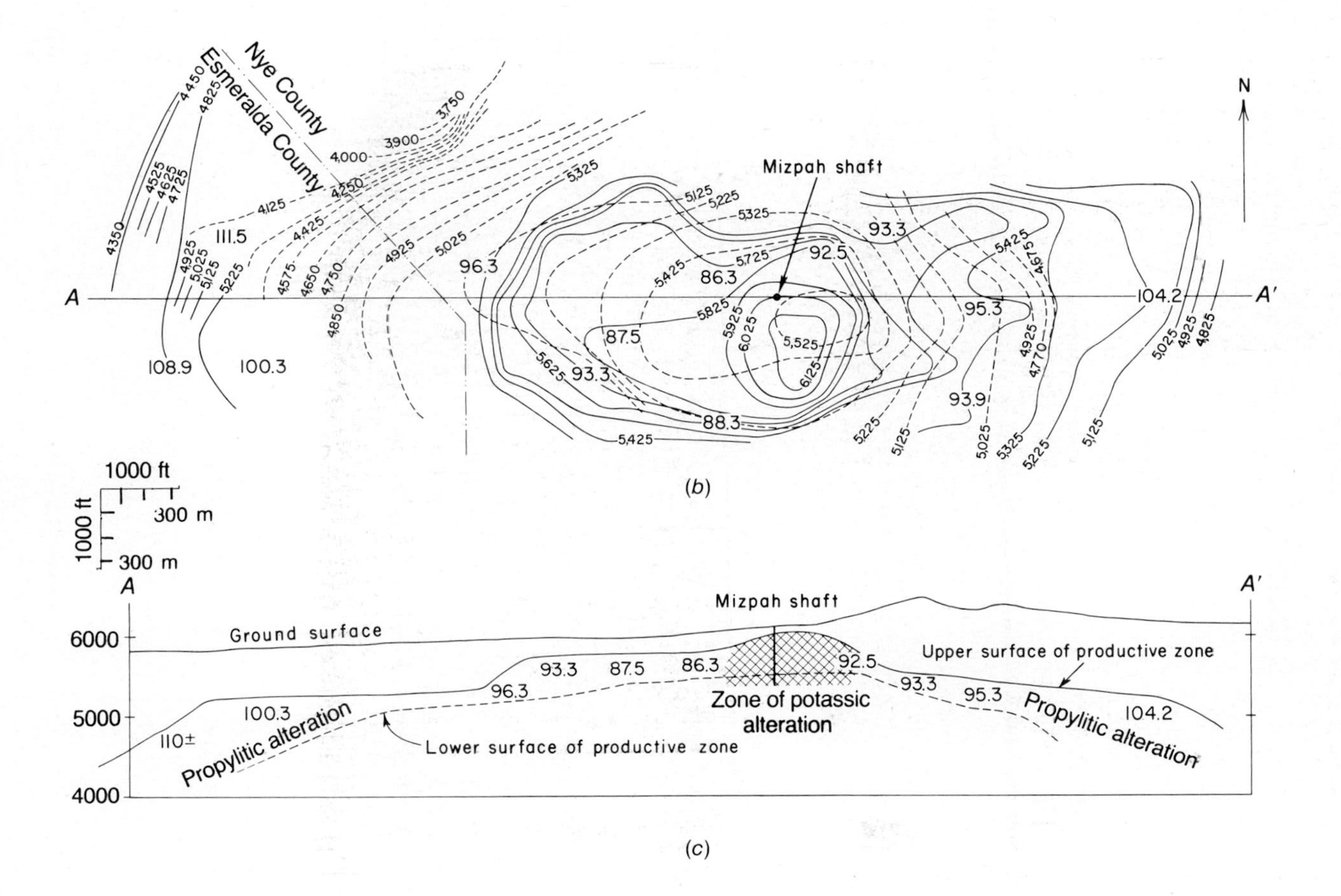

N
A
A'
Nye County
Esmeralda County
Mizpah shaft
(b)
1000 ft
1000 ft
300 m
300 m
A
A'
Mizpah shaft
Ground surface
Upper surface of productive zone
Zone of potassic alteration
Lower surface of productive zone
Propylitic alteration
Propylitic alteration
6000
5000
4000
(c)

▶ AN EXAMPLE OF OREBODY ZONING— THE RED MOUNTAIN, COLORADO, SILVER-LEAD MINE

Some of the best examples of orebody zoning are among deposits formed at shallow depths, where temperatures and pressures apparently changed rapidly. Many of these deposits are in regions of comparatively recent volcanic activity. They are typically of Tertiary age and, in general, were formed at depths only slightly below their present outcrop levels.

The Red Mountain district, in the Silverton volcanic series of the San Juan Mountains between Ouray and Silverton, Colorado, contains excellent examples of orebody zoning. The volcanic rocks of Red Mountain are intruded by many plugs and pipes of breccia, porphyritic latite, and porphyritic rhyolite. Fracturing in and around these pipes is extensive, and the area contains many weakly mineralized fissures. Large volumes of the rocks are impregnated with finely divided pyrite and enargite. Hydrothermal alteration products include the advanced argillic assemblage of kaolinite, diaspore, alunite, pyrophyllite, zunyite, and abundant quartz.

Active mining in the Red Mountain district peaked in the last decades of the nineteenth century. Typical orebodies were vertical chimneylike masses of roughly elliptical outline bordered by fractures and fault planes. Within the main brecciated and mineralized pipes there were subsidiary chimneys, fissures, and irregular bodies of ore. Most, but not all, of the chimney orebodies were surrounded by envelopes of silicified wall rock, which in turn were enclosed within highly argillized zones (Ransome, 1901; Burbank, 1941, 1947).

Orebody zoning was especially well developed in the Guston and Yankee Girl mines (Ransome, 1901), although it was recognized in many other properties of the Red Mountain district. Just below weathered outcrop, the main orebody of the Guston mine contained galena and some tetrahedrite-tennantite, which assayed 50 to 60% Pb, over 100 grams (3 to 4 troy oz) per ton of silver, and only a trace of gold. This ore continued to a depth of about 40 meters, where high-grade stromeyerite (CuAgS) abruptly increased silver values. Appreciable amounts of galena were found to a depth of about 100 meters, but below this level lead content gradually diminished. With increasing depth, stromeyerite and ruby silvers associated with pyrite and chalcopyrite characterized the richest ore of the mine. The best and most abundant ore (at 1900's prices) was recovered between the fifth and sixth levels (88 and 115 meters below the surface); select carloads assayed up to 25% (230 kg, 7400 troy oz per short ton!) silver, and the general ore ran 12% Cu and from 3 to 100 grams of gold per ton. At a depth of about 160 meters the orebody was cut by a fault; below the fault the content of silver was appreciably lower and the ore was harder and more compact. Mineralization down to the lowest level, at a depth of over 400 meters, consisted mostly of low-grade silver-bearing pyrite and some bornite, enargite, chalcopyrite, and barite; stromeyerite was uncommon. Some of the best ore in this fault block was found on the ninth level, about 230 meters beneath the surface. It consisted of bornite ore carrying 4.25 to 15.3 kg of silver and 8 to 16 grams of gold per ton, as well as 25 to 50% Cu; chalcopyrite ore with 0.5 to 2.5 kg of silver per ton, up to 35 grams of gold per ton, and 8 to 15% Cu; and massive pyrite that assayed up to 0.5 kg of silver per ton. The most striking changes below the fault were a gradual increase in the ratio of pyrite to copper and silver minerals and an increase in the gold content. Ore on the ninth level (230 meters) and below carried up to 1 kg of free gold per ton; much of the gold occurred with barite. The deepest levels contained large masses of low-grade pyrite with sporadic nodules of bornite, chalcopyrite, and barite.

In general, the zoning pattern was one of gradual and overlapping, telescoping changes downward from galena to stromeyerite, to bornite, to chalcopyrite, and finally to pyrite, as the most prominent mineral in each zone. With depth, lead dropped from 50 or 60% to nil; silver increased to uncommonly high values, then diminished; gold increased gradually to remarkable values, then dropped; and copper peaked at high to intermediate amounts and then declined. These zones did not grade smoothly from one to the next, however, and complications were not uncommon. Pyrite had the greatest vertical range; chalcopyrite extended above bornite but was important only within or below the bornite zone; the upper limit of stromeyerite was abrupt rather than gradational; and the highest values of silver and gold were associated with the greatest concentrations of copper. Nevertheless, vertical orebody zoning in the Guston mine was clearly evident. Although the vertical changes in mineralogy were indisputable, there was lateral zoning as well—between the fifth and sixth levels, the main ore was stromeyerite and galena, laterally enveloped in a low-grade pyrite-chalcopyrite aureole. The zoning is broadly consistent with Emmon's reconstructed vein, with base metals dominated by copper and iron at depth and by lead near the surface, and with silver values above the copper-iron zone. But the overall zoning is clearly complicated by the abrupt physicochemical changes and gradients that must have prevailed.

Bibliography

Barnes, H. L., 1962. Mechanisms of mineral zoning. *Econ. Geol.* 57:30–37.

——, 1979. Solubilities of ore minerals, pp. 404–460 in H. L. Barnes, Ed., *Geochemistry of Hydrothermal Ore Deposits*, 2nd ed. New York: Wiley, 798 pp.

Bastin, E. S., and F. B. Laney, 1918. The genesis of the ores at Tonopah, Nevada. USGS Prof. Pap. 104.

Beche, H. T. de la, 1839. Report on the Geology of Cornwall, Devon, and West Somerset, pp. 283–289 in *Mem. Geol. Surv.* London: Longman, 648 pp.

Borchert, H., 1951. Die Zonengliederung der Mineralparagenesen in der Erdkruste. *Geol. Rundsche.* 39(1):81–94.

Bott, M. H. P., A. A. Day, and D. Masson-Smith, 1958. The geological interpretation of gravity and magnetic surveys in Devon and Cornwall. *Phil. Trans. Roy. Soc. London* 251:161–191.

Brimhall, G. H., Jr., 1979. Lithologic determination of mass transfer mechanisms of multiple-stage porphyry copper mineralization at Butte, Montana; vein formation by hypogene leaching and enrichment of potassium-silicate protore. *Econ. Geol.* 74:556–589.

Burbank, W. S., 1941. Structural control of ore deposition in the Red Mountain, Sneffels, and Telluride districts of the San Juan Mountains, Colorado. *Colo. Sci. Soc. Proc.* 14(5):178–209.

——, 1947. Red Mountain district, Ouray County, in J. W. Vanderbilt, Ed., *Mineral Resources of Colorado.* State of Colo. Miner. Res. Bd.

Burgess, J. A., 1909. The geology of the producing part of the Tonopah mining district. *Econ. Geol.* 4:681–712.

Burnham, C. W., 1959. Metallogenic provinces of the southwestern U.S. and northern Mexico. N. Mex. Bur. Mines Bull. 65, 76 pp.

Collins, J. H., 1902. Notes on the principal lead-bearing lodes of the west of England. *Roy. Geol. Soc. Cornwall Trans.* 12:713.

Cornwall, H. R., H. W. Lakin, H. M. Nakagawa, and H. K. Stager, 1967. Silver and mercury geochemical anomalies in the Comstock, Tonopah, and Silver Reef districts, Nevada-Utah. USGS Prof. Pap. 575-B, pp. B10-B20.

Corsini, F., G. Cortecci, G. Leone, and G. Tanelli, 1980. Sulphur isotope study of the skarn (Cu-Pb-Zn) sulfide deposit of Valle del Temperino, Campiglia Marittima, Tuscany, Italy. *Econ. Geol.* 75:83–96.

Damon, P. E., M. Shafiqullah, and K. F. Clark, 1981. Age trends of igneous activity in relation to metallogenesis in the Southern Cordillera, pp. 137–154 in W. R. Dickinson and W. D. Payne, Eds., *Relations of Tectonics to Ore Deposits in the Southern Cordillera.* Tucson: Ariz. Geol. Soc. Dig. XIV, 288 pp.

Davison, E. H., 1921. The primary zoning of Cornish lodes. *Geol. Mag. London* 58:505–512.

——, 1926. *Handbook of Cornish Geology.* Roy. Geol. Soc. Cornwall, pp. 65–77.

——, 1927. Recent evidence confirming the zonal arrangement of minerals in the Cornish lodes. *Econ. Geol.* 22:475–479.

de Launay, L., 1900. Les variations des filons métallifères en profondeur. *Rev. Gén. Sci. Pures Appl.* 11:575–588.

Dewey, H., 1925. The mineral zones of Cornwall. *Geol. Assoc. Proc.* 36:107–135.

——, 1935. Copper ores of Great Britain, in *Copper Resources of the World*, 16th Int. Geol. Congr.

Dines, H. G., 1933. The lateral extent of ore shoots in the primary depth zones of Cornwall. *Roy. Geol. Soc. Cornwall Trans.* 16:279–296.

——, 1956. *The Metalliferous Mining Region of Southwest England.* Gt. Brit. Geol. Surv. Mem.

Edwards, A. B., 1952. The ore minerals and their textures. *Roy. Soc. N.S.W.J. and Proc.* 85:26–46.

Emmons, W. H., 1936. Hypogene zoning in metalliferous lodes. 16th Int. Geol. Congr. Rept., pt. 1, pp. 417–432.

Garnett, R. H. T., 1966. Relationship between tin content and structure of lodes at Geevor mine, Cornwall. *Inst. Min. Metall. Trans.* 75:B1–B22.

Gustafson, L. B., 1963. Phase equilibria in the system Cu-Fe-As-S. *Econ. Geol.* 58:667–701.

Henwood, W. J., 1843. The metalliferous deposits of Cornwall and Devon. *Roy. Geol. Soc. Cornwall Trans.* 5:213–224.

Holland, H. D., 1967. Gangue minerals in hydrothermal deposits, pp. 382–436 in H. L. Barnes, Ed., *Geochemistry of Hydrothermal Ore Deposits*, New York: Holt, Rinehart and Winston, 670 pp.

—— and S. D. Malinin, 1979. The solubility and occurrence of non-ore minerals, pp. 461–508 in H. L. Barnes, Ed., *Geochemistry of Hydrothermal Ore Deposits*, 2nd ed. New York: Wiley, 798 pp.

Hosking, K. F. G., 1951. Primary ore deposition in Cornwall. *Roy. Geol. Soc. Cornwall Trans.* 18:309–356.

——, 1964. Permo-carboniferous and later primary mineralization of Cornwall and southwest Devon, in K. F. G. Hosking and G. J. Shrimpton, Eds., *Present Views of Some Aspects of the Geology of Cornwall and Devon.* Penzance, Roy. Geol. Soc. Cornwall.

—— and G. J. Shrimpton, Eds., 1964. *Present Views of Some Aspects of the Geology of Cornwall and Devon.* Penzance, Roy. Geol. Soc. Cornwall.

Jackson, N. J., 1979. Geology of the Cornubian tin field: "A review", pp. 209–238 in C. H. Yeap, Ed., *Geology of Tin Deposits.* Kuala Lumpur: Geol. Soc. Malaysia Bull. 11, 392 pp.

James, H. L., 1954. Sedimentary facies of iron-formation. *Econ. Geol.* 49:235–293.

John, E. C., 1978. Mineral zones in the Utah Copper ore body. *Econ. Geol.* 73:1250–1259.

Jones, W. R., 1931. Discussion on "Genesis of ores in relation to petrographic processes." Brit. Assoc. Advan. Sci. Rept. Centenary Mtg., pp. 387–389.

Keith, S. B., 1983. Distribution of fossil metallogenic systems and magma geochemical belts within the Great Basin and vicinity from 145 million years ago to present, pp. 285–286 in *The Role of Heat in the Development of Energy and Mineral Resources in the Northern Basin and Range Province*. Davis, Calif., Geothermal Resources, Spec. Rept. 13.

Kutina, J., 1957. A contribution to the classification of zoning in ore veins. *Univ. Carolina Geol.* 3(3):197–225.

——, Ed., 1963. *Problems of Postmagmatic Ore Deposition, a Symposium,* vol.1. Prague: Geol. Surv. Czech., 588 pp.

——, C. F. Park Jr., and V. I. Smirnov, 1965. On the definition of zoning and on the relation between zoning and paragenesis, in *Symposium on the Problems of Postmagmatic Ore Deposition,* vol. 2,Prague: Czech Acad. Sci.

Langton, J. M., 1973. Ore genesis in the Morenci-Metcalf district. *AIME Trans.* 254:247–256.

Lilley, E. R., 1932. Geology and economics of tin mining in Cornwall, England. AIME Tech. Pub., class I, no. 41, pp. 5–10.

Locke, A., 1912. The geology of the Tonopah mining district. *AIME Trans.* 43:157–166.

MacAlister, D. A., 1908. Geological aspect of the lodes of Cornwall. *Econ. Geol.* 3:363–380.

McKinstry, H., 1963. Mineral assemblages in sulfide ores: the system Cu-Fe-As-S. *Econ. Geol.* 58:483–505.

Meyer, C., E. P. Shea, C. C. Goddard, Jr., and Staff, 1968. Ore deposits at Butte, Montana, pp. 1373–1416 in J. D. Ridge, Ed., *Ore Deposits of the United States 1933/1967,* Graton-Sales Vols. New York: AIME, 1880 pp.

Moore, W. J., and J. T. Nash, 1974. Alteration and fluid inclusion studies of the porphyry copper ore body at Bingham, Utah. *Econ. Geol.* 69:631–645.

Newell, R. A., 1975. Exploration geology and geochemistry of the Tombstone-Charleston area. Cochise County, Arizona. Unpub. Ph.D. dissert., Stanford Univ., 280 pp.

Noble, J. A., 1970. Metal provinces of the western United States. *Geol. Soc. Amer. Bull.* 81:1607–1624.

Nolan, T. B., 1935. The underground geology of the Tonopah mining district, Nevada. Univ. Nev. Bull. 29(5).

Pardee, J. T., and C. F. Park, Jr., 1948. Gold deposits of the Southern Piedmont. USGS Prof. Pap. 213.

Park, C. F., Jr., 1955. The zonal theory of ore deposits. *Econ. Geol. 50th Anniv. Vol.*, pp. 226–248.

—— and R. A. MacDiarmid, 1975. *Ore Deposits,* 3rd ed. New York: Freeman, 530 pp.

Peacock, H., 1948. An outline of the geology of the Bingham district. *Min. Metall.* 29:533–534.

Petrascheck, W. E., 1968. Kontinentalverschiebung und Erzprovinzen. *Mineral. Deposita* 3:56–65.

Phillips, C. H., N. A. Gambell, and D. S. Fountain, 1974. Hydrothermal alteration, mineralization, and zoning in the Ray deposit. *Econ. Geol.* 69:1237–1250.

Pouba, Z., and M. Stemprok, Eds., 1970. *Problems of Hydrothermal Ore Deposition*, IUGS Ser. A, no. 2. Stuttgart, 393 pp.

Pryor, T., 1923. The underground geology of the Kolar gold field. *Inst. Min. Metall. Trans.* 33:95–135.

Ransome, F. L., 1901. Economic geology of the Silverton quadrangle, Colorado. USGS Bull. 182.

Riley, L. B., 1936. Ore-body zoning. *Econ. Geol.* 31:170–184.

Sales, R. H., and C. Meyer, 1949. Results from preliminary studies of vein formation at Butte, Montana. *Econ. Geol.* 44:465–484.

Sampson, E., 1936. Zonal distribution of ore deposits. 16th Int. Geol. Congr. Rept., pt. 1, p. 461.

Schilling, J. H., 1957. The Gabbs magnesite-brucite deposit, Nye County, Nevada, pp. 1607–1622 in J. D. Ridge, Ed., *Ore Deposits of the United States, 1933/1967*, Graton-Sales Vols., New York: AIME, 1880 p.

Schneider-Scherbina, A., 1962. Uber metallogenenetisches Epochen Boliviens und den hybriden Charakter der sogenannten Zinn-Silber Formation. *Geol. Jahrb.* 81:157–170.

Spurr, J. E., 1905. Geology of the Tonopah mining district, Nevada. USGS Prof. Pap. 42.

——, 1907. A theory of ore deposition. *Econ. Geol.* 2:781–795.

——, 1915. Ore deposition at Tonopah, Nevada. *Econ. Geol.* 10:713–769.

——, 1923. *The Ore Magmas*. New York: McGraw-Hill, 915 pp.

Takeuchi, T., and H. Abe, 1970. Regularities of hydrothermal alteration of sulphur and iron-sulphide ore deposits in Japan, pp. 381–383 in Z. Pouba and M. Stemprok, Eds., *Problems of Hydrothermal Ore Deposition*, IUGS Ser. A, no. 2. Stuttgart, 393 pp.

Taylor, H. P., Jr., 1974. The application of oxygen and hydrogen isotope studies to problems of hydrothermal alteration and ore deposition. *Econ. Geol.* 69:843–883.

Taylor, R. G., 1966. Distribution and deposition of cassiterite at South Crofty mine, Cornwall. *Inst. Min. Metall. Trans.* 75:B35–B49.

Waller, G. A., 1904. Report on the Zeehan silver-lead mining field. Tasmania Geol. Surv. Bull., p. 24.

Westra, G., and S. B. Keith, 1981. Classification and genesis of stockwork molybdenum deposits. *Econ. Geol.* 76:844–873.

Worthington, J. E., and I. T. Kiff, 1970. A suggested volcanigenic origin for certain gold deposits in the slate belt of the North Carolina Piedmont. *Econ. Geol.* 65:529–537.

Yan, M. Z., and K. Hu, 1980. Geological characteristics of the Dexing porphyry copper deposits, Jiangxi, China, pp. 197–204 in S. Ishihara and S. Takenouchi, Eds., *Granitic Magmatism and Related Mineralization.* Soc. Min. Geol. Jap., Spec. Issue 8, 247 pp.

SEVEN

Geothermometry, Geobarometry, and Isotope Studies

The temperatures and pressures at which ores are deposited range from very high at depth to atmospheric at the surface. Placer deposits and most chemical-sediment-hosted ores form under atmospheric conditions. Veins, pegmatites, and magmatic segregation deposits may form at depths of as much as several kilometers and at temperatures above 500°C, some of them even above 1000°C. Lindgren used temperature and pressure of formation as the basis of his classification of ore deposits; modifications to it have depended upon the acquisition of massive amounts of precise and accurate temperatures and pressures of formation of individual minerals, parageneses, and paragenetic sequences. As indicated by studies of paragenesis and zoning, mineral assemblages can themselves be approximate indicators of temperatures and pressures of deposition. However, mineral assemblages by themselves are only indicative; in order to establish more precise limits of deposition, more accurate information is needed. Absolute knowledge of the temperature and pressure environment of mineral deposition can only be based upon laboratory studies of ores and accompanying gangue minerals. Several methods of study are in common use (Ingerson, 1955a, 1955b; Kullerud, 1959; Roedder, 1967; Kelly and Turneaure, 1970; Roedder, 1979; Roedder and Bodnar, 1980), but all must be used with great care and caution. Most methods have proven difficult to apply or are of limited usefulness and accuracy, one of the principal difficulties being the correlation of temperature and pressure findings with an established paragenesis. The study of fluid inclusions is proving to be the best method of determining temperatures of deposition. In conjunction with the study of isotopes, it is beginning to provide a clear and cohesive picture of the nature, conditions, and sequencing of fluids at the time of ore deposition.

There are practical as well as theoretical reasons for studying the temperature of ore deposition and the character of the depositing fluids. Aside from the fact that the genesis of a mineral deposit cannot be understood until the conditions of deposition and the natures of the fluids involved are determined, it is important for exploration purposes to know whether the mineralizing fluid was hot or cold, and if hot, whether or not it cooled quickly or slowly. A hydrothermal deposit precipitated at high temperature is more likely to be persistent at depth than one formed at low temperatures near the surface, because if the depositing solutions cooled slowly over gentle gradients, the vertical range of ore may be greater than where the fluids cooled rapidly. As a logical generalization, it might be stated that deep, high-pressure thermal gradients are more gradual than shallow, lower pressure ones. Consequently, compared to shallow deposits, ores deposited under deep-seated, high-pressure conditions will show less change over long distances. Geothermometric information and geobarometric information are both useful.

Similarly, a knowledge of pH, Eh, and other geochemical properties relating to the character of the ore-bearing fluids is helpful in determining environments favorable for deposition. As we shall see, fluid inclusions show evidence of high temperatures and high salinity directly above the mineralized portions of porphyry base-metal deposits consistently enough that examination of those inclusions is an increasingly useful exploration tool. Their value in determining the anatomy of a vein deposit will also become apparent.

▸ GEOTHERMOMETRY

Fluid Inclusion Studies

Fluid inclusions have unquestionably received more study in recent years than any other method for determining the temperature of deposition of ore minerals. The results of such studies are proving to be consistent and reasonable, and therefore are becoming more and more useful. In addition to giving temperature and pressure of deposition, fluid inclusions furnish considerable information as to the chemical character of the fluid. Many ore deposits have recently been subjected to exhaustive studies using fluid inclusion techniques, sometimes in conjunction with isotope studies (Kelly and Turneaure, 1970; Nash and Theodore, 1971; Takenouchi and Imai, 1971; Sillitoe and Sawkins, 1971; Roedder, 1971, 1979; Robinson and Ohmoto, 1973; Grant et al., 1980; Ahmad and Rose, 1980; Wilson et al.,

1980; Eastoe, 1982). Major developers of the fluid inclusion technique have been Edwin Roedder of the U.S.G.S. in Washington-Reston and Bernard Poty and Alain Weisbrod at the Centre de Recherches Pétrographiques et Géochimiques in Nancy, France. So many researchers now apply fluid inclusion techniques that an annual publication of abstracts of papers involving FI, or "flinc" research, as it is also known, not only has proven highly successful but also has grown to several hundred pages per year. *Fluid Inclusion Research*, the proceedings of COFFI (the Commission on Ore-Forming Fluids in Inclusions), a commission of the International Association of the Genesis of Ore Deposits (IAGOD), is edited by Roedder. The theory behind fluid inclusion studies is simple (Newhouse, 1933; Yermakov, 1957; Roedder, 1967; Roedder and Skinner, 1968). As we learn more and more about high-temperature aqueous solution geochemistry, the use of inclusions becomes more and more complex, but the results are increasingly informative and meaningful (Roedder, 1979; Roedder and Bodnar, 1980).

As crystals grow, they may trap some of the gases or liquids from which they crystallize. If an imperfection or dislocation is formed as a quartz or sphalerite crystal grows, a cavity forms which ultimately closes over as growth continues. But if the crystal is bathed in fluid, a microaliquot of that fluid is encapsulated as the inclusion is sealed. It is assumed that vacuoles in the host minerals were formed when the minerals were precipitating and that the contents of these vacuoles were true samples of the original solutions. The fluids in these vacuoles are thought to have been homogeneous before cooling. If the entrapment occurs at elevated temperatures, the aqueous fluid will probably separate into a gas and a liquid at lower temperatures as the fluid contracts and becomes "stretched," or depressurized. If the solubility of solutes also decreases with falling temperature, separate crystals called *daughter minerals* may nucleate and grow within the cavity (Figure 7-1). Clearly, the more abundant and varied the daughter minerals are, the more complex was the fluid. Common daughter minerals (Figure 7-2) include halite, sylvite, hematite, magnetite, anhydrite, and chalcopyrite; several other salts are known. They can be identified by optical-physical properties; by bursting them for scanning electron microscope (Figure 7-3) and X-ray analysis; by laser-Raman spectrometry; or by chemical analysis of the inclusion contents—by washing, crushing, rewashing, and analyzing the second "wash water" for microamounts of solute (Roedder, Ingram, and Hall, 1963). William C. Kelly at the University of Michigan is attempting to measure the potassium in sylvite in fluid inclusions and the argon gas developed from it by radioactive decay since microencapsulation, thereby dating actual ore-gangue formation.

The assumptions are made that the contents of the vacuoles were deposited from the original ore-bearing fluid, that there were no particular surface effects that caused an atypical fluid to be trapped, and that there has been no subsequent leakage either in or out. If these assumptions are met, the fluids as we see them now can offer clues to the composition and

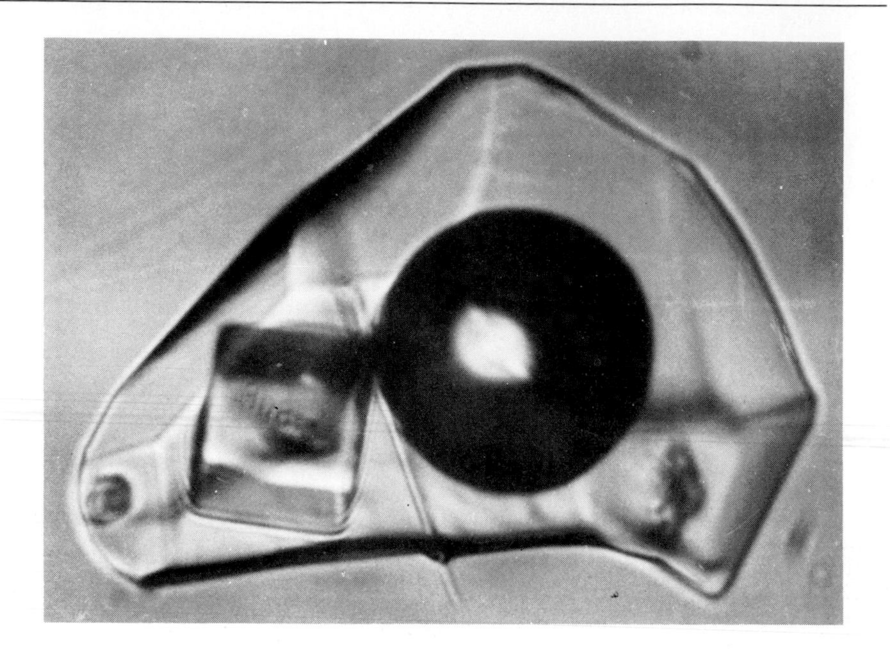

Figure 7-1. Type III (see definitions of types, Figure 7-9) primary fluid inclusion in quartz with a large halite cube, a vapor bubble, and two unidentified daughter salts in the lower corners. The cavity was filled with a homogeneous hydrothermal fluid at the temperature of formation of the quartz. As it cooled in the vein—a cassiterite vein in Bolivia—the vapor and crystals separated and grew. The vapor bubble appears black only because of index of refraction differences. When such fluid inclusions are heated—this one to 430°C—they totally rehomogenize. (*From* Kelly and Turneaure, 1970.)

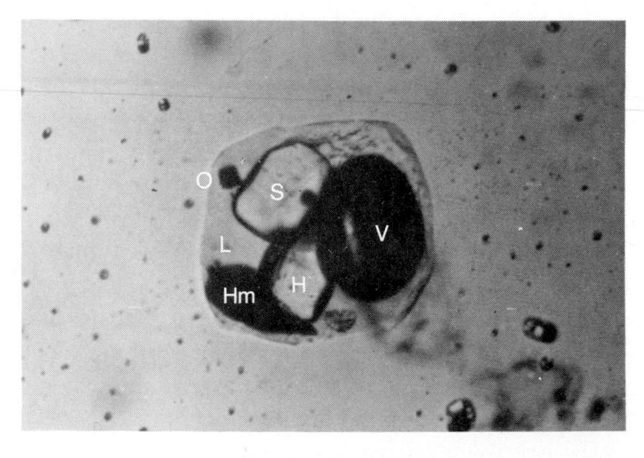

Figure 7-2. Type III inclusion from the Granisle porphyry copper system in British Columbia that is so crowded with daughter minerals that it must have been strongly saline. V = vapor; H = halite; S = sylvite; Hm = hematite; O = opaque, probably magnetite or chalcopyrite; L = liquid. The inclusion is 50 μm across. 600×. (*From* Wilson et al., 1980.)

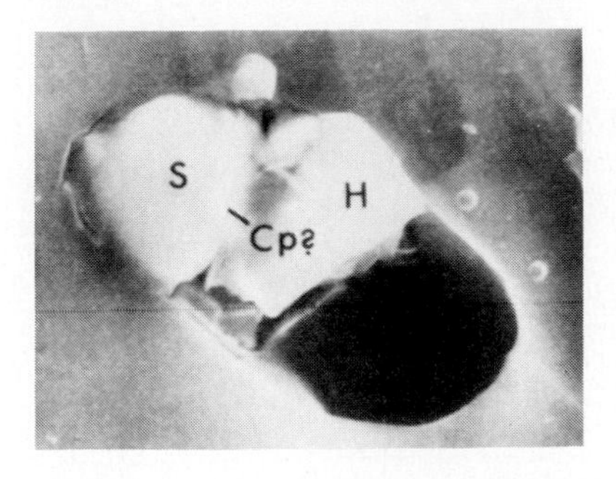

Figure 7-3. SEM (scanning electron microscope) view of an opened fluid inclusion similar to that in Figure 7-2 and from the same location. The cube-form is halite. The vapor and liquid have evaporated, of course, but the shape of the cavity can be seen. Size, scale, and symbols as in Figure 7-2. (*From* Wilson et al., 1980.)

physical state of the original ore solutions. By heating the mineral until the liquid expands to fill the vacuole—that is, until a single phase is restored—the *minimum temperature* at the time of entrapment of the original fluid can be estimated. To determine the composition of the ore-bearing fluids, the minute vacuoles are opened and direct tests are made on the liquid contents. As early as 1932, Newhouse found that sodium and chlorine are the most abundant dissolved substances in liquid inclusions. Many later studies have confirmed this conclusion. If the inclusions truly represent the ore-bearing fluids, as thought, then halides must play the large role in ore genesis that is currently attributed to them. The dissolved salts, as much as 70% of the liquids by weight, are primarily chlorides, sulfates, and carbonates of sodium, potassium, magnesium, and calcium.

Three genetic types of fluid inclusions have been distinguished: *primary*, *pseudosecondary*, and *secondary*. Many careful studies have developed criteria for recognizing the three types, and the dependability of the results of fluid inclusion studies rests directly upon the successful determination of the type of inclusion. Primary inclusions are those dispersed through a mineral with no clear relationship to any structure that would permit the escape or entry of either gas or liquid. They must be known to occur in association with primary crystallization features of the host mineral, such as growth zones, and are thus demonstrably sampling devices of the fluid that bathed the crystal while it grew. Secondary inclusions are those that form by any process after the primary crystallization of the host is complete. For example, if a formed crystal is fractured late in its hydrothermal history, fluid will fill the crack and dissolve and reprecipitate the host until the new surface is minimized and the fracture healed. Inclusions form (Figure 7-4), but they yield information pertinent only to this later hydrothermal stage. The information may be useful, but it does not establish the P-T conditions at the time of the crystal's growth.

Primary inclusions thus form in growing crystals, and "secondaries" at some later time. *Pseudosecondary inclusions* fill the overlap—they *look* secondary, but may not be if the host crystal was fractured essentially as the crystal was growing. "Pseudosecondaries" could be formed in a fracture

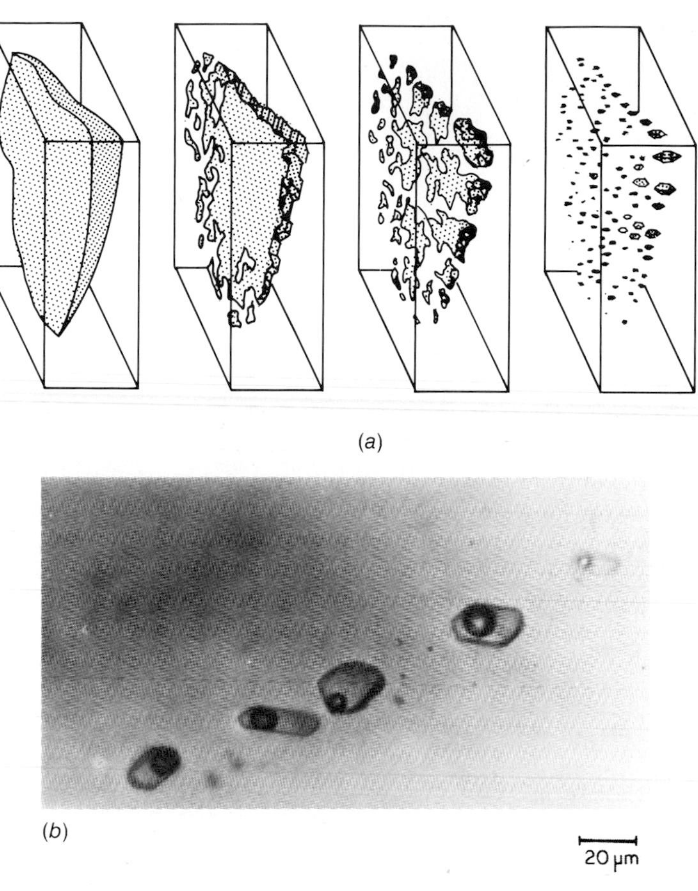

Figure 7-4. Secondary fluid inclusions. (*a*) Progressive healing of a fracture in an already formed quartz crystal. Secondary inclusions can be recognized in part because they lie on planes that cut across primary growth rings. (*From* Roedder, 1979.) (*b*) Planar occurrence of Type III inclusions from a tin porphyry in Bolivia. (*From* Grant et al., 1980.)

in a crystal while "primaries" are growing a millimeter away in a true growth ring. It is often difficult to be sure which type is being dealt with, but it can be done. Roedder (1979) provides a three-page table of textural criteria.

Fluids can, of course, be microencapsulated in both transparent and opaque minerals, but only in transparent or translucent minerals can inclusions be studied optically. Thin wafers up to 1 mm thick are polished on both sides to reduce light scatter, and these "thick sections" can be placed on a heating stage (Figure 7-5) and watched microscopically or by closed-circuit television while they are heated or cooled. Fluid inclusions can be found in almost any milky or bull quartz with any petrographic microscope by focusing into the rock slice of a thin section at moderately high magni-

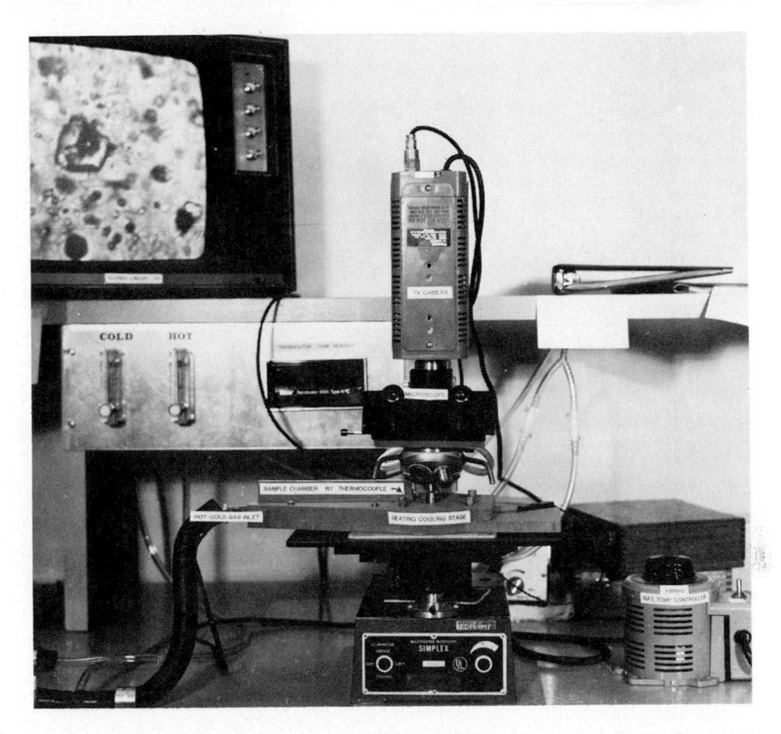

Figure 7-5. Fluid inclusion laboratory in the Laboratories of Economic Geology at the University of Arizona. The Variac (right) controls the temperature of gases that enter the gray heating stage through the insulated tube (left) and bathe the sample chip in the optical path of the microscope. The sample can be watched directly on the closed-circuit television monitor.

fications. They range from infinitesimal to several millimeters; most usable ones are 10 to 100 μm, a hundredth to a tenth the size of the period after this word.

On a "flinc" stage, the wafer and its contained fluid inclusions can be heated or cooled. In a heating experiment, the temperature is raised until the inclusion is restored to a single homogeneous fluid (Figures 7-6 and 7-7). This *filling temperature* is a minimum temperature of formation. It is subject to pressure correction depending upon its density and salinity (Roedder, 1979; Roedder and Bodnar, 1980; Cochran, 1982). The degree of salinity of the fluid in vapor-plus-liquid two-phase inclusions may be determined by placing a chip containing inclusions on a freezing stage, lowering the temperature until the liquid phase freezes, and then observing the melting temperature of ice upon rewarming (Roedder, 1963). The degree of salinity is then calculated from freezing-point depression tables, normally assuming the salinity to be caused by halite (NaCl). Freezing-point depression measurement is not necessary if solid halite is present; the volume of

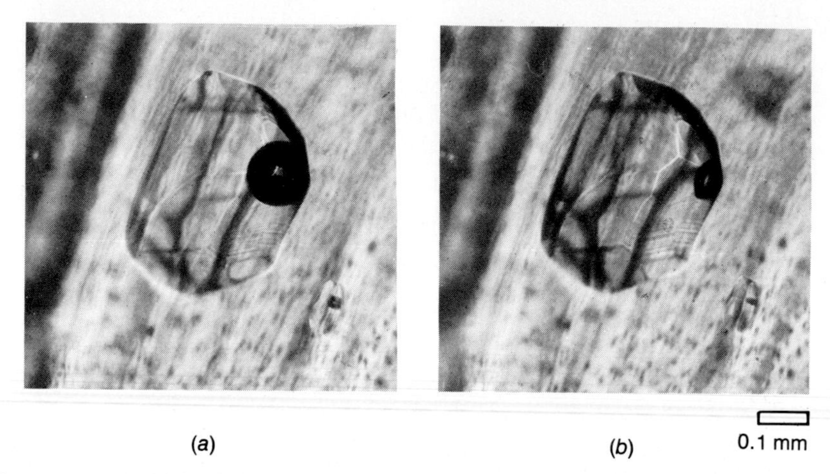

Figure 7-6. Two-phase Type I fluid inclusion (*a*) at room temperature and (*b*) at 130°C. The bubble within the liquid is much smaller at the higher temperature and disappeared completely into the liquid-fluid phase at 140°C, the temperature of homogenization in this case. Fluorite from the Deardorff mine, Cave-in-Rock district, Illinois. 70×. (*Photos by* B. M. Harrison.)

the NaCl cube and of the cavity can be measured and salinity calculated on the basis of the amount of halite and salt-saturated liquid present. Halite is the most widely distributed, abundant, and easily determined of the solid minerals in inclusions. Other phase transitions are observed if carbon dioxide, methane, or other compounds are present, and useful information can be obtained from the relevant temperatures.

Boiling of fluids can be established if primary inclusion populations include two types (Figure 7-8), one that is vapor-rich and homogenizes to a vapor upon heating, another that has a small vapor bubble, which dissolves by homogenizing to a liquid. Clearly, two fluids were trapped, a vapor and a liquid, and the hydrothermal fluid must itself have been two-phase, a condition unique to the boiling curve (Figures 2-9 and 2-10).

Most fluid inclusion investigators follow Nash and Theodore (1971) in describing fluid inclusions in terms of five types (Figure 7-9):

- I liquid with small vapor bubble, no daughter mineral(s)
- II liquid with large vapor bubble, no daughter mineral(s)
- III polyphase essential inclusions with small vapor bubble
- IV two liquids with vapor and daughter mineral(s)
- V liquid CO_2 with vapor

Ahmad and Rose (1980) used a slightly different grouping (Figure 7-9). Several subtypes have been discussed, but Nash and Theodore's classification serves most purposes, partly because their bubble types seem to occur in characteristic positions in exploration systems. For example, Type

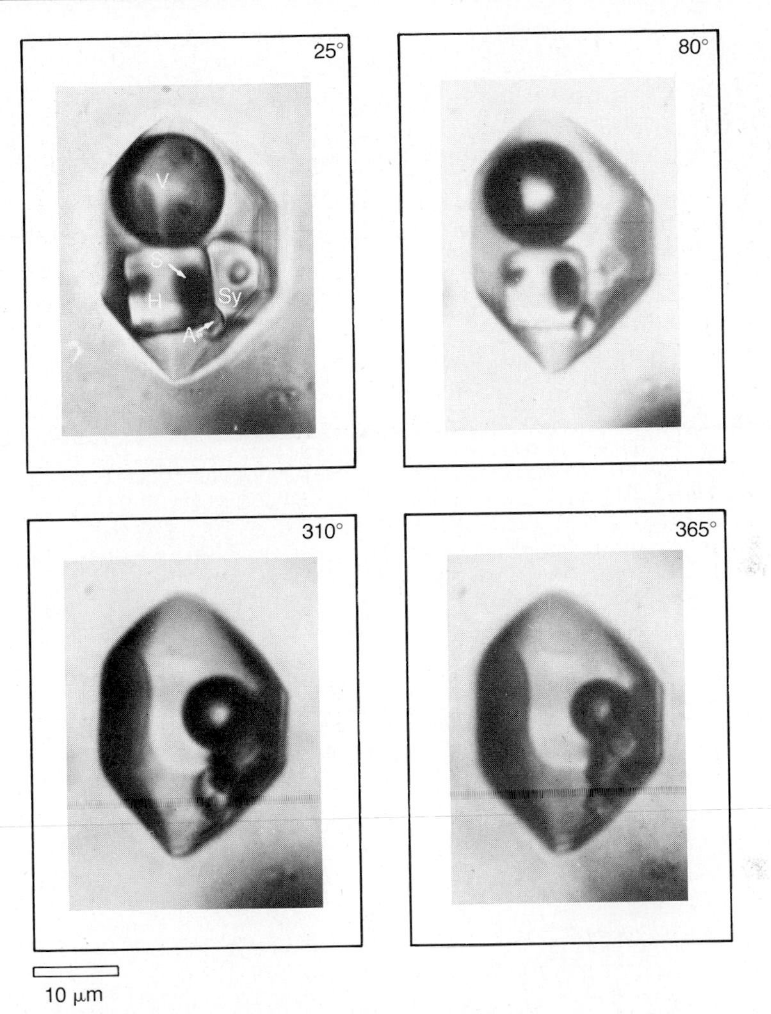

Figure 7-7. Multiphase Type III inclusion from the Bingham Canyon porphyry copper deposit at successively increased temperatures on a heating stage. Notice how the vapor bubble (V) shrinks; it disappeared at 400°C. Sylvite (Sy) dissolved at 80°C; most of the halite (H) is gone at 310°C, but anhydrite (?) (A) and hematite (?) (S) remain to 365°C. Both might dissolve at higher temperature. The inclusion is 35 μm long. (*From* Roedder, 1971.)

III inclusions are most common in immediate association with porphyry and Cordilleran Vein mineralization.

Although skepticism concerning the value of fluid inclusion studies is still expressed, so much work has been done with such consistent and reasonable results that most scientists now accept the results. Ratios of liquids to gases and solids are ordinarily recorded during observation, and these ratios are found to be comparatively constant through wide ranges of inclusions in a given deposit. This constancy would not prevail if leakage took place, in which case a wide variation in composition would be expected.

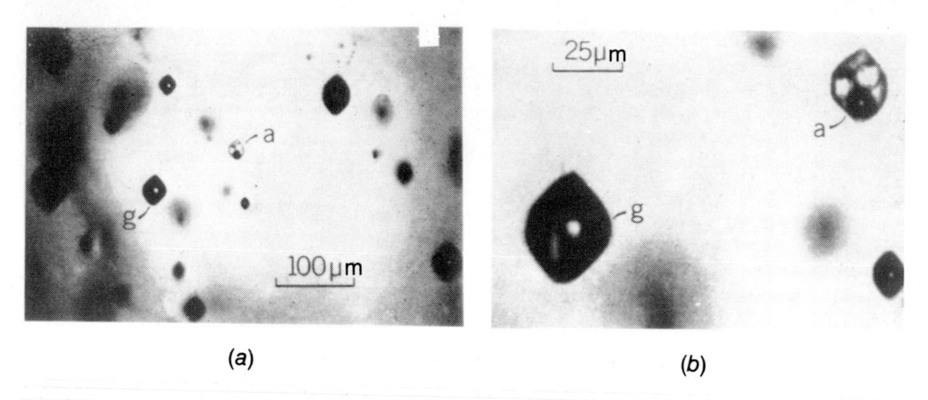

Figure 7-8. Type II and Type III inclusions in quartz from Oruro, Bolivia, showing almost certain boiling conditions. Both types of inclusions homogenize between 464 and 470°C. One is gas-rich (g) and fills to vapor upon heating, the other (a) is liquid- and crystal-rich and fills to a liquid. (*b*) Enlargement of (*a*). (*From* Kelly and Turneaure, 1970.)

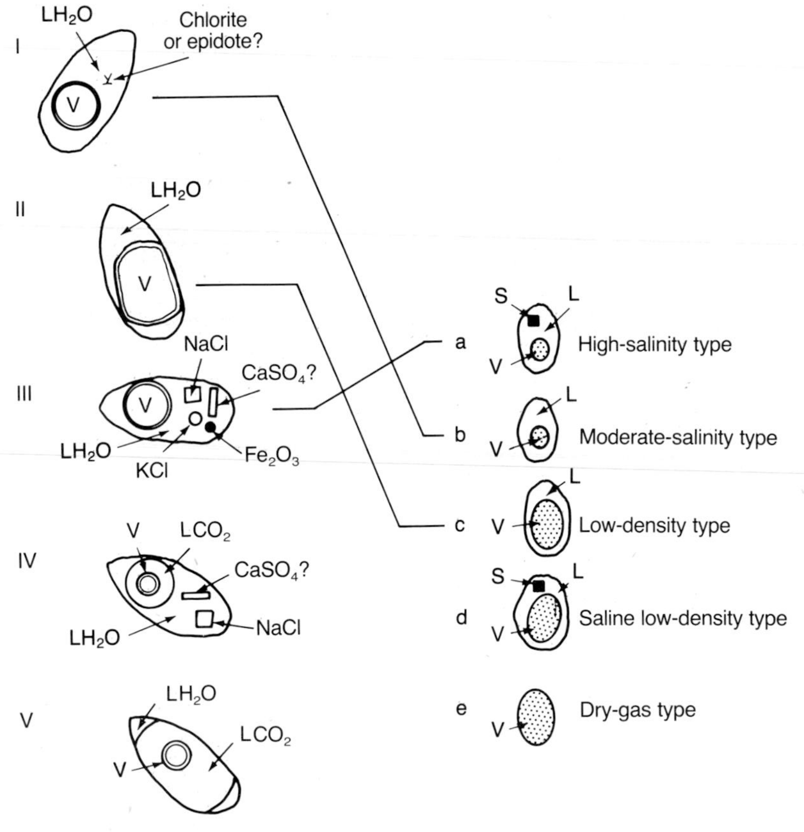

Figure 7-9. Sketches of the most important types of fluid inclusion: Types I to V from Nash and Theodore (1971) and Types *a* to *e* from Ahmad and Rose (1980), with tie lines connecting pairs where applicable. L = liquid; V = vapor.

Secondary inclusions and rare leakage can ordinarily be detected by careful study (Smith, 1954; Smith et al., 1950; Ingerson, 1955a; Richter and Ingerson, 1954; Roedder, 1960; Roedder and Skinner, 1968).

The decrepitation method applied to fluid inclusions was formerly used widely, but has now been largely abandoned. The principle of this method was to heat a wafer or chip containing inclusions until the vacuoles exploded, or *decrepitated*. This technique, monitored by microphones, was supposed to give an idea of the upper limit of the temperature of formation, but determinations are unreliable because the temperature of decrepitation depends to a great extent upon the strength of the containing mineral.

So many studies involving the successful application of fluid inclusion techniques are included in Chapters 11 to 20 that no specific examples need be given here. The enormous leverage of this technique is of course in combination with other approaches. Fluid inclusion data collected and evaluated in the context of successive generations of mineralized and altered fractures, both in a temporal, paragenetic sequence sense and with regard to distance from a center of hydrothermal activity, can reveal crucial information about the P-T-X evolution of fluids through space and time that can be gathered in no other way. Research by Richard E. Beane, Spencer R. Titley, and their University of Arizona students Preece, Reynolds, and Haynes (Chapter 11) has been productive in terms of elucidating fracture propagation and fluid evolution in porphyry copper systems; William C. Kelly and Stephen Kesler and their group at the University of Michigan have been productive in a variety of geologic systems; and Edwin Roedder and his colleagues continue to earn the attention of geologists worldwide for their overall contributions to the understanding of ore-forming processes. The discipline is established, but is in a stage of amazingly productive infancy. Electron probe studies of inclusion contents, and even of potassium-argon dating of specific inclusions and thus of ore assemblages by virtue of the potassium in their brines, are being evaluated. The future for fluid inclusion studies is bright for both scientific and exploration returns.

Other Methods

There are many other ways of measuring, directly or indirectly, the temperatures of formation of mineralization-alteration systems. Most of them are either imprecise or difficult to apply, or both, but some may be brought to bear on a particular problem. They were brought together in previous editions of this text, and the third edition (Park and McDiarmid, 1975) remains an excellent reference. These techniques include the following.

Synthesis of Minerals The preparation of phase diagrams by synthetic melting or hydrothermal studies of pertinent assemblages is carried out by placing measured amounts of elements in controlled atmospheres to produce quantifiable assemblages of reaction products. These results can be compared with natural systems, the latter almost always being far more complex. In many instances the resulting products are similar to the natural minerals. But unfortunately, natural minerals are rarely chemically pure substances; their structures contain small amounts of nonessential, substitutional elements, the presence of which almost certainly has influenced their conditions of formation. Interpreting the results of synthetic studies is further complicated by the fact that the conditions of mineral genesis in nature depend in part upon the presence of fugitive constituents such as water, and the presence of the fluxes or other compounds that may never appear as components of the synthetic studies. Consequently most geologists are reluctant to depend on experimental work based on synthetic minerals, especially if the minerals were formed in simple systems.

Determination of Melting Points The temperature at which a mineral melts, normally at atmospheric pressure, is assumed to mark the upper limit of stability for that mineral (Birch et al., 1942). Actually, melting points are helpful in establishing the maximum temperatures of mineral genesis, but they are of little value in determining the actual conditions of formation. When the melting point of a mineral is low, it can be a meaningful clue to the temperature of formation; when it is high, it may be hundreds of degrees higher than the mineral's temperature of formation in nature. The presence of small amounts of extraneous elements in solid solutions alters the behavior of both synthetic and natural crystals. The melting temperature of individual minerals is even less meaningful than the melting temperature of combinations of minerals. Mineral combinations melt at lower temperatures than any of their component minerals. The *eutectic point* is the minimum melting point for a mineral pair or trio, and is useful at magmatic temperatures but cannot be successfully applied to hydrothermal systems.

Determination of Inversion Points and Stability Ranges An extension of the determination of melting points is the study of multicomponent systems such as the Cu-Fe-As-S system mentioned earlier (Figures 6-1 and 6-2). Also, some minerals undergo internal crystallographic changes at definite temperatures and pressures. Such changes or inversions—such as from high quartz to low quartz at 573°C—are commonly recognizable and have been used in the study of ores (Ramdohr, 1931). The temperature or the combination of temperature and pressure that causes the inversion from one polymorph to another should mark the minimum and maximum conditions of formation. The inversion temperatures for some polymorphous minerals are well established or vary within predictable limits, but the stability ranges for many polymorphs are poorly known and may be influenced by impurities (Buerger, 1961). Inconsistencies in laboratory results suggest that inversion temperatures in nature may be kinetically sluggish, that is, metastable. Inversion points also depend upon pressure since polymorphous inversions involve density through lattice-volume changes. In fact, Kelly and Turneaure (1970) showed that inversion temperatures are of little practical value in geothermometry. For example, tremendous effort has been expended to determine the fact that Cu_2S has three forms: cubic above 430 ± 10°C, hexagonal between 430 and 102°C, and orthorhombic below 102°C (Kullerud, 1964; Kullerud and Yund, 1960; Miromoto and Kullerud, 1966). However, the use of Cu-S minerals as geothermometers is not as reliable as was thought because of impurities and variations in metal and volatile ratios in natural systems, and many past evaluations are in doubt. Chalcopyrite inverts from the isometric form to the tetragonal form at 525°C, with a slight subtractive pressure correction. Twinned chalcopyrite usually results from this inversion and can be assumed, with fair confidence, to represent high-temperature deposition (Kullerud, 1959), but this twinning is difficult to detect.

The nonmetallic minerals have also been used to determine inversion temperatures. Quartz has been studied as much as any gangue mineral. Upon cooling, it changes at 573°C from the high-temperature "high-quartz" form to the low-temperature "low-quartz" form, with no change in the habit of the original polymorph. The inversion temperature can be lowered by the chemical environment (Keith and Tuttle, 1952), but with reservation it can be used to estimate the temperatures at which pegmatites and high-temperature veins are deposited. Above 870°C the stable form of silica is tridymite, and above 1470°C it is cristobalite. But these transition temperatures are also pressure dependent, and the presence of water vapor changes the entire configuration.

All in all, stability ranges and inversion temperatures of individual minerals are so subject to their environment, to solid solution effects, and to fluxing—freezing-point depression—by water and its solutes as to be of little real geothermometric use.

Determination of Exsolution Points As mentioned earlier, high temperatures tend to promote disorder in mineral structures. One of the results is that nonessential elements are more readily incorporated and retained as minerals grow at high temperatures. Upon cooling and with the development of more ordered structures, the extraneous materials can be forced out. The expelled phases tend to accumulate as small blebs and blades along cleavage surfaces or crystallographic planes. This development of intergrowth structures is called *exsolution* or unmixing, and has been used to a limited extent to indicate temperature points on the geologic thermometer as applied to ore deposits. Unfortunately, wide variations in results have been reported, and determinations based upon exsolution phenomena must be viewed with skepticism.

According to Brett (1964), exsolution is effected by nucleation and subsequent growth of exsolved particles in previously homogeneous phases. The mechanism is complex and is influenced by temperature, pressure, concentration, purity, structure, and grain size of the solid solution, the coefficients of diffusion, and stress distribution within the crystal. Lamellar, lens-shaped, and other varied types of myrmekitic, mutual-boundary, and rimming textures have all been ascribed to exsolution processes. Of these, the lamellar textures are considered by Brett to be the most indicative of exsolution. Some phase studies, for example in the Cu-Fe-Ni-S and the Au-Ag-Te systems, have been more constraining than others, but by and large the precise assignment of particular exsolution textures and processes to specific temperatures has been inexact. Exsolution in sulfide minerals is influenced by sulfur vapor pressure, and as pointed out by Kelly and Turneaure (1970), wide variations in apparent temperatures of exsolution of sulfides are possibly the result of failure to consider the effects of geologic time.

Studies of Mineral Textures and Habits Mineral textures and habits have been used by Edwards (1954) to indicate the temperatures of mineral deposition. He likened these textures to similar textures found in smelter products, and pointed out that they disappear upon annealing or recrystallization. As useful as they are for determining reaction processes and sequences, too few specific textures are so geothermometrically constrained as to be useful.

Determination of Electrical Conductivity of Minerals Assuming that minerals formed at high temperature have fewer structural defects than those formed at low temperature, and that crystallographic imperfections reduce electrical conductivity, Smith (1947) concluded that the relative electrical properties of conductive minerals should be a measure of their crystallization temperatures. Accordingly, Smith calibrated pyrite as a geothermometer by measuring its conductivity-resistivity characteristics over the hydrothermal temperature range. Applications of this technique have given inconsistent and conflicting results, suggesting that the perfection of crystal growth does not depend upon temperature alone (Smith, 1948; Ingerson, 1955a; Lovering, 1958). Future studies may resolve the discrepancies and establish the electrical properties of pyrite, or some other common sulfide, as reliable geologic thermometers, but appropriate studies are not now being carried out.

Thermoluminescence *Thermoluminescence* is the property of a substance to emit visible light when it is heated; it is also known as "frozen-in phosphorescence." Visible-wavelength light is produced when energy is released as electrons escape from high-energy lattice positions and "fall" into more stable, lower energy positions or "holes." Different minerals will give characteristic "glow curves," which show the intensity and the spectral distribution of luminescence as a function of temperature, and are recorded as a sample is heated slowly while it is "watched" by a photomultiplier tube. Once a sample has been heated, its thermoluminescence is drained permanently because the electrons will remain at the lower energy levels; the sample will not glow again when reheated, unless it has been reenergized by radioactive bombardment (Daniels et al., 1953; Ingerson, 1955b). Lovering (1958) found that the limestones at Gilman, Colorado, gave glow peaks at 235 and 330°C. The low-temperature peak was generally absent from dolomitized materials, but the 330°C peak was preserved everywhere except close to the ore. The thermoluminescence data there seem to indicate that the wall rocks were heated above 235°C but not above 330°C—figures that agree with temperatures determined by fluid inclusion thermometry.

Actually, the thermoluminescence of a sample is destroyed even if the sample is heated to a temperature below that at which maximum luminescence is produced. Consequently the absence of the 235°C peak at Gilman need not imply that the wall rocks were heated above 235°C; this peak could have been removed at temperatures as low as 150°C (MacDiarmid, 1963). Complications at Gilman, even where geology was well understood, were so great that thermoluminescence was abandoned as a thermometer. The technique is useful in determining the distribution of uranium in certain deposits, but has little geothermometric promise in the 1980s.

An Example of Applied Geothermometry

The tin and tungsten deposits of the eastern Andes of Bolivia have been studied for many years by Turneaure, Kelly, and their associates (Turneaure, 1935, 1960, 1971; Turneaure and Welker, 1947; Kelly and Turneaure, 1970). Most of the metals produced in the region have come from vein systems associated with porphyritic stocks of intermediate composition or with altered sedimentary rocks. In addition to the wide veins and vein systems, many of which are complex, many narrow veins, sheeted zones, stockworks, and breccia lenses have been productive.

Hydrothermal alteration of the vein walls accompanied emplacement of the ores in the veins. Chlorite, sericite, kaolinite, alunite, and quartz are the most common gangue and alteration products; tourmaline and quartz occur both in tungsten and silver-tin veins and in the wall rocks. Ore minerals include cassiterite, bismuthinite, native bismuth, pyrrhotite, stannite, marcasite, pyrite, siderite, wolframite, franckeite, teallite, and many others.

Field mapping indicated that mineralization involved several parageneses and that an unusual range of temperatures was involved. Temperatures were thought to vary from 400 to 500°C in the early stages to about 100°C during the later stages. The deposits were also described by early workers as having formed in a near-surface environment characterized by rapid changes in both temperature and pressure.

In order to check the character of the ore-bearing fluids and the temperature of deposition of the ores against the background of extensive field work, a careful laboratory study of the minerals of tin and tungsten from the eastern Andes of Bolivia, and the gangue minerals associated with them, was undertaken by Kelly and Turneaure (1970). Turneaure—"Mr. Tin"—was a professor of economic geology at the University of Michigan, Kelly his Ph.D. graduate student and colleague. Many methods of study were intentionally attempted, and the results show clearly both the useful contributions that can be made by thorough studies of fluid inclusions and the weaknesses of other methods.

The Bolivian tin and tungsten ores contain at least minor amounts of most of the sulfide minerals that have been used previously as geothermometers. Kelly and Turneaure applied several of the standard techniques in an attempt to evaluate temperature as a factor in zoning and paragenesis. After a great deal of effort, they decided that most of the methods attempted did not reflect initial depositional temperatures and pressures. For example, sphalerite that displayed polytypism (sphalerite-wurtzite) could not be correlated with variables of geologic interest. The iron content of the sphalerite reflected equilibrium with pyrrhotite or pyrite, but the significance of this variation in terms of the temperature was questionable.

Other geothermometers, such as sulfide invariant points, high-low quartz inversion point, melting of native bismuth, and sulfide exsolution temperatures, were closely examined. These methods were unsatisfactory when used alone, but they did serve to check results from the fluid-inclusion studies, which were by far the most productive and reliable of the methods tried.

The studies of fluid inclusions showed highly systematic trends of salinity and depositional temperatures during formation of the tin and tungsten ores (Figure 7-10). Tin-bearing fluids of the early stage were complex sodium chloride brines with low carbon dioxide content. Salinities reached values as high as 46 wt% during precipitation of cassiterite and early quartz, but the later, cooler ore fluids were more dilute and gradually approached fresh water by meteoric fluid dilution in the closing stages of mineralization. Temperatures of deposition increased from about 300 to 530°C in the early stages but later declined to less than 70°C. The bulk of the sulfide minerals were deposited as temperatures declined from about 400 to 260°C. The hypogene mineral sequence is attributed chiefly to the cooling of vein fluids, but little is known of the wall-rock alteration and its influence on paragenesis. Boiling in the early vein fluids was indicated by the fluid inclusions in quartz and cassiterite (Figure 7-8); it increased the salt content and forced CO_2 into the vapor phase. Early boiling favored precipitation of the quartz and cassiterite and helps explain the restricted vertical range of very-high-grade tin ores in several shallow deposits. Evidence of boiling suggests that the ores were deposited at depths no greater than 3 km, a limit that is compatible with that based upon stratigraphic reconstruction.

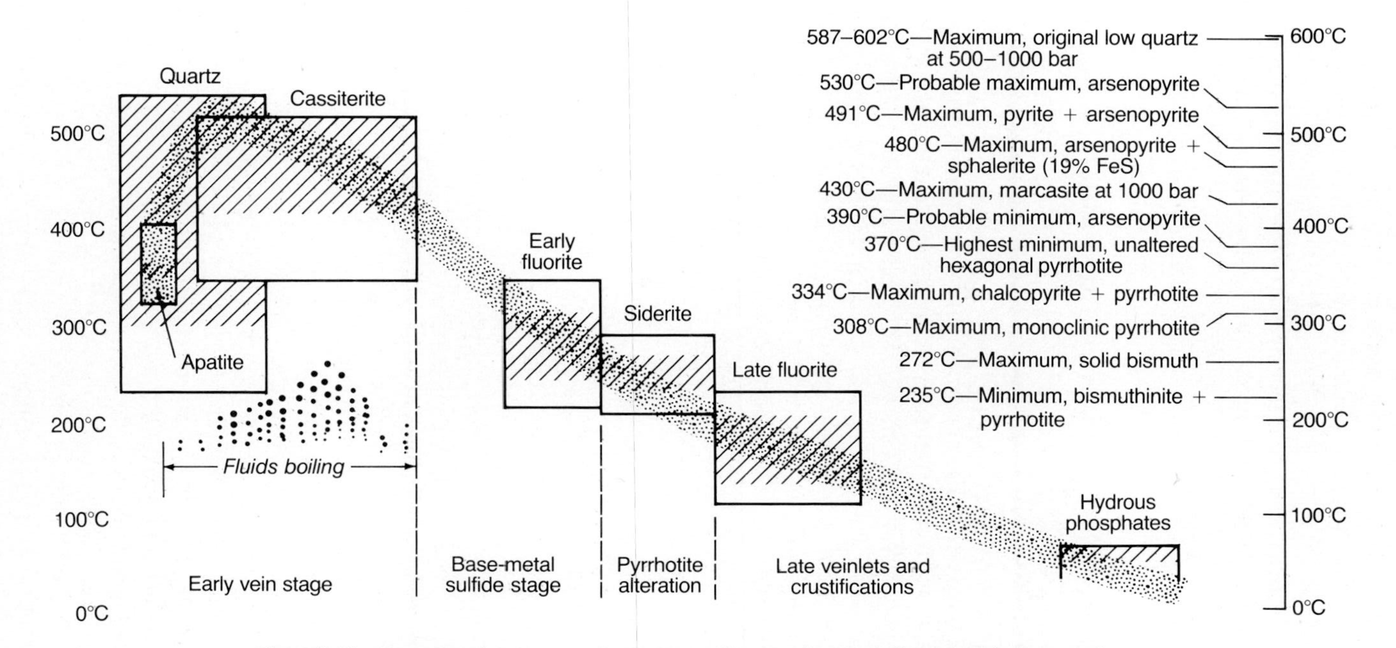

Figure 7-10. Paragenesis-paragenetic sequence diagram (horizontal) related to temperature of formation (vertical) as defined by pressure-corrected fluid inclusion filling temperatures, with other geothermometers indicated. Note the boiling zone above about 400°C revealed by fluid inclusions (see Figure 7-8) and the poor bracketing, or definition, of temperatures provided by the other techniques. (*From* Kelly and Turneaure, 1970.)

The data obtained are consistent with a single prolonged period of mineralization for the tin and tungsten deposits throughout Bolivia. A magmatic source is indicated for the mineralizing NaCl brines and their contained metals, but the continued cooling and dilution of these brines is thought to reflect gradual influx of meteoric water into the magmatic hydrothermal system. Their results are important to Bolivian tin, but they also reverberate through laboratories of economic geology by showing how powerful the techniques of fluid inclusion analysis can be.

▶ GEOBAROMETRY

It appears that however hard it may be to measure temperatures of ore-deposit formation accurately, it is even more difficult to acquire reliable pressure data. The only positive aspect of that statement is that temperature variation appears to have much greater significance to phase composition than does pressure variation, and that accurate pressure measurement is perhaps, therefore, less crucial. Still, if there were independent ways to determine what the depth of formation of a Mississippi Valley deposit has been, or whether one were standing on the root zone or the roof zone of a given deposit, it would be exceedingly useful. Overall, however, geobarometers—pressure gauges of geologic environments—have proven unreliable.

Rather than dwell on past procedures that have proven unreliable, we can describe four geobarometers that are promising in the 1980s. One is a derivative of metamorphic petrology, the stability ranges of common silicate-oxide mineral assemblages. As data accumulate on P-T regions of silicate-oxide parageneses (Figure 18-1), the feedback loops to economic geology become usable in ore-deposit analysis. Many alteration assemblages are sensitive enough to dehydration and pressure-temperature reequilibration that pressure-specific data can be applied to ore deposits if the temperature is known independently, as from "flinc" studies. We know now that many minerals earlier assigned to the high temperatures and high pressures of Lindgren's "hypothermal" class could not have existed at high pressures, and that others can also exist at low pressures and low temperatures. Certainly, increasing leverage from metamorphic petrology (Boettcher, 1976; Ganguly and Saxena, 1985) will be felt in ore-deposit interpretation.

Second, some specific silicate assemblages offer pressure-data potential. Coupled substitution of silica-alumina and alkalies in white mica sericite-phengite lattices appears to be pressure controlled (Velde, 1967; Sassi and Scolari, 1974), and X-ray techniques applied to regionally metamorphosed sedimentary rocks (Guidotti, 1973; Sassi and Scolari, 1974) indicate that the b_0 cell dimension in white micas may vary with the pressure of formation. If a sericite geobarometer were forthcoming, the obvious advantages of being able to determine the depth of formation, for example, of a given phyllic zone sample from a porphyry copper system (Figure 11-4), are clear. Sensitivity at least to 0.5 kbar will be necessary.

Third, the most portentous pressure gauge in mid-1982 is the Cu-Fe-Zn-S geobarometer (Hutchison and Scott, 1981). That system has been probed for decades in search of a composition variation that might be independent of $f(S_2)$, a_{Fe}, temperature, and so on. Early hopes for a sphalerite geothermometer, widely heralded in the 1950s, were completely dashed in the 1960s, and any reference in the literature to the "sphalerite geothermometer" must be disregarded. But in the process of such detailed investigation, the pressure dependence of the composition of sphalerite in equilibrium with pyrite and pyrrhotite with or without the presence of an intermediate Cu-Fe-S solid solution was discovered. It can be expressed in the temperature-independent portion of the Cu-Fe-Zn-S system as $P = 42.30 - 32.10 \log$ mole percent (mol %) FeS in ZnS (Figure 7-11). The composition of sphalerite can be determined from electron microprobe data, and textural-paragenetic equilibrium with pyrite-pyrrhotite determined in polished surface. Pressure data are still scant and on the high side, but the system offers promise.

Finally Roedder and Bodnar (1980) considered the uses of fluid inclusions as geobarometers. They conclude that single inclusions can yield only data that are functions of both temperature and pressure, as is implicit from Figures 2-9 and 2-10; but pairs of inclusions trapped at the same time from immiscible fluids such as brine and steam can yield both temperature *and* pressure data that are as good as our knowledge of P-T-V-X relationships in appropriate portions of the system H_2O-NaCl. Of the several requirements to permit accurate results, insufficient P-T-V-X data are the most serious weakness. The serious fluid inclusion microscopist should nonetheless consult this article for geobarometric procedures.

▸ ISOTOPE STUDIES

In recent years a great deal of effort has been devoted to the study of isotopes, including those of hydrogen, carbon, oxygen, sulfur, strontium, and lead (Hoefs, 1973; Mason and Moore, 1982). The object of these studies has been to determine the sources and compositions of ore-bearing fluids, the sources of the ore-forming components, the age of mineral deposits, to help determine temperature and other conditions of deposition, and to establish the degrees of bacteriological involvement in certain ore-forming processes. The value of isotope geochronologic studies has been enormous and is growing. Barren and productive stocks may be distinguished in part by their strontium isotopes, and Doe and Stacy (1974) list a number of exploration uses of lead isotope studies. Taylor (1967) and Krauskopf (1967) pointed out that one of the most promising ways to study the true origin

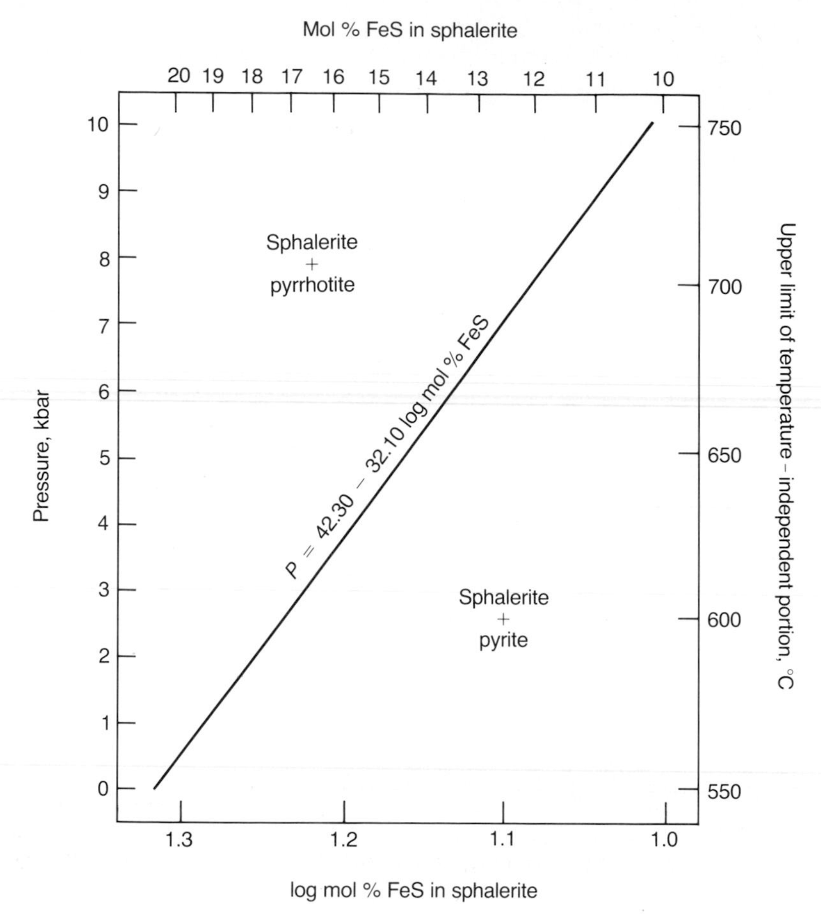

Figure 7-11. Curve of pressure versus log mol % FeS in sphalerite in stable paragenesis with pyrrhotite and pyrite. If the other conditions are met, a mol % FeS determination (top abscissa) gives a pressure readout (left ordinate). (*From* Hutchinson and Scott, 1981.)

of hot spring waters was to compare the ratios of the heavy isotopes of oxygen and hydrogen. Only seven years later, Hall et al. (1974) stated that "analyses of $^{18}O/^{16}O$ and D/H ratios of host rocks, ore, gangue minerals, and fluid inclusions have provided the basis of great advances in the understanding of hydrothermal systems involved in the formation of mineral deposits." Although isotope data have as yet been of little value in exploration, the approach is increasingly useful as more information accumulates.

Isotope analyses and studies can conveniently be placed into the two categories of *stable isotopes* and *radioisotopes*. Stable isotope variation depends upon the slightly different chemical behavior of isotopes of the same element with the built-in assumption that abundance ratios of the isotopes of a particular element are cosmically fixed and essentially unchanged in recent geologic time. Other isotopes, such as those of lead and strontium, are the end products of radioactive decay of elements with long decay times. Studies of these radiogenic isotopes depend upon the slow, constant decay of a radioactive parent element to radiogenic daughter products. These daughter isotopes are chemically distinctive, and may themselves be radioactive or stable radiogenic end products of decay series. As will be seen, the geochemistry of the parent radioelements, the time aspects of decay, and the distributions and geochemistries of the daughter products are all involved in their scientific and exploration usefulness.

The most useful pairs of stable isotopes are hydrogen-deuterium (Z = 1, ^{1}H = 99.985%, and ^{2}H = 0.015%), carbon (Z = 6, ^{12}C = 98.89%, and ^{13}C = 1.11%), oxygen (Z = 8, ^{16}O = 99.756%, and ^{18}O =0.205%), and sulfur (Z = 16, the useful isotopes being ^{32}S = 95.05% and ^{34}S = 4.21%). The most useful decay relationships among the radioisotopes are as follows, the radioactive parent given first: ^{14}C to ^{14}N, ^{40}K to ^{40}Ar, ^{87}Rb to ^{87}Sr, ^{232}Th to ^{208}Pb, ^{235}U to ^{207}Pb, and ^{238}U to ^{206}Pb. Recent studies of samarium (^{147}Sm) decay to neodymium (^{143}Nd) and of lutetium (^{176}Lu) to hafnium (^{176}Hf) are promising.

Stable Isotope Studies

Variations in the geologic distribution of the stable isotopes of the elements depend upon their slight differences in physical and chemical behavior in natural environments. These differences in turn depend on the percentage mass difference between the isotopes, differences that diminish with increasing mass numbers so that natural isotope separation of the stable isotopes of elements much heavier than Z = 20 (calcium, atomic mass = 40) is too small to be useful. It can be hoped that instrumental improvements in mass spectrometer and other techniques can raise that threshold in the future. How beneficial it would be if we could evaluate the isotope separation of chlorine (Z = 17, mass = 35.453) in brines and fluid inclusions, iron (Z = 26, mass = 55.847) in silicates, oxides, and sulfides, and copper (Z = 29, mass = 63.546), to name but three!

As stated by Faure (1977), stable isotope separation, or fractionation, stems from the fact that some thermodynamic properties of ions, radicals, and molecules depend on the masses of the atoms of which they are composed. If atomic masses vary even slightly, translational, vibrational, and rotational properties of the atoms and ions in a molecule will vary, and hence the thermodynamic properties of the ions or molecules will be affected. Molecules containing different isotopes in equivalent positions will have different chemical properties to the extent that those properties are mass dependent. "Light bicarbonate" ions $(^{1}H^{12}C^{16}O_3)^{-1}$ will behave significantly differently than "heavy bicarbonate" ions $(^{2}H^{13}C^{18}O_3)^{-1}$, to postulate two extremes of mass difference (13%). In general, bonds formed by lighter ions are weaker, so the lighter isotope ions are more mobile and reactive than equivalent heavy-isotope units.

The real utility of stable isotope studies lies in understanding how isotope ratios of an element are influenced by natural processes so that by measuring isotope ratios of oxygen, sulfur, or some other element in rocks and minerals, we can gain insight into how those rocks and minerals formed. For example, vapor evaporated from seawater has lower $^{18}O/^{16}O$ and deuterium (^{2}H)/hydrogen (^{1}H) ratios because water molecules containing the light isotopes can escape from liquid to vapor—that is, evaporate—preferentially. The values are -9 per mil for $^{18}O/^{16}O$ and -72 per mil for $^{2}H/^{1}H$, where per mil (‰) means per thousand (10‰ = 1%). The first rainfall condensing from that vapor will be richer in ^{18}O and deuterium than the vapor, but not as rich as the original seawater, and the resulting rainfall—the meteoric fluid of Chapter 2 and Figure 2-7—will be isotopically distinct both from seawater and presumably from juvenile fluids. With certain assumptions, the degree to which surficial waters may be involved even in deep-seated ore-forming processes can then be evaluated by examining isotope relationships in the ores, gangue minerals, and wall rocks.

Processes that bring about stable isotope fractionation include oxidation-reduction, evaporation and condensation as in the rainfall dynamics just described, reversible dissolution-precipitation from aqueous solution, adsorption-desorption reactions, microbially catalyzed reactions such as in sedimentary-diagenetic environments, and diffusion. Fractionation produced by differences in reaction rates of light and heavy molecules occurs at low temperatures, especially in microbial and photosynthetic reactions. Rocks and minerals precipitating from solution will retain a "memory" of the oxygen, sulfur, and carbon characteristics of the fluid from which they formed by precipitating ions of those isotopes as the stuff of minerals, especially below about 250°C. There is evidence (Sakai, 1968) that isotope fractionation of sulfur isotopes between different sulfide minerals precipitating from aqueous solution at equilibrium is in part temperature dependent, and that it also varies with pH and fugacities of oxygen and sulfur. Kinetic effects may also be involved. A complicating result of the failure of sulfate-sulfide pairs to equilibrate as temperature falls below 300°C is probably related to $SO_4^{=} - S^{=}$ reaction kinetics, pH, total sulfur in the system, and other variables. The rocks and minerals mentioned above may also have their memory of the original conditions changed by postdepositional exposure to fluids that are isotopically different. Clearly, stable isotope measurements must be interpreted and applied with great care.

Oxygen and Hydrogen Isotopes We have already mentioned that water vapor is isotopically lighter than the liquid water from which it evaporated. Clearly, some standard against which to measure isotopic shifts is required. The isotopic near-uniformity of the oceans makes seawater a logical choice, and so, indeed, SMOW (standard mean ocean water) was selected and is the reference point on diagrams of isotope variation (δD versus $\delta^{18}O$, in per mil). Vapor evaporated from SMOW is always isotopically lighter. Fortunately, for simplicity's sake, condensation of rain is isotopically almost an equilibrium process with direct proportionality between D/H and $^{18}O/^{16}O$ fractionation. Consider that the earliest rainfall from a given landward-moving air mass is enriched in ^{18}O and D relative to the vapor (but not to SMOW), and it will be apparent that successive rainfalls must be isotopically progressively lighter. The enrichment results from a reversal of the same fractionation process that enriches vapor in lighter isotopes in the first place. As the air mass progresses inland across a continent, or to higher latitudes or elevations, it becomes steadily depleted in the heavier ^{18}O and D. Since the fractionations are proportional, a straight-line plot whose equation is

$$\delta D = 8\delta^{18}O + 10 \qquad \text{per mil} \tag{7-1}$$

is generated. Although there are minor seasonal and local variations, overall contour maps of modern δD and $\delta^{18}O$ in meteoric waters—rain—have been made (Figure 7-12). These variations, described by the slope of the line of Eq. (7-1), were incorporated by Hugh P. Taylor, Jr., of the California Institute of Technology into a plot of δD versus $\delta^{18}O$, which has become familiar to readers of economic geology journals (Figures 7-13 to 7-15; a similar plot is shown in Figure 2-11). It is crucial to realize that the straight line labeled "meteoric waters" from near 0 and 0 to -166 and -24 on the δD versus $\delta^{18}O$ diagram of Figure 7-13 and others describes the landward and poleward isotope evolution of rainfall–snowfall. The isotope composition of the sea is thus basic to the isotope composition of meteoric fluids on the land masses.

Taylor (1974) stated that each of the waters of the Earth's crust—which he classified as magmatic, metamorphic, oceanic, connate, and meteoric—has different values of "heavy oxygen" and "heavy hydrogen." These differences result partly from interactions between these waters and the host rocks and minerals with which they come into contact. Silicates are normally much richer in oxygen than hydrogen, so we would expect that reacting a meteoric water with a rock of different isotope character would cause an excursion of the isotopic composition of the fluid essentially parallel to the ^{18}O abscissa. This phenomenon is readily demonstrated by the variation in $^{18}O/^{16}O$, which is much greater than δD echoed in lines *A* and *B* in Figure 7-13. Measurements of oxygen and hydrogen isotopes in alterationally pristine igneous rocks show that they range narrowly from -50 to -86 for δD and from $+5$ to $+10$ for $\delta^{18}O$. Calculation of the isotope composition of water in equilibrium with these magmatic rocks at elevated temperatures generates a "primary magmatic water" area in terms of δD and $\delta^{18}O$. By measurement and calculation, the various waters discussed in Chapter 2 have also been typed (Figures 2-11 and 7-13), and the power of oxygen-hydrogen isotope measurements in determining how a vein, rock, or ore formed becomes apparent. Near-surface clay minerals, which contain both

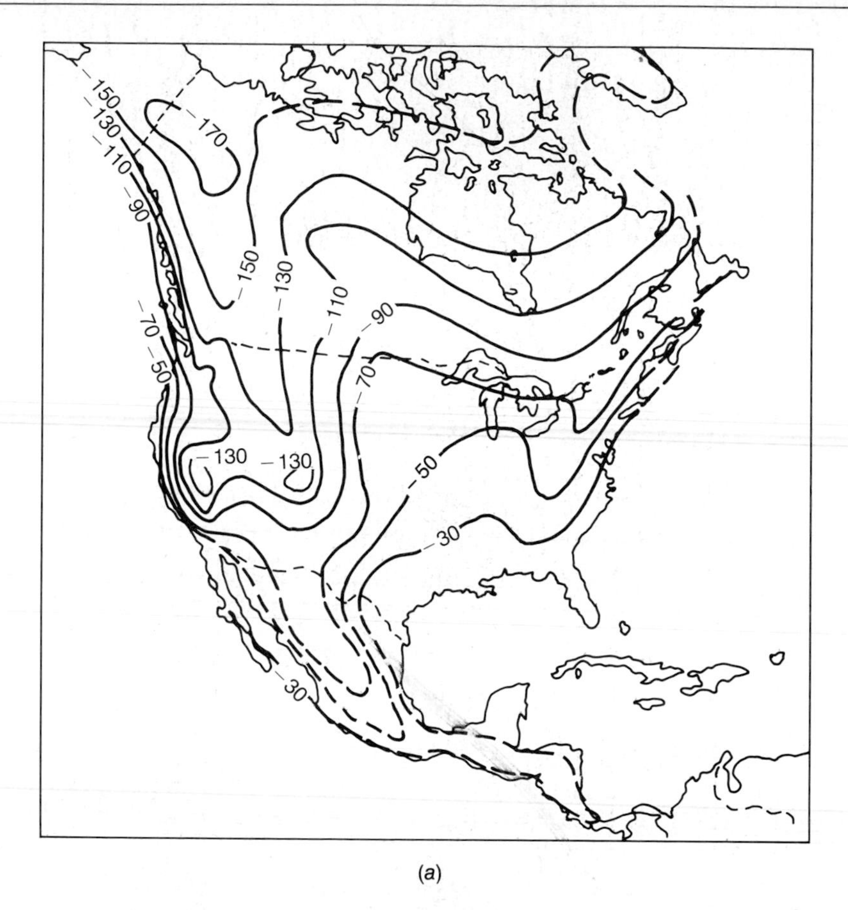

(*a*)

Figure 7-12. Maps of North America showing contours of the approximate average δD values (*a*) and $\delta^{18}O$ values (*b*) (*facing page*) of meteoric surface waters. Similar maps have been prepared for all continents. Note that the values decrease landward, especially near coastal mountains, and poleward. (*After* Taylor, 1974.)

oxygen and hydrogen, appear to form in isotope equilibrium with meteoric waters, but their isotope fractionation of O and H displaces them toward the heavier values. "Heavier" clays may also have formed at higher temperature. Interlayer water is routinely removed in analysis, so the oxygen and hydrogen that the analyses report are in true lattice positions.

Analysis of oxygen-bearing rock minerals, fluid inclusion waters, and alteration assemblages thus can now determine, in a way not available before the 1980s, the nature of the waters involved in ore-deposit formation. In general, although magmatic water signatures are found in many deep ore-forming environments associated with intrusive rocks, the extent of involvement of meteoric water has surprised many geologists. Many environments are overwhelmed throughout by surficial waters, many others

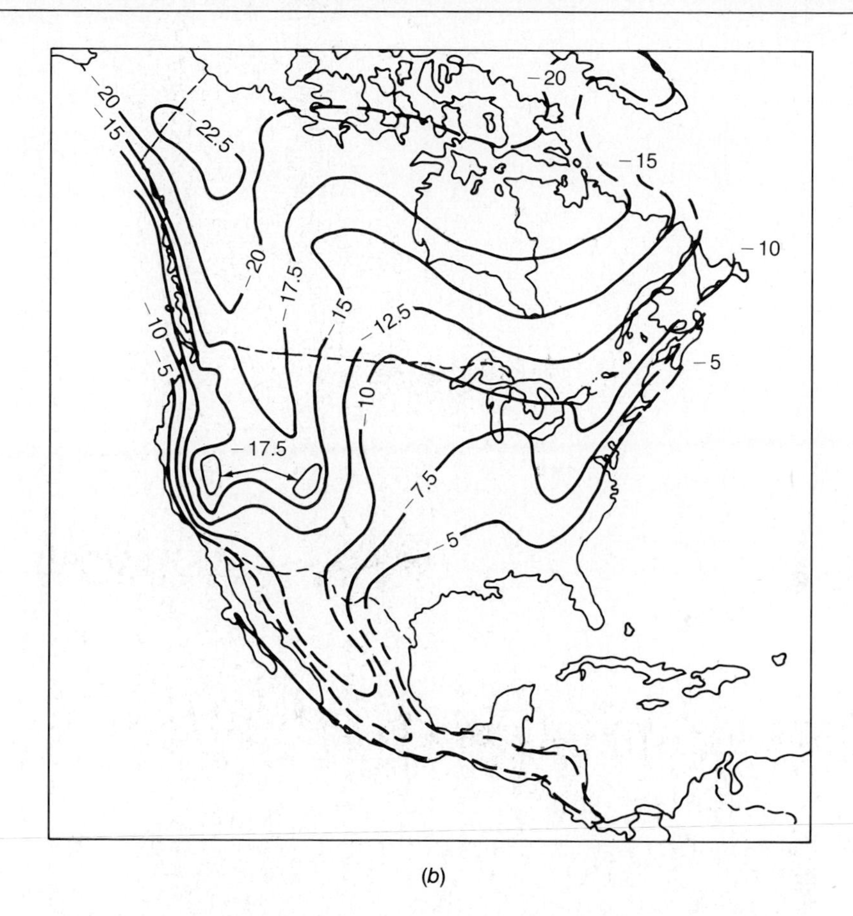

(*b*)

show a time relationship, typically with late incursions of meteoric waters, and even some deep-seated systems show strong isotope lightening away from the "magmatic water" box. Specific examples are not described here, but the results of $\delta D - \delta^{18}O$ isotope studies are included in several of the deposit descriptions to follow (Chapters 9 to 20). The student is urged to consult Taylor (1974, 1979) and Sheppard (1977) for more details in the methodology and its application.

Sulfur Isotopes One of the active areas of isotope research relative to ore-forming processes involves sulfur. Only two of the four naturally occurring stable sulfur isotopes are generally used, the lighter ^{32}S (95.02%) and the heavier ^{34}S (4.21%). ^{33}S and ^{36}S account for the remaining 0.77%. Fractionation occurs in nature either inorganically or organically. The sulfur isotope standard is that of the sulfur found in troilite (FeS) in the fragments of Arizona's Canyon Diablo meteorite, which is taken to be both cosmically primitive and unaffected by organic processes on the Earth's surface. The

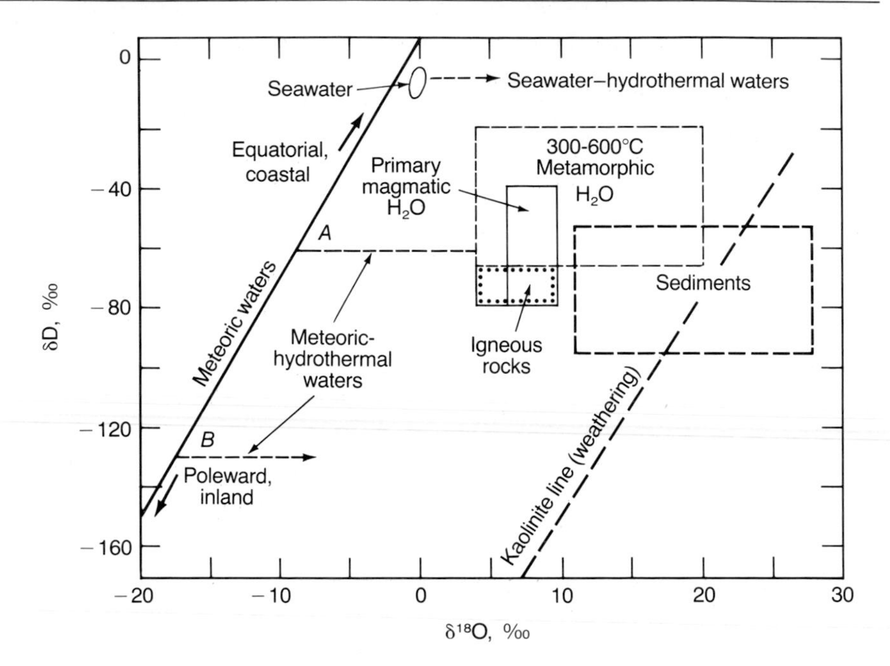

Figure 7-13. Summary diagram of isotope composition of waters of different origins. Trends of ^{18}O shift due to water-rock interaction and exchange are shown for seawater and meteoric waters of compositions A and B. (*After* Taylor, 1967.)

pertinent measured isotope ratio of $^{32/34}S$ in that meteorite is 22.22. $\delta^{34}S$ of another material is then defined as

$$\delta^{34}S = \frac{{}^{34}S/{}^{32}S_{\text{sample}} - {}^{34}S/{}^{32}S_{\text{standard}}}{{}^{34}S/{}^{32}S_{\text{standard}}} \tag{7-2}$$

The Canyon Diablo troilite $\delta^{34}S$ is taken as zero, and many other meteorites, igneous rocks, and ore sulfides have values near zero.

As stated, fractionation of sulfur isotopes in nature can be controlled either organically or inorganically. Organic separation stems from the fact that three types of anaerobic bacteria consume $(SO_4)^=$ in rivers, lakes, and oceans, break the sulfur and oxygen bonds, and excrete gaseous H_2S. The bacteria are *Desulfovibrio desulfuricans*, the most widely dispersed and active one; *Desulfovibrio orientis*, a medium-temperature species; and *Clostridium nigrificans*, a high-temperature form. Bacteria can break ^{32}S-O bonds more easily than the stronger ^{34}S-O bonds, so the $(SO_4)^=$ consumed and the H_2S produced are preferentially isotopically lighter. They may be lightened to $\delta^{34}S = -75‰$ relative to sulfate at room temperature; lightening to values of -20 to $-30‰$ is common. The H_2S then may react with iron or other elements, such as copper, lead, and zinc, if they are present,

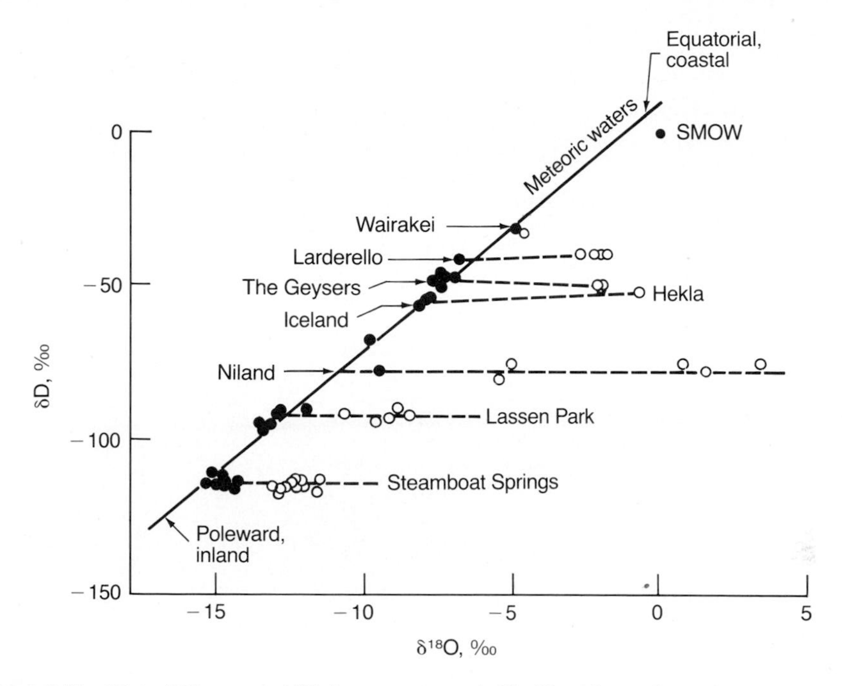

Figure 7-14. Plot of δD versus $\delta^{18}O$ for near-neutral chloride-rich geothermal waters (open circles) and local rain or snow (dark symbols) from a variety of hot spring areas throughout the world. Localities: Wairakei, New Zealand; Larderello, Italy; Niland (Salton Sea), the Geysers, and Lassen Park, California; Hekla, Iceland; and Steamboat Springs, Nevada. Notice that the precipitating waters get isotopically lighter both inland and poleward. The open circles define the degree to which the spring waters are diluted with magmatic water or equilibrated with isotopically heavier wall rocks by alteration and isotopic exchange. (*From* Taylor, 1974, Table 2-2.)

to form sedimentary sulfides. These sulfides are also isotopically light and are thereafter "labeled" as bacteriogenic. The net reaction is

$$(^{32}SO_4)^= + H_2{}^{34}S \rightleftarrows (^{34}SO_4)^= + H_2{}^{32}S \qquad K = 1.075 \text{ at } 25°C \qquad (7\text{-}3)$$

The value of K slightly greater than 1 requires that the reaction proceed to the right. Indeed, sulfides attributable to this reduction-exchange process now being deposited on the floor of the Red Sea have $\delta^{34}S$ values of −25 to −35‰. Other values of K are between 1.000 and 1.075, in part according to the rate of supply of $(SO_4)^=$. Isotope "lightening" is less at lower K, but is still large and measurable.

Marine sulfate has not remained isotopically constant during geologic time. It has ranged from $\delta^{34}S$ = +10 to about +30, and is now at about +20‰. Nonetheless, the lightening effects of bacterial catalysis have always been so marked that the technique has been of value in recognition

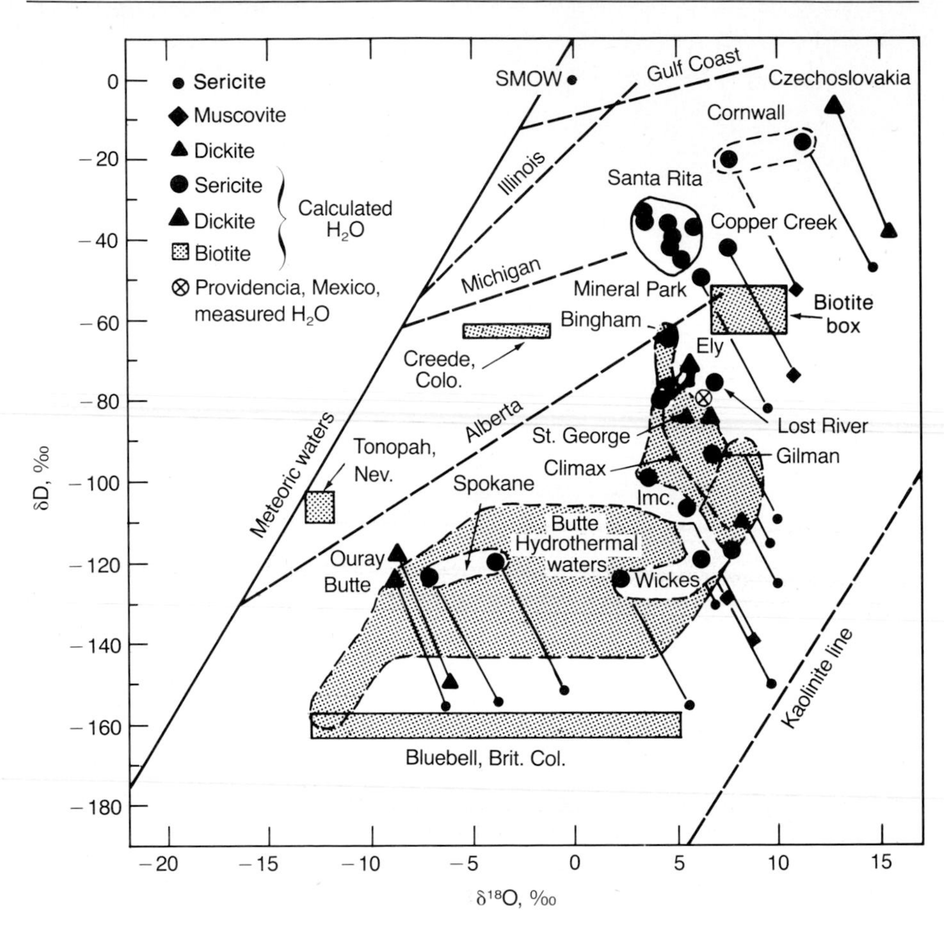

Figure 7-15. Plot of δD versus $\delta^{18}O$ for calculated hydrothermal waters from a variety of ore deposits, mainly from western North America. The stippled biotite "box" represents the "magmatic water box" of Figure 2-11 and describes waters that would have coexisted in equilibrium with hydrothermal biotites from Ely, Bingham, and Santa Rita at 650°C. Also shown are the trend lines for oil-field formation waters from various sedimentary basins in the midcontinent of North America. See also Figures 7-12, 7-13, 7-14, and 2-11. (*From* Taylor, 1974.)

of ore-forming environments of the seafloor, even, for example, in dominantly volcanic settings.

The second major control of sulfur isotope distribution involves inorganic fractionation. It is both more complex and more subtle. ^{34}S is enriched in more oxidized ions such as sulfate, and in sites of greatest bond strength in minerals. Furthermore, the isotope compositions of sulfides precipitated from hydrothermal fluids prove, as with the carbon isotopes discussed later, to be importantly controlled by the physical chemistry of the ore fluids themselves. The ion species of sulfur that are important at appropriate temperatures (25 to 500°C) include H_2S, HS^-, $S^=$, $SO_4{}^=$, HSO_4^-, $NaSO_4^-$,

KSO_4^- (Faure, 1977), and probably several more complex ions involving sulfur. As was learned in Chapter 2, the activities of these ions are pH dependent through a series of association-dissociation equilibria involving HCl, NaCl, KCl, water, and other molecules. pH largely controls relative proportions of H_2S, HS^-, and $S^=$, while $f(O_2)$ controls $(SO_4)^=$ relative to reduced species. Sakai (1968) showed that fractionation between $(SO_4)^=$ and the reduced species $S^=$ strongly favors $(^{34}SO_4)^=$, especially at lower temperatures.

The amounts of heavy isotope ^{34}S present may also vary with the paragenetic sequence of metallic sulfide species in a deposit. For example, at a given temperature the heavy isotope is most abundant in pyrite, with amounts decreasing progressively in sphalerite, chalcopyrite, and galena (Stanton, 1972). Thus pH, temperature, $f(O_2)$, $f(S_2)$, the degree of attainment of isotope disequilibrium, the starting isotope ratios, total sulfur activity, ionic speciation, and the partitioning of sulfur isotopes between sulfide minerals at various temperatures need to be known before sensitive interpretation can be made. Seldom can all these parameters be evaluated in a particular environment. Only as a result of isotope, fluid inclusion, and chemical studies were Heyl and his associates (1973) able to conclude that ore fluids in some Mississippi Valley lead deposits were heated oil field brines, and that both lead and sulfur came largely from a crustal source. Rye and Ohmoto (1974) give a fuller treatment of these relationships.

It has developed that the sulfur in hydrothermal deposits departs widely from the 0‰ of the upper mantle that might generally be expected, although 0‰ is clearly the mode (Figure 7-16). According to Field (1973), sulfides from hydrothermal deposits of the western American Cordillera are generally isotopically similar to meteoritic sulfur, regardless of the type of deposit, geologic age, or location. Field concluded that the sulfide sulfur in these deposits was derived from a deep-seated source and that contributions of heavy sulfur to it from seawater or evaporites were negligible. The isotope profile of sulfur introduced as SO_2, $SO_4^=$, or in other combinations is not known. Sedimentary deposits associated with marine conditions may be isotopically simpler because of marine isotope mass at any given geological time. Finally, metamorphism can modify initial sulfur isotope ratios since, as you might predict, ^{32}S bonds are more readily broken and ^{32}S diffuses more rapidly than ^{34}S. An application of this prediction will be considered in Chapter 13. Many workers are studying sulfur isotopes in confidence that when the many controls on isotope fractionation and distribution are understood, the usefulness of isotope studies in determining system chemistry and the source of sulfur in ore deposits will be established.

Carbon Isotopes Considering the importance of carbon and its ions and compounds in the biosphere, lithosphere, hydrosphere, and atmosphere, relatively little use has been made of stable carbon isotopes in ore deposit studies. Like sulfur, carbon occurs in reduced and oxidized species that

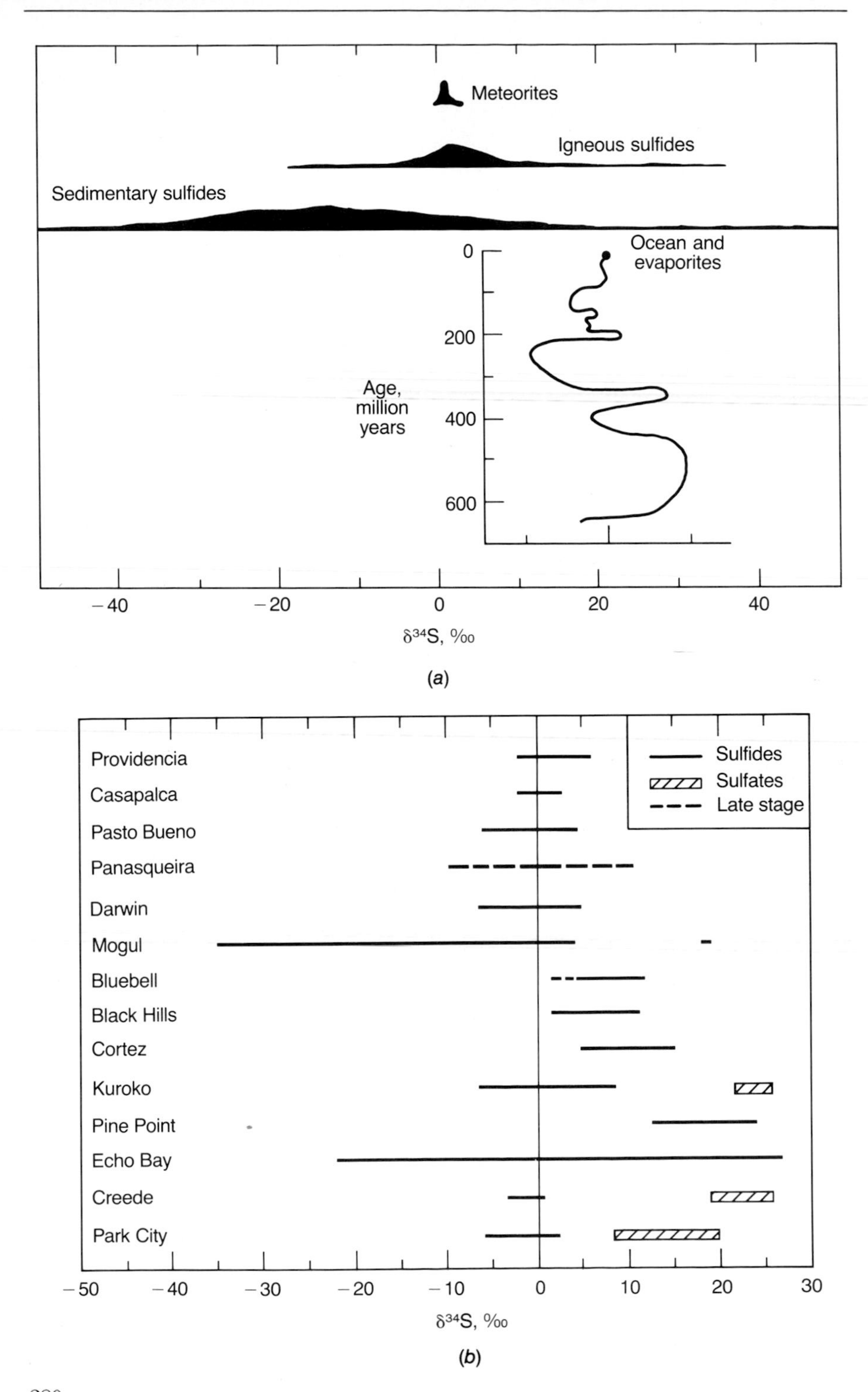
Meteorites
Igneous sulfides
Sedimentary sulfides
Ocean and evaporites
Age, million years
0
200
400
600
−40
−20
0
20
40
δ³⁴S, ‰
(a)
Providencia
Casapalca
Pasto Bueno
Panasqueira
Darwin
Mogul
Bluebell
Black Hills
Cortez
Kuroko
Pine Point
Echo Bay
Creede
Park City
Sulfides
Sulfates
Late stage
−50
−40
−30
−20
−10
0
10
20
30
δ³⁴S, ‰
(b)

behave differently. The two stable isotopes ^{12}C (98.89%) and ^{13}C (1.11%) are found as reduced species in graphite, diamond, methane (CH_4), and fossil fuels such as coal and oil shale, as intermediate species such as carbon monoxide, and as the oxidized carbon dioxide, carbonate ion, and bicarbonate ion. The standard value of $\delta^{13}C$ is that of carbon in CO_2 evolved by phosphoric acid digestion of a particular belemnite fossil. ^{12}C has always been enriched in compounds formed by photosynthesis, so reactions or processes reflecting interaction with such materials are also enriched in the light isotope. Fossil fuels such as coal, petroleum, and natural gas vary widely in $\delta^{13}C$. The carbon in carbonaceous matter—the kerogen associated with Precambrian conglomeratic gold-uranium deposits (Chapter 16) and the graphite in banded iron formations (Chapters 13 and 15)—has been identified by light carbon isotopes as organic-photosynthetic rather than inorganic, an important point in understanding these deposits. Carbon isotopes in carbonatites and diamonds vary widely, from -2 to $-10‰$ $\delta^{13}C$, in part in response to thermal fractionation, but possibly according to mantle carbon inhomogeneity. Carbon appears in hydrothermal ore deposits as CO, CH_4, and CO_2 in fluid inclusions and hot spring-fumarolic systems, as Ca-Fe-Mg-Mn carbonate gangue minerals, and occasionally as liquid hydrocarbon oils, especially in Mississippi Valley type deposits. The principal sources of that carbon (Faure, 1977) include sedimentary limestones ($\delta^{13}C \cong 0‰$), deep-seated or "average crustal" sources ($\delta^{13}C \cong -7‰$), and biogenic carbon ($\delta^{13}C \cong -25‰$). Hydrothermal carbon values include -4 to $-12‰$ $\delta^{13}C$ in CO_2 in fluid inclusions, -6 to $-9‰$ $\delta^{13}C$ in early formed carbonates, and higher values, even to positive per mills, in late formed carbonates. Information is scant thus far, but deep-seated sources of carbon are generally indicated.

Carbon isotope information appears to be of limited value in hydrothermal systems. The ultimate isotope composition of a carbon compound precipitating from a fluid depends not only on the carbon isotope ratio of the fluid but also upon pH (which affects speciation between CO, CO_2, H_2CO_3, HCO_3^-, $CO_3^=$, and CH_4, each of which affects $^{13/12}C$ distribution), the fugacity of oxygen, temperature, the ionic strength of the fluid, and on total carbon activity. Since CH_4 is strongly enriched in ^{12}C, for example, low oxygen fugacities that reduce carbon and favor CH_4 drastically modify carbon isotope profiles.

Research continues on all aspects of the problem, but the complex isotopic effects of the interplay of the independent variables discourages hope for an early solution that will be valuable to economic geologists.

Figure 7-16. (*Facing page*) (*a*) Sulfur isotope variations in nature (*From* Ohmoto and Rye, 1979.) (*b*) Range of $\delta^{34}S$ values of sulfides and sulfates from various hydrothermal ore deposits (*From* Rye and Ohmoto, 1974.)

An Example of Stable Isotope Studies—Salt Dome Sulfur

One clear example of stable isotope fractionation yields conclusive genetic information about a major ore-deposit type. The isotopes are sulfur and carbon; they involve the huge deposits of "salt dome sulfur" which are found along the coasts of the Gulf of Mexico in the United States and Mexico. Some 40 occurrences in the United States and 10 in Mexico have collectively produced a quarter billion tons of sulfur and will remain productive for many years. Structurally, the domes consist of piercement domes or diapirs of lighter salt (specific gravity = 2.2) through a heavier younger sediment cover (Figure 7-17). The principal rock type of the domes is salt with about

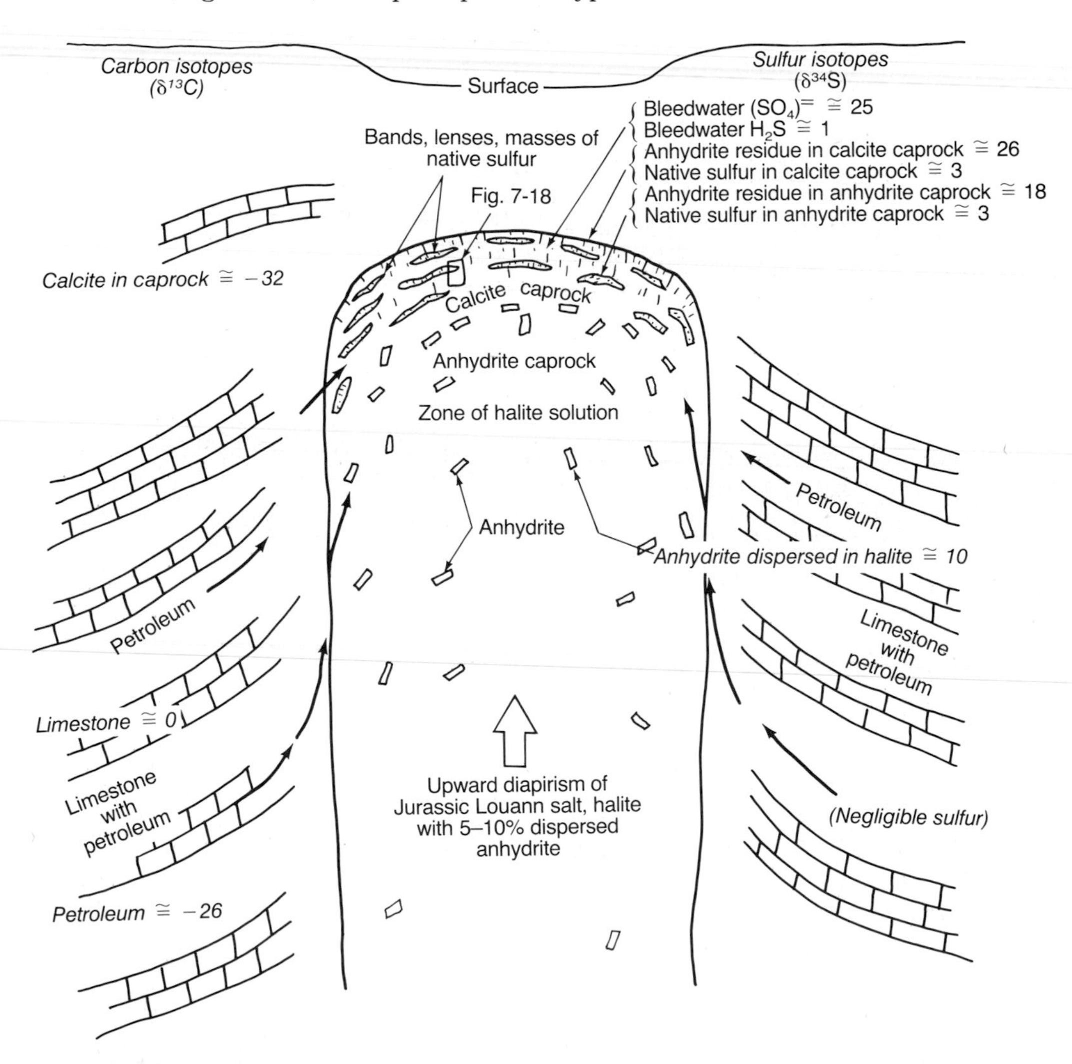

Figure 7-17. Sketch of geologic and isotope relations in salt dome systems. See also Figure 7-19. (*Isotope data from* Feely and Kulp, 1957.)

5% anhydrite. Originally the domes were drilled for the petroleum known to be trapped against their flanks in upturned drag folds, but the oil-sulfur association was soon noted. Salt diapirs through nonpetroliferous sediments are known in many parts of the world, especially in the Mediterranean Basin and the Near East, but they do not appear to generate sulfur. It was also noted early that a brecciated, cavernous, vuggy calcite "caprock" typically forms a capping originally considered to be sedimentary (Figures 7-17 and 7-18). But the true picture developed after Zobell's discovery in 1946 of sulfur-metabolizing anaerobic bacteria in these rocks and Feely and Kulp's (1957) isotope study of the salt dome systems.

Geologic, chemical, and isotope data are summarized in Figure 7-19. The carbon in calcite in the caprock proved isotopically lighter than that in

Figure 7-18. Drill core specimen of cavernous, vuggy salt dome caprock. The gray material behind the bar scale is residual anhydrite. The white to light gray banded and vug-lining material is canary yellow native sulfur and white calcite, locally with clay and insoluble residue "dirt." Both the sulfur and the calcite have grown in place, and the vug space has resulted from the solution of halite. The bar scale is in centimeters. See also Figure 7-17. (*Photo by* W. K. Bilodeau.)

sedimentary limestone, and even slightly lighter than the carbon in the petroleum. It appeared clear that the fractionation of ^{12}C into photosynthesized petroleum produced on continental margins carried through into the caprock calcite, which was therefore not the result of an inorganic sedimentary process, but must have evolved organically and derived carbon dioxide from the petroleum. Further, sulfur in the primary anhydrite ($\delta^{34}S \cong 16$) became progressively heavier as dissolution of the halite-anhydrite mix proceeded so that anhydrite in the caprock ranged from +18 to +26. Native sulfur in the same caprock was also dramatically lighter, reflecting the $SO_4^{=} - H_2S$ fractionation catalyzed by bacteria according to a reaction such as that given by Equation (7-2). It is worth noting here that the reactions of Figure 7-19 have equilibrium constants greater than 1, but they are kinetically slow. The reactions could occur naturally, but without the catalysis of bacteria deriving energy from petroleum they would advance at geologically negligible rates, and there would be no isotopic fractionation of carbon and sulfur. The case for bacterial involvement in the development of salt dome sulfur therefore appears well documented. Oxygen and hydrogen isotopes in salt dome environments would so closely approach SMOW values as to be valueless (Figure 7-13).

Radioisotope Studies

Although more than a dozen radioactive decay series have been evaluated for geologic usefulness in either source or geochronologic studies, only four are in wide enough use to warrant coverage here. They are rubidium-strontium (^{87}Rb-^{87}Sr), uranium-thorium-lead (^{235}U-^{207}Pb, ^{238}U-^{206}Pb, and ^{232}Th-^{208}Pb), potassium-argon (^{40}K-^{40}Ar), and carbon-nitrogen (^{14}C-^{14}N). Rubidium-strontium and uranium-thorium-lead systematics can be used both geochemically, as for stable isotopes, and geochronologically, so they could be considered in either category. As noted earlier, samarium-neodymium and lutetium-hafnium studies are gaining strength. The potassium-argon decay is useful for isotope dating across the geologic time scale; radiocarbon dating is useful only to about 60,000 years ago, so it is of little value in economic geology overall.

Undoubtedly the most significant equation concerning radioisotope studies is the fundamental $N = N_0 e^{-\lambda t}$, a rate equation that states that N, the number of nuclei of a radioactive element remaining from an original number N_0, is exponentially related to the decay constant λ times time t, in a negative sense, because the number of original nuclei progressively diminishes. This relationship can be expressed as $-\ln N/N_0 = \lambda t$, whereby it becomes even more apparent that radioactivity is a function of *time and the decay constant only*, and that no other factors—temperature, pressure, chemical mass, or activities—affect the rate of decay. Therefore neither metamorphism nor intrusion, neither biological activity nor deep burial will affect the rate of decay of a given population of radionuclei, so long as they

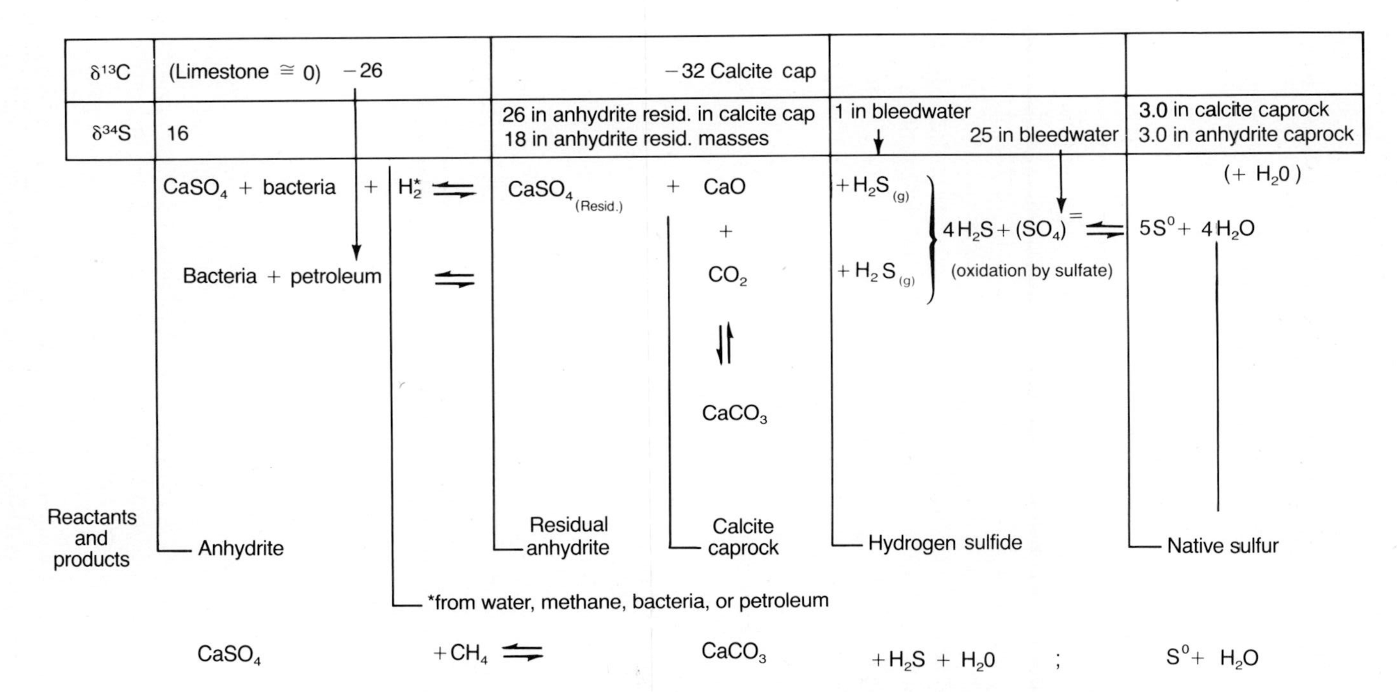

Figure 7-19. Stable isotope systematics in salt dome sulfur genesis. See also Figures 7-17 and 7-18.

remain together. Radioactivity studies can therefore be used geochronologically or to study the geochemistries of element separations, that is, source considerations.

Rubidium-Strontium A pair of elements that serves to clarify this thought is rubidium-strontium. Rubidium is an alkali element of so large an ionic radius that it cannot be accommodated in most minerals. It is almost invariably largely supplanted by smaller, more abundant alkali ions like Na^+ and K^+. It thus was almost quantitatively (totally) excluded from the mantle when the Earth formed, having been "rejected" into the sialic crust. Strontium, on the other hand, is an alkali-earth element of only slightly larger ionic radius than calcium (1.10 versus 1.00 Å) which geochemically it resembles strongly. Strontium finds many stable crystal lattice sites in the plagioclases, pyroxenes, and garnets of the mantle, and is not purged from them. It follows that the strontium isotope ratios of *mantle* materials have undergone almost no change by the radio decay of rubidium since a negligibly low value of Rb/Sr was "locked in" at the beginning of Earth history. So an intrusion emplaced in the Earth's crust, let's say 100 million years ago, should be able to give up some secrets. If it came from a zone of melting *in the mantle*, its strontium isotope ratio $^{87/86}Sr$ corrected to 100 million years should closely approach the primitive value of just over 0.700; if it penetrated and assimilated crustal rocks containing rubidium that had been producing radiogenic ^{87}Sr since their formation, their $^{87/86}Sr$ should be higher at perhaps 0.710 to 0.750; and if it formed by melting of crustal rocks alone, its $^{87/86}Sr$ should reflect no primitive mantle influence and might run as high as 0.780 to 0.800. If the Rb content and the true age of the intrusive unit are known, the postintrusion isotope shifts can be determined, and the degree of involvement of nonmantle rocks—the initial $^{87/86}Sr$ ratio—can be calculated. So mineral deposits immediately associated with intrusions can be interpreted in terms of mantle-crust source, a powerful insight (see the end of Chapter 3).

Isotope studies can, of course, become complicated if mixing or multiple intrusion is involved, and two-stage, three-stage, and even more complex mixing models have been proposed to accommodate such situations (Faure, 1977). As will be mentioned below, strontium isotopes reveal that, for example, plutons associated with many porphyry base-metal deposits have relatively low initial $^{87/86}Sr$, in the 0.705 to 0.709 range.

As Hedge (1974) pointed out, strontium isotopes can be used as tracers of the source or sources of rocks related to ore deposits, or they may be used more directly to study ore alteration and gangue minerals themselves, as in alteration feldspars and calcium or barium sulfates. Such studies have really only just begun, and results are still sparse.

Uranium-Thorium-Lead Probably the most extensively studied radioactive and stable isotope system is this one. As with Rb/Sr, it is useful both as an isotopic clock and, by geochemical separation of parent daughter elements, as a device to examine ore deposit source problems. We here deal with elements that are themselves important ore components, namely, lead and uranium, so we can study ore systems directly.

We must deal with two isotopes of uranium, one of thorium, and four of lead. ^{204}Pb is a stable isotope, possibly the decay product of a primordial radioelement now unknown. Abundances of ^{206}Pb, ^{207}Pb, and ^{208}Pb have been augmented since the formation of the Earth, ^{206}Pb having doubled and ^{207}Pb and ^{208}Pb having increased by one-half and one-third, respectively (Doe and Stacey, 1974). Since ^{204}Pb has remained unchanged, isotope statements are usually linked to it by the ratios $^{206/204}Pb$, $^{207/204}Pb$, or $^{208/204}Pb$, commonly as plots of two of these ratios (Figure 7-20). Original, primitive, primeval lead isotope ratios have been measured using as totally uranium-free, lead-rich phases as can be found. The same Canyon Diablo meteorite

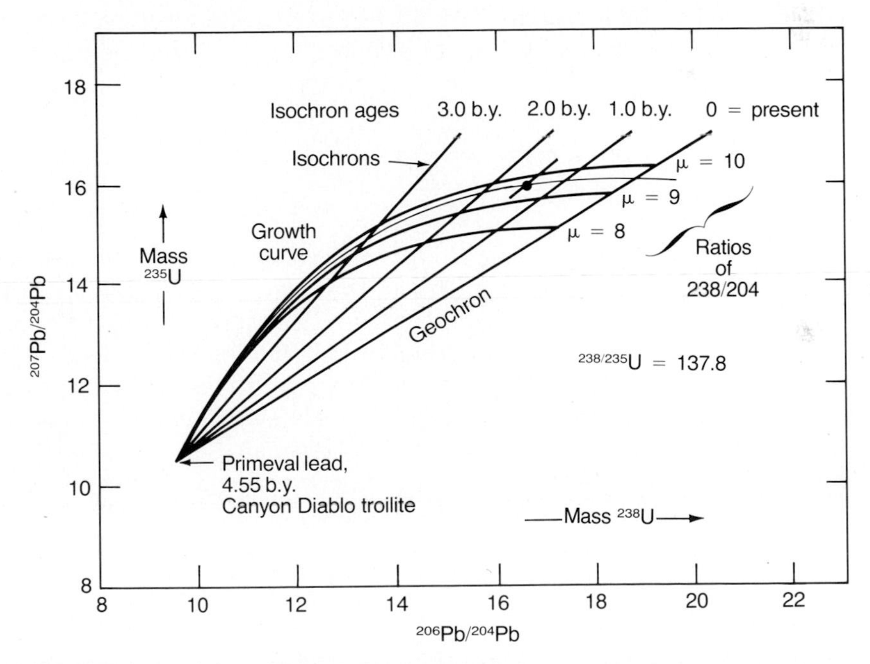

Figure 7-20. Growth curves of single-stage leads. The curved lines represent isotope ratio paths through geologic time of starting materials of different $^{238}U/^{204}Pb$ ratios μ. *Isochrons* connect points of equal geologic age. The straight-line geochron is a line connecting current isotope profiles. Note that 238/204 ratios can exceed 20 only if they are anomalous. A lead separated from uranium by withdrawal from the reservoir at 1.5 billion years with a 238/204 ratio $\mu = 9.6$ would have its isotope ratios "frozen" at the heavy dot in the figure. A lead evolved from 4.55 to 1.5 billion years at a 238/204 ratio $\mu = 9.6$ but then separated from uranium by withdrawal from the mantle reservoir at 1.5 billion years would even today have the isotope ratios represented by the dot, and therefore give temporal and geochemical evidence of its origin. The dot represents the ores of Broken Hill, Australia, in Figure 7-21, formed in the Proterozoic. (*After* Faure, 1977.)

troilite that is the sulfur isotope standard also contains traces of lead, but it contains no U or Th which might have produced ^{206}Pb, ^{207}Pb, or ^{208}Pb to "contaminate" it.

A useful aspect of lead isotope studies is that the percent difference between isotope masses is so small that the natural physicochemical fractionation so important in H, C, O, and even S isotopes can be disregarded. Oxidation-reduction of lead has no isotopic significance, and the isotopes do not fractionate between sulfides, regardless of temperature. The simplest lead isotope studies depend on several assumptions (Richards, 1971).

1. At the time of the formation of the Earth there was a single set of lead isotope ratios throughout the Earth mass, the so-called primeval lead.
2. Since then, all lead has been held in one or more closed systems, the mantle, the crust, and so on, with proportions of uranium, thorium, and lead changed only by radioactive decay.
3. From time to time "batches" of lead-bearing material have been removed from the source to form ore deposits, with negligible effect on what was left.
4. Ore leads so removed were not contaminated by other leads while traveling to the site of deposition; they reflect "frozen" original isotope abundances and ratios.

Ore leads so formed are called *single-stage* or *model-age* leads because their early separation from U and Th and their fixation in sulfide minerals like galena provide the analyst with isotope ratios that plot precisely upon growth curves which describe the overall evolution of radiogenic leads from uranium (Figure 7-20). Stated differently, model-lead ages can be measured when lead is removed from a single-stage evolving source and deposited elsewhere in a phase, like galena, in which U/Pb, and hence isotopic evolution, is infinitesimal. Lead isotope evolution ceases, and isotope ratios appropriate to the time of isolation are "locked in." Equations of the form

$$\left(\frac{^{206}\mathrm{Pb}}{^{204}\mathrm{Pb}}\right)_{\mathrm{present}} = \left(\frac{^{206}\mathrm{Pb}}{^{204}\mathrm{Pb}}\right)_{\mathrm{primeval}} + \left(\frac{^{238}\mathrm{U}}{^{204}\mathrm{Pb}}\right)_{\mathrm{present}} (e^{\lambda t - 1}) \tag{7-4}$$

where λ is the decay constant for ^{238}U and t is the age of the Earth, describe part of the isotope evolution of a closed system. The more ^{238}U, the higher the present $^{206/204}$Pb. Now if we consider Figure 7-20, we note that plotting ^{207}Pb (from ^{235}U) and ^{206}Pb (from ^{238}U), each with respect to the unchanging reference of ^{204}Pb, demands that a *family* of curves be generated if uranium abundance is not everywhere the same. If the radioactive clock has been ticking for 4.55 billion years in a given source area of low ^{238}U, the ratio of $^{206/204}$Pb will be lower than it is in a high ^{238}U source. The ratio of $^{235/238}$U appears to be nearly everywhere the same at a value of 1:137.8, so the series of growth curves will be symmetrically curved from the primeval lead origin point. The reason for this brief discussion is to stress that the ore deposit leads shown in Figure 7-20 all fall on *one growth curve* of $^{238/204} = 9.6$. The implication is that these major ore-deposit leads came from the same U-Pb-homogeneous source. That source has long been thought to be the upper mantle, but Richards (1971) and later Doe and Stacey (1974) also note that the gathering of leads from large volumes of frequently mixed, heterogeneous source materials through which U, Th, and Pb continuously diffused would have a similar stabilizing influence. If the simpler homogeneous-mantle theory is correct, all ore leads derived from a single-stage source should fall in precise chronological order along the appropriate growth curve, a condition generally but not totally met (Richards, 1971). Further, if that single-stage source is deep or isolated enough to have avoided oxidation of uranium and its geochemical separation from thorium, the same ore leads should plot chronologically on a single growth curve through a U-Th/Pb plot like $^{208}Pb/^{204}Pb$-$^{206}Pb/^{204}Pb$. The fit is surprisingly close (Doe and Stacey, 1974, Fig. 3).

But many ore leads depart greatly from plots of single-stage growth curve versus geologic age, so that more complex, anomalous origins are indicated. How do these anomalous leads develop? Let us consider a two-stage model, a two-step sequence of events. If erosion of a Precambrian orogenic belt produced a prism of sediments in a basin, those sediments would have passed through a weathering cycle. Original U-Th-Pb ratios would almost certainly have been modified (Chapter 20). Later ore fluids stemming from those rocks, or chemically interacting with them in transit to a newer ore-forming environment, would not precipitate model-age leads, but rather leads reflecting sources of more than one $^{238}U/^{204}Pb$. These anomalous leads might be two-stage, even three-stage leads, and still be manageable with existing formulas, which are extensions of Equation (7-4) (Doe and Stacey, 1974, pp. 774-775). In these calculations, the primeval U/Pb ratios are used to describe the system until the end of the first stage, for example, the erosion referred to above. Then the new U/Pb ratios are substituted into the equations and the age of the beginning of the second stage is substituted for "the age of the Earth." Using such techniques, unscrambling of three-stage leads and even more complex interrelationships can be attempted.

Using these methods, for example, Doe and his associates have shown (Doe and Zartman, 1979) that ore leads in the western San Juan caldera complex in Colorado were transported there from underlying 1700 to 1800 million-year-old Precambrian rocks, or detritus derived from them; that leads in the ores of the Coeur d'Alene district of Idaho are of Precambrian age and derivation, not of Tertiary age; and that the Homestake gold deposits of South Dakota are of approximately 2.5 billion-year-old age, with later redistribution at 1.6 billion years and in the Middle Tertiary. Perhaps most significantly, the leads of Mississippi Valley type lead-zinc deposits have highly radiogenic, two-stage, so-called *Joplin-type* anomalous leads which produce *future* model ages and which were probably scavenged from Cambrian aquifers through which ore-transporting brines moved.

Finally, Zartman (1974), Doe and Stacey (1974), and Doe and Zartman (1979) have drawn some provocative conclusions concerning the use of lead isotopes in exploration. They include the facts that *major* ore deposits tend to have single, uniform isotope ratios (Figure 7-21) and that local prospects whose leads depart from those values are likely to be subeconomic; that most major economic deposits have single-stage leads while "noneconomic showings" have leads of more complex origins; that lead isotopes may be "zoned" across a large district if central igneous and peripheral nonigneous rocks have provided leads; and that leads fall into regional provinces in the western United States, at least with ore leads in major provinces generally determined by the basement rocks.

We have ignored many complexities in lead isotope systematics, such as the use of concordia curves, instantaneous versus continuous growth models, and others. The student is urged to consult the abundant literature (Hoefs, 1973; Faure, 1977; and the many articles in *Economic Geology*) for further uses and explanations.

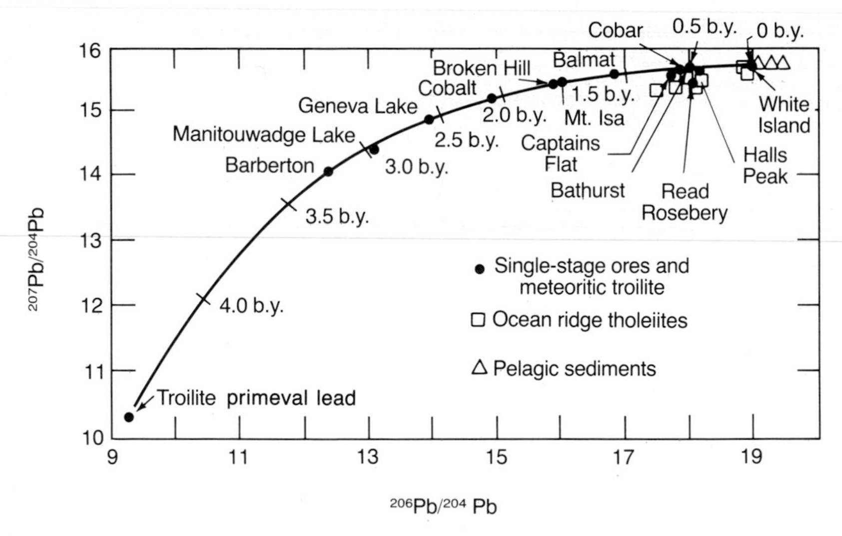

Figure 7-21. Single-stage growth curve and lead isotope ratios for leads from selected ores, ocean-ridge tholeiites, and pelagic sediments. Note how closely geologic age plots on the growth curve $\mu = 9.6$ of Figure 7-20. If the ore deposits were not isotopically single-stage, they would plot off the growth curve. A general homogeneous mantle source has therefore been suggested by Richards (1977) for those deposits the lead isotope ratios of which plot on the $\mu = 9.6$ isochron. (*From* Doe and Stacey, 1974.)

Bibliography

Ahmad, S. N., and A. W. Rose, 1980. Fluid inclusions in porphyry and skarn ore at Santa Rita, New Mexico. *Econ. Geol.* 75:229–250.

Birch, F., J. F. Schairer, and H. C. Spicer, 1942. *Handbook of Physical Constants.* Geol. Soc. Amer. Spec. Pap. 36.

Boettcher, A. L., Ed., 1976. Geothermometry-geobarometry. *Amer. Miner.*, Spec. Issues 7 and 8, 61:549–816.

Brett, R., 1964. Experimental data from the system Cu-Fe-S and their bearing on exsolution textures in ores. *Econ. Geol.* 59:1241–1269.

Buerger, M. J., 1961. Polymorphism and phase transformations. *Fortschr. Mineral.* 39(1):9–24.

Cochran, A., 1982. Fluid inclusion populations in quartz-rich gold ores from the Barberton Greenstone Belt, Eastern Transvaal, South Africa. Unpub. M.S. thesis, Tucson, Univ. Ariz., 208 pp.

Daniels, F., C. A. Boyd, and D. F. Saunders, 1953. Thermoluminescence as a research tool. *Science* 117:343–349.

Doe, B. R., and J. S. Stacey, 1974. The application of lead isotopes to the problems of ore genesis and ore prospect evaluation: a review. *Econ. Geol.* 69:757–776.

—— and R. E. Zartman, 1979. Plumbotectonics, the phanerozoic, pp. 22–70 in H. L. Barnes, Ed., *Geochemistry of Hydrothermal Ore Deposits*, 2nd ed. New York: Wiley, 798 pp.

Eastoe, C. J., 1982. The physics and chemistry of the hydrothermal systems in the Panguna porphyry copper deposit, Bougainville, Papua New Guinea. *Econ. Geol.* 77:127–153.

Edwards, A. B., 1954. *Textures of the Ore Minerals and Their Significance.* Melbourne: Australas. Inst. Min. Metall., pp. 1–31.

Faure, G., 1977. *Isotope Geology.* New York: Wiley, 464 pp.

Feely, H. W., and J. L. Kulp, 1957. Origin of Gulf Coast salt-dome sulfur deposits. *Amer. Assoc. Petrol. Geol. Bull.* 41:1802–1853.

Field, C. W., 1973. Sulfur isotope abundances in hydrothermal sulfate-sulfide assemblages of the American Cordillera (abstr.). *Econ. Geol.* 68:1206.

Ganguly, J., and S. Saxena, 1985. *Crystalline Solutions and Multicomponent Equilibria in Geologic Systems.* New York: Springer, in press.

Grant, J. N., C. Halls, S. M. F. Sheppard, and W. Avila, 1980. Evolution of the porphyry tin deposits of Bolivia, pp. 151–174 in S. Ishihara and S. Takenouchi, Eds., *Granitic Magmatism and Related Mineralization.* Soc. Min. Geol. Japan, Spec. Issue 8, 247 pp.

Guidotti, C. V., 1973. Compositional variation of muscovite as a function of metamorphic grade and assemblage in metapelites from NW Maine. *Contr. Mineral. Petr.* 42:33–42.

Hall, W. E., I. Friedman, and J. T. Nash, 1974. Fluid inclusion and light stable isotope study of the Climax molybdenum deposits, Colorado. *Econ. Geol.* 69:884–901.

Hedge, C. E., 1974. Strontium isotopes in economic geology. *Econ. Geol.* 69:823–825.

Heyl, A. V., G. P. Landis, and R. E. Zartman, 1973. Isotopic evidence for the origin of Mississippi Valley-type mineral deposits: a review (abstr.). *Econ. Geol.* 68:1207.

Hoefs, J., 1973. *Stable Isotope Geochemistry*. New York: Springer.

Hutchinson, M. N., and S. D. Scott, 1981. Sphalerite geobarometry in the Cu-Fe-Zn-S system. *Econ. Geol.* 76:143–153.

Ingerson, E., 1955a. Methods and problems of geologic thermometry. *Econ. Geol. 50th Anniv. Vol.*, pp. 341–410.

——, 1955b. Geologic thermometry. *Geol. Soc. Amer. Spec. Pap. 62*, pp. 465–488.

Keith, M. L., and O. F. Tuttle, 1952. Significance of variation in the high-low inversion of quartz. *Amer. J. Sci.* (Bowen Vol.), pp. 203–280.

Kelly, W. C., and F. S. Turneaure, 1970. Mineralogy, paragenesis and geothermometry of the tin and tungsten deposits of the eastern Andes, Bolivia. *Econ. Geol.* 65:609–680.

Krauskopf, K. B., 1967. *Introduction to Geochemistry*. New York: McGraw-Hill, 721 pp.

Kullerud, G., 1959. Sulfide systems as geological thermometers, pp. 301–335 in P. H. Abelson, Ed., *Researches in Geochemistry*. New York, Wiley.

——, 1964. The Cu-Fe-S system. *Carnegie Inst. Wash. Yearbook* 63 (1963-64): 200–202.

—— and R. Yund, 1960. Cu-S system. *Geol. Soc. Amer. Bull.* 71:1911–1912.

Lovering, T. G., 1958. Temperature and depth of formation of sulfide ore deposits at Gilman, Colorado. *Econ. Geol.* 53:689–707.

MacDiarmid, R. A., 1963. The application of thermoluminescence to geothermometry. *Econ. Geol.* 58:1218–1228.

Mason, B., and C. B. Moore, 1982. *Principles of Geochemistry*. New York: Wiley, 344 pp.

Miromoto, N., and G. Kullerud, 1966. Polymorphism on the Cu_5FeS_4-Cu_9S_5 join. *Z. Kristall.* 123:235–254.

Nash, J. T., and T. Theodore, 1971. Ore fluids in the porphyry copper deposit at Copper Canyon, Nevada. *Econ. Geol.* 66:385–399.

Newhouse, W. H., 1933. The temperature of formation of the Mississippi Valley lead-zinc deposits. *Econ. Geol.* 28:744–750.

Ohmoto, H. and R. O. Rye, 1979. Isotopes of sulfur and carbon, pp. 509–567 in H. L. Barnes ed., *Geochemistry of Hydrothermal Ore Deposits*, 2nd ed. New York: Wiley, 798 pp.

Park, C. F., Jr., and R. A. MacDiarmid, 1975. *Ore Deposits*. San Francisco: Freeman, 530 pp.

Ramdohr, P., 1931. Neue Beobachtungen über die Verwendbarkeit opaker Erze als "geologische Thermometer." *Prakt. Geol.* 39:65–73, 89–91.

Richards, J. R., 1971. Major lead orebodies—mantle origin? *Econ. Geol.* 66:425–434.

Richter, D. H., and E. Ingerson, 1954. Discussion—some considerations regarding liquid inclusions as geologic thermometers. *Econ. Geol.* 49:786–789.

Robinson, B. W., and H. Ohmoto, 1973. Mineralogy, fluid inclusions, and stable isotopes of the Echo Bay U-Ni-Ag-Cu deposits, Northwest Territories, Canada. *Econ. Geol.* 68:635–656.

Roedder, E., 1960. Fluid inclusions as examples of the ore-forming fluids. 21st Int. Geol. Congr. Rept., pt. 16, pp. 218–229.

——, 1963. Studies of fluid inclusions II: freezing data and their interpretation. *Econ. Geol.* 58:167–211.

——, 1967. Fluid inclusions as samples of ore fluids, pp. 515–574 in H. L. Barnes, Ed., *Geochemistry of Hydrothermal Ore Deposits*. New York: Holt, Rinehart and Winston, 670 pp.

——, 1971. Fluid inclusion studies on the porphyry-type ore deposits at Bingham, Utah, Butte, Montana, and Climax, Colorado. *Econ. Geol.* 66:98–120.

——, 1977. Fluid inclusion studies of ore deposits in the Viburnum Trend, southeast Missouri. *Econ. Geol.* 72:474–479.

——, 1979. Fluid inclusions as samples of ore fluids, pp. 684–737 in H. L. Barnes, Ed., *Geochemistry of Hydrothermal Ore Deposits*, 2nd ed. New York: Wiley, 798 pp.

—— and R. J. Bodnar, 1980. Geologic pressure determinations from fluid inclusion studies. *Ann. Rev. Earth Planet. Sci.* 8:263–301.

——, B. Ingram, and W. E. Hall, 1963. Studies of fluid inclusions III: extraction and quantitative analysis of inclusions in the milligram range. *Econ. Geol.* 58:353–374.

—— and B. J. Skinner, 1968. Experimental evidence that fluid inclusions do not leak. *Econ. Geol.* 63:715–730.

Rye, R. O., and H. Ohmoto, 1974. Sulfur and carbon isotopes and ore genesis: a review. *Econ. Geol.* 69:826–842.

Sakai, H., 1968. Isotopic properties of sulphur compounds in hydrothermal processes. *Geochem. J.* 2:29–40.

Sassi, F. P., and A. Scolari, 1974. The b_0 value of the potassic white micas as a barometric indicator in low grade metamorphism of pelitic schists. *Contr. Min. Petrol.* 45:143–152.

Sheppard, S. M. F., 1977. Identification of the origin of ore-forming solutions by the use of stable isotopes, pp. 25–41 in *Volcanic Processes in Ore Deposits*. London: Inst. Min. Met. Metall. Trans., 188 pp.

Sillitoe, R. H., and F. J. Sawkins, 1971. Geologic, mineralogic, and fluid inclusion studies relating to the origin of copper-bearing tourmaline breccia pipes, Chile. *Econ. Geol.* 66:1028–1041.

Smith, F. G., 1947. The pyrite geo-thermometer. *Econ. Geol.* 42:515–523.

——, 1948. The ore deposition temperature and pressure at the McIntyre mine, Ontario. *Econ. Geol.* 43:627–636.

——, 1954. Discussion—some considerations regarding liquid inclusions as geologic thermometers. *Econ. Geol.* 49:331–332.

——, P. A. Peach, H. S. Scott, A. D. Mutch, G. D. Springer, R. W. Boyle, and W. M. M. Ogden, 1950. "Pneumatolysis" and the liquid inclusion method of geologic thermometry: a reply. *Econ. Geol.* 45:582–587.

Stanton, R. L., 1972. *Ore Petrology.* New York: McGraw-Hill, 690 pp.

Takenouchi, S., and H. Imai, 1971. Fluid inclusion study of some tungsten-quartz veins in Japan. Soc. Min. Geol. Japan, Spec. Issue 3, pp. 345–350.

Taylor, H. P., Jr., 1967. Oxygen isotope studies of hydrothermal mineral deposits, pp. 109–142, in H. L. Barnes, Ed., *Geochemistry of Hydrothermal Ore Deposits.* New York: Holt, Rinehart and Winston, 670 pp.

——, 1974. The application of oxygen and hydrogen isotope studies to problems of hydrothermal alteration and ore deposition. *Econ. Geol.* 69:843–883.

——, 1979. Oxygen and hydrogen isotope relationships in hydrothermal mineral deposits, pp. 236–277 in H. L. Barnes, Ed., *Geochemistry of Hydrothermal Ore Deposits*, 2nd ed. New York: Wiley, 798 pp.

Turneaure, F. S., 1935. The tin deposits of Llallagua, Bolivia. *Econ. Geol.* 30:14–60, 170–190.

——, 1960. A comparative study of major ore deposits of central Bolivia. *Econ. Geol.* 55:217–254, 574–606.

——, 1971. The Bolivian tin-silver province. *Econ. Geol.* 66:215–225.

—— and K. K. Welker, 1947. The ore deposits of the eastern Andes of Bolivia: the Cordillera Real. *Econ. Geol.* 42:595–625.

Velde, B., 1967. Si^{+4} content of natural phengites. *Contr. Min. Petrol.* 14:250–258.

Wilson, J. W. J., S. E. Kesler, P. L. Cloke, and W. C. Kelly, 1980. Fluid inclusion geochemistry of the Granisle and Bell porphyry copper deposits, British Columbia. *Econ. Geol.* 75:45–61.

Yermakov, N. P., 1957. Importance of inclusions in minerals to the theory of ore genesis and study of the mineral forming medium. Trans. by E. A. Alexandrov, 1961. *Int. Geol. Rev.* 3(7):575–585.

Zartman, R. E., 1974. Lead isotopic provinces in the Cordillera of the western United States and their geologic significance. *Econ. Geol.* 69:792–805.

EIGHT

The Classification of Ore Deposits

The purpose of any classification is to group similar objects into classes or sets either for convenience, organization, or access, as in a collection, a library, or a computer data base, or for the purpose of learning more about the items being classified. The study of ore deposits through the last century has required—and still requires—the examination of great numbers and many types of mineral districts and the recording of their similarities and differences. Grouping together deposits with similar characteristics facilitates description, permits generalizations concerning genesis and ore controls and locations, and improves our collective abilities for exploration.

To be most useful, a classification of something as complex as ore deposits must be as true, correct, and simple as possible. Many attempts have been made to classify ore deposits since Agricola's first rough efforts, but most of those classifications have been abandoned for any of three reasons: they were not correct with regard to natural science, they were simply too cumbersome or restrictive for general use, or, most commonly, they were based on insignificant or nonuseful variants. Ultimately it must be admitted that it is difficult to classify a group of things that you do not understand, and so it has been with ore deposits. The most successful classifications have been based on purely descriptive aspects, such as by commodity produced, as in copper, nickel, or titanium deposits. But such classifications cannot satisfy geologists because there are many *genetic* types and settings of copper deposits and many deposits that contain copper as well as other metals in a multiplicity of combinations and proportions. So a commodity-based classification gives neither genetic nor exploration insight. Another type of classification that can be geologically relevant and correct is a purely descriptive one, one that classifies on the basis of wall rock type, or shape and form, or the type of structural control. Several classifications of this type have been proposed and used. For example, Bateman's textbook *Economic Mineral Deposits* (1950) was organized principally on the basis of structural controls of ores, with separate groupings of ore deposits that occur in faults, in folds, along igneous contacts, as disseminations, and so forth. But here again, some deposits are in faulted folds, or disseminated along contacts; and otherwise grossly different types of ores may occur in faults, for example—so many combinations occur that no clear-cut valid groupings emerge. All is not lost, however, because although we perceive that whether an ore occurs along a fault or in the axis of a fold is not fundamentally important, it remains true that the more purely *descriptive* a classification is, the easier is the approach to correctness and the less is the chance for genetically based misclassification.

Many geologists continue to urge that genetic aspects of classification are foolish until we really understand how nearly all ore deposits form, and perhaps even then. J. D. Ridge, J. C. Griffiths, and a group at Pennsylvania State University—and several other individuals—insist that exploration could be effectively carried out on either a purely descriptive basis—drilling dolomitized reefal limestones for sphalerite-galena deposits simply because of the observed Mississippi Valley deposit type association—or on a statistical one—selecting drill-site locations from a random number table or drilling at grid intersections regardless of geology.

But any scientist knows that it is human nature to try to figure things out. Rereading Chapter 1 reveals that from the first stirrings of our science, people have always been interested in determining *how* ores form, even more than describing them in terms of rocks, structures, and associations. Most of the articles in the latest issues of *Economic Geology* are focused not on what orebodies are, but on what processes created them. But obviously, the problems of genetic classification have stemmed from the fact that until the 1950s we really knew very little about the origin of most deposits. With the recognition of the role of volcanism in ore-deposit formation and with the improved understanding of lithogenesis and lithotectonics that modern research and the plate tectonics revolution have fostered, we can come far closer than ever before to a realistic genetic classification. Genetic classification has another drawback unrelated to ignorance, in that some ore deposits or districts may span more than one environment. For example, in many mining districts such as Butte, Montana, mineralization in the outer parts of the district was deposited at relatively low temperatures and pressures, while mineralization in the center of the district was formed at higher temperatures and pressures. And we noted in Chapter 5 that Butte's vein ores argue its inclusion in a Cordilleran

Vein category, but as the district became known at depth—and as our understanding of porphyry base-metal deposits improved—it became clear that Butte is also a porphyry copper deposit. It is impossible genetically to classify the district under only one category; a district is usually placed in the category that applies to the bulk of the ore.

So it is that nineteenth and early twentieth century classifications emphasized form, texture, and the mineral content and associations of ore deposits, while later classifications have been mainly developed around theories of genesis and environment of deposition. It was recognized early that a clear distinction could be made between certain types of sedimentary ores and others associated with igneous processes. Further breakdown into types has been difficult because geologists did not know the origin of many deposits. Even yet, no universally acceptable classification of ore deposits has been proposed, although several are in general use and some new or revised ones have been proposed. We will consider several—one by Niggli, one by Schneiderhöhn, and the Lindgren classification and its modification to be used in this book. A classification that responds to insights provided by plate tectonic revelation will be given in Chapter 21 after the geologic aspects of ore deposits have been studied. Europeans have favored Niggli's volcanic-plutonic classification and Schneiderhöhn's ore-association classification. The most widely used scheme in the United States—and in North and South America in general—has been Lindgren's depth-temperature classification. Each of these three classifications was developed during the early part of this century when vein types of ore deposits were most productive. Such deposits as the massive sulfides associated with volcanic piles, disseminated copper and molybdenum deposits, and stratiform deposits of the Mississippi Valley type, while known, were of far less relative economic value than they are now. For some reason, at the same time, they appear to have been of less scientific interest. Modern studies have made a great deal of information available and necessitate revision and modernization of the earlier classifications.

Niggli (1929) grouped the epigenetic ores into *volcanic*, or near-surface, and *plutonic*, or deep-seated (Table 8-1). The plutonic deposits were divided into hydrothermal, pegmatitic-pneumatolytic, and orthomagmatic subgroups, depending upon whether the ores formed from liquids or gases or as direct crystallization products within the magma. Subclasses within those genetic groups were based on commodity and ore-mineral associations. It is broad enough that it is usable in part, but it is too nonspecific to be generally useful in exploration. C1 would presumably include porphyry coppers, Cordilleran Veins, Carlin-type gold, and the Iron Springs, Utah, district, an unmanageable diversity of deposit types to be under one heading. Fundamentally, this classification differs little from Lindgren's, and most criticisms applied to Niggli's classification can also be applied to Lindgren's. Also since high-pressure fluids above the critical point are neither gases nor liquids, the pneumatolytic-hydrothermal distinction—the

TABLE 8-1. Niggli's Classification of Ore Deposits

I. Plutonic, or intrusive
  A. Orthomagmatic
    1. Diamond, platinum-chromium
    2. Titanium-iron-nickel-copper
  B. Pneumatolytic to pegmatitic
    1. Heavy metals, alkaline earths, phosphorus-titanium
    2. Silicon-alkali-fluorine-boron-tin-molybdenum-tungsten
    3. Tourmaline-quartz association
  C. Hydrothermal
    1. Iron-copper-gold-arsenic
    2. Lead-zinc-silver
    3. Nickel-cobalt-arsenic-silver
    4. Carbonates, oxides, sulfates, fluorides
II. Volcanic, or extrusive
  A. Tin-silver-bismuth
  B. Heavy metals
  C. Gold-silver
  D. Antimony-mercury
  E. Native copper
  F. Subaquatic-volcanic and biochemical deposits

question of whether ore components are transported as true gases or vapors as opposed to liquids—has been obviated. The classification defies field application because a mineral deposit that is formed from gaseous-transported materials cannot be distinguished from one formed from liquid-transported materials; in the modern view they are both fluids, and Niggli's B and C should be merged together.

Schneiderhöhn (1941) classified ore deposits according to (1) the nature of the ore fluid; (2) mineral associations; (3) a distinction between deep-seated and near-surface deposition; and (4) the type of deposition, host, or gangue. The significant category in the classification (abbreviated in Table

TABLE 8-2. Schneiderhöhn's Classification of Ore Deposits

I. Intrusive and liquid-magmatic deposits
II. Pneumatolytic deposits
  A. Pegmatitic veins
  B. Pneumatolytic veins and impregnations
  C. Contact pneumatolytic replacements
III. Hydrothermal deposits
  A. Gold and silver associations
  B. Pyrite and copper associations
  C. Lead-silver-zinc associations
  D. Silver-cobalt-nickel-bismuth-uranium associations
  E. Tin-silver-tungsten-bismuth associations
  F. Antimony-mercury-arsenic-selenium associations
  G. Nonsulfide associations
  H. Nonmetallic associations
IV. Exhalation deposits

8-2) is the metal-mineral associations. Schneiderhöhn proposed a detailed list of typical mineral associations, categorized according to the types of ore, host, and gangue found in each. It was therefore genetic with descriptive subcategories. Schneiderhöhn's system is popular in Europe and is advocated by many Americans. Noble (1955) argues that it is the best genetic classification because mineral associations represent metal associations in the ore-forming fluids. Although the schemes of Schneiderhöhn and Lindgren have fundamental similarities, they differ in emphasis. Under Schneiderhöhn's system, a deposit that does not fit one of the given ore-mineral associations or its subdivisions is readily categorized merely by formulating a new group or subdivision. The success of this system for field use, however, is inversely proportional to the number of major groups needed to accommodate all ore deposits; that is, each new category needed weakens the classification. A more detailed example of Schneiderhöhn's system (for group IIIA) is as follows:

III. Hydrothermal deposits
   A. Gold and silver associations
      1. Hypabyssal suite (deep-seated)
         a. Katathermal (high temperature and pressure) gold-quartz veins
         b. Gold-bearing impregnation deposits in silicate rocks
         c. Gold-bearing replacement deposits in carbonate rock
         d. Mesothermal gold-lead-selenium deposits
      2. Subvolcanic suite (near-surface)
         a. Epithermal propylitic gold-quartz veins and silver-gold veins
         b. Epithermal gold-tellurium veins
         c. Epithermal gold-selenium veins
         d. Alunitic gold deposits
         e. Epithermal silver deposits

The student should ask whether a gold-impregnation deposit in silicate rocks really differs from one in carbonates. The system, like Niggli's, is acceptable because a given ore deposit can be removed from one pigeon hole and placed in another if a new fact about it is discovered. But such a shift would not reveal much about ore formation in general, and the classification is considered quite mechanical.

Waldemar Lindgren (Figure 1-3) introduced his classification in 1913; it evolved through the next 20 years of its author's changing perspectives and ore-deposit research to the form in Table 8-3 (Lindgren, 1933). Terms such as *telethermal* (Graton, 1933) and *xenothermal* (Buddington, 1935) were added, but the deposits they were meant to accommodate have since been reclassified. Ridge (1968) recognized the need for revision, though he retained Lindgren's basic principles.

TABLE 8-3. Lindgren's Classification of Ore Deposits

I. Deposits produced by chemical processes of concentration. Temperatures and pressures vary between wide limits.
  A. In magmas, by processes of differentiation
    1. Magmatic deposits proper, magmatic segregation deposits, injection deposits. Temperature 700 to 1500°C; pressure very high
    2. Pegmatites. Temperature very high to moderate; pressure very high
  B. In bodies of rocks
    1. Concentration effected by introduction of substances foreign to rock (epigenetic)
      a. Origin dependent upon the eruption of igneous rocks
        i. Volcanogenic; deposits associated usually with volcanic piles. Temperature 100 to 600°C; pressure moderate to atmospheric
        ii. From effusive bodies; sublimates, fumaroles. Temperature 100 to 600° C; pressure moderate to atmospheric
        iii. From intrusive bodies; igneous metamorphic deposits. Temperature probably 500 to 800°C; pressure very high
      b. By hot ascending waters of magmatic origin
        i. Hypothermal deposits; deposition and concentration at great depths or at high temperature and pressure. Temperature 300 to 500°C; pressure very high
        ii. Mesothermal deposits; deposition and concentration at intermediate depths. Temperature 200 to 300°C; pressure high
        iii. Epithermal deposits; deposition and concentration at slight depth. Temperature 50 to 200°C; pressure moderate
        iv. Telethermal deposits; deposition from nearly spent solutions. Temperature and pressure low; upper terminus of hydrothermal range
        v. Xenothermal deposits; deposition and concentration at shallow depths, but at high temperature. Temperature high to low; pressure moderate to atmospheric
      c. Origin by circulating meteoric waters at moderate or slight depth. Temperature to 100°C; pressure moderate
    2. By concentration of substances contained in the geologic body itself
      a. Concentration by dynamic and regional metamorphism. Temperature to 400°C; pressure high
      b. Concentration by groundwater of deeper circulation. Temperature 0 to 100°C; pressure moderate
      c. Concentration by rock decay and residual weathering near surface. Temperature 0 to 100°C; pressure moderate to atmospheric
  C. In bodies of water
    1. Volcanogenic; underwater springs associated with volcanism. Temperature high to moderate; pressure low to moderate
    2. By interaction of solutions. Temperature 0 to 70°C; pressure moderate
      a. Inorganic reactions
      b. Organic reactions
    3. By evaporation of solvents

II. Deposits produced by mechanical processes of concentration. Temperature and pressure moderate to low

Source: Lindgren (1933) modified by Graton (1933) and Buddington (1935).

Lindgren's system was attractive in the Americas of the first half of the twentieth century because it seemed especially apt for the vein ores that then dominated the mining scene. Although the temperature-of-formation aspects could hardly be evaluated in the field, it was then—and still is—workable in most of its aspects as a field tool. Lindgren believed strongly in direct magmatic contributions to mineralizing fluids; his heading of section B1b reads "by hot ascending waters of magmatic origin"; that should certainly be changed to "by hot ascending waters of magmatic, metamorphic, oceanic, connate, or meteoric origin". Such a change in the system of classification conforms well with the findings of fluid inclusion and isotopic studies.

The temperature and pressure designations in Lindgren's scheme were at best only approximate, and certainly he recognized that they were subject to constant modification. For example, although most metallization in mesothermal deposits takes place between 400 and 250°C, early and late ore deposition may transgress these limits. Fluid inclusion data of the 1970s have blurred the temperature assignments of the various Lindgren categories. Riggs (1981) found that pressure-temperature plots of published fluid inclusion data from deposits earlier assigned to "Lindgren categories" give a virtual "scattergram", with extensive overlap of the different types. Lindgren's zones were based in large part upon minerals present in them. Little systematic phase equilibrium study had been generated then, and many minerals thought to be exclusively high-temperature phases were used as criteria. Pyrrhotite and arsenopyrite were regarded solely as high-temperature minerals. Yet we now know that pyrrhotite may exist in high-temperature hexagonal and low-temperatuare monoclinic forms, and that conditions for its deposition range from 25 to 870°C. Arsenopyrite is stable from room temperature up to 700°C (Clark, 1960). So many minerals that Lindgren considered to form at only high temperature are now known to form also at lower temperature. As data on phase equilibria and the stability of minerals have accumulated and as geothermometry has developed, the mineral assemblages typical of a given Lindgren depth zone have proven to be of too broad a thermal range. Schmitt (1950) proposed a classification chart with the ordinate and the abscissa defined by depth and temperature with Lindgren-type categories plotted in that P-T space. This method, although more precise, requires information not readily obtainable in the field. It has developed that all of the deposits in Lindgren's "hypothermal" category (B1bi) have been reliably reassigned, mostly into the volcanogenic group, his C1. Although the specific temperature assignments have largely broken down, many of Lindgren's uses of descriptive terms have remained appropriate. For that reason, the organization of this text retains parts of the Lindgren classification.

Previous editions of this book adhered closely to the Lindgren format with chapters headed "hypothermal", "mesothermal", "epithermal", "telethermal", and "xenothermal", among others. Only "epithermal" is retained in this edition because it is the only term to have survived the progress of the last 20 years essentially intact. The classification used here has been designed to reflect current thought. The tendency in exploration, in the design of research projects and the articles reporting them, and in the formats of national and international meeting and conference programs has been to place deposits in process-related, kindred groups. Porphyry base-metal deposits have formed a recognized group of deposits since the 1930s. In the 1970s that set has been validly subdivided to include porphyry coppers, porphyry molybdenums, porphyry tin, and climax type molybdenum deposits, each of which has different characteristics and even different tectonic settings. Table 8-4, then, is a statement of how Lindgren might have revised his own classification. It is the one used in the design of this book.

TABLE 8-4. Lindgren's Classification of Ore Deposits Modified for 1985

I. Deposits produced by chemical processes of concentration. Temperatures and pressures vary between wide limits.
  A. In magmas, by process of differentiation
    1. Magmatic segregation, injection, layered mafic intrusion. Temperature 700 to 1500°C; pressure very high
    2. Carbonatities, kimberlites. Temperature 700 to 1500°C; pressure very high
    3. Anorthosites, gabbros. Temperature 700 to 1500°C; pressure very high
    4. Porphyry base-metal deposits in part (see B1bi). Temperature moderate; pressure moderate
    5. Pegmatites. Temperature high to moderate; pressure high
  B. In bodies of rocks
    1. Concentration effected by introduction of components-epigenetic
      a. Origin dependent upon the eruption of igneous rocks
        i. Subaerial volcanogenic, associated with volcanic piles. Temperature 100 to 1200°C; pressure atmospheric to moderate
        ii. Sublimates, fumaroles. Temperature 100 to 600°C; pressure atmospheric
      b. Origin dependent upon ascending hydrothermal fluids of magmatic or meteoric origin
        i. Porphyry base-metal deposits (see A4). Temperature 200 to 800°C; pressure moderate
        ii. Cordilleran Veins; intermediate to shallow depths
        iii. Igneous metamorphic. Temperature 300 to 800°C; pressure low to moderate
        iv. Epithermal deposits; shallow to intermediate depths. Temperature 50 to 300°C; pressure low
      c. Origin dependent upon solution-remobilization, circulating meteoric waters
        i. Mississippi Valley deposits. Temperature 25 to 200°C; pressure low
        ii. Western states uranium. Temperature 25 to 75°C; pressure low
      d. Origin dependent upon circulating seawater
        i. Oceanic crust deposits, smokers, Red Sea. Temperature 25 to 350°C; pressure low
        ii. Volcanic exhalites in part
    2. By concentration of substances contained in the geologic body itself, epigenetic or syngenetic
      a. Concentration by dynamic and regional metamorphism. Temperature 25 to 600°C; pressure high
      b. Concentration by groundwater of deeper circulation. Temperature 0 to 150°C; pressure moderate
        i. Athabasca uranium
      c. Concentration by rock decay and residual weathering near surface. Temperature 25 to 50°C; pressure atmospheric

TABLE 8-4. (*Continued*)

C. In bodies of water; syngenetic
   1. Volcanogenic; underwater springs associated with volcanism. Temperature 25 to 350°C; pressure hydrospheric; oceanic crust deposits
      a. Massive sulfides-Cyprus
      b. Manganese-nickel-copper nodules
   2. Volcanogenic; underwater springs associated with sediments. Temperature 25 to 75°C; pressure hydrospheric
      a. Black shale hosted (?)
   3. By interaction of solutions. Temperature 0 to 70°C; pressure moderate
      a. Inorganic reactions
      b. Organic reactions
   4. By evaporation of solvents
      a. Evaporites. Temperature 25 to 75°; pressure low, atmospheric
   5. By chemical sedimentation. Temperature 25 to 75°C; pressure low
      a. Base metals
      b. Phosphates

II. Deposits produced by mechanical processes of concentration. Temperature and pressure low, surficial
   1. Alluvial placers
   2. Marine placers

III. Deposits produced following meteorite impact

Lindgren's genetic classification modified for 1985 is adventurous because some deposits fit into more than one category and because some deposits will need to be moved subject to future findings.

Charles Meyer (1981) reviewed ore-forming processes by classifying types of deposits and entering them in his Table 1, the abscissa of which is geologic time, from 4 billion years to the present. His classification (Table 8-5) is pragmatic, based on geologic association, genesis, and commodity. Meyer did not intend his classification to be complete, and there is some overlap in it, as for example "gold in iron formations" and "gold-quartz veins". It is presented as an eminently workable grouping that reflects modern process and genetic thought, as does the outline used in this text.

Other classifications are useful for special purposes, and specific terms apply without being parts of major classifications. Lovering (1963) introduced the terms *diplogenic* and *lithogenic*. Diplogenic was proposed for deposits where the components are partly syngenetic and partly epigenetic. Although the term refers primarily to time, in a sense it also refers to space; it carries no implication of the source of the epigenetic constituent or the conditions of formation. An example of a diplogenic deposit given by Lovering is one in which the syngenetic cation calcium unites with the epigenetic anion fluorine to replace limestone with fluorite. Lovering's second term, lithogenic, is applied to the mobilization of elements from one rock and their transportation elsewhere to be lodged in another. Sangster (1976) coined a similar term in "sedimentogenic", a word that he used to describe lead-zinc deposits formed by transport of Pb and Zn ions from a sedimentary source into a new host to form Mississippi Valley type deposits

TABLE 8-5. Meyer's Classification of Ore Deposits

Ores in mafic igneous rocks
- Chromite
  - Stratiform in layered complexes
  - Pods in Alpine peridotites
- Nickel-sulfide ores
  - Kambalda type
  - In amphibolites
  - Sudbury type
  - Insizwa type
- Titanium with anorthosite
  - Stratiform in layered complexes
  - Ilmenite in massifs

Volcanogenic massive sulfides in volcanic assemblages
- Cyprus-type in ophiolite suites
- Noranda-type in andesite-rhyolite suites
- Kuroko and allied types

Ores in sediments
- Sediment-hosted sulfide deposits
  - Copper in shales and sandstone
  - Lead-zinc in clastic sediments
  - Mississippi Valley type
- Iron ores
  - Banded iron formations
  - Clinton-Minette ores

Stratabound deposits
- Uranium deposits
  - Unconformity vein type
  - Sandstone and calcrete type
- Gold ores
  - Gold in iron formations
  - Gold-quartz veins
  - Gold-uranium conglomerates

Granodiorite-quartz monzonite, hydrothermal
- Porphyry coppers
- Tin-tungsten deposits

A pragmatic classification of ore deposits by geologic association, genetic type, and commodity.

Source: From Meyer (1981).

(Chapter 20). Lithogenic deposits in Lovering's sense could be derived from any rock type through the action of magmatic, metamorphic, or meteoric fluids.

Another mode of classification that has developed since 1968 is to assign ore-deposit types to plate tectonic-lithotectonic settings. It is increasingly evident that many of the forces and dynamics that underlie plate motions are immediately involved with lithogenesis, and that many rocks therefore have plate tectonic "niches". If that is true, and if ore deposits are in fact "extensions" of the rocks in which they occur, then ore-deposit analysis should reveal an order that is relatable to plate environments and that might be expected to help solve some of the problems of plate tectonic models. Several classifications have been proposed (Guild, 1971; Mitchell and Garson, 1972; Guilbert, 1981; Sawkins, 1984), many individuals have considered parts of the problem (Sillitoe, 1972; Sawkins, 1972; Solomon and Griffith, 1974), and several volumes have now used the plate tectonic framework to discuss ore deposits (Strong, 1976; Walker, 1976; Windley, 1977; Wright, 1977; Strangway, 1980; the Arizona Geologic Society Digest 14, 1981, Sawkins, 1984). The skeleton of a plate tectonic classification (Guilbert, 1981) is presented as Table 21-1 in Chapter 21 on metallogeny, the study of mineral belts and epochs. Its influence will be seen in Chapters 9 to 20. As those chapters unfold, the plate tectonic approach will offer insights into problems of regional ore-deposit genesis—also known as metallogeny—and provide explorationally and scientifically provocative groupings to the ore deposit geologist. It will be more understandable after the "geologic chapters" have been mastered, but it will help those who are familiar with plate tectonics to refer to it from time to time.

In conclusion, the problems of ore deposit classification are by no means easy or solved. It is tempting to say that we understand ore deposit genesis so well in the mid-1980s that the fundamental organization of this book will not change, and that no major reassignments of districts will need to be made. But such an assertion would require that the stunning progress of the "first decade of plate tectonics" and of the recent insights into volcanism's role in ore genesis be closed history. Such an assertion has no basis in recent events, and further elucidation of ore types and ore genesis are both inevitable and desirable. The Table of Contents of this book, especially for Chapters 9 to 20, is itself a classification that groups similar deposits into currently useful and informative subsets, and it should be referred to constantly for perspective, organization, and a "road map" while the following chapters are digested.

Bibliography

Bateman, A. M., 1950. *Economic Mineral Deposits.* New York; Wiley, 961 pp.

Buddington, A. F., 1935. High-temperature mineral associations at shallow to moderate depths. *Econ. Geol.* 30:205–222.

Clark, L. A., 1960. The Fe-As-S system: phase relations and applications. *Econ. Geol.* 55:1345–1381; 1631–1652.

Graton, L. C., 1933. The depth-zones in ore deposition. *Econ. Geol.* 28:513–555.

Guilbert, J. M., 1981. A plate tectonic-lithotectonic classification of ore deposits, pp. 1–10 in W. R. Dickinson and W. D. Payne, Eds., *Relations of Tectonics to Ore Deposits in the Southern Cordillera.* Tucson: Ariz. Geol. Soc. Dig. XIV, 288 pp.

Guild, P. W., 1971. Metallogeny: a key to exploration. *Mining Eng.* 23:69–72.

Lindgren, W., 1933. *Mineral Deposits*, 4th ed. New York: McGraw-Hill, 930 pp.

Lovering, T. S., 1963. Epigenetic, diplogenetic, syngenetic, and lithogene deposits. *Econ. Geol.* 58:315–331.

Meyer, C., 1981. Ore-forming processes in geologic history. *Econ. Geol. 75th Anniv. Vol.*, pp. 6–41.

Mitchell, A. H. G., and M. S. Garson, 1972. Relationship of porphyry copper and circum-Pacific tin deposits to paleo-Benioff zones. *Inst. Min. Metall.* 81:B10–25.

Niggli, P., 1929. *Ore Deposits of Magmatic Origin.* Transl. by H. C. Boydell London: Thomas Murby.

Noble, J. A., 1955. The classification of ore deposits. *Econ. Geol. 50th Anniv. Vol.*, pp. 155–169.

Ridge, J. D., 1968. Changes and developments in concepts of ore genesis—1933 to 1967 in J. D. Ridge, Ed., *Ore Deposits of the United States, 1933/1967*, Graton-Sales Vols., New York: AIME, 1880 pp.

Riggs, N. R., 1981. Fluid inclusion T-P data of major districts plotted by Lindgren categories. Unpub. paper, avail. from J. M. Guilbert, Univ. Ariz., 25 pp.

Sangster, D. F., 1976. Carbonate-hosted lead-zinc deposits, pp. 447–456 in K. H. Wolf, Ed., *Handbook of Stratabound and Stratiform Deposits*, vol. 6. New York: Elsevier.

Sawkins, F. J., 1972. Sulfide ore deposits in relation to plate tectonics. *J. Geol.* 80:377–396.

——, 1984. *Metal Deposits in Relation to Plate Tectonics.* New York: Springer, 325 pp.

Schmitt, H. A., 1950. The genetic classification of the bed rock hypogene mineral deposits. *Econ. Geol.* 45:671–680.

Schneiderhöhn, H., 1941. *Lehrbuch der Erzlagerstättenkunde.* Jena: Gustav Fischer.

Sillitoe, R. H., 1972. A plate tectonic model for the origin of porphyry copper deposits. *Econ. Geol.* 67:184–197.

Solomon, M., and J. R. Griffiths, 1974. Aspects of the early history of the southern Tasmanian orogenic zone, pp. 29–46 in A. K. Denmead, G. W. Tweedale and A. F. Wilson, Eds., *The Tasman Geosyncline—A Symposium.* Brisbane: Geol. Soc. Aust., Queensland Div.

Strangway, D. W., Ed., 1980. *The Continental Crust and Its Mineral Deposits.* Geol. Assoc. Can. Spec. Pap. 20, 804 pp.

Strong, D. F., Ed., 1976. *Metallogeny and Plate Tectonics.* Geol. Assoc. Can. Spec. Pap. 14, 660 p.

Walker, W., Ed., 1976. *Metallogeny and Global Tectonics.* New York: Wiley, 413 pp.

Windley, B. F., 1977. *The Evolving Continents.* New York: Wiley, 385 pp.

Wright, J. B., Ed., 1977. *Mineral Deposits, Continental Drift, and Plate Tectonics.* New York: Hutchinson and Ross, Benchmark Pap. in Geol. 44, 417 pp.

NINE

Deposits Related to Mafic Igneous Rocks

The deposits to be treated in this chapter are specifically related to mafic rocks. Those igneous rocks range from among the largest, most extensive igneous petrologic systems in the world—such as the Bushveld Complex—down to moderate-sized bodies like carbonatites. In each of them, the ore minerals are hosted by, and are therefore parts of, the igneous rocks themselves. With this in mind, the need to understand this realm of economic geology legitimizes a subdiscipline known as *economic petrology*, the bringing to bear of the tools and approaches of the petrologist to problems until recently explained by more general economic geologists. The following descriptions bear heavily on thin-section petrology, polished-surface mineragraphy, and geochemistry, with the premise that we cannot hope to understand the genesis and occurrence characteristics of specific minerals like chromite unless we also consider the genesis of the kindred rocks formed with them. The scale of consideration is thus enlarged to include entire ore-forming petrologic systems.

Ore deposits formed during fractional crystallization of magmas were recognized and named before Lindgren developed his classification system (Vogt, 1894). The term *magmatic segregation deposit* is now applied to all ore deposits that are direct crystallization products of a magma except for pegmatites, porphyry base-metal deposits, and others that involve hydrothermal transport. They usually form in the magma chamber, and thus are deep-seated intrusive bodies, but differentiated or immiscible melts and crystal mushes can be driven into magma chamber walls or roofs to form orebodies that are dikes, sills, and even extrusive flows.

A magmatic segregation deposit may constitute an entire intrusive rock mass or a single compositional layer within such a body, or it may be defined by the presence of valuable accessory minerals in an otherwise normal igneous rock. The ore minerals may be early or late fractionation products concentrated by gravitative settling of crystals or liquids, liquid immiscibility, or filter pressing; and they may remain in place or be injected as an ore magma into a previously solidified pluton or the surrounding country rock. The possibility that separation of immiscible magmatic liquids, such as a sulfide or oxide liquid from a silicate melt, has been important in magmatic segregation ore formation was reemphasized by Fischer (1950), Hawley (1962), McDonald (1967), Philpotts (1967), MacLean (1969), and MacLean and Shimazaki (1976). Fischer and Philpotts dealt with magnetite-apatite fluids as immiscible fractions of a silicate melt and MacLean considered silicate-sulfide liquid immiscibility.

Certain ore minerals are characteristic of specific igneous rocks, although others show no consistent affiliations. Ores commonly found with mafic rocks include chromite, ilmenite, apatite, diamonds, nickel, copper, and platinum group elements; those with igneous rocks of intermediate composition are magnetite, hematite, ilmenite, and vanadium; and those associated with siliceous rocks are magnetite, hematite, and such accessories as zircon, monazite, uraninite, and cassiterite. Many associations are even more restrictive; for example, chromite is closely associated with peridotite and dunite, or with serpentine derived from these ultramafic rocks, in Alpine peridotite deposits (Thayer, 1946, 1969). The tendency toward a specific ore–host-rock association (Buddington, 1933) is one of the strongest lines of evidence advanced by proponents of magmatic segregation as an ore-forming process.

Deposits to be considered in this chapter are those that form as parts of the mafic portion of igneous rock systems and in intrusive settings. They are in general lodged in cratonic masses or at least in continental crust. Deposits formed in oceanic crust—and there are many similarities—will be considered in Chapter 10. Those to be dealt with now include the largest ore-forming magmatic systems known, the layered mafic intrusions, or LMIs, of which the Bushveld Igneous Complex of South Africa, the Great Dike of Zimbabwe, the Sudbury Complex of Ontario, Canada, and the Stillwater Complex and the Duluth Complex of the United States are the best known examples. Although they are huge, LMIs such as the Jurassic Dufek Massif in Antarctica are still being found and explored (Ford, 1976). As more terranes like the Amazon Basin and the Australian interior become known, more LMIs will doubtlessly be discovered. LMIs constitute major sources of chromium, nickel, copper, the platinum metals, titanium, iron, vanadium, tin, and by-product sulfur.

Also to be treated in this chapter are the anorthosites, large nearly monomineralic plagioclase bodies which contain the world's most significant igneous titanium orebodies as rutile, ilmenite, and titanomagnetite. Examples are the Sanford Lake deposit in upstate New York and the Allard Lake occurrences in Quebec, Canada. Finally, an intriguing igneous family, the kimberlite-carbonatite group, will be described. This stem, perhaps the least well understood from the petrologists' point of view, reaches from igneous rocks dominated by olivines, phlogopite, minor carbonate—and accessory diamond—through those that are essentially pure igneous dolomite-calcite-siderite carbonate. The kimberlites are indeed known for their diamond content; occurrences in the type area of Kimberley, South Africa, will be described. Carbonatites are more varied in both the petrologists' and the economic geologists' view. Petrologically they are divided into two types. One is associated with other ultramafic rocks like dunite, peridotite, and pyroxenite; Palabora, South Africa, with its ore-grade copper, vermiculite, apatite, iron-titanium, zirconium, and uranium-thorium, will be described in detail. The second type, which appears to be more differentiated, is associated with alkalic-subsilicic rocks like nepheline and alkali syenites. These carbonatites contain deposits of rare earth elements, as at Mountain Pass, California; of apatite phosphate as at Cargill, Ontario; and of a broad spectrum of elements including titanium, zirconium, molybdenum, iron, vanadium, and rare earths (lanthanides), as at Magnet Cove, Arkansas. Only a brief treatment of Mountain Pass and Magnet Cove will be given. Finally, copper-nickel-sulfide deposits in ultramafic submarine volcanic flows will be described.

Deposits related to mafic intrusive systems have been interpreted in a plate tectonic context since about 1970, and improvements in their understanding are substantial. Rock types discussed in this chapter are considered to have formed in cratonic settings, but in cratons that were undergoing rifting or proto rifting. The association of layered mafic intrusions with cratonic-scale major tectonic separations leads to some remarkable results. For example, the Great Dike of Zimbabwe is an LMI with a length-to-width ratio of 100; it stretches in a straight line (see Figure 9-4) some 500 km north-northeast across Zimbabwe, is only 3 to 12 km wide, and is a major source of metallurgical and chemical-grade chromite.

Anorthosite intrusions were perceived by Herz (1969) to have been intruded in a narrow time window at around 1.4 billion years and along what appears to represent a then-major continent-scale linear which reached from Wyoming through what is now the St. Lawrence River valley area of Canada perhaps through what is now Scandinavia into the U.S.S.R. The province of kimberlites, carbonatites, and alkalic rocks is one of cratons so deeply rifted that upper mantle partial melting is tapped or generated. These igneous rocks are uncommon in orogenic belts, occurring only along lineaments within stable continental interiors.

The textures of minerals in many magmatic segregation ore deposits are essentially the same as textures in the enclosing parent igneous rocks (Newhouse, 1936; Ramdohr, 1940; Uytenbogaardt, 1954). For example, chromite in peridotite, even where it is abundant enough to constitute an ore-grade deposit, normally displays textures similar to those in peridotites with only a few percent chromite. Grains of magnetite in a fine-grained syenite are smaller than those in a coarser grained syenite, but are otherwise similar. In these cases, minerals may be anhedral (Figure 9-1) to euhedral (Figure 9-2) and interstitial to silicate minerals. Even where overgrowths and crystallization from immiscible liquids are involved, textural relationships in magmatic segregation deposits are in general familiar to the petrographer, and those textures and interpretations from them can be used successfully in exploration, evaluation, and production situations.

Minerals formed early in magmatic differentiation will not be in complete equilibrium with the melt during its later stages. Many of these early formed minerals become partly resorbed and show rounded faces and other effects of late magmatic and deuteric activity. Figure 9-3 shows clusters of chromite grains that were crowded against coarser olivine crystals when both were deposited in the Bushveld Complex magma chamber. Most of the olivine reacted postdepositionally with melt to form bronzite, leaving

Figure 9-1. Thin section of ilmenite-anorthosite from Allard Lake, Canada. Note the narrow reaction rim around the ilmenite (left of center) and the textural normalcy of plagioclase and anhedral interstitial ilmenite. Increased titanium content would produce a shift in relative amounts of plagioclase and ilmenite, with textures remaining similar. See also Figure 9-2. 15×, crossed nicols.

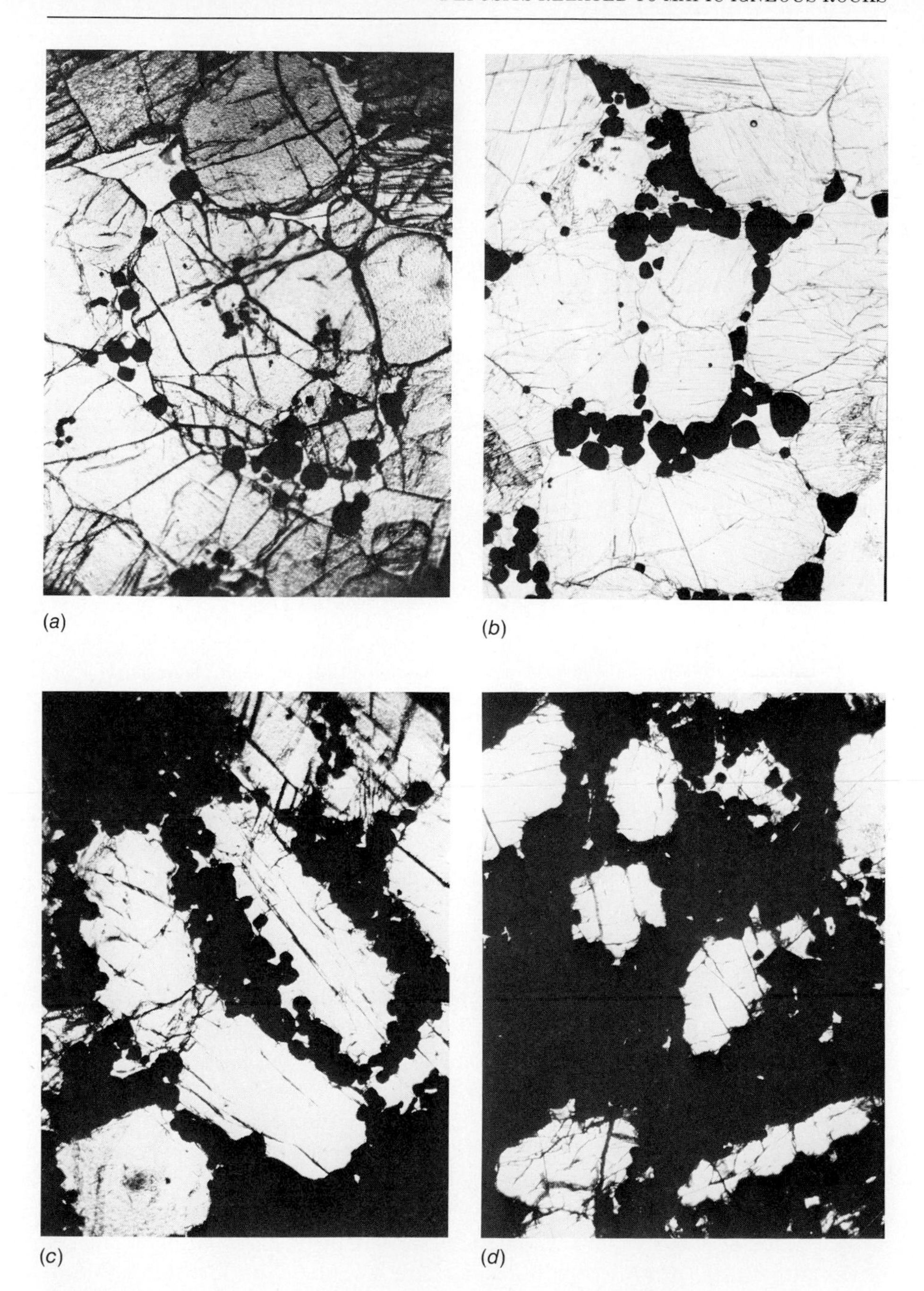

Figure 9-2. (*a*) Norite with accessory chromite; (*b*) chromitiferous norite; (*c*) bronzite-chromitite; (*d*) chromitite, all from the BIC. Notice the euhedral character of the black chromite grains in (*a*) and (*b*) and that qualitative aspects of the chromite-bronzite textures change but little while chromite content increases from 5 to 80 vol %. Each photo covers an area of 4 by 5 mm. Plane light.

Figure 9-3. Reaction rim showing an olivine grain (light gray) in a rim of bronzite (dark gray) surrounded by chromite (black). Note preservation of the outline of the original olivine grain by chromite and the parallelism of the chromite front with the remaining olivine core. From an olivine chromitite in the BIC harzburgite subzone. The original olivine was 2 by 3 mm. Crossed nicols.

only the central olivine remnants. The postreaction bronzite-olivine boundary shapes were predisposed by the original chromite-olivine boundary shapes (Figure 9-3). Such reaction-corrosion effects may obscure the primary textures of magmatic segregation deposits, making it difficult to distinguish between these deposits and some deposits of hydrothermal origin. Thus a debate has arisen over whether certain deposits were produced by magmatic segregation or by hydrothermal processes. In reality the two processes probably grade into one another, but only where differentiation becomes more advanced. Singewald (1917) concluded, for example, that many iron ores, especially titaniferous iron ores, are late-stage magmatic differentiates whose concentration is directly related to the action of mobile fluids. The process advocated by Singewald had been proposed earlier by de Launay, who suggested that as crystallization abates, crystal fractionation and hydrothermal activity must overlap, with an increase in the relative importance of aqueous fluids. Thus there would be the gradation from the magmatic stage to the hydrothermal stage as suggested in Figure 2-7. If de Launay and Singewald are correct, as is now commonly assumed, then such corrosive effects as those shown in Figure 9-3 are to be expected in magmatic segregation deposits, especially in those formed during later stages of differentiation where water activity is higher. Mineragraphic studies of magmatic sulfide ores indicate that the more mobile later fluids play an important role in magmatic segregation.

The gradation between magmatic segregation and hydrothermal activity is also manifested by wall-rock alteration effects, which are absent from most early magmatic differentiates, but may be clearly developed around the products of late fractionation. It should be noted, however, that wall-rock alteration is not characteristic of magmatic segregation deposits associated with mafic rocks. The presence of abundant fluids—as with late magmatic products—may not entail much alteration if there is little difference between ore and wall-rock pressure-temperature conditions. Norton and Taylor (1979) have documented evidence of extensive stable isotope exchange in rocks in and near the Skaergaard Intrusion in Greenland, but these oxygen-hydrogen exchanges were not accompanied by visible hydrolysis or hydration of original magmatic minerals. Among the main problems to be dealt with in LMIs is the nature of their layering and the means of accumulation of the oxide and sulfide ore phases.

▸ LAYERED MAFIC INTRUSIONS

Bushveld Igneous Complex, South Africa—Chromium-Platinum

Two of the most amazing ore systems in the world are the Bushveld Igneous Complex, or BIC, and the Sudbury Complex, or Sudbury Irruptive. The BIC holds the Earth's greatest reserves of chromium and the platinum metals platinum, palladium, osmium, iridium, rhodium, and ruthenium, and huge reserves of iron, titanium, vanadium, and tin. The Sudbury district boasts the world's largest production of nickel, a similar amount of copper, and hundred-year reserves of those metals, lesser platinoids, and sulfur. Both complexes are LMIs, the BIC the more strongly layered of the two. Both have been attributed to meteor-triggered magmatism, of which more will be said later.

The BIC, first of all, is large, measuring 375 km east-west by 300 km north-south. It extends over 67,000 km^2 at present levels of exposure (Table 9-1) and is thus roughly the size of Maine or Ireland. It has been the object of extensive ongoing study from Hall's 1932 memoir through Daly, Hess, and Wagner to the contemporary studies of the University of Wisconsin's Eugene Cameron and co-authors (1959, 1963, 1964, 1969, 1975, 1978, 1982), and Cousins (1969), Willemse (1969b), Vermaak (1970), Molyneux (1972), and Hunter (1976) in South Africa. Continuing research by the Economic Geology Research Unit of the University of the Witwatersrand, Johannesburg Consolidated Industries, and others is being coordinated by Von Gruenewaldt at the Bushveld Research Institute of the University of Pretoria.

A full volume of new studies will be published (Viljoen, 1985) as part of a five-volume compendium on South African orebodies. The best geologic map available is the provisional tectonic map of the Bushveld Complex (Hunter and Gomes, 1975), and the best single review article on the BIC is by Hunter (1976).

The BIC lies north of Pretoria in the northeastern portion of the Republic of South Africa (Figures 9-4 and 9-5). Broadly, it is a quadrilobate body, somewhat wider on its east side than its west, with a hook-shaped northward protrusion on its north side, the Potgietersrus Lobe. The complex is thought to be truly intrusive, with metavolcanic and metasedimentary older units called the Rooiberg Felsites and associated granophyric rocks above and older Transvaal Supergroup metasediments below, most prominently the Magaliesburg Quartzite (Figures 9-5 and 9-6). Hunter (1976) considered it intrusive, but shallowly so; others have speculated that it may even have been a thick flow rather than a sill-like body. For many years it was thought that the BIC was a single, continuous, essentially lopolithic,

TABLE 9-1. Areas and Diameters of Layered Mafic Intrusive Complexes

		Equivalent Circular Diameter	
	km^2	km	miles
Kapalagula Complex, Tanzania	23	5.4	3.4
Tabankulu Complex, South Africa	30	6.1	3.8
Rhum Complex, Scotland	30	6.1	3.8
Ardnamurchan Complex, Scotland	62	8.9	5.6
Skye Complex, Scotland	73	9.6	6.0
Ingeli Complex, South Africa	83	10.3	6.3
Skaergaard Complex, Greenland	104	11.5	7.2
*Stillwater Complex, United States	194	15.7	9.8
*Muskox Intrusion, Canada	350	20.1	12.5
Dore Lake Complex, Canada	470	24.5	15.3
Colony Complex, Sierra Leone	492	25.0	15.6
*Insizwa Complex, South Africa	544	26.3	16.4
†Sudbury Complex, Canada	1,342	41.3	25.8
Usushwana Complex, Swaziland–South Africa	1,650	45.8	28.6
†Great Dike, Zimbabwe	3,265	64.4	40.3
*Duluth Complex, United States	4,715	77.5	48.4
Kunene Complex, Angola-Namibia	7,770	99.5	62.2
Dufek Intrusion, Antarctica	+8,000	100.9	63.1
†Bushveld Complex, South Africa	67,340	292.8	183.0

Outcrop areas and equivalent circular diameters of most of the world's layered mafic intrusions. Equivalent circular diameter is only a device—most are circular, elliptical, or lobate, but some, the Great Dike and Muskox, for example, are dike-form. Notice the relative enormity of the Bushveld Complex.

*Subeconomic deposits known.
†Cr, Ni, Cu, or Pt mined.

Source: Original data from Willemse (1969b).

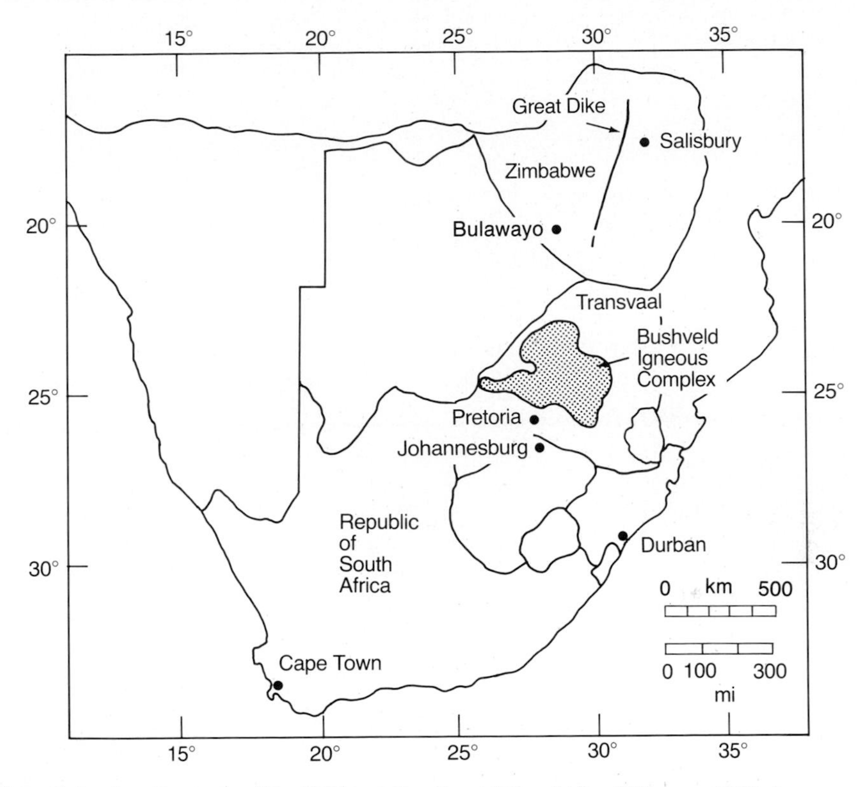

Figure 9-4. Locality map of the BIC and the Great Dike. (*After* Willemse, 1969b.)

funnel-shaped body consisting of a lower, older layered mafic phase and a younger, central granitic phase differentiated from the same magma. The Bushveld Red Granites are now known to be younger than and intrusive into the layered mafic rocks, but the term *BIC* is still generally applied to both groups of rocks. It now appears (Hunter, 1976) that the complex consists of four lobes (Figure 9-5). Using azimuths, the eastern lobe extends from 20 to 120°; a southeastern lobe produces a gravity high from 120 to 200°, which is largely buried by younger Permian Karroo-age sediments; the western lobe swings from 220° at Pretoria to 300°; and the northern, or Potgietersrus, limb protrudes to the north. A cross section of *just the eastern lobe* (Figure 9-6, *A-A'* on Figure 9-5, no vertical exaggeration) shows the gently, centrally directed dips on the east side that describe the Eastern Norite Belt units. Older sections would have shown those dips continuously westward dipping across the diagram, probably with a fault block on the west side. Each of the four lobes appears to have been formed at the same time. They must all have been centers of contemporaneous magmatic activity and in chemical communication, if not equilibrium; there is strong

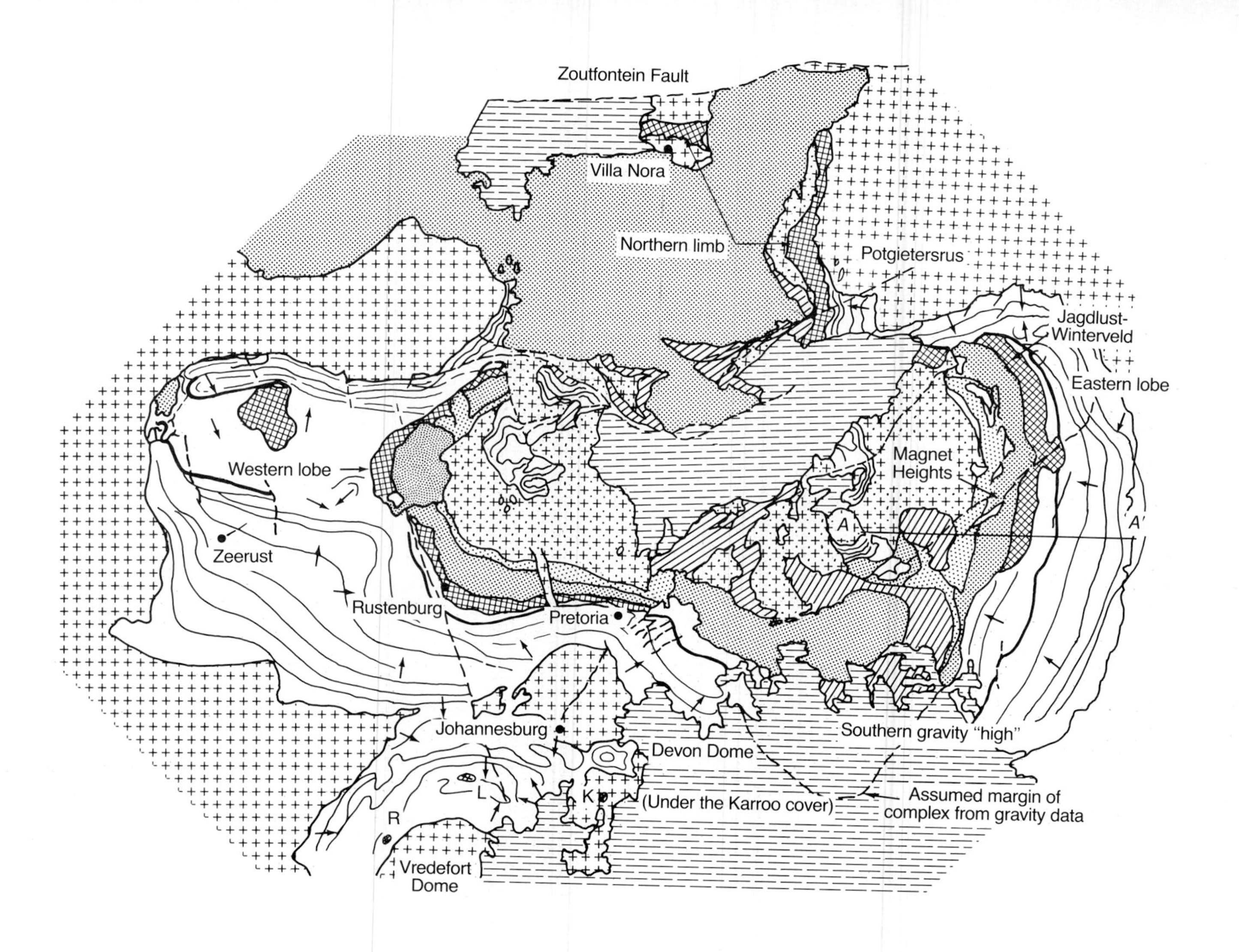

Zoutfontein Fault
Villa Nora
Northern limb
Potgietersrus
Jagdlust-
Winterveld
Eastern lobe
Western lobe
Magnet
Heights
A
A'
Zeerust
Rustenburg
Pretoria
Johannesburg
Southern gravity "high"
Devon Dome
L
K
(Under the Karroo cover)
Assumed margin of
complex from gravity data
R
Vredefort
Dome

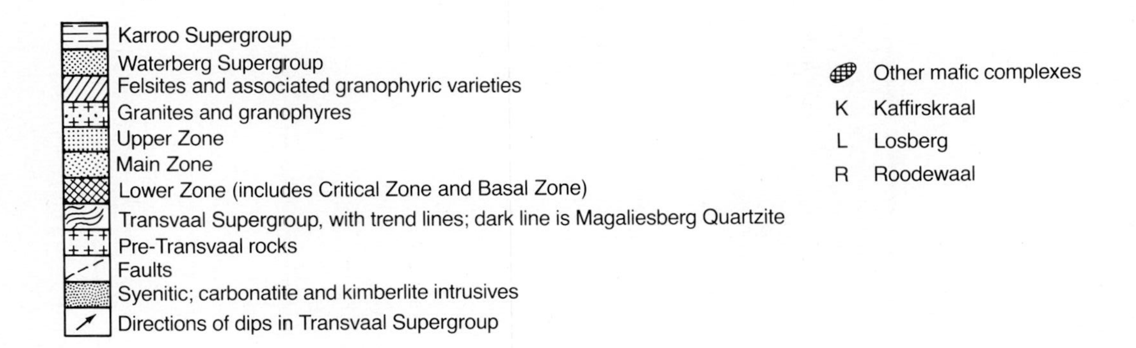

Figure 9-5. Simplified geological map of the BIC. The principal chromite reserves occur in the upper "critical" portion of the Lower Zone, and the world's largest source of platinum metals, the Merensky Reef, forms the boundary between the Lower Zone and the Main Zone in the eastern and western lobes especially near Rustenburg. *A–A′* locates the cross section of Figure 9-6. (*After* Hunter, 1976.)

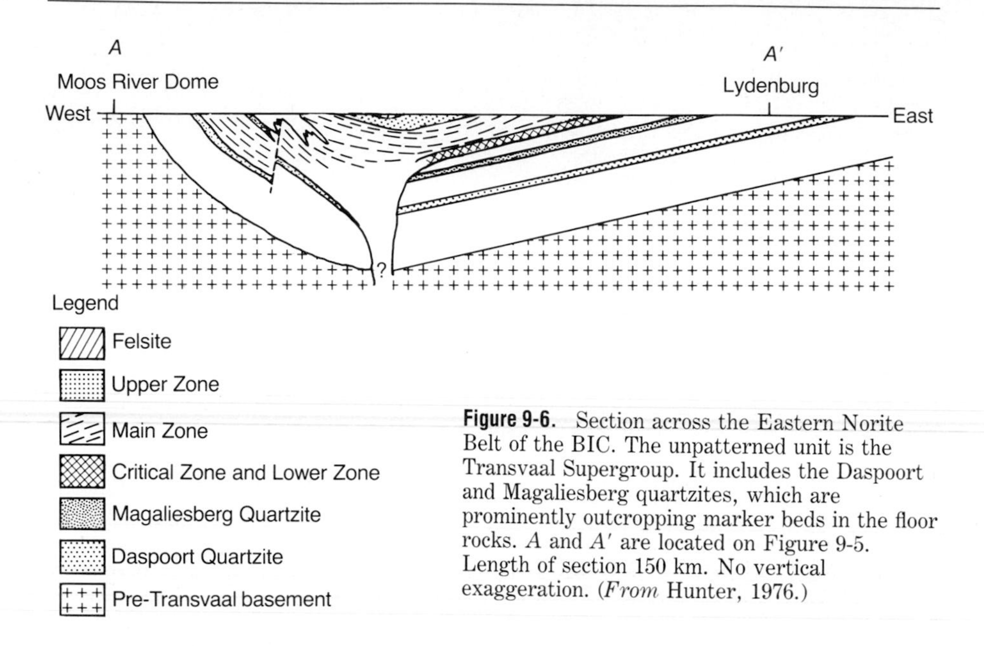

Figure 9-6. Section across the Eastern Norite Belt of the BIC. The unpatterned unit is the Transvaal Supergroup. It includes the Daspoort and Magaliesberg quartzites, which are prominently outcropping marker beds in the floor rocks. *A* and *A'* are located on Figure 9-5. Length of section 150 km. No vertical exaggeration. (*From* Hunter, 1976.)

similarity between the layers of the eastern and western lobes, and the Merensky Reef and the UG-1 chromitite are present in both. Similarities are less well developed in the southern and northern lobes.

In the vertical sense, Hall (1932), Cameron (1959 et seq.), Willemse (1969b), and others have divided the complex into several intervals. Divisions recently recognized in the Eastern Norite Belt (Cameron, 1978) are the Lower Zone, the Critical Zone, the Main Zone, the Upper Zone, and the Bushveld Red Granites and Granophyres (Table 9-2). In general, the complex consists of layered igneous rocks with the more mafic magnesium-chromium-rich rocks at the base of the pile. Layering may be relatively inconspicuous, or dramatic (Figures 9-7 and 9-8); it may occur on a scale of tens to hundreds of meters, or on a scale of centimeters (Figure 9-9); and it may yield knife-sharp millimeterwide contacts, or gradational ones, typically over several centimeters or tens of centimeters. Layering is best developed in the Critical Zone (Hall, 1932; Cameron and Emerson, 1959), as shown in Figure 9-10. Two of the most astonishing lithotypes in the complex are the Steelpoort Main chromite Seam (SMS), with the Leader chromite Seam 0.4 meter above it, and the Merensky Reef (MR). The SMS is near the top of the Lower Critical Zone pyroxenites. It is a chromitite, consisting of over 90% sand-sized chromite grains surrounding sparse platy inclusions of pyroxenite. It can be walked along strike for 65 km, and over most of that distance is essentially constant in thickness at 110 $\pm$ 8 cm. It is also nearly constant in composition (Table 9-3). The Leader Seam is similarly constant in composition (Table 9-3), thickness (47 $\pm$ 8 cm), and persistence along strike. Contacts of chromitites are generally sharp at the

TABLE 9-2. Thicknesses and Lithologies of Zones in the Bushveld Igneous Complex

Older Zone Names	New Zone Names*	Subzone Designations*	Thickness, Meters	Structure	Rock Types
Upper Zone	Upper Zone		2130	Massive to layered	Gabbro, Fe-olivine gabbro, magnetites
Main Zone	Main Zone		2780	Massive	Gabbro, norite, anorthosite
Critical Zone	(Upper) Critical Zone	Anorthosite Series	900	Strongly layered	Norite, anorthosite, chromitites
	(Lower) Critical Zone	Pyroxenite Series	530	Strongly layered	Bronzitites, norites, peridotites, chromitites
Transition Zone	Lower Zone	Upper Bronzitite	260		Harzburgites with dunite and bronzitite layers
		Harzburgite Subzone	530	Laminated	
		Lower Bronzitite Subzone	408	Vaguely laminated	Bronzitite
Chill (Basal) Zone		Basal Subzone	250	Vaguely laminated	Bronzitite with interstitial plagioclase, harzburgites
			7780		

"Stratigraphic" units in the Eastern Belt of the BIC. These units are cut by the Bushveld Red Granite. See also Figure 9-10.

*Cameron, 1978.

(a)

(b)

Figure 9-7. (*a*) Amazingly sharply defined layering of UG-1 chromite (dark) in anorthosite (white to gray) at the Dwars Rivier Geologic Reserve near Steelpoort, South Africa. The layers, not simply superposed, merge and split, always opening toward the west (left). (*b*) Similar layering at Red Mountain Deposit No. 12, Cook Inlet, Alaska. (*Photo by* P. W. Guild.)

base, especially at the base of the Leader Seam, but gradational upward. The SMS and the Leader Seam, or layers, dip gently centripetally at 10 to 35°, that is, toward the center of the BIC, as do the enclosing units, a sequence of layered, laminated bronzitites with disseminated chromite. The SMS has been mined downdip more than 1 km in places, its regularity of thickness and dip greatly facilitating mining procedures. Its status as one of the world's great chemical-grade chromium reserves has scarcely been damaged by mining; 97% of the world's chromium resources of over 30 billion tons are listed in southern Africa, 70% in the Republic of South

Figure 9-8. Fine centimeter-scale banding of bronzite, olivine, and chromite in the lower member of the C unit (Figure 9-10) above the Winterveld 343 adit, but below the SMS of the BIC. The white vertical streaks are magnesite ($MgCO_3$), a weathering product of olivine. The layers dip gently westward into the complex.

Figure 9-9. Persistent, smooth layering of mineralogically contrasting rock types in the BIC. The light lower layer is anorthosite, the gray upper layers are norite, with dark continuous millimeter-centimeter black bands of chromite and bronzite. This outcrop marks the contact between units F (below) and G (above) (Figure 9-10). The dark lense-shaped bodies are discoid pyroxenite inclusions emplaced at the time of accumulation of the rocks.

Zone	Subzone / series	Zone	Division	Unit	Thickness	Lithology	
Main Zone		Main Norite Zone				Norite interlayered with anorthosite (a)	
						Bastard Merensky Reef	MR
Critical Zone	Anorthosite series	Critical Zone	Upper			Merensky Reef	
				X	1190′	Norite, anorthosite (a), and feldspathic bronzitite (b)	
					48′	Feldspathic bronzitite	UG-2
				W	155′	Gabbro (above) and norite (below)	
				R	115′	Feldspathic bronzitite	
				O, M	655′	Anorthosite, norite (n), gabbro (g), and feldspathic bronzitite (b), with thin chromitites in upper part	UG-1, Fig. 9-7
				L	120′	Feldspathic bronzitite with thin chromitites	
				K	18′	Anorthositic norite	
	Pyroxenite series			J	430′	Norite	
				H	11′	Anorthosite, norite, and bronzitite	
				G	101′	Feldspathic bronzitite and norite	
				F	16′	Anorthosite, norite, and bronzitite	Fig. 9-9
			Lower	E	340′	Feldspathic bronzitite interlayered with chromitites	
				D		Steelpoort and Leader chromite seams	SMS
					175′	Feldspathic bronzitite interlayered with chromitites	
				C	460′	Feldspathic dunite and harzburgite (dn) and feldspathic bronzitites, with thin chromitites	Fig. 9-8
				B	895′	Feldspathic bronzitite with thin chromitites	
Lower Zone	Up. Bz.	Transition Zone	Upper		290′	Bronzitite	
	Harzburgite Subzone				1775′	Harzburgite interlayered with dunite (d) and bronzitite (b)	
	Lower Bronzitite		Lower		1355′	Bronzitite	
					25′	Interlayered anorthositic norite and norite	
					45′	Bronzitite	
	Basal Subzone	Lower Zone				Feldspathic bronzitite	

TABLE 9-3. Compositions of Chromites from Progressively Higher Chromitites in the BIC

	Steelpoort Main Seam	Leader Seam (2 m)*	UG-1 Dwars Rivier (637 m)*	Merensky Reef (974 m)*
Cr_2O_3	47.42	47.31	45.74	44.77
Al_2O_3	14.83	14.49	16.67	13.69
Fe_2O_3	7.35	7.44	5.89	8.41
FeO	19.22	19.03	22.20	23.86
MgO	10.20	10.07	8.47	7.45
Cr/Fe	1.62	1.62	1.43	1.22

Composition of chromites from the Steelpoort Main Seam and Leader Seam chromitites upward to the Merensky Reef. Values represent percentage by weight. Note that chromite composition can vary widely, with Mg^{+2} and Cr^{+3} concentrated in higher temperature, magmatically earlier, stratigraphically lower phases. Fe^{+2} and Fe^{+3} (approaching magnetite) are enriched in later, higher, lower temperature assemblages).

*Above the Steelpoort Main Seam.

Source: From Cameron (1978).

Africa's Bushveld Complex and 30% in Zimbabwe's Great Dike (Vermaak, 1979). The SMS is not the only one mined. Other seams, some of higher Cr/Fe ratio, have been worked in the western lobe, and metallurgical-grade chromite (Cr:Fe > 1.8) has been produced in the Potgietersrus lobe. Another significant source of refractory and chemical-grade chromite is a 1.2-meter-thick chromitite in Lower Zone rocks near the easternmost point of the BIC on Figure 9-5.

Above the SMS, the Merensky Reef lies about 100 meters below the contact between the Critical Zone and the Main Zone of the complex. The reef, or layer, has been traced in outcrop—originally by prospector Hans Merensky himself in the 1920s—for 150 km along the Eastern Norite Belt (Schwellnus, Hiemstra, and Gasparrini, 1976), for 250 km around the Western Norite Belt from Pretoria through Rustenburg (Vermaak, 1976; Cousins and Vermaak, 1976) around to near Thabazimbi (Wagner, 1929). A similar, but not identical, unit, the Platreef, extends for substantial distances in the northern limb. The reef is an integral part of the Bushveld layered sequence. It too dips gently inward, at about 10°, and is conformable to layers above and below. UG-1 chromitite layers in anorthosite, which

Figure 9-10. (*Facing page*) Generalized stratigraphy of the lower portion of the Eastern Norite Belt in the northeast sector of the BIC (Cameron, 1971) with revised nomenclature for zones and subzones (Cameron, 1978). Terminology of the Critical Zone and the Main Zone is unchanged from earlier publications. SMS flags the Steelpoort Main Seam, MR the Merensky Reef. The positions of the chromitites UG-1 (Dwars Rivier) and UG-2 are also shown. Layering common in both is more pronounced and more rapidly variable in the Critical Zone. The suffix –ic denotes postcumulus crystallization of plagioclase; the lowest important cumulus plagioclase is at the base of the Anorthosite Series.

are essentially identical to those at Dwars Rivier in the Eastern Norite Belt (Figure 9-7), are in similar stratigraphic positions; they occur 140 meters below the Merensky Reef near Rustenburg in the Western Belt, but nearly 500 meters below it in the Eastern Belt. Another chromitite layer called the UG-2, also containing platinum-group metal values as Pt-Fe alloys, occurs about 110 meters stratigraphically below the Merensky Reef in the west, 400 meters below it in the east. Both are located in Figure 9-10. The Merensky Reef itself is sandwiched between footwall norite below and a hanging-wall succession of 1 meter of pyroxenite, 5 meters of norite, and 5 meters of anorthosite mottled with orthopyroxene clots above. It is best described as a thin mafic pegmatitic norite layer, consisting as it does of intergrown 1-to-5-cm grains of plagioclase and bronzite. It may carry accessory olivine, and it typically has a thin chromite-rich band near or at its base where sulfides and platinum-metal values are highest. Chromite and orthopyroxene are cumulates, and plagioclase and clinopyroxene are postcumulus. The average stope height in the mines is only 75 cm (30 in), but the reef may be as much as 90 cm or as little as 30 cm thick in places. It contains from about 9 ppm (Wagner, 1929) to as much as 40 ppm platinum metals with an average near 10 ppm, or 10 grams per ton. About 60% of these values occur in solid solution in the base-metal sulfides pyrrhotite (high Pt, no Pd), pentlandite (high Pd, Rh), and pyrite (all platinum-group elements, but minor amounts) (Vermaak and Hendriks, 1976). The remaining 40% of the platinum-metal values occur as the discrete platinoid-metal minerals cooperite (PtS), laurite (RuS_2), braggite (Pt, Pd, Ni)S, and Pt-Fe alloys or their intergrowths (Vermaak and Hendriks, 1976). Many other precious-metal minerals and alloys have been described, but they are of minor value. Some precious-metal ions may proxy in chromite; more data are being sought. The platinum-metal values are universally associated with chromite, typically in the centimeter-or-so-thick chromite band which is characteristically near the base of the norite pegmatoid of the reef. The reef is mined for some 30-km strike length in the Rustenburg area alone. The Impala mines and the Union area to the north and west add to this total, making these the most extensive mines in the southern hemisphere. Mining extends from outcrop to points 1 km downdip. Published reserves are nearly 6 billion grams; unproven resources are nearly twice that much more (Hunter, 1976), and are clearly the largest platinoid resources in the world. Large additional resources have been identified in UG-2, and other chromitites in the BIC and elsewhere are being evaluated.

The origin of the BIC and its components is too complicated to be covered here in more than a superficial way. The sequence of events is generally agreed to have included (Hunter, 1976):

1. Deposition of the Early Proterozoic Transvaal Supergroup sediments, with at least three volcanic BIC-precursor events

2. Injection of a series of diabase sill sheets into the uppermost Pretoria group of the Transvaal Supergroup sediments
3. Extrusion of epicrustal Rooiberg felsites and granophyres, with interbedded arkoses and grits forming at the same time
4. Intrusion above the Transvaal sediments and beneath Rooiberg Felsites of the main plutonic phase and formation of layered ultramafic, mafic, and intermediate rocks at about 1.9 billion years ago
5. Intrusion of the late felsic plutonic phase represented by the Bushveld Red Granites (Figure 9-5) through all earlier rocks

Within that framework, however, many vexing problems remain. Gravity measurements require that the BIC cannot now be a continuous sill-like sheet extending under cover from the Eastern Belt to the Western Belt along a chordlike geoidal plane. But were preexisting, continuous layers obliterated or upturned (Figure 9-6) by intrusion of the central granite phases? Does each of the lobes represent a separate but contemporaneous cognate intrusion? Were the lobes interconnected, as the continuity of layers suggests, or only similar in differentiation-evolution? Was the whole Bushveld "intrusive" event triggered by the impact of a single meteorite, or perhaps (Rhodes, 1975) by a cluster of four? Are the red granites, with their tin-fluorite deposits as at Zaaiplaats (Lenthall, 1974, Lenthall and Hunter, 1977) differentiates of the more mafic older units, or do they represent a separate later magmatic event?

As significant as these problems are, the problem that has provoked the liveliest discussion is that of the genesis of the layering and the layered units themselves. There remains little argument that they are fundamentally magmatic, although metasomatic alternatives are still provocatively suggested (Gilmour, 1980). Cameron and his coworkers have presented abundant geochemical-mineralogical data (1959 et seq.) that indicate an irregular but progressive differentiation from mafic, magnesium-chromium-enriched units near the base to intermediate, FeO-enriched, more sialic phases near the top, conforming to conventional tholeiitic differentiation. If crystallization proceeded by simple crystallization and the raining-down of cumulus crystals in a static melt, the rocks would not be so layered nor so profoundly mineralogically contrasting from layer to layer, even as from nearly pure chromite to nearly pure pyroxene in bands only a few grains thick (Figure 9-8). The layering has been ascribed (1) to magma mixing upon frequent multiple intrusions; (2) to reversals in the thermal or chemical regime resulting in shifts in crystallization ratios near the magma chamber roof so that different "crops" of crystals fell to the floor at successive times; (3) to variations in the fugacity of oxygen in the melt and expansion-contraction of the chromite stability field; (4) to convective circulation as suggested at the Skaergaard Complex in Greenland; (5) to differential magmatic sedimentation and the winnowing of minerals of varying shapes, sizes, and specific gravities from overcooled, convecting magmatic currents moving first outward and downward along the sides, then centrally across the chamber floor; (6) to sheeted, turbiditelike spreadings of crystal mushes banked along the walls outward and downward across the floor; (7) to the evolution and gravitative separation of immiscible silicate and oxide magmatic liquids; (8) to varied bottom crystallization controlled by interaction of melt composition, true pressure-temperature gradients, and the adiabatic gradient; (9) to differential flotation of lighter phases like plagioclase during sedimentation of denser solids like pyroxene and chromite; (10) to variations in total pressure; and, of course, to combinations of all of these explanations. Differential magmatic sedimentation appears to explain many of the physical and chemical requirements (Cameron and Emerson, 1959), but the thinness and the lateral extent of individual layers and of the layered sequence as a unit, at 8 km thick versus at least 80 km horizontally in the northeast portion of the Eastern Norite Belt, appear to rule out convection as we know it in other liquids. McBirney and Noyes (1979) point out that settling rates and viscosities do not permit "winnowing" and suggest that Jackson's (1959) ideas of variable bottom crystallization as proposed for the Stillwater Complex of Montana can perhaps be more generally applied. Bottom crystallization could account only with difficulty for many of the physical-mechanical features of the BIC layered rocks, especially the amazingly fine-scale layering shown in Figure 9-8. Vermaak (1976, 1980) thinks that the Merensky Reef formed a few meters below a "ceiling" that was a mush of plagioclase tablets suspended in magma, with the settling of cobblelike "boulders" of plagioclase with interstitially growing mafic minerals, of chromite grains, and of the precious metal minerals listed earlier. Others believe the Merensky Reef to represent injection of a new pulse of more mafic magma with a "whiff" of immiscible sulfide liquid and chromite. Research into the answers to these questions and to the testing of proposed models continues.

Certainly one of the requirements of any solution to these problems is that it should be consonant with the fact that many layered mafic intrusive bodies exist. They range in age from the 2.5 billion years of the Great Dike through the 1.9 billion years of the Bushveld (600 million years younger than the Great Dike, but descriptively almost identical to it) and the Jurassic age of the Dufek Massif to the Middle Tertiary age of the Skaergaard Complex (55 million years). Many Precambrian intrusions are dominated by accumulations of chromite, magnetite, or both, such as the BIC, the Stillwater Complex, and the Great Dike; younger ones do not appear to contain chromitites. Others, such as the Sudbury Complex next to be described and the Nizhny Tagil and N'orilsk Complex in the Urals of the U.S.S.R., generated sulfide accumulations. Still others like the Duluth Complex of Minnesota, the Muskox Intrusion of the Northwest Territories,

Canada, and perhaps the Stillwater Complex, may have spawned both. All of these bodies are layered mafic intrusive complexes.

Another significant resource in LMIs is that of layered magnetite-rich units. They are exemplified by the Magnet Heights occurrence in the Bushveld Igneous Complex (Molyneux, 1970) near Lydenburg (Figure 9-5); similar units occur in the Stillwater and Duluth complexes, the Skaergaard Complex, and elsewhere. The magnetite-rich layers are found well above the generally basal magnesio-chromite layers, magnesium and chromium being high-melting components and both FeO and Fe_2O_3 generally increasing with differentiation, and therefore being found in younger, higher members of LMIs. Similarly, the magnesium-rich olivines and orthopyroxenes yield upward to iron-rich clinopyroxenes, and labradorite yields to andesine and oligoclase. In the BIC, monomineralic rocks near the magnetites include anorthosite and pyroxenite; all three of these occur as specialized layers within dioritic, gabbroic, and noritic units which dip gently to the west. More than 30 magnetite layers of thickness from a few centimeters to several meters have been mapped. They are all measured from, and related to, the Main Magnetite, which occurs in the Upper Zone (Figures 9-5 and 9-6; Table 9-2), the uppermost 2 km of the layered sequence of the BIC. Most of the magnetite layers have sharp lower and gradational upper contacts, but all combinations of "sharp" and "gradational" have been observed. They consist of 80 to 100% magnetite, 1 to 10% sporadically distributed coarse exsolved granular ilmenite, less than 1% sulfides, some trace exsolved oxide phases, with the balance mostly plagioclase and lesser pyroxene. The magnetite contains a few percent titanium and 0 to 2% V_2O_5. Titanium increases upward. The V_2O_5 content is significant as a vanadium resource. Huge tonnages of vanadiferous magnetite are present even although V_2O_5 decreases regularly from 2% in the lower seams 350 meters below the Main Seam through 1.5% in the Main Seam itself to almost none in the highest magnetite layers 1300 meters above the Main Seam. V_2O_5 is being looked for in other LMIs, especially the Stillwater Complex (Dietrich, 1985).

Molyneux (1970) described several postdepositional mineralogic changes in the ore related to postmagmatic oxidation, exsolution, and weathering, and listed several critical properties of the Magnet Heights BIC occurrences:

1. Plagioclase and magnetite are intimately associated in the magnetite seams and their transitional contacts; pyroxene and olivine are scarce.

2. Plagioclase and magnetite started to crystallize before later, interstitial clinopyroxene and fayalite, at least in the seams.

3. Magnetite contacts are typically gradational upward into anorthosite, and sharp downward against it.

4. Anorthosite with disseminated magnetite dominates the sections below the groups of magnetite seams.

5. The magnetite bands do not cut from one level to another, although they may thicken or thin laterally.
6. Platy crystals and xenoliths are oriented in the plane of the layering.
7. There is "extroardinary" lateral persistence of even the thinnest, millimeter-scale bands.
8. V_2O_5 decreases steadily upward.

Willemse (1969a) suggested that the magnetite-rich layers are cumulus rocks formed by crystal setting of magnetite octahedra in a feldspathic melt, with extensive post depositional overgrowth of magnetite. Molyneux (1970) stated that the extreme lateral persistence and regularity of the units implies tranquil rhythmic crystallization near the then high-level magma chamber floor, but he offered no explanation for the phase or cryptic layering. These magnetite-rich rocks have been mined for iron ore and steelmaking, but the TiO_2 and V_2O_5 contents are metallurgically troublesome. They represent enormous reserves of iron, and means of treating them are advancing. As indicated, they may represent the most extensive resource of vanadium, itself a ferro alloy in stainless and high-performance steels. Molyneux noted that a sequence in the western lobe of the BIC, 400 km distant, is stratigraphically similar, with the main seam "the same thickness, lithology, and V_2O_5 content in both places." The first four upper seams, the lower seams, and the seams of Subzone D are similar in thickness, in position in the column, and in V_2O_5 content in both places.

Sudbury Complex, Ontario—Copper-Nickel-Platinum

The Sudbury district, Ontario, Canada, has been the world's most productive nickel deposit. It has yielded about 8 million tons of this metal, about the same amount of copper, and minor quantities of the platinum metals, cobalt, iron, sulfur, gold, silver, selenium, and tellurium. The geology of the area has been studied and restudied and the literature is voluminous (Coleman, 1905, 1913, 1926; Yates, 1938, 1948; Hawley, 1962; Souch et al., 1969; Naldrett el al., 1970, 1972; Peredery and Naldrett, 1975; Dietz, 1972, and the collection of *New Developments* papers in which it appears; Pattison, 1979; and Fleet, 1977, 1979). The Sudbury Complex has been the center of many geologic controversies, and its origin is still hotly discussed. As will be shown, debate now centers largely upon whether or not magmatic activity was triggered by a meteorite colliding with a Precambrian surface. A recent paper (Pattison, 1979) essentially presupposes this astrobleme, or bolide, concept as central to irruptive activity; most of the geologists of Inco Ltd. agree. Many other geologists with Falconbridge Nickel, another major company mining the complex, and "outsiders" remain skeptical. An interesting observation is that only one group or the other can be correct; the "middle ground" of a "partial impact" cannot exist. It is also noteworthy that most published reports since Robert Dietz's original suggestion in 1964 that the form and size of the complex are consistent with those of a deformed meteorite scar have added more evidence that it indeed is one (Guy-Bray et al., 1966; Guy-Bray and Peredery, 1971; Dence, 1972; French, 1968, 1972; Brocoum and Dalziel, 1974; Pattison, 1979). Although contrary arguments have been presented (Card and Hutchinson, 1972; Cantin and Walker, 1972; Fleet, 1977, 1979), the pendulum has swung toward the meteoriticists in recent years. Much of the evidence will be presented below. The sublayer rocks and ore deposits will be described first, then the rest of the complex, and finally the underlying and remarkable overlying rocks will be treated.

The Sudbury basin is a large elliptical depression about 60 km long by 30 km wide, with its long axis trending east-northeast (Figures 9-11 and 9-12). Lodged as a complex, lopolithic, generally funnel-shaped mass below the roof of Whitewater Group sediments and above the older basement complex is the nickel irruptive. The *irruptive* (in British usage synonymous with intrusive) complex involves at least three major units, from bottom to top called the *sublayer*, the *norite*, and the *micropegmatite*. The copper-nickel ores are uniquely associated with the lowermost sublayer, especially near its basal contact with older footwall rocks below. Pattison (1979) described the sublayer as composed of two major variants, an igneous gabbro-norite-diorite rock and a group of mafic to felsic, generally leucocratic breccias. Both rock types may contain Fe-Ni-Cu sulfides, and virtually all of the sulfides are contained in these sublayer materials. The sublayer (Figure 9-12) occurs as gently dipping sheets and irregular lenses along the floor of the complex (Figure 9-13), as small bodies in depressions or troughs in the complex floor called *embayments* (Figure 9-14), and as steeply dipping to nearly vertical dikelike bodies called *offsets*, which project outward into the footwall (Figure 9-15). The sublayer is typically gabbroic (augite-bronzite-plagioclase) to noritic (bronzite-plagioclase) where it is the floor unit of the irruptive and in the embayments, but in the offsets it is a hornblende-actinolite-biotite quartz diorite, which Pattison (1979) considers also to have been gabbro-norite prior to assimilation of wall rocks or to deuteric (?) alteration. The Foy Offset contains some unreacted gabbronorite within 1.5 km of the irruptive contact.

Souch and his colleagues (1969) recognized three orebody types exemplified by the North Range, the South Range, and the offsets, representing, respectively, normal footwall contact, embayment, and offset environments. The Creighton mine on the south limb of the basin, the Levack mine on the north limb, and the Frood-Stobie Offset deposit north of the town of Sudbury will used here as examples of those three environments of ore deposition.

The Creighton ore zone includes a series of individual sheetlike orebodies that have been outlined to a depth of approximately 3 km (Yates, 1948; Souch et al., 1969) and that continue downdip an unknown distance.

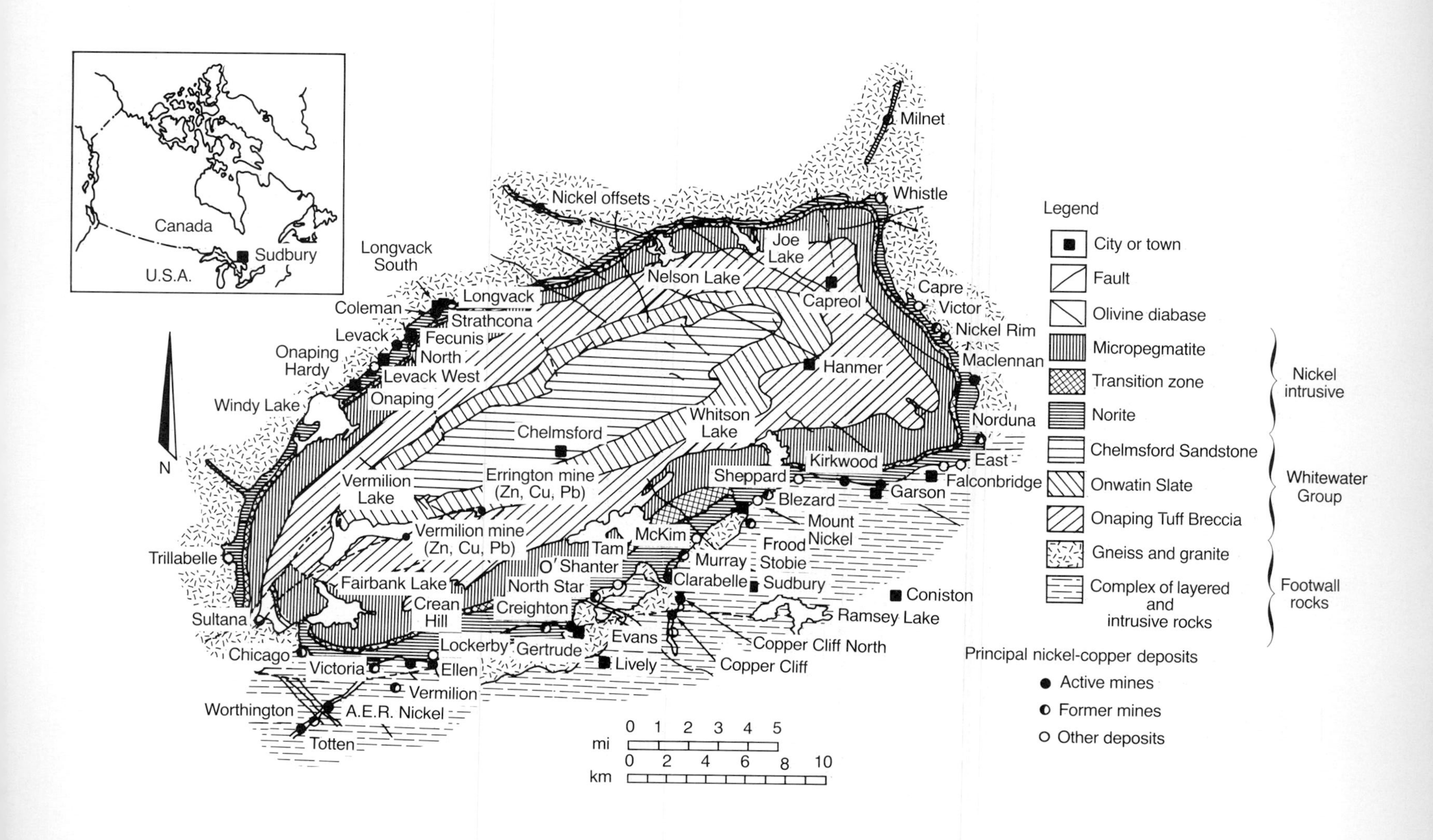

Canada
U.S.A.
Sudbury
N
Milnet
Whistle
Nickel offsets
Joe Lake
Nelson Lake
Capreol
Capre
Victor
Nickel Rim
Maclennan
Hanmer
Norduna
East
Falconbridge
Garson
Kirkwood
Sheppard
Blezard
Mount Nickel
Frood
Stobie
Murray
Sudbury
Coniston
Ramsey Lake
Copper Cliff North
Copper Cliff
Clarabelle
McKim
Whitson Lake
Chelmsford
Errington mine (Zn, Cu, Pb)
Vermilion mine (Zn, Cu, Pb)
Vermilion Lake
Tam
O'Shanter
North Star
Creighton
Evans
Lively
Gertrude
Lockerby
Crean Hill
Fairbank Lake
Ellen
Vermilion
Victoria
Chicago
A.E.R. Nickel
Worthington
Totten
Sultana
Trillabelle
Windy Lake
Onaping
Levack West
Onaping Hardy
North
Levack
Fecunis
Strathcona
Coleman
Longvack
Longvack South
Legend
City or town
Fault
Olivine diabase
Micropegmatite
Transition zone
Norite
Chelmsford Sandstone
Onwatin Slate
Onaping Tuff Breccia
Gneiss and granite
Complex of layered and intrusive rocks
Nickel intrusive
Whitewater Group
Footwall rocks
Principal nickel-copper deposits
Active mines
Former mines
Other deposits
mi
0 1 2 3 4 5
km
0 2 4 6 8 10

Figure 9-11. (*Facing page*) Geological and locality map of the Sudbury district, Ontario. Note the close association between open and closed dots, which are Cu-Ni-Pt orebodies, the complex floor, and the dikeform offshoots. The rock type at the floor of the complex is called *sublayer*. It and the offshoots are shown more clearly in Figure 9-12. (*From* Boldt and Queneau *via* Stanton, 1972.)

The contact of sulfides and sublayer rocks downward with the older granite and gneiss of the footwall is sharp, but upward against the norite the contact is obscure and poorly defined. The sublayer materials are cut by a series of faults that parallel each other and are concentrated close to the contact; slices of ore sulfides have been structurally inserted into the footwall rocks well after the ores crystallized. Large "rolls" are recognized in the faults, both along strike and downdip. In most places the faults lie close to the footwall and control the position of the ore. Pieces of massive sulfide occur along the faults and in the brecciated footwall rock adjacent to the faults.

The ore itself takes several forms described in detail by Hawley (1962) and Souch et al. (1969). The most massive material is composed of more than 95% sulfide; it is ore of this type that has characterized many of the contact deposits of the South Range. Sulfides also occur as disseminations

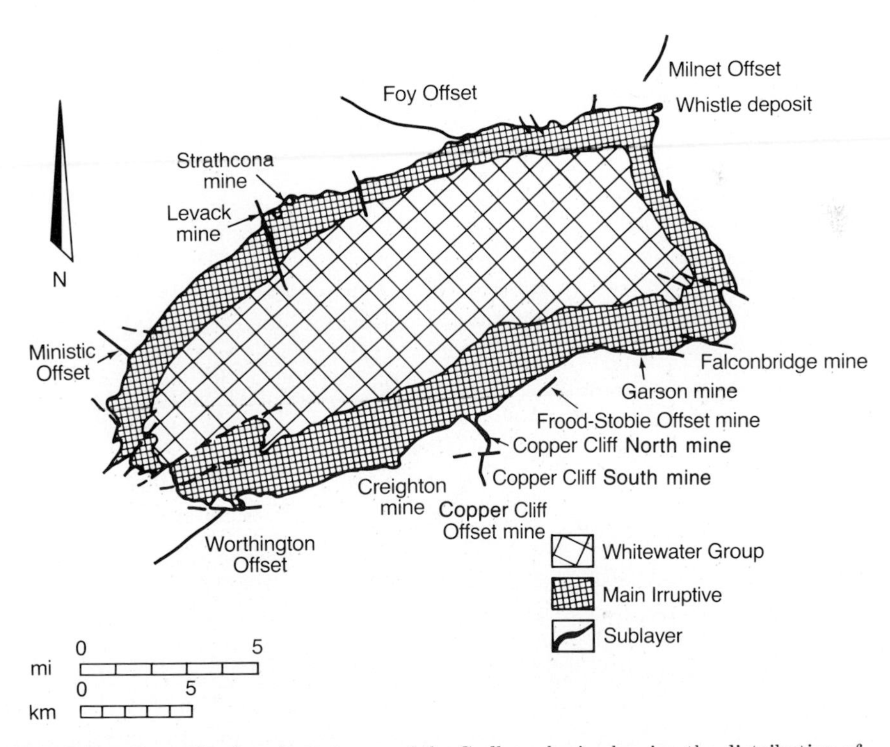

Figure 9-12. Generalized geological map of the Sudbury basin showing the distribution of major occurrences of sublayer material (thickness somewhat exaggerated) and principal offset dikes. (*After* Pattison, 1979.)

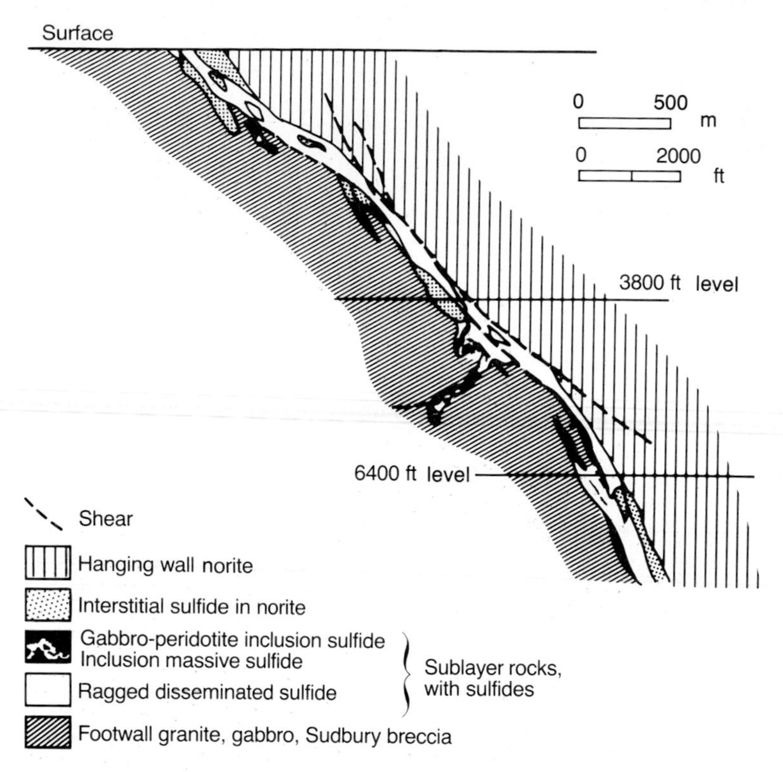

Figure 9-13. Generalized section through the Creighton ore zone, looking west. This array is typical of most Sudbury Complex ore, with sulfides on or slightly penetrating the footwall contact. (*From* Souch et al., 1969.)

of beadlike or dropletlike blebs in silicate rock (Fleet, 1977), as irregular grains interstitial to the gabbro-norite textural fabric, as disseminations in deformed, contorted schist-gneiss wall-rock fragments, and in sublayer gabbro which also contains peridotite inclusions, both in normal contact settings and in offsets. Low-grade material such as disseminated sulfide in sublayer rock is volumetrically by far the most abundant ore type at Sudbury. Semimassive and massive ore constitutes a lesser proportion. Another important ore type, especially in the embayments, is that of sulfide as matrix interstitial to breccia fragments of the granite and gneiss of the floor rocks. Figure 9-14 shows at least three varieties of breccia, all of which are locally sulfide-cemented; sulfidic footwall breccias have recently been discovered and are being mined more than 100 meters into the basement floor of the North Range (Abel et al., 1979).

Mineralogically, the sulfide material is essentially the same wherever it occurs. Its bulk composition was set when an early high-temperature sulfide liquid crystallized to form nearly homogeneous sulfide solid solution, perhaps containing small amounts of platinum-metal mineral microcrystals. During cooling from temperatures probably near 1000°C to normal geo-

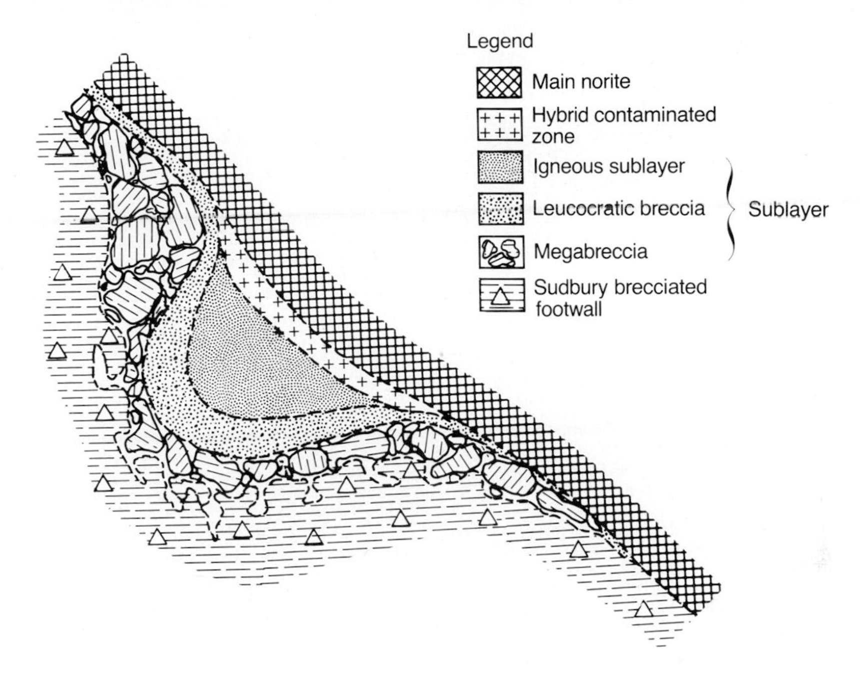

Figure 9-14. Cross section through an idealized sublayer occurrence on the North Range of the Sudbury structure near Levack. Everything between the Main Norite and the locally brecciated footwall is sublayer material, typically containing sulfides. (*From* Pattison, 1979.)

thermal gradient temperatures, the solid solution reorganized itself, first with the exsolution separation of chalcopyrite ($CuFeS_2$), then of pentlandite, the principal Sudbury nickel mineral ($[Fe, Ni]_9S_8$, Ni $\cong$ 35%). The sulfide minerals are now dominantly pyrrhotite, pentlandite, and chalcopyrite with minor amounts of pyrite and cubanite, although cubanite exceeds chalcopyrite in places in the massive sulfides. Accessory magnetite is common. In microscopic detail, however, the mineralogy is varied; minor amounts of at least 64 species are recognized. Hawley (1962) listed 10 secondary or supergene minerals and 18 nonmetallic and gangue minerals. Cabri and Laflamme (1976) and Hoffman et al. (1979) report further electron microprobe and mineragraphic studies. Typical ore contains about 1% Cu, 1% Ni, and a few parts per million platinum metals.

The southward-dipping Levack and Strathcona mines on the North Range (Figure 9-11) consist of sulfide- and inclusion-bearing sublayer norite, sulfide-bearing granitic breccia, and massive sulfides with rounded to angular basement rock or peridotite inclusions. The ore is in an embayment that follows and dimples the regional trend of the irruptive contact (see Figure 9-14). Sulfides in the norite are sparse but abundant enough locally to constitute ore. A second type of ore in the granite breccia is one of sulfides in blebs in the matrix, which coalesce locally to form minable pods. A third type of ore is one of sulfides disseminated in granite breccia intruded

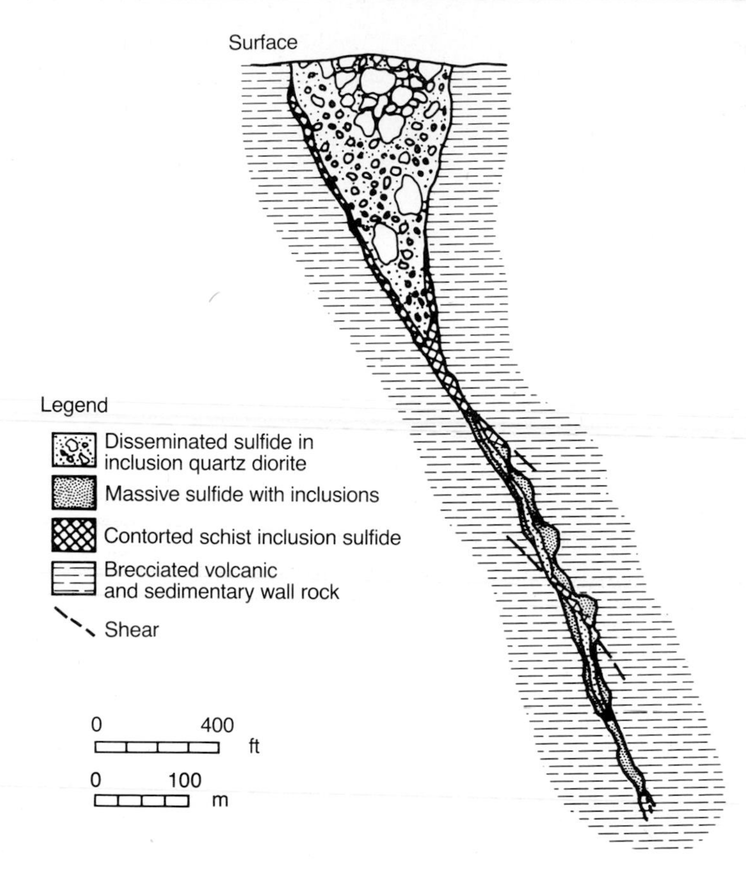

Figure 9-15. Generalized section through the Frood Offset orebody, looking southwest. All but the granite-gneiss wall rock is actually sublayer material. Note the size of the inclusions, many of which are peridotite, relative to the narrowness of the offset. Sulfides occur from top to bottom. (*After* Souch et al., 1969.)

along the norite contact and in shattered areas in the underlying granitic gneiss. The controlling structures are obscure, but faulting similar to that at the Creighton mine is absent. What happened at the interface at the base of the complex appears critical.

The Frood-Stobie Offset is a mineralized dikelike body of quartz diorite sublayer rock that cuts at low angles across the northeasterly trend of steeply dipping volcanic and sedimentary floor rocks. The composition of the quartz diorite in it varies considerably, but it is dominantly a fine- to medium-grained rock containing plagioclase, quartz, hornblende, and biotite, with lesser amounts of sulfides and magnetite. Much of the quartz diorite is the peculiar breccia known as inclusion quartz diorite (Souch et al., 1969). Breccia also forms a discontinuous envelope around much of the orebody, and was once considered to be an intrusive expression of the activity that produced the Onaping volcanics that lie above the main intrusive mass, to be considered below. Hawley concluded that the offset material had been forced downward into cracks in the floor of the complex. The presence of huge inclusions of igneous rocks in some of the offsets (Figure 9-15) is consistent with that interpretation.

A cross section of the Frood-Stobie mine is shown in Figure 9-15. Zurbrigg and his colleagues (1957) reported that the orebodies were in and along a downward-directed dike; other offset deposits appear more like radial dike fillings. Two types of ore distinguished in the offsets are "disseminated" and "nondisseminated". They include massive sulfides, breccia sulfides, a breccia containing mineralized schist fragments in a sulfide matrix, stringer ore, and a low-grade siliceous ore. Although the predominant minerals are pyrrhotite, chalcopyrite, pentlandite, pyrite, and cubanite, Zurbrigg lists accessory minerals such as niccolite, maucherite, galena, sperrylite, hessite, and several minerals of the platinum group. The rather sharp local variations in pyrrhotite, pentlandite, and chalcopyrite ratios led Zurbrigg to describe zoning in the Frood-Stobie deposit. Three zones were discussed by Zurbrigg and later by Hawley: disseminated ore, massive ore, and siliceous ore at depth (Zurbrigg et al., 1957; Hawley, 1962). According to Hawley, this zoning resulted from gravitative settling of droplets of sulfide through a silicate melt, which formed the disseminated and massive zones at the base. Because the most mobile, texturally youngest minerals are concentrated at the base and were fugitive downward, Hawley described the zoning as "upside down," or downward directed, as opposed to more normal, upward and outward directed zoning in most other deposit types. Fleet (1977) noted that the blebs of sulfide in silicate that have been interpreted as evidence of a preexisting sulfide liquid are curiously variable in mineralogic ratios of the sulfides, and discounts the importance of immiscible liquid separation.

That the floor, or footwall, rocks are generally Archean granites and Huronian (Aphebian) volcanic and sedimentary rocks has been noted. Card and Hutchinson (1972) emphasized that the only evidence of volcanism in that portion of the southern province of Canada is a tholeiitic to calc-alkaline volcanic pile restricted to the immediate Sudbury area, which is temporally and spatially appropriate to have represented early stages of Sudbury igneous activity. The floor rocks contain at least three features that have provoked much debate—the Sudbury breccias, the pseudotachylite breccias, and the shatter cones. The *Sudbury breccias* are so called because of a singular absence of lithologic mixing. The breccias themselves may be dikelike, sheetlike, pipelike, irregular, or large diffuse zones ranging in scale from millimeters to hundreds of meters, with varying degrees of fragment rotation. The matrix and the fragments are typically composed only of the rock type in which the breccia occurs; the compositions of matrix and fragments change at formation contacts, with almost no transport or mixing across contacts. *Pseudotachylite breccias* are remarkable for the same reason, but they also show microscopic evidence that their dark-colored, fine-grained to conchoidal groundmass material was once fused and glassy, that is, molten, presumably by strong frictional heating on shear planes. *Shatter cones* are structures that may also crosscut lithologic contacts. They are nested stacks of nose-cone-like, coaxial, fluted conical shear surfaces, apices generally upward, and centimeters to a few meters in height. All three of these features have been widely attributed to shock metamorphism resulting from meteorite impact, although some say that pseudotachylite brecciation (Wenk, 1978) and shatter cones (Fleet, 1979) can both be caused by seismic tectonic stresses. Fleet in fact concluded that the shatter cones near Sudbury formed geologically much later than did the complex, although that conclusion is disputed.

Immediately above the sublayer units and their contained orebodies are the two other units of the main Sudbury Irruptive. Lying like a huge funnel-shaped, vaguely layered lopolithic mass below the Whitewater Group and above the older basement complex and sublayer rocks are the norites and gabbros of the main body, the finely cross-hatched portion of Figure 9-12. The norites range considerably in composition, and much of the lopolith is altered. The feldspars are dusty with sericite and epidote, and pyroxenes are commonly changed to amphiboles. Mafic norite—plagioclase and orthopyroxene—at the base grades upward to felsic norite by decrease in pyroxene, increased Na/Ca in plagioclase, and increased quartz. Chemical trends in the pyroxenes are similar to those in the Bushveld (Figure 9-16), generally showing decreased Mg/Fe upward. The norite is characterized by coarse poikilitic feldspar enclosing pyroxene, as much as 50% hypersthene, and minor amounts of augite and quartz. Felsic norite grades into quartz gabbro by increase in both augite and quartz. The gabbro includes a distinctive magnetite-apatite-rich layer. At the top of the intrusive mass is the micropegmatite, a felsic igneous layer consisting of subhedral sodic feldspars in a matrix of quartz-K-feldspar granophyre and granular quartz (Stevenson, 1963; Souch et al., 1969; Naldrett et al., 1970). The micropegmatite reportedly grades into the underlying norite in many places, though rubidium-strontium age dating by Fairbairn and his colleagues (1968) indicates an age of about 1.7 billion years for the micropegmatite and about 2 billion years for the underlying norite. Peredery and Naldrett (1975) concluded that the micropegmatite is younger than the norites and that it is not part of a simple layered differentiate. Kuo and Crocket (1979) analyzed rare earth element patterns in Sudbury Complex rocks and concluded that the three major rock units are comagmatic. They indicate that the norites represent cumulate phases from a liquid that formed the main mass of quartz norite–quartz gabbro, with the micropegmatites crystallizing highest and last from a residual melt.

This controversy notwithstanding, another key surface is the upper contact of the irruptive's felsic phase (the micropegmatite) and the overlying Whitewater Group rocks. At the base of the Whitewater is the Onaping Formation, a complex of breccias about 2 km thick. Fragments in this brec-

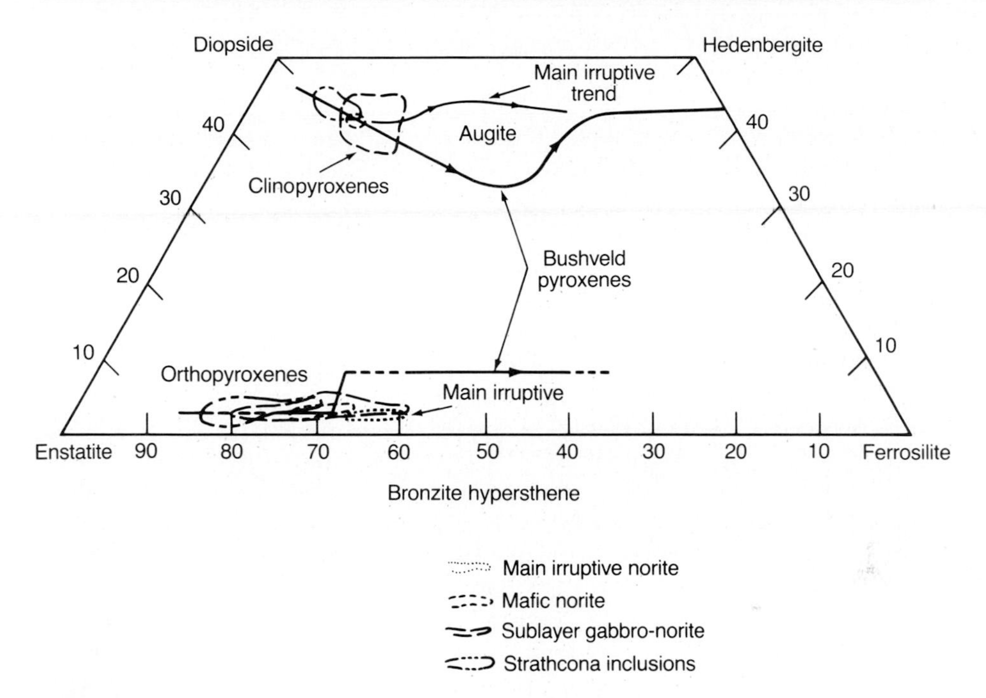

Figure 9-16. Pyroxene quadrilateral with compositional fields for pyroxenes from the sublayer, the main Sudbury Irruptive, and the exotic inclusions. Bushveld pyroxene trends are shown for comparison. Geologically lower, older, more magnesian layers fall at the left; they progress to higher, younger, more ferruginous layers that plot at the right. (*From* Pattison, 1979.)

cia are said to be somewhat coarser near the base than they are toward the top, though the formation is generally unsorted. The fragments are composed of various types of country rock cemented by a matrix of dark ashlike material and devitrified glass. Fragments of sulfides, especially pyrrhotite and sphalerite, are common, and pieces of identifiable basement rocks are found. The unit as a whole and the pyrrhotite contained in fragments within it are anomalously nickel-rich. The origin of this breccia has been the subject of debate and is still not clear. The explanation that formerly was widely held is that the breccia is a tuff or ignimbrite (Speers, 1957; Hawley, 1962). Williams (1956) and Thomson and Williams (1959) referred to it as having been deposited from an ignimbritelike "glowing avalanche." In the past few years, many authors have accepted the thesis, first advanced by Dietz (1964), that the Onaping is a fall-back breccia resulting from the impact of a large meteorite (Guy-Bray et al., 1966; French, 1968; Guy-Bray and Peredery, 1971; Pattison, 1979). This hypothesis is supported by the presence of shock metamorphic features in clasts in the breccia.

The formations above the Onaping are unbrecciated. Directly above the Onaping is the Onwatin Slate, locally more than 300 meters thick, of fine-grained, carbonaceous, pyritic argillite that in places contains beds of limestone and chert. The Onwatin Slate in turn is overlain by the Chelmsford Sandstone, or graywacke, which fills the central part of the basin. Extensive geological mapping throughout the region has failed to correlate the members of the Whitewater Group with strata outside of the Sudbury basin. Their absolute geologic age, which would dictate whether they were older roof rocks intruded by the irruptive or of syn- or postirruptive age, has not been established.

The structure of the basin is far from simple. The northern contact dips about 41° south and the southern contact dips 65° north, as indicated by observation and magnetic studies (Souch et al., 1969). Two systems of faulting are recognized. The first trends nearly east-west and is typified by the Cameron Creek Fault on the west and the Airport Fault on the east. The Cameron Creek fault system dips to the south and seems to have a vertical offset of at least 5 km, with the south side up relative to the north. The second prominent fault system trends north-northwest; individual faults dip steeply west or are vertical. The whole basin has been intruded by a regional swarm of narrow, vertical, olivine-diabase dikes.

Finally the genesis of the complex and its ores must be considered. It should be noted, however, that regardless of the triggering mechanism at Sudbury, crystallization of the intrusive mass appears to have proceeded more or less normally so that Sudbury can be compared with other LMIs, whether or not they too have unusual geologic histories.

Clearly, the foregoing sections have presented evidence that has been applied differently by different investigators. There is some doubt that a meteorite slammed into the Precambrian surface at all (Wenk, 1978; Fleet, 1979). Card and Hutchinson (1972) pointed out that it is fortuitous in the extreme that a meteor happened to hit not only at the junction of four tectonic provinces, but also in a single area within thousands of square kilometers where volcanic activity was already established. Others suggest that Lake Wanapitei, a younger subcircular lake basin at the northeast end of the Sudbury basin, was the point of impact that formed the shatter cones and some of the breccias. As stated earlier, many of the geologists who know the area best are convinced that a meteorite impact did indeed trigger melting, intrusive activity, and the formation of the ores, and that the originally circular form of the basin was deformed to its present elliptical shape by Grenville-age stresses. Pattison presented a schematic of a high-velocity impact at the Sudbury area (Figure 9-17) and concluded that only two processes explain the high-energy intrusive nature of the sublayer and its various ores (Pattison, 1979, p. 257): (1) segregation of a sulfide-rich

Figure 9-17. (*Facing page*) Development of the Sudbury structure as the result of a hypervelocity meteorite impact 2 billion years ago. (*a*) Prebolide geology and extent of cratering at the instant of impact. (*b*) Scene at moments to minutes after collision. (*c*) Hours to days later. (*d*) As the norites begin to crystallize. The Cu-Ni ores were emplaced between (*c*) and (*d*). (*From* Pattison, 1979.)

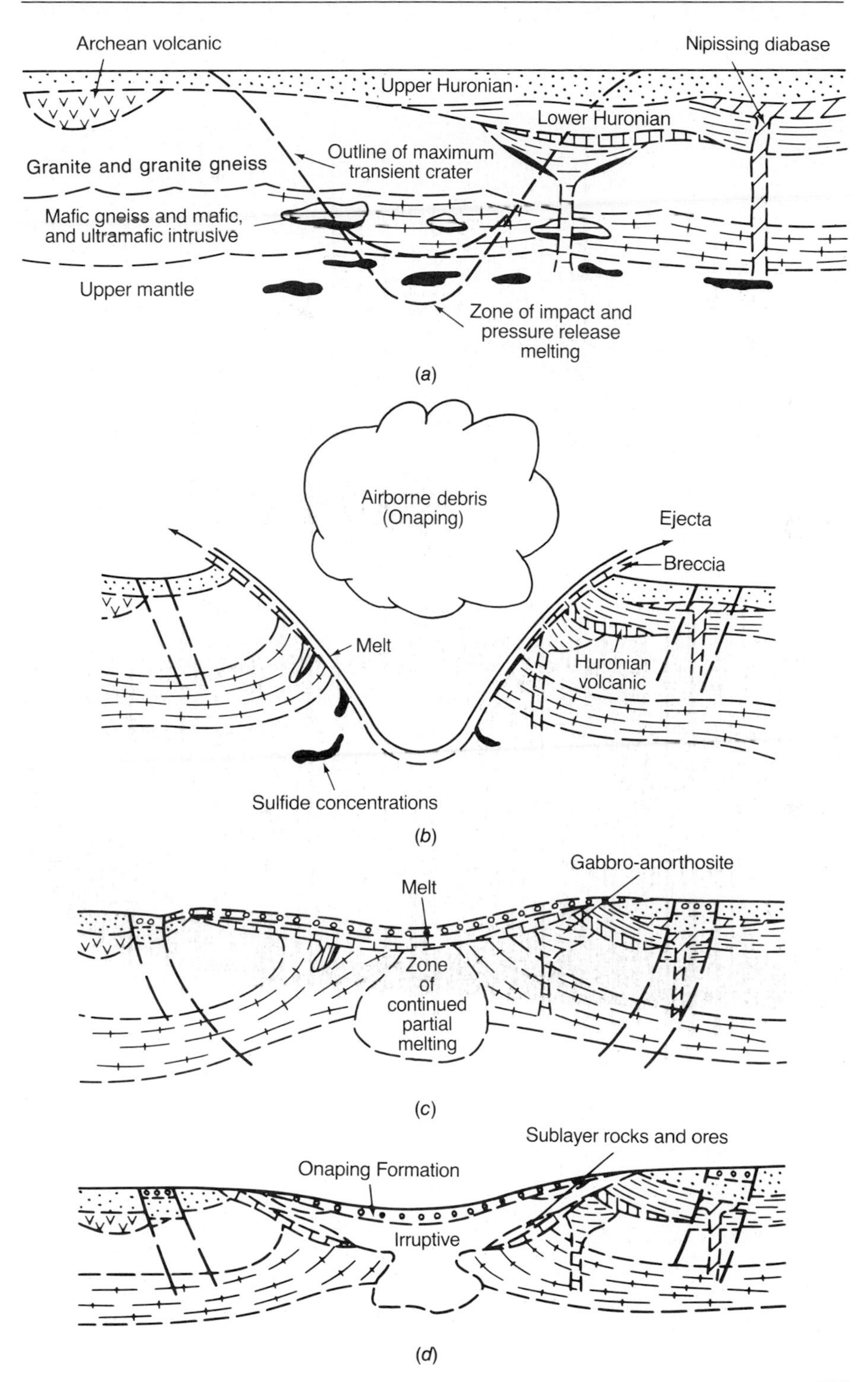

Archean volcanic
Nipissing diabase
Upper Huronian
Lower Huronian
Granite and granite gneiss
Outline of maximum transient crater
Mafic gneiss and mafic, and ultramafic intrusive
Upper mantle
Zone of impact and pressure release melting
(a)
Airborne debris (Onaping)
Ejecta
Breccia
Melt
Huronian volcanic
Sulfide concentrations
(b)
Gabbro-anorthosite
Melt
Zone of continued partial melting
(c)
Sublayer rocks and ores
Onaping Formation
Irruptive
(d)

magmatic differentiate in the lower levels of an impact-triggered magma and its early intrusion along the contact between brecciated footwall rocks and overlying Onaping Formation; or (2) direct emplacement of sulfide-enriched impact melt along the walls of the crater, the sulfides being derived from previous concentrations in mafic magmatic country rocks. In either case, the leucocratic breccias were formed by mechanical attrition of brecciated footwall rocks as the igneous sublayer was intruded. The peridotite inclusions in sublayer rocks are interpreted as having been brought from greater depths and intruded with the sublayer and its sulfides; the sulfidic breccias result from injection and mixing of sulfide liquid and shattered country rocks; and the offsets are seen as resulting from radial and downward injection-intrusion of sulfide-rich melt into impact-induced planes of weakness. Earlier arguments that focused upon hydrothermal activity and upon simple sinking and puddling of an emulsion of droplets of immiscible sulfide liquid to the bottom of the lopolithic magma chamber appear inadequate, although the application of liquid immiscibility at greater depths to produce the sulfide and silicate portions of the intruded sublayer units is viable. Efforts are continuing to determine the impact of meteorite collision with the Precambrian Sudbury surface, while the district continues to be one of the worlds greatest sources of nickel and copper, and one of the largest ore-bearing systems known.

It also appears that Fe-Cu-Ni-S liquids compositionally similar to those that existed at the Sudbury Complex have been involved with ultramafic volcanism, and have been erupted to form pyrrhotite-chalcopyrite-pentlandite Cu-Ni ores with basalts, basaltic komatiites, and komatiites. These deposits will be briefly considered at the end of this chapter.

▸ ANORTHOSITES—TITANIUM

The expanded use of titanium in high-performance light-weight structural metals and as TiO_2 in pigments since World War II has encouraged the development of titaniferous ores. These ores may be composed of titanium-bearing magnetite and hematite, intimate exsolution mixtures of these minerals with ilmenite, ilmenite itself ($FeTiO_3$), or rutile (TiO_2). Most are thought to be of magmatic segregation origin.

Virtually all economically significant concentrations of titanium minerals are associated either with magmatic anorthosites or with marine placer deposits derived from them (Chapter 16). As Stanton (1972) and Herz (1976) pointed out, there are at least two types of anorthosites with which deposits of titanium minerals are associated. First, anorthosite is a monomineralic rock composed of 90% or more of intermediate to calcic plagioclase. Anorthosite should mean "90% or more of anorthite, itself a plagioclase of $An_{90\text{-}100}$," but instead it has been applied to nearly monomineralic plagioclase rocks of An_{35} or more. There are at least two kinds of these rocks: (1) the layered rocks near the upper portions of LMIs which formed after the mafic minerals had crystallized and sunk, or by the floating of plagioclase crystals within the magma chamber and (2) the so-called anorthosite massifs, plutons typically containing plagioclase that is andesine or labradorite (An_{35-65}). The former are somewhat better understood, the Anorthosite Series in the upper portion of the Critical Zone of the Bushveld Igneous Complex being an excellent example. They develop by gravity stratification in ultramafic to mafic complexes, are characterized by rhythmic layering, show many cumulate textures, are free of megacrysts, are not typically sheared, and are composed mostly of plagioclase in the An_{70-100} range (Herz, 1976).

The anorthosite massifs still pose serious problems. Herz (1969) noted that accumulated data concerning them indicate that they are both geochronologically and tectonically constrained. For reasons still not well understood, these mafic calcium-rich rocks are all Proterozoic in age at 1.3 ± 0.2 billion years. They occur along a belt (Figure 9-18), which is diffuse at its western end west of the Laramie Range, Wyoming, but which includes dozens of deposits from the Great Lakes along the St. Lawrence River of Canada to Labrador and by plate reconstruction across to northern

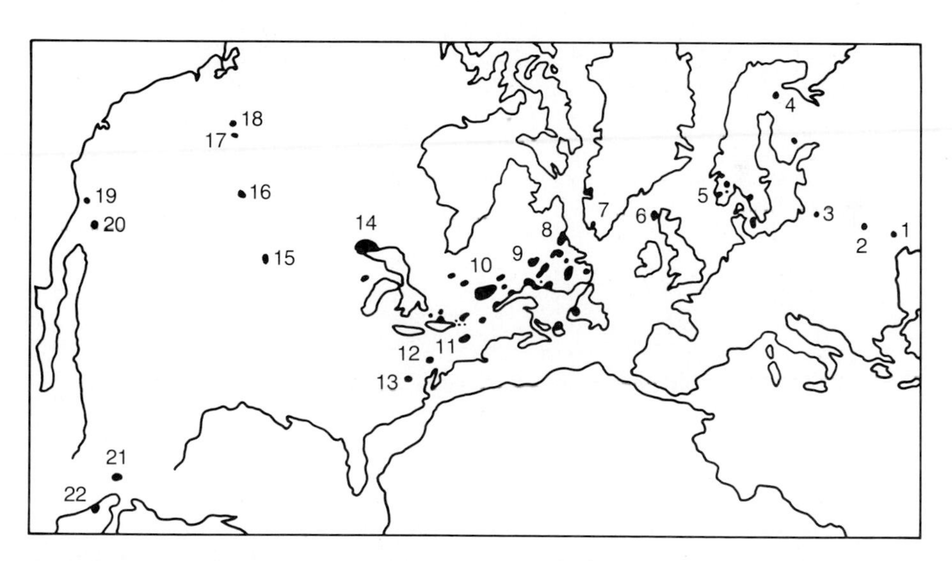

Figure 9-18. Anorthosites of the northern hemisphere plotted on Bullard's North America reconstruction. Anorthosites are: 1, Korosten, Ukraine; 2, Korsun-Novomir-gorod, Ukraine; 3, Suwalki, Poland; 4, Utsjoki, Finland; 5, southern Norway; 6, South Harris, Outer Hebrides; 7, Gardar, Greenland; 8, Kiglapait-Nain, Labrador; 9, Michigamau, Labrador; 10, Lac St. Jean, Quebec, including the Allard Lake district and its Lac Tio occurrence; 11, Lake Sanford, Adirondacks, New York; 12, Honeybrook, Pennsylvania; 13, Roseland, Virginia; 14, Duluth, Minnesota; 15, Cambridge Arch, Nebraska; 16, Laramie Range, Wyoming; 17, Bitterroot Range, Montana; 18, Boehls Butte, St. Joe, Idaho; 19, San Gabriel Range, California; 20, Orocopia Range, California; 21, Pluma Hidalgo, Oaxaca, Mexico; and 22, Sierra de Santa Marta, Colombia. (*From* Herz, 1969.)

Britain, southern Scandinavia, and into the Ukraine. Emslie (1978) portrays two clusters of anorthosite intrusive events, a 1.4 to 1.5 billion-year age group in Labrador, Canada, and in southwestern to midcontinent United States, and a younger group at 1.1 billion years along the St. Lawrence–Grenville Province in Canada. For both age groups Emslie stresses their anorogenic intracontinental rifting or incipient rifting tectonic setting in a broad belt across North America, Britain, and Scandinavia. Simmons and Hanson (1978) indicate that anorthosites can result from partial melting of tholeiitic compositions, rather than deep eclogite ones, at fairly shallow depths in the asthenosphere and again probably in response to shallow rifting.

According to Herz (1976), nearly all large igneous deposits of titanium are associated with anorthosite massifs. The largest ilmenite deposit in the United States is at Lake Sanford, New York (number 11 in Figure 9-18), and the largest rutile deposit is in the Roseland anorthosite in Virginia (number 13). The world's largest known ilmenite deposit is in Canada at Lac Tio (number 10) in the Allard Lake anorthosite, Quebec, Canada, part of the belt noted above. Large rutile deposits are found (Herz, 1976, p. D1) in the

> *. . . Pluma Hidalgo anorthosite of Oaxaca, Mexico, and the St. Urbain anorthosite of Quebec, Canada. Large ilmenite deposits occur in anorthosite at Egersund-Sogndal, Norway, and in the Ukraine, USSR. Noritic gabbro, an end member of common anorthositic assemblages, contains ilmenite deposits, such as those at Otanmaki, Finland. Deposits in granulite facies wall-rock gneisses may be closely related to those in adjacent anorthosite, as they are at Roseland.*
>
> *Some massif anorthosites, such as those in Nain and Michigamau, Labrador, show gravity layering, but in most complexes gravity layering is generally poorly developed or absent. Ages determined to date on all massif anorthosites are Precambrian, about 1300 ± 200 m.y. (Herz, 1969). These rocks crystallized under slow cooling conditions in the upper mantle or lower crust; the resulting megacrysts are commonly tens of centimeters in length. The feldspars are andesine, in places antiperthitic, or labradorite; pyroxenes are the most common accessory minerals, but olivine may be present, as it is in the Laramie anorthosite of Wyoming. Some of the minerals show kink banding, and plagioclase twin lamellae may be bent or broken.*

Herz (1976) also said that titanium-rich minerals crystallize late in the magmatic history of anorthositic complexes, as shown in Figure 9-1. The behavior of titanium during the crystallization of a magma is controlled by several factors, including the initial abundance of titanium, the changing chemical activities of silicon, aluminum, and iron, the variable partial pressure of oxygen, the possible formation of titanium-rich immiscible liquids, and temperatures of crystallization (Verhoogen, 1962). In a normally crystallizing tholeiitic magma having sufficiently high oxygen fugacity, titanium and iron will form oxide minerals early in magmatic history. These oxide minerals may segregate out and accumulate by gravity settling because of their high specific gravity as they did in the titanomagnetite layers in the Anorthosite Series of the Bushveld Complex. Fe-Ti-rich oxide liquids may be the normal residue of mafic magmatic differentiation, concentrated in late melts by exclusion of Fe and Ti from plagioclase so that the oxides crystallize interstitially and late in anorthosite genesis (Figure 9-1). Finally, an immiscible liquid that has the composition of iron and titanium oxides with phosphate as apatite may separate from a predominantly silicate liquid within a magma chamber (Philpotts, 1967). The formation of some ilmenite- and titaniferous magnetite-apatite orebodies has been ascribed to this phenomenon (Kolker, 1982).

The early crystallization history of anorthositic plutons is a much debated topic (Yoder, 1968; Green, 1969), but it is generally agreed that rocks rich in iron-titanium oxide and mafic minerals are emplaced late in the history of the anorthosites themselves (Isachsen, 1969). There is abundant evidence at Lake Sanford of younger titano-magnetite-rich anorthosites and anorthosite gabbros intruding low titania anorthosites with which they appear to be continuously related. If those slightly younger rocks have crystallized from the same magma, it must have been progressively enriched in iron and titanium either by processes of differentiation or by formation of an immiscible liquid so that young titaniferous phases intrude cognate older ones.

Anorthosite massifs are divisible into two subgroups (Herz, 1976, p. D2)

> *. . . based on their plagioclase and oxide compositions (Anderson and Morin, 1968). At St. Urbain, Quebec, one type can be seen intruding the other, but this relationship is exceptional; any one massif generally consists of only one type. The two types recognized are (1) labradorite anorthosite massifs characterized by plagioclase of* $An_{68\text{-}45}$ *composition and either titano-magnetite or its oxidized equivalent magnetite-ilmenite, and (2) andesine anorthosite massifs containing plagioclase* $An_{48\text{-}25}$*, hemo-ilmenite, and En:An ratios (enstatite content of orthopyroxene and anorthite content of plagioclase) greater than one. The latter is generally called "Adirondack type" (Buddington, 1931) and is associated with the world's principal ilmenite deposits. An unusual type of andesine anorthosite massif, found in Roseland, Virginia, Pluma Hidalgo, Mexico, and St. Urbain, Quebec, Canada, contains feldspar megacrysts that are strongly antiperthitic and have a* K_2O *content of 3 to 4 percent, which is about double the normal amount. This type has been called "alkalic anorthosite" (Herz, 1968) and has associated deposits of rutile as well as ilmenite.*

The distinction between ilmenite-rich andesine-type and magnetite-ilmenite-rich labradorite-type anorthosites is further substantiated by major and trace element analyses by Duchesne and Demaiffe (1978) who conclude that the andesine type may be either more shallowly derived by partial melting or contaminated by lower continental crust materials. Some typical labradorite anorthosite massifs include the Michigamau of Labrador (Emslie, 1970) and the Duluth Gabbro Complex of Minnesota (Taylor, 1964); andesine anorthosites include massifs of the Adirondack Mountains in New York (Buddington, 1931), Allard Lake in Quebec (Hargraves, 1962), and Roseland, Virginia (Herz, 1968).

Several careful studies of the magnetite-ilmenite deposits near Lake Sanford, New York, indicate that these ores are differentiates of an anorthosite-gabbro magma (Buddington, 1939; Balsley, 1943; Stephenson, 1948, Gross, 1968). The ores are thought to have formed from segregations of a magnetite-ilmenite liquid, part of which was trapped and crystallized in interstices between crystals of plagioclase and augite, but much of which was tectonically squeezed through and out of such crystal meshes in the process called filter pressing. This Fe-Ti-oxide liquid was injected into adjacent, locally still-plastic gabbros and anorthosites, locally forming high-grade Fe-Ti-oxide ores. Material grading from gabbro containing less than 10% magnetite-ilmenite to almost pure magnetite-ilmenite rock is found. The magmatic segregation interpretation was challenged by Gillson (1956); he argued that the ores are related to pneumatolytic replacement of the anorthosite-gabbro. He offered evidence that rock types in the area were produced by alteration of anorthosite along faults and fracture zones, and he concluded that the metallization was a late phase of this alteration. His arguments were convincingly refuted by Gross (1968) by citation of evidence closely linking Fe-Ti occurrences with magmatic features.

Unusually large ilmenite-hematite deposits have been found and developed at Allard Lake, Quebec, Canada. The deposits have been interpreted by several geologists as magmatic segregations (Deardon, 1958; Hammond, 1951; Rose, 1969; Bergeron, 1972; Herz, 1976). They are in a large mass of anorthosite and anorthositic-gabbro that invades Precambrian garnet-amphibole metamorphic rocks. The anorthosites were in turn invaded by granite. Coarse-grained ilmenite containing exsolved hematite makes up the ore, which forms irregular lenses, narrow dikes, large sill-like masses, and various combinations of these forms. Most of the ore contains 32 to 35% TiO_2. Typical analyses of ore from the Allard Lake district and the Lac Tio deposit are given in Table 9-4.

The titanium ores throughout the region are confined to light-gray medium- to coarse-grained anorthosites. The Lac Tio deposit, discovered in 1945, is said to be the largest body of titanium ore of its type in the world. This somewhat triangular-shaped tabular orebody is roughly 1 km wide and across and 100 meters thick, and contains an estimated 125 million tons or more of ore (Hammond, 1952). A system of fractures composed of two steeply dipping sets of joints and faults that trend approximately at right angles to each other forms a prominent regional pattern throughout much of the anorthosite surrounding the Lac Tio deposit. The more conspicuous set of normal faults strikes nearly north-south. The deposits are displaced by the faults, but the ore appears to be independent of these structures. The anorthosite shows strong shearing, or cataclasis, which predates these structures and is typical of this ore type.

The Lac Tio ore contains crystal aggregates of thick, tabular ilmenite grains up to 10 mm across and 2 mm thick. Minor amounts, generally about 5%, of plagioclase, pyroxene, biotite, pyrite, pyrrhotite, and chalcopyrite make up the interstitial material. Microscopic examination reveals an intimate exsolution texture developed during progressive cooling. Hematite laths and lenses in ilmenite are repeated at finer scales, with exsolution lamellae of ilmenite in hematite blebs and finer shreds of hematite from that ilmenite, and so on. The ilmenite contains up to 25% of intergrown hematite, and the ilmenite-hematite mixture is so fine grained that grinding cannot effectively separate the two minerals.

Widespread accessory ilmenite and commercial concentrations of ilmenite in the anorthosite indicate a genetic relationship between ilmenite and anorthosite. Dikes and veinlets of ilmenite in anorthosite, as well as inclusions of anorthosite in the orebodies, attest to the younger age of the ore. Both Hammond (1952) and Dearden (1958) suggest that the ilmenite and the anorthosite are differentiates of the same parent magma, the large deposits of commercial ore representing a late segregation. The immiscible Fe-Ti-PO_4 liquid must have had the composition of the ore (Table 9-4) and must have been injected into dilational shear zones in the hot precrystalline

TABLE 9-4. Analyses of Anorthosite Titanium Ores (In Percent) from Allard Lake and Lac Tio, Quebec

	Allard Lake			
Constituent	Sample 1	Sample 2	Sample 3	Lac Tio
TiO_2	34.8	36.0	34.4	34.3
Fe (Fe^{+2} + Fe^{+3})	38.0	42.8	39.1	37.1
S	0.36	0.40	0.39	0.3
P_2O_5	0.004	0.010	0.012	0.015
Cu	0.037	0.12	0.14	—
V	0.22	0.21	0.21	0.27
Mn	0.08	0.08	0.10	0.12
Ni	0.03	0.01	0.02	—
Co	0.014	0.013	0.019	—

Values represent percentage of ore by weight.

Source: From Hammond (1952) and Bergeron (1972).

anorthosite to form tabular bodies as Lac Tio, dikelike bodies at the Puizzalen area, and lensoid masses at the Mills locale (Rose, 1969).

Small amounts of vanadium, generally 0.1 or 0.2%, are found in both the Lake Sanford and Allard Lake deposits. Vanadium is common in many iron and titanium ores, and appears especially in ores of magmatic affiliation. Vanadium is seldom recovered from these ores, but they may be a source of large amounts in the future. During World War II, vanadium was recovered from slag as a by-product of the German iron-and-steel industry (Fischer, 1946). According to Vaasjoki (1947), vanadium is concentrated with magnetite rather than ilmenite, probably because vanadium and ferric iron ions are nearly the same size. However, Hutton (1945) stated that the differences in ionic size between V^{+3} or Fe^{+3} and Ti^{+4} are not large enough to cause such a selective concentration of vanadium. He favored the hypothesis that the vanadium becomes concentrated in residual magmatic liquids after ilmenite crystallizes and consequently is more available during the later growth of magnetite. It is presently being extracted from titanian magnetites of the Bushveld Complex and has been reported in amounts up to 2–3% (Dietrich, 1985) in layered magnetites of the Stillwater Complex, Montana.

Exploration for titanium deposits must first focus upon locating anorthosites or alkalic rocks and carbonatites from which minor amounts of titanium concentrates have been mined (Herz, 1967). Once an anorthosite has been located, one would examine heavy minerals in watersheds for ilmenite, hematite, or rutile; test for high-gravity, magnetic, and induced polarization anomalies; examine peripheral and upper marginal volumes of the massif and adjacent wall rocks; seek out areas of high apatite and sheared, cataclasized plagioclase megacrysts; and follow out directional increases in ilmenite and rutile.

Magmatic liquids consisting essentially of nearly pure magnetite-hematite-apatite have been intruded near the surface, and even extruded to form flows. These bodies, which have been mined in Chile, Mexico, and Missouri, are late magmatic, typically low in titanium, and associated with calc-alkaline rocks along Cordilleran plate margins. They do not appear to be tectonically related to higher titanium-iron liquids, although the processes of differentiation-concentration may have been similar. They are discussed in Chapter 11.

▸ KIMBERLITES—DIAMOND

The type area for kimberlites is Kimberley, South Africa, the site in the 1870s of the first and probably most dramatic diamond rush. The area has yielded more than 200 million carats of diamond since its discovery. Although alluvial diamonds in the Vaal and Orange rivers and marine placer diamonds both on and offshore along the South Atlantic coastline from South Africa north to Zaire have at times been more valuable, Kimberley has become one of the principal centers for the study of the igneous occurrence of the mineral. Although 90% of the world's diamonds have been extracted from placer deposits, it is thought that most placer diamonds were originally released by weathering from kimberlites. Kimberlites, kimberlitelike rocks, and diamonds have been found on all continents except Antarctica, and in the 1980s exploration for them is active in the United States, South America, Asia, Africa, and Australia. Interest is focused especially upon southeastern Australia where 14 clusters of more than a score of kimberlitelike pipes have been discovered (Ferguson and Sheraton, 1979), upon western Australia where kimberlites and diamonds in lamprophyres have been reported, and upon the western United States, Michigan, and Canada, especially the so-called "State Line" diamondiferous kimberlite pipes along the Colorado-Wyoming border (McCallum and Mabarak, 1976, McCallum, Mabarak, and Coopersmith, 1979; Smith et al., 1979). South Africa has recently produced 9.5 million carats (1600 kg) a year, 80% from kimberlite and 20% from placers, and 48% gem quality and 52% industrial quality. Zaire annually ships 12.5 million carats, mostly from placers and nearly all of industrial grade, and the USSR produces 12 million carats from pipes and placers, 80% of industrial grade, and sends a large proportion of its gem diamonds to European markets. De Beers Consolidated and its Central Selling Organization has for years dominated in the negotiated marketing rights for all this production, but in the early 1980s the cartel is being challenged by Zaire and others.

The rock called *kimberlite* is unusual both in chemistry-petrology and in occurrence. Early definitions were simple, as, for example, "pipes or dikes of mafic igneous rock containing mantle phases, commonly including diamond." But such definitions also include kindred carbonatites (see next section), peridotites, and other mafic intrusive rocks, so the definition has been expanded. It currently includes features that are mineralogical (Skinner and Clement, 1979) and textural (Clement and Skinner, 1979). Compositionally, kimberlite is a volatile-rich, potassic ultramafic igneous rock dominated by olivine, with subordinate minerals of mantle derivation. Kimberlites all contain olivine, so their mineralogical classification is based on less universal minerals. Different kimberlites are characterized by the dominance of monticellite (calcium olivine, Ca_2SiO_4), phlogopite (magnesian biotite), diopside (Ca-Mg clinopyroxene), serpentine, or calcite, with minor and variable amounts of apatite, magnetite, chromite, garnet, diamond, and other high-pressure, high-temperature upper mantle minerals.

However, composition is not the whole story. Inclusions of upper mantle rock xenoliths such as garnet lherzolite, eclogite, and harzburgite (Boyd and Meyer, 1979, II) and the occasional presence of diamond itself indicate deep derivation of these rocks. Also, until recently they were viewed as having drilled their ways to the surface in fairly rapid, straightforward fashion (Kennedy and Nordlie, 1968), but it is now recognized that the process is complex.

Hawthorne (1975) diagrammed a composite kimberlite pipe (Figure 9-19) and Smith et al. (1979) depicted their interpretation of the State Line pipes (Figure 9-20). It has been noted by many workers (Kennedy and

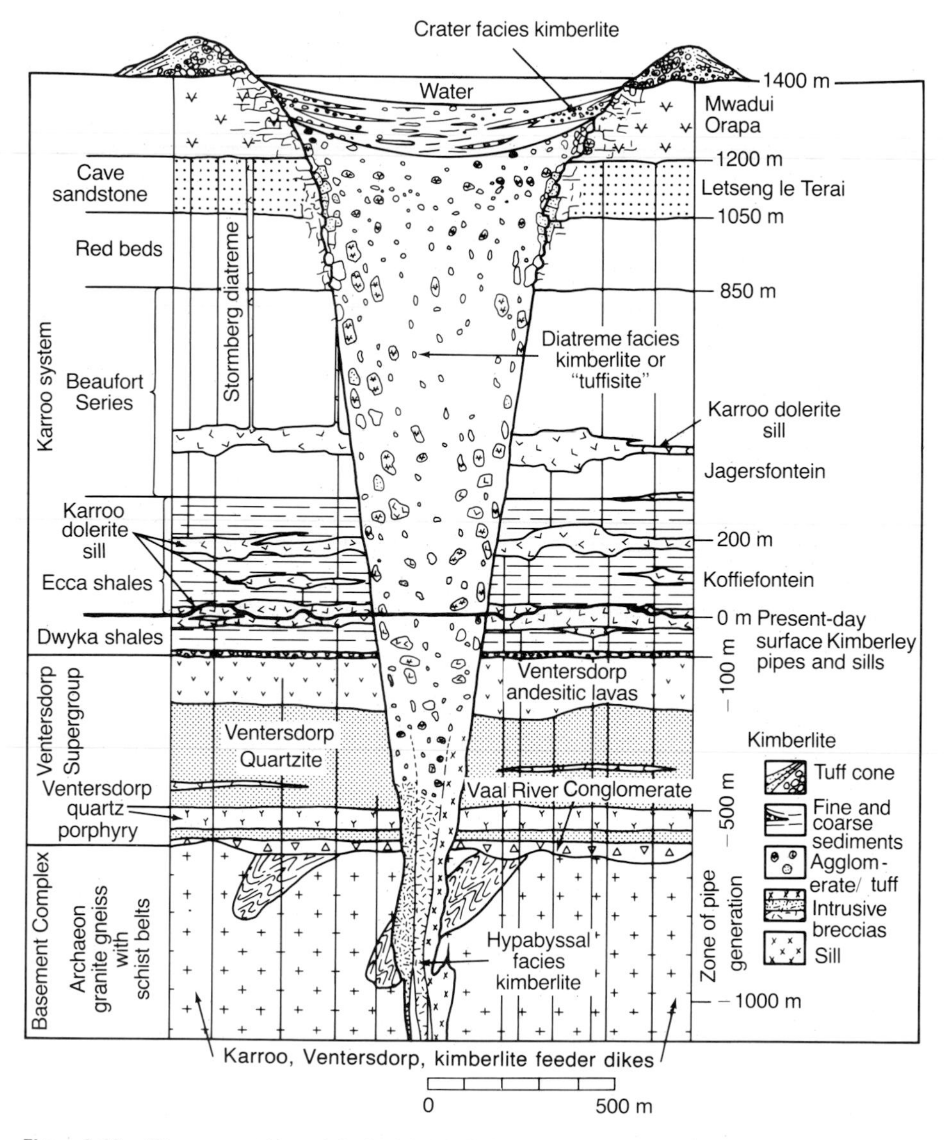

Figure 9-19. Diagrammatic model of a kimberlite pipe shortly after emplacement. The African names at the right designate outcrop erosion levels of specific pipes. (*After* Hawthorne, 1975.)

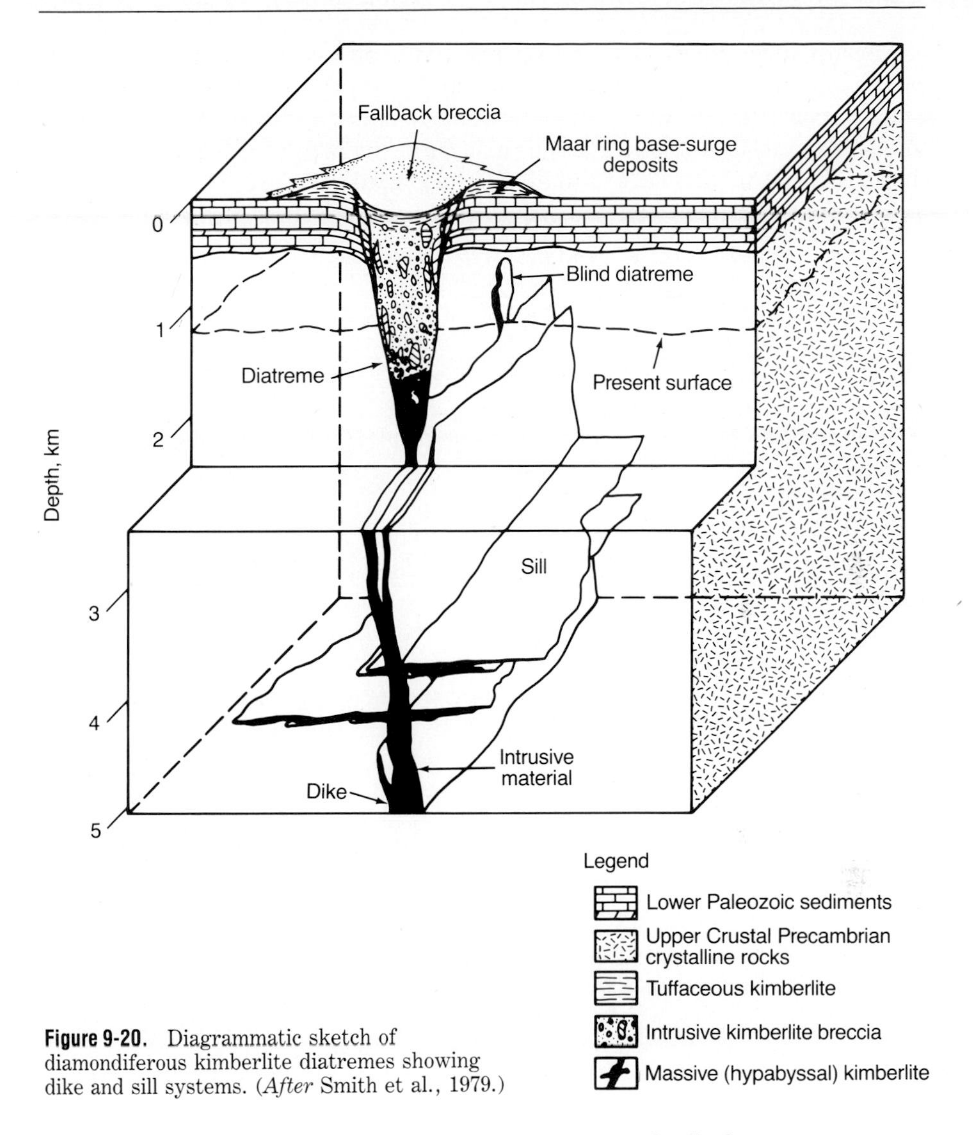

Figure 9-20. Diagrammatic sketch of diamondiferous kimberlite diatremes showing dike and sill systems. (*After* Smith et al., 1979.)

Nordlie, 1968) that these carrot-shaped bodies gradually became narrower and more irregular at depth, some ultimately becoming dikeform. It has been thought that they represent a rapid, violent upward rush of deep-mantle material from the asthenosphere, perhaps 200 km below the surface, in the form of a diatreme with explosive surface expression, fallback, and base-surge phenomena, and probably violent CO-CO_2-H_2-H_2O degassing. Expansion of these gases might have chilled the intrusive units such that diamond, formed at 50 to 70 kbar pressure and 1400 to 1600°C, was metastably preserved. Not all kimberlite pipes contain diamonds at levels currently exposed. Of about 100 occurrences in the Kimberley area, only 20 have been productive, and this 20:100 ratio of productive to barren is exceptionally high. Pipes near the Orange River alluvial paleoplacer deposits are deeply eroded and barren of diamonds; perhaps their higher reaches were diamondiferous and supplied resistate diamonds to the river systems. In that regard, it is significant that modern gravels in the Vaal and Orange rivers contain only reworked diamonds, not new arrivals.

Current thought regarding kimberlite emplacement is that intrusive advance was rapid at 100 to 150 km depths, with magma wedges or lenses, like balloons ascending in a viscous medium; that the rate slowed as the surface was approached, with stoping becoming more significant; that the slowest advance rates were realized just below surface where a "gas cap" formed, ultimately to explode upward and outward to form fallback tuffisites and chaotic breccias within and around a flared, conical crater, with brecciated wall rocks (Skinner, Clement, and Hawthorne, 1980). It now appears that the typical kimberlite may "pinch out" at depth, and thus be shaped like an inverted teardrop rather than as a conventional volcanic pipe. Most kimberlite occurrences involve multiple intrusive events which form a cluster of several discrete intrusions. Clement and Skinner (1979) presented a textural classification that reflects that fact. They suggest (see Figure 9-19) that three texturally distinctive kimberlite facies should be recognized: the hypabyssal, diatreme, and crater facies. Hypabyssal kimberlites are deeper seated, are porphyritic, and result from crystallization of kimberlite magma in the "teardrop" beneath the diatreme. Diatreme-facies kimberlites represent the bulk of the intrusive body shown in Figure 9-19. They contain mantle- and crustal-derived rock fragments, with kimberlite lapilli, kimberlite fragments, and discrete kimberlite minerals. The upper portions are dominated by tuffisite kimberlites and tuffisitic kimberlite breccias, *tuffisite* being a term used by Clement and Skinner to describe a rock that looks like a tuff and has the characteristics of one, but that was intruded into position. Larger blocks in the pipes are always stratigraphically lower than their original level by as much as 1 km. Small fragments may have moved up or down. In the Kimberley pipe, mined out in 1914 at a mining depth of 1079 meters, mixing of xenoliths representing 3 km of stratigraphy was measured. Crater facies may be pyroclastic fallback breccias or epiclastic water-lain material. Kimberlites, then, are clearly ultramafic epicrustal intrusive systems; their details are still being studied. Geologic ages range from middle Precambrian (the Premier Pipe near Pretoria, Republic of South Africa) at least through the Mesozoic; most of those near Kimberley are Cretaceous in age, as is the Murfreesboro, Arkansas, occurrence.

Diamond ores are perhaps the "lowest grade" mined anywhere. The open-pit mine at the town of Koffiefontein 80 km southeast of Kimberley profitably mines tuffisite kimberlite breccia containing about 25 carats of diamond per 100 cubic meters, or about 10 carats per 100 tons. 0.2 gram make 1 carat, so 5 carats are 1 gram. At a specific gravity of 3.5 g/cm^3 1

cm^3 weighs 16 carats. So a huge 100-ton ore truck laboring upgrade out of this pit statistically carries just over half a cubic centimeter of economic mineral! Figured another way, the ore contains only 0.02 ppm, or 0.000002 wt%, of diamond. And only 35% of that is gem quality, the remainder being industrial grade. Still, the average 1980 price was approximately $150 per carat for the raw stones, so extraction is profitable. In 1980 the Koffiefontein mine produced about 12,000 tons of ore, and thus 1200 carats of diamond, per day from a pipe approximately 350 meters across. It is planned to convert completely to underground operations when the pit gets too deep. Stones over 100 carats—say, a 3-cm octahedron—are rare, but 10 to 30 carat stones are recovered at a rate of about 8 or 10 per week. Diamonds are recovered from the ore by heavy-liquid separation in ferrosilicon (specific gravity 3.1), by the property unique to diamond that it adheres to a greased surface, and by the Sortex Method, jets of air activated by photocells that perceive visible light fluorescence induced in diamond by an X-ray beam as concentrates move along a V belt, the air jets popping the diamonds and adjacent particles off the belt.

In prospecting and evaluation it has been learned that sampling and recovery problems are so great that as much as a tenth of a kimberlite pipe may be mined before its economic viability is truly established. Evaluation of newly found pipes, prospecting for new ones, and identifying diamondiferous members of a cluster of pipes are all difficult. The best guide, other than the presence of diamond itself, is mineralogical; chromian diopside, magnesian ilmenite, and high-chromium, low-calcium Mg-Al (pyrope) garnets almost always occur in kimberlite with diamond. Other heavy minerals are avidly studied, and compositional tie lines between clinopyroxenes, olivines, garnets, spinels, and other phases that might prove unique to productive assemblages are being studied by electron probe microanalysis and other methods. Exploration geochemistry is so far useless. Cr, Ni, Ti, and other anomalies exist, but for less than 100 meters beyond pipe limits. Mantle mineral haloes may extend 500 meters, rarely more. Geophysical detection of pipes has been only moderately successful, with gravity, aeromagnetics, and EM most useful.

It has been demonstrated that the largest pipes in a cluster tend to contain the largest numbers of large, gem-quality, best grade diamonds. Diamond size distribution in a pipe is log normal, with nodes at different dimensions in different pipes. The final screening is at 1.2 mm at many mines; smaller stones are discarded. Within the pipes, a further variation is wall-rock dilution, especially the amount of near-surface material incorporated in the kimberlites.

Boyd and Finger (1976) showed that the ultramafic xenoliths found in various kimberlites are derived from different depths in the mantle. A vertical "stratigraphy" of inclusion type has been established, so the depth of origin of a given kimberlite can be assessed. Kimberlites can also occur as small diatremes, dikes, or even sills. Surface expression of kimberlites can be circular and pluglike, but some, such as the Main Fissure at Bellsbank, 80 km northwest of Kimberley, are sharply dikeform. The Main Fissure kimberlite was cut locally by almost perfectly circular pipes 50 to 100 meters across. The Bellsbank dikes may be only 1 or 2 meters wide, and are only locally minable for diamonds (Figure 9-21).

Perhaps the most significant advance in the understanding of diamond geology is the recognition and clarification of the lithologic-mineralogic spectrum of kimberlites described above. Many rocks around the world previously called minettes, lamprophyres, and even diabases may in fact be kimberlites that have been dismissed uncritically and prematurely as nondiamondiferous, and the reverse is also true, that many rocks have mistakenly been called kimberlites. Continued research and clarification of the petrogeny of kimberlites, and of the occurrence of diamond in them, are essential.

▸ CARBONATITES

Another mafic rock that appears to be closely related to kimberlites is that of the carbonatite. Literally hundreds of carbonatite occurrences have been found since 1940; only scores were known earlier. Many of them are cylindrical, pipelike, and prominently concentrically zoned, such as the Palabora occurrence described below. A few are irregularly shaped, unzoned bodies like that at Mountain Pass, California, probably the greatest rare earth (lanthanide) deposit in the world. All carbonatites contain calcite, dolomite, or siderite which has been concluded to be truly igneous. Still, an *igneous rock* composed of flow-banded calcite with barite phenocrysts takes some getting used to; they occur at Mountain Pass, California; Magnet Cove, Arkansas; and at many other localities.

Beyond the essential carbonate component, carbonatites are of several as yet poorly understood classes. Many appear to be magnetite-apatite-rich, others rare-earth-element, fluorine-barium-rich. They have also been classified geochemically into alkali, ferric iron, and zirconium-rich (agpaitic) and alkali-poor, FeO-CaO-MgO-rich, and zirconium-poor (miascitic) types. Heinrich (1966) used a structural-occurrence classification with four entries: (1) with alkalic ring complexes, such as Palabora and Magnet Cove; (2) with alkalic complexes not of the ring type, such as Mountain Pass, California; (3) not associated with alkalic rocks; and (4) as flows and pyroclastic rocks. Note here that indeed not all carbonatites are associated with other mafic igneous rocks.

Carbonatites are exploited for a wide variety of metals and minerals singularly concentrated within them, including rare earths, niobium-tantalum, zirconium-hafnium, iron-titanium-vanadium, uranium-thorium, and industrial minerals such as apatite, vermiculite, and barite. Some appear

(*a*)

Figure 9-21. (*a*) Kimberlite dike at Bellsbank in the Cape Province, Republic of South Africa. The diamondiferous dikes, largely mined out, are only a few meters wide. The plane of the dike is a surface along which columnar diapirlike kimberlite plugs have intruded, producing smoothly fluted chimneylike bodies (*b*) in dolomite that have also been mined for their contained diamonds.

(*b*)

to be hightly differentiated rocks, occurring near and with syenites and other alkalic lithologies, but others appear to be undifferentiated, occurring with curious diopside-olivine pyroxenite-harzburgite intrusives. At least geologically young ones appear to be related to rift environments; they occur with kimberlite associates, generally within rifted cratonic interiors and with the suggestion of deep crustal or upper mantle processes involving carbonatitic liquid separation and injection, or some other form of mantle degassing. The kimberlite-carbonatite kinship can be debated. Kimberlites invariably contain at least accessory carbonates, but carbonatites are not known to contain diamond, mantle phases such as chrome garnets, or mantle xenoliths. Although a kinship seems reasonable, either mode of origin or depth of source must distinguish them. Most carbonatite outcrops are of low topographic relief, and many have been discovered to be the floors of subcircular to circular lakes, for example on the Canadian Shield. They range in geologic age from early Precambrian—Palabora is 1.9 billion years, Mountain Pass 1.2—quite evenly through Phanerozoic ages to recent times. Some carbonatitic volcanic flows at Oldoinyo Lengai and elsewhere in the East African Rift may result from modern carbonatite magmatism.

Two standout examples of the different types of carbonatites are the magnetite-apatite-rich Palabora occurrence to be described here and the rare-earth-element-rich Mountain Pass body, of which an excellent account is given by Olson et al. (1954). Erickson and Blade (1963) provide a clear and thorough description of the Palabora-like Magnet Cove, Arkansas, occurrence.

Palabora Carbonatite, South Africa—Copper-Phosphate-Iron

One of the most fascinating ore deposits developed in recent years is the one at the village of Phalaborwa, immediately west of Kruger National Park in northeastern Transvaal, Republic of South Africa (Lombaard et al., 1964; Palabora Staff, 1976). There a large composite alkalic carbonatite pipe, known as the Palabora Complex, approximately 7 km long by 3 km wide, vertically invades Archean granitic basement rocks (Figure 9-22). Copper, apatite, and vermiculite are mined from the northern two of the trio of pipes, and co-product magnetite, uranium in uranothorianite, cobalt in linnaeite, and zirconium-hafnium in baddeleyite (ZrO_2) are recovered, as are trace amounts of nickel, gold, silver, and platinum metals. The hill in the center named Loolekop, or "sitting hill" in Bantu, has long since disappeared into a highly mechanized, computer-controlled, 75,000-ton-per-day open-pit mine (Figure 9-23). It is the largest in Africa and a striking contrast to the primitive African wilds maintained as near as 5 km to the east in Kruger Park.

The Palabora Complex is as unusual and intriguing a petrologic terrane as it is an orebody. The flat Lowveld peneplane around the town of Phalaborwa is studded with scores of alkalic silicate plugs, now expressed

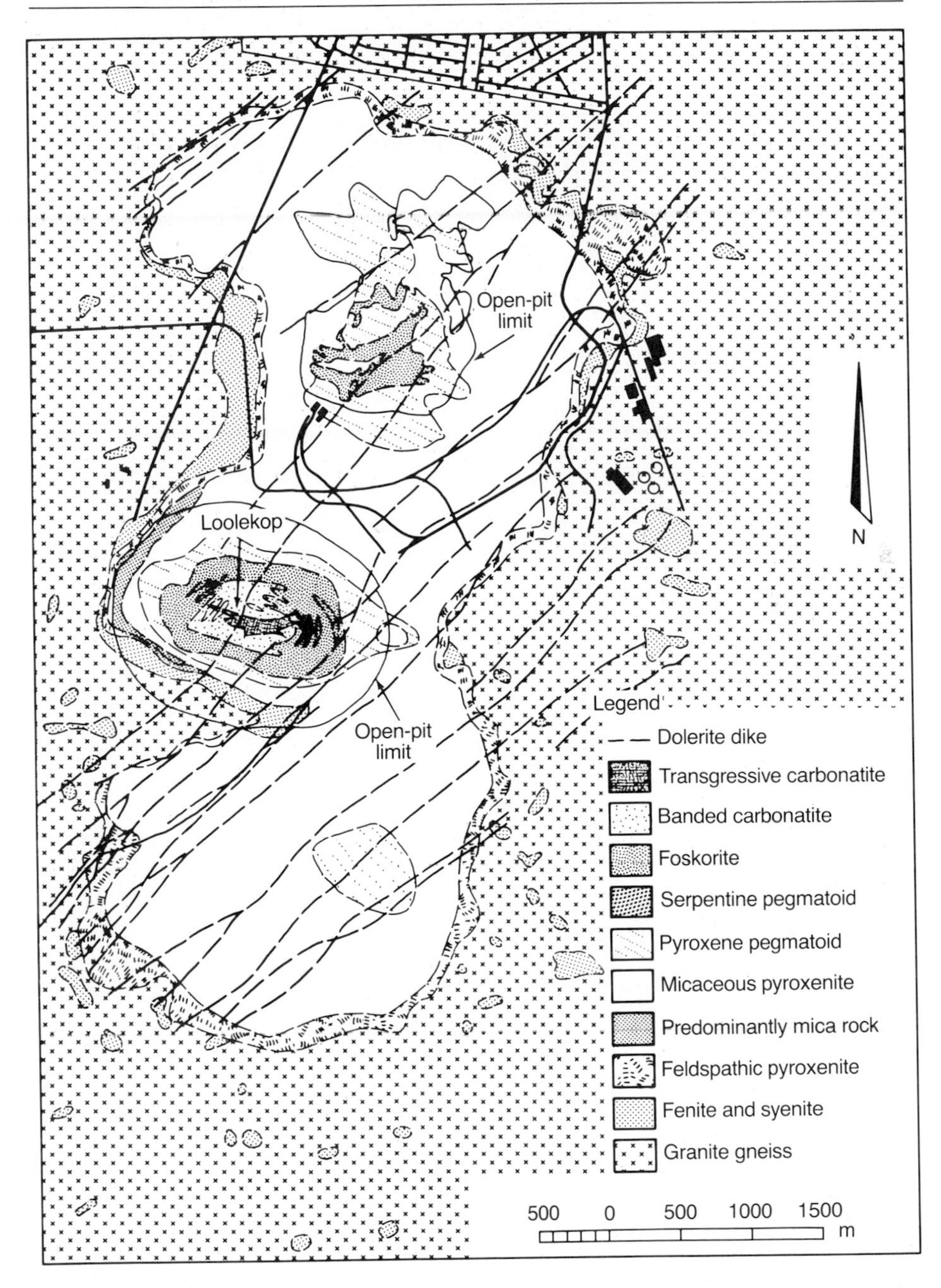

Figure 9-22. Palabora Complex, Transvaal, Republic of South Africa. The northern sector is mined for vermiculite and apatite-phosphate, the central portion for copper. (*From* Palabora Mine Staff, 1976.)

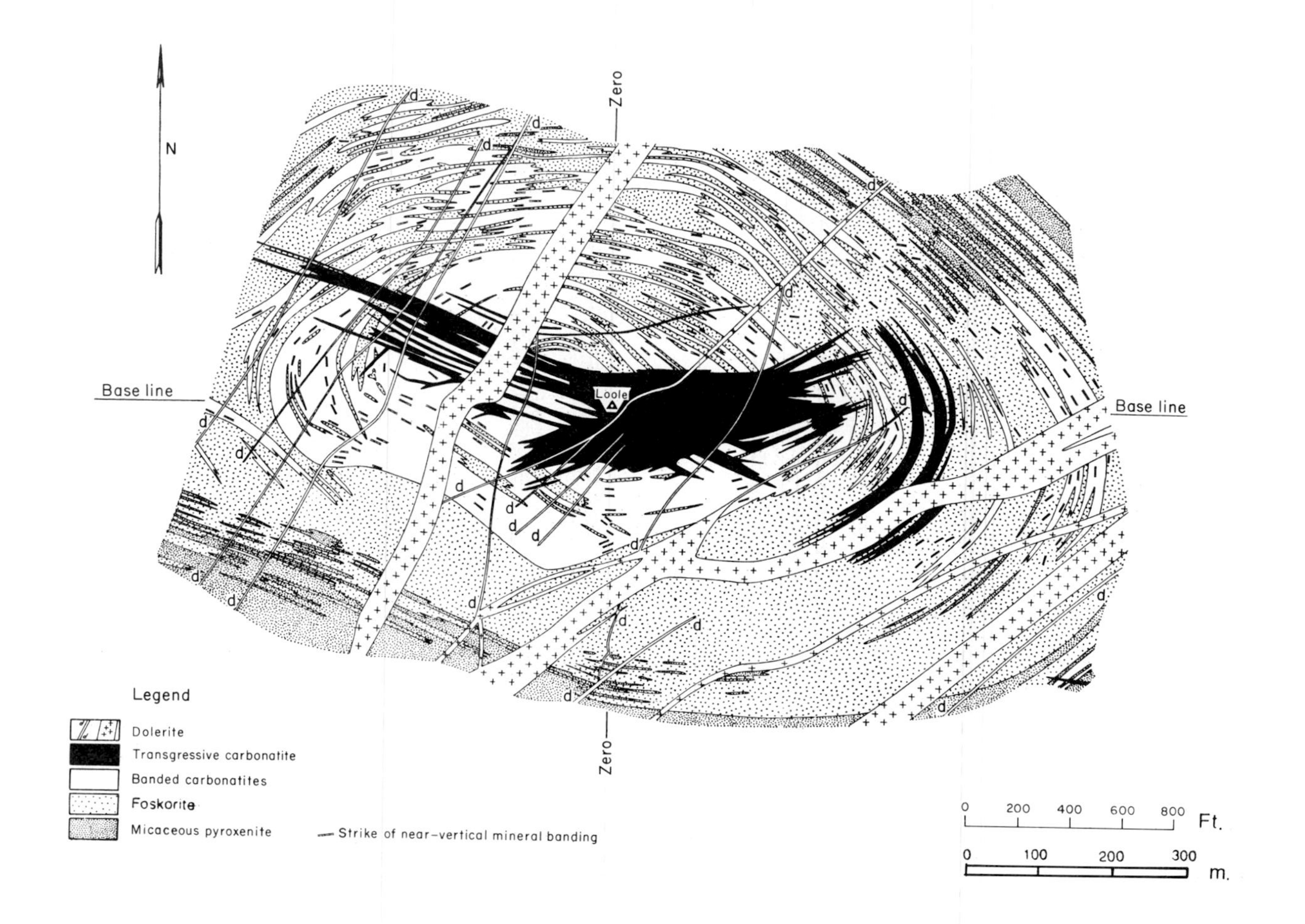

N
Zero
Zero
Base line
Base line
Loole
d
Legend
Dolerite
Transgressive carbonatite
Banded carbonatites
Foskorite
Micaceous pyroxenite
Strike of near-vertical mineral banding
0 200 400 600 800 Ft.
0 100 200 300 m.

as conical syenite hills, but only the Palabora Complex itself shows the mineralogic-lithologic variability described below. The complex has been isotopically dated at 1960 million years, about 1 billion years younger than the granitic cratonic crust in which it occurs, and almost the same age as the Bushveld Igneous Complex far to the southwest. As Figure 9-22 shows, the complex consists primarily of micaceous pyroxenite, a diopside-phlogopite (biotite)-apatite rock in which apatite constitutes about 15 vol %. The narrow rind between granitic gneiss country rocks and the pyroxenite is a contact zone of hybrid rocks involving feldspathic pyroxenite, syenite, and fenite, a sodic subsilicic hydrothermal alteration assemblage found around many carbonatite pipes.

The overall form of the complex is that of three adjacent composite pipes, a northern, a central, and a southern component, each of which is steep-walled and pipelike in form (Figure 9-23) and each of which maintains its carrotlike shape and individuality to the deepest drill intercepts at nearly 2 km down. The northern body is mafic pegmatitic rock composed largely of serpentine and hydrated phlogopite (vermiculite); it has been mined for vermiculite since 1946. The southern pipe is also pyroxene pegmatoid. It contains 7 to 10% PO_4 as apatite and minor vermiculite, but it is not currently being mined. The central, or Loolekop, pipe is of greatest economic interest. As shown in Figure 9-22 and in detail in Figures 9-24 and 9-25, the pipe has a strongly coaxial sheeted structure. Progressing inward from the micaceous pyroxenite main mass are "shells" of coarser grained pyroxene pegmatoid, then an unusual girdle of magnetite-olivine-apatite rock called *foskorite*, and finally a central, concentrically sheeted pipe of *banded carbonatite* cut by a younger, largely crosscutting carbonatite called the *transgressive carbonatite*. Contacts between all of these rock types suggest progressive, transitional inward changes, the carbonatite apparently having been intruded in two stages. The older is a medium- to coarse-grained rock with conspicuous banding roughly parallel to the walls of the pipe. This banded carbonatite, as well as much of the younger transgressive material, consists of calcite intergrown with dolomite and accessory chondrodite, olivine, phlogopite, and biotite (Lombaard et. al., 1964). Banding results from the alignment of magnetite, which makes up as much as 25% of the rock, and silicate grains. Contacts between the older carbonatite and foskorite are commonly gradational; toward the borders of the pipe the two rocks are intimately interlayered.

The younger carbonatite, called transgressive carbonatite because it crosscuts other rocks, is especially abundant near the center of the pipe and tails out to the east and west. Figure 9-23, a plan of the 122-meter level, shows this relationship very well; part of the younger carbonatite is

Figure 9-23. (*Facing page*) Geological map of the Palabora Complex at the 400-ft level. (*From* Lombaard et al., 1964.)

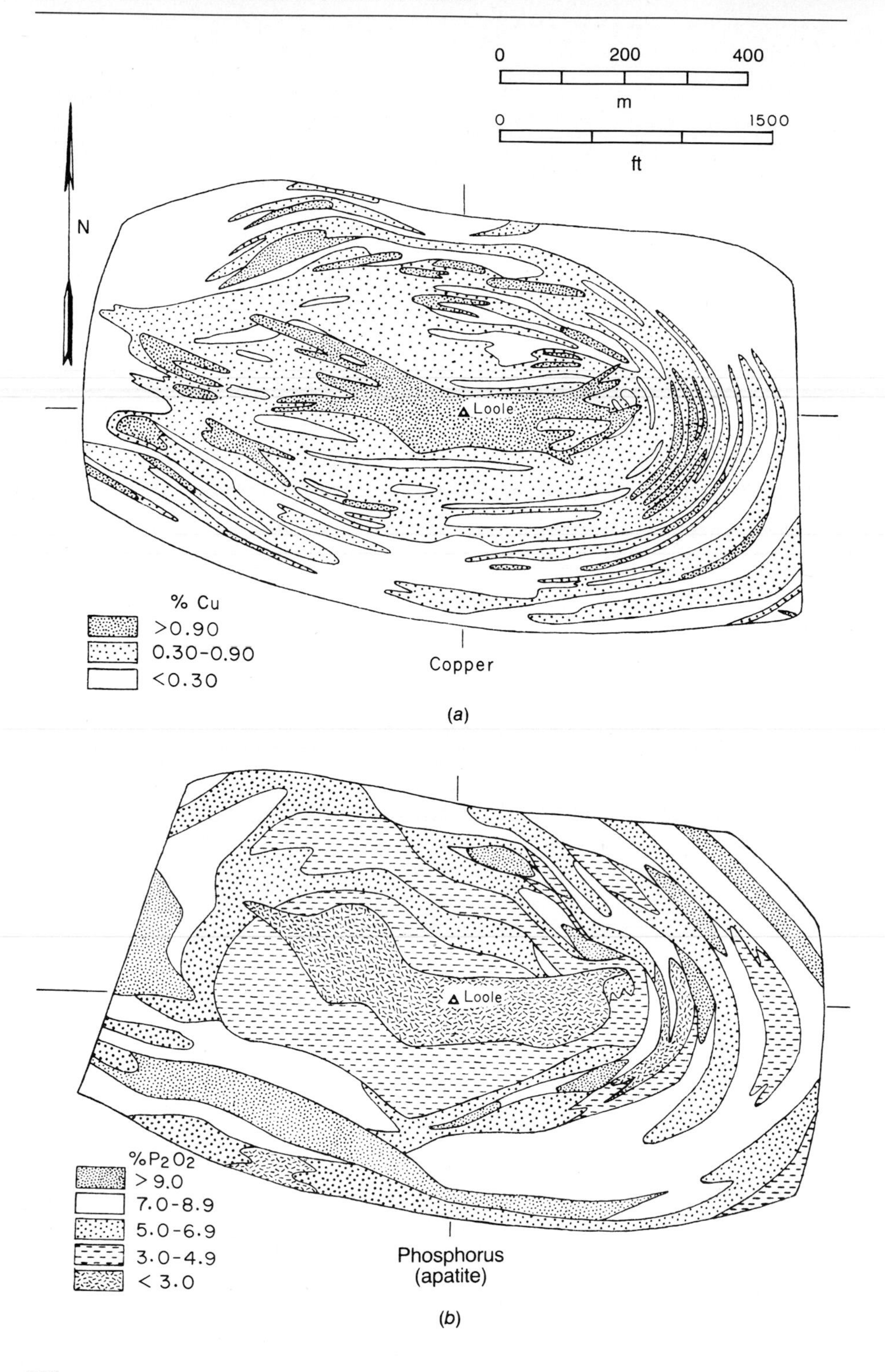
0
200
400
m
0
1500
ft
N
Loole
% Cu
>0.90
0.30-0.90
<0.30
Copper
Loole
%P2O2
> 9.0
7.0-8.9
5.0-6.9
3.0-4.9
< 3.0
Phosphorus
(apatite)

(*a*)

(*b*)

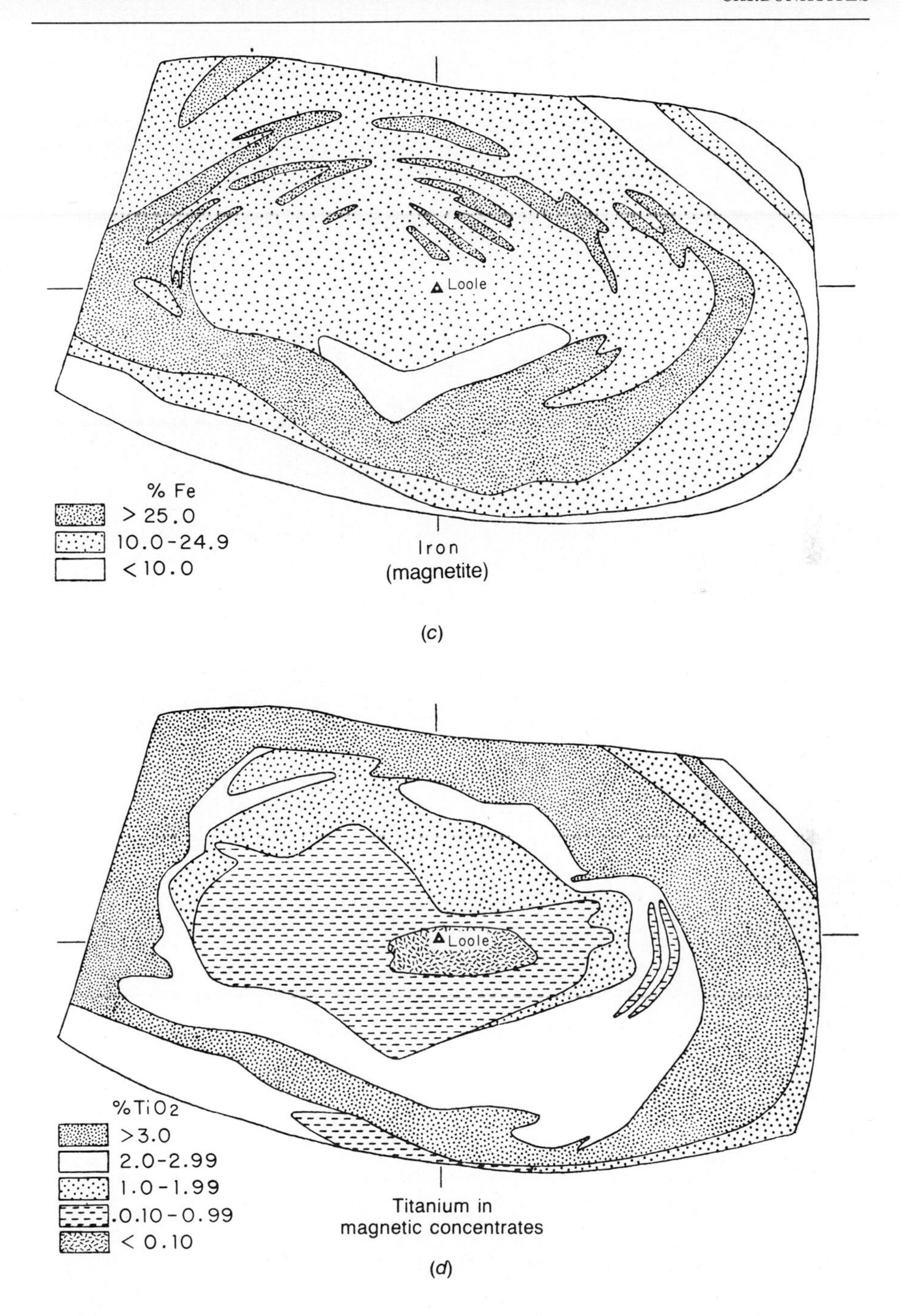

Figure 9-24. Distribution of copper, phosphorus, iron, and titanium at the 400-ft level, Palabora Complex. (*From* Lombaard et al., 1964.)

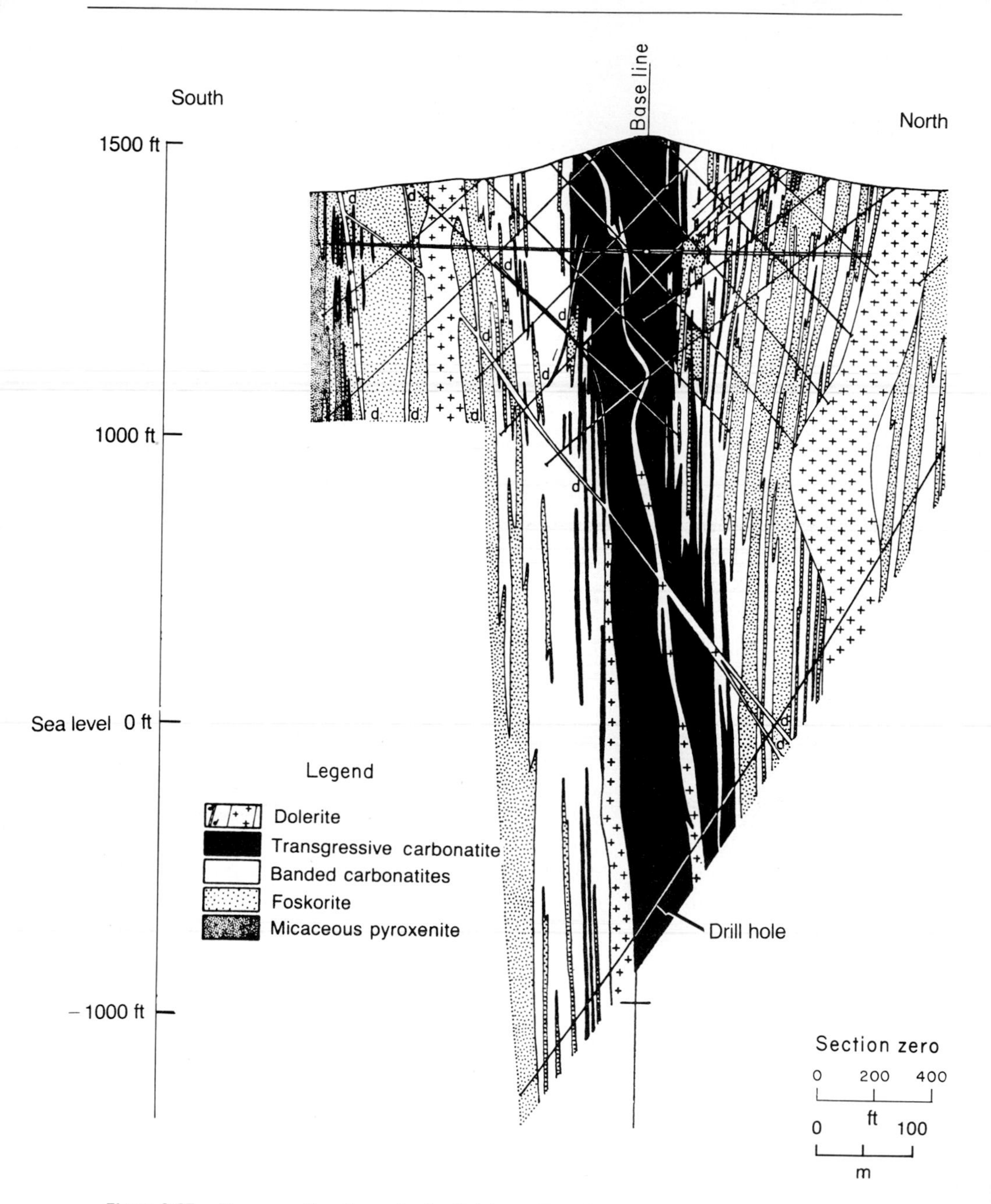

Figure 9-25. Cross section through the Palabora Complex pipe along the zero plane on Figure 9-23. (*From* Lombaard et al., 1964.)

fine-grained and is locally banded like the older rock, thus making it difficult to distinguish the two.

Copper mineralization pervades the carbonatites, and the distribution of ore minerals corresponds generally with the rock type (Figure 9-24). Copper occurs as chalcopyrite in transgressive carbonatite at grades averaging nearly 1% near the center of the pipe; that rock type is low in phosphorus, titanium, and iron (Figure 9-24*b* to *d*). Banded carbonatite contains intermediate copper values averaging 0.5% Cu, primarily as bornite. Foskorite contains only 0.2 to 0.3% Cu as bornite with minor chalcocite, but phosphorus, titanium, and iron contents are higher here. Copper minerals are disseminated in the foskorite and banded carbonatite and are progressively more veinlet-controlled in the younger transgressive carbonatite. The copper mineralization appears to be an integral part of the emplacement of the complex, rather than an epigenetic overprint, because such a large fraction of the copper minerals is disseminated in and integral to the host rocks. The dolerite (diabase) dikes which cut but do not otherwise affect the complex and its ores may be as old as 1880 million years, leaving little time for truly epigenetic mineralization. Further, several carbonatites, chiefly in Uganda, Finland, and elsewhere in the Transvaal, contain significant copper (Marais, 1980). The close and unique association of the ore with intrusive carbonatites, and the fact that copper was introduced in two stages before and during emplacement of the transgressive carbonatite, argues for close genetic associations between the carbonatite itself and the metals mined. Likewise, the presence of apatite, baddeleyite, and magnetite suggests magmatic affiliations.

The copper orebody, about 0.6 by 1 km in plan view, is said to contain about 700 million tons of ore averaging 0.68 % Cu. Chalcopyrite is the principal sulfide, with bornite, chalcocite, and locally minor amounts of other copper minerals. Late valleriite coatings on magnetite and other sulfides cause losses of copper during metallurgical extraction. Magnetite is a major constituent of the ore and of value as a by-product, and titanium is present in much of the magnetite in the carbonatites. Around the periphery of the ores, the magnetite contains from 0.6 to 2.8% TiO_2, which inhibits its use by the steel industry. The transgressive carbonatite in the center of the plug contains only minor amounts of apatite, but elsewhere, especially in the foskorite, apatite is abundant. It is a valuable by-product, mined for use in fertilizers. Sulfur smelted from the sulfides is oxidized to sulfuric acid and is used in the local production of liquid phosphate fertilizers. Gypsum produced in those reactions is used to produce cement and sheetrock wallboard.

As far as they have been probed, at least to 2 km (Figure 9-25), the pipes continue unchanged downward. The three pipes do not merge, grades are unchanged, and the orebody and complex are "open" at depth. They thus constitute another major African–South African resource.

▸ ULTRAMAFIC VOLCANIC ROCK ASSOCIATIONS—COPPER-NICKEL

Recognition of the common association of nickel sulfides with both plutonic and volcanic ultramafic rocks (Naldrett, 1973) opened new exploration avenues in the 1970s, which will yield discoveries in the 1980s. The discovery of the Thompson orebodies in northern Manitoba, Canada, of the Kambalda deposits in Western Australia (Woodall and Travis, 1970), and the recognition in recent years that there are chalcopyrite-pentlandite-pyrrhotite ores associated with ultramafic, high MgO volcanic flows called *komatiites* in South Africa, Australia, and Canada has led to a reevaluation of basaltic volcanic Precambrian terranes. The komatiites are uniquely characterized by *spinifex textures* (Pyke, Naldrett, and Eckstrand, 1973), a texture produced by intergrowth and interpenetration of long, skeletal quench crystals of olivine and pyroxene which resemble a mat of an Australian grass called "spinifex grass" (Figures 9-26 and 9-27). Compositionally they are indeed ultramafic at from 20 to 45% MgO with correspondingly high modal and normal forsteritic olivine. They are thus the extrusive equivalents of peridotites, harzburgites, and even dunites. Canadian deposits associated with ultramafic flows characterized by komatiite compositions and spinifex textures include the Alexo, the Marbridge, and, near Timmins, Ontario, the Langmuir; in southern Africa, the Damba prospect and Shangani mine (Viljoen et al., 1976) and several others in Zimbabwe are recent discoveries (Williams, 1979); and in Australia, the Kambalda copper-nickel deposit is now being mined (Marston and Kay, 1980). All of these deposits involve lenses or sheets a few meters thick of pyrite-pyrrhotite-chalcopyrite-pentlandite ore which are stratabound within sections of komatiite flows, commonly associated in space and time with the volcanogenic deposits to be described in Chapter 13. They are included in this chapter because it is increasingly apparent that although they are of submarine origin, these Cu-Ni-S deposits did not involve submarine exhalation of hydrothermal fluids and are probably of direct magmatic-extrusive origin. Extensive remapping of primitive volcanic piles is continuing in many parts of the world.

The Kambalda deposits are in Archean rocks of the Western Australian Precambrian shield. They were discovered in January, 1966, following the recognition of a nickeliferous limonitic gossan. Geologic mapping shows that this gossan is at the lower contact of an ultramafic komatiite body that lies between two metabasalts. The lower contact of the ultramafic body crops out as a broad domal structure with a strike length of 20 km (Figure 9-28). Induced polarization, magnetometer, and geochemical soil surveys were used to prospect the contact zone. Most of the ore is at the base of the komatiite in contact with metabasalt, though some is in lenses within the ultramafic flows. Structural depressions in the basal contact appear to

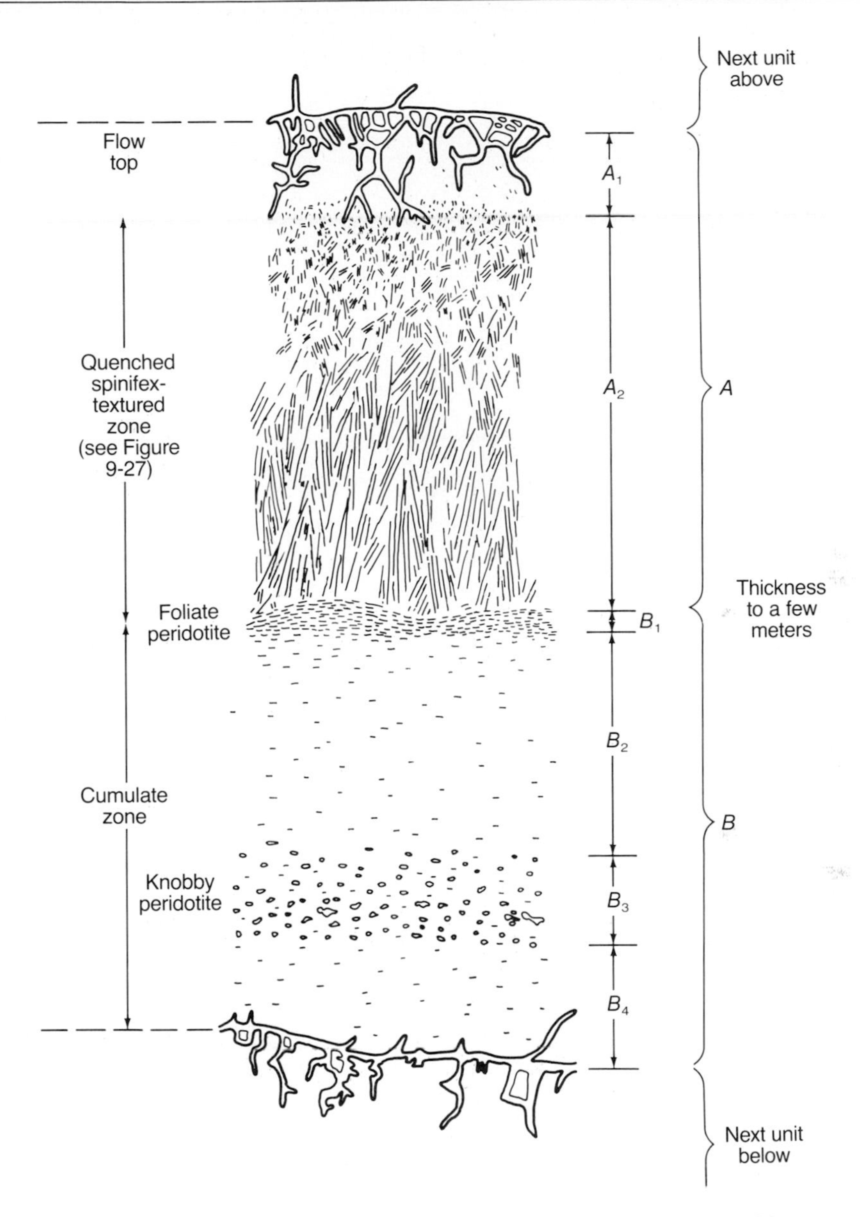

Figure 9-26. Cross section through a typical spinifex-textured komatiite flow in Munro Township, Ontario. The komatiites are ultramafic olivine-pyroxene flows first described from near the Komati River in South Africa. *Spinifex* refers to the texture of zone A_2, the extraordinary skeletal, lathlike olivine crystals which resemble mats of an Australian grass of that name. These rocks are unique hosts to Cu-Ni massive sulfide ores. See also Figure 9-27. (*After* Pyke, Naldrett, and Eckstrand, 1973.)

Figure 9-27. Two pieces of spinifex-textured komatiite showing the characteristic lathlike skeletal crystals of olivine. (*a*) Specimen from the Komati River type area in the Republic of South Africa shows a weathered surface on which the grasslike texture is emphasized. (*b*) Piece of diamond drill core from the Langmuir mine area, Timmins, Ontario, shows that the olivine crystals are truly platy and interpenetrating. (*Photo by* W. K. Bilodeau.)

constitute local ore controls, perhaps where the sulfide matte extruded and "puddled" (Figure 9-29). The sulfide liquids sank into depressions in the newly formed floor rocks, and olivine crystals froze in place while floating to the sulfide surface. They left a *net texture* of adjacent olivine tablets with interstitial sulfides first described by Naldrett (1973) at the Alexo mine in Ontario, Canada (Figure 9-30). Seven orebodies at Kambalda on the "puddled" contact, and Marston and Kay (1980) report similar new major discoveries near the Juan deposits to the northwest. These sulfide layers are a few meters thick and hundreds of meters across, and involve a few million

Figure 9-28. (*Facing page*) (*a*) Simplified geological map of the Kambalda dome, Western Australia, and (*b*) cross section (along line *A-A'*) of the Lunnon ore shoot. Orebody outlines have been projected to the surface for illustrative purposes; all rocks are Archean in age. Note that the orebodies are strataform and conformable to the flows in which they occur, and that they occur only at the base of and within ultramafic, locally komatiitic flows. See also Figure 9-29. (*After* Naldrett, 1981.)

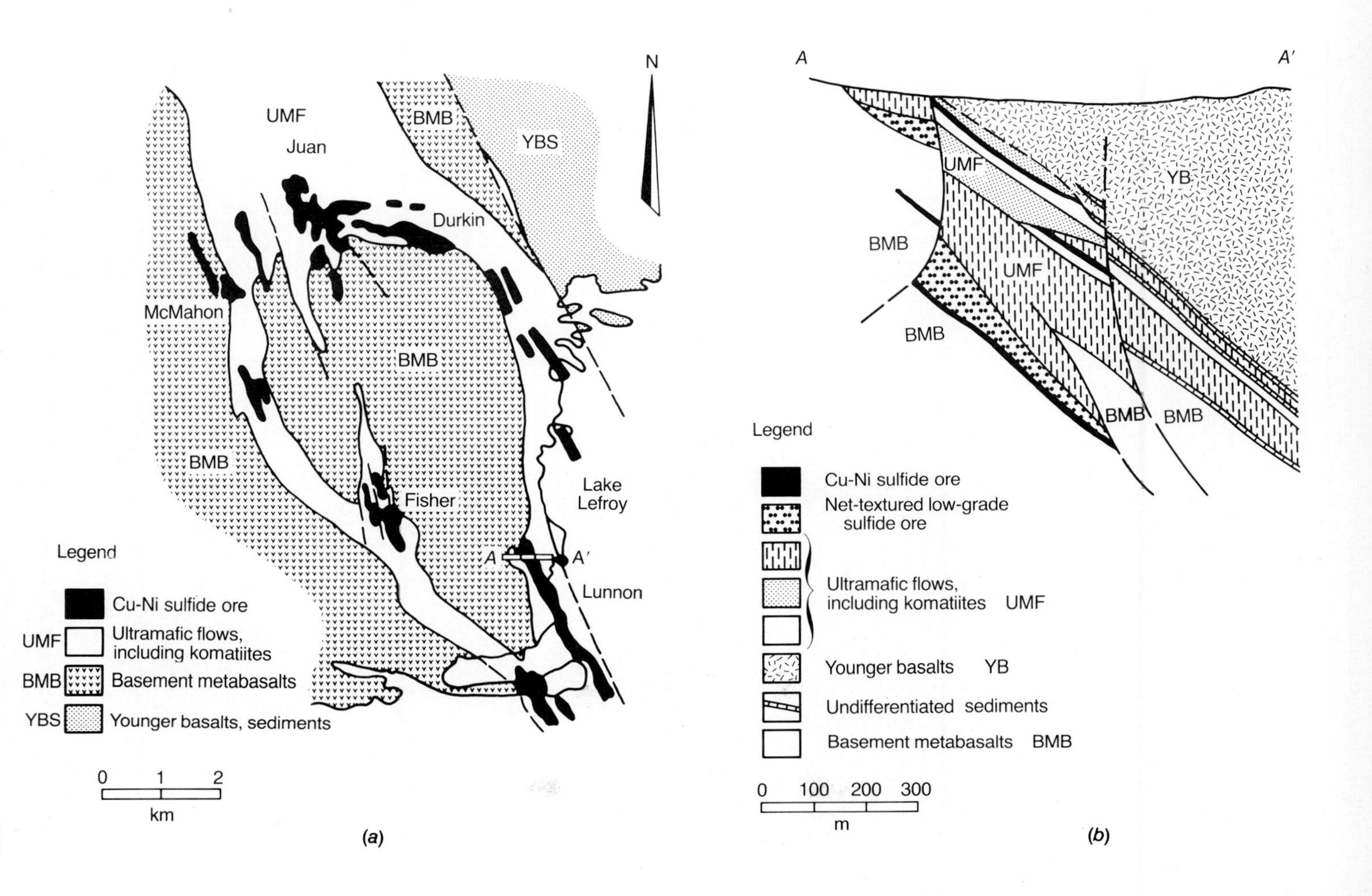
N
UMF
Juan
BMB
YBS
Durkin
McMahon
BMB
BMB
Fisher
Lake
Lefroy
A
A′
Lunnon
Legend
Cu-Ni sulfide ore
UMF
Ultramafic flows, including komatiites
BMB
Basement metabasalts
YBS
Younger basalts, sediments
0 1 2
km
(a)
A
A′
UMF
YB
BMB
UMF
BMB
BMB
BMB
Legend
Cu-Ni sulfide ore
Net-textured low-grade sulfide ore
Ultramafic flows, including komatiites UMF
Younger basalts YB
Undifferentiated sediments
Basement metabasalts BMB
0 100 200 300
m
(b)

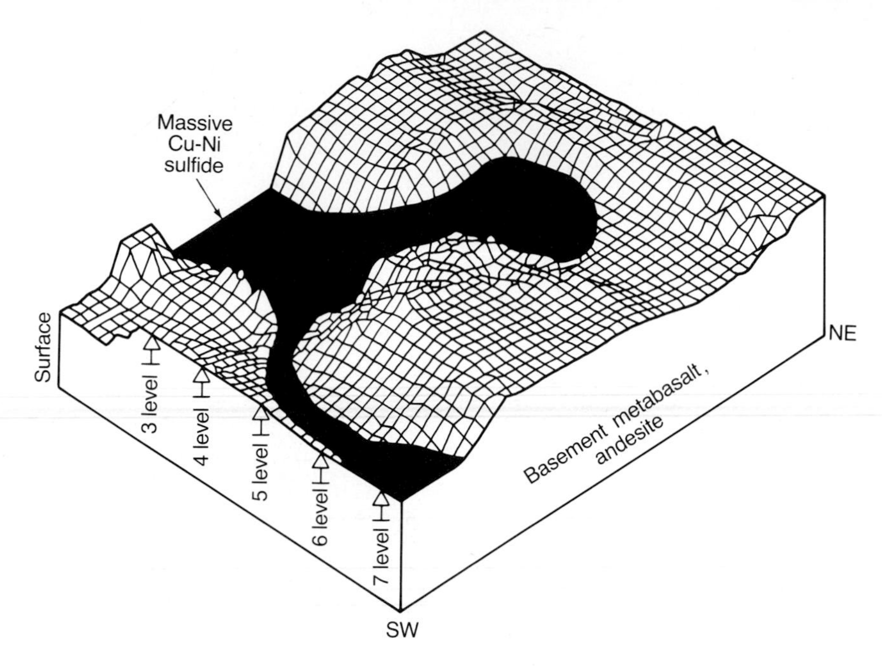

Figure 9-29. Three-dimensional view of the Langmuir 2 Cu-Ni sulfide ore occurrence rotated to a horizontal, original flow surface position and showing how the massive pyrrhotite-chalcopyrite-pentlandite liquid accumulated, or "puddled," in the low parts of the irregular upper contact of footwall andesites and ultramafic flows. (*After* Naldrett, 1981.)

tons of ore at best. At Kambalda, the orebodies occur in a 3-km-thick sequence of Archean ultramafic-mafic volcanics in the Yilgarn Craton. Post-consolidation metamorphism thickened some ores and mobilized some into faults and along previously barren contacts, but created no new sulfides.

According to Ewers and Hudson (1972), the profitable Lunnon Shoot in the Kambalda district has a footwall of pillow basalt. Successively above the footwall is a narrow zone of magnetite-rimmed euhedral chromite crystals in silicates, massive ore with 95% sulfides, matrix or disseminated ore with 20 to 60% sulfides, and hydrated and carbonated ultramafic rocks in the hangingwall. Ewers and Hudson considered that the data were consistent with the ore having been emplaced at elevated temperatures as an ultramafic mush containing olivine and chromite crystals, magnesia-rich silicate melt, and an immiscible sulfide-oxide liquid. Gravity separation produced the ore zone, and the hanging wall represents the original crystal mush with minor amounts of sulfides. Pyrite, concentrated near the top of the massive ore, was the first subsolidus sulfide mineral to exsolve, at about 400°C, and cobalt was preferentially concentrated in this pyrite.

The ore at Kambalda and Juan consists of both massive and disseminated sulfides. Where both are present, the disseminated mineralization overlies the massive. The primary mineralization is a pyrrhotite-pentlandite

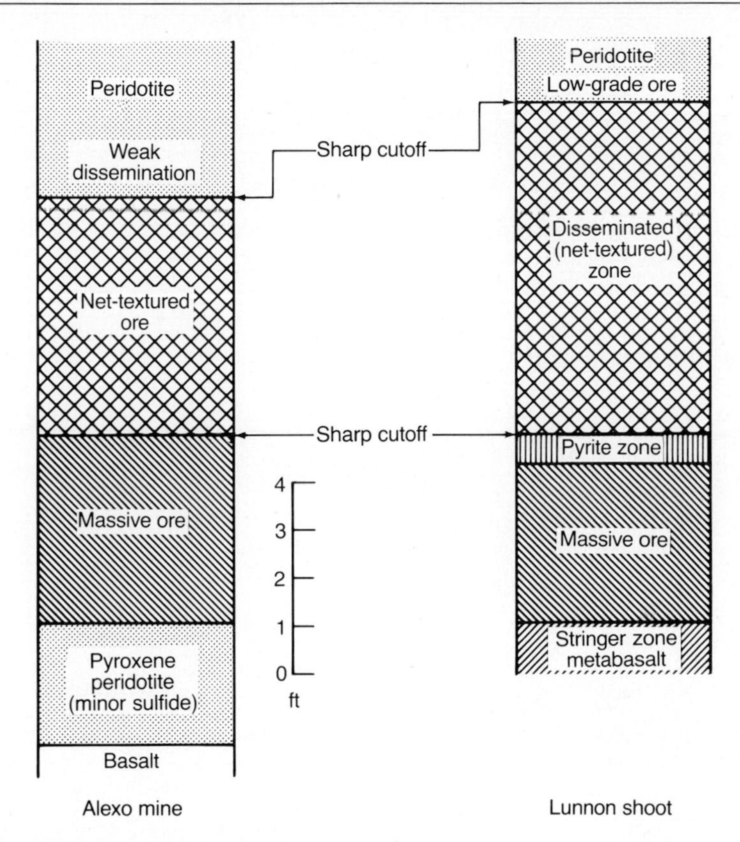

Figure 9-30. Comparison of the ore zone—disseminated, net-textured, low-grade sulfide zones at the Alexo mine near Timmins, Ontario, and the Lunnon ore shoot in the Kambalda district of western Australia. *Net texture* refers to a texture formed by the freezing of sulfide liquid while olivine crystals float upward within it and form a loose meshwork when immobilized. (*After* Naldrett, 1981.)

assemblage with subordinate cobaltiferous pyrite and minor amounts of chalcopyrite. Platinum metal values are present but low. A supergene zone characterized by violarite, a purplish supergene mineral of composition $(Ni,Fe)_3S_4$, lies between the leached gossan and the primary sulfides and is a useful guide to ore.

Canadian deposits are also associated with spinifex-textured komatiite flows, including the classic outcrops in Munro Township described in Pyke, Naldrett, and Eckstrand (1973) and Naldrett and Cabri (1976). They occur in the lower ultramafic-mafic platform basalts that underlie tholeiite-calcalkaline terranes that host exhalite Cu-Zn ores, and are ultimately part of their geologic setting. The Langmuir Cu-Ni sulfide mine was found in spinifex komatiites only a few kilometers south of, and in the basement rocks related to, the Kidd Creek Cu-Zn-Ag orebody near Timmins, Ontario (Chapter 13).

Bibliography

Abel, M. K., R. Buchan, C. J. A. Coats, and M. E. Penstone, 1979. Copper mineralization in the footwall complex, Strathcona Mine, Sudbury, Ontario. *Can. Mineral.* 17:275–285.

Anderson, A. T., Jr., and M. Morin, 1968. Two types of massif anorthosites and their implications regarding the thermal history of the crust, pp. 57–69 Y. W. Isachsen, Ed., *Origin of Anorthosite and Related Rocks*. N.Y. State Mus. Sci. Serv. Mem. 18, 466 pp.

Balsley, J. R., Jr., 1943. Vanadium-titanium-iron ores, Lake Sanford, New York. USGS Bull. 940D.

Bergeron, M., 1972. Quebec Iron and Titanium Corporation ore deposits at Lac Tio, Quebec. 24th Int. Geol. Congr. Guidebook B09, 8 pp.

Boyd, F. R., and L. W. Finger, 1976. Homogeneity of minerals in mantle rocks from Lesotho. Ann. Rept. Dir. Geophys. Lab., 1975–76, pp. 519–528.

—— and H. O. A. Meyer, Eds., 1979. *Kimberlites, Diatremes, and Diamonds: (I) Their Geology, Petrology, and Geochemistry; (II) The Mantle Sample: Inclusions in Kimberlites and Other Volcanics*. Washington, DC: Amer. Geophys. Union, I:400 pp.; II:424 pp.

Brocoum, S. J., and I. W. D. Dalziel, 1974. The Sudbury Basin, the Southern Province, the Grenville Front, and the Penokean Orogeny. *Geol. Soc. Amer. Bull.* 85:1571–1580.

Buddington, A. F., 1931. The Adirondack magmatic stem. *J. Geol.* 39:240–263.

——, 1933. Correlation of kinds of igneous rocks with kinds of mineralization, pp. 350–385 in *Ore Deposits of the Western States*, Lindgren Vol., New York: AIME, 797 pp.

——, 1939. Adirondack igneous rocks and their metamorphism. Geol. Soc. Amer. Mem. 7, pp. 64–67

Cabri, L. J., and J. H. G. Laflamme, 1976. The mineralogy of the platinum group elements from some copper-nickel deposits of the Sudbury area, Ontario. *Econ. Geol.* 72:1449–1458.

Cameron, E. N., 1963. Structure and rock sequences of the Critical Zone of the Eastern Bushveld Complex. Mineral. Soc. Amer. Spec. Pap. 1, pp. 93–107.

——, 1969. Postcumulus changes in the Eastern Bushveld Complex. *Amer. Mineral.* 54:754–799.

——, 1971. Problems of the Eastern Bushveld Complex. *Fortschr. Miner.* 48:86–108.

——, 1975. Postcumulus and subsolidus equilibration of chromite and coexisting silicates in the Eastern Bushveld Complex. *Geochim. et Cosmochim. Acta* 39:1021–1033.

——, 1978. The lower zone of the Eastern Bushveld Complex in the Olifants River Trough. *J. Petrol.* 19:437–462.

——, 1982. The upper Critical Zone of the Eastern Bushveld Complex; precursor to the Merensky Reef. *Econ. Geol.* 77:1307–1327.

—— and G. A. Desborough, 1964. Origin of certain magnetite-bearing pegmatites in the eastern part of the Bushveld Complex, South Africa. *Econ. Geol.* 59:197–225.

—— and M. E. Emerson, 1959. The origin of certain chromite deposits of the eastern part of the Bushveld Complex. *Econ. Geol.* 54:1151–1213.

Cantin, R. D., and R. G. Walker, 1972. Was the Sudbury Basin circular during deposition of the Chelmsford Formation? pp. 93–101 in J. V. Guy-Bray, Ed., *New Developments in Sudbury Geology*, Geol. Assoc. Can. Spec. Pap. 10.

Card, K. D., and R. W. Hutchinson, 1972. The Sudbury structure: its regional geological setting. pp. 67–79 in J. V. Guy-Bray, Ed., *New Developments in Sudbury Geology*, Geol. Assoc. Can. Spec. Pap. 10.

Clement, C. R., and E. M. W. Skinner, 1979. A textural-genetic classification of kimberlite rocks. Kimberlite Symp III, Univ. Cambridge, 4 pp.

Coleman, A. P., 1905. The Sudbury nickel field. Ontario Bur. Mines Ann. Rept. 14, pt. 3.

——, 1913. The nickel industry. Dept. of Mines, Canada, no. 170.

——, 1926. The magmatic origin of the Sudbury nickel ores. *Geol. Mag.* 63:108–112.

Cousins, C. A., 1969. The Merensky Reef of the Bushveld Igneous Complex. Econ. Geol. Mon. 4, pp. 239–251.

—— and C. F. Vermaak, 1976. The contribution of Southern African ore deposits to the geochemistry of the platinum group metals. *Econ. Geol.* 71:287–305.

Deardon, E. O., 1958. Lac Tio ilmenite deposit. AIME Preprint 5818A4.

Dence, M. R., 1972. Meteorite impact craters and the structure of the Sudbury basin, pp. 7–18 in J. V. Guy-Bray, Ed., *New Developments in Sudbury Geology*, Geol. Assoc. Can. Spec. Pap. 10.

Dietrich, D. R., 1985. Vanadium distribution in magnetite-rich layers of the Stillwater Complex, Montana. Unpub. M. S. Thesis, Univ. Ariz.

Dietz, R. S., 1964. Sudbury structure as an astrobleme. *J. Geol.* 72:412–434.

——, 1972. Sudbury astrobleme, splash-emplaced sub-layer and possible cosmogenic ores, pp. 29–40 in J. V. Guy-Bray, Ed., *New Developments in Sudbury Geology*, Geol. Assoc. Can. Spec. Pap. 10.

Duchesne, J. C., and D. Demaiffe, 1978. Trace elements and anorthosite genesis. *Earth and Plan. Sci. Lett.* 38:249–272.

Emslie, R. F., 1970. The geology of the Michikamau intrusion, Labrador, Can. Geol. Surv. Pap. 68–57, 85 pp.

——, 1975. Nature and origin of anorthositic suites. *Geosci. Can.* 2:99–116.

——, 1978. Anorthosite massifs, Rapakivi granites, and late Proterozoic rifting of North America. *Precambrian Res.* 7:61–98.

Erickson, R. L., and L. V. Blade, 1963. Geochemistry and petrology of the alkalic igneous complex at Magnet Cove, Arkansas. USGS Prof. Pap. 425.

Ewers, W. E., and D. R. Hudson, 1972. An interpretative study of a nickel-iron sulfide ore intersection, Lunnon Shoot, Kambalda, Western Australia. *Econ. Geol.* 67:1075–1091.

Fairbairn, H. W., P. M. Hurley, and W. H. Pinson, 1968. Rb-Sr whole-rock age of the Sudbury lopolith and basin sediments. *Can. J. Earth Sci.* 5:707–714.

Ferguson, J., and J. W. Sheraton, 1979. Petrogenesis of kimberlitic rocks and associated xenoliths of southeastern Australia, pp. 140–160 in F. R. Boyd and H. O. A. Meyer, Eds., *Kimberlites, Diatremes, and Diamonds: (I) Their Geology, Petrology, and Geochemistry.* Washington, DC: Amer. Geophys. Union, 400 pp.

Fischer, R. P., 1946. German iron ores yield vanadium. AIME Tech. Pub. 2070.

——, 1950. Entmischungen in Schwermetalloxyden, Silikaten und Phosphaten. *Neues Jahrb. Mineral. Abh.* 81:315–364.

Fleet, M. E., 1977. Origin of disseminated copper-nickel sulfide ore at Frood, Sudbury, Ontario. *Econ. Geol.* 72:1449–1458.

——, 1979. Partitioning of Fe, Co, Ni, and Cu between sulfide liquid and basaltic melts and the composition of Ni-Cu sulfide deposits; discussion. *Econ. Geol.* 74:1517–1519.

Ford, A. B., 1976. Stratigraphy of the layered gabbroic Dufek intrusion, Antarctica. USGS Bull. 1405-D, 36 pp.

French, B. M., 1968. Sudbury structure, Ontario, Canada: some petrographic evidence for an origin by meteoritic impact, in B. M. French and M. N. Short, Eds., *Shock Metamorphism of Natural Materials.* Baltimore: Mono Book Corp.

Gillson, J. L., 1956. Genesis of titaniferous magnetites and associated rocks of the Lake Sanford district, New York. AIME *Trans.* 205:296–301.

Gilmour, P., 1980. Pers. Commun.

Green, T. H., 1969. High-pressure experimental studies on the origin of anorthosite. *Can. J. Earth Sci.* 6:427–440.

Gross, S. O., 1968. Titaniferous ores of the Sanford Lake district, New York, pp. 140–153 in J. D. Ridge, Ed., *Ore Deposits of the United States 1933/1967,* Graton-Sales Vols., New York: AIME, 1880 pp.

Guy-Bray, J. V., and Geological Staff, 1966. Shatter cones at Sudbury, Canada. *J. Geol.* 74:243–245.

—— and W. V. Peredery, 1971. Guide notes, Sudbury excursion, May, 1971. Geol. Assoc. Can.

Hall, A. L., 1932. The Bushveld Igneous Complex of the central Transvaal. So. Afr. Geol. Surv. Mem. 28, 554 pp.

Hammond, P., 1951. Allard Lake ilmenite deposits. *Geol. Soc. Amer. Bull.* 62:1448.

——, 1952. Allard Lake ilmenite deposits. *Econ. Geol.* 47:634–649.

Hargraves, R. B., 1962. Petrology of the Allard Lake anorthosite suite, Quebec, pp. 163–189 in *Petrologic Studies.* Geol. Soc. Amer. Buddington Vol., 660 pp.

Hawley, J. E., 1962. The Sudbury ores: their mineralogy and origin. *Can. Mineral.* 7(pt.1):1–29.

Hawthorne, J. B., 1975. Model of a kimberlite pipe. *Phys. Chem. Earth* 9:1–15.

Heinrich, E. W., 1966. *The Geology of Carbonatites.* Chicago: Rand McNally, 550 pp.

Herz, N. L., 1967. Titanium deposits in anorthosite massifs. USGS Prof. Pap. 959 D, pp. D1-D6.

——, 1968. The Roseland alkalic anorthosite massif, Virginia, pp. 13–22 in Y. W. Isachsen, Ed., *Origin of Anorthosite and Related Rocks.* NY State Mus. Sci. Serv. Mem. 18, 466 pp.

——, 1969. Anorthosite belts, continental drift, and the Anorthosite Event. *Science* 164:944–946.

——, 1976. Geology and resources of titanium. USGS Prof. Pap. 959-A,B,C,D,E,F. 49 pp.

Hoffman, E. L., A. J. Naldrett, R. A. Allcock, and R. G. V. Hancock, 1979. The noble-metal content of ore in the Levack West and Little Stobie mines, Ontario. *Can. Mineral.* 17:437–451.

Hunter, D. R., 1976. Some enigmas of the Bushveld Complex. *Econ. Geol.* 71:229–248.

—— and N. A. De N. C. Gomes, 1975. Provisional tectonic map of the Bushveld complex. Econ. Geol. Res. Unit, Johannesburg, Univ. Witwatersrand.

Hutton, C. O., 1945. Vanadium in the Taranaki titaniferous iron-ores. *New Zealand J. Sci. Tech.* 27:15–16.

Isachsen, Y. W., 1969. Origin of Anorthosite and Related Rocks—a summarization, pp. 435–445 in *Origin of Anorthosite and Related Rocks.* NY State Mus. Sci. Serv. Mem. 18, 466 pp.

Jackson, E. D., 1959. The chromite deposits of the Stillwater Complex, Montana, pp. 1495–1510 in J. D. Ridge, Ed., *Ore Deposits of the United States 1933/1967*, Graton-Sales Vols. New York: AIME, 1880 pp.

Kennedy, G. C., and B. E. Nordlie, 1968. The genesis of diamond deposits. *Econ. Geol.* 63:495–503.

Kolker, A., 1982. Mineralogy and geochemistry of Fe-Ti oxide and apatite (nelson-ite) deposits and evaluation of the liquid immiscibility hypothesis. *Econ. Geol.* 77:1146–1158.

Kuo, H. Y., and J. H. Crocket, 1979. Rare earth elements in the Sudbury Nickel Irruptive; comparison with layered gabbros and implications for nickel irruptive petrogenesis. *Econ. Geol.* 74:590–605.

Lenthall, D. H., 1974. Tin production from the Bushveld Complex. Econ. Geol. Res. Unit Circ. 93, Johannesburg, Univ. Witwatersrand.

—— and D. R. Hunter, 1977. The geology, petrology and geochemistry of the Bushveld granites and felsites in the Petgeitersrus tin field. Econ. Geol. Res. Unit Circ. 110, Johannesburg, Univ. Witwatersrand, 91 pp.

Lombaard, A. F., N. M. Ward-Able, and R. W. Bruce, 1964. The exploration and main geological features of the copper deposit in carbonatite at Loolekop, Palabora Complex, in S. H. Haughton, Ed., *The Geology of Some Ore Deposits in Southern Africa*. Johannesburg: Geol. Soc. So. Afr., 739 pp.

MacLean, W. H., 1969. Liquidus phase relations in the FeS-FeO-Fe_3O_4-SiO_2 system, and their application in geology. *Econ. Geol.* 64:865–884.

—— and H. Shimazaki, 1976. The partition of Co, Ni, Cu, and Zn between sulfide and silicate liquids. *Econ. Geol.* 71:1049–1057.

Marais, P., 1980. Pers. Commun.

Marston, R. J., and B. D. Kay, 1980. The distribution, petrology, and genesis of nickel ores at the Juan Complex, Kambalda, Western Australia. *Econ. Geol.* 75:546–565.

McBirney, A. R., and R. M. Noyes, 1979. Crystallization and layering of the Skaergaard Intrusion. *J. Petrol.* 20:487–554.

McCallum, M. E., and C. D. Mabarak, 1976. Diamond in kimberlite diatremes of northern Colorado. *Geology* 4:467–469.

——, C. D. Mabarak, and H. G. Coopersmith, 1979. Diamonds from kimberlites in the Colorado-Wyoming state line district, pp. 42–58 in F. R Boyd, and H. O. A. Meyer, Eds., *Kimberlites, Diatremes, and Diamonds: (I) Their Geology, Petrology, and Geochemistry*. Washington, DC: Amer. Geophys. Union, 400 pp.

McDonald, J. A., 1967. Evolution of part of the Lower Critical Zone, Farm Ringhoek, Western Bushveld. *J. Petrol.* 8:165–209.

Molyneux, T. G., 1970. The geology of the area in the vicinity of Magnet Heights, eastern Transvaal, with special reference to the magnetic iron ore, pp. 228–241 in *Symposium on the Bushveld Complex and Other Layered Intrusions*. Johannesburg: Geol. Soc. So. Afr. Spec. Pub. 1.

——, 1972. A survey of the mineral deposits related to the BIC, South Africa. 24th Int. Geol. Congr. Proc., Sec. 4, pp. 225–232.

Naldrett, A. J., 1973. Nickel sulphide deposits—their classification and genesis with special emphasis on deposits of volcanic association. *Can. Inst. Min. Metall. Trans.* 76:183–201.

——, 1981. Nickel sulfide deposits: classification, composition and genesis. *Econ. Geol. 75th Anniv. Vol.*, pp. 628–685.

——, Guy-Bray, E. L. Gasparrini, T. Podolsky, and J. C. Rucklidge, 1970. Cryptic variation and the petrology of the Sudbury nickel irruptive. *Econ. Geol.* 65:122–155.

—— and L. J. Cabri, 1976. Ultramafic and related mafic rocks: their classification and genesis with special reference to the concentration of nickel sulfides and platinum-group elements. *Econ. Geol.* 71:1131–1158.

——, L. Greenman, and R. H. Hewins, 1972. The main irruptive and the sub-layer at Sudbury, Ontario. 24th Int. Geol. Congr. Proc., Sec. 4, pp. 206–214.

Newhouse, W. H., 1936. Opaque oxides and sulfides in common igneous rocks. *Geol. Soc. Amer. Bull.* 47:1–52.

Norton, D., and H. P. Taylor, Jr., 1979. Quantitative simulation of the hydrothermal systems of crystallizing magmas on the basis of transport theory and oxygen isotope data; an analysis of the Skaergaard intrusion. *J. Petrol.* 20:421–486.

Olson, J. C., D. R. Shawe, L. C. Pray, and W. N. Sharp, 1954. Rare earth mineral deposits of the Mt. Pass district, San Bernardino County, California. USGS Prof. Pap. 261.

Palabora Geological and Mineralogical Staff, 1976. The geology and economic deposits of copper, iron, and vermiculite in the Palabora Complex: a brief review. *Econ. Geol.* 71:177–192.

Pattison, E. F., 1979. The Sudbury sublayer. *Can. Mineral.* 17:257–274.

Peredery, W. V., and A. J. Naldrett, 1975. Petrology of the upper irruptive and the sub-layer at Sudbury, Ontario. *Econ. Geol.* 70:164–175.

Philpotts, A. A., 1967. Origin of certain iron-titanium oxide and apatite rocks. *Econ. Geol.* 62:303–315.

Pyke, D. R., A. J. Naldrett, and O. R. Eckstrand, 1973. Archean ultramafic flows in Munro township, Ontario. *Geol. Soc. Amer. Bull.* 84:955–988.

Ramdohr, P., 1940. Die Erzmineralien in gewöhnlichen magmatischen Gesteinen. Preuss. Akad. Wiss. Abh. 2.

Rhodes, R. C., 1975. New evidence for impact origin of the Bushveld Complex, South Africa. *Geology* 3:549–554.

Rose, E. R., 1969. Geology of titanium and titaniferous deposits of Canada. Geol. Surv. Can. Econ. Geol. Rpt. 25.

Schwellnus, J. S. I., S. A. Hiemstra, and E. L. Gasparrini, 1976. The Merensky Reef at the Atok platinum mine and its environs. *Econ. Geol.* 71:249–260.

Simmons, E. C., and G. N. Hanson, 1978. Geochemistry and origin of massif-type anorthosites. *Contr. Mineral. Petrol.* 66:119–135.

Singewald, J. T., 1917. The role of mineralizers in ore segregations in basic igneous rocks. Johns Hopkins Univ. Contr. Geol. (Mar.), pp. 24–35.

Skinner, E. M. W., and C. R. Clement, 1979. Mineralogical classification of southern African kimberlites, pp. 129–139 in F. R. Boyd and H. O. A. Meyer, Eds., *Kimberlites, Diatremes, and Diamonds: (I) Their Geology, Petrology, and Geochemistry*. Washington, DC: Amer. Geophys. Union, 400 pp.

——, C. R. Clements, and J. B. Hawthorne, 1980. Pers. Commun.

Smith, C. B., M. E. McCallum, H. G. Coopersmith, and D. H. Eggler, 1979. Petrochemistry and structure of kimberlites in the Front Range and Laramie Range, Colorado-Wyoming, pp. 178–189 in F. R. Boyd and H. O. A. Meyer, Eds., *Kimberlites, Diatremes, and Diamonds: (I) Their Geology, Petrology, and Geochemistry*. Washington, DC: Amer. Geophys. Union, 400 pp.

Souch, B. E., T. Podolsky, and Geological Staff, 1969. The sulfide ores of Sudbury: their particular relationship to a distinctive inclusion-bearing facies of the nickel irruptive. Econ. Geol. Mon. 4, pp. 252–261.

Speers, E. C., 1957. The age relation and origin of common Sudbury breccia. *J. Geol.* 65:497–514.

Stanton, R. L., 1972. *Ore Petrology*. New York: McGraw-Hill, 690 pp.

Stephenson, R. C., 1948. Titaniferous magnetite deposits of the Lake Sanford area, New York. *AIME Trans.* 178:397–421.

Stevenson, J. S., 1963. The upper contact phase of the Sudbury micropegmatite. *Can. Mineral.* 7:413–419.

Taylor, R. B., 1964. Geology of the Duluth Gabbro Complex near Duluth, Minnesota. Minn. Geol. Surv. Bull. 44, 63 pp.

Thayer, T. P., 1946. Preliminary chemical correlation of chromite with the containing rocks. *Econ. Geol.* 41:202–217.

——, 1969. Gravity differentiation and magmatic re-emplacement of podiform chromite deposits. Econ. Geol. Mon. 4, pp. 132–146.

Thomson, J. E., 1956. Geology of the Sudbury Basin. Ontario Dept. Mines Ann. Rept. 65 (pt. 3), pp. 1–56.

Thomson, J. E., and H. Williams, 1959. The myth of the Sudbury lopolith. *Can. Min. J.* 80:3–8.

Uytenbogaardt, W., 1954. On the opaque mineral constituents in a series of amphibolitic rocks from Norra Storfjallet, Vasterbotten, Sweden. *Ark. Mineral. Geol.* 1(5–6):527–543.

Vaasjoki, O., 1947. On the microstructure of titaniferous iron ore at Otanmaki. Comm. Geol. Finlande Bull. 140, P. Eskola Vol., pp. 104–114.

Verhoogen, J. 1962. Oxidation of Fe-Ti oxides in igneous rocks. *J. Geol.* 70:168–181.

Vermaak, C. F., 1970. The geology of the lower portion of the Bushveld Complex and its relationship to the floor rocks in the area west of the Pilanesberg, western Transvaal., pp. 242–265 in *Bushveld Igneous Complex and Other Layered Intrusions*. Johannesburg: Symp. Geol. Soc. So. Afr. Spec. Pub. 1.

——, 1976. The Merensky Reef—Thoughts on its environment and genesis. *Econ. Geol.* 71:1270–1298.

——, 1979. The global status of the South African minerals economy and data summaries of its key commodities. Geol. Soc. So. Afr. Rev. Pap. 1, 57 pp.

——, 1980. Pers. Commun.

—— and L. P. Hendriks, 1976. A review of the mineralogy of the Merensky Reef, with specific reference to new data on the precious metal mineralogy. *Econ. Geol.* 71:1244–1269.

Viljoen, M. J., A. Bernasconi, M. van Coller, E. Kinloch, and R. P. Viljoen, 1976. The geology of the Shangani nickel deposit, Rhodesia. *Econ. Geol.* 71:76–95.

——, Ed., 1985. *The Geology and Petrology of the Bushveld Igneous Complex* (approx. title), 5 vols. Johannesburg: Geol. Soc. So. Afr., in press.

Vogt, J. H. L., 1894. Beiträge zur genetischen Classification der durch magmatische Differentiationsprocesse und der durch Pneumatolyse entstandenen Erzvorkommen. *Z. Prakt. Geol.* 2:381–399.

Wager, L. R., and W. A. Deer, 1939. The petrology of the Skaergaard intrusion, Kangerdlugssuag, east Greenland. *Medd. Grönland*, 105(4).

Wagner, P. A., 1929. *The Platinum Deposits and Mines of South Africa*. London: Oliver and Boyd.

Wenk, H. R., 1978. Are pseudotachylites products of fracture or fusion? *Geology* 6:507–511.

Willemse, J., 1969a. The vanadiferous magnetite iron ore of the Bushveld igneous complex. Econ. Geol. Mon. 4, pp. 187–208.

——, 1969b. The geology of the Bushveld igneous complex, the largest repository of magmatic ore deposits in the world. Econ. Geol. Mon. 4, p. 1–22.

Williams, H., 1956. Glowing avalanche deposits of the Sudbury Basin. Ontario Dept. Mines Ann. Rept. 65 (pt. 3), pp. 57–89.

Williams, N., 1979. Studies of the base metal sulfide deposits at McArthur River, Northern Territory, Australia; II, The sulfide-S and organic-C relationships of the concordant deposits and their significance; replies. *Econ. Geol.* 74:1695–1697; 1699–1702.

Woodall, R., and G. A. Travis, 1970. The Kambalda nickel deposits, Western Australia, pp. 517–533 in *Mining and Petroleum Geology*, 9th Commonw. Min. Met. Congr. Proc., vol. 2.

Yates, A. B., 1938. The Sudbury intrusive. *Roy. Soc. Can. Trans.*, Ser. 3., 32 (sec. 4):151–172.

——, 1948. Properties of International Nickel Company of Canada, in *Structural Geology of Canadian Ore Deposits*. Montreal: Can. Inst. Min. Metall., 948 pp.

Yoder, H. S., Jr., 1968. Experimental studies bearing on the origin of anorthosite, pp. 13–22 in Y. W. Isachsen, Ed., *Origin of Anorthosite and Related Rocks*. NY Mus. Sci. Serv. Mem. 18, 466 pp.

Zurbrigg, H. F., and Geological Staff, 1957. The Frood-Stobie mine, in *Structural Geology of Canadian Ore Deposits*, vol. 2. Montreal: Can. Inst. Min. Metall. 524 pp.

TEN

Deposits Related to Oceanic Crust

The world's modern technology uses up to 3 million tons a year of chromium in chemicals, in steel alloys, as a rust-resistant plating, and in refractory brick and other heat-resistant materials. The United States imports all of its 1 million tons of chromium consumed per year from Turkey, the Philippines, the Republic of South Africa, Zimbabwe, and the U.S.S.R. Chromium ores occur in two types of mafic and ultramafic igneous rock systems, the layered mafic intrusive bodies described in the preceding chapter and the ophiolites-Alpine peridotites to be described here. Both ophiolites and Alpine peridotites involve the tectonic emplacement of slices of upper mantle and oceanic crust—mafic and ultramafic rocks formed at oceanic spreading centers—into or onto continental crust. The oceanic crust rocks can be deformed in the process but may be nearly undeformed; the upper mantle floor on which they formed always appears squeezed, deformed, and smeared, and almost certainly was so before emplacement. We shall first consider the formation of oceanic crust, then the emplacement and the characteristics of ophiolites and Alpine peridotites, and then ore deposits associated with them. Cyprus-type massive copper-zinc sulfide deposits, which may be found on top of ophiolites, are illustrated here but described in Chapter 13.

The ocean floor has been studied intensively during recent years by the dredging of grab samples from its surface, by drilling from research vessels, by marine geophysical sensing, and by geologists in submersible vessels making direct observations and collecting seafloor samples. We have been able to use "on-land" techniques to study what we think are pieces of ocean floor, the ophiolites, that have been tectonically emplaced in subaerial environments where they can be mapped and studied handily, as in Cyprus and Oman. Geophysical measurements and geologic studies have shown that oceanic crust is composed of three major layers, although more detailed petrographic study has revealed several sublayers. From the top down, the first one is a layer of deep-sea sediments, a few centimeters to a few meters of fine-grained laminated siliceous or ocherous rocks commonly containing radiolarian fossils and other biotic debris. Layer 2 beneath the sediments is composed of several hundred to several thousand meters of variably altered, distinctly layered, commonly columnar-jointed, pillowed basalts cut by diabase dikes. The diabase dikes increase in width and frequency with depth. Finally Layer 3 is as much as 10 km thick and is thought to be composed of vaguely layered, medium- to coarse-grained, variably serpentinized dunites, peridotites, harzburgites, gabbros, norites, troctolites, and chromitites. Rock assemblages repeating this three-fold association of cherts on pillow lavas above serpentinized units were described from a score of separate localities around the globe by Steinmann in 1905 and 1927. They were dubbed "Steinmann's trinity," and remained often observed but largely unexplained until the 1970s.

We can now offer an explanation of Steinmann's trinity. The associations that Steinmann mapped are rafts or slices of the three layers of oceanic crust which are part of the assemblage that we now collectively call *ophiolites*. The lowermost unit (Layer 3) represents plutonic crystallization of deep-seated melts derived by partial melting of asthenospheric upper mantle material beneath midoceanic or back-arc basin spreading centers. The magma chambers are thought to be about 15 to 20 km across (East Pacific Rise Study Group, 1981), and crystallization regimens probably not unlike those described in Chapter 9 for the Bushveld Complex prevail. They have existed for hundreds of millions of years much as they exist as you read this paragraph. The chambers are probably broad-beam hull-shaped (Figure 10-1), with the floor slowly retreating by gravity in both directions, downward and away from the spreading center, thus with layering dipping gently inward. The middle layer (Layer 2) represents outflow of some of the liquids crystallizing in Layer 3. They are the basalts that spill out onto the submarine floor at the midocean rise. As they move laterally, they become the surficial crust over the subjacent pluton, the deck of the hull in the analogy above. Dikes slashing upward along and near the rise reinvade older dike swarms and deliver basaltic magma to the rock-water interface, where submarine pillowed flows and masses form and move out laterally. The upper, later rocks clearly have fewer dikes than the lower ones. These upper rocks are chloritized, serpentinized, and spilitized (albitized), that is, chemically altered by seawater circulated through them by convection (Figure 10-1 and Chapter 13). Finally as Layer 2 and Layer 3 are pushed or pulled, or slide down gravity gradients from oceanic rises to become oceanic crust

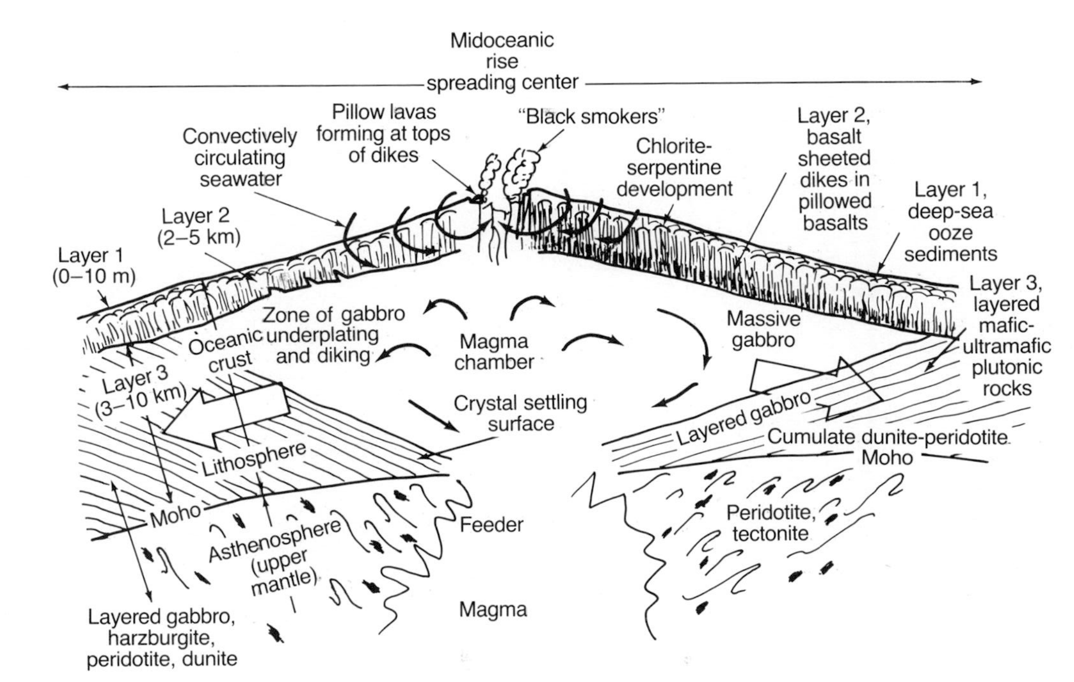

Figure 10-1. Sketch of the dynamics of oceanic crust formation along midoceanic rises, with implications for ophiolites, Alpine peridotites, and Cyprus-type massive sulfides. Some ophiolites involve only oceanic crust; others include a thickness of the upper mantle peridotite tectonite.

between ridge and continent, Layer 1 is precipitated as a skim of siliceous diatoms, radiolarian debris, limey and cherty precipitates, and iron-manganese hydroxides that constitute dirt and fish scales on the decks in the earlier analogy. So Layers 1, 2, and 3, which we now perceive to have been Steinmann's trinity, are a logical geologic result of mechanisms that create oceanic crust.

Most oceanic crust is ultimately subducted and consumed, but a small percentage—perhaps 0.001% (Coleman, 1977)—is not smoothly subducted but is lodged as slivers and slabs in continental margins by *obduction*, that is, for various "accidental" tectonic reasons involved with arc or continental collision, rammed edgewise into the crust rather than being carried down the subduction zone. Rafts of oceanic crust that become lodged in or are thrust up onto the continental or island-arc margin are assemblages called *ophiolites*. They may or may not include basal portions of the deformed upper mantle material called *tectonized peridotite* on which the oceanic crust formed, the asthenosphere of Figure 10-1. Figure 10-2 shows a section of the Samail ophiolite in Oman (Hopson et al., 1981), which includes a basal slab of that mantle material.

Alpine complexes—named for occurrences along the Alps of southern central Europe—are important modes of occurrence of the ophiolite assemblage. They were perceived by Harry Hess as long ago as the 1940s and 1950s as occurring in belts along many of the mountain ranges of the world (Figure 10-3). Geologically, Alpine complexes consist of what appear to be squeezed, kneaded, smeared masses of dunite, peridotite, harzburgite, lherzolite, olivine gabbro, and gabbro, all showing a tectonized fabric of gneissic, contorted foliation of lenticular, discordant, tightly folded layers. Lherzolite is like peridotite, but its pyroxene content is more than half the clinopyroxene augite. The most mafic of those complexes are called *Alpine peridotites*; they are dunites, peridotites, and harzburgites consisting of olivine and orthopyroxene with disseminations or lenses of chromite. These podiform smeared-out sacklike bodies of chromite, or *chromitite* where the spinel content is more than 90 vol %, suggest the earlier existence of chromite lenses or layers in original plutonic dunite-harzburgite sandwiches. They are found only in Alpine peridotite, not in other Alpine complex rocks. Alpine complexes, and of course the subset Alpine peridotites, thus appear to be slices of the asthenospheric upper mantle substrate to oceanic crust shown in Figure 10-1. Where ophiolite assemblages have been as little disturbed internally during obduction as they were at Oman (Figure 10-2), recognition of the geologic relationship between the lithospheric oceanic crust part and the asthenospheric tectonite peridotite part is easy. However, the ophiolite assemblage is readily dismembered so that bodies of Layers 1, 2, and 3 and bodies of upper mantle materials can be emplaced separately. Alpine peridotite contacts against rocks in which they have been lodged show nil to slight metamorphism; many are fault-bounded, as might

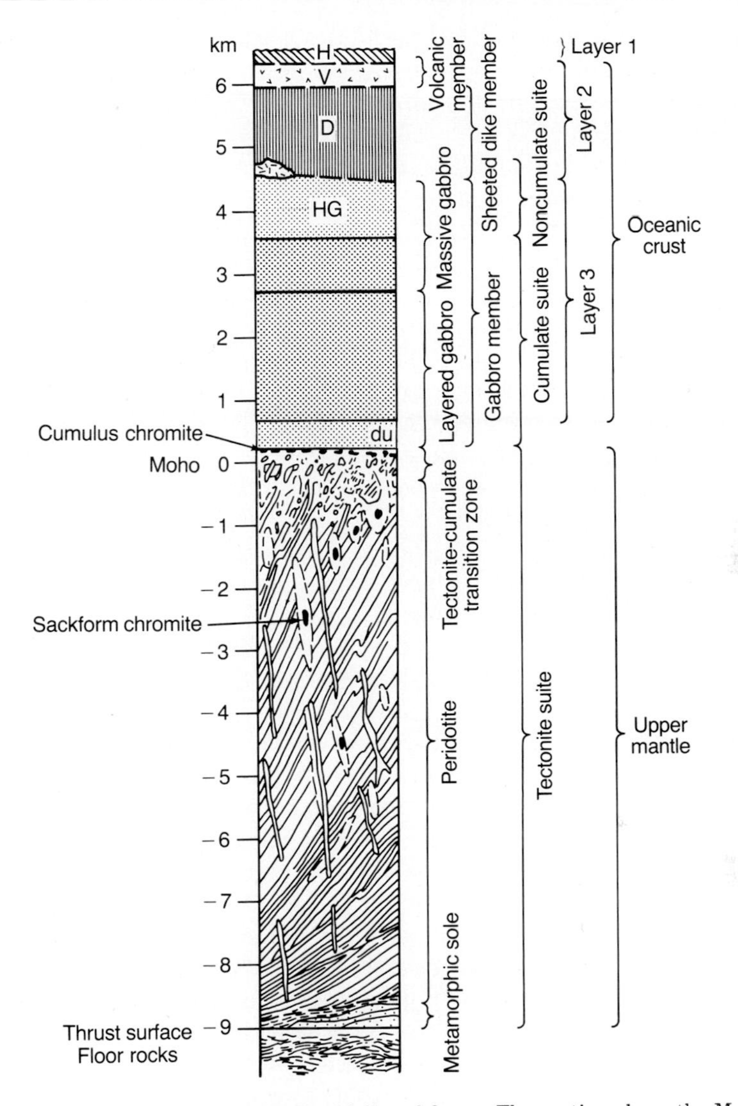

Figure 10-2. Cross section of the Samail ophiolite of Oman. The portion above the Moho was oceanic crust, the portion below it peridotite of the asthenosphere, or upper mantle. H is Layer 1 sediment. V is massive and pillowed basalt over D, a sheeted diabase dike member, V and D comprising Layer 2. HG (high-level noncumulus gabbro) overlies other massive, then layered gabbros of Layer 3, with cumulate olivine and chromite (du) at the base. Below that, smeared and deformed "tectonized peridotite" continues to the thrust surface below the "metamorphic sole." Chromite also occurs as sackform masses in this unit. The entire package is over 15 km thick, and was thrust on top of older metamorphic and sedimentary rocks 65 to 90 million years ago. (*From* Hopson et al., 1981.)

be expected. Although smeared peridotitic rocks are easily recognized and established as Alpine complexes, it may be impossible to relate them to specific ophiolite packages if originally contiguous oceanic crust layers have been dismembered and displaced elsewhere. Many Alpine peridotite bodies such as that at Burro Mountain in California (Burch, 1968) show only the upper mantle type assemblages.

Ophiolites have been described for some decades, although their significance has been recognized only recently (Moores and MacGregor, 1972; Coleman 1977, 1984; Moores, 1982). Although their parentage has been contested (Miyashiro, 1975), their oceanic crust origin and tectonic significance are now well established. They can be internally flat-faulted so that portions of the original crust may be doubled up or attenuated, but they normally appear to have moved monolithically with a great range of intensity of internal deformation. These rafted masses are most abundant around the perimeters of modern ocean basins and are found, for example, at scattered localities all around the Pacific basin. They are most common in Paleozoic and younger island-arc fore-arc settings, but they are now being identified in some geologically older suture zones within continental crust, especially in Precambrian terranes in Egypt, Arabia, and the Urals of the U.S.S.R. They are of many sizes from tens of hundreds or even thousands of square kilometers in plan view, and from less than 1 to 15 km thick, as at Oman.

As Figure 10-2 suggests, chromite in ophiolite assemblages can occur in at least two modes, described long ago (Sampson, 1942) as sack- or pod-shaped masses in the tectonized peridotite upper mantle material, or as discontinuous to quasicontinuous layers in the highest temperature, most mafic basal cumulates of Layer 3. The former are typically smeared-out schlieren-like lensoid bodies in dunite jackets; they can be seen to be part of a layered sequence, but the layering is contorted and deformed. The Layer 3 chromitites are normally more discernibly stratiform, regular, laterally persistent, and with well-preserved cumulate textures. Moutte (1982), in describing the Tiébaghi chromite deposits on the island of New Caledonia (Figure 17-4), referred to two occurrence types of chromitite, one as schlieren, pods, nodules, and lenses in what is probably tectonized peridotite, and the other as "the Alpha and Vieille Montagne 1 deposits situated higher in the series (which) show chromite-rich layers of regular thickness, with large open folds and with well-preserved cumulate textures," including chain textures and overgrowths (Figure 9-2). These latter are probably cumulate materials of Layer 3 indicated in Figures 10-1 and 10-2. Moutte shows that Mg/Fe^{+2} and Cr/Fe^{+3} decrease upward, conforming to progressive magmatic differentiation trends. Disseminated chromite also occurs in peridotite, locally in amounts warranting extraction. Both the massive and the disseminated chromites are typically rich in chromium and magnesium and low in iron within the variable $(Fe,Mg)^{+2}(Cr,Al,Fe)_2^{+3}O_4$ formula of chromite. That tendency makes Alpine peridotite chromite with its Cr/Fe ratios near 2.0 highly desirable for metallurgical and refractory applications. Values as high as 3.3 (55 wt % Cr_2O_3) are normal at the Tiébaghi mine on New Caledonia (Moutte, 1982).

▸ ALPINE PERIDOTITE CHROMITE

Moa, Cuba

The chromite deposits of Moa, Cuba, constitute a well-known example of ophiolite-Alpine peridotite chromite and are described here, but the scholar should also read descriptions of the Masinloc deposit, Luzon, the Philippines (Stoll, 1958), the Turkish occurrences (Thayer, 1964), those in Greece (Zachos, 1969), the Indian examples (Srinivasachari, 1976), and the Tiébaghi deposits of New Caledonia (Moutte, 1982). A brief description of the Cyprus ophiolite chromites (Greenbaum, 1977; Panayiotou, 1978) will be followed by some even briefer exploration suggestions. Generally chromitiferous Alpine peridotites are common along the Tethyan suture zone from Cyprus and Greece to Turkey, along the Ural Mountain zone in the U.S.S.R, from Japan through the Philippines to New Zealand, along the Appalachians of the United States, and around the Caribbean (Jackson and Thayer, 1972). They are virtually all less than 1200 million years old. The specific origins of a particular ophiolite or group of ophiolites with respect to whether the package of oceanic crust formed at a midoceanic spreading center, such as the East Pacific Rise, or at a back-arc or inter-arc spreading center are the subject of current study. Coleman (1984) presents petrologic and tectonostratigraphic evidence that the North American Cordilleran ophiolites and Alpine peridotites all formed at a spreading center in a back-arc or inter-arc basin between the North American plate and a Mesozoic island arc that stood offshore to the west. The ophiolites were obducted, and the oceanic crust was subducted, when these basins were closed; ophiolites commonly constitute contact zones between accreted fragments, microplates, and the continent.

In the Moa district of northeastern Cuba (Figures 10-3 and 10-4) are several Alpine peridotite-ophiolite chromite deposits that are clear-cut tectonized periodotite-oceanic crust occurrences. These refractory chromite deposits have been studied in detail by Guild (1947), who described them as sacklike bodies of massive, nearly pure chromite in serpentinized ultramafic rocks. The orebodies are in an extensive ultramafic complex which is a broadly domed, laterite-covered slope along the northern range of Oriente Province. Several chromite-bearing districts continue along an east-west belt of serpentinized ultramafics; only Moa is described here because it is better understood and was less complexly deformed during emplacement than the other districts. Rapid headward erosion of short vigorous streams has cut deep V-shaped valleys, destroying part of the lateritic surface and

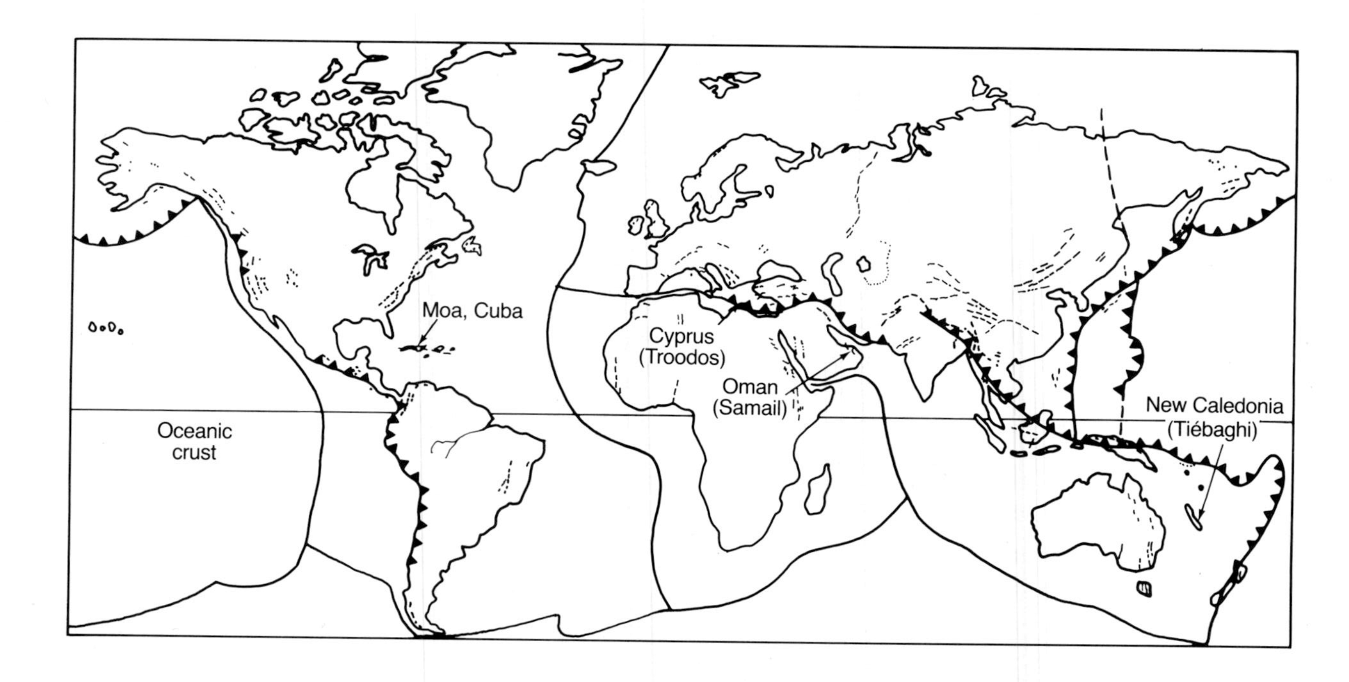

Figure 10-3. Major tectonic elements of the Earth, with ophiolite-Alpine peridotite belts. Modern subduction zone and mid-oceanic rise positions are shown. The broken and dotted lines on the continental plates are paleo-suture zones marked by ophiolite-Alpine periodic belts. Oman; Moa, Cuba; Tiébaghi; and Cyprus are discussed in the text. (*After* Gass, 1982.)

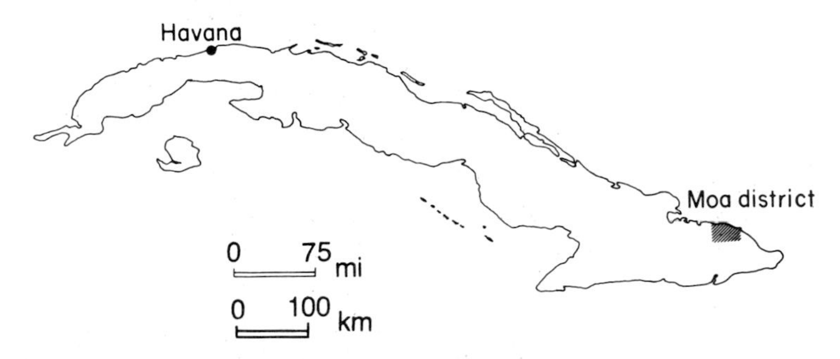

Figure 10-4. Map of Cuba, showing the location of the Moa district.

exposing bedrock. Between the streams, the well-preserved old surface rises at a slope of 3 or 4° from sea level on the north to an altitude of nearly 1 km near the crest of the Cuchillas de Toar range (Figure 10-5). The climate is humid, and thick vegetation considerably handicaps geologic mapping. Chromite bodies project from the canyon walls at a few places, supplying abundant float to the stream beds. Although the deposits have been known since late in the nineteenth century, no exploration was done until World War I, which motivated a small amount of prospecting. Real development began in 1940; during the next seven years more than 500,000 tons of refractory chromite ore averaging about 33% Cr_2O_3 were produced (Guild, 1947).

The subjacent country rock complex is made up of chlorite schist, metamorphosed volcanic tuffs and lavas, thin impure marbles, and diorite exposed south of the Moa district and thought to represent the base onto which the ultramafics were thrust. The serpentines show no trace of the intense metamorphism that affected the older rocks. Typical of Alpine peridotites, no metamorphism of the underlying host rocks is attributable to the emplacement of the Moa body. Overlying the ultramafics to the west are Upper Cretaceous conglomerates that contain cobbles of serpentine, and Tertiary deposits (Figure 10-5).

The serpentine massif originally consisted of harzburgite with subordinate dunite, chromitite, and pyroxenite. It contains some less mafic facies, such as gabbro, troctolite, and anorthosite which reflect variable proportions of olivine, pyroxene, plagioclase, and chromite. With the exception of pyroxene and chromite, all combinations and gradations of these minerals are found. The olivine component of the ultramafic rocks is serpentinized, and the feldspathic rocks are altered to unctuous aggregates of hydrous minerals, including antigorite (?), zoisite, edenite, chlorite, and zeolites.

The chromite concentration in the ore varies. Chromite is subordinate to the silicate minerals in some deposits, but the best deposits contain only a small percentage of gangue minerals, and average ore carries about 5% silicates by weight. The most common gangue minerals are serpentinized olivine-pyroxene and minor alteration products of feldspars. Low-grade ores

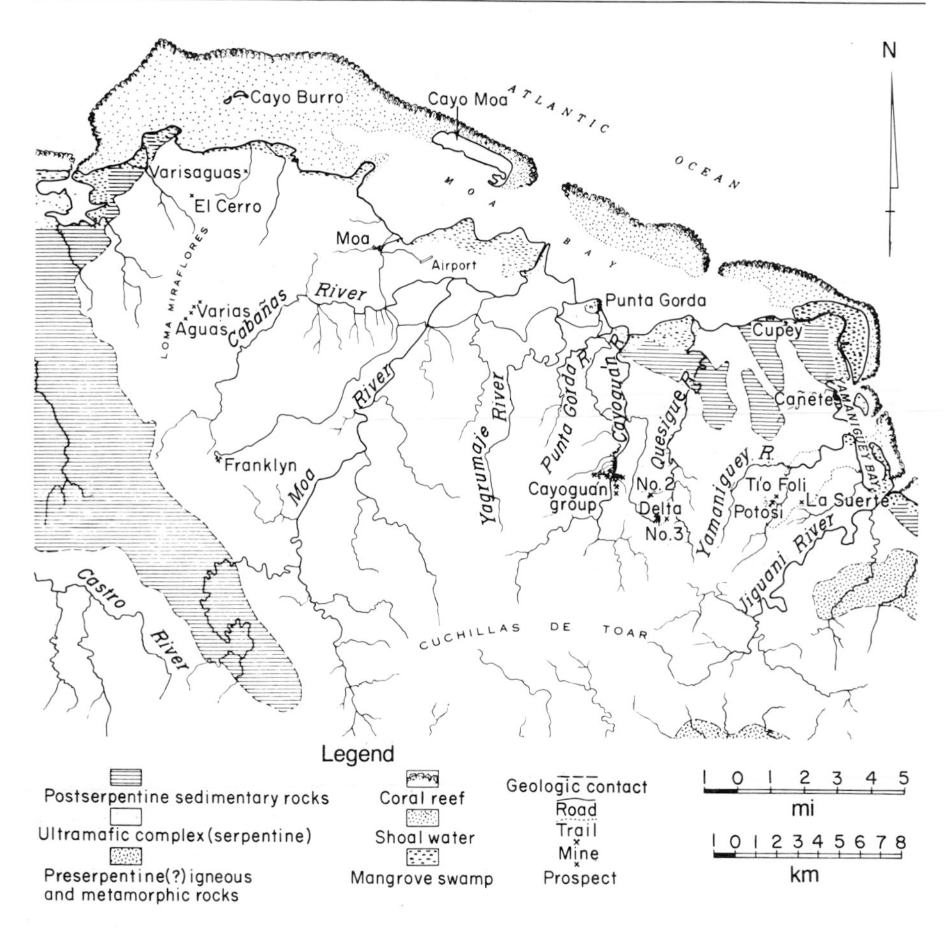

Figure 10-5. Map of the Moa chromite district, Oriente Province, Cuba. See Figure 10-3 for location. Figure 10-6 focuses on the Cayogúan group. (*From* Guild, 1947, Fig. 2.)

contain individual grains of anhedral chromite disseminated in dunite, whereas the high-grade ores, which constitute most of the commercial deposits, appear almost massive and contain only minor amounts of interstitial silicates. Most of the chromite is fine- to medium-grained with a seriate texture, but a few grains are measured in centimeters, and crystals up to 15 cm across have been found.

The chromite deposits are tabular to lenticular, and range from thin streaks 1 or 2 meters long to much larger masses; some are elongated downdip and others along strike (Figure 10-6). Guild (1947) described one deposit as 500 meters downdip, 250 meters along strike, and 30 meters thick; it caps a hillslope between A' and A'' in Figure 10-6. Its contacts are generally sharp, but other orebodies grade outward through alternating parallel layers of variable serpentine content in which the proportion of chromite decreases outward from the deposit into the country rock. Still other deposits are composed of interlayered olivine and chromite and have no central concentration of chromite. Each chromite deposit is surrounded by a dunite jacket ranging in thickness from half a meter or less to tens of meters. The chromite-dunite pods are encased in peridotite, hence there is a general sequence from chromite to dunite to peridotite in all orebodies. At any single place the chromite-dunite and dunite-peridotite contacts are virtually parallel. Guild noted in 1947 that fault contacts are more common than undisturbed igneous contacts, and that the association of chromite with dunite is so well established that a fault was plotted where drilling encountered peridotite instead of dunite adjacent to ore (Guild, 1947). This distribution, and the features described in this paragraph, suggest that chromite was originally deposited in layers or lenses in olivine-rich rocks, the contacts having later been smeared, deformed, and faulted. It is noteworthy that chromite bodies in Alpine peridotite-ophiolite association are consistent throughout the world. They are always within dunite layers, even in variegated mafic–ultramafic rocks, having been deposited early with olivine, as indicated in Figure 10-1, or occurring in tectonized mantle ultramafic rocks, as shown in Figure 10-2.

Fortunately the internal structures of the ultramafic complex have been modified but not obliterated by later deformation so that the rock fabric still reflects the processes of crystallization. Compositional layering and platy mineral grains oriented parallel to the layering give the rocks a planar fabric. The most prominent layering is that of the chromite bodies themselves. Feldspathic lenses and streaks or layers of olivine in the chromite are all parallel to the contacts. Another planar structure is formed by the orientation of flat pyroxene grains in the peridotite, generally but not everywhere parallel to the compositional layering. It is especially noticeable on weathered outcrops.

The ultramafic complex is intricately jointed and faulted, with joints encountered every few centimeters or meters. Among the myriad fractures is a class of openings called *pull-aparts*. They consist of blocks or slabs of chromitite that are cut by veinlike, dikelike dunite, now serpentinized. Such pull-apart structures, filled by mechanical flow during emplacement, are common in Alpine peridotites. The dikes are more numerous and more coarse-grained in the brittle chromite masses than in the other rocks. Their contents were apparently cold injections into dilatant fractures formed during emplacement. Dunite dikes in chromitite masses were noted by Thayer (1964) in Turkey. These dikelets reflect tension-separation of the massive, probably more brittle chromitite masses and cold intrusion by syntectonic recrystallization and physical rotation and flow of olivine granules (Carter and Ave'Lallemant, 1970).

The orebodies and the ultramafic complex at Moa are clustered in a narrow vertical zone consistent with emplacement of a slab after solidifi-

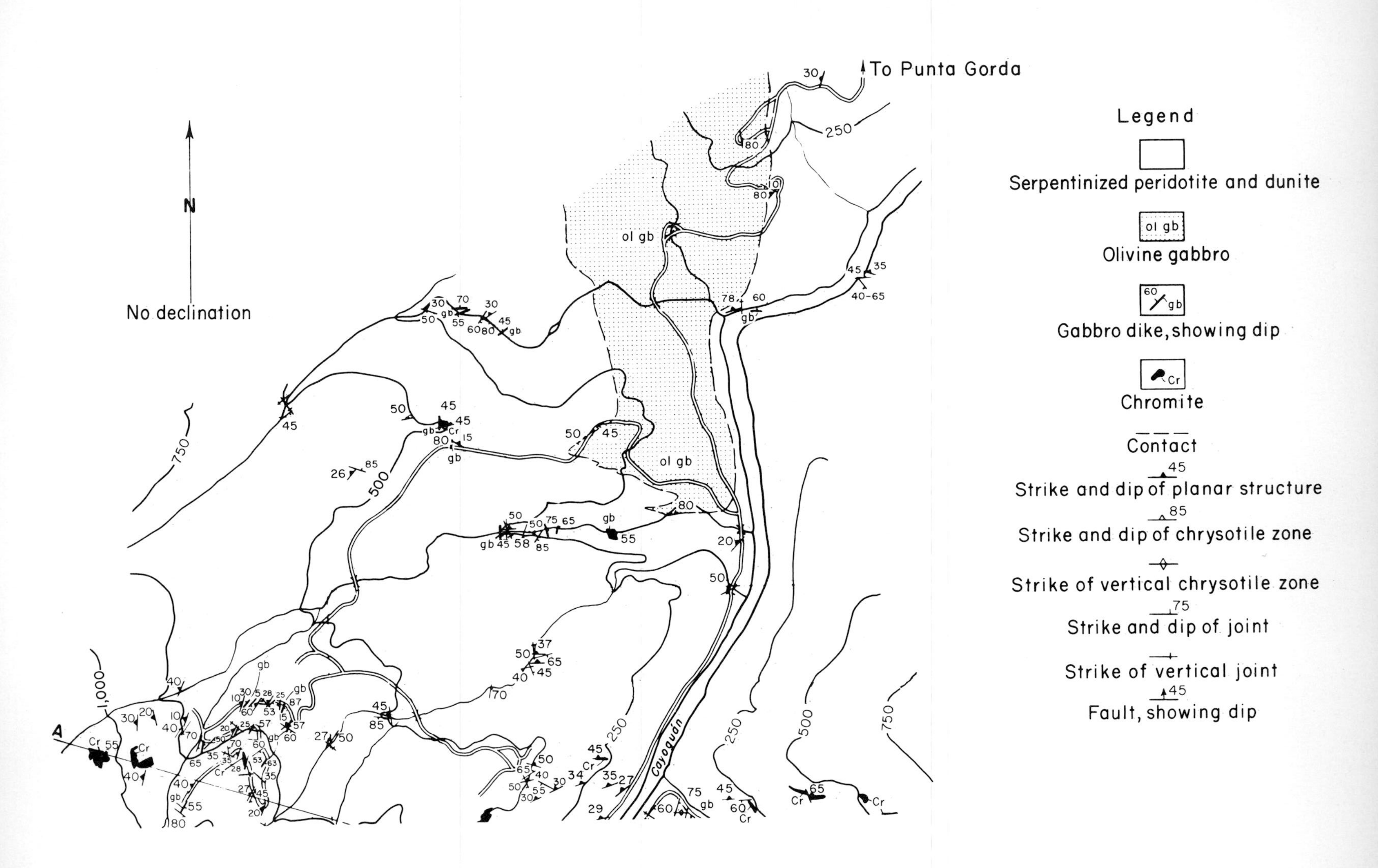

To Punta Gorda
No declination
N
Cayoguán
ol gb
Legend
Serpentinized peridotite and dunite
Olivine gabbro
Gabbro dike, showing dip
Chromite
Contact
Strike and dip of planar structure
Strike and dip of chrysotile zone
Strike of vertical chrysotile zone
Strike and dip of joint
Strike of vertical joint
Fault, showing dip

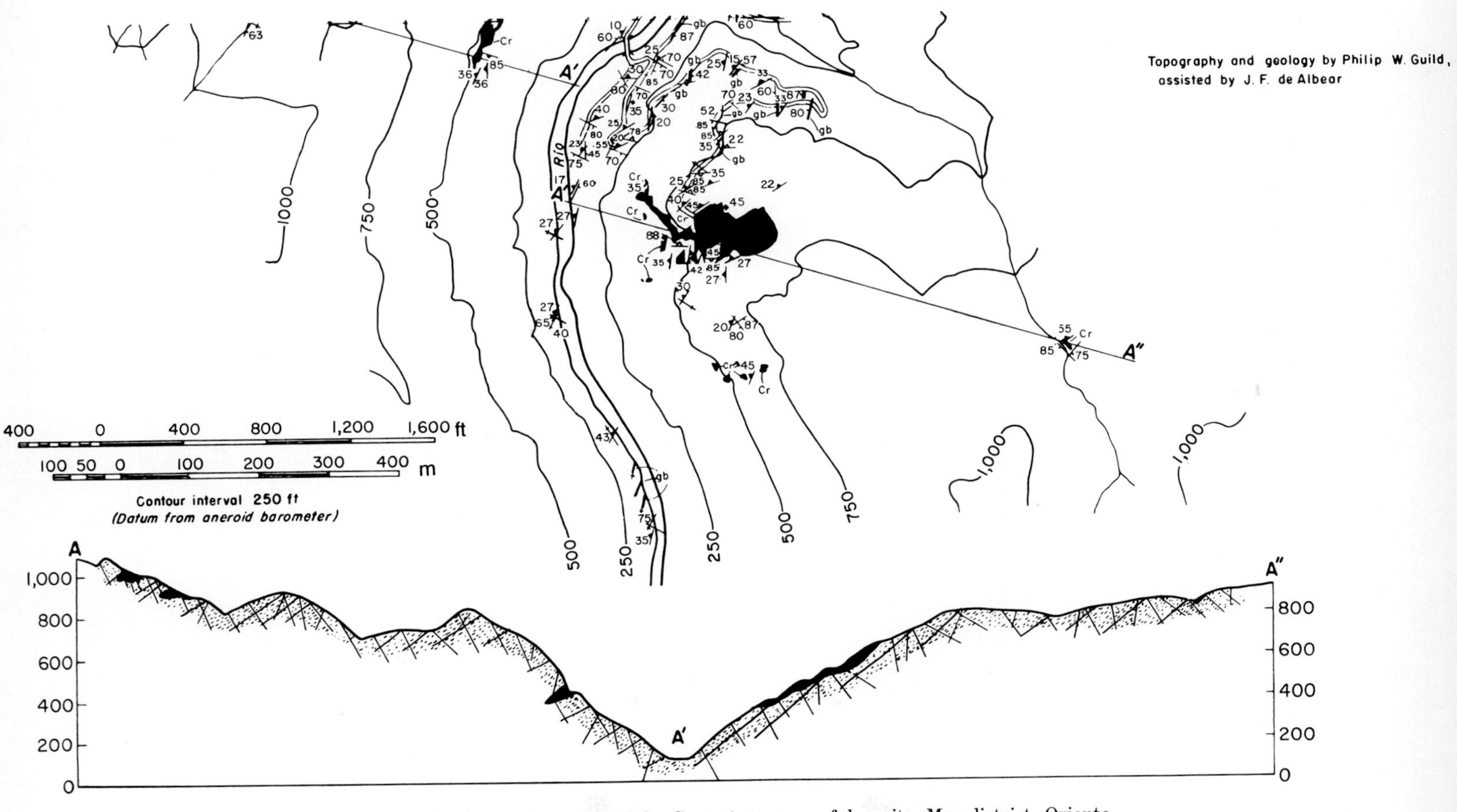

Figure 10-6. Map and section of the Cayogúan group of deposits, Moa district, Oriente Province, Cuba. See also Figure 10-5. The chromitite bodies dip to the west parallel to silicate foliation and lamination. (*From* Guild, 1947, Fig. 6.)

cation. Both the parallelism of the orebodies and their layered structures could be said to be similar to relationships in crustal LMIs. Two of the larger orebodies, the Cayoguán and the Narciso-Cromita deposits, have similar dimensions and almost identical compositions (see Figure 10-6 and Table 10-1), a distinction that cannot be made for any other local pair of deposits. This similarity implies that they may be fragments of a single mass that was broken and separated during emplacement, either in basal Layer 3 or as original units in what is now tectonized peridotite. The higher percentages of the "high-melting" components MgO and Cr_2O_3 in these Alpine peridotite chromites are typical and confirm their consanguinity with high-temperature more mafic rocks.

Guild (1947) concluded that the ore was a product of early magmatic differentiation that formed discrete lodes before the ultramafic units reached their present position. Today that chromite-dunite association would be considered to have occurred when the rocks, now serpentinized, formed as Layer 3 of oceanic crust or tectonized peridotite along a midoceanic ridge (Figure 10-1). Indeed Guild argued that there could not have been sufficient time for compositional layering and coarse crystals to develop after intrusion at the present site. Moreover, pyroxene crystals were oriented during growth, and the paragenesis of chromite, olivine, and pyroxene implies that

TABLE 10-1. Analyses of Chromites from Cuban Alpine Peridotite Chromium Ores

Component	Cayoguán	Narciso	Potosi
Cr_2O_3	38.03	38.26	39.62
Al_2O_3	30.96	30.30	24.96
Fe_2O_3	2.59	2.55	5.21
FeO	10.36	10.84	14.00
MgO	17.43	17.22	14.60
MnO	0.03	0.09	0.15
CaO	0.02	0.30	0.20
TiO_2	0.06	0.19	0.44
SiO_2	0.24	0.52	0.80
H_{20}	0.10	0.12	0.20
P_2O_5	0.06	0.06	—
S	none	trace	—
Totals	99.88	100.45	100.06
Cr/Fe	2.64	2.56	1.86

Composition of chromite mineral from the Cayoguán, Narciso, and Potosi deposits, Moa district, Cuba. Values represent percentage by weight. Notice the high Cr/Fe ratio and high MgO compared with that of LMI chromite in Table 9-3.

Source: From Guild (1974).

the ore should have been solid at temperatures of Alpine peridotite-ophiolite emplacement.

Weathering has had little effect on the chromite ores. In one of the small deposits, a pale green chlorite replaces euhedral chromite, leaving only skeletal relict crystals of the chromite grains. This chlorite is not chromiferous, suggesting that chromium ions can be mobile under certain weathering conditions and capable of diffusing away from a reaction front. The pink chrome-bearing chlorite kammererite is also encountered in many of the deposits, but for the most part the chromitites are merely oxidized to lateritic limonites at the outcrop.

Troodos, Cyprus

The island of Cyprus has attracted attention in recent years because it contains a mafic intrusive complex that is thought to represent an excellent example of an ophiolite slice (Wilson, 1959; Moores and Vine, 1971; Greenbaum, 1977). The ophiolite complex, known as the Troodos Massif, was emplaced during the Cretaceous period. It consists of pillow basalts overlying a sheeted dike complex, and then an intrusive complex of gabbro that grades downward into olivine gabbro, then into an ultramafic body of harzburgite and dunite (Figure 10-7, 10-8). Some of the harzburgites are serpentinized, and small amounts of asbestos have been recovered. A generalized geological map of the area is shown in Figure 10-7. The upper parts of the gabbro and the lower basalts are cut by many closely spaced diabase dikes that form conspicuous sheeted masses. They are overlain by fine-grained ferruginous, siliceous, sulfidic sediments called *umbers* (Figure 10-9). The complex thus includes all three layers of oceanic crust, the most important here being a Layer 3 plutonic coarse-grained assemblage of lower cumulate ultramafic rocks and upper gabbros underlying Layer 2 basaltic volcanic rocks. The underlying deformed tectonized harzburgites contain scores of podiform chromite deposits—64 are counted in a 16-km^2 area. They occur as isolated pods and layers in dunite near harzburgite, in harzburgite within dunite lenses, or in narrow dunite envelopes, or jackets, exactly as at Moa, Cuba. Chromite-silicate textures include primary ones such as the cumulate textures of Figure 9-2 and silicate overgrowths over chromite, and secondary ones such as younger schlieren, or sheared, "strung out" textures produced by deformation. The ore is in isolated pods and discontinuous layers of variable size and grade. The pods do not seem to fit a pattern, but disseminated chromite expectably shows decreased chromium and increased iron upward in the evolution of silicate rocks from dunite through olivine-clinopyroxene peridotite (wehrlite) to olivine-clinopyroxene-biotite peridotite. Chromitite in both Layer 3 cumulate layered rocks and underlying tectonized peridotite is recognized on Cyprus. Greenbaum (1977, p. 1175) concludes that the upper part of

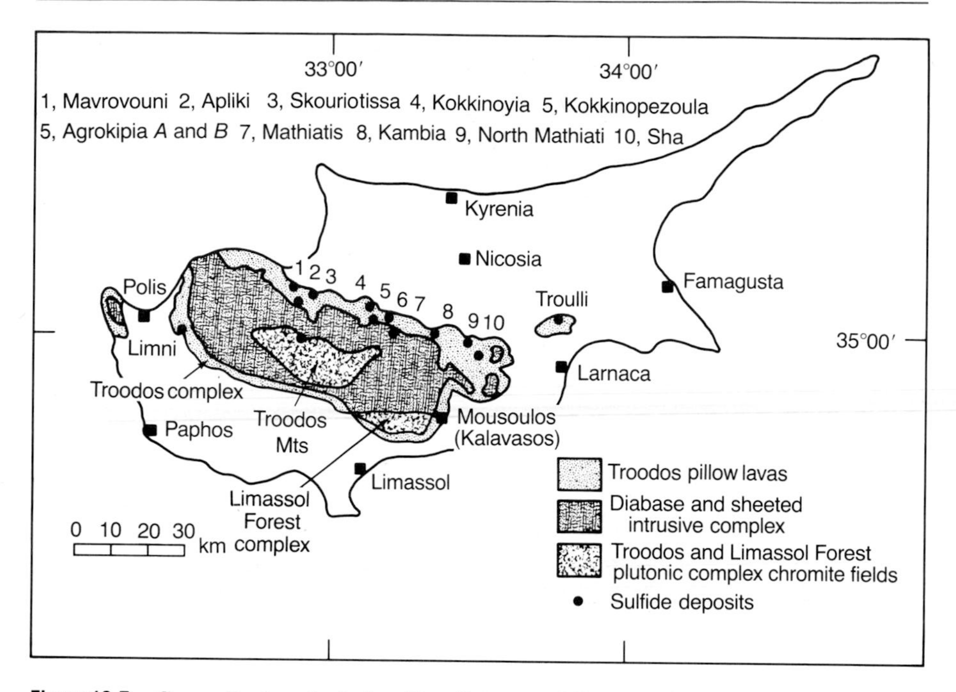

Figure 10-7. Generalized geological and location map of Cyprus. The numbered dots designate massive sulfide chalcopyrite-pyrite deposits named above. The textured units dip to the northeast, so the sulfides are in the uppermost layers. See also Figure 10-8. (*After* Searle, 1972.)

> *. . . the plutonic complex evolved as a differentiated body of magmatic cumulates laid down upon a basement of depleted-mantle harzburgite. The chromitites originated as isolated magmatic segregation ore deposits near the base of the cumulate dunite during episodic crystallization at localized centers within the magma. Postcumulus mobilization and accompanying penetrative deformation of the lower ultramafic rocks caused the tectonic overprint of schlieren structures on earlier magmatic textures, and associated large-scale rock flowage resulted in deep infolding between lithologic units; outliers of chromite-bearing dunite were formed in this way.*

The conformity of this description with aspects of ophiolite generation and emplacement is clear.

As mentioned earlier, oceanic crust is considered again in Chapter 13 in connection with sulfide deposits that occur as lenses and layers formed in Layer 1 and exposed to sunlight by obduction. It is further considered in Chapter 17 on deposits related to weathering because Layer 3 and upper mantle rocks are relatively rich in nickel substituting for iron and magnesium in olivine and orthopyroxene. When Ni-rich harzburgite dunites of ophiolites and Alpine peridotites are exposed to weathering as at Riddle,

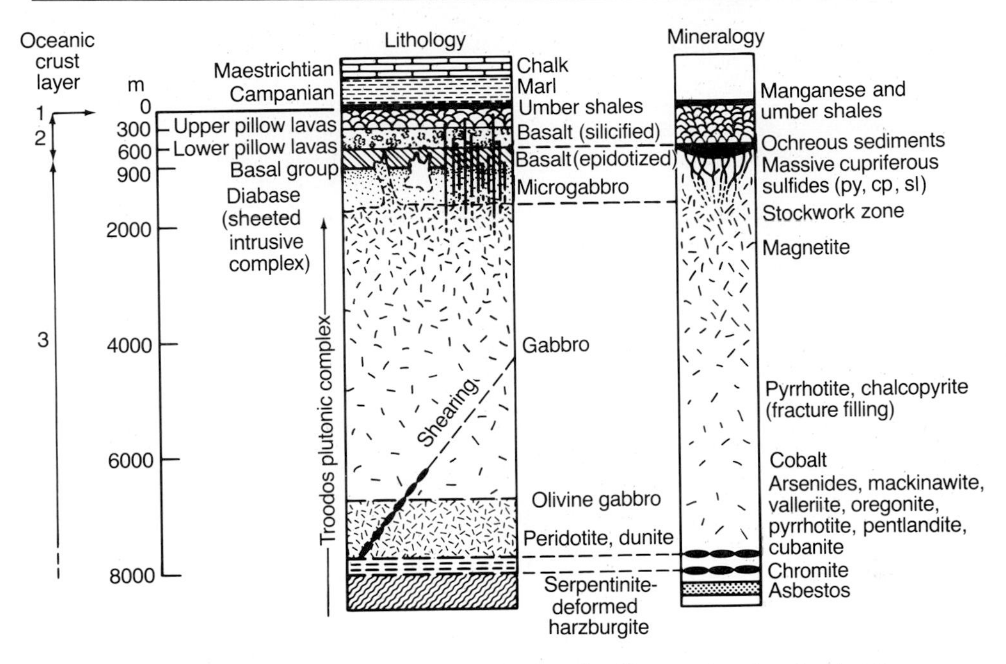

Figure 10-8. Diagrammatic sections through the Troodos Igneous Complex showing the distribution of rock types and associated mineralization. (*After* Searle, 1972.)

Oregon, and on the island of New Caledonia, nickel-rich serpentines called *laterites* form. These laterites are an important commercial source of nickel and a minor source of by-products chromite and cobalt.

Exploration for chromitites in Alpine peridotites is difficult, but can best be directed by a structural geologist-petrologist. For example, there are, to be sure, hundreds of Alpine peridotites along the Cordillera from Mexico through the Aleutians and also in the Appalachians. Placer chromite and platinum have been mined for decades at variable rates and amounts in California, British Columbia, and Alaska. Prospecting within a given Alpine peridotite body would have to involve stress and strain analysis of emplacement mechanisms to evaluate both the deformation of original chromitite layers and the magnitude and importance of separations, pull aparts, and so on. The petrologist would bring an understanding of crystallization chemistry, the directionality of Mg/Fe and Cr/Fe variations (Moutte, 1982), and the significance of rock type, texture, and primary and secondary mineral distributions.

Finally it should be noted that almost all of the world's production of chrysotile asbestos has come from terranes that are now recognized as ophiolite-Alpine peridotite settings. Thetford, Quebec, Canada, the single most productive locale, is characterized by veins and veinlets of cross-fiber asbestos—asbestos perpendicular to vein walls—that have resulted from

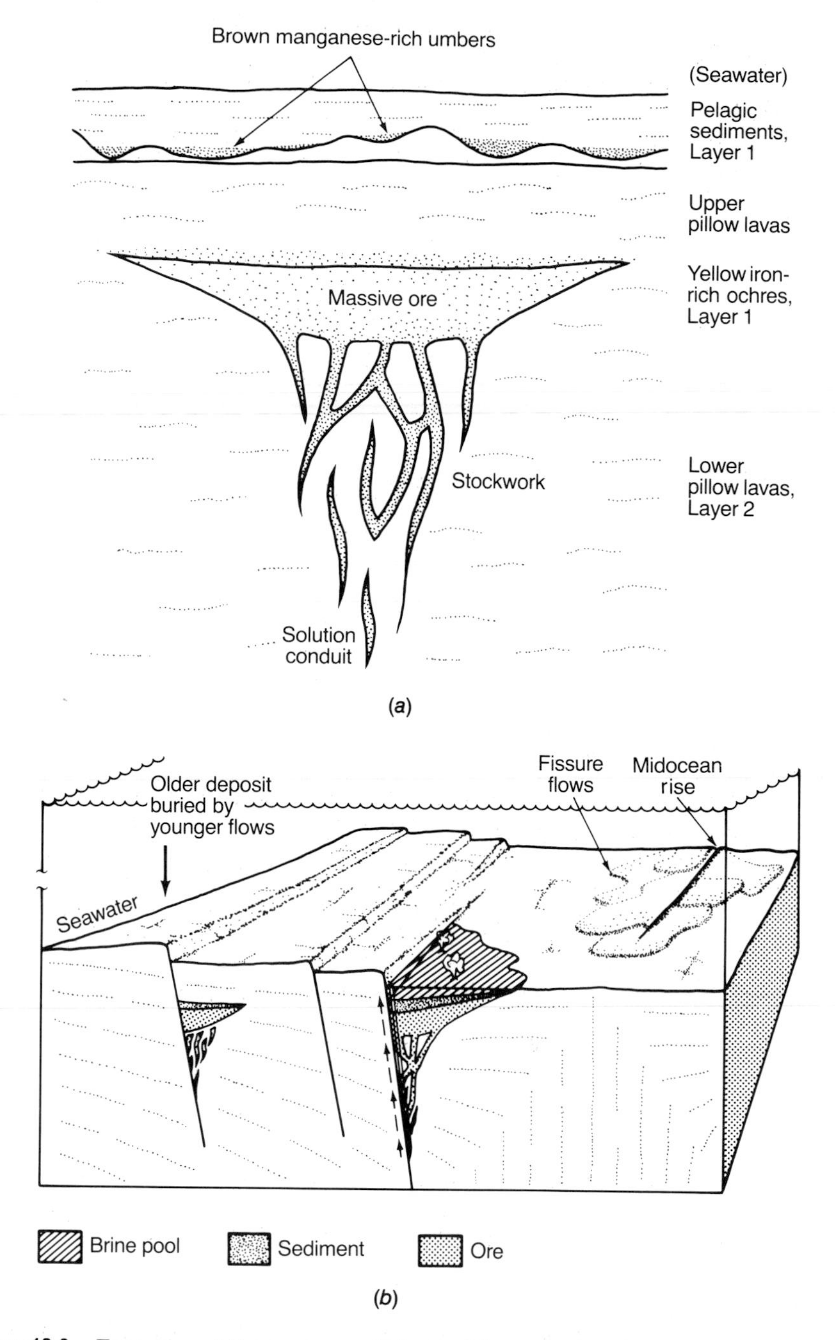

Figure 10-9. Formation of Cyprus-type sulfide precipitates at the seafloor along midoceanic rises. (*a*) Idealized cross-sectional view of the top of the oceanic crust as shown in Figure 10-8. (*b*) A more generalized view relates Figure 10-8 and (*a*) to submarine rise structural-lithologic settings. (*From* Tarling, 1981.)

hydration of olivine-pyroxene rocks in an early Paleozoic ophiolite slab. The Cassiar, Yukon Territory, Canada, deposits are similar, with asbestos in dilatant veins and veinlets, but the host rock is apparently tectonized harzburgite-type Alpine peridotite.

Bibliography

Ave'Lallemant, H. G., and N. Carter, 1970. Syntectonic recrystallization of olivine and modes of flow in the upper mantle. *Geol. Soc. Amer. Bull.* 81:2203–2220.

Burch, S. H., 1968. Tectonic emplacement of the Burro Mt. ultramafic body, Santa Lucia Range, California. *Geol. Soc. Amer. Bull.* 79:527–544.

Carter, N. L., and Ave'Lallemant, H. G., 1970. High temperature flow of dunite and peridotite. *Geol. Soc. Amer. Bull.* 81:2181–2202.

Coleman, R. G., 1977. *Ophiolites: Ancient oceanic Lithosphere?* New York: Springer, Minerals-Rocks Ser. 12, 240 pp.

——, 1984. The diversity of ophiolites. *Geol. en. Mijnbouw,* in press.

East Pacific Rise Study Group, 1981. Crustal processes of the mid-ocean ridge. *Science* 213:31–40.

Gass, I. G., 1982. Ophiolites. *Sci. Amer.* 247(2):122–131.

Greenbaum, D., 1977. The chromitiferous rocks of the Troodos Ophiolite Complex, Cyprus. *Econ. Geol.* 72:1175–1194.

Guild, P. W., 1947. Petrology and structure of the Moa district, Oriente Province, Cuba. *Amer. Geophys. Union Trans.* 28:218–246.

Hopson, C. A., R. G. Coleman, R. T. Gregory, J. S. Pallister, and E. H. Bailey, 1981. Geologic section through the Samail ophiolite and associated rocks along a Muscat-Ibra transect, southeastern Oman Mountains. *J. Geophys. Res.* 86(B4):2527–2544.

Jackson, E. D., and T. P. Thayer, 1972. Some criteria for distinguishing between stratiform, concentric, and Alpine peridotite-gabbro complexes. 24th Int. Geol. Congr. Proc., Sec. 2, pp. 289–296.

Miyashiro, A., 1975. Classification, characteristics, and origin of ophiolites. *J. Geol.* 83:249–281.

Moores, E. M., 1982. Origin and emplacement of ophiolites. *Revs. Geophys. Space Phys.* 20(4):735–760.

—— and I. D. MacGregor, 1972. Types of alpine ultramafic rocks and their implications for fossil plate interactions. Geol. Soc. Amer. Mem. 132, pp. 209–223.

—— and F. J. Vine, 1971. The Troodos massif, Cyprus, and other ophiolites as oceanic crust: evaluation and implications *Roy. Soc. London, Phil. Trans.,* Ser. A, 268:443–466.

Moutte, J., 1982. Chromite deposits of the Tiébaghi ultramafic massif, New Caledonia. *Econ. Geol.* 77:576–591.

Panayiotou, A., 1978. The mineralogy and chemistry of podiform chromite deposits in the serpentinites of the Limassol Forest, Cyprus. *Mineral. Deposita* 13:259–274.

Sampson, E., 1942. Chromite deposits, pp. 110–125 in W. H. Newhouse, Ed., *Ore Deposits as Related to Structural Features.* Princeton, NJ: Princeton Univ. Press, 280 pp.

Searle, D. L., 1972. Mode of occurrence of the cupriferous pyrite deposits of Cyprus. *Inst. Min. Metall. Trans.* 81:B189–B197.

Srinivasachari, K., 1976. Chromite deposits of India. *Indian Miner.* 30:16–21.

Steinmann, G., 1905. Geologische Beobachtungen in der Alpen, 2. Die Schardt'sche Überfaltungstheorie und die geologische Bedeutung der Tiefseeabsätze und der ophiolithischen Massargesteine. *Ber. Nat. Ges. Freiburg,* 16:44–65.

——, 1927. Die ophiolitischen Zonen in den Mediterranen Kettengebergen. 14th Int. Geol. Congr., Proc. Sec. 2, pp. 638–667.

Stoll, W. C., 1958. Geology and Petrology of the Masinloc chromite deposit, Zambales, Luzon, P.I. *Geol. Soc. Amer. Bull.* 69:419–448.

Tarling, D. H., Ed., 1981. *Economic Geology and Geotectonics.* New York: Wiley, 213 pp.

Thayer, T. P., 1964. Principal features and origin of podiform chromite deposits, and some observations on the Guleman-Soridag district, Turkey. *Econ. Geol.* 59:1497–1524.

Wilson, R. A. M., 1959. The geology of the Xeros-Troodos area, Cyprus. Geol. Surv. Cyprus Memo. I, 184 pp.

Zachos, K., 1969. The chromite mineralization of Vourinos Ophiolitic Complex, Northern Greece. Econ. Geol. Mon. 4, pp. 147–153.

ELEVEN

Deposits Related to Intermediate to Felsic Intrusions

Many of the great deposits of base and precious metals in the world fall into this category; much of the turbulent and romantic history of geology and mining is associated with districts like Butte, Montana; Coeur d'Alene, Idaho; Cornwall, England; and Cerro de Pasco, Peru. Deposits to be considered here are associated with the great calc-alkaline orogenic belts of the world, with their diorite-monzonite-granite intrusive rocks, their andesite-latite-rhyolite extrusive rocks, and porphyritic equivalents of both. Deposit types to be discussed and described in separate sections include what are known as porphyry base-metal deposits, large complex petrologic systems with subsets of porphyry copper, porphyry molybdenum, Climax-type molybdenum, and porphyry tin deposits. Igneous metamorphic-metasomatic deposits are commonly found where porphyry systems cut sedimentary sections that are mainly carbonates of Paleozoic age; these skarn or tactite deposits are described. Cordilleran Vein type deposits involving both replacement and open-space filling in fault structures generally associated with intrusive activity, pegmatites, and tin granites and tin-tungsten veins are all described. Annual production (Miller, 1976) from deposits of these types makes them avidly sought after as an exploration category—they provide over 40% of total metal-mine value output around the world.

The relationship between mountain building and ore deposits has of course long been recognized, and mountainous terrains have attracted prospectors and geologists from our emergence from the stone age into the ages of bronze, iron, steel, and now energy. More specifically, a relationship between intrusive rocks and certain types and compositions of ores has been perceived, and a relationship between geologic structure, intrusive activity, and ore occurrence has been abundantly observed. During the last two decades we have been able to discern some of the processes and mechanisms that are the reasons behind those associations. Although undeniable first-order links between plate tectonic mechanisms, ore deposits, and the rocks that contain them have not yet been established, the circumstantial evidence grows stronger each year. Few scientists still argue that the concept of subduction and subduction zones is essentially fallacious. If the assertion that subduction operates and has operated for the last 200 or 300 million years generally as the geophysicists and petrologists insist that it has, then certainly most of the major ore deposits of the American Cordilleras formed at least while subduction was progressing "beneath them." The question then becomes one of how plate tectonic mechanisms are related to and have affected ore deposition.

If we look back through geologic time, we find that the great mountain belts involving calc-alkaline rocks are generally young. Certainly, calc-alkaline rocks even in systems resembling island arcs were no strangers to Proterozoic and Paleozoic lithotectonics, but the rapid increase of the mass ratio of calc-alkaline rocks to other igneous types—komatiites, tholeiites, and alkalic rocks—in the last 200 to 300 million years is well established. It is also established that island arcs and orogenic belts are the "homes" of calc-alkaline igneous compositions, and of the ore deposits listed earlier in this section. Finally it is noteworthy that the single most important ore type in this exploration group is the porphyry base-metal deposit. Except for a few subeconomic, relatively ill-defined occurrences in the Archean, a large well-developed but uneconomically low-grade 1.8-billion-year-old Proterozoic porphyry copper deposit called the Haib in Namibia, and a few Paleozoic occurrences as in Russia (Laznicka, 1976), Australia (Ambler and Facer, 1975), and the Appalachian United States and Canada (Allcock, 1982), the porphyry ore systems and the ascendence of the calc-alkaline rocks which spawned them are the products of Mesozoic-Cenozoic orogeny. It must also be significant that the association of even the oldest of these deposits is with calc-alkaline complexes, presumably but by no means assuredly relatable to early, proto-plate-tectonic phenomena.

Finally, a global map of porphyry ore deposits (Figure 11-1) is also a map of primarily late Mesozoic-Cenozoic mountain building, intrusion, and orogeny. Points on the map designating pre-200-million-year-old minable occurrences are nearly absent. So also, deposits related to porphyry systems, such as many of the skarn and Cordilleran Vein districts, are similarly biased toward younger calc-alkaline, probably plate-tectonic-related provinces.

▸ IGNEOUS IRON DEPOSITS

Most of the nearly 1 billion tons per year of the world's iron is mined from volcano-sedimentary or sedimentary deposits (Chapters 13 and 15). A small but important 1 to 2% is mined from intrusive magmatic segregation ores, slightly more from extrusive volcanic iron ores. Both of these types owe their existence to the segregation of a magmatic fluid rich in iron, normally with 4 to 5% phosphorus. The distinction between the two is only whether this ferruginous melt is injected into country rocks as intrusive, podlike masses of magnetite, or whether it vents to the surface to be laid down as stratiform, generally stratabound iron-rich flows and tuffs with layered volcanic rocks, typically andesites and latites, as footwall, hanging wall, or both. Examples of the intrusive type are not uncommon, as at Pea Ridge and Iron Mountain, Missouri (Emery, 1968); Grängesberg, Sweden; Larap in the Philippine Islands; and elsewhere. Examples of volcanic iron deposits are rare but significant—El Laco, Chile; Cerro de Mercado, Durango, Mexico; and the Iron Mountain district of Missouri (Snyder, 1969) are but a few. Several others, perhaps even including Pea Ridge, Missouri, and Savage River, Tasmania, were originally described as intrusive but are now seen as perhaps having formed in an extrusive setting. The most important of these is the Kiirunavaara mine in Kiruna, Sweden, which for years has been cited as the type example of intrusive iron ore. Because the final determination has not yet been made, the description of Kiruna is reproduced here largely as it was given in the preceding edition of this book (Park and MacDiarmid, 1975). Note that it can be read to be consistent with either theory. Parak (1975) made a case for an extrusive setting, but was refuted by Frietsch (1978). If Parak is correct, these deposits could be treated in Chapter 12; Cerro de Mercado, Mexico, is in fact described there.

Before we describe Kiruna, a few comments about iron-rich melts are called for. Geochemically the differentiation that produces calc-alkaline rocks affects iron enrichment in the same melts. Normally Mg/Fe ratios decrease and total iron increases along with other differentiation indices. But special conditions must prevail to produce a melt that is nearly pure iron oxide. There are several, perhaps working in concert. Calling upon extraordinarily high-iron starting melts is unattractive, because the high-iron subvolcanic melts appear to be part of a regional zoning pattern that depends more upon melt differentiation than upon source rock geochemistry. There are four other explanations: (1) high sodium content leading to the "alkali-iron effect," the formation of sodium-iron-oxygen complex ions in late melts, which both increase the melt's iron content and flux it; (2) low early fugacity of oxygen, which restricts Fe^{+3} activity in the melt and prevents the progressive extraction of iron by steady magnetite ($FeO \cdot Fe_2O_3$) separation; (3) high phosphorus contents, which flux the melt and permit late low-temperature mobility of normally high-temperature phases; and (4) the combina-

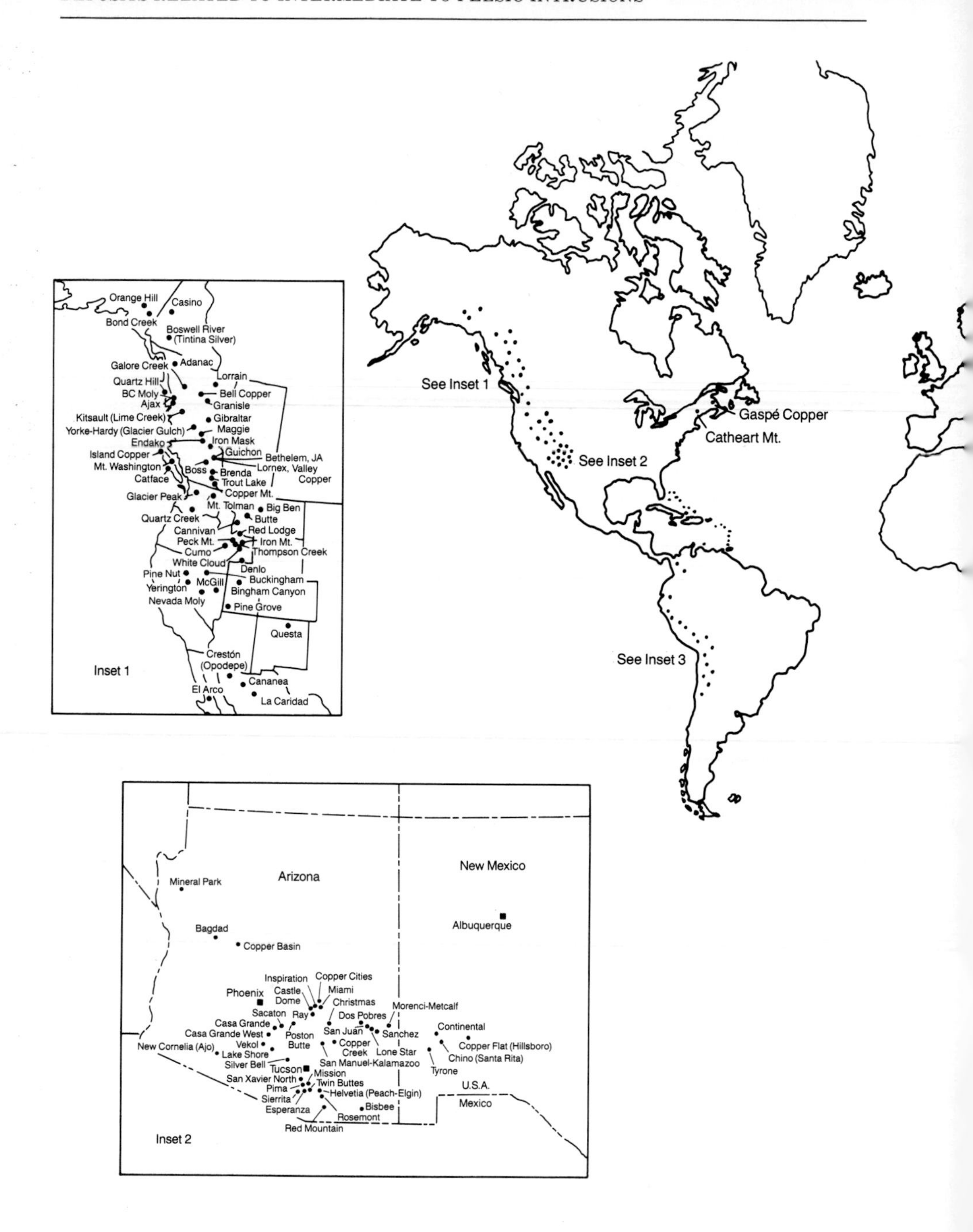

See Inset 1
See Inset 2
See Inset 3
Gaspé Copper
Catheart Mt.
Inset 1
Orange Hill
Casino
Bond Creek
Boswell River (Tintina Silver)
Galore Creek
Adanac
Lorrain
Quartz Hill
BC Moly
Ajax
Bell Copper
Granisle
Kitsault (Lime Creek)
Gibraltar
Yorke-Hardy (Glacier Gulch)
Maggie
Endako
Iron Mask
Island Copper
Guichon
Bethelem, JA
Mt. Washington
Lornex, Valley Copper
Catface
Boss
Brenda
Trout Lake
Glacier Peak
Copper Mt.
Mt. Tolman
Big Ben
Quartz Creek
Butte
Cannivan
Red Lodge
Peck Mt.
Iron Mt.
Cumo
Thompson Creek
White Cloud
Pine Nut
Denlo
Buckingham
Yerington
McGill
Bingham Canyon
Nevada Moly
Pine Grove
Questa
Crestón (Opodepe)
Cananea
El Arco
La Caridad
Inset 2
Arizona
New Mexico
Mineral Park
Albuquerque
Bagdad
Copper Basin
Inspiration
Copper Cities
Phoenix
Castle Dome
Miami
Christmas
Sacaton
Ray
Morenci-Metcalf
Dos Pobres
Casa Grande
Continental
Casa Grande West
Poston Butte
San Juan
Sanchez
New Cornelia (Ajo)
Vekol
Copper Creek
Lone Star
Copper Flat (Hillsboro)
Lake Shore
Chino (Santa Rita)
Silver Bell
Tucson
San Manuel-Kalamazoo
Tyrone
Mission
San Xavier North
Twin Buttes
U.S.A.
Pima
Helvetia (Peach-Elgin)
Mexico
Sierrita
Esperanza
Bisbee
Rosemont
Red Mountain

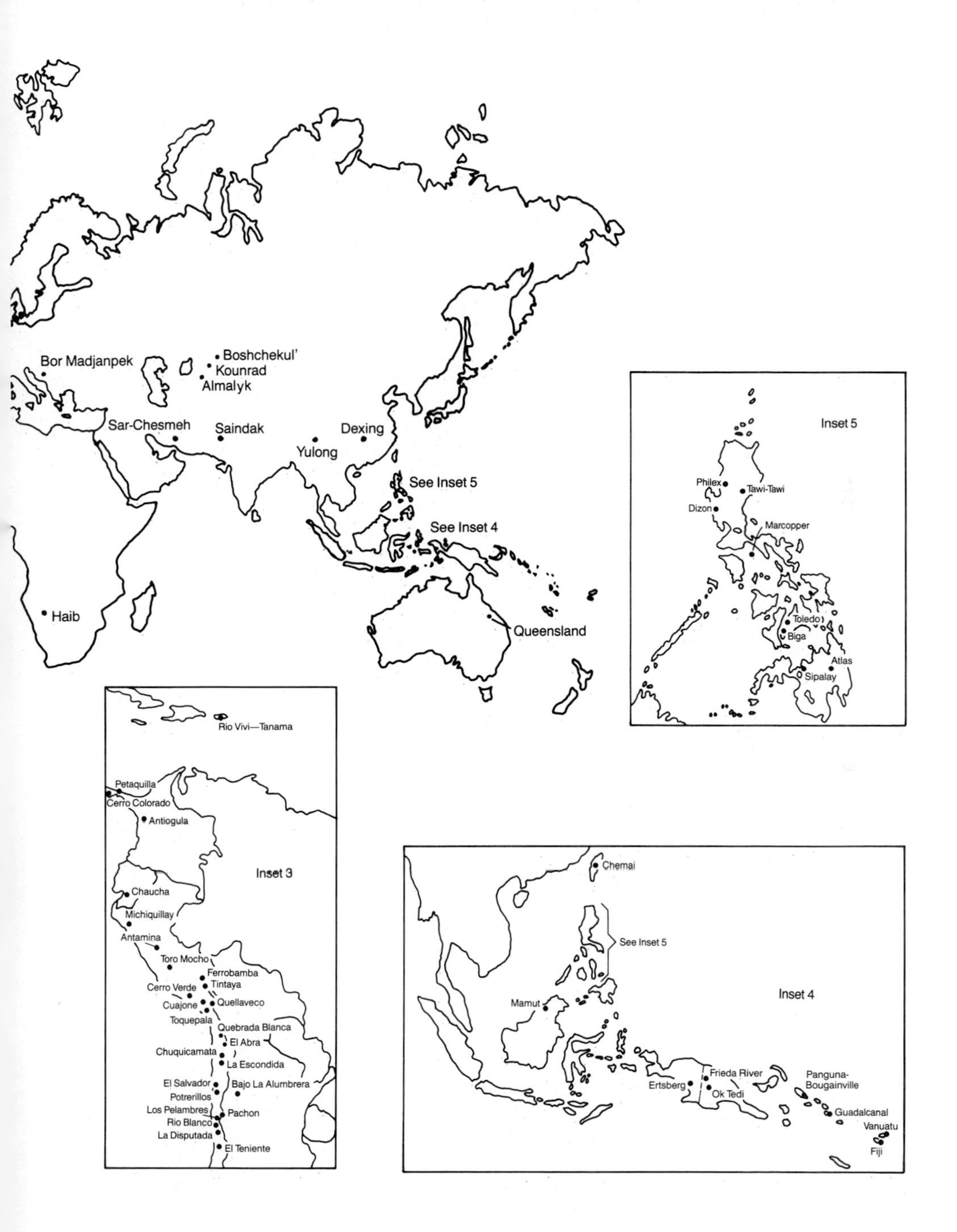

Figure 11-1. Distribution of porphyry copper-molybdenum deposits in the world. Notice the coincidence of these deposits with Mesozoic-Cenozoic orogenic belts and calc-alkaline volcanism.

tion of two or more of these effects to move the melt into a temperature-composition field of two-liquid immiscibility wherein a silica-rich and an alkali-phosphorus-magnetite-(hematite)-rich liquid separate from one another, the latter being separately intruded or extruded.

Kiruna, Sweden

Probably the best known and most productive iron deposit in the world explained as a product of magmatic segregation is that of the high-grade ores at Kiruna in northern Sweden (Figure 11-2). The deposit has been studied intensively over a period of about 75 years, and most geologists who have examined the area have agreed that the ores are magmatic segregations (Stutzer, 1907; Geijer, 1910, 1931a, 1931b, 1960; Geijer and Magnusson, 1952; Schneiderhöhn, 1958; Frietsch, 1973, 1978; Parak, 1975). The Kiruna district contains several iron deposits, the largest of which is Kiirunavaara. Kiirunavaara is a sill-like orebody between Precambrian syenite porphyry in the footwall and quartz porphyry in the hanging wall (Figure 11-3). Overlying the quartz-bearing porphyry is the Lower Hauki complex, consisting of highly altered flows and silicified tuffs (?), and the Vakko Series (or Upper Hauki complex) of sediments. All of these rock units strike north or northeast and dip 50 to 60° east. The orebody appears to be intrusive between the syenite and quartz-porphyry, extending for over 5 km along strike. It averages about 90 meters thick, except at its narrow northern end, which extends over 900 meters under a lake. Production began in 1903, and by 1980 it had totaled nearly 400 million metric tons averaging 62% iron and estimated to be about one-fourth of the available ore. It is

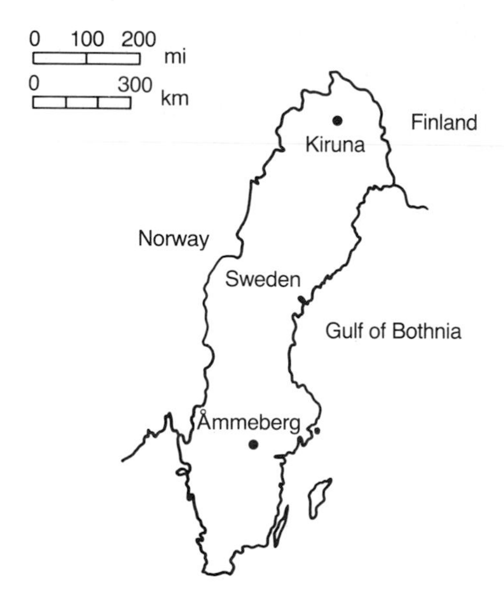

Figure 11-2. Map of Sweden, showing the location of the Kiruna district.

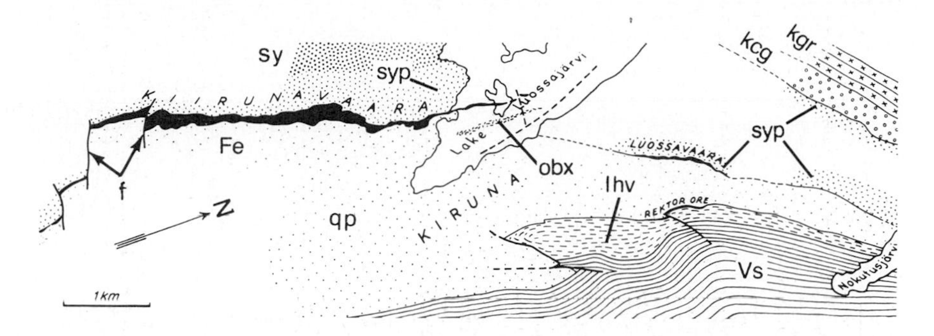

Figure 11-3. Map of the Kiruna district, Sweden. Fe = iron ore; obx = zone containing ore veins (ore breccia); kgr = Kiruna greenstones; kcg = Kurravarra conglomerate; sy = syenite; syp = syenite porphyry; qp = quartz-bearing porphyry; lhv = lower Hauki volcanic complex; Vs = Vakko sedimentary series; f = fault. (*From* Geijer, 1960.)

the principal source of ore used to produce the world-famous Swedish steel and iron products.

The ore consists of fine-grained magnetite or, more rarely, hematite, and contains variable amounts of fluorapatite. The phosphorus content is generally high at greater than 2%, due to the apatite; in fact, the term *Kiruna type ore* universally refers to high-phosphorus iron ores. After apatite, actinolite and diopside are the most abundant nonmetallic minerals. In places, prisms of apatite exhibit a trachytic arrangement, and workers have even observed streaked out or "stratified" alternations of pure apatite and magnetite-apatite laminae. At Kiirunavaara both the apatite and the iron minerals are fine-grained, and a well-defined layering or banding is developed which Geijer attributed to flow banding.

Contacts between ore and wall rocks are sharp. A thin amphibole skarn is found in places, and small veins or apophyses of ore branch out into both the footwall and the hanging wall. Such separate systems of ore veins are locally referred to as ore breccia. Porphyry dikes, apparently intermediate in composition and texture between the footwall and hanging-wall porphyries, intrude the footwall and spread out beneath the hanging wall. The dikes, in turn, are broken and engulfed in the ore, though one such dike intrudes the magnetite. These relations imply that the ore was emplaced after most of the porphyry dikes were intruded, but before the igneous activity had completely stopped. The ore is offset by faults with predominantly strike-slip movement. Granophyre dikes apparently unrelated to the ore and older igneous rocks were intruded during the faulting.

Geijer (1931a, 1931b, 1960, 1967) concluded that the Kiirunavaara deposit resulted from the intrusion of highly mobile ore magmas, presumably before the country rocks were tilted. The magnetite-apatite fluid became concentrated during magmatic differentiation as an immiscible fraction within the mother magma. Separation of the two immiscible fractions took place deep within the Earth's crust, prior to intrusion. This interpretation is directly supported by Fischer's (1950) and Philpotts' (1967) confirmation that an immiscible magnetite-apatite liquid can exist. Geijer (1967), Philpotts (1967), Parak (1975), and Frietsch (1978) give excellent reviews of the mineralogical and textural properties of the magnetite-hematite ores of magmatic affiliation, especially those with a high percentage of apatite. Although he emphasizes the magmatic character of the ores, Geijer stated emphatically that the consolidation of these iron ores is not the same as that of a normal igneous rock. He attributed the differences to the role of volatile substances, and said that the form of emplacement in many places is forceful injection rather than simple filling of preexisting open fissures.

Many magnetite deposits other than at Kiruna contain coarse-grained asparagus-colored apatite, of which the crystals vary in length up to several centimeters. The apatite may be dispersed in massive magnetite-hematite masses, and crystals commonly stand perpendicular to the walls of fractures, though they may be separated from the walls by thin layers of magnetite. Where fragments of country rock are present in magnetite, apatite prisms are likely to be normal to the boundaries of the fragments (Geijer, 1967). Scapolite has been reported from the ore at El Laco, Chile (Ruiz et al., 1965, pp. 221–247) and is present in several other deposits of magmatic affiliation. According to Geijer, the presence of normally hydrothermal scapolite does not mean that the ores cannot be magmatic in origin. Hornblende, actinolite, and in a few places tremolite are common companions of magnetite orebodies.

Other lines of evidence support Geijer's hypothesis. The orebodies invade the porphyries and show liquid flowage by their trachytoidal texture. The contact zones underwent slight metasomatism, indicating the presence of volatiles. These volatile constituents were presumably active in separating the two immiscible magmatic fractions within the mother magma. Furthermore, all of the minerals of the orebody are also in the associated igneous rocks. Other orebodies in the district are clearly related in mineralogy, structure, and chronology to the Kiirunavaara deposit, indicating a closely related genesis; all types of ore, from magmatic to hydrothermal, can be seen in these deposits, and it is clear that some formed at lower temperatures than the Kiirunavaara ores. In the entire district, then, there is evidence of closely related magmatic and hydrothermal ore deposition, wherein the early, high-temperature ores were products of direct magmatic segregation and the later ores were influenced by greater concentrations of volatiles.

The preceding paragraphs and the accompanying geological map (Figure 11-3) permit at least two interpretations. Parak (1975) suggests that the syenite porphyry and quartz porphyry on either side of the massive magnetite-apatite ores shown in black in Figure 11-3 were in fact extrusive rocks, as was the ore itself. Syenite porphyries are associated with high-iron systems elsewhere, and are the footwall rocks here. Note also that the orebody is distinctly sheetlike, shows flow banding, lamination, and brecciation. The only facts at clear odds with a volcanic origin are that veins of oxides are described as intrusive into both the floor and the roof rocks, and skarnification may affect both. Studies of the district are continuing.

▸ PORPHYRY BASE-METAL DEPOSITS

This ore deposit group is unusual in view of its overall economic impact and the recency of its definition. *Porphyries* were so called as early as the 1920s (Parsons, 1933, 1957), but it remained for introductory notes and a collection of papers edited by Titley and Hicks (1966), articles in the Graton-Sales volumes edited by Ridge (1968), and several papers early in the 1970s (Lowell and Guilbert, 1970; Rose, 1970; James 1971) to establish many of the geologic, mineralogic, geochemical, and zonal characteristics of the group. Since the early 1970s an avalanche of descriptive, conceptual, and experimental papers have clarified the deposit type, although many problems concerning genetic models, geochemistry of the hydrothermal processes, and even exploration signatures remain. An overriding feature of the porphyries is that they are gigantic hydrothermal-petrogenetic systems that influence cubic kilometers around and including the intrusive stock or dike systems that cause their emplacement. It took geologists many years to shift perspective from outcrop or vein-stope scales to the cubic-kilometer scale, a change in exploration approach that spread during the 1950s, 1960s, and 1970s from the southwestern United States to British Columbia, Central and South America, Australia, and finally the South Pacific.

One of the quirks of geologic taxonomy is that the word *porphyry* has survived as an integral part of the label of these deposits. Although the word is not an easy one for most people, the environmental connotations and constraints of the terms *porphyry* and *porphyritic texture* are uniquely apt to this deposit type. Porphyritic rocks—generally the epizonal or hypabyssal dacite, latite, quartz latite, and rhyolite porphyries but also including their plutonic, phaneritic equivalents quartz diorite, monzonite, quartz monzonite, and granite—have been identified as significant phases in *all* of the deposits which would otherwise be identified with the group. Porphyritic textural development is the product of chemical, thermal, barometric, and temporal processes (Whitney, 1977; Burnham, 1979) that occur in hypabyssal conditions with ranges of 1 to 2 kbar pressure, 1.5 to 4 km depth in the crust, and at temperatures of 750 to 850°C. Petrologists and economic geologists agree that this epizonal environment is the hearth of the porphyry base-metal deposit, which is emplaced at similar pressures and temperatures above 250 but below 500°C, rarely to 600 or 700°C.

Porphyry Copper Deposits

San Manuel–Kalamazoo, Arizona No definition of *porphyry copper* has been universally accepted. Mining geologists prefer to emphasize that porphyry coppers are "large, low-grade copper deposits regardless of their genesis and host rock type that are amenable to mass mining methods." That definition fits porphyry copper deposits, but it does not exclude deposits of sedimentary rock, carbonatite, or LMI affiliation. The geological engineer might specify rock mechanics and joint spacings, but again these properties are not specific to the geologic essence of the group. Parsons in 1933 and 1957 listed a series of characteristics which included response to weathering, size, shape, and a large number of descriptive variables. Kirkham (1971) provided an adequate short geologic definition with "large, low to medium grade deposits in which hypogene sulfides are primarily structurally controlled, and which are spatially related to felsic or intermediate porphyritic intrusions."

Titley (1966, 1972) properly placed great importance upon the relative amounts of porphyritic intrusive rocks and the nature of wall rocks involved. He described "simple" porphyry deposits as being either all in the homogeneous igneous rocks that spawned the ores, or nearly all in external lithologies, and "complex" ones as lying athwart these chemical-mineralogical heterogeneities. Lowell (1974) said that a porphyry copper deposit is any disseminated pyrite-chalcopyrite-molybdenite deposit exhibiting good concentric zoning. However, not all porphyries have disseminated values, and not all show the "good" zoning required.

We will return to problems of definition, because they become problems of classification, to end this section. An appropriately specific definition which will be used here is that a porphyry copper deposit, or PCD, is a large low- to medium-grade deposit, primarily of chalcopyrite and molybdenite, in which hypogene sulfide and silicate zoning spans potassic-propylitic alkali metasomatic and phyllic-argillic hydrolitic alteration, and which is temporally and spatially related to an epizonal calc-alkaline porphyritic intrusion.

We now briefly examine the several components of that definition in order to describe what a porphyry deposit is. We will then discuss some variables in the temporal and chemical aspects of porphyry systems. The first descriptor is the adjective "large." Anyone who has visited the open-pit mines at Butte, Montana; Bingham Canyon, Utah; Morenci, Arizona; Cananea, Mexico; Toquepala, Peru; and Chuquicamata, Chile, and the two largest underground copper mines in the world at El Teniente, Chile, and San Manuel, Arizona, has seen a total of perhaps 15 billion tons of copper ore. The average tonnage of porphyry copper deposits is much less at perhaps 200 million tons, with a minimum of about 100 million tons and an unknown maximum of perhaps 2 to 3 billion tons at Chuquicamata and El Teniente. In area, they are typically 1 or 2 km^2 or so in outcrop, although alteration effects to be described later extend to 10 or 20 times the area of the ore zone. Clearly, however, they are large petrogenetic hydrothermal systems, each influencing many cubic kilometers of the Earth's crust wherever it occurs.

"Low to medium grade" refers in porphyries to the value sum of several elements that may be concentrated, typically copper, molybdenum, silver, gold, lead, zinc, and manganese. Other elements like traces of arsenic, bismuth, tin, tungsten, uranium, rhenium, and platinum are recovered but are seldom included as contributors to ore value. In porphyry copper deposits, the values of by-product elements are converted to "copper equivalents" to calculate grade or tenor; most porphyry deposits provide values in the 0.6 to 0.9% Cu equivalent range. A few are higher in grade, the 1 to 2% Cu ores of El Teniente, Bingham Canyon, and Chuquicamata being highest; many are lower, with the ores of the Sierrita orebody near Tucson, Arizona, being the lowest at 0.35% Cu equivalent to sustain mining operations. The porphyry molybdenum deposit at Quartz Hill, Alaska, lists 1.3 billion tons of only 0.17% MoS_2 as ore. Shifting metal prices dictate that several traditional porphyry copper deposits may derive more revenue from co-products than from copper, as for molybdenum at Sierrita and Kingman, Arizona, gold at Panguna on Bougainville Island and Bingham Canyon, and silver at Pima-Mission, Arizona. Clearly, then, grades vary widely, but they are typically below 1% Cu, or 10 kg Cu per metric ton; enormous quantities of ore—in many instances from 75,000 to 100,000 tons per day—must be handled to ensure profits.

Chalcopyrite is by far the dominant hypogene Cu-Fe-S mineral of economic interest in porphyries. Bornite, demanding higher Cu^{+2}/Fe^{+2} activity ratios and higher metal-to-sulfur ratios in the ore-forming systems, is rare; many deposits have none at all, in others it is common, and in a few it is abundant. Hypogene chalcocite is even rarer. Other Cu-Fe-S sulfosalts like valleriite and cubanite are nearly unreported. Some porphyry systems contain enough arsenic and antimony that tennantite-tetrahedrite and enargite are important; such systems are more common in South America than elsewhere. Norton and Knight (1977) provided thermodynamic basis to the observation that tennantite is most common in the upper, lower temperature reaches of porphyries, as will be described for El Salvador and Chuquicamata. It should further be noted that many porphyries are ringed by outer zones of lead, zinc, manganese, silver, and gold mineralization. Where minerals of these metals are deposited in veins (Figure 11-4*b* and *c*), they may be minable; many porphyry copper open-pit mines are ringed by the old head frames of underground vein mines. Such vein systems may be parts of the Cordilleran Vein type described later in the chapter. The mines around Butte, Montana's, porphyry "core" constituted major vein mines of sphalerite and galena, and Butte's peripheral rhodochrosite veins made it the most important manganese district in the United States. Silver and gold in solution in many of these minerals, and as separate sulfosalt species, was also mined there. Molybdenum, present only as molybdenite (MoS_2), is typically centrally and deeply located, but it can be external to copper.

The next important aspect of the definition involves sulfide-silicate zoning. Lowell (1968) reported his success in mounting an exploration effort to find the mammoth Kalamazoo porphyry in Arizona on the then-novel basis of coaxial alteration zoning symmetry. Zoning in porphyries was recognized at Butte by Sales (1914) and by Sales and Meyer (1949, 1950), and was to be classically described by Meyer et al. (1968). It was known to be applicable at many other deposit locales (Titley and Hicks, 1966), but the unifying internal characteristics that zoning analysis offered to understanding porphyry deposit economic geology as a whole had not become common experience in the mid to late 1960s. Since concept-oriented exploration had been successful, Lowell and Guilbert (1970) used the San Manuel–Kalamazoo deposit as a framework to describe characteristics of sulfide and silicate zoning at 25 other porphyries. DeGeoffroy and Wignall (1972) extended this analysis to involve 58 deposits. The zoning pattern at San Manuel can be separated into the three superimposable aspects of alteration, mineralization, and sulfide occurrence (Figure 11-4). It should be stressed here that this representation is not necessarily typical, although the geology described has proven to be widely applicable; and that it is a static portrayal of the final distribution of phases and components, with no process, kinetic, or physicochemical implications. Many aspects of porphyry copper deposits are evident in the schematics of the San Manuel–Kalamazoo deposit (Figure 11-4). Translation of these aspects to a composite description indicates that the "typical" Cordilleran porphyry copper deposit is emplaced in late Cretaceous sediments and metasediments and is associated with a 65-million-year Laramide quartz monzonite stock. Its host intrusive rock is elongate irregular, 1.3 by 2 km in outcrop, and is progressively differentiated from quartz diorite to quartz monzonite inwardly. The host is more like a stock than a dike and is controlled by regional-scale faulting. The orebody is an oval 1 by 2 km in plan view, is a thick-walled pipelike body in three dimensions, and has gradational boundaries. Of the 200 million tons of ore in this composite orebody 70% is in the igneous host rocks, 30% in preore rocks. Metal values include 0.45% hypogene Cu with 0.35% supergene Cu and 0.015% Mo. Alteration is zoned from potassic (quartz-K-feldspar-biotite) at the core, and earliest, outward through phyllic (quartz-sericite-pyrite), argillic (quartz-kaolin-montmorillonite), and propylitic (epidote-calcite-chlorite), the propylitic zone extending 1 km outward beyond the copper ore zone. Over the same interval, sulfide species vary from chalcopyrite-molybdenite-pyrite through successive assemblages of pyrite-chalcopyrite-molybdenite, pyrite-chalcopyrite, and pyrite to an assemblage of galena-sphalerite-pyrite with minor gold and silver values in solid solutions, as metals, and as sulfosalts. Characteristics shift from disseminations through respective zones of microveinlets (crackle fillings), veinlets, veins, and fi-

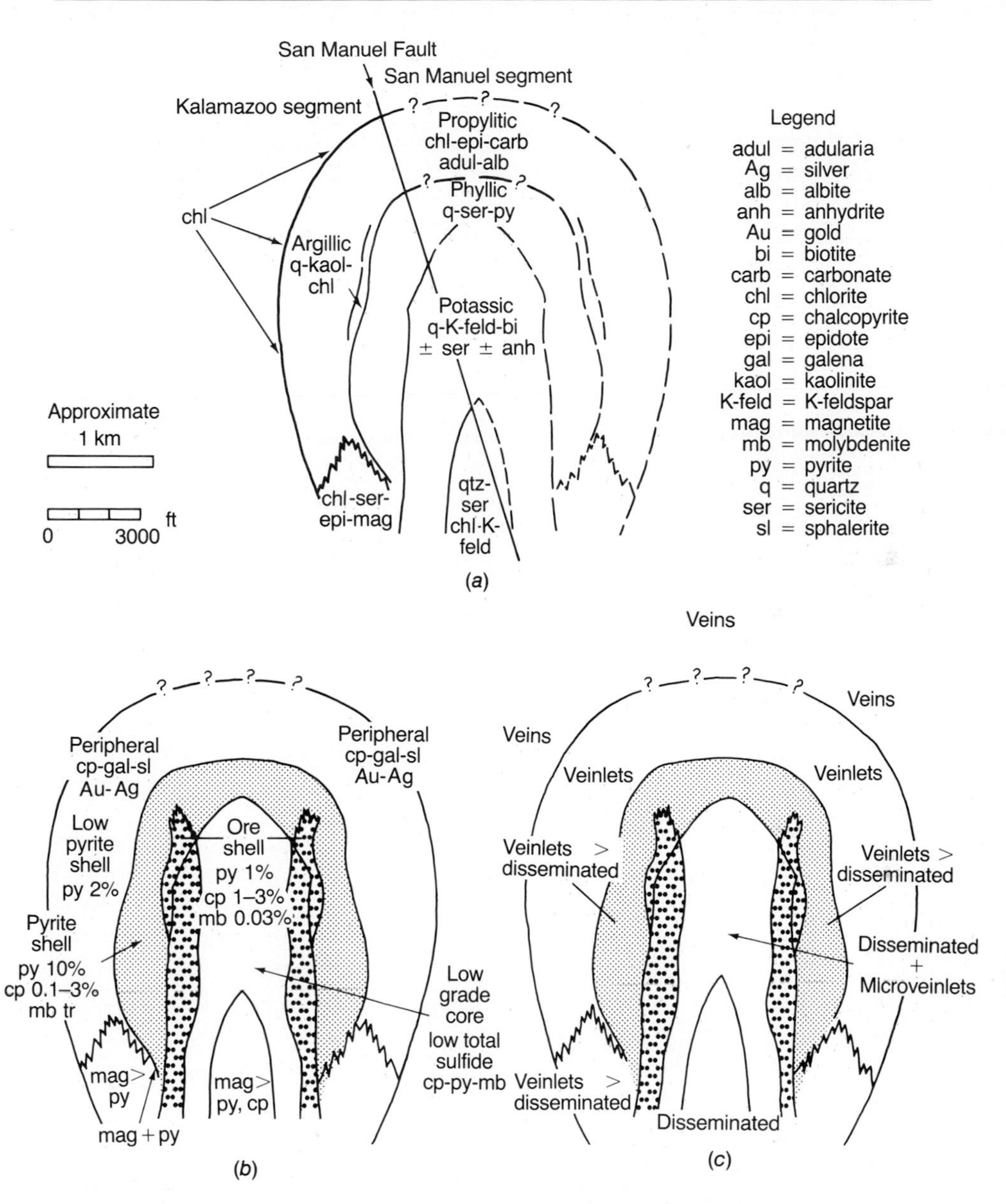

Figure 11-4. Schematic of concentric alteration-mineralization zones at San Manuel–Kalamazoo. (*a*) Alteration zones. Broken lines on the Kalamazoo side indicate uncertain continuity or location, and on the San Manuel side extrapolation from Kalamazoo. (*b*) Mineralization zones. (*c*) Occurrence of sulfides.

nally to individual structures on the periphery which may contain high-grade mineralization. Breccia pipes with attendant crackle zones are common. Quartz-sericite-pyrite zones are commonly laced with stockwork fracturing.

The actual individual expression of zoning is affected by level of exposure, structural and compositional homogeneity, wall-rock composition, size, tectonic setting, and postore faulting or intrusive activity. Vertical dimensions can exceed 3 km, with the upper reaches of the porphyry environment perhaps only at subvolcanic depths of 1 km or so. The vertical and lateral zoning described is repeated with sufficient constancy that depth of exposure at many deposits can probably be evaluated.

The last aspect of the porphyry deposit definition refers to the association with epizonal, felsic, porphyritic intrusions. Figure 11-5 is a sketch

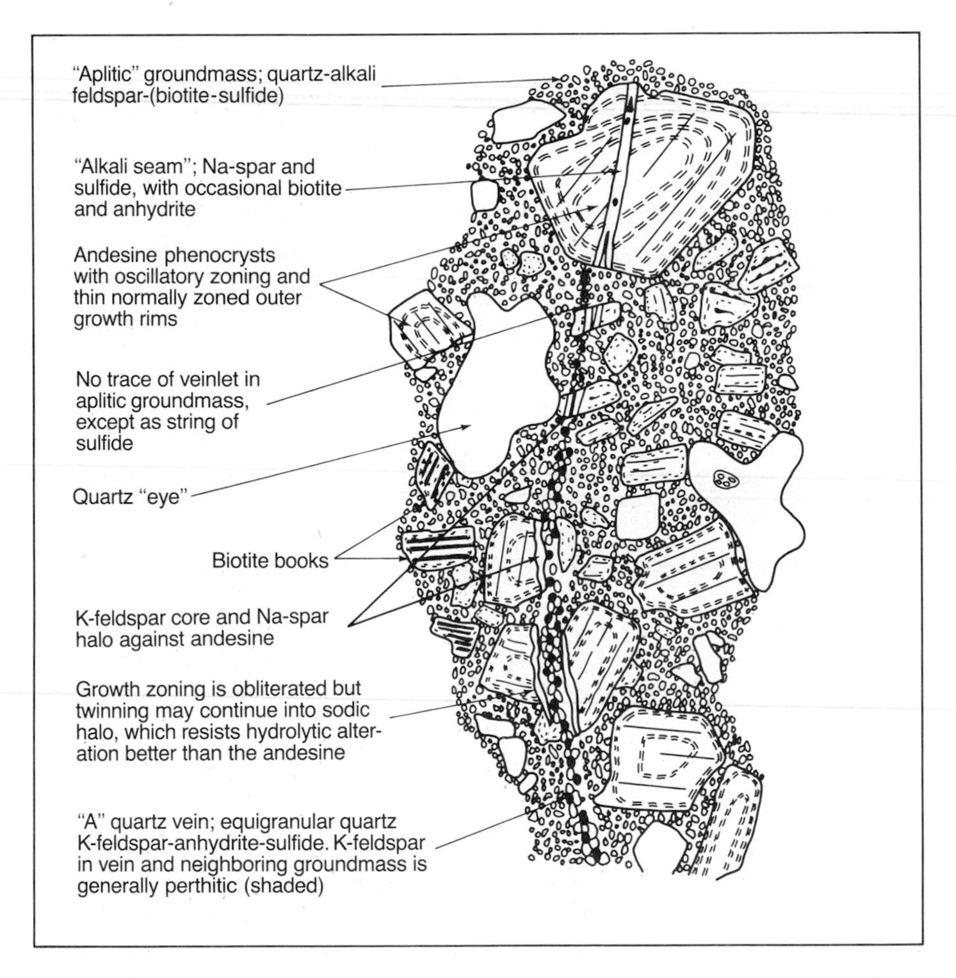

Figure 11-5. Sketch of a portion of a quartz latite porphyry thin section in plane light. The height of the diagram is 1 cm. Plagioclase with or without K-feldspar phenocrysts in a sucrose-aplitic groundmass of quartz and K-feldspar is typical. Quartz eyes always show lobate outlines thought to bespeak partial resorption. This specimen from El Salvador, Chile, is cut by an early stage A veinlet (potassic alteration) described at the bottom of the figure. The rock has not been sericitized. (*After a drawing by* Langerfeld *in* Gustafson and Hunt, 1975.)

of the characteristics of a typical quartz porphyry, or quartz-eye porphyry, as it is also known. They are typically rhyodacite (granodiorite equivalent), latite (monzonite), or quartz latite (quartz monzonite) porphyry. The textures shown and described in the side notes of Figure 11-5 are classic. The drawing has been generalized from one of an El Salvador, Chile, porphyry (Gustafson and Hunt, 1975); a similar sketch could be made of porphyries from any of scores of deposits. Significantly, porphyritic rocks like these are thought to be associated with shallow-seated epizonal intrusive activity, the hearth of porphyry base-metal ores (Whitney, 1977).

Specific aspects of San Manuel–Kalamazoo not necessarily clear from Figure 11-4 and the above "composite deposit" description are that the copper grade of its hypogene chalcopyrite ores is a relatively constant 0.75% Cu with 0.015% Mo as molybdenite. As Figure 11-6 shows, the orebody and its alteration envelopes were formed in an upright bullet-shaped configuration by the intrusion of Laramide monzonite porphyry dike swarms into Precambrian quartz monzonite, with about 50% of the ore values dispersed in each rock type, and with well-developed copper, zinc, and other elements dispersed as halos extending outward into the wall rocks beyond the orebody itself (Lowell, 1968; Chaffee, 1975). The orebody was almost exposed by early Tertiary erosion, then tilted, faulted, and offset to show the two orebody segments in Figure 11-6*b*. San Manuel, developed under the commodity pressures of World War II, was found first and was in production when the wall-rock alteration and structural geologic analyses described by Lowell (1968) led to the discovery in 1965 of the deeper seated Kalamazoo down-faulted segment.

Other Areas It would be impossible to provide geologic descriptions of the scores of porphyry copper deposits now in production, or even of the dozen most important ones. In one of the single most important papers on porphyry geology, Gustafson and Hunt (1975) reported 80 man-years of geologic research on the El Salvador, Chile, deposit, contributing to description and to petrologic, geochemical, and genetic concepts. El Salvador is described briefly at the end of this section. A series of geologic-geochemical-explorational papers on the Bingham Canyon deposit was published as one issue of *Economic Geology* in 1978 (pages 1215–1365); another on porphyry deposits of Australia and the Southwest Pacific comprised pages 608–985 of *Economic Geology* of that same year, and several books such as Sutherland-Brown (1976) and Hollister (1978) provide excellent recent descriptions and data. Titley (1982) presented a further description of southwestern North American porphyry copper deposit geology and systematics.

Before we turn to Chuquicamata to learn from its geology, a few paragraphs on recent developments in porphyry deposit studies will aid in the comprehension of the deposit type. Progress has been made in the area of fluid inclusion studies. Built upon the work of Edwin Roedder (Roedder,

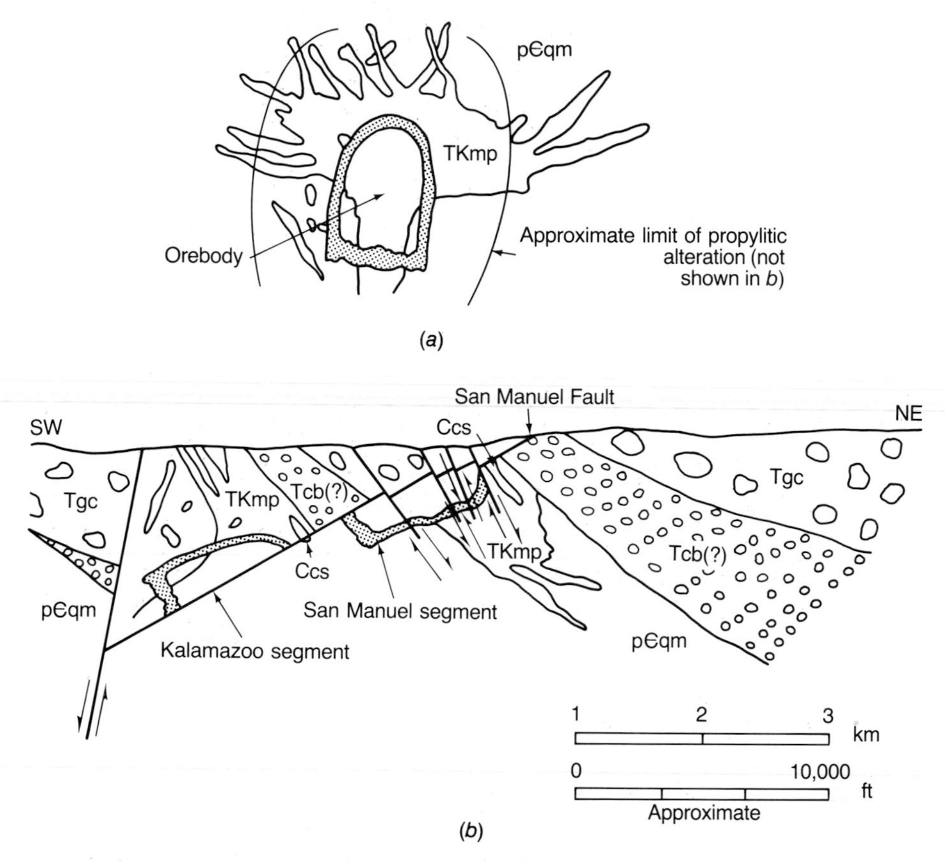

Figure 11-6. Structural history of the San Manuel–Kalamazoo deposit. (*a*) At time of emplacement and (*b*) at present. Note the umbrellalike flare of the dike swarm, and the trivial chalcocite enrichment zone (CCs). p€qm = Oracle Granite, actually quartz monzonite; TKmp = monzonite porphyry; Tcb = Cloudburst Formation; Tgc = Gila Conglomerate. (*From* Lowell and Guilbert, 1970.)

1967, 1971; Roedder and Skinner, 1968) reported in Chapter 7, many researchers have studied variations in fluid temperature, pressure, salinity, ion complexing, dissolved species, and physical state in space and time in porphyry systems. Through studies by Moore and Nash (1974), Gustafson and Hunt (1975), Chivas and Wilkins (1977), Moore and Moore (1979), Wilson et al. (1980), Ahmad and Rose (1980), Beane and Titley (1981), and others, porphyry base-metal deposits have been found to be truly mesothermal, with ore deposition from strongly saline fluids up to 30 to 60% NaCl ranging from 250 to 500°C, rarely 600 to 700°C, commonly boiling. But the greatest utility of fluid inclusion analysis will come from its combination with structural studies and alteration petrology in the 1980s. Moore and Nash (1974), Preece and Beane (1982), Haynes and Titley (1980), and Ahmad and Rose (1980) studied the Bingham Canyon, Utah; Sierrita, Arizona; and Santa Rita, New Mexico, porphyry deposits in novel ways. Preece and Beane (1982) studied fluid inclusion characteristics, alteration assemblages, and the relative ages of both by veinlet interactions, and were able to discern at Sierrita the presence of both a higher temperature, more saline early fluid and a later, cooler, less saline one, as described by Moore and Nash (1974) at Bingham. Haynes (1980) measured the total length per square half-meter of outcrop of each of a number of intersecting, sequential alteration-mineralization veinlet types at the same deposit. By selecting such squares over a 1-km distance from the center to the perimeter of Sierrita, and by combining alteration, petrographic, and fluid inclusion analyses, he was able to depict the evolution of the fluid and fracture characteristics in space and time (Figure 11-7). Ahmad and Rose (1980) performed similar fluid inclusion analyses at Santa Rita, New Mexico.

Another approach to porphyry base-metal deposit studies has been through stable isotope, petrologic, and geochemical studies. $^{18/16}O$ and D/H

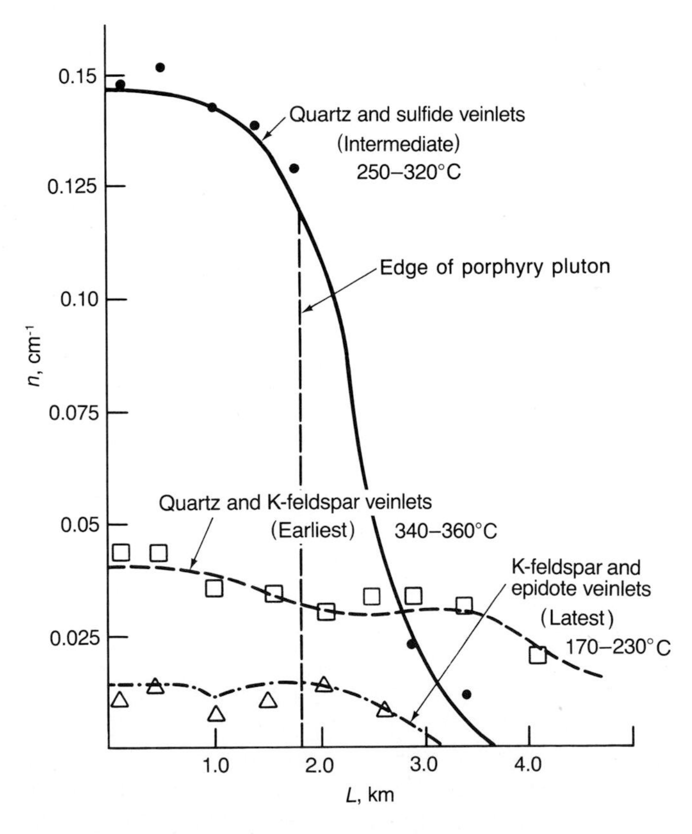

Figure 11-7. Diagram relating fracture-veinlet abundance n (centimeters of length of each type per square centimeter) to distance L from the center of the Sierrita porphyry system by earliest to latest veinlet type. The temperatures determined from fluid inclusion filling temperatures for each assemblage are high, then falling. (*After* Haynes and Titley, 1980.)

studies by Sheppard, Nielsen, and Taylor (1971), Hall, Friedman, and Nash (1974), Sheppard and Taylor (1974), Taylor (1974), Sheppard and Gustafson (1976), and Batchelder (1977) all argue that meteoric waters enter the circulation systems of porphyry deposits (Figure 11-4*b*) and that the ratio of meteoric to juvenile fluids appears to increase upward, laterally, and with time. Geochemical studies such as those by Rose (1970), Field (1975), Chaffee (1975, 1976), Olade and Fletcher (1975, 1976), Putman (1975), Olade (1976), Brimhall (1979, 1980), and Korkowski (1982) indicate additive outward dispersion halos of most of the ore-forming elements at the levels of observation. Studies of compositions, structures, trace element contents, and order-disorder relationships of biotite (Beane, 1974; Kesler et al., 1975; Jacobs and Parry, 1976, 1979; Banks, 1973; Graybeal, 1973; Mason, 1978), muscovite (Thompson, 1974; Montoya and Hemley, 1975; Guilbert and Schafer, 1978), K-feldspar (Lehtinen, 1974; Olade, 1980; Bird and Helgeson, 1980), and even rutile and apatite (Williams and Cesbron, 1977), considered with petrographic studies, are further illuminating pressure-temperature-composition relationships. Geochemistry and petrography together are adding new dimensions to studies of stable assemblages, paragenetic relations, and mineralogy in thin sections by using activity-activity diagrams (Chapter 7) and experimental mineral equilibrium studies (Hemley et al., 1980) to track fluid-composition–wall-rock responses through the time and space continua of ore-deposit formation. Meyer and Hemley (1967) studied alteration by heating pulverized quartz monzonite bathed in aqueous solutions of varying K^+, Na^+, H^+, and Cl^- at various temperatures. They published a diagram that has been added to most recently by Burnham and Ohmoto (1980) (Figure 11-8) who superimposed studies of fluid-melt relations at high temperatures onto lower temperature phenomena reported by Meyer and Hemley. Figure 11-8 depicts many aspects of porphyry base-metal deposit alteration environments. Chapters by Burnham on magmas and hydrothermal fluids and by Meyer and Hemley and Rose and Burt on hydrothermal alteration in Barnes (1967, 1979) constitute major advances.

Finally, another approach is that of computer analysis and simulation of ore-forming processes. Papers in 1977 by Cathles, Norton and Knapp, Norton and Knight, and Villas and Norton and in 1978 by Norton and by Fehn, Cathles, and Holland have established that pore fluids, surface-related meteoric waters, and groundwaters will convect in the vicinity of an intruded stock which can be viewed as a heat source. Their data are in accord with stable isotope studies mentioned above. The geochemical-geothermal implications and requirements of this convection in terms of the source of metals, of sulfur, of alkalies and halogens in ore-forming fluids, and of equilibration between fluids and buffering wall rocks are being studied. The problems are enormous. For example, the true compositions of biotites and their variations in porphyries are not at all well known; even if they were, unknowns concerning the thermodynamics of solid solution in

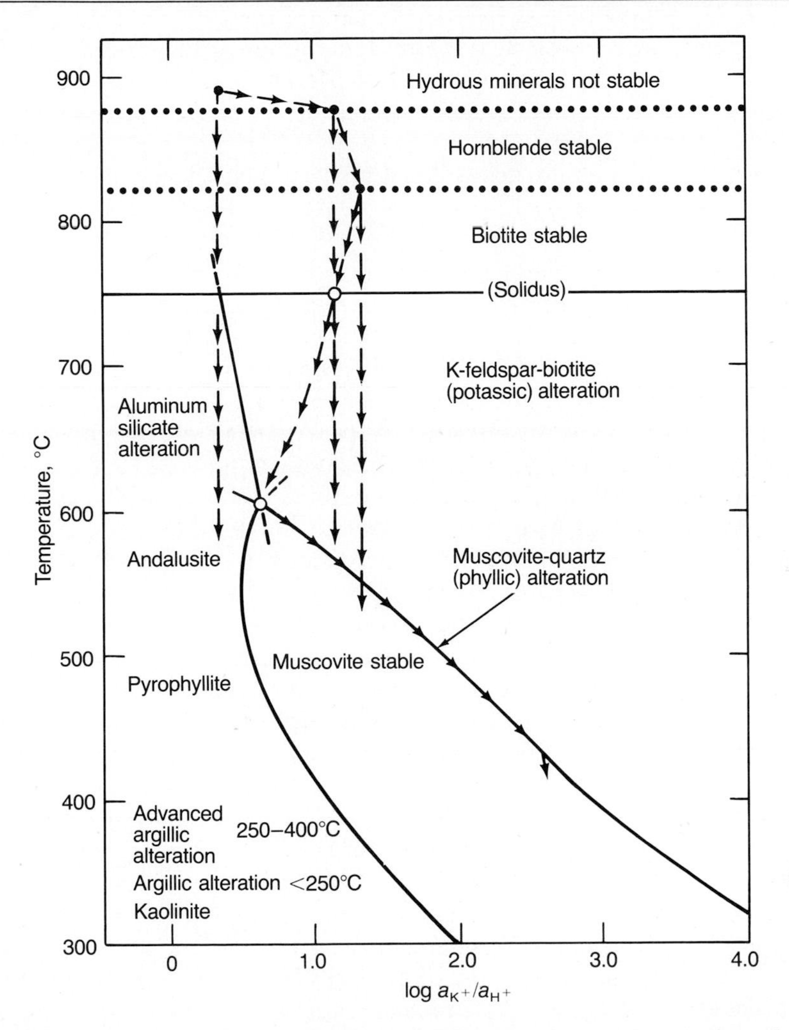

Figure 11-8. Log a_{K^+}/a_{H^+} versus temperature at 1 kbar showing variations in the ratio of KCl to HCl of magmatic aqueous chloride solutions as a function of phase assemblage, and possible nonequilibrium cooling paths of these fluids in a porphyry fracture system. Solid circles represent compositions of aqueous chloride solutions in equilibrium with granodiorite magmas and represent the "starting points" of hydrothermal fluid evolution. (*After* Burnham and Ohmoto, 1980.) See also Figure 5-3.

layer lattice silicates are almost prohibitively ill-known and complex to allow their incorporation into existing computer programs. Nonetheless, the continued approach of computer-derived geologic constructions to understanding ore-forming processes and reaction products as they can be seen and studied in field and laboratory supports both activities.

A result of the studies listed above is the recognition that porphyry copper deposit genesis is a multistaged, continual evolution that is almost certainly not exclusively magmatic in its affinities. The static diagrams of Lowell and Guilbert (Figure 11-4) are not process descriptions, nor do they include any aspects of time other than that of the final result.

El Salvador, Chile The approaches listed above support ideas on the genesis of the porphyry deposit at El Salvador, Chile, expressed in Figure 11-9 by Gustafson and Hunt (1975). El Salvador is a medium-sized but high-grade deposit that originally contained 300 million tons of 1.6% Cu. It is in north central Chile (Figures 11-1 and 11-10). It has an enriched blanket of chalcocite-digenite (Chapter 17) about 1.5 km across and 200 meters thick, with minable chalcopyrite-bornite-molybdenite hypogene ore extending below. Wall rocks include Cretaceous andesitic sediments and thick Tertiary andesites and subordinate rhyolites which were intruded by a complex series of granodiorite and rhyodacite porphyry stocks and dikes. Their emplacement generated chalcopyrite-bornite ores in a coaxially zoned alteration-mineralization pattern. As in many deposits, a low-total-sulfide, potassically altered core is surrounded successively outward by annular rings of chalcopyrite-bornite, chalcopyrite-pyrite, and pyrite-chalcopyrite, with a "pyrite halo" beyond the ore. Silicates are also zoned from K-silicate centrally through a sericite-kaolin zone to marginal propylitization, but sericite is generally supplanted by biotite in response to higher Fe-Mg levels of the andesite host rocks than is generally met with in more felsic hosts (Guilbert and Lowell, 1974). Potassic alteration is rich in biotite, K-feldspar, and anhydrite, and that which might be a normal phyllic zone is overwhelmed by biotite. Near the top of the orebody is a thin blanketlike zone dominated by the quartz, pyrophyllite, diaspore, corundum, tennanite, and enargite alteration-mineralization of the advanced argillic lithocap assemblage.

As shown in Figure 11-9, Gustafson and Hunt postulate that early mineralization (*b*) stemmed from the stock with minor addition of groundwater, with early establishment of K-silicate alteration grading outward into propylitic, not phyllic, assemblages in an environment of rising temperatures. Later an invasion of convecting groundwater and a retrograde temperature regimen gave rise to an assemblage characterized by sericite and pyrite, which invaded and crosscut earlier potassic-propylitic alteration in a relationship that has been called *phyllic overprint* at many deposits. The very late stage (*d*) is dominated by meteoric hydrothermal solutions cooled to hot spring temperatures. The cooler fluids produced the high-level pyrophyllite-diaspore argillic assemblage. Gustafson and Hunt generalized the El Salvador story (page 859) to a genetic model

Figure 11-9. (*Facing page*) Geologic and process history of the formation of the porphyry copper deposit at El Salvador, Chile. (*a*) Schematic geologic cross section before tilting and erosion. (*b*) Early alteration and mineralization before intrusion of L porphyry. (*c*) Main period of late alteration and mineralization after intrusion of L porphyry. (*d*) Very late postmineral stage of hot springs and intrusion of latites.

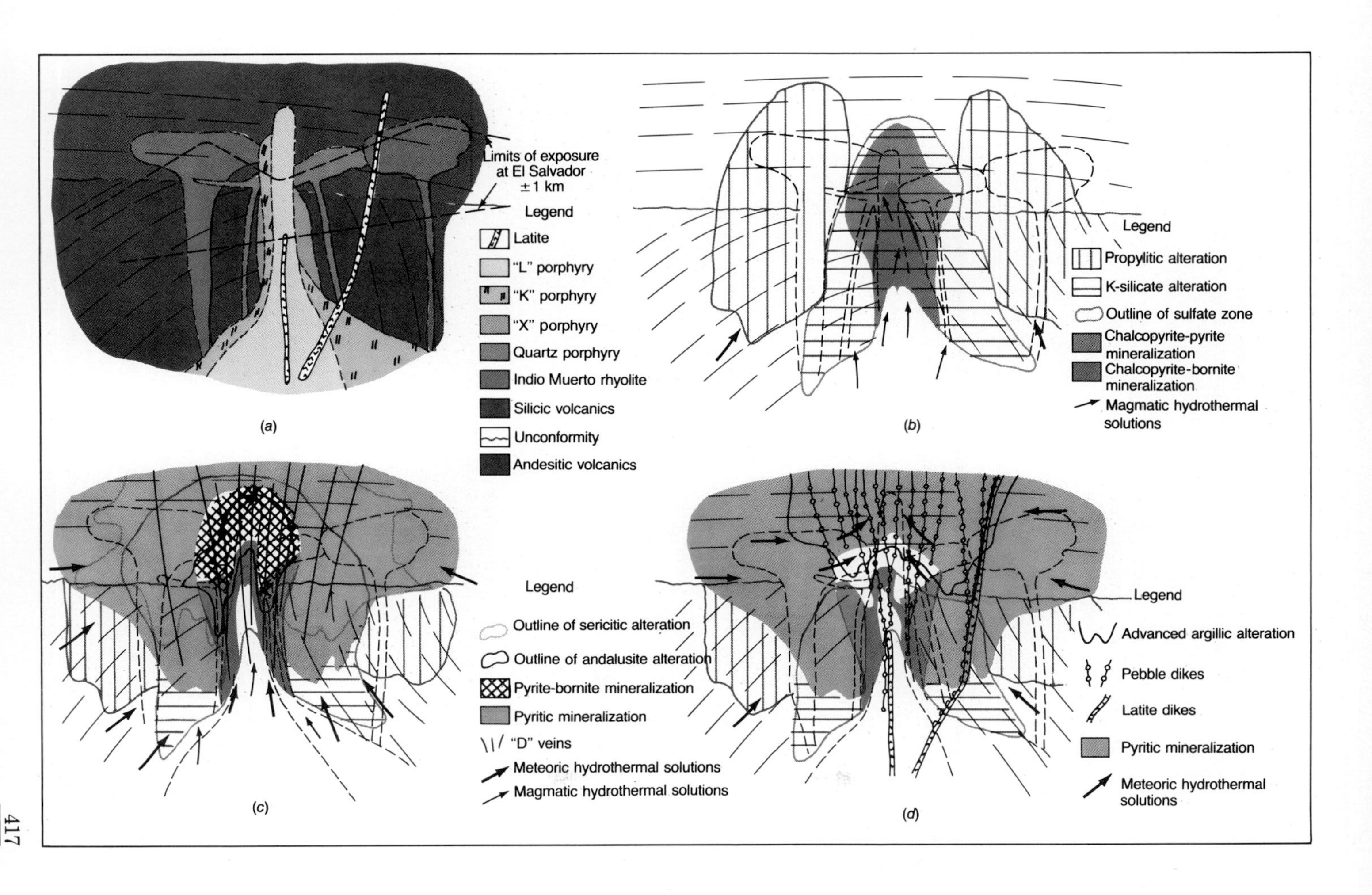
Limits of exposure at El Salvador ±1 km
Legend
Latite
"L" porphyry
"K" porphyry
"X" porphyry
Quartz porphyry
Indio Muerto rhyolite
Silicic volcanics
Unconformity
Andesitic volcanics
(a)
Legend
Propylitic alteration
K-silicate alteration
Outline of sulfate zone
Chalcopyrite-pyrite mineralization
Chalcopyrite-bornite mineralization
Magmatic hydrothermal solutions
(b)
Legend
Outline of sericitic alteration
Outline of andalusite alteration
Pyrite-bornite mineralization
Pyritic mineralization
"D" veins
Meteoric hydrothermal solutions
Magmatic hydrothermal solutions
(c)
Legend
Advanced argillic alteration
Pebble dikes
Latite dikes
Pyritic mineralization
Meteoric hydrothermal solutions
(d)

> *. . . proposed for the emplacement and deposition of porphyry copper deposits in general. Essential elements of this genetic model are (1) shallow emplacement of a complex series of porphyritic dikes or stocks in and above the cupola zone of a calc-alkaline batholith; (2) separation of magmatic fluids and simultaneous metasomatic introduction of copper, other metals, sulfur, and alkalies into both the porphyries and wall rocks; and (3) the establishment and inward collapse of a convective groundwater system which reacts with the cooling mineralized rocks. The well-known similarities of porphyry copper deposits from many parts of the world are variations on a common theme. The differences and unique features exhibited by individual deposits reflect the imprint of local variables upon the basic model. The local variables include depth of emplacement, availability of groundwater, volume and timing of successive magma advances, and the concentration of metals, sulfur, and other volatiles in the magmas, as well as depth of exposure.*

Guilbert and Lowell (1974) would add original wall-rock composition as another important variable.

The "phyllic overprint" story significantly modifies earlier, fundamentally prograde magmatic concepts. Current research is in part aimed at evaluating the timing and origin of quartz-sericite-pyrite assemblages at specific deposits. Phyllic alteration and stockwork structures are essentially absent at some deposits, as at Sierrita, Arizona, and El Arco, Baja California, Mexico, so phyllic alteration is not essential. It is therefore tempting to say that phyllic alteration *in general* is an overprint phenomenon. However, at both Butte, Montana, and San Manuel–Kalamazoo, Arizona, several polytypes, compositions, colors, and occurrence styles of sericite reach depths to 3 km below present surface and are unlikely to be solely of "overprint" origin. Burnham and Ohmoto (1980) include sericite as an orthomagmatic extension of potassic alteration (Figure 13-8). The overprint concept is clearly a valid one, not only with regard to exploration tactics but also with regard to the economic petrology of porphyry systems; its evaluation continues (Piekenbrock, 1983).

Chuquicamata, Chile " 'Awesome' is perhaps the best general description of Chuquicamata" (Cook, 1978, page 321). "Chuqui," in one of the driest parts of the Atacama Desert of northern Chile (Figure 11-10), is among the five largest copper deposits known (Lopez, 1939, 1942; Renzetti, 1957; Ambrus, 1979). The mine is 220 km northeast of Antofagasta and about 150 km from the coast, at an elevation of 3000 meters in the foothills of the Andes mountains.

Chuquicamata is an excellent example of a porphyry copper deposit, and along with other deposits of this type, is classed as mesothermal. Copper ornaments found in Inca graves indicate that the deposits were worked before the Spanish conquistadors arrived (Taylor, 1935). Small amounts of mining, confined principally to narrow but rich high-level veins, was done chiefly by English and Chilean companies between 1879 and 1912. The early operations were consolidated by the Guggenheim interests, then purchased by an Anaconda Company subsidiary in 1914 for 20 million dollars. Anaconda operated it for 60 odd years until it was nationalized in 1972. Since then it has been capably operated by Corporacion del Cobre (CODELCO-

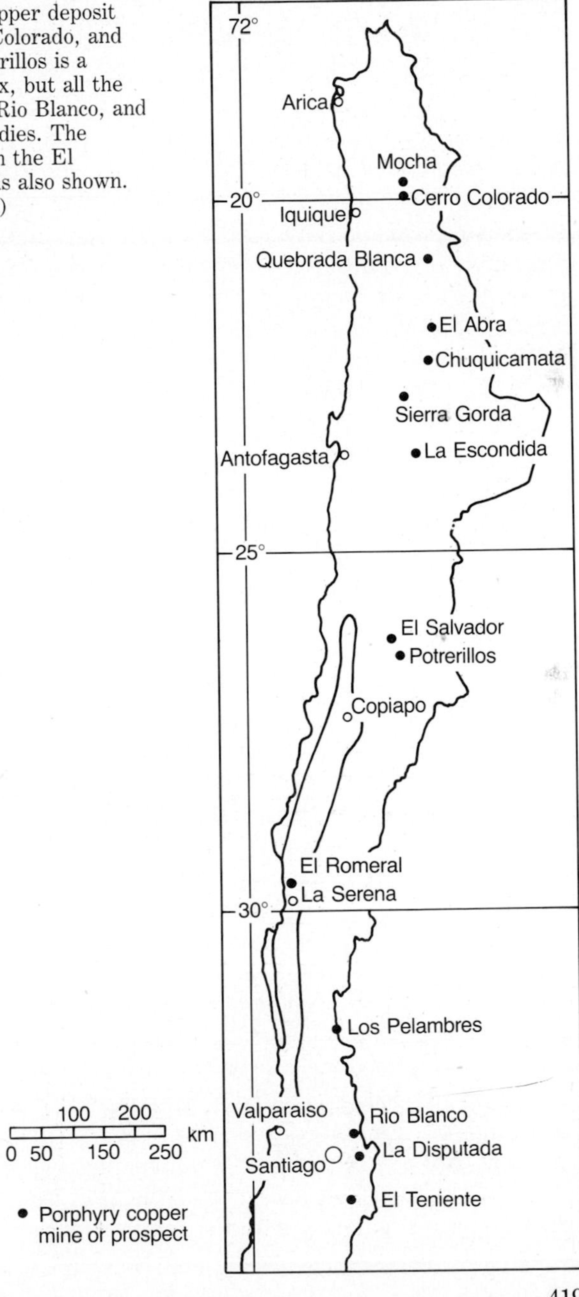

Figure 11-10. Map of porphyry copper deposit locations in Chile. Mocha, Cerro Colorado, and Sierra Gorda are prospects, Potrerillos is a mined-out porphyry-skarn complex, but all the rest are minable. Chuquicamata, Rio Blanco, and El Teniente are world-class orebodies. The Chilean iron province, centered on the El Romeral deposit described later, is also shown. (*From* Gustafson and Hunt, 1975.)

CHUQUI), a Chilean government arm. It is the largest producer of copper in the world—a sign there proudly claims "La Mina de cobre mas grande del mundo." It is the largest copper open pit at 1.8 by 3.7 km, with 500 million tons of 1.75% Cu material removed, and it *still* contains 1.5 billion tons of 1% copper ore above 400-meter depth. Scattered drill holes to depths of 1 km show that the orebody continues downward. Much of the following information has been derived from visits and by personal communication (1975, 1978) with Chief Geologist J. Ambrus.

The mineralization is related to a period of faulting, folding, and igneous activity accompanying intrusion of the Andean batholith porphyritic rocks during the Late Mesozoic and Early Tertiary ages. Older Jurassic sediments and metamorphic rocks were first gently folded and covered with a series of volcanic flows and then strongly deformed as the Andean batholith was intruded. After considerable erosion had exposed the plutonic rocks, the area was again uplifted and more volcanics were deposited. Slight warping and volcanism have continued into recent time (Taylor, 1935; Jarrell, 1944). Figure 11-11 shows a Chuqui pit map which is old but still adequate; the pit bench lines would now be greatly expanded and deepened. Figure 11-12 shows alteration in the same area, from Taylor (1935), Lopez (1939), Sillitoe (1973), Ambrus, and by observation. The oldest rocks are the wall rocks to the southeast and the Jurassic Elena Granodiorite dated at 122 million years east and northwest of the pit. The Fortuna Granodiorite (38 million years) lies west of the West Fissure, a major fault with kilometers of strike-slip displacement and with regional significance—the deposits of Pampa Norte, El Abra (Ambrus, 1977), Quebrada Blanca, and Copaquire lie along it to the north (Ambrus, 1979). Between the West Fissure and the Elena Granodiorite is an elongate wedge of 24-million-year-old complexly intruded quartz porphyries, probably originally quartz latite porphyry. Three facies can be distinguished by groundmass and phenocryst distinctions, the East Side, the West Side, and the Banco porphyries. Contacts are difficult to map, but discernible; they are not differentiated on Figure 11-11, but rather are lumped under "Chuquicamata porphyry." Hypogene protore copper in these rocks averaged 0.8 to 0.9% Cu, with values from 0.5 to 1.0% Cu. These hypogene sulfides occur as (1) K-feldspar-stable, chalcopyrite-bornite-molybdenite "magmatic" or potassic veinlets, (2) an early hydrothermal chalcopyrite-molybdenite-sericite set, and (3) a late hydrothermal pyrite-chalcopyrite enargite-major sericite group. All of these vein sets lace the porphyries, as they do at El Salvador (Gustafson and Hunt, 1975). Here the veinlet sets are centered on the Chuqui porphyries.

Two strongly fractured stockwork zones define the western and eastern limits of the mineralized zone. Shears vary in strike from N10°E to as much as N76°E, forming a violin-shaped orebody with the neck pointed south. Between the eastern and western shear zones there are several intermediate shear zones and a complex system of tension fractures, most

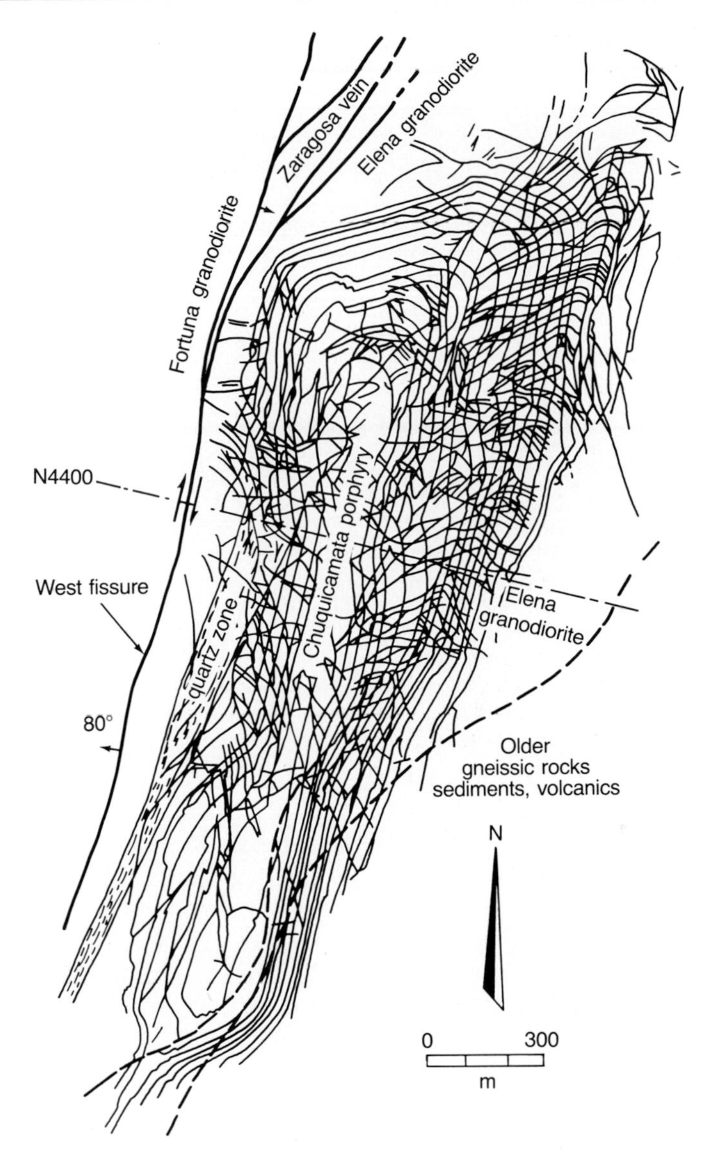

Figure 11-11. Geologic plan of Chuquicamata, Chile. The nested lines are pit bench outlines, the branching and heavier ones faults or fractures. (*From* Perry, 1952.)

of which form a horsetail pattern branching northeastward from the shears (Figure 11-11). Lopez (1939, 1942) suggested that the shears and tension fractures are all related to a simple shearing couple acting horizontally in a north-south direction. Given this, the maximum shear stresses would have been oriented northeast-southwest, parallel to the horsetail tension frac-

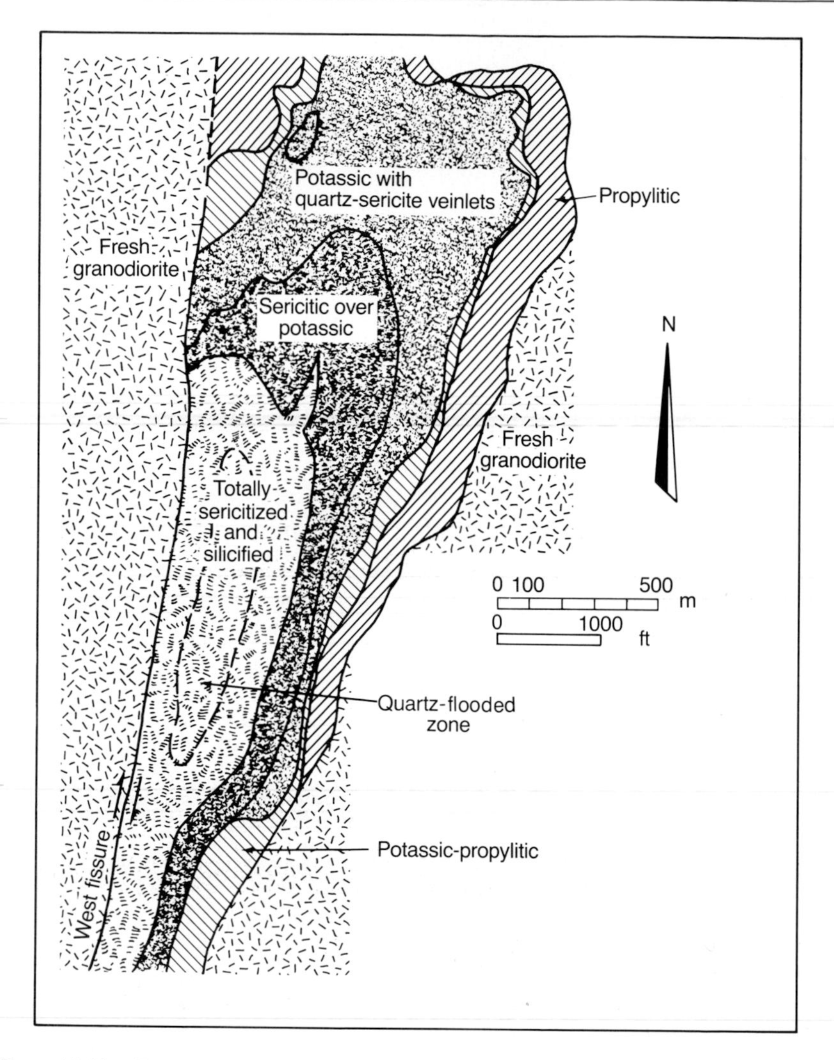

Figure 11-12. Hydrothermal alteration at Chuquicamata, Chile, based on Figure 11-11.

tures. Another prominent set of tension fractures strikes northwest, roughly normal to the horsetail fissures. The many components of the combined stockwork-fracture-shear zones locally brecciated the porphyries, and in places "make" into veins. Sulfides permeated these fractures.

Hydrothermal alteration patterns (Figure 11-12) are complex. Most striking is a zone of total pervasive conversion to quartz-sericite-pyrite in the heart of the pit and centered on the Chuqui porphyries. The altered rocks of the benches are essentially snow-white cliffs of mica and quartz (Figure 11-13). The next zone outward represents a drop in the extensiveness of sericitization; the rocks are less white, and relict feldspars can be seen with a hand lens. External to that volume is one shown with a speckled

(a)

Figure 11-13. (*a*) Extensive phyllic alteration of the Chuqui porphyry on the east side (to the left of the sample pick) of the nearly vertical West Fissure in the pit at Chuquicamata, Chile. The originally tan quartz latite porphyry at the left has been totally converted to snow-white masses of quartz-sericite-pyrite. The mass effect is apparent in (*b*), a panorama of the pit. The gray upper material is iron-oxide-stained by weathering and leaching (Chapter 17), the lower benches showing the white pervasive alteration effect of conversion to dominant sericite.

(b)

pattern in Figure 11-12 in which Chuqui porphyry and Elena Granodiorite, veined and impregnated with orthoclase and biotite, are cut and veined with quartz-sericite veinlets. Is this zone "sericite overprint"? A thin zone of potassic alteration with epidote and chlorite then yields outward to the classic propylite association of chlorite-calcite-epidote-K-feldspar which merges outward with fresh granodiorite.

Now if sericite and silicification are stripped from Figure 11-12, the remaining pattern shows potassic alteration merging outward with propylitic alteration, which in turn merges outward with fresh rock. The quartz and sulfides of magmatic and early hydrothermal veinlets are immune to sericitization; they are still found in the center of the pit. But that volume of high fracture density is now completely sericitized, and age relationships suggest younger, postore sericitization. Considering the San Manuel–Kalamazoo patterns (Figure 11-4*a*), however, alteration appears inside out, with potassic assemblages *external* to phyllic. More reasonably, earlier potassic alteration interfaced outward with propylitic. Phyllic alteration was superimposed on both, but the center of its activity was controlled by relatively high porosity-permeability within the central ellipse of high fracture density and high-volume fluid flow. In short, field relations argue that the bulk of Chuquicamata sericite is of overprint affinity, and that its lateral extension, perhaps because of ease of vertical flow, was not as great as it is at most porphyry coppers. So the interface between potassic and propylitic alteration was extraordinarily preserved. Alteration *appears* to be "inside out," but is in fact a result of the distribution of permeability and timing.

Following mineralization, a remarkable series of low-temperature supergene phenomena converted one of the largest and highest grade hypogene porphyry copper deposits into the richest deposit in its class. Chuquicamata has experienced a complex history of oxidation, leaching, and supergene enrichment. During the Late Tertiary uplift of the area, the climate changed from humid to arid, and the water table fluctuated radically. Copper leached from the oxidation zone was carried down to the water table, where it was reprecipitated as a sulfide enrichment of the primary ore. As the land surface was eroded, the water table and its associated supergene ores moved downward through the hypogene ores. An apparently sudden, geologically recent change to an arid climate abruptly lowered the water table and left the supergene zone high and dry above the newly established deep water table. Subsequent leaching and oxidation have, in general, been slow, but where the ore contained sufficient pyrite to react with water and form sulfuric acid, and where there was very little feldspathic material to neutralize this acid, copper was leached. But over 200 million tons of easily treated, high-level oxidized ore containing 1% Cu, of which less than 20 million tons remain, have been mined in the past. The oxidized zone contains many rare minerals, many of which are water-soluble but are preserved in the extreme aridity of the Atacama Desert. Atacamite and antlerite were especially abundant and economically important. Sulfates, oxides, and carbonates characterize the oxidized zone. The copper minerals include cuprite (Cu_2O), tenorite and its variety melaconite (CuO), native copper, brochantite [$Cu_4SO_4(OH)_6$], antlerite [$Cu_3SO_4(OH)_4$], kroehnkite [$Na_2Cu(SO)_4)_2 \cdot 2H_2O$], chalcanthite [$CuSO_4 \cdot 5H_2O$], natrochalcite [$NaCu_2(SO_4)_2OH \cdot H_2O$], cuprocopiapite [$CuFe_4(SO_4)_6(OH)_2 \cdot nH_2O$], salesite ($CuIO_3OH$), turquoise, leightonite [$K_2Ca_2Cu(SO_4)_4 \cdot 2H_2O$], lindgrenite [$Cu_3(MoO_4)_2(OH)_2$], chenevixite [$Cu_2Fe_2(AsO_4)_2(OH)_4 \cdot H_2O$], bellingerite [$3Cu(IO_3)_2 \cdot 2H_2O$], atacamite [$Cu_2Cl(OH)_3$], and numerous other species (Bandy, 1938; Palache and Jarrell, 1939; Jarrell, 1944). Sulfates of iron are also abundant in the oxidized zone.

Cook (1978) listed 114 verified mineral species at Chuqui. The presence of chlorine and iodine in some of the compounds has been attributed both to wind transport of these halides from the ocean (Jarrell, 1944) and to recent volcanism (Bandy, 1938). Mixed oxide and sulfide ores are also abundant. Where the porphyry was thoroughly silicified, it locally became relatively impermeable, and shielded patches of hypogene ore from the leaching action of descending waters (Jarrell, 1944). The sulfide-enriched blanket that was developed before final uplift contained less pyrite than hypogene ore, because supergene copper was deposited at the expense of iron in pyrite. As a result, when this supergene sulfide zone was stranded above the water table, it could not generate enough sulfuric acid to dissolve copper minerals and carry them down to the new zone of enrichment. Instead, copper minerals were merely oxidized in place to sulfates, forming oxidized and mixed oxide-sulfide orebodies. Another oddity is that the hypogene ore that had previously underlain the supergene blanket contained sufficient pyrite that its breakdown after the water table dropped led to the leaching of copper; as a result, high-level oxide ore is separated in places from a supergene-enriched sulfide blanket by a layer of iron-oxide waste. Furthermore, the old supergene-enriched zone was not uplifted uniformly; it was tilted, leaving its south end lower than its north end. The present-day water table truncates this surface, isolating some sulfides in the north and submerging part of the oxidized zone in the south. Overall, enrichment upgraded the supergene blanket to unusual degrees—trainloads of 15 to 20% Cu ore are not uncommon, and chalcocite-covellite supergene ores account for most of Chuqui's reserves.

As if Chuqui were not phenomenal enough, yet another mode of ore is found there. In 1957 a churn drill hole quite accidentally penetrated "green-copper"-cemented gravels a few kilometers south of the main occurrences at Chuqui. It was formed from lateral runoff of copper-bearing waters when Chuqui proper was being oxidized and enriched. Called *La Exotica*, it is a sheet at least 1 by 3 km in size, up to 100 meters thick; it totals 180 million tons that average 1.6 to 1.8% Cu, and is composed of chrysocolla, atacamite, melaconite, cupiferous wad, and libethenite. One-third of the copper minerals cement the gravels, and two-thirds occur as impregnations, fracture fillings, and even replacements in the propylitized granodiorite and metamorphic rocks beneath the Miocene gravels. La Exotica has been mined intermittently since 1970.

Overall, the Chuquicamata ores owe their concentration to several factors. It seems inescapable that the Chuqui porphyries were derived from a batholith that delivered and localized transition elements along the West Fissure tectonic zone. The West Fissure must have tapped this activity and influenced the localization of the Chuqui porphyries, although it is not intruded or mineralized itself (Ambrus, 1979). Alternating periods of silicification and fracturing prepared the rock for ore deposition; fluids appear to have ascended through fractured zones along the west side of the main pit area, spreading eastward and westward with alteration and mineralization, first with potassic-propylitic, later with phyllic assemblages. Finally copper was leached and redeposited by meteoric waters, concentrating what was otherwise a more dispersed zone of mineralization.

Recent developments at Chuquicamata include a drainage tunnel driven 400 meters below the original surface; this tunnel lowered the water table, allowing the zone of supergene enrichment to be mined. As a result, mining is now in the sulfide zone below the oxidized ores, and Chuquicamata promises to be a major producer of copper for many years.

Several explanations and classifications of subtypes of porphyry copper deposits have been offered to account for variations within the deposit type and departures from the "typical" porphyry. Guilbert and Lowell (1974) discussed six influences on such departures, emphasizing the importance of intruded wall-rock composition in predisposing alteration mineralogy and zoning in a given deposit. Hollister (1975) noted that porphyries with abundant alteration biotite and dioritic intrusive phases are common in island arc settings and added a plate tectonic component in suggesting the "diorite model" as distinct from quartz monzonite types. This distinction should be made with great care, and only as Hollister intended it. For example, the Safford, Arizona, and El Salvador, Chile, deposits have been called examples of the diorite model, although they occur in tectonic settings which are identical to those of their neighboring quartz-monzonite-affiliated Morenci and Chuquicamata, respectively. The differences are probably the results of variations controlled by local environments of emplacement. Similarly, examples of the "quartz monzonite model" occur beside those of the "diorite model" in island arcs in New Guinea and the Philippines. In recent years a group of porphyry coppers associated with sodic rocks and albitic alteration have been described from New Guinea and from a belt in eastern British Columbia. They should be looked for elsewhere; they may have been emplaced in zones of tectonic extension rather than compression. Sutherland-Brown (1976) identified three temperature-pressure clusters of porphyries and proposed a threefold classification of plutonic, phallic, and volcanic. Efforts to classify will doubtlessly continue as barometric, thermometric, chemical, structural, and tectonic data build.

Porphyry Molybdenum, Climax-Type Molybdenum, and Porphyry Tin Deposits

Three important subsets of the porphyry base-metal deposit type are the porphyry molybdenum deposit, or *porphyry moly*, the *Climax-type moly*, and the *porphyry tin* deposit. It might be thought, in view of the joint occurrence of molybdenite and chalcopyrite in porphyry coppers, that there are continuous populations of porphyry base-metal deposits along a binary plot of copper and molybdenum, and perhaps even within a ternary plot of copper, molybdenum, and tin. In fact, the critical nature of the competency of hydrothermal fluids is emphasized by the fact that there is *not* a continuum. Figure 11-14 is a binary plot of the copper and molybdenum ratios in 57 porphyry copper-moly deposits. Note the cluster of points near the copper end, the gap between Cu_{74}-Mo_{16} and Cu_5-Mo_{95}, and the cluster at the molybdenum end. There are too few tin content values to permit a diagram, but there would be a third discrete cluster at the tin apex of a ternary diagram; porphyry tin deposits contain little copper and only traces of molybdenum. The schisms are real. For example, the 1974

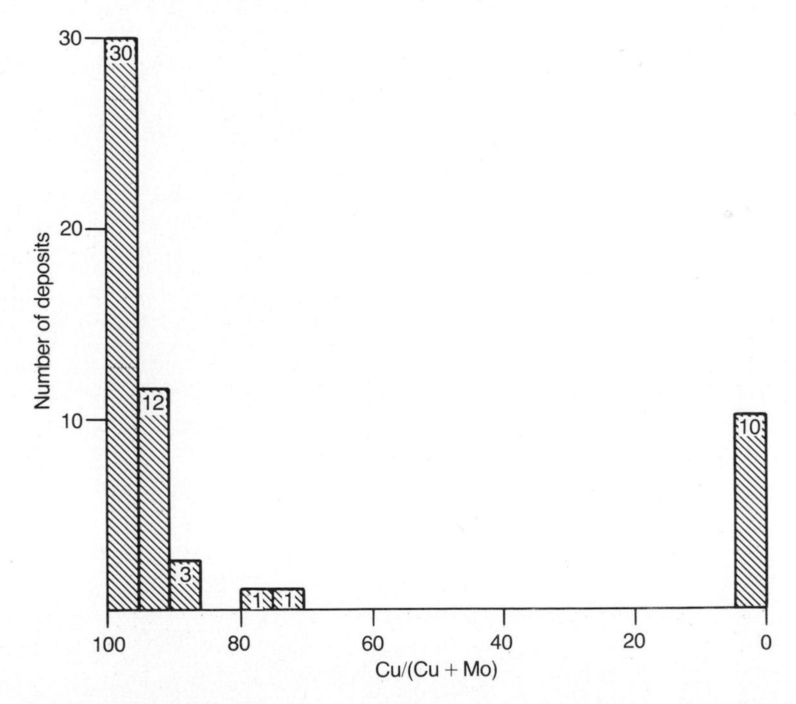

Figure 11-14. Histogram of the copper and molybdenum contents of 57 porphyry deposits, five of which are Climax-type moly deposits. 15 of the 57 contain no molybdenum; the five porphyry molys and the five Climax-type molys contain no copper. Note the breadth of the gap between Brenda and Nogal Peak, British Columbia, at 0.77 and 0.74, to the 0.00 of the Quartz Hill deposit, and others. See text for discussion of the high Mo end. (*From* Lainé, 1974, *with Quartz Hill, Alaska, added.*)

discovery by U.S. Borax of what has proven to be an orebody of 1.3 billion tons of 0.10% Mo (0.17% MoS_2) at Quartz Hill near Ketchikan, Alaska, revealed a deposit that contains only 30 to 60 ppm Cu. The highest value so far recorded in thousands of assays is only 0.014% Cu, or 140 ppm; chalcopyrite has not yet been logged in 15 km of diamond drill core! Neither are Ag, Au, Sn, or W values appreciable; the plot of Quartz Hill on Figure 11-14 sits squarely at the molybdenum end on top of other deposits such as Climax, Urad-Henderson, and Mt. Emmons, Colorado, and B.C. Moly and Endako, British Columbia. When this diagram was first prepared (Lainé, 1974), no distinction was made between porphyry moly and Climax-type moly deposits. It has now been perceived (Westra and Keith, 1981; Bookstrom, 1981; Mutschler et al., 1981; White et al., 1981) that a distinction between molybdenum-rich porphyry and Climax-type occurrences should be made. Porphyry moly deposits are end members of the porphyry copper-moly family. Different groups call them quartz monzonite type, granodiorite type, and calc-alkaline type; for simplicity's sake, and to emphasize their continuity with porphyry coppers, they are here called *porphyry moly type.* They have similar fluorine-to-chlorine ratios, occur along compressive Cordilleran continental margins in I-type granodiorite to quartz monzonite host rocks, and differ mainly in Cu/Mo ratio. They include Quartz Hill, Alaska; Thompson Creek and White Cloud, Idaho; Endako, British Columbia; and others.

Climax-type moly deposits differ markedly. They are relatively high-grade at 0.3 to 0.5% MoS_2 (= 0.2 to 0.3% Mo; MoS_2 = 61% Mo, 39% S), occur in atectonic to tensional rift environments with A-type (atectonic or anorogenic) high-silica porphyritic granitic alkalic intrusive igneous rocks with K_2O higher than Na_2O, show active intrusion structures like radial and ring dikes, have variable initial $^{87/86}Sr$ ratios (0.705 to 0.760) and high F/Cl ratios, and have low copper contents. The best examples are Climax, Urad-Henderson, and Mt. Emmons, Colorado; Questa, New Mexico; and Pine Grove, Utah. White et al. (1981) provide a comprehensive detailed description of Climax-type systems and deposits.

Since porphyry coppers have been described and differ little from porphyry moly occurrences, the latter need little further treatment. Quartz Hill, Alaska, is an excellent example. The ores are in an I-type biotite-amphibole quartz monzonite, quartz latite porphyry, and aplite complex of Miocene age 80 km east of Ketchikan, Alaska. The only ore mineral is molybdenite. It occurs in quartz-molybdenite veinlets in the "standard format" of white quartz veinlets 1 cm across with smoky films of molybdenite on either wall, as veinlets of moly without quartz, called *moly paint,* and as disseminations. Alteration is quite inconspicuous. It is dominated by potassic assemblages, especially K-feldspathization; but the rocks overall are so low in mafic phases and so felsic that they record neither peripheral propylitization nor extensive potassic or phyllic alteration. Veinlets typically show minor recrystallized or augmented K-feldspar, but the alteration is not flamboyant. Fluorite is absent; the principal gangue mineral is quartz. The molybdenum occurs in spiderweblike stockworks of quartz moly veinlets, in a breccia dike called "superbreccia," and as disseminations. Ore values appear to diminish below a geochemical "floor" roughly 300 meters below the present surface—quartz-filled fractures and other structural openings continue downward, but MoS_2 values drop markedly. The orebody thus appears to be sheetlike, or tabular. As mentioned above, copper is rare and molybdenum values are substantially lower than they are in the Climax-type deposits described next. Hudson, Arth, and Muth (1981) suggest that the Quartz Hill environment was atectonic at the Early Miocene time of intrusion. Quartz Hill was discovered in 1974 by a coastal exploration search program mounted from a launch with on-board heliport and analytical laboratories. The deposit was inconspicuous from the air, but anomalous stream sediment values of 35 and, farther upstream, 168 ppm Mo brought U.S. Borax geologists J. Stephens and R. Kistler to the area for reconnaissance. They found the cobbles of the river bed of White Creek to be blue-gray with molybdenite-coated fracture surfaces. The deposit will be "brought in," with care to minimize adverse environmental impact, in the 1980s.

Climax-type moly deposits, as stated earlier, appear to be atectonic or rift-related, and thus may be restricted to extensional environments such as the Rio Grande rift, the Basin and Range province, and similar structural settings. No Climax-type moly occurrences are known from South America; perhaps the tectonics of western North America, where a continental plate has "snuffed out" a midoceanic rise, are uniquely suitable among recent geologic settings. Bookstrom (1981) suggested that the tectonic habitat of Climax-type deposits is the hinterland inland from magmatic arcs with shallowly dipping subduction zones, and that they form as the environment shifts from compressional through atectonic to tensional when subduction ceases. All but one of the North American deposits are less than 30 million years old; Big Ben in Montana is 50 million years. The two best known Climax-type moly deposits—Climax and Urad-Henderson—are only 50 km NE-SW apart in the Colorado Mineral Belt of the Rocky Mountains, the latter 80 km due west of Denver near Berthoud Pass. Studies of the Climax deposit under Stewart Wallace in the 1960s were so thorough and skillful that geologic relationships educed there (Wallace et al., 1968) contributed integrally to the discovery of 300 million tons of 0.4% MoS_2 of the Henderson deposit in 1968 by Wallace and his staff (Wallace et al., 1968; White and MacKenzie, 1973; Wallace, 1974; and Wallace, MacKenzie, Blair, and Muncaster, 1978). The Henderson ores are described in the latter paper. They are similar enough to those at Climax that either could be presented as an example.

Climax, Colorado, Molybdenum Relevant igneous activity at Climax, 100 km southwest of Denver, Colorado, occurred between 33 and 24 million years ago when a series of four granitic stocks intruded Precambrian gneisses, the Precambrian Silver Plume granite, and one another. Each of the four intrusive events of the composite Climax stock (Figure 11-15) followed its predecessor closely in time, and each reached only to a successively deeper level. The intrusives are more felsic than their equivalents in porphyry copper systems: the first and second are rhyolite porphyries, the third is an aplite, a fine-grained sugary intergrowth of quartz and orthoclase, and the fourth is a porphyritic granite. They are all relatively silica-rich and alkali-rich with high K^+/Na^+ ratios. K_2O averages 5.5%, Na_2O about 3%. Each stage generated a separate orebody, each with its own similar morphology, its own element zoning, and its own alteration patterns. This repetitive intrusion and mineralization is another distinctive characteristic of Climax-type molys; many show multiple composite intrusion-mineralization histories. The orebodies each have the form of an inverted bowl, or shield, or umbrella, a draped layer about 200 meters thick, 1 to 1.3 km across, and with perhaps half a kilometer of overall height (Figure 11-16). Their trace in plan, then, is nearly circular. Overall grade averages about 0.25% Mo over some 500 million tons of ore. The single ore mineral is molybdenite, 95% of which is in quartz moly veinlets, 1-cm-wide white quartz veinlets with steel gray to smoky gray molybdenite films along the walls. Minor tin as cassiterite (SnO_2) and tungsten as huebnerite [$(Fe,Mn)WO_4$] are recovered. Figure 11-15 shows the zoning of the Upper orebody, the second orebody of Figure 11-16*b*, a zoning that characterizes each of the hoods of mineralization, the fourth being barren. Two silicified zones—one with huebnerite and topaz above the molybdenite zone, one a "high-silica" zone below it—are clearly parts of nearly identical, consecutive alteration zoning patterns that are repeated and complexly overlapping. It involves early potassic alteration, with abundant K-feldspar, quartz, and some sericite. The Henderson deposit shows clearer alteration zoning (MacKenzie, 1970; Wallace et al., 1978; White et al., 1981). Figure 11-17 shows that the deposit progresses from a high-silica zone, which is at the apex of a cupola and within a deep central K-feldspar potassic molybdenite ore zone, upward and outward through higher level zones of quartz-magnetite-topaz, sericite-quartz-pyrite, quartz-kaolin, high-level garnet, and finally a peripheral chlorite-calcite-epidote propylitic zone of complete biotite destruction that covers a 2-by-3-km elliptical area at the surface, with discernible propylitization extending even farther. Copper, lead, zinc, barium, and several other element geochemical anomalies are formed in halos around the orebodies. The Henderson and Climax systems are typical of the Climax-type moly deposit as contrasted with porphyry coppers or porphyry molys in that they appear to have formed at slightly higher temperatures, near 500°C (Hall, Friedman, and Nash, 1974); their fluids, gangue minerals, and alteration silicates, like sericite, are characterized by traces of fluorine (F^-) rather than chlorine (Cl^-); they are nearly devoid of copper; they are geologically younger at Oligocene-Miocene rather than Late Cretaceous to Eocene; and they lie

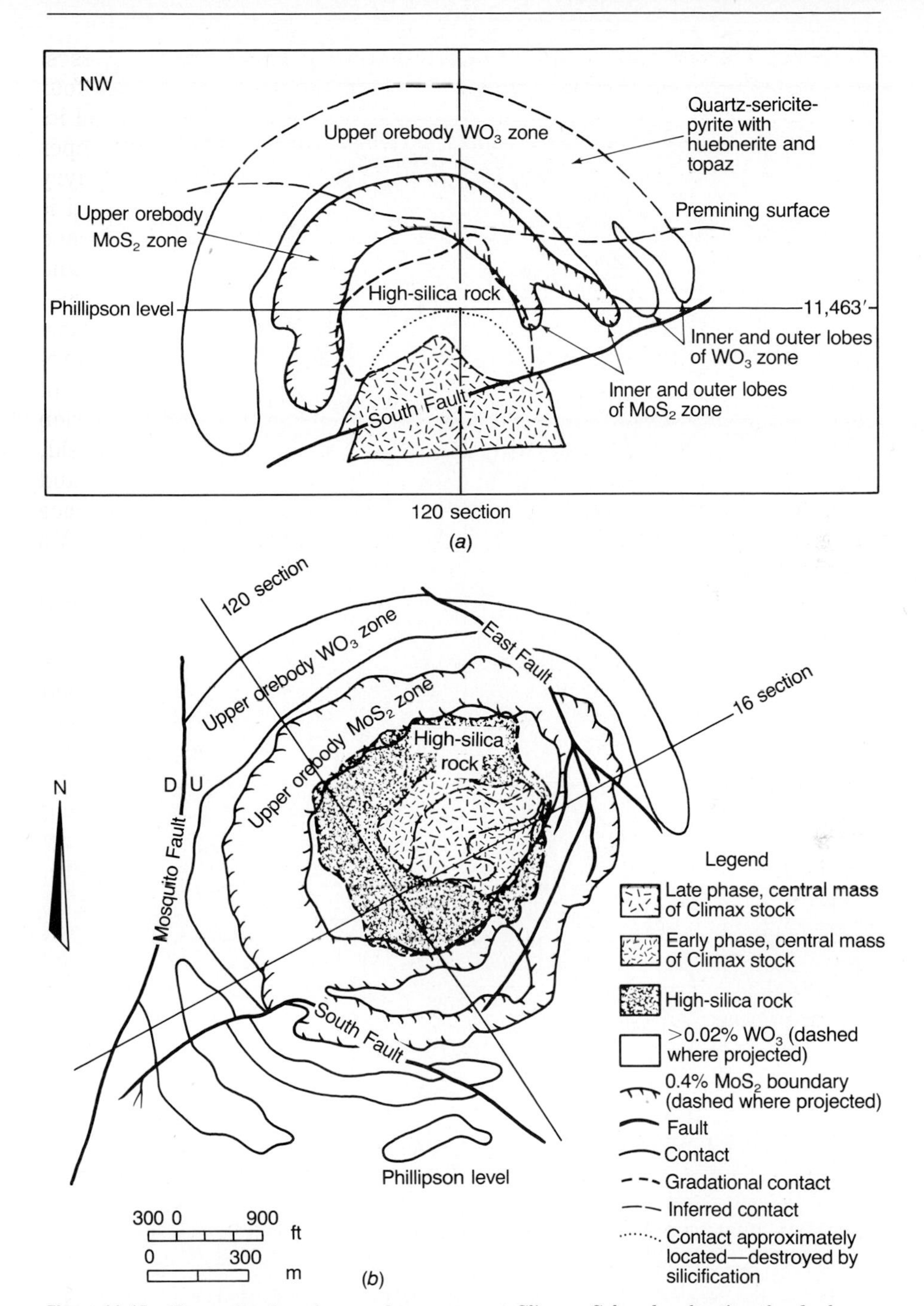

Figure 11-15. Generalized geology and ore zones at Climax, Colorado, showing the dual nature of the Upper orebody and its vertical zoning. (*a*) Cross section and (*b*) map or plan view on the Phillipson level, 3496-meter elevation. Compare closely with Figure 11-16*b*. (*From* Wallace et al., 1968.)

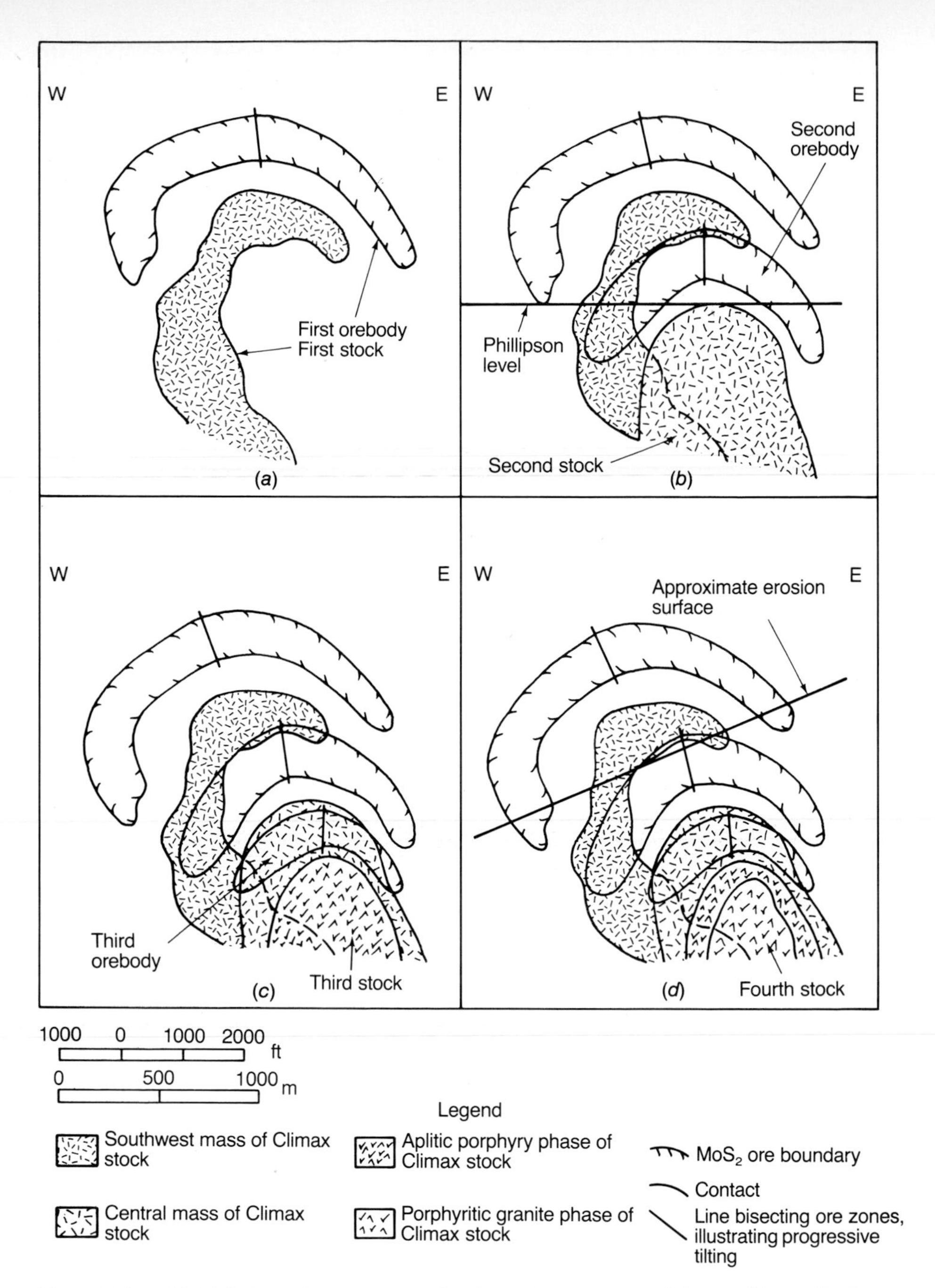

Figure 11-16. Multiple intrusion and mineralization and progressive tilting at Climax. Each of the umbrella-shaped orebodies is circular in plan view, as seen in Figure 11-15*b*, and each is itself vertically zoned (Figure 11-15*a*). Only molybdenite mineralization is shown here. Labels cite relative ages and are not names. (*a*) Southwest mass of Climax stock and Ceresco orebody. (*b*) Central mass of Climax stock and Upper orebody (see Figure 11-15). (*c*) Aplitic porphyry phase of Climax stock and Lower orebody. (*d*) Porphyritic granite phase of Climax stock (late "barren" stage mineralization not shown). (*From* Wallace et al., 1968.)

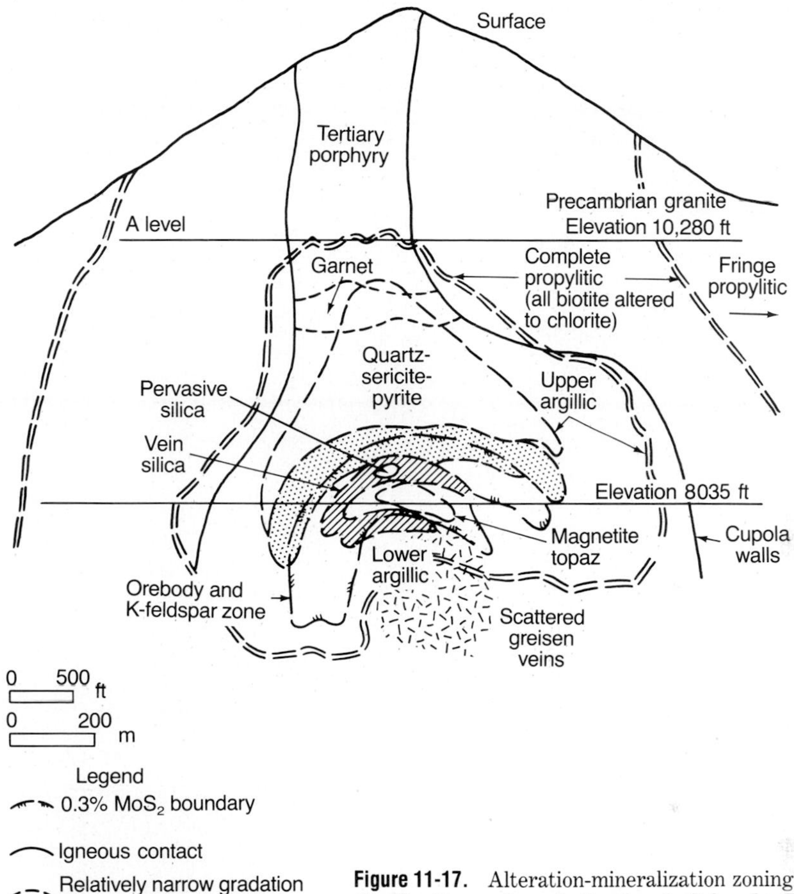

Figure 11-17. Alteration-mineralization zoning at the Henderson deposit, Berthoud, Colorado. The array shown here is typical of that of Climax-type moly deposits. (*After* White et al., 1981.)

inland from continental margins. Mitchell and Garson (1972), Oyarzun and Frutos (1974), and others have suggested that chlorine-dominated copper-rich intermediate intrusive systems are generated along subduction zones at shallow low-temperature levels close to the trench, and that the fluorine-dominated tin-tungsten-molydenum silicic intrusions are generated at deeper, hotter, more inboard levels. Mutschler et al. (1981) present evidence that Climax-type moly deposits are distinct from porphyry moly ones and that the former represent high-level batholith-related cupola concentrations of large-ion lithophile elements, fluorine, and the other components listed above. Several considerations, including initial strontium isotope ratios, suggest that assimilation of continental crust may be more important in Climax-type moly systems.

That the *porphyry tin* deposits of Bolivia are a subgroup was first suggested by Sillitoe, Halls, and Grant (1975). Those deposits, described by Turneaure (1971) and Kelly and Turneaure (1970), were previously considered to be simply extensive igneous-related hydrothermal vein systems, part of the scale consideration mentioned earlier in connection with recognition of porphyry coppers. They may also represent cupolas over larger bodies of intrusive alkalic leucogranite rather than the calc-alkalic quartz monzonites that characterize PCDs (Bookstrom, 1983).

Llallagua, Bolivia, Tin Sillitoe, Halls, and Grant (1975) described Catavi, the largest single tin deposit, at Llallagua, Bolivia. A brief study of Figures 11-18 and 11-19 shows general similarity with porphyry copper and moly deposits, except that mineralized veins are greatly more important. Other porphyry tins at Oruro, Potosi, and Chorolque, all in Bolivia, are similar. These authors drew 12 points of similarity and three of dissimilarity in comparing porphyry tin and copper moly deposits.

Points of similarity summarized from their paper serve to describe Llallagua: mineralization is localized in 1 to 2 km^2 calc-alkaline latite porphyry stocks; the environment is shallow substratovolcanic; porphyry contacts with wall rocks are sharp but complex; stocks were passively emplaced; metamorphism is inconspicuous; complicated histories of multiple brecciation before and during mineralization are apparent, with cassiterite-cemented breccias; central phyllic alteration at the surface is ringed peripherally with widespread propylitization, with deeper quartz-tourmaline common; phyllic alteration occupies a stockwork; alteration and disseminated mineralization occupy oval to cylindrical coaxial columns more than 1 km in vertical extent; there are peripheral veins of Pb, Zn, Ag, and Ba;

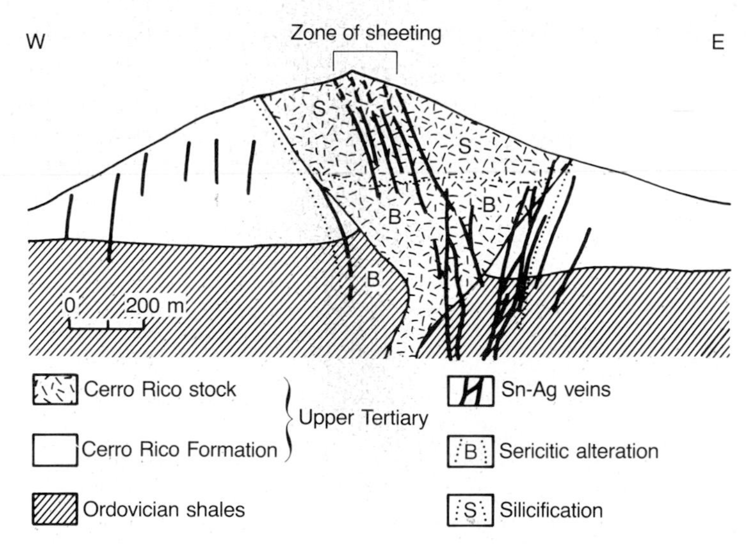

Figure 11-18. Generalized east-west cross section of the Cerro Rico stock at Potosi, Bolivia, showing the zone of sheeting (*after* Turneaure, 1960) and the approximate positions of silicification and sericitic alteration. (*From* Sillitoe, Halls, and Grant, 1975.)

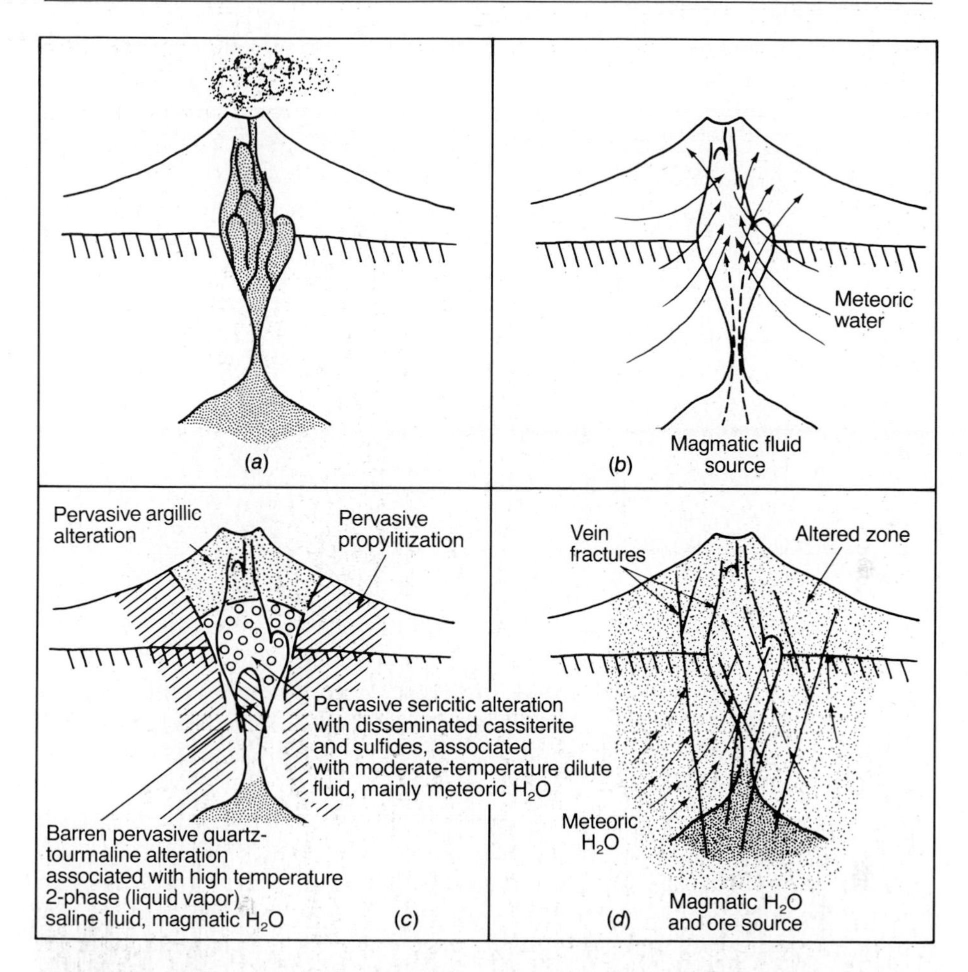

Figure 11-19. Schematic evolution of porphyry tin deposits. (*a*) A composite porphyry and breccia stock is emplaced in the vent of a felsic stratovolcano. (*b*) A magmatic-meteoric hydrothermal system centered on the high-level stock develops. (*c*) Early pervasive alteration and disseminated cassiterite mineralization are set. (*d*) As temperature and hydrothermal activity in the upper regions decline, structural readjustments within the underlying pluton and reimposition of regional tectonic stress produce a major fault system which taps a continuing supply of mineralizing fluid in the deep magmatic source. Vein ores form that crosscut the earlier altered and mineralized zones. See also Figure 11-9. (*From* Grant et al., 1980, *with propylitization added from* Sillitoe et al., 1975, Figure 7.)

some of the metals occur in through-going veins within the stock, as at Butte and Endako, although this feature is generally anomalous; and fluid inclusions (Kelly and Turneaure, 1970) are mesothermal in filling temperatures and contain high-salinity Na-Ca-Cl brines. Points of dissimilarity are: supergene enrichment is absent; potassic alteration does not occur at present shallow levels of development and exploration; and host stocks seem to narrow rather than flare with depth. Grant, Halls, Sheppard, and Avila (1980) reaffirm and add to Kelly and Turneaure's (1970) data. The latter found that pervasive Llallagua sericitization was produced early by high-temperature, 400 to 450°C, boiling saline fluids, and that early quartz-cassiterite veins followed at temperatures which fell from 450 to 300°C, with late sulfides precipitated at lower temperatures. Grant et al. also showed $^{18/16}O$ and D/H values indicating that the major mineralization directly involved isotopically magmatic fluids. Figure 11-19 presents their concept of the evolution of porphyry tin systems; compare it with Figures 11-9 and 11-17. Finally Halliday (1980) reported evidence that tin mineralization at Cornwall, England, appears geochemically, temporally, and thermally (300 to 400°C) related to porphyry dike injections, rather than to crystallization of the granites themselves. He also reports the existence of potassic alteration which is generally postgreisenization, or phyllic alteration, and may precede or accompany tin mineralization.

Skarn Deposits

Rocks intruded by igneous masses are commonly recrystallized, altered and mineralized, and replaced, especially near the intrusive contact. These changes, caused by heat and by fluids emanating from or activated by the intrusives, have been collectively labeled *igneous metamorphism*, *pyrometamorphism*, *pyrometasomatism*, and contact *metamorphism*. Although each of these terms means roughly the same thing, igneous metamorphism is less restrictive than the others, and many geologists prefer it. Pyrometamorphism refers only to thermal effects, and pyrometasomatism focuses upon replacement activity close to an igneous contact, although many deposits in metamorphic rocks are found at considerable distance from any known intrusive mass. Pyrometasomatism may also refer (Titley, 1973) to an alteration assemblage including skarn minerals that may be found at a contact, but the same minerals may also be found along veins or at considerable distance from a contact (Figure 11-20), not automatically implying high-temperature, high-pressure conditions. Igneous metamorphism can refer to all forms of alteration associated with the intrusion of igneous rocks, and thus is the preferred general term (Figure 11-21). However, because the deposits to be described involve dominant metasomatic effects in the presence of metamorphic ones, none of these terms is really apt. Einaudi (1982) has recommended that the nongenetic term *skarn* be adopted.

Skarns are best developed around the borders of small- to moderate-sized, discordant intrusive masses of intermediate composition, such as monzonites and granodiorites. Lesser metasomatic effects are also found around other intrusions, whether silicic or mafic (Edwards and Baker, 1953), although effects adjacent to mafic plutons are more thermal than hydrothermal. Skarn ore deposits adjacent to, and as extensions of, porphyry base-metal deposits are common (Figure 11-21).

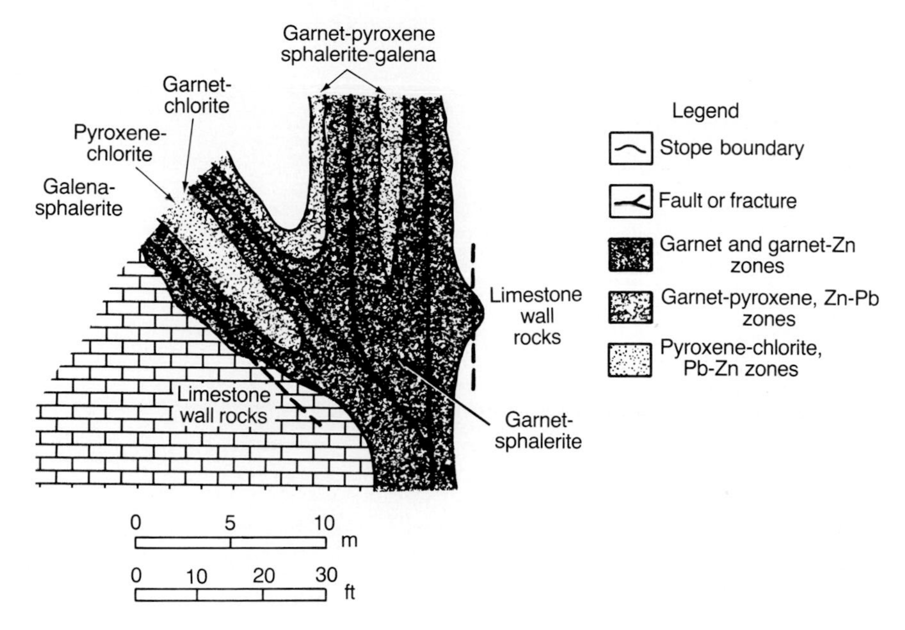

Figure 11-20. Plan map of a stope in the Linchburg mine near Magdalena, New Mexico, showing fault control of skarn minerals. The relationships demonstrate first that not all skarn assemblages are immediately contact-related, and second that skarnlike assemblages can be a hydrothermal alteration type. The stope was excavated to extract several meters of garnet-sphalerite zinc ore. (*From* Titley, 1963.)

Ore deposits of igneous metamorphic-metasomatic origin have traditionally been assumed to have formed at high temperature and pressure, usually deep within the Earth, and to have been exposed at the surface only after appreciable uplift and erosion. Recent studies (Titley, 1973; Kwak, 1978a, 1978b; Einaudi 1982) indicate that not all skarns and contact zones are formed at high temperature. Reactions in the contact zone depend upon the nature of the intruded rock, emanations produced by the intrusive, fluids activated in the wall rocks, and the overall pressure-temperature regime (Titley, 1973; Guilbert and Lowell, 1974). Relatively nonreactive rocks such as quartzites may be unchanged, even at the contact; others such as carbonates may be altered up to several kilometers from the pluton. Still others, such as andesites or shales, may be altered more than quartzites but less than carbonates. Two types of alteration are recognized: (1) recrystallization to coarser grains of the same mineral, or local isochemical rearrangement of the constituents already in the rocks to form new mineral assemblages, and (2) metasomatism, requiring the addition of materials. Most igneous metamorphic aureoles show both. Considerable discussion has arisen as to the source of metasomatic materials because the small intrusives are altered by the same processes as the bordering rocks so that the direction of material flux becomes difficult to establish. Added materials

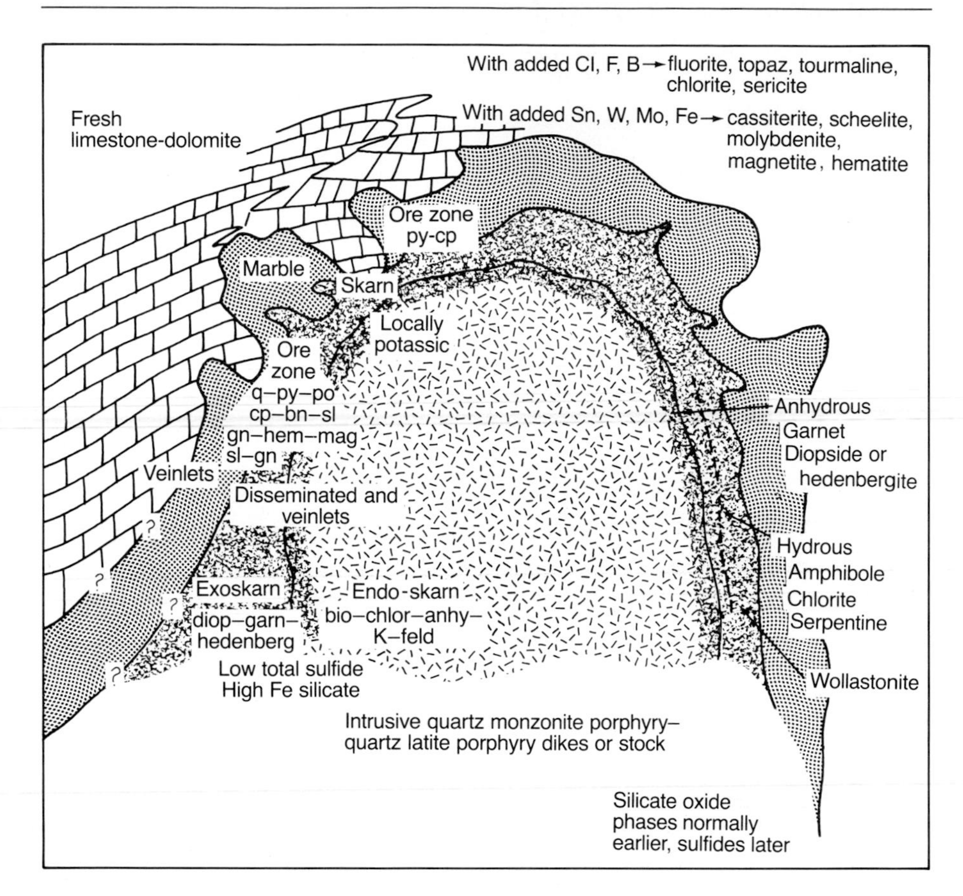

Figure 11-21. Stock with a skarn contact zone. The skarn zone is subdivided on the right side, the ore zone generalized on the left. (*After* Guilbert and Lowell, 1974.)

may have been derived locally from the intrusive mass, they may have migrated toward the stock from adjacent sediments, or they may have been carried from volumes at depth in the parent magma chamber, in which case the exposed intrusion will be only a small appendage of a large pluton, that is, a cupola in which the more mobile constituents have accumulated. If the small intrusive body solidified before the metamorphic processes were completed, the igneous rock itself will probably have been altered or endomorphosed (Lindgren, 1905; Perry, 1969).

The most striking metamorphic aureoles are developed in carbonate rocks in which narrow or broad zones of tactite or skarn are commonly developed. *Skarn* was defined originally by Swedish miners as the amphibole contact rock with which the magnetite ores of Sweden are associated. The term *tactite* was introduced by Hess (1919) to include all metasomatic products of igneous metamorphism. In present-day usage, skarn is the favored term used for a silicate rock of complex metamorphic-metasomatic mineralogy formed in carbonate rocks in a contact metamorphic aureole. Shales and most mafic igneous rocks form "hornfels," amphibolites, or metaigneous rocks; sandstones and siltstones form quartzites.

The assemblage of alteration products normally depends upon the character of the invaded rocks and the composition of the metasomatizing fluid, although where large amounts of materials have been added, for example, in ore deposits, the minerals formed may bear little relation to the invaded rock. In general the minerals of skarn zones are both diagnostic and conspicuous. In limestones, for example, skarn is characterized by lime-rich minerals such as grossularite or andradite, wollastonite, tremolite, epidote, and the Ca-Fe pyroxenes hedenbergite or salite. Dolomites develop serpentine, diopside, the humite-chondrodite group, and other minerals comparatively rich in magnesium. Skarns in carbonate hosts are also likely to contain other minerals from the garnet and scapolite families. Unusual minerals like ilvaite [$CaFe_2^{+2}Fe^{+3}(SiO_4)_2 \cdot OH$], the manganese-rich epidote piemontite [$Ca_2(Mn,Fe,Al)_3^{+3}Si_3O_{12} \cdot OH$], and members of the Ca-Mn-Fe pyroxene-pyroxenoid series like johansennite and bustamite [$(Ca,Mn,Fe)^{+2}SiO_3$] are widely but sparsely reported.

Pure carbonate rocks near contacts or skarn zones are generally coarsely marbleized, and many of them are white. Sandstones may be recrystallized to quartzite, but generally they have been considered to be unreactive and relatively unaltered except where they are argillaceous or calcareous. However Quick (1976) showed that quartzites and siltstones adjacent to porphyry copper intrusions may be significantly mineralized by open-space filling as in the Precambrian Apache Group quartzites and siltstones at Ray, Arizona. Atkinson and Einaudi (1978) reported similar effects at Bingham Canyon. In clastic sequences, metamorphism is likely to be widespread where an intrusive mass cuts shale beds which become baked and hardened or recrystallized into a dense, sugary rock known as hornfels. Most hornfels are apparently products of simple thermal metamorphism, although in places additions and subtractions especially of silica have been made. Shaly rocks in igneous metamorphic aureoles commonly display knotty or nodular textures, the small knots consisting of clusters around grains of garnet, cordierite, biotite, amphibole, or other porphyroblasts. Likewise, many volcanics are silicified and form hard, dense hornfelslike aureoles. Calcium-bearing volcanics may alter to calc-silicate suites similar to skarns, with prograde diopside, garnet, and plagioclase followed by retrograde actinolite, scapolite, epidote, and chlorite (Bookstrom, 1977).

As might be expected, thermal metamorphic products in shales are high in alumina, the characteristic minerals including biotite, ottrelite [$H_2(Fe,Mn)Al_2SiO_7$] and other micas, andalusite, sillimanite, hornblende, actinolite, garnets, scapolite, cordierite, and many others, as predicted from activity-activity diagrams in appropriate systems (Slaughter, Kerrick, and Wall, 1975; Einaudi, 1982). Minerals of skarns and hornfels are commonly distributed in more or less well-defined and regular zones around centers of igneous activity. Their composition and distribution depend not only upon the nature of fluids and rocks, but also upon the temperatures and pressures of the reaction environment.

Silica and iron are probably the most abundant elements added to igneous metamorphic zones. Silica may enter into one or more silicate minerals (silication), or it may silicify the host completely (silicification). The silica sometimes forms small inconspicuous isolated quartz crystals, and elsewhere produces massive deposits of cryptocrystalline jasperoid. Silicification is common in both carbonate rocks and shales.

In contrast to the complex silicate mineralogy of skarns, ore minerals in igneous metamorphic deposits are usually simple sulfides and oxides. The sulfides include commonly iron-rich sphalerite, galena, chalcopyrite, bornite, and in places molybdenite. A few igneous metamorphic zones are comparatively iron-free, but most contain abundant pyrite or large amounts of the oxides magnetite and hematite.

Recent chemical and petrographic studies of skarns related to ore deposits have started to show some unifying characteristics (Beane, 1980). When a stock at 800 to 1000°C invades a "cold" stratigraphic section at 50°C, conduction theory demands contact temperatures of half the difference, or about 450°C. Each half-stockwidth outward sees temperatures lower by 50%. Isotherms expand outward, with temperatures at any point increasing, so long as the stock continues to supply heat, a regimen called *prograde*. When it begins to cool, isotherms will similarly retreat, and *retrograde* conditions result. When a stock invades a pure limestone, simple recrystallization to marble, commonly with a few needles of wollastonite or tremolite, is developed. Sandstones and siltstones are similarly recrystallized to quartzite. Shales, as already described, become hornfels. Impure limestones behave more interestingly. As isotherms move out, so does an expanding front of marbleization. It is typically followed by an expanding front of anhydrous minerals, especially garnet-diopside-pyroxene. The garnets are typically *grandites*, that is *gr*ossularite ($Ca_3Al_2Si_3O_{12}$)-*an*dradite ($Ca_3Fe_2Si_3O_{12}$), resulting from a combination of calcium from limestone with kaolin and with indigenous or added iron and silica:

$$\underset{\text{limestone}}{3CaCO_3} + \underset{\text{kaolinite}}{Al_2Si_2O_5(OH)_4} + SiO_2 \rightleftarrows \underset{\text{grossularite}}{Ca_3Al_2Si_3O_{12}} + 3CO_2 + 2H_2O \tag{11-1}$$

$$3CaCO_3 + Fe_2O_3 + SiO_2 \rightleftarrows \underset{\text{andradite}}{Ca_3Fe_2Si_3O_{12}} + 3CO_2 \tag{11-2}$$

Grossularite has also been found in pure limestone in tin and tungsten skarns; aluminum must have been added by hydrothermal fluids (Einaudi et al., 1981).

Pyroxenes in skarns are typically magnesian, and are thus diopsidic. Many anhydrous skarns, then, contain andradite [(Ca,Fe^{+2} >

$Mg)_3Fe_2{}^{+3}Si_3O_{12}]$ and diopside $[Ca(Mg_{0.8}Fe_{0.2})Si_2O_6]$ in an environment involving silica, iron, magnesium, and aluminum metasomatism in a medium of high original calcium with or without magnesium. Dolomite reactant rocks of course favor higher magnesia minerals like diopside, periclase, forsterite, and even pyrope. A band of wollastonite ($CaSiO_3$) is normally found between garnet-pyroxene skarn and external marble. Intermediate volcanic rocks, with calcium and magnesium, may alter to metamorphic suites containing diopside, plagioclase, and garnet.

Generally this anhydrous thermal event is followed in both time and space by an event involving veinlets and masses of hydrous equivalents of the Ca-Mg-Fe minerals just discussed, including talc $[Mg_3Si_4O_{10}(OH)_2]$, tremolite $[Ca_2Mg_5Si_8O_{22}(OH)_2]$, epidote $[Ca_2Al_3Si_3O_{12}(OH)_2]$, clinochlore $[Mg_5(Al,Fe)_2Si_3O_{10}(OH)_8]$, calcite, and quartz. Sulfides are commonly introduced at this later stage; they are almost never found replacing limestone or marble, but rather are in veins cutting and replacing garnet-diopside skarn. Figure 11-24, part of the description of the Silver City district which follows, is a skarn paragenetic sequence diagram that shows the anhydrous–hydrous-plus-sulfide sequence. Although it is tempting to suppose that anhydrous skarns are equivalent to thermal metamorphism generated "on contact" with an intruding melt, and that superimposed hydrous skarnification and metallization is of later hydrothermal affinity, both anhydrous and hydrous assemblages are formed by hydrothermal metasomatism. In many porphyry-related skarn systems, potassic alteration in silicate igneous rocks occurs in contact with anhydrous skarnification of adjacent sediments, and the hydrous phases in the metamorphic aureoles are later and time-equivalent to phyllic overprint assemblages in the porphyries.

The terms *endoskarn* and *exoskarn* (Perry, 1969) refer to igneous metamorphic-pyrometasomatic effects on either side of the actual contact between stock and host rocks. Endoskarn involves hybridization mixing effects on the igneous side, and exoskarn refers to effects at and beyond the contact. Endoskarns normally include garnet or epidote, reflecting the high Ca-Fe environments nearby.

Metamorphic petrology and processes in ore-forming systems can be represented in terms of fronts expanding and retreating past a given point, an assemblage focus nicknamed "the Anaconda approach" because of development and adherence to that concept by Einaudi (1982), Atkinson and Einaudi (1978), and other Anaconda personnel at Carr Fork, an Anaconda property. An alternative also used is to track rock-fluid interactions through essentially isothermal activity-activity diagrams. Plotting $a_{Ca}/(a_{H^+})^2$ against $a_{Mg}/(a_{H^+})^2$ or $a_{Fe}/(a_{H^+})^2$ with variable pCO_2 permits depiction of most of the phases described. Neither approach is of course the whole answer.

Discussion and research in the early 1980s has been focusing upon the direction of mass movement of both water and contained ions under metamorphic conditions. Expositors of convection hold that thermal gradients are sufficient to drive the circulation of meteoric groundwaters such that nothing need be contributed by the stock except heat. Others suggest that water undersaturation of the melt causes Fe, Mg, Cu, Zn, and S to be drawn along with water into the melt, with deposition achieved at the contact. And classical magmatic hydrothermalists see most of the chemical readjustments as emanating from the stock outward across the endoskarn-exoskarn interface. Almost certainly the relative importance of these three alternatives varies from site to site.

Examples of skarn will clarify the foregoing discussion. A resurgence of economic interest in skarn deposits such as Carr Fork, Utah, is currently making a great deal of theoretical information available. Although many skarn deposits are highly irregular and difficult to mine (Lehman, 1980), some have been highly profitable.

Large (1972) published a quantitative discussion of both the endoskarns and the exoskarns of the Bold Head scheelite deposits on King Island, Australia. Here an aureole formed in sediments, mostly in dolomitic limestone members and in volcanics, around a stock of Devonian granodiorite. Tungsten in scheelite occurs with Ca-Fe (hedenbergite) pyroxene skarns in horizons that were originally limestones. Kwak (1978a, 1978b) showed that recrystallization and the growth of zones of hornblende, diopside, grossularite, vesuvianite, wollastonite, and scapolite at Bold Head began in a prograde regime of rising temperature at the low temperature of 150 to 200°C. At 330 to 370°C anhydrous pyrometasomatism involving andradite garnet, hedenbergite, Mo-rich scheelite, pyrrhotite, sphene, quartz, and calcite skarns formed, with the introduction of Fe, W, Mo, S, and Ti. Increased temperatures—490 to 530°C—resulted in anhydrous skarn recrystallization with increased Fe and Al and decreased Mg and Mn in pyroxenes and increased fugacity of sulfur so that Mo-rich scheelite went to molybdenite and normal scheelite, and pyrite was deposited. Temperatures then fell, with development of more sulfides and some retrograde hydrous minerals. Note the tendency here from early low-temperature "stewing" through high-temperature dehydration and oxide mineralization to retrograde rehydration and sulfide deposition.

Lehman (1980) described igneous metamorphism-pyrometasomatism in the Washington Camp district in the Patagonia Mountains of extreme south-central Arizona where 100,000 tons of high-grade lead-zinc-silver-copper ores have been mined from erratically distributed orebodies at the contact of the 64-million-year-old Patagonia Granodiorite with Paleozoic carbonate rocks. Endoskarns involved abundant K-feldspar, common secondary biotite, and sparse plagioclase and garnet rocks, nearly a potassic alteration assemblage. Skarns in the upper Paleozoic carbonates were an anhydrous assemblage of andradite-diopside-wollastonite, an intermediate vesuvianite-epidote-sulfide stage, and a later hydrous chlorite-talc-antigorite-montmorillonite event. Lehman states that sulfides occur in skarns only in garnet-diopside-bearing rocks opposite potassic alteration in the stock, even though the sulfides are texturally slightly younger. They are patchy in their distribution and thus costly to explore for, find, and mine. Dolomitic units produced enstatite-periclase-diopside-forsterite assemblages.

Gale (1965), Kinnison (1966), Himes (1973), and Langlois (1978) discussed the geology of the Mission-Pima deposit, Arizona, where pyrometasomatic skarns occur with disseminated chalcopyrite which developed at low temperature at distances as much as hundreds of meters from dikes of Laramide porphyry. Both andradite and diopside were formed, probably at temperatures as low as those reported by Kwak at King Island. Hydrous minerals are associated with the sulfides.

Central District, New Mexico The Central district in southwestern New Mexico, also known as the Santa Rita area, is 25 km east of Silver City near the town of Santa Rita. It includes iron and zinc ores of igneous-metamorphic origin, a major disseminated porphyry copper deposit, sphalerite replacements in limestone, and zinc-lead-copper veins. The district has been well studied and is unusually well known (Lindgren et al., 1910; Paige, 1916; Schmitt, 1933; Hernon and Jones, 1968; Rose and Baltosser, 1966; Nielsen, 1970; and Ahmad and Rose, 1980).

The Central district contains Precambrian granites, gneisses, schists, and hornfels, Paleozoic limestones and shales underlain by a basal unit of sandstone, and Upper Cretaceous to Tertiary clastic and igneous rocks. The igneous rocks include diorite sills, andesite intrusives and associated pyroclastics, quartz diorite sills and laccoliths, composite stocks of granodiorite and quartz monzonite, dikes of quartz monzonite and granodiorite, quartz latite and rhyolite pyroclastics, and basalt flows (Lasky and Hoagland, 1950; Hernon et al., 1953). Jones, Hernon, and Moore (1967) list a score of igneous rocks that intrude as sills, laccoliths, stocks, and dikes of Late Cretaceous–Early Tertiary age.

The Late Cretaceous–Early Tertiary metallization is related to granodiorite-quartz monzonite stocks and dikes, although minor mineralization also accompanied Late Tertiary igneous activity. There are three major stocks in the Central district: the Hanover-Fierro stock, the Santa Rita stock, and the Copper Flat stock (Figure 11-22). The Pinos Altos stock farther to the northwest is the site of a late 1970s discovery of a large skarn-porphyry copper deposit by Exxon Minerals, Inc. The regional geology is complex; the district lies on the northeast limb of a broad, shallow syncline modified by minor folds and local domes associated with the stocks and laccoliths. Faulting in the region is also complex; it spans at least six periods, ranging from before the stocks to after the last lava flows. Recurrent movement took place along many faults during several or all of these episodes. The mineralized area is in a horst, bounded on the northeast by the Mimbres Fault and on the southwest by the Silver City Fault, within which the rocks are highly faulted and mineralized and structural details are complex (Figure 11-22). The most conspicuous faults trend northeast; many are filled with dikes and some are bordered by skarns.

108°07'30"
108°00'
32° 52' 30"
32° 45'
Pinos Altos Range
Pinos Altos
TKv
TKi
BARRINGER
SANTA RITA
HORST
FAULT
ARCH
Fierro
HANOVER-FIERRO PLUTON
MIMBRES FAULT
BAYARD
Hanover
Copper Flat
MIRROR FAULT
Santa Rita
HORNET FAULT
NANCY FAULT
FORT
Central
SLATE FAULT
LOBO FAULT
SAN JOSE FAULT
GROUNDHOG FAULT
Bayard
Chino mine dump
SANTA RITA STOCK
SANTA RITA QUADRANGLE
Cobre Mountains
Lone Mountain
RIO DE ARENAS FAULT
SILVER CITY FAULT
Hurley
0 6 km
0 1 2 3 4 5 mi
NEW MEXICO
Edge of Miocene(?) and younger rocks
Discordant plutons
TKi, mafic plutons
TKv, mafic volcanic rocks
Concordant plutons
Cretaceous to Tertiary
Lower Paleozoic to Upper Mesozoic sedimentary rocks including limestones
Contact, showing dip
Dashed where approximately located
Major normal fault
Dashed where approximately located; dotted where concealed. Bar and ball on downthrown side
Minor normal faults and fractures
Anticline, approximately located
Showing trace of axial plane and bearing and plunge of axis
Syncline, approximately located
Showing trace of axial plane
Strike and dip of beds

The Hanover-Fierro stock is the largest of the plutons. It is somewhat oval, extending about 4 km in a north-south direction and averaging 1 km or so in width. The towns of Hanover and Fierro lie at the south and north ends, respectively. The Santa Rita stock is about 1.5 km southeast of Hanover; the Copper Flat stock, the smallest of the three stocks, is about 3 km southwest of Hanover. Each stock is surrounded by a metamorphic aureole, chiefly manifested by replacement skarns in carbonate rocks and shales. As shown in the illustrations, differences in the compositions of intruded rocks caused considerable local differences in the alteration products formed in the metamorphic zones. For example, higher magnesium content in the older Paleozoic rocks along the northern end of the Hanover-Fierro stock and the higher calcium content of the younger rocks near the south end of the stock caused striking differences in the metamorphic products at the two extremities. Near Fierro, the skarn zone contains abundant Mg-Fe minerals serpentine, tremolite, and magnetite, with some wollastonite; at Hanover the mineral assemblage is Ca-Fe-rich garnet, epidote, hedenbergite, tremolite, ilvaite, and sphalerite. Some metamorphic minerals are restricted to certain host rocks; epidote ($CaFe_3{}^{+3}Si_3O_{12}\cdot OH$), for example, not surprisingly selects ferruginous and aluminous rocks such as shales and argillaceous limestones, and sphalerite favors the pure limestones (Schmitt, 1939). Furthermore, the skarn aureole itself is zoned with anhydrous garnet-pyroxene ore near the intrusives; this zone is in turn separated by a band of coarse marble from an outer hydrous zone of actinolite-tremolite skarn (Lasky and Hoagland, 1950).

The deposition of the skarn ores is governed by proximity to the intrusive bodies and by local lithology and structural geology. The ore-bearing fluids appear to have ascended along fractures, and the skarn and ore zones are most extensive near them. Most of the metamorphic aureole is within about 1 km of the intrusives, but apophyses along faults and dikes extend much farther. Even in massive garnet skarns, bedding is accentuated by differences in texture and size of mineral grains. The metamorphic zones make a single continuous band between the Hanover and Santa Rita stocks (Figure 11-23).

Zinc and iron ores form separate skarn deposits in the Central district. The largest and richest zinc sulfide orebodies near Hanover—the Empire mine near Zinc Hill, for example—are in the upper crinoidal part of the Lake Valley Limestone of Mississippian age. Lesser amounts of sphalerite replace other limestones, but the Lake Valley Limestone contains a 6-meter-thick shale bed, the Parting Shale, which appears to have efficiently dammed ascending ore fluids. Much of the zinc ore in the Pewabic and

Figure 11-22. (*Facing page*) Major structural features in the Santa Rita quadrangle, New Mexico. The skarns discussed in this section in general surround the stocks in Paleozoic limey clastic and limestone sedimentary rocks (clear on the map). (*From* Jones, Hernon, and Moore, 1967.)

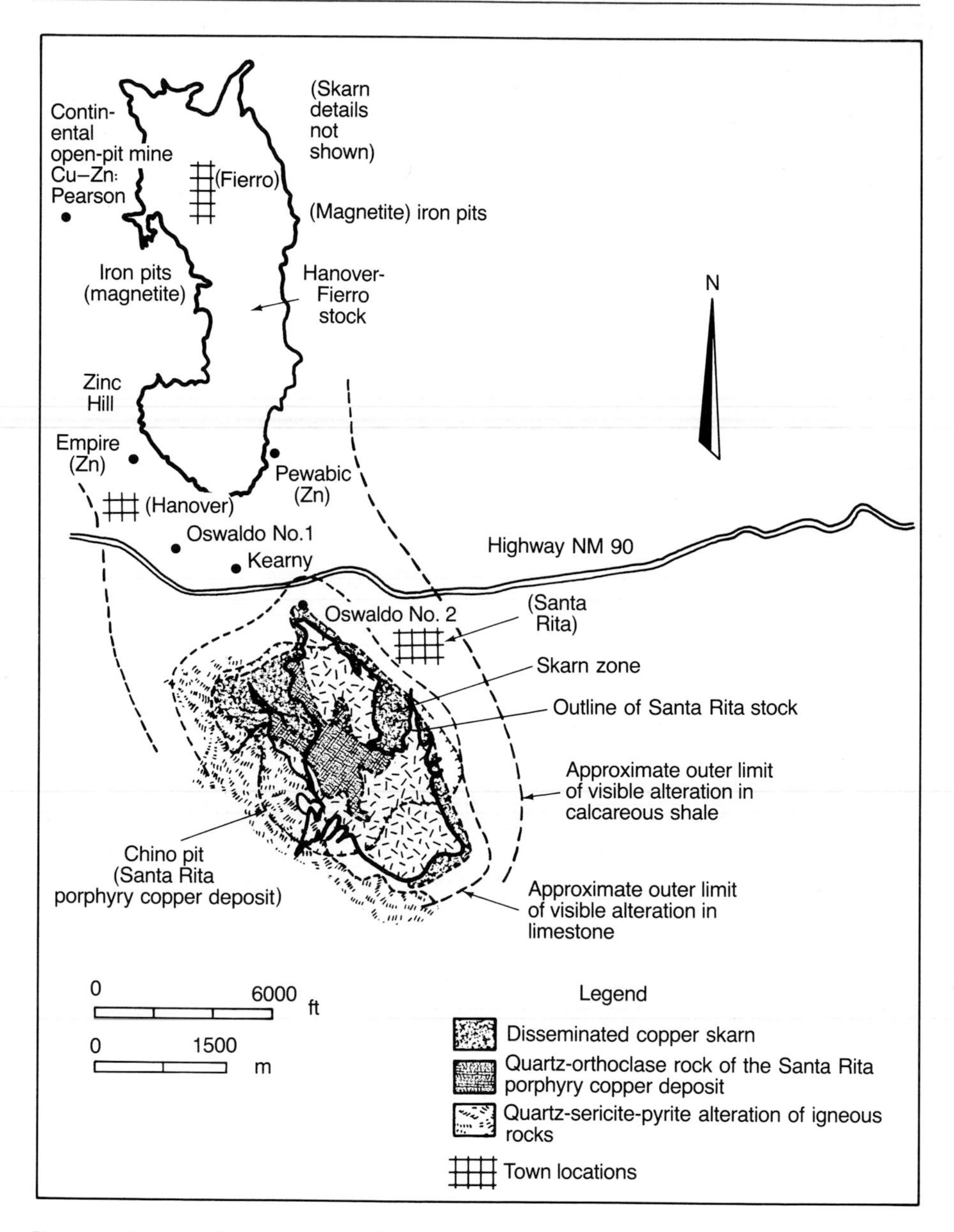

Figure 11-23. Distribution of mineralized skarn (stippled) and discernible skarnification in the limestone and shale (inner and outer dashed lines) around the Santa Rita stock, New Mexico. The mineralogies of the skarns are compared in Table 11-2 and their paragenetic sequences diagrammed in Figure 11-25. Gains and losses of components are shown in Figure 11-24. (*After* Nielsen, 1970.)

Empire mines is concentrated in the limestone immediately below the Parting Shale; locally, however, mineralization extends into the shale and, in places, above it into the limestone.

In effect, the skarn zone is localized beneath the Parting Shale, and skarn distribution controlled the deposition of sulfide ore minerals. Thus the igneous metamorphic deposits are generally in the skarn, which they replace. Lasky and Hoagland (1950) observed that the orebodies near the Copper Flat stock lie well within the garnet-pyroxene zone, but near Hanover they are at the outermost edge of this zone, against marble but still replacing skarn. Consequently most of the orebodies near Hanover have at least one marble wall. Sphalerite is commonly associated with hedenbergite rather than with the nearby garnet. In fact, most of the hedenbergite and salite there contains several percent zinc sulfide as replacements, and zinc replaces iron in the hedenbergite lattice. Atypically, zinc ores also formed in limestone at considerable distances from skarn, beyond the skarn zone; these lower temperature ores are limestone replacement deposits.

Some of the sphalerite deposits in the Central district are relatively large. A typical orebody in the Pewabic mine, near the southeast edge of the Hanover-Fierro stock (Figure 11-23), was localized along the intersection of a thrust fault and a vertical fracture zone, forming a cigar-shaped mass about 15 meters in diameter and 200 meters long (Schmitt, 1939). Its parageneses are given in Figure 11-24. An analysis of the zinc ore mined early in the history of the Hanover mine is given in Table 11-1. As in most mining districts, the earliest ore was higher grade than later shipments; although the mineralogic species have not varied, the most profitable, highest grade material was mined first. In the past, the ore brought a premium price because of its extremely low lead content. As the deposits have been worked farther and farther from the intrusive contacts, however, the amount of galena has increased, with the result that lead is now an important ore constituent.

Skarn iron ore deposits have also been mined in the Central district. In contrast to sphalerite, which occurs near the outer fringe of the metamorphic aureole, magnetite is concentrated close to the igneous contacts (Spencer and Paige, 1935). Although magnetite is widespread throughout the district, the principal commercial deposit is near Fierro, especially where the contact of the stock parallels bedding in the sedimentary rocks. The best ores are concentrated in the Lower Ordovician El Paso Limestone; lesser amounts are in the Cambrian Bliss Sandstone, itself ferruginous. The magnetite orebodies are roughly tabular, conforming to the stratification of the rocks, and alternating with layers of serpentine, wollastonite, and lesser amounts of garnet, hedenbergite, tremolite, and epidote. Small veinlets of pyrite and chalcopyrite in magnetite signal a slightly later phase of mineralization. Schmitt (1939) found that the silicates and oxides of the Pewabic ores preceded the sulfides and that ferric iron predominated during the early silicate-oxide phase, whereas ferrous iron predominated during the sulfide phase (Figure 11-21). Specular hematite, the first ore mineral, was deposited at the same time as the silicates, but most metallization followed the formation of skarn.

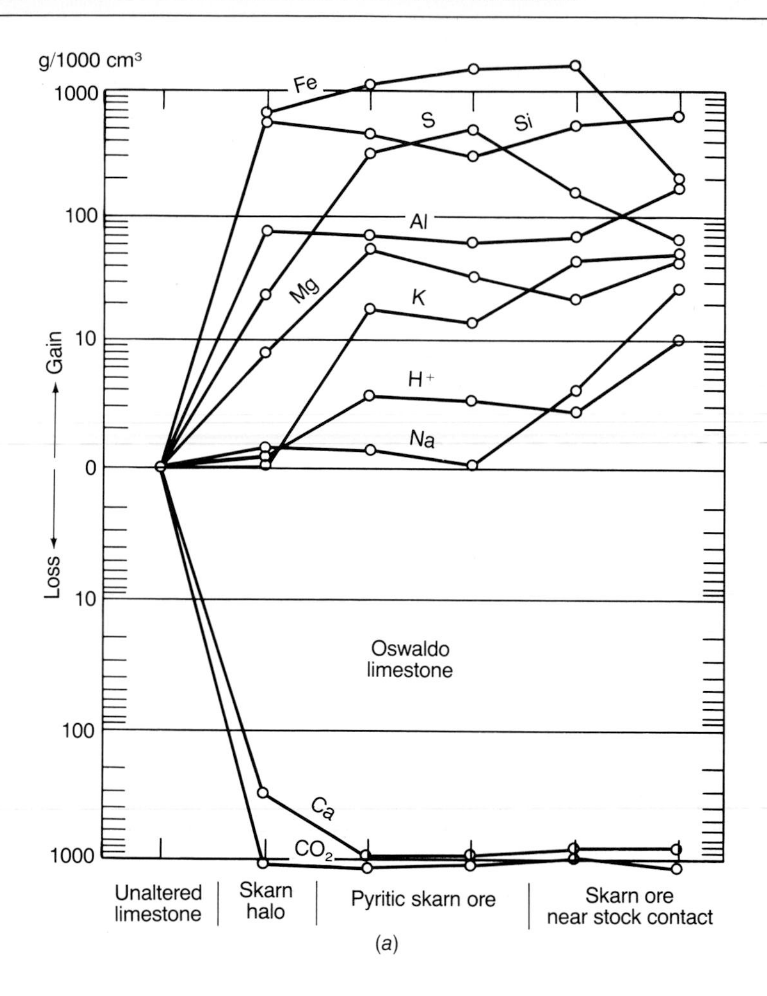

(*a*)

TABLE 11-1. Analysis of Early Zinc Ore from Hanover, New Mexico

Constituent	Percent
Zn	17.49
Pb	0.13
Fe	6.19
Mn	0.94
CaO	8.09
MgO	1.81
CO_2	3.37
S	10.19
Insoluble	49.77
Total	97.98

Source: From Spencer and Paige (1935).

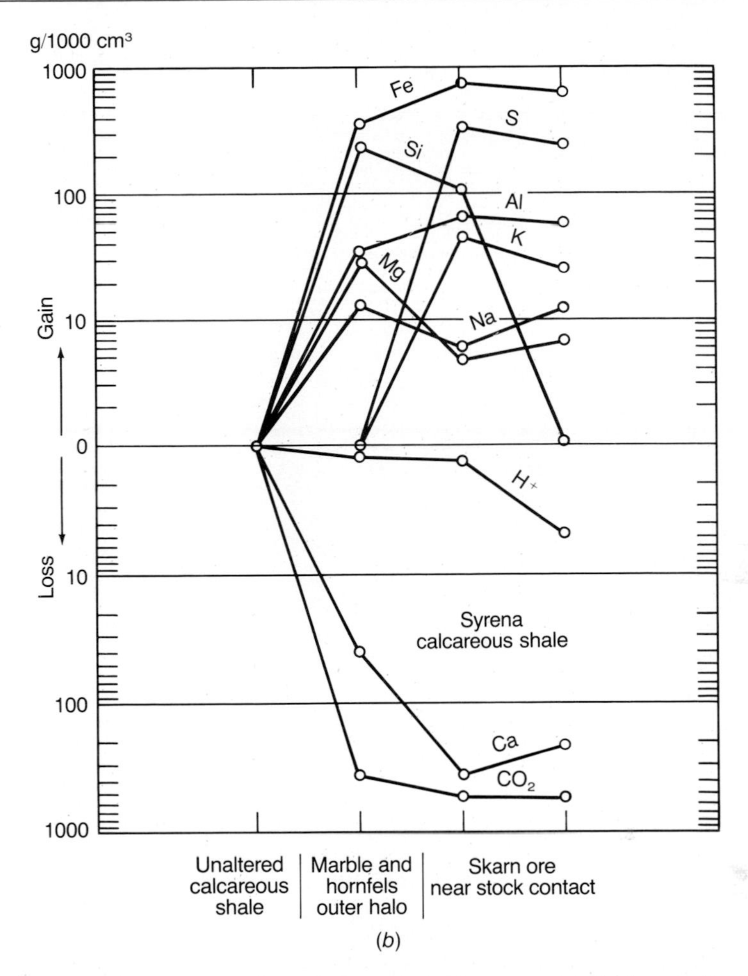

Figure 11-24. Gain-loss diagrams from fresh unaltered limestone (*a*) (*facing page*) and shale (*b*) to skarns against the Santa Rita stock. Addition of rock-forming silicate constituents and sulfur, and destruction of calcite, are evident. See Table 11-2. (*From* Nielsen, 1970.)

The skarn and ore deposits are part of a complex interaction of sediments and the stocks; the ores do not appear to have been abstracted from the immediately adjacent igneous rocks, since the latter are commonly quite fresh against mineralized skarns. East of the Hanover-Fierro stock are Paleozoic sediments striking westward toward it. Lower Paleozoic dolomitic sediments beneath Devonian Percha Shale are dominated by magnetite-serpentine. Upper Paleozoic units above the Percha are garnet-pyroxene-sphalerite-chalcopyrite skarns in limestone. It almost seems that the Percha separated two circulation and thermal systems. The adjacent Hanover stock is fresh, but only 8 km southeast intrusive rock is characterized by potassic-phyllic-propylitic alteration with significant chalcopyrite-pyrite—the Santa Rita porphyry copper—and its Paleozoic wall rocks are garnet-diopside-

TABLE 11-2. Skarn, Sulfide Mineral, and Alteration Zoning Around the Santa Rita Stock

Oswaldo Formation Limestone*				
Disseminated Copper Ore About 60 to 600 m from Stock Contact		Visible Alteration Halo Approx. 60 to 1000 m from Stock Contact		Background > 2000 m from Stock Contact
"Normal" ore, py/cp < 10/1	Pyrite-rich, ore py/cp > 10/1	Inner	Outer	
MAGNETITE (2–40 wt %)	MAGNETITE (20–75 wt %)	ANDRADITE	CALCITE	CALCITE
PYRITE (3–15 wt %)	PYRITE (5–25 wt %)	EPIDOTE	QUARTZ	QUARTZ
CHALCOPYRITE (1–3 wt %)	CHALCOPYRITE (1–3 wt %)	DIOPSIDE	TREMOLITE	kaolinite
QUARTZ	QUARTZ	quartz	kaolinite	illite
ANDRADITE	Fe-Rich ACTINOLITE	magnetite	diopside	chlorite
Fe-RICH EPIDOTE	MONTMORILLONITE	actinolite	andradite	
andradite	Fe-RICH EPIDOTE	pyrite	epidote	
biotite	CHLORITE		pyrite	
siderite	biotite		sphalerite	
orthoclase	orthoclase			
montmorillonite	siderite			
	zeolite			

Syrena Formation Calcareous Shales*

Disseminated Ore Zone up to About 600 m from Stock Contact	Visible Alteration Halo Approx. 60 to 1800 m from Stock Contact		Background > 1000 m from Stock Contact
	Inner	Outer	
QUARTZ	EPIDOTE	CALCITE	CALCITE
EPIDOTE	QUARTZ	QUARTZ	QUARTZ
PYRITE (2–20 wt %)	ANDRADITE	PLAGIOCLASE†	ILLITE
ORTHOCLASE†	DIOPSIDE	ACTINOLITE	CHLORITE
BIOTITE†	PYRITE (2–10 wt %)	epidote	kaolinite
MAGNETITE	MAGNETITE	garnet	
MONTMORILLONITE†	Fe-RICH CHLORITE†	chlorite	
CHALCOPYRITE (1–2 wt %)	plagioclase†	pyrite	
chlorite			
actinolite			
diopside			
specularite			
calcite			
andradite			

Skarn mineral zoning in Oswaldo Formation limestone (above) and Syrena Formation calcareous shales (below) around the Santa Rita stock, New Mexico. The geography of the halos is shown in Figure 11-23. Note that there are generally more silica-alumina-rich skarn minerals in the shale-derived equivalent, and more Ca-Mg-Fe phases in the carbonate skarn.

*Capital letters = major phases; lowercase letters = minor phases.
†Minerals emphasized in shaley lithology.

Source: From Nielsen (1970).

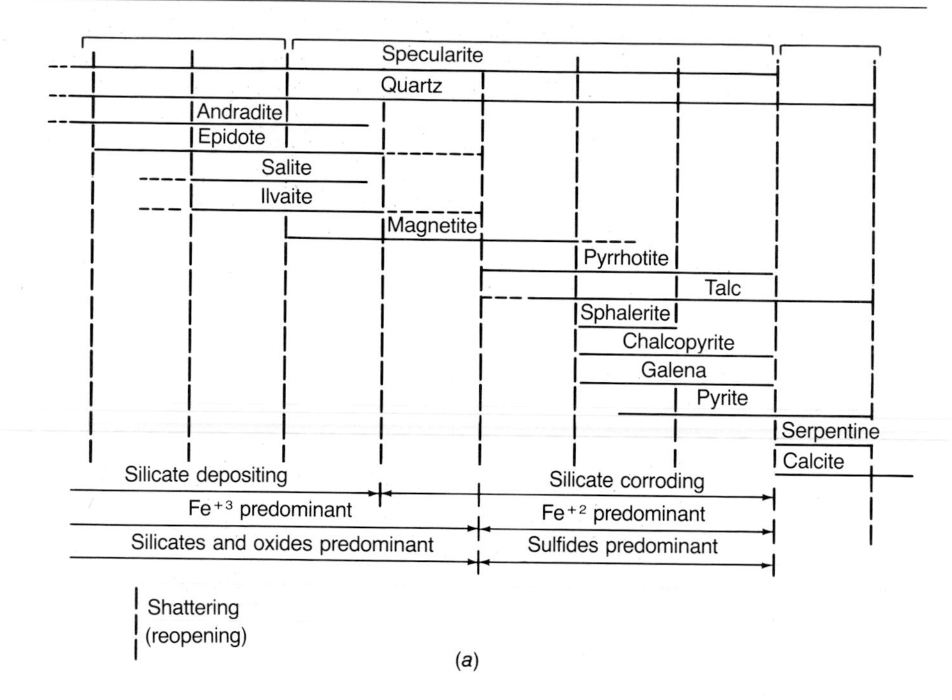

(*a*)

epidote-magnetite-pyrite skarn with chlorite-chalcopyrite-sphalerite, a classic skarn-ore deposit assemblage. Nielsen (1970) and Ahmad and Rose (1980) studied these skarns, which are well developed up to 1 km beyond the contact, and discernible as much as 2 km away (Figure 11-23 and Table 11-2). Ahmad and Rose reported (page 231) that

> . . . *intense alteration extends to 3000 ft (1 km) from the stock; traces of alteration are visible to 5000 ft (1.8 km) in the limestone. The limestone and calcareous shales near the contact are replaced by massive andradite, epidote, diopside, quartz, magnetite, actinolite, and pyrite, while farther from the contact actinolite, tremolite, quartz, and calcite become more abundant. These massive replacements are cut by veins containing quartz, magnetite, pyrite, and chalcopyrite bordered by hydrous silicates* [Fig. 11-25]. *This early garnet-diopside period with massive andradite, diopside, epidote, quartz, and magnetite, correlated with the early stage, is succeeded by a later hydrous silicate and sulfide period with actinolite, siderite, epidote, feldspar, biotite, magnetite, pyrite, and chalcopyrite. Minerals of this later period occur largely in and adjacent to veins and fractures. Based on the presence of orthoclase, biotite, and magnetite, the quartz-sulfide veins and their alteration envelopes of hydrous silicates probably formed during the later part of the early stage, although some of these veins may extend into the late stage.*

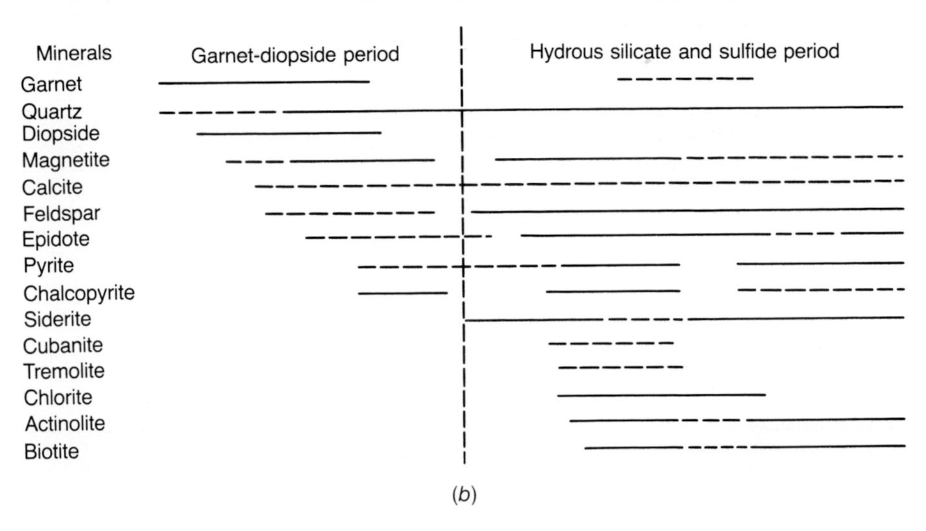

Figure 11-25. Paragenetic sequence of (*a*) (*facing page*) the Pewabic ores, Central district, New Mexico, as given by Schmitt (1939) and (*b*) the Santa Rita skarns by Ahmad and Rose (1980), with time advancing to the right. Notice that anhydrous phases are early, that "silicate corroding" is contemporaneous with sulfide deposition, and that the hydrous phases talc and serpentine were recognized as late in the sequence.

Their Figure 1, with its figure number changed to 11-25 to conform to this text, corroborates Nielsen's summary as given in Table 11-2 and the gain-loss diagrams of Figure 11-24.

In addition to progress in skarn studies in Japan (Samabe and Kuwahata, 1979; Shimazaki, 1980) and Italy (Verkaeren and Bartholomé, 1979), two other skarn deposits have recently been described. They are the Mason Valley ores near the Yerington, Nevada, porphyry copper deposit (Einaudi, 1977) and Carr Fork, at depth adjacent to the Bingham Canyon porphyry copper stock near Salt Lake City, Utah (Reid, 1978, and Atkinson and Einaudi, 1978). Three of Atkinson's and Einaudi's figures are especially apt. Figure 11-26 shows zoning around both the igneous and the metamorphic portions of the porphyry deposit centered on the Bingham Stock, in which the integrated system is seen to extend across a 6-km span. Figure 11-27 shows the same volume in cross section with actual metal ratios showing Pb/Cu expectably increasing outward; Figure 11-28 shows how lithology affects local skarn mineralogy.

▸ HYDROTHERMAL IRON DEPOSITS

Another type of deposit associated with calc-alkaline plutonism are the magnetite-rich iron deposits. They are commercially far less important as global sources of iron than the banded iron formations (Chapter 13) and ferrugi-

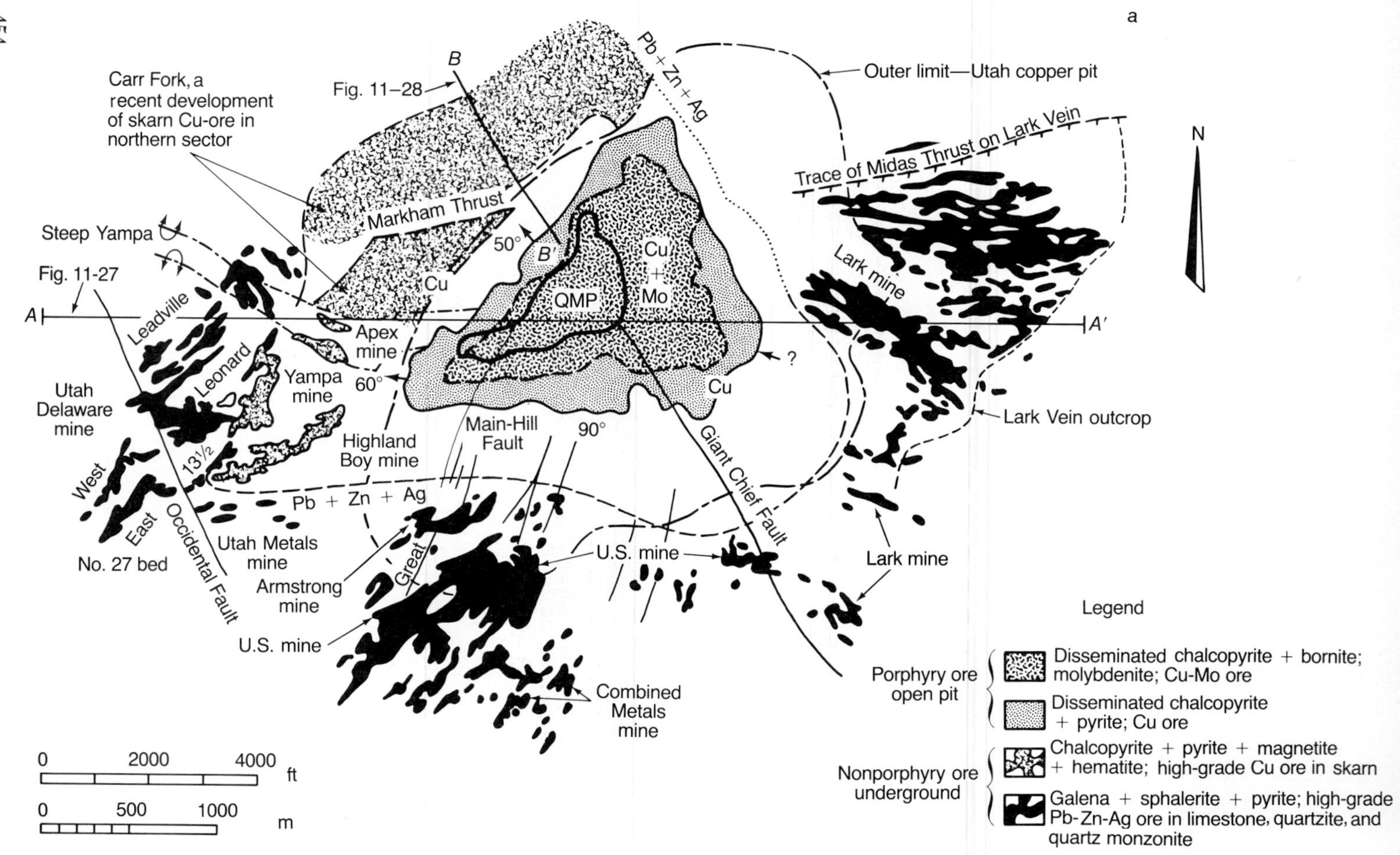
a
N
Outer limit—Utah copper pit
Pb + Zn + Ag
Trace of Midas Thrust on Lark Vein
B
Fig. 11–28
Carr Fork, a recent development of skarn Cu-ore in northern sector
Markham Thrust
Steep Yampa
Fig. 11-27
A
A′
50°
B′
Cu
Cu + Mo
QMP
Lark mine
Leadville
Apex mine
Leonard
Yampa mine
60°
?
Cu
Lark Vein outcrop
Utah Delaware mine
Main-Hill Fault
90°
Highland Boy mine
Giant Chief Fault
West
13½
Pb + Zn + Ag
East
Occidental Fault
Utah Metals mine
No. 27 bed
Great
U.S. mine
Lark mine
Armstrong mine
U.S. mine
Legend
Combined Metals mine
Porphyry ore open pit
Disseminated chalcopyrite + bornite; molybdenite; Cu-Mo ore
Disseminated chalcopyrite + pyrite; Cu ore
0
2000
4000
ft
0
500
1000
m
Nonporphyry ore underground
Chalcopyrite + pyrite + magnetite + hematite; high-grade Cu ore in skarn
Galena + sphalerite + pyrite; high-grade Pb-Zn-Ag ore in limestone, quartzite, and quartz monzonite

nous sediments (Chapter 15), but many countries without sedimentary ores depend upon them. The hydrothermal magnetite deposits associated with intermediate composition intrusive units are found at many locations around the Pacific basin, most notably in Chile (El Romeral, Las Truchas, Peña Colorada), Peru (Marcona and Acari), central America, Australia, and Japan (Park, 1972). They are manifest as magnetite skarns such as that in Triassic carbonates in contact with a Jurassic monzonite stock at Pumpkin Hollow, Nevada, and as replacements in noncarbonate wall rocks as in the Chilean iron province. That district runs for 600 km north-south and is 30 km wide (Figure 11-10); it includes at least four operating deposits (Ruiz, 1965), one of which—El Romeral—will be described. We will then examine the hydrothermal magnetite deposits of Iron Springs, Utah.

El Romeral Magnetite Deposits, Chile

The magnetite deposits of El Romeral are well known through the studies of several Chilean geologists (Ruiz, 1965; Ruiz et al., 1965, 1968), and an article by Bookstrom (1977), on which this description is largely based. The magnetite deposits are north-south elongate masses of 20 to 70% iron oxide emplaced in andesite porphyry and metasediments centered on a stock of the Romeral diorite (Figure 11-29). The main orebody (Figure 11-30) is lenticular. It is 850 meters long, 250 meters wide, and extends 400 meters in depth. Magnetite contains 72% Fe, so the high-grade portions are nearly pure magnetite. On close inspection, the magnetite is seen to contain small intergrowths of actinolite and discontinuous veinlets of apatite, calcite, chlorite, andesine, and other gangue minerals. Microscopically the ore consists of spongy intergrowths of sievelike, skeletal magnetite grains and actinolite prisms with granules of clinozoisite, sphene, chlorapatite, andesine, scapolite, chlorite, and quartz. The massive magnetite of the orebodies yields to abundant, highly visible fine veinlets on the margins; the presence of veinlets in both ore and walls suggests that small fractures were important distributors. Unmatching, irregular veinlet walls, and actinolite-altered wall rocks suggested progressive wall-rock replacement to Bookstrom (1977). The main orebody is cut by dikes, some of which are also mineralized; the diking and mineralization overlapped.

The several orebodies (Figure 11-29) are younger than the Romeral diorite and are in fault-bounded slices of preintrusive altered and metamorphosed volcanic rocks—the Jurassic La Liga andesite porphyry—and older Paleozoic-Mesozoic metasedimentary schists, phyllites, and quartz-

Figure 11-26. (*Facing page*) Metal, mine, mineral, and deposit-type zoning around the Bingham Canyon stock near Salt Lake City, Utah. Porphyry type Cu-Mo-Au ores are central and mined from the Utah Copper pit (Figure I-2). Skarn ores are next outward near the stock contact with Paleozoic sediments, and lead-zinc replacements in limestone lie farther out. QMP = quartz monzonite porphyry. See also Figures 11-27 (*A–A′*) and 11-28 (*B–B′*). (*From* Atkinson and Einaudi, 1978.)

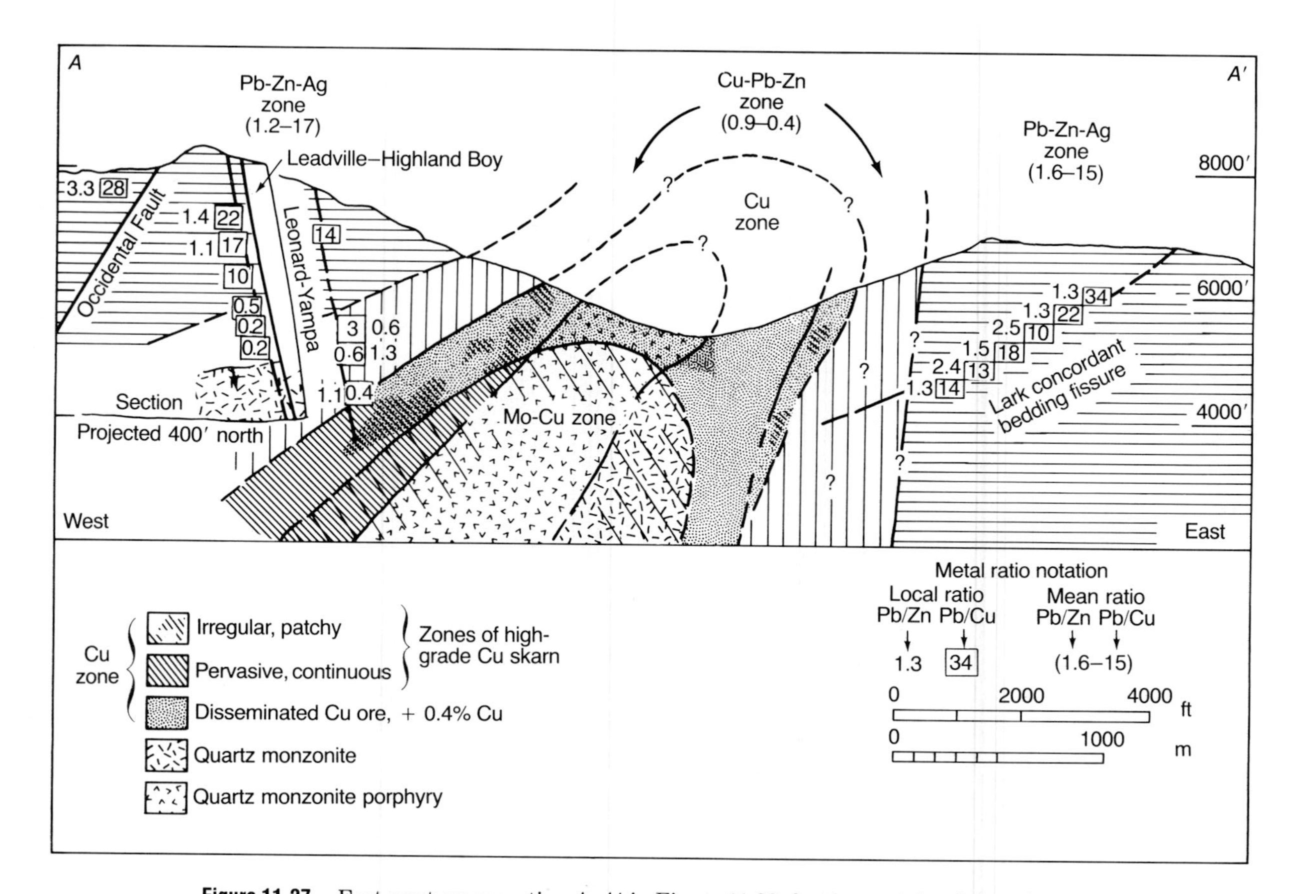

Figure 11-27. East-west cross section A–A' in Figure 11-26 showing metal and deposit-type zoning in the Bingham, Utah, district. Note the development of skarn ores on both sides of the quartz monzonitic intrusion. (*After* Atkinson and Einaudi, 1978.)

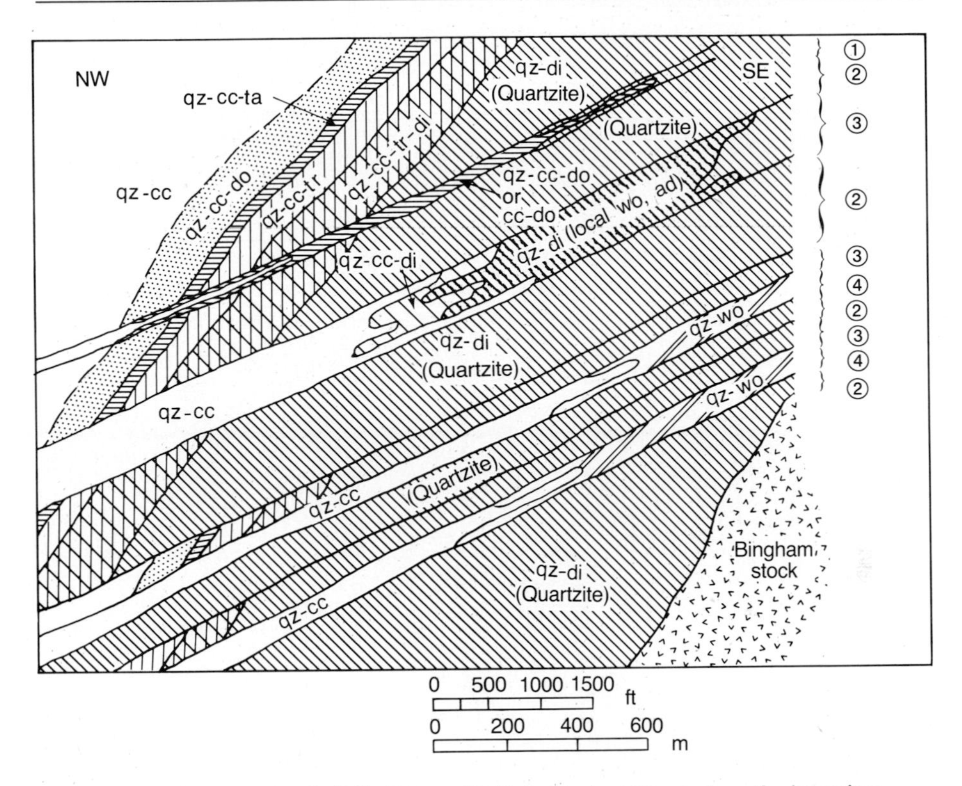

Figure 11-28. Cross section *B–B′* in Figure 11-26, looking northeast along the intrusive contact of the Bingham stock. Note the generalized 1200-meter-wide quartz-diopside zone in quartzite. The crosscutting bands represent special compositions of original sedimentary lithology: ① thin limestone; ② quartzite and calcareous quartzite; ③ interbedded calcareous siltstone, silty limestone, limestone; ④ thick cherty limestone. Note how composition and permeability telescopes and modifies metasomatic zoning. qz = quartz; di = diopside; wo = wollastonite; ad = andradite garnet; tr = tremolite; ta = talc; cc = calcite; do = dolomite. The diagram is a "time lapse" one depicting zoning before andradite-magnetite-chalcopyrite replaced the wollastonite zone in the limestones and actinolite-biotite-chalcopyrite-pyrite replaced the diopsidic quartzites. (*From* Atkinson and Einaudi, 1978.)

ites. These rocks were intruded and metamorphosed by the Cretaceous Romeral diorite, producing excellent examples of skarn-equivalent metamorphic assemblages in noncarbonate rocks. Figure 11-30 gives a larger scale view of the main orebody. It shows the metamorphism-alteration response of the metasediments and andesites as occurring around the main orebody and against the Romeral stock. Magnetite, clinozoisite, plagioclase, diopside, garnet, cordierite, sphene, and chlorapatite are developed as anhydrous phases, and actinolite, scapolite, tourmaline, chlorite, epidote, micas, and clays are hydrous ones. Pyrite and calcite are sparse, chalcopyrite is rare, and molybdenite is not reported. Bookstrom describes "dioritization," the generation of plagioclase-hornblende-diopside assemblages in quartzite and andesite porphyry. Cordierite porphyroblasts grew in phyllite

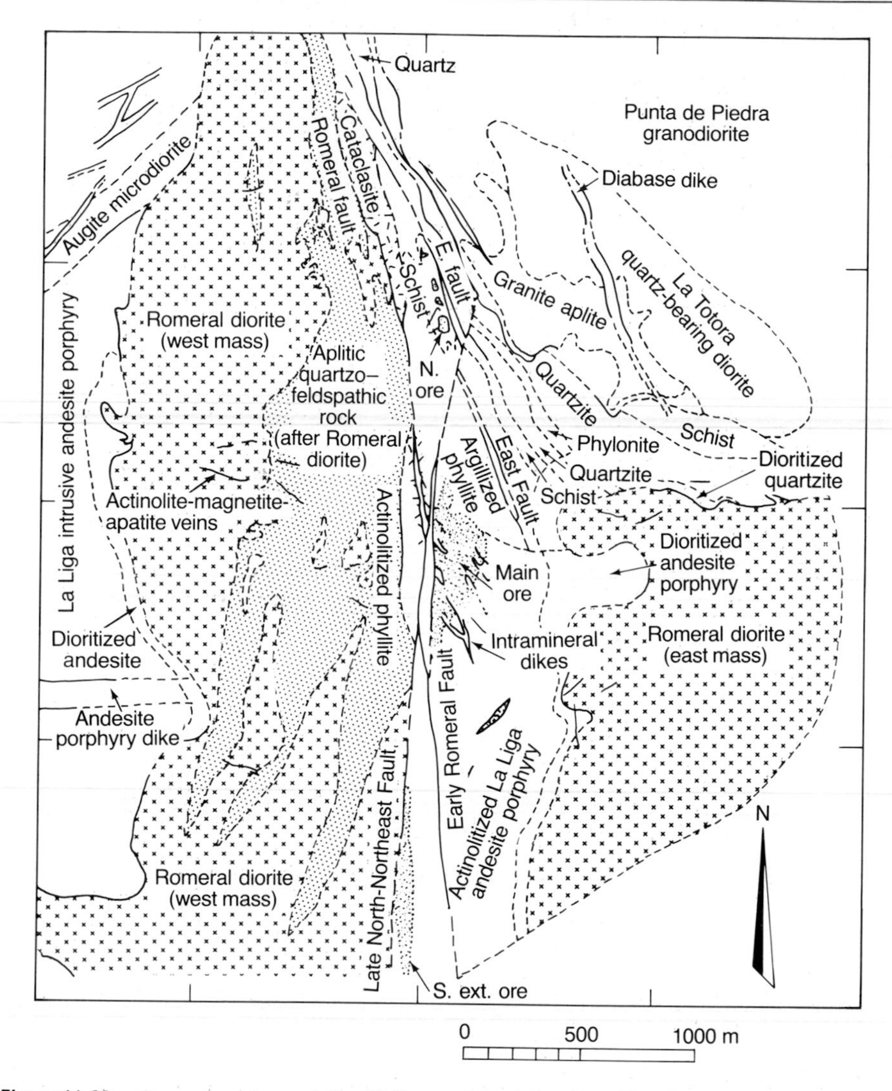

Figure 11-29. General geology of the El Romeral iron deposits. The stock (plus signs) generated hydrothermal fluids that appear to have emplaced the magnetite orebodies in older La Liga andesites and metasedimentary rocks. The aplite is younger and probably not importantly related. (*From* Bookstrom, 1977.)

and schist, and diopside and oligoclase grew in the andesites. Chloritization followed emplacement of the magnetite, as did the introduction of felsic components to form the aplitelike body shown in Figure 11-29. Bookstrom (1983) considers the calc-silicate alteration of the calcium-bearing andesites and diorites to comprise a prograde anhydrous assemblage of diopside-plagioclase-garnet and a retrograde hydrous one of actinolite-plagioclase-scapolite-epidote. The amphibolite grade metamorphic assemblages that accompanied magnetite orebody and veinlet emplacement permit assignment

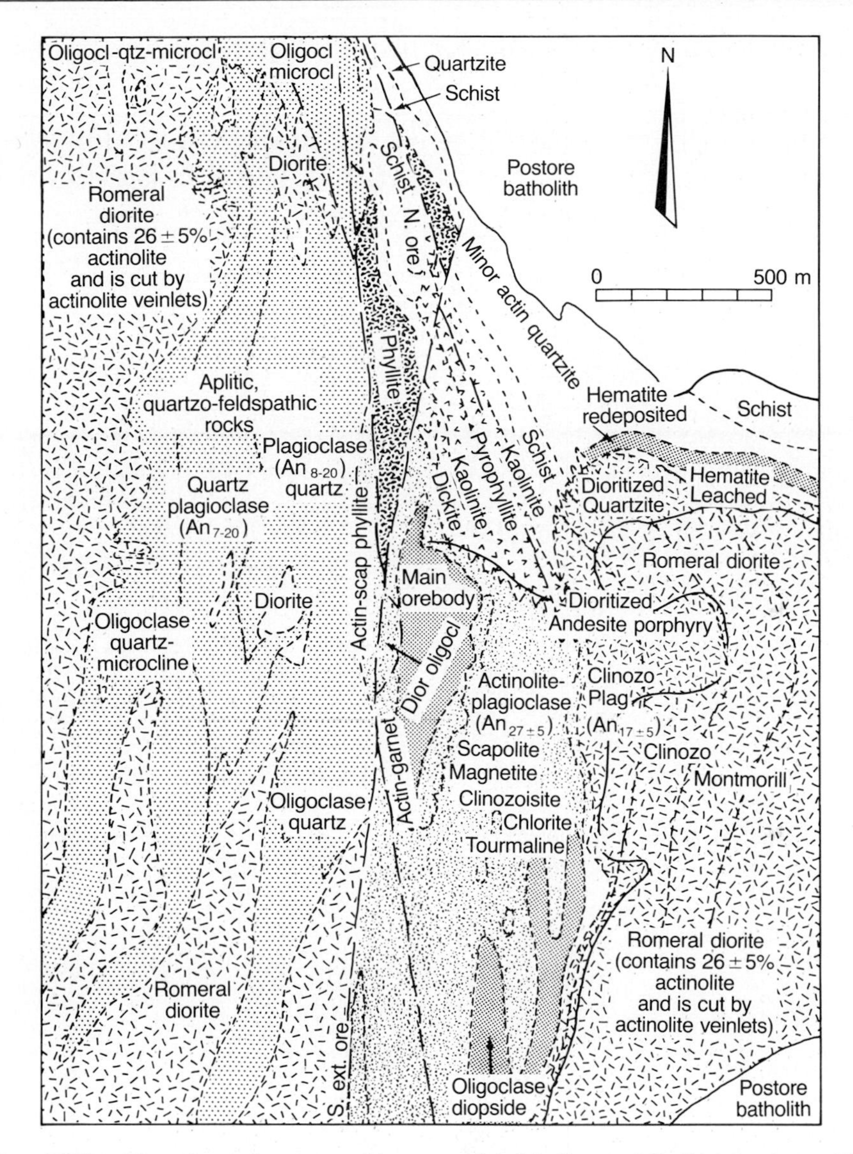

Figure 11-30. Alteration and metamorphic zones related to Romeral diorite intrusion and El Romeral magnetite orebody formation. (*From* Bookstrom, 1977.)

of pressure-temperature values for magnetite mineralization at about 475 to 550°C and 2kbar (7-km depth), followed by retrograde transition with decreasing temperatures from amphibolite-greenschist-facies to greenschist-facies conditions (Figure 18-1).

The emplacement of the magnetite was a hydrothermal metasomatic effect that accompanied metamorphism. The silicified and feldspathized portions of the diorite pluton are now iron-depleted, but the four stages of intramineral dikes appear to represent iron- and volatile-enriched residual

Romeral diorite magma, as indicated by their repeated emplacement into the main mass of Romeral diorite and the ore itself while the magnetite bodies formed.

Bookstrom (1977) concluded that the Romeral pluton invaded the Paleozoic-Mesozoic wall rocks and metamorphosed and hydrothermally mineralized them. Most of the iron came from iron- and volatile-enriched residual magmas evolved from the Romeral magma. The Romeral magma appears to have remained active at depth while magnetitic hydrothermal replacement-metamorphism-alteration proceeded above it; periodic pulses of dike-filling magma cut the ore.

This style of iron mineralization is common in general aspects around the Pacific margin and in association—as Bookstrom describes it—with subduction mechanisms and calc-alkaline plutonism. Another style of magnetite orebody development, as exemplified by Iron Springs, Utah, is far less representative of iron deposits in the quantitative sense, but it is illustrative of processes that merit description.

Iron Springs Magnetite Deposits, Utah

The magnetite deposits of the Iron Springs district west of Cedar City in southwestern Utah (Figure 11-31) furnish an example of a different style of igneous metamorphic deposition (Leith and Harder, 1908; Young, 1948, Mackin, 1947, 1968; Mackin and Ingerson, 1960). This district was mentioned in Chapter 2 in connection with deuteric leaching as a source of metals; here the manner of ore deposition is also of interest. A detailed description is presented because fewer examples of deuteric leaching and concentration are well described than are suspected to exist.

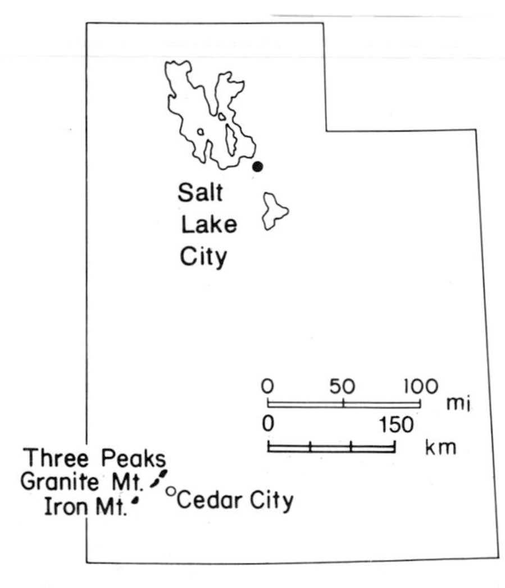

Figure 11-31. Location map of the Iron Springs district, Utah.

Ore deposition in the Iron Springs district is associated with three intrusive masses: Iron Mountain, Granite Mountain, and Three Peaks (Figure 11-31). As shown, the intrusives are aligned in a northeast direction; each is oval in plan, 5 to 8 km long, with its long axis parallel to the regional alignment. In addition to replacement bodies along the borders of the intrusives, vertical veins of magnetite traverse the plutons. Much more ore has been obtained from Iron Mountain and Granite Mountain than from Three Peaks; Granite Mountain is emphasized in subsequent diagrams. These deposits differ from the magnetite bodies at Fierro, New Mexico, just described, and from most hydrothermal magnetite deposits, in that most of the contact-related ores at Iron Springs replaced limestone directly, rather than replacing an earlier skarn. In fact, the calc-silicate minerals that are typical of skarns are virtually absent from the ore zone. The Iron Springs district is considered here because it is intimately stock-related, as are skarn-type ores.

The lowest stratigraphic unit exposed in the district is a massive, 60-to-100-meter-thick blue and gray limestone of Jurassic age known as the Homestake Formation. Overlying the Homestake Formation is the Late Jurassic Entrada shale and sandstone formation. It varies in thickness from 20 to 70 meters due to erosion, and is covered disconformably by a Cretaceous sequence of variegated freshwater limestones, shales, sandstones, and conglomerates of the Iron Springs Formation. The Iron Springs Formation was folded, then eroded and overlain unconformably by younger Cretaceous sediments. The folding and erosion left the Iron Springs Formation with a variable thickness of 300 to 2000 meters; rarely is the full 2000-meter section preserved.

The three intrusive bodies are all laccolithic; they intruded along the base of the Homestake Formation, inflated that contact, and bowed the overlying sediments upward. The three laccoliths consist principally of quartz monzonite (Mackin, 1954) or granodiorite porphyry (Mackin and Ingerson, 1960), with fine-grained chilled borders. Mackin (1947) delineated three zones in the laccoliths, the Peripheral, Interior, and Selvage Joint zones. Each intrusive is bordered by a peripheral shell of fine-grained quartz monzonite 30 to 70 meters thick (Figure 11-32), an erosion-resistant rock that forms ridges, is finer grained, less friable, and of higher magnetite content than the quartz monzonite of the interior region, and contains fresh biotite and hornblende phenocrysts. The Interior Zone of the quartz monzonite is coarser grained, and plagioclase, biotite, and hornblende have been altered to clays and chlorite. Because of this alteration, the rock weathers readily and generally forms low crumbly knobs or flat barren stretches. Unlike the Peripheral Zone, very little fresh rock crops out in the Interior Zone. Both the interior and the peripheral shells of each laccolith are jointed along radial and concentric planes that are spaced from 1 meter to a few tens of meters apart. Most radial joints strike at nearly right angles to the concentric joints; neither the radial nor the concentric joints show any altera-

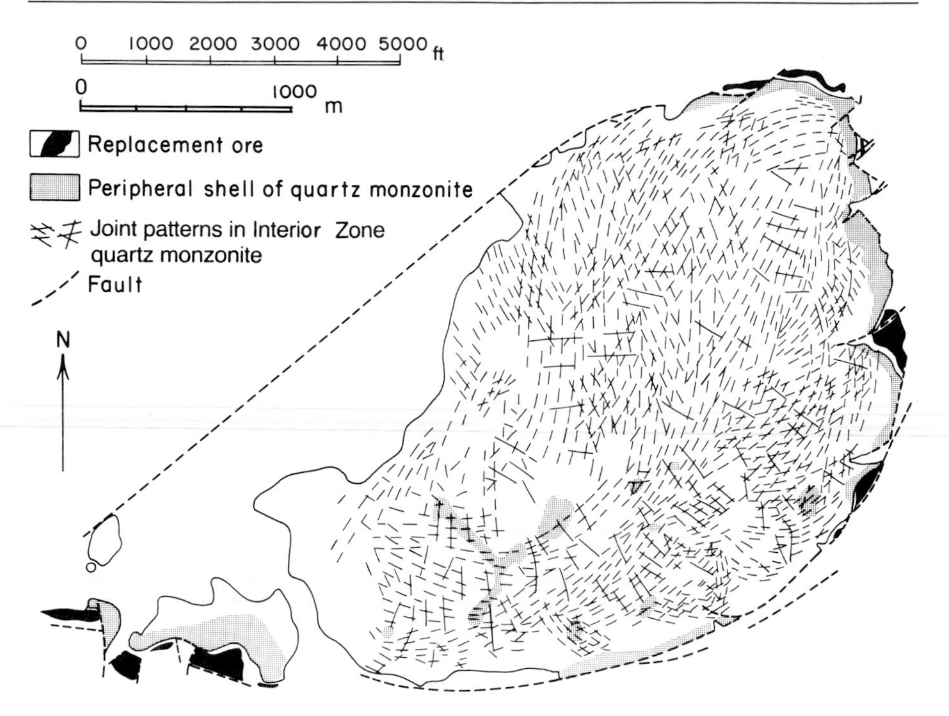

Figure 11-32. Map of quartz monzonite outcrops of the Granite Mountain laccolith, Utah, showing the joint pattern and areas of Peripheral Zone facies. Note the contiguousness of replacement orebodies and Peripheral shell facies. See Figure 11-34 for the relationships of the quartz monzonite and ore to sedimentary rocks. (*From* Mackin, 1954.)

tion selvages in this Interior Zone—they are simple, clean fractures, shown schematically in Figure 11-32.

The third zone lies between the Peripheral Zone and the Interior Zone, and has a rugged topography. Mackin called it the Zone of Selvage Joints because the rock on both sides of individual joints is bleached in a selvage, or envelope, that ranges from 1 cm to more than 10 cm in width (Figure 11-33). About 20% of the rock volume in this Zone of Selvage Joints has been bleached. Differential weathering in the Zone of Selvage Joints produces a ribbed surface because each selvage forms a hard surface crust that stands above the surrounding quartz monzonite. This bleached crust grades within 1 cm into the soft, deuterically altered quartz monzonite characteristic of the Interior Zone. Mafic minerals in the selvage zones have been bleached to light green or white chlorites and sericites. Differences between the three laccolithic zones are due to cooling hsitory and the nature and degree of deuteric alteration rather than to any differences in original rock type. The Zone of Selvage Joints is a subzone of the Interior Zone at its outer edge, and is not distinguished on the maps of Figures 11-32 and 11-34*a*. It is best shown on the cross section *A-B* in Figure 11-34*b*. Selvage joints are radial, concentric, and oblique. The radial set is a continuation of the radial joints in the Interior Zone, is normal to the intrusive contacts,

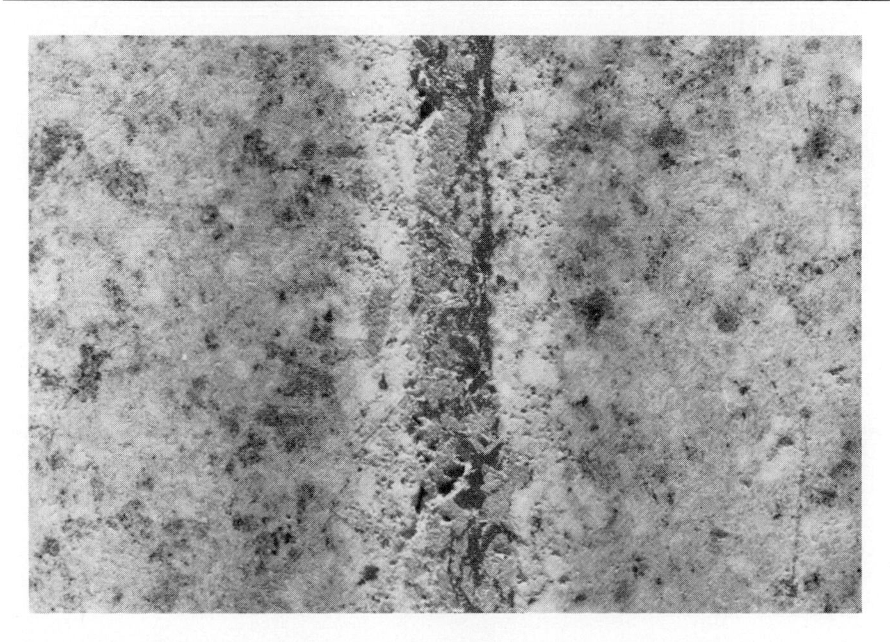

Figure 11-33. Selvage joint through the Three Peaks quartz monzonite showing the bleached, iron-depleted selvage bordering magnetite (gray) in a veinlet. Magnetite once in the quartz monzonite is interpreted to have been dissolved, to have diffused into the veinlet, and to have been reprecipitated there. 4×. (*Specimen collected by* A. S. Radtke; *photo by* W. J. Crook.)

and dips vertically. By contrast, the concentric joints parallel the igneous contacts and dip into the intrusives, perpendicular to the contacts (see Figure 11-34*b*). Some 30 to 50% of the selvage joints contain magnetite with accessory pyroxene, apatite, calcite, and hematite. Comb structures and vugs indicate that the fissures are open-space fillings rather than replacement veinlets. Thousands are less than 10 cm wide, scores are 2 meters wide, and some are over 3 meters wide, mostly in radial joints. In general the radial veins are wedge-shaped, thickening toward the margins of the laccoliths both in plan and in cross section.

Chemical analyses indicate that the iron deposited in fissures and as replacement of limestone could have been derived from quartz monzonite in the Zone of Selvage Joints. The deuterically altered Interior Zone contains 3% Fe, about as much as the fresh Peripheral Zone rocks, but the bleached selvage zones are all relatively deficient in iron. Calculations show that the volume of rock leached could have supplied more than enough iron to account for the deposits in and around laccoliths.

Detailed mapping shows that the ore deposits are associated with bulges in the laccolith roofs (Mackin, 1947, 1954). Where the intrusive was not arched strongly upward, neither joints nor the surrounding limestones were mineralized. There was apparently a critical period during crystallization of the magmas when fracturing of the roof permitted egress of vol-

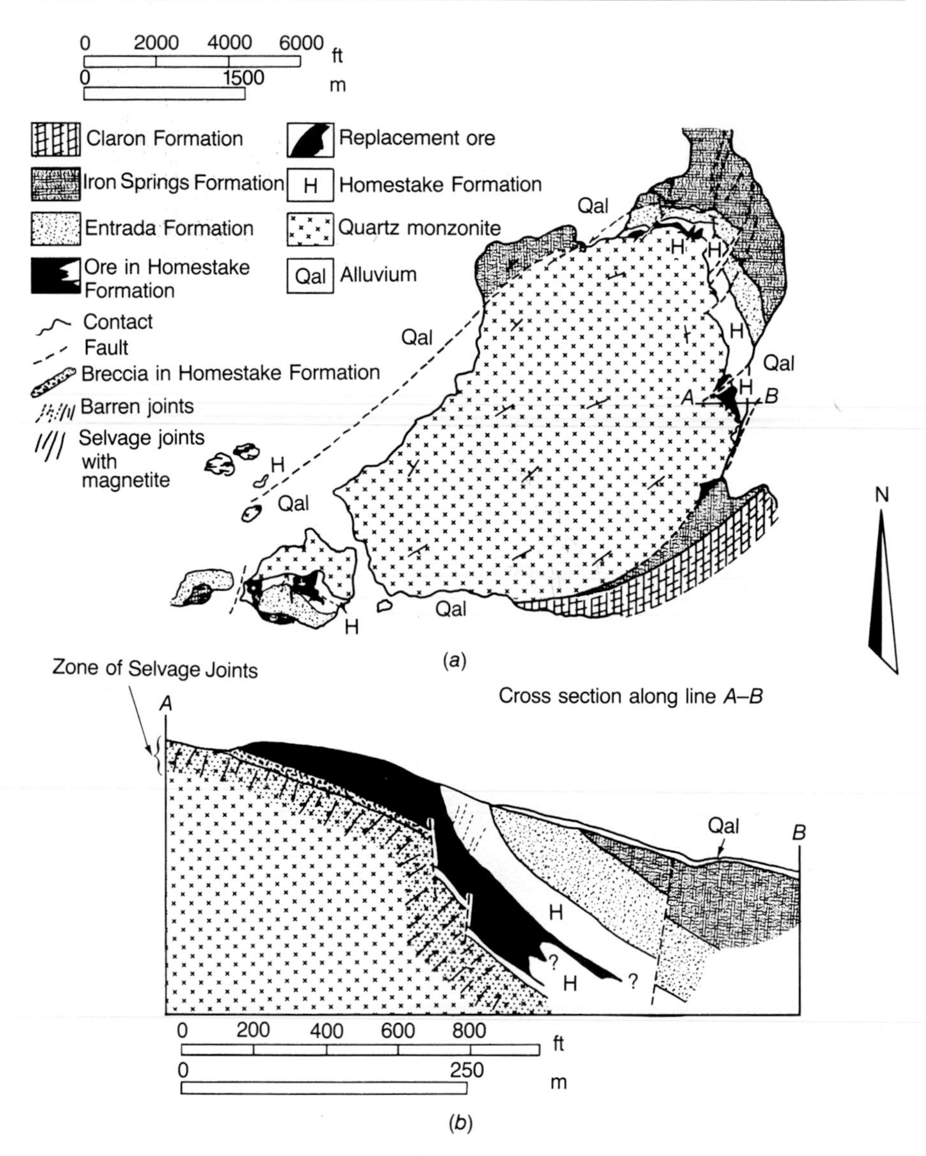

Figure 11-34. (*a*) Geologic map and (*b*) cross section of the Granite Mountain laccolith, Utah. See also Figure 11-32. (*From* Mackin, 1947, Figure 12.)

atiles. The Three Peaks laccolith has a relatively flat roof and did not provide joints for escaping mineralizers; consequently this intrusion has only a thin Zone of Selvage Joints and is unproductive. Only where late-intrusion tension cracks and breccia zones tapped the selvage zone was ore deposited beyond the laccolithic contact; it appears that iron-rich fluids capable of moving from the Zone of Selvage Joints through the Peripheral Zone left replacement deposits in the Homestake Limestone.

Replacement orebodies in the Homestake Limestone measure as much as 300 meters in dip and strike dimensions and are as much as 100 meters thick in places. They are pod-shaped bodies scattered irregularly along the igneous contact. Their outer limits conform to bedding only where the entire thickness of limestone was replaced (Figure 11-13). Calculations by Mackin (1968) indicate that for every mole of iron added, only 0.08 mol Si, 0.05 mol Mg, and 0.02 mol Al were added to the limestone, surprisingly little when compared to the millions of tons of iron that were introduced. Phosphorus in apatite and fluorine in phlogopite and apatite are troublesome in the ore because they can cause difficulties in smelting.

Ore deposition was not preceded by the development of skarn, though traces of mica, quartz, garnet, and a few lime silicates are found. Some silicification took place, but it did not localize the ore. Copper and iron sulfides, minor constituents of the replacement ores, are found farthest from the intrusive contacts.

In summary, Mackin (1947, 1954) believed that the Iron Springs ores originated locally within the laccoliths. They were leached by hydrothermal fluids from ferromagnesian minerals during late stages of igneous activity, leaving behind areas of bleaching and alteration. Before the interiors of the laccoliths were completely crystallized, but after the outer zones had solidified enough to support tension cracks, upward surges of the interior crystal mush developed bulges in the laccolith roofs, causing distension and local jointing. Wherever the peripheral shell was breached by joints or breccia zones, iron-bearing hydrothermal fluids ascended into the overlying limestones, forming replacement deposits (Figure 11-34*b*). If the ore-bearing fluids could have moved freely out of the laccoliths, a continuous blanket of iron ore might have developed. The actual occurrence of ore is sporadic, however, reflecting the fact that irregularly distributed avenues guided the mineralizers. The ore solutions by-passed a bed of siltstone and selectively replaced only limestone, implying a chemical control of ore deposition. If the limestone had not been available, the ore-bearing fluids presumably would have migrated some distance from the laccoliths before precipitating the iron, probably dispersing it in a large volume of rock.

▸ CORDILLERAN VEIN TYPE DEPOSITS

Many of the major base- and precious-metal deposits of the world occur as open-space filling or replacement polymetallic vein deposits which may not be obviously related to adjacent igneous activity. More specifically, they do not appear to be related to porphyry systems or other recognizable ore-forming environments. Many of these deposits are of major economic and natural resource significance because some of the principal ore districts of the world are included. Although the Cordillera of the Americas is the type area, it will be seen that vein systems of that type occur in compressive margin settings around the world; the term *Cordilleran Vein* is no more geographically restrictive than is *Mississippi Valley type* or any other type area label. A partial listing includes deposits like the Ontario-Tintic-Mayflower silver mines of Utah, the Magma copper deposits at Superior, Arizona, the gold-silver telluride deposits of British Columbia, eastern Washington, and the Front Range of Colorado, the silver-lead deposits of Coeur d'Alene, Idaho, and the limestone-dolomite replacement zinc mines of Gilman and the silver mines of Leadville, Colorado. The "home" of these deposits, however, is central Peru, where the deposits of Quiruvilca, San Cristóbal, Casapalca, Cerro de Pasco, Julcani, Morococha, and others have produced rich ores for decades. Gold deposits in Japan and the Philippines, base- and alloy-metal mines in Taiwan, the Peoples Republic of China, and southeast Asia, and a band of deposits reaching through the Himalayas, Alps, and Hercynides into western Europe—including some of the oldest mines in the world, like the silver mines at Laurium in Greece which funded the Athenian city states for centuries—are of this class. Interestingly and significantly, many of the deposits that might be placed in this group have recently been reclassified into other deposit types. The great veins of Butte, Montana, and Morococha, Peru, are parts of porphyry systems, and the tin and tin-silver vein systems of Bolivia have recently been described (Figures 11-18 and 11-19) as parts of porphyry tin deposits.

Sawkins (1972) first established the identity and collective characteristics of the Cordilleran Vein type. He noted that these deposits had individually been called *postmagmatic* or *magmatic hydrothermal*, terms that are too nonspecific. He cited as main features (1) association in space and time with calc-alkaline igneous activity, (2) hydrothermally transported ore components deposited at clearly epigenetic stages from solutions in fractures and fault veins, (3) ore minerals as typically open-space fillings or coprecipitates in silicate host rocks, and as replacements in carbonate hosts, (4) deposition within about 1 km of the surface, (5) well-developed metal zonation in the planes of veins, and (6) sulfide sulfur isotope ratios close to zero per mil, the juvenile-magmatic standard. He could have added that (7) they are structurally controlled and yield in exploration and development to structural geologic methods, (8) they display typically well-developed bilaterally symmetrical wall-rock alteration, be it alkali metasomatic, hydrolytic, or skarn, and (9) they normally contain the same suite of elements and ore minerals as porphyry coppers within their pronounced zonation from tin (cassiterite)-tungsten (huebnerite-wolframite)-molybdenum (molybdenite) through copper-zinc to zinc-lead-manganese-silver. Sawkins further noted that these high-level deposits are vulnerable to erosional destruction and are thus not found in deeply eroded terranes, orogenic root areas, and so forth, and that available oxygen-hydrogen data indicate that early epizonal-plutonic-magmatic hydrothermal fluids were progressively mixed with meteoric waters, but that mixing was restricted to late stages of vein deposition. Post-1972 findings generally concur (Rye and Sawkins, 1974; Sawkins, O'Neil, and Thompson, 1979); Casapalca involved no demonstrable meteoric admixture during ore-formation stages (Rye and Sawkins, 1974).

An unusual aspect of this deposit type is its conceptual vulnerability—as porphyry ore deposits become better understood, many of these deposits can be related to them and considered part of them, as at Morococha, Butte, and Llallagua. As Sawkins points out, they are indeed related to the kinds of epizonal intrusive rocks that spawn porphyries. But even if the relationship is a close one, even if deep drilling reveals a deep-seated porphyrylike environment beneath the Magma Vein system or Casapalca, the veins that would then be considered as "porphyry flares" (Figure 11-4*b*) would still deserve a separate classification status. It is thus true that much can be learned from Butte about Magma—in many respects they are similar, although most economic geologists consider Butte a porphyry system and Magma a Cordilleran Vein.

We will examine the Magma deposit at Superior, Arizona, the vein deposits at Casapalca, and the lead-silver ores of Coeur d'Alene, Idaho. In each case, ask yourself about "the porphyry connection."

Magma Mine, Arizona

The Magma mine at Superior, Arizona (Figure 11-35), has provided both vein and replacement mesothermal ores. Copper, silver, gold, and zinc have been produced from the mine (Ransome, 1912; Short and Ettlinger, 1927; Short et al., 1943; Gustafson, 1961; Hammer and Peterson, 1968). Until 1950 operations were restricted to a system of rich veins, but in 1948 the first of several extensive manto replacement orebodies was discovered

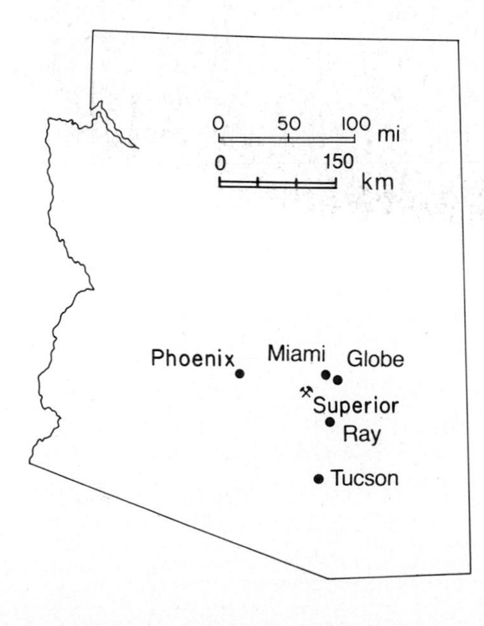

Figure 11-35. Map of Arizona showing the location of the Magma mine at Superior.

at depth along the eastward extension of the Magma Vein system. Although mining of the Magma Vein ended in 1965, the replacement ores are still being mined. As will be seen, the Cu-Fe-As-S vein contents are similar to those at Butte, Montana.

Well-exposed rocks of Precambrian and Paleozoic age, overlain to the east with angular discordance by Tertiary Whitetail conglomerates and dacitic volcanics, make up the geologic column in the Superior district (Figure 11-36). The Precambrian rocks include the Pinal Schist and a younger sequence of Apache Group shales, conglomerates, quartzites, and some limestones. The youngest Precambrian unit is the Troy Quartzite, 0 to 220 meters thick, which unconformably overlies the older Precambrian sediments. Younger Precambrian diabase sills with aggregate thickness of more than 1 km invaded the Upper Precambrian rocks, inflating the section. Intrusion of the diabase was followed by deformation and erosion that produced a widespread unconformity; in the Magma mine area the Precambrian and Paleozoic layered rocks are almost parallel, but the latter were deposited on a surface of substantial erosional relief.

The Cambrian is represented by a quartzite that is similar to the Troy and with which it has been confused. This Cambrian quartzite is 0 to 100 meters thick in the mine area (Hammer and Peterson, 1968) and is probably equivalent to the Bolsa Quartzite of southern Arizona (Krieger, 1961; Peterson, 1963). Other Paleozoic units include 110 to 140 meters of Devonian Martin Limestone, 150 meters of Mississippian Escabrosa Limestone, and up to 400 meters of Pennsylvania Horquilla and Earp (Naco) Limestone (Hammer and Peterson, 1968). Dikes of Laramide dacite porphyry and Late Cenozoic basalt are also found in the mine area. The dacite porphyry is mineralized wherever it is cut by the veins, but basalt dikes are younger than the ores and are unmineralized.

The sediments were faulted and tilted during the Laramide orogeny, or probably earlier; a few east-west faults were invaded by dacite porphyry magma, and some were to act as passageways for ascending ore fluids. Apparently faulting continued into the period of ore deposition—owing to these movements, the dacite porphyry dikes were brecciated, and repeated opening of fault zones guided ore-bearing solutions upward presumably along declining pressure-temperature gradients. The dacite melt and ore metals may both have originated in a deep plutonic mass of batholithic dimensions (Ettlinger, 1982). A quartz diorite stock near the Silver King copper-silver-zinc-productive stockwork crops out less than 3 km north of the Magma mine (Short et al., 1943; Puckett, 1970) and may be part of this batholith (Figure 11-36). Postore faulting associated with the Miocene Apache Leap dacitic volcanic activity reactivated ancient structures that trend predominantly north to northwest. Further tilting of the strata to the east may have resulted from this deformation.

The Magma Cordilleran Vein deposit is in a remarkably mineralized terrain. Figure 11-36 locates many mines in the area shown, but two facts

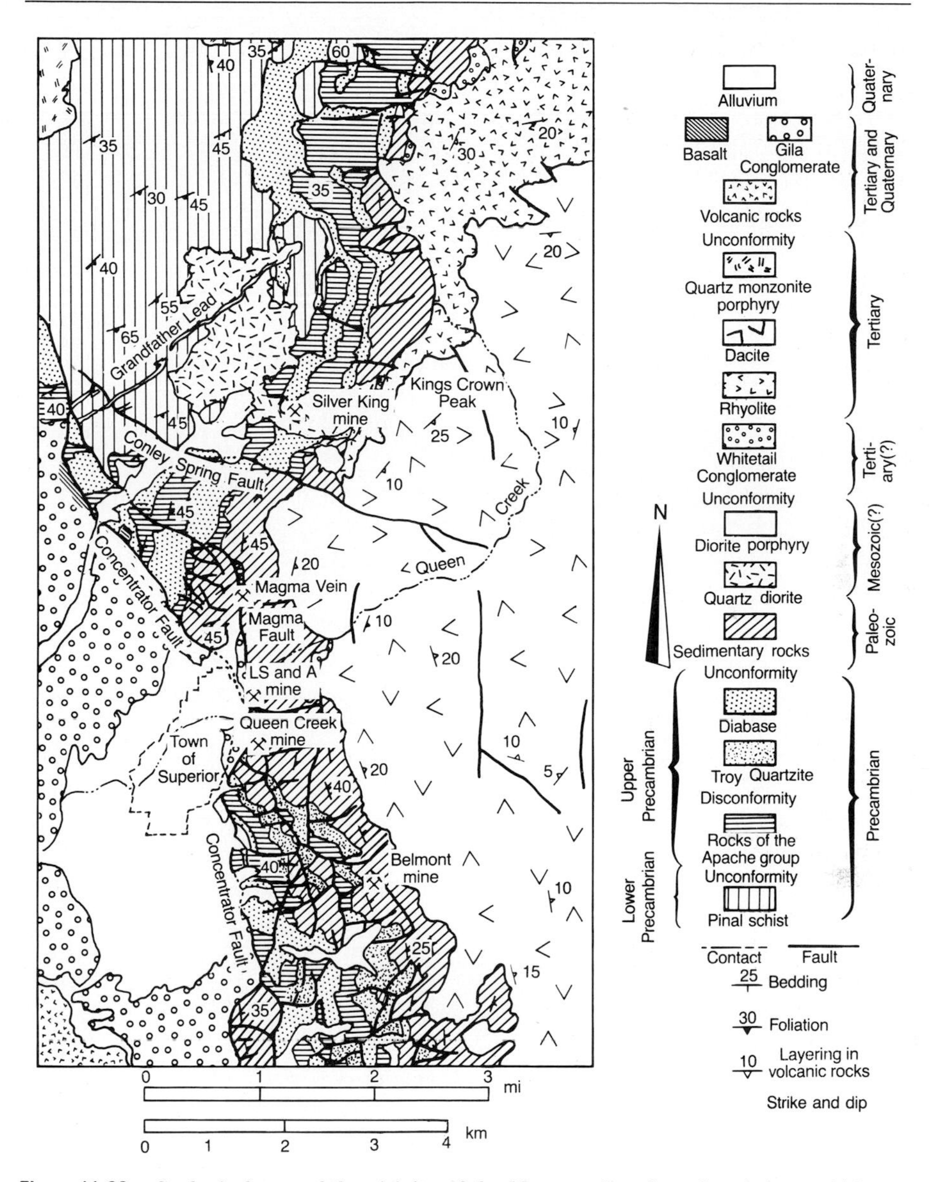

Figure 11-36. Geological map of the vicinity of the Magma mine, Superior, Arizona. (*After* Hammer and Peterson, 1968.)

are noteworthy. The first is that Apache Leap Dacite on the east side of the map and the Gila Conglomerate to the southwest are both younger and may cover many other nearby lodes. The second is that the Silver King is a hydrothermal silver deposit, that the Grandfather Lead is a sericitized vein dike, that the porphyry copper districts of Globe-Miami, Ray, and Christmas, Arizona, lie respectively only 27 km northeast, 16 km southeast, and 40 km southeast of Superior. Might there be an undetected porphyry system near, probably below, the Magma veins?

Three subparallel east-trending faults contain most of the vein ore at Magma. The bulk of the ore has been recovered from the Magma Vein, a normal fault zone with vertical displacement that varies from 120 meters at the west end to 150 meters at the east end of the mine, and with slight right-lateral strike offset (Hammer and Peterson, 1968). In surface outcrops, the fault strikes east and dips about 65°N, but below about the 900-ft level it dips 78°S. It shifts in strike from east to N80°E eastward below postmineral cover (Hammer, 1983). Most of the ore is in a large continuous shoot that plunges steeply to the west, subparallel to stratigraphy, but there are also similar oreshoots that plunge gently eastward along the vein fault (Figure 11-37). Mineralization is found along 3 km of strike and to a depth of 1.5 km. On individual levels, the main shoot extends as much as 1 km along the vein fault. The ore filled open spaces along the fault, which ranges from less than a quarter of a meter to over 20 meters in width (Wilson, 1950), but replacement of breccia and sheared material accounts for the bulk of mineralization (Wilson, 1950; Short et al., 1943). Other mineralized faults are branch veins known as the Koerner Vein south of the Magma Vein and the North Branch Vein in the deeper western part of the mine. These veins are subparallel and identical in ore mineralogy to the Magma Vein.

Early workers at Magma placed great emphasis upon the role of wall-rock composition as an ore control, a concept embodied in the older literature. As the mine was deepened and extended, and as data accumulated, it became apparent that there exists a correlation between vein thickness and grade and structurally induced permeability. The width of ore in a given vein segment depends upon net dilational opening and upon permeability developed by both crossing structures and changes in strike and dip of the vein fault. Width has little to do with specific wall-rock type or age. Wide, rich shoots are found where ore fluids migrating along the vein—generally upward and laterally—were dammed and channeled by cross faults and where flow was compounded at structural intersections. The thickest and richest orebodies are found where channelways joined and the volume of ore-fluid flow presumably increased.

The lower part of the Martin Limestone does contain a unit that is especially favorable for replacement. Above the 2550 level, this favored horizon—elsewhere called the O'Carroll bed—was almost 20 meters thick. Where the Magma Vein cut it, sphalerite-rich replacement mineralization expanded to as much as 10 meters on either side, affording some rich stopes (Wilson, 1950). Also this horizon is the host for an extensive stratiform manto of ore on the east side of the district, as shown on the right side of Figure 11-38 and in Figure 11-39. The lower Martin manto deposit, a planar body sheeting laterally from the vein as an aircraft stabilizer juts outward from the rudder, is localized along both the Magma Vein and the South

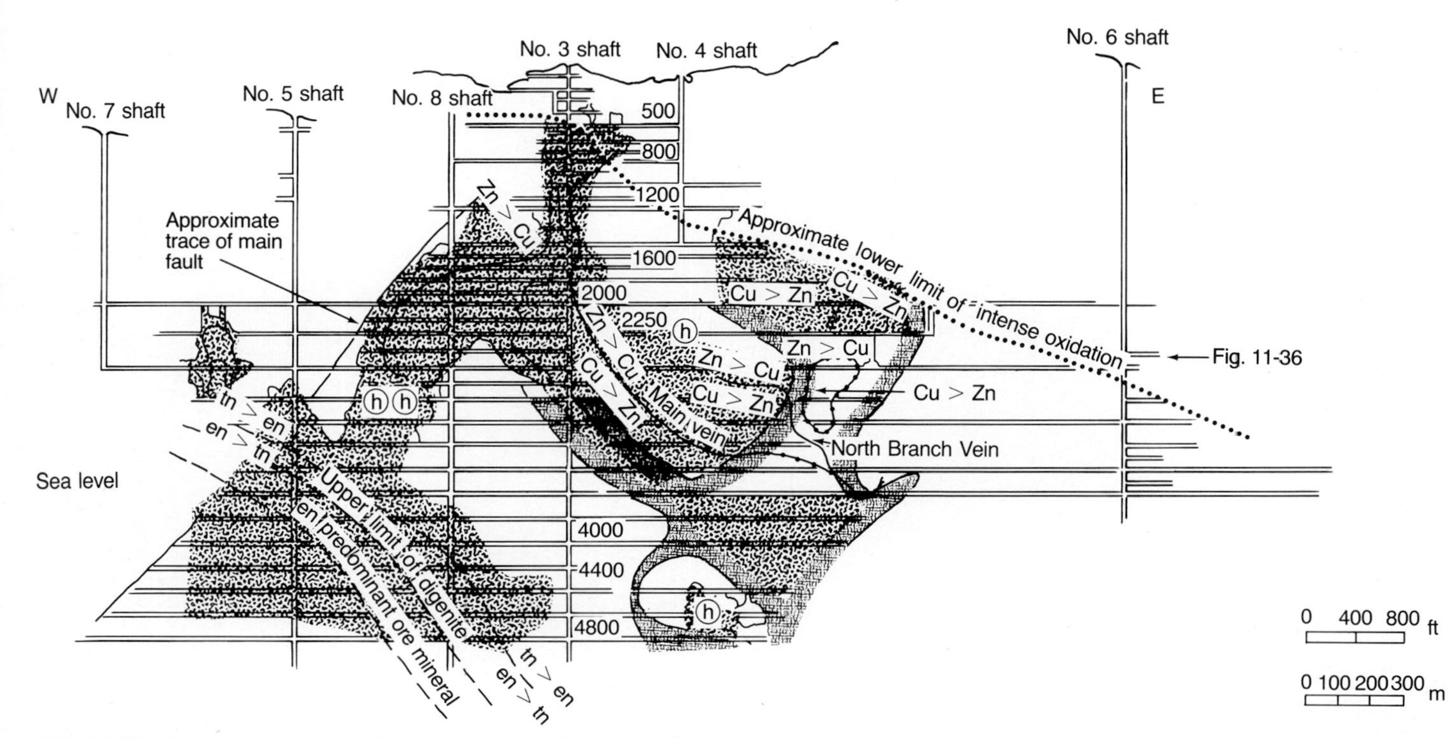

Figure 11-37. Longitudinal section facing north along the Magma Vein, Superior, Arizona, in the plane of the main ore shoot. See also Figures 11-38 and 11-39 for a three-dimensional composite view. Notice that Zn > Cu contours flank both sides of the main orebody in the vicinity of the 1400 level, indicating mineral zoning relative to the ore fluid channelway and not to bedding. Also, the arching of the Tn > En and the "upper limit of digenite" lines suggests that mineral zoning is only fortuitously related to bedding and may be related to deeper, district-scale temperature and composition gradients. (*From* Hammer and Peterson, 1968.)

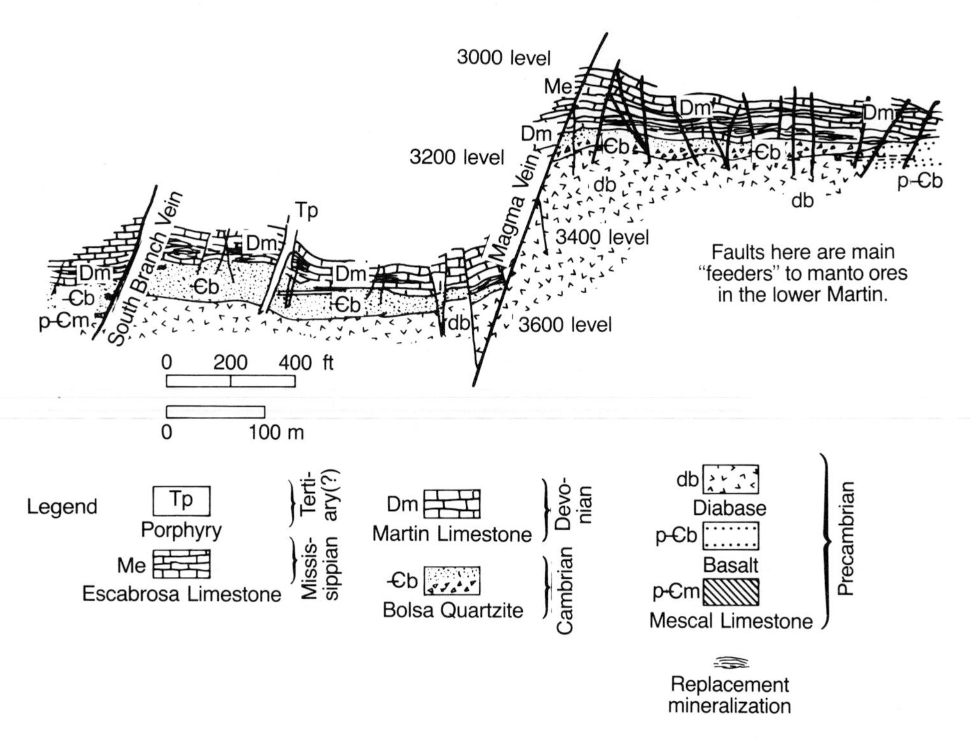

Figure 11-38. Vertical north-trending section (looking west) through the limestone-replacement deposit, Magma mine, Arizona. Some replacement ore shoots show direct connection with veins, but others are isolated. See text and Figure 11-36 for a description of the replacement geometry. This west-facing plane is perpendicular to the Main Fault Vein, as indicated. See also Figure 11-39. (*From* Hammer and Peterson, 1968.)

Branch Vein, a feeding structure which strikes east-southeast from its junction with the Magma Vein (Figure 11-39). The replaced lower Martin beds extend from about 6 meters above the base of the formation to as much as 21 meters above it, and the manto averages about 7 meters thick. Replacement in the plane of the beds extends laterally a maximum of 275 meters. Development indicates that the manto fingers out updip at about the 2000-ft level and extends more than 1.2 km easterly down the gentle dip of the bedding. In addition to the lower Martin replacement manto, significant replacement ore is also mined from the Precambrian Mescal Formation Limestone, a horizon in the upper Martin, the lower and upper parts of the Mississippian Escabrosa Limestone, and in basal units of the Pennsylvania Horquilla Formation limestone units (Hammer, 1983).

Both the veins and the replacement mantos consist of sulfide-rich ores that contain pyrite, chalcopyrite, bornite, sphalerite, tennantite, enargite, chalcocite, digenite, galena, and stromeyerite, in decreasing order. These minerals are zoned through the deposit. Marked differences in ore mineralogy define zones subparallel to the bedding (Figure 11-37); as a result, the zonal distribution of ore forms a pattern that appears to be related to

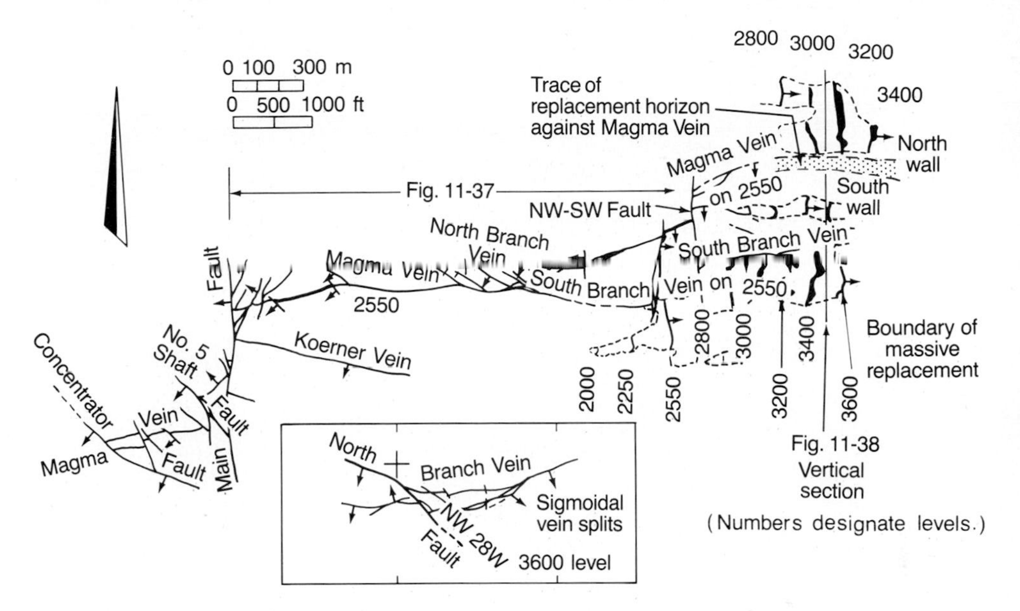

Figure 11-39. Structural plan map of the 2550-ft level and part of the 3600 level of the Magma mine, Superior, Arizona, showing its major branches, significant subsidiary fractures, offsets by major faults of the north-striking set, and the vein forms. The outline of the east-dipping limestone-replacement orebodies in the Martin Limestone is projected to the diagram. The inset shows a structural map of part of the 3600 level. Most of the veins contained ore, with thickest and richest sections at vein branchings and fault and vein intersections. The vertical section referred to at right is Figure 11-38. (*From* Hammer and Peterson, 1968.)

depth but is gently inclined to the east. However, the mineralogical zones do not conform to individual lithologic wall-rock units. In a general way, the upper eastern zone is characterized by pyrite-sphalerite-chalcopyrite; the intermediate orebodies are pyrite-chalcopyrite-bornite-chalcocite-tennantite; and the lower western zone is pyrite-chalcopyrite-enargite-bornite-chalcocite-digenite (Short and Wilson, 1938; Webster, 1958; Gustafson, 1961). In general the mantos contain specularite, pyrite, chalcopyrite, bornite, and minor amounts of chalcocite, sphalerite, quartz, magnetite, siderite, and barite. The overall zoning seems independent of the stratigraphy. If the zones were controlled by wall-rock chemistry, they would correspond to single lithologic units. In the Magma Vein, however, zone boundaries fall within stratigraphic units—in the diabase, for example—rather than along contacts. In general the mineral assemblages that are typical of the deeper enargite-digenite ores formed at higher temperatures than did the shallow sphalerite-rich ores, suggesting that the zoning was produced by pressure and temperature gradients related to a heat source at depth beneath the western or central part of the mine. The copper-iron-sulfur minerals, however, do not appear to be zoned with respect to a deep "hot source"; rather, they are zoned "locally" around channelways of greater presumed flow of hydrothermal fluids. These channelways are the centers of the stronger

Figure 11-40. Pyrite replacements of earlier hematite in manto ore from the Magma mine. The black and dark gray are quartz. The brighter white generally at upper left center is pyrite, which has replaced and inherited the acicular form of medium-gray hematite at lower center and right. 240×, plane light.

oreshoots and are marked by pods, lenses, and sheets of massive bornite-chalcocite or enargite-tennantite outward to pyrite-chalcopyrite along the outer fringes.

Paragenetic studies indicate that pyrite was the earliest sulfide deposited; some pyrite clearly replaces hematite (Figure 11-40), such that an early pyrite-hematite assemblage is indicated. Sphalerite deposition preceded that of copper minerals. Galena and stromeyerite were deposited later, probably at about the same time as the principal copper minerals enargite-chalcopyrite, bornite-tennantite, and chalcocite-digenite (Gustafson, 1961). The paragenesis progressed from relatively low sulfur pyrite-hematite and then chalcopyrite-bornite-tennantite through progressively more copper- and sulfur-rich parageneses, including chalcopyrite-bornite-chalcocite, tennantite-enargite, and enargite-tennantite-bornite-chalcocite-digenite, both with time and geometrically inward toward main channels of mineralization (Gustafson, 1961). This general trend upward and outward in a vein, and through time, is common in Cordilleran Veins (McKinstry, 1963; Sales and Meyer, 1949). The student should now reexamine Figure 6-1 and especially Figure 6-2, which was drawn with Magma in mind.

Wall-rock alteration—sericitization and silicification—accompanied the deposition of ores. It was most extensively developed in the diabase sills, but it also affected the schists and clastic sediments (Short et al., 1943). Unreplaced equivalents of carbonate beds that are host to the mantos are extensively recrystallized; under advanced attack, these units become vuggy and porous (Webster, 1958), but most carbonates are not obviously altered.

The ores were oxidized near the surface, and where pyrite was available to produce ferric sulfate and sulfuric acid, meteoric waters leached some of the copper and carried it down to form a supergene enrichment zone (Chapter 17). The oxidized zone dips eastward, roughly parallel to the sediments, further evidence that tilting was postore (Short and Ettlinger, 1927; Wilson, 1950).

Casapalca, Peru

The highest altitude, continuously occupied communities in the world are the mining towns of Peru and Bolivia. In this group, Casapalca (Figure 11-41) is only moderately high at 4000 meters, or 13,100 ft. Cerro de Pasco is at 4400 meters, Morococha ranges from 4400 to 4500 meters, and San Cristóbal is at 4840 meters, or 15,900 feet. The brief discussion of Casapalca here is based upon reports by Petersen (1965), Rye and Sawkins (1974), Wu and Petersen (1977), and field notes.

Mineralization at Casapalca is exclusively in a series of several veins. The principal C vein (Figure 11-42) strikes N45°E and dips 70 to 80° to the northwest. Other veins are subparallel, with a few splays off to the east like the P and S veins. The veins cut a series of broadly closed-folded north-south anticlines and synclines in underlying Cretaceous limestone, 1500 meters of Tertiary subaerial sediments of the Casapalca Formation, 1500 meters of andesitic tuffs, flows, and volcanoclastic sediments of the Carlos Francisco Formation, and about the same thickness of sediments and volcanics that closed out sedimentation before Miocene folding and dioritic intrusive activity. One of the major intrusions is cut by mineralized faults; age relationships between mineralization and these stocks are not clear, but the mine staff expresses confidence that a responsible stock or cupola will be found not far below present mining levels and the 12-km-long Graton drainage tunnel 1 to 1.5 km below the surface. The only skarn in the district, a tremolite-grossularite-quartz-epidote assemblage in Cretaceous limestones, occurs in that tunnel under the orebody. The veins are along shear planes responding to north-south-directed principal stresses, extend about 6 km along strike, are 2 km high, and show examples of such structural ore controls as cymoid loops and fault plane separations (Figure 11-43).

Mineralization timing, temperature, mineralogy, and zonation can best be shown in Figures 11-44 and 11-45 and Table 11-3. Figure 11-44 shows that sphalerite-galena-huebnerite-arsenopyrite ores were generally earliest, at highest mesothermal temperatures of 370°C; huebnerite [$(Fe,Mn)WO_4$] and arsenopyrite ($FeAsS$) remain centralized, but sphalerite and galena extend outward to become widely distributed through the veins. Pyrite and quartz also accompany sphalerite and galena in these early veins and vein wall linings. Late sulfides and sulfosalts are strongly zoned, as shown in Figure 11-45 and Table 11-3. Not only are mineral species themselves zoned,

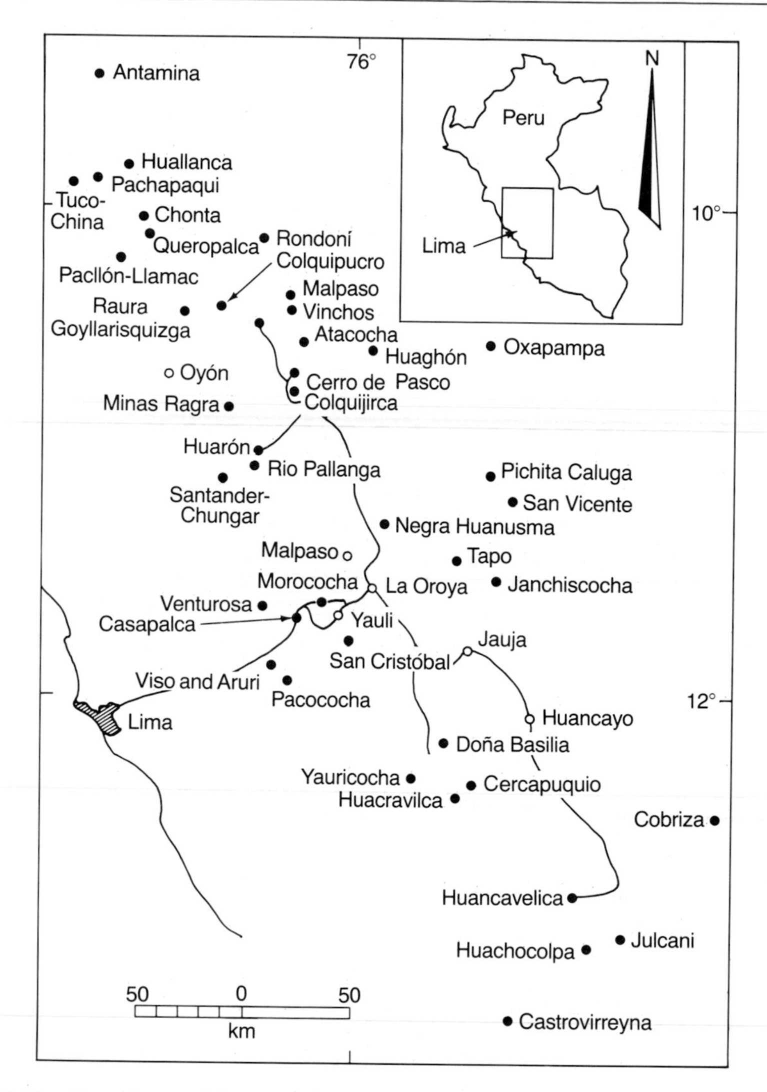

Figure 11-41. Map of central Peru showing the principal mines (black dots), most of which are Cordilleran Vein type deposits. (*From* Petersen, 1965.)

but several change in composition across zonal boundaries; for example, tennantite-tetrahedrite $[(Cu,Ag)_{12}(As,Sb)_4S_{13}]$ is higher in Cu and As centrally, and higher peripherally in Ag and Sb (Wu and Petersen, 1977). Chalcopyrite is abundant in Zone 1 (Figure 11-45); Sb-As-Bi sulfosalts take over peripherally. Alteration is extensive and of high intensity centrally, but becomes less intense and more restricted outward (Table 11-3). Solutions ranged erratically from 4 to 40% NaCl during main-stage deposition, and thermal gradients were real but gentle (Rye and Sawkins, 1974).

Average ore grade sent to the mill has been 2% Pb, 3% Zn, 0.3% Cu, and 10 oz per ton (310 ppm = 0.03%) Ag, but single assays not uncommonly

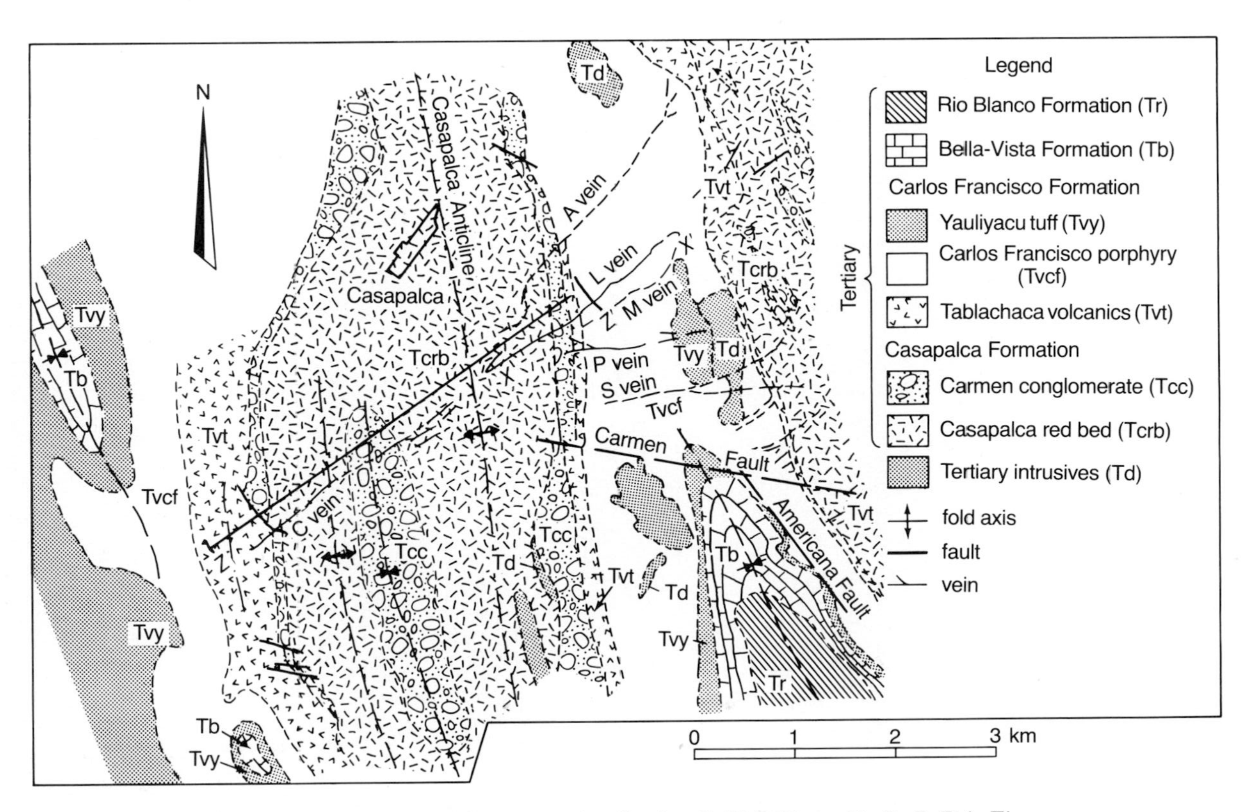

Figure 11-42. Surface geology of Casapalca district. X–X' is Figure 11-43; Z–Z' is Figure 11-45. (*From* Wu and Petersen, 1977.)

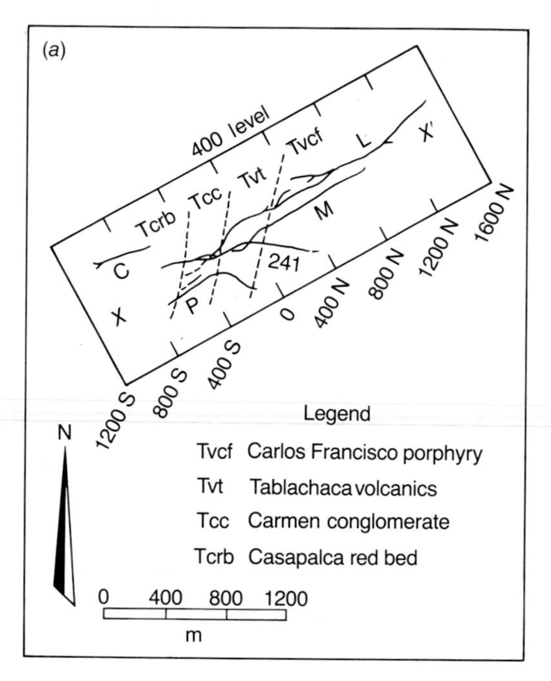

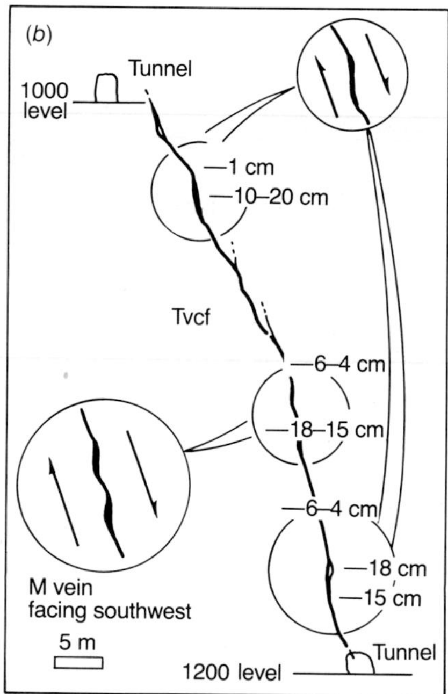

Figure 11-43. (*a*) Plan view and (*b*) cross section of veins at Casapalca selected to show classic structural effects. The plan view shows branching, anastomosing veins with several partial or complete cymoid loops; *X–X′* is located on Figure 11-32. The cross section shows by sketches and with measurements how throw develops thick ore sections in irregularly planar open-space-filled veins. (*After* Wu and Petersen, 1977.)

reach highs of 15 to 20% Pb or Zn, 5% Cu, and 3100 ppm (100 oz per ton = 0.3%) Ag. The veins are not extraordinarily wide; they average about 1 meter wide, with abundant flanking veinlet zones and replacements, es-

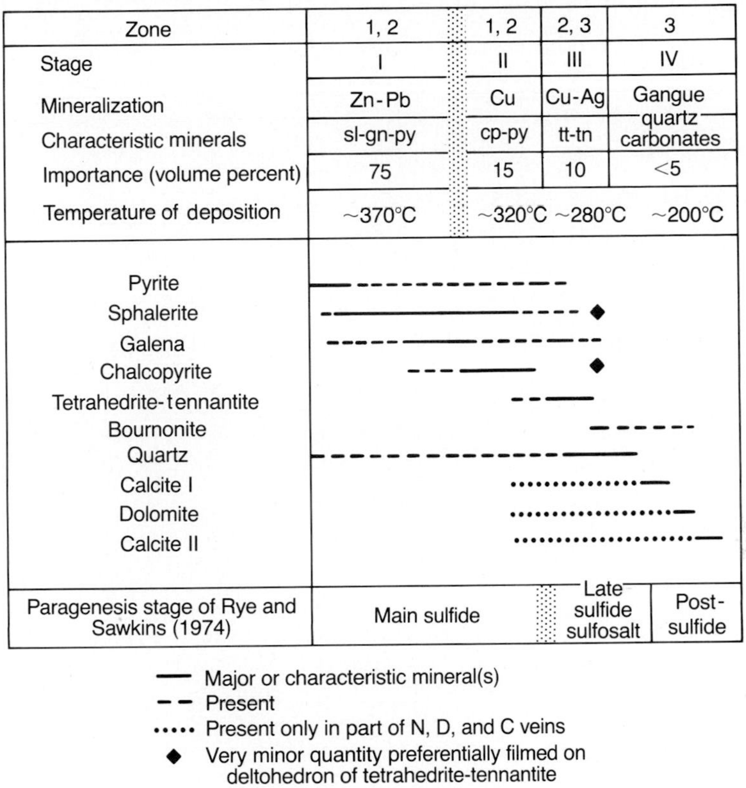

Zone	1, 2	1, 2	2, 3	3
Stage	I	II	III	IV
Mineralization	Zn-Pb	Cu	Cu-Ag	Gangue quartz
Characteristic minerals	sl-gn-py	cp-py	tt-tn	carbonates
Importance (volume percent)	75	15	10	<5
Temperature of deposition	∼370°C	∼320°C	∼280°C	∼200°C

Pyrite
Sphalerite
Galena
Chalcopyrite
Tetrahedrite-tennantite
Bournonite
Quartz
Calcite I
Dolomite
Calcite II

Paragenesis stage of Rye and Sawkins (1974)	Main sulfide	Late sulfide sulfosalt	Post-sulfide

— Major or characteristic mineral(s)
-- Present
..... Present only in part of N, D, and C veins
◆ Very minor quantity preferentially filmed on deltohedron of tetrahedrite-tennantite

Figure 11-44. General paragenetic sequence in veins at Casapalca, especially the northern C vein. sl = sphalerite; gn = galena; py = pyrite; cp = chalcopyrite; tt-tn = tetrahedrite-tennantite. Zones refer to Figure 11-45. Temperature data are from Rye and Sawkins (1974).

pecially in sedimentary hosts. A typical assay on the 1700 level (Figure 11-45) showed 15% Zn, 5% Pb, 0.5% Cu, and 10 to 20 oz Ag in the vein, with an adjacent 2 meters of disseminations and veinlets in wall rock which contain 6% Zn, 3% Pb, 0.4% Cu, and 60 ppm (2 oz) Ag. Bedding-plane control of replacement so well developed at Magma is minimal here, but is visible at the meter scale. Reserves at 2500 tons mined per day are in sight for about 20 years. Tungsten is not recovered, but an early, deep, huebnerite stage is profitably mined at nearby San Cristóbal and elsewhere.

Coeur D'Alene, Idaho

The Coeur d'Alene district in northern Idaho (Figure 11-46), one of the largest silver districts in the world, is an excellent example of Cordilleran Vein deposits. It has produced nearly 3 billion dollars worth of ore since it was discovered in 1879, has been part of the romantic and turbulent mining and labor movement history of the northwestern United States, and

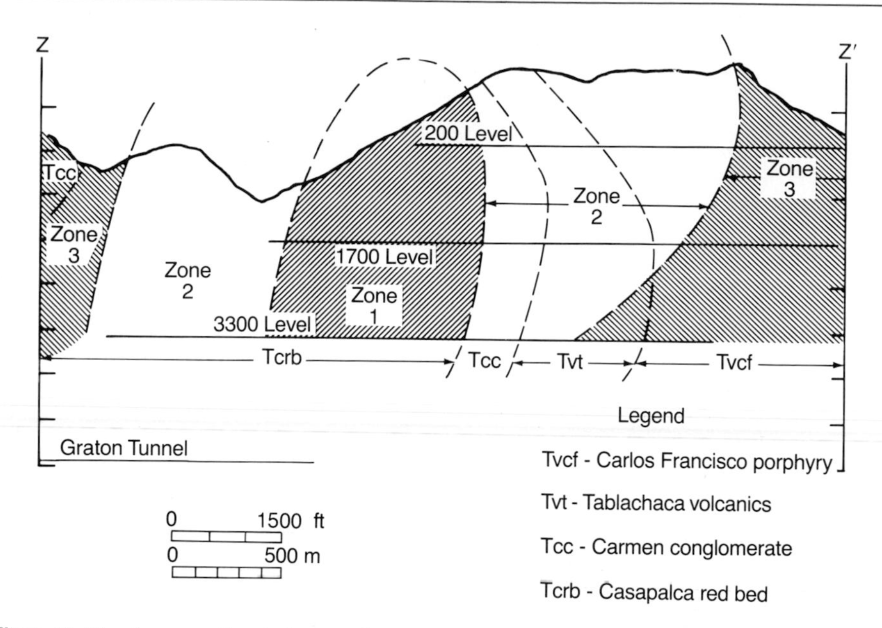

Figure 11-45. Long section facing northwest showing zoning in the C and L veins at Casapalca, Peru. Wall-rock alteration, ore mineral, and ore metal characteristics are given in Table 11-3; more information, including temperatures of deposition, is given in Figure 11-44. *Z–Z′* is located on Figure 11-42. (*After* Wu and Petersen, 1977.)

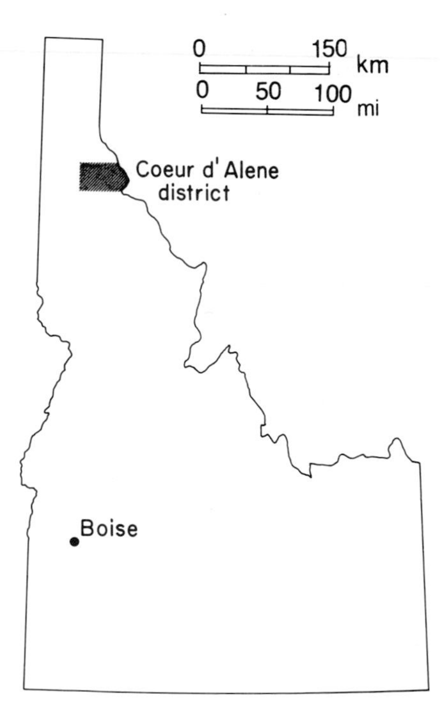

Figure 11-46. Map of Idaho showing the location of the Coeur d'Alene district.

TABLE 11-3. Wall-Rock Alteration, Ore-Mineral, and Ore-Metal Zoning at Casapalca, Peru

	Wall-Rock Alteration: Silicification, Sericitization, Pyritization, and Propylitization	Characteristic Minerals Other Than Pyrite, Sphalerite, and Galena	Metal Distribution
Zone 1	Most intense as observed in both surface and underground; rock altered to dense, fine-grained silica-rich rock with disseminated minute pyrite cubes and epidote nodules; alteration extends several hundred meters from the vein as sericitization-silicification.	Chalcopyrite, As-rich tetrahedrite, central huebnerite-arsenopyrite, high sphalerite.	Cu, W, As, Zn > Pb
Zone 2	Less intense alteration and bleaching; width varies from 30 meters or more adjacent to Zone 1 to a few centimeters or less in Zone 2; sericitization less extensive, minor carbonates; pyritohedral pyrite.	Sb-rich tetrahedrite, bournonite, chalcopyrite with sphalerite, increasing galena, tetrahedrite.	Pb > Zn, Ag, Cu, Sb
Zone 3	Maximum alteration zone is less than a few centimeters wide, generally propylitization, chlorite-epidote-carbonates.	Stibnite, realgar, orpiment, jamesonite.	Spotty Ag Sb, Pb, As
Deep Graton Tunnel	Skarn of tremolite-quartz-grossularite-epidote.		

See cross section, Figure 11-45.
Source: Revised from Wu and Petersen (1977).

continues to be a major producer and employer. The area has yielded 43,000 metric tons—1 billion oz—of silver, 10 million metric tons of lead, 3 million tons of zinc, and 200,000 tons of copper. In the past, most of the ore was lead-zinc-silver, but the recent developments of high-tetrahedrite deposits in the southern part of the district have added copper and antimony to the metals produced in quantity. Several good descriptions of the deposits are available (Ransome and Calkins, 1908; Umpleby and Jones, 1923; McKinstry and Svendsen, 1942; Sorenson, 1947, 1951; Shenon, 1948; Fryklund, 1964; Hobbs et al., 1965; Hobbs and Fryklund, 1968; Reid et al., 1975; Gott and Cathrall, 1980).

The Coeur d'Alene mines extend over an area spanning 50 km from east to west and 20 km from north to south (Figure 11-47). At the northern extreme the mines were worked for gold in the early days. Near the center the principal commodities recovered are lead and zinc, the southern edge of the district contains copper-silver ore, and the Lucky Friday mine in the east produces silver. Coeur d'Alene is thus an example of district zoning.

Three groups of rocks underlie the district: The Upper Precambrian sediments of the Belt Supergroup, Cretaceous igneous rocks that intrude the Belt, and unconsolidated Tertiary and Quaternary clastics. The Belt rocks consist principally of quartzites, argillites, and calcareous rocks, reaching a maximum thickness of 6 km. Abundant mud cracks, ripple marks, crossbedding, and stromatolites indicate that most of this sequence is of shallow-water origin. The Belt has been divided into six lithologic units. They are, in chronologic order, the Prichard Formation, 3.6 km of black argillite and argillaceous quartzite; the Burke Formation, a 550-to-700 meter sequence of gray to white quartzite, some of which is argillaceous; the Revett Formation, 640 to 1000 meters of massive gray to white quartzite with some interbedded argillaceous quartzite; the St. Regis Formation of thin-bedded purple to purplish gray argillite and quartzite; the Wallace Formation, 1300 to 2000 meters of gray black calcareous argillite and white to gray calcareous quartzite; and the Striped Peak Formation, which comprises 700 meters of purple, pink, and green quartzite and calcareous argillite, and apparently contains no ore (Sorenson, 1951).

These strata were folded, faulted, and trivially metamorphosed in the Proterozoic and again at the close of the Mesozoic era. During this latter Laramide orogeny, monzonite stocks and accompanying lamprophyre dikes and sills (McKinstry and Svendsen, 1942) were intruded after folding. The stocks, which vary from diorite and quartz monzonite to syenite, appear to be satellitic to the Idaho batholith (Umpleby and Jones, 1923). Diabase sills in the district predate the monzonite and folding and are probably Precambrian (Reid et al., 1981), but the lamprophyre dikes are younger than the monzonite. In spite of their age and the degree of folding, the Belt rocks have undergone surprisingly little metamorphism. The argillites have a slaty cleavage, and the arenites are mostly quartzitic, but in general the metamorphic effects are minor. By contrast, sediments around the borders

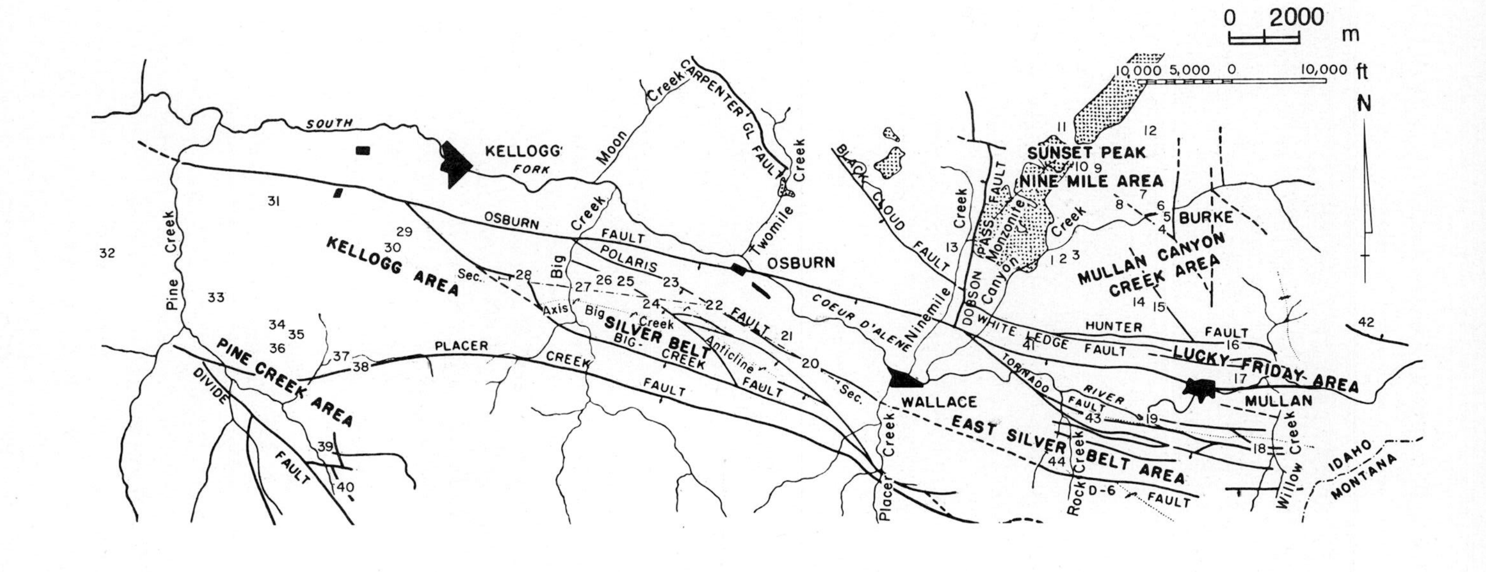

1. Gem
2. Frisco
3. Black Bear
4. Hecla
5. Poorman
6. Tiger
7. Sherman
8. Standard Mammoth
9. Custer
10. Tamarack
11. Interstate
12. Hercules
13. Day Rock
14. Star
15. Morning
16. Gold Hunter
17. Lucky Friday
18. Atlas
19. Moe
20. Galena
21. Argentine
22. Coeur d'Alene mines
23. Nellie
24. Silver Summit
25. Chester
26. Polaris
27. Sunshine
28. Crescent
29. Last Chance
30. Bunker Hill
31. Page
32. Hypotheek
33. Liberal King
34. Denver
35. Sidney
36. Pittsburg
37. Nevada Stewart
38. Highland Surprise
39. Douglas
40. Constitution
41. Golconda
42. Snowstorm
43. Gem State
44. Rock Creek

Figure 11-47. Coeur d'Alene mining district, Idaho, showing the mines and major faults. (*From* Sorenson, 1951, Figure 1.)

of some monzonite stocks show igneous metamorphic alteration in the form of bleaching, sericitization, and even granitization (Anderson, 1949).

As shown in Figure 11-47, the district is cut by several west-northwesterly faults, the most pronounced being the great Osburn Fault. This structure can be traced hundreds of kilometers beyond the Coeur d'Alene district. It dips 55 to 65° south and has 4.5 km of normal dip-slip offset (Shenon, 1948) and 25 km of right lateral strike-slip displacement (Hobbs and Fryklund, 1968). It is the principal locus of Coeur d'Alene mineralization, but it extends 300 km to the east into Montana, and is an integral part of the tectonic history of the northwestern United States (Hobbs, 1968; Meyer et al., 1968). It is a recurrent fault that has been intermittently active since Precambrian times. The actual fault zone is 30 to 60 meters wide and consists of sheared, brecciated, and pulped rock. South of the Osburn Fault are many parallel and oblique faults that appear to be second-order structures caused by the Osburn faulting. All the second-order faults dip south. Tensional stresses set up by movements along these major structures developed local openings through which mineralizing fluids ascended (Sorenson, 1951). The major structural elements differ on opposite sides of the Osburn Fault. To the south, fold axes trend east-west, or parallel to the faults, but on the north side both the folds and the faults strike north to northwest parallel to a broad anticlinal arch heading toward the Idaho batholith.

Most of the ore deposits lie along fractures and shear zones in the Belt Supergroup rocks. Minor permeable fractures that contain only small amounts of gouge were more favorable for mineralization than were the faults with large displacements. Minor mineralization in the monzonite stocks and diabase dikes proves their preore age, but lamprophyre dikes were emplaced after the ore. Concentrations of ore lie along areas of strong structural deformation and in the brittle, brecciated host rock units (McKinstry and Svendsen, 1942; Sorenson, 1951). Where the rocks were brittle and the bedding planes were oblique to the direction of shear, open-fissure zones were produced; here the quartzites are more favorable hosts than the argillites. For example, in the Polaris mine many fissures are mineralized in the brittle quartzites of the St. Regis Formation, but their continuations in the less brittle argillite of the Wallace Formation are not (Figure 11-48).

Coeur d'Alene ore minerals include galena, sphalerite, tetrahedrite, chalcopyrite, pyrrhotite, magnetite, arsenopyrite, and minor amounts of bornite, chalcocite, stibnite, the Cu-Sb-S sulfosalts boulangerite and bournonite, gersdorffite (NiAsS), scheelite, and uraninite. Arsenopyrite is an indicator of economically favorable areas because it forms envelopes around ore shoots (Mitcham, 1952). Quartz, siderite, other carbonates, pyrite, and, locally, barite are the principal gangue minerals. These ore and gangue minerals form a typical mesothermal assemblage. Geothermometric studies indicate mesothermal temperature ranges (Fryklund and Fletcher, 1956).

The lead-zinc ore averages 10% combined lead and zinc and about 31

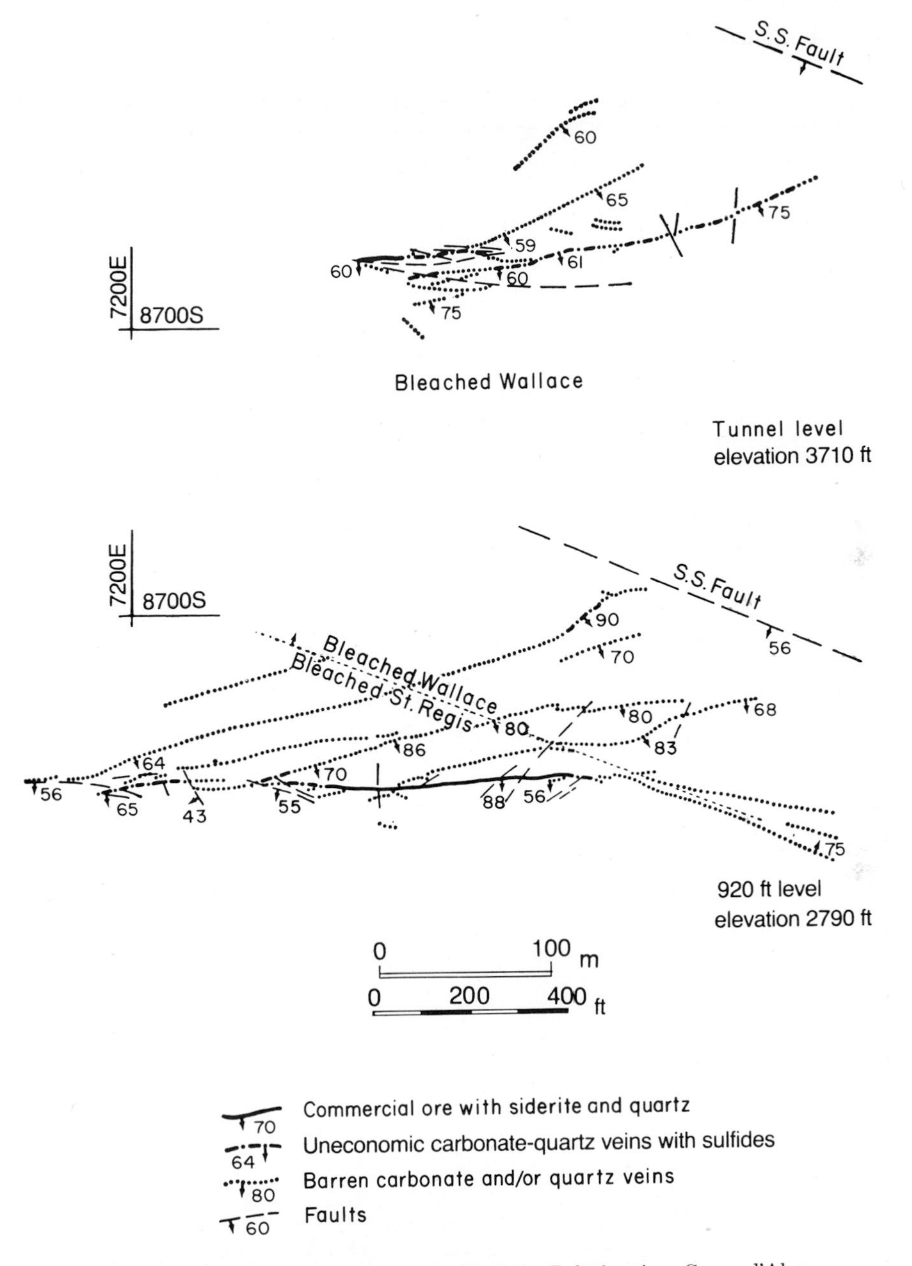

Figure 11-48. Plan of the 920-ft and tunnel levels of the Polaris mine, Coeur d'Alene district, Idaho. The St. Regis is brittle quartzite, the Wallace a less brittle argillite. Mineralization is generally stronger (heavy lines) in the St. Regis, with fractures dying out into the Wallace. (*From* Sorenson, 1951, Figure 3.)

ppm (1 oz) silver per unit of lead (Shenon, 1948). The silver is not in the galena but in tetrahedrite, which in turn tends to be associated with galena (Warren, 1934). Grade in silver appears to increase slightly with depth at the Lucky Friday mine, and the Lucky Friday, like many units, is first and foremost a silver producer. Six separate periods of mineralization ranging from Precambrian to Tertiary have been described by Fryklund (1964), but Hobbs and Fryklund (1968) suggest that most of the mineralization is late Cretaceous in age. Geologic mapping has not yet solved the enigma of the timing of ore deposition. Some workers believe that the principal lead-silver mineralization is of the same age as or later than the monzonite stocks of Cretaceous age. Gott and Cathrall (1980) describe halos of major elements arrayed around the Gem stock that constitute a strong argument that the stocks are either the source or the mobilizer of the metals. The isotopic age of the lead is Precambrian (Doe, 1973); the lead must have been derived from ancient basement or Belt rocks, not from a Cretaceous magmatic source. Many investigators have expressed the opinion that the lead and other ore elements have metamorphically derived from the Belt Supergroup rocks (Landis and Leach, 1983). Similar "old" leads emplaced by younger rocks have been found at many places, including Butte, Montana. Finally there are areas, for example in the Hercules mine, where veins are folded and contact-metamorphosed within the aureole of the Gem stock, but planar and unmetamorphosed, with normal district sulfides, beyond it. Apparently at least some mineralization, perhaps much of it, is therefore of late Precambrian East Kootenay age (1100 to 750 million years ago).

Mineral paragenesis is difficult to establish for the district as a whole. In general, the gangue minerals were deposited early, and most of them continued to be deposited throughout the sulfide phase. Magnetite, pyrrhotite, uraninite, and arsenopyrite were among the earliest ore minerals, and may be Precambrian in age; tetrahedrite, galena, and sphalerite are of intermediate textural age; and chalcopyrite is one of the last minerals deposited (Ransome and Calkins, 1908; Anderson, 1940; Hosterman, 1956). The accessory minerals have not been included in this paragenetic sequence because their order of deposition is not clear; indeed, the paragenesis of even the principal minerals is questionable because of the modification of primary fabrics by postore faulting. A few of the accessory minerals apparently formed as reaction products between two other compounds. For example, bournonite ($CuPbSbS_3$) seems to represent an intermediate step in the replacement of tetrahedrite ($Cu_{12}Sb_4S_{13}$) by galena (PbS) (Anderson, 1940). Vertical zoning is inconspicuous; pyrrhotite increases with depth, and in a few veins galena is more abundant near the surface and sphalerite is richer at depth. Lateral zoning has already been described as real but not as well developed at Magma, Casapalca, San Cristóbal, and Butte.

Much of the mineralization lies in the bleached, altered country rock that trends parallel to anticlinal axes and faults and has generally been a good guide to ore. The bleaching locally results from the development of sericite replacing chlorite, glauconite, and other minerals in the Belt sediments. This alteration is not universal, however; some unaltered argillites contain as much sericite as the bleached rocks. Besides the variable sericite, the bleached zones contain disseminated pyrite, quartz, carbonates, and in places chlorite. Sericite is one of the few persistently stable minerals in the alteration halo, reflecting the mesothermal conditions of the environment. The metallization took place late in the hydrothermal activity, and the intrusion of lamprophyre dikes marks the geologic close of the igneous period.

This complex vein deposit system, one of the greatest silver districts in the world, is still problematical in genesis, but its extent and many descriptive characteristics are now well known. Studies by U.S.G.S. personnel and by Reid, Hall, and others at the University of Idaho continue.

▸ PEGMATITES

Although several geologic types of pegmatites have been recognized, commercially interesting pegmatites have clearly been associated with felsic plutonic activity from Precambrian times to the present. In the broadest definition, pegmatites are just unusually coarse-grained igneous or metamorphic rocks. Metamorphic pegmatites are formed when the more mobile quartzofeldspathic constituents of a rock are concentrated in dilation openings during metamorphic differentiation. They are mineralogically simple, unzoned, virtually never of economic interest, and need not be considered further here. Jahns (1955) defined pegmatites as "holocrystalline rocks that are at least in part very coarse-grained, whose major constituents include minerals typically found in ordinary igneous rocks, and in which extreme textural variations, especially in grain size, are characteristic." Mafic pegmatites—clots or lenses of olivine, pyroxenes, and plagioclase in peridotites, gabbros, and other mafic rocks—are fairly common, but again they seldom contain valuable minerals except in exceptional circumstances such as in the Merensky Reef of the Bushveld Complex (Chapter 9); they also need not be considered further here. Zoned syenite pegmatites in the Bancroft, Ontario, area have yielded uranium ores, but most syenite pegmatites are not economically noteworthy. Simple coarse-grained quartz-K-feldspar clots, knots, and veins are also common late residues of felsic melts but neither are these of economic interest.

We can focus, then, on what are known as granitic pegmatites. Most of them consist of quartz, perthite, and albite, with subordinate to accessory muscovite, biotite, or both. They can be mineralogically complex with respect to the LIL (large ion lithophile) elements like lithium, potassium, sodium, rubidium, and beryllium. They may be either homogeneous and unzoned, or heterogeneous and zoned. Unzoned ones may have the same bulk composition as zoned ones. Heterogeneous pegmatites are normally strikingly zoned from earlier outer to younger inner envelope zones. The best known pegmatites are intermediate to silicic in composition, and are highly colored, conspicuous bodies whose coarse textures and unusual mineral compositions make them favorite mineral collecting grounds.

Our knowledge of the pegmatites stems from long-term studies by petrologists from Brögger (1890) and Crosby and Fuller (1897) to Burnham (Jahns and Burnham, 1969), which were greatly stimulated and broadened by shortages of critical metals and minerals during World War II to include economic geologists such as Cameron, Jahns, McNair, Page, and others. Intensive field and laboratory studies begun then, coupled with the petrologic and mineralogic appeal of the pegmatites, have led to a rich and abundant literature; Jahns (1955), in describing the study of pegmatites to that date, listed 698 *selected* references! Uses for the rare minerals and metals found in pegmatites greatly expanded with the electronics revolution after World War II, and many of these metals are in short supply. Increased demand has periodically encouraged an intensified search, which has resulted in considerable information about pegmatite genesis (Cameron et al., 1949; Jahns, 1946, 1951, 1953, 1955; Beus, 1966; Jahns and Burnham, 1969; Gallagher, 1975; Putman and Sullivan, 1979). Pegmatites represent the final water-rich, siliceous melts of intermediate to silicic igneous magmas, and can generally be thought of as final residual melts rich in silica, alumina, water, halogens, alkalies, and lithophile elements not readily accommodated in the common igneous "Bowen's reaction series" minerals. They have thus been—and to some extent still are—important sources of beryllium, lithium, rubidium, cesium, tantalum, niobium, lesser sources of uranium, thorium, rare earth elements, molybdenum, tin, and tungsten, and major sources of muscovite mica, perthite-quartz for glass manufacture, high-purity silica, and a variety of gem stones and salable mineral specimens, including emeralds, other beryls, topaz, tourmalines, and many more.

Although pegmatites can be found in almost any shape, they are most commonly dikelike or lensoid. Most pegmatites are small, but dimensions can vary from a few meters to hundreds of meters in the longest dimension and from 1 cm to as much as 200 meters in width. They rarely form extensive and continuous tabular bodies in the form of deep-seated veins; only a few exceptional deposits over 2 km long are known. Since igneous pegmatites characteristically solidify late in igneous activity, they tend to be associated with plutonic or hypabyssal intrusions from which the volatile fractions could not readily escape. The great majority of pegmatites developed in deep-seated high-pressure environments. They are rare in unmetamorphosed sediments or shallow intrusives, lavas, or tuffs. Pegmatites rarely develop conspicuous alteration halos, although some hydrolysis and silicification may occur in the walls.

Geothermometric studies indicate that pegmatites form over a wide range of temperatures. Much of the older literature reports temperatures of about 575°C, on the assumption that the quartz formed near the alpha-beta inversion point. However, fluid inclusion studies have indicated temperatures ranging from as low as about 150 to 500°C for many pegmatite zones (Ingerson, 1947), and other geothermometers record temperatures up to 700°C. The consensus is that most pegmatites are formed at the upper end of the range between 700 and 250°C (Jahns, 1955). Jahns and Burnham suggest separation and crystallization temperatures of pegmatite melts over a range from 1300°C down to 650°C, rarely to 600°C. Mohon (1975) used several techniques to find that the Monte Cristo pegmatite in Yavapai County, Arizona, crystallized from a water-rich melt containing 5 to 7% water at around 650°C with later stage hydrothermal activity extending down to about 200°C. Babu (1969) applied several geothermometers to the pegmatites of India and reported temperatures of 700°C for the pegmatite magma, with descending temperatures of 760 to 500°C for pegmatite biotite, 760 to 450°C for tourmaline, 600 to 300°C for feldspar, 500 to 400°C for beryl, 500 to 435°C for muscovite, 500 to 400°C for milky quartz, and 300°C for clear quartz in the core.

Homogeneous Pegmatites

The vast majority of pegmatites are mineralogically simple, unzoned ones. They consist mostly of coarse-grained quartz and feldspars with subordinate mica, and are ordinarily uniform in both composition and texture from wall to wall. Except where they can be mined for lithium, feldspars, or micas, they have little economic value. However, homogeneous pegmatites have become economically important in this century. Most of the world's pegmatite lithium reserves are in the homogeneous pegmatites of the productive Kings Mountain district, North Carolina (Kesler, 1961; Kunasz, 1982), and at Manono-Kitole, Zaire. In the United States, commercial potash-soda feldspar has been produced largely from homogeneous pegmatites in New England and North Carolina. Since flotation was introduced into the processing of pegmatite ores, there has been little interest in zoned pegmatites as sources of feldspar. Minor sheet mica production is still recorded in India.

Simple pegmatites can result from metamorphic differentiation or from one comparatively short period of igneous activity. The palingenesis of sediments may produce an S-type melt corresponding chemically to a mixture of quartz, K-feldspars, and muscovite; it is probable that many simple pegmatites are of this local-melting origin. Most of the larger commercial or potentially commercial homogeneous pegmatites are of igneous origin, whereas the pegmatites formed during metamorphism are typically small and irregular. Lit-par-lit pegmatites intertongued with metamorphic rocks are common, in places appearing to grade into country rock.

Heterogeneous Pegmatites

Heterogeneous pegmatites result from igneous processes rather than from recrystallization or palingenesis associated with metamorphism, although it may develop that most igneous pegmatites are associated with

S-type rocks. They are thought to have formed as a result of one period of crystallization, after injection of a water-rich siliceous late-stage melt, during which the first-formed minerals react with a progressively changing residual magmatic fluid from the outer to the inner zones with time. Hunt (1871) was probably first to recognize that zoning is a common aspect of pegmatites with a zoning order revealed by geologists of the U.S.G.S. in the 1940s (Cameron et al., 1949). Their studies detailed the fact that zoning was not everywhere identical in its completeness, but that zones were always present in *the same order* if they were present at all. They recognized four major types of zones, the border, wall, intermediate, and core zones (Figure 11-49). A zone in a pegmatite is a lithologic unit defined on the basis

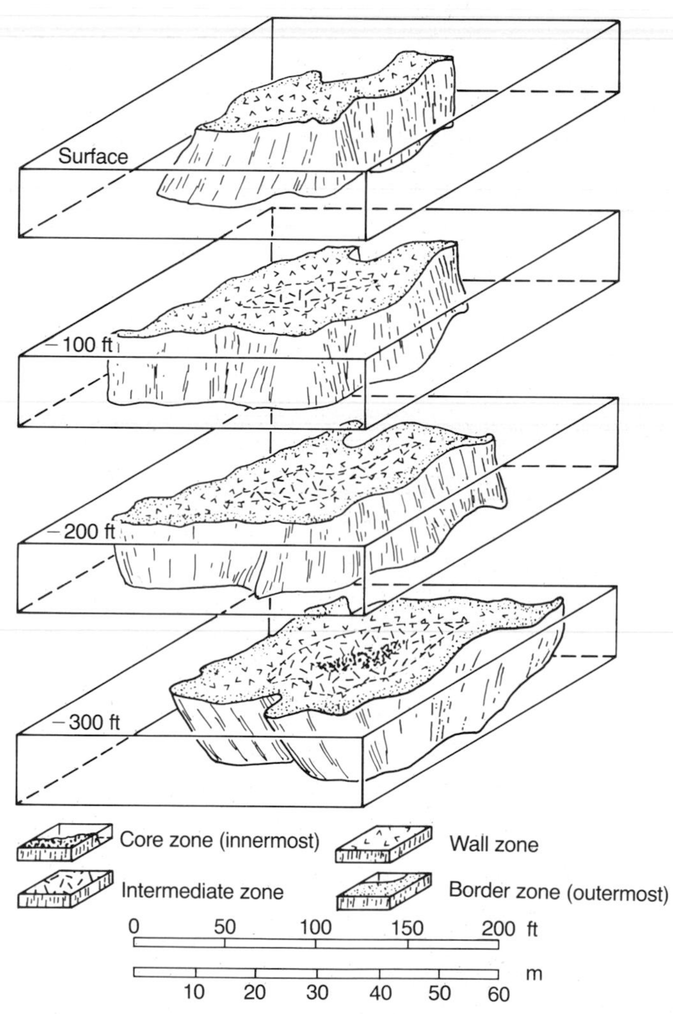

Figure 11-49. Idealized diagram of a zoned pegmatite, showing appearance of the "nested" border, wall, intermediate, and core zones at successively different levels. (*From* Cameron et al., 1949.)

of mineralogy *and* texture. Each zone is composed of a particular mineral assemblage, but the same mineral assemblage may make up two or three zones that differ in texture, mineral proportions, or both. The border zone is thin, averaging a few centimeters in most pegmatites. Although in some deposits it cannot be recognized, it is likely to be the most continuous of the zones. It is a selvage, or transition, between the inner, more characteristic pegmatitic materials and the wall rocks, and is typically aplitic in texture. The most common minerals in border zones are fine-grained feldspars, quartz, and muscovite; accessory minerals include garnet, tourmaline, beryl, or even some of the rare species. Valuable metallic constituents are absent from this zone. Some border zones may be the chilled margins of pegmatitic masses, as suggested by an overall chemical composition similar to that of the pegmatite as a whole; others differ distinctly from the bulk composition of the entire pegmatite (Cameron et al., 1949).

The wall zone is well developed in many pegmatites and absent in others. In general the wall and border zones contain the same minerals, though the proportions may differ, but the wall zone is characteristically more coarsely textured and thicker than the border zone. It marks the advent of genuinely pegmatitic coarse crystallization, although most of the truly giant crystals occur farther in toward the core. The essential minerals of most wall zones include plagioclase, perthite, quartz, and muscovite; subordinate but appreciable amounts of tourmaline, biotite, apatite, beryl, and garnet may be present. Metallic constituents may also be present; in a few deposits they have been of economic value. Mica and beryl are the principal commercial minerals recovered from wall zones, and the wall zone commonly constitutes the innermost site of garnet, the "endoskarn analogue" of skarn terminology.

Intermediate zones, where present, may contain a variety of minable minerals. Although some pegmatites pass directly from their border or wall zones into the core and do not have an intermediate zone, most complex granitic pegmatites carry as many as five or six intermediate zones, and it is here that the detailed mapping of Cameron, Jahns, Page, McNair, and other U.S.G.S. geologists was most revealing. They discovered that in zoned granitic pegmatites there is a definite sequence of mineral assemblages from the walls of the pegmatites inward. The order is the order from 1 to 11 in Table 11-4. If, for example, the wall and border zones of a pegmatite consist of assemblage 4, no zone consisting of assemblages 1, 2, or 3 will be found in the pegmatite. On the other hand, one or more of assemblages 5, 6, 7, and 8 may be present, but if so, they will occur in a zone or zones inside the wall zone composed of assemblage 4. Very few pegmatites contain all the assemblages of Table 11-4, but the succession of those assemblages that are present will be that of the table. Many pegmatites contain only assemblages 1, 3, 4, and 11.

The sequence of mineral assemblages of Table 11-4 was found to hold for more than 100 pegmatites mapped in New England, 39 in North Car-

TABLE 11-4. Mineral Assemblages of the Intermediate Zones of Major U.S. Pegmatite Districts

Essential Mineral Assemblages	Southeastern States	New England	Petaca, New Mexico	Black Hills, South Dakota	Accessory Minerals, Products
1 Plagioclase-quartz-muscovite	×	×		×	Sheet mica, high BeO beryl, schorlite, niobite-tantalite, cassiterite; Be, Nb, Ta, Sn
2 Plagioclase-quartz	×	×		×	Niobite-tantalite; Nb-Ta
3 Quartz-perthite-plagioclase, with or without muscovite, with or without biotite	×	×	×	×	Beryl, sheet mica, perthite, niobite-tantalite
4 Perthite-quartz	×	×	×	×	Beryl, quartz-K-feldspar, K-feldspar, niobite-tantalite; Be, Nb, Ta
5 Perthite-quartz-plagioclase-amblygonite-spodumene		×		×	Amblygonite, spodumene, beryl; Li, Be
6 Plagioclase-quartz-spodumene	×	×		×	Spodumene, cassiterite; Li, Sn
7 Quartz-spodumene	×	×		×	Spodumene, gem minerals; Li
8 Lepidolite-plagioclase-quartz		×		×	Lepidolite, microlite; Li, Ta, Nb
9 Quartz-microcline (rare)				×	
10 Microcline-plagioclase-lithia micas-quartz (rare)		×		×	Tantalite, uraninite, cassiterite; Ta, U, Sn
11 Quartz core zone	×	×	×	×	Quartz

×, mineral assemblage observed.

Source: Adapted from Cameron et al. (1949).

olina, 73 in the Petaca district, New Mexico, and 79 in the Black Hills, South Dakota, a total of almost 300 occurrences. Mapping in India (Babu, 1969), Zimbabwe (Rijks and Van der Veen, 1972; Gallagher, 1975), Namibia (Cameron, 1955; Roering, 1966), Tanzania (Solesbury, 1967), and South America (Herrera, 1968) has confirmed the early findings and added to this list of conforming pegmatite bodies. The significance of Table 11-4 is that it enormously simplifies exploration (Kesler, 1961) and development (Norton and Page, 1956) of pegmatites. Immediately obvious is that beryl is found in outer subzones; that the perthite-quartz graphic granite eutectic composition of assemblage 4 is ideal for the glassmaker, especially with a flux to aid melting; that if lithium is to be present at all it will lie between this graphic granite and the quartz core; that the lithium content declines inward; and that high-purity silica will be found at the center of symmetry if at all. Not so obvious is that the constituent minerals *also* shift toward more alkalic, felsic, differentiated compositions inward across the zoning (Figure 11-50). The intermediate zones, then, span from assemblage 1 to assemblage 11, the latter also constituting the quartz core.

It should be stressed that an intermediate zone is any one between the wall zone and the core. The term implies absolutely nothing as to mineral composition. Suppose, in a given pegmatite, that there are three inter-

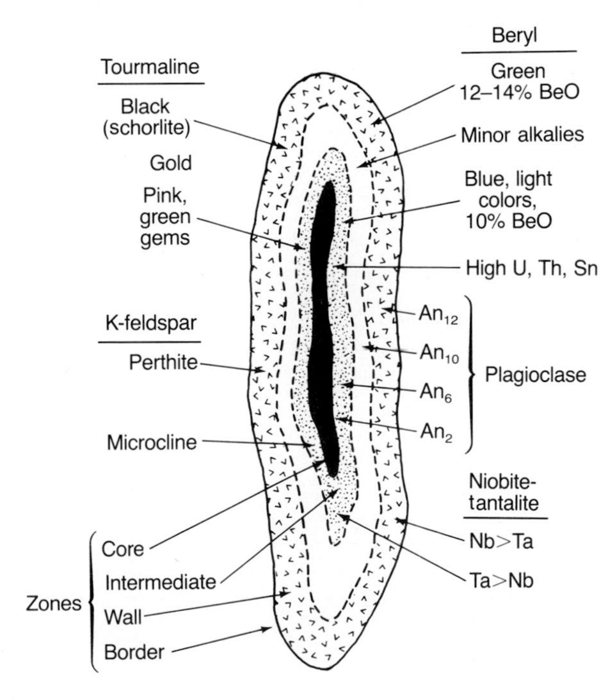

Figure 11-50. Cryptic zoning from wall zone to core in individual pegmatite mineral species. (*Composited from* Cameron et al., 1949.)

mediate zones. They should be designated as outer, middle, and inner intermediate zones, or as intermediate zones 1, 2, and 3. Each of these is a zone in its own right, not a subdivision of a zone. Two adjacent intermediate zones may have the same mineral assemblage, differing in texture or in mineral properties, or they may have different mineral assemblages. As one of many important examples that could be cited, consider the Deep Creek No. 1 pegmatite, Bryson City district, North Carolina. Mineral assemblage 4 (Table 11-4) makes up three distinct intermediate zones. The outer one consists of graphic granite (Figures 11-51 and 11-52), the middle one of giant massive perthite crystals with interstitial quartz, and the inner one of quartz with numerous scattered giant crystals of perthite. Inside the latter is a quartz core.

A special feature of some zoned pegmatites is the presence of giant crystals within the inner zones. Abnormally large crystals of quartz, feldspars, micas, beryl, apatite, tourmaline, and other pegmatite minerals have been reported. Individual crystals are measured in meters, even to tens of meters (Figure 11-53). Jahns (1953) contends that they must have crystal-

Figure 11-51. Specimen of the cunieform, hieroglyphiclike texture that gives "graphic granite" its name. The white is perthitic orthoclase, the dark is quartz intergrown with it in a texture thought to have formed at the quartz-K-feldspar eutectic. It is from a zone composed of assemblage 4 of a pegmatite at Auburn, Maine. (*Photo by* E. S. Bastin.)

Figure 11-52. Contact between assemblages 3 and 4 in a pegmatite at Fisher's Quarry, 5 km northeast of Brunswick, Maine. The blocky fracture at the left is principally perthite-plagioclase cleavage; the graphic granite at the right is similar to but coarser than that of Figure 11-51. Width of photograph is 3 meters. (*Photo by* E. S. Bastin.)

Figure 11-53. Giant crystals in intermediate assemblage 6 on the south side of the Etta pegmatite mine in the Black Hills district, South Dakota, showing the enormous single crystals occasionally encountered. The man is almost 2 meters tall; the albite (cleavelandite) crystals at the upper left are over half a meter long, and the spodumene logs across the center are over 10 meters long and 2 meters across! (*Photo by* W. T. Schaller.)

lized directly from a volatile-rich pegmatitic liquid under delicately balanced thermal and chemical conditions; that is, the giant crystals must have grown rapidly, and the ions must have been able to diffuse readily through the pegmatitic fluid. "Logs" of beryl and spodumene as much as 15 meters long and tapering from centimeter-scale seed crystals at their bases to several meters width at their inward-facing tops (Figures 11-53 and 11-54) have been described from the Etta and Keystone pegmatites in the Black Hills of South Dakota (Page et al., 1953).

Many commodities and minerals have been extracted from pegmatites. Coarse sheet muscovite mica is most abundant in assemblages 1 and 3, as is BeO-rich beryl. Niobium-rich niobite-tantalite commonly occurs in assemblages 3 and 4. Masses of graphic granite, the eutectic mixture and "glassmaker's dream," constitute assemblage 4 (Figures 11-51 and 11-52); assemblage 4 zones of coarse, nongraphic perthite and quartz have been the sole source of high-grade K–feldspar. Numbers 5 through 8 contain the lithium phosphates and silicates amblygonite and spodumene, with progressively

Figure 11-54. Specimen of quartz, K-feldspar, and beryl from the Monte Cristo pegmatite, Yavapai County, Arizona. The beryl crystals grew markedly in cross section as they grew lengthwise toward the center of the pegmatite (upward in the photo). Such crystals, called "logs" and measured in tens of meters, have been found in many pegmatites. The largest crystal known is a log of beryl 18 meters long and 3.5 meters across from a pegmatite in the Malagasy Republic (Rickwood, 1981). (*Photo by* W. K. Bilodeau.)

less lithium inward toward petalite-zinnwaldite-lepidolite volumes in the inner subzones. These minor zones also contain most of the gem quality topaz, tourmaline, and other precious to semiprecious minerals as well as relatively tantalum- and uranium-rich oxide assemblages. The quartz core's high purity is attractive to ceramists, glassmakers, steelmakers, and others.

The core zone in pegmatites is commonly a solid mass of barren white quartz, coarse-grained quartz with feldspar, or quartz with large, euhedral crystals of tourmaline or spodumene. The core is generally near the center of a pegmatitic body and may form discontinuous pods along a central axis. It is ordinarily barren of other minerals, though a few exceptions are known.

In addition to the zonal construction of pegmatites, which normally constitutes as much as 95% of the volume of a pegmatite, two other structural-genetic modes are regularly encountered, namely, fracture fillings and replacement bodies (Figure 11-55). Fracture fillings (Figure 11-55*a*) may be confined to a single zone or subzone, may occupy contacts between zones, or they may crosscut older, outer units in orthogonal or radial patterns. They are filled with pegmatitic material characteristic of the zone that was crystallizing when the fractures formed, so the vein dikes never cut inward toward the core. Replacement bodies (Figure 11-55*b*) may replace preexisting zoned pegmatite material, or they may occur in the walls of fractures in the settings just described or in massive irregular diffuse bodies generally near the core. Both fracture fillings and replacement bodies reflect ongoing structural and chemical activity during pegmatite crystallization. Relations between the zones and replacement pegmatite in a mapped occurrence are shown in Figures 11-55 and 11-56.

Mining of zoned pegmatites may be costly because the intermediate zones, fracture fillings, and replacement bodies are discontinuous and variable in dimension, even though they are relatively orderly in internal distribution. The huge homogeneous lithium-rich pegmatites of the King's Mountain district, North Carolina (Kesler, 1961), can be mined by open-pit methods, but most zoned pegmatite mining involves careful, highly selective excavation, hand sorting of minerals, and expensive benficiation procedures. Pegmatite mining in the 1980s is at relatively low activity levels, but interest in them continues at high levels as the various elements contained, like uranium and tin, rise and fall in demand and price.

Pegmatites range in age from the earliest Archean of Kenya-Tanzania-Zimbabwe in Africa at more than 3.5 billion years and in Manitoba, Canada, to the Jurassic Pala pegmatite near San Diego, California, at 147 million years. They occur on all continents, typically where orogenic zone roots have been exposed by relatively deep erosion. A belt of Caledonian-Hercynian, mid-to-late Paleozoic occurrences extends from Canada's northeastern provinces through Maine, New Hampshire, and the other New England States down the Appalachians through the Carolinas into Georgia and Alabama. Pegmatites in the western United States are generally Precambrian in the erosional window that exposes Archean rocks in the Black

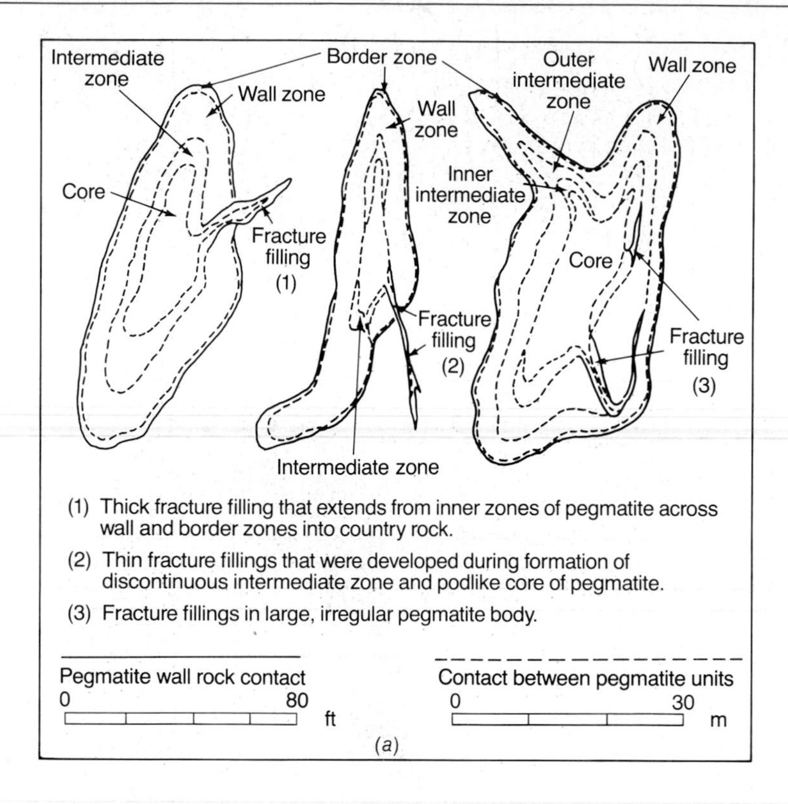

(*a*)

Hills, South Dakota, area and in uplifted dissected Proterozoic rocks of Colorado, New Mexico, and Arizona. Pegmatites in these districts, and others, occur in association with granitic plutons in middle- to high-rank metamorphic rocks, but the occurrence of individual pegmatites may not be tied to particular plutons—they are part of a regional plutonism and metamorphism package. Elsewhere in the world they are generally in Precambrian shield areas in Brazil, Africa, India, and the U.S.S.R., and in tectonically more recently active areas such as Scandinavia and Namibia. The majority of igneous pegmatites lie near the borders of plutons, either within them or in the nearby country rock. The position and the size of these pegmatites have been controlled to a large extent by the structure and fracture pattern in the border areas. Russian and Canadian geologists in particular are studying regional zoning patterns of pegmatites (Cerny, 1982).

In spite of the great amount of work that has been done on pegmatites, very little is actually known about methods of their emplacement. According to Gevers (1936), the pegmatites of Namaqualand, in the northwestern Republic of South Africa, possess not only internal zoning, but the bodies themselves are distributed in zones in and around an associated granitic

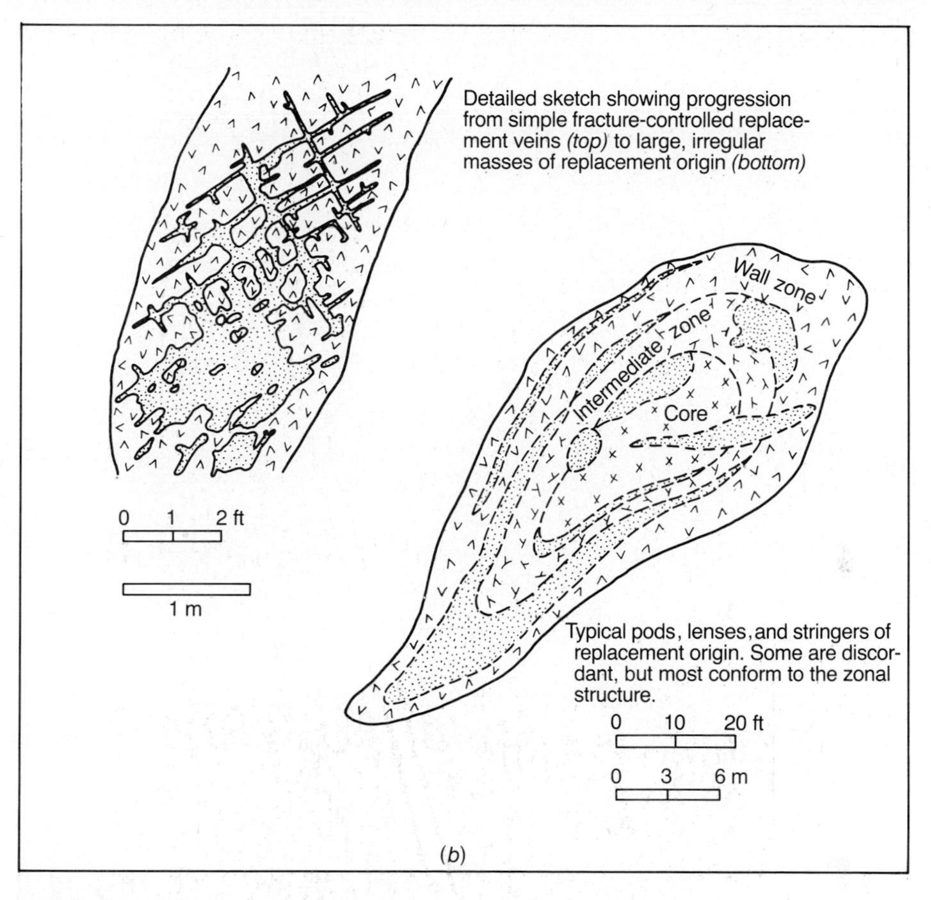

Figure 11-55. (*a*) (*facing page*) Idealized plans of pegmatites, showing typical relations of fracture fillings in pegmatities. (*b*) Relations of replacement bodies (stippled) to host pegmatite. (*From* Cameron et al., 1949.)

batholith. Gevers classed the Namaqualand pegmatites in southern Africa as interior or core pegmatites, marginal or hood pegmatites, and exterior or roof pegmatites. The interior pegmatites are small, scattered, and generally barren of economic minerals. They become increasingly scarce with depth into the batholith. Marginal pegmatites are abundant and large in the upper or outer shell of the batholith. The most abundant and most economically rewarding are the exterior pegmatites, found outside the parent granite body, but grading into the marginal pegmatites. The exterior pegmatites are distributed over a zone several kilometers wide, depending upon the size of the parent batholith, and many contain valuable mineralized zones and subzones.

The relationships pointed out by Gevers fit into a generalization stated by Emmons (1940). Emmons noted that the valuable pegmatites are in and near the tops of batholiths—within the intrusives or injected into their

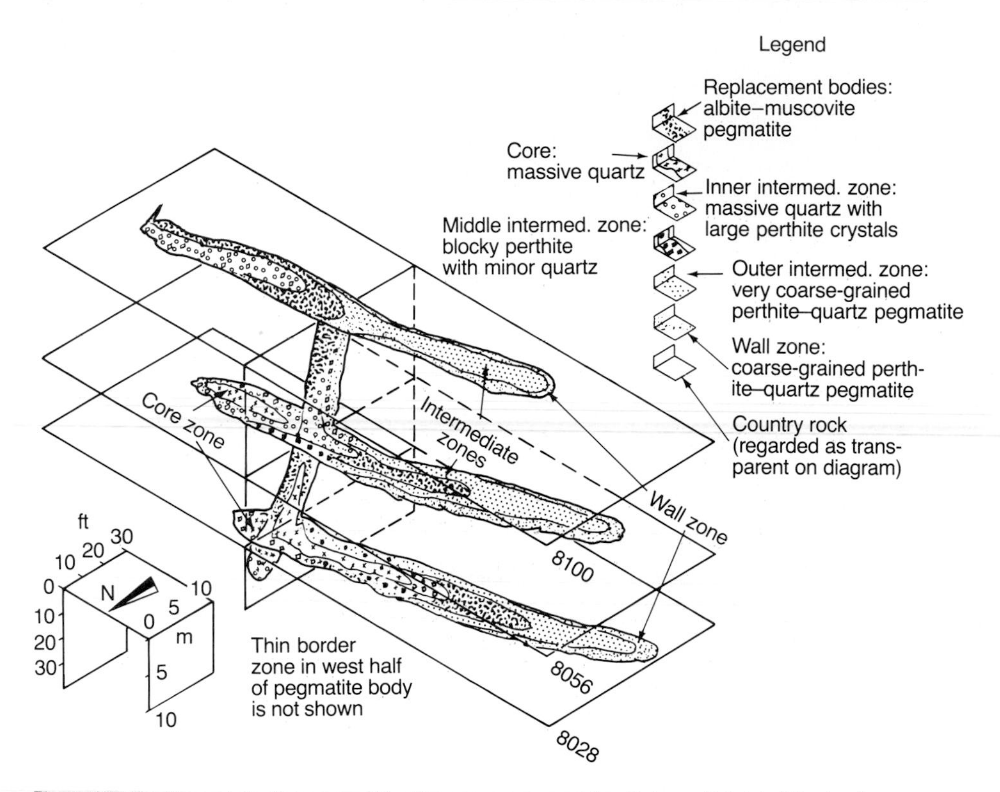

Figure 11-56. Isometric diagram of the Lonesome pegmatite, Petaca district, Rio Arriba County, New Mexico, showing distribution of zones and replacement bodies. (*From* Cameron et al., 1949.)

roofs. This observation seems to agree with the latest field data. Further, Heinrich (1963) found that homogeneous pegmatites tend to be near or within the batholithic source, and the heterogeneous pegmatites form from the latest pegmatitic fluids, which are more mobile and have a greater rare element content than the fluids producing simple pegmatites. As with many generalizations in geology, however, there are exceptions. Tôrre de Assuncão (1944) pointed out such an exception in Portugal, where the valuable pegmatites are concentrated in the core of a nepheline syenite pluton. Nevertheless these observations of regional zoning conform to ideas concerning pegmatite genesis (Jahns and Burnham, 1969, and others) to be considered next.

Pegmatite Genesis

Any theory of zoning in granitic pegmatites must take into account certain fundamental relationships, among which are the arrangements of the zones, gradational to sharp contacts, the transection or replacement of outer zones by inner zones (but not vice versa), the progressive sodium enrichment of plagioclases toward the inner zones, and the paragenetic consistency of zones and mineral assemblages from one pegmatite to another (Figures 11-49 and 11-50; Table 11-4).

Of the several explanations that have been advanced, one is that the zonal structure is formed by fractional crystallization of a pod of melt in place under disequilibrium conditions. Thus the reaction between crystals and the remaining liquid would be continuous and evolving, creating successive layers of contrasting composition (Brögger, 1890). Crystallization in a partially closed system that undergoes repeated pressure releases and consequent resurgent boiling may account for the occasional outward transgressions of younger zones upon older ones. This hypothesis is supported both in theory and in mineralogical detail by Bowen's reaction principle. It is certainly a plausible explanation for those pegmatites that corrrespond to Bowen's reaction series from the border to the core. The pegmatite from the San Gabriel Mountains (Figure 11-57) is a good example; it is a mafic pegmatite in coarse-grained norite. The deposit consists of a wall zone with augite-labradorite, an outer intermediate zone of hornblende-labradorite, an inner intermediate zone of andesine-hornblende, and a core of perthite-quartz-albite-epidote. Grain sizes increase gradually from the

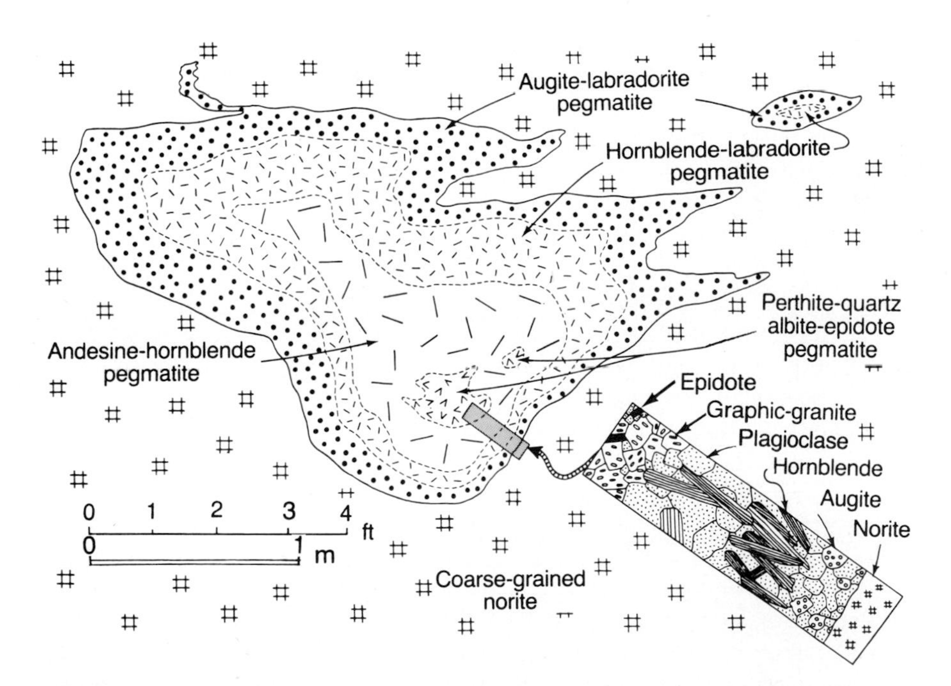

Figure 11-57. Zoning of minerals in a gabbroic pegmatite from the western San Gabriel Mountains, California. The sequence of zones and crystallization events recapitulates Bowen's reaction series, as is normal in zoned granitic pegmatites. See also Figures 11-49 to 11-56. (*From* Jahns, 1954.)

outer zone through the intermediate zone, and the core is a relatively coarse mixture of graphic granite, albite, and epidote (Jahns, 1954). According to Bowen's series, this sequence implies both fractional crystallization and a gradual increase in the volatile content, or at least mobility, of the rest melt.

A second explanation of zoning is that progressively changing solutions deposit materials along the walls of open veinlike channels (Hunt, 1871). This hypothesis does not depend upon local in situ crystal fractionation and conditions of disequilibrium. The pegmatitic fluids would be expected to vary in composition for any of a number of reasons, such as magmatic differentiation at the source and contamination with wall rocks or other fluids in transit. The observed intermediate zone sequencing and cryptic zoning would be difficult to explain with this degree of openness.

A third hypothesis is that complex pegmatites are developed in two stages: (1) formation of a simple pegmatite by the direct crystallization in a closed system of a pegmatitic fluid, and (2) partial or complete replacement of the pegmatite as hydrothermal solutions pass through it in a partially open system (Schaller, 1933; Hanley et al., 1950; Cotelo Neiva, 1954).

Most modern workers agree with either the first or the third hypothesis, or with some combination of the two (Landes, 1933; Derry, 1931; Cameron et al., 1949). An argument against the possibility that zones develop along open channels—as proposed in the second hypothesis—lies in the fact that many interior zones are completely enclosed in outer zones. Since the zones become progressively younger toward the center, it would be difficult for this repetitive arrangement to develop in truly open systems, though crustified zoned veins would form in such open systems. The third hypothesis—the two-stage system—has been objected to on the evidence that replacement would also be likely to have developed from the outside of a pegmatite toward its core. The zones would represent progressively thinner shells of replacement formed from fluids of changing composition, in which circumstance the paragenetic sequence would be that replacement veins in through-going systems would develop outward from a central fissure. If pegmatite zones do result from replacement reactions in fluids of variable composition, the universal similarity of zonal sequences would seem to be fortuitous rather than a necessary result of the process involved (Cameron et al., 1949). Thus the general concept of a pegmatitic system is one in which the zones develop from the walls inward, within a restricted magmatic-hydrothermal system. The role of a gaseous phase or a supercritical volatile-rich fluid may be critical in this model. Late in the pegmatitic process, hydrothermal fluids may travel through fractures and react with the older minerals. This modern concept of the origin of complex pegmatites is precisely what Brögger proposed in 1890, and was reinforced by experimental studies by Jahns and Burnham (1969). They showed that pegmatites crystallize from a rest silicate melt in the presence of an exsolved siliceous, volatile-rich aqueous fluid of much lower viscosity (Figure 11-58), an im-

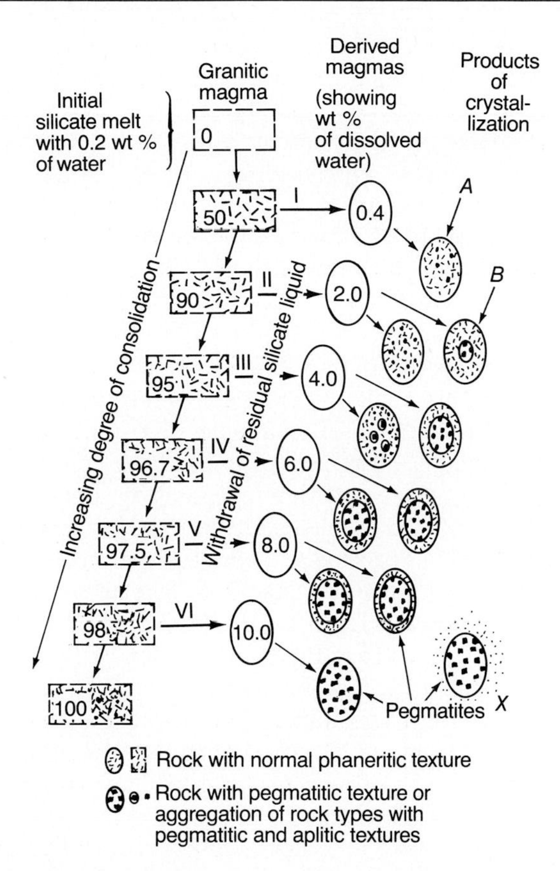

Figure 11-58. Crystallization of granitic magma and the withdrawal of residual silicate liquid at six different stages to form bodies of granite and pegmatite with contrasting lithologies. The magma is assumed to crystallize under constant confining pressure corresponding to a water solubility limit of 10 wt %. Figures in boxes of the left-hand column indicate the percentages of original silicate liquid used up, and those in ellipses of the adjacent column indicate contents of dissolved water in bodies of magma derived at the various stages. Consolidation of these bodies is illustrated in the right-hand column, and each pair of bodies designated *A* and *B* represents contrasting types of segregation. Body *X*, at lower right, is surrounded by a zone of metasomatically altered country rock. (*From* Jahns and Burnham, 1969.)

miscibility of two fluids probably resulting from second boiling at some stage after crystallization had begun. They state (page 843) that

> *. . . partitioning of constituents between melt and aqueous fluid, rapid diffusion of constituents through the aqueous phase, and gravitational rising of this fluid through the system contribute to formation of pods, zones, and other rock units of unusual composition and texture.*

Their diagram (Figure 11-58) shows ratios of pegmatitic to normal, phaneritically grained granite produced by the separation of the aqueous fluid at varying stages of crystallization of a parent melt starting with 0.2 wt % water. If separation is not effected until stage VI when the parent melt is 98% crystallized, the pegmatite melt will contain 10% water and will crystallize entirely to zoned granitic pegmatite material.

Pegmatites are studied for reasons other than their mineralogy. In many systems they represent transitions between ordinary intrusive, granitoid masses and the veins and mineral deposits derived from hydrothermal fluids. As indicated above, with the exception of the pegmatites formed by metamorphic differentiation and related processes, most pegmatites are thought to be rest magmas, the residual volatile-rich fractions left after the main magmatic mass has solidified. The pegmatite residues crystallize in place or are squeezed into available openings where they solidify. Theoretically it would seem that pegmatites should grade into hydrothermal deposits, especially into higher temperature, deep-seated quartz veins. Many writers state that transitions between pegmatites and quartz veins are known or even common (Schneiderhöhn, 1941; Bateman, 1950; Lutton, 1959). Actual transitions between typical individual pegmatite bodies and typical quartz veins are few; most involve swarms of veins and dikes (Shand, 1943); many are like the Passagem Lode in Minas Gerais, Brazil, and to some extent the veins in the Southeast Piedmont, United States. In these areas the veins are predominantly quartz, but they contain minor amounts of scattered garnet, kyanite, tourmaline, apatite, muscovite, phlogopite, biotite, feldspars, and a few other high-temperature minerals. The Passagem Lode is considered to be a deep-seated vein, though an early description classified it as a pegmatite because of its mineral composition and coarse texture (Hussak, 1898). Minerals other than quartz are not abundant, however, nor have variations in the vein material been noted either at depth or laterally. Deposits of the southern Piedmont are similar; most are not typical pegmatites, but at the Old Franklin pit in Clay County, Alabama, a zone of quartz stringers was mined for gold, and one narrow vein grades into what seems to be a characteristic homogeneous pegmatite. The composition of veins in the Southern Piedmont is like that of the enclosing rocks (Ross, 1935), apparently due in part to contamination from materials mobilized metamorphically from within the country rock. Lutton (1959) asserted that a granite-pegmatite-quartz molybdenite veinlet continuum can be shown around many porphyry molybdenum systems, and the general presence of quartzofeldspathic "alteration" and pegmatoid development indeed argues for an environmental-genetic link. Sillitoe (1973) used the presence of pegmatites with porphyry copper mineralization to indicate that both had been developed at relatively deep epizonal levels. Halliday (1980) reports that pegmatites appear to participate in a temporal and geological continuum from granite through pegmatite and porphyry development to tin ore development at Cornwall, England, and thus appear to be part of the transition from igneous to hydrothermal conditions.

Petaca District, New Mexico Both economically and otherwise, the pegmatites of the Petaca district in northern New Mexico are ideal examples for study. The Petaca district is an area about 25 km long and 7 km wide, trending north-south in Rio Arriba County north of Santa Fe. The pegmatites there lie within Precambrian rocks south of the San Juan Mountains. Studies of the Petaca pegmatites made during World War II were reported in detail by Jahns (1946). His description forms the basis for the following summary.

The pegmatites intrude Precambrian quartzites, quartz-mica schists, granite, and subordinate amphibole schists, andalusite schists, staurolite schists, and metamorphosed rhyolites. The Tusas Granite (Just, 1937) invades the metamorphic rocks and appears to be the source of the pegmatitic fluids. Nearly all the commercial pegmatites are in the schists and quartzites, beyond the granite contact. Small irregular pegmatites in the granite itself are of little or no economic value. The commercial pegmatites reach dimensions of 100 by 500 meters in plan, but most are 100 meters long and 10 to 20 meters wide.

More than 50 pegmatites are exposed in the Petaca district, including simple pegmatites in the Tusas Granite and simple sills and dikes as well as irregularly shaped cross-cutting complex pegmatites in the metamorphics. Only the latter are of economic value. Folds and minor structures controlled the emplacement of many. Associated with the pegmatites are swarms of quartz veins. Both veins and pegmatites appear to have a common origin in the granite, but most of the veins are younger. Some quartz veins cut through the outer zones of pegmatites and merge with their cores; a few veins gradually change along strike from pure quartz to a quartz vein with feldspathic walls. Figure 11-59 was chosen to illustrate both a typical pegmatite and one of the Petaca pegmatites. Figure 11-59 shows a typical narrow zone of pegmatite impregnation of its wall rocks at their contact.

The pegmatites are primarily of value for their mica, but they also contain significant concentrations of beryllium, niobium, tantalum, bismuth, uranium, thorium, and the lanthanide rare earth metals. Mineralogically they consist of microcline, quartz, albite, and muscovite, with accessory biotite, spessartite ($Mn_3Al_2Si_3O_{12}$, a garnet), fluorite, beryl, columbite-tantalite [$(Fe,Mn)(Nb,Ta)_2O_6$], samarskite [$(Fe,Ca,UO_2)(Ce,Y)(Nb,Ta)_6O_2$], monazite, ilmenite, magnetite, uraninite, lepidolite, tourmaline, and copper sulfides along with rare grains of apatite, native bismuth, bismuthinite, microlite [$(Ca,Na)_2Ta_2O_6(O,OH,F)$], cassiterite, fergusonite [$(Y,Er,Ce)(Nb,Ta)O_4$], gadolinite ($Be_2FeY_2Si_2O_{10}$), galena, phenakite (Be_2SiO_4), phlogopite, pyrite, scheelite, topaz, and many secondary alterations of these minerals.

Most commercial pegmatites in the Petaca district include border, wall, intermediate, and core zones and display superimposed replacement minerals. The border zones generally consist of fine- to medium-grained microcline and quartz with subordinate mica and, in places, garnet, fluorite,

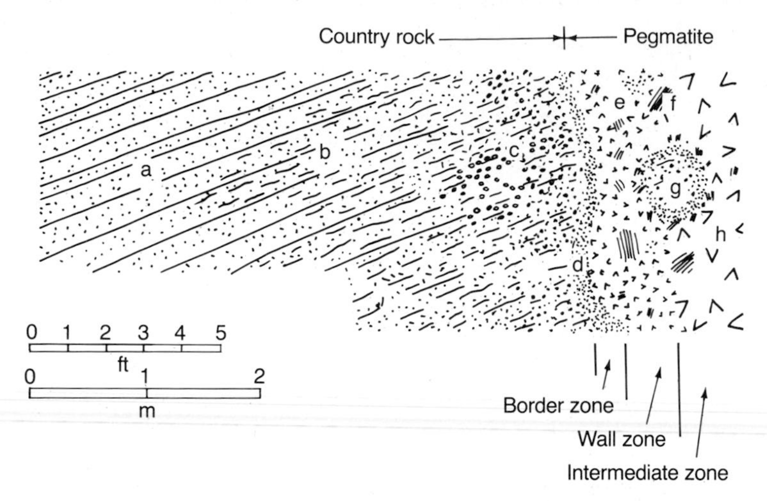

Figure 11-59. Gradational relations between a pegmatite and its country rock in the Petaca district, New Mexico. a = slabby micaceous quartzite; b = quartzite with muscovite-rich partings and disseminated small flakes of muscovite; c = mica-impregnated quartzite with metacrysts of microcline and albite-oligoclase; d = fine-grained, mica-rich border zone of pegmatite; e = wall zone medium-grained perthite-quartz-albite-oligoclase-muscovite pegmatite, assemblage 3; f = large block of muscovite; g = inclusion of altered quartzite lithologically similar to that in c; h = coarse-grained perthite-quartz pegmatite of the intermediate zones, assemblage 4. (*From* Jahns, 1946.)

and beryl. They range from less than 1 cm to several meters in thickness. Locally the pegmatitic fluids reacted with the country rocks developing gradational mica-rich border zones. Wall zones of coarse perthite and quartz (assemblage 4), with accessory mica, garnet, fluorite, and beryl, are as much as a few meters thick. In some pegmatites the wall zones form complete shells, but in others they are only partly developed or were partly removed by reaction with fluids that formed the intermediate zones. Garnet, generally rare in pegmatites, is restricted to the border and wall zones. As many as three intermediate zones can be mapped in the Petaca pegmatites, although some have no intermediate zone at all. The intermediate zones characteristically form hoodlike units along the crests of plunging pegmatite bodies and seldom encompass the whole core, so that these zones vary greatly in thickness. The cores are near the troughs of the plunging pegmatites and consist of massive quartz, quartz containing large microcline crystals, or, in some cases, coarse-grained microcline-quartz pegmatite. They are normally in the thickest part of the pegmatite and they consist of single ellipsoidal or pipelike masses or series of disconnected pods. The outer intermediate zone is commonly coarse graphic granite (assemblage 4); the middle intermediate zone is coarse, blocky microcline; and the inner intermediate zone is generally massive quartz locally containing giant microcline crystals (assemblage 9).

The mineral paragenesis indicates that the zones developed progressively from border to core. Minerals of the outer zones were corroded and replaced by later pegmatitic fluids with which they were no longer in equilibrium; veins of the late fractions cut the peripheral zones. In general, perthite, garnet, and beryl formed early; albite, muscovite, monazite, columbite-tantalite, bismuth, and the sulfides formed later; quartz developed during all stages. The latest minerals to form were samarskite, uraninite, and smoky quartz.

Some of the pegmatites were essentially restricted to fractures, but others permeated the walls and left noticeable alteration zones along gradational contacts. In places the wall rock was replaced by pegmatite, leaving relict oriented stringers or ghosts of schist and quartzite. Figure 11-59 demonstrates the typical effect of pegmatitic saturation of the quartzite. The alteration zone averages only 1 or 2 meters thick and projects along permeable parting planes in the quartzite.

Jahns (1946) concluded that the Petaca pegmatites formed in two stages, an early magmatic stage and a later hydrothermal stage. The early stage produced zoned pegmatites of relatively simple mineralogy, the hydrothermal stage most of the albite, muscovite, and rare minerals. These residual fluids must have been rich in soda, silica, alumina, and significant quantities of niobium, tantalum, beryllium, thorium, uranium, fluorine, bismuth, copper, sulfur, and the rare earths. The hydrothermal fluids traveled along fractures and reacted with the previously formed pegmatitic minerals, leaving behind new minerals of a replacement origin. It was during this hydrothermal stage that most of the metals of economic interest were produced. Jahns' conclusions are completely consistent with, even corroborative of, those experimentally based conclusions of Jahns with Burnham 23 years later.

▸ GRANITIC TIN AND URANIUM DEPOSITS

Two more ore types clearly related to the intrusion of intermediate to felsic plutons are known as *tin granites* and *granitic uranium deposits*. Tin granites account for over 80% of the free world's present tin production, chiefly from Malaysia, South Africa, and Australia, and they constitute an important resource elsewhere, as in Tasmania, Alaska, and Nigeria. Granitic uranium occurrences are many and include the largest "hardrock" uranium mine in the world, at Rössing, South-West Africa (Namibia). Many ore deposits doubtlessly remain, some to be discovered and some to be created by changing economics.

It has become apparent since the publication of a paper by Chappel and White (1974) that at least two distinct major suites of granitic rocks occur throughout the world's orogenic belts, the *I-type* and the *S-type*. Porphyry copper, molybdenum, and tin deposits are associated with I-type granitoid rocks, and tin, tungsten, tantalum, molybdenum, beryllium, niobium, bismuth, and uranium "granite" deposits are found only in S-type granites, most of which are also *two-mica granites*. Two-mica granites contain quartz, twinned microcline, muscovite, biotite, and commonly accessory garnet, allanite, monazite, magnetite, zircon, and apatite. I-type granites are composed of quartz, mostly untwinned microcline or orthoclase, hornblende, biotite, and similar accessories, except that sphene is present and garnet absent. Two-mica granites are more felsic and much lighter in color index in hues of red and yellow; they are generally nonporphyritic, and show extensive internal variation; they are more commonly foliated, and are associated with metamorphic rocks. Their initial $^{87/86}Sr$ is high at 0.708 to 0.710, and they appear to have been intruded at high tectonic levels. Ishihara (1977) finds that I-type plutons are inclined to carry accessory magnetite, S-types only the more reduced phase ilmenite, although the fit is not perfect. Curiously, rocks that fit the S-type description occur in orogenic belts, such as in Malaya, Cornwall, France, and Tasmania, but others also are found in anorogenic (nonorogenic) settings such as that of the Bushveld Red Granites of South Africa. It has been suggested (Chappel and White, 1974; Plimer, 1978) that two-mica S-granites are derived by partial melting of continental crust or sedimentary rocks, and that I-granites are principally of upper mantle subduction zone origin. Groves and McCarthy (1978) point out that descriptive aspects of tin granites, which are relevant to exploration, are similar in the orogenic and anorogenic settings of the Blue Tier batholith of Tasmania and the Bushveld Red Granite of South Africa, respectively. To one who has visited the Zaaiplaats deposits in the Bobbejaankop and Lease Granites of South Africa, the Blue Tier descriptions to be cited below are strikingly similar. Groves and McCarthy emphasize (1978, page 11) that a

> *. . . comparison between tin-bearing granitoids in an anorogenic setting (Bushveld Complex) and an orogenic setting (Blue Tier Batholith, Tasmania) reveals a number of genetically important similarities. These include: in situ fractional crystallization of late melt in a sheet-like form near a roof zone; the association of barren pegmatites overlying the ore, and of aplites; and the occurrence of conformable tin-bearing sheets, often exhibiting greisenization.*

Groves and McCarthy consider tin-bearing granites to be specialized, advanced differentiates of crustally derived granitoid melts, and note that highly variable but low Sr and Ba contents and anomalously high levels of Li and Rb will aid the search for tin. Olade (1980) studied Jurassic tin granites in Nigeria and found Sn values over 25 ppm, low K/Rb, low Ba/Rb, and high Rb/Zr ratios to be valuable pathfinders to zones in granites in which late magmatic alkali-volatile phases concentrated tin. Plimer (1977) studied six Sn-W-Mo-Bi deposits in Paleozoic granitic rocks in eastern Australia, and reported that they are 1-to-10-meter-wide pipelike bodies of quartz, fluorite, calcite, cassiterite, and other ore minerals enclosed in subcircular greisen envelopes, with solution temperatures and salinities decreasing outward and with time from about 500 to 250°C and from 30 to 10% NaCl. Trace element patterns are similar to those reported by Olade, and the granites are S-granites. Ishihara et al. (1980) report that S-granites prevail in the tin granite districts of Thailand. Tantalum deposits recently identified in Egypt are in two-mica granites, as are disseminated uranium-beryllium-tungsten deposits in China, Korea, France, and elsewhere.

Blue Tier Tinfield, Tasmania

The Blue Tier tinfield and batholith deserves description. It lies at the extreme northeastern corner of Tasmania (Figure 11-60). Tin was discovered in the district in 1874, and alluvial mining started soon thereafter. The wash ores were patchy, local, and soon exhausted. Lode mining began in 1895 at the Anchor mine, which has produced about 80% of the lode output of the district.

The productive area is underlain entirely by late orogenic S-type granitic rocks of Upper Devonian age intruded into Silurian-Ordovician sediments. A few remnant patches of Permian sedimentary rocks are also present. Considerable Tertiary olivine basalt lies west-southwest of the district, and narrow diabase dikes of Mesozoic age cut northeast across the granitic rocks (Reid and Henderson, 1928). Figure 11-60 is a generalized map of the district after Groves (1972).

The granite is a composite tabular body 45 by 75 km across and about 12 km thick. It is generally separated into two principal types. The older of these ranges from hornblende-biotite granodiorite to coarse porphyritic granodiorite-quartz monzonite with phenocrysts of oligoclase up to 10 cm long (Thomas, 1953), which may be I-granites. The roof zones of these older steep-sided intrusive masses are cut by younger flat-topped sheetlike units of granite, with feeder dikes in places. These granites are the youngest parts of the batholith, contain 75 to 77% SiO_2, and are known locally as the "tin granite." They are the distinctively patterned masses in Figure 11-60. Groves and McCarthy (1978, page 12) said that

> *. . . all muscovite-biotite granites are even grained, reddish to yellow-brown rocks containing less than 5% total mica and commonly with miarolytic cavities. They contain large flakes of apparently primary muscovite, and rocks from the upper parts of the sheets contain accessory, apparently primary topaz, tourmaline, fluorite, and cassiterite. Biotites are typically Li- and F-rich but Cl-poor (Groves, 1974). The granites occupy less than 10% of the surface area of the batholith, and probably represent an even smaller volume, perhaps 2%, of the batholith.*

These tin granites, the biotite-muscovite units of Figure 11-60, sound like typical S-granites, although the authors propose a model of in situ fractional

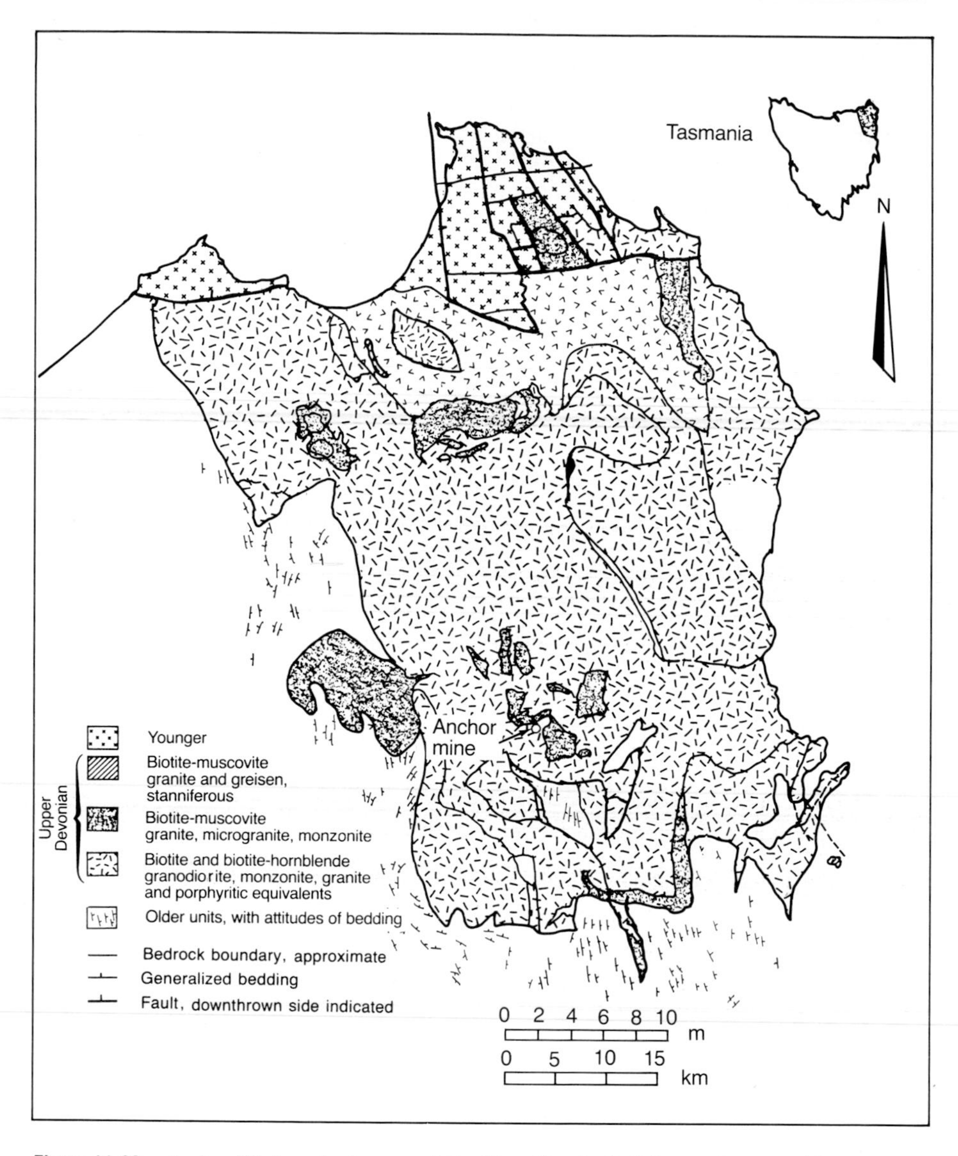

Figure 11-60. A simplified geologic map of the Blue Tier batholith, northeastern Tasmania, showing the distribution of granitic rocks. The tin mines and deposits, such as the anchor mine, are almost all in the sheetlike, youngest, two-mica biotite-muscovite granites and greisens that constitute 10% of the outcrop area, 2% of the volume of the Blue Tier batholith. (*After* Groves, 1972.)

crystallization to account for them. Clearly, the whole story of I- and S-granites and their contained mineralization is not yet known. New studies involving ratios of samarium and its radiogenic decay product neodymium (Farmer and De Paolo, 1983), rubidium and its radiogenic product strontium, and lutetium and its isotope hafnium (Patchett, 1983), hold great promise to explain the origins of these rock systems. Tin is particularly rich in structural irregularities along the upper contacts of the sheeted granites. Greisenization is common and was described by Groves and Taylor (1973) as having been produced by ascending fluids containing tin, fluorine, lithium, and rubidium. These fluids reacted with feldspars and biotite to form muscovite, topaz, fluorite, and cassiterite. The most productive tin ore bodies are flat-lying bodies known locally as "floors." Cassiterite is found as disseminated grains in tin granite and greisen, pegmatites and granitic dikes, and in quartz and greisen veins. Reid and Henderson (1928) stated that the "floor' deposits really are closely spaced, nearly vertical greisen veinlets that extend to considerable depths, but that they contain cassiterite only in their upper 70 meters. Jack (1965) stated that the "floors" appeared to be flat joints penetrated by mineralizing fluids. At the Anchor mine, the contact between muscovite-biotite granite and the overlying porphyritic biotite granodiorite-quartz monzonite is commonly masked by pegmatitic lenses. Below the contact are irregular arched patches, many of crescent shape, of muscovite-biotite granite and greisen, leucogranite, aplite, and pegmatite that may contain tin in economic quantities, generally 0.2% Sn or better. Relatively thin greisen veinlets cut this sequence.

Groves (1974) and Groves and McCarthy (1978) stated that the tin is genetically related to the sheetlike intrusions of muscovite-biotite granite. Variations throughout the granite and changes during greisenization are consistent with contemporaneous metasomatism rather than with postmagmatic alteration. Ore in the greisenized granite contains topaz, abundant fluorite, and minor amounts of scheelite, wolframite, and galena in addition to the cassiterite. In the veins, the ore commonly contains cassiterite, chalcopyrite, wolframite, and molybdenite.

This description is remarkably similar to that of the Zaaiplaats tin district in the Red Granites of the Bushveld Complex of South Africa, even to the presence there of pegmatite phases, of sheetlike zones of two-mica granites with disseminated cassiterite, fluorite, and topaz, and of rich veins and pipes which have been mined downward into dissemination zones. Pipelike bodies almost identical to those described by Plimer in Australia are common there, and geochemical patterns mimic those already described.

Tin granites therefore appear to be a deposit type; a volume edited by Yeap (1981) and published by the Geological Survey of Malaysia entitled *The Geology of Tin* provides modern descriptions and discussions of the world-class Malaysian deposits. It is valuable to assess their similarities with respect to deposits described in the foregoing paragraphs. Recent information from Cornwall (Stone, 1980) depicts the intrusives there as two-mica S-types.

Tin has also been found in anomalous amounts in rhyolites in Mexico (Lee-Moreno, 1972) and at several localities in the Black Mountains of New Mexico. Search for tin in porphyry environments, high-level granites, and felsic volcanic rocks continues.

Granitic Uranium Deposits

The complex economic geology of uranium includes many deposits in and associated with felsic plutonic to epizonal igneous bodies. Deposits in sediments are described in Chapters 16 and 20, and deposits related to unconformities are considered in Chapter 20. Uranium in igneous rocks is found in important quantities in granites and in lesser to trivial amounts in syenites, kimberlites, and carbonatites (Chapter 9). In "granites," it is found in normal I-granites and alkali-rich granites, anatectic S-granites, pegmatites, and in veins related to each of these three subtypes. Pegmatite and vein settings have been discussed (Figure 11-50 and Table 11-3); we focus here on the granite-alkali granite occurrences.

The uranium ions U^{+4} and U^{+6} are both too large and too highly charged to participate in main-line silicate fractional crystallization phases. Like tin and its companions, it is therefore concentrated increasingly in siliceous late melts. Uranium values increase steadily with fractional crystallization from 0.00*X* ppm in ultramafic rocks to 2 to 20 ppm in "average granites." Toward the end of crystallization of either I-type or S-type granitoid rocks, it may precipitate as the oxide uraninite in disseminations; it may partition increasingly into accessory minerals like allanite, titanite, zircon, and monazite; it may form uranothorianite $(U,Th)(SiO_4)_2$; or it may accompany other oxides into late crystallization zones such as the sheeted zones of tin granites just described, there to be deposited as veinlets, veins, or disseminations of uranium minerals, including the silicates coffinite and brannerite. A few parts per million uranium is normal in porphyry copper deposits (Davis and Guilbert, 1973), and uranium was being extracted from several, as at Twin Buttes, Arizona, in 1980. Typically uranium-thorium values are highest in the most differentiated rocks, be they granites or rhyolites. Uranium values are commonly in late veinlets or in a mineralization mode clearly related to but younger than the igneous body in which they occur—texturally epigenetic, as is chalcopyrite in veinlets in a porphyry copper deposit. Indeed the term "porphyry uranium deposit" has been suggested to describe disseminated or dispersed veinlet occurrences, but it has been rejected because proposed deposits need show but little of the alteration, fluid flow, scale, and geologic process characteristics of other porphyry base-metal deposits.

Disseminated uranium deposits may occur in either I- or S-granites, although they are both more common and more important in the latter. Since S-granites seem to involve anatectic melting of sedimentary rocks or continental crust, it is not surprising that S-granites appear to form a continuum with metamorphic complexes, metamorphic core complexes (Crittenden, Coney, and Davis, 1980), and terranes that include pegmatites, migmatites, lit-par-lit and boudinage phenomena, and other evidences of partial melting of crustal rocks. Uranium concentration of possible economic significance has been identified in the Kettle Dome metamorphic core complex rocks in the state of Washington, and tungsten mineralization is now seen as being related in part to such complexes (Silver, 1980). The greatest single deposit of uraninite now known in the Earth's crust is the Rössing deposit near Swakopmund, South-West Africa (Namibia), where major uraninite and minor davidite [$(Fe,U,Ca)(Ti,Fe)_3O_7$] and oxidized phases occur in high-silica alaskites, pegmatites, and migmatites. As described in detail in Chapter 18, they outcrop as an elliptical anticline-syncline folded complex of the Damaran orogeny of 500 to 600 million years (Berning et al., 1976). Grade runs from 30 to 1000 ppm (0.003 to 0.1%) U and reserves of several hundred million tons of 500 ppm (0.05%) U to 500-meter depth make this major deposit of crucial geopolitical significance. The rocks are typical S-type granites. Uraninite is in biotite- and zircon-rich zones in alaskite that cut gneiss. Similar deposits occur in Nigeria (Bowden and Turner, 1974; Olade, 1980) and in granite-metamorphic complexes elsewhere in the world. Deposits of this type, as well as disseminations in high-level I-type rocks, will be increasingly sought.

The Ross-Adams deposit at Bokan Mountain, Alaska, is an excellent example of the latter type. Bokan Mountain is 60 km southwest of Ketchikan in southeastern Alaska and only 100 km west of the 1974 molybdenite discovery at Quartz Hill, Alaska. The deposit consists of veins, pods, and pipelike bodies of uranothorite, uraninite, and coffinite as disseminations and in randomly dispersed microveinlets with fluorite, hematite, pyrite, galena, molybdenite, quartz, and calcite. These minerals, and hence the ore deposits, are in an early Jurassic peralkaline granite which shows 12 distinct intrusive episodes of sodic granites, porphyritic units, and pegmatites in a ring dike-stock, I-type granite intrusive complex (MacKevett, 1963; Lanphere, MacKevett and Stern, 1964; Thompson, Pierson, and Lyttle, 1982). Uranium minerals are the last magmatic-hydrothermal manifestations; they constitute disseminations, veinlets, and segregations in the granite and in related aplites and pegmatites, especially where high sodium, zirconium, and rubidium values are noted. The Ross-Adams yielded some 10,000 to 20,000 tons of ore in 1957 from a diffuse zone 10 by 55 meters in plan and 5 to 15 meters thick. Grades ranged from 30,000 ppm (3%) U through considerable tonnage at 10,000 ppm (1%) U to a 4000-ppm (0.4%) average. Other bodies will probably be found. Although the area is not now in production, similar deposits should be sought in orogenic belts worldwide.

Kelly and Rye (1979) portrayed a consanguineous relationship between a Hercynian (Silurian) S-type granite cupola environment, horizontal hydraulically dilated veins, and quartz-tin-tungsten hydrothermal mineralization in them at Panasqueira, Portugal. Temperatures declined during metallization from 360 to 120°C, salinities ranged from 5 to 10% NaCl downward, and pressures were relatively low at 600-to-1300-meter depth.

Miller and Bradfish (1980) describe in a short paper a belt of Mesozoic and Cenozoic S-type granitoid intrusive plutons which stretches down the Cordillera of the United States east of a belt of I-type granitoids. The plutons are known to contain tin and molybdenum; this melding of isolated data into a regional pattern will doubtlessly affect tin-tungsten-molybdenum exploration.

Bibliography

Ahmad, S. N., and A. W. Rose, 1980. Fluid inclusions in porphyry and skarn ore at Santa Rita, New Mexico. *Econ. Geol.* 75:229–250.

Allcock, J. B., 1982. Skarn and porphyry copper mineralization at Mines Gaspé, Murdochville, Quebec. *Econ. Geol.* 77:971–999.

Ambler, E. A., and R. A. Facer, 1975. A Silurian porphyry copper prospect near Yeoval, N.S.W. *J. Geol. Soc. Austral.* 22:229–242.

Ambrus, W., J., 1977. Geology of the El Abra porphyry copper deposit, Chile. *Econ. Geol.* 72:1062–1085.

——, 1979. Emplazamiento y mineralización de los porfidos cupriferos de Chile. Ph.D. Dissert., Univ. Salamanca, Spain, 313 pp.

Anderson, A. L., 1949. Monzonite intrusion and mineralization in the Coeur d'Alene district, Idaho. *Econ. Geol.* 44:169–185.

Anderson, R. J., 1940. Microscopic features of ore from the Sunshine Mine. *Econ. Geol.* 35:659–667.

Atkinson, W. W., Jr., and M. T. Einaudi, 1978. Skarn formation and mineralization in the contact aureole at Carr Fork, Bingham, Utah. *Econ. Geol.* 73:1326–1365.

Babu, V. R. R. M., 1969. Temperatures of formation of pegmatites of the Nellore mica belt, Andra Pradesh, India. *Econ. Geol.* 64:66–71.

Bandy, M. C., 1938. Mineralogy of three sulfide deposits of northern Chile. *Amer. Mineral.* 23:669–760.

Banks, N. G., 1973. Biotite as a source of some of the sulfur in porphyry copper deposits. *Econ. Geol.* 68:697–708.

—— and J. S. Stuckless, 1973. Chronology of intrusion and ore deposition at Ray, Arizona, pt. II: fission track ages. *Econ. Geol.* 68:657–664.

Barnes, H. L., Ed., 1967. *Geochemistry of Hydrothermal Ore Deposits.* New York: Holt, Rinehart and Winston, 670 pp.

——, Ed., 1979. *Geochemistry of Hydrothermal Ore Deposits.* 2nd ed. New York: Wiley, 798 pp.

Batchelder, J., 1977. Light stable isotope and fluid inclusion study of the porphyry copper deposit at Copper Canyon, Nevada. *Econ. Geol.* 72:60–70.

Bateman, A. M., 1950. *Economic Mineral Deposits.* New York: Wiley, 916 pp.

Beane, R. E., 1974. Biotite stability in a porphyry copper environment. *Econ. Geol.* 69:241–256.

——, 1980. Pers. Commun.

——, 1982. Hydrothermal alteration in silicate rocks, pp. 117–138 in S. R. Titley, Ed., *Advances in Geology of the Porphyry Copper Deposits, Southwestern North America.* Tucson: Univ. Ariz. Press, 560 pp.

Beane R. E., and S. R. Titley, 1981. Porphyry copper deposits: hydrothermal alteration and mineralization. *Econ. Geol. 75th Anniv. Vol.*, pp. 235–269.

Berning, J., R. Cooke, S. A. Hiemstra, and U. Hoffman, 1976. The Rössing uranium deposit, South-West Africa. *Econ. Geol.* 71:351–368.

Beus, A. A., 1966. *Geochemistry of Beryllium and Genetic Types of Beryllium Deposits.* New York: Freeman, 401 pp.

Bird, D. K., and H. C. Helgeson, 1980. Chemical interaction of aqueous solutions with epidote-feldspar mineral assemblages in geologic systems. 1: Thermodynamic analysis of phase relations in the system CaO-FeO-Fe_2O_3-Al_2O_3-SiO_2-H_2O-CO_2. *Amer. J. Sci.*, 280:907–941.

Bookstrom, A. A., 1977. The magnetite deposits of El Romeral, Chile. *Econ. Geol.* 72:1101–1130.

——, 1981. Tectonic setting and generation of Rocky Mountain porphyry molybdenum deposits, pp. 215–226 in W. R. Dickinson and W. D. Payne, Eds., *Relations of Tectonics to Ore Deposits in the Southern Cordillera.* Tucson: *Ariz. Geol. Soc. Dig.* 14, 288 pp.

——, 1983. Pers. Commun.

Bowden, P., and D. C. Turner, 1974. Peralkaline and associated ring complexes in the Nigeria-Niger Province, West Africa, in H. Sorenson, Ed., *The Alkaline Rocks.* New York: Wiley.

Brimhall, G. H., Jr., 1979. Lithologic determination of mass transfer mechanisms of multiple-stage porphyry copper mineralization at Butte, Montana; vein formation by hypogene leaching and enrichment of potassium-silicate protore. *Econ. Geol.* 74:556–589.

——, 1980. Deep hypogene oxidation of porphyry copper potassium-silicate protore at Butte, Montana; a theoretical evaluation of the copper remobilization hypothesis. *Econ. Geol.* 75:384–409.

Brögger, W. C., 1890. Die Mineralien der Syenitpegmatitgänge der Südnorwegischen Augit- und Nephelinsyenite. *Kristall. Mineral.* 16:215–235.

Burnham, C. W., 1979. Magmas and hydrothermal fluids, pp. 71–136 in H. L. Barnes, Ed., *Geochemistry of Hydrothermal Ore Deposits*, 2nd ed. New York: Wiley, 798 pp.

—— and H. Ohmoto, 1980. Late stage processes of felsic magmatism, pp. 1–11 in S. Ishihara and S. Takenouchi, Eds., *Granitic Magmatism and Related Mineralization*. Soc. Min. Geol. Jap., Spec. Issue 8, 247 pp.

Cameron, E. N., 1955. Concepts of the internal zoning of granitic pegmatites and their application to certain pegmatites of South-West Africa. *Trans. Geol. Soc. So. Afr.* 38:45–70.

——, R. H. Jahns, A. H. McNair, and L. R. Page, 1949. Internal structure of granitic pegmatites. Econ. Geol. Mon. 2.

Cathles, L. M., 1977. An analysis of the cooling of intrusives by groundwater convection which includes boiling. *Econ. Geol.* 72:804–826.

Cerny, P., 1982. Anatomy and classification of granitic pegmatites, pp. 1–40, *Mineral. Assoc. Can.*, in P. Cerny, Ed., *Short Course in Granitic Pegmatites in Science and Industry*. Winnipeg, Man. 555 pp.

Chaffee, M. A., 1975. Geochemical exploration techniques applicable in the search for copper deposits. USGS Prof. Pap. 907B, pp. B1–B26.

——, 1976. The zonal distribution of selected elements above the Kalamazoo porphyry copper deposit, San Manuel district, Pinal County, Arizona. *J. Geochem. Explor.* 5:145–165.

Chappel, B. W., and A. J. R. White, 1974. Two contrasting granite types. *Pac. Geol.* 8:173–174.

Chivas, A. R., and R. W. T. Wilkins, 1977. Fluid inclusion studies in relation to hydrothermal alteration and mineralization at the Koloula porphyry copper prospect, Guadalcanal. *Econ. Geol.* 72:153–169.

Cook, R. B., 1978. Famous mineral localities: Chuquicamata, Chile. *Mineral Rec.* 9:321–333.

Cotelo Neiva, J. M., 1954. Pegmatitos com cassiterite e tantalite-columbite da Cabracão (Pontedo Lima-Serra de Arga). Univ. Coimbra, Portugal, Mem. 36.

Crittenden, M. D., Jr., P. J. Coney, and G. H. Davis, 1980. Cordilleran metamorphic core complexes. Geol. Soc. Amer. Mem. 153, 490 pp.

Crosby, W. O., and M. L. Fuller, 1897. Origin of pegmatite. *Amer. Geol.* 19:147–180.

Davis, J. D., and J. M. Guilbert, 1973. Distribution of the radioelements potassium, uranium, and thorium in selected porphyry copper deposits. *Econ. Geol.* 68:145–160.

DeGeoffroy, J., and T. K. Wignall, 1972. A statistical study of geological characteristics of porphyry copper-molybdenum deposits in the Cordilleran belt—application to the rating of porphyry prospects. *Econ. Geol.* 67:656–668.

Derry, D. R., 1931. The genetic relationships of pegmatites, aplites, and tin veins. *Geol. Mag. Gt. Brit.* 68:454–475.

Doe, B. R., 1973. Lead isotopes, ore genesis, and ore prospect evaluation. *Econ. Geol.* 68:1206.

Edwards, A. B., and G. Baker, 1953. Scapolitization in the Cloncurry district of northwestern Queensland. *Geol. Soc. J.* 1:1–33.

Einaudi, M. T., 1977. Petrogenesis of the copper-bearing skarn at the Mason Valley Mine, Yerington district, Nevada. *Econ. Geol.* 72:769–795.

——, 1982. Skarns associated with porphyry plutons. I: Description of deposits, southwestern North America; II: General features and origin, pp. 139–183 in S. R. Titley, Ed., *Advances in Geology of the Porphyry Copper Deposits of Southwestern North America.* Tucson: Univ. Ariz. Press, 560 pp.

——, L. D. Meinert, and R. J. Newberry, 1981. Skarn deposits. *Econ. Geol. 75th Anniv. Vol.*, pp. 317–391.

Emery, J. A., 1968. Geology of the Pea Ridge iron ore body, Missouri, pp. 359–369 in J. D. Ridge, Ed., *Ore Deposits of the United States 1933/1967*, Graton-Sales Vols. New York: AIME, 1880 pp.

Emmons, W. H., 1940. *Principles of Economic Geology.* New York: McGraw-Hill.

Ettlinger, I. A., 1928. Ore deposits support hypothesis of a central Arizona batholith. AIME Tech. Pub. 63.

Farmer, G. L., and D. J. DePaolo, 1983. Origin of Mesozoic and Tertiary granite in the western United States and implications for pre Mesozoic crystal structure, 1, Nd and Sr isotopic studies in the geocline of the northern Great Basin. *J. Geophys. Res.* 88, B4: 3379–3402.

Fehn, U., L. M. Cathles, and H. D. Holland, 1978. Hydrothermal convection and uranium deposits in abnormally radioactive plutons. *Econ. Geol.* 73:1556–1566.

Field, C. W., et al., 1975. Chemical trends in Mesozoic plutons associated with porphyry-type mineralization of the Pacific Northwest. AIME Preprint 75-L-359, 25 pp.

Fischer, R., 1950. Entmischungen in Schmelzen aus Schwermetalloxyden, Silicaten, und Phosphaten. *Neues Jahrb. Mineral. Abh.* 81:315–364.

Frietsch, R., 1973. The origin of the Kiruna iron ores. *Geol. Foren. Stockholm Forh.* 95:375–380.

——, 1978. On the magmatic origin of the iron ores of the Kiruna type. *Econ. Geol.* 73:478–485.

Fryklund, V. C., Jr., 1964. Ore deposits of the Coeur d'Alene district, Shoshone County, Idaho. USGS Prof. Pap. 445.

—— and J. D. Fletcher, 1956. Geochemistry of sphalerite from the Star mine, Coeur d'Alene district, Idaho. *Econ. Geol.* 51:223–247.

Gale, R. E., 1965. The geology of the Mission copper mine, Pima mining district, Arizona. Unpub. M.S. Thesis, Stanford Univ., 162 pp.

Gallagher, M. J., 1975. Composition of some Rhodesian lithium-beryllium pegmatites. *Trans. Geol. Soc. So. Afr.* 78:35–70.

Geijer, P., 1910. Igneous rocks and iron ores at Kiirunavaara, Lousavaara, and Toulluvaara, Sweden. *Econ. Geol.* 5:699–718.

——, 1931a. The iron ores of the Kiruna type. Sver. Geol. Unders. Arsb. Ser. C Avh. Uppsatser 367.

——, 1931b. Pre-Cambrian geology of the iron-bearing region of Kiruna-Gallivare-Pajala. Sver. Geol. Arsb. Ser. C Avh. Uppsatser 366, pp. 185–221.

——, 1960. The Kiruna iron ores. 21st Int. Geol. Congr. Guide Excursions A25 and C20, pp. 3–17.

——, 1967. Internal features of the apatite-bearing magnetite ores. Sver. Geol. Unders. Arsb. Ser. C Avh. Uppsatser 624.

—— and N. H. Magnusson, 1952. The iron ores of Sweden, in *Symposium sur les Gisements de Fer du Monde*, 19th Int. Geol. Congr., vol. 2.

Gevers, T. W., 1936. Phases of mineralization in Namaqualand pegmatites. *Geol. Soc. So. Afr. Trans.* 39:331–378.

Gott, G. B., and J. B. Cathrall, 1980. Geochemical exploration studies in the Coeur d'Alene district, Idaho and Montana. USGS Prof. Pap. 1116.

Grant, J. N., C. Halls, W. Avila, and G. Avila, 1977. Igneous geology and the evolution of hydrothermal systems in some subvolcanic tin deposits of Bolivia, pp. 117–126 in *Volcanic Processes in Ore Genesis*. Inst. Min. Metall. Spec. Pap. 7, 188 pp.

——, C. Halls, S. M. F. Sheppard, and W. Avila, 1980. Evolution of the porphyry tin deposits of Bolivia, pp. 151–174 in S. Ishihara and S. Takenouchi, Eds., *Granitic Magmatism and Related Mineralization*. Soc. Min. Geol. Jap., Spec. Issue 8, 247 pp.

Graybeal, F. T., 1973. Copper, manganese, and zinc in coexisting mafic minerals from Laramide intrusive rocks in Arizona, *Econ. Geol.* 68:785–798.

Groves, D. I., 1972. The geochemical evolution of tin-bearing granites in the Blue Tier batholith, Tasmania. *Econ. Geol.* 67:445–457.

——, 1974. Geochemical variation within tin-bearing granites, Blue Tier batholith, Northeast Tasmania, pp. 154–158 in M. Stemprok, Ed., *Metallization Associated with Acid Magmatism*, Prague: Ustredni Ustav Geol. Czech.

—— and T. S. McCarthy, 1978. Fractional crystallization and the origin of tin deposits in granitoids. *Mineral. Deposita* 13:11–26.

—— and R. G. Taylor, 1973. Greisenization and mineralization at Anchor Tin mine, northeast Tasmania. *Inst. Min. Metall. Trans.* 82:135–146.

Guilbert, J. M., 1981. A plate tectonic-lithotectonic classification of ore deposits, pp. 1–10 in W. R. Dickinson and W. D. Payne, Eds., *Relations of Tectonics to Ore Deposits in the Southern Cordillera.* Tucson: Ariz. Geol. Soc. Dig. XIV, 288 pp.

—— and J. D. Lowell, 1974. Variations in zoning patterns in porphyry copper deposits, *Can. Inst. Min. Metall. Bull.*, 67(Feb.):99–109.

—— and R. W. Schafer, 1978. Preliminary geochemical characterization of muscovites in porphyry base metal alteration assemblages. *5th IAGOD Quad. Symp. Proc.*, pp. 57–68.

Gustafson, L. B., 1961. Paragenesis and hypogene zoning at the Magma mine, Superior, Arizona. Unpub. Ph.D. Dissert., Harvard Univ.

—— and J. P. Hunt, 1975. The porphyry copper deposit at El Salvador, Chile. *Econ. Geol.* 70:857–912.

Hall, W. E., I. Friedman, and J. T. Nash, 1974. Fluid inclusion and light stable isotope study of the Climax molybdenum deposits, Colorado. *Econ. Geol.* 69:884–901.

Halliday, A. N., 1980. The timing of early and main stage ore mineralization in southwest Cornwall. *Econ. Geol.* 75:752–759.

Hammer, D. F., 1983. Pers. Commun.

—— and D. W. Peterson, 1968. Geology of the Magma mine area, Arizona, pp. 1282–1310 in J. D. Ridge, Ed., *Ore Deposits of the United States 1933/1967*, Graton-Sales Vols. New York: AIME, 1880 pp.

Hanley, J. B., E. W. Heinrich, and L. R. Page, 1950. Pegmatite investigations in Colorado, Wyoming, and Utah, 1942–44. USGS Prof. Pap. 227.

Haynes, F. M., 1980. The evolution of fracture-related permeability within the Ruby Star granodiorite, Sierrita porphyry copper deposit, Pima County, Arizona. M.S. Prepub. Manus., Univ. Ariz., Tucson, 48 pp.

—— and S. R. Titley, 1980. The evolution of fracture-related permeability within the Ruby Star granodiorite, Sierrita porphyry copper deposit, Pima County, Arizona. *Econ. Geol.* 75:673–683.

Heinrich, E. W., 1963. Zoning in pegmatite districts. *Amer. Mineral.* 38:68–87.

Hemley, J. J., J. W. Montoya, J. W. Marinenko, and R. W. Luce, 1980. Equilibria in the system Al_2O_3-SiO_2-H_2O and some general implications for alteration/mineralization processes. *Econ. Geol.* 75:210–228.

Hernon, R. M., W. R. Jones, and S. L. Moore, 1953. Some geological features of the Santa Rita Quadrangle, New Mexico, in *Guidebook of Southwestern New Mexico*, 4th Field Conf. Socorro, N. Mex., N. Mex. Geol. Soc.

—— and W. R. Jones, 1968. Ore deposits of the Central Mining District, Grant County, New Mexico, pp. 1211–1237 in J. D. Ridge, Ed., *Ore Deposits of the United States 1933/1967*, Graton-Sales Vols. New York: AIME, 1880 pp.

Herrera, A. O., 1968. Geochemical evolution of zoned pegmatites in Argentina. *Econ. Geol.* 63:13–29.

Hess, F. L., 1919. Tactite, the product of contact metamorphism. *Amer. J. Sci.* 198:377–378.

Himes, D., 1973. Mineralization and alteration at Pima mine—a complex porphyry copper deposit. AIME *Trans.* 254:166–176.

Hobbs, S. W., 1968. Geologic setting of metallic ore deposits in the northern Rocky Mountains and adjacent areas, pp. 1351–1372 in J. D. Ridge, Ed., *Ore Deposits of the United States 1933/1967*, Graton-Sales Vols. New York: AIME, 1880 pp.

—— and V. C. Fryklund, 1968. The Coeur d'Alene district, Idaho, pp. 1417–1435 in J. D. Ridge, Ed., *Ore Deposits of the United States 1933/1967*, Graton-Sales Vols. New York: AIME, 1880 pp.

——, A. B. Griggs, R. E. Wallace, and A. B. Campbell, 1965. Geology of the Coeur d'Alene district, Shoshone County, Idaho. USGS Prof. Paper 478.

Hollister, V. F., 1975. The porphyry molybdenum deposit of Compaccha, Peru, and its geologic setting. *Mineral. Deposita* 10:141–152.

——, 1978. *Geology of the Porphyry Copper Deposits of the Western Hemisphere.* New York: AIME, 219 pp.

Hosterman, J. W., 1956. Geology of the Murray area, Shoshone County, Idaho. USGS Bull. 1027-P.

Hudson, T., J. G. Arth, and K. G. Muth, 1981. Geochemistry of intrusive rocks associated with molybdenum deposits, Ketchikan Quadrangle, southeastern Alaska. *Econ. Geol.* 76:1225–1232.

Hunt, T. S., 1871. Notes on granitic rocks. *Amer. J. Sci.* 101:82–89, 182–191.

Hussak, E., 1898. Der goldführende, kiesige Quartzlagergang von Passagem in Minas Geraes, Brasilien. *Z. prakt. Geol.* 6:345–357.

Ingerson, E., 1947. Liquid inclusions in geologic thermometry. *Amer. Mineral.* 32:375–388.

Ishihara, S., 1977. The magnetite-series and ilmenite-series granitic rocks. *Min. Geol.* 27:293–305.

——, H. Sawata, K. Shibata, S. Terashima, S. Arrykul, and K. Sato, 1980. Granites and Sn-W deposits of peninsular Thailand, pp. 223–241 in S. Ishihara and S. Takenouchi, Eds., *Granitic Magmatism and Related Mineralization.* Soc. Min. Geol. Jap., Spec. Issue 8, 247 pp.

Jack, R., 1965. Tin deposits of north-east Tasmania, pp. 497–500 in J. McAndrew, Ed., *Geology of Australian Ore Deposits*, vol. 1. 8th Commonw. Australas. Min. Metall. Congr.

Jackson, N. J., 1981. Geology of the Cornubian tin field—a review, pp. 209–238 in C. H. Yeap, Ed., *Geology of Tin Deposits*, Kuala Lumpur, Malay.: Bull. Geol. Soc. Malaysia 11, 392 pp.

Jacobs, D. C., and W. T. Parry, 1976. Comparison of the geochemistry of biotite from some Basin and Range stocks. *Econ. Geol.* 71:1029–1035.

—— and ——, 1979. Geochemistry of biotite in the Santa Rita copper deposit, New Mexico. *Econ. Geol.* 74:860–887.

Jahns, R. H., 1946. Mica deposits of the Petaca district, Rio Arriba County, New Mexico. N. Mex. Bur. Mines Mineral Res. Bull. 25.

——, 1951. Geology, mining, and uses of strategic pegmatites. AIME *Trans.* 190:45–59.

——, 1953. The genesis of pegmatites. *Amer. Mineral.* 38:563–598; 1078–1112.

——, 1954. Pegmatites of Southern California. *Calif. Div. Mines Bull.* 170:37–50.

——, 1955. The study of pegmatites. *Econ. Geol. 50th Ann. Vol.*, pp. 1025–1130.

—— and C. W. Burnham, 1969. Experimental studies of pegmatite genesis: I. A model for the derivation and crystallization of granitic pegmatites. *Econ. Geol.* 64:843–864.

James, A. H., 1971. Hypothetical diagrams of several porphyry copper deposits. *Econ. Geol.* 66:43–47.

Jarrell, O. W., 1944. Oxidation at Chuquicamata, Chile. *Econ. Geol.* 39:215–286.

Jones, W. R., R. M. Hernon, and S. L. Moore, 1967. General geology of Santa Rita quadrangle, Grant County, New Mexico. USGS Prof. Pap. 555.

Just, E., 1937. Geology and economic features of the pegmatites of Taos and Rio Arriba Counties, New Mexico. N. Mex. Bur. Mines Mineral Res. Bull. 13.

Kelly, W. C., and R. O. Rye, 1979. Geologic, fluid inclusion, and stable isotope studies of the tin-tungsten deposits of Panasqueira, Portugal. *Econ. Geol.* 74:1721–1822.

—— and F. S. Turneaure, 1970. Mineralogy, paragenesis, and geothermometry of the tin and tungsten deposits of the eastern Andes, Bolivia. *Econ. Geol.* 65:609–680.

Kesler, S. E., L. M. Jones, and R. L. Walker, 1975. Intrusive rocks associated with porphyry copper mineralization in island arc areas. *Econ. Geol.* 70:515–526.

Kesler, T. L., 1961. Exploration of the Kings Mountain pegmatites. *Min. Eng.* 13:1062–1068.

Kinnison, J. E., 1966. The Mission copper deposit, pp. 281–287 in S. R. Titley and C. L. Hicks, Eds., *Geology of the Porphyry Copper Deposits, Southwestern North America*. Tucson: Univ. Ariz. Press, 287 pp.

Kirkham, R. V., 1971. Intermineral intrusives and their bearing on the origin of porphyry copper and molybdenum deposits. *Econ. Geol.* 66:1244–1249.

Korkowski, B. J., 1982. Trace element dispersion patterns around the North Silver Bell Extension, Silver Bell porphyry copper deposit, Pima County, Arizona. Unpub. M.S. Thesis. Univ. Ariz., Tucson, 82 pp.

Krieger, M. H., 1961. Troy quartzite (younger Precambrian) and Bolsa and Abrigo formations (Cambrian), northern Galiuro Mountains, southeastern Arizona. USGS Prof. Pap. 424-C, pp. 160–164.

Kunasz, I., 1982. Foote Mineral Company's Kings Mountain operation. pp. 505–511 in P. Cerny, Ed., *Short Course in Granitic Pegmatites in Science and Industry.* Winnipeg, Man.: Mineral. Assoc. Can., 555 pp.

Kwak, A. P., 1978a. Mass balance relationships and skarn-forming processes at the King Island scheelite deposit, King Island, Tasmania, Australia. *Amer. J. Sci.* 278:943–968.

——, 1978b. The conditions of formation of the King Island scheelite contact skarn, King Island, Tasmania, Australia. *Amer. J. Sci.* 278:969–999.

Lainé, R. P., 1974. Geological-geochemical relationships between porphyry copper and porphyry molybdenum ore deposits. Unpub. Ph.D. Dissert., Univ. Ariz., Tucson, 326 pp.

Landes, K. K., 1933. Origin and classification of pegmatites. *Amer. Mineral.* 18:33–56, 95–103.

Landis, G. P., and D. L. Leach, 1983. Silver-base-metal mineralization as a product of metamorphism; Coeur d'Alene district, Shoshone County, Idaho. *Geol. Soc. Amer.*, 36th Ann. Mtg., Rocky Mt. Sect., Abs. with Prog. 15:327.

Langlois, J. D., 1978. Geology of the Cyprus-Pima mine, Pima County, Arizona. *Ariz. Geol. Soc. Dig.* 11:103–113.

Lanphere, M. A., E. M. MacKevett, Jr., and T. W. Stern, 1964. Potassium-argon and lead alpha ages of plutonic rocks, Bokan Mountain, Alaska. *Science* 145:705–707.

Large, P. R., 1972. Metasomatism and scheelite mineralization at Bold Head, King Island. Austral. Inst. Min. Metall. Proc. 238, pp. 31–45.

Lasky, S. G., and A. D. Hoagland, 1950. Central mining district, New Mexico. 18th Int. Geol. Congr. Rept., pt. 7, pp. 97–110.

Laznicka, P., 1976. Porphyry copper and molybdenum deposits of the U.S.S.R. and their plate tectonic setting. *Inst. Min. Metall. Trans.* 85:B13–B32.

Lee-Moreno, J. L., 1972. Geological and geochemical exploration characteristics of Mexican tin deposits in rhyolitic rocks. Unpub. Ph.D. Dissert., Univ. Ariz., Tucson, 108 pp.

Lehman, N. E., 1980. Pyrometasomatic deposits of the Washington Camp-Duquesne district, Santa Cruz County, Arizona. *Min. Eng.* 32:181–188.

Lehtinen, M., 1974. Degree of Al-Si order in K-feldspars, a combination of X-ray and infrared data. *Contr. Mineral. Pet.* 47:223–230.

Leith, C. K., and E. C. Harder, 1908. The iron ores of the Iron Springs district, southern Utah. USGS Bull. 338.

Lindgren, W., 1905. The copper deposits of the Clifton-Morenci district, Arizona, USGS Prof. Pap. 43.

——, L. C. Graton, and C. H. Gordon, 1910. The ore deposits of New Mexico. USGS Prof. Pap. 68, pp. 305–318.

Lopez, V. M., 1939. The primary mineralization at Chuquicamata, Chile. *Econ. Geol.* 34:674–711.

——, 1942. Chuquicamata, Chile, pp. 126–128 in W. H. Newhouse, Ed., *Ore Deposits as Related to Structural Features*, Princeton, NJ: Princeton Univ. Press, 280 pp.

Lowell, J. D., 1968. Geology of the Kalamazoo orebody, San Manuel district, Arizona. *Econ. Geol.* 63:645–654.

——, 1974. Regional characteristics of porphyry copper deposits of the Southwest. *Econ. Geol.* 69:601–617.

—— and J. M. Guilbert, 1970. Lateral and vertical alteration-mineralization zoning in porphyry ore deposits. *Econ. Geol.* 65:373–408.

Lutton, R. J., 1959. Pegmatites as a link between magma and copper molybdenum ore. *The Mines Mag.*, Dec., pp. 15–24.

MacKenzie, W. B., 1970. Hydrothermal alteration associated with the Urad and Henderson molybdenite deposits, Clear Creek County, Colorado. Ph.D. Dissert., Univ. Mich., 208 pp.

MacKevett, E. A., 1963. Geology and ore deposits of the Bokan Mt. uranium-thorium area, southeastern Alaska. USGS Bull. 1154.

Mackin, J. H., 1947. Some structural features of the intrusions in the Iron Springs district, Utah, in *Guidebook to the Geology of Utah*, No. 2. Salt Lake City: Utah Geol. Surv.

——, 1954. Geology and iron ore deposits of the Granite Mountain area, Iron County, Utah. USGS Map MF 14.

——, 1968. Iron ore deposits of the Iron Springs district, Southwest Utah, pp. 992–1019 in J. D. Ridge, Ed., *Ore Deposits of the United States 1933/1967*, Graton-Sales Vols. New York: AIME, 1880 pp.

—— and E. Ingerson, 1960. An hypothesis for the origin of ore-forming fluid. USGS Prof. Pap. 400B.

Mason, D. R., 1978. Compositional variations in ferromagnesian minerals from porphyry-copper-generating and barren intrusions of the Western Highlands, Papua, New Guinea. *Econ. Geol.* 73:878–890.

McKinstry, H., 1963. Mineral assemblages in sulfide ores: the system Cu-Fe-As-S. *Econ. Geol.* 58:483–505.

—— and R. H. Svendsen, 1942. Control of ore by rock structure in a Coeur d'Alene mine. *Econ. Geol.* 37:215–230.

Meyer, C., and J. J. Hemley, 1967. Wall rock alteration, pp. 166–235 in H. L. Barnes, Ed., *Geochemistry of Hydrothermal Ore Deposits.* New York: Holt, Rinehart and Winston, 670 pp.

——, E. P. Shea, C. C. Goddard, Jr., and Staff, 1968. Ore deposits at Butte, Montana, pp. 1373–1427 in J. D. Ridge, Ed., *Ore Deposits of the United States, 1933/1967,* Graton-Sales Vols. New York: AIME, 1880 pp.

Miller, C.F., and L. J. Bradfish, 1980. An inner Cordilleran belt of muscovite-bearing plutons. *Geology* 8:412–416.

Miller, L. J., 1976. Corporations, ore discovery, and the geologist. *Econ. Geol.* 71:836–847.

Mitcham, T. W., 1952. Indicator minerals, Coeur d'Alene silver belt. *Econ. Geol.* 47:414–450.

Mitchell, A. H. G., and M. S. Garson, 1972. Relationships of porphyry copper and circum-Pacific tin deposits to paleo-Benioff zones. *Inst. Min. Metall. Trans.* 81:B10–B25.

Mohon, J. P., 1975. Comparative geothermometry for the Monte Cristo pegmatite, Yavapai County, Arizona. Unpub. M.S. Thesis, Univ. Ariz., 90 pp.

Montoya, J. W., and J. J. Hemley, 1975. Activity relations and stabilities in alkali feldspar and mica alteration reactions. *Econ. Geol.* 70:577–582.

Moore, F., and D. J. Moore, 1979. Fluid inclusion study of mineralization at St. Michael's Mount, Cornwall. *Inst. Min. Metall. Trans.* 88:B57–B64.

Moore, W. H., and J. T. Nash, 1974. Alteration and fluid inclusion studies of the porphyry copper ore body at Bingham, Utah. *Econ. Geol.* 69:631–645.

Mutschler, F. E., E. G. Wright, S. Ludington, and J. T. Abbott, 1981. Granite molybdenite systems. *Econ. Geol.* 76:874–897.

Nielson, R. L., 1970. Mineralization and alteration in calcareous rocks near the Santa Rita stock, New Mexico, in *Geol. Soc. New Mex. Guidebook,* 21st field Conf., pp. 133–139.

Norton, D., 1978. Sourcelines, sourceregions, and pathlines for fluids in hydrothermal systems relating to cooling plutons. *Econ. Geol.* 73:21–28.

—— and R. B. Knapp, 1977. Transport phenomena in hydrothermal systems: the nature of porosity. *Amer. J. Sci.* 277:913–936.

—— and J. E. Knight, 1977. Transport phenomena in hydrothermal systems: cooling plutons. *Amer. J. Sci.* 277:937–981.

Norton, J. J., and L. R. Page, 1956. Methods used to determine grade and reserves of pegmatites. *Min. Eng.* 8:401–414.

Olade, M. A., 1976. Geochemical evolution of copper-bearing granitic rocks of the Guichon Creek Batholith, British Columbia, Canada. *Can. J. Earth Sci.* 13:199–209.

——, 1980. Geochemical characteristics of tin-bearing and tin-barren granites, northern Nigeria. *Econ. Geol.* 75:71–82.

—— and W. K. Fletcher, 1975. Primary dispersion of rubidium and strontium around porphyry copper deposits, Highland Valley, British Columbia. *Econ. Geol.* 70:15–21.

—— and ——, 1976. Trace element geochemistry of the Highland Valley and Guichon Creek batholith in relation to porphyry copper mineralization. *Econ. Geol.* 71:733–748.

Oyarzun, J. M., and J. J. Frutos, 1974. Sobre el posible control tectonico de los yacimientos de cobre porfidico en la Cuenca Andina Chilena. *Phys. Earth Plan. Inst.* 9:5.

Page, L. R., J. W. Adams, M. P. Erickson, W. E. Hall, J. B. Hanley, P. Joralemon, J. J. Norton, L. C. Pray, T. A. Steven, W. C. Stoll, and R. F. Stopper, 1953. Pegmatite investigations, 1942–1945, Black Hills, South Dakota. USGS Prof. Pap. 247.

Paige, S., 1916. The Silver City (New Mexico) Folio. USGS Folio 199.

Palache, C., and O. W. Jarrell, 1939. Salesite, a new mineral from Chuquicamata, Chile. *Amer. Mineral.* 24:388–392.

Parak, T., 1975. Kiruna ores are not "intrusive-magmatic ore of the Kiruna type." *Econ. Geol.* 70:1242–1258.

Park, C. F., Jr., 1972. The iron ores of the Pacific Basin. *Econ. Geol.* 67:339–349.

—— and R. A. MacDiarmid, 1975. *Ore Deposits*, San Francisco, Freeman, 530 pp.

Parsons, A. B., 1933. *The Porphyry Coppers*. New York: AIME, 581 pp.

——, 1957. *The Porphyry Coppers in 1956*. New York: AIME, 270 pp.

Patchett, P. J., 1983. Importance of the Lu-Hf isotopic system in studies of planetary chronology and chemical evolution. *Geochim. Cosmochim. Acta* 47:81–91.

Perry, D. V., 1969. Skarn genesis at the Christmas mine, Gila County, Arizona. *Econ. Geol.* 64:255–271.

Perry, V. D., 1952. Geology of the Chuquicamata orebody. *Min. Eng.* 4:1166–1168.

Petersen, U., 1965. Regional geology and major ore deposits of central Perú. *Econ. Geol.* 60:407–476.

Peterson, D. W., 1963. Pers. Commun.

Philpotts, A. A., 1967. Origin of certain iron-titanium and apatite rocks. *Econ. Geol.* 62:303–315.

Piekenbrock, J. R., 1983. The structural and chemical evolution of phyllic alteration at North Silver Bell, Pima County, Arizona. Unpub. M.S. Thesis, Univ. Ariz., Tucson, 95 pp.

Plimer, I. R., 1977. Alteration and ore deposition associated with high level S-type granites in eastern Australia. *2nd Int. Sympos. Water-Rock Interactions Proc.*, pp. 11130–11138.

——, 1978. Proximal and distal stratabound ore deposits. *Mineral. Deposita* 13:345–353.

Preece, R. K., III, and R. E. Beane, 1982. Contrasting evolution of hydrothermal alteration in quartz monzonite and quartz diorite wall rocks at the Sierrita porphyry copper deposit, Arizona. *Econ. Geol.* 77:1621–1641.

Puckett, J. C., Jr., 1970. Petrographic study of a quartz diorite stock near Superior, Arizona. Unpub. M.S. Thesis, Univ. Ariz., Tucson, 48 pp.

Putman, G. W., 1975. Base metal distribution in granitic rocks. II: Three-dimensional variation in the Lights Creek stock, California. *Econ. Geol.* 70:1225–1241.

—— and J. W. Sullivan, 1979. Granitic pegmatites as estimators of crustal pressures—a test in the eastern Adirondacks, New York. *Geology* 7:549–553.

Quick, J. D., 1976. Contact alteration and mineralization of stratified rocks in southeastern Arizona. Unpub. M.S. Thesis, Univ. Ariz., Tucson, 183 pp.

Ransome, F. L., 1912. Copper deposits near Superior, Arizona. USGS Bull. 540, pt. 1, pp. 139–158.

—— and F. C. Calkins, 1908. The geology and ore deposits of the Coeur d'Alene district, Idaho. USGS Prof. Pap. 62.

Reid, A. M., and Q. J. Henderson, 1928. The Blue Tier tinfield. Tasmania Geol. Surv. Bull. 38.

Reid, J., 1978. Skarn alteration of the Commercial Limestone, Carr Fork area, Bingham, Utah. *Econ. Geol.* 73:1315–1325.

Reid, R. R., S. W. Caddey, and J. W. Rankin, 1975. Primary refraction control of ore shoots, with examples from the Coeur d'Alene district, Idaho. *Econ. Geol.* 70:1050–1061.

——, W. R. Greenwood, and G. L. Nord, Jr., 1981. Metamorphic petrology and structure, St. Joe area, Idaho. Geol. Soc. Amer. Bull., pt. 2, 92:94–205.

Renzetti, B. L., 1957. Geology and petrogenesis at Chuquicamata, Chile. Unpub. Ph.D. Dissert., Univ. Indiana.

Rickwood, P. C., 1981. The largest crystals. *Amer. Mineral.* 66:885–907.

Ridge, J. D., Ed., 1968. *Ore Deposits of the United States, 1933/1967*, Graton-Sales Vols. New York: AIME, 1880 pp.

Rijks, H. R. P., and A. H. Van der Veen, 1972. The geology of the tin-bearing pegmatites in the eastern part of the Kamativi district, Rhodesia. *Mineral. Deposita* 7:383–395.

Roedder, E., 1967. Fluid inclusions as samples of the ore-forming fluid, pp. 515–574 in H. L. Barnes, Ed., *Geochemistry of Hydrothermal Ore Deposits.* New York: Holt, Rinehart and Winston, 670 pp.

——, 1971. Fluid inclusion studies on the porphyry-type ore deposits at Bingham, Utah, Butte, Montana, and Climax, Colorado. *Econ. Geol.* 66:98–120.

—— and B. J. Skinner, 1968. Experimental evidence that fluid inclusions do not leak. *Econ. Geol.* 63:715–730.

Roering, C., 1966. Aspects of the genesis and crystallization sequence of the Karibib pegmatites, South-West Africa. *Econ. Geol.* 61:1064–1089.

Rose, A. W., 1970. Zonal reactions of wallrock alteration and sulfide distribution at porphyry copper deposits. *Econ. Geol.* 65:920–936.

—— and W. W. Baltosser, 1966. The porphyry copper deposit at Santa Rita, New Mexico, pp. 205–220 in S. R. Titley and C. L. Hicks, Eds., *Geology of the Porphyry Copper Deposits, Southwestern North America.* Tucson: Univ. Ariz. Press, 560 pp.

Ross, C. S., 1935. Origin of the copper deposits of the Ducktown type in the Southern Appalachian region. USGS Prof. Pap. 179.

Ruiz-F., C., 1965. The volcanic iron ores of Chile, pp. 221–247 in C. Ruiz-F. et al., *Geología y Yacimientos Metaliferos de Chile.* Santiago: Inst. Investigaciones Geológicas, 305 pp.

——, L. Aguirre, J. Corvalán, C. Klohn, E. Klohn, and B. Levi, 1965. *Geología y Yacimientos Metaliferos de Chile.* Santiago: Inst. Investigaciones Geológicas, 305 pp.

——, F. Ortiz, A. Moraga, and A. Aguilar, 1968. Genesis of the Chilean iron ore deposits of Mesozoic age. *Proc. 23rd Int. Geol. Congr.*, vol. 7, pp. 323–338.

Rye, R. O. and F. J. Sawkins, 1974. Fluid inclusion and stable isotope studies on the Casapalca Ag-Pb-Zn-Cu deposit, central Andes, Peru. *Econ. Geol.* 69:181–205.

Sales, R. H., 1914. Ore deposits at Butte, Montana. *AIME Trans.* 46:4–106.

—— and C. Meyer, 1949. Results of preliminary studies of vein formation at Butte, Montana. *Econ. Geol.* 44:465–484.

—— and ——, 1950. Interpretation of wall-rock alteration at Butte, Montana. *Colo. School Mines Quart.* 45:261–273.

Samabe, S., and K. Kuwahata, 1979. Geology, skarnization, and mineralization of the Nippo Ore Deposit, Kamaishi Mine, Iwate Prefecture. *Min. Geol.* 29:161–174.

Sawkins, F. J., 1972. Sulfide ore deposits in relation to plate tectonics. *J. Geol.* 80:377–396.

——, J. R. O'Neil, and J. M. Thompson, 1979. Fluid inclusion and geochemical studies of vein gold deposits, Baguio District, Philippines. *Econ. Geol.* 74:1420–1434.

Schaller, W. T., 1933. Pegmatites, in *Ore Deposits of the Western States,* Lindgren Vol. New York: AIME, 797 pp.

Schmitt, H. A., 1933. The Central mining district, New Mexico. AIME Contrib. 39 (class 1).

——, 1939. The Pewabic Mine. *Geol. Soc. Amer. Bull.* 50:777–818.

Schneiderhöhn, H., 1941. *Lehrbuch der Erzlagerstättenkunde.* Jena: Gustav Fischer.

——, 1958. *Die Erzlagerstätten der Erde* (On magmatic oxide ore deposits), vol. 1. Stuttgart: Gustav Fischer, 315 pp.

Shand, S. J., 1943. *Eruptive rocks: Their Genesis, Composition, and Classification* New York: Wiley.

Shenon, P. J., 1948. Lead and zinc deposits of the Coeur d'Alene district, Idaho. *Proc. 18th Int. Geol. Congr. Symp. on the Geology, Paragenesis, and Reserves of Ores of Lead and Zinc,* pp. 78–80.

Sheppard, S. M. F., and L. B. Gustafson, 1976. Oxygen and hydrogen isotopes in the porphyry copper deposit at El Salvador, Chile. *Econ. Geol.* 71:1549–1559.

——, R. L. Nielsen, and H. P. Taylor, 1971. Hydrogen and oxygen isotope ratios in minerals from porphyry copper deposits. *Econ. Geol.* 66:515–542.

—— and H. P. Taylor, Jr., 1974. Hydrogen and oxygen isotope evidence for the origins of water in the Boulder batholith and the Butte ore deposits, Montana. *Econ. Geol.* 69:926–946.

Shimazaki, H., 1980. Characteristics of skarn deposits and related acid magmatism in Japan. *Econ. Geol.* 75:173–183.

Short, M. N., and I. A. Ettlinger, 1927. Ore deposition and enrichment at the Magma mine, Superior, Arizona. *AIME Trans.* 74:174–222.

——, F. W. Galbraith, E. N. Harshman, T. H. Kuhn, and E. D. Wilson, 1943. Geology and ore deposits of the Superior mining area, Arizona, Ariz. Bur. Mines Geol. Ser. 16, Bull. 151.

—— and E. D. Wilson, 1938. Magma mine area, Superior, pp. 90–98 in E. D. Wilson, Ed., *Some Arizona Ore Deposits*, Ariz. Bur. Mines Geol. Ser. 12, Bull. 145.

Sillitoe, R. H., 1973. Geology of the Los Pelambres porphyry copper deposit, Chile. *Econ. Geol.* 68:1–10.

——, C. Halls, and J. N. Grant, 1975. Porphyry tin deposits in Bolivia. *Econ. Geol.* 70:913–927.

Silver, D. B., 1980. The distribution of tungsten in limestone contact environments, Silver Bell mine, Dos Cabezas Mountains, Arizona. Unpub. M.S. Thesis. Univ. Ariz., Tucson, 58 pp.

Slaughter, J., D. M. Kerrick, and V. J. Wall, 1975. Experimental and thermodynamic study of equilibria in the system CaO-MgO-SiO_2-H_2O-CO_2. *Amer. J. Sci.* 275:143–152.

Snyder, F. G., 1969. Precambrian iron deposits in Missouri, pp. 231–238 in

H. D. B. Wilson, Ed., *Magmatic Ore Deposits, a Symposium*, New Haven, Conn.: Econ. Geol. Mono. 4, 366 pp.

Solesbury, F. W., 1967. Gem corundum pegmatites in NE Tanganyika. *Econ. Geol.* 62:983–991.

Sorenson, R. E., 1947. Deep discoveries intensify Coeur d'Alene activities. *Eng. Min. J.* 148:70–78.

——, 1951. Shallow expression of Silver Belt ore shoots, Coeur d'Alene district, Idaho. *AIME Trans.* 190:605–611.

Spencer, A. C., and S. Paige, 1935. Geology of the Santa Rita mining area, New Mexico, USGS Bull. 859.

Stone, M. C., 1980. Pers. Commun.

Stutzer, O., 1907. Geologie und Genesis der lappländischen Eisenerzlagerstätten. *Neues Jahrb. Mineral. Geol. Paläontol.* 24 (supp.):548–675.

Sutherland-Brown, A., Ed., 1976. *Porphyry deposits of the Canadian Cordillera.* Can. Inst. Min. Metall. Spec. Vol. 15, 510 pp.

Taylor, A. V., Jr., 1935. Ore deposits at Chuquicamata, Chile, in *Copper Resources of the World*, 16th Int. Geol. Congr., Vol. 2.

Taylor, H. P., Jr., 1974. The application of oxygen and hydrogen isotope studies to problems of hydrothermal alteration and ore deposition. *Econ. Geol.* 69:843–883.

Thomas, D. E., 1953. The Blue Tier tinfield. *Proc. 5th Empire Min. Metall. Congr.*, 1:1213–1221.

Thompson, A. B., 1974. Calculation of muscovite-paragonite-alkali feldspar phase relations. *Contr. Mineral. Petrol.* 44:173–194.

Thompson, T. B., J. R. Pierson, and T. Lyttle, 1982. Petrology and petrogenesis of the Bokan granite complex, southeastern Alaska. *Geol. Soc. Amer. Bull.* 93:898–908.

Titley, S. R., 1963. Lateral zoning as the result of monoascendant hydrothermal processes in the Linchburg Mine, New Mexico, pp. 312–316 in J. Kutina, Ed., *Problems of Postmagmatic Ore Deposition, a Symposium*, vol. 1. Prague: Geol. Surv. Czech., 588 pp.

——, 1966. Preface, pp. ix–x in S. R. Titley and C. Hicks, Eds., *Geology of the Porphyry Copper Deposits, Southwestern North America.* Tucson: Univ. Arizona Press, 287 pp.

——, 1972. Some geological criteria applicable to the search for southwestern North American porphyry copper deposits. Min. Metall. Inst. Jap.–AIME Mtgs., Tokyo, Japan, 16 pp.

——, 1973. Pyrometasomatic—an alteration type. *Econ. Geol.* 68:1326–1328.

——, Ed., 1982. *Advances in Geology of the Porphyry Copper Deposits, Southwestern North America.* Tucson: Univ. Ariz. Press, 560 pp.

—— and L. B. Gustafson, Eds., 1978. Papers on Ok tedi, Panguna, other PCD's of S. W. Pacific. *Econ. Geol.*, Spec. Issue, 73:597–985.

—— and C. Hicks, Eds., 1966. *Geology of the Porphyry Copper Deposits, Southwestern North America.* Tucson: Univ. Ariz. Press, 287 pp.

Tôrre de Assuncão, C. F., 1944. Algumas observacoes petrologicas nas Caldas de Monchique. Lisboa Univ. Mus. Lab. Mineral. Geol. Bull., Ser. 4, no. 11–12, pp. 55–66.

Turneaure, F. S., 1960. A comparative study of major ore deposits of Central Bolivia. *Econ. Geol.* 55:217–254; 574–606.

——, 1971. The Bolivian tin-silver province. *Econ. Geol.* 66:215–225.

Umpleby, J. B., and E. L. Jones, Jr., 1923. Geology and ore deposits of Shoshone County, Idaho. USGS Bull. 732.

Verkaeren, J., and P. Bartholomé, 1979. Petrology of the San Leone magnetic skarn deposit, S. W. Sardinia. *Econ. Geol.* 74:53–66.

Villas, R. N., and D. Norton, 1977. Irreversible mass transfer between circulating hydrothermal fluids and the Mayflower stock. *Econ. Geol.* 72:1471–1504.

Wallace, S. R., 1974. The Henderson ore body—elements of discovery, reflections. *AIME Trans.* 256:216–227.

——, W. B. MacKenzie, R. G. Blair, and N. K. Muncaster, 1978. Geology of the Urad and Henderson molybdenite deposits, Clear Creek County, Colorado. *Econ. Geol.* 73:325–368.

——, N. K. Muncaster, D. C. Jonson, W. B. MacKenzie, A. A. Bookstrom, and V. E. Surface, 1968. Multiple intrusion and mineralization at Climax, Colorado, pp. 605–640 in J. D. Ridge, Ed., *Ore Deposits of the United States 1933/1967*, Graton-Sales Vols. New York: AIME, 1880 pp.

Warren, H. V., 1934. Silver-tetrahedrite relationship in the Coeur d'Alene district, Idaho. *Econ. Geol.* 29:691–696.

Webster, R. N., 1958. Exploration extends Magma's future. *Min. Eng.* 10:1062–1065.

Westra, G., and S. B. Keith, 1981. Classification and genesis of stockwork molybdenum deposits. *Econ. Geol.* 76:844–873.

White, W. H., A. A. Bookstrom, R. J. Kamilli, M. W. Ganster, R. P. Smith, D. E. Ranta, and R. C. Steininger, 1981. Character and origin of Climax-type molybdenum deposits. *Econ. Geol. 75th Anniv. Vol.*, pp. 270–316.

—— and W. B. MacKenzie, 1973. Hydrothermal alteration associated with the Henderson molybdenite deposit, Colorado. *Econ. Geol.* 68:142.

Whitney, J. A., 1977. A synthetic model for vapor generation in tonalite magmas and its economic ramifications. *Econ. Geol.* 72:686–690.

Williams, S. A., and F. P. Cesbron, 1977. Rutile and apatite: useful prospecting guides for porphyry copper deposits. *Mineral. Mag.* 41:288.

Wilson, E. D., 1950. Superior Area, pp. 84–98 in Arizona zinc and lead deposits, pt. 1, Ariz. Bur. Mines Geol. Ser. Bull. 18, 156 pp.

Wilson, J. W. J., S. E. Kesler, P. L. Cloke, and W. C. Kelly, 1980. Fluid inclusion geochemistry of the Granisle and Bell porphyry copper deposits, British Columbia. *Econ. Geol.* 75:45–61.

Wu, I., and U. Petersen, 1977. Geochemistry of tetrahedrite and mineral zoning at Casapalca, Peru. *Econ. Geol.* 72:993–1016.

Yeap, C. H., Ed., 1981. *The Geology of Tin.* Kuala Lumpur, Malay., Bull. Geol. Soc. Malaysia 11, 392 pp.

Young, E. B., 1948. The Pioche district. *Proc. 18th Int. Geol. Congr. Symp. on the Geology, Paragenesis, and Reserves of Ores of Lead and Zinc*, pt. 7, pp. 98–106.

TWELVE

▼

Deposits Related to Subaerial Volcanism

Many of the world's most spectacular and historically significant deposits are shallowly buried or crop out in or near subaerial, commonly intermediate to silicic volcanic rocks. They are generally related to volcanic processes and to the emplacement, cooling, and consolidation of the volcanic rocks that contain them. They have a tectonic position, as might be expected. Many of them result from extravasation of the intermediate to felsic igneous systems discussed in the last chapter; they occur along convergent plate boundaries and are therefore related to orogenic belts much as are porphyry, skarn, and granitic deposits and Cordilleran Vein and pegmatite deposits. Associated rock types are basalts, dacites, rhyodacites, latites, and rhyolites in pyroclastic, welded pyroclastic, and flow regimes associated with ignimbrites, cauldron and caldera complexes, and other volcanic symptoms of the volcanic belts of the world.

Deposits considered in this chapter include the classic epithermal type, essentially as defined by Burbank, Nolan, and Lindgren (1933), Wisser (1966), and others before them. Epithermal deposits include the bonanza precious metal districts for which Europe and the Americas are so renowned; in North America, old names like Comstock Lode–Virginia City, Tonopah, Creede, Cripple Creek, Guanajuato, Pachuca, and Potosi are being joined by new ones like Round Mountain, McDermitt, Delamar, Summitville, and Oatman and Creede reawakened. We will also consider a relatively recently recognized subtype, the so-called "invisible gold" deposits of the Carlin-Cortez, Nevada, type, and develop the idea of "bulk low-grade" epithermal precious-metal gold or silver deposits—broad, altered-mineralized volumes of volcanic pyroclastic rocks and even associated lake-bed volcaniclastic sediments. Waterloo, California, and Candelaria, Nevada (Watson, 1977), are examples of those deposits. Although they do not involve hydrothermal transport, we will also examine the volcanic magnetite flows of Durango, Mexico; El Laco, Chile; and elsewhere, occurrences presumably related to magnetite deposits like those on the Kola Peninsula of Fennoscandia-U.S.S.R. The tin-rhyolite association mentioned at the close of Chapter 11 could be considered here. In a review article, Sillitoe (1977) included stratiform "manto" occurrences of native copper-chalcocite-bornite in flow-top vesicles and ash flow tuffs in some South American calc-alkaline volcanic rocks in this group. Similar occurrences are known in Sonora and near Arivaca, Arizona. Most economic geologists would also include the famous deposits of native copper in propylitized basalts and interbedded conglomerates of the Keweenaw Peninsula in northwestern Michigan as epithermal. Lastly, the Senator antimony deposit in west central Turkey (Bernasconi, Glover, and Viljoen, 1980) is probably epithermal.

▸ EPITHERMAL SILVER-GOLD DEPOSITS

Epithermal deposits as originally defined are products of volcanism-related hydrothermal activity at shallow depths and low temperatures. Deposition normally takes place within about 1 km of the surface in the temperature range of 50 to 200°C, although temperatures to 300°C are now known to be common. Most deposits are in the form of siliceous vein fillings, irregular branching fissures, stockworks, breccia pipes, vesicle fillings, and disseminations. Replacement textures are recognized in many of the ores, but open-space fillings are common and in most deposits are the dominant form of emplacement. Drusy cavities, comb structures, crustifications, and symmetrical banding are generally conspicuous. Colloform, agatelike textures also characteristic of epithermal environments presumably reflect moderate temperatures and free hydrothermal fluid circulation. The fissures have a direct connection with the surface, which allowed the ore-bearing fluids to flow with comparative ease; in fact, some modern hot springs and steam vents are almost certainly surface expressions of underlying epithermal systems (White, 1955; White, Muffler, and Truesdell, 1971; Bernasconi, Glover, and Viljoen, 1980). Barton, Bethke, and Toulmin (1971) deduced flow rates to have been at the rapid rate of 0.2 to 1 cm/second in the OH vein at Creede, Colorado. They found flecks of hematite on the upper surfaces of quartz crystals protruding into veins, and calculated the rate of flow necessary to carry them upward.

A few epithermal deposits can be related directly to deep-seated intrusive bodies, but this relationship is demonstrable only where especially deep erosion has occurred. Many epithermal deposits have no observable association with plutonic rocks. Most ores are in or near areas of Tertiary volcanism, especially near volcanic necks and other structures that tap underlying source materials and reservoirs. Because these deposits are formed near the surface and in tectonically rising areas, they are susceptible to destruction by almost immediately subsequent erosion. They occur, then, in young volcanic rocks, and are nearly unknown in pre-Cenozoic rocks. Wisser (1960) emphasized that district-scale gentle doming is almost universal in epithermal districts. The volcanic environment and continuing hot spring activity engender hot waters in some mines; for example, caustic hot waters were encountered at depth in mines in the Comstock Lode of Nevada and in several of the mercury mines of California.

The country rocks near epithermal veins commonly are extensively altered, even though the vein walls may be sharply defined. Relatively high porosity and open-channel permeability allow fluids to circulate in the wall rocks for great distances, and favorable temperature gradients promote reactions between cool host rocks and warm to hot invading solutions. As a result, wall-rock alteration is both widespread and conspicuous, as is shown in Figure 5-11. Among the principal alteration products are chlorite, sericite, alunite, zeolites, clays, adularia, silica, and pyrite. Chlorite is probably the most common alteration mineral in this zone. Propylitization is the dominant alteration process, *propylite* being an aggregate of secondary chlorite, pyrite, epidote, sericite, carbonates, and albite or adularia in mafic to intermediate volcanics like basalts, andesites, and dacites (Chapter 5). Studies in the late 1970s revealed widespread inconspicuous potassium feldspathization associated with epithermal districts (Howell, 1977); sericitization, silicification, and K-feldspathization are the most common alteration products flanking veins in the more felsic rhyodacite, latite, and rhyolite host lithologies. The silica, sericite, chlorite, and pyrite of epithermal alteration halos are generally fine-grained. Carbonate minerals, especially calcite, dolomite, ankerite, and rhodochrosite, are also alteration products. Furthermore, kaolin and montmorillonite clay minerals may be abundant and conspicuous (Sudo, 1954), forming zones of different colors parallel to the walls of veins. The gangue minerals in epithermal veins include white, clear, greenish, or amethystine quartz, chalcedony, adularia, calcite, dolomite, ankerite, rhodochrosite, barite, and fluorite. Typical "high-temperature" minerals such as tourmaline, topaz, and garnet are absent.

Sulfosalt ore minerals are characteristic of epithermal deposits. They include the silver sulfantimonides and sulfarsenides polybasite, stephanite, pearceite, pyrargyrite, proustite, and others, all combinations of Ag, As, Sb, and S; the gold and silver tellurides petzite [$(Ag,Au)_2Te$], sylvanite [$(Au,Ag)Te_2$], krennerite [$(Au,Ag)Te_2$], calaverite ($AuTe_2$), hessite (Ag_2Te), and so on; and stibnite, acanthite, cinnabar, and native mercury. Some of the world's richest concentrations of native gold and electrum, the natural gold-silver alloy, were deposited under epithermal conditions; the famous bonanza deposits at Goldfield, Nevada; Cripple Creek, Colorado; and Hauraki, New Zealand, are examples. Most epithermal deposits have "bottoms" which must be physicochemical in nature; the structures and gangue mineral fillings such as quartz and carbonates continue downward, but ore values drop sharply below a given level, which may be related to a zone of boiling. In some deposits, however, typical epithermal mineralogies merge downward with galena, sphalerite, chalcopyrite, and other sulfides commonly found in mesothermal or Cordilleran Vein deposits. This interdigitation is found at Creede and Silverton in Colorado, at Pachuca, Mexico, and elsewhere. Graton (1933) proposed the term *leptothermal* to describe ores representing the transition between moderate and low temperatures of formation, the prefix *lepto* meaning small or weak. Recent fluid inclusion studies have so blurred thermal distinctions between epithermal, leptothermal, and mesothermal, however, that leptothermal is now applied generally just to base-metal mineralization and gangue minerals that continue below epithermal precious-metal occurrences.

It is not uncommon to find large, highly colored supergene gossans, or iron oxide cappings, covering epithermal ores. During weathering, the wide-spread pyrite in the altered wall rock is oxidized to limonite—goethite and hematite—forming a conspicuous guide to ore deposits. Supergene enrichment to the extent found in copper deposits is absent, although minor enrichment in silver and gold is reported. Just as epithermal deposits have a discrete bottom or base, they also appear to have upper assay limits which may parallel the lower ones and lie at or a few meters below the surface. The tops almost certainly also represent chemical-physical thresholds related to surface effects with or without surface-related leaching.

Epithermal deposit articles by Schmitt (1950), Wisser (1966), White (1955, 1967), and Sillitoe (1977) have added to definitions and understandings developed decades ago when many epithermal districts now considered "worked out" were still active. They include the observations that silver and gold minerals are texturally younger than base-metal sulfides, with gold normally youngest; that veins close to the subvolcanic basement are base-metal- and silver-rich, lower in quartz, and narrow, while veins higher in the average system are gold-rich, quartz-rich, and wide; that fluid inclusion filling temperatures range from less than 100 to 330°C with an average more nearly 250°C than 200°C; that salinities are typically low at less than 2% NaCl equivalent; and that low $\delta^{18}O$ at -16 to $+4$ ‰ and low D/H values of from -90 to -140 ‰ suggest dominance of meteoric fluids, with magmatic fluid input discerned at several deposits (O'Neill and Silberman, 1974). It appears that near-surface boiling is a trigger of precipitation, generally in the 230 to 260°C range, in systems that may be periodically sealed by precipitation of quartz, then reopened by increased pressure or seismic activity.

The genesis of epithermal precious-metal deposits has been a subject of lengthy discussion. The elements of that debate can be summarized best in Figures 12-1 to 12-3. Zies pointed out in 1929 that epithermal mineralization and broad-scale, fumarolic or solfataric kaolinite-alunite-pyrophyllite alteration can occur even in flows that are removed from a source conduit. He described geothermal springs in a series of volcanic rocks that flowed into a U-shaped valley at Katmai, Alaska, to produce the renowned Valley of Ten Thousand Smokes, with thousands of "rootless" epithermal fumaroles. Schmitt (1950) generalized Zies' ideas with the suggestion that epithermal deposits might be generated totally by shallow circulation of meteoric fluids even without a magmatic component (Figures 12-1 and 12-2). On the other hand, it cannot be controverted that in some districts the base and precious metals are essentially all delivered by magmatic fluids (Figure 12-3) with variable degrees of dilution, precipitation, and redistribution

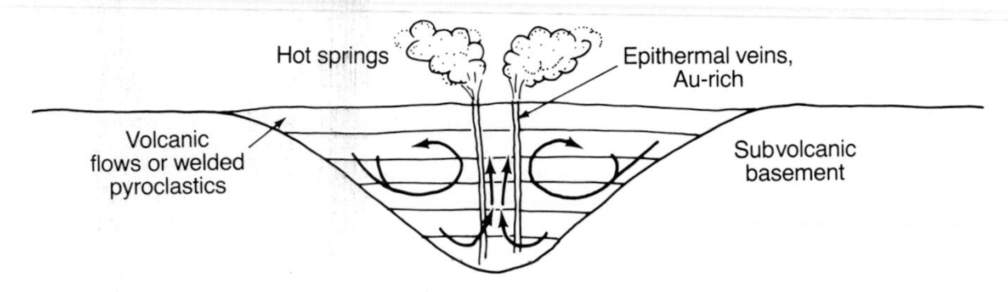

Figure 12-1. Meteoric waters convecting through a cooling volcanic pile dissolve silica, alkalies, halogens, and precious metals and deposit them in veins beneath active hot springs or fumaroles. Patterned after Katmai, Alaska (Zies, 1929), and ideas of Schmitt (1950).

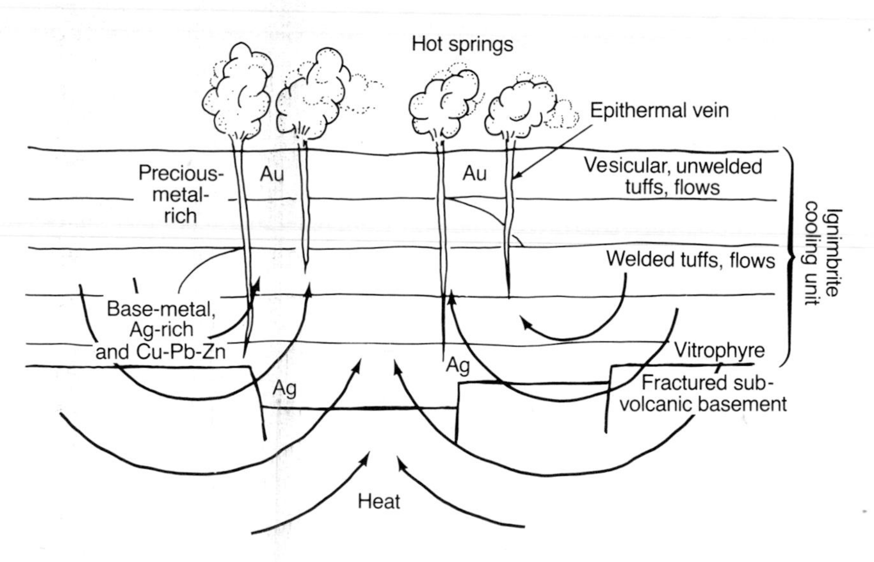

Figure 12-2. Deep circulation of meteoric fluids down through an ignimbrite cooling unit 1 km thick into subvolcanic basement rocks. Leaching of silica, alkalies, base metals, and precious metals supplies epithermal veins above. Similar circulation could be imposed upon a stratovolcanic edifice with an anhydrous subjacent stock. Patterned on possible relationships at Creede, Colorado.

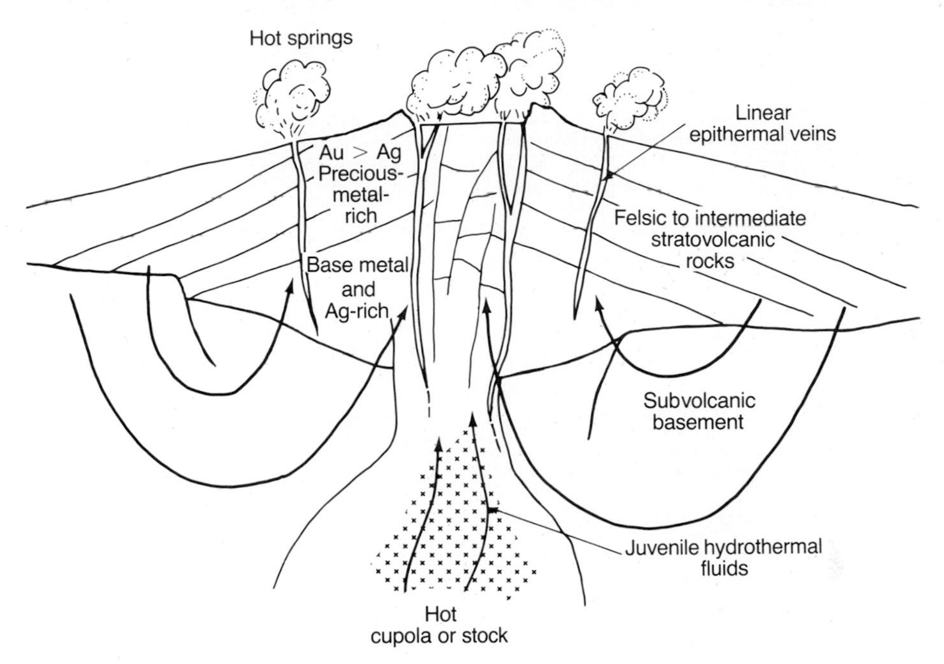

Figure 12-3. Combined juvenile (magmatic) and meteoric fluids. The magmatic component may be either early, as at the Comstock Lode, or late, as in several Peruvian deposits. The relative amounts of juvenile versus leached metal values is thought to vary widely, as from 100% at Tayoltita, Mexico, to low at Pachuca (Dreier, 1976) and Guanajuato (Gross, 1975). It is suggested by some that no meteoric component is necessary and by others that no juvenile component is required. Methods of defining the ratio for particular districts or mines are still impressive.

influenced by concurrent or late incursions of meteoric waters. It is certain that meteoric waters could hardly be excluded from the well-developed fault-vein systems that intersected the surface at the time of epithermal activity.

Pachuca–Real del Monte, Mexico

The famous bonanza silver districts of Pachuca and Real del Monte are 100 km north-northeast of Mexico City (Figure 12-4); they are typical epithermal deposits. They would represent a single district were they not effectively separated by the crest of the Sierra de Pachuca. Pachuca lies along the west flank of the mountain range; Real del Monte is on the east flank, only about 1 km away. Because Real del Monte receives about twice as much rainfall as Pachuca, their physical geographies differ considerably. But geologically they represent two ends of a single mineralized volume (Ordoñez, 1902; Wisser, 1937, 1942, 1966; Winchell, 1922; Geyne, 1956; Geyne et al., 1963; Dreier, 1976).

The Sierra de Pachuca is made up of Tertiary volcanic rocks overlying Cretaceous sedimentary rocks. Thick andesite flows and associated tuffs

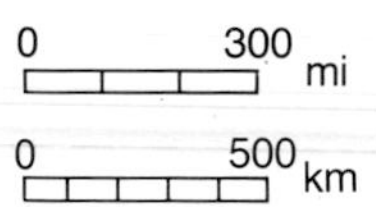

Figure 12-4. Map of Mexico showing the location of some epithermal deposits discussed in the text, namely Pachuca–Real del Monte, Guanajuato, and Tayoltita. Cerro de Mercado is not epithermal, but is related to subaerial volcanism and is also discussed in this chapter.

and breccias constitute the bulk of the older pre-Tertiary subvolcanic basement. They are overlain by Oligocene-Pliocene rhyolites, dacites, and local late basalt flows which total about 200 meters in thickness. Premineral dikes of andesite, dacite, and rhyolite strike generally N75°W across the district (Figure 12-5). The most abundant dikes are rhyolitic and are the last phase of igneous activity before the ore-bearing fluids were introduced (Geyne, 1956). Fossil plants in the volcanic rocks make it possible to date the periods of igneous activity. The andesites are Oligocene to Miocene, the rhyolites Late Miocene or Early Pliocene, and the younger basalt lavas are Pliocene. The age of the ore, then, is Early Pliocene, slightly younger than the rhyolites.

The structural geology has been studied in detail because the ore deposits are localized along faults. Regionally the volcanic rocks describe a broad, shallow syncline, but in the immediate mining district the bedding varies in dip from horizontal to 40 or 50° in all directions. East-west and northwest-southeast faults dominate the structural pattern, which is emphasized further by the northwest-southeast dike system (Figure 12-5). In addition to the east-west faults, the eastern part of the district, Real del Monte, has a well-developed set of mineralized north-south faults. The parallel fracture zones apparently represent regional faulting. The dikes ascended along the earliest fractures, but faulting continued throughout the period of mineralization. Nearly all of the east-west faults have steep dips and normal downthrown displacements. The north-south faults formed near the end of the period of east-west faulting, just at the time of ore deposition. These late faults are small, but recurrent movement and timing made them especially favorable as avenues for ascending ore fluids (Wisser, 1942; Geyne, 1956; Dreier, 1976).

The ore deposits are in quartz veins along the faults, forming shoots where local conditions favored development of breccias, wide shattered zones, or broad fault openings. Paragenetically the hypogene sulfide sequence is

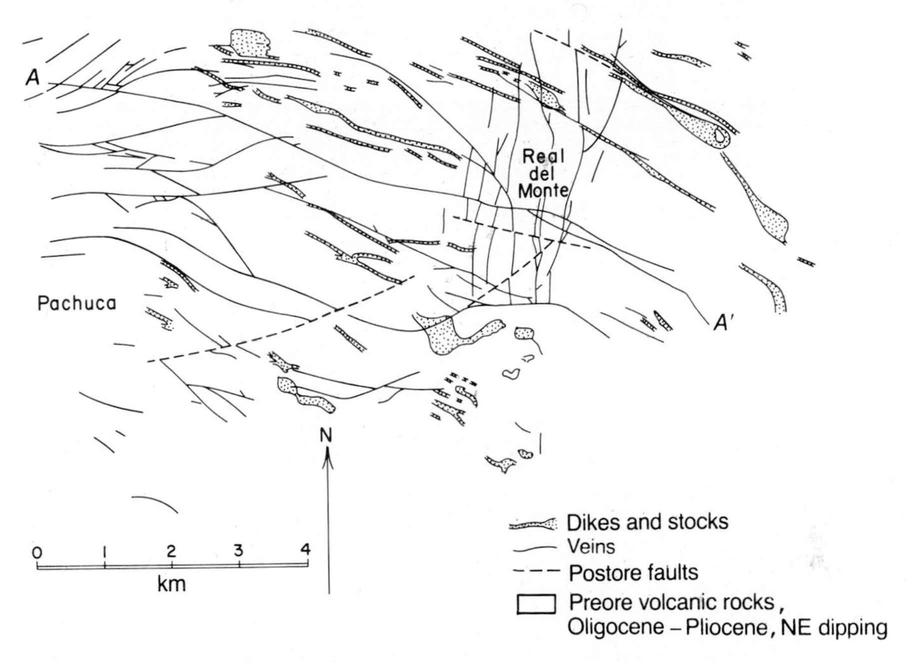

Figure 12-5. Veins, faults, and intrusive rocks in the Pachuca–Real del Monte district, Mexico. The paired lines and spotted areas are preore dikes and other intrusives, mostly rhyolite but including some dacite; the solid lines are veins along faults; and the broken lines are postore faults. Intrusives are as mapped on the surface; veins and faults are projected to the common level. *A-A'* refers to Figure 12-6 and Table 12-1. (*From* Geyne, 1956.)

pyrite, sphalerite, galena, chalcopyrite, acanthite, polybasite, and stephanite (Bastin, 1948: Geyne, 1956; Dreier, 1976). Chalcocite, covellite, and bornite occur locally, but they represent spotty secondary enrichment. By far the most abundant compound of silver is acanthite (Ag_2S), a product of both hypogene and supergene processes. Other secondary minerals include native silver, silver chlorides and bromides, malachite, azurite, anglesite, and oxides of manganese and iron. Besides quartz, the veins contain calcite, dolomite, rhodochrosite, rhodonite, bustamite, and barite.

Propylitization is the most widespread wall-rock alteration. If affects extensive volumes of the volcanics. Minor amounts of sericitized, kaolinized, pyritized, and silicified wall rocks serve as valuable guides to ore (Thornburg, 1952; Geyne, 1956). Dreier suggests that potassic alteration flanked by more pervasive propylitization may be more common flanking deep veins, and that phyllic alteration generally occurs at shallow levels.

Although the district as a whole stretches 10 km north-south by 12 km east-west, and single structures extend 8 to 10 km, individual minable ore shoots in the veins seldom reach more than 1 km in length. Vertical dimensions are generally shorter (Figure 12-6), with veins rarely extending more than 700 meters down. As a result of leaching by meteoric waters, most orebodies do not reach the surface. Except where exposed in deep

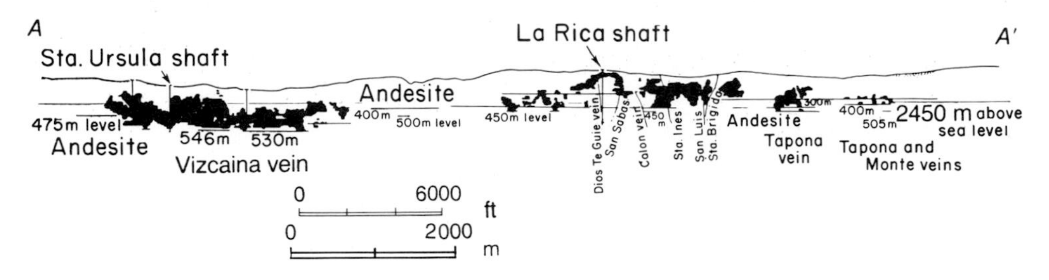

Figure 12-6. Longitudinal section (looking northerly) along the Vizcaina-Tapona vein, Pachuca–Real del Monte district, Mexico. See also Figures 12-4 and 12-5. The blacked out areas denote minable ore which bottomed at 450 to 550 meters and did not crop out. An almost identical profile could be drawn for the Oatman district and for Creede, Colorado, discussed later. (*From* Thornburg, 1952.)

Pliocene-Pleistocene canyons, the tops of the ore deposits are anywhere from 20 meters beneath the surface in the east-west fault system to more than 250 meters beneath the surface in the north-south faults, and the ore shoots continue 300 to 700 meters down. The reasons for localization of ore along the veins cannot always be ascertained. Some shoots occupy wide fault intersections; others are in open zones where the faults changed attitude or formed branches, loop structures, and feather joints. In some places ore was deposited along the contacts between veins and rhyolite dikes (Wisser, 1942; Geyne, 1956; Thornburg, 1945). Dreier found a general correlation between higher ore grade and greater vein width. Veins average 1 to 2 meters wide with essentially no wall-rock disseminations (Dreier, 1976). Thornburg (1952) noted that silver and gold values typically extend into the wall rocks as stringers that trend parallel to the veins. Consequently, the average stopes were 2 to 3 meters wide.

The evidence for shallow low-temperature deposition of ore minerals is abundant. Repeated seismic movements along the faults caused multiple, repeated brecciation of early quartz, allowing later quartz, adularia, and ore minerals to enter. Banded veins, vugs, and replacement textures are all common. Some of the quartz is even chalcedonic, indicating possible deposition at low temperature and colloidal size ranges. At depth the veins branch, the quartz content diminishes, and the ore values decrease abruptly. Geyne and his colleagues (1963) estimated that the zone of ore deposition was between 300 and 1000 meters below the original ground surface. Table 12-1 shows the paragenesis of a typical vein, the Vizcaina-Tapona-Monte. This paragenesis is characteristic of many epithermal vein deposits.

Dreier (1976) reported that his studies showed no mineralogic zoning over the 700-meter vertical ore interval, although fluids in the upper 100 to 200 meters were boiling during deposition; that the fluids averaged 0.5 wt % NaCl equivalent; that pH at sulfide mineralization temperatures of 200 to 250°C was 6.0 ± 0.2; that $f(O_2)$ was 10^{-38} atm; and that silver concentrations in the fluids were low at about 0.01 ppm. Hydrothermal fluids

TABLE 12-1. Paragenetic Sequence at Pachuca–Real del Monte, Mexico

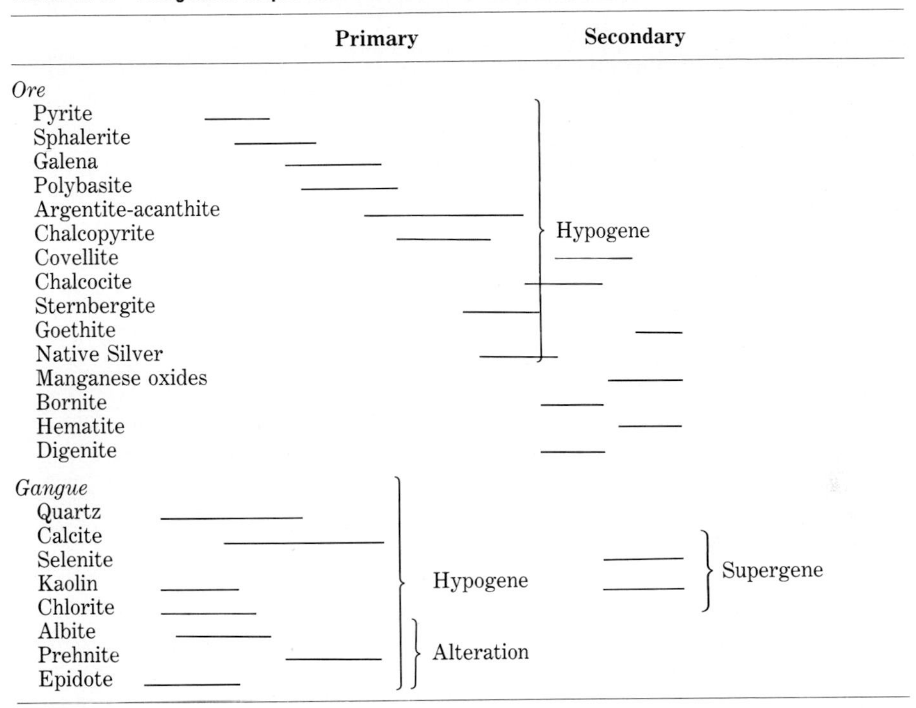

Paragenetic sequence of the minerals in the Vizcaina-Tapona-Monte vein, Pachuca, Mexico. The minerals are listed by "amount present," which decreases downward. See Figures 12-5 and 12-6 for location.

Source: From Geyne *et al.* (1963).

apparently did not issue at the surface as hot springs, and Dreier postulated a deep circulation system driven by a buried pluton perhaps 2 to 3 km beneath the area that concentrated silver at the top of the system by convective overturn almost analogous to the physical chemistry of a percolator coffee pot (Figures 12-1 and 12-3). Here also, leptothermal base-metal sulfides continue downward below precious-metal cutoffs. Metallization thus appears related to an underlying magmatic source expressed at the surface as rhyolite dikes and small stocks or plugs. Some of the tectonic features are related to these stocks; the contemporaneity of fault movement, stock-dike emplacement, and ore deposition implies their close interrelationship.

Oxidation and supergene leaching appear to have removed ore minerals from the upper parts of veins that reach the surface. Much of the silver that was dissolved from the zone of oxidation was reprecipitated beneath the water table at the expense of iron, copper, and sulfur. This supergene-enriched zone has accounted for some of the most valuable ore in the district.

Since about 1526, when mining began in the Pachuca–Real del Monte area, the district has produced more than 1.2 billion troy oz (37 million kg, or 37,000 metric tons) of fine silver and 6.2 million troy oz (200,000 kg, or 200 metric tons) of fine gold, the combined output being worth over $1.3 billion at 1960 prices (Geyne et al., 1963), or almost $28 billion at 1980 prices. This amounts to about 6% of the world's silver production for this period, most of which has been achieved during the present century. As evidence of this productivity, the district is honeycombed with over 2000 km of underground workings. Recent exploration and development have found new veins, lateral extensions and unmined portions of old veins, and minor low-grade ores. The district will remain productive for decades.

Oatman, Arizona

Another district that serves to illustrate epithermal deposit characteristics and the exploration excitement that they evoke is in the area around the town of Oatman in northwestern Arizona. The town was never large, and the present ghost town belies the richness of the veins that supported its growth. The Oatman area consists of a veneer of 2 km of mid-Tertiary subsilicic, intermediate, and silicic volcanic rocks on Precambrian schists and gneisses. Thorson (1971) described, from base to top, the Eocene Miocene Alcyone trachyte crystal welded tuff, with interbedded, inward-thinning sedimentary tuff breccias followed by quartz latite outpourings. These units, probably caldera-form, were intruded by the Times granophyric latite porphyry 22.6 ± 1.8 million years ago. A second cycle of activity produced a broad latite-quartz latite shield volcano and a basin-filling volcanic pile of Esperanza Trachyte, Oatman Latite, and Gold Road Latite, a kilometer thickness of flows and tuffs intruded 10.4 ± 0.5 million years ago by the subvolcanic Moss porphyry stock which produced them. All of these rocks were rent by a swarm of normal northwest-southeast faults which swing to east-west toward the south. Vertical displacements as great as 100 to 200 meters are uncommon, and throw is generally minor. Branching and widening of veins at changes in dip and strike are abundant, especially where fault blocks moved laterally (Lausen, 1931; Clifton, Buchanan, and Durning, 1980). Discovery of the Moss Vein in 1863 and of many others between 1900 and 1920 led to the establishment of several fortunes. The generally low-sulfide ores are in wide, originally open, now quartz-filled fault fissures. Visible particles of gold-rich electrum of Au_{80}-Ag_{20} composition are common; it was the principal ore mineral. Negligible chalcopyrite, pyrite, and marcasite have been found, and supergene minium (Pb_3O_4), wulfenite ($PbMoO_4$), and iron and manganese oxides are reported. Gangue minerals are major quartz, minor adularia and calcite, and trace fluorite and gypsum. Five successive stages of mineralization can be correlated throughout the district. They show classic examples of epithermal textures (Figures 12-7 to 12-10, Table 12-1). The stages are variably mineralized, contain variable amounts of partially replaced platy calcite crystals, and are generally symmetrically banded, commonly with adularia. Some veins show all five stages; others record only those stages of deposition during which they were open (Figure 12-8). Waxy green quartz vein fillings are common in the high-grade ores of stage 4. Values are generally in ore shoots rather than in continuous stretches. Although banded quartz veins cross the area, ore values seldom crop out and have not been found more than 400 meters below the surface.

Recent studies, principally at Oatman (Clifton, Buchanan, and Durning, 1980), have been based on four exploration precepts which further illuminate epithermal deposit characteristics. First, irregularities in vein aperture—and thus thickness of ore—can best be seen by contouring the distance from the footwall of a given vein to an inclined smooth data plane, as was done at Slocan, B.C. (Figure 3-37). Wide zones occupying north-concave bends in strike-slip fault veins have proven to be the sites of ore shoots in dilatant openings extending downdip as "flat pipes." Second, these "flat pipes" are also the loci of altered zones which flair outward where veins are thickest and highest in grade. Wide zones of quartz-sericite phyllic alteration flank veins only where they are mineralized, argillic and propylitic alteration is more general, and flanking, quartz-filled veins and veinlets are ubiquitous. Third, fluid inclusions indicate mineralization temperatures from 200 to 240°C and that fluids were boiling *below* levels at which ore values were precipitated, thus providing geothermal gradient information that is useful in exploration drilling. Finally, contouring of vein tops also indicates that the upper and lower extremities of ore formation conform to a domical form, much as at Tonopah. Buchanan (1979) found similar relationships at a much larger scale at Guanajuato, Mexico, except that ore there was deposited in zones of boiling in the veins, and that argillization-sericitization formed a cap environment over the veins, as it does at Creede and Pachuca. Salinities were 0.5% NaCl with temperatures of 200 to 300°C.

Another mode of occurrence of epithermal deposits is in association with *calderas* or *caldera complexes*, examples of which are now being vigorously prospected at McDermitt, Nevada-Oregon, and at Creede, Colorado (Figures 12-11 to 12-13). A major gold deposit now being mined at Round Mountain, Nevada, occurs in a caldera setting. Caldera environments differ in three respects from the elongate fault zones that typify the Comstock Lode, Oatman, and Pachuca, for example. First, ores may be emplaced virtually as volcanism continues, and hence are vulnerable to burial; second, mineralization may be developed in the subsided, central "race-track" or moat portions of the caldera and be either involved with lake sedimentation or buried by it; and third, the structural patterns are likely to be more varied by involving radial and ring fractures that are part of caldera development, zones of weakness between adjacent contemporaneous or older calderas, and the many other complications of an active volcanic field.

Before describing Creede, let us examine the anatomy of the caldera environment. Calderas are broad pie-pan-shaped volcanic depressions that

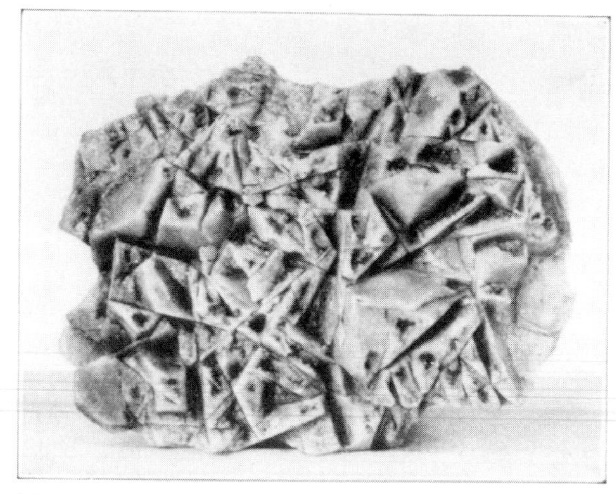

(*a*)

(*b*)

are normally circular in plan view and have centrally located vents. At their margins they have gently outward dipping volcanic flanks. Inward they show steep rampartlike walls that face inward toward the moat and central spire, if one exists. The basin and moat, sometimes filled with an inner lake, are formed by subsidence of the central portions of the volcanic complex, presumably into the partly emptied magma chamber below. Incomplete subsidence or another period of resurgence may produce a central spire, a typically conical volcanic neck at the bull's eye. Its presence may leave an

(*c*)

(*d*)

Figure 12-7. Examples of epithermal textures from the Oatman district, Arizona. (*a*) (*facing page*) Quartz of the second stage of deposition, showing a pseudomorphic replacement of calcite of a platy habit common in epithermal veins (Figure 12-9). (*b*) (*facing page*) Banded quartz of the third stage, crustified with the youngest vein filling at the top. (*c*), (*d*) Dark greenish quartz and white adularia of the fifth stage facing upward in (*c*) and closing in a centerline of adularia in (*d*). (*From* Lausen, 1931.)

annular flat-floored depression, or racetrack, between spire and walls, which upon filling with lake water, volcaniclastic sediments, or late volcanic flows resembles the moat around a castle. Lakes in such depressions are called *caldera lakes* or *crater lakes*. Ring and radial faults abundantly developed by stresses during volcanism may produce many different sites of mineralization. We have described a subsidence caldera; they can also form by collapse, by explosion, or rarely, by erosion. The caldera complex involves dikes, sills, and subjacent stocks; vent breccias and volcanic necks; flows

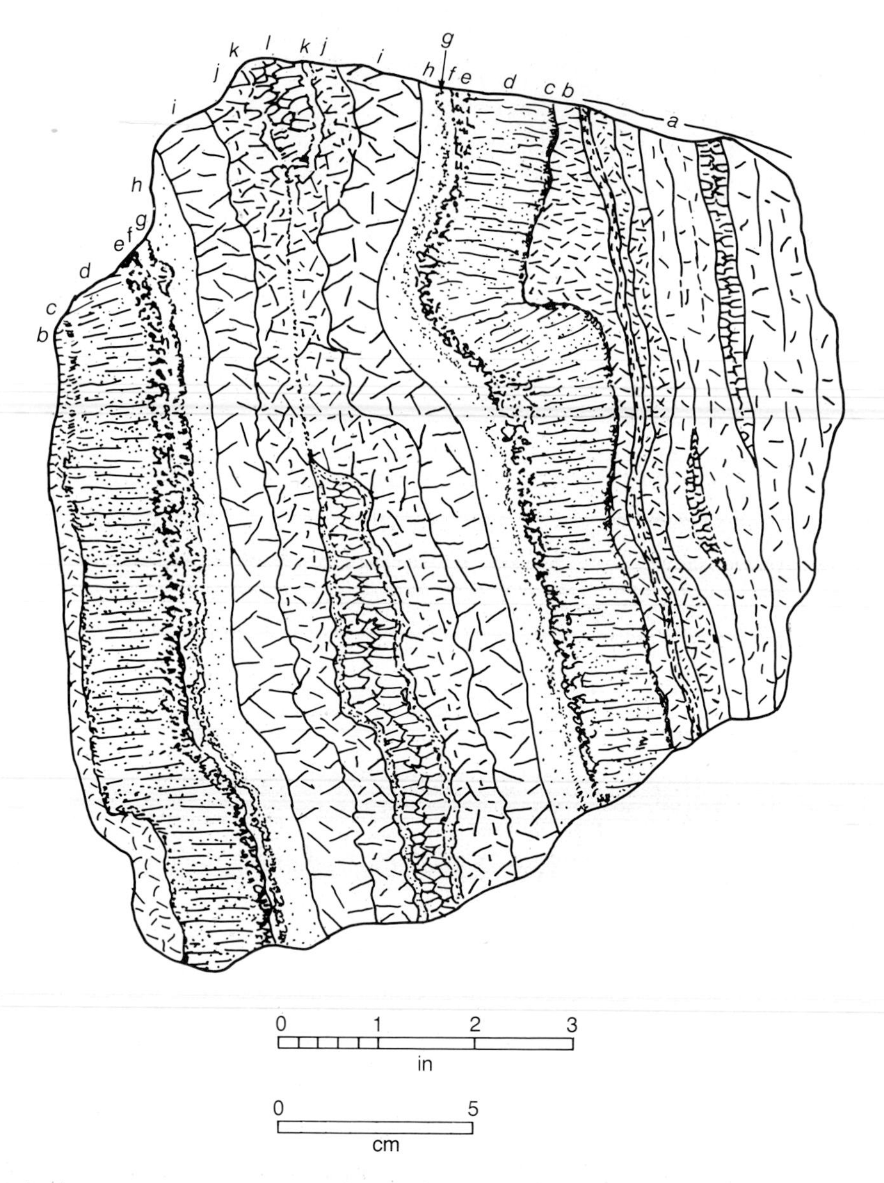

Figure 12-8. Banded ore from the stope above the 600 level of the United Eastern mine, Oatman, Arizona, showing ideal inward-facing cockade and comb structure quartz about a centerline through layer *l*, all in stage 5. *a* = brown quartz in thin crusts; *b* = rather finely crystalline white milky quartz; *c* = thin layer of honey-yellow oily looking quartz and probably also microscopic adularia; *d* = clear quartz intimately associated with the honey-yellow variety and with calcite, probably also with microscopic adularia; *e* = a layer of dark calcite, quartz, and probably adularia; *f* = pale greenish-white fine-grained quartz; *g* = thin layer of honey-yellow oily looking quartz; *h* = like *f*; *i* = rather coarsely crystalline milk-white quartz; *k* = oily yellow quartz; *l* = clear comb quartz. (*From* Lausen, 1931.)

Figure 12-9. Typical epithermal vein texture; from Bodie, California. The bladed mineral is calcite of a habit characteristic of these occurrences. It has been replaced by quartz, crystals of which also project into vugs. See also Figures 4-21 and 12-7*a*. (*Photo by* A. Knopf.)

and pyroclastic rocks that drape the flanks and partially fill the moat; and talus slopes against the inner walls and spire that may consist of tuff, cinder, ignimbrites, and agglomerates. Rocks deposited against the rampart become thinner inward, toward the center, while other units thin outward. Consider that the whole complex may be the site of vigorous hot spring and fumarole activity, and the many possible loci of mineralization become apparent. For example, mercury mineralization has been known for years in the volcaniclastic lake sediments in the McDermitt caldera, but only recently have explorationists probed the ring fractures that project downward from the inner caldera walls to find more minable mercury ores. Uranium has also been found in the sediments, and the area is presently being reevaluated.

Creede, Colorado

The Creede district has been the subject of a long-sustained geologic study, mainly by geologists of the U.S.G.S., which has resulted in many papers, most recently Lipman, Doe, Hedge, and Steven (1978), Barton,

Figure 12-10. The banded quartz of, for example, Figures 12-7*b* and 12-8 as it looks in outcrop. The comb structure and open-space filling aspects of epithermal veins is clear. This vein is at the Nevada Original Bullfrog mine in the Bullfrog district, Nye County, Nevada. (*Photo by* F. L. Ransome.)

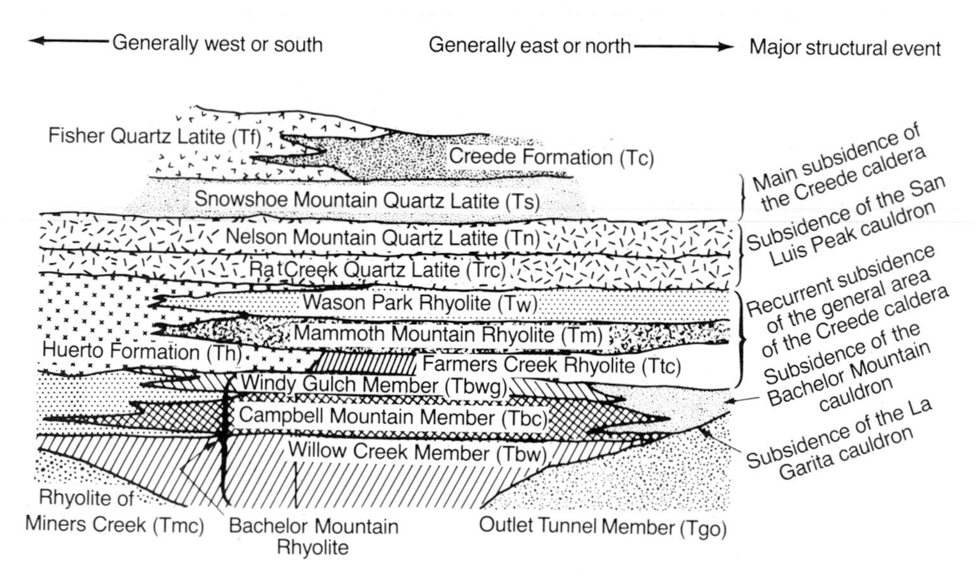

Figure 12-11. General relationships of rock units in the Creede district, Colorado. The ores are in the caldera-form Bachelor Mountain Formation shown at the bottom. Study the "major structural event" notes—the Bachelor Mountain units were spewed onto the floor of a broad caldera and were faulted later. The Creede Formation fills the present-day moat that formed late in caldera evolution. (*After* Steven and Ratté, 1965.)

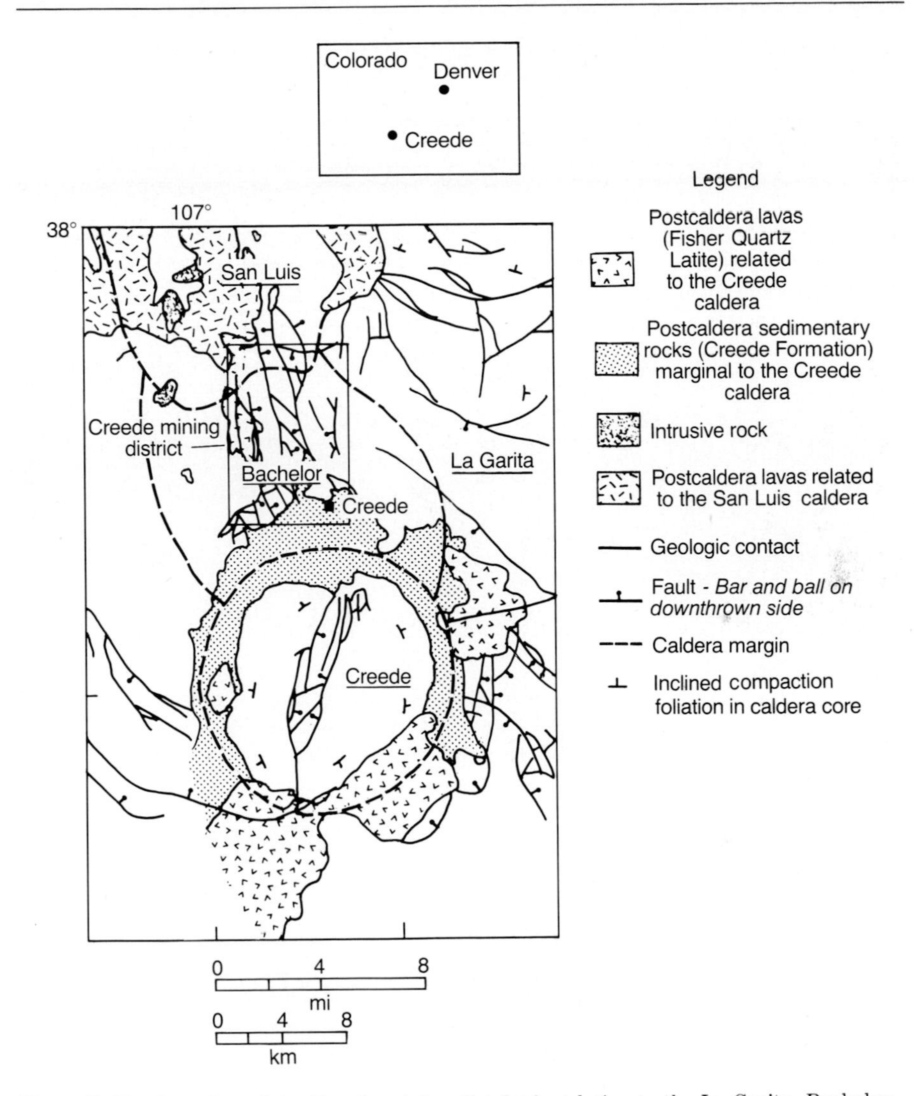

Figure 12-12. Location of the Creede mining district in relation to the La Garita, Bachelor, San Luis, and Creede calderas, central San Juan volcanic field, Colorado. The area in the rectangle is enlarged in Figure 12-13. (*After* Steven and Eaton, 1975.)

Bethke, and Roedder (1977), and Bethke and Rye (1979). Giudice (1980) focused on fracture-related and disseminated mineralization there. It is in the central San Juan Mountains of southwestern Colorado in the basin of Willow Creek, a tributary of the Rio Grande. The district was discovered in 1883 (Emmons and Larsen, 1923) and since that time is said to have produced metal valued at about $100 million, mostly in silver.

The rocks of the area, flows, tuffs, and sediments derived from volcanic materials, are all of volcanic origin (Larsen, 1930). All are of rhyolitic or latitic composition; Figure 12-11, after Steven and Ratté (1965), gives

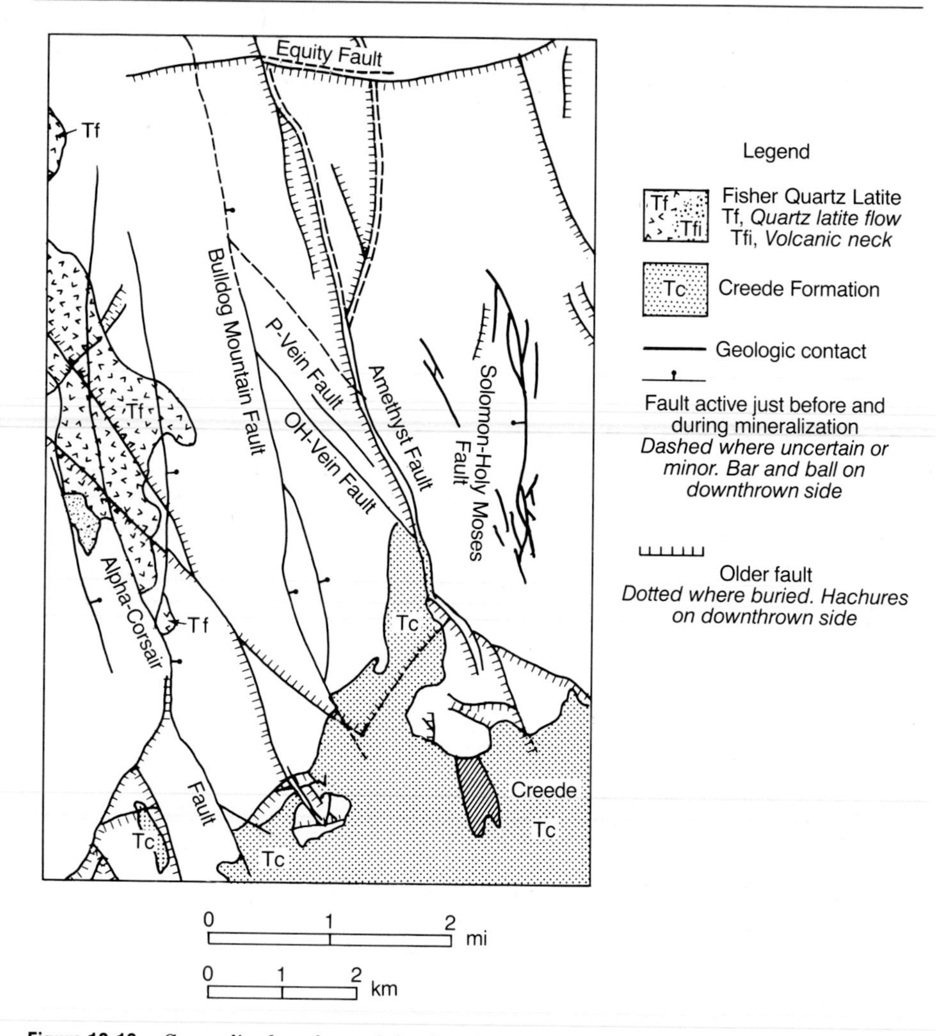

Figure 12-13. Generalized geology of the Creede mining district. See also Figures 12-11 and 12-12. The town of Creede is on the Creede Formation, a sedimentary unit in the moat of the Creede caldera. (*From* Steven and Eaton, 1975.)

the stratigraphic column and shows the general relationships among the formations. Subsidence took place after most of the eruptions and resulted in the formation of calderas, four of which—La Garita, San Luis, Bachelor, and Creede—are recognized in the area (Figure 12-12, after Steven and Eaton, 1975). As a result of the subsidence, the formations in general are highly brecciated and faulted. Intrusive rocks are scarce. Four types have been recognized—rhyolite porphyry, quartz latite porphyry, basalt, and hornblende quartz latite porphyry (Steven and Ratté, 1965). Rhyolite porphyry is exposed south of Bulldog Mountain in irregular sill-like intrusions.

Quartz latite porphyry is confined to the southeast corner of the area and is in dikes striking west of north. These dikes are as much as 100 to 200 meters across. Basalt was found in only one narrow dike, and hornblende quartz latite porphyry was seen only underground.

The Creede caldera is just one of a cluster of perhaps a score of circular caldera structures that dimple the surface of the San Juan volcanic field of southwestern Colorado (Steven and Lipman, 1976). Lipman considers the volcanic field to be a skin of ignimbrites and related flows and pyroclastic rocks above a subvolcanic batholith that might resemble the more deeply eroded and exposed Boulder batholith of Montana. Were there epithermal deposits above the Boulder batholith? Are there porphyry systems like Butte under the San Juan field? There are many ore deposits dominantly of epithermal affinity in the Tertiary San Juans—names like Silverton, Summitville, Ouray, and Creede itself are part of western U.S. history. The Creede caldera (Figure 12-12) was the last of a series of caldera events in the area between 28 and 25 million years ago, and its caldera geometry is still topographically clear as a circular form which cuts older ones (Figure 12-12). Rhyolitic ignimbrites of the Bachelor Mountain and Wasson Park formations were then cut by a swarm of radial faults, some of which strike northwest from the Creede caldera toward the San Luis caldera (Figures 12-12 and 12-13). Among them are the fabled Amethyst, OH, and P veins.

Ore is restricted to veins and fractures, but some mineralization has been recorded locally as disseminations into the Creede Formation. The principal veins fill fissures along normal faults; only the Equity Fault, in the northern part of the district, is a reverse fault. Subordinate fissures commonly split from the principal veins. Splits are particularly well developed in the hanging walls; at places they rejoin the principal fissures at depth. Hanging walls apparently were shattered as they were dragged down by gravity along the footwalls. Offsets along the major faults are considerable; throw on the Amethyst Fault is estimated to be more than 500 meters and the displacement of the reverse Equity Fault is thought to be about 400 meters.

Most of the faults strike northwest although the Equity Fault trends nearly due east. Dips are generally steep at 60 to 70°; in the eastern part of the district the dips are toward the west, but west of the Amethyst Fault the dips are eastward (Figure 12-13). These veins contained classic open-space fillings of both coarsely crystalline and chalcedonic-colloform banded quartz, amethyst, adularia, and carbonate gangue minerals with acanthite (Ag_2S), native silver, and minor silver sulfosalts. The ores were locally so rich as to warrant fully the use of the term *bonanza*. Jasper was recorded, and the presence of waxy green quartz as at Oatman was common. This "green quartz" is reported from almost all epithermal camps. It is known to miners as a pathfinder material indicating that ore is nearby, as it usually is, but the actual cause of the green color is unknown. Giudice (1980) was unable, even by electron microprobe, to identify whatever pigment may occur. It is also true that the absence of "green quartz" does not preclude nearby ore grades.

Most production at the present time comes from the Bulldog vein, but this is a recent development and the Amethyst vein system is much better known and has produced most of the ore in the district. Both the Amethyst and the Bulldog veins show evidence of extensive and repeated brecciation and rehealing during vein deposition. The veins are cut by many fractures that formed after ore deposition and are characterized by slickensides and granulated rock and ore. A few veins, the OH for example (Figure 12-13), show only minor breaking and rehealing during ore deposition.

The main Amethyst vein is largely a mineralized fault breccia, fragments of silicified country rock set in a matrix of clay, chlorite, amethystine and white quartz, and sulfide minerals. Widths of the vein have ranged from 1 meter to more than 10 meters. According to Emmons and Larsen (1923), barite was present in all of the ores from the Amethyst vein; they averaged between 10 and 20% barium sulfate. Gold content in the southern part of the Amethyst system is said to have been negligible, but farther north the ore averaged about 0.1 oz per ton, or 3 ppm. Additional minerals in the ore are galena, acanthite, stephanite, sphalerite, pyrite and marcasite, chalcopyrite, and their oxidation products cerargyrite, native silver, anglesite, cerussite, smithsonite, jarosite, malachite, chrysocolla, limonite, and manganese oxides. Gangue minerals include white and amethystine quartz, chlorite, barite and fluorite with minor rhodochrosite, hematite, siderite, and calcite. Part of the ore is thought to have been deposited in open spaces and part by replacement of the wall rocks. Wall rocks generally are silicified and sericite has been identified. Adularia has been noted near the Amethyst vein. Ribbon quartz and crustification are abundant in the vein.

Significantly, the bonanza ores at Creede display most of the epithermal open-space filling and mineralogic characteristics described in this chapter and in Chapter 4. The Amethyst and OH veins were systematically stoped from wall to wall in the early twentieth century, sealed up, and forgotten. So were many epithermal deposits. But with the spectacular growth of silver prices in the 1970s, many such deposits are being reevaluated. Giudice (1980) studied the distribution and occurrence of silver values along veins that had been so long ignored at Creede and found that silver mineralization had soaked irregularly into the walls near the intersection of the major Amethyst and OH veins, there to be fixed in microveinlets, as vug fillings, in pumice fragments, as disseminated pyrite replacements, and as solid solutions in secondary sulfide minerals. The distribution is not simply a function of induced permeability adjacent to the major structures, but is rather a complex interaction of original intrinsic lithologic porosity-permeability, alteration, and structural penetration and its timing. The mineralization is zoned, but irregularly so. The body of ore values is but one of those that has prompted many explorationists to return to the turn-of-the-century bonanza districts around the world for another evaluation.

▸ CHINESE ANTIMONY DEPOSITS

The world's most productive single antimony source is the stibnite-rich submarine volcanogenic massive sulfide district named Consolidated Murchison in the Murchison greenstone belt of northeastern Transvaal, Republic of South Africa, but the world's principal resources and reserves of antimony lie in the southern and southwestern parts of the Peoples' Republic of China in the provinces of Hunan, Kweichow (Guizhou), Kwangsi (Guangxi), Kwangtung (Guangdong), and Yunnan (Figure 12-14). The deposits lie in three crudely defined east-west belts associated with Mesozoic (160 to 120 million years) plutonic rocks, principally granodiorites and quartz

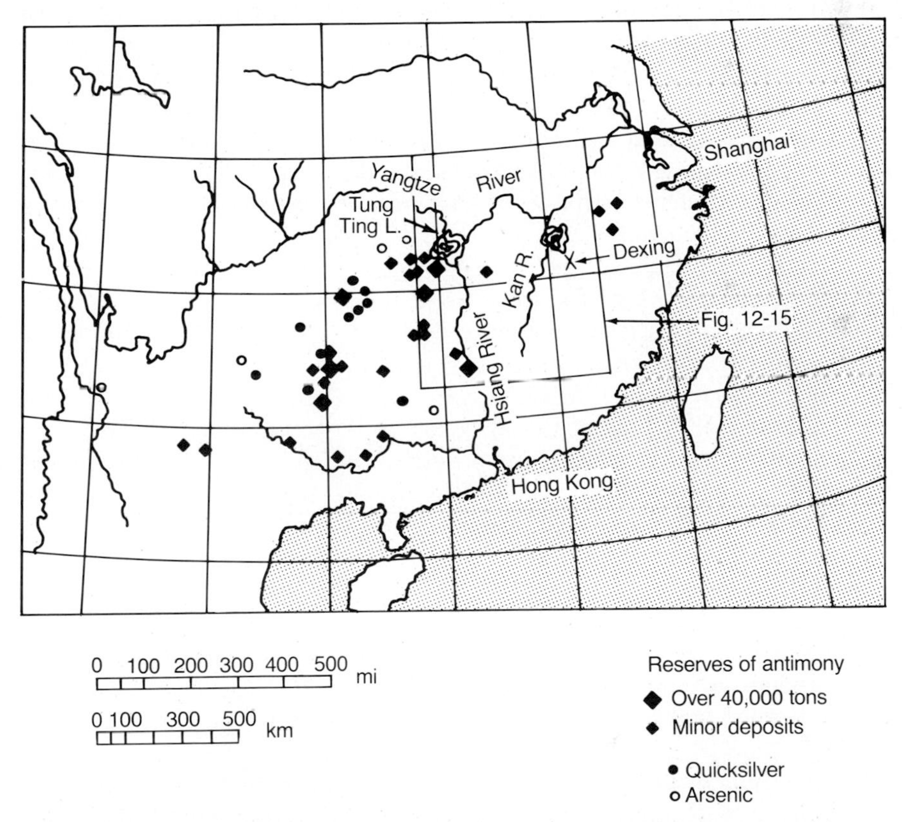

Figure 12-14. Map of China showing the location of antimony, quicksilver, and arsenic deposits. The Dexing porphyry copper deposit (Yan and Hu, 1980) is shown only for reference. The Pan Hsi antimony mines (see text) are immediately southwest of Tung Ting Lake. (*From* Juan, 1946, Plate 7.)

diorites. The northern belt, extending from northeastern Guizhou through northern Hunan, is the richest; the southern belt, which reaches from southern Yunnan to central Guangxi, is the least productive; the central belt follows a mountain range along the Guizhou-Guangxi and Hunan-Guangdong borders. Hunan, the chief source, contains the richest deposits and about 88% of the reserves (Tegengren, 1921; Juan, 1946; Wang, 1977; Rathjen, 1979). Several recently recognized porphyry copper-molybdenum deposits lie generally toward the coastline; the first target for production is Dexing (Yan and Hu, 1980). As in the Americas, the porphyries are geologically older than associated epithermal mineralization. The Chinese antimony deposits are atypical of epithermal deposits in general in that they all occur in sedimentary rocks. There are veins in shales, slates, sandstones, and quartzites, and there are replacement lodes in limestones and dolomites. The ores are in Precambrian to Mesozoic rocks. The Guizhou ores are in Precambrian slates, some small deposits in Yunnan are in Triassic rocks, and the Hunan ores are in Devonian clastics and Carboniferous carbonates. They are similar in many respects to epithermal antimony deposits in Turkey (Bernasconi, Glover, and Viljoen, 1980).

There are two principal kinds of antimony deposits in this extensive region: stibnite-quartz veins in clastic host rocks and stibnite-galena-arsenopyrite replacement deposits in carbonate rocks. Although the most productive mines include replacement ores, the dominant ore is stibnite-quartz open-space vein filling. Stibnite (Sb_2S_3) is easily recognized by its striated lathlike prismatic form. Where anhedral, it is identifiable by its color and by the fact that saturated KOH instantly stains stibnite canary yellow.

The veins are restricted to zones of shearing or brecciation. In many places, individuals fissures open into stockworks or irregularly distributed veinlets; simple structural geology is the best guide to ore. Apparently the slates faulted to favor continuous veins and the quartzites shattered to favor stockwork deposits. Few of the veins are uniform in their content of stibnite, which occurs sporadically in pockets or bunches. Quartz, which forms crystal-lined vugs in many places, is generally the only gangue mineral. Cinnabar and pyrite are found in some of the stibnite veins in subordinate amounts. Because of differences in host rocks, stockworks predominate in value in Hunan and veins constitute the ore in Guizhou. The average vein is about half a meter wide; widths of more than 1 meter are exceptional. Stibnite-quartz ore ranges from 6 to 25% stibnite; it is hand-sorted to much higher grades. Because no effort is known to have been made to mine or explore beyond 200 meters in depth, the vertical extent of the vein ore is unknown.

Replacement deposits of stibnite in carbonate country rocks form irregular pods, with associated galena, arsenopyrite, and pyrite. Relatively continuous and uniform veins carrying the same minerals as the replacement deposits are also found in limestones. The stibnite content of limestone deposits is generally higher than that of stibnite-quartz veins; it ranges from 20 to 60% (Juan, 1946).

In mountainous country along the Zi River in central Hunan, there are several quartz-stibnite vein deposits known as the Banxi (Panshin) mines (Figure 12-14). Tilted slates, shales, and quartzites have been invaded by granites which caused widespread jointing but only slight igneous metamorphism. Vertical veins cut obliquely through the eastward striking sediments. The veins are very narrow, seldom exceeding 40 cm in width, and consist of stibnite in a gangue of quartz or interlaminated slivers of sericitized-silicified wall rock. Antimony ore is restricted to short shoots in narrow footwall seams, which, in turn, are accompanied by kaolinic alteration. The deepest of the mines extend only about 200 meters. The lower levels show that the ore shoots decrease in length, even though well-defined quartz-filled fissures continue to greater depths. They thus meet many of the criteria of epithermal deposits.

Not far from the Banxi mines, in rugged country between the Zi and Yuan rivers, stibnite deposits are associated with scheelite and gold-bearing quartz. These are the Wuxi tungsten-antimony-gold mines. The deposits are veins in folded slates, shales, and a few interbedded quartzites. Originally the mines were worked for gold and the stibnite and scheelite were discarded as waste; even the quartz "gangue" contains 7 to 10 grams of gold per ton. Quartz veins with sharply defined, regular footwalls and indistinct hanging walls carry both gold and stibnite, which is generally concentrated along the sharp footwall side. The principal veins are persistent in both strike and dip, the deepest workings being at the water table 165 meters from the surface. In places the veins are as much as 2 meters wide, and handsorted ore from these wide sections carries 20 to 30% antimony. At depth, the shoots contract sharply, and stibnite gives way to pyrite.

The Xikuangshan (Hsikeng Mountain) district 120 km to the south in central Hunan (Figure 12-15) contains the largest and richest structure in the entire region (Tegengren, 1921). Gently folded dolomitic limestone and a few beds of shale, sandstone, and low-grade bituminous coal overlie a quartzitic sandstone that contains the ore deposits. The orebodies are concentrated along the crest of a gentle anticline, the folding of which caused tension faulting, jointing, and brecciation in brittle quartzites sandwiched between shales. Veins, large pockets of cockscomb quartz, and stibnite ore extend for about 2 km. The most favorable bed for ore is a quartzite along the upthrown side of a prominent axial plane fault. This bed, more than 50 meters thick, contains several favorable strata; the deposits are scattered through it, where ground preparation structures intersect the favorable beds. Most stibnite is in veins, irregular veinlets, or lenticular bodies; veins of pure stibnite more than 1 meter wide are not uncommon. Open-space filling was dominant (Tegengren, 1921). A small deposit follows the same fault 10 km south of Xikuangshan. The ore consists almost entirely of long,

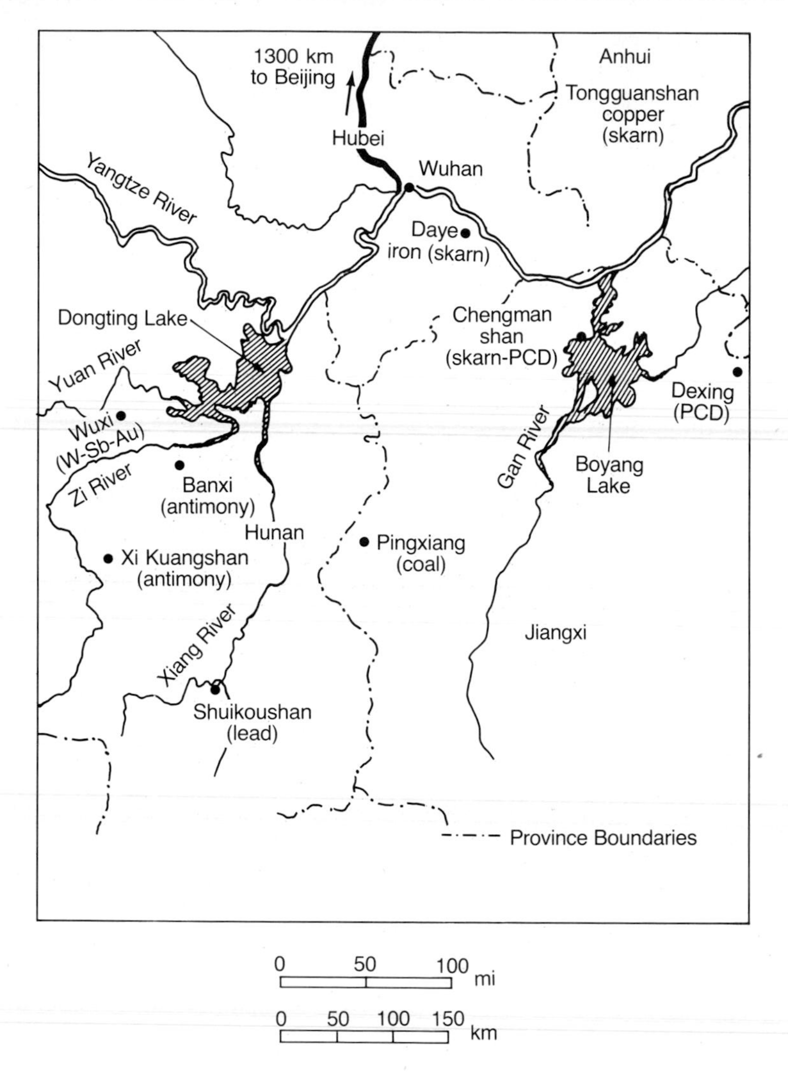

Figure 12-15. Major mineral districts in the Lower Yangtze Valley area, Peoples' Republic of China. (*Revised by* Ren Qijiang, 1983.) Dongting, Xiang, and Gan are new alliteration forms of similar, older words in Figure 12-14.

radiating, or columnar crystals of stibnite, oxidized in places to stibiconite ($Sb_3O_6 \cdot OH$), which contain only a fraction of 1% of combined arsenic, lead, and copper.

The Chinese antimony deposits, especially those of Hunan Province, are said to have been mined since the sixteenth century, yet most of the workings are confined to the zone above the water table. Large-scale systematic mining was started in 1927, but hand sorting and hand labor are still used. The available resources are said to be adequate for many years.

These deposits are similar to epithermal occurrences elsewhere, except that they occur in sediments. Stibnite is a common accessory mineral in many epithermal districts, and sulfantimonides have already been said to be common in this deposit type.

▸ BULK LOW-GRADE SILVER-GOLD DEPOSITS

Before we conclude the discussion of epithermal deposition, a recent trend in precious-metal exploration and development must be briefly covered. Two trends have made a new deposit type economic. First, the prices of silver and gold have been freed from currency and, for better or worse, the metals increased in relative value such that these previously unattractive exploration and production commodities are now relatively attractive. Short supply and strong demand have generally raised prices in the 1980s. Second, the diversities of geologic occurrence of silver and gold have been shown by recent discoveries to be greater than was heretofore perceived and more varied exploration ideas are proving successful.

The Delamar mine in Idaho is the first modern open-pit silver mine. It currently produces 2000 tons per day of ore containing 3.5 oz (110 grams) of silver and 0.04 oz (1.3 grams) of gold per short ton. The principal ore mineral is naumannite (Ag_2Se), which occurs in a swarm of quartz veins and as impregnations dispersed in a silicified rhyolitic ash-flow tuff in a Miocene caldera complex. The Sam Goosley silver-copper-antimony-gold deposit in British Columbia is another open-pit mine. Watson (1977) and Berger and Eimon (1982) describe the Candelaria and Rochester silver deposits in Nevada, the Waterloo in southeastern California, the Hog Heaven in Montana, the Hardshell in Arizona, and parts of the Creede district in Colorado as examples of bulk, dispersed low-grade ores. None of these areas features prominent individual vein structures, but the deposits meet many of the criteria of epithermal districts. They also represent the fact that either special circumstances of porosity and permeability of host rocks adjacent to veins or the unusual outflow of hot springs, producing a commingling of epithermal vein fluids and caldera lake sediments, can generate large volumes of low-grade mineralization. In many circumstances these values may be "invisible" to the extent that they either are fine-grained or consist of inconspicuous minerals. The important aspect is that neither mineralization nor alteration in epithermal environments need be restricted to the veins, but rather, values may be dispersed in host rocks or involved with either normal lakebed sediments or caldera-lake beds like the Creede Formation. "Old epithermal" districts, and new ones, must be carefully evaluated, and entirely new concepts, such as mobilization and reprecipitation of gold along thrust fault and décollement structures in the southwestern United States, are revealing new, potentially important habitats of gold ore.

▸ CARLIN-TYPE GOLD DEPOSITS

The "invisible" gold deposits are another subtype of the epithermal deposit group. They have not long been recognized; Carlin began production in 1965, Cortez, currently closed, began in 1968, and Jerritt Canyon in 1980. Other similar deposits named Getchell and Gold Acres, Nevada, had longer intermittent histories, but were lower in grade. They are now being reevaluated, several new deposits like Maggie Creek and Alligator Ridge are being brought in, and old districts like Getchell and Cortez will probably be reactivated. An aspect of these deposits, many also known as the *Carlin type*, is that the dominantly metallic gold particles are so small that they are indeed invisible to the unaided eye and thus detectable only by assay rather than by inspection or panning. Intensive geologic studies to establish the geologic characteristics of the "invisible" gold ores were begun in the Lynn mining district and its Carlin orebody (Figure 12-16) in 1967 by R. J. Roberts, then by Newmont Mining Corporation's Carlin Gold Mining Company, and, after production began, by U.S.G.S. personnel led by A. S. Radtke. Most of the following description is from Radtke, Rye, and Dickson's (1980) report on the geology and stable isotope studies of Carlin, one of the largest of the known epithermal disseminated bodies. Cortez contained 5 million tons of 7 ppm (0.2 oz per ton) gold and Carlin about 12 million tons of 10 ppm (0.3 oz per ton) gold, and the Jerritt Canyon deposit in the Independence Range of northeastern Nevada has been announced by Freeport Gold at a conservative 12 million tons of 7 ppm gold ore. The most recently announced discovery at Gold Quarry, 26 km from Carlin, may be the best deposit of them all, containing as much as 20 million ounces (620 metric tons) of gold. High gold prices greatly increased the tempo of exploration in Nevada in 1980, and several new deposits were found. Maggie Creek, 14 miles south of Carlin, contains 5 million tons of 3 ppm gold ore, 2.3 million tons of which are of 5 ppm gold grade and will soon be producing.

The name "Carlin type" was first defined by Roberts, Radtke, and Coats (1971) to include key aspects such as dissemination, spatial relation to the Mississippian age Roberts Mountain thrust fault, domed and fractured carbonate host rocks, fine-grained gold, arsenic-sulfur minerals, and others. In 1974 Radtke altered the earlier definition by eliminating any direct dependence on the Roberts Mountain thrust, thereby including the larger number of deposits on Figure 12-16. It is accepted now that the thrust fault has no genetic relation to mineralization. The Carlin type is characterized (Radtke, 1974) as involving invisible metallic gold,

> *. . . fine-grained, simple, and relatively uniform minerals, high concentrations of arsenic, antimony, and mercury, low contents of base metals, and paucity of quartz veins . . . with late-stage carbonate and sulfate gangue minerals . . . Carbonate host rocks commonly contain as much as 0.5 weight % carbon.*

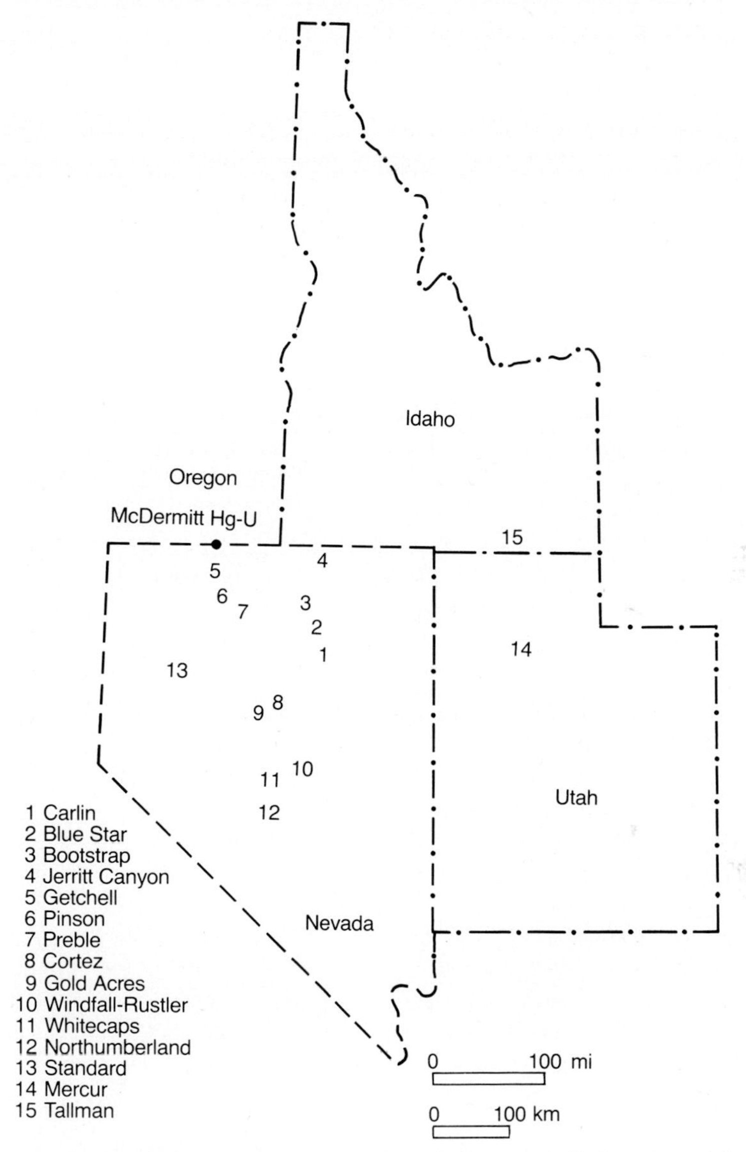

Figure 12-16. Index map showing the location of the Carlin-type gold deposits and the McDermitt caldera mercury-uranium mines.

At Carlin the disseminated gold deposits are spatially related to the regional-scale Roberts Mountain thrust fault, having formed in the thrust zone or in 100 to 200 meters of the carbonate sediments beneath it. They have been found especially where the carbonates and the thrust plane have both been bowed upward and erosionally breached to provide a "window" into the lower plate. The doming may be part of a regional control, but Roberts (1966) insisted that other requisites include a specific source of the gold, "ground preparation" fracturing to provide access for gold-bearing fluids, and the presence of carbonate rocks and organic carbon as precipitants of it.

At Carlin the late Tertiary orebodies formed by replacement of calcite and dolomite in impure dolomite units in the upper 250 meters of the Silurian-Devonian Roberts Mountains Formation. Early fluids silicified the

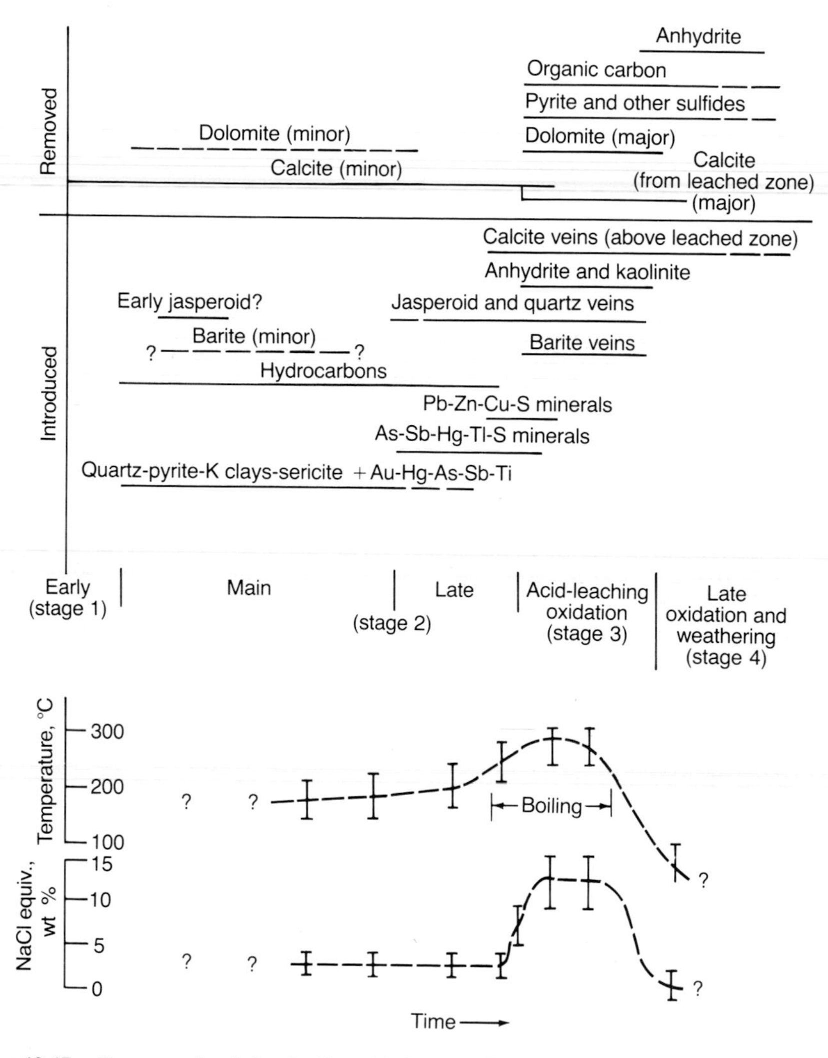

Figure 12-17. Paragenesis of the Carlin gold deposit. The acid-leaching and the accompanying oxidation (stage 3) were superimposed on the late hydrothermal (stage 2) events in the upper part of the deposit. The extent to which main- and late-stage events (including mineralization) continued at depth during acid leaching in the upper levels is not known. Also shown are average temperature and salinity trends of hydrothermal fluids as indicated by fluid inclusion data. The curves are highly schematic. (*From* Radtke, Rye, and Dickson, 1980.)

carbonate units. Main-stage mineralization at 175 to 200°C and 2 to 4% NaCl equivalent salinity introduced Si, Al, K, Ba, Fe, Au, Tl, Hg, S, As, Sb, and organic materials (Figure 12-17); quartz, pyrite, and sericite minerals were deposited, as was the ultrafine 5 to 30 μm gold, microparticles of which can best be seen by electron microscope. These minerals and gold ores were formed in Zone 3 of Figure 12-18 and in the stippled portion in the schematic of Figure 12-19, on which it is indicated that fluids rose from a presumed shallow stock along faults and along an older dike system to spread out through the Roberts Mountains limestone under the Devonian Popovich limestone. Minor base-metal minerals were deposited after the gold.

Temperatures rose to as high as 275 to 300°C toward the end of the mineralization event (Figure 12-17). Boiling began; it boiled off the volatiles H_2O, CO_2, and H_2S, increased nonvolatile salinity to 10 to 15% NaCl, and it lead to the generation of H_2SO_4 at high levels. This sulfuric acid stage 3 (Figure 12-17) produced intense acid leaching and oxidation of ore and rocks near the surface in Zone 1 of Figure 12-18. Calcite and dolomite were con-

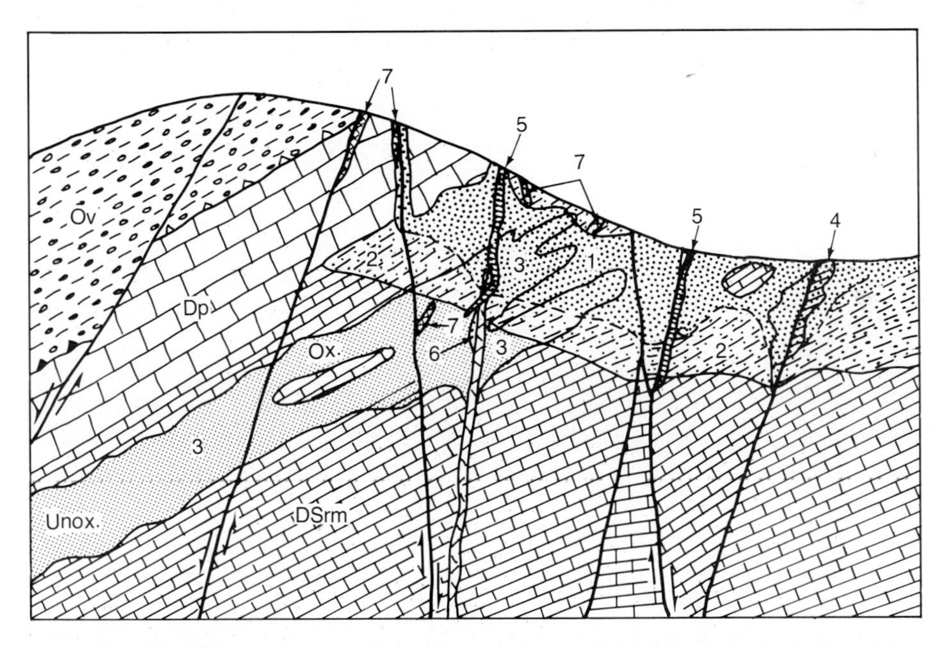

Figure 12-18. Schematic north-south cross section through the main ore zone and Popovich Hill showing major geologic features at Carlin. Lithologic units include the Vinini Formation (Ov), the Popovich Formation (Dp), and the Roberts Mountains Formation (DSrm). 1 = zone of leaching-alteration (heavy dots); 2 = zone of late supergene alteration (dots and dashes) extending from the surface downward through and below acid-leaching zones; 3 = main ore zone includes lower unoxidized ores and upper oxidized ores (fine dots); 4 = jasperoid bodies in solid black dot and line pattern; 5 = barite veins in horizontal bar pattern; 6 = quartz veins in solid black dots; 7 = calcite veins in cross-hatched pattern. A dike intruded along the fault near the center of the figure. (*From* Radtke, Rye, and Dickson, 1980.)

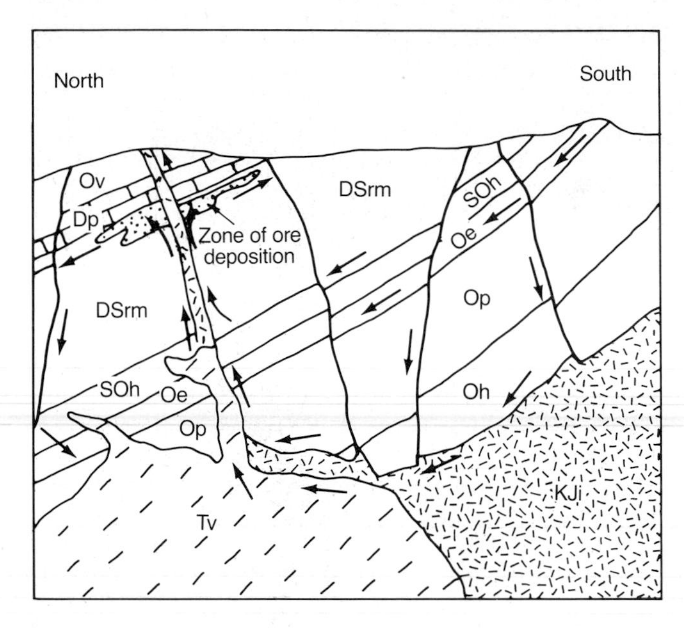

Figure 12-19. Hydrothermal system and solution paths (large arrows) inferred for the formation of the Carlin gold deposit. See Figure 12-18 for explanation of symbols; also, Oh = Ordovician Hamburg dolomite; Op = Ordovician Pogonip limestone; SOh = Siluro-Ordovician Hanson Creek dolomite; KJi = 125-million-year Mesozoic granodiorite; Tv = 14-million-year rhyolite-rhyodacite volcanic rocks, the probable heat source that drove the fluids. (*From* Radtke, Rye, and Dickson, 1980.)

sumed, sulfides and organic compounds were oxidized, kaolinite and anhydrite were formed, and silica was dumped in that same Zone 1. Temperatures and salinities then fell (Figure 12-17), hydrothermalism ended, and late supergene oxidation (Zone 2, Figure 12-18) penetrated well below Zone 1, affecting both the previously oxidized Zone 1 and the orebody Zone 3. Oxygen and hydrogen isotopes indicate that most of the hypogene and supergene waters were meteoric and had high $^{34/32}S$ ratios similar to those in the underlying Ordovician sediments; they suggest the circulation pattern of Figure 12-19. The source of the metals is not known. Their collective similarity to the epithermal suite of elements is strongly suggestive of the influence of a shallow stock, although Radtke, Rye, and Dickson (1980) favor a chemical model involving leaching of metals from adjacent and subjacent carbonate rocks. The normal ores contain 8 ppm Au, 25 ppm Hg, 100 ppm Sb, 400 ppm As, and 10 to 50 ppm Tl. Silver is absent. The gold occurs with mercury, antimony, and arsenic as coatings on pyrite and fracture fillings in that mineral. Many classifications of ore have been established, namely, normal, siliceous, pyritic, carbonaceous, arsenical, oxidized, and leached-oxidized. As stated, the Carlin ores and processes are similar in many respects to those prevailing at other disseminated gold deposits in the Basin and Range states. The ore type is being sought elsewhere; nothing known about them indicates that Carlin-type deposits should be found only in Nevada, although they are known principally there to this date. The Tallman-Duval in Idaho and the Mercur deposit in Utah (Figure 12-16) are similar, and others are being sought elsewhere.

▶ OTHER DEPOSITS RELATED TO SUBAERIAL VOLCANISM

Other deposit types that can be placed in this category include the volcanic iron deposits and their related phosphatic ores, volcanic tin deposits, volcanic manto copper deposits, some manganese oxide ores in felsic volcanics as at the Lucifer deposit, Baja California, and Keweenaw-type basalt-native copper settings. Kiruna iron ores were described in Chapter 11 as related to intermediate to felsic intrusion; the subaerial extrusive volcanic iron ores of Cerro de Mercado, Mexico, are described here. Tin in extrusive felsic rocks, especially the rhyolite stocks and flows of Mexico, New Mexico, and Bolivia-Argentina, were also mentioned in Chapter 11.

Cerro de Mercado, Durango, Mexico, Iron Deposits

In the last decade it has been established that massive magnetite of magmatic segregation origin can either be intruded as huge lenses of iron ore as at Pea Ridge, Missouri (Emery, 1968) or at Kiruna, Sweden, according to Frietsch (1978); or it can be extruded where it is found much as are other flow units, interlayered with felsic volcanic rocks, as at El Laco in Chile, Cerro de Mercado in Durango, Mexico (Swanson et al., 1978), or at Kiruna, according to Parak (1975). Neither of these modes of formation, which probably involve both fractional crystallization and liquid immiscibility at intermediate evolutionary stages of igneous differentation, should be confused with the gravitative formation of layered magnetites in LMIs, such as the titaniferous magnetites high in the Main Zone of the Bushveld Igneous Complex (Chapter 9). Magmatic trends that would produce an iron-rich residual melt, or an immiscible oxide liquid, were discussed in Chapter 11.

Cerro de Mercado, about 10 km north of the city of Durango, Mexico, was discovered in 1552 by Giné Vasquez de Mercado, for whom the district was named. Since 1881, when systematic production of iron ore was first attempted, the deposit has supplied a large part of the iron and steel used in Mexico. The ores are part of a volcanic caldera complex (Figure 12-20), and are generally stratiform layers of hematite-magnetite-apatite tuffs, pyroclastics, agglomerates, flows, and dikes.

The country rock near the iron ore is mainly quartz latite, rhyolite, and rhyolitic tuff, commonly colored red from finely divided hematite. The stratigraphy of the area was described by Swanson et al. (1978); a quartz latite porphyry intrusion was also recognized by Ortiz Asiain (1956). The ore is in thick, outwardly dipping wedge-shaped layers in fault blocks within the margins of a large caldera (Lyons and Clabaugh, 1973; Swanson et al., 1978) in the Cacaria Formation (Figure 12-20), section *A-A'*). It consists mainly of massive, at places brecciated hematite, hematite replacing magnetite, and magnetite, and is associated with rhyolite domes, flows, and air-fall tuffs.

A thick lava flow that preceded the collapse of the caldera, the Aguila rhyolite, lies below the ore. This rhyolite is cut by magnetite-hematite dikes with asparagus-colored, almost gem quality apatite crystals. At their upper ends the dikes became vents, and magnetite was mostly oxidized to hematite as it was erupted as flows, as thoroughly oxidized ash and pyroclastic material, and as agglomerates with fragments of magnetite and partially oxidized magnetite-hematite mixtures in a fine-grained hematite fragmental matrix. Specimens of coarse-grained martite—magnetite octahedrons several centimeters on a side replaced by hematite—are common. A thin layer of hematite powder, probably ash, was deposited on the existing erosion surface of several hundred square kilometers adjacent to the iron ore. Above the ore is finely divided powder of crystalline magnetite that is laminated except where disrupted by an overlying flow of rhyolite; a rhyolite dike cuts upward through the orebody and transforms vertically into a flow that caps the orebody. The iron ores thus appear to be highly specialized igneous rocks participating in an unusual caldera-form volcanic edifice. K-Ar determinations indicate that the underlying and overlying ash flows were emplaced about 30 million years ago (Lyons and Clabaugh, 1973).

At Peña Morada, a small inactive mine about 7 km west of Cerro de Mercado (Figure 12-20), a layer of nearly pure hematite lies upon an ashlike hematite deposit that contains angular rhyolite fragments. These rhyolite fragments have been partially melted and streaked out immediately below the hematite layer, which appears to have been emplaced as a very hot, volatile-rich flow. Small amounts of hedenbergite-diopside have been recognized, and montmorillonite and silica are locally abundant. The beautiful crystals of pale greenish-yellow apatite known as asparagus stone are widely distributed in the iron ores and are sought as a semiprecious material.

Until recently the origin of the ores has been widely debated. Weidner reported in 1858 that the ores resulted from volcanic eruptions produced by isolated cones. Others have considered the ore to have formed as dikelike bodies, and still other geologists describe the deposits as of contact metamorphic origin (Salazar et al., 1923). Gonzales Reyna (1956) thought the ores were probably magmatic segregation deposits, and Foshag (1929) thought that the ore resulted from replacement by ascending hydrothermal solutions. The preferred origin is that proposed by Lyons and Clabaugh (1973) and by Swanson and his colleagues (1978). They interpret the deposits as flows of iron oxides, largely hematite and martite, and the overlying magnetite and hematite powder as having been deposited by ash falls and from fumaroles.

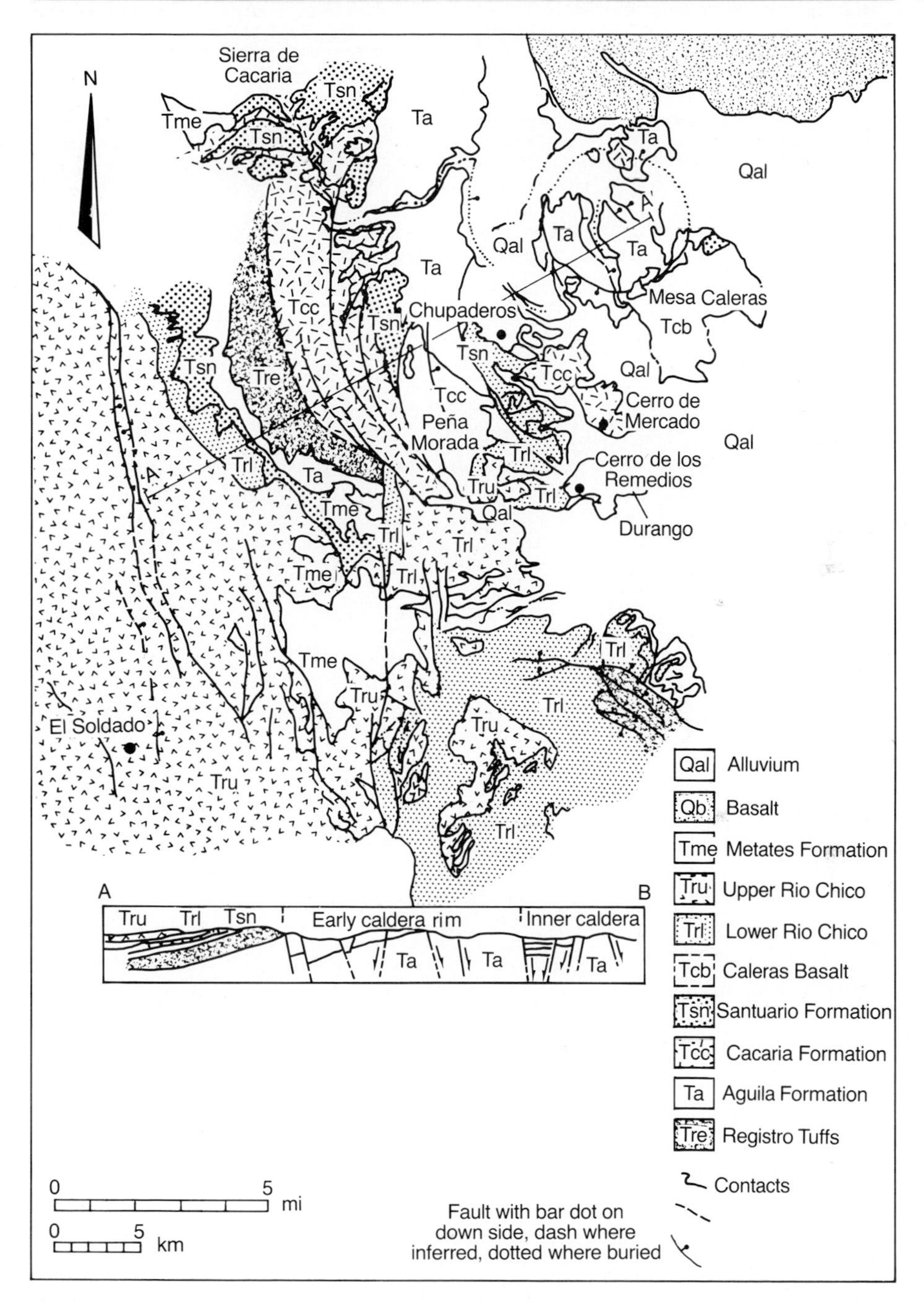

Figure 12-20. Generalized geological map of the Durango and the Cerro de Mercado mine districts, Mexico. See Figure 12-4 for Mexico location map. (*After* Swanson et al., 1978.)

Basalt-Andesite Copper Deposits

Michigan Copper Deposits The manto andesite copper deposits of Peru, Chile (Sillitoe, 1977), and Arizona may be related to Michigan copper ores that occur in basalt and conglomerate. The copper ores of the Calumet-Hecla district of the Keweenaw Peninsula in northern Michigan are examples of low-temperature hydrothermal deposits that have been channeled into their present positions through permeable conglomerates and the broken tops of gently dipping lava flows (Butler and Burbank, 1929; White, 1968). Concentrations of ore are found as cement in conglomerate beds between the basalt flows, in the fragmental, vesicular amygdaloidal flow top layers of individual flows, and in sparse fissures which cut both. These sites were favorable for ore deposition because of extremely high permeabilities; the ores are therefore essentially tabular and stratabound. The basalts, of late Proterozoic Precambrian age, are continental tholeiite basalts that poured out in a basin that was also receiving the conglomerate sediments. The resulting 2-to-3-km-thick pile of 1-to-200-meter-thick basalt flows with about 3% interleaved conglomerates, such as the Calumet and Kingston units, is called the Portage Lake series. The deposits contain almost no sulfur, so native copper is the predominant ore mineral, with rare chalcocite (Cu_2S) and digenite ($Cu_{1.8}S$), both of which have high metal-to-sulfur ratios. Native copper is commonly associated with native silver, another indicator of low sulfur fugacities, and metallic nuggets called *half breeds*, which contain both native copper and silver, are common in old mineral collections. Although most of the native copper is measured in millimeters or centimeters, masses of lacy interconnected amygdule and conglomerate matrix fillings of copper have been found, some weighing hundreds of kilograms. Although the fissures provide only a small percentage of the ore, large masses of native copper up to several hundred metric tons have been found in them. Some of these masses were so large and so obdurate to drills, explosives, cutting torches, wire saws, and other conventional tools that they were either skirted and left behind in mining or painstakingly reduced with hacksaws. The native copper deposits are associated with albitized wall rock and amygdule fillings of calcite, chlorite, epidote, quartz, prehnite, pumpellyite, and the zeolite laumontite ($CaAl_2Si_4O_{12} \cdot 4H_2O$). Open-space filling vastly predominates over replacement, and mineralogy, textures, and the absence of alteration indicate exceedingly low pressures, and temperatures perhaps less than 100°C. Stoiber and Davidson (1959) and White (1968) presented field and laboratory data in support of the argument that metamorphism of lavas at depth altered the basalt to pumpellyite-epidote-prehnite-chlorite, and released copper and other chemical constituents. Low-temperature fluids carried the copper updip along zones of high permeability to fill open spaces at near-surface levels. The district has supplied nearly 6 million tons of copper since the 1860s. As the first major copper mining district in North America, it attracted waves of miners and engineers from Europe and Scandinavia to the United States, craftsmen who then dispersed to other fledging districts like Butte, Montana, and Bisbee, Arizona, and contributed materially to their culture, history, and technology.

Andesite Copper Desposits Veins, veinlets, vesicle- and amygdule-filling blebs, and disseminations of native copper, native silver, chalcocite, bornite, and rare chalcopyrite are also found in calc-alkaline volcanic rocks of mid-Mesozoic to Pliocene age in the American Cordilleras. They have been ore-bodies only in Chile and Bolivia (at Coro Coro), but as Sillitoe (1977) points out, it seems reasonable to conclude that economic copper concentrations of this type will be found in calc-alkaline belts elsewhere in the world as exploration continues. It is tempting to speculate that they may be I-type extrusive equivalents of porphyry coppers while the tin rhyolites are S-type extrusives of tin-rich plutons. These manto coppers are stratiform and stratabound, tens of meters thick, and they extend for kilometers along strike. Andesitic, dacitic, and latitic flow tops, breccias, and ignimbrites are preferred hosts. Since porosity and permeability decrease downward, so also does copper grade; deposits typically have sharp hanging walls and gradational footwalls. Gangue minerals and alteration other than scant silicification, carbonatization, and rare pyritization are rare. Ruiz et al. (1971) described the Buena Esperanza deposit in the Atacama Desert near Antofagasta, Chile. Twenty-eight horizons 2 to 25 meters thick and mineralized with chalcocite and minor bornite are intercalated in a 300-meter-thick stack of vesicular-topped, locally flow-brecciated Jurassic andesites. Some sediments occur, and the presence of mineralized fragments attests to the fact that mineralization had commenced while volcanism was still active. Fluid inclusions in quartz and calcite fill at 112 to 195°C. Sillitoe suggests a Valley of Ten Thousand Smokes analogy (Figure 12-1), possibly with involvement of lagoonal or lacustrine waters. Occurrences similar to that at Buena Esperanza are common in Chile and Peru, are reported in andesites near Arivaca, Arizona, and should be sought elsewhere.

Bibliography

Barton, P. B., Jr., P. M. Bethke, and E. Roedder, 1977. Environment of ore deposition in the Creede mining district, San Juan Mountains, Colorado, Pt. III: Progress toward interpretation of the chemistry of the ore-forming fluid for the OH vein. *Econ. Geol.* 72:1–24.

——, ——, and P. S. Toulmin, III, 1971. An attempt to determine the vertical component of flow rate of ore-forming solutions in the OH vein, Creede, Colorado. Soc. Min. Geol. Jap., Spec. Pap. 2, pp. 132–136.

Bastin, E. S., 1948. Mineral relationships in the ores of Pachuca and Real del Monte, Hidalgo, Mexico. *Econ. Geol.* 43:53–65.

Bernasconi, A., N. Glover, and R. P. Viljoen, 1980. The geology and geochemistry of the Senator antimony deposit, Turkey. *Mineral. Deposita* 15:259–274.

Bethke, P. M., and R. O. Rye, 1979. Environment of ore deposition in the Creede mining district, San Juan Mountains, Colorado. Pt. IV: sources of fluids from oxygen, hydrogen, and carbon isotope studies. *Econ. Geol.* 74:1832–1851.

Berger, B. R., and P. I. Eimon, 1982. Comparative models of epithermal silver-gold deposits. AIME Preprint 82–13, 30 pp.

Buchanan, L. J., 1979. Pers. Commun.

Burbank, W. S., 1933. Base metals, pp. 641–651 in *Ore Deposits of the Western States*, Lindgren Vol., Pt. VI: Epithermal deposits. New York: AIME, 797 pp.

Butler, B. S., W. S. Burbank, et al., 1929. The copper deposits of Michigan. USGS Prof. Pap. 144.

Clifton, C. G., L. J. Buchanan, and W. P. Durning, 1980. Exploration procedures and controls of mineralization in the Oatman mining district, Oatman, Arizona. AIME Preprint 80–143, 17 pp.

Dreier, J. E., 1976. The geochemical environment of ore deposition in the Pachuca-Real del Monte district, Hidalgo, Mexico. Unpub. Ph.D. Dissert., Univ. Ariz., Tucson, 115 pp.

Emery, J. A., 1968. Geology of the Pea Ridge Iron Ore Body, pp. 359–369 in J. D. Ridge, Ed., *Ore Deposits of the United States, 1933/1967*, Graton-Sales Vols. New York: AIME, 1880 pp.

Emmons, W. H., and E. S. Larsen, 1923. Geology and ore deposits of the Creede district, Colorado. USGS Bull. 718, 189 pp.

Foshag, W. F., 1929. Mineralogy and geology of Cerro de Mercado, Durango, Mexico. U.S. Natl. Mus. Proc., vol. 74, 27 pp.

Frietsch, R., 1978. On the magmatic origin of iron ores of the Kiruna type. *Econ. Geol.* 73:478–485.

Geyne, A. R., 1956. Las rocas volcanicas y los yacimientos argentiferos del distrito minero de Pachuca-Real del Monte, Estado de Hidalgo. 20th Int. Geol. Congr., Excursions A-3 and C-1, pp. 47–57.

——, C. Fries, Jr., K. Segerstrom, R. F. Black, I. F. Wilson, and A. Probert, 1963. Geology and mineral deposits of the Pachuca-Real del Monte district, State of Hidalgo, Mexico. Mex. D. F., Consejo Recursos Naturales no Renovables Pub. 5E.

Giudice, P., 1980. Mineralization at the convergence of the Amethyst and OH fault systems, Creede district, Mineral County, Colorado. Unpub. M.S. thesis, Univ. Ariz., Tucson.

Gonzales Reyna, J., 1956. *Riqueza Minera y Yacimientos Minerales de Mexico.* Banco de Mexico Pub., pp. 229–251.

Graton, L. C., 1933. The depth-zones in ore deposition. *Econ. Geol.* 28:513–555.

Gross, W. H., 1975. New ore discovery and source of silver-gold veins, Guanajuato, Mexico. *Econ. Geol.* 70:1175–1189.

Howell, K. K., 1977. Geology and alteration of the Commonwealth mine, Cochise County, Arizona. Unpub. M.S. thesis, Univ. Ariz., Tucson, 225 pp.

Juan, V. C., 1946. Mineral resources of China. *Econ. Geol.* 41:399–474.

Larsen, E. S., 1930. Recent mining developments in the Creede district, Colorado. USGS Bull. 811B, pp. 89–112.

Lausen, C., 1931. Geology and ore deposits of the Oatman and Katherine districts, Arizona. Ariz. Bur. Mines Bull. 131, 126 pp.

Lindgren, W., 1933. *Mineral Deposits*, 4th ed. New York: McGraw-Hill, 930 pp., esp. 444–513.

Lipman, P. W., B. R. Doe, C. E. Hedge, and T. A. Steven, 1978. Petrologic evolution of the San Juan volcanic field, southwestern Colorado: Pb and Sr isotope evidence. *Geol. Soc. Amer. Bull* 89:59–82.

Lyons, J. L., Jr., and S. E. Clabaugh, 1973. Pyroclastic and extrusive iron ore at Durango, Mexico. *Econ. Geol.* 68:1216–1217.

Nolan, T. B., 1933. Precious metals, pp. 623–640 in *Ore Deposits of the Western States*, Lindgren Vol., Pt. IV, Epithermal deposits. New York: AIME, 797 pp.

O'Neil, J. R., and M. L. Silberman, 1974. Some mineralogical characteristics of the silver deposits in the Guanajuato mining district, Mexico. *Econ. Geol.* 69:1178–1185.

Ordoñez, E., 1902. The mining district of Pachuca, Mexico. *AIME Trans.* 32:224–241.

Ortiz Asiain, R., 1956. Notas sobre el Cerro de Mercado. 20th Int. Geol. Congr., Excursions A2-A5, pp. 119–124.

Parak, T., 1975. Kiruna ores are not "intrusive magmatic ores of the Kiruna type." *Econ. Geol.* 70:1242–1258.

Radtke, A. S., 1974. Genesis and vertical position of fine-grained disseminated replacement-type gold deposits in Nevada and Utah, U.S.A. *4th IAGOD Sympos. Proc.*, Varna, Bulgaria, vol. 1, pp. 71–77.

——, R. O. Rye, and F. W. Dickson, 1980. Geology and stable isotope studies of the Carlin gold deposit, Nevada. *Econ. Geol.* 75:641–672.

Rathjen, J. A., 1979. Antimony mineral commodity profiles. U.S. Bur. Mines, 13 pp.

Ren, Qijiang, 1983. Pers. Commun.

Roberts, R. J., 1966. Metallogenic provinces and mineral belts in Nevada. Nev. Bur. Mines Rept. 13, pt. A, pp. 47–72.

——, A. S. Radtke, and R. R. Coates, 1971. Gold-bearing deposits in north-central Nevada and southwestern Idaho. *Econ. Geol.* 66:14–33.

Ruiz, C. F., A. Aguilar, E. Egert, W. Espinosa, F. Peebles, R. Quezada, and M. Serrano, 1971. Strata-bound copper sulphide deposits of Chile, pp. 252–260 in Joint Symp. Vol. IMA-IAGOD Mtgs., 1970, Tokyo, Soc. Min. Geol. Jap., Spec. Issue 3, 500 pp.

Salazar, J., L. Salinas, P. Gonzales, M. Santillan, A. Aceveda, and A. R. Martinez Q., 1923. El Cerro de Mercado, Durango, Mexico. Inst. Geol. Mex. Bol. 44.

Schmitt, H. A., 1950. The fumarolic-hot spring and "epithermal" mineral deposit environment. *Quarterly Colo. Sch. Mines* 45,1B:209–229.

Sillitoe, R. H., 1977. Metallic mineralization affiliated to subaerial volcanism: a review, pp. 99–116 in *Volcanic Processes in Ore Genesis*. London: Inst. Min. Metall., Spec. Pap. 7, 188 pp.

Steven, T. A., and G. P. Eaton, 1975. Environment of ore deposition in the Creede mining district, San Juan Mountains, Colorado. I: Geologic, hydrologic and geophysical setting. *Econ. Geol.* 70:1023–1037.

—— and P. W. Lipman, 1976. Calderas of the San Juan volcanic field, southwestern Colorado. USGS Prof. Pap. 958, 35 pp.

—— and J. L. Ratté, 1965. Geology and structural control of ore deposition in the Creede district, San Juan, Colorado. USGS Prof. Pap. 487, 87 pp.

Stoiber, R. E., and E. S. Davidson, 1959. Amygdule mineral zoning in the Portage Lava series, Michigan Copper district. *Econ. Geol.* 54:1250–1277; 1444–1460.

Sudo, T., 1954. Types of clay minerals closely associated with metalliferous ores of the epithermal type. Tokyo Kyoiku Daigaku Sci. Rept. 3(23):173–197.

Swanson, E. R., R. P. Keizer, J. L. Lyons, and S. E. Clabaugh, 1978. Tertiary volcanism and caldera development near Durango City, Sierra Madre Occidental, Mexico. *Geol. Soc. Amer. Bull.* 89:1000–1012.

Tegengren, F. R., 1921. The Hsi-K'uang-Shan antimony mining fields, Hsin-Hua district, Hunan. Geol. Surv. China Bull. 3, pp. 1–26.

Thornburg, C. L., 1945. Some applications of structural geology to mining in the Pachuca-Real del Monte area, Pachuca silver district, Mexico. *Econ. Geol.* 40:283–297.

——, 1952. The surface expression of veins in the Pachuca silver district of Mexico. *AIME Trans.* 193:594–600.

Thorson, J. P., 1971. Igneous petrology of the Oatman district, Mohave County, Arizona. Unpub. Ph.D. Dissert., Univ. Calif. Sta. Barbara, 189 pp.

Wang, K. P., 1977. Far East and South Asia. Mineral Perspectives Series MP-1. U.S. Bur. Mines, 92 pp.

Watson, B. N., 1977. Large low grade silver deposits in North America. *World Min.*, Mar., pp. 44–49.

White, D. E., 1955. Thermal springs and epithermal ore deposits. *Econ. Geol. 50th Anniv. Vol.*, pp. 99–154.

——, 1967. Mercury and base-metal deposits with associated thermal and mineral waters, pp. 575–631 in H. L. Barnes, Ed., *Geochemistry of Hydrothermal Ore Deposits*. New York: Holt, Rinehart and Winston, 670 pp.

——, 1968. Environments of generation of some base metal ore deposits. *Econ. Geol.* 63:301–335.

——, L. J. P. Muffler, and A. H. Truesdell, 1971. Vapor-dominated hydrothermal systems compared with hot-water systems. *Econ. Geol.* 66:75–97.

White, W. S., 1968. The native copper deposits of northern Michigan, pp. 303–326 in J. D. Ridge, Ed., *Ore Deposits of the United States, 1933/1967*, Graton-Sales Vols. New York: AIME, 1880 pp.

Winchell, H. V., 1922. Geology of Pachuca and El Oro, Mexico. *AIME Trans.* 66:27–41.

Wisser, E., 1937. Formation of the north-south fractures of the Real del Monte area, Pachuca silver district, Mexico. *AIME Trans.* 126:442–487.

——, 1942. The Pachuca silver district, Mexico, pp. 229–235 in W. H. Newhouse, Ed., *Ore Deposits as Related to Structural Features*. Princeton, NJ: Princeton Univ. Press, 280 pp.

——, 1960. Relation of ore deposition to doming in the North American Cordillera. Geol. Soc. Amer. Mem. 77.

——, 1966. The epithermal precious-metal province of northwest Mexico. Nev. Bur. Mines Rept. 13, pt. C., pp. 63–92.

Yan, M. Z., and K. Hu, 1980. Geological characteristics of the Dexing porphyry copper deposits, Jiangxi, China, pp. 197–204 in S. Ishihara and S. Takenouchi, Eds., *Granitic Magmatism and Related Mineralization*. Soc. Min. Geol. Jap., Spec. Issue 8, 247 pp.

Zies, E. G., 1929. The Valley of Ten Thousand Smokes. Nat. Geog. Soc. Contr. Tech. Pap., Katmai Ser. 1(4).

THIRTEEN

Deposits Related to Submarine Volcanism

It is remarkable that so little was known about the seafloor and submarine ore-forming processes as recently as 1960. Most world maps and globes then were colored three-fourths light blue, and little attention was, or could have been, paid to ore-forming processes on the seafloor before that time. That situation has changed dramatically in the last 25 years, however, and submarine volcanism and volcanically driven hydrothermal activity are now recognized as having been important from the development of the first Archean crust to this very moment. Sulfide deposits of copper-nickel, copper-zinc, zinc-lead-silver, iron, gold, tungsten, tin, antimony, and mercury are now attributed to this environment, as are banded iron formations, manganese and barite deposits, and others. Their elucidation has led to the resolution of new lithotectonic relationships. These deposits were described by Derry (1973) as having formed on a *contemporaneous surface*, a surface constituting part of a conformable, continuous rock-orebody-rock sequence in which the rock components are volcanic and the orebody components involve volcanism, exhalation of volcanic fluids, and sedimentation. Among the remarkable developments of the period has been the appreciation of what have become known as volcanogenic massive sulfide deposits.

It is difficult to define and describe *massive sulfide deposits* for several reasons. First, they vary in several important geologic aspects according to geologic age, from earliest Archean to the present (Hutchinson, 1965). Second, although they are recognized as having specific plate tectonic settings (Sillitoe, 1972; Pearce and Gale, 1980; Guilbert, 1981; Sawkins, 1984), they were perhaps best developed in Archean and Proterozoic systems at least 2 billion years before plate tectonics as presently expressed had developed. Third, they involve the many commodities listed in the preceding paragraph in deposits that do not appear to be truly transitional one to another, so to describe the class thoroughly could require inclusion of descriptions of major deposits of each metal. Fourth, the several types of environment of deposition vary according to both water depth and lateral distance between the submarine hydrothermal source and the site of deposition—near deposits are called *proximal*, and laterally removed ones are termed *distal*. Fifth and last, many of the wall-rock alteration-mineralization systematics, including the sources of components, their transportation, the behavior of fluids when they enter the seawater medium, and the actual precipitation of proximal massive sulfides and distal sulfide, oxides, carbonates, and metals, are still imperfectly understood.

The term *volcanogenic*, or *volcanogenic-exhalative*, refers to stratabound mineral deposits that have been formed by volcanic processes and the activities of thermal springs at the bottoms of bodies of water (Oftedahl, 1958). Although many types of deposits are of this origin, recent authors have tended to narrow the term volcanogenic to the massive sulfide bodies deposited from submarine thermal springs during periods of volcanic activity. Mineralizing fluids issued from vents and either poured out upon the seafloor or rose to various levels in the sea to permit the dispersal of ore-forming components in the seawater medium prior to their precipitation. Some fluids appear to have permeated and replaced shallow layers of volcanic tuffs and associated seafloor sediments. This chapter will stay with a broader definition of "volcanogenic," which includes oxides, carbonates, sulfates, and so on, of a variety metals, not simply massive sulfides.

The idea that an ore deposit may form by deposition from volcanic and thermal springs pouring out on the seafloor is not a new one, but rather has been held for many years. In particular it has been used to describe banded iron formation and layered manganese oxide deposits (Van Hise and Leith, 1911; Park, 1942; Hewett, 1966; Gilmour, 1965). Only in recent years, however, has the concept been found to apply widely to accumulations of massive sulfides and related ores associated with thick volcanic piles. Many massive sulfide deposits are now thought to be stratabound and closely related to the processes of volcanism, especially to the formation of rhyolites and rhyolite domes (Kinkel, 1962, 1966; King, 1958; Griffitts et al., 1972; Ridler, 1970, 1976; Ishihara, 1971; Sangster and Scott, 1976).

Gilmour (1971) pointed out that one of the most important influences on the geology of a deposit in this category is measured in terms of proximal and distal attributes (Jambor, 1979; Plimer, 1978). Proximal deposits logically are dominated by immediate volcanic processes and volcanic products and occur typically as parts of the volcanic pile that produced them; they are exemplified by the Canadian deposits of the Abitibi and Matagami greenstone belts. Progressively more distal deposits result from lateral transmission of ore-forming components through seawater, and may involve greater and greater components of nonvolcanic sediments being shed into the basin and mixing with volcanic sediments. Deposits of this type—such as Broken Hill, New South Wales, the Mount Isa–McArthur River deposits in Australia, and the ones at Aggeneys and Gamsberg, Republic of South Africa—are described in Chapter 14.

The theory of volcanogenesis in massive sulfides is supported by the discovery of the metal-bearing deposits now being formed by hot brines in the Red Sea and especially by descriptions of 380°C metal-charged hydrothermal fluids pouring into the sea from fissures in the East Pacific Rise off the coasts of British Columbia, Mexico, and Ecuador. Oceanographic studies in the Red Sea were first reported by Miller and later by other scientists (Miller, 1964; Degens and Ross, 1969; Shanks and Bischoff, 1980). Hot concentrated saline waters or metals are found here in as many as 18 separate deeps. In one of the largest of the basins, the Atlantis II deep, the temperature is about 56°C and the salinity is about 255 parts per thousand, or 25.5%. The deeps are aligned along what appears to be a rift zone, and submarine volcanic activity is involved in the generation of the heated and saline waters. Iron, copper, zinc, and silver are present in appreciable amounts. Further observations of the Red Sea deposits should prove to be of great interest in the study of ore genesis. But even more exciting are descriptions of "black smokers," 2-meter-high chimneylike encrustations of sphalerite-chalcopyrite jetting turbid columns of 380 ± 20°C hydrothermal brines, blackened by entrained sulfide precipitates, along the Rivera Rise, near the mouth of the Gulf of California (Spiess et al., 1980). These "smokers" must closely resemble the submarine sources of deposits to be described. They have been visited and photographed by the ALVIN submersible, and their future study should answer many of the continuing problems mentioned earlier.

A central concept in considering these ore deposits is that of "exhalation" of ore-forming substances. Oftedahl (1958) used the term *exhalative* to describe ore-related materials debouched from submarine hot springs or fumaroles into seawater. The environment is not fully volcanic, but is clearly related to volcanism. It involves volcanic heat which drives both juvenile magmatic waters and convecting, ingested seawater. Ore components inserted from subjacent igneous environments, leached from volumes of seafloor rocks, swept by downward, inward, then upward convecting seawater (Figure 10-2), and provided by the seawater itself are heated and jetted out into the sea along fractures or through conduits precisely analogous to submarine fumaroles. The process is called *exhalation*. Ridler (1970) called chemically precipitated sediments formed of exhaled components *exhalites*, just as rocks formed by evaporative processes are called evaporites. Properly and precisely used, the term is a good one, but one must be careful to apply it only to stratiform units known beyond reasonable doubt to have formed through exhalative processes. Clearly, such exhalites are widespread, stratiform, stratabound conformable units that normally constitute time-stratigraphic units. They may locally contain components that are ore by themselves and they almost always grade distally to geochemically distinctive but uneconomically thin strata. Even if such units are not economic, their exploration significance is clear (see Figures 13-1 to 13-5). Close to the source, those compounds with the lowest solubility products will precipitate from the hot spring fluids. Those compounds are typically the sulfides, and their early precipitation accounts for the fact that the "black smoker" chimneys are made up of sulfide encrustations. Massive sulfide components are thus likely to precipitate proximally. Iron and manganese oxides and hydroxides stay in solution or form colloids which may be mechanically dispersed in the sea and precipitate distally. Noble metals which form no sulfides—especially gold—may be carried with these oxides, and so banded iron formations with or without gold values are common lateral facies of massive sulfide deposits. Sato (1977) has shown that temperature-density relations of the exhaled fluids may dictate their behavior in the sea. Temperature affects density-buoyancy, as do salinity and the presence or absence of dissolved sulfides or entrained sulfide flecks. Dense aqueous fluids may spread out immediately and puddle in seafloor depressions. Less dense solutions may rise to a level of gravitative equilibrium and then spread laterally (Solomon and Walshe, 1979), and least dense fluids may simply disperse into the sea. Excellent photographs of models of fluid behavior in the sea were presented by Turner and Gustafson (1978). It is important to recognize that massive sulfide ores as most geologists define them result from deposition from aqueous submarine solutions (see Figure 13-11*c*), not from submarine eruption of liquid sulfide melts.

Among the most important of the world's proximal sulfide deposits are the Cu-Zn orebodies of the Noranda district, Quebec, described here, the Kidd Creek ores near Timmins, Ontario, and the Kuroko deposits of Japan. A longer list of proximal orebodies would have to include some of the New Brunswick–Bathhurst–Newcastle and Maine orebodies such as the Blackbird; the Rio Tinto–Tharsis–Aznalcollar ores of the Silurian-age Pyrite, Belt of Spain; several Scandinavian districts such as Skellefte and Falun, Sweden, and Outokoumpu and Vihanti, Finland; the Prieska deposit in South Africa; and Mons Cupri in Australia and Mt. Lyell in Tasmania. Dozens more are known, for example in Arizona, and scores more will be found.

We examine here the Cu-Zn, Cu-Zn-Pb-Ag, and Cu-Ni proximal deposits of Canada, the Kuroko copper-zinc deposits of Japan, and the Cyprus-type massive sulfides. Subsequent sections treat the more distal banded iron formations and gold deposits, and Chapter 14 covers truly distal base-metal occurrences.

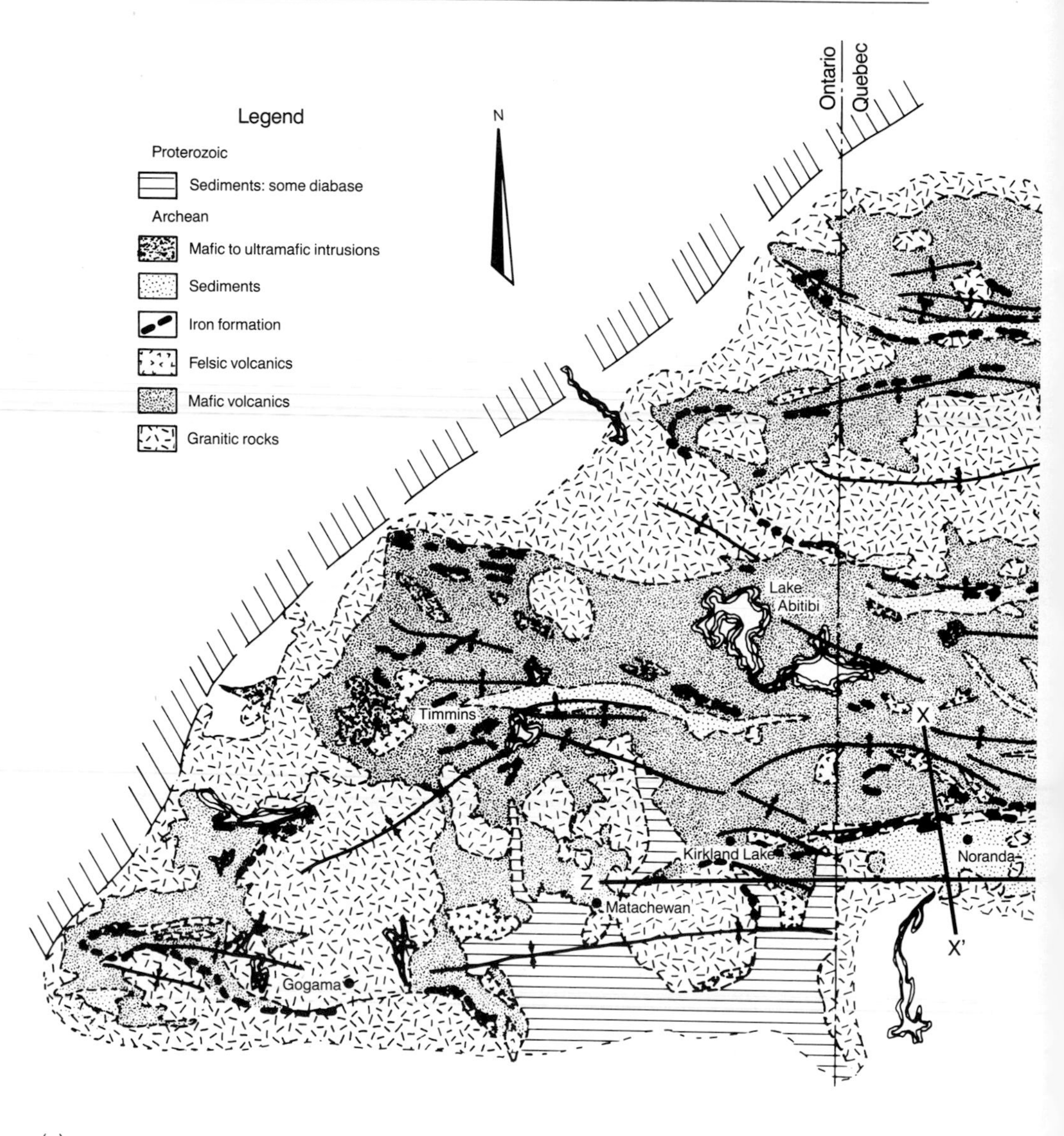
Legend
Proterozoic
Sediments: some diabase
Archean
Mafic to ultramafic intrusions
Sediments
Iron formation
Felsic volcanics
Mafic volcanics
Granitic rocks
N
Ontario
Quebec
Lake Abitibi
Timmins
X
X'
Z
Kirkland Lake
Noranda
Matachewan
Gogama

(*a*)

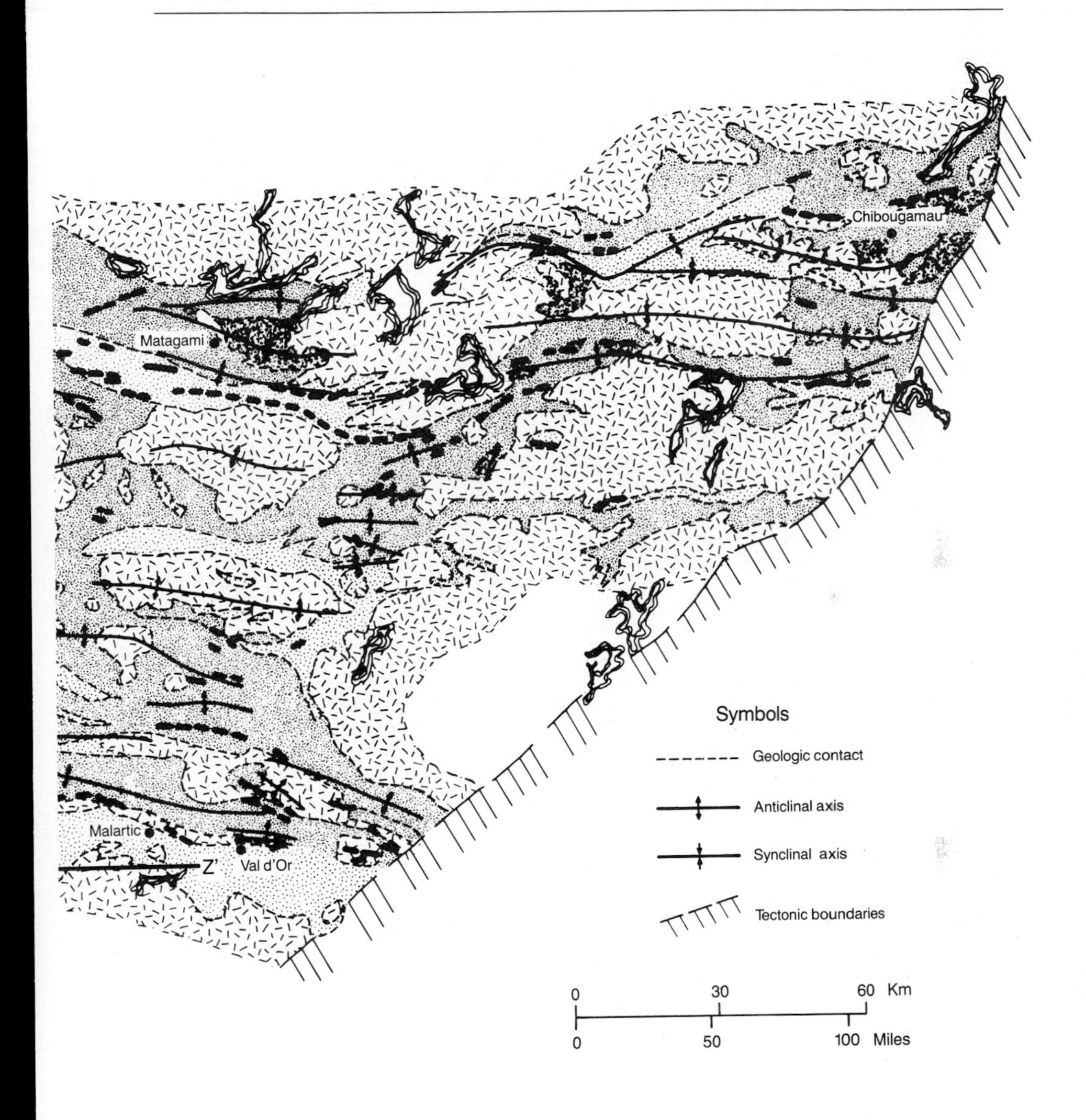

Figure 13-1. (*a*) Archean geology and structures of the Abitibi Greenstone Belt of Ontario and Quebec, Canada, a series of related volcanic centers on a granitic basement. (*b, next page*) Volcanic centers and mineralized zones of the Abitibi Belt. Any one of these centers looks like that of Figure 13-2, which might as well represent line X–X'. The plan view of facies changes in the purported exhalite horizon (Figure 19-12) runs from Z to Z'. Figure 13-3 illustrates the Abitibi Belt and outlines of other greenstone belts of Canada's Precambrian shield terrane to show how common they are. Cratonic areas on the other continents contain similar greenstone belts. (*After* Goodwin and Ridler, 1970.)

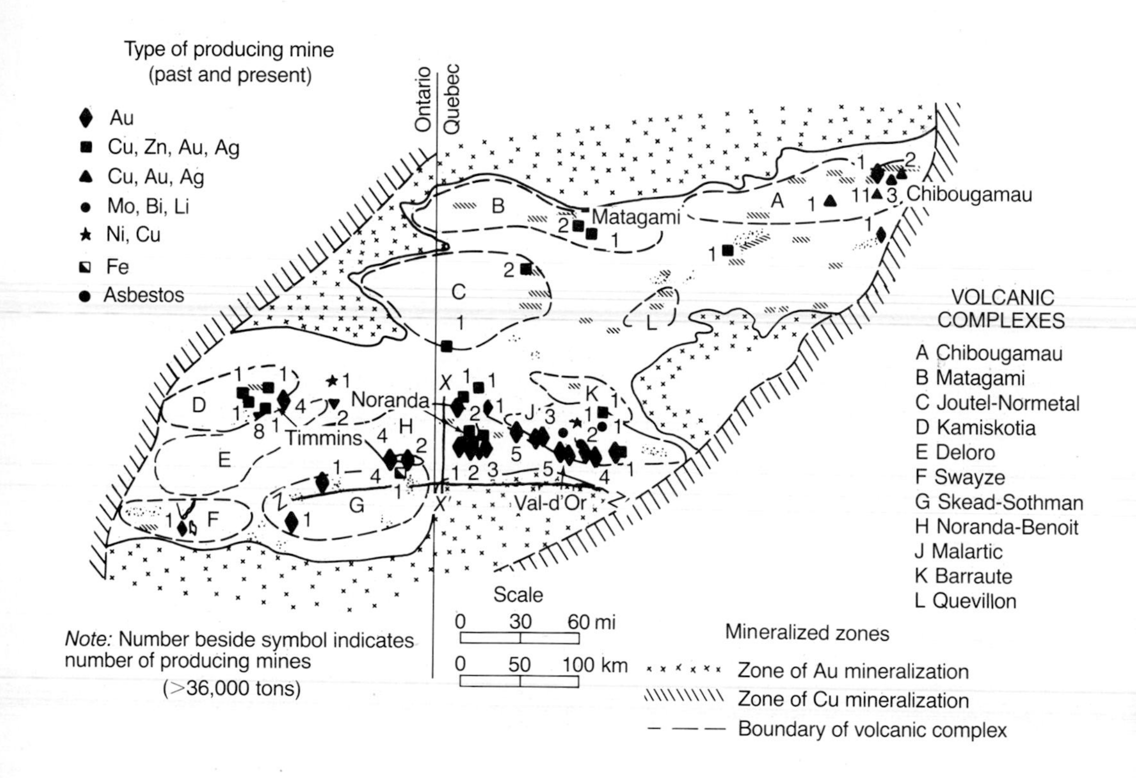

(*b*)

Throughout this book, the terms *stratabound* and *stratiform* are carefully used. *Stratabound* refers to ore deposit components that are contained within a single stratigraphic unit or layer but that can be either concordant or discordant (cross-cutting) within that layer. *Stratiform* means truly bedded in detail; it implies that the finer- and larger-scale elements of a layered occurrence are of similar age and mode of origin. All stratiform ore deposits are stratabound, but the reverse is not true. The infilling of void spaces in a sedimentary breccia by sulfides would be stratabound but not stratiform, as would tongues of replacement sulfides at various angles to true bedding but confined within specific bedding planes.

▸ VOLCANOGENIC MASSIVE SULFIDE DEPOSITS

Abitibi Copper-Zinc-Silver Deposits, Canada

Sangster and Scott (1976) and others consider that the first and lowermost stage in the process of formation of the Canadian massive sulfide deposits was the accumulation in Archean times, 2.4 to 2.7 billion years ago, of a complex, thick sequence of submarine flows of pillowed and vesicular basalts. Portions of these basal platform flows are the ultramafic spinifex-textured high-magnesia komatiites and komatiitic basalts described in Chapter 9, most of them tholeiitic; they collectively represent the oldest mafic crustal elements in the *greenstone belts.* Above these basement rocks are younger flows, flow breccias, and tuffs, mainly of basalt to andesite composition. The uppermost materials of the volcanic section are dacitic to rhyolitic, massive, and almost textureless flows and pyroclastic layers. Repeated cycles built huge volcanic edifices, perhaps as much as 10 km high. What we now call greenstone belts (Figure 13-1) are thus composed of older broad, flat shields of mafic or ultramafic flows with scattered subshields and stratovolcanic complexes on them, the whole mildly metamorphosed to greenschist facies. Greenstone belts typically are underlain and surrounded by older "crystalline basement" rocks, and commonly have complexly intruded, metamorphosed, and tectonized margins. The Abitibi Belt is shown in Figure 13-2; similar belts occur in Archean cratons in Africa, South America, the U.S.S.R., and elsewhere.

The Noranda sequence of the Abitibi Belt—called the Blake River Group—included at least five mafic to felsic cycles (Spence and de Rosen-Spence, 1975) and many interlaminated time-stratigraphic levels of submarine exhalite sulfide precipitation. As the rhyolitic rocks at the end of each cycle were extruded, a hiatus in active volcanism was recorded in places by explosive volcanism, which produced breccias called *mill rock,* by exhaled massive sulfides, banded iron formation, and quartz-carbonate-gold ore deposit components and by graywacke-type sediments (Figures 13-2 and 13-3). It is also characteristic of the ore-containing volcanic piles that they are intruded by dikes and irregular masses as compositionally diverse as the extrusives. The massive sulfide deposits are thus distributed along contemporaneous time-stratigraphic surfaces marked by variable amounts of sulfides, cherts, tuffs, banded iron formations and ferruginous carbonate rocks that are explorationally highly significant. Tuffs, also called *tuffisites,* and cherty rocks or tuffs containing anomalous silica and iron and manganese oxides are thus persistent tens to hundreds of kilometers beyond the sulfide ores, but in physical continuity with them. Sangster noticed in the early 1970s that silicified rhyolite breccias are so common in the footwall rocks near proximal massive sulfide orebodies that an operating mine's concentrator mill can commonly be heard running nearby when one stands on one of them. Dubbed mill rocks, these breccias are another exploration

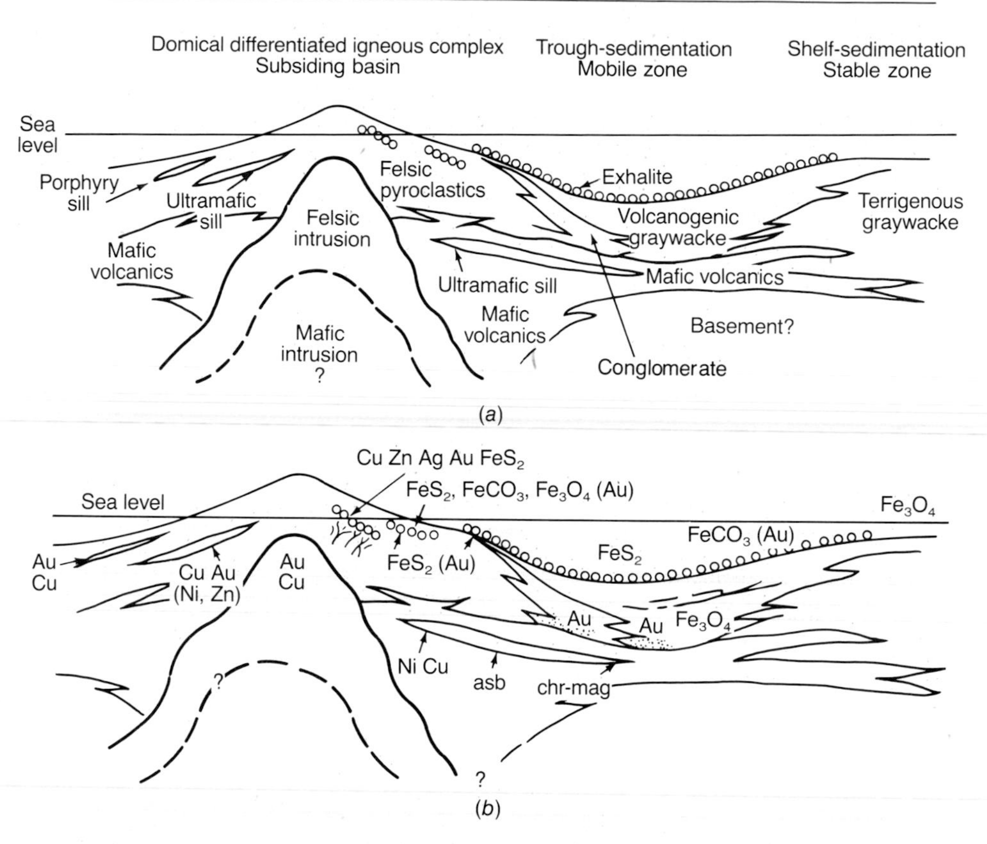

Figure 13-2. (*a*) Idealized, generally submarine volcanic edifice of Archean age showing rock-type–sediment-type interaction between volcanic and continental regimes. (*b*) Locations of various ore-deposit types related to these Archean volcanic sources. See also Figure 13-11. (*From* Hutchinson, Ridler, and Suffel, 1971.)

hallmark (Lajoie, 1977). They normally represent true explosive brecciation that just precedes ore formation; fragments of sulfides may be found in them, but sparsely. Many of the features described here are depicted in Figure 13-4 from Simmons (1973), which shows the stepwise development of the Millenbach deposit; Figure 13-5 depicts a generalized proximal volcanogenic massive sulfide system.

Another footwall characteristic is the so-called *alteration pipe*, a tap-root-shaped pathway of upward circulating ore-bearing fluids. Gilmour (1965) described an example at the Lake DuFault deposit in the Noranda district of western Quebec, Canada. These hydrothermal fluid conduits may be circular to subcircular in plan view; at Jerome, Arizona, the conduit to the United Verde massive sulfide layer is fissure-form, or planar (Anderson and Nash, 1972). The cores of the alteration pipes in Canada and elsewhere are generally composed of fine-grained massive "black chlorite" with disseminated sulfides and magnetite. Chlorite content decreases outward in the

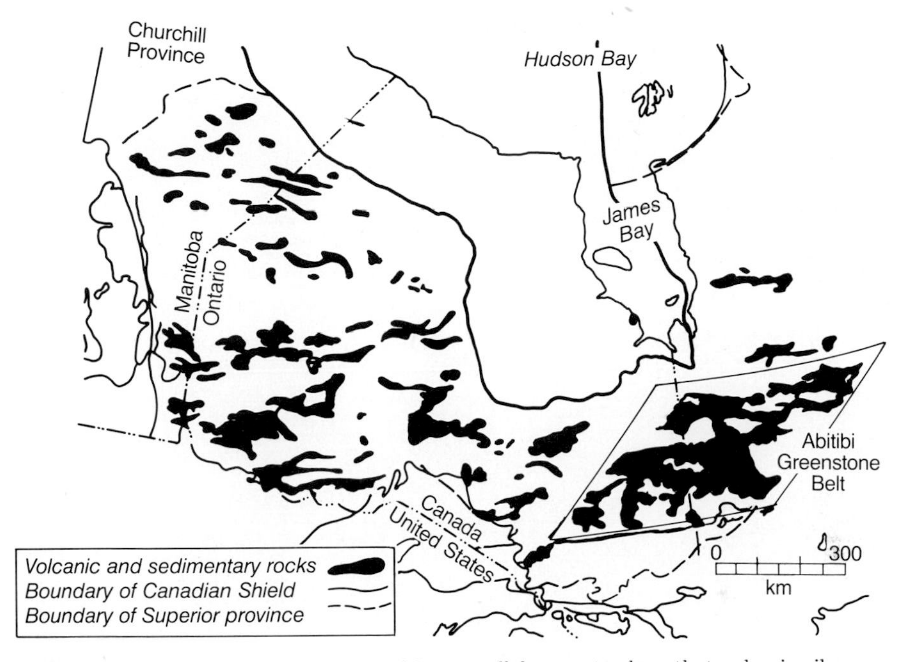

Figure 13-3. Area of Figure 13-1 outlined in a parallelogram to show that volcanic piles, such as those cross-sectioned in Figure 13-2, extend across the Precambrian shield of Canada. They can be found in shield areas in continental cratonic nuclei in all continents. (*After* Goodwin and Ridler, 1970.)

pipes, and the borders are normally gradational and poorly defined. Sericite, widely distributed and most easily recognized close to the outer borders of the chlorite-rich core, represents potassium buffering of felsic wall rocks against Fe-Mg-rich solutions. The chlorite-sericite rock grades outward into massive siliceous rock, particularly into rhyolite. This chlorite-sericite alteration has been described in the Matagami district by Roberts and Reardon (1978) and at the Millenbach mine in the Noranda area by Riverin and Hodgson (1980). In the "black chlorite" zone beneath the massive sulfide is a stockwork zone, a three-dimensional network of chalcopyrite-pyrite veinlets which lace the chloritized rock and which represent upward-migrating solution pathways of the exhaling fluids. Veinlets 1 or 2 cm across, isolating irregular blocks the size of a fist, are well developed at Kidd Creek (Simmons, 1973; Walker et al., 1975) and Delbridge (Boldy, 1968) in Canada, the United Verde and Bruce orebodies in Arizona (Baker and Clayton, 1968; Anderson and Nash, 1972), and elsewhere.

But it is the massive sulfide bodies themselves that are most astounding. The ores commonly contain 50% sulfide, frequently reaching 100%, at which grade a block only 50 cm on a side weighs a metric ton. The ores are mineralogically simple, consisting primarily of pyrite-pyrrhotite-chalcopyrite-sphalerite, some with galena. Plimer and Finlow-Bates (1978) report an increase in the pyrite-to-pyrrhotite ratio in younger deposits; Hutchinson

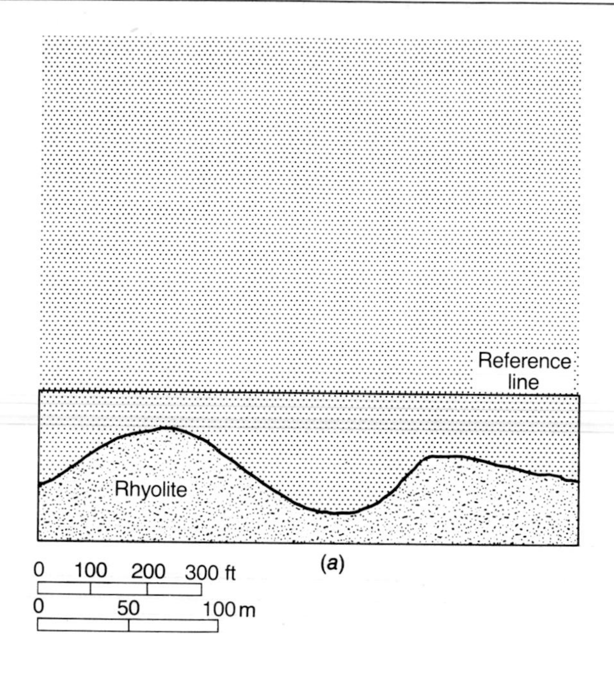

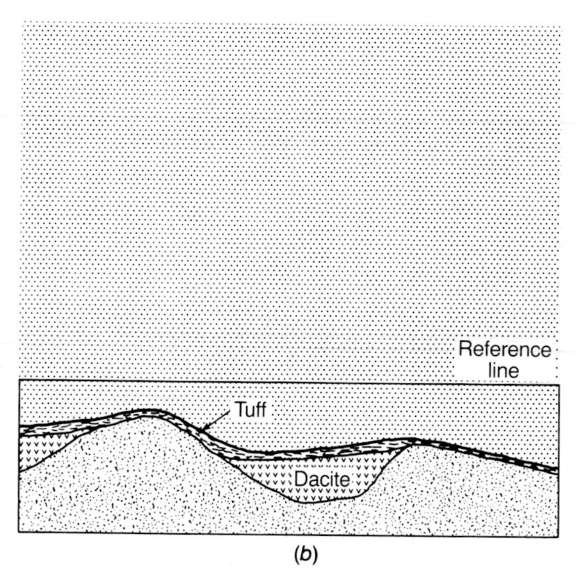

(1965) noted that the lead content increases markedly from Archean through Proterozoic to Paleozoic and younger ores; and Gilmour (1971) noted that the copper-to-lead ratio diminishes greatly from proximal to distal, the latter dominated by lead and zinc. Kidd Creek, the largest massive volcanogenic orebody yet discovered, contains a greater variety of minor minerals

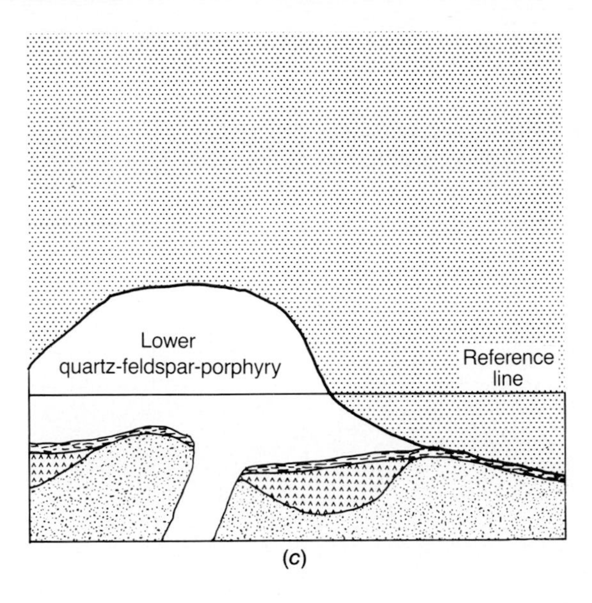

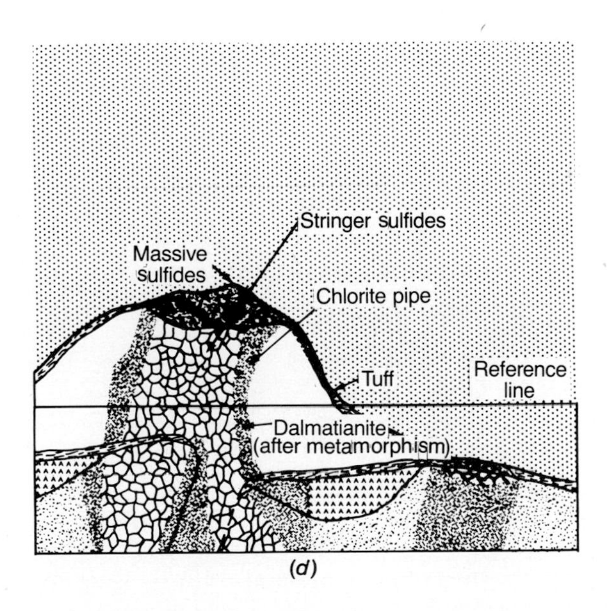

Figure 13-4. Sequence of geologic events leading to the development of the Millenbach mine ores, Noranda district, Quebec, Canada. (*a*) Rhyolite basement. (*b*) Dacite extrusion, tuff deposition. (*c*) Lower quartz-feldspar-porphyry extrusion. (*d*) Sulfide emplacement, tuff deposition. (*e*) (*next page*) Upper quartz-feldspar-porphyry, acid dike intrusion. (*f*) Andesite extrusion, faulting, tilting. Notice the commonality of tuff and ore sulfides, and the exploration significance of the tuff. See also Figures 13-19 and 19-16. (*After* Simmons, 1973.)

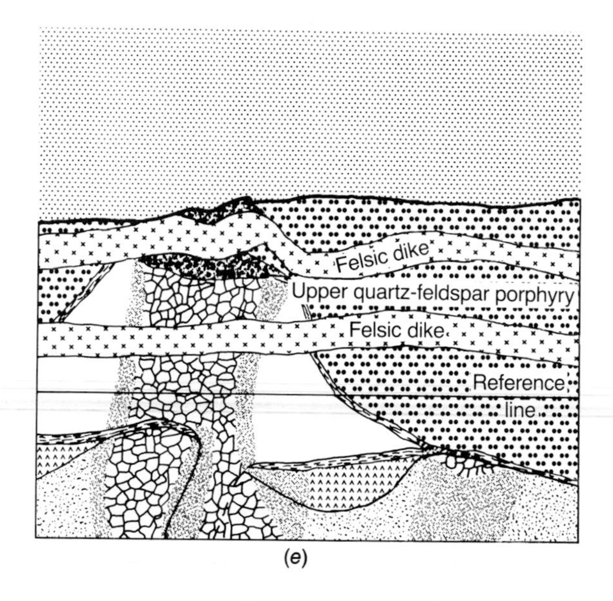

(*e*)

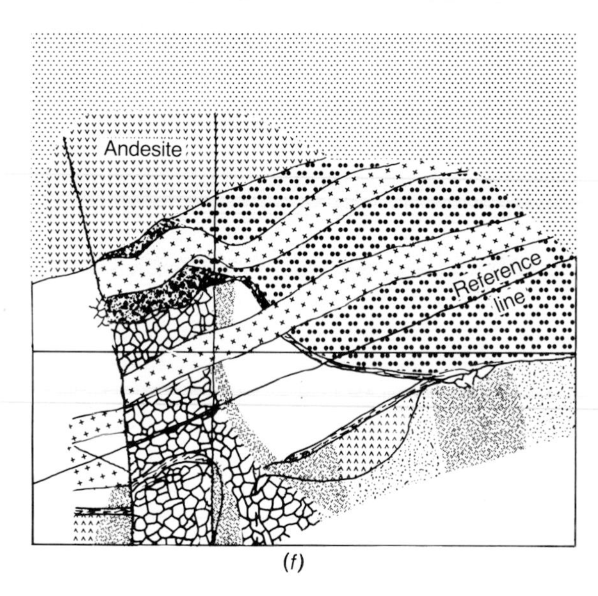

(*f*)

than most massive sulfides (Walker et al., 1975). The ores are vaguely banded in the attitude of bedding, and they are zoned from high chalcopyrite-to-sphalerite ratios at the base to high sphalerite-to-chalcopyrite ratios at the top; the ores thus appear to be monocyclic. The origin of this zoning may reflect evolution with time from high copper to high zinc from the source; or it may be due to differential "storage" in and sulfide sedimentation rates from the sea, zinc sulfide precipitation being slower (Figure 13-11*c*); or it may be due to a postdepositional chemical or electromotive

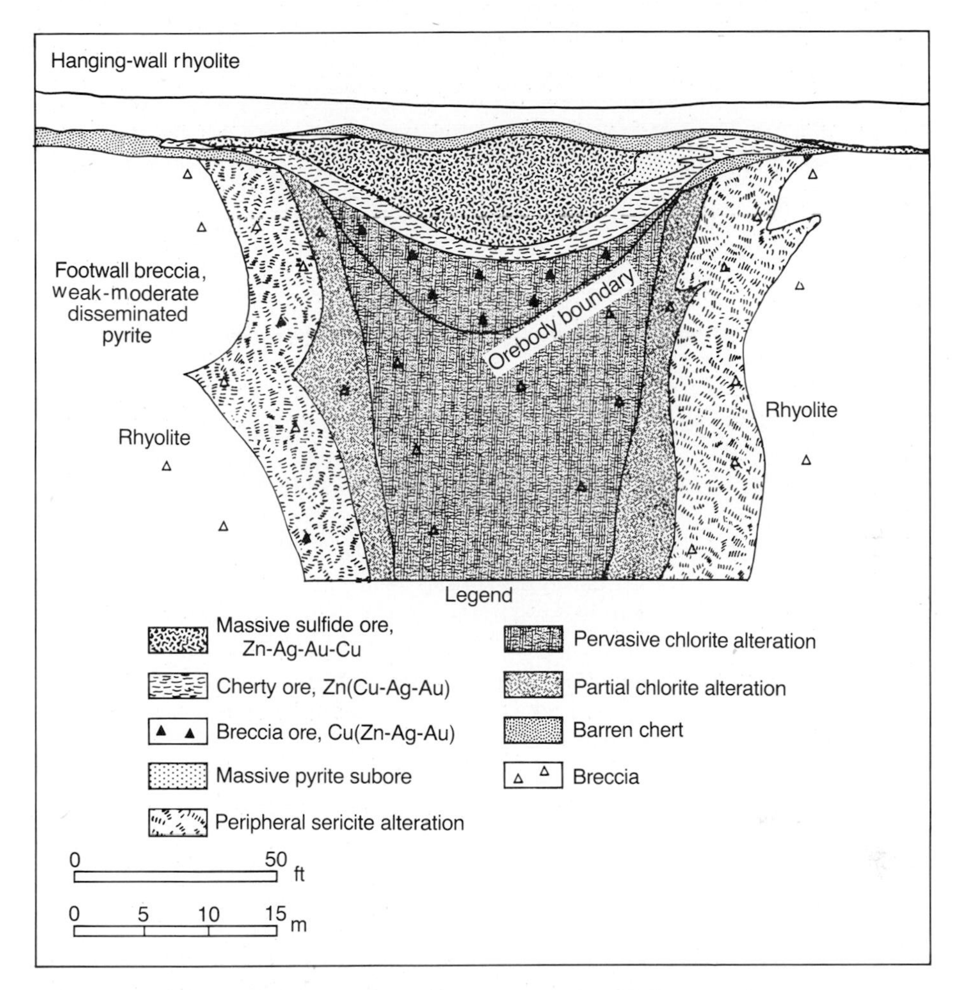

Figure 13-5. Plan of a typical ore shoot, the B Lens at the 850-ft level of Delbridge mine, Noranda district, Quebec, Canada. (*After* Boldy, 1968.)

galvanic effect. Graphite or carbonaceous material is common either within or near the upper portions of the sulfide layer. Colloform and framboidal structures are abundant; they are especially common where the carbon content of the rock is high (Sangster, 1972; Simmons, 1973). Sangster (1971) showed that sulfur isotope ratios in massive sulfides demand that sulfur either came ultimately from the seas or equilibrated with seawater sulfur.

According to Sangster (1972), most volcanogenic massive sulfide deposits are zoned in consistent patterns of four kinds. *Morphologic zoning* exists where both massive ores and stringer ores are known; the stringer ores are everywhere below the massive sulfides. *Mineralogic zoning* in an ideal ore would consist of massive pyrite-chalcopyrite underlain by stringer ore of pyrrhotite-chalcopyrite and overlain by pyrite-sphalerite-galena. In massive ore, galena and sphalerite are more abundant in the upper parts of the orebodies. Chalcopyrite increases toward the footwall and grades into stringer ore. *Textural zoning* is defined by the fact that the sphalerite-rich upper parts are crudely banded, expressed as monomineralic layers of pyrite and sphalerite. The chalcopyrite-rich parts of the orebodies seldom show banding. *Compositional zoning* parallels the distribution of the three major sulfides, sphalerite, chalcopyrite, and galena, but is usually more quantitative and can be detected on routine assay charts. Sphalerite, because of its wide variation in color, is difficult to recognize, but shows up clearly on assay plots. The student should mentally superimpose these varieties of zoning onto Figure 13-5.

Greenschist metamorphism in the Canadian Shield has mildly recrystallized some of the orebodies, without disruption of most primary textures. Where subvolcanic stocks have intruded near the ore systems, black chlorite in the feeder pipes has been metamorphosed to a magnesium-aluminum-rich spotted intergrowth of cordierite and anthophyllite called *dalmatianite* (de Rosen-Spence, 1969; Figure 13-14*d*). Metamorphism, tilting, shearing, and faulting have obscured enough primary relationships that many of the deposits were placed in Lindgren's hypothermal category until their geology became understood in the period 1965–1975. Continued studies of geochemistry and geothermometry and interpretation of outward gradations from proximal sulfide to distal carbonate-oxide environments are proving explorationally and scientifically valuable.

Plimer (1978) considered differences between proximal and distal deposits. Figure 13-6 schematically shows variations from proximal at the left to distal at the right. He concluded (page 345) that

> *. . . with increasing distance in time and space from a volcanic center, the proportions of volcanics to sediments and intensity of hydrothermal alteration decrease. Footwall alteration pipes, stringer and disseminated ore zones dominated by quartz-, sericite- and Mg-rich assemblages are spatially associated with proximal stratiform base metal deposits, whereas slight alteration to quartz-, sericite- and Fe(Mn)-bearing assemblages are associated with distal stratiform base metal deposits. With increasing distance in time and space from the volcanic center, the Fe, Cu, and S content of base metal deposits decreases and the Zn, Pb, Ag, Mn, Ba, and F content increases. The trend from copper-dominated to zinc-lead-dominated iron sulfide ores is probably a result of the increasing mixing of seawater with metal-bearing hydrothermal fluid with increasing distance from the volcanic center.*

Jambor (1979) suggested that the structural terms *autochthonous* and *allochthonous* might be used in the sense that proximal autochthonous massive sulfides formed in place over the pipe or conduit from which they issued, while proximal allochthonous ones were fragmental and suggestive of downslope movement.

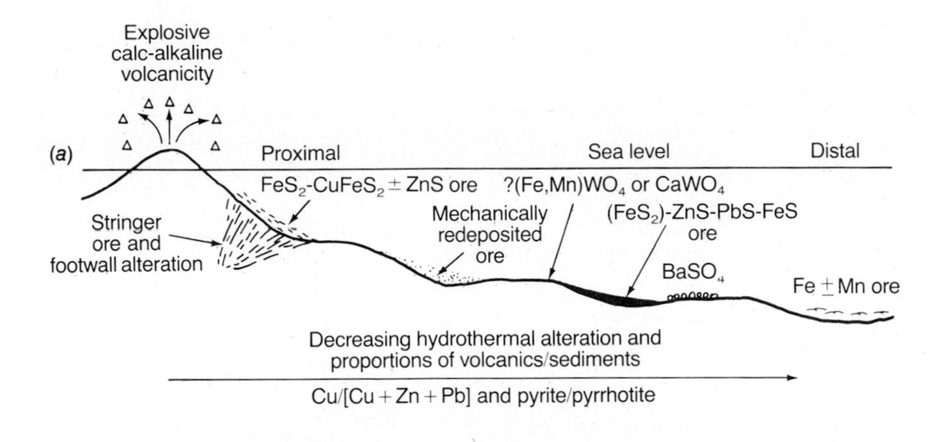

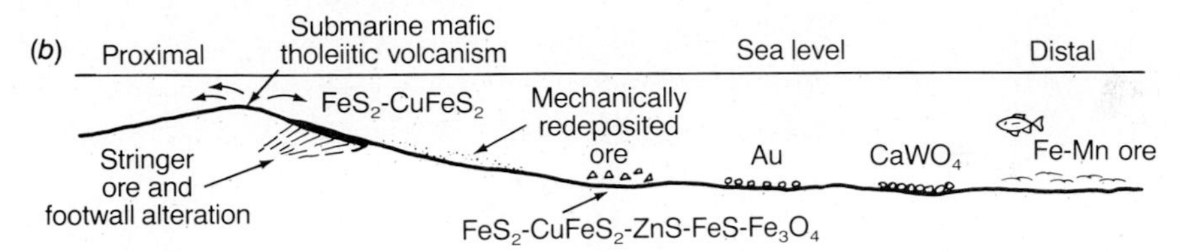

Figure 13-6. Plimer's schematic of proximal versus distal volcanogenic massive sulfide deposit characteristics. Other authors would greatly extend the auriferous sediment exhalite, especially distally. The presence of the fish is not intended to preclude Precambrian applicability. Note also that Large (1977), Jambor (1979), and other authors would argue that pyrrhotite does not increase distally relative to pyrite. The mechanically redeposited material would be Jambor's "allochthonous ore." (*From* Plimer, 1978.)

Detailed descriptions of proximal Cu-Zn and Cu-Zn-Ag deposits abound in recent literature. Kidd Creek near Timmins, Ontario, is the largest massive sulfide yet discovered. It did not outcrop, but two distinctive outcrops and a geophysical anomaly led to its finding (Miller, 1976). One of the outcrops was well-developed mill rock rhyolite breccia (Figure 13-7), the other was fresh hanging-wall andesite, and between them a considerable EM anomaly was discerned. The 300-million-ton orebody—one of the great accumulations of copper, zinc, lead, and silver in the Earth's crust—is lensoid in form, overlies a chalcopyrite stringer zone, shifts from massive chalcopyrite-pyrrhotite-pyrite at its base to massive sphalerite-pyrite at its top, and is capped by pyrite-carbonaceous rock breccias and ferruginous cherts and ultimately by barren andesites. The massive sulfide lens is approximately 1200 meters long, about 100 meters thick, and has been probed to more than 1500 meters in depth, the dimension of width before tilting. It was first mined from an astonishingly productive open pit, and is now being mined by underground caving techniques. The student is urged to read more detailed accounts of Kidd Creek by Walker et al. (1975), of the Millenbach deposit by Simmons (1973), and of the Mattagami Lake deposits

(*a*)

(*b*)

Figure 13-7. (*a*) Knob of the mill rock rhyolite breccia outcrop of the Kidd Creek mine near Timmins, Ontario, and (*b*) close-up of the sheared silicified nature of the breccia. Only two bedrock areas were exposed through glacial drift, including this most distinctive and portentous one.

by Roberts (1975) and Roberts and Reardon (1978). The United Verde deposit at Jerome, Arizona, was described by Anderson and Nash (1972). Arizona Proterozoic massive sulfide deposit characteristics were summarized by Anderson and Guilbert (1978).

Kuroko Copper-Zinc Deposits, Japan

Gilmour (1971) observed that massive sulfide deposits of the Canadian type appear to have formed in the early stages of orogeny at any given location, at a time of eugeosynclinal environments associated with island arcs and consuming continental margins, an observation reinforced and well expressed by Hutchinson (1979). Kuroko-type deposits appear to be related to more felsic intrusions (Pearce and Gale, 1980), and may represent orogenically later deposits more closely related to porphyry base-metal deposits than to eugeosynclinal massive sulfides. The Kuroko occurrences resemble Precambrian massive sulfide ore deposits, but they occur in modern island arc or interplate environments which are compatible with synorogenic conditions. They have been deposited in volcanic rocks over Paleozoic basements, and might therefore be considered similar to Paleozoic-age massive sulfide environments in the Appalachian United States (Ducktown, Tennessee), the Sierra Nevada Foothills Copper Belt deposits (Copperopolis and the Penn mine, California), and deposits in the Tasman geosyncline along the eastern parts of Australia and in Tasmania (Hutchinson, 1980). Although the discussion of Kuroko ores is generally limited to the deposits in Japan, similar deposits also occur in the Philippine Islands (Bryner, 1967), in Fiji (Frenzel and Ottemann, 1967), and possibly in Turkey (Griffitts, Albers, and Oner, 1972). Comprehensive cooperative studies of Kuroko environments by Japanese and American geologists are being carried out in the early 1980s (Ohmoto and Skinner, 1983).

The Kuroko (Kuromono) ores of Japan are examples of volcanogenic massive sulfide deposits (Kinoshita, 1924, 1929, 1931; Griggs, 1947; Collins, 1950; Horikoshi and Sato, 1970; Ohmoto, 1972). The Kuroko ores have been the subject of debate for many years. Among geologists who have studied them, they are now accepted as unmetamorphosed to weakly metamorphosed stratabound polymetallic deposits genetically related to submarine volcanic activity during the Miocene period only 13 to 13.5 million years ago. Japanese geologists were among the first to perceive the importance of submarine volcanism in ore genesis, and their post–World War II studies of this ore type were precedent-setting.

The Japanese word "Kuroko" means "black ore." The term *Kuroko type* is commonly applied to six mineralogical categories of ore. Siliceous ores (keiko) contain sulfides, particularly chalcopyrite, disseminated in silicified rock. Yellow ores (oko) are primarily pyrite with minor amounts of chalcopyrite and quartz. The black ores (kuroko) are intimate mixtures of dark iron-rich sphalerite, galena, barite, and minor quantities of pyrite and chalcopyrite; wurtzite, enargite, tetrahedrite, marcasite, and numerous other minerals are found locally in small amounts. Veins and large masses of gypsum (sekkoko) are in related but separate bodies, and large discrete bodies of barite are found. Chalcopyrite-rich stringer zones in pipes beneath the ores are called "ryukoko." Finally, the sulfide ores are overlain by ferruginous cherts (tetsusekiei beds). Five of the six ore types have equivalents in Archean Canadian deposits; Archean sulfates, gypsum and barite, are rare.

Kuroko-type orebodies, especially common in northeastern Japan, occupy stratigraphic horizons characterized by the accumulation of sandy and muddy sediments that contain abundant molluscan fossils of warm-water species. Deposition is thought to have taken place in relatively quiet, shallow, and isolated caldera lake basins, which may have been near sea level and open to the ocean. Pertinent volcanic activity began with the accumulation of a floor of pyroclastic debris which constitutes the Miocene Motoyama dacite formation. The Motoyama units were in part reworked by turbidity currents during slightly later formation of lava domes and flows. These volcanic materials with which the ores are usually associated are silicic rhyolites and dacites that intruded the Motoyama dacite. These intrusions triggered explosive volcanism when intruding melts encountered seawater either several tens of meters beneath the Motoyama surface, at the seafloor (Figure 13-8*a* and *c*), or upon the extrusion of blisterlike domes (Figure 13-8*b* and *c*), as at the Millenbach deposit (Figure 13-4*c*). When the domes exploded, they produced breccia-agglomerate sheets.

Solutions carrying sulfur and base metals moved upward through and around the rhyolite domes, forming a network of chalcopyrite veins and chlorite zones in the relatively brittle footwall volcanic rocks. Where the solutions reached the unconsolidated saltwater-saturated tuffs and breccias over and around the domes, without reaching the seafloor, they reacted quickly. Massive sulfides, together with silica, gypsum, and barite, extensively replaced the brecciated materials near the surface and were deposited at the footwall-water surface as well (Sato, 1977). Changes in volume and density by alteration of the floor rocks caused slumping and the incorporation of blocks of different kinds of partially replaced material in the ore. Various solutions were involved. Massive bodies of gypsum, barite, sulfides, or mixed materials of several types were produced where fluids of different base-metal and sulfur-species compositions poured out onto this submarine surface (Figure 13-8*d*). After the main period of mineralization, the deposits were covered with pyroclastics and flows of dacitic composition. Waning fluid movement from the subvolcanic source caused mild alteration of the higher, younger volcanic beds (Jenks, 1966, 1971).

Watanabe (1970) states that many of the Kuroko-type ores are sedimentary and localized at specific horizons, and prospecting has been suc-

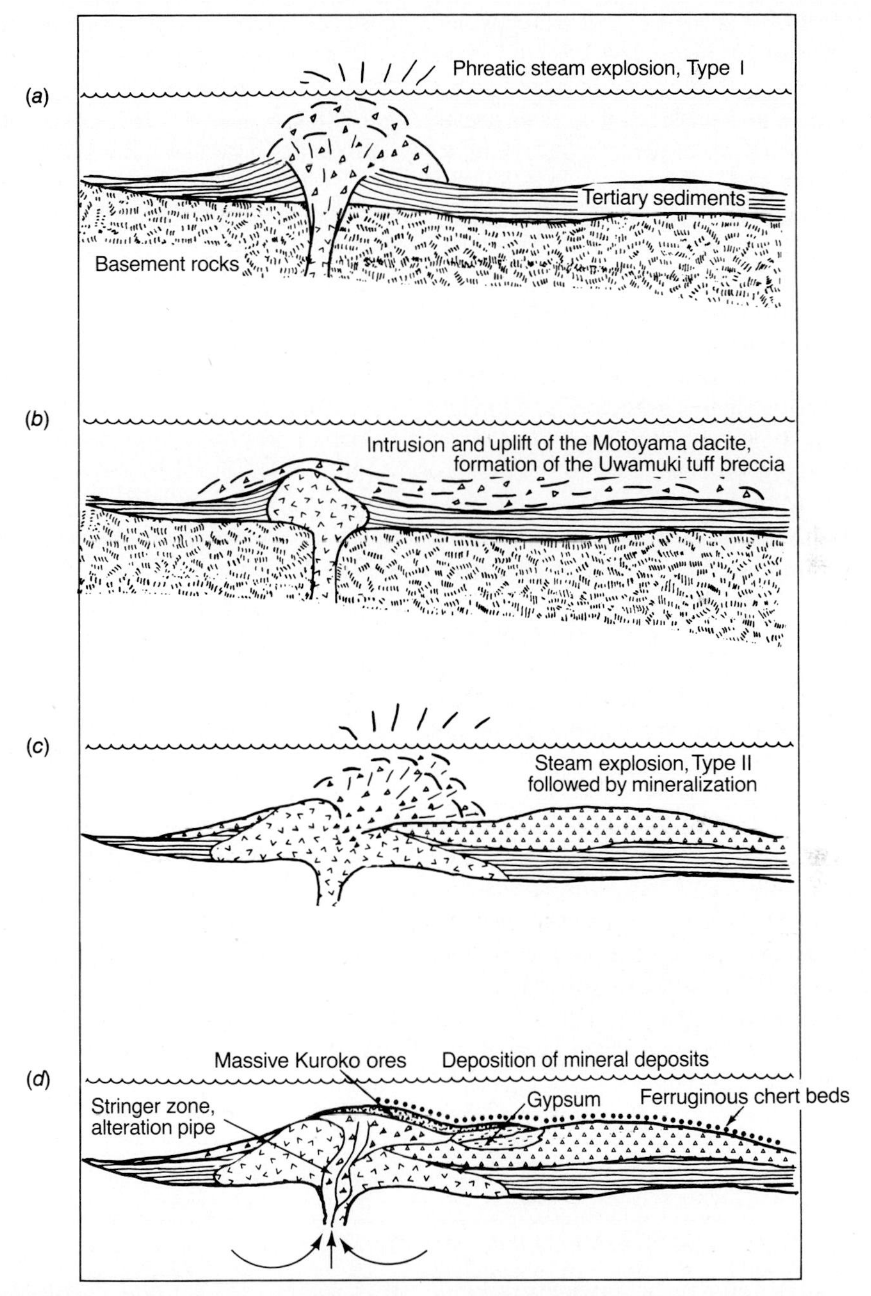

Figure 13-8. Steam explosions leading to the development of Kuroko ores at Kosaka, Japan. Type I steam explosion occurs when hot rock meets cold seawater. Type II steam explosion, followed by mineralization, is more laterally directed. Gypsum beds are common in Kuroko deposits, uncommon in older ore deposits, perhaps representing relatively high Eh in recent seas. (*After* Horikoshi, 1969.) See north end of Figure 13-9.

cessful where this criterion has been used. Both sedimentation and replacement at shallow depths, probably within a few meters or tens of meters of the surface, were active processes; Kuroko-type ores may be either bedded or irregular massive replacement bodies. Below many of the ores are stringer ores consisting of chalcopyrite veinlets in silicified, chloritized volcanics and sediments. The best evidence for volcano-sedimentary origin includes the stratabound character of the deposits, the extremely fine but well-formed laminations, false bedding, graded bedding, and suprajacent ferruginous cherts. Direct precipitation on the seafloor seems to be the most reasonable explanation for the genesis of the majority of the ore.

Kuroko orebodies range in size from small nodules to irregular masses as much as 800 meters long by 300 meters wide by 100 meters thick. Many are flat lenticular bodies. The Kuroko-type ores are fine-grained. They commonly show framboidal textures, concentric banding, nodules, and colloform structures. In fact, these textures are so abundant that Kinoshita (1924, 1929) thought that the deposits had precipitated from colloidal solutions. The presence of veins of silica gel, probably of "diagenetic" timing, was noted in the upper parts of the massive sulfide deposits of Turkey by Griffitts, Albers, and Oner (1972).

Four zones of alteration are recognized near the Kuroko-type ores of Japan: (1) strong silicification in the footwalls of the orebodies, commonly accompanied by small amounts of sericite and chlorite; (2) sericite, chlorite, and quartz intimately associated with the ores; (3) sericite, chlorite, and pyrite above the ores; and (4) montmorillonite and zeolite mineralization that grades outward and upward into unaltered rock (Matsukuma and Horikoshi, 1970). Sudo (1954) reported montmorillonite, iron-montmorillonite, sericite, and chlorite as alteration products. District-scale alteration is from a mordenite zeolite fringe alteration inward through sericite to chlorite. The Kuroko-type ores are unmetamorphosed, or at most have been only slightly affected by thermal metamorphism. Regional dynamic metamorphism is absent. In older volcanic provinces in other parts of the world, even where dynamic greenschist metamorphism has been active, it is not uncommon to find massive sulfide bodies that have many features resembling the Kuroko ores.

Sangster (1972) lists points of similarity and dissimilarity between the metamorphosed Archean Canadian massive sulfide deposits and the Miocene Kuroko ores of Japan. The following are points of similarity.

1. Both involve calc-alkaline submarine volcanic rocks.
2. Both tend to be in clusters or districts related to centers of volcanic activity.
3. Both show strong spatial correlation with acidic, explosive phases of volcanism.

4. Both consist of two main types of ore, massive sulfides and stringer ore. The massive ore in both areas is essentially conformable with surrounding rocks, but the stringer ore clearly cuts the stratigraphy. The massive ores are banded.

5. Both are commonly capped by a layer of ferruginous chert-hematite in Japan and magnetite in Canada.

6. Both show compositional zoning relative to stratigraphy, with lead-zinc increasing and copper decreasing upward.

7. Both are underlain by alteration materials that enclose stringer ores.

The following are points of dissimilarity.

1. Footwall alteration in the Kuroko deposits is mainly silicification, whereas silica is removed during alteration and magnesium is increased in the Canadian deposits. Scott (1948) and Gibson, Watkinson, and Comba (1983) described siliceous alteration in rhyolites beneath the Quemont orebodies in the Noranda district, Canada, but magnesium metasomatism is far more common.

2. Postore alteration of hanging-wall rocks, which can be directly related to mineralization, is common above the Kuroko ores, but is recognized in Canada in only a few districts which show "stacked" orebodies.

3. The anhydrite-gypsum ores of Japan have no known equivalent in Canada, probably because Precambrian oceans were considerably more reduced than those of Tertiary times.

4. Differences in mineralogy exist. In the Kuroko ores bornite and tetrahedrite-tennantite are major constituents, but in Canada they are seldom more than minor accessories. Galena is abundant in Japanese ores, but is seldom encountered in Canada.

The Hanoaka-Matsumine district in northern Honshu is typical of Kuroko deposits. According to Ogura (1972), the ores from this district are zoned with siliceous rock at the base and gypsum above, followed by siliceous iron sulfide (keiko) ores and then by yellow (oko) ores; the black (kuroko) materials are on top. Colloform and spherulitic textures are abundantly developed, especially in the massive ores. The Hanoaka-Matsumine orebodies are considered to have formed as submarine hydrothermal sediments in a caldera in the upper part of a pyroclastic rock sequence interlayered with and overlain by mudstones (Figure 13-9). Sedimentary structures, such as stratification and fine laminations, are widely recognized. The introduction of the ore followed intrusion of a rhyolite dome, with solutions permeating through and around the dome. The ores were depos-

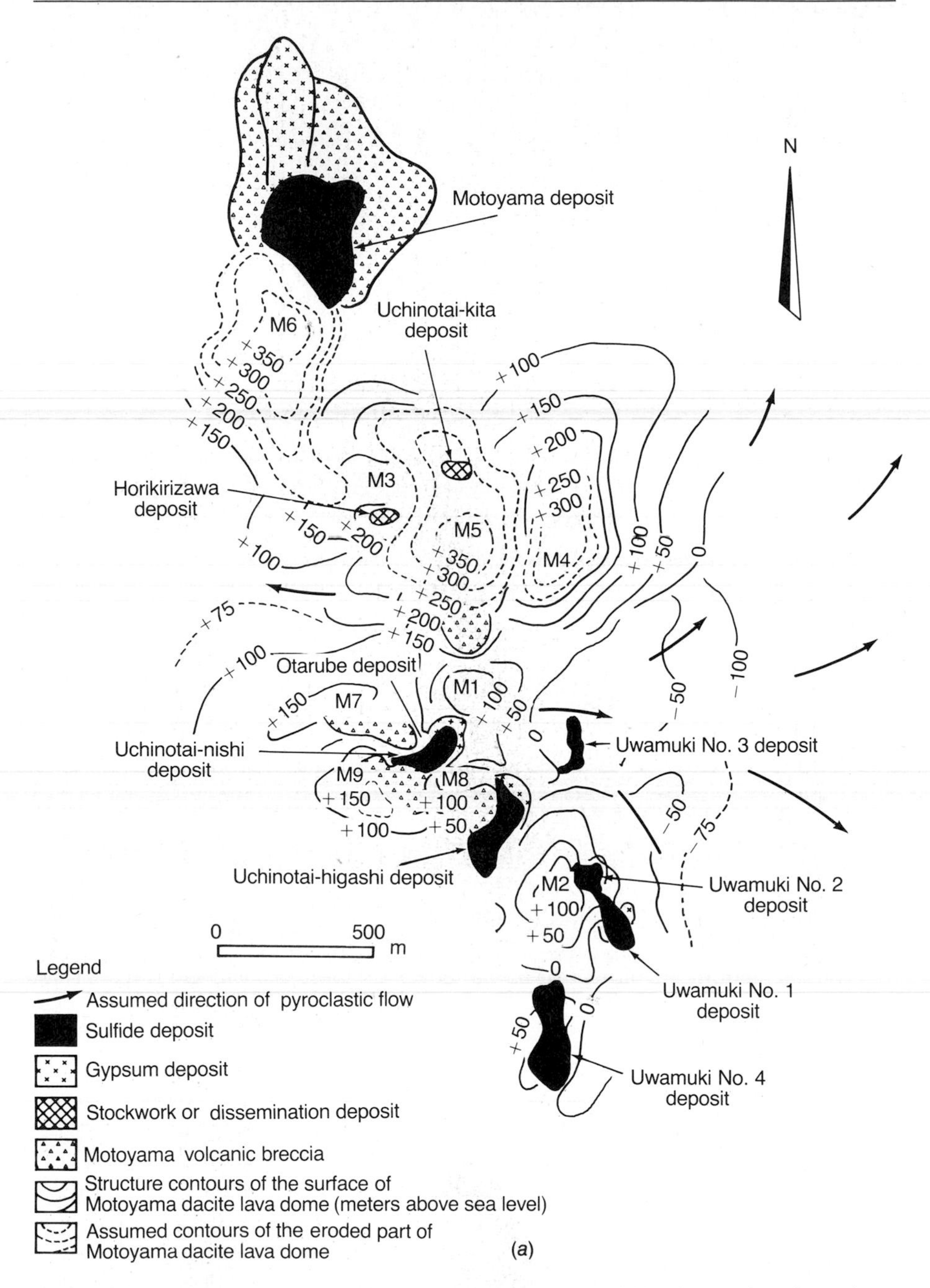

Figure 13-9. (*a*) Plan view and (*b*) (*facing page*) section view of Kuroko ore deposit settings at the Kosaka mine, Japan. (*b*) is a north-south cross section through the Uwamuki No. 4 deposit at the south edge of the cluster on (*a*). It shows the relations between rhyolite doming, breccia sheet formation, mineralization, and stockwork pipes. The plan view also shows the importance of doming and how the Kuroko orebodies are clustered. Figure 13-8 is a north-south section through the Motoyama deposit. (*From* Urabe and Sato, 1978.)

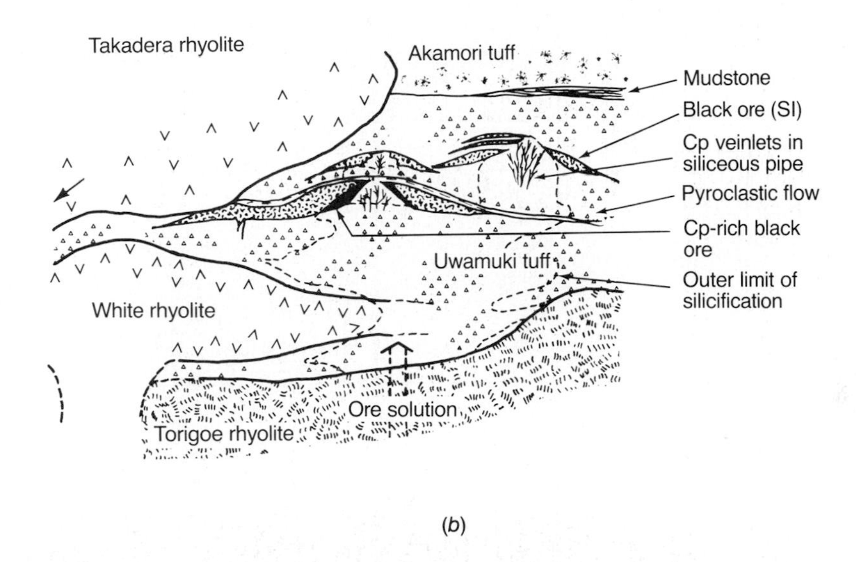

(*b*)

ited in bordering sediments or, where the thermal springs reached the sea, were spread out on the seafloor. Ogura (1972) determined the temperature of formation of the ore at about 200°C. Recent studies (Sato, 1977) indicate that the Kuroko sulfides were precipitated on the Miocene seafloor from weakly acidic, reducing brines at temperatures of 200 to 250°C and as a result of falling temperatures. Oxygen-hydrogen and sulfur isotope ratios demand seawater as a predominant medium of transportation, and lead isotope data demand a magmatic component.

Read-Rosebery Lead-Zinc-Copper Deposits, Tasmania

The Rosebery mine in western Tasmania lies in the west-central portion of that shield-shaped island about 100 km southwest of the coastal town of Burnie and in a cluster of deposits which have individually been, and are now collectively, interpreted as volcanogenic. The nearest neighbors include the Hercules and Mt. Lyell copper-gold ores and the Mt. Farrell lead-zinc-copper-silver deposits, both a few kilometers to the southeast, and several other small deposits (Green, Solomon, and Walshe, 1981).

They occur in eugeosynclinal volcanic environments that are exemplary of the term. The deposits are in the Cambrian Dundas Trough eugeosynclinal volcanic and volcaniclastic rocks of the Mt. Read and Rosebery groups overlain by miogeoclinal open-water sediments, the former marking the onset of early orogenic activity. Hall et al. (1953, 1965), Braithwaite

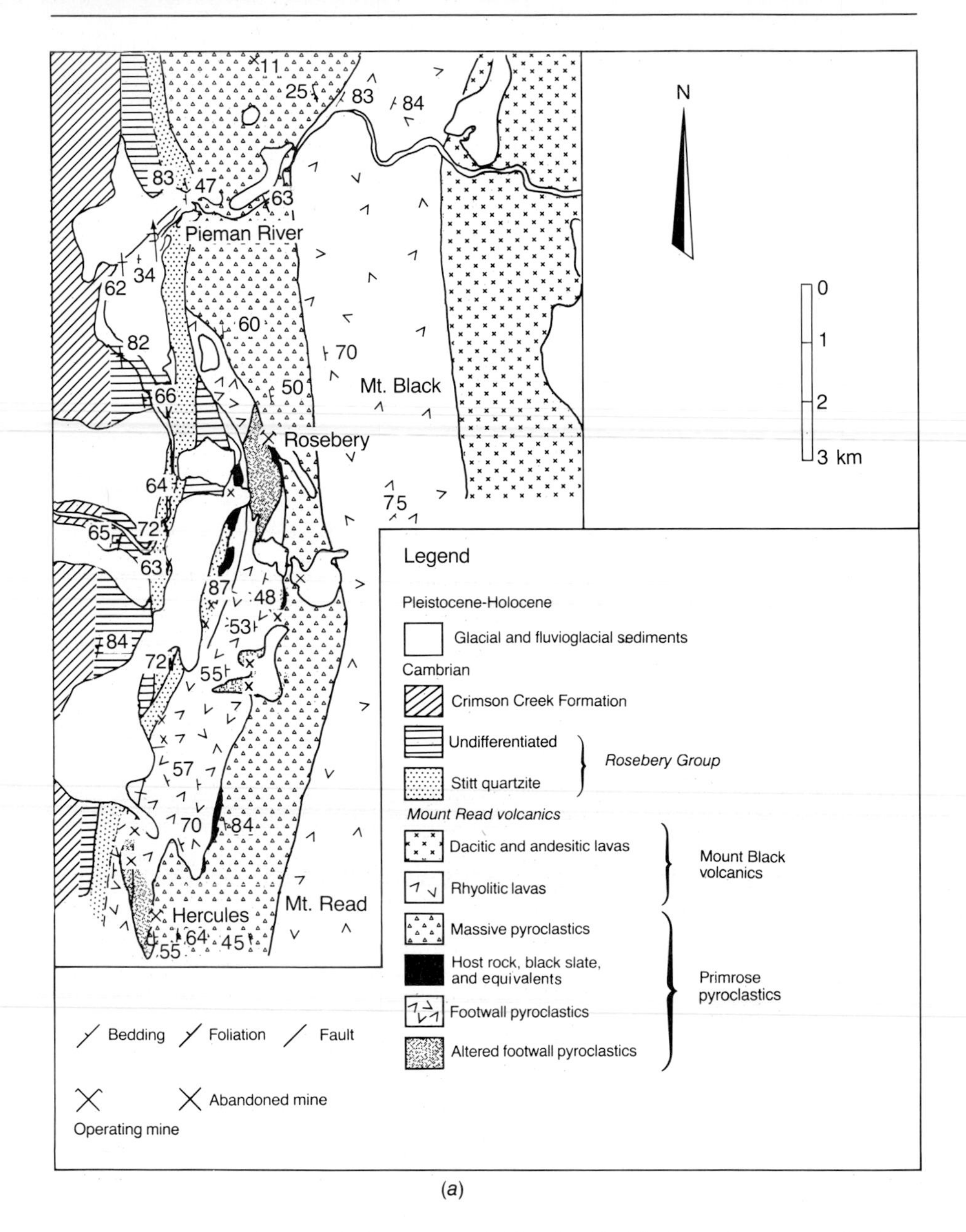

(*a*)

(1971, 1974), and Green, Solomon, and Walshe (1981) describe the orebodies at Rosebery as massive pyrite, galena, and sphalerite deposited in a marine environment in a fine-grained tuffaceous black slate member in a sequence of pyroclastic rocks including welded and nonwelded crystal tuffs. Minerals in the orebody are principally pyrite, sphalerite, and galena, with lesser amounts of chalcopyrite, bournonite ($PbCuSbS_3$), arsenopyrite, pyrrhotite, tetrahedrite, pyrargyrite (Ag_3SbS_3), and gold. The ore is finely banded and

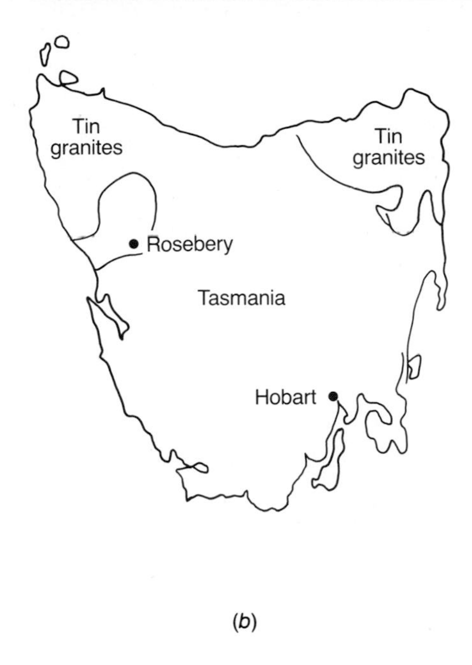

Figure 13-10. (*a*) (*facing page*) Geologic and (*b*) location map of the Rosebery deposits. The area underlain by tin-bearing granites—the "tin fields" of Tasmania discussed in Chapter 11—is also shown on (*b*). (*After* Green, Solomon, and Walshe, 1981.)

stratiform tabular. Alteration products, abundantly developed in the footwall layers beneath the ore, consist of quartz, sericite, chlorite, and kaolinite, are marked by Na and Sr depletion and K, Mg, Mn, Rb, and H_2O addition, and are inferred to represent a feeder zone.

The geologic structure of the area is complex, and the rocks at the Rosebery mine are folded, faulted, and metamorphosed, but they retain their stratiform configuration (Figure 13-11) and still show soft-sediment deformation. The shales are sheared so much that bedding is largely destroyed, and locally the black slate ore horizon between the underlying tuffs and the overlying pyroclastics is missing. Faulting may be responsible for the lack of shale, although Hall and his colleagues (1965) think an unconformity may explain the absence. The most prominent structure in the area is a large overturned fold that, combined with faulting, resulted in shearing and development of shear cleavage, particularly in the rocks below the massive pyroclastics. It may represent a soft-sediment fold complicated by younger tectonic events. Figure 13-11 shows a geologic plan of the 13th level of the Rosebery mine. Close inspection of the figure is informative. If the folding is "stretched out" and restored to volcano-sedimentologic attitudes as was done by Green, Solomon, and Walshe (1981) (Figure 13-11*b*), the high Fe-Cu ore is seen to be centrally disposed, with the Zn-Pb ores extending outward to distal manganiferous carbonates. Barite is central and in the hanging wall, a standard stratigraphic array.

Sulfur isotope data are complex. Green, Solomon, and Walshe (1981) report that Fe- and Cu-rich phases were deposited earlier from buoyant saline fluids, but that the Pb-Zn phases represent later reversed buoyancy

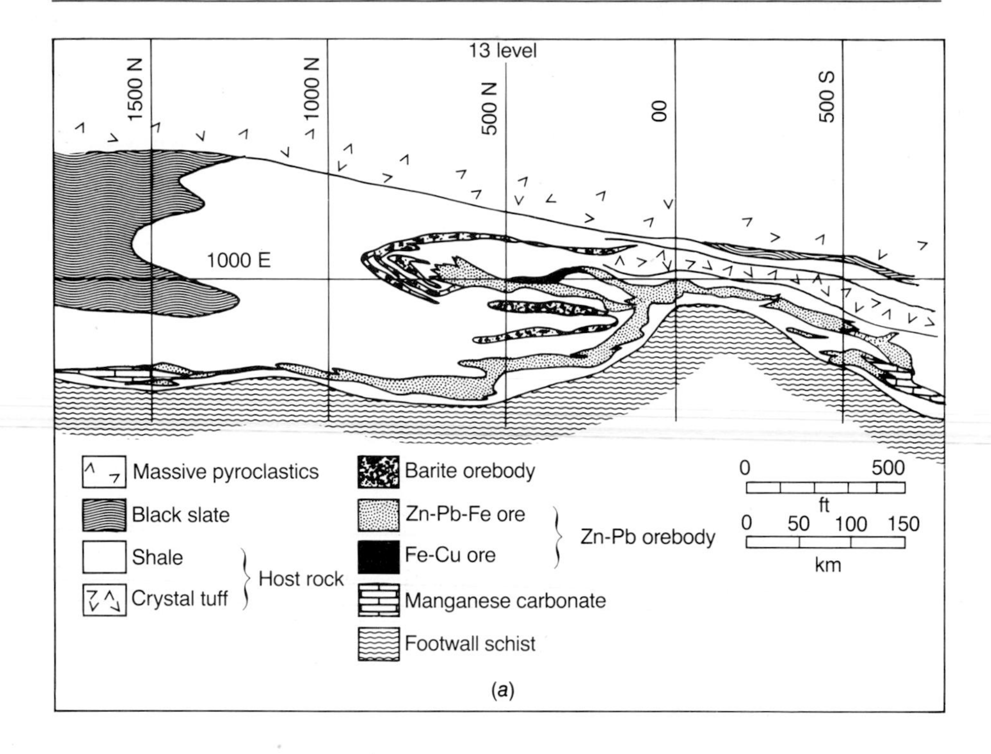

(*a*)

of the fluids (Figure 13-11*b*). Temperatures in the 300°C range are indicated. The ore fluid is thought to have been generated by convective circulation of seawater in a volcanic pile heated by a Cambrian granitoid pluton.

The massive sulfide ores of the Tasman geosyncline resemble the Kuroko deposits of Japan closely enough that they have been considered to be a type intermediate between the Japanese deposits characterized by explosive volcanism with sulfate-sulfide deposition and the massive sulfide ores of Canada (Sangster, 1972).

Cyprus-Type Copper-Zinc Deposits

Chapter 10 ended with a discussion of the Troodos Igneous Complex of the Island of Cyprus, a slice of ophiolitic oceanic crust that contains podiform chromite deposits in the peridotite-dunite of its Layer 3 and tectonized harzburgite. This ophiolite also contains copper-zinc-sulfide deposits as part of Layer 1 sedimentary lithology. They are considered to be the "type area" of exhalative massive sulfide deposits formed on basaltic, oceanic crust seafloor, presumably near midoceanic rises. As such, they are a special case of submarine massive sulfide deposits. Deposits of the Cyprus type found in New Brunswick and Newfoundland in Canada include Tilt Cove,

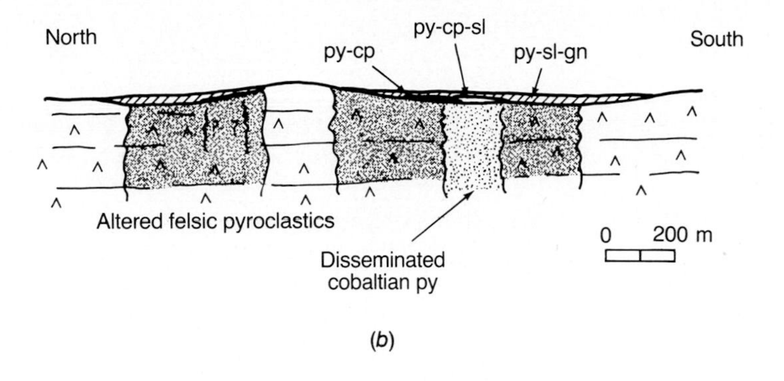

(*b*)

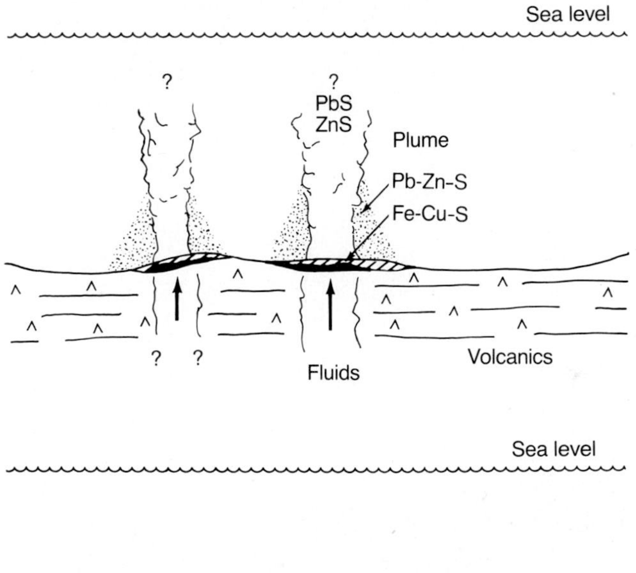

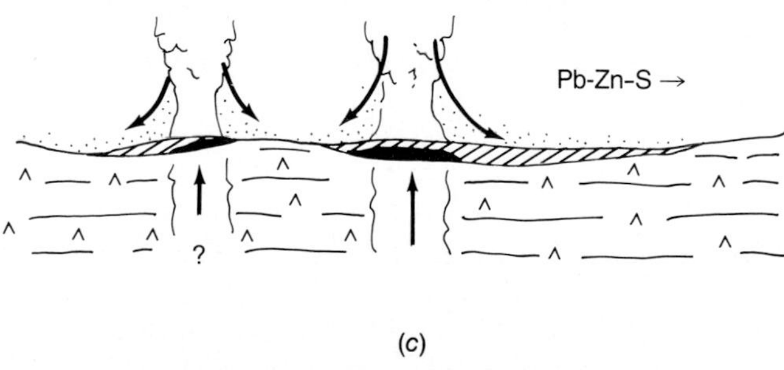

(*c*)

Figure 13-11. (*a*) (*facing page*) Geologic plan of the 13th level of the Rosebery mine, Tasmania. See text for interpretation. (*From* Brathwaite, 1974.) (*b*) Reconstruction of the orebody showing primary zoning and prefolded form. (*c*) Sketch to illustrate how Green, Solomon, and Walshe (1981) think the deposit zoning may have been formed.

Lush's Bight, Little Bay, Whaleback, and Bett's Cove, all of Ordovician age. They are also found in Paleozoic ophiolites in Japan and the Philippines; the Ural Mountains of the U.S.S.R. have revealed over 100 Cyprus-type occurrences; and scores of prospects and mines are known in Paleozoic and Mesozoic units in Turkey, Greece, and Italy. Similar deposits of various ages are known from Mexico, Central America (especially Guatemala and Costa Rica), Cuba, and Ecuador-Colombia. Younger ones are described from Sumatra-Java-Timor in the southwest Pacific. In short, there are hundreds of known occurrences. It is an ore deposit type of considerable importance, with millions of tons of past, present, and future annual production.

Cyprus-type deposits were first emphasized in modern economic geologic context when Hutchinson (1965) compared and contrasted them with Archean and Proterozoic massive sulfide deposits. Hutchinson and others noted that these deposits have a number of common denominators. They occur on footwall rocks that in general consist of albitized pillowed basalts called spilites, which are draped with thin-bedded silica, iron oxide, and manganese oxide-rich oceanic sediments commonly described as *Fe-Mn cherts* or *silica-hematites*. Locally the sediments contain sulfides, dominantly chalcopyrite, sphalerite, and pyrite; minor barite, psilomelane-pyrolusite, and silver-gold occur in some of the beds or lensoid sulfidic portions of the sediments. Basaltic breccias or water-fragmented tuffs and flows called *hyaloclastites* are commonly associated rock types, along with agglomerates, argillites, graywackes, mafic flows, and deep-water limestones and silica-rich sediments as hanging walls.

A description of the setting of the Cyprus deposits, with color photographs and sections, is provided by Rona (1973); the deposits themselves are described by Hutchinson (1965), Hutchinson and Searle (1971), Constantinou and Govett (1973), and Solomon (1976). The reader should also see the text and Figures 10-7, 10-8, and 10-9 at the end of Chapter 10, where Cyprus was described. The Troodos Complex is a belt of rocks across the girdle of the Island of Cyprus (see Chapter 10). They dip moderately to the north and east, and are composed of a 7-to-8-km-thick slab of basal ultramafic and mafic cumulate rocks like the dunite-peridotite-chromite assemblage described earlier. These plutonic rocks yield upward to a layer of extrusive pillow basalts about 5 km thick, which are cut by basalt dikes toward the base in such a fashion as to suggest repeated faulting, uprush and outpouring of basaltic magma, and consolidation of surface layers so that dikes are more common with increasing depth, exactly the mechanisms expected along a spreading oceanic ridge. As the uppermost pillowed basalts (Layer 2*a*), their diked substrata basalts (Layer 2*b*), and the gabbro-peridotite "basement" migrate away from the ridge (Figure 10-2), sediments accumulate on their "roof." On Cyprus the uppermost 0.5 km of pillowed basalts is called the "Lower Pillow Lavas" (Figures 13-12 and

10-8). They are overlain by locally sulfidic layers called the Ochre group, which include manganese-poor, iron- and barium-rich sediments, commonly with sulfide bands and fragments, and locally with tuff, bedded chert, and limestones. These rocks are well named; they strongly resemble the artist's ochre, a yellowish-brown pulverulent pigment. Some orebodies, such as the Kokkinoyia deposit pictured in Figure 13-12, occur at this level. The ochres and sulfidic sediments are overlain by 200 to 300 meters of mafic basaltic "Upper Pillow Lavas," which in turn are the floor for the largest sulfide accumulations, including the orebodies at Skouriotissa, Mavrovouni, and Mathiati. The sulfide masses are then overlain by a thick section of sediments of the Parapedhi Formation, which constitute the younger hanging-wall limits of the Troodos Complex. Significantly, this unit is called Umber, suggesting a darker, richer colored, but still ocherous unit. It is a thin-bedded manganese- and iron-rich, locally pyritic bentonitic shale with ra-

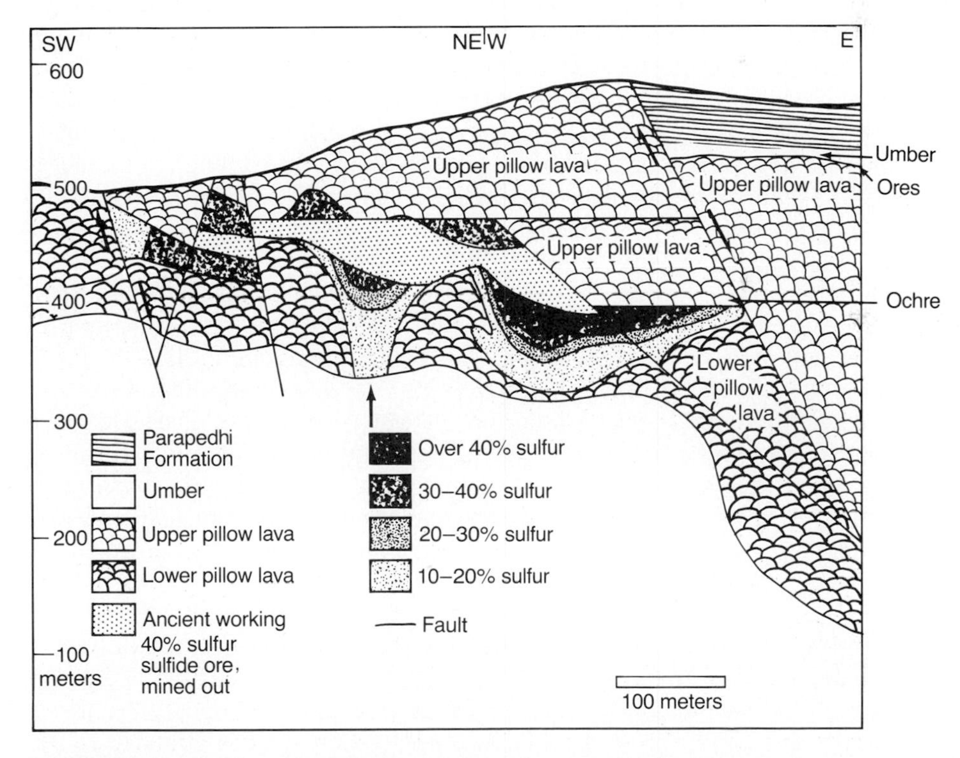

Figure 13-12. Ores and Ochre mark the boundary between the LPL and UPL (Lower and Upper Pillow Lavas). Other major orebodies, such as the Skouriotissa and the Mavrovouni, occur at the Umber level on top of the UPL, here marked "Ores." Note the apparent pipe-zone conduit leading to the million-ton faulted ore lens. (*After* Constantinou and Govett, 1972.)

diolarian jasperoidal cherts; Hutchinson described it in 1965 as a rock "rich in Si, Mn, and Fe oxides, pyrite, and tuffaceous material." It marked the end of the volcanic record; the Troodos Complex is overlain by Cretaceous to Pliocene marls, chalks, and limestones.

Thus the Cyprus ores lie along a narrow band along the northern side of the Troodos Complex terrane, at its contact with younger rocks which were deposited on top of it. Tectonically, Cyprus is an island in the Mediterranean Sea which lies south of the Anatolian Plateau of Turkey, a buttress against which the African foreland has collided. The deeps of the Levantine Sea are to the south of Cyprus, which is itself a highly positive gravity and magnetic anomaly, with thin or absent continental crust. It is thus considered to be an upwarped or upbuckled, broken slice of the seafloor, probably the result of Cretaceous-age thrusts, with a skin of younger sediments washed onto it from the continent before its emergence.

The orebodies are typically small, ranging from 15 or 20 million down to a few thousand tons. They are irregular to lenticular to podlike to gutter-shaped in form; many of the roughly lenticular ones are convex downward as though they formed in shallow basins. They are conformable to underlying layered volcanic and sedimentary rocks. Rona (1973) described them as roughly elliptical, 200 by 300 meters across, and up to 250 meters thick. Constantinou and Govett (1972) emphasized stockworklike mineralized zones beneath the massive sulfides which provoke descriptors like "mushroom-shaped" (Figures 13-12 and 10-9). Sulfur in the lavas—the background values—is well under 10%. In mineralized lavas, sulfur reaches 10 to 20%; in the stockwork mineralized brecciated conduits, 20 to 30% sulfur is common. In the ore horizon, low-grade materials reach 30 to 40% and massive sulfide consistently tops 40% sulfur; for reference, pure pyrite contains 53% sulfur, chalcopyrite 35%, and sphalerite 33%. Typical ores contain 4% Cu, 0.5% Zn, 8 ppm Ag, 8 ppm Au, 48% S, 43% Fe, 5% SiO_2, and minor CaO and MgO. Major minerals are pyrite and chalcopyrite; minor phases include sphalerite and marcasite, and traces of pyrrhotite, bornite, and a series of oxidation-related phases are found. Some of the ores have been oxidized to limonite, jarosite, and natrojarosite, an oxidation resulting largely from the interaction of newly deposited sulfides with oxidizing seawater before burial occurred. Trace elements include Ni and Co, but in small amounts. There is no alteration discerned in the hanging-wall Parapedhi rocks, but extensive propylitization, chloritization, and serpentinization in the footwall rocks is consistent with deeply circulating, thermally driven convective flow of seawater (Parmentier and Spooner, 1978; Heaton and Sheppard, 1977; Mottl, Holland, and Corr, 1979). Bachinski (1977) reported that sulfides of the Newfoundland deposits contain $^{34/32}S$ somewhat higher than juvenile sulfur at + 5 to 13‰. Upon correction for increasing $f(O_2)$ during deposition, the values fit Ordovician seawater values and conform to Cyprus values, suggesting that seawater circulation was indeed important. It should be appreciated that there is no proof that Cyprus-type massive sulfides were formed specifically at a spreading center midoceanic rise or along a medial graben-rift zone, but the evidence that they formed as part of oceanic crust development and as an integral part of Steinmann's trinity becomes better documented each year. It is not known whether Cyprus-type and East Pacific Rise "black smoker" deposition are synonymous, but analogies can be drawn (East Pacific Rise Study Group, 1981).

With new ophiolite sections being discovered as geologic mapping progresses in many sectors of the globe, recognition of the ore type and its characteristics is increasingly strategic.

▶ BANDED IRON FORMATIONS

The world of the 1980s produces and consumes nearly 1 billion tons of iron ore each year. The vast majority of that—on the order of 90%—is extracted from Precambrian-age cherty *banded iron formations*, or *BIFs*. These BIFs are thin- to medium-bedded interlaminations of iron oxide, iron carbonate, or iron silicate materials with chert or jasper. The terms *iron formation* or *banded iron formation* are used widely by geologists, though not always consistently. Deposits of iron formation generally have features in common, but differences between individual deposits and types of deposits (Gunderson, 1960) have meant that considerable confusion and misunderstanding have arisen over the definition of the term. James (1954) attempted to clarify the matter by generalizing the definition. He defined *iron formation* as "a chemical sediment, typically thin-bedded or laminated, containing 15% or more iron of sedimentary origin, commonly but not necessarily containing layers of chert." Since this definition is broad enough to embrace all so-called iron formation, it is used in this book. Most economic iron formation contains between 25 and 35% iron. The presence of thin layers or nodules of chert is considered by many an essential characteristic of iron formation (Figures 13-13 and 13-14). Kimberley (1978) proposed that "ironstone" be used to define chemical sediments with more than 15% Fe, and that "iron formation" be used to describe a mappable rock unit or "package" dominated by ironstone and with ironstone layers defining its top and bottom.

It must be noted at the outset of this section that two broad classes of BIF units are described here, only one of which—the Algoma type of Archean age—is specifically relatable to submarine volcanic processes. The other class—the Superior type of early Proterozoic age—may include some volcanic input, but need not. The two are discussed here for what will be seen to be obvious reasons, but the differences between them will be noted later and should be kept in mind.

The iron ores of the early days of the Lake Superior iron district of the United States and Canada were oxidized and leached near surface to locally nearly pure massive to pulverulent hematite, but the primary unoxidized iron formations beneath and lateral to them, now called *taconites*,

Figure 13-13. Banded iron formation in outcrop. Jaspilite, forming the upper part of the Negaunee Iron Formation of the Marquette district, Michigan, consists of interlayered reddish jasper and specular hematite. Jasper Knob, Negaunee, Michigan. (*Photo by* H. L. James, 1973.)

are commonly mineralogically complex. Iron in the BIFs of the world is contained in several compounds summarized in Figure 13-15 and Table 13-1, and typically including granular magnetite (Fe_3O_4), hematite (Fe_2O_3) as granular black material, as crystalline specularite, as powdery "soft red hematite," or as reniform "kidney ore," and massive limonite [$Fe_2O_3 \cdot nH_2O$, $Fe(OH)_3$, or $FeO \cdot OH$]. Siderite ($FeCO_3$), chlorite [$Fe_6Si_4O_{10}(OH)_8$], greenalite [the Fe^{+2} analog of kaolinite, $(Fe,Mg)_6Si_4O_{10}(OH)_8$], chamosite [the Fe^{+3} analog of kaolinite or greenalite, $Fe_4Si_4O_{10}(OH)_8$], minnesotaite [the Fe^{+2} analog of talc, $(Fe,Mg)_3Si_4O_{10}(OH)_2$], grunerite [the amphibole, $(Fe,Mg)_7Si_8O_{22}(OH)_2$], stilpnomelane [$(Ca,Na,K)_{0.5}(Fe^{+2},Mg,Al)_4Si_4O_{10}(OH)_2 \cdot 2H_2O$], the olivine fayalite (Fe_2SiO_4), the ferruginous chert "jasper," and the sulfides pyrite (FeS_2) and pyrrhotite ($Fe_{1-x}S$) vary in abundance. The term *taconite* is a name now applied worldwide to a BIF dominated by magnetite, iron silicates, and chert, sometimes with hematite and siderite.

Figure 13-14. Outcrop of Archean-age Algoma-type banded iron formation of the Boston Formation south of Kirkland Lake, Ontario, and north of the Adams mine. The dark gray is magnetite; the white layers are quartz recrystallized from chert. The millimeter- to centimeter-scale bedding has been deformed by soft-sediment flowage, not by tectonic forces. Compare with Figure 13-13. See also Figure 13-16.

TABLE 13-1. Original BIF Precipitates and Metamorphic Equivalents

Compound	Inferred Initial Precipitate	Now Observed	
SiO_2	Amorphous	Chert	
Fe_2O_3	Amorphous $Fe_2O_3 \cdot nH_2O$	Hematite	
Fe_3O_4	$Fe_3O_4 \cdot nH_2O$ hydromagnetite	Magnetite	
$FeCO_3$	Siderite	Siderite	
$Fe_3Si_2O_5(OH)_4$	Amorphous ferrous silicate	$Fe_3Si_2O_5(OH)_4$ greenalite $Fe_7Si_8O_{22}(OH)_2$ grunerite	→ $Fe_3Si_4O_{10}(OH)_2$ minnesotaite Fe_2SiO_4 fayalite
Fe sulfide	FeS	FeS_2	
Na-Fe silicate	Na-Fe silicate gels	Riebeckite Stilpnomelane Chert	

Source: From Garrels, Perry, and Mackenzie (1973).

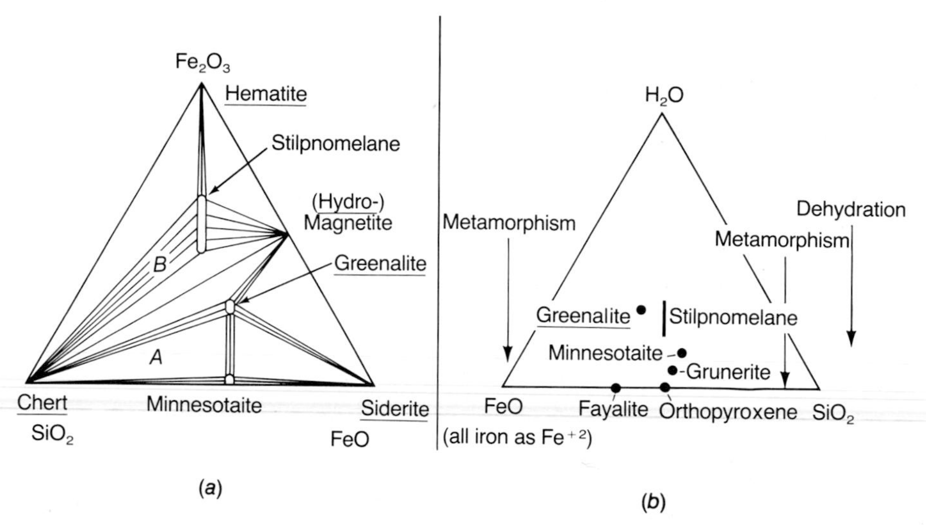

Figure 13-15. Two representations of mineralogy in BIF and its metamorphic equivalent taconite. Table 13-1 gives precipitation products and what is now seen. (*a*) Ternary plot shows the compositions of minerals in low-rank or unmetamorphosed BIF, in the presence of CO_2 and H_2O. Phases that can be deposited as sediments are underlined. Tie lines and triangles enclose stable assemblages according to where a given bulk composition plots: a composition plotting at point *A* would consist of chert, minnesotaite, and greenalite, but more Fe_2O_3-rich *B* would involve only chert and stilpnomelane. (*b*) Some of the minerals in (*a*), but shown in terms of ferrous iron, silica, and water. Metamorphism moves in the direction of dehydration, with sedimentary greenalite beginning an approximate sequence from the kaolin analog greenalite through the micas stilpnomelane and minnesotaite and the amphibole grunerite and the pyroxene ferrosilite, the olivine fayalite being only rarely produced. (*From* Klein, 1973.)

Iron formation consisting predominantly of granular quartz-hematite mixtures like those in Brazil are called *itabirites*. The ferruginous chert jasper is common in the Australian *jaspilite* BIFs and in many other Precambrian iron formations.

Iron formations are widely distributed in the world, but the largest and most abundant deposits are around the Atlantic Ocean and the Indian Ocean. Deposits of iron formation are few and of little consequence around the Pacific Basin. Major districts are present in Brazil, Venezuela, the Lake Superior region of the United States, the Labrador region of Canada, in South Africa, along the west coast of Africa, particularly in Gabon, Liberia, and Mauritania, in several districts in Russia, in India, in Manchuria, and in western Australia, where large deposits were discovered in the 1960s. The most recent major discovery, made in 1967, is of more than 16 billion tons of 67% Fe in the Carajas Range, 700 km south of the mouth of the Amazon River in Brazil.

Iron formations as seen in the field are distinctive. They are normally thin-bedded, with strong color contrast between dark bands of hematite or magnetite, red bands of jasper, and light to white bands of chert (Figure 13-13). Figure 13-16 is a close-up of a polished slab of BIF, a monument near the Sherman mine discovery site near Temagami, Ontario. Note the thin banding and the abundance of pull-apart and apparent slump features—such soft-sediment deformation is a common feature in BIFs. The layers and nodules of silica are chert, though in many places—especially where it is metamorphosed and later weathered—the silica resembles quartzite or sandstone and is frequently so called. Examination of many specimens has led to the conclusion that the material is predominantly the result of chemical sedimentation, locally with detrital admixture. In some districts silica has a mosaic texture (Tyler, 1949) indicative of crystallization in situ, and iron formation in general is notably lacking in the heavy accessory minerals expected in a clastic sediment. As in any sequence of sedimentary rocks, local littoral accumulations of mechanically deposited grains grade laterally and basinward into chemical precipitates. Nevertheless, the bulk of the silica in iron formation is chemically precipitated chert.

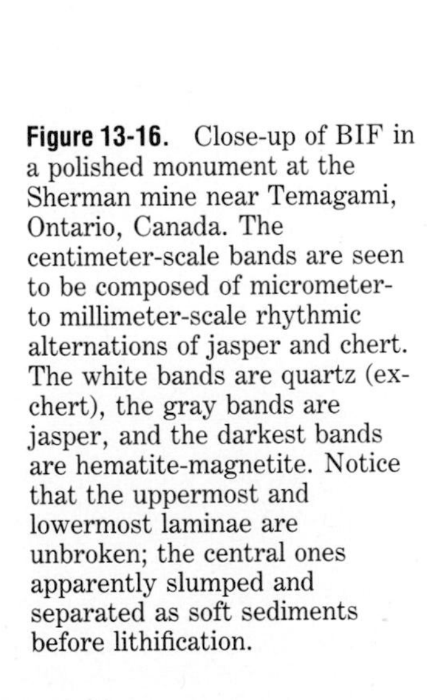

Figure 13-16. Close-up of BIF in a polished monument at the Sherman mine near Temagami, Ontario, Canada. The centimeter-scale bands are seen to be composed of micrometer- to millimeter-scale rhythmic alternations of jasper and chert. The white bands are quartz (ex-chert), the gray bands are jasper, and the darkest bands are hematite-magnetite. Notice that the uppermost and lowermost laminae are unbroken; the central ones apparently slumped and separated as soft sediments before lithification.

Bacteria may have been important in the deposition of iron and silica in iron formation, but it is not possible to determine their importance since the role of bacteria is not fully understood. Iron-secreting bacteria exist now in modern environments and were apparently active in Precambrian times as well (Harder, 1919; Gruner, 1922; Beerstecher, 1954; Baarghorn and Tyler, 1965; Cloud, 1973; Margulis, 1981). Under favorable conditions, these microorganisms may have been a major factor in the deposition of iron by catalyzing an inorganic reaction that would produce the same results if given enough time. Both processes—the organic and the inorganic—were undoubtedly active; the question of which process fixed the most iron in a given deposit is as yet unanswerable, although blue-green algal remains and algal textures are locally common. LaBerge (1973) described biologically produced 30-μm spheroids of silica as abundant in iron formations of the world. More will be said of depositional chemistries later; Figure 13-17 shows the Eh-pH variations possible.

BIFs of the oxide facies are normally highly distinctive in the field, both observationally and geophysically. Soft-sediment deformation has in many places produced chaotic crenulation, commonly with pull-apart features, nonpenetrative random textures, and abrupt local thickenings and thinnings (Figures 13-13, 13-14, and 13-16, and 13-18). Obvious bedding is still preserved, but it is commonly wavy to crenulated in outcrop. Further tectonic deformation normally accompanying greenschist metamorphism has produced more organized, through-going chevron folding and systematic crenulation (Figure 13-19). BIFs rich in silica normally stand out prominently in outcrop owing to their resistance to erosion. Those rich in magnetite can be closely followed with a magnetometer, even when they lie under tens of meters of muskeg, swamp, tundra, or glacial debris, a fact that enhances their value in terrane mapping and exploration. Their exploration value is further appreciated when it is understood that they can be "walked into" massive sulfide deposit areas in proximal settings, such as in the Abitibi Greenstone Belt (Figure 13-2).

The role of metamorphism of BIFs has been shown over the years to be great. Hematitic ores are plums of high-grade, silica-leached, oxidized BIF that grade laterally, downdip, and stratigraphically downward into beneficiable taconite that is essentially mildly metamorphosed BIF. Expectably, low-temperature oxides, hydrous oxides, carbonates, and silicates like limonite, siderite, and greenalite originally deposited on the seafloor were dehydrated and upgraded by sedimentary compaction and regional metamorphism and locally by contact metamorphism around major plutonic systems such as the Duluth Gabbro to produce the higher rank layer lattice silicates minnesotaite and stilpnomelane, the amphiboles grunerite and cummingtonite, the pyroxenes hedenbergite and ferrohypersthene, magnetite, and other minerals (Bayley and James, 1973; Klein, 1973).

BIF has been considered to be ferruginous sedimentary rock, which resulted from the weathering, transportation, and precipitation of iron de-

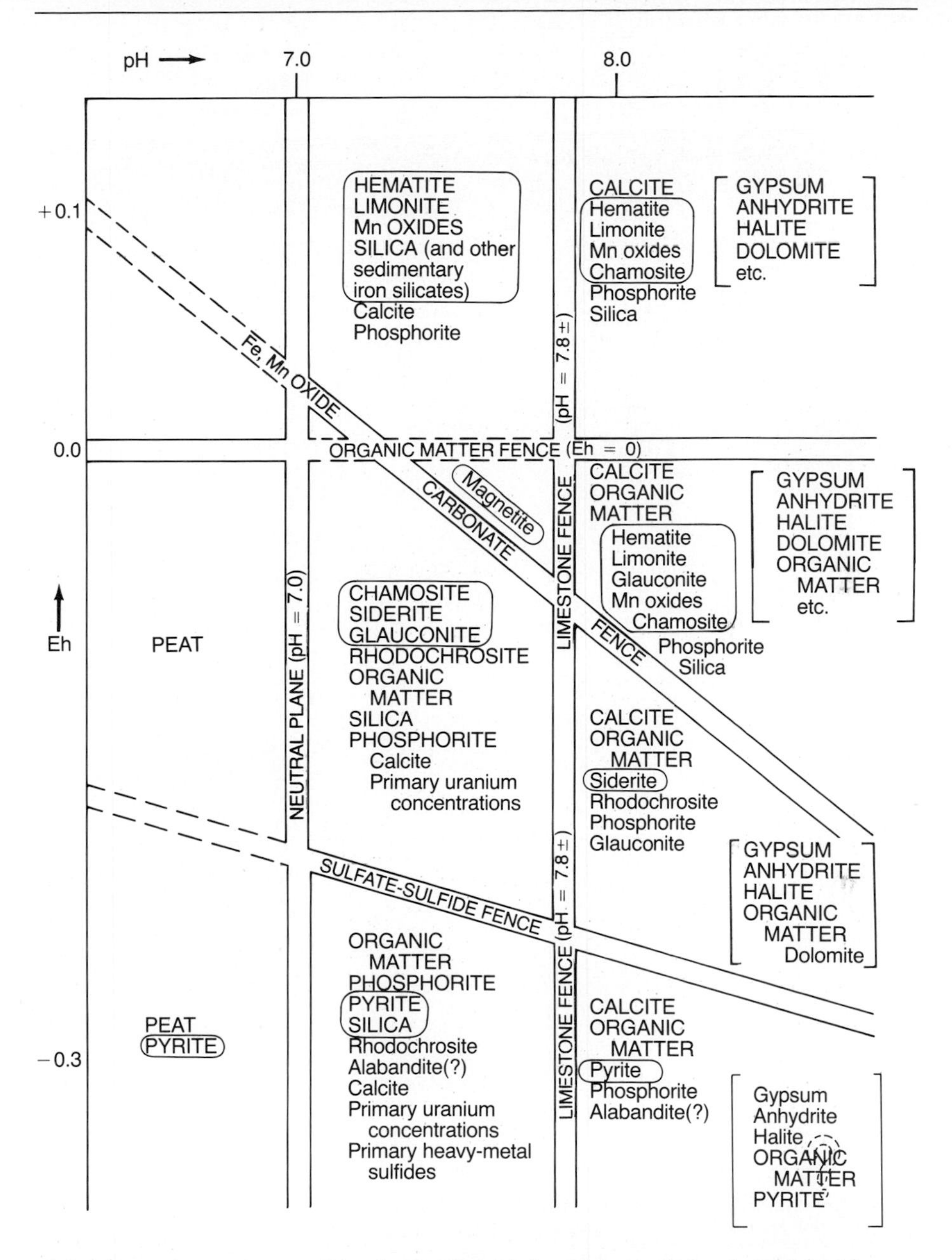

Figure 13-17. Fence diagram showing Eh-pH fields in which chemical end members of nonclastic sediments are formed under normal seawater conditions. Associations in brackets are for hypersaline conditions (salinity >200 ‰). James' sulfide facies lies below the sulfate-sulfide fence and at lower pH than the limestone fence; the oxide facies is above the oxide-carbonate fence. Iron minerals pertinent to this discussion have been circled, and magnetite has been added immediately above the oxide-carbonate fence. (*From* Krumbein and Garrels, 1952.)

Figure 13-18. Split core of iron formation from the Bomi Hills, Liberia. Notice the fine banding and the abrupt random thickening and thinning characteristic of soft sediment deformation. Three-fourths natural size.

Figure 13-19. Outcrop of crenulated chevron-folded magnetite-quartz (ex-chert) BIF in the Kirkland Lake, Ontario, area. The individual bands are only millimeters thick but are laterally continuous for tens of meters. The entire unit has been tectonically folded.

rived from landmasses by erosion, with deposition of iron as oxides, hydroxides, and hydrous oxide-silicate minerals in sedimentary systems. The role of volcanism was appreciated by some early geologists (Van Hise and Leith, 1911; Royce, 1942; Trendall, 1965, 1968). The debate is still lively, and at least six theories to account for BIF are discussed in current literature:

1. Silica and iron associated with volcanism were poured out on the seafloor from springs of magmatic origin (Van Hise and Leith, 1911; Trendall, 1965, 1968; Gross, 1980).

2. Iron and silica carried in true solution from nearby landmasses were rhythmically deposited as sediments in water, probably in response to seasonal variations in the composition of the water. Various explanations have been offered for the mode of deposition, most of which involve direct inorganic precipitation of silica and iron or one of several biochemical processes (Baarghorn and Tyler, 1965; Eugster and I-Ming, 1973).
3. Iron formation beds were originally deposited as more thickly bedded fine-grained ferruginous tuffs and other iron-rich sediments that were diagenetically oxidized and silicified under the influence of solutions that were partly volcanic in origin. The silicification caused separation to finer beds of banded cherts and jaspers that alternate with more iron-rich layers (Dunn, 1935, 1941).
4. Iron formations were deposited as end members, or final products, of carbonate sedimentary cycles (Button, 1976).
5. Deposition of iron formation resulted after buildup of iron concentration in the sea. Jolliffe (1966) envisioned a primitive Archean acidic sea with a pH of 6 or less, an Eh of about 0, and with seawater in equilibrium with an atmosphere rich in CO_2. Under these conditions, iron released by erosion and by volcanism would remain as ferrous iron in the sea. As time progressed, the CO_2 of the atmosphere was gradually depleted, and an increase in the pH of the sea resulted by removal of H_2CO_3. A point of saturation was ultimately reached, and $FeCO_3$ and Fe_3O_4 started to precipitate. The gradual buildup of oxygen and the depletion of CO_2 in the atmosphere eventually led to the wholesale precipitation of iron in seawater as magnetite, hematite, and siderite.
6. Iron formation resulted from the upwelling of cold, deep seawater onto a warm continental shelf. The cold water would be saturated in $CO_3^{=}$, Ca^{+2}, and Fe^{+2}. A few milligrams of Fe^{+2} per liter—3 ppm—would precipitate as ferric iron oxides, hydroxides, or silicates (Figure 13-15 and Table 13-1) with increased temperature, oxidation, and CO_2 loss (Holland, 1973). The same waters would be supersaturated in amorphous silica and would precipitate chert.

There were very few means of determining which mechanisms actually were applicable in particular deposits until Gross (1965a, b, 1980) and Kimberley (1978) attempted to differentiate essentially volcanic from sedimentary iron formations. While one group including Gross, James, Bayley, and Sims was struggling to subdivide various types of iron-rich fundamentally sedimentary rocks and unravel the complexities of their formation, another group including A. M. Goodwin at the University of Toronto, Canada, R. L. Stanton of the University of New England, New South Wales, Australia, and others was perceiving that one of the persistent lithostrati-

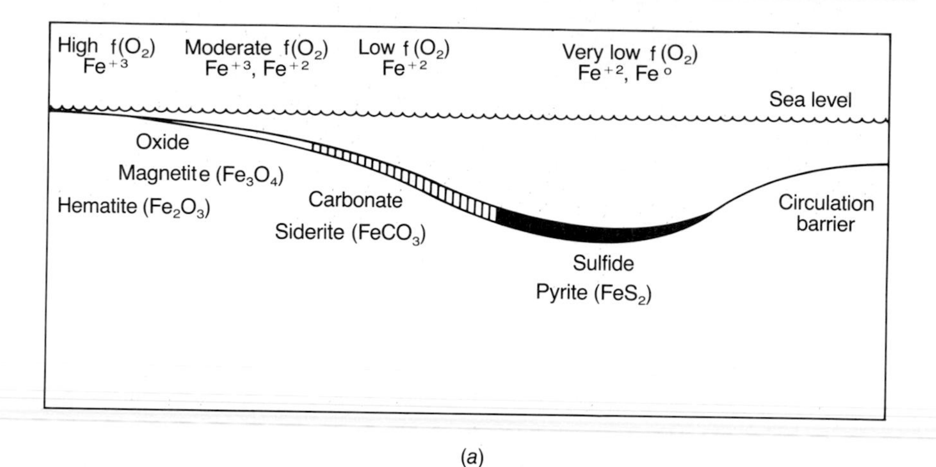

(*a*)

Figure 13-20. (*a*) Modification of James' diagram of depositional zones in a hypothetical basin in which iron compounds are being precipitated. (*From* James, 1954; Figure 3.) (*b*) (*facing page*) Its application in the form of a reconstructed stratigraphic section of the Michipicoten basin by Goodwin (1973). The oxide-carbonate facies transition marks the slope edge of the shelf upon which conglomerate-bearing sediments rest. The carbonate-sulfide facies transition overlies felsic pyroclastic piles within the margin of the basin, with sulfide facies in the deeper water direction.

graphic units associated with Archean volcanism in greenstone belts was the BIF. These workers discerned that BIF rocks constituted time-stratigraphic units that thickened and thinned and changed in mineralogy from place to place, basin to basin, and environment to environment, and that they could be used for correlation and as mapping aids in deciphering the complex Archean volcanic greenstone belts (Goodwin, 1965, 1973). They also perceived real differences between Archean-age units (3.5 to 2.5 billion years) and younger Precambrian ones (2.5 to 1.9 billion years). Other geologists like King in Australia, the Viljoens and Anhaeusser in South Africa, and Tatsumi in Japan were perceiving the same associations between volcanism and BIF. Amidst this growing recognition expressed by Kinkel, Stanton, Gilmour, Hutchinson, Ridler, and the others mentioned, two papers emerged as milestones. One was by the Norwegian Oftedahl (1958), who saw the importance of the BIFs as part of a volcanic ore-forming process that he described in terms of exhalation; the other was by Harold James (1954), who articulated the facies concept of iron formations. James recognized (Figure 13-20) that high-energy littoral environments were relatively well oxygenated and characterized by hematite stability and magnetite metastability, and that progressively basinward, deeper water, lower energy, more reducing environments were characterized by magnetite, siderite, and pyrite-pyrrhotite stability; this facies approach developed into a new means of interpreting ferruginous sediments and volcaniclastic rocks. Finally a new rubric on the association of exhaled gold and iron formation has helped our understanding of both BIF and gold deposits.

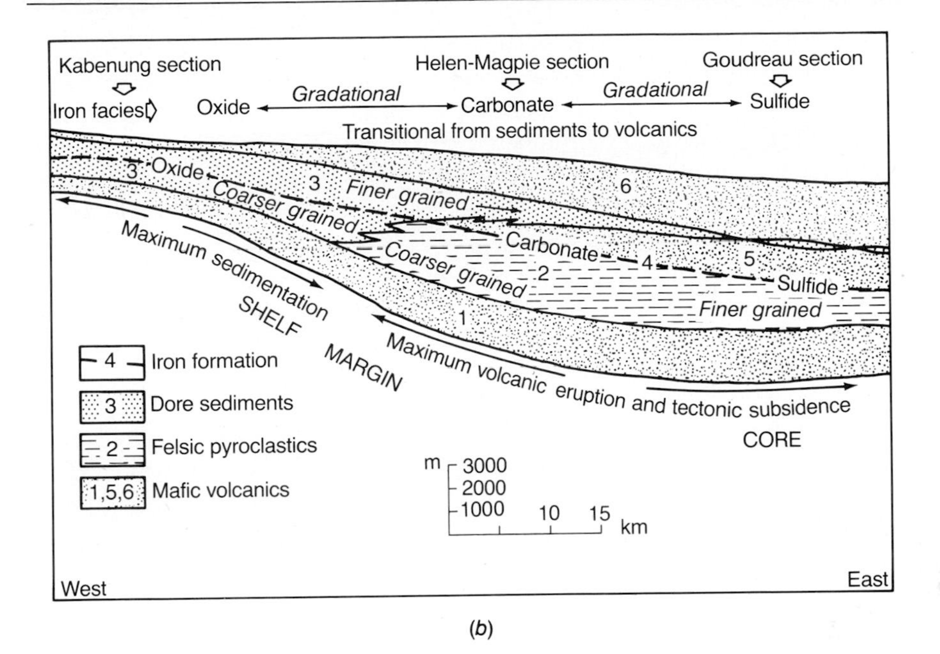

(*b*)

James' ideas were born in observations like these. In iron formations in general, two subfacies of the oxide zone are found, one characterized by hematite and the other by magnetite, both as primary sediments. The hematite rock normally consists of fine-grained hematite interlayered with chert or jasper, with common oolitic structures. Evidently the hematite facies accumulated in a strongly oxidizing near-shore environment like that in which younger iron-rich rocks, such as those of the Clinton Formation of the eastern United States, were deposited. The silicate facies contains abundant hydrous ferrous silicates—greenalite, stilpnomelane, or minnesotatite—and is most commonly associated with either carbonate- or magnetite-bearing rocks, suggesting that the optimum conditions for deposition ranged from slightly oxidizing to slightly reducing, probably with mild post-depositional metamorphism. The magnetite facies, very common in the Lake Superior region, consists of magnetite interlayered with silica, carbonates, iron silicates, or some combination of these minerals; its mineralogy and associations suggest weakly oxidizing to moderately reducing conditions. When uncontaminated by detrital grains, the carbonate facies consists of interbedded siderite, or iron-rich ankerite, and chert. It is a product of an environment in which the oxygen concentration was high enough to destroy most organic material but not high enough to permit ferric compounds to form. Lake Superior's sulfide facies is represented by black carbonaceous slates containing as much as 40% pyrite. The free carbon content of these slates typically ranges from 5 to 15%, indicating that ultrastagnation conditions prevailed during deposition. Individual crystals of pyrite can be microscopic, and the iron content is not conspicuous in hand specimens.

Chemical relationships between these assemblages (Figure 13-17) had already been assessed by Krumbein and Garrels (1952) and were later to be studied by Eugster and I-Ming (1973). James noted that these relations could best be explained on the basis of depositional-environmental facies as most simply stated in Figure 13-20. James and other scientists like Gastil and Knowles (1960) found that the concept could be applied to explain similar relationships described elsewhere, even in Proterozoic deposits such as at Wabush Lake near the Quebec-Labrador border in Canada (Figure 13-21), and the idea found ready acceptance. One of the areas where it was successfully applied was in the Abitibi Greenstone Belt and related terranes in Canada by Goodwin (1973) and his associates. Parts of the abstract of that paper are presented verbatim here (page 915) and can be used also to establish some of the differences between Archean (Algoma) and Proterozoic (Superior) iron formation types.

Iron-formation of Archean age (older than 2500 million years) is widely distributed in supracrustal assemblages of the Canadian Shield. Although individual iron-formations are comparatively small, the total ore reserve is 35,000 million tons or 25 percent of the established iron ore resources in Canada. Archean iron-formations form an intimate part of volcanic-rich greenstone belts or segments. Individual iron-formations are commonly associated with upper pyroclastic phases of predominantly tholeiitic to calc-alkaline, mafic to felsic volcanic sequences and nearby turbidite assemblages. The iron-formations are readily attributed to volcanic processes in terms of the source of the chemical components.

The distribution of Archean iron facies conforms in the main to the common worldwide depositional pattern of iron. Thus oxide, carbonate, and sulfide facies are present in that shallow-to-deep order upon Archean paleoslopes which are components of Archean basins. Iron facies thereby provide an insight to Archean basin analysis. In general, oxide facies is the most common and readily recognized form of Archean iron-formation. Sulfide and carbonate facies iron-formations, although widely distributed, are comparatively thin, discontinuous, and inconspicuous. Most if not all Archean iron-formation is attributed to a volcanic exhalative source (exhalite).

A plot of known Archean iron-formation by facies within the greenstone belts of the Canadian Shield reveals the presence of a number of large Archean basins. Each basin, although presently elliptical in outline as a result of structural deformation, may originally have been more nearly circular in outline and 750 to 1000 km in diameter. The margin of a basin features a triple lithofacies association of (1) oxide-carbonate-sulfide BIF transition, (2) arc-type felsic volcanic rocks, and (3) proximal volcanic conglomerates.

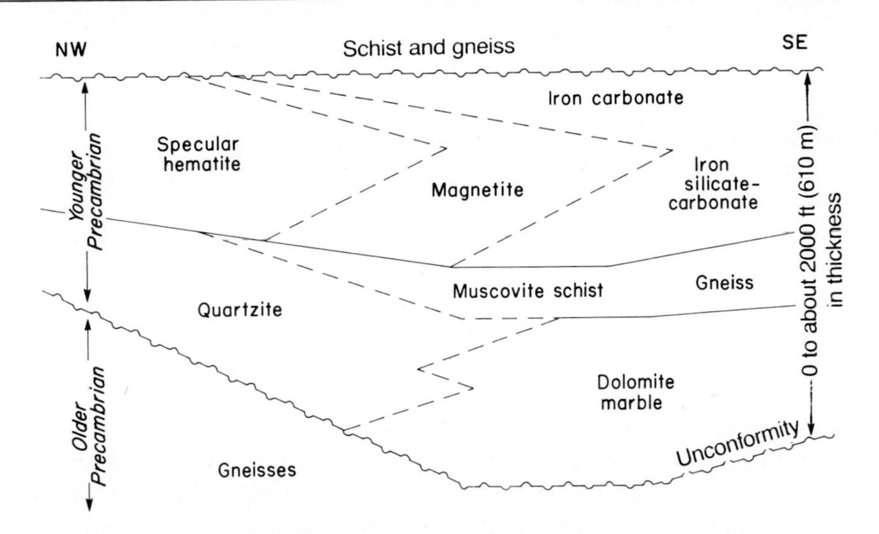

Figure 13-21. Diagrammatic section across the Wabush Lake area on the Quebec-Labrador border in eastern Canada, showing the distribution of sedimentary facies. The ferruginous upper portions conform well to James' facies concept (Figure 13-20), implying deeper water to the southeast and early and late northwest-transgressive seas. (*From* Gastil and Knowles, 1960; Figure 2.)

Deep-water or volcano-slope proximal depressions feature sulfide facies iron formation, with stratigraphically persistent carbonate or oxide iron formation formed in shelf environments. Although they are not of the same age, perhaps we can think of these Abitibi Algoman-SVOP as proximal and of the Superior-MECS as distal, in an expanded sense of the words as they are used in massive sulfide environment description. The MECS of Kimberley contain less of the proximal-type directly volcanic trace elements and typically less pyrite.

Kimberley (1978) and Bayley and James (1973) have provided excellent basic studies of iron formations. Bayley and James noted that there have been three great periods of Precambrian iron formation development, namely, older than 2.6 billion years, from 2.6 to 1.9 billion years, and from 1.9 to 0.6 billion years. The oldest group, called the Algoma type, represents several periods of orogeny, volcanism, and plutonism, and the deposits are generally enclosed in and closely related to volcanic rocks. Kimberley called these units SVOP-IF, or shallow volcanic platform iron formations. They include the Temagami deposits at the Adams and Sherman mines in the Abitibi Greenstone Belt in Ontario, the Helen and Soudan deposits in Minnesota, units in the Wind River, Wyoming, and Beartooth, Montana, areas, and many BIF units in South Africa and Zimbabwe. These accumulations are seldom more than 50 meters thick, and they contain much more magnetite than hematite, with oxide and carbonate facies iron minerals interlaminated with chert, jasper, and finely granular quartz. Silicate facies, including greenalite-chamosite-minnesotaite, are sparse; carbonaceous sulfide facies are common. Compositions are remarkably constant at 40 to 55% SiO_2 and 28 to 37% Fe. Trace elements that might be considered consistent with volcanic exhalation are found—manganese to several percent, lesser barium, a few to a hundred or more parts per million of Co, Ni, Cu, Cr, As, and Sr, and gold measured in parts per billion. Adjacent rocks include graywackes with prominent volcanically derived components and eugeosynclinal pyroclastic and flow units ranging from deeply subjacent basalts to more abundant intermediate to felsic lithologies. Immediate footwall units may include altered-fractured, even brecciated zones suggestive of pipes or solution conduits.

Variations on the theme of iron formations related to volcanic processes—especially with regard to Archean submarine hot springs—have received geochemical, geologic, and even observational support in the last decade. Bostrom and Peterson (1966) collected seafloor sediment samples over a 6000-km traverse perpendicular to the East Pacific Rise. When the sediments were redispersed in seawater, the murkiness of samples collected near the rise, which yielded to clarity away from it, was attributed to colloidal iron and manganese hydroxides originally suspended in seawater and presumably stemming from hot springs issuing from the East Pacific Rise. Cu, Cr, Ni, Pb, and Ba were also anomalously high. Atwater (1980) and others have witnessed dense clouds of material that includes iron hydrolysates, issuing from "black smokers" along the East Pacific Rise. Iron compounds are abundantly present in the thermal brines of the Red Sea and the Salton Sea.

Vastly more abundant and economically more significant are the Superior-type iron formation units of 1.9-to-2.6-billion-year age (Bayley and James, 1973). These cherty BIFs are typically thick, commonly to several hundred meters. Hematite is slightly more abundant; magnetite still predominates among oxides. But more iron is probably present in magnetite-iron silicate or steel-blue hematite-iron silicate assemblages than as simple oxide-silicate units. These deposits—of which the Biwabik Formation of the Mesabi Range of northern Minnesota is an example—were classified by Kimberley as MECS-IF, an acronym for metazoan-poor, extensive, chemical-sediment-rich, shallow-sea iron formation. These Superior-type BIFs are distinctive petrographically, compositionally, and in host rock settings from Algoma types, including the fact that close affinity to volcanism is not demonstrable. These Superior-type BIF units are the fundamental iron formation connoted by the term as used by James (1954), by Marsden (1968), by Bayley and James (1973), and by the profession at large, and are discussed below in greater detail. Morro do Urucum in western Brazil, the Mesabi Range and the Gogebic-Iron River-Menominee Range in the Lake Superior area, and the Hamersley Range in Australia are examples of large districts in which volcanism seems to have been absent. In these latter areas, which are also post-Archean early to mid-Proterozoic in age, non-volcanic, perhaps deep marine or terrigenous sedimentary affinity is indicated. Marsden (1983) considers that portions of the Mesabi may have been a paleolittoral environment.

It has been demonstrated by Lepp and Goldich (1959), Cloud (1973), and Garrels, Perry, and Mackenzie (1973) that the Earth's atmosphere appears to have become markedly more oxidizing at about 2 billion years ago, which makes attractive the concept that the major BIFs accumulated as magnetite-chert, siderite-chert, or carbonate-chert chemical sediments in reducing Precambrian environments. Most iron formation contains no fossils other than blue-green algae and nondiagnostic and questionable organisms. Shallow-water stromatolite clumps and algal reefs and colonies have been reported, but are rare. O'Rourke (1961) presented evidence, however, that there are BIFs that resemble Precambrian ores and have been assumed to be that old, but which are in fact post-Precambrian in age. Kimberley (1978) stated that the evolution of tectonic and magmatic environments was more important than biologic or atmospheric-hydrologic evolution. He notes that eugeosynclinal-type volcanic environments prevailed in the Archean, with continental shelf environments and inland sea environments prevailing in the Middle Precambrian and Phanerozoic, respectively, with all three represented in Late Precambrian times. Other sedimentary iron ores that formed under oxidizing conditions—the oolites such as the Clinton ores of the eastern United States, the minettes of Alsace-Lorraine, bog iron and iron carbonate beds otherwise known as "black band" ores—will be described in Chapter 15.

Adams-Sherman Algoma-Type BIF Deposits, Ontario, Canada

Two Algoma-type concentrating-grade BIF deposits are the Adams mine south of Kirkland Lake, Ontario, and the Sherman deposit in the Temagami district of Ontario. Both are currently being mined, both are part of the Archean Abitibi Greenstone Belt, and both are stratigraphically part of the ferruginous sedimentary system shown in Figures 13-1, 13-22, and 19-12, and schematically in Figure 13-2. The Adams orebody is composed of thinly laminated magnetite-quartz beds of the Boston iron formation in a layer that varies in thickness to up to several hundred meters and that extends for some 10 km along strike (Figure 13-14). The BIF, averaging 27% Fe in the ground, is magnetically upgraded to provide pellets of 62 to 63% Fe content, 1 million tons a year of which is shipped 2000 km to Pittsburgh, Pennsylvania, the longest iron ore rail haul in the world. The ores contain locally troublesome amounts of pyrite and minor hematite and jasper; they have been mildly contorted, although soft-sediment slump structures are most common. Minor amounts of "exhalite elements"—Ni, Cr, Mn, Ba, and even traces of Au—are reported.

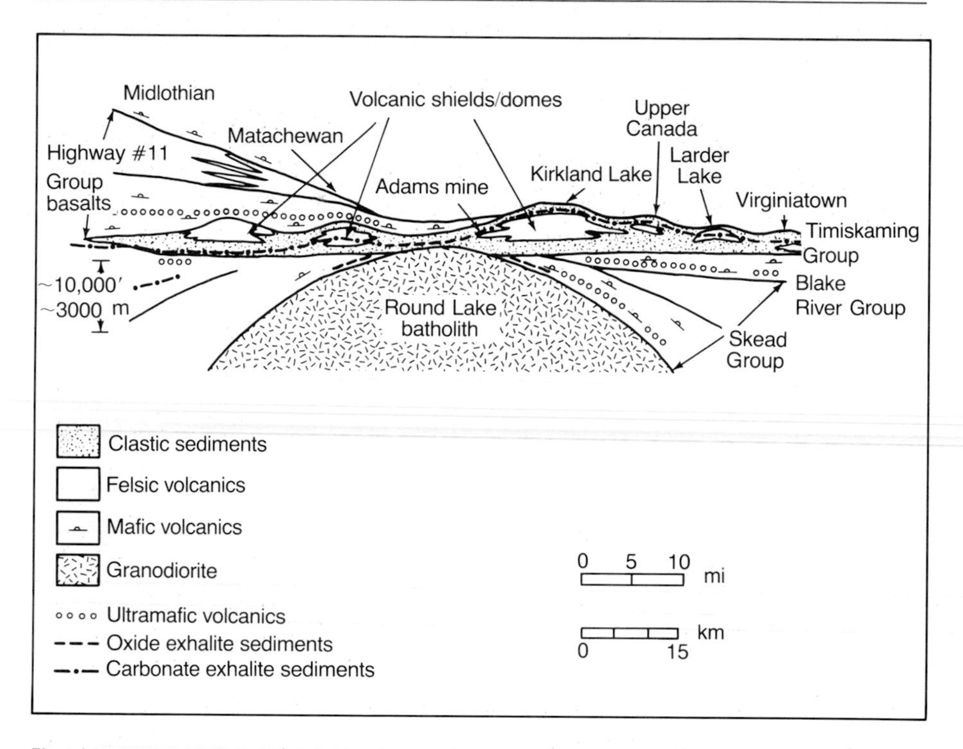

Figure 13-22. Stratigraphic relationships within the Abitibi Greenstone Belt. Notice the volcanic centers and the persistence of the exhalite iron formation layers of submarine chemical sedimentation-exhalation origin. See also Figures 13-1 and 19-12. (*From* Ridler, 1976.)

The lowermost cycle of iron formation at the Adams mine is underlain by massive flows of tholeiite basalt containing pillows; small lenses of lean iron formation are interbedded in these flows. Well-bedded felsic tuffs lie above a sharp contact with the mafic volcanics. These tuffs may be absent or range up to tens of meters thick; they commonly contain interbedded iron formation. A layer of nodular and fine-grained disseminated sulfides from several centimeters to 10 meters thick rests on top of the tuff. Above the sulfide layer, lean, cherty, low-iron ironstone sediments begin. Finely laminated chert and magnetite are interbedded, and subsequent metamorphism has destroyed little of the fine layering. Going up in the sequence, the amount of magnetite with respect to chert increases, and magnetite layers thicken. Minor amounts of other minerals are found in the iron formation. Hematite occurs in magnetite bands, giving them a reddish coloration, and garnet, tremolite, actinolite (a bluish amphibole), pyrite, and chlorite are also present. Above the ore unit, lean iron formation with iron content less than 5% reappears. On the BIF, a horizon of graphitic, sulfidic tuff that may be up to 3 meters thick is present; it is overlain by 200 to 300 meters of chert-derived quartzite. Sedimentary deposition was ended by fissure volcanism that produced mafic and ultramafic flows unrelated to the previous felsic activity. The iron formation units are overlain by komatiites with spinifex texture.

The Peria Pit is one of the eight iron bodies outlined at the Adams mine. Most of the iron formation horizons measure only 30 to 50 meters thick, but regional folding and brecciation due to slumping during and just after deposition (Figure 13-14) has thickened the units considerably. The largest orebody, the South Pit, is 1 km long and 200 meters thick. The Adams ores are intimately related to proximal submarine volcanism.

Sherman mine ores at Temagami (Boyum and Hartviksen, 1970) are also magnetite-quartz-rich, mainly well-banded gray to white magnetite-chert rocks, but they contain more jasper and less pyrite and show only mild greenschist-grade metamorphic effects (Figure 13-15). The iron formation averages about 60 meters thick for 2 km of strike length. Maximum thickness is 100 meters. Interbedded tuffs a few centimeters thick are now chlorite and stilpnomelane. The rocks have been folded synclinally, and both limbs, now dipping 75° toward each other, are mined. The ore, of which 1 million tons a year is shipped, averages 25 to 30% Fe and permits pellets of 65% Fe. At both pits the BIF units lie stratigraphically between layered flows and volcaniclastic rocks; the basal units include an andesite unit that has been fractured to the verge of brecciation and silicified by hot spring fluids presumed to have been entering the basin at the time of deposition. The finely laminar BIF units show soft sediment deformation (Figure 13-16) and are exemplary of Algoma-type ores.

Lake Superior Region Superior-Type BIF Deposits, Minnesota-Wisconsin

The Lake Superior region of the north-central United States and southern Canada (Figure 13-23) through 1970 was the most productive iron ore province in the world. From the first production in Michigan in 1855 and in Minnesota in 1884 through 1980, the mines have yielded more than 6 billion tons of ore. Although most of the accessible and highest quality "natural" ores have already been mined, the region can produce taconite iron ore pellets for many years; known reserves are adequate for at least a century. The future of the region is assured by the development of economical methods of handling the hard magnetic taconites from the Mesabi Range and the "jasper" ores of the Marquette Range.

There are six principal districts—or ranges, as they are called—within the United States sector of the Lake Superior region. These ranges (Figure 13-23 and Table 13-2) are the Mesabi, the Vermilion, and the Cuyuna ranges in Minnesota; the Penokee-Gogebic across the Wisconsin-Michigan border; and the Marquette and Menominee ranges in Michigan. The Menominee Range is further divided into the Iron River, Crystal Falls, and Iron Mountain districts. Of the six ranges, the Mesabi has been by far the most pro-

TABLE 13-2. Lake Superior Region Precambrian Stratigraphic Correlations

Series		Cuyuna	Mesabi	Vermilion	Gogebic	Marquette	Menominee
Keweenawan		Mafic intrusives Silicic intrusives	Duluth Gabbro	Duluth Gabbro	Unconformity Presque Isle Granite Mafic intrusives Sandstones, shales, and conglomerates Silicic flows Mafic flows Quartzite and conglomerate		
Huronian	Upper Middle	Virginia Slate Upper slates *Deerwood* Lower slates	Virginia Slate (1660 m.y.)	Rove Slate	Unconformity Granite Tyler Slate	Silicic intrusives Mafic intrusives Michigamme Slate Upper slates *Bijiki* Lower slates Clarksburg volcanics *Greenwood* Goodrich Quartzite	Silicic intrusives Mafic intrusives *Michigamme Slate* South belt of Quinnesec greenstone (position uncertain)
			?	?	Unconformity		
			Biwabik (1900–2200 m.y.) Pokegama Quartzite	*Gunflint* (1700 m.y.)	Mafic intrusives and extrusives *Ironwood* Palms Quartzite	*Negaunee* Siamo Slate Ajibik Quartzite	*Vulcan* Greenstones (Lake Antoine region)

Lower	Dolomite			Unconformity		
				Bad River dolomite Sunday quartzite	Wewe Slate Kona dolomite Mesnard quartzite	Randville Dolomite Sturgeon Quartzite
		Unconformity Giants Range Granite (2670 m.y.)	Vermilion Granite (2680 m.y.)	Granite		Granite
		Knife Lake Slate	Knife Lake Slate Slate (2700–2750 m.y.) Ogishke conglomerate member			
Keewatin			Granite	Unconformity Granite and granitoid gneiss	Granite, syenite, peridotite Palmer Gneiss	Granites and gneisses
		Greenstones, schists, and porphyries	Ely Greenstone (2660 m.y.)	Greenstones and green schists	Kitchi Schist and Mona Schist	

Comparative Precambrian stratigraphic columns from the principal districts of the Lake Superior region from west to east. Cuyuna, Mesabi, Vermilion are in Minnesota, the others in Wisconsin and Michigan. Units of or containing iron formation are italicized.

Source: Table from Leith, Lund, and Leith (1935); dates from Goldich (1973).

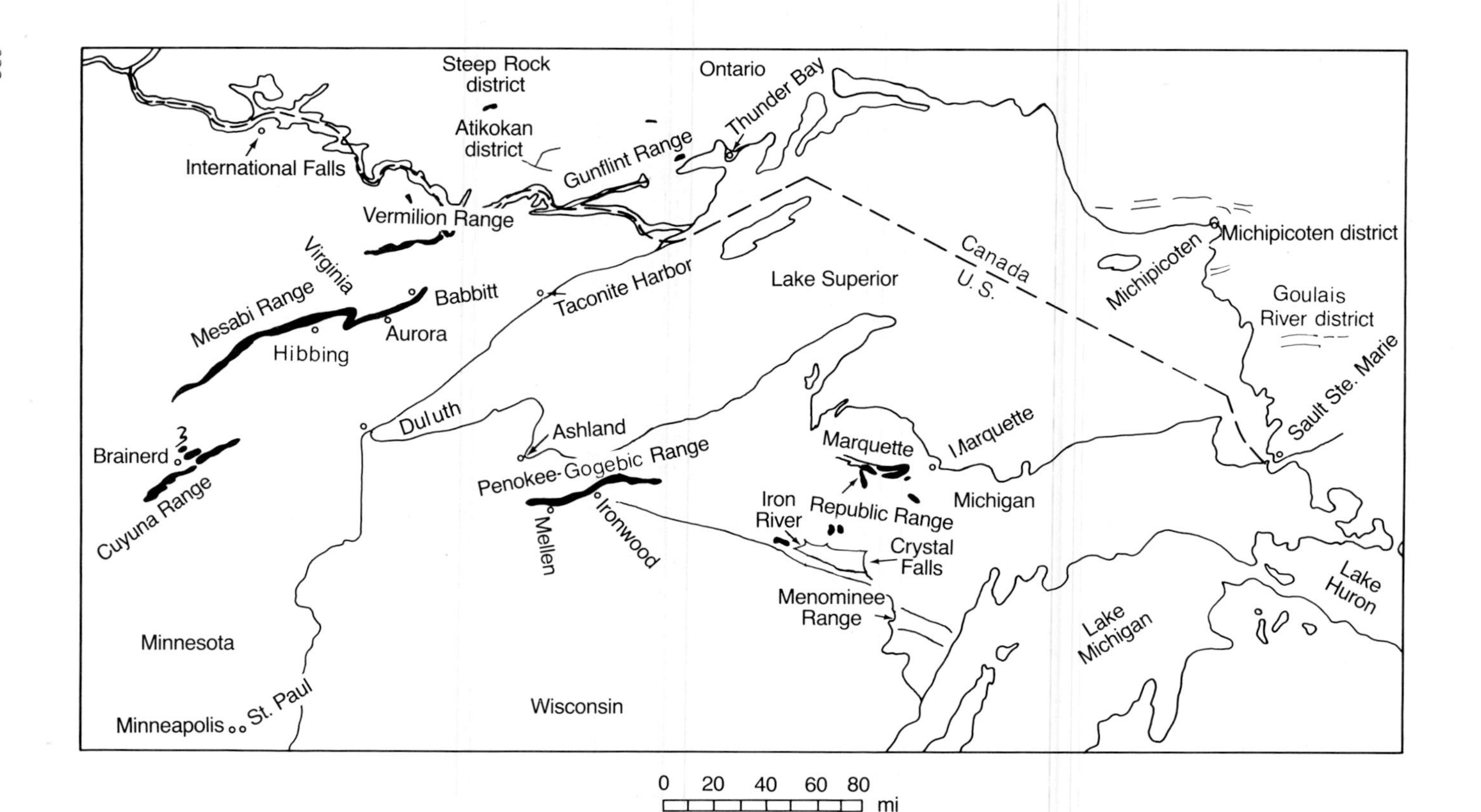

Figure 13-23. Map of the Lake Superior region showing the location of major iron ore districts. (*From* Lake Superior Iron Ore Association, 1952.)

ductive (Figure 13-24). Many subordinate deposits such as the Gunflint Range, the Republic trough, the Amasa district, the Baraboo Range, and the Florence district, have contributed small amounts of ore. In Canada the two most productive deposits are the Michipicoten Range and the Steep Rock Lake district.

The literature concerning the Lake Superior deposits is voluminous; many excellent discussions of both the region and individual districts have been published (Van Hise and Leith, 1911; Gruner, 1930, 1946; Royce, 1938; Tyler, 1949; Dutton, 1952; James, 1954; Grout and Wolff, 1955; Marsden, 1968; Sims, 1972; Sims and Morey, 1972; Bayley and James, 1973). One whole issue of *Economic Geology* (1973, 68:913–1179) focused attention upon BIFs.

Much of the Lake Superior region is covered with glacial drift and vegetation that handicaps geological studies, so that correlation between the ranges is difficult. The possibility of finding new ore beneath the drift and of extending the known ranges into covered territory has been assessed—much exploration has been done and some has been successful. In addition to geological

Figure 13-24. Aerial view of one of the large open-pit mines in iron ore of the Mississippi Group, Mesabi Range, Minnesota. When the photo was taken, iron formation was being removed at the left center, and waste rock was being dumped in the pit where ore has been removed at the right. (*Courtesy of* Hanna Mining Co.)

studies, extensive geophysical work has been done. Magnetic methods using the dip needle and the magnetometer have been productive, and most of the entire region has now been mapped with airborne magnetic equipment with follow-up on the ground. Magnetic techniques have helped in the mapping of bedrock structure beneath the drift. Because most high-grade ores are related to structures—especially to folds—magnetic methods have contributed directly to the discovery of considerable amounts of ore.

The transition in the mid-1950s from the mining of hematite-goethite ores to the production of concentrating-grade BIF taconite ores processed to pellets has had profound effects in the iron ranges. The observation that "the waste of one generation is the ore of the next" is well exemplified by the story of iron formation. Until about the end of World War II, iron formation was not ore; the iron content was considered too low and the silica content too high. Iron formation was mined as ore only where the silica had been leached by weathering or other processes and the iron oxidized to ferric oxides, principally hematite. The ores were called "natural" or "direct shipping" ore, and the standard was called "fifty-one fifty"—51.5% Fe. It was typically a red, soft material. With the development of jet-piercing drills and economic methods of concentrating and pelletizing the iron, low-grade magnetite "taconites"—plain BIF—became ore; low-grade direct-shipping ores of less than 50% iron now find it increasingly difficult to compete, and indeed are consumed only internally by large companies. However, iron formation is not ore unless it is situated so that it can be cheaply and easily handled, that is, it must be at or near the surface, near transportation, and readily amenable to concentration. The geologic relationship between Biwabik iron formation taconite ores and natural hematite ores is shown in Figure 13-25. The most significant changes (Ohle, 1983) are that the high-grade "natural" ores were situated like small splotches on the 130-kilometer strike length of, for example, the Mesabi; now much of the Biwabik iron formation is potential ore. Geophysical exploration was reoriented because the old ores were high-hematite and thus magnetic *lows*; the taconites are magnetite-rich and thus magnetic *highs*. Also, grain size, grindability, mineralogy, and impurities—geologic aspects of the rocks—are overridingly important in the taconites. Many old, worked out direct-shipping ore pits were found to bottom cleanly in taconite, and so they were reopened. Others show a transition from "natural ore" to "oxidized ore" through "oxidized taconite" to "taconite." All of the Lake Superior plants now process taconite by crushing it, magnetically separating magnetite, adding 1% bentonite as a binder, rolling the mix to 1-to-2-cm pellets, and

Figure 13-25. (*Facing page*) Cross section of a typical Mesabi orebody as it might have been found in 1900. The volume contained within the orebody is composed of "natural" hematite ores. Their derivation by weathering and their stratigraphic and spatial relation to today's ores—the adjacent taconites—is clear. (*From a* Lake Superior Iron Ore Association, 1952, *diagram first published by* Wolff, 1917.)

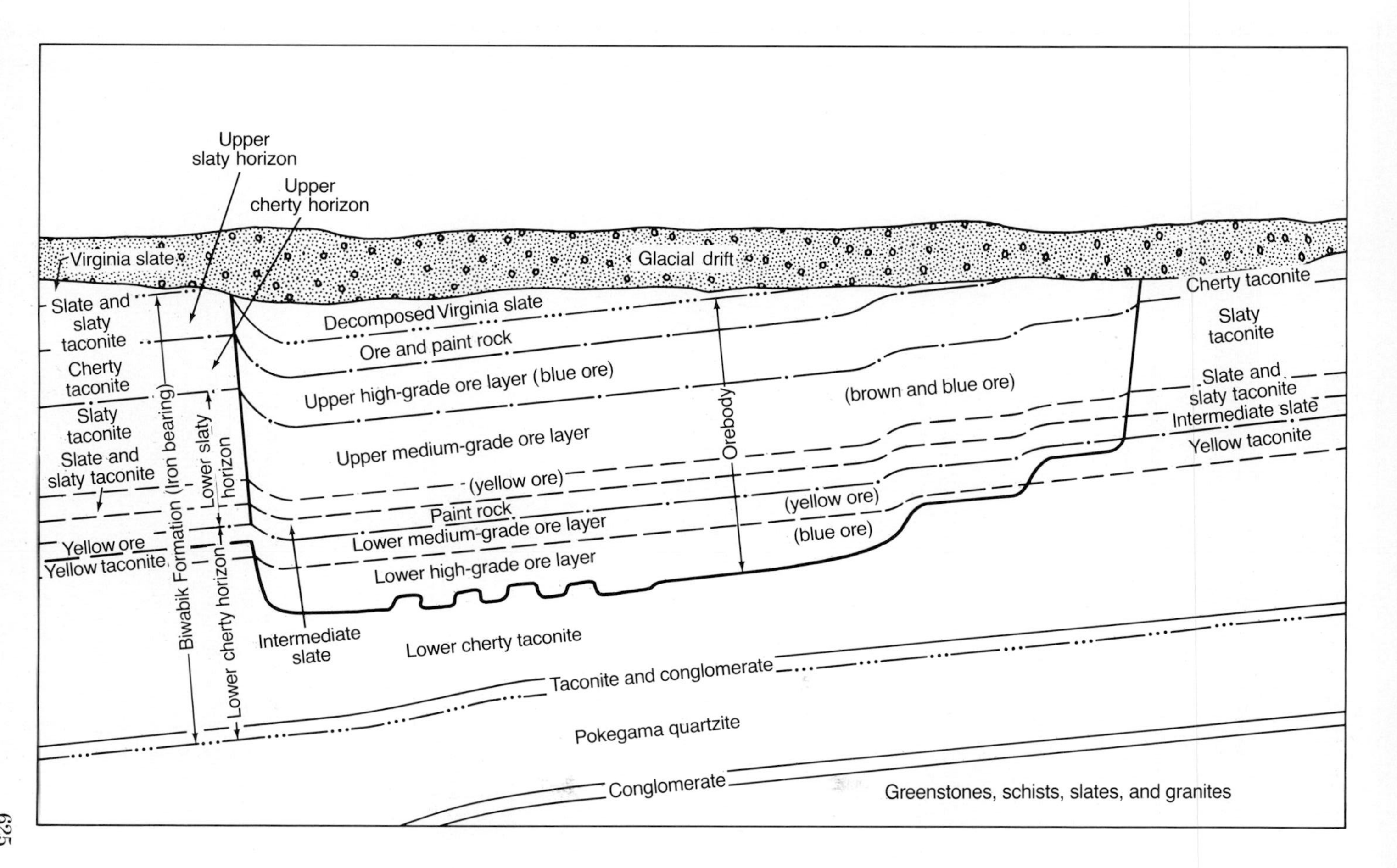

Upper slaty horizon
Upper cherty horizon
Virginia slate
Glacial drift
Slate and slaty taconite
Cherty taconite
Slaty taconite
Slate and slaty taconite
Yellow ore
Yellow taconite
Biwabik Formation (Iron bearing)
Lower slaty horizon
Lower cherty horizon
Decomposed Virginia slate
Ore and paint rock
Upper high-grade ore layer (blue ore)
Upper medium-grade ore layer
(yellow ore)
Paint rock
Lower medium-grade ore layer
Lower high-grade ore layer
Intermediate slate
Orebody
(brown and blue ore)
(yellow ore)
(blue ore)
Lower cherty taconite
Taconite and conglomerate
Pokegama quartzite
Conglomerate
Greenstones, schists, slates, and granites
Cherty taconite
Slaty taconite
Slate and slaty taconite
Intermediate slate
Yellow taconite

briefly baking them for hardening. Potential ore deposits of BIF elsewhere in the world must be evaluated in terms of this modern technology.

As described by Marsden (1968), the Mesabi Range (Figure 13-23) has been by far the leading producer in the region. Most of the ore shipped was a mass of soft reddish-brown hematite and limonite from which most of the silica and other elements had been leached. In general the mineralogy of the Lake Superior "natural" ores was simple. Hematite and goethite were the principal minerals, either the soft reddish to brown variety of the Mesabi Range, or the locally hard, massive variety of the Marquette and Vermilion ranges. On the Menominee and Cuyuna ranges the ore was composed of soft yellow to brown limonite with some earthy hematite. Several beds in the Cuyuna Range carried as much as 8% manganese, which enhanced their usefulness for steel making in the United States. The Mesabi Range is structurally simple. Except for a Z-shaped bend near the center of the outcrop area, the orebody strikes east-northeast and dips 4 to 7° southeast in the western part of the area and 6 to 12° southeast in the eastern part. In contrast to the generally simple structure of the Mesabi Range, most of the other districts are structurally complex. The iron formation is intricately folded and faulted, and as a result, detailed knowledge of both the stratigraphy and the structure is essential to exploration (Schmidt and Dutton, 1957; Pettijohn, 1946; Dutton et al., 1945; Marsden, 1968).

Throughout most of the region, the iron formation contains between 25 and 35% iron. It is hard, tough rock, difficult to break and expensive to beneficiate. The eastern end of the range consists of hard magnetite taconite, metamorphosed by gabbros of the Duluth Complex (French, 1968). The western end consists of weathered taconite in which most of the silica is retained. The oxidized western taconite was also mined, and was beneficiated to an economic grade by removing the silica by washing and gravity methods. On the Marquette Range a pelletizing process has been developed for hard jasper-hematites, but here the iron is concentrated by flotation or flocculation and flotation.

The Precambrian stratigraphy of the Lake Superior region varies considerably from one district to another. Probably the most comprehensive correlations were made by Leith and his coworkers in 1935; a modified version of their correlation chart is reproduced in Table 13-2. The great abundance of slate or argillite associated with the iron formation is a conspicuous feature; many of these rocks are lithologically similar to one another and are not easy to distinguish in the field. They are also intricately deformed in some places. Most of the slates and argillites are uniformly gray or black, but a few show gradational bedding and contain appreciable amounts of finely divided carbon and pyrite. In the Menominee Range, a carbonaceous pyrite slate ignites spontaneously when exposed to the air, and for this reason is avoided in the mines wherever possible.

Volcanic materials are widely distributed, particularly in the Huronian and Keweenawan series. Greenstones—mildly metamorphosed calc-alkaline volcanic units—are the most abundant igneous rocks, though intrusive bodies are also common. Many mafic dikes are present, and in the Gogebic Range the best "natural" ores were concentrated along these intrusives which acted as guides to oxidizing surface waters. The Duluth Gabbro borders the northwestern shore of Lake Superior and cuts off the eastern extension of the Mesabi Range. Granitic rocks (Giants Range granite) are also widespread north of the Mesabi Range. Many of the slates described in the last paragraph have bulk compositions which plot along calc-alkaline volcanic rock trends in AFM or QFM diagrams, and are probably metamorphosed volcaniclastic materials.

The stratigraphic succession which includes the Mesabi ores also bears out the facies concept. Examination of Figures 13-25 to 13-27 tells the story well, and relates the stratigraphy to the ore types. Figure 13-26 shows the stratigraphy of the 200-meter-thick Biwabik Formation, the Mesabi Range's only horizon. Figure 13-27 illustrates its regional correlations and economic significance. The entire section is iron formation, except for the approximately 10 meters of ferruginous slate at the base of the lower slaty unit—virtually the entire section is called taconite. The stratigraphic section of Figure 13-26 is given as it would appear in the drill hole of Figure 13-27, which shows how the units conform to facies concepts. A progression from shore to deeper water trends from high-energy clastic quartzites through fine-grained argillites or chert-magnetite chemical sediments to ultrafine-grained iron silicate rocks of the lower slaty. Even BIFs are involved in this progression from clastic to chemical. The figure also shows how transgression and regression analysis is useful in exploration—the lower and upper cherts are actually facies-continuous through time. The right-hand third of Figure 13-27 depicts the stratigraphy of the outcrop area of the Mesabi. Figure 13-25 relates the units of the Biwabik Formation as the stratigrapher would describe them (lower cherty, and so on) to their equivalents as the mine geologists and mining engineers would get to know them (lower cherty taconite, and so on), and shows how the now exhausted or essentially defunct "natural" ores were related through oxidation and silica leaching to the taconites.

The evaluation of taconite ores is complex and basically geological and mineralogical (Ohle, 1972). Grade is of course important, but not so much total iron or soluble iron as magnetic iron, determined by the Davis Tube method of pouring a slurry through a coil and measuring the amount of magnetite retained. Mineralogy is vital (Figure 13-15) with high magnetite the prime consideration, and grain size is critical, with 100-to-200-mesh liberation of magnetite most desirable. The waste component is also evaluated: more than 6% silica is undesirable, but less than 3% SiO_2 prevents slag formation in blast furnaces and is likewise undesirable. Thus gangue mineralogy, metamorphic rank, and the nature of intergrowth of minerals all come into the evaluation. Chemically the low levels of sodium and potassium in taconites are excellent because those elements are anathema to

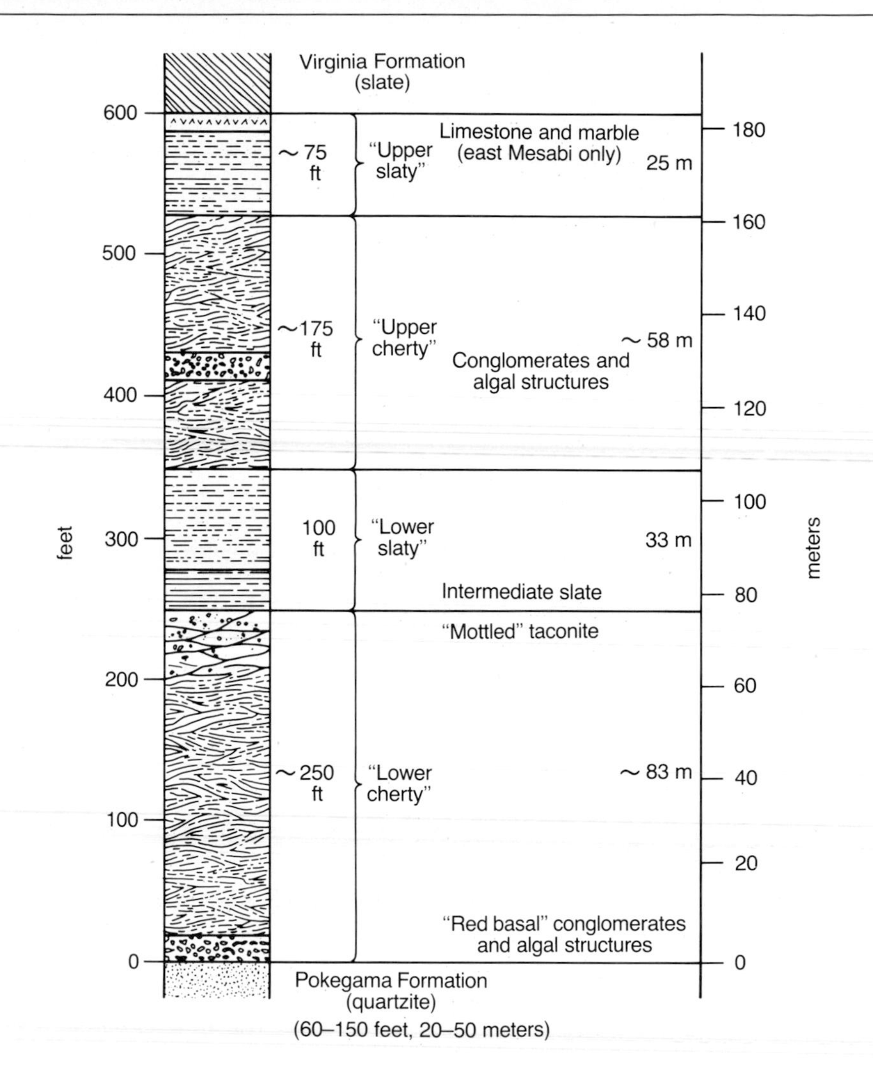

Figure 13-26. Stratigraphic section of the Biwabik Formation as the drill in Figure 13-27 would see it. See also Figures 13-25 and 13-27. (*From* French, 1963.)

blast furnace linings; more than 0.7% Ti produces a pasty slag and poor separation, and phosphorus affects the steel-making properties of the ore. Nongeologic factors include transportation costs, infrastructure costs, and of course overall energy costs per ton of finished product.

Natural ores are still important in many parts of the world, and will typically be mined until their subjacent or laterally adjacent taconites become more economic. The Carajas deposits in Brazil, discovered in 1967,

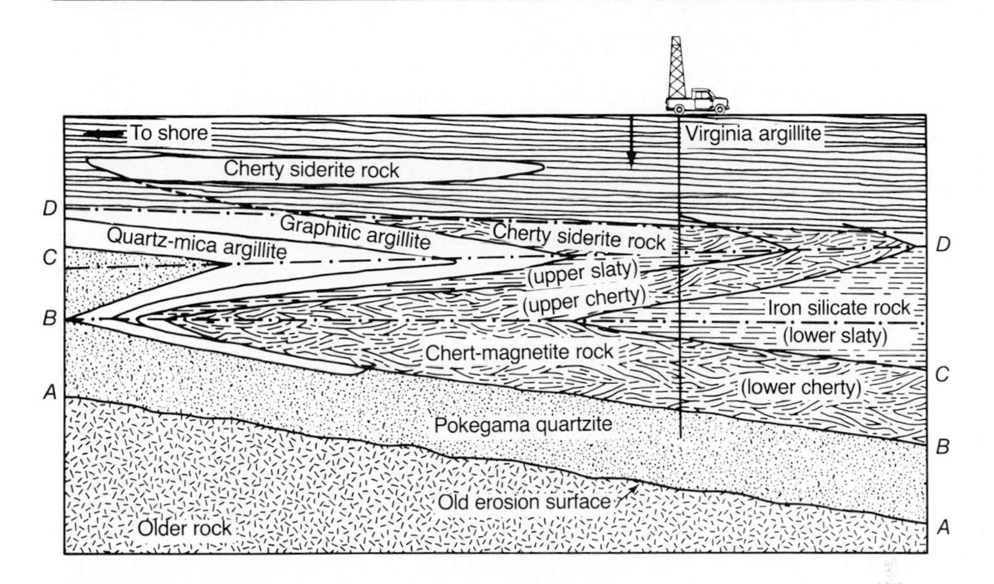

Figure 13-27. Reconstruction of depositional cycles of Biwabik Formation. *A*/*B* first transgression; *B*–*C* first regression; *C*–*D* second transgression. Arrow at top right indicates approximate position of present outcrop of western Mesabi Range. The lines *AA*, *BB*, *CC*, and *DD* are moments in time. The drill hole is figurative and refers to Figure 13-26. See also Figure 13-25. (*From* White, 1954.)

contain 16 billion tons of material containing 66.5% Fe, 2.2% Al_2O_3 + SiO_2, and 1.1% P_2O_5. Pure hematite contains 70% Fe; these ores are described as "hard, semihard, mill, and soft," terms similar to those used for years in the Mesabi. Material called *canga* is also abundant at Carajas and elsewhere in Brazil. Canga is defined as a ferruginous breccia or conglomerate composed of fragments of hematite and itabirite cemented by limonite or hematite, and sometimes by other lateritic constituents. The term originated in Brazil and is widely used for iron-rich laterites that usually accumulate over taconites, itabirites, or in boulder piles at the bases of slopes. Silica is leached from many cangas, and where the silica is entirely removed and the iron oxides are recrystallized and compacted, hard blue hematite masses may result. Many districts underlain by iron formation contain pods, lenses, and layers of high-grade hard hematitic ore formed by the leaching of silica resulting from weathering. The hard blue hematite in Gabon may be directly related to the transformation of canga by meteoric waters (Park, 1959). In many places where the underlying rock is siliceous itabirite or taconite, only the high-grade "natural ore" bodies may now be mined profitably.

▸ EXHALITE GOLD DEPOSITS

According to many economic geologists it is but a short conceptual step from banded iron formations to many of the world's large gold deposits. As just considered, most of the banded iron formations appear to be chemical sediments resulting from the issuance of metal-rich hot springs associated with seafloor volcanism and of a type now known to be issuing from mid-oceanic rise structures. Many of these iron formations, particularly those of the Algoman type, contain small amounts of gold. A considerable exploration boom that developed in the 1970s continues into the 1980s, when it was realized that gold may occur in oxide, carbonate, or sulfide facies iron formations. These rocks may be mined essentially as they were deposited, as at the Kerr-Addison mine near Larder Lake, Ontario, or the gold may have been mobilized with quartz, carbonates, and sulfides to form much richer structurally controlled quartz-carbonate-gold-arsenopyrite veins, as at Kirkland Lake, Ontario. Such veins are typical within, or within 1 km of, the presumed source rocks. Many ores once thought to be of hydrothermal replacement origin, especially the supposed hypothermal ores of the Homestake, South Dakota, type, are now considered by many geologists to have been generated as gold-rich, iron-rich sediments related to volcanism. Iron formation at Itabiri, Brazil, was first mined locally as a gold ore.

Again, credit goes to Oftedahl (1958) for bringing attention to the place of ferruginous sediments in ore-forming systems, but among the early proponents of the idea were Park (1942), Ridler (1970, 1976), and Sawkins and Rye (1974). Ridler noticed that most of the banded iron formation units in the Abitibi Greenstone Belt in Canada contain anomalously high amounts of gold. Figure 13-28 shows two histograms of gold abundances in such rocks. The Kaminak Group materials show "anomalous" gold values in 44% of the samples, the Boston iron formation samples in 67%. In researching the idea of gold in exhalites, Ridler (1976) noted that many deposits now considered to be of exhalative origin, such as the Horne mine at Noranda, contain significant gold, but he noted that the presence of gold in some exhalative units does not prove ferruginous exhalites necessarily to be characterized by high gold contents. As a test, he sampled random iron formations in the relatively unexplored Rankin-Ennadi Inlet greenstone belt and produced the data of the Kaminak Group in Figure 13-28. He exclaimed that 4% of the banded iron formation samples from the test area contained gold values that were either near or within conventional economic gold mine limits, and that the concept of gold's chemical participation in the Algoman banded iron formation was demonstrated.

In light of these discoveries, it is all the more remarkable that some lithologic discontinuities that might have been called major faults in many districts were actually called "breaks" by early day geologists. An excellent example is the Larder Lake Break in eastern Ontario. There exists a band

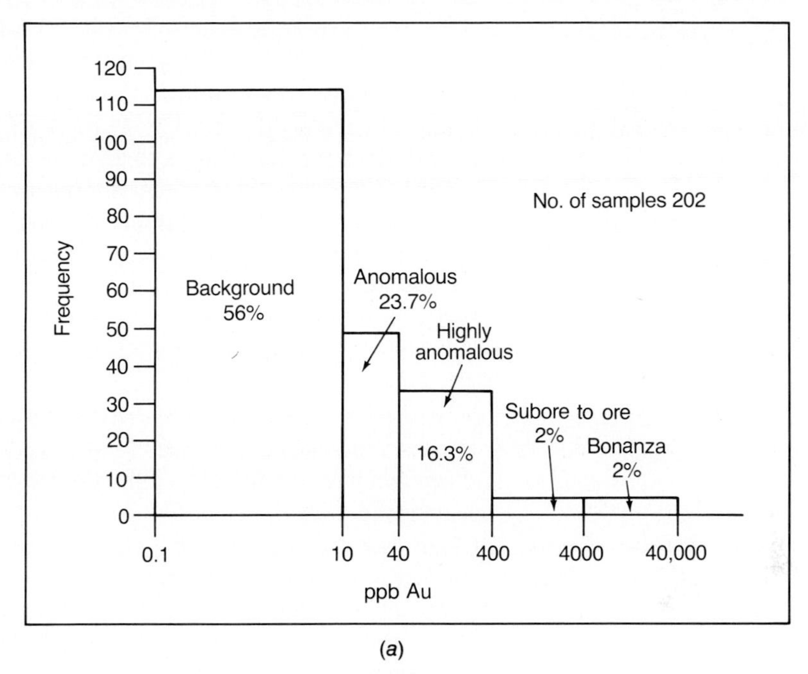

(*a*)

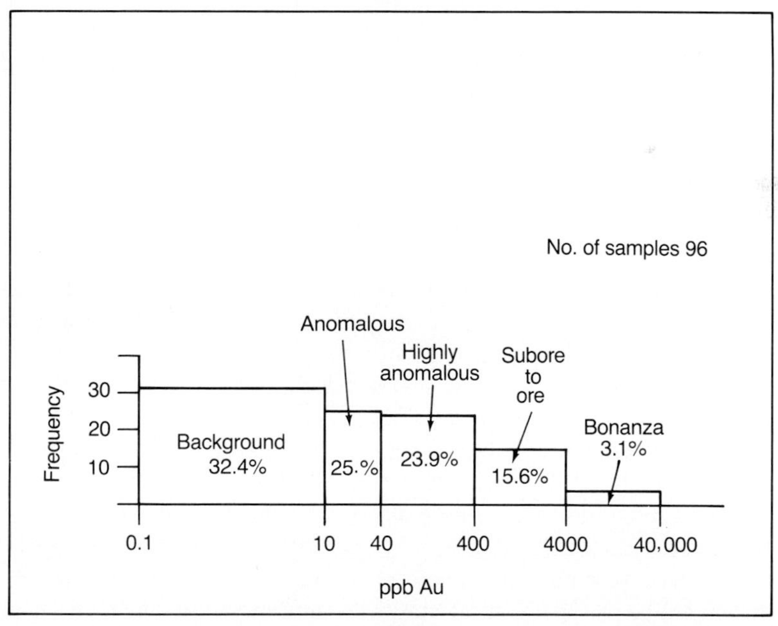

(*b*)

Figure 13-28. Histograms showing amounts of gold in ferruginous exhalite sediments of (*a*) the Kaminak Group on the west shores of Hudson Bay and (*b*) the Boston Iron Formation near Kirkland Lake, Ontario. Note the logarithmic scale. 40,000 ppb = 40 ppm = 1.28 troy oz per ton. (*From* Ridler, 1976.)

of what appear to be sedimentary rocks extending from east of Noranda, Quebec, westward into Ontario (Figure 13-1) and well beyond the city of Kirkland Lake, a mining town whose shafts were sunk more than 1 km along nearly vertical quartz-carbonate-native gold veins. The band of sediments appears stratigraphically discontinuous with the volcanic rocks on either side of it, and major faulting was long suspected of having disrupted whatever sequence existed earlier. It is now claimed by many district geologists that what might have been perceived as fault displacement is actually a true sedimentologic nonconformity; that auriferous, ferruginous banded sediments of oxide, carbonate, and sulfide facies were deposited on an Archean basin floor of felsic volcanic rocks and then buried during later volcanism (Figure 13-2 and 13-22). Much of the gold remained in situ and constitutes modern gold ore grade, as Figure 13-28 suggests. In some places structural dilation coupled with greenschist metamorphism caused local redistribution of gold, quartz, and carbonates from the sedimentary rocks into veins, veins which later became the targets of active mining. Some areas along the belt of apparent sediments contain what is called *flow ore* material, as at Thompson-Bousquet in Quebec and in parts of the Kerr-Addison mine near Larder Lake, Ontario. Flow ore is siliceous sinterlike gold-bearing material which may represent the actual vents from which exhalation occurred. All these relationships appear to apply in eastern and western Canada, the Appalachian United States, Brazil, southern Africa, western Australia, Zimbabwe, and the U.S.S.R. They are perhaps more general than has been realized. Fripp (1976) noted that stratabound mineral deposits, particularly of gold and associated sulfides, are distributed within beds of banded iron formation in the Archean greenstone belts of Zimbabwe. In the following paragraph, paraphrased from his abstract, Fripp notes that at the Vubachikwe mine in Zimbabwe the orebodies are in several thin beds of banded iron formation which are interlayered with mafic and felsic water-lain tuffs. These rocks form part of a mafic-ultramafic volcanic assemblage characteristic of the Sebakwian Group and are overlain by basaltic and andesitic lavas and tuffs characteristic of the younger Bulawayan Group. The orebodies are stratiform and are confined to beds of sulfide and mixed sulfide-carbonate facies banded and nonbanded iron formation, which consist of layers alternately rich in chert, arsenopyrite, pyrrhotite, and ankerite. The micrometer-size gold occurs in arsenopyrite grains, and the average grade of 11 ppm constitutes ore. Petrographic and structural evidence indicates that the gold and sulfides were deposited before metamorphism and before deformative events in the area. Fripp noted in all the deposits he studied that the gold, the sulfides, and the iron formation were consanguineous. Experimental evidence on gold solubility and data from active fumarolic-stage geothermal areas support the view that thermal brines are capable of transporting significant quantities of gold and other metals to the surface as soluble complex ions. A volcanogenic model for the origin of banded iron formation, the sulfides, and the gold is favored. Such deposits are considered to represent submarine chemical precipitates deposited from solutions extruded from subaqueous (and possibly subaerial) active fumaroles. He further suggested that circulating fluids might leach gold from banded iron formation units shortly after deposition, relocating it either in nearby veins or reissuing it onto the seafloor. In any event, Fripp recognized the relationship of gold and banded iron formations, as did Fryer et al. (1979) and Kerrick and Fryer (1979) at the Dome mine in the Timmins district, Ontario (Chapter 19).

Although the discussions cited above stress Archean phenomena, the association carried into the Proterozoic. Exhalite units interlaminated in the Proterozoic Yavapai greenschist belt of Arizona include abundant examples of sulfide facies base-metal ores (Anderson and Guilbert, 1978) and many siliceous, cherty, ferruginous sediments, some of which contain gold. Swan, Hausen, and Newell (1982) described an occurrence of quartz-gold material in sulfide-facies banded iron formation, which grades laterally into carbonate facies and ultimately onlaps a small rhyolite dome 6 km to the north near Mayer, Arizona.

Although many of these gold ores appear best explained as exhalites, there are still many unknowns concerning both the process and the geology. Altered flows, some showing relict-spinifex textures, are stratigraphically close to some layers identified as exhalites, and the widespread presence of a green chromian mica called fuchsite in exhalites is more a problem to the proponents of exhalation than to those who suggest that many "exhalites" may be talcose, carbonated ultramafic flows.

Homestake Gold District, South Dakota

A development in the systematics of gold geology was the suggestion in 1970 that the ores that had been called the "Homestake type" are not hypothermal hydrothermal replacement deposits, but that they stem from the original deposition of a gold-rich carbonate facies iron formation. Economic geologists have long remarked on the geologic similarities of the Homestake deposit at Lead, South Dakota, the United States' most productive gold district, the Champion lodes of the Kolar district in India, the Morro Velho mines in Brazil, and the recently discovered and developed Contwoyto Lake area (Figure 19-1) in Northwest Territories, Canada. Their common denominators are that each deposit occurs in a terrane of regionally metamorphosed folds of Archean greenschist rocks, that each involves relatively low but uniform grade in gold, that wall-rock alteration is unrecognized, and that the host rock is an otherwise unusual ferruginous carbonate-rich stratum of cummingtonite schist. Persistent, uniform grades, lack of alteration, and the presence of minerals like pyrrhotite and arsenopyrite once thought to form only at high temperatures and pressures prompted early workers to place these deposits in Lindgren's hypothermal category. But since the work of Rye (1972) and Sawkins and Rye (1974), and with the maturity of volcanogenic models of ferruginous-auriferous exhalite rocks, the Homestake-type deposits have been reassigned to an exhalation-related class of ore deposits originally formed at low temperatures and seafloor pressures.

The Homestake ores were discovered in 1874 and have been productive nearly continuously since 1876; the Kolar and Morro Velho deposits are equally long-lived. A 1980 announcement by IU International heralds the development at Contwoyto Lake of a minimum of 2.7 million tons of similar ore above 200 meters depth. It averages 0.38 troy oz (12 grams) per ton of gold; Homestake's average over the years has been almost exactly the same 12 grams per ton; economics permit present production to be at about 6 grams.

The Homestake mine is in the northern Black Hills of western South Dakota. The district is the largest producer of gold and gold ore in the western hemisphere and is also one of the deepest mines, operating at depths as much as 2.5 km below the surface. The district has been described many times, and its geology is well known (Paige, 1924; McLaughlin, 1931; Gustafson, 1933; Wright, 1937; Noble, 1950; Slaughter, 1968; Sawkins and Rye, 1974; Rye, Doe and Delavaux, 1974).

The area is underlain by a sequence of schists and gneisses of Precambrian age (Figure 13-29), overlain unconformably by Lower Paleozoic and Tertiary clastic sediments. The metamorphosed units were predominantly shales, silts, and probably tuffs and other volcanic rocks, locally with layers and lenses of quartzite and other materials. Original lithologic features have been almost obliterated by regional metamorphism, but they appear to have been eugeosynclinal or shelf greenstones and graywackes. These host rocks are presumably volcanic in origin, and are pre-2.5-billion-year Archean in age. Bayley and James (1973) classify them as Algoma-type rocks. In the basal footwall Poorman Formation, the bottom of which is not exposed, ankerite is the dominant constituent after quartz. The rock is a finely laminated quartz-ankerite-illite-graphite-hematite-magnetite phyllite that may have been of dacitic composition. The Homestake Formation itself is the only unit of economic interest. It overlies the Poorman and consisted of a high-iron low-magnesium carbonate rock called *sideroplesite,* with abundant quartz and minor sulfides to be listed below. Biotite-to-garnet-rank metamorphism has converted it to a rock called *cummingtonite schist,* the rock unique to most of the gold districts listed earlier, which consists of the high-iron low-magnesium amphibole cummingtonite, ankerite, garnet, biotite, and graphite, with pyrite, gold, auriferous arsenopyrite, magnetite, and sphalerite. The Homestake Formation is of special significance as virtually all of the ore in the mine has been recovered from it; none has been found in the Poorman, and only traces are known in the overlying Ellison Formation. Where less metamorphosed, the Homestake Formation still consists of fine-grained sideroplesite and lenses and thin layers of white quartz that are considered to have been originally chert in an iron formation con-

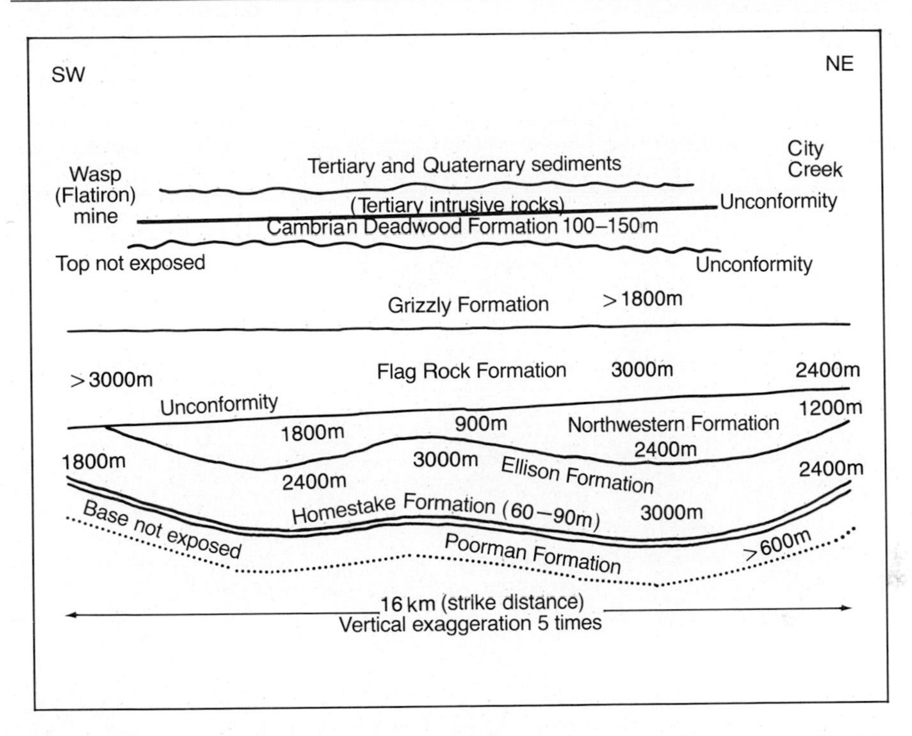

Figure 13-29. Stratigraphic column of Precambrian and younger rocks in the Lead district, South Dakota. The Homestake Formation contains all of the Precambrian-age gold; its undeformed thickness is more like 20 meters than 60 to 100 meters as shown. The beds are restored to their nearly prefolded positions. (*After* Noble and Harder, 1948.)

figuration. The assemblage is identical to what Fripp (1976) might have predicted from his Zimbabwean work involving Archean fumarolic discharges enriched in Au-As-Fe-CO_2-S-SiO_2 in far less metamorphosed terranes. The 1-to-2-km-thick Ellison Formation above consists of more felsic rocks, phyllites and quartzites of quartz-illite-biotite-ankerite-tourmaline-chlorite. Were these rocks rhyolites or rhyolite pyroclastics? Precambrian intrusive rocks, thought to have been originally gabbroic in character, are now metamorphosed to amphibolites. The sequence of Precambrian rocks is estimated to be about 7 km thick, but deformation has been so intense that thickness cannot be determined with certainty. The Precambrian section is cut by intrusive rocks of Tertiary age that consist of granites and rhyolitic, syenitic, and monzonitic porphyries.

Most outcrops of the Homestake Formation are red, reddish brown, or dark brown due to the presence of iron, but greenish brown to green colors may be caused by chlorite. Pods of light gray to white crystalline quartz are common. Lesser amounts of biotite, chlorite, magnetite, and a little graphite are recognized. Banding resulting from concentrations of

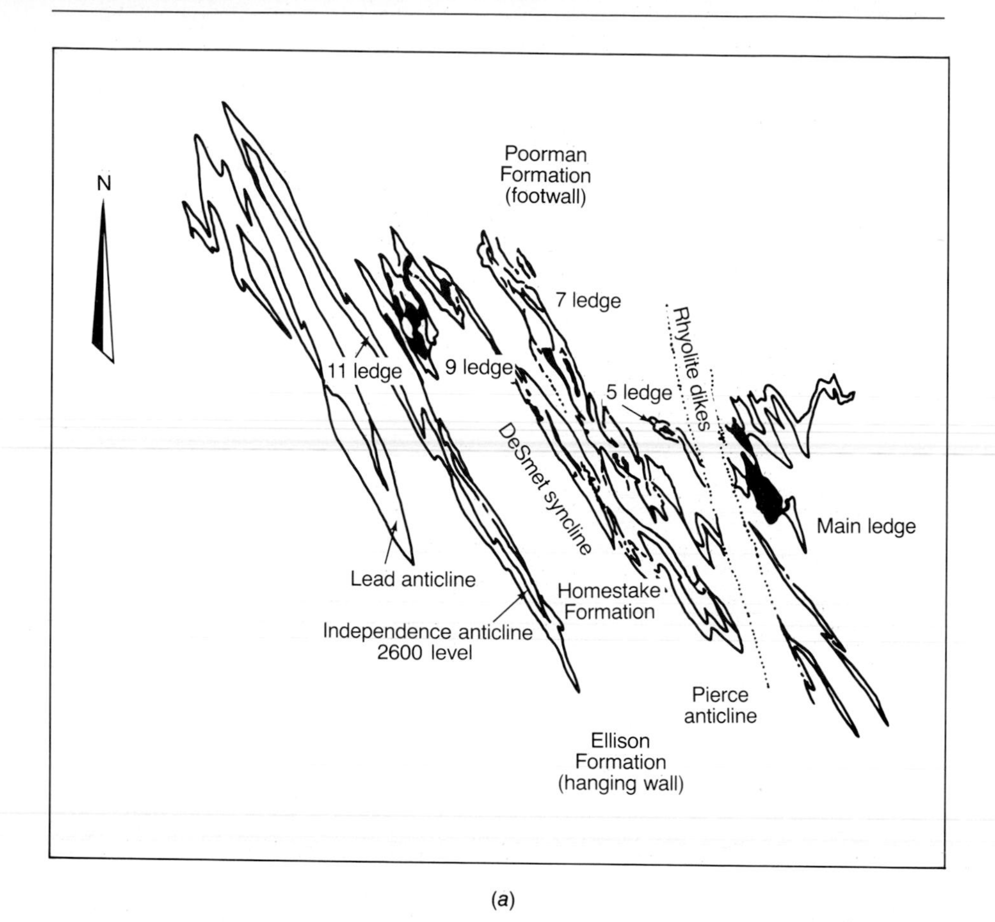

(*a*)

Figure 13-30. Outline of the Homestake Formation on (*a*) the 2600 level and (*b*) (*facing page*) the 4100 level of the Main Orebody of the Homestake mine in an area slightly larger than the rectangle on Figure 13-31. The contortions of the folds bring about a marked foreshortening of total strike length of the Homestake Formation. Ore is solid black, dikes are dotted. (*From* Slaughter, 1968.)

magnetite and graphite is seen at a few places. The Homestake Formation was originally on the order of 20 meters thick, but it is now so greatly distorted that locally it is many times the normal thickness or it may be missing entirely (Figures 13-29 and 13-30).

The dominant structure of the region is that a broad dome of contorted layers of rocks, with the Cutting Stock of Tertiary age occupying a central position (Slaughter, 1968). The layered Precambrian rocks that comprise this dome (Figure 13-29) are highly deformed into a sequence of anticlines and synclines (Figures 13-30 to 13-32). Southwest of the Homestake mine, one of the best developed structures is the complex Poorman anticline (Figure 13-31), the axial plane of which strikes a little west of north and dips

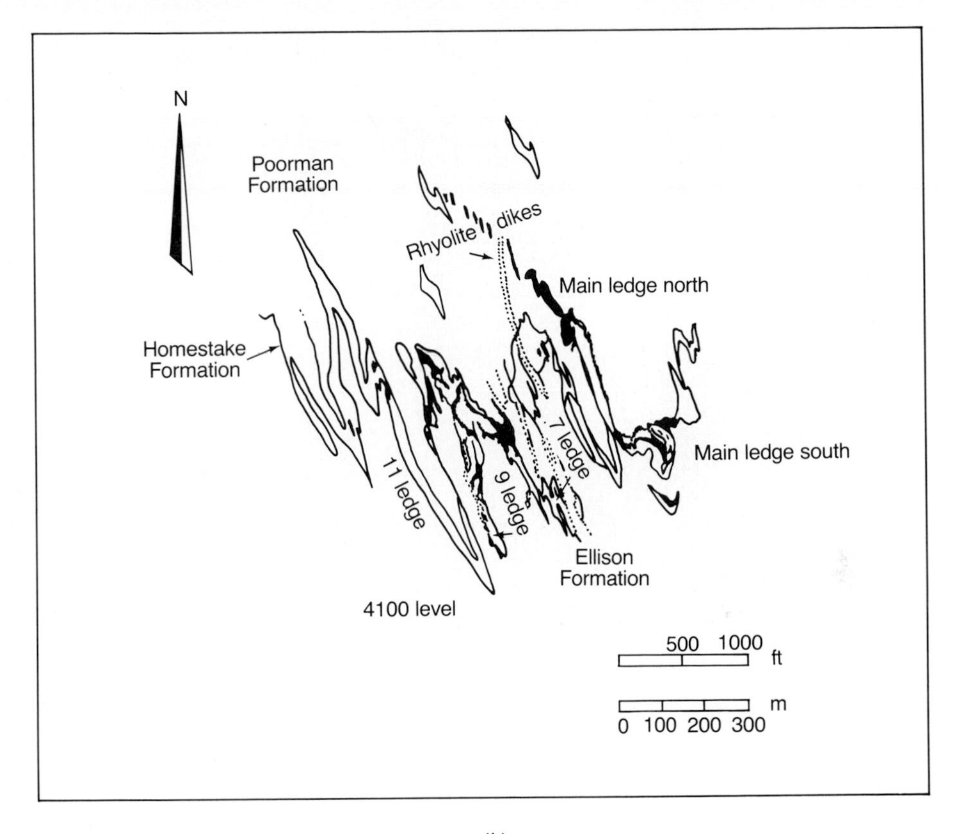

(*b*)

east, a geometry repeated in the ore-bearing Pierce anticline (Figure 13-32) and many others. The axial line plunges about 25° southeast. Many lesser folds are recognized, both east and west of the Poorman anticline. Precambrian rocks around the Homestake mine show the effects of extreme isoclinal folding on which has been superimposed an apparent shear folding that crosses the isoclinal planes at small angles (Slaughter, 1968; Figure 13-30). Many of the isoclinal folds are long and narrow, and considerable slipping has taken place along closely spaced shear planes. The appearance of apparently warped or refolded folds may result as much from shearing as from cross folding. The older rocks have been intensely metamorphosed, the grade of metamorphism increasing from the southwest toward the intrusive body in the northeast (Figure 13-31; Noble and Harder, 1948). Three metamorphic zones are recognized, a biotite zone, an almandite garnet zone, and a staurolite zone. Amphibolites were developed from the original gabbros (Dodge, 1942). Sideroplesite, the essential mineral of the original Homestake Formation, has been converted to cummingtonite; cummingtonite rank conforms closely to garnet rank metamorphism. Mineralogical changes indicate that potash has migrated from zones of increasing meta-

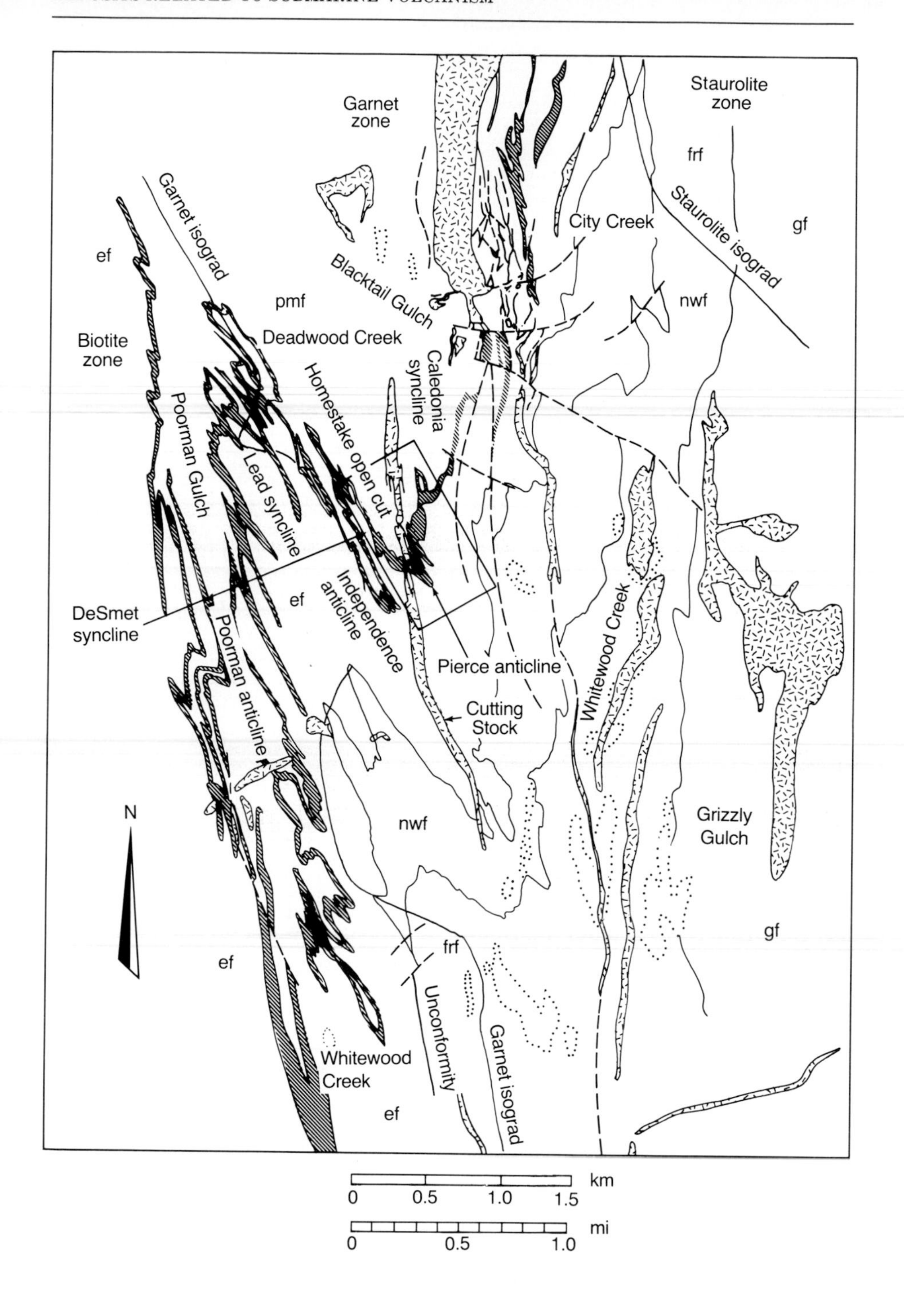
Garnet zone
Staurolite zone
frf
Garnet isograd
City Creek
Staurolite isograd
gf
ef
Blacktail Gulch
nwf
pmf
Deadwood Creek
Biotite zone
Caledonia syncline
Homestake open cut
Poorman Gulch
Lead syncline
ef
Independence anticline
DeSmet syncline
Poorman anticline
Pierce anticline
Whitewood Creek
Cutting Stock
N
nwf
Grizzly Gulch
gf
frf
ef
Unconformity
Garnet isograd
Whitewood Creek
ef
km
0
0.5
1.0
1.5
mi
0
0.5
1.0

Figure 13-31. (*Facing page*) Geologic map of the structure and distribution of Precambrian metasediments and Tertiary intrusive rocks of the Homestake district. Regional metamorphic isograds and zones are shown. The rectangle encloses the approximate area of Figure 13-30. Notice that your eye can track the Homestake Formation along strike through hundreds of anticline-syncline pairs and several major fold structures like the Pierce anticline. Tertiary intrusives are in line stipple. Amphibolites are dotted. Faults are dashed lines. Homestake Formation is cross-ruled. pmf = Poorman Formation; ef = Ellison Formation; nwf = Northwestern Formation; frf = Flag Rock Formation; gf = Grizzly Formation. (*After* Slaughter, 1968.)

morphic grade ahead of the northeastward-advancing isothermal surfaces. The Precambrian rocks are cut at the surface by many Tertiary dikes of rhyolitic, latitic, and syenitic compositions. The dikes are generally less than 3 meters thick and are common in the upper levels of the Homestake mine, though they are scarce or absent on the lower levels. They have controlled scattered gold-bearing veinlets as mentioned below.

The ore at Homestake occurs in spindle-shaped or pipelike bodies that follow fold axes (Figures 13-30 and 13-32). Some ore has been recovered from anticlines, but by far the most has been mined from synclines, gen-

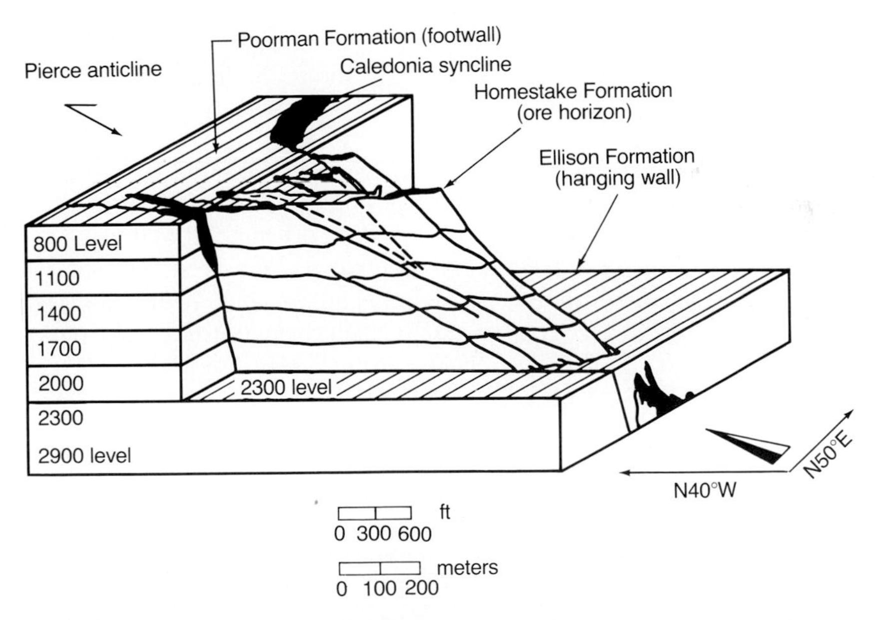

Figure 13-32. Isometric representation of the Pierce anticline in the Homestake district. The walls of the block are the rectangle on Figure 13-31. Note the thickening of the Homestake Formation at troughs and crests of folds, and the thinning along the limbs. Fold noses thus become prime mining targets. They are also slightly richer in gold, as explained in the text. Since this diagram was drawn by Noble (1950), mining has progressed down the plunge of the folds to below the 5000 level. A rhyolite dike that cuts the ore is not shown. See also Figures 13-29, 13-30, and 13-31.

erally not far from their axial areas. Replacement textures are common. Localization of the gold values is believed due both to induced permeability in the schists in fold axial lines and to the commonly observed propensity of plastic materials to thin on fold limbs and thicken in their axes. At least four observations require not only simple plastic deformational thickening but also preferential local mobilization and redeposition of sulfur and gold. First, gold assay values are slightly higher near axial planes; simple "squeezing" should not involve grade enhancement. Second, the gold occurs in discontinuous gold-quartz and gold-arsenopyrite-quartz stringers that are texturally late. Noble mapped and identified four structurally and temporally distinct stages of mineralization (Noble, 1950; Slaughter, 1968): (1) quartz, chlorite, and auriferous arsenopyrite; (2) quartz, ankerite, and pyrrhotite; (3) pyrrhotite, and (4) (after intrusion of Tertiary dikes) pyrite and calcite. The same assemblage and association occurs over more than 2 km vertically and over several square kilometers in plan. Gold is closely associated as blebs and fracture fillings in the arsenopyrite of stage 1, and is thus stratabound but not necessarily stratiform. Quartz-chlorite-arsenopyrite replaces minerals in the cummingtonite schist, but remains conformable to it. Third, Rye and Rye (1974) followed Rye, Doe, and Delavaux's (1974) demonstration that lead isotope relations demand an ultimate Precambrian age for the mineralization with the conclusion that ^{32}S is enriched in sulfides in the axial nodes with respect to sulfides in the limbs, a shift in $^{34/32}S$ ratios consistent with preferred migration of the lighter isotope along metamorphically derived gradients. Finally, stable isotope studies indicate a sedimentary origin of the sulfur in the deposit; most of the sulfur is considered to have been derived originally from seawater sulfate. The sulfur and lead isotope data suggest that the gold and other constituents of the ore deposit were indigenous to the Homestake Formation and were probably of syngenetic origin in a ferruginous carbonate facies sulfidic sediment. Redistribution of susceptible elements (including gold) from the original syngenetic carbonate facies sideroplesite iron formation into ore deposits in dilatant zones along fold axes during metamorphism appears an attractive hypothesis.

The age of the ore bodies has been of particular interest because gold has been recovered in the past not only from the Homestake Formation of Precambrian age, but also in minor amounts from gold-calcite-pyrite veins that cut the Tertiary dikes. This distribution has led some geologists to think that all of the mineralization is Tertiary in age; others have believed that gold was introduced during both Precambrian and Tertiary times; and still others have considered that the gold is Precambrian and that migration during Tertiary intrusion has concentrated some in the late veins. Placer gold also has been reported at the base of the Tertiary sediments, but confirmation of this is lacking. Minor detrital gold is found in the Cambrian Deadwood quartzite and on the weathered unconformable surface below it. The question of the source of the gold in the Tertiary deposits remains unclear, but it may well have accompanied intrusion of the Tertiary rhyolite dikes. Rye and Rye (1974) summarized their conclusions approximately as follows: According to Noble (1950), a Tertiary origin is likely because "(1) the ores have been deposited in a sequence of interrelated stages, and the latest stage is later than the Tertiary dikes," and (2) "the cross folds in which the Homestake ores are localized were caused by the intrusion of the Tertiary igneous rocks." A strong but missing piece of evidence for a Tertiary origin for the Homestake mineralization would be the areal distribution of demonstrably Tertiary orebodies. Chinn (1970) has shown that the cross folds to which the orebodies are restricted were probably not formed by Tertiary intrusives. Noble pointed out that the first three stages of mineralization as he defined them cannot be directly related to his fourth stage, and only the fourth stage can be shown to be Tertiary.

The lines of evidence for a Precambrian origin, including those of Irving et al. (1904), Gustafson (1933), and Rye (1972), are: (1) fossil placer gold occurs in the overlying stream-channel conglomerate of the Upper Cambrian Deadwood Formation; (2) the ore is restricted to one stratigraphic horizon; (3) the textures observed in the ores can be interpreted as having a metamorphic origin; (4) similar mineralization is recognized many kilometers away where Tertiary activity is conspicuously absent; and (5) the Homestake mineralization is similar to mineralization in auriferous Precambrian iron-rich formations all over the world, such as Morro Velho of Brazil and the Kirkland-Larder Lakes area of Canada. Rye and Rye further suggest, as many geologists affirm, that

> . . . *the genetic model for the Homestake gold ores (most) in keeping with the geologic data and observed isotope systematics is one in which gold, arsenic, sulfur, and silica were introduced into the depositional environment by thermal spring processes, together with iron, magnesium, and carbon during accumulation of the original sedimentary unit. Later, the syngenetic gold, arsenic, sulfur, carbon, and silica migrated into dilatant zones formed during metamorphism and complex folding of the Homestake formation and adjacent units, forming the mineral assemblages emplaced prior to Tertiary rhyolite dikes. Tertiary plutonism produced hydrothermal fluids which redistributed some of the Precambrian ore components, forming the Tertiary assemblages recognized in the Homestake mine and the overlying Tertiary orebodies.*

The original exhalite was almost certainly a massive unbanded unit, such as that shown in Figure 19-13, not the finely laminated oxide facies banded iron formations of Figures 13-13 and 13-14.

Homestake Mining Company geologists and many others engaged in regional and district exploration studies are recognizing the fact that Archean sediments, which might contain similar syngenetic units, are known to crop out over hundreds of thousands of square kilometers of Precambrian terrain in western and northwestern North America (Goodwin, 1973) as well as in Brazil, Asia, and Australia. They also note that exposures of Archean rocks through the erosional window of the Black Hills are in proportion to a porthole in the side of a battleship; how many Homestakes are shallowly covered under Great Plains sediments and rocks analogous to them elsewhere? As noted some years ago (Sawkins and Rye, 1974), this relationship between gold, sulfides, and the carbonate facies of early Precambrian iron formations should provide an important key to the exploration for gold deposits in shield areas.

It was mentioned early in this chapter that the so-called volcanogenic deposit group includes the classic Cu-Zn-Ag, Cu-Pb-Zn, and Fe-Au ores and a puzzling group of almost monometallic ores. Among those are the antimony deposits of the Murchison Greenstone Belt, Republic of South Africa, which are almost monomineralic stibnite, and the mercury ores of Almadén. Although a great deal remains to be learned about Almadén, Spanish geologists are rapidly deciphering many aspects of that district. It is described here as an example of a compositional "end member" type of massive sulfide deposit.

▸ OTHER DEPOSITS RELATED TO SUBMARINE VOLCANISM

Almadén Mercury Deposits, Spain

The Almadén mine and district about 200 km southwest of Madrid in the western part of the Ciudad Real province (Figure 13-33) is the world's richest mercury mine. It has been so for as long as mercury or cinnabar has been sought, and still lists enormous reserves. The deposits were well known to ancient Moorish and Arabic peoples, and to the Romans who used the cinnabar as a paint pigment. According to Bennett (1948), the cinnabar was used as far back as the fourth century B.C., but the tenth century

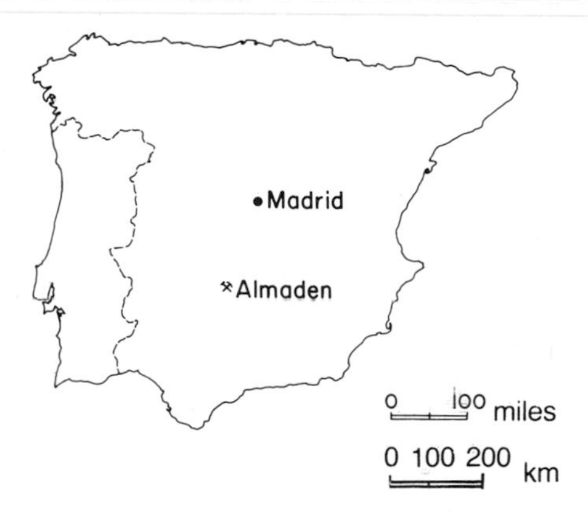

Figure 13-33. Map of Spain and Portugal showing the location of the Almadén district.

Moors were the first to distill the mercury commercially. In recent years the deposits have been a fruitful source of income to the Spanish government, which owns and operates the mines (DeKalb, 1921; Hernandez Sampelayo, 1926; Schuette, 1931, Menendez and Puget, 1949; Saupé, 1973; Arnold, Maucher, and Saupé, 1973; Maucher, 1976). Geologic studies and exploration have been greatly accelerated since Francisco Franco's death in 1975, and the Almadén district will continue to earn its world status. The deposit had long been considered to be epithermal, presumably epigenetic, and related to volcanically derived fluids circulating in sedimentary and volcanic rocks, but recent studies—first credited to Saupé—have shown it to belong in this category of deposits related to submarine volcanism and sedimentation.

The rocks near Almadén consist of lower to middle Paleozoic clastic rocks intruded by Silurian-Devonian (Hercynian) diabases and substantially younger silicic porphyries (Almela Samper, 1959; Almela Samper et al., 1962; Maucher, 1976). Basement rocks consisting of Precambrian slates are overlain by a thick preore epicontinental series of Ordovician shales, quartzites, and flysch deposits, including a Lower Ordovician massive crossedded quartzite of the Armorican Formation (Figure 13-34). It dips vertically, is about 300 meters thick, and forms a prominent ridge trending eastwest just south of the town of Almadén. Upper Ordovician to lower Silurian sandstones, slates, and quartzites overlie this ridge-forming quartzite unit, and Silurian graptolitic shales, sandstone, and quartzite follow in sequence. One of these units is a quartzite called the Criadero quartzite, which contains all of the mercury ores. It is only 50 meters thick in the mines area,

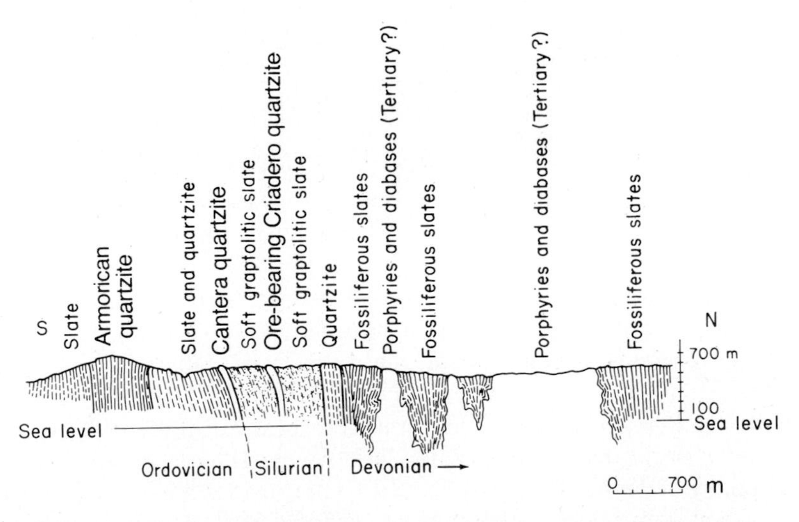

Figure 13-34. Cross section, approximately north-south, through the Almadén district, Spain. Scale does not permit inclusion of volcanic components and the Piedra Frailesca unit associated with the ore-bearing Criadero quartzite. (*From* Almela Samper, 1959.)

and it thins gradually outward. Its immediate footwall is a basaltic flow (Maucher, 1976), which heralded a brief but locally vigorous mafic volcanic event with many centers, throats, flows, and sills. Overlying the mineralized strata are Silurian-Devonian quartzites, sandstones, a few beds of fossiliferous limestone, and interlayered basalt flows. Just north of Almadén, extrusive flows and shallow intrusives of diabase and various types of late porphyries are mapped in the Devonian rocks; one small quartz porphyry mass near the ore deposits seems to intrude the Silurian rocks conformably.

All the strata are either vertical or dip steeply to the north (Figure 13-34). The igneous rocks, although poorly exposed, seem to parallel the sediments, which strike generally east-west. It is therefore difficult to determine whether a particular sill or flow was actually an intrusive or extrusive unit. But it is clear that the Almadén area was a mafic volcanic center at the time that clastic sediments—the quartz grains of what is now the Criadero quartzite—were being washed into a district-scale basin. Volcanism and sedimentation continued through the Silurian-Devonian Hercynian orogeny, with Carboniferous folding, that affected Spain, Portugal, and most of western Europe, and that also saw the deposition of the Rio Tinto, Tharsis, and Aznalcollar massive volcanogenic sulfide deposits in the "Pyrite Belt" in southwestern Spain.

The nature of the igneous activity associated with ore deposition at Almadén has been controversial. Some geologists concluded that there are no igneous rocks near the ore deposits; others insist that there exists an intimate association between the cinnabar and volcanic rocks (DeKalb, 1921; Ransome, 1921; Van der Veen, 1924). Much of the controversy resulted from inadequate field study combined with the fact that the igneous history is complex, with both a variety of occurrence modes and a variety of ages and compositions. It has proven difficult to assign responsibility for mineralization to any one of the many igneous rocks. The present geologic staff sees no relationship between the felsic rocks and the ores, and assigns full responsibility to the diabases.

The problem is further confused by the presence of a strange breccia near the mercury deposits. Locally, this rock is known as Piedra Frailesca (friarlike rock) because of its textural resemblance to the irregular, coarse, open weave of the robes of Franciscan monks. This rock has been at the center of debate over the presence or absence of igneous activity. The Piedra Frailesca is a fine-grained granulated rock containing fragments of basalt, with minor quartzite, slate, and limestone (Almela Samper, 1959). It seemed until recently to be restricted to a single stratigraphic position (Almela Samper and Febrel, 1960), but it is now known also to cut across rocks beneath its main stratigraphic level. Near the principal workings at Almadén, the Piedra Frailesca forms a cone or wedge-shaped structure some 300 meters wide at the surface but absent or inconspicuous in the deepest workings. Petrographically the matrix of Piedra Frailesca was reported to be similar to a crushed quartzite (Ransome, 1921), so it has long been considered to be a tectonic breccia. Almela Samper and Febrel (1960) concluded that it is a submarine basaltic tuff, a hyaloclastite formed as a late development of the associated lava flows during and after the deposition of the Criadero quartzite. It contains quartzite fragments and blocks, some of which are mineralized with cinnabar, proving that at least some of the mercury had already been deposited when the Piedra Frailesca formed.

The ore at Almadén consists predominantly of cinnabar, native mercury, and pyrite (Van der Veen, 1924; Raynaud, 1941). Two closely spaced quartzite strata—one a layer, one lens-shaped—contain the orebodies; they dip from 70° north near the surface to nearly vertical at depth (Figures 13-35 and 13-36). The lodes are named, in stratigraphic order from base to top and from south to north, the San Pedro–San Diego, San Francisco, and San Nicolas; they are, respectively, 5 to 7 meters wide, 3 meters wide, and 3.3 meters wide on the intermediate levels, where all of them are about 400–500 meters long (DeKalb, 1921). The "vein" material is sheared, jointed, and brecciated quartzite cemented with quartz, the mercury minerals, and pyrite. The sulfides and native mercury are stratabound (Figure 13-35),

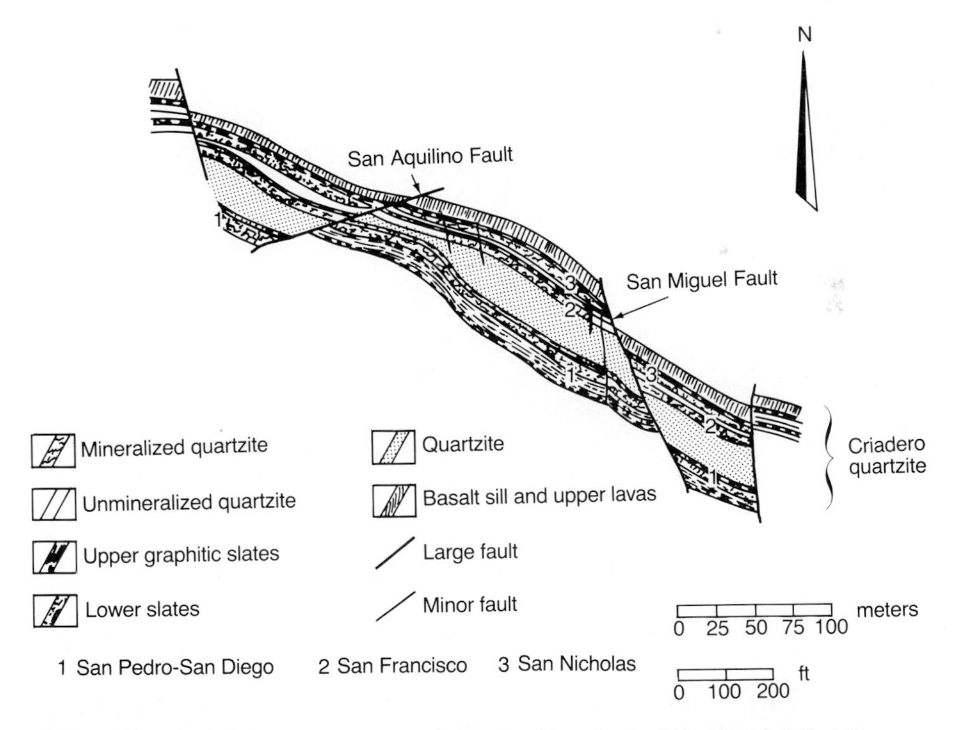

Figure 13-35. Plan view of the 14th level, Almadén mine, Spain. Remembering that the sediments are nearly vertically dipping, this plan view is also a cross section of the Criadero quartzite. Numbers 1, 2, and 3 identify the three stratiform ore horizons. The quartzite contains a few beds of shale and slate. Along strike to the west (left) the quartzite and ore are cut off by the Piedra Frailesca. Values increase to the west, decrease to the east. See also Figure 13-36. (*From* Almela Samper et al., 1962.)

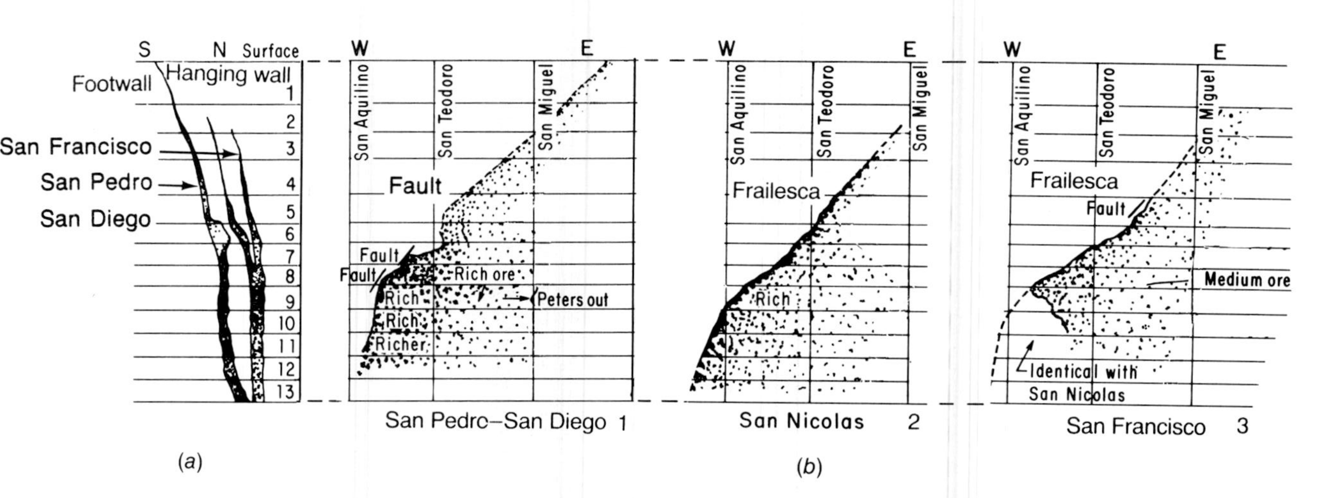

Figure 13-36. (*a*) Cross section and (*b*) long sections of upper levels of the ore horizons at Almadén. The cross section faces west, as does Figure 13-34. The San Pedro–San Diego orebody is near the base of the Criadero quartzite, the San Francisco layer near the top, so the cross section is essentially a cross section of the quartzite. See also Figure 13-35. The three long sections face north. Ore values are in general higher to the west and in upper portions, and diminish gradually to the east. The western limit is a combination of fault surfaces, breccia zones, and sections of Piedra Frailesca, which locally crosscut and terminate the ore. Mining now extends down to the 15 level, where the three horizons are still as distinct as they appear in Figure 13-35. (*After* Schuette, 1931.)

commonly with beads or droplets of the native metal and subhedral to euhedral cinnabar in vugs and pores in the quartzite. The native mercury can often be shaken free of a specimen by a snap of the wrist. Pyrite is widely distributed, even in the adjacent slates. Weathering of the pyritized slate produces conspicuous rusty outcrops that distinguish this slate from nonpyritiferous slate. There are places where mud and mud flow layers in the quartzite are mercury-rich, with mercury in ripple marks and load casts. In beds in the quartzite, the mercury minerals locally appear to replace quartz grains (Figure 13-37), to be intergrown with quartz overgrowths in granular nucleii, and to occupy microveinlets and veinlets. Intergranular fillings of cinnabar and native mercury are abundant. Typical ore now contains 1% Hg, but meter thicknesses of 70 to 80% Hg were encountered. Cinnabar is intimately intergrown with diagenetic framboidal pyrite in quartzites and mudstones (Arnold, Maucher, and Saupé, 1973). Neither gangue nor alteration minerals occur with the ore assemblages, except for pyrite; the ores contain traces of Ni and several hundred ppm Ga, but no consistently detectable Sb, As, Au, or Ag. Boron is absent. Preliminary data indicate that sulfur is isotopically heavy, and probably of sedimentary origin.

Rich zones 1 to 3 meters wide are restricted to the central planes of the lodes, and the values decrease gradually toward the walls. Until 1928 mining was confined to the area between two complex cross faults that interrupt the orebodies on the east and west sides, but underground exploration beyond these faults has proven the extension of ore. In long section, the ore layers were longest at intermediate levels. Deep development shows both length and mercury content of the ore layers to be diminishing.

The following account of how the deposit formed is that of Saupé (1967), the Almadén staff (Soler, 1978), and from observation. They describe the following series of events as consistent with a combined syngenetic volcanic-sedimentary origin, with later redistribution of some of the mercury. First, basaltic sills, dikes, flows, and the Piedra Frailesca, with pieces of these rocks interbedded with the Criadero, indicate that quartzose sediment accumulated in an environment of submarine volcanism, which was now quiescent, now explosive. Maar rings (lateral blowout, or base-surge deposits), the Frailesca, and mapped structures suggest phreatic explosions, quite reasonably with caldera structures like that at McDermitt, Nevada-Oregon, and in the San Juan Mountains of Colorado (Chapter 12). Next, seafloor solutions associated with the basaltic volcanism flowed out into the Criadero sedimentary environment, almost as if into a caldera-form trap. They locally followed submarine paleochannels and primary sedimentologic permeabilities and spread out three times to form the three successive mineralized horizons. There is no ore in the diabases themselves. Finally, continued volcanism included the probably phreatic explosion that emplaced the Piedra Frailesca with its entrained mineralized and unmineralized sedimentary blocks, and a family of postore dikes and sills that locally bake the ore. The Frailesca pipe contains only fragments, ash, and clay, but the sills, dikes, and a pronounced gravity high suggest that magmatism was closely subjacent.

The continued volcanism or the folding events may have driven leaching and redistribution of the important amounts of mercury which are found as veins and veinlets in the now vertical plane of the ore horizons. These veinlets show mercury-depleted envelopes suggesting hydrothermal leaching, show minor flanking silicification, sericitization, and chloritization, and contain minor amounts of the only gangue minerals in the district—quartz, barite, dolomite, pyrite, and the zeolite natrolite.

The Almadén mine is unique among the quicksilver deposits of the world, although its association with volcanism, possible calderas, and either marine or lacustrine submergence starts to suggest a deposit class that

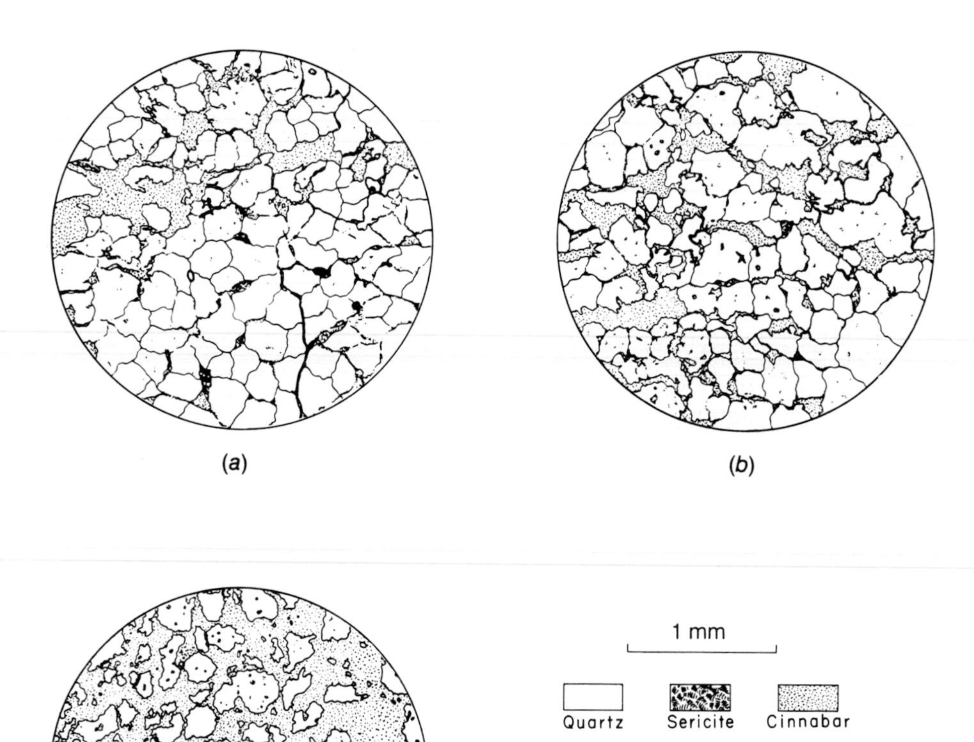

Figure 13-37. *Camera lucida* drawings of the Almadén ore, showing stages in the replacement of quartz grains by interstitial cinnabar. (*a*) Early; (*b*) intermediate; (*c*) advanced. (*From* Ransome, 1921.)

would include the Cenozoic McDermitt, Nevada-Oregon, ores (Speer, 1977) and the major Idria, Yugoslavia, Mesozoic deposits (Mlakar and Drovenik, 1971, in Maucher, 1976). Other volcanic centers should be sought, both in the Almadén district and elsewhere. Curiously, apparently genetically similar deposits associated with ultramafic flows in the Murchison Greenstone Belt in northeastern South Africa contain antimony, but essentially no mercury or arsenic. These Consolidated Murchison mines show no evidence of explosive volcanism, but they are similar in most other ways.

Although mineralization is traceable on the surface near Almadén for more than 20 km with numerous scattered showings of cinnabar, most of the mining has been confined to the immediate area of Almadén. At least eight other volcanic centers have been discovered in the district. Recent drilling at one of these centers, which was plumbed for 70 meters by a broad shaft sunk with ladders, picks, and shovels by Arabic miners 1000 years ago, has revealed another huge mercury orebody. That orebody, Entredicho, contains at least 30,000 metric tons of mercury down to only 100 meters and is a major producer in 1985. Since 1499, the date of the earliest records, the total production from Almadén has been about 250,000 metric tons of mercury. Reserves at Almadén and Entredicho are adequate for at least another century of continuous production.

Bibliography

Almela Samper, A., 1959. Esquema geológico de la zona de Almadén, Ciudad Real. *Inst. Geol. Minero España Bol.* 70:315–330.

——, M. Alvarado, J. Coma, C. Felgueroso, and I. Quintero, 1962. Estudio geológico de la región de Almadén. *Inst. Geol. Minero España Bol.* 73:197–327.

——, and T. Febrel, 1960. La roca frailesca de Almadén: un episodio tobaceo en una formación basáltica del Siluriano superior. Inst. Geol. Minero España Nota Comun. 59, pp. 41–72.

Anderson, C. A., and J. T. Nash, 1972. Geology of the massive sulfide deposits at Jerome, Arizona—a reinterpretation. *Econ. Geol.* 67:845–863.

Anderson, P. A., and J. M. Guilbert, 1978. The Precambrian massive sulfide deposits of Arizona—a distinct metallogenic epoch and province. *IAGOD 5th Quadr. Symp. Proc.* 2:39–48.

Arnold, M., A. Maucher, and F. Saupé, 1973. Diagenetic pyrite and associated sulfides at the Almadén mercury mines, Spain, pp. 7–20 in G. C. Amstutz and A. J. Bernard, Eds., *Ores in Sediments.* New York: Springer, 350 pp.

Bachinski, D. J., 1977. Sulfur isotopic composition of ophiolitic cupriferous iron sulfide deposits, Notre Dame Bay, Newfoundland. *Econ. Geol.* 72:243–257.

Baker, A., III, and R. L. Clayton, 1968. Massive sulfide deposits of the Bagdad district, Yavapai County, Arizona, pp. 1311–1327 in J. D. Ridge, Ed., *Ore Deposits of the United States, 1933/1967*, Graton-Sales Vols. New York: AIME, 1880 pp.

Barghoorn, E. S., and S. A. Tyler, 1965. Microorganisms from the Gunflint Chert. *Science* 147:563–577.

Bayley, R. W., and H. L. James, 1973. Precambrian iron-formations of the United States. *Econ. Geol.* 68:934–959.

Beerstecher, E., Jr., 1954. *Petroleum Microbiology.* New York: Elsevier.

Bennett, E., 1948. Almadén, world's greatest mercury mine. *Min. Metall.* 29:6–9.

Boldy, J., 1968. Geological observations on the Delbridge massive sulfide deposit. *Can. Inst. Min. Metall. Trans.* 71:247–256.

Bostrom, K., and M. N. A. Peterson, 1966. Precipitates from hydrothermal exhalations of the East Pacific Rise. *Econ. Geol.* 61:1258–1265.

Boyum, B. H., and R. C. Hartviksen, 1970. General geology and ore-grade control at the Sherman mine, Temagami, Ontario. *Can. Inst. Min. Metall. Trans.* 73:267–276.

Brathwaite, R. L., 1971. Structure of the Rosebery ore deposit. *Australas. Inst. Min. Metall. Proc.* 241:1–13.

——, 1974. The geology and origin of the Rosebery ore deposit. *Econ. Geol.* 69:1086–1101.

Bryner, L., 1967. Geology of the Barlo mine and vicinity, Dasol, Pangasinan Province, Luzon, Philippine Islands. Phil. Is. Bur. Mines Rept. Invest. 60.

Button, A., 1976. Iron-formation as an end-member in carbonate sedimentary cycles in the Transvaal Supergroup, South Africa. *Econ. Geol.* 71:193–201.

Chinn, W., 1970. Structural and mineralogical studies at the Homestake mine, Lead, South Dakota. Unpub. Ph.D. Dissert., Univ. Calif., Berkeley.

Cloud, P., 1973. Paleoecological significance of the banded iron formation. *Econ. Geol.* 68:1135–1143.

Collins, J. J., 1950. Summary of Kinoshita's Kuroko deposits of Japan. *Econ. Geol.* 45:363–376.

Constantinou, G., and G. J. S. Govett, 1972. Genesis of sulphide deposits, ochre and umber of Cyprus. *Inst. Min. Metall. Trans.* 8:B34–B46.

——, 1973. Geology, geochemistry, and genesis of Cyprus sulfide deposits. *Econ. Geol.* 68:843–858.

Degens, E. T., and D. A. Ross, Eds., 1969. *Hot Brines and Recent Heavy Metal Deposits in the Red Sea.* New York: Springer.

DeKalb, C., 1921. The Almadén quicksilver mine. *Econ. Geol.* 16:301–312.

de Rosen-Spence, A. F., 1969. Genèse des roches à cordièrite-anthophyllite des gisements cupro-zincifères de la région de Rouyn-Noranda, Quebec, Canada. *Can. J. Earth Sci.* 6:1339–1345.

Derry, D. R., 1973. Ore deposition and contemporaneous surfaces. *Econ. Geol.* 68:1374–1380.

Dodge, T. A., 1942. Amphibolites of the Lead area, South Dakota. *Geol Soc. Amer. Bull.* 53:561–584.

Dunn, J. A., 1935. The origin of iron ores in Singhbhum, India. *Econ. Geol.* 30:643–654.

——, 1941. The origin of banded hematite ores in India. *Econ. Geol.* 36:355–370.

Dutton, C. E., 1952. Memorandum on iron deposits in the United States of America, in *Symposium sur les Gisements de Fer du Monde,* 19th Int. Geol. Congr., vol. 1.

——, C. F. Park, Jr., and J. R. Balsley, Jr., 1945. General character and succession of tentative divisions in the stratigraphy of the Mineral Hills district. U.S.G.S. Prelim. Rept.

East Pacific Rise Study Group, 1981. Crustal processes of the mid-ocean ridge. *Science* 213:31–40.

Eugster, H. P., and C. I-Ming, 1973. The depositional environments of Precambrian banded iron-formations. *Econ. Geol.* 68:1144–1168.

French, B. M., 1963. Progressive contact metamorphism of the Biwabik iron-formation, Mesabi Range, Minnesota. Univ. Minn. Press Bull. 45.

Frenzel, G., and J. Ottemann, 1967. Eine Sulfid-Paragenese mit kupferhaltigem Zonarpyrit von Nukundamu/Fiji. *Mineral. Deposita* 1:307–316.

Fripp, R. E. P., 1976. Stratabound gold deposits in Archean banded iron formations, Rhodesia. *Econ. Geol.* 71:58–75.

Fryer, B. J., R. Kerrich, R. W. Hutchinson, M. G. Peirce, and D. S. Rogers, 1979. Archean precious-metal hydrothermal systems, Dome mine, Abitibi greenstone belt. I. Patterns of alteration and metal distribution. *Can. J. Earth Sci.* 16:421–439.

Garrels, R. M., E. A. Perry, Jr., and F. T. Mackenzie, 1973. Genesis of Precambrian iron-formations and the development of atmospheric oxygen. *Econ. Geol.* 68:1173–1179.

Gastil, G., and D. M. Knowles, 1960. Geology of the Wabush Lake area, southwestern Labrador and eastern Quebec, Canada. *Geol. Soc. Amer. Bull.* 71:1243–1254.

Gibson, H. L., D. H. Watkinson, and C. D. A. Comba, 1983. Silicification: hydrothermal alteration in an Archean geothermal system within the Amulet rhyolite formation, Noranda, Quebec. *Econ. Geol.* 78:954–971.

Gilmour, P., 1965. The origin of the massive sulfide mineralization in the Noranda district, northwestern Quebec, Canada. *Geol. Assoc. Can. Proc.* 16:63–81.

——, 1971. Stratabound massive pyritic sulfide deposits—a review. *Econ. Geol.* 66:1239–1244.

Goldich, S. S., 1973. Ages of Precambrian banded iron-formations. *Econ. Geol.* 68:1126–1134.

Goodwin, A. M., 1956. Facies relations in the Gunflint iron formation. *Econ. Geol.* 51:565–595.

——, 1965. Mineralized volcanic complexes in the Porcupine Kirkland Lake-Noranda region, Canada. *Econ. Geol.* 60:955–971.

——, 1973. Archean iron formations and tectonic basins of the Canadian Shield. *Econ. Geol.* 68:915–934.

—— and R. H. Ridler, 1970. The Abitibi orogenic belt, pp. 1–30 in A. J. Baer, Ed., *Basins and Geosynclines of the Canadian Shield*, Geol. Surv. Can. Pap. 70-40, 265 p.

Green, G. R., M. Solomon, and J. L. Walshe, 1981. The formation of the volcanic hosted massive sulfide ore deposit at Rosebery, Tasmania. *Econ. Geol.* 76:304–338.

Griffitts, W. R., J. P. Albers, and O. Oner, 1972. Massive sulfide copper deposits of the Ergani-Maden area, southeastern Turkey. *Econ. Geol.* 67:701–716.

Griggs, A. B., 1947. Zinc-lead resources of Japan. Supreme Command Allied Powers Nat. Res. Sec. Rept. 65, pp. 23–26.

Gross, G. A., 1965a. Geology of iron deposits in Canada, vol. 1: General geology and evaluation of iron deposits. Can. Geol. Surv. Econ. Geol. Rept. 22.

——, 1965b. Origin of Precambrian iron formations, discussion. *Econ. Geol.* 60:1063–1065.

——, 1980. A classification of iron formations based on depositional environments. *Can. Mineral.* 18:215–222.

Grout, F. F., and J. F. Wolff, Sr., 1955. The geology of the Cuyuna district, Minnesota, a progress report. Minn. Geol. Surv. Bull. 36.

Gruner, J. W., 1922. The origin of sedimentary iron formations: the Biwabik Formation of the Mesabi Range. *Econ. Geol.* 17:407–460.

——, 1930. Hydrothermal oxidation and leaching experiments: their bearing on the origin of Lake Superior hematite-limonite ores. *Econ. Geol.* 25:697–719; 837–867.

——, 1946. Mineralogy and geology of the taconites and iron ores of the Mesabi Range, Minnesota. St. Paul, Minn.: Iron Range Res. and Rehab. Rept.

Guilbert, J. M., 1981. A plate tectonic-lithotectonic classification of ore deposits, in W. R. Dickinson and W. D. Payne, Eds., Relations of tectonics to ore deposits in the Southern Cordillera. Tucson: *Ariz. Geol. Soc. Dig.* 14:1–10.

Gunderson, J. N., 1960. Lithologic classification of taconite from the type locality. *Econ. Geol.* 55:563–573.

Gustafson, J. K., 1933. Metamorphism and hydrothermal alteration of the Homestake gold-bearing formation. *Econ. Geol.* 28:123–162.

Hall, G., V. M. Cottle, P. B. Rosenhain, and R. R. McGhie, 1953. The lead-zinc deposits of Read-Rosebery and Mount Farrell, in A. B. Edwards, Ed., *Geology of Australian Ore Deposits.* Melbourne: Australas. Inst. Min. Metall.

——, ——, ——, ——, and J. G. Druett, 1965. Lead-zinc ore deposits of Read-Rosebery (Tasmania), in *Geology of Australian Ore Deposits.* 8th Commonw. Min. Metall. Congr., vol. 1.

Harder, E. C., 1919. Iron-depositing bacteria and their geologic relations. U.S.G.S. Prof. Pap. 113.

Heaton, T. H. E., and S. M. F. Sheppard, 1977. Hydrogen and oxygen isotope evidence for sea-water hydrothermal alteration and ore deposition, Troodos Complex, Cyprus, pp. 42–57 in *Volcanic Processes in Ore Genesis.* Inst. Min. Metall. Spec. Pub. 7, 188 pp.

Hernandez Sampelayo, P., 1926. Minas de Almadén. 14th Int. Geol. Congr. Guidebook, Excursion B-1.

Hewett, D. F., 1966. Stratified deposits of the oxides and carbonates of manganese. *Econ. Geol.* 61:431–461.

Holland, H. D., 1973. The oceans: a possible source of iron in iron-formations. *Econ. Geol.* 68:1170–1172.

Horikoshi, E., 1969. Volcanic activity related to the formation of the kuroko-type deposits in the Kosaka district, Japan. *Mineral. Deposita* 4:321–345.

—— and T. Sato, 1970. Volcanic activity and ore deposition in the Kosaka mine, pp. 181–196 in *Volcanism and Ore Genesis,* Watanabe Vol. Tokyo: Tokyo Univ. Press, 448 pp.

Hutchinson, R. W., 1965. Genesis of Canadian massive sulfide deposits reconsidered by comparison to Cyprus deposits. Symposium on Stratabound Sulfides. *Can. Min. Metall. Bull.* 58:286–300.

——, 1979. Evidence of exhalative origin for Tasmanian tin deposits, *Can. Inst. Min. Metall. Bull.*, Aug., 72:90–104.

——, 1980. Massive base metal sulfide deposits as guides to tectonic evolution; pp. 659–684 in D. W. Strangway, Ed., *The Continental Crust and Its Mineral Deposits.* Toronto: Geol. Assoc. Can. Spec. Pap. 20.

——, R. H. Ridler, and G. G. Suffel, 1971. A model for Archean metallogeny. *Can. Inst. Min. Metall. Bull.*, Apr., 64:48–57.

——, and D. L. Searle, 1971. Stratabound pyrite deposits in Cyprus and relations to other sulphide ores, pp. 198–205 in S. Ishihara, Ed., *Geology of Kuroko Deposits.* Soc. Min. Geol. Japan Spec. Issue 3, 435 pp.

Irving, J. D., S. F. Emmons, and T. A. Jaggar, Jr., 1904. Economic resources of the northern Black Hills, USGS Prof. Pap. 26.

Ishihara, S., Ed., 1971. *Geology of Kuroko Deposits*. Soc. Min. Geol. Japan Spec. Issue 3, 435 pp.

Jambor, J. L., 1979. Mineralogical evaluation of proximal-distal features in New Brunswick massive-sulfide deposits. *Can. Mineral.* 17:649–665.

James, H. L., 1954. Sedimentary facies of iron formation. *Econ. Geol.* 49:235–293.

Jenks, W. F., 1966. Some relations between Cenozoic volcanism and ore deposition in northern Japan. *N.Y. Acad. Sci. Trans.* 28:463–474.

——, 1971. Tectonic transport of massive sulfide deposits in submarine volcanic and sedimentary host rocks. *Econ. Geol.* 66:1215–1224.

Jolliffe, A. W., 1955. Geology and iron ores of Steeprock Lake. *Econ. Geol.* 50:373–398.

——, 1966. Stratigraphy of the Steeprock Group, Steeprock Lake, Ontario, in *The Relationship of Mineralization to Precambrian Stratigraphy in Certain Mining Areas of Ontario and Quebec*. Toronto: Geol. Assoc. Canada.

Kerrich, R., and B. J. Fryer, 1979. Archean precious-metal hydrothermal systems, Dome Mine, Abitibi greenstone belt, II REE and oxygen isotope relations. *Can. J. Earth Sci.* 16:440–458.

Kimberly, M. M., 1978. Paleoenvironmental classification of iron formation. *Econ. Geol.* 73:215–229.

King, H. F., 1958. Notes on ore occurrences in highly metamorphosed Pre-Cambrian rock. *Australas. Inst. Min. Metall., Stillwell Anniv. Vol.*, pp. 143–167.

Kinkel, A. R., Jr., 1962. Observations on the pyrite deposits of the Huelva district, Spain, and their relation to volcanism. *Econ. Geol.* 57:1071–1080.

——, 1966. Massive sulfide deposits related to volcanism and possible methods of emplacement. *Econ. Geol.* 61:673–694.

Kinoshita, K., 1924. Colloidal solutions as the mineralizing solutions of the "kuromono" deposits. Tohoku Imper. Univ. Sci. Rept., 3rd Ser. 2:23–30.

——, 1929. On the genesis of the "kuromono" deposits. *15th Int. Geol. Congr., Comptes Rendus*, vol. 2, pp. 454–474.

——, 1931. On the "kuroko" (black ore) deposits. *Jap. J. Geol. Geogr.* 8:281–352.

Klein, C., Jr., 1973. Change in mineral assemblages with metamorphism of some banded Precambrian iron-formations. *Econ. Geol.* 68:1075–1088.

Krumbein, W. C., and R. M. Garrels, 1952. Origin and classification of chemical sediments in terms of pH and oxidation-reduction potentials. *J. Geol.* 60:1–33.

LaBerge, G. L., 1973. Possible biological origin of Precambrian iron-formations. *Econ. Geol.* 68:1098–1109.

Lajoie, J., 1977. Sedimentology: a tool for mapping "mill rock." *Geosci. Can.* 4:119–122.

Lake Superior Iron Ore Assn., 1952. *Lake Superior Iron Ores*, 2nd ed. Cleveland: Lake Superior Iron Ore Assn.

Large, R. L., 1977. Chemical evolution and zonation of massive sulfide deposits in volcanic terrains. *Econ. Geol.* 72:549–572.

Leith, C. K., R. J. Lund, and A. Leith, 1935. Pre-Cambrian rocks of the Lake Superior region. USGS Prof. Pap. 184.

Lepp, H., and S. S. Goldich, 1959. The chemistry and origin of iron formations. *Econ. Geol.* 54:1348–1349.

Margulis, L., 1981. *Symbiosis in Cell Evolution; Life and Its Environment on the Early Earth.* San Francisco: Freeman, 419 pp.

Marsden, R. J., and B. D. Kay, 1980. The distribution, petrology, and genesis of nickel ores at the Juan Complex, Kambalda, Western Australia. *Econ. Geol.* 75:546–565.

Marsden, R. W., 1968. Geology of the iron ores of the Lake Superior region in the United States, pp. 489–506 in J. D. Ridge, Ed., *Ore Deposits of the United States 1933/1967,* Graton-Sales Vols. New York: AIME, 1880 pp.

——, 1983. Pers. Commun.

Matsukuma, T., and E. Horikoshi, 1970. Kuroko deposits in Japan: a review, pp. 153–180 in T. Tatsumi, Ed., *Volcanism and Ore Genesis.* Tokyo: Univ. Tokyo Press, 448 pp.

Maucher, A., 1976. The stratabound cinnabar-stibnite-scheelite deposits, pp. 477–500 in K. H. Wolf, Ed., *Handbook of Stratabound and Stratiform Ore Deposits,* Vol. 7. New York: Elsevier, 656 pp.

McLaughlin, D. H., 1931. The Homestake enterprise, ore genesis and structure. *Eng. Min. J.* 132:640–645.

Menendez, A., and L. Puget, 1949. The riches of Almadén. *Min. World* 11(7):34–36; 11(8):38–41; 11(9):35–37.

Miller, A. R., 1964. High salinity in sea water. *Nature* 203:590–594.

Miller, L. J., 1976. Corporations, ore discovery and the geologist. *Econ. Geol.* 71:836–847.

Mottl, M. J., H. D. Holland, and R. F. Corr, 1979. Chemical exchange during hydrothermal alteration of basalts by seawater—II. Experimental results for Fe, Mn, and sulfur species. *Geochim. Cosmochim. Acta* 43:869–884.

Noble, J. A., 1950. Ore mineralization in the Homestake gold mine, Lead, South Dakota. *Geol. Soc. Amer. Bull.* 61:221–252.

—— and J. O. Harder, 1948. Stratigraphy and metamorphism in a part of the northern Black Hills and the Homestake mine, Lead, South Dakota. *Geol. Soc. Amer. Bull.* 59:941–976.

——, ——, and A. L. Slaughter, 1949. Structure of a part of the northern Black Hills and the Homestake mine, Lead, South Dakota. *Geol. Soc. Amer. Bull.* 60:321–352.

Oftedahl, C., 1958. A theory of exhalative-sedimentary ores. *Geol. Foren. Stockholm. Forh.* 80:1–19.

Ogura, H., 1972. Geology and "kuroko" ore deposits of the Hanaoka-Matsumine mine, northern Japan. 24th Int. Geol. Congr., Sec. 4, Mineral Deposits, pp. 318–325.

Ohle, E. L., 1972. Evaluation of iron ore deposits. *Econ. Geol.* 67:953–964.

——, 1983. Pers. Commun.

Ohmoto, H., 1972. Origin of hydrothermal fluids responsible for the kuroko deposits in Japan (abs.). *Min. Eng.* 24(12):74.

—— and B. J. Skinner, Eds., 1983. *The Kuroko and Related Volcanogenic Massive Sulfide Deposits.* New Haven, Conn.: Econ. Geol. Mon. 5, 604 pp.

O'Rourke, J. E., 1961. Paleozoic banded iron-formations. *Econ. Geol.* 56:331–361.

Paige, S., 1924. Geology of the region around Lead, South Dakota. USGS Bull. 765.

Park, C. F., Jr., 1942. Manganese deposits of Cuba. USGS Bull. 935-B, pp. 75–97.

——, 1959. The origin of hard hematite in itabirite. *Econ. Geol.* 54:573–587.

Parmentier, E. M., and E. T. C. Spooner, 1978. A theoretical study of hydrothermal convection and the origin of the ophiolitic sulfide ore deposits of Cyprus. *Earth Planet. Sci. Lett.* 40:33–44.

Pearce, J. A., and G. H. Gale, 1980. Identification of ore-deposition environments from trace-element geochemistry of associated igneous host rocks, pp. 14–24 in *Volcanic Processes in Ore Genesis.* Inst. Min. Metall. Spec. Pub. 7, 188 pp.

Pettijohn, F. J., 1946. Geology of the Crystal Falls-Alpha iron-bearing district, Iron County, Michigan. USGS Strateg. Mins. Invest., Prelim. Map 3–181.

Plimer, I. R., 1978. Proximal and distal stratabound ore deposits. *Mineral. Deposita* 13:345–354.

—— and T. Finlow-Bates, 1978. Relationship between primary iron sulphide species, sulphur source, depth of formation, and age of submarine exhalative sulphide deposits. *Mineral. Deposita* 13:399–410.

Ransome, F. L., 1921. The ore of the Almadén mine. *Econ. Geol.* 16:313–321.

Raynaud, J., 1941. Le minerai de la mine d'Almaden, Espagne. *Soc. Geol. Belg. Bull.* 64:226–237.

Ridler, R. H., 1970. Relationship of mineralization to volcanic stratigraphy in the Kirkland-Larder Lakes area, Ontario. *Geol. Assoc. Can. Proc.* 21:33–42.

——, 1976. Stratigraphic keys to the gold metallogeny of the Abitibi Belt. *Can. Min. J.* 97(6):81–88.

Riverin, G., and C. J. Hodgson, 1980. Wall-rock alteration at the Millenbach Cu-Zn Mine, Noranda, Quebec. *Econ. Geol.* 75:424–444.

Roberts, R. G., 1975. The geological setting of the Mattagami Lake Mine, Quebec: a volcanogenic massive sulfide deposit. *Econ. Geol.* 70:115–129.

—— and E. J. Reardon, 1978. Alteration and ore-forming processes at Mattagami Lake mine, Quebec. *Can. J. Earth Sci.* 15:1–21.

Roberts, H. M., and M. W. Bartley, 1943a. Replacement hematite deposits, Steep Rock Lake, Ontario. AIME Tech. Pub. 1543.

—— and ——, 1943b. Hydrothermal replacement in deep-seated iron ore deposits of the Lake Superior region. *Econ. Geol.* 38:1–24.

Rona, P. A., 1973. Plate tectonics and mineral resources. *Sci. Amer.* 229:86–95.

——, 1978. Criteria for recognition of hydrothermal mineral deposits in oceanic crust. *Econ. Geol.* 73:135–160.

Royce, S., 1938. Geology of the Iron Ranges, in *Lake Superior Iron Ores.* Cleveland: Lake Superior Iron Ore Assn.

——, 1942. Iron ranges of the Lake Superior district, pp. 54–63 in W. H. Newhouse, Ed., *Ore Deposits as Related to Structural Features.* Princeton, NJ: Princeton Univ. Press, 280 pp.

Rye, D. M., 1972. The stable and lead isotopes of part of the northern Black Hills: age and origin of the Homestake and surrounding ore bodies. Unpub Ph.D. Dissert., Univ. Minn., 119 pp.

——, B. R. Doe, and M. H. Delavaux, 1974. Homestake gold mine, South Dakota—II. Lead isotopes, mineralization ages, and sources of lead in ores of the northern Black Hills. *Econ. Geol.* 69:814–822.

—— and R. O. Rye, 1974. Homestake gold mine, South Dakota: I stable isotope studies. *Econ. Geol.* 69:293–317.

Sangster, D. F., 1971. Sulphur isotopes, stratabound sulphide deposits, and ancient seas, pp. 295–299 in S. Ishihara, Ed., *Geology of Kuroko Deposits.* Soc. Min. Geol. Japan, Spec. Issue 3, 435 pp.

——, 1972. Precambrian volcanogenic massive sulfide deposits in Canada: a review. Geol. Surv. Can. Pap. 72-22.

—— and S. D. Scott, 1976. Precambrian stratabound massive Cu-Zn-Pb sulfide ores of North America, pp. 129–222 in K. H. Wolf, Ed., *Handbook of Stratabound and Stratiform Ore Deposits,* vol. 6. New York: Elsevier, 585 pp.

Sato, T., 1977. Kuroko deposits, their geology, geochemistry, and origin, pp. 153–161 in *Volcanic Processes in Ore Genesis.* Inst. Min. Metall. Spec. Pap. 7.

Saupé, F., 1967. Note préliminaire concernant la genèse du gisement de mercure d'Almadén. *Mineral. Deposita* 2:26–33.

——1973. La géologie du gisement de mercure d'Almaden. Sci. Terre Mem. 29.

Sawkins, F. J., 1984. *Metal Deposits in Relation to Plate Tectonics.* New York: Springer, 325 pp.

—— and D. M. Rye, 1974. Relationship of Homestake-type gold deposits to iron-rich Precambrian sedimentary rocks. *Inst. Min. Metall. London Trans.* 83:B56–B60.

Schmidt, R. G., and C. E. Dutton, 1957. Bedrock geology of the south-central part of the North range, Cuyuna district, Minnesota. Sheets 1–3, USGS Mineral Invest. Field Stud. Map MF 99.

Schuette, C. N., 1931. Occurrence of quicksilver ore bodies. *AIME Trans.*, pp. 403–488.

Scott, J. S., 1948. Quemont mine, pp. 773–776 in *Structural Geology of Canadian Ore Deposits*. Montreal: Can. Inst. Min. Metall.

Shanks, W. C., III, and J. L. Bischoff, 1980. Geochemistry, sulfur isotope composition, and accumulation rates of Red Sea geothermal deposits. *Econ. Geol.* 75:445–459.

Sillitoe, R. H., 1972. A plate tectonic model for the origin of porphyry copper deposits. *Econ. Geol.* 67:184–197.

Simmons, B. D., 1973. Geology of the Millenbach massive sulfide deposit, Noranda, Quebec. *Can. Inst. Min. Metall. Bull.*, Nov., 66(736):67–78.

Sims, P. K., 1972. Banded iron formations in the Vermilion district, pp. 79–81 in P. K. Sims and G. B. Morey, Eds., *Geology of Minnesota*, Schwartz Vol., Minn. Geol. Surv. 100th Anniv., 632 pp.

—— and G. B. Morey, Eds., 1972. *Geology of Minnesota*, Schwartz Vol., Minn. Geol. Surv. 100th Anniv., 632 pp.

Slaughter, A. L., 1968. The Homestake mine, pp. 1436–1459 in *Ore Deposits of the United States, 1933/1967*, Graton-Sales Vols., New York: AIME, 1880 pp.

Soler, M., 1978. Pers. Commun.

Solomon, M., 1976. "Volcanic" massive sulfide deposits and their host rocks—a review and their explanation, pp. 21–54 in K. H. Wolf, Ed., *Handbook of Stratabound and Stratiform Ore Deposits*, vol. 6. New York: Elsevier, 585 pp.

—— and J. L. Walshe, 1979. The formation of massive sulfide deposits on the sea floor. *Econ. Geol.* 74:797–813.

Speer, W. E., 1977. Geology of the McDermitt Mine area, Humboldt County, Nevada. Unpub. M.S. thesis, Univ. Ariz., 65 pp.

Spence, C. D., and A. F. de Rosen-Spence, 1975. The place of sulfide mineralization in the volcanic sequence at Noranda, Quebec. *Econ. Geol.* 70:90–101.

Spiess, F. N., et al., 1980. Hot springs and geophysical experiments on the East Pacific Rise. *Science* 207:1421–1433.

Stanton, R. L., 1959. Mineralogical features and possible mode of emplacement of the Brunswick Mining and Smelting ore bodies, Gloucester County, New Brunswick. *Can. Inst. Min. Metall. Bull.* 52:631–642.

Sudo, T., 1954. Types of clay minerals closely associated with metalliferous ores of epithermal types. Tokyo Kyoiku Daigaku Sci. Rept., ser. C, no. 23, pp. 186–192.

Swan, M. M., D. M. Hausen, and R. A. Newell, 1982. Lithological, structural, chemical, and mineralogical patterns in a Precambrian stratiform gold occurrence, Yavapai County, Arizona, pp. 143–157 in D. M. Hausen and W. C. Parks, Eds., *Process Metallurgy*. New York: AIME.

Trendall, A. F., 1965. Origin of Precambrian iron formations, discussion. *Econ. Geol.* 60:1065–1069.

——, 1968. Three great basins of Precambrian banded iron formation deposition: a systematic comparison. *Geol. Soc. Amer. Bull.* 79:1527–1544.

Turner, J. S., and L. B. Gustafson, 1978. The flow of hot saline solutions from vents in the sea floor—some implications for exhalative massive sulfide and other ore deposits. *Econ. Geol.* 73:1082–1100.

Tyler, S. A., 1949. Development of Lake Superior soft ores from metamorphosed iron formation. *Geol. Soc. Amer. Bull.* 60:1101–1124.

Urabe, T., and T. Sato, 1978. Kuroko deposits of the Kosaka mine, northeast Honshu, Japan—products of submarine hot springs on Miocene sea floor. *Econ. Geol.* 73:161–179.

Van der Veen, R. W., 1924. The Almadén mercury ores and their connection with igneous rocks. *Econ. Geol.* 19:146–156.

Van Hise, C. R., and C. K. Leith, 1911. The geology of the Lake Superior iron region. USGS Mon. 52.

Walker, R. R., A. Matulich, A. C. Amos, J. J. Watkins, and G. W. Mannard, 1975. The geology of the Kidd Creek Mine. *Econ. Geol.* 70:80–89.

Watanabe, T., 1970. Volcanism and ore genesis, pp. 423–432 in T. Tatsumi, Ed., *Volcanism and Ore Genesis*. Tokyo: Univ. Tokyo Press, 448 pp.

White, D. A., 1954. The stratigraphy and structure of the Mesabi Range, Minnesota. Minn. Geol. Surv. Bull. 38, 92 pp.

Wright, L. B., 1937. Gold deposition in the Black Hills of South Dakota and Wyoming. *AIME Trans.* 126:390–415.

FOURTEEN

Deposits Related to Submarine Volcanism and Sedimentation

In Chapter 13 we dealt with a group of deposits that are directly related to submarine volcanism. Those deposits are *proximal* within the sense of "near to," and *distal,* referring to "farther from," with regard to the actual point of venting of volcanically derived fluids into a marine eugeosynclinal basin. As we withdraw in concept from those hot spring sites of active volcanism and volcanosedimentary mineralization, we encounter a group of ore deposits called *black-shale-hosted* ores—and some geologic difficulties. On the one hand, it is attractive to think of the precipitation of banded iron formations and iron- and gold-rich stratabound stratiform units of the Algoma type yielding to Biwabik-like shelf sediments of the Superior type either with distance from the source or out onto a shelf. On the other hand, it is also appealing to think of a progressive shift from proximal to distal as involving progressive dilution of volcanic components by terrestrial, terrigenous sediments as though one traveled from the volcanic-arc side to the continental side of a back-arc basin, as though crossing the Sea of Japan (Figure 21-3). Gilmour (1971) proposed such a descriptive relationship between many of the stratiform deposits then known, and many economic geologists still use his classification, updated in 1976 (Figure 14-1). It suggests that one can arrange many of the major deposits in a sequence trending from Cu-Zn-Ag massive sulfide deposits in clearly volcanic lithotectonic environments such as Kidd Creek (Chapter 13), to Pb-Zn-dominated stratiform ores in more tranquil, continentally derived clastic-detritus-rich or

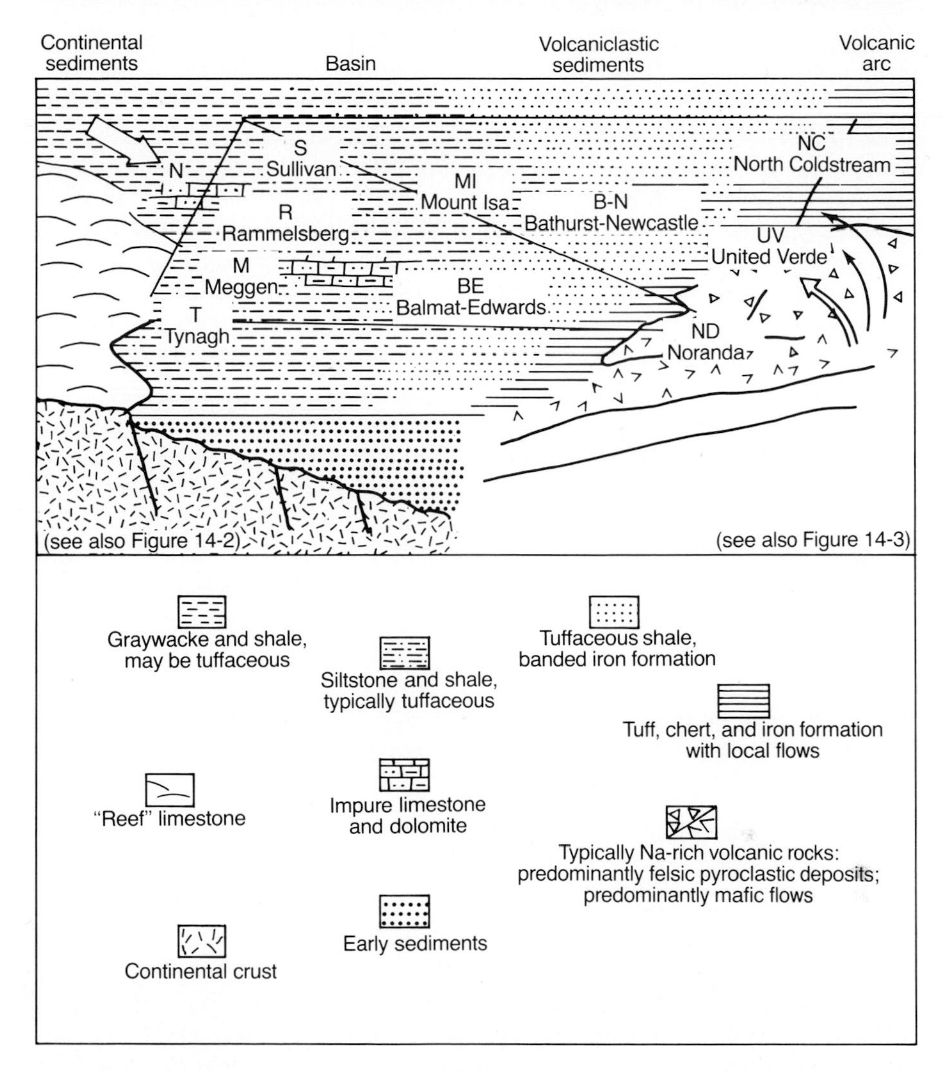

Figure 14-1. Diagram of the spectrum of deposits considered in this chapter. Chapter 13 dealt with the proximal volcanogenic deposits at the right-hand side. The legend overlays the sketch and identifies progressively more "distal" environments to the left. Horizontal scale is hundreds of kilometers. (*From* Gilmour, 1976.)

carbonate reef settings. If the former is in close association with the volcanism of an island arc setting, the latter might be envisioned as occurring on the continent side of a back-arc basin, as for example on the right side of Figure 13-2*b*. Implicit in such a geometry are the assumptions that the metals, and presumably sulfur, travel distances of tens or hundreds of kilometers in the seawater medium, then to be concentrated by precipitation, and that there might be little evidence of hot spring or fumarolic activity near the deposits.

Here come some difficulties. The first implication is dubious in large part by virtue of the dilution-dispersion, rather than transport-concentration, that must be expected; the second is dispatched by the fact that altered-mineralized conduits are found in the footwall sediments of many, although by no means all, of the black-shale-hosted deposits. The columnar conduit beneath the Sullivan, British Columbia, lead-zinc orebodies produced tourmaline-chlorite-silica alteration, and probably was an active hot spring, as pipes at Rammelsberg, Germany, and Bleïda, Morocco, are thought to have been. Other "pipes" may simply have delivered warm, metal-laden brines from adjacent deep basins or from deep fracture systems (Figure 14-2) to sediment-rich basins like that in which the Belt Supergroup accumulated or to sediment-starved basins as at Bleïda, and not represent active, even remotely distal volcanism at all. Leblanc and Billaud (1978) envision active volcanism inserting base metals into a shale-depositing environment (Figure 14-3) on the northwest edge of the African craton in the late Proterozoic 760 to 623 million years ago, but such environments are difficult to identify in Mesozoic-Cenozoic tectonic settings such as ensialic back-arc basins. Their diagram shows the volcanism-related "end-member" of black-shale-hosted deposit settings. Figure 14-2 shows brine insertion of metals and marine delivery from distant volcanic sources. Clearly, the final determination of tectonic settings of this deposit class is yet to be made. Many do lie on the continentward side of eugeosynclinal basins far from silicic volcanic centers, as shown in Figure 13-2*b*, but others may have

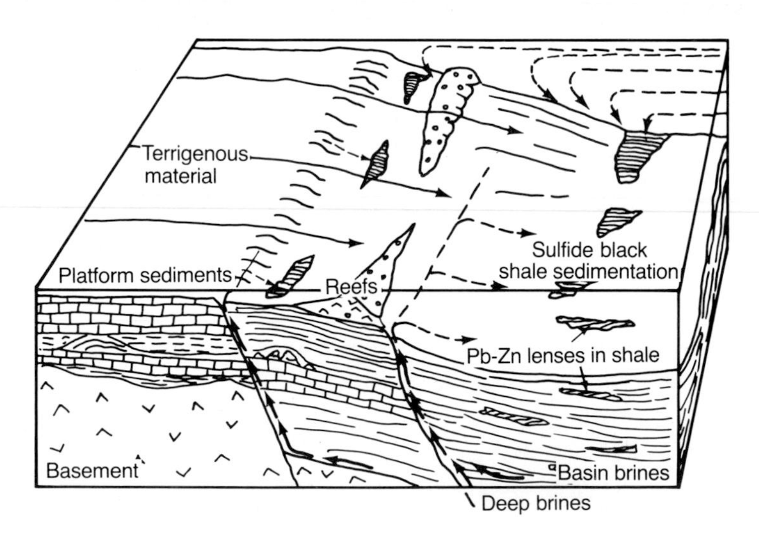

Figure 14-2. Black-shale-hosted environments on the continent side of a back-arc basin. Such platform-basin margins or ensialic back-arc basins involve basin brines (forward section) or distal marine delivery of base metals (back section) for deposition with shale. See also Figure 14-3.

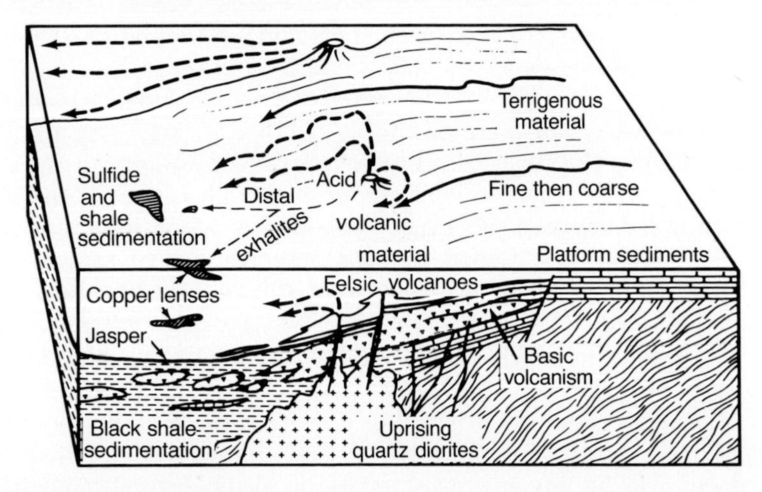

Figure 14-3. Paleogeographic and metallogenic reconstruction of the Bleïda area, Morocco, on the volcanic arc side of a back-arc basin. Volcanism is importantly involved, with dilution by terrigenous sediments. This diagram could easily have been drawn showing a full-blown volcanic arc with a subduction zone dipping under diagram far to the right. See also Figure 14-2. (*After* Leblanc and Billaud, 1978.)

formed in more active volcanic settings into which terrigenous sediments were being dumped, as for example in an intracratonic rift basin or aulacogen such as the Red Sea. It would be a mistaken oversimplification to apply the terms proximal and distal in the strict sense of their use in Chapter 13; as mentioned, many of the black-shale-hosted deposits formed directly over their vent conduits.

A list of deposits to be considered here includes districts which contain a substantial part of the world's lead-zinc reserves, especially with the addition of some substantial recent discoveries in Alaska, the Yukon–Northwest Territories of Canada, the northeastern United States, and South Africa. Deposits with apparent volcanic association are the newly found Broken Hill–Black Mountain–Gamsberg–Aggeneys occurrence near the Orange River in the northwest Cape Province of South Africa and the Broken Hill mines in New South Wales, Australia (Stanton 1972, 1976a–d 1978; Laing, Marjoribanks, and Rutland, 1978). Both of these deposits are mid-Proterozoic in age, both are metamorphosed to garnet-granulite rank, and both originally contained about 200 million tons of lead-zinc ore. The Carpentaria-age rocks that host Broken Hill, New South Wales, may run under younger cover rocks and be continuous with outcrops far to the north that enclose the Mt. Isa ores described below, still with manifestations of volcanism, and the McArthur River Pb-Zn deposits in evaporite-related pyritic-dolomitic-gyspiferous-carbonaceous black siltstones. Other Precambrian deposits include the Iron King Pb-Zn district in Arizona (Gilmour and Still, 1968), Rosh Pinah (Page and Watson, 1976) in Namibia (South-West

Africa), Bleïda in Morocco (Leblanc and Billaud, 1978), and the 200-million-ton finely laminated ores of the Sullivan district, British Columbia (Campbell, Ethier, and Krause, 1980). Paleozoic districts include Ordovician ores in New Brunswick, which are part of a belt that includes recently discovered deposits like the Blue Hills–Blackhawk in Maine, Tynagh and Tara in Ireland, and old districts which include the Great Gossan Lead in Virginia, Ducktown in Tennessee, and many Scandinavian deposits. Devonian deposits at Rammelsberg, Ramsbeck, and Meggen, West Germany, Upper Paleozoic districts in the Western Brooks Range, Alaska, the Howard's Pass district in the Selwyn Basin of Yukon–Northwest Territories, Canada, and possibly the Kupferschiefer districts in East Germany and Poland are also of this type.

Before we look in detail at the Mt. Isa and Broken Hill ores as representative, a listing of some of the characteristics of black-shale-hosted occurrences will be advantageous, much of the information having been assembled by Nieman (1976). These deposits are typically stratiform and stratabound at all scales, being composed of lensoid or wedge-shaped aggregate piles of thin but remarkably continuous beds of galena-sphalerite-pyrite. The sulfidic laminae are generally millimeters to a few centimeters thick, rarely more than 1 meter. They may be nearly monomineralic or simple mixtures of a few minerals, but individual compositions and metal ratios may be maintained laterally for hundreds of meters, even kilometers, as at Broken Hill, New South Wales. Single laminae or beds may extend for tens to hundreds of meters, the resultant textures (Figures 14-4 and 14-5) resembling the fine and variable bedding of BIF (Figure 13-13). The sulfide bands are typically intercalated with thin beds of tuff, volcaniclastic materials, or silty or shaley terrigenous sediments suggesting tranquil deposition. The orebodies are sheetlike (Figure 14-6) or they comprise variably offset stacked sheets (Figure 14-7); they cannot be bulbous or domical. Boundaries between ore bands and between ore and waste bands are typically sharp in the vertical direction, but the orebodies are laterally zoned, with overall values and Pb/Zn ratios decreasing outward. The dominant ore-mineral sulfides by far are galena and sphalerite, with major pyrite and pyrrhotite. Lesser amounts of chalcopyrite, bornite, covellite, chalcocite, marcasite ($FeS_{2-x}H_x$), mackinawite (Fe_9S_8), the thiospinel griegite (Fe_3S_4), and arsenopyrite have been reported. The predominant iron sulfide—normally pyrite or pyrrhotite—usually extends far beyond the limits of economic Pb-Zn mineralization in carbonaceous shale units. Gangue minerals are rare because veins are essentially absent; where present in layers, gangue minerals include the sedimentary minerals, especially quartz, calcite, chlorite, and barite. Alteration is absent except in the altered pipelike bodies beneath the ore lenses at several deposits, where tourmaline (Sullivan), barite (Rammelsberg), silica (Rammelsberg, Alaska, New Brunswick), and chlorite (Alaska, New Brunswick) are common. Barite and the manganese phases rhodonite and rhodochrosite are common both beneath

Figure 14-4. Example of fine bedding and banding in sediment-hosted sulfide layers. This layer of ore—here about 1 meter thick—is one of several in the Sullivan, British Columbia, Canada, deposit; massive argillite with essentially no sulfides is seen above and below the contacts emphasized with chalk. This fine banding is similar to that at Rammelsberg, McArthur River, and many other deposits of this type. The light-colored material here is pyrite, with sphalerite and galena in the darker layers. See also Figures 13-13, 13-14, and 13-15 of BIF. (*From* Freeze, 1966.)

and lateral to the ore lenses (Meggen and Rammelsberg, West Germany, Broken Hill, New South Wales; Broken Hill, Republic of South Africa).

Textures and structures perceived in unmetamorphosed members of the shale-hosted group collectively indicate low temperatures of formation, probably with important biologic influences. Grain sizes of both sulfide and silicate-gangue particles are exceedingly fine—indeed, the difficulty of economical separation of 5-μm sphalerite and galena grains from one another and from silicates and other minerals has delayed production (New Brunswick) or indefinitely postponed it (McArthur River). Delicate lamination, fine-graded bedding, the presence of sulfide overgrowths, framboidal pyrite, atoll structures, colloform banding, concretions, and soft sediment deformation textures all are consistent with low-temperature, stagnant, biologically influenced environments. They appear (Vaughn, 1976) to represent near neutral or slightly acid seafloor environments with pH 6 to 8, low Eh at -0.1 mV, low $f(O_2)$, and moderate $f(S_2)$.

(*a*)

(*b*)

Figure 14-5. (*a*) Rough and (*b*) polished hand sample showing the fine continuous bedding of sulfide minerals—here galena, sphalerite, and pyrite—that characterizes black-shale-hosted deposits. The specimen is from Rammelsberg, West Germany (Figure 14-7), but could as well have come from Sullivan (Figures 14-4 and 14-6) or many others.

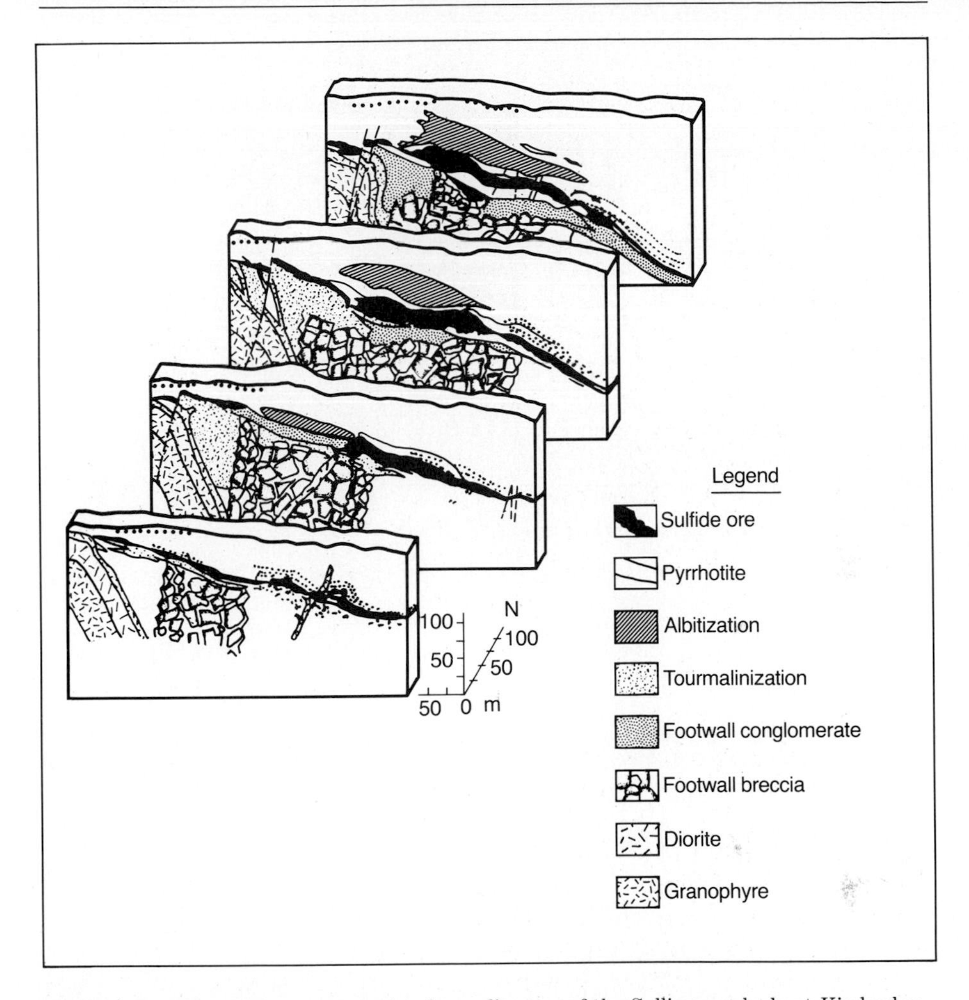

Figure 14-6. North-facing cross-section fence diagram of the Sullivan orebody at Kimberley, British Columbia, Canada, an example of a single-sheet sediment-hosted deposit. The sulfides (black) extend for several kilometers but are essentially one horizon. See also Figure 14-4. (*From* Campbell, Ethier, and Krause, 1980.)

▸ MT. ISA LEAD-ZINC-SILVER-COPPER DEPOSITS, QUEENSLAND, AUSTRALIA

The base-metal deposits at Mt. Isa, Queensland, Australia (Figure 14-8), form one of the largest concordant accumulations of base metals in the world. Silver, lead, zinc, and copper are recovered. The deposits were not discovered until 1923, with operations starting in 1928 (Blanchard and Hall, 1942; Carter, 1953; Fisher, 1960; Carter, Brooks, and Walker, 1961; Bennett, 1965; Mathias and Clark, 1975; Mount Isa Mines, Ltd., 1977).

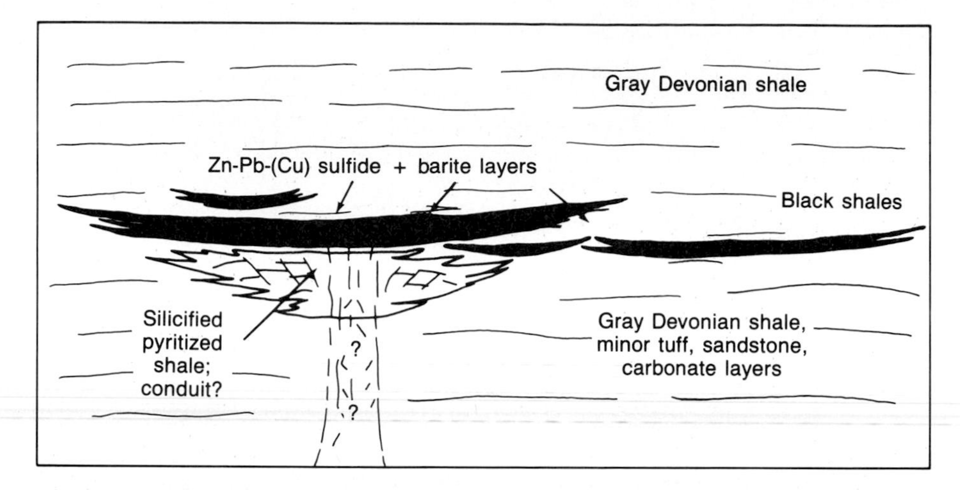

Figure 14-7. Depiction of variably offset stacked sheets of ore in the black-shale-hosted deposit at Rammelsberg, West Germany. The lower layers, temporally equivalent to Devonian-age Wissenbacher Shale that extends laterally from all units shown, are silicified, pyritized, and contain copper and zinc sulfides near the ore lenses. The massive sulfide bodies, about 30 meters thick and up to 500 meters wide (black), contain variable amounts of finely bedded sphalerite and galena with barite. See also Figures 14-5 and 14-11. (*After* Anger, Nielsen, Puchelt, and Ricke, 1966.)

The valuable deposits are confined to Lower Proterozoic Carpentaria Group strata of the Australian Precambrian shield. The general stratigraphy is shown in Figure 14-9 (Bennett, 1965). The Mount Isa Group, more than 5000 meters thick, is a sequence of steeply dipping shales, limey shales, and siltstones of sedimentary and volcanic origin. Prominent outcrops trend north-south, to the north about 30 km to the Hilton mine and beyond and to the south for about 30 km, where they go under younger cover. Within the Mount Isa Group is the Urquhart Shale, which contains all of the economically valuable minerals. The Urquhart Shale, about 1000 meters thick,

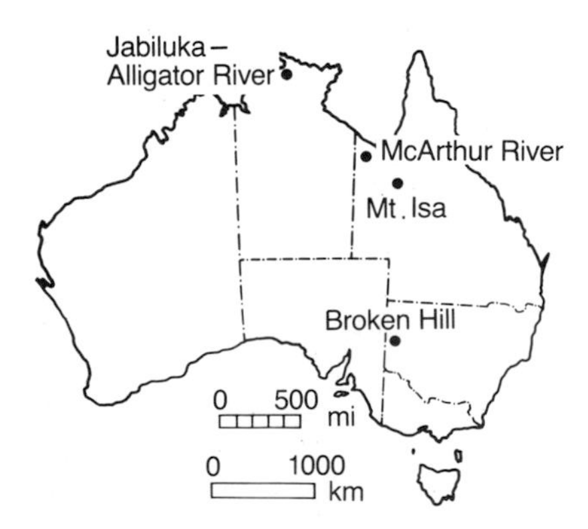

Figure 14-8. Map of Australia showing the locations of the Broken Hill district, New South Wales, and Mt. Isa and McArthur River, Queensland. The Jabiluka–Alligator River uranium deposits are mentioned in Chapter 20.

is a light to dark gray, banded dolomitic and volcanic material, transitional along strike into a fine-grained thin-bedded dolomitic and pyritic shale with well-defined bedding planes separating laminae ranging from 1 mm to 30 cm thick. Abundant carbonaceous materials are present in the shale, and the unit is laterally pyrite-rich at the stratigraphic level of the orebodies within it. The Urquhart contains scores of thin, widespread distinctive marker beds composed of about 80% K-feldspar, 15–20% dolomite, and about 5% silica. These "TMs," or "tuff marker" beds, average 3 cm in thickness, ranging from 1 to 120 cm, and are remarkably persistent. They are useful in deciphering structural and stratigraphic details.

The Mount Isa Group crops out on the western limb of a north-plunging anticline, the axis of which lies east of the district and trends nearly due north. The axial plane of the fold dips about 60 to 65° west. Minor folding is common and results in many structural complications. The area has been extensively faulted. Most of the faults are of the conjugate strike-slip type, but there are both normal faults and high-angle reverse faults, especially west of Mt. Isa (Blanchard, 1968). About 3 km west of the mine area, the anticline has been sliced off by the Mt. Isa shear fault, which has been traced in a north-south direction for over 60 km (Bennett, 1965). The fault dips west from 50 to 70°, and the west side has been moved northward and upward an unknown distance. The fault is the sum of a series of many small displacements rather than the result of a single movement or a single fault plane. A similar fault zone, the Racecourse Shear, marks the eastern limit of the district.

Age	Formation		Thickness	
			Feet	Meters
Lower Proterozoic		Western volcanics	?	
		unconformity?		
	Mount Isa Group	Magazine Shale	700	215
	Mount Isa Group	Kennedy Siltstone	1020	310
	Mount Isa Group	Spear Siltstone	560	170
	Mount Isa Group	Urquhart Shale	3000	915
	Mount Isa Group	Native Bee Siltstone	2600	795
	Mount Isa Group	Breakaway Shale	3400	1040
	Mount Isa Group	Moondarra Siltstone	4000+	1200+
		Judenan Beds		
		Myally Beds		
		Eastern Creek Volcanics	15,000+	4500+
		Mount Guide Quartzite	8000	2440
		Argylla Formation	—	

Figure 14-9. Stratigraphy and thicknesses of units in the Mt. Isa area. (*From* Bennett, 1965).

Intrusive rocks are absent in the immediate vicinity of the mines, but about 8 km west of Mt. Isa and west of the Mt. Isa fault is an extensive exposure of granite. No relationship between the granite and the mineralization is known.

Deposits of silver-lead-zinc were discovered first, and during the early years were the only sources of ore (Figure 14-10). Copper ores were found later on the west side, and these now constitute the principal ores of the district. All of the ores are in the Urquhart Shale, but the copper orebodies are stratigraphically separated from those of the other metals. Copper ores

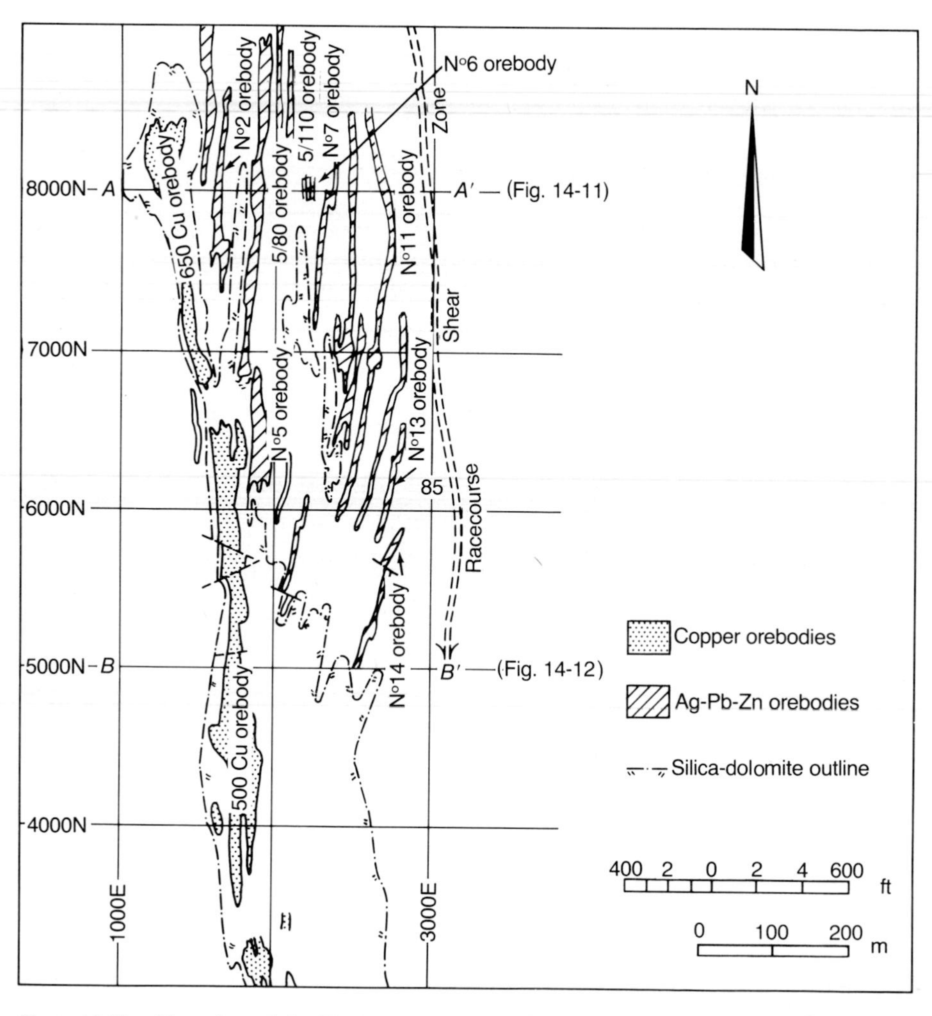

Figure 14-10. Plan view of the Mt. Isa mines 11 level, 500 meters below surface. The north-south Mt. Isa fault is to the west, the host rock here being the Urquhart Shale. The beds and ore zones dip 50 to 60° to the west. Section *A–A′* is Figure 14-11, *B–B′* is Figure 14-12. (*After* Mount Isa Mines, Ltd., 1977.)

are in the hanging wall above the silver-lead-zinc; the two types are not intermixed. They are also generally surrounded by an envelope of massive siliceous dolomite called *silica-dolomite.*

The silver-lead-zinc ores are in layers of shale, with pyrite, which are distributed throughout the Urquhart Shale. Where the layers of mineralized materials are closely spaced, they constitute ore. From north to south, the ore changes gradually from zinc-rich to silver-lead-zinc-rich; pyrite becomes increasingly abundant and other sulfides less plentiful. Silver and lead dominate in the southern parts of the silver-lead-zinc zone, but they decrease northward as sphalerite and pyrite become more abundant, and they yield to chalcopyrite in the extreme south. Figures 14-11 and 14-12 are cross sections through both the silver-lead-zinc and the copper orebodies (Mount Isa Mines, Ltd., 1977). These figures show clearly that lobelike bodies of silica-dolomite rock are on the updip or higher levels of the mine. The silica-dolomite widens in depth to a maximum development at the contact with the "greenstone" of the basement (Figures 14-10 and 14-11).

The original outcrops of the silver-lead-zinc-bearing shales were thin bands of cerussite ($PbCO_3$) with some anglesite ($PbSO_4$) and pyromorphite [$Pb_5(PO_4)_3Cl$], commonly colored with a bluish coating of iron and manganese oxides which enabled ready recognition and facilitated mapping. Part of the surface ore assayed as much as 50% lead and hundreds of grams of silver per ton. Outcrops generally were on the flanks of erosion-resistant jasper ridges that have been traced for kilometers. Vegetation on the outcrops was scarce or absent. The silver-lead-zinc ores were oxidized to a depth of about 50 to 60 meters, and along faults and fissures to as much as 200 meters. Below the oxidized zone was a transition zone about 20 meters thick. Silver was largely removed from the top 30 meters of the oxidized zone, but was redeposited about 25 meters above the primary ore (Blanchard and Hall, 1942). Zinc was removed from the upper 30 meters of the oxidized ores, and supergene smithsonite ($ZnCO_3$) was deposited in the transition zone where the grade of ore was 15 to 20% higher in zinc than was the primary ore. The primary silver-lead-zinc ores contain galena, high-iron sphalerite, and tetrahedrite, all fine-grained and intimately mixed. Pyrite, pyrrhotite, quartz, carbonates, and graphite are also common. Much lesser amounts of arsenopyrite, marcasite, chalcopyrite, valleriite (a complex Cu-Fe-S-Mg-Al-OH mineral), proustite (Ag_3AsS_3), polybasite [$(Ag,Cu)_{16}Sb_2S_{11}$], and argentite (acanthite) have been reported. These minerals occur in thin laminae as described in the introductory paragraphs of this chapter, interlaminated with one another in high-grade ore or with dark carbonaceous Urquhart Shale in lower grade rock.

Similar to the silver-lead-zinc ores, the copper orebodies are distributed in en-echelon tabular bodies trending north-south, but they are confined to "silica-dolomite" rock within the Urquhart Shale. This rock is extensively brecciated, although the breccia may not be of tectonic origin

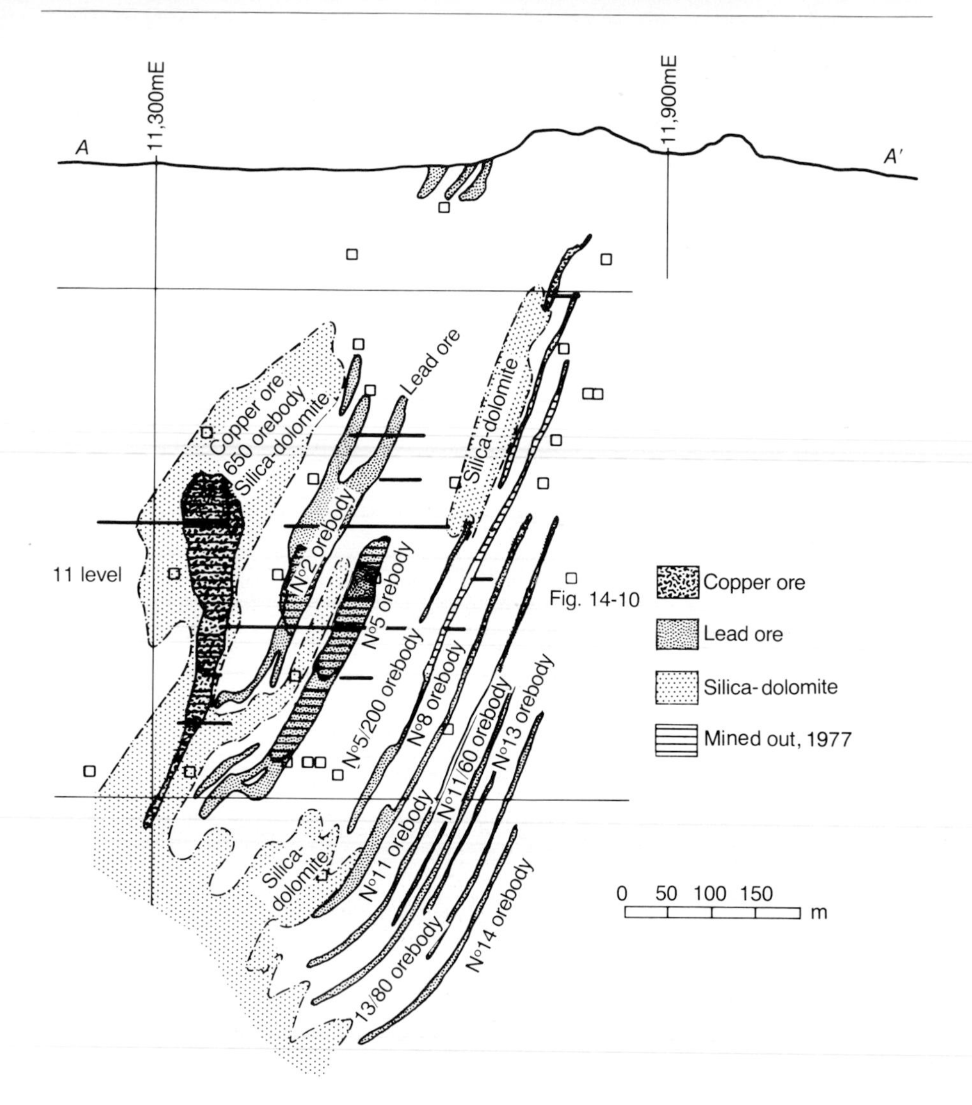

Figure 14-11. Cross section through the silver-lead-zinc ores of Mt. Isa along section *A–A′* on Figure 14-10. See also *B–B′* on Figure 14-12. (*After* Mount Isa Mines, Ltd., 1977.)

(Carter et al., 1961). Neither silica-dolomite nor copper ore crop out. Primary minerals in the copper ores include chalcopyrite, pyrrhotite, and arsenopyrite; marcasite, cobaltite [(Co,Fe)AsS], valleriite [$(Cu,Fe)_4S_4(Mg,Al)_3(OH)_6$], chalcostibite ($CuSbS_2$), galena, dyscrasite (Ag,Sb), jamesonite ($Pb_4FeSb_6S_{14}$), magnetite, and gold have been detected in polished sections (Carter et al., 1961). Oxidized copper minerals include the copper carbonates malachite and azurite, cuprite, chrysocolla, and minor native copper. Other supergene minerals reported are chalcocite, tenorite, covellite, chalcanthite ($CuSO_4{\cdot}5H_2O$), brochantite [$CuSO_4(OH)_6$], tetrahedrite ($Cu_{12}Sb_3S_{13}$),

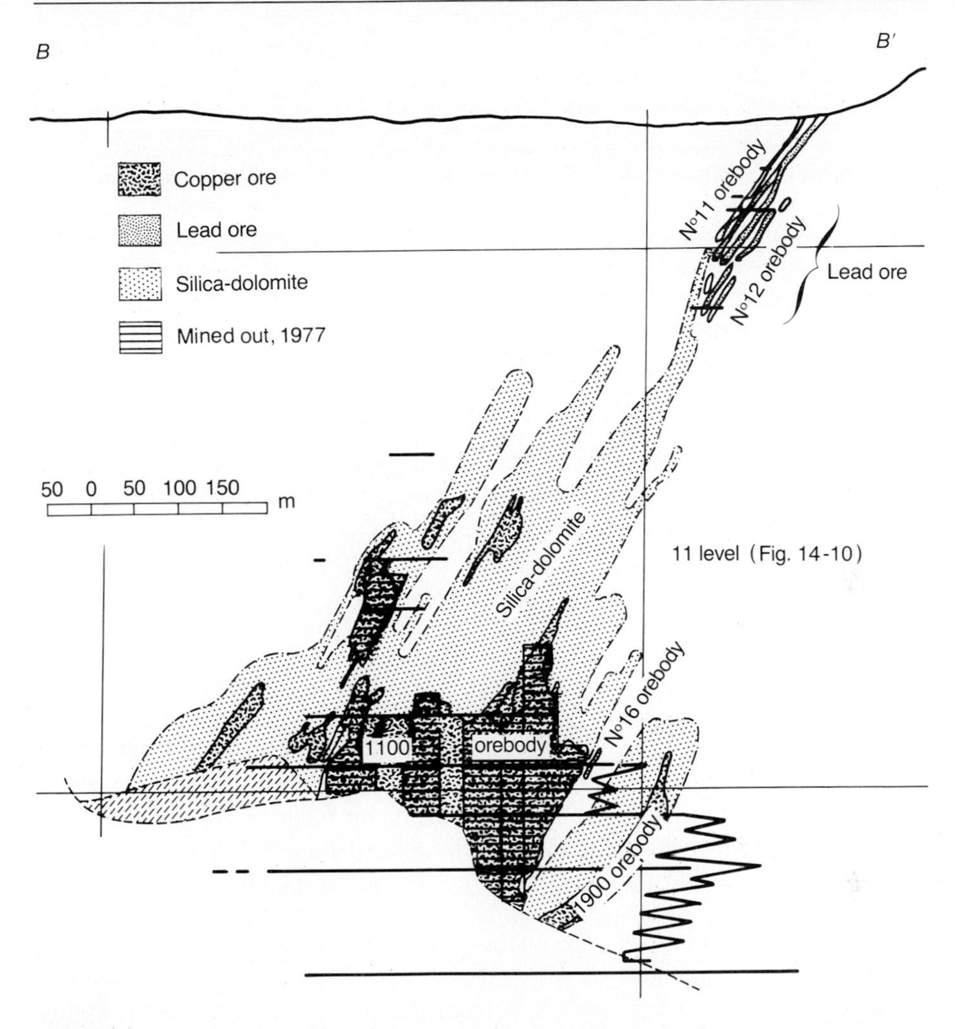

Figure 14-12. Cross section through the predominantly copper orebodies of the Mt. Isa district along section *B–B'* on Figure 14-10. See also *A–A'* on Figure 14-11. (*After* Mount Isa Mines, Ltd., 1977.)

atacamite [$Cu_2Cl(OH)_3$], and dioptase [$CuSiO_2(OH)_2$]. The primary ores are predominantly chalcopyrite and pyrite. Gangue minerals in the copper ores include quartz, calcite, and dolomite, locally with siderite, kaolin, tremolite, actinolite, scapolite, magnetite, diopside, augite, selenite, tourmaline, fluorite, apatite, psilomelane, barite, epidote, garnet, and sphene (Carter et al., 1961).

The orebodies of both types are arranged in en-echelon patterns (Figures 14-11 and 14-12) (Blanchard, 1968). A single stratiform orebody may extend over 1 km along strike and more than 0.7 km downdip. The widths of the individual orebodies depend mainly upon economics, as mineral values gradually decrease laterally in the plane of the lamination. In one of the best known zones, the Black Star, the ore is up to 50 meters wide (or thick), but widths elsewhere are much less. Bennett (1965) stated that mineralization is always associated with concentrations of spherical framboidal pyrite grains. A coarse euhedral form of pyrite is also scattered through the shale and is concentrated along fractures and faults. The complexly folded beds show much coarser grained sulfides along their axial planes than do the undisturbed beds. These coarse-grained sulfides include galena, sphalerite, pyrite, and pyrrhotite, with quartz, carbonates, and graphite. The ores change character close to the silica-dolomite, and coarse-grained galena, pyrrhotite, carbonates, sphalerite, chlorite, chalcopyrite, and albite are present. Most silver is said to be in the form of argentite (acanthite), in tetrahedrite, or in the lattice of galena.

The origin of the lead-zinc-silver ore deposits was hotly debated in the 1950s and 1960s, but the orebodies so closely follow the stratigraphy and the minerals so faithfully appear to be part of original sedimentary material that the ores are now generally considered to be syngenetic, possibly of volcanic origin. No regional structural control has been demonstrated during mapping and sampling, but widespread lead-zinc anomalies are known in Carpentaria-age sedimentary units elsewhere. These ores may be correlative with the generally similar McArthur River deposits to the northwest, and may even be relatable to the Broken Hill ores far to the south (Figure 14-8). Sedimentary or volcano-sedimentary origins are generally favored for all three deposits, although hydrothermal activity at McArthur River has recently been convincingly postulated by Williams (1978a,b). Renfro (1974) proposed an arid coastline sabkha origin (Chapter 16) for these ores, but his ideas have not been adopted by Mt. Isa geologists.

The origin of the copper-rich ores at Mt. Isa has been more controversial. Carter et al. (1961) state that the closeness of the copper ores to the Mt. Isa fault may not be accidental, but may have a genetic significance. Copper ores are also said to be concentrated at the intersections of faults, at changes of pitch of fold and fold axes, and in flexures and fault planes. They also are described as locally cutting across bedding planes in apparent epigenetic hydrothermal relationships. Other geologists disagree with the explanation of these features, saying that it seems peculiar that the distribution of two types of ore with different geneses would be so closely associated. One explanation is that both the silver-lead-zinc and the copper ores are syngenetic or volcanogenic in origin but that the copper ores have been remobilized during mild metamorphism and formation of the silica-dolomite. Another is that the silica-dolomite bodies were actually reef structures which grew in shallow Precambrian seas after the deposition of deeper water euxinic shales, with "delivery" of lead-zinc-silver generally preceding copper. The two ore types would thus be related in genesis, time, and overall form, but be stratigraphically separate. Mathias and Clark (1975) support this interpretation, suggesting that the metals are ultimately of distal volcanic origin, perhaps related to the distant volcanism that produced the TM horizons that punctuate the Urquhart shales. Gilmour would probably concur (Figure 14-1).

▸ BROKEN HILL LEAD-ZINC-SILVER DEPOSITS, NEW SOUTH WALES, AUSTRALIA

Another type of stratified sediment-hosted deposit is illustrated by the Broken Hill district in westernmost New South Wales (Figure 14-8). Originally staked in 1883 as a tin claim although no tin was present, it is one of the principal lead-zinc-silver producing areas of the world, having exceeded the billion-dollar mark in production more than 15 years ago. The district produces more than 1 million tons of ore annually and has maintained a steady, developed ore reserve of 12 or 13 million tons. Average ore yields about 15% Pb, 12% Zn, and 5 oz (150 ppm) Ag per ton (King and O'Driscoll, 1953). In addition to these metals, the smelting of Broken Hill ores annually produces about 60,000 tons of sulfuric acid, about 200 tons of cadmium, and smaller amounts of gold, antimony, copper, and cobalt. There still remain more than 100 million tons of "reasonably high grade lead-silver-zinc ore" at Broken Hill (King, 1983).

The Broken Hill district has been studied by many able geologists, and excellent reports concerning all phases of the geology are available (Jaquet, 1894; Andrews, 1922; Gustafson, Burrell, and Garretty, 1950; King and Thomson, 1953; King and O'Driscoll, 1953; Stillwell, 1953; King, 1958; a series of papers by Hodgson and coworkers, 1974, 1975a, b, 1977; Stanton, 1972, 1976a–d, 1978; Both and Smith, 1974; Both and Rutland, 1976; Johnson and Klingner, 1975; and most recently by Laing, Marjoribanks, and Rutland, 1978). The district is in a semiarid region of sweeping deeply weathered plains and rough rocky ridges, one of which was a north-south whaleback "broken hill" (Figure 14-13), now removed by mining. Highly contorted Precambrian rocks that underlie the entire district are divided into two series, the Willyama and the Torrowangee, which are separated by a widely recognized unconformity. The lower, older, probably Upper Precambrian Willyama Series that hosts the ore layers was originally a thick sequence of sandstones and shales, possibly with volcanic components including banded iron formations in the footwall units. Both series have been regionally metamorphosed to amphibole-garnet, locally granulite rank assemblages including sillimanite-biotite-garnet gneisses, itabirites, granulites, quartzites, augen gneisses, amphibolites, and sericite, andalusite, hornblende, and staurolite schists, which will be interpreted below. Metamorphic and igneous (?) pegmatites are common. The pegmatites range from large masses to irregular veinlets and lit-par-lit replacements or migmatitic injections (Gustafson et al., 1950; King and Thomson, 1953). Andrews (1948) noted that the ore-bearing zone coincides with the centrolineal belt of maximum metamorphic intensity. Several authors, especially King (1958), have noted that the Willyama units are similar in every respect but geography and the paucity of carbonates to the Grenville Province of Canada, probably were metamorphosed at the same time, and contain similar ore deposits—Montauban-les-Mines and Calumet, Quebec, and Balmat-Edwards, New York, all resemble Broken Hill.

Unconformably overlying the Willyama Series is a thick section of Proterozoic sediments that includes argillites, quartzites, shales, limestones, conglomerates, and several glacial deposits and constitutes the Torrowangee Series. Locally, near its base, the Torrowangee Series is metamorphosed, but overall post-Willyama metamorphic effects are minor.

Structural interpretation of the region has changed radically in recent years from a concept involving irregular complex synclinal basins separated by anticlinal zones of intense plastic flowage and shearing (Andrews, 1922, Figure 14-14*a*) through one that the orebodies lie in anticlines (Gustafson, Burrell, and Garretty, 1950, Figure 14-14*b*) or occupy a drag fold on the eastern flank of a major syncline that extends throughout the mineral field, a relationship proposed by Lewis, Forward, and Roberts (1965) and shown in Figure 14-14*c* with sections from earlier interpretations. The overall plunge of what appears to be a major syncline is at a low angle to the south, out of the page toward you. The dominant structural features are zones of isoclinal folding complicated by intense shearing along bedding planes and faults. Faults, although found in many places, play no role except in offsetting mineralized beds. Deformation was such that the rocks generally yielded by plastic flow rather than by rupture. The folding was essentially one of shearing adjustments and plastic flow, but without the development of axial cleavage. Folding is also en echelon on both hand specimen and regional scales.

Laing, Marjoribanks, and Rutland revised earlier interpretations (Figure 14-14*d*). They concluded that the entire metasediment-structural volume

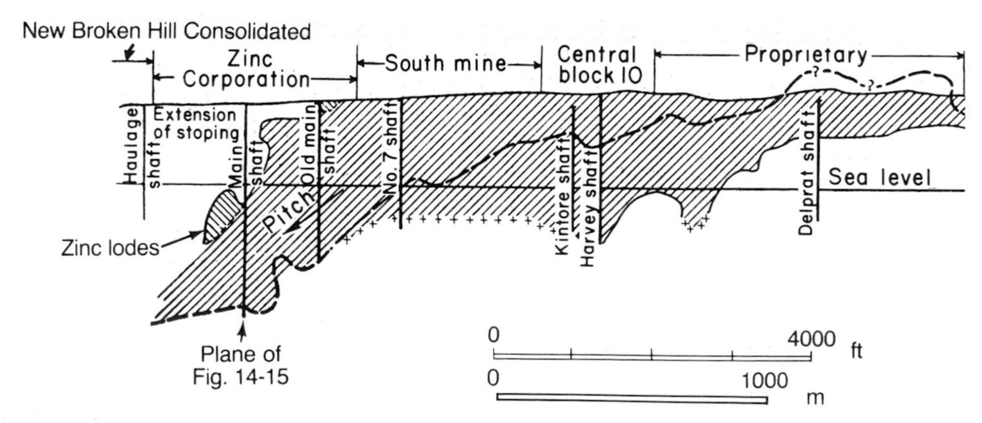

Figure 14-13. Longitudinal projection along the Broken Hill lode, New South Wales, facing west. (*From* Gustafson et al., 1950.)

has been overturned so that the Main Syncline of Lewis, Forward, and Roberts (1965) is actually an *inverted* syncline; it would thus now be called the Main Synform, or even the Main Anticline, and a major "anticline" east of the ore zones is the Broken Hill Antiform, an isoclinal syncline rather than an anticline. In the discussions that follow, the mine terminology of hanging wall (up) and footwall (down) is retained, but remember that the spatial sense is inverted from geologic relationships.

The Broken Hill lode strikes north-northeast, parallel to the foliation and trend of the metamorphosed Willyama sediments. The mineralized zone crops out for about 10 km along strike and plunges downward at each end. The ore-bearing zone is actually a composite lode of two or more metal-rich beds of contorted gneiss, nestled one above the other (Figure 14-15) as though they were originally flat-lying intercalated sheets. Each lode, or layer, is distinguished by details of its gangue mineralogy and metal ratios. The number 2 lens (now the upper) is characterized by calcite, bustamite [$(Mn,Ca)SiO_3$], wollastonite, and both rhodonite and hedenbergite. By contrast, the Number 3 lens (the lower) contains fluorite, Fe-Mg-garnet, rhodonite, and pyroxmangite [$(Mn,Fe)SiO_3$]. Both lodes contain other minerals in common, including Mg-Al garnets, quartz, K-feldspar, apatite, rutile, and rarer species. The lower horizon has relatively high zinc/lead and silver/lead ratios compared to the upper horizon, and rhodonite is much more abundant in the lower lens than in the upper. In the southern part of the district, in the mines of the Zinc Corporation and the New Broken Hill Consolidated, as many as six layers of ore have been developed. The upper (stratigraphically lower) four ore layers are known as the zinc lodes (Carruthers, 1965). These zinc lodes are shown in Figure 14-15 in relation to the No. 3 and No. 2 ore horizons. The No. 2 and No. 3 are shown at a larger scale in Figure 14-16.

Broken Hill ore is mined principally for sphalerite and galena, but it also contains tetrahedrite, pyrrhotite, marcasite, chalcopyrite, arseno-

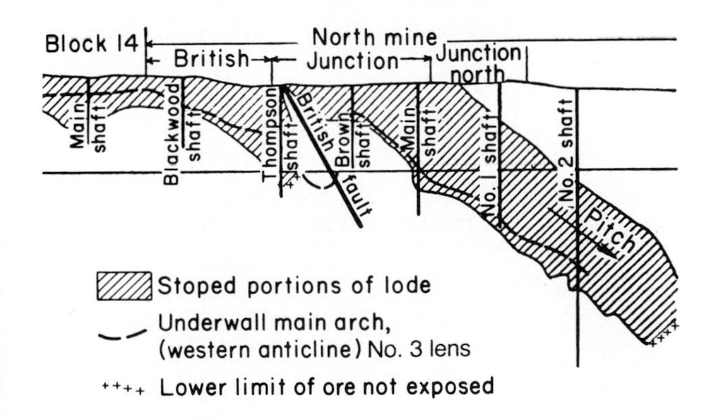

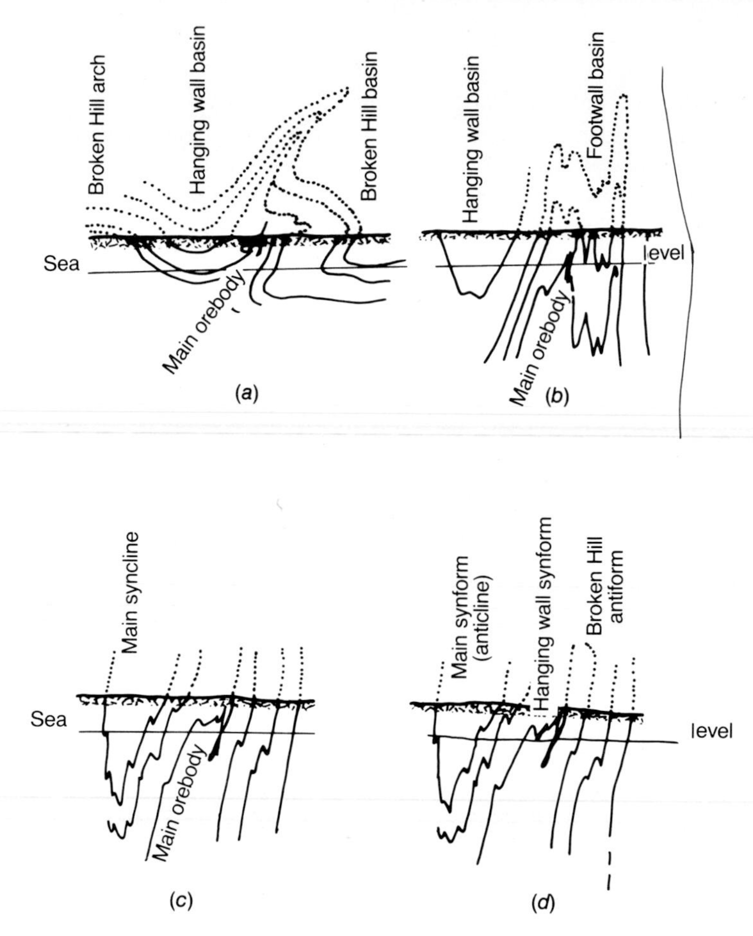

Figure 14-14. Four successive structural interpretations of the Broken Hill, New South Wales, district dated 1922, 1950, 1965, and 1978. Each faces north. See text for explanation of *a*, *b*, and *c*; *d* incorporates Laing, Marjoribanks, and Rutland's finding that the entire sequence has been inverted.

pyrite, loellingite ($FeAS_2$), dyscrasite (Ag_3Sb), pyrargyrite, gudmundite (FeSbS), and cubanite ($CuFe_2S_3$). Stillwell (1953) reported small amounts of many other minerals, including wolframite, scheelite, molybdenite, cobaltite, pyrite, stannite, breithauptite (NiSb), niccolite, bornite, jamesonite, native gold, and native antimony. This mineral assemblage appeared to hydrothermal theorists to represent formation over a perplexingly broad temperature range. Arsenopyrite, pyrrhotite, wolframite, and cobaltite were considered typical high-temperature "hypothermal" minerals; tetrahedrite, bornite, and jamesonite are usually found in mesothermal deposits; and pyrargyrite and marcasite typify the epithermal environment. Most of these constituent elements are now known from exhalation-related deposits, and resident and "outside" geologists alike generally accept that origin.

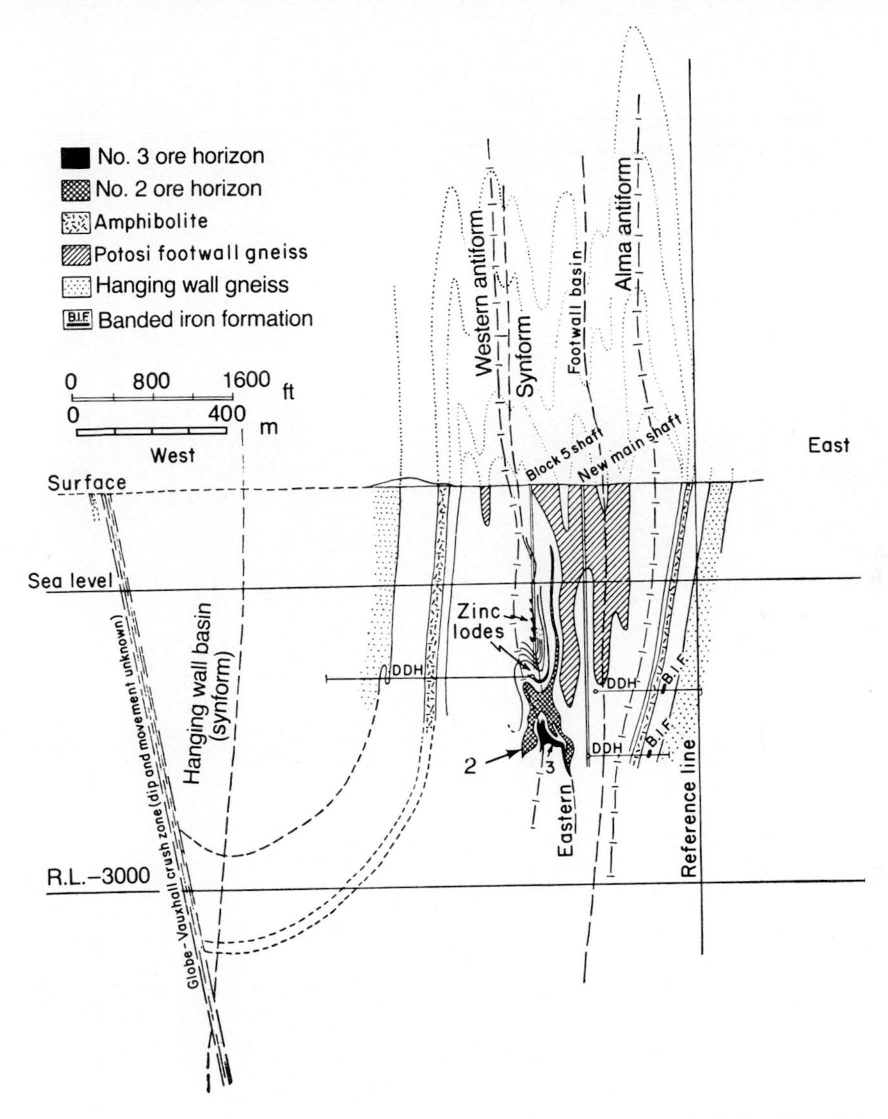

Figure 14-15. Cross section of the southern Broken Hill lode, New South Wales. See also Figure 14-13, where the zinc lodes are shown; its main shaft is shown here as New main shaft. (*From* King and O'Driscoll, 1953, Figure 3.)

Although each lens or layer has its own characteristic mineral assemblage, the ore layers are in general remarkably uniform throughout the entire length of the lode. Differences between the lenses seem to be related to original variations in layer composition rather than to district-scale zoning effects. Zonation is reflected in slightly greater concentrations of pyrrhotite, arsenopyrite, and sphalerite at the south end of the lode. The paragenesis of the minerals is obscure (King and O'Driscoll, 1953). Petrographic

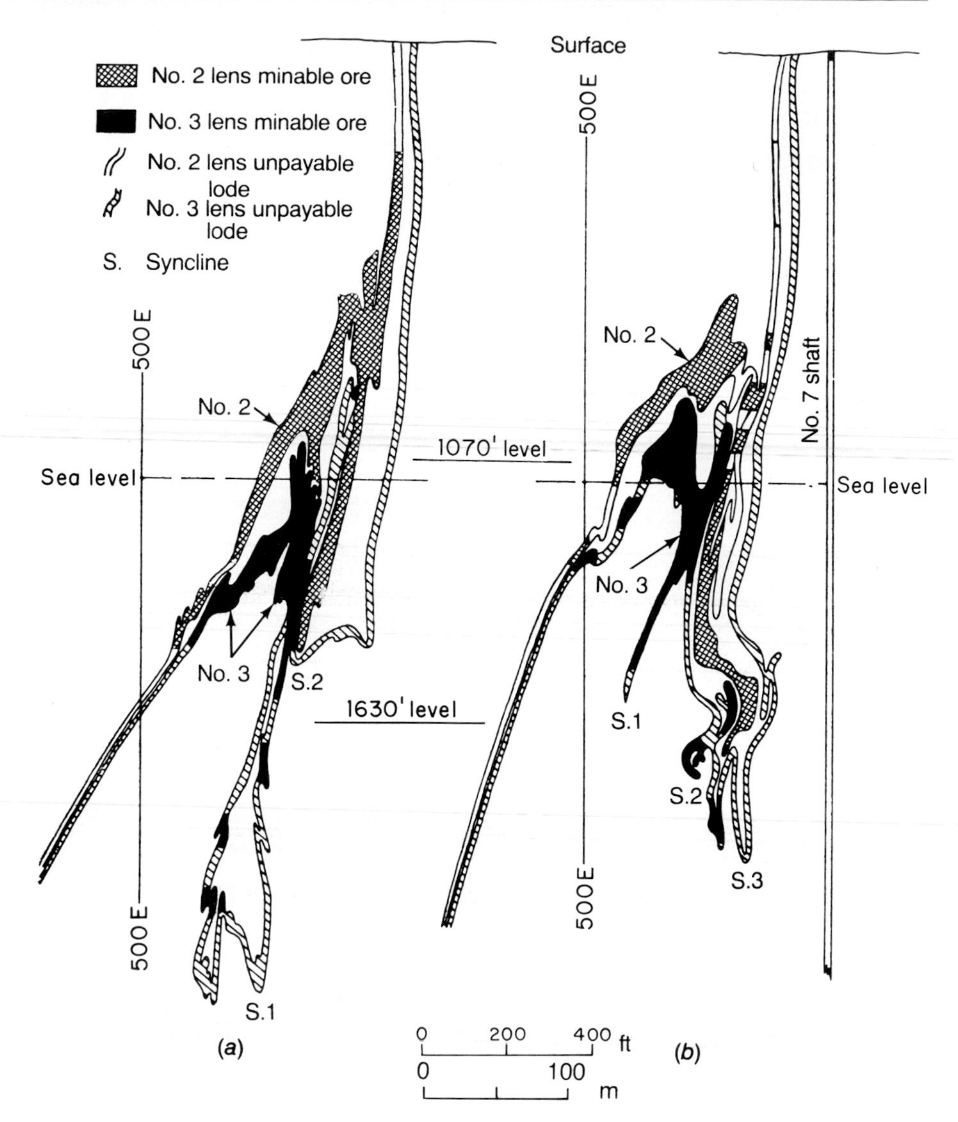

Figure 14-16. Cross sections (looking north) of the Main Lode, Broken Hill district, New South Wales. (*a*) At coordinate 2200S. (*b*) At coordinate 1800S. (*From* Lewis, Forward, and Roberts, 1965.)

studies indicate that the sulfides followed the principal gangue minerals and that pyrrhotite and arsenopyrite were deposited before zinc, lead, and silver minerals, but Ramdohr (1950, 1953) suggested that the mineral relationships are due to postdepositional recrystallization during the period of metamorphism. If Ramdohr is correct, the ores may represent a metamorphosed epigenetic deposit or, as King and Thomson (1953) first advocated, a metamorphosed syngenetic deposit. Geothermometry cannot solve this controversy because the ore temperatures would reflect the temperatures of granulite metamorphism.

Wall-rock alteration is not widespread, nor is it uncommon to find ore against unaltered gneiss; however, some of the gneiss adjacent to the lodes has been silicified, garnetized, or sericitized. Bleached zones are left in the walls where biotite and sillimanite are intergrown with sericite. The most extensive apparent alteration is replacement by silica, and where silicification has been thorough, the gneisses look like fine-grained quartzites (Gustafson et al., 1950).

As has clearly been alluded to, controversy between a hypogene hydrothermal epigenetic origin and a syngenetic sedimentary or volcano-sedimentary origin has swirled around the Broken Hill deposits. According to the hydrothermal theory, the ore was emplaced by hot fluids that moved up dips and along the crests of folds, replacing up to six favorable stratigraphic horizons; folding and metamorphism are believed to have channeled and localized, and thus to have antedated, the ore. The alternative hypothesis is that the ores were deposited as black-shale-hosted chemical sediments that were subsequently folded and recrystallized during metamorphism. Even the syngenetic ore would show some replacement textures because metamorphism would recrystallize it, and would force it to flow in plastic fashion from the limbs to the crests of the folds. Both hypotheses are supported by careful geologists who are thoroughly acquainted with the area, and neither hypothesis can answer all the questions.

The syngenetic hypothesis first proposed by King and Thomson (1953) was novel for the Broken Hill lodes and profoundly important to the evolution of geologic thought concerning stratiform ores. Whether advocating igneous or metamorphic processes, sponsors of earlier theories had all argued that the ore was genetically related to the granites and pegmatites. But King and Thomson were impressed by the persistence of the No. 2 and No. 3 lenses, which contain dissimilar mineral assemblages in spite of being stratigraphically close, even locally in contact. They suggested that the selective replacement of two favorable strata by two different ore fluids, as reflected by persistent but different Zn/Ag/Pb ratios along 10 km of highly contorted rocks—and four additional ore layers in the southern portion of the district (King, 1983)—could not be so perfect. The physical problems of the transmission of ore-bearing fluids along the lodes present a further problem, in that the mechanism of transport for such large quantities of solutions along a 10-km replacement lode is difficult to envisage. Consequently, King and Thomson (1953) suggested that the ores were deposited with the Willyama sediments—some 12 distinct units that present characteristics and problems common to those of the ores—and were later folded, metamorphosed, and granitized. This interpretation has been supported by radiometric age determinations based upon the lead isotope ratios of the galenas, which indicate that most of the lead was formed 1.4 to 1.6 billion years ago (Figure 7-21) and did not experience a complex geochemical history prior to deposition (Russell and Hawley, 1957; Russell and Farquhar, 1960). Some of the ores, however, give anomalous ages, and hence support the hypothesis that at least part of the mineralization is epigenetic.

Most advocates of syngenesis are impressed by the continuity of individual layers and the fact that compositions of these layers do not change. Richards (1966) clearly states his thesis that geometrical and compositional relationships between the orebodies, garnetiferous and manganiferous quartzites, and itabirites "suggest that the three rock types are cogenetic." He states further that since two of these types, the garnet- and manganese-bearing quartzites and the itabirites, are metamorphosed sedimentary beds, then the third type, which is the ore bed, is also likely to be of sedimentary origin. Advocates of the syngenetic hypothesis maintain that metamorphism and deformation have obscured many of the features that support the sedimentary hypothesis. These arguments have taken on much more convincing dimensions in recent years. Stanton (1972), for example, discussed the relationship of the ore to the iron formation that seems everywhere to accompany it. He assumes as a working hypothesis that both the ores and the iron formation are of volcanogenic origin. In five papers published in 1976 and 1978 he demonstrated that metamorphism even of granulite rank does not importantly redistribute base metals; that original chemical differences of the No. 2 and No. 3 horizons "survived" that metamorphism; that exhalite, BIF-like units virtually everywhere at Broken Hill are part of the ore "package"; that the whole rock geochemistry of the enclosing siliceous sediments suggests at least some volcanic contribution; that the garnet-quartzite rocks (the "garcues" of local vernacular) are probably metamorphosed Mg-Fe-chlorite; that the sillimanite and sillimanite-anthophyllite units are probably metamorphosed aluminous (kaolinitic) altered conduit rocks; and that the high iron, manganese, barium, phosphorus, and base-metal strata can be seen in unmetamorphosed deposits like Rammelsberg and Sullivan to be associated with syngenetic stratabound deposits.

The advocates of a hydrothermal origin believe that ore deposition was controlled by wall-rock chemistry as well as structure and that sensitive, selective replacement of slightly different original favorable beds is indicated. Attenuation along the crests of folds having left these zones relatively permeable, they acted as conduits for ore-bearing fluids migrating up the plunge. Further localization was caused during the period of ore deposition by repeated fracturing in the favorable strata (Gustafson et al., 1950). Crumpling of the weak beds between layers of stronger rocks produced irregular saddle-shaped structures that nearly parallel the original bedding. According to the hydrothermal theory, the ores are replacements of the favorable beds along this zone of maximum deformation.

Notwithstanding these arguments, the syngenetic arguments are so strong that the latest published statement (Laing, Marjoribanks, and Rutland, 1978) indicates (page 1112) that *the orebodies lie close to the interface between a lower, partly volcanogenic sequence and a higher, unvolcanogenic sequence. The lead lodes lie stratigraphically above the zinc lodes . . . as stratiform, premetamorphic lenses . . . originally deposited . . . within a limited stratigraphic interval.*

The length of the discussion devoted to Broken Hill, New South Wales, geology and genesis might seem disproportionate were it not for the facts that several similar deposits are known on as many continents (King, 1958) and that a truly remarkably similar deposit in almost identically composed and metamorphosed terrane has been discovered in just the last decade. It is the Broken Hill–Black Mountain–Aggeneys deposit in South Africa, which Stanton has described (1976e) as so similar to the New South Wales occurrence as to demand similar genetic controls. And with the discovery of Aggeneys, are there not more to be sought in the Australian, African, Brazilian, Canadian, and Asian shields? The black-shale-hosted deposits are actually and conceptually an increasingly important subclass of base- and precious-metal ore deposits, and the modern explorationist should be prepared to deal with both unmetamorphosed and metamorphosed occurrences.

Bibliography

Anger, G., H. Nielsen, H. Puchelt, and W. Ricke, 1966. Sulfur isotopes in the Rammelsberg ore deposit, Germany. *Econ. Geol.* 61:511–536.

Andrews, E. C., 1922. The geology of the Broken Hill district. N.S.W. Mem. Geol. 8.

——, 1948. Geology of Broken Hill, New South Wales. *Proc. 18th Int. Geol. Congr.*, pt. 7, pp. 117–122.

Bennett, E. M., 1965. Lead-zinc-silver and copper deposits of Mount Isa. *8th Commonw. Min. Metall. Congr. Trans.*, vol. 1, pp. 233–246.

Blanchard, R., 1968. Interpretation of leached outcrops. Nev. Bur. Mines Bull. 66.

—— and G. Hall, 1942. Rock deformation and mineralization at Mount Isa. *Australas. Inst. Min. Metall. Proc.* 125:1–60.

Both, R. A., and R. W. R. Rutland, 1976. The problem of identifying and interpreting stratiform ore bodies in highly metamorphosed terrains, the Broken Hill example, pp. 261–326 in K. H. Wolf, Ed., *Handbook of Stratabound and Stratiform Ore Deposits*, vol. 4. New York: Elsevier.

—— and J. W. Smith, 1975. A sulfur isotope study of base-metal mineralization in the Willyama Complex, Western N.S.W., Austral. *Econ. Geol.* 70:308–318.

Campbell, F. A., V. G. Ethier, and H. R. Krause, 1980. The massive sulfide zone: Sullivan orebody. *Econ. Geol.* 75:916–926.

Carter, E. K., J. H. Brooks, and K. R. Walker, 1961. The Precambrian mineral belt of north-western Queensland. Australas. Bur. Miner. Res. Geol. Geophys. Bull. 51.

Carter, S. R., 1953. Mount Isa mines, in A. B. Edwards, Ed., *Geology of Australian Ore Deposits*. Melbourne: Australas. Inst. Min. Metall.

Carruthers, D. S., 1965. An environmental view of Broken Hill ore occurrences. *8th Commonw. Mining Congr. Trans.*, vol. 1, pp. 339–351.

Fisher, N. H., 1960. Review of evidence and genesis of Mount Isa orebodies. *Proc. 21st Int. Geol. Congr.*, vol. 16, pp. 99–111.

Freeze, A. C., 1966. On the origin of the Sullivan orebody. Kimberley, B.C. Can. Inst. Min. Metall. Spec. Vol. 8, pp. 263–294.

Gilmour, P., 1971. Stratabound massive pyritic deposits—a review. *Econ. Geol.* 66:1239–1244.

——, 1976. Some transitional types of mineral deposits in volcanic and sedimentary rocks, pp. 111–160 in K. H. Wolf, Ed., *Handbook of Stratabound and Stratiform Ore Deposits*, vol. 1. New York: Elsevier.

—— and A. R. Still, 1968. The geology of the Iron King mine (Arizona), pp. 1239–1257 in J. D. Ridge, Ed., *Ore Deposits of the United States 1933/1967*, Graton-Sales Vols. New York: AIME, 1880 pp.

Gustafson, J. K., H. C. Burrell, and M. D. Garretty, 1950. Geology of the Broken Hill ore deposit, N.S.W., Australia. *Geol. Soc. Amer. Bull.* 61:1369–1437.

Hodgson, C. J., 1974. The geology and geological development of the Broken Hill lode in the New Broken Hill Consolidated mine, Australia. Part I: structural geology. *J. Geol. Soc. Austral.* 21:413.

——, 1975a. The geology and geological development of the Broken Hill lode in the New Broken Hill Consolidated mine, Australia. Part II: petrology and petrogenesis. *J. Geol. Soc. Austral.* 22:35–56.

——, 1975b. The geological development of the Broken Hill lode in the New Broken Hill Consolidated mine, Australia. Part III: petrology and petrogenesis. *J. Geol. Soc. Austral.* 22:195–228.

—— and J. W. Lydon, 1977. Geological setting of volcanogenic massive sulfide deposits and active hydrothermal systems: some implications for exploration. *Can. Min. Metall. Bull.* 70:360–366.

Jaquet, J. B., 1894. Geology of the Broken Hill lode and Barrier Ranges mineral field. Geol. Surv. N.S.W. Mem. 5.

Johnson, I. R., and G. D. Klingner, 1975. Broken Hill ore deposit and its environment, in C. L. Knight, Ed., *Economic Geology of Australia and Papua New Guinea: I, Metals*. Australas. Inst. Min. Metall. Mon. 5, pp. 476–491.

King, H. F., 1958. Notes on ore occurrences in highly metamorphosed Precambrian rocks. Australas. Inst. Min. Metall., Stillwell Anniv. Vol., pp. 143–167.

——, 1983. Pers. Commun.

—— and E. S. O'Driscoll, 1953. The Broken Hill lode, in A. B. Edwards, Ed., *Geology of Australian Ore Deposits*. Melbourne: Australas. Inst. Min. Metall.

—— and B. P. Thomson, 1953. The geology of the Broken Hill district, in A. B. Edwards, Ed., *Geology of Australian Ore Deposits*. Melbourne: Australas. Inst. Min. Metall.

Laing, W. P., R. W. Majoribanks, and R. W. R. Rutland, 1978. Structure of the Broken Hill mine area and its significance for the genesis of the orebodies. *Econ. Geol.* 73:1112–1136.

Lambert, I. B., 1976. The McArthur River Zn-Pb-Ag deposit: features, metallogenesis, and comparisons with some other stratiform ores, pp. 535–585 in K. H. Wolf, Ed., *Handbook of Stratabound and Stratiform Ore Deposits*, vol. 6. New York: Elsevier.

Leblanc, M., and P. Billaud, 1978. A volcano-sedimentary copper deposit on a continental margin of upper Proterozoic age: Bleïda, Anti-Atlas, Morocco. *Econ. Geol.* 73:1101–1111.

Lewis, B. R., P. S. Forward, and J. B. Roberts, 1965. Geology of the Broken Hill lode, reinterpreted. *8th Commonw. Min. Metall. Congr. Trans.*, vol. 1, pp. 319–332.

Mathias, B. V., and G. J. Clark, 1975. Mount Isa copper and silver-lead-zinc orebodies—Isa and Hilton Mines, in C. L. Knight, Ed., *Economic Geology of Australia and Papua New Guinea: I, Metals*. Australas. Inst. Min. Metall. Mon. 5, pp. 351–371.

Mount Isa Mines, Ltd., 1977. *Operations at Mount Isa*, revised ed., 70 pp.

Nieman, W., 1976. Pers. Commun.

Page, D. C., and M. D. Watson, 1976. The Pb-Zn deposit of Rosh Pinah mine, South West Africa. *Econ. Geol.* 71:306–327.

Ramdohr, P., 1950. Die Lagerstätte von Broken Hill im New South Wales im Lichte der neuen geologischen Erkenntnisse und erzmikroskopischer Untersuchungen. *Heidelberger Beitr. Min. Pet.* 2:291–333.

——, 1953. Über Metamorphose und sekundäre Mobilisierung. *Geol. Rundsch.* 42:11–19.

Renfro, A. R., 1974. Genesis of evaporite-associated stratiform metalliferous deposits—a sabkha process. *Econ. Geol.* 69:33–45.

Richards, S. M., 1966. The banded iron formations at Broken Hill, Australia, and their relationships to the lead-zinc orebodies. *Econ. Geol.* 61:72–96; 257–274.

Russell, R. D., and R. M. Farquhar, 1960. Dating galenas by means of their isotopic constitutions—II. *Cosmochim. Geochim. Acta* 19:41–52.

—— and J. E. Hawley, 1957. Isotopic analyses of leads from Broken Hill, Australia, with spectrographic analyses. *Amer. Geophys. Union Trans.* 38:557–565.

Stanton, R. L., 1972. A preliminary account of chemical relationships between sulfide lode and "banded iron formation" at Broken Hill, New South Wales. *Econ. Geol.* 67:1128–1145.

——, 1976a. Petrochemical studies of the ore environment, Broken Hill, New South Wales, Australia: 1. Constitution of "banded iron formations." *Inst. Min. Metall. Trans.* 85:B33–B46.

——, 1976b. Petrochemical studies of the ore environment at Broken Hill, New South Wales: 2. Regional metamorphism of banded iron formations and their immediate associates. *Inst. Min. Metall. Trans.* 86:B118–B131.

——, 1976c. Petrochemical studies of the ore environment at Broken Hill, New South Wales: 3. Banded iron formations and sulfide orebodies: constitutional and genetic ties. *Inst. Min. Metall. Trans.*, 86:B132–B141.

——, 1976d. Petrochemical studies of the ore environment at Broken Hill, New South Wales: 4. Environment synthesis. *Inst. Mining Metall. Trans.* 86:B221–B233.

——, 1976e. Pers. Commun.

——, 1978. Tectonic deformations at Broken Hill, New South Wales, Australia, and their significance for interpretation of ore environments: discussion and contributions. *Inst. Min. Metall. Trans.*, 87:B176–180.

Stillwell, F. L., 1953. Mineralogy of the Broken Hill lode, in A. B. Edwards, Ed., *Geology of Australian Ore Deposits.* Melbourne: Australas. Inst. Min. Metall.

Vaughn, D. J., 1976. Sedimentary geochemistry and mineralogy of the sulfides of lead, zinc, copper, and iron and their occurrence in sedimentary ore deposits, pp. 317–363 in K. H. Wolf, Ed., *Handbook of Stratabound and Stratiform Ore Deposits*, vol. 2. New York: Elsevier.

Williams, N., 1978a. Studies of the base metal sulfide deposits at McArthur River, Northern Territory, Australia. I: The Cooley and Ridge deposits. *Econ. Geol.* 73:1005–1035.

——, 1978b. Studies of the base metal sulfide deposits at McArthur River, Northern Territory, Australia. II: The sulfide-S and organic-C relationships of the concordant deposits and their significance. *Econ. Geol.* 73:1036–1056.

FIFTEEN

Deposits Related to Chemical Sedimentation

Vincent McKelvey said in 1950 that "principles of sedimentation are little used in the production of sedimentary ore and rock materials, and such principles as *are* used are elementary in nature" (page 501). His statement was valid then, and has remained so through the 1970s. As the 1980s progress, we are still somewhat primitive in the application of sedimentology to ore deposition, but as Eric Cheney of the University of Washington has said, "the eighties will be the decade of ore deposits in sedimentary rocks." Heightened awareness of the role of sedimentologic processes has developed because of improved understanding of the chemical and mechanical dynamics of sedimentation, of the sedimentary petrology of depositional and diagenetic environments, and of sedimentary lithotectonics. Many deposits—like those described in the preceding two chapters—are now recognized as having formed in sedimentary basins. In this chapter we focus on deposits considered to have formed by chemical sedimentologic precipitation of ore-forming constituents, an origin commonly shared by the immediate or adjacent host rocks. Use of the term *sedimentary ores* in this chapter and elsewhere refers strictly to those ore components localized by processes of sedimentation or diagenesis, not to all ores that occur in sedimentary rocks, some of which have been introduced epigenetically. We examine a range of environments from littoral-continental to deep sea, for such is the siting of these ores. Several major sulfide deposits such as the Kupferschiefer district of Poland and the German Democratic Republic and the Zambian Copperbelt are thought to have formed in near-shore environments, at the boundary between terrestrial-fluvial and marine conditions. Several large manganese deposits and accumulations of oolitic iron ores, such as those of Clinton, Alabama, are believed to have formed in shallow marine waters in geosynclines or on miogeoclines or continental shelves. Several other important commodities form there—limestone, "cement rock," and dolostone, as well as the world's great deposits of phosphates. Phosphate—used principally as a fertilizer—has formed, and is forming today, by chemical precipitation of the calcium phosphate apatite from cold, landward-directed, up-welling seawater on continental shelves in-board from the continental slopes and deep seas. Landlocked lakes and bays and tectonic arms of the seas in arid environments may produce concentrations of salts by evaporation. So important are all of these deposits that an entire issue of *Economic Geology* was recently devoted to them (Howard, 1979). And finally, the deep-sea floor is one of the loci of the nucleation and growth of nodular manganese-copper-nickel accretions called *manganese nodules*. We will not examine mechanical sedimentation and its part in forming clastic, or detrital, placer-type deposits until Chapter 16.

Chemical sedimentation takes many forms and produces deposits formed in a wide range of Eh-pH composition environments, although the basic condition of solubility contrast in low-temperature aqueous solution is common to all of them. By "solubility contrast" is meant the concept that a given component must be soluble as a simple or complex ion in a provenance terrane and be transported in solution or as a colloid into an environment in which solubility products are exceeded or complex ions destabilized, or both. Commonly a transition from one anion to another is involved, as from carbonate or oxide to sulfide. It is generally assumed that transport from landward to marine environments is required, but the sea itself as a reservoir of metals may be underrated. A perennial and valid argument is that if the sea alone were the source of sedimentary ores of iron, manganese, and copper, then sedimentary deposits of those metals should be more widespread and conventional lithologic entities. But sedimentary ore deposits do represent "abnormal" accumulations, although they are environmentally interpretable and transitional to normal environments at their perimeters.

Almost certainly, then, specialized chemical sediments that are ore deposits involve delivery of unusual amounts of a metal into an environment in which the delivery mode is chemically disrupted and an insoluble compound formed. Chemical precipitation results. But a further requirement is that concentration must happen either by virtue of abundant chemical precipitation or by moderate precipitation in a diluent-starved, clastic-detrituspoor depositional basin. Even if chalcocitelike mineraloids are being deposited in a stagnant bay, the resulting sediment cannot reach ore grade if the system is overwhelmed with either an influx of clastic components or the deposition of gangue compounds. Thus is becomes clear that mechanical and chemical weathering supply ore materials to basins of deposition just as they supply quartz, clays, and dissolved solids for the production of clastic and nonclastic sediments. Some metals could have been transported clastically and mobilized and reprecipitated during diagenesis. Under favorable conditions of transportation, sorting, and deposition, some of the ore materials become sufficiently concentrated to constitute economic deposits. These sedimentary ores are generally classified first as either chemical precipitates or mechanical accumulations and second according to their chemical or mineralogical composition. Whether chemically or mechanically derived, sedimentary ores are syngenetic deposits, *syngenetic* meaning "formed contemporaneously with the enclosing rocks."

Many large and valuable sedimentary deposits have been exploited, and additional discoveries are to be expected as our knowledge of sedimentary processes grows. The minerals industry increasingly appreciates the value of studies of sedimentary deposits and of the processes that produced them.

Deposits of many of the metals have been precipitated as primary sediments from surface waters by chemical and biochemical processes (McKelvey, 1950; Mason, 1958; Garrels and Mackenzie, 1971; Vaughn, 1976, in Wolf's ten-volume set, 1976). Although most elements are amenable to chemical sedimentary precipitation in one form or another, only a few have formed large deposits. Principal ores of this type include the oxides, silicates, and carbonates of iron and manganese, such as the Clinton-type iron ores and the oolitic manganese sediments at Chiaturi in the Urals of Russia. Base-metal deposits such as the Kupferschiefer copper-zinc-lead strata at Mansfeld, Germany, appear to be sediments that have accumulated under unusual conditions. The origin of other bedded sulfides is considered by many geologists to be either syngenetic sedimentary or volcanogenic (Dunham, 1964; Sangster, 1972). Low-grade uranium, vanadium, and rare element deposits, particularly those associated with marine black shales and phosphorites, also constitute chemical sediments. It should be clear that many of the exhalite ores treated in Chapter 13 are also chemical sediments, but that they involve a volcanic source. We here focus on a nonvolcanic terrestrial or marine source.

The chemical precipitation of sediments is controlled by many factors, chief among which are the availability of the ions in question and the pH and Eh of the environment (Figure 13-18). Oxidation-reduction potentials are different for each element. The oxidation state of an element in a particular sedimentary environment is largely controlled by the oxygen content of the water, which is in turn ordinarily a function of depth and nearness to atmospheric mixing at the shoreline. Most dissolved solids are supplied to the depositional basins by streams and rivers, so mineral deposits, if they result, are likely to be thickest and best developed along the shoreline. Reefform or barform deposits are thus more common than broad equidimensional sheets. Oxygen levels are set by interaction of the sea with atmospheric oxygen, by equilibrium between oxide ions in solution, by biologic processes, and by the activities of carbonate and sulfide species. In general, stagnation of water—its removal from high-energy sedimentologic environments—leads to low Eh and relatively low pH.

The role of biochemical processes in the precipitation of ore deposits has been a subject of long-standing debate. These processes remain underevaluated and probably underrated (Hem, 1960; Temple, 1964; Baas-Becking and Moore, 1961; Roberts, 1967; Trudinger, Lambert, and Skyring, 1972; Margulis, 1981a). Certain bacteria and algae can cause the precipitation of oxide compounds by acting as catalysts for oxidation reactions. Furthermore, anaerobic bacteria are able to reduce sulfates and produce H_2S or $(HS)^-$, which in turn may cause the precipitation of ore minerals. Sulfide deposits of the base metals may form where complex metal ions encounter H_2S or $(HS)^-$, and native sulfur may form through the oxidation of H_2S by inorganic processes or by aerobic bacteria as the reduced species enter an aerated zone (Dessau et al., 1962). Iron- and manganese-fixing bacteria have been recognized since the nineteenth century, but their actual role in the formation of ore deposits is unknown. Since most of the reactions they cause may also take place without their involvement, although at a slower rate, the physical and chemical environment may be the true controlling factor. The presence of sulfides in carbonaceous shale facies of basin sediments supports the hypothesis of a biogenic involvement, but oxide, carbonate, and silicate facies generally lack evidence of organic activity. The problem is clouded by the similarity between the conditions under which microorganisms cause iron and manganese to precipitate and those conditions under which these metals will precipitate inorganically (Beerstecher, 1954). Margulis (1981b) has marshaled chemical, petrographic, and descriptive evidence showing that the hydrosphere has teemed with bacterial and algal life since 3.5 billion years ago, and that virtually all sedimentological processes have proceeded in the presence of the biomass and were probably influenced by it.

A visual distinction between hand samples of sedimentary ores formed as chemical precipitates of terrigenous metals and those formed by hydrothermal or exhalative processes may be megascopically difficult or nearly impossible to make, although they should be chemically distinctive. The problem is generally a matter of distinguishing pseudomorphous bedding from original bedding or interstitial replacement material from diagenetic cement. Fine-grained banded sulfide ores like those from the Rammelsberg district, Germany (Figure 14-5) are especially difficult to interpret. Recent interpretations have been heavily weighted in favor of primary depositional features.

▸ SEDIMENTARY BASE-METAL DEPOSITS

Wherever there is a source of metal ions flowing into a basin with redox conditions favorable for precipitation, an ore deposit has a chance to form. Such conditions may be met in basins of deposition where decaying organic debris or bacterial action generate an exceptional bottom-mud reducing environment, and where the accumulation of clastics is practically nil. Several large deposits in the world seem to belong in this category, and in each case doubt about whether or not the ore minerals are actually syngenetic-diagenetic diminishes annually. Perhaps the principal objection to this mechanism of ore deposition is the mysterious source of metals; normal seawater does not carry an appreciable supply. Given the base-metal ions, however, there is no chemical objection to precipitation in a reducing sedimentary environment.

Geologists have long advocated a syngenetic chemical sedimentary origin for the base-metal deposits of the Kupferschiefer near Mansfeld, Germany (Pompecki, 1920; Schneiderhöhn, 1923; Trask, 1925; Jung and Knitzschke, 1976). According to Dunham (1964), the Kupferschiefer of northern Europe is a bituminous-calcareous shale that lies near the base of the Zechstein formation of Middle Permian age; the word translates to "copper shale," or "cupriferous shale." The Kupferschiefer sedimentary unit is only about 1 meter thick, but it extends from northern England eastward through the Netherlands, across Germany, and into Poland. It appears to follow a long shallow arm of a Permian sea. It is only locally cupriferous enough to be mined. The Kupferschiefer overlies a thin conglomerate and is overlain by limestones. It contains marine animal and plant remains, but they do not represent an in situ ecologic assemblage. Apparently the environment was similar to that of the present Black Sea—shallow stagnant waters that were not amenable to a flourishing fauna.

The ore minerals of the Kupferschiefer are sulfides, principally bornite, chalcocite, chalcopyrite, galena, sphalerite, tetrahedrite, and pyrite. Minor amounts of silver, nickel, cobalt, selenium, vanadium, and molybdenum have also been recovered. Copper is the most important product; over wide areas it averages as much as 3% of the deposit. The unusual nature of the deposit is apparent in the east end of the Harz Mountains of Germany, where the shale, which is only 22 cm thick, has been worked over an area of about 140 km^2. According to Dunham (1964), at least 6000 km^2 are underlain by shale that contains more than 1% Zn; another almost equal area has similar concentrations of lead. The total area underlain by metalliferous shale is at least 20,000 km^2. It should be stressed that most of this carbonaceous shale is anomalous in Cu, Pb, and Zn. Ratios shift such that one metal or the other predominates locally.

Much discussion has focused on the source of the metals. Jung and Knitzschke (1976) favor a sedimentary syngenetic origin with metals delivered from now eroded landmasses which were adjacent to the shallow Permian ensialic sea and with H_2S sulfur coming from the decomposition of organic matter and reduction of sulfates, an origin confirmed by sulfur isotopic studies. The Cu-Pb-Zn ions were deposited as the sea transgressed, and were derived from the continent. An alternative to this hypothesis is an exhalative sedimentary origin, with metals contributed by submarine volcanic springs and with precipitation where stagnant bottom conditions prevailed.

Another important basin of deposition of base metals is that of the Belt Supergroup in Montana, Idaho, British Columbia, and Alberta. Scores of stratiform, presumably sedimentary deposits of copper and copper-silver have been recognized there for years, but the only economically viable deposit at present is ASARCO's Spar Lake deposit near Troy, Montana (Balla, 1980). The Spar Lake deposit is a 20-meter-thick, 70-million-ton layerlike subunit in the Revett Formation—the ore quartzite—that contains an average of 7500 ppm (0.75%) Cu and 40 ppm Ag. Unfaulted boundaries are gradational, and the ores are vertically zoned from chalcopyrite upward into bornite-chalcocite, then back to chalcopyrite and pyrite (Balla, 1982), so highest metal-to-sulfur ratios occur in the middle strata. The best grades are in shales to the west, the poorest grades in silty portions to the east,

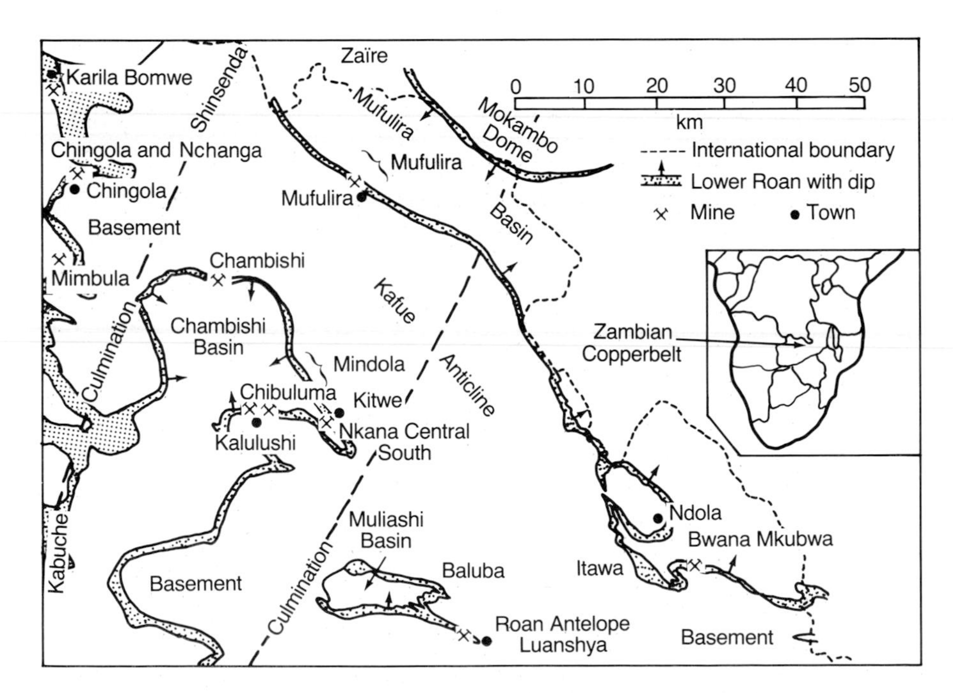

Figure 15-1. Location map of major features of the geology and geography of the Zambian Copperbelt, including an inset map of southern Africa. The Roan-Muliashi basin at the bottom of this map is enlarged as Figure 15-4.

although sulfides elsewhere are generally confined to quartzite-siltite members in the Revett. Cross bedding, uniform gentle dips, shallow-water and deltaic structures, and beach and bar phenomena suggest near-shore environments. Copper sulfides are most commonly disseminated through the quartzite, or occur as bedding plane concentrations in either normal or cross-bedding arrays with the heavy minerals rutile, zircon, and tourmaline. Clark (1971) described sulfide participation in graded bedding, cross bedding, scour and cut-and-fill structures, and regular bedding. The primary sulfides chalcopyrite, galena, and bornite have been in part metamorphically redistributed into narrow discontinuous veinlets. Pyrite extends well beyond the ore zone in the cupriferous layers. Several features suggest that the sulfides are closely related to sedimentation or diagenesis as chemical rather than as detrital sediments; they are geographically and temporally restricted; sediments and sulfides have the same isotopic ages; sedimentary structures and processes dictate the details of sulfide distribution; there is apparent diagenetic local redistribution of values; wall-rock alteration is absent; and there are no recognized local sources or ingress avenues for epigenetic mineralizing fluids. Lange and Sherry (1983) suggest that the metals may be introduced immediately postdepositionally along regional-scale syndepositionally active faults, but most authors stress a syngenetic or diagenetic origin. Not surprisingly, there are several exploration groups working in the area.

Zambian Copperbelt

One of the world-class copper deposits is a district discovered and developed in the 1920s in what was then Northern Rhodesia and is now Zambia. The deposits were first owned by British and American interests. They were nationalized by emergent Zambia in 1970, and their production has dwindled through the 1970s as a result of their remoteness and problems connected with the self-determination of several central African countries. The mines lie in central Africa, 1500 km from both the Indian Ocean and the South Atlantic Ocean (Figure 15-1).

The ores of Zambia occur in 900-million-year-old younger Precambrian sedimentary rocks called the Roan Group of the Katanga Supergroup, or Mine Series. The principal host rock is the "ore shale," a carbonaceous, sulfide-rich, reduced euxinic sediment in which copper sulfides precipitated. But sulfides also were laid down in siltstones, arkoses, and quartzites, which were deposited by long-shore currents in sunlit seas, but which were anoxic and chemically reducing immediately below the water-sediment interface. The floor to the sediments was an Archean granitic and gneissic basement terrane, possibly with local copper-enriched areas that provided copper to their drainages. Topographically, the inundated coastal Precambrian terrane was one of knobs and small hills that formed shallow ridges and islands; this relief created eddies and currents that in turn controlled clastic sediment particle behavior and distribution and bottom chemistry. Early studies re-

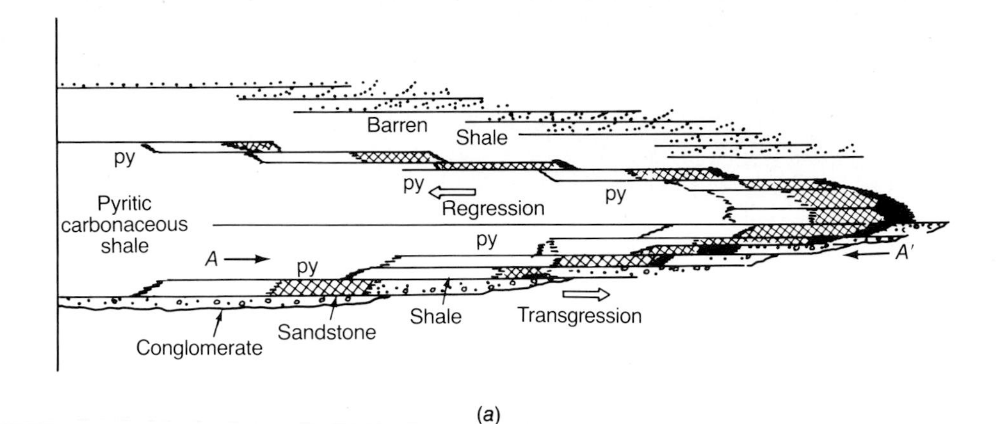

(*a*)

vealed a distinctive zoning of both metal distribution and mineralogy that seemed to be consistent with transgressive-regressive facies distribution, as shown in Figure 15-2. By applying standard sedimentology to these rocks, it appeared that at any given instant, near-shore sediments were barren. In shallow bottom muds, sediments including chalcocite (Cu_2S) with its high metal-to-sulfur ratio were deposited. Next seaward were bornite (Cu_5FeS_4) muds, and with decreasing Cu-Fe and metal-sulfur ratios away from shore, the succession went chalcocite-bornite-chalcopyrite-pyrite. Carrolite ($CuCo_2S_4$) was deposited sparingly with chalcopyrite, with linnaeite (Co_3S_4) deposited beyond copper in the pyrite zone farther seaward. This facies shift was preserved as a lateral and vertical mineralogical-chemical facies variation. These minerals could have been directly deposited, or they could have formed from original gelatinous, odiferous muds of appropriate composition and metal and metal-sulfur ratios during diagenesis, compaction, or mild metamorphism of sediments sometime after deposition.

The sulfides and oxides are in sedimentary rocks that appear to have been deposited in arid coastline environments with anhydrite and quartz sand dunes, as at Mufulira (Figure 15-2), and in littoral tide-flat environments, as at Chambishi and N'Changa, relationships sketched in Figure 15-2. The sediments were then variably metamorphosed and down-folded into the older, granitic floor rocks. They exist today in steep-walled, downwarped, complexly folded synclines (Figures 15-3 and 15-4). As shown in Figure 15-5, the cupriferous units have been mined downdip as far as possible by open-pit methods, and will generally be mined by underground techniques in the future.

Mineralized sediment thicknesses range from a few meters to tens of meters. They are restricted to a narrow sediment interval on or a few meters above the basement, although the Mine Series is kilometers thick. The ores average from 3 to 6% Cu from mine to mine, exceptionally to 15 or 20% Cu. In many mines, especially at N'Changa and Chambishi (Figure 15-6), sulfides at the modern surfaces have been oxidized in situ to cuprite,

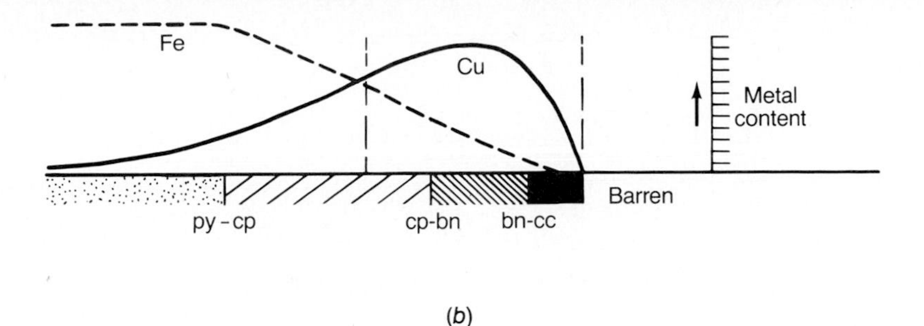

(*b*)

Figure 15-2. (*a*) (*facing page*) Zoning of minerals and sedimentary facies in the Roan Series at several Zambian copper occurrences. As basal conglomerate, sandy, and shaley faces advance landward (to the right), pyritic (py) and cupriferous facies (black) advance with them. The package retreats, without conglomerate, as the cycle reverses. (*From* Fleischer, Garlick, and Haldane, 1976.) (*b*) Metal and mineral zoning at any given level or time, as for example along line *A–A′* in (*a*). Notice that not only the copper content but also the copper-sulfur ratio increases shoreward. (*From* Shaner, 1982.)

Figure 15-3. Chambishi open pit in the Zambian Copperbelt. The dark ore shale lies in the bottom of the pit. The light-colored folds in the far wall are like those diagrammed in Figure 15-4 and shown in Figure 15-7.

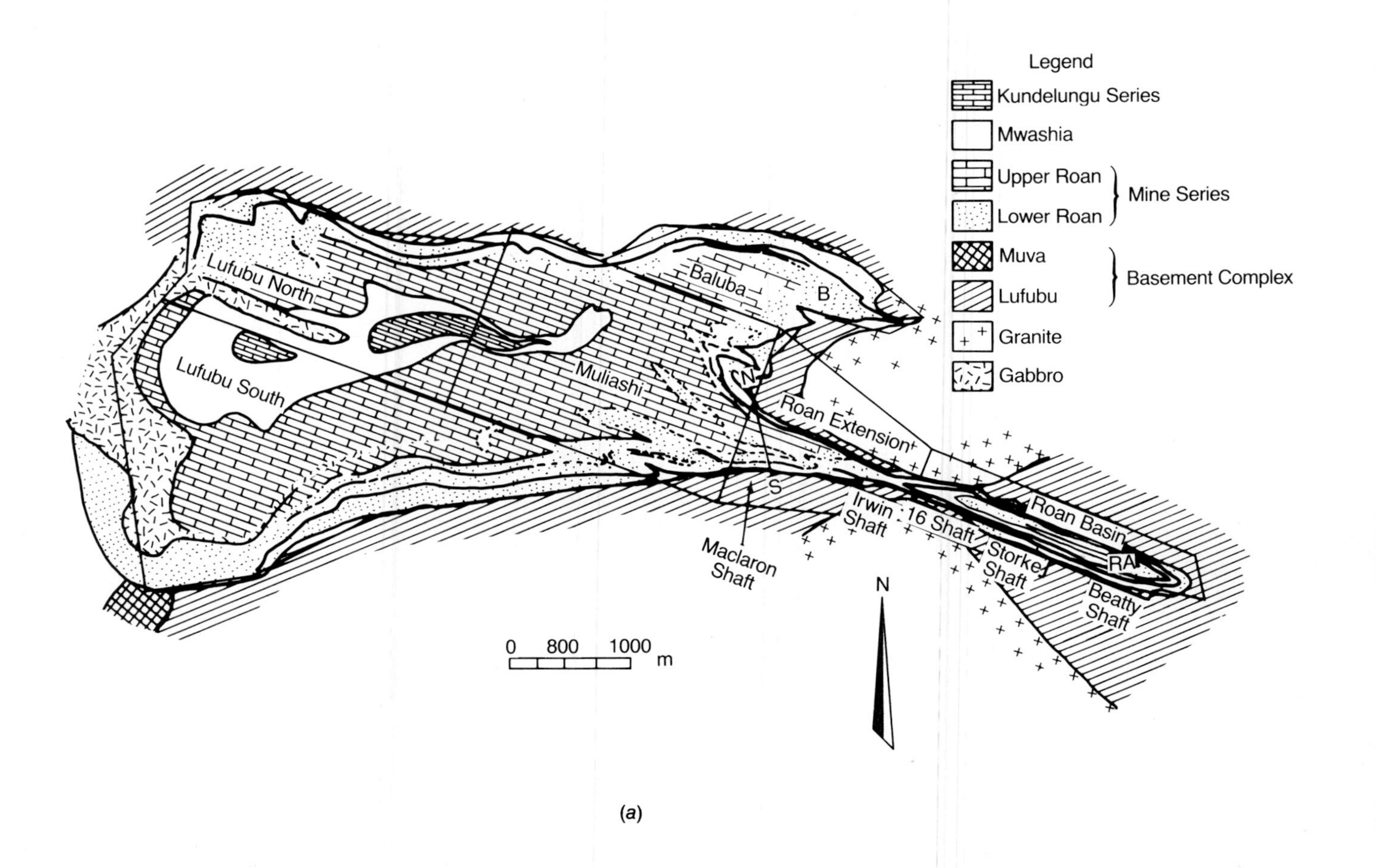

Legend
Kundelungu Series
Mwashia
Upper Roan
Lower Roan
Mine Series
Muva
Lufubu
Basement Complex
Granite
Gabbro
Lufubu North
Lufubu South
Baluba
B
Muliashi
N
Roan Extension
S
Irwin Shaft
16 Shaft
Maclaron Shaft
Storke Shaft
Roan Basin
RA
Beatty Shaft
N
0
800
1000
m

(a)

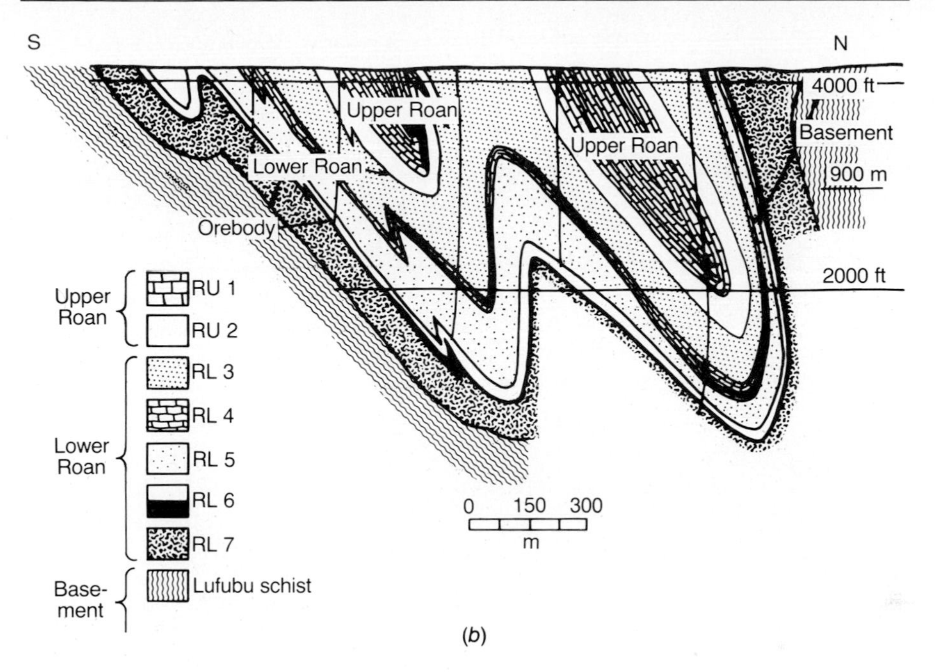

Figure 15-4. (*a*) (*facing page*) Surface map and (*b*) cross section showing the form and downfolding of the Roan Muliashi basin, one portion of the Zambian Copperbelt. (See also Figure 15-1.) The Baluba mine is at B, the fabled Roan Antelope at RA. The cross section shows how the ore shale was deformed, and how it must now be followed by mining to great depths. The ore shale is shown in black in both diagrams. NS is the section line. (*From* Fleischer, Garlick, and Haldane, 1976.)

tenorite, malachite, and azurite, without enrichment. Ore reserves still total hundreds of millions of tons. Metamorphism accompanying postore folding has destroyed whatever fine-scale primary textures might have existed. The copper sulfides are fine-grained where the rocks are fine, coarser where they are coarse. Metamorphically derived gash veinlets contain quartz, calcite, and sulfides. Copper minerals as integral parts of graded bedding and cross bedding are common, especially at Mufulira (Figure 15-1); indeed, several local geologists note that the copper value distribution in sedimentary structures is consistent with a detrital origin and emplacement of some of the copper-iron sulfides.

The mode of origin has long been debated, but resident geologists have been convinced of their chemical sedimentary origin for decades. A compendium by Mendelsohn (1961), a book by Bowen and Gunatilaka (1977), and an article by Fleischer, Garlick, and Haldane (1976) describe the deposits thoroughly, each concluding that sedimentology and chemical sedimentation provide the dominant genetic constraints. The ore deposits appear to have formed when a presumably iron-, copper-, cobalt-, and uranium-rich river system delivered waters into a near-shore reducing environment

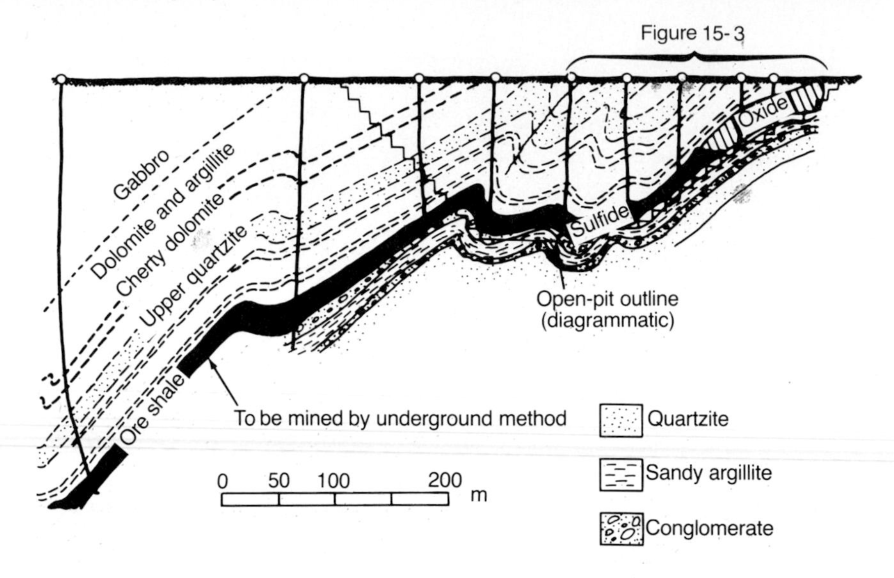

Figure 15-5. Folding and stratigraphy at Chambishi. The Upper Quartzite and younger units have been stripped to permit extraction of the ore shale, which since 1976 has been mined mainly by underground techniques. The essentially vertical lines represent drill holes. (*After* Mendelsohn, 1961.)

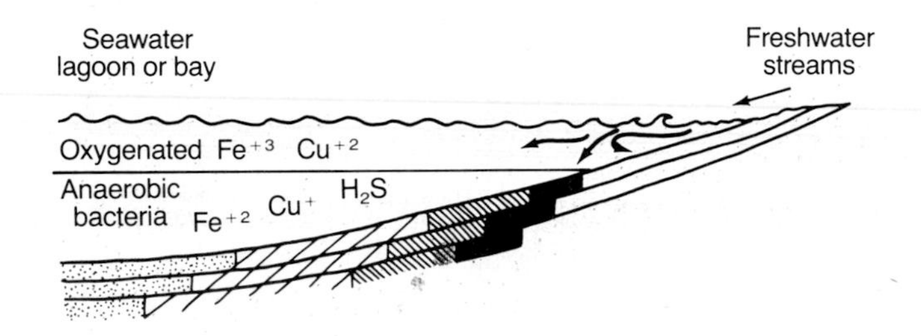

Figure 15-6. Relationships of environments of deposition interpreted in a context of river delivery of metals to a bay or lagoonal site of deposition. Compare with Figure 15-2 and consider with Figure 15-7. (*After* Shaner, 1982.) Decoration as in Figure 15-2.

of tidal flats, estuaries, and bays (Figures 15-6 and 15-7). Sulfide distribution is consistent with geochemical precepts that one might predict on the basis of Eh and pH, especially with regard to sulfide speciation. The deposits appear to owe their existence in part to the facts that clastic sedimentation was relatively slow so that chemically deposited base-metal compounds could collect to reasonable percentages without being overwhelmed by particulate dilution. Arguments concerning whether or not these deposits were real sedimentary accumulations or hydrothermal epigenetic deposits were stilled when it was predicted that uranium should have participated in the sedimentary geochemistry, and concentrations of UO_2 were

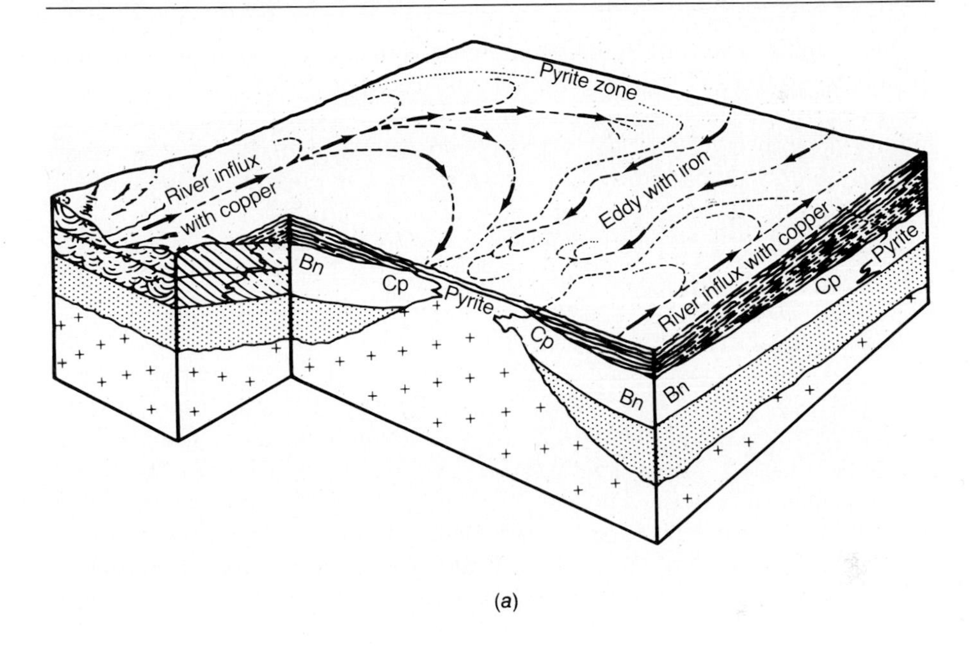

(*a*)

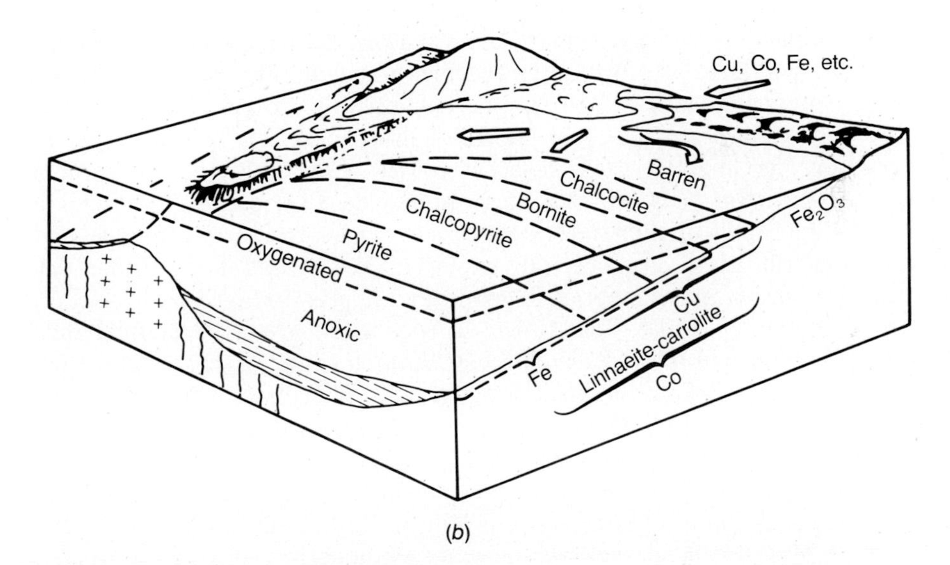

(*b*)

Figure 15-7. Two perspective schematic sketches of Zambian Copperbelt depositional environments. (*a*) River flowing from the cross-bedded aeolian dune terrain like that near the Mufulira deposit out into shallow seawater. Shoals over a granite high cause eddy currents, which interrupt with pyrite what would otherwise now be ore segments. (*From* Mendelsohn, 1961.) (*b*) Similar setting from the seaward side. It shows copper, cobalt, iron, and other metals in solution being carried into a sea. Copper is precipitated in a sulfidic sludge only in the near shore feather edge of the lower layer, while iron protosulfide is precipitated increasingly in depth; after diagenesis or metamorphism this gives the sedimentary zoning of the minerals as shown. (*From* Fleischer, Garlick, and Haldane, 1976.)

in fact found where they should be in more reduced, deeper water portions of the ore systems (Fleischer, Garlick and Haldane 1976).

Renfro (1974) proposed a mode of formation of base-metal-rich sediments like those of Zambia, a mode generally called the *sabkha model* (Figure 15-8). It applies only to desert coastlines and involves subterranean terrestrial groundwaters flowing down subsurface gradients to the coastline and impinging upon reduced conditions, algal mats, and euxinic muds at the shoreline. Metal ions in the groundwaters replace pyrite or would be precipitated by hydrogen sulfide, and metalliferous sediments would be created. The idea appears sound, and was applied to the Mufulira ores, although the local geologists consider that belt of ore too large both geographically and quantitatively to have been so generated. A test of the idea in presumably ideal circumstances of the mud flats near Puerto Penasco, Mexico, was negative (Shaner, 1982). Smith (1976) applied the concept to stratiform shaley copper ores in the Permian Flowerpot Shale of Texas, stating (page 441) that "evaporative pumping of copper-bearing groundwater through the sabhka surface (occurred). As groundwater passed through algal mats buried beneath the sabhka sediments, copper was stripped out by H_2S and $(HS)^-$ generated by sulfate-reducing anaerobic bacteria" and also replaced earlier formed pyrite.

The Cu-Co-U ores of the Kolwezi and Shinkolobwe districts of Zaire are in similar but slightly younger units, and with carbonate involvement. The high grades of 5 to 6% Cu as chalcocite and the high tonnages of the Tenge Fungurume area in Zaire might well affect world prices and supplies were the resource to be developed, and the U.S.S.R.'s keen interest in controlling the huge Cu-Co-U resources of the Kolwezi-Kitwe and Zambia areas is a matter of record. Although no new discoveries have been made in the Copperbelt of Zambia in recent years, vast tracts to the west and southwest remain to be thoroughly explored. The potential of sedimentary rocks as sources of base metals is generally underestimated.

A further example of stratiform sulfides is furnished by the fahlbands in the Precambrian of southern Norway. *Fahlbands* are sparse sulfide disseminations that follow the strike of gneisses and schists. They have been traced along strike for many kilometers, and their widths range from a few centimeters to a few hundred meters. These layered beds, containing mostly small amounts of pyrite, chalcopyrite, sphalerite, and galena, have been described as being of hydrothermal origin, as impregnations related to the injection of basic dikes, or as sedimentary syngenetic deposits.

Gammon (1966) favors a syngenetic origin of the sulfides. He envisions their deposition in a near-shore deltaic environment, where conditions were such that organic materials were deposited and covered before decomposition was completed. Probably sulfate-reducing bacteria liberated hydrogen sulfide, which fixed the iron and copper and lead to the production of pyrite and chalcopyrite. During later metamorphism, the sulfides were re-

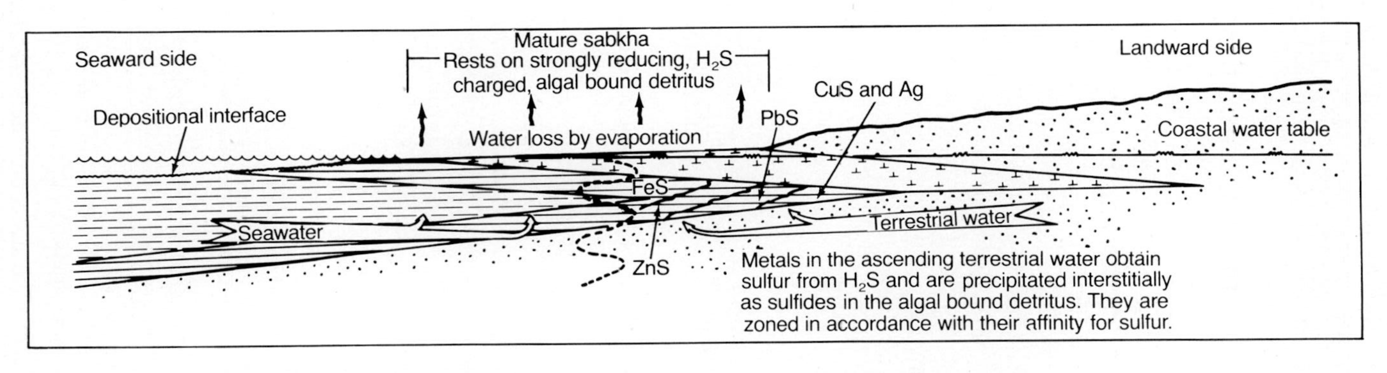

Figure 15-8. Coastal sabkha environment, showing impingment of metal-bearing groundwaters from the landward side upon algal mats and other reducing metal-sulfide-precipitating chemistries. (*After* Renfro, 1974.)

crystallized and lost all trace of their original organic origin and detailed textures. Gammon based his conclusions primarily upon textures visible in spite of metamorphic effects. Sulfides are interstitial to rock-forming silicates, and lenses of sulfides do not cut bedding foliation (Gammon, 1966).

▸ OTHER CHEMICAL PRECIPITATES

Carbonaceous shales, phosphatic shales, and many similar marine sediments contain minor amounts of uranium, vanadium, silver, arsenic, gold, molybdenum, and other metallic elements (Krauskopf, 1955, 1956b; Mercer, 1976). Our knowledge of the metal contents of sediments is fragmentary, largely because the amounts of metals are so small that most deposits are not commercial and have not received much attention. Shales appear to contain slightly higher percentages of metals than the other sedimentary rocks, but sampling and distribution studies have not been adequate. It has been repeatedly shown, however, that some metals are enriched in black, carbonaceous marine shales. Krauskopf (1955) found that several elements are enriched more than a thousandfold in selected organic sediments. These elements are concentrated by chemical precipitation in the reducing environment created by rotting organic sediments; by adsorption on clay particles, colloidal gels, and organic debris; and by organic processes such as bacterial action.

Krauskopf (1956b) studied the concentration of zinc, copper, lead, bismuth, cadmium, nickel, cobalt, mercury, silver, chromium, molybdenum, tungsten, and vanadium in seawater and concluded that the seas are greatly undersaturated in each of these metals. Accordingly, direct chemical precipitation of these elements cannot be responsible for the routinely observed concentrations in sediments. Local precipitation of sulfides may remove some of the metals from seawater, but cannot be the chief control; the concentrations in solution are unrelated to the sulfide solubilities, and some of the metals do not form stable sulfides. Again, it was concluded that adsorption and organic precipitation must be significant factors in removing many of the dissolved metals present in seawater (Krauskopf, 1956b).

Marine sediments have also attracted recent attention because certain types contain appreciable, though as yet noncommercial, amounts of uranium and vanadium (McKelvey and Nelson, 1950; McKelvey et al., 1955). The uranium-bearing black shales are typically rich in organic matter and sulfides. Many black shales and phosphorites contain 0.01 to 0.02% U, and diagenetic nodules in the Alum Shale of Sweden carry as much as 0.5% U. The character of the uranium-bearing mineral in the black shales is not known, but these shales are characteristically more phosphatic than others, so the uranium may occur as a phosphate or as adsorbed ions on organic matter (McKelvey et al., 1955). In the phosphorites the uranium content varies roughly with the phosphate content, thereby giving evidence that the uranium is contained in a phosphatic mineral. The depositional environment of these uranium-bearing formations—both the black shales and the phosphorites—is characteristic of low-lying, stable areas, where the influx of clastic material is small. Many of these deposits are associated with diastems or minor unconformities; as far as is known, nonmarine black shales are not uraniferous. The precipitation of uranium in carbonaceous black shales may be brought about by its chemical adsorption on apatite, on living or dead plankton (McKelvey et al., 1955), or by reduction of U^{+6} ions to the less soluble U^{+4} ions through the action of biochemically generated H_2S (Goldschmidt, 1954; Jensen, 1958). Quadrivalent uranium ions and divalent calcium ions are approximately the same size, which permits uranium to substitute for calcium in the apatite structure of phosphorites. Consequently uranium ions compete for positions with calcium ions, and only the calcite-poor varieties of shale or phosphorite are able to absorb appreciable quantities of uranium.

Vanadium is also concentrated in shales and organic sediments (Goldschmidt, 1954). It is typically associated with sedimentary uranium in reducing environments, but unlike uranium it forms stable sulfides. Both solid and liquid hydrocarbons have been known to contain highly abnormal concentrations of vanadium, perhaps a biochemical precipitate or a product of reduction in sapropelic muds. The patronite (VS_4) of Peru was probably formed by enrichment of vanadium and sulfur in vanadiferous hydrocarbons through natural fractionation in an oil seep. Phosphatic rocks, such as the Phosphoria Formation of the northwestern United States, have yielded concentrations of vanadium that are apparently related to organic matter in the sediments rather than to phosphate minerals (Jacob et al., 1933). The presence of both vanadium and uranium in these carbonaceous sediments is thought to be due to reduction from higher, more soluble valences rather than from direct biologic concentration in plant and animal tissues.

Mercer (1976) found that Co/Ni ratios in pyrite of sedimentary origin are generally less than unity, and that trace and minor element populations in "normal" shales are commonly astonishingly high and are but poorly understood.

▸ SEDIMENTARY IRON DEPOSITS

Several types of iron ore deposits form directly as chemical precipitates. Probably the most widespread are the *iron formations*, also known as *taconite* in the Lake Superior district, as *itabirite* in Brazil, as *jaspilite* in Australia, and as *banded ironstone* elsewhere in the British Commonwealth. Two major types—the volcanism-related Algoma type and the less obviously affiliated Superior type—were described in Chapter 13 as being the result of volcanic hot spring effusion and distal marine sedimentation. But there are several other iron ore types that do not appear to be related to volcanism. These iron ore types include oolitic ferruginous deposits, such as the Clinton ores of the southeastern United States and the minette ores of the Alsace-Lorraine, bog iron ores, and iron carbonate beds called *black band* ores.

Many geologists believe that ordinary river waters can dissolve and transport enough iron to account for the immense amounts in iron formations, a hypothesis supported by field and laboratory evidence. First, the relationship between iron deposits and volcanic activity is not found everywhere; the Clinton district to be described is an example of a large district in which volcanism seems to have been absent. Moore and Maynard (1929) discovered that abundant organic matter in a weathering regime allows cold water to extract, chelate, or complex, and transport enough iron and silica to build up large deposits of iron formation. They said that iron and silica travel and precipitate as colloids protected by organic matter. Their experiments showed that a mixture of these sols will form a banded sediment with an iron-rich layer at the base and nearly pure silica at the top. Krauskopf (1956a) has cast doubt on the colloidal nature of the silica, but the process may operate as well without true colloidal activity.

It has been suggested that nonvolcanogenic iron formations are epicontinental sediments formed as chemical precipitates from river waters that entered saline lakes, geosynclines, or other perhaps closed basins. The paucity of clastic diluents generally means that the adjacent land surface had been reduced to the limit of peneplanation. The combination of closed basins and peneplanation is rare, and the precipitation of iron formation would have been restricted to periods of exceptionally perfect base leveling (Woolnough, 1941). Studies in Manchuria by Sakamoto (1950) led to the conclusion that sedimentation in shallow but variable-depth lakes, coupled with a monsoonlike climate, was responsible for the banded iron deposits there. Periodic fluctuations in the pH of the lake water resulted in cyclic deposition of iron oxides and silica in alternating bands. Sakamoto attributed the iron and silica to mature weathering in neighboring areas and noted that the minerals must have been transported in ways that varied according to the seasons, such as the development of distinct soil horizons in wet and dry climates. During wet seasons, when the waters were acidic, iron was carried to lakes; arid seasons would have caused precipitation of the iron and concomitant transportation of silica because of a sharp rise in pH. Accordingly, a uniform climate could not have produced the banded iron formations.

The deposits that are discussed below are generally post-Precambrian in age and part of a sedimentary package that includes geosynclinal deposition. To the extent that these ores are chemical rather than detrital, it is required that iron was leached from adjacent, presumably peneplaned source rocks, carried as complex ions or hydrolysates such as $FeCl_2 \cdot nH_2O$ or $Fe(OH)_3 \cdot nH_2O$ as the ferrous or ferric ion, and precipitated as a flocculate, as a biochemical precipitate, by agglomeration and settling, or as a direct chemical precipitate. The first products of sedimentation were probably limonites, either goethite or hematite, which were diagenetically dehydrated and compacted. If biogeochemical or mechanical processes which formed oolites intervened, then oolitic deposits resulted. Further compaction or metamorphism completed the dewatering process.

Kimberley (1978) described this group as *s*andy, *c*layey, and *o*ölitic *s*hallow-inland-sea *i*ron *f*ormation (SCOS-IF). Kimberley (1980) reported the discovery of small deposits in Colombia and Venezuela, and Lunar and Amoros (1979) described several in Spain. Kimberley (1980) described the Eocene Paz de Rio district, Colombia, as typical, it being an extensive 5-meter-thick iron-rich, aluminous, chert-poor, phosphatic sediment laid down in a sandy argillaceous oolitic environment. Garrels and MacKenzie (1971) showed that pO_2 and pCO_2 in most modern soil atmospheres are so deep within the hematite-water stability field that virtually all iron in weathering regimes must be oxidized (Figure 15-9). Garrels, Perry, and MacKenzie (1973) stated that perhaps the major difference between Precambrian and Phanerozoic iron sedimentary geochemistry is that the reduced hydrospheric conditions of the early and middle Precambrian allowed iron to be transported as the soluble ferrous ion, but that only less soluble ferric iron is stable now, so that iron is carried only in the suspended load of streams. It would therefore enter sedimentary systems as colloidal or coarser detritus direct from the weathering of normal rocks and soils or from the weathering of iron formations, and could then, according to local Eh, pH, and ion activities, be involved in diagenetic reactions. There is certainly abundant evidence that iron is not as insoluble and immovable on a local scale as Garrels, Perry, and MacKenzie would imply; as discussed below and shown in Figure 15-9, acidic waters can carry ferrous iron ions in solution. Experiments by Castaño and Garrels (1950) showed that even aerated rivers with pH less than 7 could carry significant ferrous iron. Dimroth (1976) indicates that Phanerozoic iron has been transported (1) as detritus ferric hydroxide sols and as oxide films on clay and silt detritus and (2) as ferrous iron in reduced unaerated waters.

Oolitic Iron Ores

Iron ores of chemical sedimentary origin, but distinct from the classic BIFs of Chapter 13, are mined in many places throughout the world. Of these ores, the oolitic deposits are probably the most valuable. Oolitic ores vary considerably; in some the oolites abound and make up most of the rock but in others they are scattered throughout a clay or limestone matrix. The oolites also vary considerably in composition; they may consist of hematite, limonite, siderite, or chamosite, with or without calcite or chalcedonic silica.

One of the world's most extensive deposits of oolitic iron ore is in the Clinton Formation, which crops out intermittently from upper New York State southward into Alabama, where it goes under Coastal Plain sediments. The Clinton Formation, of Silurian age, is composed of thin-bedded iron-stained sandstone, shale, and oolitic hematite. Locally the beds are calcareous and grade into impure limestones (Burchard and Butts, 1910;

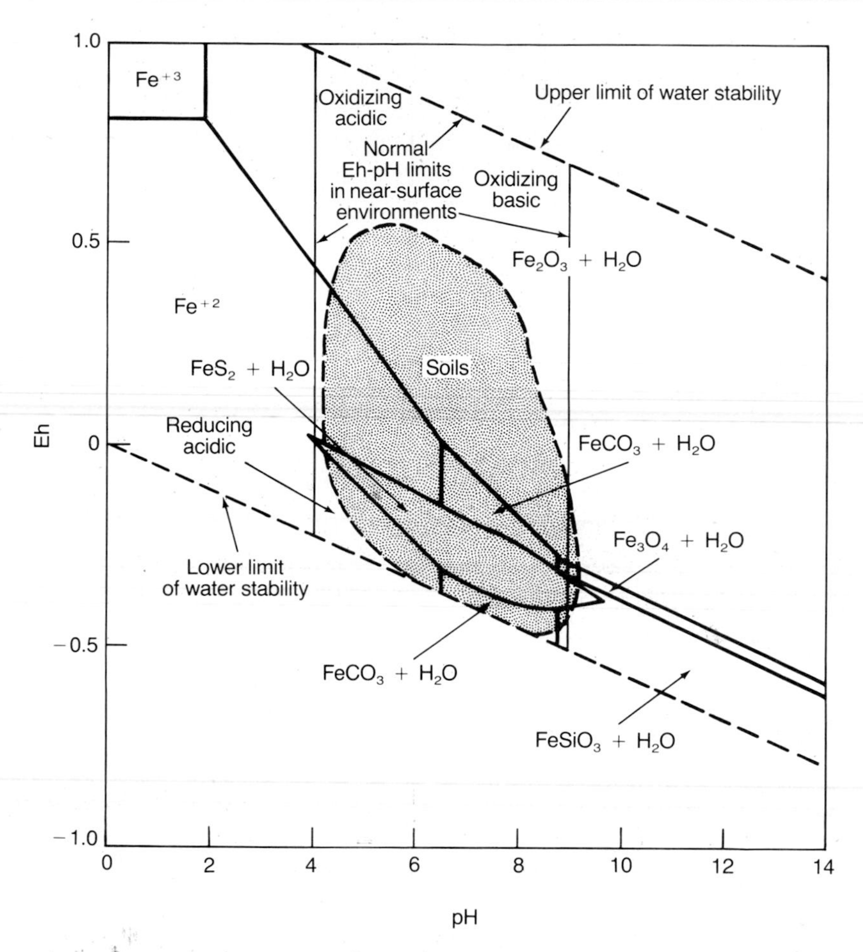

Figure 15-9. Eh-pH diagram showing the areas of dominance of dissolved iron species and the stability fields of some iron minerals at 25°C and 1 atm total pressure. The fields of the dissolved species are shown where the total activity of the ions is greater than 10^{-6} mol. The environmental conditions are total dissolved $CO_2 = 10^0$ mol, total sulfur $= 10^{-6}$ mol, and amorphous silica present. The upper and lower dashed lines are the stability boundaries of H_2O. Lines showing the limits of natural environments have been added. (*After* Garrels and MacKenzie, 1971.)

Alling, 1947; Hunter, 1970). Fossils are abundant, and many have been replaced entirely by hematite, presumably diagenetically. The hematitic beds attain their greatest development in the Birmingham district of Alabama, where they are as much as 7 meters thick.

Three types of ore have been mined in the Birmingham district: (1) oolitic, in which the hematite forms oölites generally in a hematite, calcite, or siliceous matrix; (2) flaxseed, in which the hematite forms small flat grains, or flattened oolites; and (3) fossil ore, in which hematite has replaced numerous fossils, principally mollusks and bryozoans. As a result of the supergene leaching of the calcite matrix, much of the ore in the upper parts of the mines is soft; at depth the ores are harder and more siliceous. The ores are relatively low grade, the main impurity being calcite. But since calcite is necessary as a fluxing agent in smelting and since coking coals were mined in the nearby Warrior coalfields, the Birmingham deposits were able to compete successfully with higher grade deposits found elsewhere. With depth the Birmingham ores become increasingly siliceous and more refractory to smelt. The mines were highly productive well into the twentieth century, but have been closed since World War II.

The Clinton Formation is of shallow-water, marine origin (Alling 1947; Hunter, 1970). It exhibits cross bedding, mud cracks, animal tracks, oolitic structures, and other shallow-water features, and in places it contains lenses of conglomerate. The hematite is probably of both primary and diagenetic origin, as evidenced by replaced fossils and concentric layers in calcareous oolites (Alling, 1947). Iron appears to have been carried into shallow marine basins at the same time the other sediments were accumulating. As mentioned, Castaño and Garrels (1950) demonstrated experimentally that aerated river waters with a pH of 7 or lower are able to carry significant quantities of ferrous iron in solution. If such a solution enters a marine environment where solid calcium carbonate is at equilibrium with seawater at a pH of 7 or more, iron will be precipitated as ferric oxide both in the water and as a direct replacement of calcium carbonate. This mechanism is the most credible explanation of the Clinton ores, which are considered to be early diagenetic replacements of calcareous ooze, fossil fragments, and oolites to which additional contributions were made by direct chemical precipitation, the iron having been brought in by streams and rivers.

The oolitic limonite ores of Alsace-Lorraine and Luxembourg, which have fed the steel mills of Europe for decades, are in Middle Jurassic shales, sandstones, and marls. The ores average between 30 and 35% Fe in a gangue of calcium carbonate and silica. The ooids consist dominantly of limonite, though siderite, chlorite, chamosite, and hematite are also present. According to Cayeux (1909), the iron minerals precipitated chemically and diagenetically replaced original calcitic oolites, as they did in the Clinton ores.

A third example of oolitic ore is found at Wabana, along the coast of Newfoundland. These deposits consist of beds that are a few centimeters to 10 meters thick and lie within the upper 15 meters of a sequence of Ordovician sandstones and shales. Only the three thickest beds were of economic value. Shallow-water to subaerial depositional conditions are evidenced by raindrop imprints, ooids, cross bedding, mud cracks, and the presence of shallow-water marine fossils such as brachiopods and trilobites. The oolites consist of concentric shells of hematite and chamosite, in places embedded in a siderite matrix. Both the oolites and the matrix are penetrated by algal borings, thereby giving evidence that the ores were established in their present condition before they were buried. The deposition of iron oxide apparently terminated abruptly, as evidenced by overlying graptolitic shale which contains oolitic pyrite (Hayes, 1915, 1928, 1929, 1931). These deposits are exposed along the northwest shore of Bell Island and dip about 9°NNW under Conception Bay. Mining operations have been carried on for a distance of more than 4 km downdip, placing the deepest workings well out under the bay. The ore averages 51.5% iron, 11.8% silica, 0.9% phosphorus, and 1.5% water. An area of 100 km^2 or more is available for mining, of which hardly more than one-tenth had been developed by 1955 (Lyons, 1957).

Siderite Deposits

Sedimentary beds containing siderite, commonly known as *black-band ores*, are widely distributed throughout the world. Efforts to mine these deposits have generally been unsuccessful because of low grade, but sedimentary siderite has been worked successfully in Germany and the British Isles and in Texas from the Eocene Wickes Formation. During the colonial period in the United States, small amounts of iron carbonates were extracted from Paleozoic geosynclinal marine sediments in Massachusetts, Ohio, Pennsylvania, and a few other areas.

Bog Ores and Spring Deposits

Limonitic bog iron ores occur in small low-grade deposits with much manganese, phosphorus, water, clay, and other impurities. They are of interest mainly because some are good examples of biochemical precipitation of iron minerals. The iron content of bog waters is higher than that of most other surface waters because the iron is stabilized by humic complexes and low pH (Rankama and Sahama, 1950). Bacterial action causes the precipitation of ferric oxides and hydroxides from the breakdown of humic iron complexes and ferrous bicarbonate. Supplies of iron are transported to the bog waters by streams and springs in situations that can be observed today in some northern glacial lakes. At present, bog ores are of very minor economic significance.

Special environmental conditions may account for local concentrations of both calcium carbonate and iron. An example is offered by spring deposits; travertine forms where spring waters are charged with calcium carbonate, and iron oxide forms from iron-rich waters. For example, at the Kuchan mine in Hokkaido, Japan, about 5 million tons of limonite were mined where a cold spring flowed from a hillside. The ferrous iron in solution was oxidized as soon as the water was exposed to the air, and the ensuing deposit formed a terracelike enbankment.

▸ SEDIMENTARY MANGANESE DEPOSITS

Manganese behaves chemically much as iron behaves—the two elements can accumulate in similar environments and under similar conditions. Under oxidizing conditions, pyrolusite or some other form of MnO_2 would be expected to form preferentially; at intermediate values of Eh and pH, hausmannite or the manganese carbonates or silicates should be deposited; and in strongly reducing environments, alabandite (MnS) or manganosite (MnO) should form (Figure 15-10; Krauskopf, 1957). The extremely low Eh-pH conditions necessary for alabandite and manganosite are not likely to be attained in sedimentary environments, but the other minerals are common and seem to be deposited according to their individual thermodynamic restrictions. Figure 15-10 shows the stability fields of many of the manganese minerals. Although some compounds having more complex or variable compositions are generalized by a simple formula, the figure still gives a fairly accurate picture of manganese environments and the facies to be expected under various conditions. For example, the diagram indicates that pyrolusite (MnO_2) overlaps some of the environments of manganite (Mn_2O_3) and braunite ($3Mn_2O_3 \cdot MnSiO_3$) but not the stability fields of rhodochrosite ($MnCO_3$), alabandite (MnS), or hausmannite (Mn_3O_4); the natural mineral assemblages corroborate this expectation.

Both oxides and carbonates of manganese are widely distributed throughout the world. The carbonates are generally complex, containing variable amounts of calcium, magnesium, and iron along with the manganese. Many sedimentary manganese oxide deposits are nearly pure, but other perhaps volcanogene ones contain minor amounts of cobalt, nickel, tungsten, copper, and barium, or such extraneous materials as clay, limestone, chert, and tuff. Manganese silicates ordinarily do not accumulate as sedimentary deposits, although at the San Francisco mine in Jalisco, Mexico, braunite ($Mn^{+2}Mn_6{}^{+3}SiO_{12}$) is the most abundant sedimentary mineral.

Roy (1976) divided stratiform manganese deposits into two broad groups, nonvolcanogenic and volcanogenic-sedimentary, according to the source of their contained manganese. Inevitably some deposits contain manganese from both sources, especially the ferromanganese nodules to be described below. A more comprehensive classification is as follows:

1. Volcanogenic
   a. Deposits associated with agglomerates, tuffs, and other clastic materials of subaerial or submarine hot spring affinity
   b. Deposits associated with banded iron formations of distal submarine exhalite origin

2. Nonvolcanogenic
   a. Bog ore, lacustrine, and fluvial settings
   b. Continental-terrigenous sediments in geosynclinal or shelf settings

3. Hybrid
   a. Ferromanganese nodules
   b. Sea floor sediments more remote from submarine hot springs

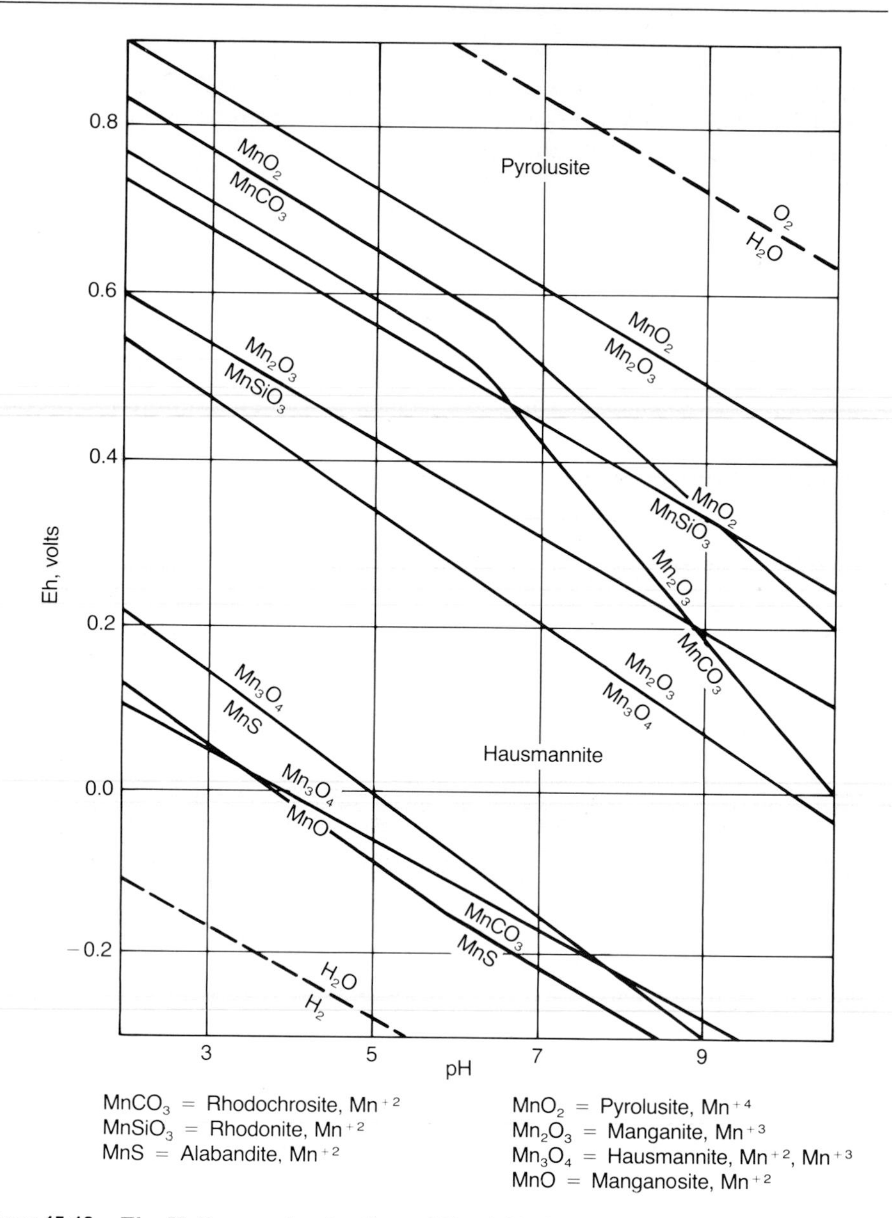

Figure 15-10. Eh-pH diagram showing the stability fields for common manganese minerals. Pyrolusite (MnO_2) is stable in the shaded area at high Eh, hausmannite (Mn_3O_4) at moderate Eh (also shaded), and divalent Mn minerals at low Eh. (*From* Krauskopf, 1957.)

One of the many excellent examples of the first type is the Lucifer manganese deposit near Santa Rosalia, Baja California, Mexico. The deposits formed when Miocene manganiferous hot spring waters percolated through and cemented volcanic breccias and agglomerates and flowed out subaerially or into a bay or lake to form stratiform bodies of pyrolusite (Freiberg, 1983). Hot springs can thus transport manganese to a lake or ocean floor where part of the manganese precipitates out of the water and becomes an integral part of the ordinary sediments (Zantop, 1978).

Many finely laminated manganese ores, usually consisting of oxide minerals, are interlayered with and genetically related to reddish or greenish andesitic tuffs and clastic volcanic sediments. These manganese-rich beds exhibit sedimentary features identical to those of the overlying and underlying tuffaceous beds. They are generally classified as sedimentary deposits, although where they are localized along faults they have been interpreted as hydrothermal (Wilson and Veytia, 1949). The deposits are thin-bedded, with individual layers a few centimeters thick and less. In a few places the beds resemble ordinary banded iron formation because the manganiferous layers are separated by thin beds and nodules of chert. More commonly, however, the interlayered materials are clays, altered tuffs, or clastic volcanic debris. Locally the layers between manganese strata pinch out, and the manganese becomes massive. Trace elements include Ni, Co, Cu, and Ba; they should be classified as volcanogene.

Workable deposits of manganese associated with banded iron formation (Chapter 13) are usually low grade. Banded iron formation generally contains less than 1% MnO (Bayley and James, 1973), although some iron ores contain enough manganese that it is a steel-making asset. The manganiferous beds commonly are separate from the iron-rich beds, as at Morro do Urucum, Brazil; in other deposits, such as the Cuyuna range, Minnesota, and Lagoa Grande, Minas Gerais, Brazil, manganese is distributed throughout many of the iron-oxide layers. Pipes or chimneys of high-grade manganese are encountered at several places in the Menominee and Penokee-Gogebic ranges in Wisconsin and Michigan (Figure 13-23). The voluminous literature dealing with iron formation seldom discusses the origin of the accompanying manganese. Why manganese is in separate laminae in some iron deposits but distributed throughout the iron-rich strata in others has long puzzled geologists. Presumably separation is caused either by variations in the rates of supply or by local, even seasonal variations in the environment of chemical sedimentation. Ljunggren (1955) pointed out that manganese and iron separate during precipitation in bogs and during movement of mineral-bearing waters through soils and subsoils. In both Sweden and Finland, iron and manganese have been found as separate precipitates within single lakes or inlets. The degree of separation seems to be directly dependent upon the pH of the waters: acidic waters retain high ratios of dissolved manganese to iron, but alkaline, less acidic waters precipitate the two oxides together (Ljunggren, 1955). The solubilities of both manganese and iron are functions of their oxidation states; in the reduced state they are both readily soluble, but each forms essentially insoluble oxidized ions. As Krauskopf (1957) pointed out, iron compounds in nature are uniformly less soluble than the corresponding manganese compounds, and ferrous ion is preferentially oxidized with respect to manganous ion under naturally occurring Eh-pH conditions. Consequently iron should be precipitated before manganese from any solution containing both metals, unless the Mn/Fe ratio is very high. Similarly, manganese should be more extensively distributed than iron by weathering or meteoric processes as, for example, in dendrites. The problem of iron and manganese separation, both within the laminae of iron formation and as pure minerals of each metal, has not been solved, but differential oxidation and solubility seem to be the most logical explanation.

Manganese silicates analogous to greenalite, for example, are not found in most unmetamorphosed iron formations in spite of the abundance of both manganese and silica in these rocks. Possibly the development of Mn-silicates is directly related to warm or hot waters which are essentially absent during the distal deposition of iron formation. Silicates are present in many of the small manganese deposits of Japan, and bementite and neotocite are common minerals in deposits of the Olympic Peninsula, Washington. In both Japan and Washington, the ores are in thin irregular beds interlayered with basalt and andesite flows, perhaps representing hot spring deposits.

Bacteria are known to secrete manganese as well as iron, though the extent to which they are operative is debatable. Zapffe (1931) studied manganese in groundwaters and concluded that bacteria may play an active part in its precipitation. In support of this hypothesis, the sedimentary ores of the Tennengebirge, Austria, have been attributed to bacterial action (Cornelius and Plöchinger, 1952). Algae are also known to precipitate manganese (Twenhofel, 1950). It has also been suggested that the separation of iron and manganese may be due in part to the presence of bacteria that use one of the metals but not the other (Krauskopf, 1957). As with iron, although the effects of bacteria may be considerable, the available evidence is inconclusive, and other processes appear to explain the larger accumulations of manganese satisfactorily.

Bog environments with or without hydrous iron oxides typically contain a few percent MnO_2 as pyrolusite, and nodular manganese oxide accretions similar to deep-sea nodules are abundant in Green Bay off of Lake Michigan in Wisconsin, the Finger Lakes of New York State, and thousands of bogs and lakes occupying glacial depressions in North America, Europe, Scandinavia, and the U.S.S.R. (Callender and Bowser, 1976). The manganese has been leached from nearby glacially derived sands and silts. But as Roy (1976) states, all of the large to gigantic deposits, those containing from 100 million to several billion tons of ore and the world's commercial sources, are nonvolcanogenic terrigenous sediments. Mikhalev (1946) concluded that the deposition of sedimentary manganese ores take place in near-shore basins, especially along continental margins.

Some of the world's largest manganese deposits are sediments that are apparently independent of volcanism. The famous Russian areas of Nikopol on the Dnieper River—1.7 billion tons of 15 to 35% Mn ores—and the Chiatura field on the south slope of the Caucasus mountains are of this type. At Chiatura the ore beds average 1 to 3 meters thick and cover about 150 km^2. Manganese was derived from the weathering of neighboring granites, syenites, and possibly some andesitic volcanics. The ore consists of oolitic and nodular pyrolusite in a matrix of manganese oxides, and the associated sediments are shales, marls, and sandstones (de la Sauce, 1926). Less extensive deposits of this type in the United States are the oxides of Artillery Peak, Arizona, and the carbonate beds of Arkansas, Virginia, and Maine. In general, carbonate beds are uneconomic unless they are weathered and concentrated as oxides. An exception to this generality is the extensive carbonate bed near Molango, Hidalgo, Mexico, where beds of the manganoan calcite kutnahorite and calcian rhodochrosite are as much as 60 meters thick and extend along strike for about 50 km. Some of the beds are 25% or more manganese. The ore is crushed, roasted, and then pelletized. The final product is a high lime-manganese oxide pellet, which makes an excellent furnace feed for steel mills.

A large deposit of manganese ore—more than 250 million tons—is present on Groote Eylandt in the Gulf of Carpenteria, northern Australia. The deposit is thought to be of Cretaceous age because a few fossils of probable Cretaceous age have been recovered from the ore. The mine geologists consider that the deposit is of primary sedimentary origin, but the area is deeply weathered and the deposits have many of the characteristics of typical laterite, particularly abundant pellets, most of which are less than 1 cm in diameter. Locally the pellets are recemented to form larger nodules or small irregular masses. The ore is being mined from shallow pits that are spread over an area of several square kilometers in the western part of the island. No volcanic activity has been recognized on the island or the nearby mainland.

Sedimentary ores are commonly concentrated in ribbon-shaped deposits parallel to paleoshorelines. An example of this type is in the Elqui River valley in central Chile, where a ribbon-shaped bed of manganese oxides averaging about 20 cm thick is enclosed in highly altered volcanic debris. It has been followed in the valley for 30 km along strike, but as far as can be determined, it is only 1.5 km wide across strike. Field evidence indicates that it is a reef-form deposit that parallels an ancient shoreline. In support of this contention, impure limestones are almost everywhere associated with the manganese and tuffs. Deposits similar to the Chilean manganese are known in Cuba at Charco Redondo, El Cristo, Ponupo (Figure 15-11*a*), and Taratana; in Mexico at Paridero and San Francisco in the

(*a*)

(*b*)

Figure 15-11. (*a*) Bed of powdery manganese oxides and tuff overlain by massive white limestone. Ponupo mine, Oriente Province, Cuba. (*b*) Workings along the outcrop of a bed of braunite ($3Mn_2O_3 \cdot MnSiO_3$) and pyrolusite at the San Francisco mine near Autlán, Jalisco Province, Mexico.

state of Jalisco (Figure 15-11*b*); and in the United States near Lake Mead in Nevada and Arizona (Park, 1942, 1956; Hewett and Webber, 1931; McKelvey et al., 1949).

▸ PHOSPHATE DEPOSITS

A chemical sediment of first-order importance is phosphate rock, the backbone of the fertilizer industries and agricultural capability of all nations. Phosphatic materials are of many origins. Guano—bird droppings—are rich in phosphates "inherited" from shoreline marine foodstuffs; guano is mined and 1% of our global phosphate consumption is shipped from several southwest Pacific islands, especially Nauru and Christmas. Another 23% of the world's phosphate is obtained from carbonatites (Chapter 9) and from alkalic igneous rock complexes. But by far the most productive sources at 76% are marine phosphatic chemical sediments.

Phosphate-rich sedimentary rock, or phosphorite—material containing more than 15 to 20% P_2O_5 and in general economic—forms on continental shelves. The extensive deposits of the Permian Phosphoria Formation of Montana and Idaho, the vast Moroccan ores, the Miocene phosphates of Florida and North Carolina, and phosphatic sediments forming today off the coasts of Peru, Mexico, and southern Africa, all formed in shelf environments. The process appears to be one involving upwelling of deep, cold seawater and its longshore flow across shallow, warm, sunlit shelf environments like that shown in Figure 15-15 below, or the mixing of cold and warm ocean currents. Cold water can dissolve more CO_2 and apatite [$Ca_5(PO_4)_3F$] than can warm water, so PO_4 averages 0.3 ppm in deep cold seawaters but only 0.01 to 0.003 ppm in warm shallow water (McKelvey, 1967). As deep, apatite-saturated, slightly acidic cold water is warmed, CO_2 is "degassed" both by increased temperature and by decreased pressure. Carbonic acid (H_2CO_3) is weakly dissociated to $2H^+ + CO_3^=$ and $H^+ + (HCO_3)^-$ at depth. When the water is warmed, both pairs yield to $H_2O(l) + CO_2(g)$, H^+ is consumed, CO_2 is lost, and pH increases. Apatite is less soluble in alkaline waters, even if they are warmer. The waters thus become supersaturated in calcium phosphate, which precipitates and promotes a thriving biota from algae through shellfish to vertebrates. The skeletons and debris of this biomass, with direct inorganic apatite precipitation, produce considerable thicknesses and tremendous areas of fossiliferous, oolitic, pelletal, nodular, or micritic phosphate sediments called *microsphorites*. Impurities include calcite, quartz, and clays. If upwelling is sustained, the deposits may become thick and rich enough and widespread enough to be valuable. Fine-grained microsphorite typically forms in quiet environments, pelletal masses in agitated waters. The phosphatic units commonly show extensive biologic, chemical, and mechanical reworking—

phosphate from shells, bones, plankton and chemical precipitation is redissolved and reprecipitated while being stirred and mixed by waves, currents, and burrowing animals. Longshore currents peel up precipitated microsphorite and carry it or roll it along, redepositing it in low spots on the shelf as oolites, pellets, clasts, and pebbles, commonly with rims of new phosphate. Deposits are forming today at several places, including off Sechura, Peru, where cold phosphatic Humbolt Current waters flowing north meet warmer currents flowing southward. Figure 13-17 shows Eh-pH fields of preferred deposition.

Recent reports by Cook and McElhinney (1979), Riggs (1979a, b), and Howard (1979) show that phosphorites range from Precambrian to recent in age, and that they are found on every temperate continent. They appear to have formed at various times and places within 40° of latitude north or south of the paleoequator. The marine apatite is normally fine-grained carbonate fluorapatite $Ca_5(PO_4,CO_3)_3(F,OH,Cl)$, with worldwide average trace element compositions including significant Na, Mn, Sr, Ti, V, and U, and minor Cr, Mo, F, Zn, and the rare earth elements, the lanthanides. Uranium and fluorine are extracted as by-products; other elements may be as prices and technology permit.

The most productive deposits in the United States—in the world, for that matter—have been those of central Florida. The Lee Creek deposit of Texas Gulf Sulfur in eastern North Carolina is achieving major status since its discovery in the 1960s; the Phosphoria Formation Montana-Idaho deposits are high-grade, but they must be mined far from major markets and underground at high cost, so their competitive radius is limited. The Florida deposits are in the Miocene Hawthorn and Bone Valley formations, which include the several-meter-thick principal phosphorite.

Florida Phosphate Deposits

The Florida phosphate deposits are part of a larger belt of late Tertiary phosphatic sediments that extends from Florida northward along coastal Georgia, South Carolina, and North Carolina, where the Lee Creek mine is located, into Virginia (Figure 15-12). The Florida deposits have been mined since 1888, have been the world's principal source of phosphate, and will be productive for many years to come. The description to follow has been derived from Cook and McElhinny (1979), Gurr (1979), Riggs (1979a, b), and several other sources.

Phosphate in Florida is found in three modes. The *Florida hard rock phosphate*, or *Alachua type*, found in a band shown in Figure 15-12, was composed of Miocene phosphatic clayey sands above irregular apatite replacement bodies in underlying Eocene Ocala limestone or Oligocene Suwanee Formation limestone bedrock. The replacement mantos, thought to have been formed by downward percolating phosphatic waters involved with weathering, have largely been mined out. *River pebble* deposits are

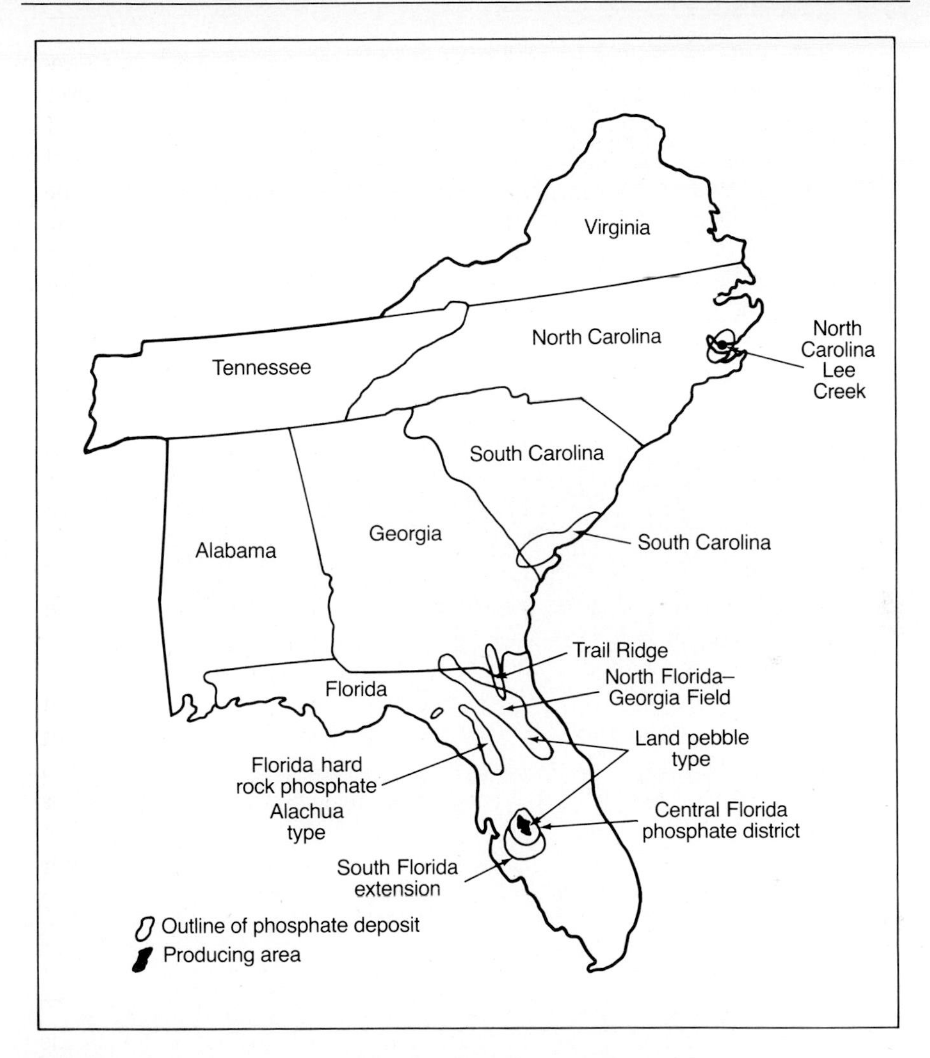

Figure 15-12. Location of major Miocene-Pliocene-Pleistocene phosphate deposits in the southeastern United States. The location of the younger Trail Ridge Fe-Ti oxide marine placer deposits discussed in Chapter 16 is also shown.

thin Holocene layers of phosphatic gravel along streams and rivers, and are economically trivial. The major occurrence mode is that of *land pebble* deposits.

Figure 15-13 shows a generalized stratigraphic section of the units pertinent to land pebble deposits. Overburden thickness ranges from 0 to 50 meters, averaging 3 to 15. Phosphates occur in the Middle Miocene Hawthorn Formation, but the land pebble ores are in the Pliocene-Pleistocene Bone Valley Formation. Its lower portion consists of coarse- to fine-

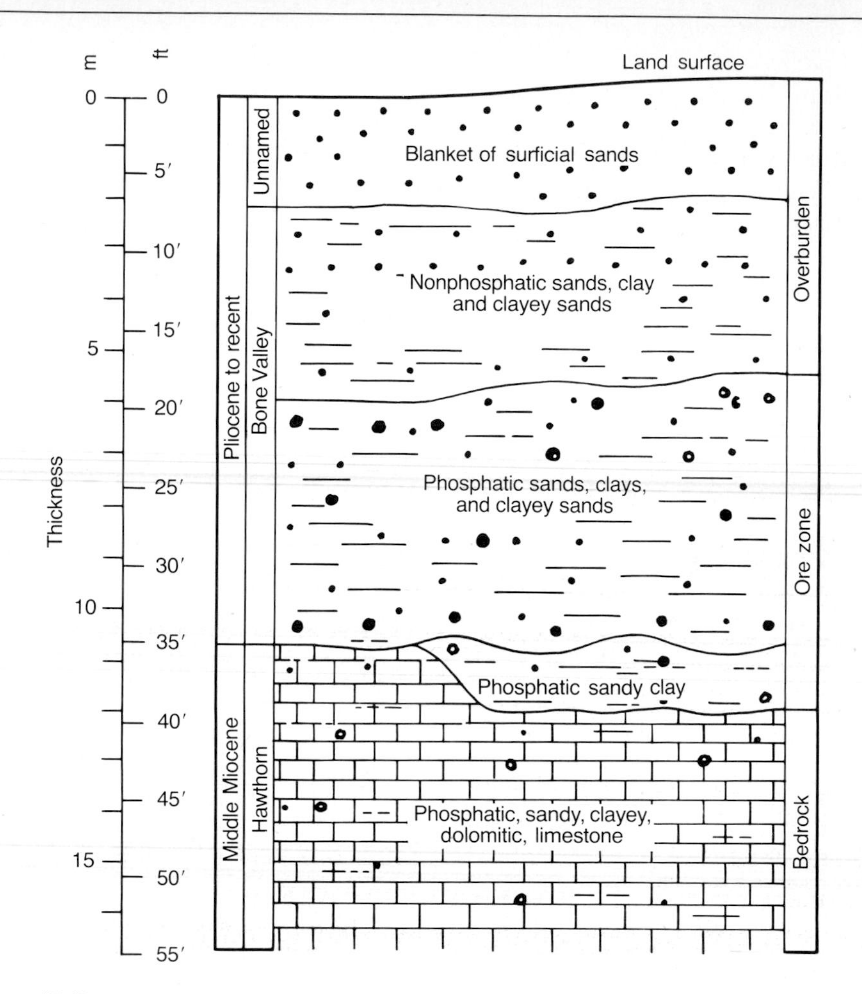

Figure 15-13. Generalized geological section in the central Florida land pebble phosphate district. (*From* Gurr, 1979.)

grained, poorly sorted phosphatic clayey sands, sandy clays, and gravels. It ranges in thickness from 1 meter to as much as 15, normally ranging from 5 to 8 meters in minable areas. It is like a blanket draped over subdued Miocene topography and underlying vast areas of Florida's flat, swampy surface. It contains from 10 to 60% phosphate pebbles, which themselves contain 30 to 32% P_2O_5; the average ore is about one-third pebble, one-third quartz sand, and one-third clay, the latter two being discarded. The units are locally crossbedded, contain interbeds of sands or clays, and thin lenses of dolomite.

Riggs (1979a) identified several petrologic phosphate components. One is a phosphatic mud, a microcrystalline material apparently directly deposited on the depositional surface. This mud can be torn up to form pellets and pebbles of rice to walnut size, it can be ingested and excreted by organisms to form pellets, or it can aggregate around other grains and pebbles. Most of the pebbles, then, are true aggregates composed of a complex mixture of mineralogical, biological, and sedimentological components. Figure 15-14 shows some pebble structures. There are also bacterialike rods, microorganism fossil hash, dolomite rhombs, terrigenous sand and clay, and

Figure 15-14. Some representative structures of phosphorites. (*a*) Sawed cross section of three layers of laminated pale orange to brown primary microsphorite. The dark grains are quartz sand. Scale bar is 5 mm. (*b*) Photomicrograph of primary microsphorite laminations in thin section. The light and dark (brown) bands are phosphatic-mud-rich and bacteria-rod-rich layers, respectively. Scale bar is 50 μm (0.05 mm). (*c*) and (*d*) Two samples of phosphate pebbles show rounding and represent the erosion, transportation, and redeposition aspects of the "phosphate machine." (*c*) Composed of microsphorite mud with sand grains (light) and primary laminated microsphorite (dark). Scale bar is 2 mm. (*d*) Composed of microsphorite pebbles and phosphatic fossil molds. Scale bar is 5 mm. (*e*) and (*f*) Photomicrographs that could be of (*c*) and (*d*), respectively. (*e*) Phosphate intraclast pebble of quartz grains (white) in brown phosphate mud and phosphate mud matrix. Scale bar is 1 mm. (*f*) Oval to spherical microsphorite pellets in a phosphate mud matrix. Scale bar is 0.2 mm. (*Photos from* Riggs, 1979a.)

fish bones, shark teeth, and other vertebrate debris. Different portions of the Bone Valley vary in their content of phosphatic material, and it is of course the highest grade intervals, both vertically and laterally, that are sought by drilling for drag-line mining.

A high percentage of the phosphate—and its associated sediment—appears to have been multiply reworked, both mechanically by wave action and chemically by solution and local redeposition. Riggs (1979b) described the environment as a "phosphate machine," an extensive coastal shallow near-shore shelf setting with cold, chemically phosphate-saturated upwellings moving across the shallow platforms into coastal environments (Figure 15-15), where phosphate muds precipitated. Chemical, biological, and physical processes in the high-energy environment broke up these microsphorites, generated more muds, granules, and pebbles, and longshore currents swept clasts, cement, and debris into adjacent entrapment basins, both along the shelf and downdip. Riggs, (1979b, page 285) wrote:

> *Thus, wherever the phosphorus sources were adequate, the physical current and geochemical systems were appropriate, and the shallow marine environments had the proper geometry, the "phosphate machine" produced and supplied clastic phosphorites to the associated "entrapment basins." The ultimate magnitude of phosphorite deposition was then dependent upon the size and extent of the structural system, the duration of the phosphogenic system through geologic time, and the volume and rate of terrigenous or carbonate diluent sedimentation. Subsequent fluvial erosion and subaerial weathering severely modified the most updip portions of the Hawthorn phosphorites following emergence.*

Florida land values and ecological constraints have greatly reduced the volume of mining in recent years, but the deposits are still important producers. Apatite or phosphorite can be crushed and applied directly as fertilizer, but solution and $(PO_4)^{=}$ availability is normally too slow. Apatite is soluble in sulfuric acid to produce phosphoric acid, a product called *triple superphosphate*, and calcium sulfate. Enormous amounts of sulfuric acid are consumed, so the phosphate and sulfur industries are symbiotic. Phosphoric acid and triple superphosphate ($[3(Ca(H_2PO_4)_2 \cdot H_2O)]$ = 48% available phosphorus) are both liquids and can be used directly as fertilizers and in the chemicals industry.

▸ EVAPORITES

Another group of chemical sediments is that package called the *evaporites*. The name is at once genetic and descriptive, covering the group of rocks formed by the processes and results of evaporation. Marine affinity is generally assumed if no modifier is used, but both marine and lacustrine evap-

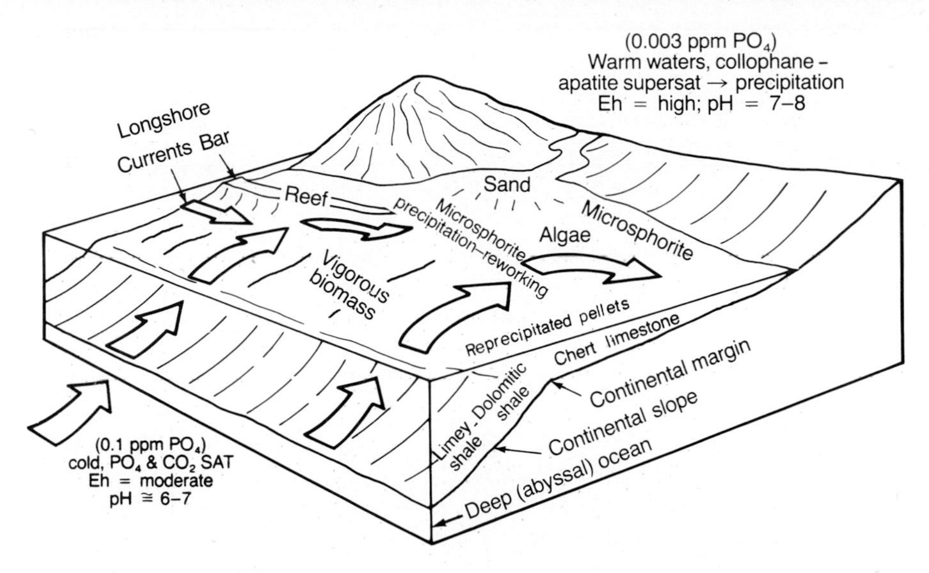

Figure 15-15. Perspective sketch showing delivery of phosphate to near-shore sites of precipitation by upwelling, cold, deep ocean waters. Reworking and further concentration of microsphorite by longshore currents on the continental shelf are significant. See text. Water movement indicated by arrows.

orites are quite common in the geologic record. Marine evaporites total over 3% of the total stratigraphic thickness of sedimentary rocks, and are known back into the Precambrian. Lacustrine evaporites are widespread in fault block and other depressions in modern desert environments, but because of their fragile, specialized nature they are not well represented in the ancient stratigraphic record.

Both types of evaporites have specific tectonic settings. Marine evaporites are now thought to have formed in arid climates in rift zones at or near continental margins; evaporite sediments commonly appear in the stratigraphic record along protorift zones where continental blocks have separated, as at today's Red Sea. They also form in foundered blocks with connections to the sea and in subsidiary basins that are part of that extensional protorift setting. They generally, then, involve seawater flowing into land-locked basins along coastal margins, basins which have no egress, with resultant evaporation of the seawater and continued concentration of the nonvolatile salts that remain as a "distillation residue" (Figure 15-16). Lacustrine evaporites also require unusual tectonic regimes. They appear to form either in extensional fault blocks at high structural levels (Chapin and Cather, 1981) or in topographic lows in rain-shadow deserts. Large lakes form geologically quickly and then undergo relatively slow evaporation, normally with periodic influxes of water and with no, or very restricted, outflow.

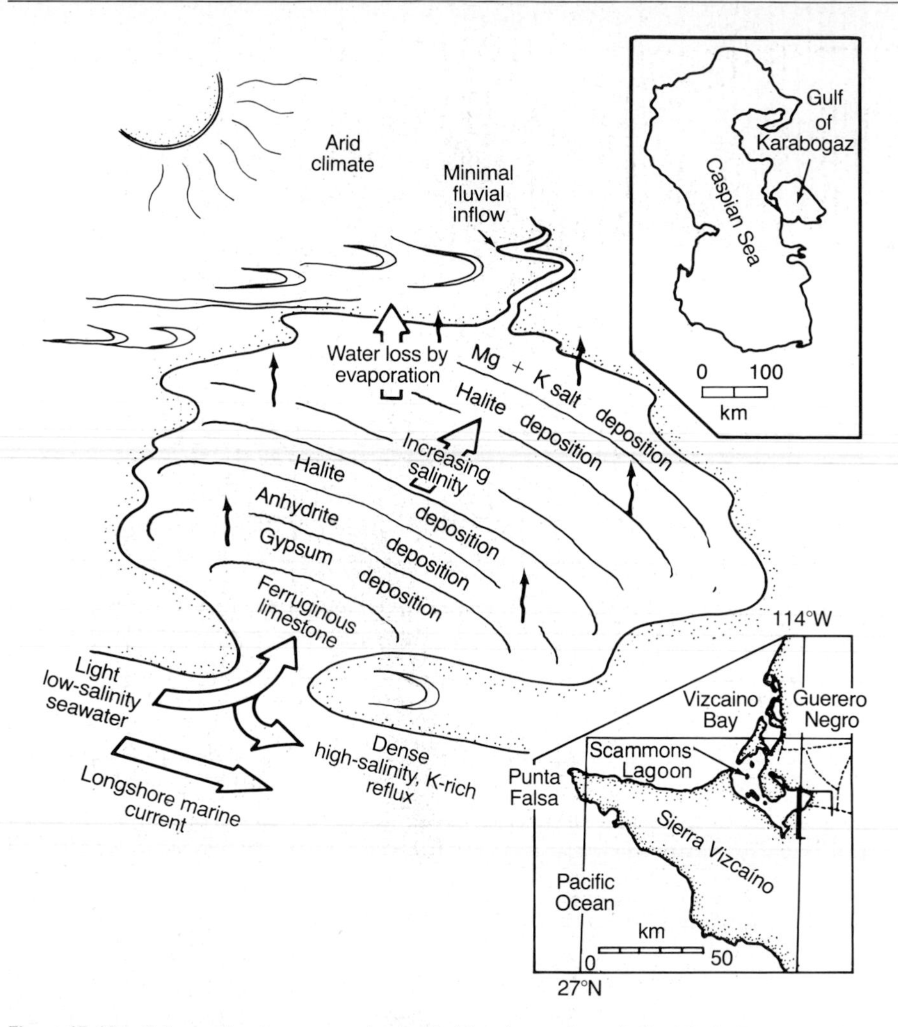

Figure 15-16. Schematic of an evaporite basin like the modern Gulf of Kara Bogaz on the east side of the Caspian Sea or Scammon's Lagoon on Baja California (insets). The curved lines are isopleths of salinity, which increases with more prolonged evaporation on the inland side. The lateral margins might elsewhere be rift faults, the basin occupying a graben.

Because marine evaporites involve progressive solar distillation of seawater, the composition of seawater (Table 15-1) constrains the mineralogy and bulk composition of marine evaporites. The principal solute components are Na^+, K^+, Ca^{+2}, Mg^{+2}, Cl^-, and $SO_4^=$. Marine evaporites constitute the world's major source of salt, and thus of both Na and Cl; of gypsum, anhydrite, and of native sulfur biogenically generated from them (Chapter 7); of K, Mg, Br, and I; and of Rb, Sr, and other minor elements. Lacustrine evaporites are more varied because they contain evaporation-concentrated

TABLE 15-1. Analysis of Seawater

Compound or Salt	wt % of Seawater	wt % of Dissolved Salts	g/kg of Seawater, or Parts Per Thousand
H_2O	96.2		
$CaCO_3$	0.0114	0.33	0.11
Fe_2O_3	0.003		
$CaSO_4$	0.1357	3.48	1.20
$SrSO_4$		0.05	0.02
NaCl	2.9424	78.03	26.90
NaF		0.01	neg.
$MgCl_2$	0.3219	9.21	3.18
$MgSO_4$	0.2477	6.53	2.25
K_2SO_4	0.0863		
NaBr	0.0556		0.03
$MgBr_2$	0.0076	0.25	0.09
KCl	0.0505	2.11	0.7
	100.00	100.00	34.48

Ion	wt % of Ions in Seawater or Seasalt	g/kg × 10^3 or ppm of Ions in Seawater
Na^+	30.61	10,556
K^+	1.10	380
Mg^{+2}	3.69	1,272
Ca^{+2}	1.15	400
Sr^{+2}	0.03	8
Cl^-	55.06	18,890
Br^-	0.19	65
HCO_3^-	0.41	140
$SO_4^=$	7.68	2,649
H_3BO_3	0.07	26
	100.00	34,386

residues of waters which are volcanic in origin, rain waters which have rinsed and equilibrated with lithologically varied, generally arid terranes, and "atomized" or aerosol salts wafted in from distant marine coast droplets in "sea air" or mists. Lacustrine salts therefore are far more diverse and "regional" in their composition. Chile has been the world's prime source of mined nitrates. The dry lakes of California, Nevada, and Turkey have produced boron and the boron salts borax, tincalconite, colemanite, and ulexite, and the Great Salt Lake of Utah provides Mg, Br, and Li. Other lakes are sources of lithium, boron, sodium bicarbonate (nahcolite), sodium carbonate (trona), sulfate, sodium aluminate (dawsonite), and oil shales. Some monomineralic saline deposits are formed by thermal springs.

Evaporite commodities are typically mined by conventional underground techniques. Room and pillar methods are used to mine Silurian salt beds under the city of Detroit, Devonian potash in Saskatchewan, and Permian Zechstein salts in England, France, and Germany. Some of the soluble salts can be dissolved and recovered by pumping and circulating brines, as is done at Searles lake in the Mojave Desert of California. Most of the third world's table salt is obtained from coastal evaporite pans (Figure 15-17), many of which are "farmed" by controlled draining and flooding. Successive "crops" of salts can be obtained by decantation as different solubility products are reached with progressive evaporation. Great Salt Lake brines are moved from one salt crystallization "pan" to the next while evaporation continues.

Marine evaporites were among the first ore-forming systems that were successfully studied and correctly analyzed. Experiments on salt species resulting from progressive artificial evaporation of Mediterranean Sea water were carried out in the mid-1800s. Results from these experiments are still quoted (Usiglio, 1849), although they have been refined (Harvie, Weare, and Hardie, 1980; Eugster, Harvie, and Weare, 1980). If we consider a "one shot" evaporation cycle, the admission of one flow of seawater into a basin such as the Salton Sea or the Dead Sea and its subsequent evaporation, then sequence and quantity or stratigraphic thickness of salts is predictable. That sequence was correctly identified by Usiglio, although it has been slightly modified in detail since. It is given as Table 15-2 with approximate thicknesses of evaporites resulting from the evaporation of a column of seawater 1 km high and is included in Figure 15-18. Early studies by the Germans Borchert, Braitsch, and Kühn were fundamental to the kind of understanding expressed in Figure 15-18. Another German—Ernst Jä-

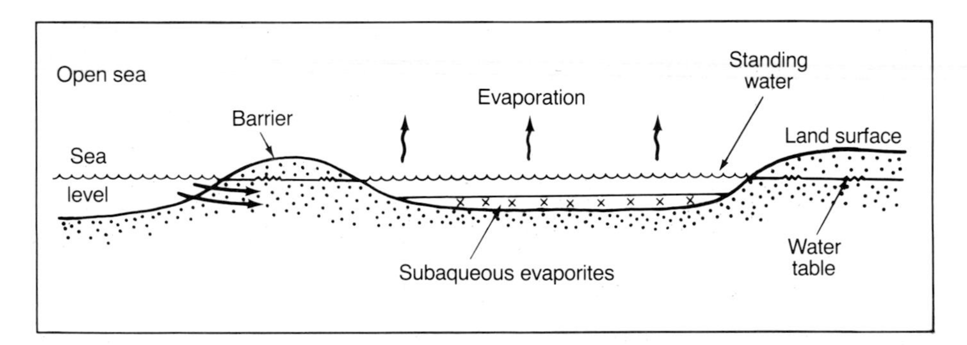

Figure 15-17. Schematic cross section of a coastal evaporite pan. Seawater seeps through the barrier into the pan where solar evaporation concentrates the salts past their solubility products, inducing precipitation. Water level in the pan may be 1 or 2 meters below sea level. Seawater can of course be artificially admitted into a managed pan by dams, gates, or pumps. (*After* Renfro, 1974.)

TABLE 15-2. Thicknesses and Precipitates of Salts from Evaporated Seawater

% of Total	Meters	Feet	
			Top of Column
2	0.1	0.3	Potassium chloride,[8] sulfates; bitterns[6,7,8]
79.7	3.7	12.0	Halite[4] with magnesium sulfates,[5] chlorides,[6] and NaBr[7]
13.3	0.6	2.0	Anhydrite[3] and halite,[4] with $MgSO_4$,[5] $MgCl_2$,[6] and NaBr[7] Gypsiferous[3] halite,[4] with $MgSO_4$[5] and $MgCl_2$[6]
5	0.2	0.7	Gypsum[3] Calcitic[2] gypsum[3]
	0.02	0.05	Ferruginous[1] limestone[2]
100.0	4.6	15.1	
			Bottom of column

		Meters	Feet
1.	Fe_2O_3	0.0002	0.0006
2.	$CaCO_3$	0.002	0.05
3.	$CaSO_4 \cdot 2H_2O$	0.17	0.56
4.	NaCl	3.60	11.8
5.	$MgSO_4$	0.30	1.00
6.	$MgCl_2$	0.41	1.34
7.	NaBr	0.05	0.16
8.	KCl	0.06	0.21
		4.59	15.12

Precipitates formed by the evaporation of 1 km^3 of seawater, in feet and meters of thickness. The upper portion could be considered a stratigraphic column; the lower portion represents the total thickness of individual compounds distributed through the strata as described above. Superscripted numbers in upper portion refer to the compounds below.

necke—experimented, drew ternary diagrams with Mg, K, and SO_4 as apices, and showed salt relationships at several temperatures toward the final drying of evaporite systems. These *Jänecke diagrams* (Stewart, 1963; Krauskopf, 1967) show enormous complexity of salt species and quantities in natural evaporites as unbonded water is finally reduced to zero. True, the opening paragraph of this section suggested that this group of sediments can be interpreted almost classically in terms of aqueous solution chemistry. It can, but there are at least four complicating factors. First, the large number of trace or minor components in the system requires that scores of mixed-salt variably hydrated minerals can occur. Second, although pressure variation is small, seasonal and diurnal temperature variation in salt pans is astonishingly high, especially as evaporation nears completion in vuggy, white layers in the hot sun. The dynamics of convection in shallow, stratified, intermixing brines and of resolution, recrystallization, and diagenesis in precipitates, again especially toward the end of an evaporation cycle, are formidable. And finally postdepositional reactions between the salts and trapped or migrant brines can obscure the original paragenetic sequence. Harvie and Weare (1980) and Eugster, Harvie, and Weare (1980) used modern electrolyte chemistry, phase equilibrium studies, and computer models to refine mineral sequence appearance predictions in the complicated Ca-Na-K-Mg-Cl-H_2O system. If back-reaction is maintained—that is, if the progressively more saline brine remains in equilibrium contact with salts successively precipitated from it—the crystallization sequence computes as gypsum—halite—glauberite—polyhalite, or $CaSO_4$-$NaCl$-$Na_2Ca(SO_4)_2$-$Ca_2K_2Mg(SO_4)_4 \cdot 2H_2O$. If precipitated salts are fractionated out of the system—that is, if early calcium-salts are buried and isolated from reaction with late brines—then the calcium-salt reservoir is removed early, and the sequence is calcium-poor at the end. The sequence is changed to gypsum-halite-polyhalite-bloedite-kainite-carnallite [$CaSO_4$-$NaCl$-$Ca_2K_2Mg(SO_4)_4 \cdot 2H_2O$-$Na_2Mg(SO_4)_2 \cdot 4H_2O$-$K_4Mg_4(Cl)SO_4 \cdot 11H_2O$-$KMgCl_3 \cdot 6H_2O$]. This sequence is the one most commonly encountered in natural evaporites (Figure 15-18).

It is no surprise, however, that the geologic-tectonic-topographic circumstances that would lead to simple "one shot" evaporation of 1-km depth of seawater are rarely encountered. In fact, when the world's evaporites are carefully studied, it is seen that some basins accumulate disproportionate thicknesses of anhydrite-gypsum, others hundreds of meters of halite with almost no sulfates, and yet others unusual complements of the most soluble late-crystallizing salts of K, Na, Mg, Cl, and Br, called *bitterns*.

Figure 15-18. (*Facing page*) Some of the systematics of marine evaporite formation. The black "teardrops" describe crystallization with time, or the progress of evaporation, evolving to the right. (*After* Strakhof, 1970a.)

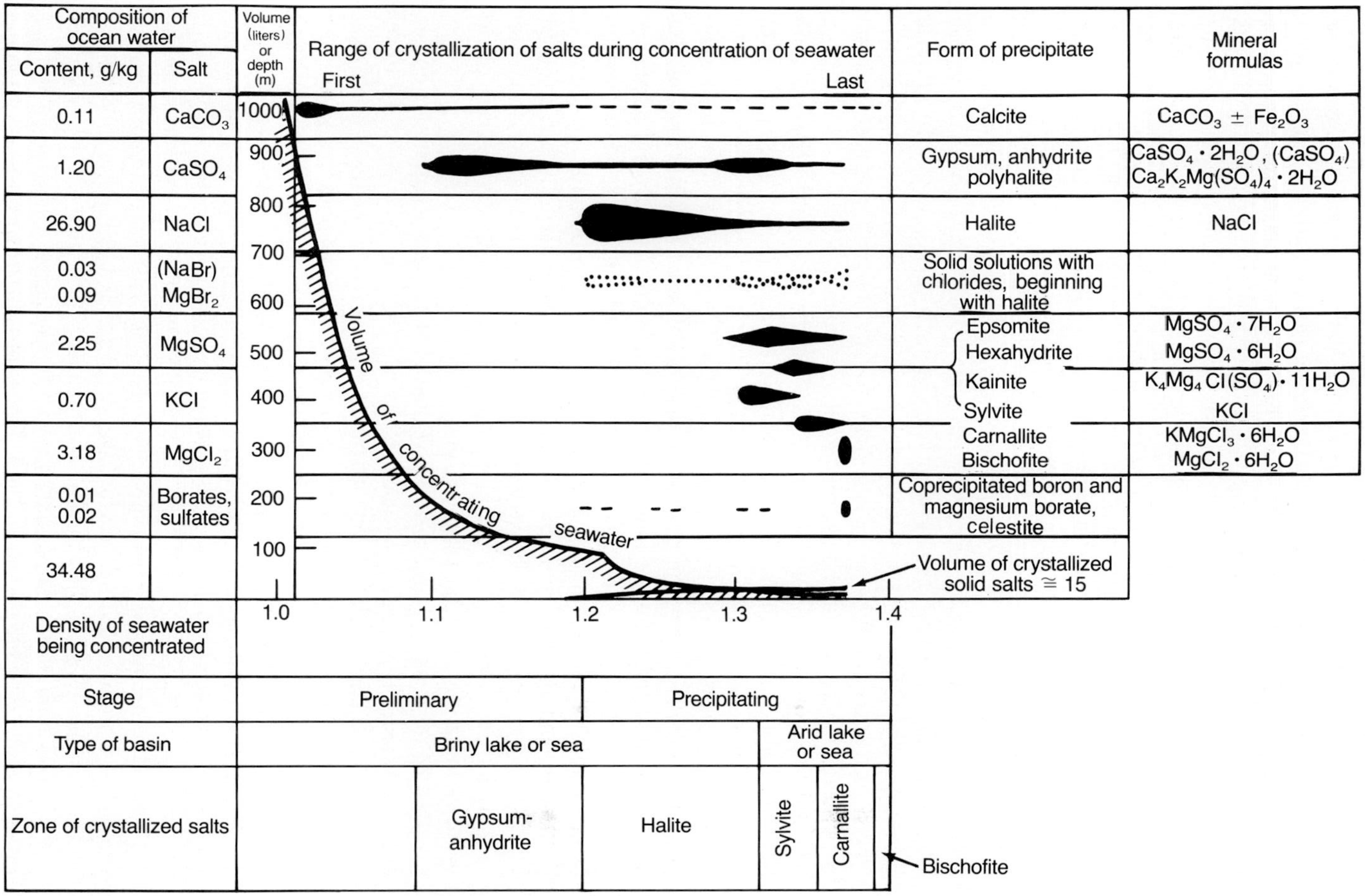

Composition of ocean water: Content, g/kg	Composition of ocean water: Salt	Volume (liters) or depth (m)	Range of crystallization of salts during concentration of seawater (First → Last)	Form of precipitate	Mineral formulas
0.11	$CaCO_3$	1000		Calcite	$CaCO_3 \pm Fe_2O_3$
1.20	$CaSO_4$	900		Gypsum, anhydrite polyhalite	$CaSO_4 \cdot 2H_2O$, $(CaSO_4)$ $Ca_2K_2Mg(SO_4)_4 \cdot 2H_2O$
26.90	NaCl	800		Halite	NaCl
0.03 0.09	(NaBr) $MgBr_2$	700		Solid solutions with chlorides, beginning with halite	
2.25	$MgSO_4$	600 500		Epsomite Hexahydrite	$MgSO_4 \cdot 7H_2O$ $MgSO_4 \cdot 6H_2O$
0.70	KCl	400		Kainite Sylvite	$K_4Mg_4Cl(SO_4) \cdot 11H_2O$ KCl
3.18	$MgCl_2$	300		Carnallite Bischofite	$KMgCl_3 \cdot 6H_2O$ $MgCl_2 \cdot 6H_2O$
0.01 0.02	Borates, sulfates	200		Coprecipitated boron and magnesium borate, celestite	
34.48		100		Volume of crystallized solid salts ≅ 15	
Density of seawater being concentrated		1.0	1.1 · 1.2 · 1.3 · 1.4		
Stage		Preliminary	Precipitating		
Type of basin		Briny lake or sea	Arid lake or sea		
Zone of crystallized salts			Gypsum-anhydrite · Halite · Sylvite · Carnallite · Bischofite		

Volume of concentrating seawater

Several explanations for these variations have been suggested. One, depicted schematically in Figures 15-16 and 15-19, is that density stratification develops as evaporation in a basin proceeds. For example, if the inlet is deep enough (Figure 15-19*a*), seawater may flow in at the surface at normal salinity and density, undergo evaporation and reach supersaturation of calcium salts as it flows inward; it may become increasingly dense through continued solvent loss while precipitating gypsum or anhydrite, sink, flow back out across the threshold and beneath inflowing seawater recharge, and blend with the sea once more, carrying sodium chloride and bittern components with it. A gypsum-anhydrite unit may thus build up in the absence of halite and the bitterns. If the inlet is shallow, the backflow does not occur, and bitterns will form within the basin (Figure 15-19*b*). Also, progressive decantation of brines into successively more landward basins, backreef or back-bar areas, or estuaries may permit anhydrite-gypsum to yield laterally rather than vertically to halite, then to the bitterns. In general, the K-Na-Mg-Cl-Br salts are indeed found in thin inboard pockets or in "pans" in halite sequences that appear to have nearly completely dried before reinundation and the start of another cycle.

Questions concerning the deposition of gypsum ($CaSO_4 \cdot 2HO$) and anhydrite ($CaSO_4$) are still poorly answerable. Gypsum seems to be the first of the two precipitated in nature, but anhydrite is by far the more common mineral in evaporite sediments. Stewart (1963), Krauskopf (1967), and others note that anhydrite precipitation is favored at higher temperatures of 40°C or more and at salinities of more than 3.35%. Since salt precipitation normally starts at higher salinities, anhydrite should be the standard precipitate, but attempts to synthesize it in simulated natural conditions have all failed. Gypsum appears to be the earlier salt, presumably followed at higher salinity or temperature by anhydrite. Gypsum is then normally dehydrated to anhydrite by burial pressures; indeed, anhydrite may be entirely a secondary "metamorphic" postdepositional mineral.

Space does not permit a detailed description of an evaporite deposit, but the student is referred to descriptions of the Permian evaporites of the Salado, Castile, and related formations of New Mexico by King (1942) and Jones (1954). These units have been major producers of potash and are now prime sources of biogenic sulfur (Ruckmick, Wimberley, and Edwards, 1979) formed in layered anhydrite-gypsum units in a fashion analogous to the salt dome sulfur production described in Chapter 7. The Permian of Poland, Germany, and western Europe is also enormously productive of potash in the Zechstein strata (Smith and Crosby, 1979). The Boulby mine in the Zechstein of northeastern England (Woods, 1979) shows a nearly complete evaporite sequence—compare the interval from the Billingham Main Anhydrite to the Upgang Formation in Figure 15-20 with Tables 15-1 and 15-2. The evaporites of Devonian age in the Williston basin of Saskatchewan-Manitoba of Canada and in the Dakotas of the United States are major suppliers of potash for the North American midcontinent agricultural belt

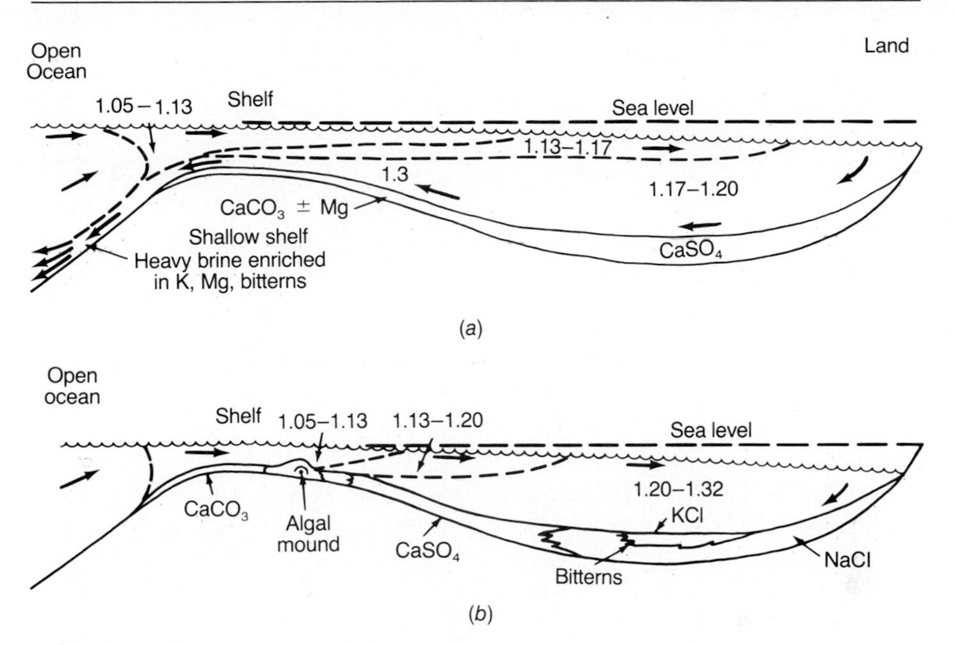

Figure 15-19. Schematics of marine evaporite formation in shallow basins, such as those of Figure 15-16. In (*a*) the sea level conditions are higher than in (*b*). The channel is deep enough to permit return of denser late-stage brines to the sea, and only limestone and gypsum-anhydrite precipitate in the basin. (*b*) No reflux occurs over the shallow threshold, and bitterns are forced to precipitate as their solubility products are exceeded. (*From* Hite, 1970.)

from a locally 200-meter-thick salt section (Anderson and Swinehart, 1979; Worsley and Fuzesy, 1979). One of several excellent descriptive reference volumes is *Marine Evaporites* (Dean and Schreiber, 1978).

Lacustrine evaporites, as indicated, are also highly specialized and localized chemical sediments. A review of the Chilean nitrate deposits has recently been presented by Ericksen (1981) and of Searles Lake, a representative of the lacustrine deposits of the Sierra–Death Valley chain of dry lakes, by Smith (1979). The remarkable Eocene Green River Wyoming-Colorado-Utah alkali-carbonate–oil shale deposits have been discussed by Bradley (1972), Eugster and Surdam (1973), and Chapin and Cather (1981).

▸ MANGANESE NODULES

We have considered chemical sediments that form in littoral environments, in bays and fault-block basins at continental margins, and in geosynclines and on the continental shelves themselves. As we progress conceptually seaward, we consider the majority of the Earth's surface, the abyssal ocean deeps. These "deeps" are the vast, cold darknesses of the ocean floors, 320 million km^2 of an average depth of 3.8 km, over 12,500 ft. The ocean floors

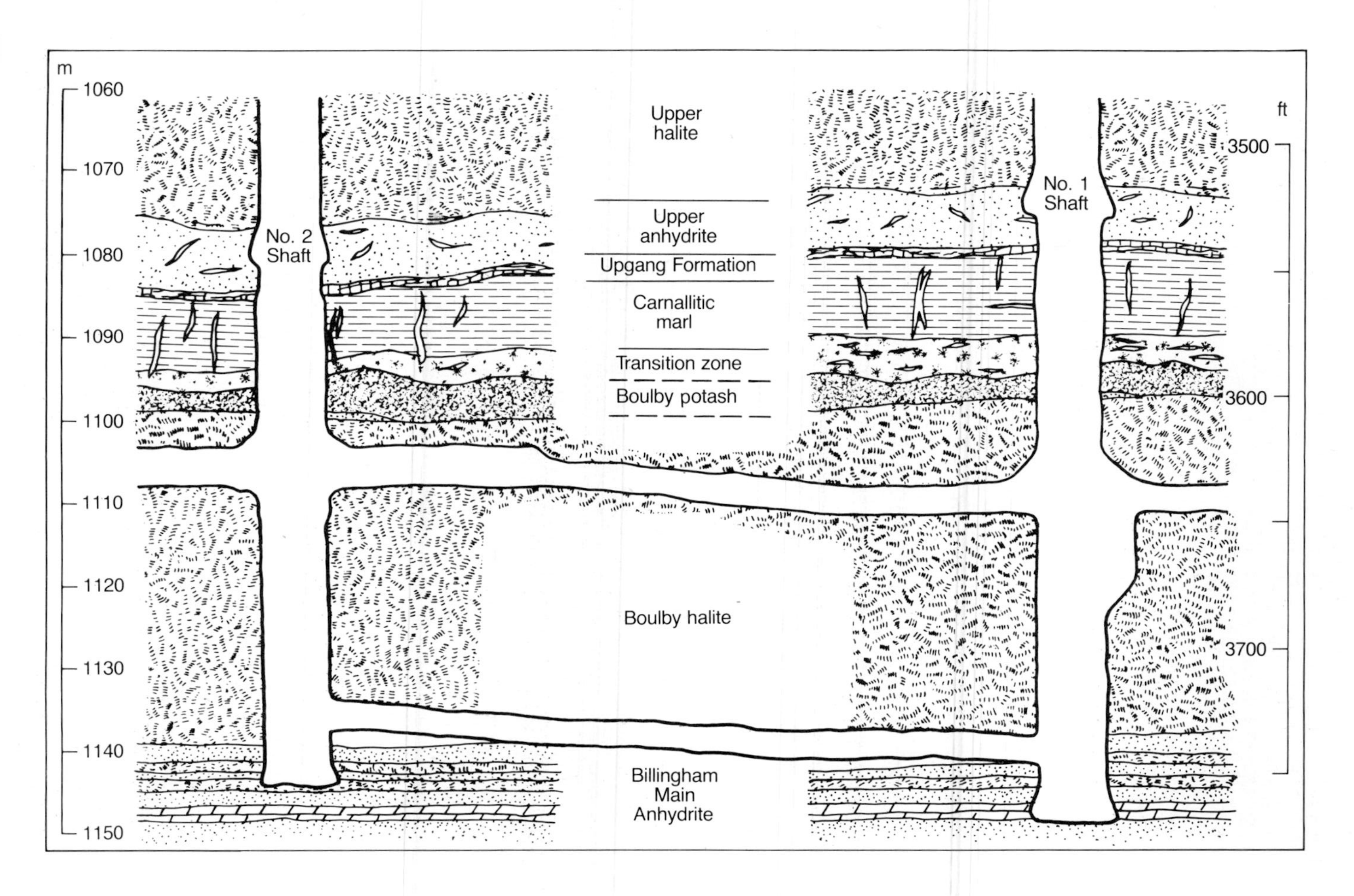

m
1060
1070
1080
1090
1100
1110
1120
1130
1140
1150
No. 2 Shaft
Upper halite
Upper anhydrite
Upgang Formation
Carnallitic marl
Transition zone
Boulby potash
Boulby halite
Billingham Main Anhydrite
No. 1 Shaft
ft
3500
3600
3700

Figure 15-20. (*Facing page*) Cross section of the evaporite strata and mine workings of the Boulby mine, England. Note that the succession from anhydrite at the base through a thick section of halite to the potash-rich beds beneath the Upgang Formation constitutes a normal marine evaporite cycle, with the start of another at the Upgang. (*From* Woods, 1979.)

were little known until submarine and nuclear detonation detection devices proliferated in the 1940s and 1950s and revealed their topography and form, also leading of course to the plate tectonics revolution. The ocean floors have now been photographed, dredged, visited by submersible craft, drilled, and geophysically detailed. The deep-ocean drilling is not of the off-shore oil-well type, which is restricted to sedimentary units on the continental shelves, but rather the piston-coring and shallow drilling of recent research programs, such as the Joint Oceanographic Institutes for Deep Earth Sampling (JOIDES) project and the Deep Sea Drilling Project (DSDP).

The deep oceans are sediment-starved. Since oceanic crust is produced at midoceanic rises and consumed in subduction zones, the average age of oceanic crust and thus of the seafloor is only a few tens of millions of years. With only a few millimeters of sedimentation per year, the sediment load on deep oceanic crust is typically thin, consisting of the few centimeters of siliceous oozes, ochres, diatoms and radiolaria, and other debris described

Figure 15-21. Manganese-rich nodules lying on the seafloor in one of the nodule-rich areas southeast of Hawaii. (*From* Heath, 1981.)

(*a*)

(*b*)

Figure 15-22. Close-up photos of typical nodules showing their globular laminated nature. (*From* Meyer, 1973.)

in Chapter 13 as Layer 1. The ooze appears to become more anomalously transition-metal-rich as a midoceanic rise is approached (Bostrom and Peterson, 1966); but one of the startling discoveries of the marine world was that a pavement of black, nodular, golf-ball-sized accretions of manganese and iron oxide composition carpets much of the ocean floor (Figure 15-21;

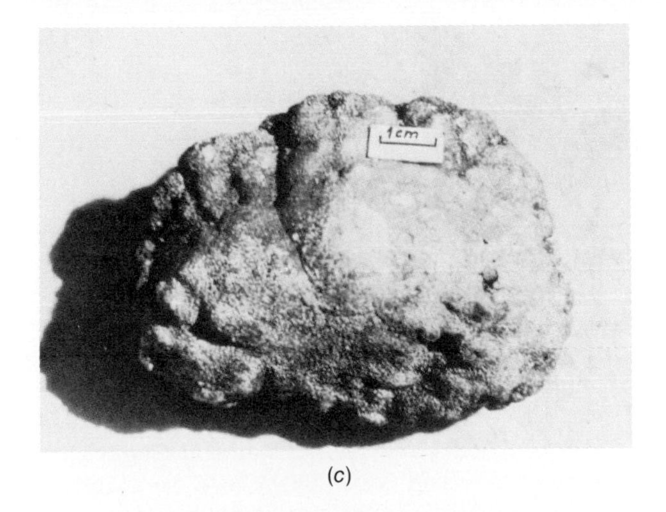

(*c*)

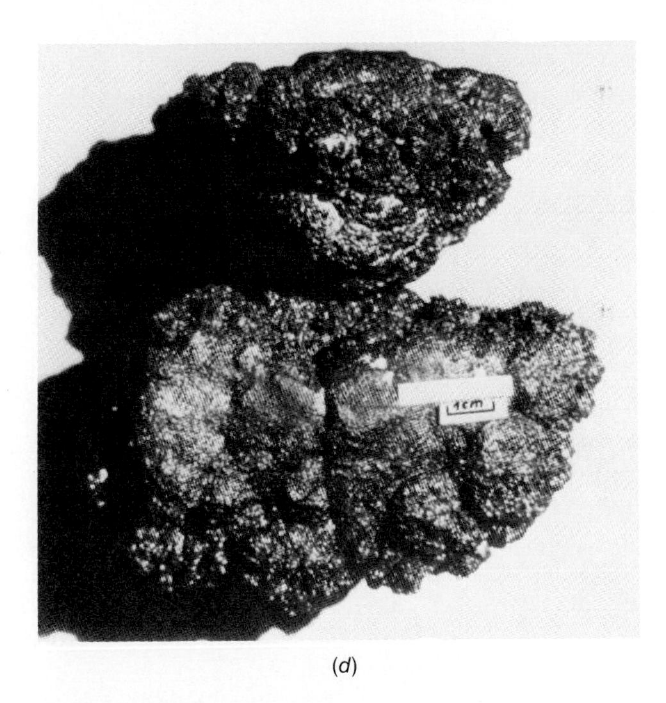

(*d*)

Sorem and Fewkes, 1979; Heath, 1981). They are widely scattered on the floors of all of the major ocean basins at depths of 15,000 ft (5000 meters) or more. The nodules range from the size of a pea to blunt slabs more than 1 meter across; most are globular and a few centimeters in diameter (Figure 15-22). The greatest concentrations and the best grades of nodules are in the Pacific ocean outside of the territorial limits claimed by any country. Estimates of tonnages of recoverable nodules range from about 50 billion to more than a trillion tons. Where they are abundant, there may be more than 100 per square meter, forming a pavement only one nodule thick. Where they are less abundant, they are in general evenly spaced one from another.

The nodules commonly contain 20 to 30% manganese oxides, with some iron. As much as 2 to 3% of combined copper, nickel, and cobalt are present. The manganese oxides pyrolusite and birnessite are deposited as concentric layers around cores of sand grains, sharks' teeth, or other material. They are abundant immediately at the ocean floor, but are also found occasionally embedded in buried sediment. They appear to nucleate and grow slowly, essentially in situ, with Fe, Mn, Cu, Co, Ni, and other metals supplied by the sedimentation of colloids and fine particles through the seawater, or from the sedimentary oozes in which they lie (Cheney and Vredenberg, 1968). The source of the metals may be volcanic exhalative, terrigenous, or even extraterrestrial cosmic dust or meteorite ash dust. Most geologists consider the nodules to be related to submarine volcanic processes and many deposits do appear to be concentrated in areas of volcanic activity (Bonatti and Nayudu, 1965). Bostrom and Peterson (1966) showed that the enrichment in iron, manganese, and other elements in Pacific ocean bottom muds is distributionally related to the East Pacific Rise for hundreds of kilometers on either side.

The origin of the manganese oxides associated with seafloor mid-oceanic rise volcanic processes is of considerable theoretical and economic interest. Hot volcanic materials ejected under water tend to become finely and thoroughly fragmented. The fragments are agitated, and because of their textures are ideally prepared for leaching by volcanically heated seawater or by hydrothermal waters contributed during volcanism. Under these conditions, ferromagnesian minerals which contain manganese are susceptible to alteration. Volcanic emanations or convectively circulating seawaters that percolate through oceanic crust basalts contribute to decomposition of the ferromagnesian minerals and may transport manganese and iron up and out (see Figure 10-9) (Bonatti, 1975; Humphris and Thompson, 1978; Kuypers and Denyer, 1979). Dissolved manganese migrates upward to the floor of the sea and is deposited on the flanks of the rises or on remote seafloor and may be regarded as exhalative.

Metallurgical studies of the nodules have been designed primarily to recover copper, nickel, and cobalt, though some efforts are also being made to save the manganese. In 1985 all of these elements are available at far lower cost from deposits on the continents. However, a great deal of time and money have been spent in efforts to develop an economic method for recovering the nodules from the great depths involved (Heath, 1981). Several groups of companies claim their recovery processes to be ready and able to start work at once on a commercial basis. They are being delayed because of low prices for copper and nickel on the world market, but primarily because of political problems. Who owns the floors of the deep ocean basins? No law of the sea yet applies to these deposits. Various groups sponsored by the United Nations have met in Geneva, but they are still reported to be far from agreement in early 1985.

Bibliography

Alling, H. L., 1947. Diagenesis of the Clinton hematite ore of New York. *Geol. Soc. Amer. Bull.* 58:991–1018.

Anderson, S. B., and R. P. Swinehart, 1979. Potash salts in the Williston Basin, U.S.A. *Econ. Geol.* 74:358–376.

Baas-Becking, L. G. M., and D. Moore, 1961. Biogenic sulfides. *Econ. Geol.* 56:259–272.

Balla, J. C., 1980. Pers. Commun.

——, 1982. Geology of the Troy deposit, northwestern Montana, p. 8 in *Symp. on the Genesis of Rocky Mountain Ore Deposits.* Denver Reg. Explor. Geol. Soc.

Bayley, R. W., and H. L. James, 1973. Precambrian iron formations of the United States. *Econ. Geol.* 68:934–959.

Beerstecher, E., Jr., 1954. *Petroleum Microbiology.* New York: Elsevier.

Bonatti, E., 1975. Metallogenesis at oceanic spreading centers, in *Annual Reviews of Earth and Planetary Sciences*, vol. 3. Palo Alto: Annual Reviews, Inc., pp. 401–431.

—— and Y. R. Nayudu, 1965. The origin of manganese nodules on the ocean floor. *Amer. J. Sci.* 263:17–39.

Bostrom, K., and M. N. A. Peterson, 1966. Precipitates from hydrothermal exhalations of the East Pacific Rise. *Econ. Geol.* 61:1258–1265.

Bowen, R. N. C., and A. Gunatilaka, 1977. *Copper, Its Geology and Economics.* London: Applied Sci. Pub., 366 pp.

Bradley, W. H., 1972. Oil shale formed in desert environment, Green River Formation, Wyoming. *Geol. Soc. Amer. Bull.* 84:1121–1154.

Burchard, E. F., and C. Butts, 1910. Iron ores, fuels, and fluxes of the Birmingham district, Alabama. USGS Bull. 400.

Callender, E., and C. J. Bowser, 1976. Freshwater ferromanganese deposits, pp. 341–394 in K. H. Wolf, Ed., *Handbook of Stratabound and Stratiform Ore Deposits*, vol. 7. New York: Elsevier.

Castaño, J. R., and R. M. Garrels, 1950. Experiments on the deposition of iron with special reference to the Clinton iron ore deposits. *Econ. Geol.* 45:755–770.

Cayeux, L., 1909. *Les Minérais de Fer Oolithique de France*, vol. 1. Paris: Imprimerie Nationale.

Chapin, C. E., and S. M. Cather, 1981. Eocene tectonics and sedimentation in the Colorado Plateau–Rocky Mountain area, pp. 173–198 in W. R. Dickinson and W. D. Payne, Eds., *Relations of Tectonics to Ore Deposits in the Southern Cordillera*. Ariz. Geol. Soc. Dig., vol. 14, 288 pp.

Cheney, E. S., and L. D. Vredenberg, 1968. The role of iron sulfide in the diagenetic formation of iron-poor manganese nodules. *J. Sed. Petrol.* 38:1363–1365.

Clark, A. L., 1971. Stratabound copper sulfides in the Precambrian Belt Supergroup, northern Idaho and northwestern Montana, pp. 261–267 in *Joint Symposium Vol., IMA-IAGOD Mtgs., 1970*. Tokyo: Soc. Min. Geol. Japan, Spec. Issue 2, 500 pp.

Cook, D. R., Ed., 1957. *Geology of the East Tintic Mountains and Ore Deposits of the Tintic Mining District*. Utah Geol. Soc. Guidebook to the Geology of Utah, no. 12, Univ. Utah, 177 pp.

Cook, P. J., and M. W. McElhinny, 1979. A reevaluation of the spatial and temporal distribution of sedimentary phosphate deposits in the light of plate tectonics. *Econ. Geol.* 74:315–330.

Cornelius, H. P., and B. Plöchinger, 1952. Der Tennengebirge-N-Rand mit seinen Manganerzen und die Berge im Bereich des Lammertales. *Austrian Geol. Bundesanstalt Jahrb.* 95:145–225.

Dean, W. T., and B. C. Schreiber, Eds., 1978. *Marine Evaporites*. Soc. Econ. Paleontol. Miner., 188 pp.

Dessau, G., M. L. Jensen, and N. Nakai, 1962. Geology and isotopic studies of Sicilian sulfur deposits. *Econ. Geol.* 57:410–438.

Dimroth, E., 1976. Aspects of the sedimentary petrology of cherty iron-formation, pp. 203–254. in K. H. Wolf, Ed., *Handbook of Stratabound and Stratiform Ore Deposits*, vol. 7. New York: Elsevier.

Dunham, K. C., 1964. Neptunist concepts in ore genesis. *Econ. Geol.* 59:1–21.

Ericksen, G. E., 1981. Geology and origin of the Chilean nitrate deposits. U.S.G.S. Prof. Pap. 1188.

Eugster, H. P., C.E. Harvie, and J. H. Weare, 1980. Mineral equilibria in a six-component seawater system, Na-K-Mg-Ca-SO_4-Cl-H_2O, at 25°C. *Geochim. Cosmochim. Acta* 44:1335–1347.

—— and R. C. Surdam, 1973. Depositional environment of the Green River Formation of Wyoming: a preliminary report. *Geol Soc. Amer. Bull.* 84:1115–1120.

—— and B. F. Jones, 1979. Behavior of major solutes during closed-basin brine evolution. *Amer. J. Sci.* 279:609–631.

Fleischer, V. D., W. G. Garlick, and R. Haldane, 1976. Geology of the Zambian Copperbelt, pp. 223–352 in K. H. Wolf, Ed., *Handbook of Stratabound and Stratiform Ore deposits*, vol. 6. New York: Elsevier.

Freiberg, D. A., 1983. Geologic setting and origin of the Lucifer manganese deposit, Baja California Sur, Mexico. *Econ. Geol.* 78:931–943.

Gammon, J. B., 1966. Fahlbands in the Precambrian of southern Norway. *Econ. Geol.* 61:174–188.

Garrels, R. M., and F. T. MacKenzie, 1971. *Evolution of Sedimentary Rocks*. New York: Norton, 397 pp.

——, E. A. Perry, Jr., and F. T. MacKenzie, 1973. Genesis of Precambrian iron-formations and the development of atmospheric oxygen. *Econ. Geol.* 68:1173–1179.

Goldschmidt, V. M., 1954. *Geochemistry*. Oxford: Clarendon.

Gurr, T. M., 1979. Geology of U. S. phosphate deposits. *Min. Eng.* 31:682–691.

Harvie, C. E., and J. H. Weare, 1980. The prediction of mineral solubilities in natural waters: the Na-K-Mg-Ca-Cl-SO_4-H_2O system from zero to high concentration at 25°C. *Geochim. Cosmochim. Acta* 44:981–997.

——, ——, and L. A. Hardie, 1980. Evaporation of seawater; calculated mineral sequences. *Science* 208:498–500.

Hayes, A. O., 1915. Wabana iron ore of Newfoundland. Can. Geol. Surv. Memo. 78.

——, 1928. Wabana iron mines and deposits, Newfoundland. *Min. Metall.* 9:361–366.

——, 1929. Further studies of the origin of the Wabana iron ore of Newfoundland. *Econ. Geol.* 24:687–690.

——, 1931. Structural study of the Conception Bay region and of the Wabana iron ore deposits of Newfoundland. *Econ. Geol.* 26:44–64.

Heath, G. R., 1981. Ferromanganese nodules of the deep sea. *Econ. Geol. 75th Anniv. Vol.*, pp. 736–765.

Hem, J. D., 1960. Some chemical relationships among sulfur species and dissolved ferrous iron. USGS Water Supply Pap. 1459–C, pp. 57–73.

Hewett, D. F., and B. N. Weber, 1931. Bedded deposits of manganese oxides near Las Vegas, Nevada. Univ. Nev. Bull. 25(6).

Hite, R. J., 1970. Shelf carbonate sedimentation controlled by salinity in the Paradox Basin, southeast Utah. *Proc. 3rd Symp. on Salt,* vol. 1, pp. 48–66.

Howard, P. F., 1979a. Phosphate. *Econ. Geol.* 74:192–194.

——, Ed., 1979b. Issue devoted to phosphate, potash, sulfur. *Econ. Geol.* 74(2):191–495.

Humphris, S. E., and G. Thompson, 1978. Trace element mobility during hydrothermal alteration of oceanic basalts. *Geochim. Cosmochim. Acta* 42:127–136.

Hunter, R. E., 1970. Facies of iron sedimentation in the Clinton Group, pp. 101–124 in G. W. Fisher, F. J. Pettijohn, J. C. Reed, Jr., and K. N. Weaver, Eds., *Studies of Appalachian Geology: Central and Southern.* New York: Wiley-Interscience, 460 pp.

Jacob, K. D., W. L. Hill, H. L. Marshall, and D. S. Reynolds, 1933. Composition and distribution of phosphate rock with special reference to the United States. US Dept. Agric. Tech. Bull. 364.

Jensen, M. L., 1958. Sulfur isotopes and the origin of sandstone-type uranium deposits. *Econ. Geol.* 53:598–616.

Jones, C. L., 1954. The occurrences and distribution of potassium minerals in southeastern New Mexico. *New Mex. Geol. Soc. Guidebook*, 5th Field Conf. eastern New Mex., pp. 107–112.

Jung, W., and G. Knitzschke, 1976. Kupferschiefer in the German Democratic Republic with reference to deposits in S. E. Harzforeland, pp. 353–406 in K. H. Wolf, Ed., *Handbook of Stratabound and Stratiform Ore Deposits*, vol. 6. New York: Elsevier.

Kimberly, M. M., 1978. Paleoenvironmental classification of iron formations. *Econ. Geol.* 73:215–229.

——, 1980. The Paz de Rio oolitic-inland-sea formation. *Econ. Geol.* 75:97–106.

King, P. B., 1942. Permian of west Texas and southeastern New Mexico, part 2, in R. K. Deford and E. R. Lloyd, Eds., *West Texas–New Mexico Symposium, Bull. Amer. Assoc. Petrol. Geol.* 26:535–763.

Krauskopf, K. B., 1955. Sedimentary deposits of rare metals. *Econ. Geol. 50th Anniv. Vol.*, pp. 411–463.

——, 1956a. Dissolution and precipitation of silica at low temperatures. *Geochim. Cosmochim. Acta* 10:1–26.

——, 1956b. Factors controlling the concentrations of thirteen rare metals in sea water. *Geochim. Cosochim. Acta* 9:1–32.

——, 1957. Separation of manganese from iron in sedimentary processes. *Geochim. Cosmochim. Acta* 12:61–84.

——, 1967. *Introduction to Geochemistry*. New York: McGraw-Hill, 721 pp.

Kuypers, E. P., and C. P. Denyer, 1979. Volcanic exhalative manganese deposits of the Nicoya ophiolitic complex, Costa Rica. *Econ. Geol.* 74:672–678.

Lange, I. M., and R. A. Sherry, 1983. Genesis of the sandstone (Revett) type of copper-silver occurrences in the Belt Supergroup of northwestern Montana and northeastern Idaho. *Geology* 11:643–646.

Ljunggren, P., 1955. Geochemistry and radioactivity of some Mn and Fe bog ores. *Geol. Foren. Stockholm Forh.* 77:33–44.

Lunar, R., and J. L. Amoros, 1979. Mineralogy of the oolitic iron deposits of the Ponferrada-Astorba zone, northwestern Spain. *Econ. Geol.* 74:751–762.

Lyons, J. C., 1957. Wabana iron ore deposits, in *Structural Geology of Canadian Ore Deposits*, vol. 2. Montreal: Can. Inst. Min. Metall. Eng., 524 pp.

Margulis, L., 1981. *Symbiosis in Cell Evolution; Life and Its Environment on the Early Earth*. San Francisco; Freeman, 419 pp.

——, 1981b. Pers. Commun.

Mason, B., 1958. *Principles of Geochemistry,* 2nd ed. New York: Wiley.

McKelvey, V. E., 1950. The field of economic geology of sedimentary mineral deposits, in P. D. Trask, Ed., *Applied Sedimentation.* New York: Wiley.

——, 1967. Phosphate deposits. USGS Bull. 1252–D.

——, D. L. Everhart, and R. M. Garrels, 1955. Origin of uranium deposits. *Econ. Geol. 50th Anniv. Vol,* pp. 464–533.

—— and J. M. Nelson, 1950. Characteristics of marine uranium-bearing sedimentary rocks. *Econ. Geol.* 45:35–53.

——, J. H. Wiese, and V. H. Johnson, 1949. Preliminary report on the bedded manganese of the Lake Mead region, Nevada and Arizona. USGS Bull., 948-D, pp. 83–101.

Mendelsohn, F., 1961. Ore genesis: summary of the evidence, pp. 130–165 in F. Mendelsohn, Ed., *The Geology of the Northern Rhodesian Copper Belt.* London: MacDonald, 523 pp.

Mercer, W., 1976. Minor elements in metal deposits in sedimentary rocks: a review of the recent literature, pp. 1–27 in K. H. Wolf, Ed., *Handbook of Stratabound and Stratiform Ore Deposits,* vol. 2. New York: Elsevier.

Meyer, K., 1973. Surface sediment and manganese nodule facies encountered on R/V Valdivia cruises 1971/73, pp. 125–130 in M. Morgenstein, Ed., *Papers on the Origin and Distribution of Manganese Nodules in the Pacific and Prospects for Exploration.* Honolulu: Hawaii Inst. Geophys.

Mikhalev, D. N., 1946. On the great and minor epochs of accumulation of manganiferous sediments. *Compt. Rend. (Dokl.) Acad. Sci. USSR.* 54:339–341.

Moore, E. S., and J. E. Maynard, 1929. Solution, transportation, and precipitation of iron and silica. *Econ. Geol.* 24:272–303; 365–402; 506–527.

Park, C. F., Jr., 1942. Manganese deposits of Cuba. USGS Bull. 935–B, pp. 75–97.

——, 1956. On the origin of manganese, in *Symp. sobre Yacimientos de Manganeso, 20th Int. Geol. Congr.,* vol. 1.

Pompecki, J. F., 1920. Kupferschiefer und Kupferschiefermeer. *Z. Deut. Geol. Ges.* 72:329–339.

Rankama, K., and T. G. Sahama, 1950. *Geochemistry.* Chicago: Univ. Chicago Press.

Renfro, A. R., 1974. Genesis of evaporite-associated stratiform metalliferous deposits—a sabkha process. *Econ. Geol.* 69:33–45.

Riggs, S. R., 1979a. Petrology of the Tertiary phosphate system of Florida. *Econ. Geol.* 74:195–220.

——, 1979b. Phosphorite sedimentation in Florida—a model phosphogenic system. *Econ. Geol.* 74:285–314.

Roberts, R. J., 1967. The genesis of disseminated and massive sulfide deposits in Saudi Arabia. Saudi Arabian Proj. Rept. 207, Prep. for Dir.-Gen. Min. Res., Minist. Petrol. Min. Res., Jiddah, Saudi Arabia, by USGS.

Roy, S., 1976. Ancient manganese deposits, pp. 395–476 in K. H. Wolf, Ed., *Handbook of Stratabound and Stratiform Ore Deposits*, vol. 7. New York: Elsevier.

Ruckmick, J. C., B. H. Wimberly, and A. F. Edwards, 1979. Classification and genesis of biogenic sulfur deposits. *Econ. Geol.* 74:469–474.

Sakamoto, T., 1950. The origin of the Pre-Cambrian banded iron ores. *Amer. J. Sci.* 248:449–474.

Sangster, D. F., 1972. Precambrian volcanogenic massive sulfide deposits in Canada: a review. Geol. Surv. Can. Pap. 72–22.

Sauce, W. B. W., de la, 1926. Beiträge zur Kenntnis der Manganerzlagerstätte von Tschiaturi im Kaukasus. Abh. Prakt. Geol. Bergwirtsch. 8.

Schneiderhöhn, H., 1923. Chalkagraphische Untersuchung des Mansfelder Kupferschiefers. *Neues Jahrb. Mineral. Geol. Paläontol.* 47(supp.):1–38.

Shaner, L. A., 1982. Reconnaissance study of metal sulfide deposition in tidal flat and sabkha-like environments, Gulf of California, Sonora, Mexico. Unpub. M.S. Thesis, Univ. Ariz., Tucson, 80 pp.

Smith, D. B., and A. Crosby, 1979. The regional and stratigraphic context of Zechstein 3 and 4 potash deposits in the British sector of the southern North Sea and adjoining land areas. *Econ. Geol.* 74:397–408.

Smith, G. E., 1976. Sabkha and tidal-flat facies control of stratiform copper deposits in north Texas, pp. 407–443 in K. H. Wolf, Ed., 1976, *Handbook of Stratabound and Stratiform Ore Deposits*, vol. 6. New York: Elsevier.

Smith, G. I., 1979. Subsurface stratigraphy and geochemistry of late Quaternary evaporites, Searles Lake, California. USGS Prof. Pap. 1043.

Sorem, R. K., and R. H. Fewkes, 1979. *Manganese Nodules: Research Data and Methods of Investigation.* New York: Plenum, 723 pp.

Stewart, F. H., 1963. Marine evaporites, pp. Y1–Y52 in *Data of Geochemistry*, 6th ed. USGS Prof. Pap. 440–Y.

Strakhof, N. M., 1970. *Principles of Lithogenesis*, vol. 3. New York: Plenum, 577 pp.

Temple, K. L., 1964. Syngenesis of sulfide ores: an evaluation of biochemical aspects. *Econ. Geol.* 59:1473–1491.

Trask, P. D., 1925. The origin of the ore of the Mansfeld Kupferschiefer, Germany: a review of the current literature. *Econ. Geol.* 20:746–761.

Trudinger, P. A., I. B. Lambert, and G. W. Skyring, 1972. Biogenic sulfide ores: a feasibility study. *Econ. Geol.* 67:1114–1127.

Twenhofel, W. H., 1950. *Principles of Sedimentation*, 2nd ed. New York: McGraw-Hill, pp. 444–451.

Usiglio, J., 1849. Etudes sur la composition de l'eau de la Méditerrannée et sur l'exploitation des sels qu'elle contient. *Ann. Chim. Phys.*, ser. 3, 27:172–191.

Vaughn, D. J., 1976. Sedimentary geochemistry and mineralogy of the sulfides of Pb, Zn, Cu, and Fe and their occurrence in sedimentary ore deposits, pp. 317–360 in K. H. Wolf, Ed., *Handbook of Stratabound and Stratiform Ore Deposits*, vol. 2. New York: Elsevier.

Wilson, I. F., and M. Veytia, 1949. Geology and manganese deposits of the Lucifer district northwest of Santa Rosalia, Baja California, Mexico. USGS Bull. 960-F, pp. 177–233.

Wolf, K. H., Ed., 1976. *Handbook of Stratabound and Stratiform Ore Deposits*, 10 vols. New York: Elsevier.

Woods, P. J. E., 1979. The geology of Boulby Mine. *Econ. Geol.* 74:409–418.

Woolnough, W. G., 1941. Origin of banded iron ore—a suggestion. *Econ. Geol.* 36:465–489.

Worsley, N., and A. Fuzesy, 1979. The potash-bearing members of the Devonian Prairie Evaporite of southeastern Saskatchewan, south of the mining area. *Econ. Geol.* 74:377–388.

Zantop, H., 1978. Geologic setting and genesis of iron oxides and manganese oxides in the San Francisco manganese deposit, Jalisco, Mexico. *Econ. Geol.* 73:1137–1149.

Zapffe, C., 1931. Deposition of manganese. *Econ. Geol.* 26:799–832.

SIXTEEN

Deposits Related to Clastic Sedimentation

In general, weathering processes acting upon rocks contribute to the overall mineral resource picture in five ways. First, weathering may produce new minerals in situ that are more useful than their progenitor minerals, such as bauxites or kaolinite produced by the hydrolysis of feldspars. Examples include bauxite deposits developed from syenite in the Little Rock area, Arkansas, the long productive kaolinite deposits of Cornwall, Devonshire, England, which are partly of weathering origin, and kaolin developed by the weathering of pegmatite feldspar at Spruce Pine, North Carolina. Second, weathering may trigger the redistribution of valuable elements such as copper and nickel in a family of processes called *supergene enrichment*, which are considered in Chapter 17. Third, fine detritus and chemical complexes may be released by weathering to become involved as precipitates in chemical sedimentation (Chapter 15) or the formation of conventional sediments. Fourth, weathering may release resistant minerals from hard rocks such as granite into softer or otherwise more workable soillike aggregates at and near the surface. One excellent example is the "freeing" of diamonds from hard, unweathered kimberlite "blue ground" into friable, easily disaggregated weathered kimberlite "yellow ground," which contained the most easily recovered diamonds at Kimberley, South Africa, in the U.S.S.R., and in Arkansas. Other examples are the release of cassiterite from tin-bearing granites into cassiterite-kaolinite "grus" material in many localities, notably Malaysia, and the release of muscovite into coproduct kaolinite at the Spruce Pine deposit just mentioned. And finally, weathering releases resistant minerals like gold, diamonds, and cassiterite to be winnowed from their weathered matrix, transported, and concentrated by sedimentologic processes into deposits called *placers*. Hails (1976) defined placers as

> *. . . surficial mineral deposits formed by mechanical concentration, commonly by alluvial but also by marine, aeolian, lacustrine, or glacial agents, of heavy mineral particles such as gold from weathered debris.*

▸ PLACER DEPOSITS

Resistate minerals, those that are chemically stable or metastable in the weathered zone of the Earth's surface and are not decomposed by weathering processes as the surrounding rocks are dissolved or disintegrated, either remain in the soil or are carried away by rain, streams, waves, or wind. The lighter particles of some resistates are readily moved and become dispersed; more brittle ones break easily along cleavage or fracture planes and become so fine-grained that they, too, are dispersed. But heavy, stable, durable minerals left as residual particles in soil may be washed into drainages. Where the slopes below the outcrops are steep, or where other conditions encourage movement, the resistant particles slide, creep, or are washed gradually down the slope until they reach a stream bed. In streams, they are separated from finer grained, hydraulically lighter clay- and silt-sized materials that are washed downstream. The heavy minerals may then be concentrated into the sands and gravels of streams and beaches. The process of elutriation, or separation by agitation of minerals in aqueous suspension in streams or at beaches, causes heavy particles to settle to the bottom with the removal of the lighter or more brittle, finer grained gangue. The result is a concentration of the heavy, tough, and chemically resistant minerals. These minerals may accumulate near the outcrops as residual concentrations; they may be washed into streams and accumulate in sand bars or in riffles and irregularities along the channel floors; or they may reach bodies of water where they are reworked by wave action and deposited in beach sands. Moving waters sort them according to their specific gravities, their shapes, their pickup velocities, the velocity of flow, bottom gradients, and other factors. Minerals with properties favorable for deposition are concentrated at the expense of lighter, brittle particles, which are broken, scattered, and transported into the deeper basins of deposition. Sutherland (1982) showed that longshore transport of diamonds reduces average grain size but improves quality because inferior stones are destroyed. The mere presence of gem diamond is commonly enough to define an orebody; hydraulic concentration is ineffective because the specific gravity of diamond is only 3.5.

Important concepts in considering placer mineral concentration are *equivalent entrainment*, *equivalent hydraulics*, and *equivalent spherical diameter*. These terms relate specific gravity, particle size, shape, and surface properties such as roughness and even electrostatic behavior to how grains are picked up by moving water and to their settling rates. It is thus possible to account for similar settling rates of well-rounded quartz sand grains, flatter, larger but lighter mica flakes, and much smaller denser particles of, for example, gold. The same particle may require higher or lower water velocity to be freed from its neighbors and put into suspension—called *pickup velocity*—than is actually required to move it along a streambed, so the actual ratios of minerals deposited can vary complexly from place to place in a streambed or marine terrace.

The most common and abundant placer minerals are the native metals, especially gold and the platinum group, and many of the heavy "inert" oxides, silicates, and other phases such as cassiterite, chromite, wolframite, rutile, magnetite, ilmenite, zircon, and many gemstones. Since sulfides readily break up and decompose in modern oxygenated environments, they seldom accumulate in placers. In a few exceptional instances, however, small amounts of relatively insoluble sulfides—for example, the cinnabar at New Almaden, California—have been recovered from geologically young placers formed near a lode deposit. Magnetite and ilmenite are among the most abundant minerals in placers, but fluvial concentrations of these are rarely sufficiently rich to be of economic interest. Magnetite or "black sand" behavior in stream meanders, rapids, and confluences can be studied casually in most of the mountain streams of the world. Some of the world's greatest tin and diamond deposits are placers; examples are the cassiterite-rutile residual and river placer deposits of Nigeria and the diamond gravels of the Vaal and Orange rivers in the Republic of South Africa.

But the world's most valuable placer commodity is gold. Placer gold varies widely in composition, depending on the character of the original mineral and the distance from the source lode to the placer resting place. Native gold in veins is generally alloyed with silver, less commonly with copper and other metals. Because both silver and copper are oxidizable and more water-soluble than gold, those metals are selectively leached from electrum during weathering, erosion, and downstream transport. Consequently, gold far removed from its source tends to become purer than the original material (Desborough, 1970). The constant pounding and abrasion that particles of placer gold receive as they travel downstream also result in a gradual reduction in grain size away from the lode. The malleability of gold leads to the production of *colors*, tiny thin flakes of gold that may actually float on water and that reflect brighter glints of light than their small mass would suggest. Incidentally, a tantalizing problem that deserves study is that of nugget formation. Few base- or precious-metal deposits contain gold as coarse as the nuggets that appear to be derived from them. How did the 6.2-kg nugget recently found in Brazil form? How was the ultrafine dispersed gold in the Butte, Montana, veins aggregated during erosion into the nuggets of gold that drew prospectors up Silver Bow Creek to the discovery area? This discovery history and gold particle size discrepancy is more the norm than the exception. Kilogram nuggets are far more common than kilogram masses of vein gold, and both physical aggregation by "impact welding" and surficial solution-redeposition of supergene gold in streams must occur.

The geologic details of placer deposit formation are still poorly understood because the behavior of heavy minerals in the various sedimentological regimes has not long been studied quantitatively. A standard reference is Bilibin's *Principles of Placer Geology* (1938). More recent works on placer deposits of Arizona (Johnson, 1972) and Alaska (Cobb, 1973) are more inventory catalogs than geologic descriptions. Other reports (Raicevic and Cabri, 1976; Foster, Foord, and Long, 1978) are primarily mineralogical. A manual on basic placer mining, published by the California Division of Mines and Geology (Anon., 1982), describes the construction of riffle devices for mining so clearly that geologic accumulation is also indirectly described, but few exploration guides are given. Averill (1946) is a comprehensive guide to fluvial placer gold mining. Broadly applicable experimental data and conclusions published by Adams, Zimpfer, and McLane (1978), Minter (1976, 1978), and Smith and Minter (1980) are discussed here in the section on the Witwatersrand. Quantitative or predictive sedimentologic studies that might be useful in the exploration of modern or fossil placers are scant, but Mosley and Schumm (1977), Slingerland (1977), Minter (1976, 1978), Pretorius (1981a), and Smith and Beukes (1983) should be consulted. McDonald (1983) emphasizes economics, and Armstrong (1981) is the best collection of articles on quartz pebble conglomerate deposits.

Relatively coarse heavy minerals can be expected to be concentrated in *riffles* in a streambed. Riffles are transverse linear features (Figure 16-1) that range in size from hairline joints *A* through fracture and fault surfaces *B* to transverse dikes *C* or contacts between resistant and nonresistant strata *D*. Heavy minerals will also be retained preferentially in plunge pools *E*, in eddy pools in rapids, and in potholes. Figure 16-2 shows some of the kinds of joints that can trap "heavies." One of the richest placer gold occurrences in mining history was a few square kilometers of potholes and karstic crevices and cavities in a limestone portion of the streambed of the Calaveras River near Columbia, California (Figure 16-3). These pockets yielded 4,000,000 troy oz (124,000 kg) of gold in 15 years, with nuggets of up to 30 kg in the coarse gravel pothole fillings. The total value would be almost $1.5 billion at today's prices.

Coarse "heavies" will also accumulate—with much coarser pebbles or

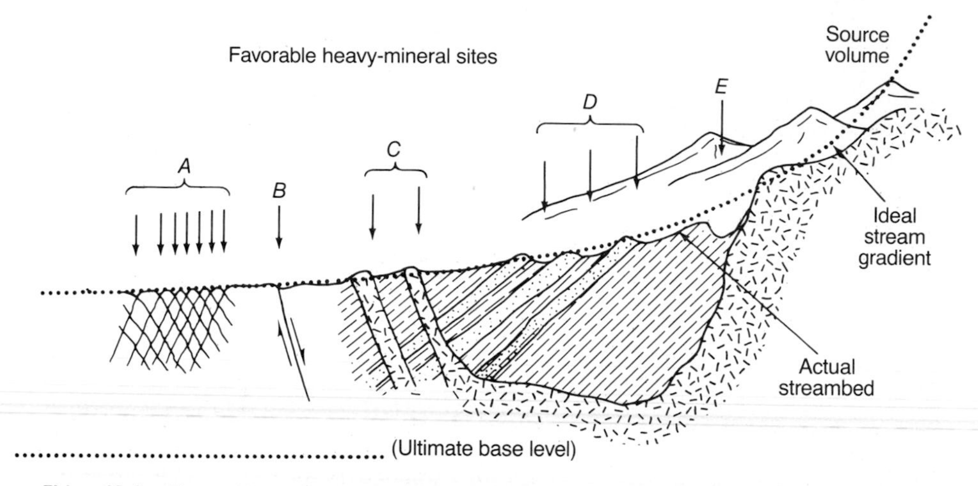

Figure 16-1. Favorable sites along a hypothetical stream profile for the accumulation of heavy minerals being carried downstream from a source volume include crevices formed by steep joints *A*, faults *B*, or outcrops of resistant dikes *C* or strata *D*, and plunge pools *E*, eddies, or potholes in rapids.

Figure 16-2. Joints across the dry rhyolite streambed of Sycamore Canyon, Arizona, constitute riffle traps for heavy minerals at times of flow (*A* in Figure 16-1.)

Figure 16-3. Karstic limestone portion of the Calaveras River streambed near Columbia, California. The crevices, potholes, and cavities were ideal sedimentologic traps for coarse, heavy gold particles. See text. (*Photos by* R. C. Loyd.)

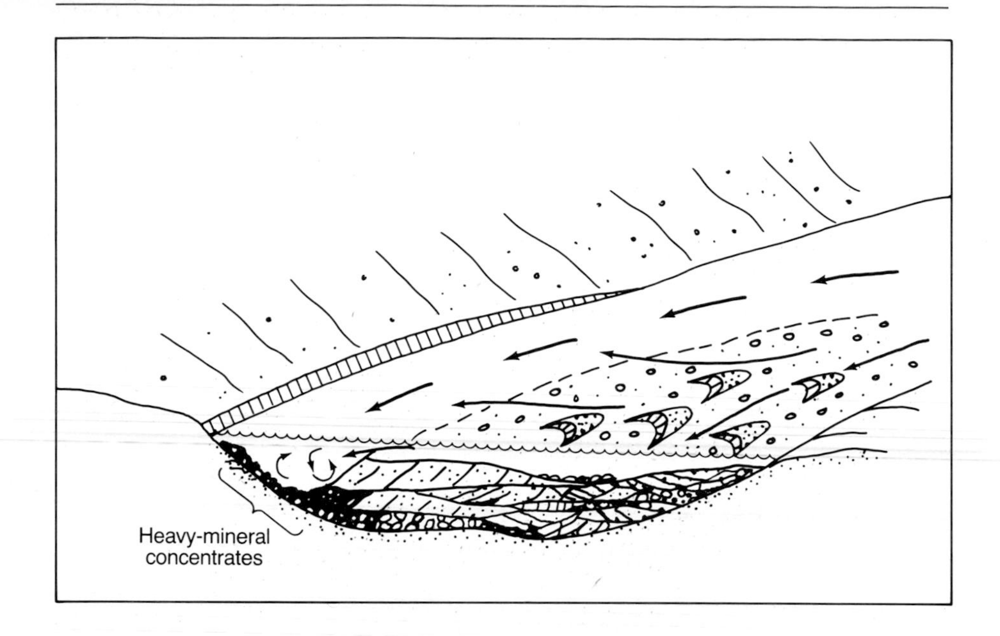

(*a*)

Figure 16-4. (*a*) Concentrations of fine-grained heavy minerals can form as "lag deposits" when clay-, silt-, and sand-sized particles of lighter minerals are winnowed away in convergent channels (see flow arrows) in slower streams. (*After* Smith and Minter, 1980.) (*b*) (*facing page*) In rapid streams, streaks and "tails" of coarser "heavies" are dropped in slack water velocity "shadows" in several characteristic locations. (*After* Boyle, 1979.)

cobbles of durable rock types—as lag material in the cut banks and channels of streams (Figure 16-4*a*). Fine-grained "heavies" associated with clays and silts may accumulate in the slack water areas on slip slopes, the inner edges of meanders (Figure 16-4*b*), or in velocity "shadows" below stream confluences.

In general, heavy mineral accumulations in fluvial, marine, or glacial environments should be carried out along geomorphic-hydraulic guidelines (Smith and Minter, 1980) by searching out areas in which the deposition of coarse silicate particles has accompanied the winnowing and removal of clay- and silt-sized particles and the retention of small dense grains of the same equivalent spherical diameter and hydraulic properties. Within each of these large-scale targets, finer scale sedimentologic features such as channel fillings, lag gravels, cross-bed foresets, and bottom-set "spoons," ripple marks, point bars, algal mats, and other primary structures are sought.

Placer deposits have formed throughout geologic time, but most exploited ones are of Cenozoic age. Most placer deposits are small and form above local base level; hence most are eventually removed by erosion and transported downstream before they can be preserved by burial. Moreover,

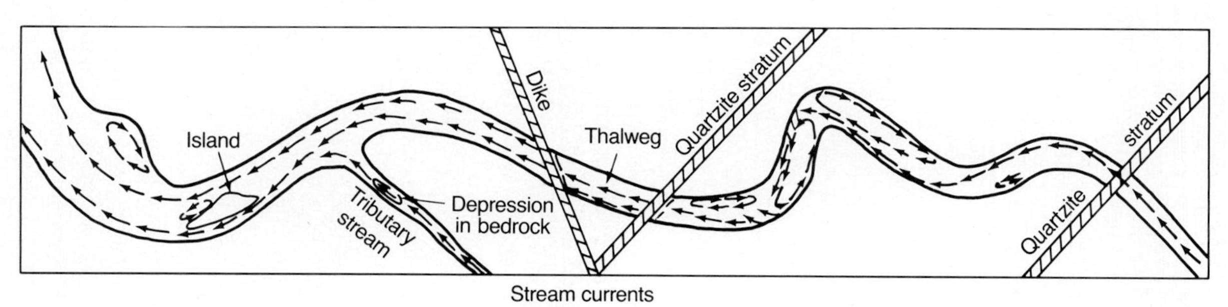

Stream currents

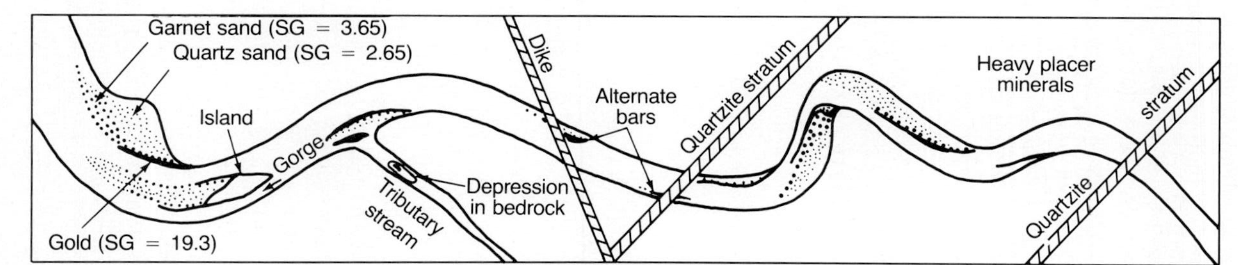

Distribution of heavy minerals

(*b*)

older fossil placer deposits are likely to be tilted, folded, and lithified, so that unless they contain unusually valuable minerals or exceptional concentrations of them, they cannot compete economically with the unconsolidated younger near-surface materials whose chief virtue may involve ease of handling, disaggregation, and value recovery.

Fossil placers, or those buried under younger rocks, are mined in some areas. In the Sierra Nevada of California and in the Victoria gold fields of Australia, for example, placer gold deposits have been shallowly to deeply buried beneath younger stream materials and lava flows. Lavas that flowed down stream valleys and covered placer deposits derived from the Mother Lode veins of California have subsequently been left as residual ridges when the terrane between the lava-filled streambeds was eroded away. As a result, the old stream valleys are easy to locate, but sporadic gold concentrations are difficult to find beneath the volcanic flows. If placer deposits were evenly distributed along stream courses, the problem of underground mining would be greatly simplified, but of course this is not the case. Most of the environments described above are transitory and discontinuous, and some, such as those in meanders, tend to migrate downstream. Finding these pay streaks is enough of a challenge in surface deposits without extrapolating the techniques to buried placers. Thus much of the placer gold covered by lava flows in the Sierra Nevada remains undiscovered. Similar examples in Cordilleran North and South America and elsewhere in the world could be cited.

Placer deposits have been divided into *eluvial*, on hill slopes; *alluvial*, on fans; *fluvial*, in streams; *lacustrine*, in lakes; *glacial;* and *marine*. Russian authors have proposed the terms *autochthonous* for eluvial or nontransported, generally coarse-grained placer accumulations and *allochthonous* for transported placers, but the terms have not become popular. Eluvial placers include portions of the tin placers of Malaysia. Small, young lacustrine placers tend to be ephemeral, as are lakes themselves, and of little overall importance. Glacial placers are similarly both ephemeral and nonsystematic, and are little exploited, except as "pathway indicators" to point to sources of diamonds, iron formations, and the like. Stream placers and marine deposits, both as offshore bar and beach-ridge accumulations, are of great significance as repositories of gold, diamonds, rare earth elements, and base and alloy metals such as iron, titanium, thorium, and zirconium. The Trail Ridge occurrence of Florida will serve as an example of marine placers, and the Witwatersrand gold deposits of South Africa have emerged as the best understood fluvio-lacustrine placers in the world.

Trail Ridge Heavy-Mineral Beach Sand Deposits, Florida

The world produces magnetite, ilmenite, rutile, zircon, and gemstones from scores of marine beach placers, many of which have been and will be mined for decades. They include Richards Bay, South Africa, the Travancore deposits of southwestern India, the east and west coasts of Australia, and the diamondiferous gravels off Namibia-Angola. Few have received more attention than the titanium deposits of eastern Florida (Martens, 1928, 1935; Creitz and McVay, 1949; Lynd, 1960; Pirkle and Yoho, 1970; Pirkle, Pirkle, and Yoho, 1974; Force, 1976). Some placers have been deposited along existing beaches; others have been left along older elevated shorelines. Two principal deposits in Florida are elevated sandbars. One is east of Jacksonville and the other, known as Trail Ridge, runs from near Starke, Florida, northward 200 km parallel to the Atlantic coastline into the state of Georgia (Figure 15-12). The average heavy-mineral content in both deposits is very low, and only the combination of low-cost dredging and beneficiation by the efficient "Humphrey spiral" centrifugal concentrator has made mining possible.

The Holocene Trail Ridge beach sand concentrates contain ilmenite, rutile, staurolite, zircon, sillimanite, tourmaline, kyanite, andalusite, pyroxene, and small amounts of corundum, gahnite, garnet, and monazite (Creitz and McVay, 1949; Carpenter et al., 1953; Pirkle, Pirkle, and Yoho, 1974). Epidote, hornblende, sphene, and other minerals have been identified elsewhere in Florida sands, but not at Trail Ridge. Most of the titanium at Trail Ridge is as leucoxene, an alteration product of ilmenite, sphene, or other titanium minerals, which consists for the most part of microcrystalline rutile, anatase, or brookite, but which is in part amorphous. This leucoxene, even as structurally variable as it is (Overholt et al., 1950), is more amenable than ilmenite to titanium extraction. "Pseudorutile," an alteration product intermediate between ilmenite and rutile, is also reported (Teufer and Temple, 1966).

Concentrations of titaniferous sands are widely distributed in Florida, but most are not of commercial grade. Studies of the modern beaches, which are similar in mineralogy and structure to the elevated bars such as Trail Ridge, have revealed the source and mode of concentration of heavy minerals. All the heavy-mineral species present in the Florida beach sands are found in the old metamorphic and intrusive rocks of the Southern Piedmont to the north. Geologists agree that the most likely source of the heavy minerals along the Florida coast is the granitic and gneissic rock basement of the Piedmont Province, especially the Appalachian-Piedmont parts of Georgia and the Carolinas (Figure 15-12). The sands were evidently transported to the coast by streams and reworked along the coast by southward-moving longshore currents (Force, 1976). Heavy sand deposits are found for long distances north of Florida, but in general they become less concentrated northward (McKelvey and Balsley, 1948). Also, the content of ilmenite and rutile in the sands decreases northward as the magnetite content increases, as is true in the basement rocks. Thirty-six modern beach placer titanium deposits are known between northern Maine and southern Florida. Two are north of North Carolina, 4 are in North Carolina, 9 each are in South Carolina and Georgia, and 12 are in Florida. Mining is current near Lakehurst, New Jersey (Puffer and Cousminer, 1982), and in Florida.

The Trail Ridge deposits consist of completely friable, free-running black sands composed of oxide, silicate, and phosphate resistate minerals listed earlier. They are fundamentally stratiform, a series of slightly offset, superposed ribbons or elongate crescents of varying percentages of moderately well-sorted, well-rounded black sands in quartz-carbonate sands (Figures 16-5 and 16-6). Individual ribbons are 1 or 2 meters thick, 10 meters to tens of meters wide, and hundreds to thousands of meters long parallel to the coastline. Locally they are composed totally of heavy minerals, and percentages from 50 to 85% are routine. Grades of 5 to 25% Fe-Ti oxides are being mined profitably in New Jersey (Puffer and Cousminer, 1982). Ratios of individual heavy minerals vary widely along strike and from layer to layer within ribbons. They occur high on modern and paleo beaches, normally along what is recognizable as the high-tide storm line. Puffer and Cousminer (1982) studied the New Jersey deposits sedimentologically and found that the highest heavy-mineral abundances are in the best sorted sands that they interpreted to be backshore beach and dune facies sands.

The Trail Ridge deposits formed, as most marine placers do, as a result of the combined effects of long-shore currents and on-shore wave-wind-storm action. Longshore currents moved both heavy and light minerals southward as a mixed sediment load according to the size, specific gravity, durability, and abundance of each mineral in the various source units. Durability is significant, and the ratio of dark, heavy, fine-grained

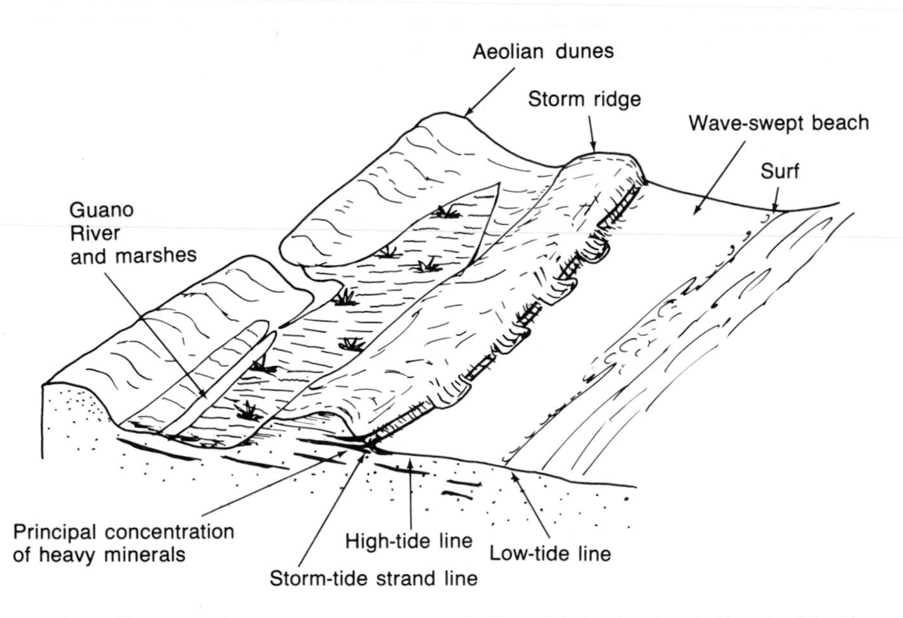

Figure 16-5. Generalized section of beach south of Mineral City, St. John's County, Florida. Vertical scale is exaggerated. (*Adapted from* Martens, 1928.)

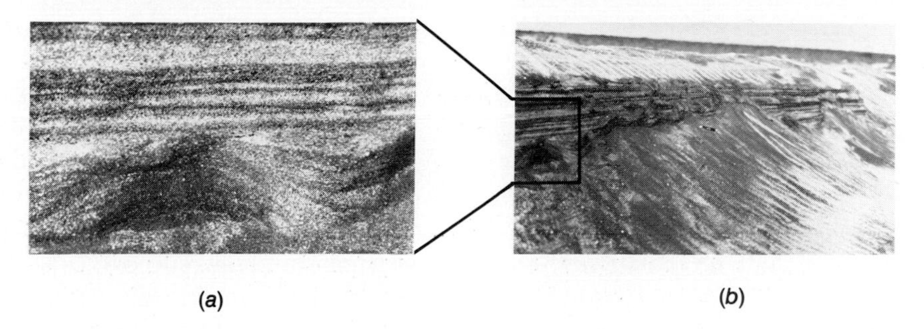

(a) (b)

Figure 16-6. Layers enriched in black, heavy ilmenite-magnetite sand grains along a storm-tide strandline near Punta Colorada, Baja California Sur, Mexico. (*a*) is detail of (*b*).

durable minerals increases southward, diluted by the input of unmodified sediment load from river mouth to river mouth along the way. When the sands are deposited on the beaches of Florida, they can be thought of as having a heavy-mineral content that varies widely from layer to layer and depositional site to depositional site, but which has been delivered—increment by increment—by the long shore currents.

It then remains for various combinations of wave action to concentrate the "heavies." Figure 16-5 shows that the heavy minerals at Trail Ridge are concentrated at the beachhead rather than in the aeolian dunes or in the beaches swept by day-to-day tidal and wave action. Wind is an ineffective concentrator of heavy minerals in general, but storm tides are highly effective. Minerals can be suspended longest and carried farthest in high-energy turbulent wave action. A storm at high tide will carry both light and heavy minerals beachward and dash them onto the storm-line sands. But flowback will carry more "lights" than "heavies," and storm-line accumulations of heavy minerals can become both sizable and rich (Figure 16-6). The size, grade, and extent of a given deposit are complex functions of the source of supply, relative coastline-mean sea level stability, and the frequency, energy, consistency of direction, and reproducibility of storms.

Pirkle and Yoho (1970) describe Trail Ridge as a broad sand ridge stretching generally north-south and in the youngest, modern sediments; they overlie units related to the phosphate deposits discussed in Chapter 15. The most widespread concentrations of heavy minerals are along a 30-km stretch at the southern end of the ridge. Evidence indicates that this part of Trail Ridge was formed along a Pleistocene shoreline bordering the northern part of the Lake Wales Ridge. Trail Ridge was localized by south-flowing currents impinging on the northern end of the Lake Wales remnant. The Lake Wales Ridge was crucial to the environment in which the Trail Ridge orebody formed because it was an older, less heavy-mineral-rich strand deposit; it underwent another cycle of wave-action enrichment of the heavy minerals now mined from Trail Ridge.

Martens (1935) emphasized the fine-grained character of the heavy sands and the presence in them of coarser fragments of shells, apatite, and glaucophane from local formations. He suggested that much detrital material in the beach sands did not come directly from the older rocks of the Piedmont, but had been held for varying lengths of time in some of the Coastal Plain formations. Variations in grain size and composition of the sands are in part inherited from local sources supplied by the direct wave erosion of the land during high tides and storms. Sand on southern Florida beaches is generally coarser than sand to the north; grain size illogically seems to increase away from the source of supply. The change from fine to coarse grains is fairly abrupt; most of the beaches to the north of Cape Canaveral are fine-grained, and most of those to the south are coarse-grained. This relationship suggests that the southward-moving longshore currents are deflected oceanward there; the southern beaches receive most of their sand from direct erosion of the south Florida landmass, some of which was in turn derived from the Piedmont region during earlier cycles of erosion.

Most important is the fact that Trail Ridge itself, and the slightly younger oceanward Boulounge and Green Cave Springs heavy-mineral deposits, are the results of the erosion of the provenance areas of the Appalachian-Piedmont provinces and times of maximum marine energy, of transgression, and of prolonged wave-action elutriation of beach materials. In the process of winnowing, they underwent yet another stage of pulverization, abrasion, and size-specific gravity separation such that only the heavier, higher equivalent-hydraulic-diameter particles were lodged in storm-generated high-tide strand lines. Similar phenomena and results dictate exploration procedures for diamonds in the surf zone offshore from Angola and Zaire, for ilmenite-rutile-zircon sands in the elevated beaches of Japan, Travancore, India, and Australia, and for beach sands around the world.

Witwatersrand Gold Deposits, Republic of South Africa

The Witwatersrand district lies in South Africa between the cities of Johannesburg and Welkom (Figure 16-7). "The Rand," as it is commonly known, has been and remains by far the most productive gold district in the world. It has yielded over 35 million kg of gold valued at nearly $60 billion and still has enormous reserves. Although the ore now mined averages only about 7 ppm = 7 grams (0.23 troy oz) of gold per ton, as much as 1.0 million kg of gold (32 million oz) have been recovered in a single year. Since the discovery that appreciable amounts of uranium can be recovered as a coproduct of the gold ore, the district has also become one of the world's principal sources of fissionable materials. Indeed, some mine units produce more uranium value than gold.

The geologic literature on the Witwatersrand is voluminous (Pretorius, 1975, 1981a; Minter, 1982). Almost every aspect of the deposits has been studied, but some fundamental issues are still debated. The Rand

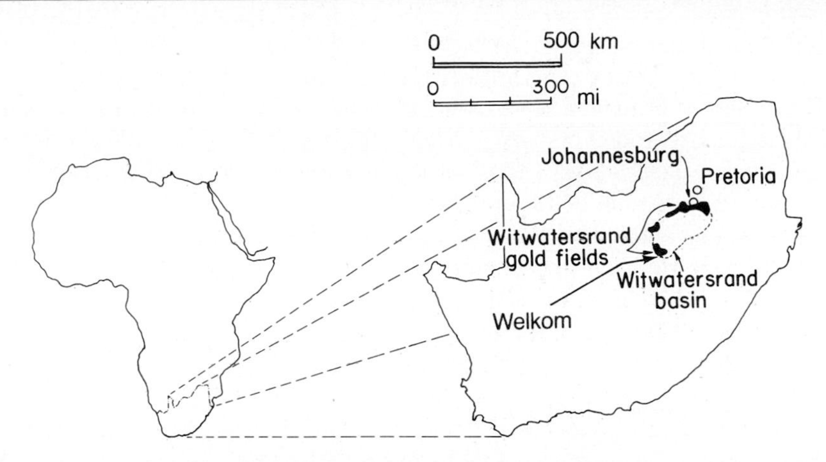

Figure 16-7. Map of the Republic of South Africa showing the location of the Witwatersrand deposits.

deposits are discussed here under placer deposits because this theory of origin is favored among South African geologists, and because almost every increment of evidence gathered during the last decade confirms a sedimentological origin. As of 1985, questions concerning the type of placer origin—whether alluvial plain, lacustrine delta, or marine environments prevailed—are discussed more than are alternative models. More will be said about the alternative models later; the reader is referred to the predecessor of this text for a more comprehensive discussion of far-ranging genetic alternatives (Park and MacDiarmid, 1975). Geologic research coordinated by D. A. Pretorius and the Economic Geology Research Unit of the University of the Witwatersrand and sponsored by many South Africa mining houses has greatly clarified our understanding of the distribution of gold in the Witwatersrand strata, and of specific environments of deposition.

It is now agreed by most workers that sediments consisting of detrital quartz grains and pebbles, sericite, gold, uraninite, pyrite, and accessory minerals were transported into the Witwatersrand basin from the north, northwest, and southwest. The source area of the sediments and their contained ore minerals appears to be the Archean granite basement domes and greenstone belts that constitute the cratonic basement rocks of 3.5-to-4.0-billion-year age and that surround the basin. Viljoen, Saager, and Viljoen (1970) showed that accessory and major minerals from the base to the top of the Witwatersrand basin sediments indicate a pattern consistent with an early denudation of felsic supracrustal rocks followed by intermediate, then deep-seated mafic rocks. It appears that successively deeper levels of a source area were exposed, weathered, and denuded, with resistant minerals being transported to the basin and deposited there in "inverted" sequence.

Later Köppel and Saager (1974) studied lead isotopes of sulfides in the gold deposits of the Precambrian greenstones of the Barberton Mountain Land in the eastern Transvaal, Republic of South Africa, and compared them with the lead isotopes of pyrite from gold deposits of the Witwatersrand basin. They concluded that the isotopic evidence supported the contention that these greenstone belts and their gold deposits contributed detritus to the Witwatersrand basin. Radiometric ages of radiogenic lead in galena and detrital monazite in the sediments are among the oldest known at 3040 and 3160 million years, respectively, and match the ages of granites and greenstones in the Archean basement rocks (Allsopp, 1961; Nicolaysen et al., 1962). It appears compelling that these Archean terranes were the source of the "mineralized" sediments.

We are relatively confident that the atmosphere and the hydrosphere were chemically less oxygenated prior to 1.8 to 1.9 billion years ago than they are now (Cloud, 1968). It appears that resistant minerals, including pyrite and uraninite that would be oxidized and destroyed today, entered stream systems in a hinterland, were carried basinward, and were then deposited in huge fanlike sedimentary bodies near the basin edges. Pretorius (1975, 1976, 1981) presented evidence that these bodies were alluvial fans deposited in large lake basins such that fluvial sediments splayed out in a deltaic lacustrine environment. Minter and his coworkers (Minter, 1976, 1978; Smith and Minter, 1980) have effected at least two major contributions. Minter studied sedimentologic detail in the Vaal Reef and described not only the placer systematics of the gold and uranium but also the hydraulics and the dynamics of their transportation and deposition.

Stratigraphy The oldest rocks of the region are Archean Precambrian rocks that make up a basement complex of granite, granitic gneiss, and greenstone terranes, the latter among the geologically oldest volcanic belts on Earth. The basement rocks occur now as gently unwarped to steeply diapiric domes, represented by B on Figure 16-8; they are also shown on Figures 16-9, 16-10, and 16-11. The greenstones include extensive areas, or volumes, of regionally chloritized rocks that contain districts of quartz-carbonate-gold veins (Fripp, 1976; Cochran, 1982) and younger potassic granite bodies with accessory uraninite. Overlying the basement in the "Wits" area is the Dominion Group, a sequence of thin conglomerates and thick lava flows that underlie the Witwatersrand Supergroup. One gold-bearing zone has been discovered in the Dominion unit (Figure 16-8), and uranium-rich strata in it are now being exploited (Vermaak, 1979). Uncon-

Figure 16-8. (*Facing page*) Map of the distribution of the Witwatersrand System beneath younger strata. Dips are gently centripetal in and downdip from the stippled Wcr outcrop band, and steep to overturned around the central Vredefort dome.

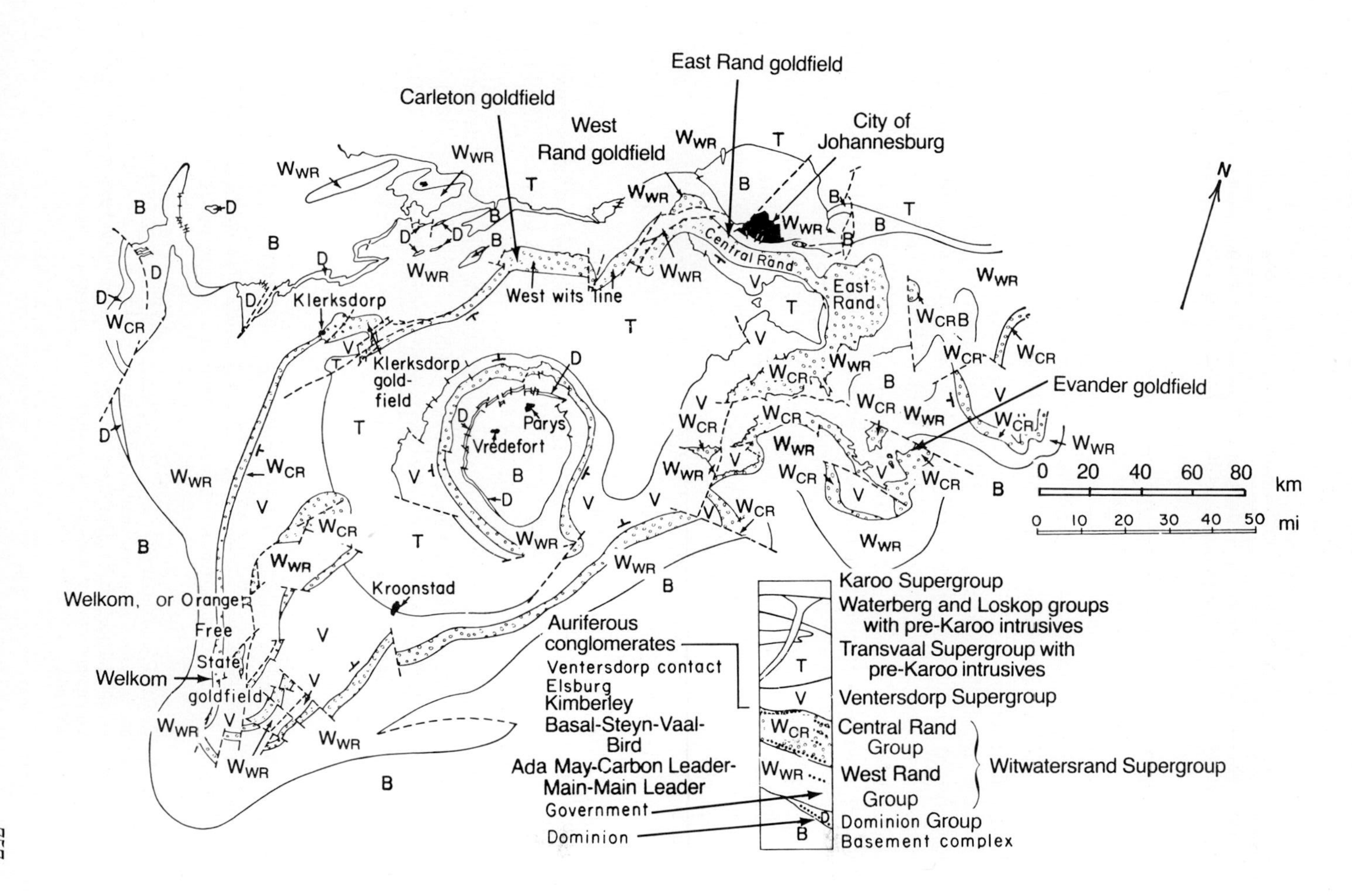
East Rand goldfield
Carleton goldfield
West
Rand goldfield
City of
Johannesburg
Central Rand
East
Rand
West wits line
Klerksdorp
Klerksdorp
gold-
field
Parys
Vredefort
Kroonstad
Evander goldfield
Welkom, or Orange
Free
State
goldfield
Welkom
N
0 20 40 60 80 km
0 10 20 30 40 50 mi
Auriferous
conglomerates
Ventersdorp contact
Elsburg
Kimberley
Basal-Steyn-Vaal-
Bird
Ada May-Carbon Leader-
Main-Main Leader
Government
Dominion
Karoo Supergroup
Waterberg and Loskop groups
with pre-Karoo intrusives
Transvaal Supergroup with
pre-Karoo intrusives
Ventersdorp Supergroup
Central Rand
Group
West Rand
Group
Witwatersrand Supergroup
Dominion Group
Basement complex

formably overlying the Dominion Group and the basement is the 7000-meter-thick Witwatersrand Supergroup, which is divided into lower and upper units, the West Rand Group and the Central Rand Group (Minter, 1982). The lower West Rand Group of the Witwatersrand (Wwr on Figure 16-8) consists of shales, quartzites, grits, and conglomerates, with one gold-producing conglomerate bed. The upper younger Central Rand Group (Wcr) is about 90% quartzites, grits, and rare shales, and contains most of the gold-bearing conglomerates, the most productive horizons being at and about 1 km above its base. A stratigraphic correlation of the entire basin and the contained ore horizons is given by Minter (1982) in Tankard et al. (1982). It must be emphasized that this 7-km-thick "Wits system" is only 6 to 12% conglomerate, but these conglomerates are key horizons for both gold and uranium. As will be shown, they and their associated sands are fluvial and lacustrine, not marine, sedimentary rocks. Tilting and erosion of the Witwatersrand Supergroup, deposited 2.5 to 2.7 billion years ago, were followed by igneous activity that formed the overlying Ventersdorp Supergroup of lavas and interbedded clastic sediments (V on Figure 16-8). The Ventersdorp varies greatly in thickness, exceeding 1 km in places. Following erosion and subsidence of the Ventersdorp, a thick sequence of clastic and dolomitic limestones known as the Transvaal Supergroup was deposited in a new basin. The end of this period was marked by Bushveld-age igneous activity followed by folding and thrusting. At the end of the Paleozoic era, the Karroo Supergroup of shallow marine, aeolian, fluvial, and glacial sediments and coal formations was deposited over the entire area; it still covers more than half of the Witwatersrand.

Tectonic Setting The Witwatersrand sedimentary rocks lie in a large synclinorium, a structural basin that measures about 400 km long and 150 km wide and is elongate in a northeasterly direction. The frame of the synclinorium was formed as part of a complex fabric of regional and district-scale interference folding described by Pretorius (1981a). This folding produced domes where anticlinal axes of more than one age or direction of folding coincided, and basins where synclinal axes intersected. The broad Witwatersrand synclinal basin was near the edge of the craton, and was filled with a presumably shallow broad intracratonic freshwater lake, the bottom of which lowered progressively to accept the Witwatersrand sediments. It may have been open to the sea to the south. Actively growing domes around the basin margin (Figures 16-9*a*, 16-10, and 16-11) tectonically and geomorphically controlled the positions of rivers that brought their sediment

Figure 16-9. (*Facing page*) (*a*) Locations of Witwatersrand goldfields within the basin and within the Central Rand Group sediments (solid isopachs). (*After* Pretorius, 1981a.) (*b*) Array of basement domes that localized the river systems, which delivered the basin-filling detritus, including the gold particles. (*After* Minter, 1982.) Arrows in both show sediment transport directions.

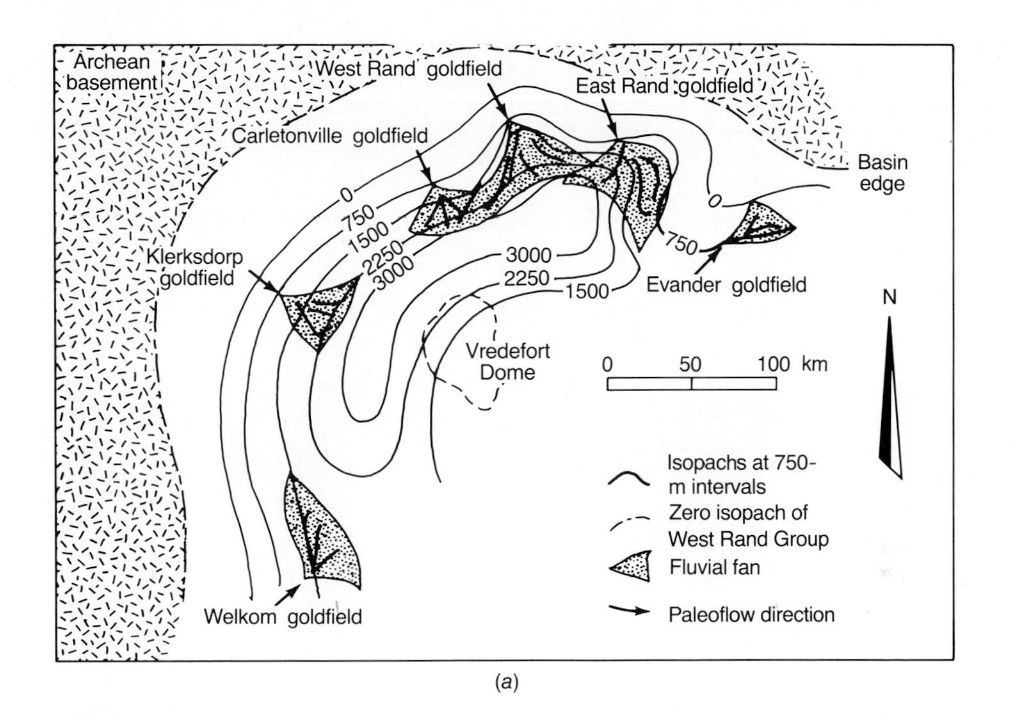
Archean basement
West Rand goldfield
East Rand goldfield
Carletonville goldfield
Basin edge
0
750
1500
2250
3000
Klerksdorp goldfield
3000
2250
1500
750
0
Evander goldfield
N
Vredefort Dome
0
50
100 km
Isopachs at 750-m intervals
Zero isopach of West Rand Group
Fluvial fan
Welkom goldfield
Paleoflow direction

(*a*)

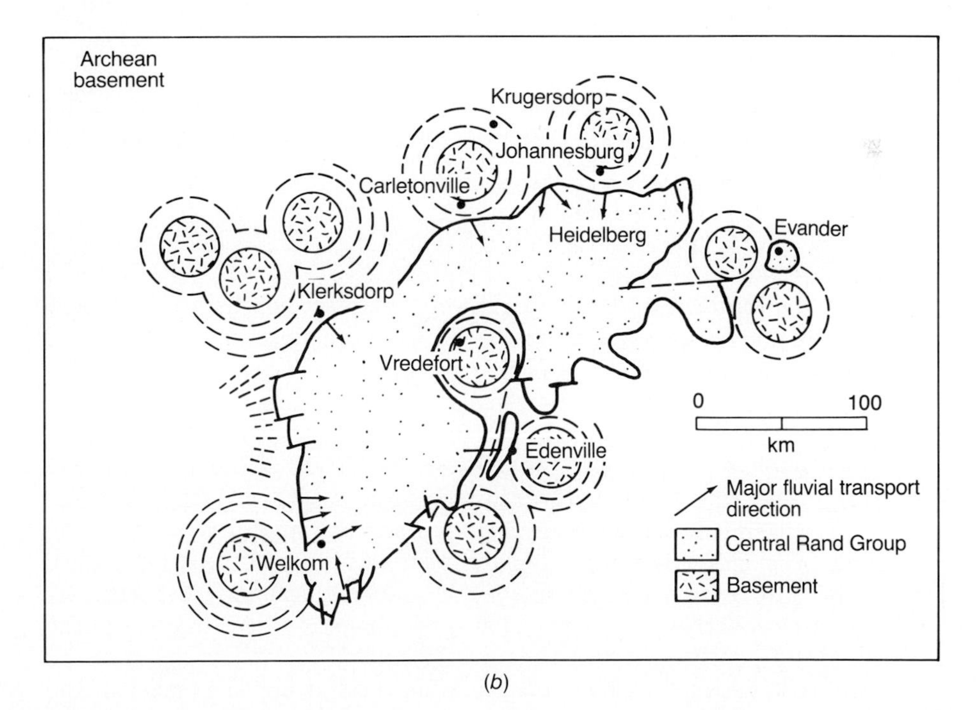
Archean basement
Krugersdorp
Johannesburg
Carletonville
Heidelberg
Evander
Klerksdorp
Vredefort
0
100
km
Edenville
Major fluvial transport direction
Central Rand Group
Basement
Welkom

(*b*)

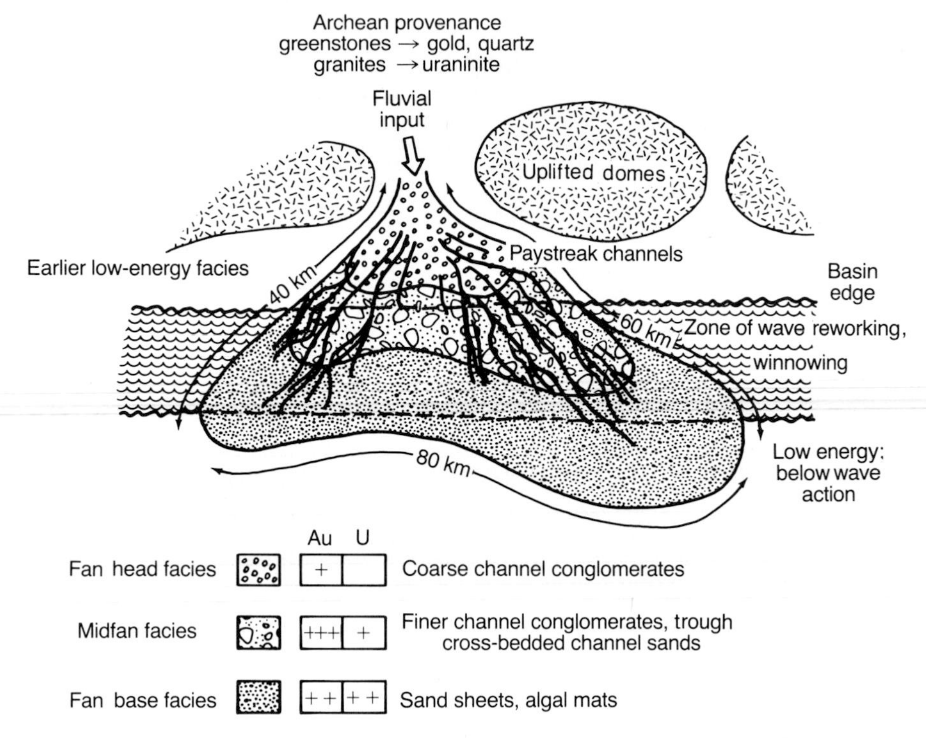

Figure 16-10. Conceptual model of a Proterozoic Witwatersrand goldfield as a fluvial fan developed at the mouth of a major river flowing from a source area in the northwest and debouching into a shallow-water intermontane intracratonic Proterozoic lake. The general geometry of the fan, the various fan facies of deposition, the locations of coarse clastics and algal mats, the arrangement of varying zones of enrichment in gold and uranium, the portion of the fan subjected to transgression and winnowing, and the approximate dimensions of a typical fan are shown. The plus signs designate relative quantities of gold and uraninite. (*After* Pretorius, 1981a.)

load from the hinterland. The locations of the river mouths are indicated by the arrows on Figure 16-9. Sand and silt were the most abundant detrital fragments washed in, but broad braided channels like those of today's Yukon or Mackenzie rivers led downstream to the deltas and fans sketched in Figures 16-9*a*, 16-10, and 16-11). The oldest, lowest sediments deposited were the Dominion Group and the West Rand Group sediments, and the principal ore-bearing conglomerates lie not at the outer basal fringes of this basin, but within it in the younger Central Rand Group sediments, thereby describing a smaller oval roughly 300 by 110 km, the stippled units in Figure 16-8. It also follows that the mines are concentrated along the northwest side and at both extremities of the oval synformal structure. In general the sediments dip gently to steeply basinward around the outer edges of the structure, but flatter dips have been encountered at depth in most of the deeper, more central mines. The synclinorium is complicated by many faults,

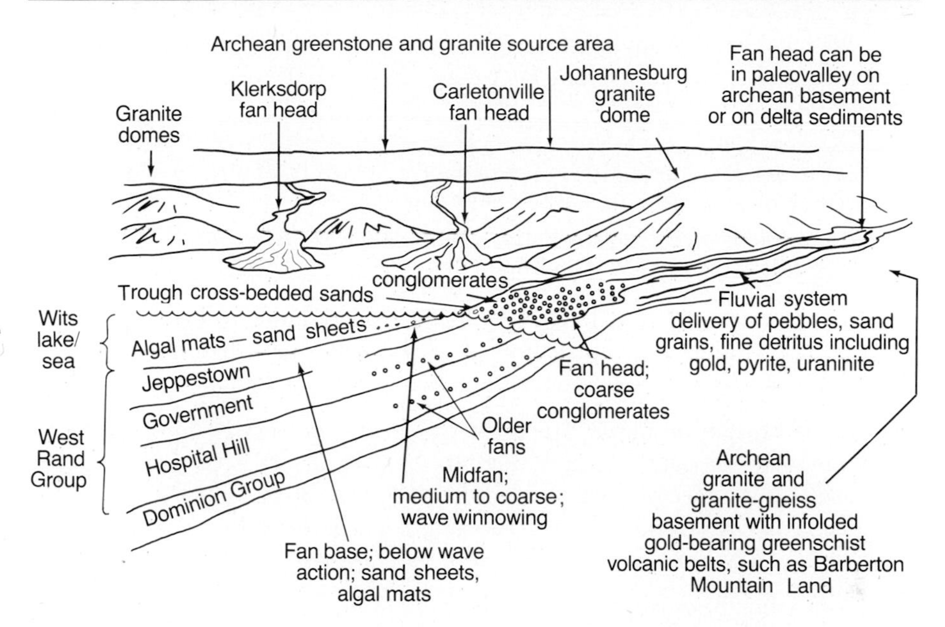

Figure 16-11. General setting of sediment delivery and sedimentation at the time of formation of the Main Leader, the Carbon Leader, and the Ada May Reef, respectively, on the West Rand, Carletonville, and Klerksdorp fans. Not to scale, greatly foreshortened horizontally.

and large-scale thrust and small normal faults are abundant in the mines. Several mine areas have had to be abandoned because faults complicated the mining operations so badly that gold could not be recovered economically.

Sedimentation As mentioned above, the goldfields constitute only a small fraction of Witwatersrand sedimentation. The term *goldfield* has developed "on the Rand" to designate these specialized shield- or fan-shaped auriferous conglomerate systems interleaved into the arenite thicknesses of the supergroup. As shown in Figure 16-9*a* and as described by Pretorius, there are six separate major goldfield mining areas in the district, each representing a wedge of sediment entry into the basin and each involving several horizons of conglomerate development, as implied in the sketch of Figure 16-11. The Welkom goldfield has 4 auriferous-uraniferous horizons, the Klerksdorp 7, the Carletonville 3, the West Rand 10, the East Rand 9, and the Evander 1. Each represents some degree of perfection of the development of the dynamics shown in Figures 16-10 and 16-11, which should be considered carefully. The Central Rand Group has been found to consist of a series of sedimentary cycles that range in thickness from 30 to 600 meters, averaging 250 meters. They are defined by basal unconformities marked by lag gravels, become finer upward, are composed of major sand–minor conglomerate, and include siltstones, sandstones, trough cross-bedded channel sands, cross-bedded sand sheets, and conglomerates.

Wits gold and uranium placer ores now extracted occur in at least four geologic sites in the goldfields (Pretorius, 1976, 1981b; Smith and Minter, 1980):

1. In the fan head or midfan: in the sandy or sandy-pebbly matrix at and near the bases of fluvial pebble-supported conglomerates in channel fills, where "openwork conglomerates" trapped and retained sand-sized heavy-mineral grains from the sediment flux (Figure 16-12)
2. In the midfan: in pyritic trough cross-bedded sands in erosion-deposition channels with gold, uraninite, and pyrite particles on foresets, in bottom-set spoons and scours, and in basal lag sands or gravels
3. In the midfan or upper fan base: in sheets of cross-bedded sands by winnowing of quartz grains, leaving thin layers of heavy minerals as lag deposits with or without pebbles (Figures 16-12 to 16-15), and thus also as the lag sands and gravels along unconformities at the bases of sedimentary units
4. In the fan base: in carbonaceous layers on unconformities, in scour pools, and in algal mats, which acted both as mechanical riffle traps and as chemical traps

The ore-bearing conglomerate strata in the Witwatersrand area are known as *reefs* or *bankets*, the latter term being a Dutch word for the almond cake that the conglomerates resemble. Early attempts to correlate the reefs and their nomenclature across the whole basin have been negated by the fact that similar appearing horizons may or may not have formed contemporaneously on the different fans. The strata are not continuous as in a layer cake, although the fans are qualitatively similar and may have intertonged distally from place to place and from time to time. For example, the Main Reef group, also called the Johannesburg Subgroup and the most important ore zone in the district, includes several ore-bearing conglomerates. In the East Rand goldfield near Johannesburg, this group includes the Main Reef, the Main Leader, and the South Reef. About 30 km to the east, in the Evander goldfield only the Main Leader persists; along the West Rand, a goldfield extending from 30 to 80 km southwest of Johannesburg, a rich gold-uranium conglomerate known as the Carbon Leader car-

Figure 16-12. (*Facing page*) (*a*) Sketch of a hand specimen of conglomerate in the Ventersdorp Contact Reef, Witwatersrand gold district, Republic of South Africa. Gold particles occur with the pyrite but are too fine-grained to be shown. The rounded quartz pebbles and "jelly beans" of pyrite both indicate stream transport. Quartz particles are larger than pyrite, which are larger than gold, in reverse order of their specific gravities and in keeping with hydraulic equivalency. (*b*) Basal lag conglomerate from the Basal Reef composed of 3-to-4-cm quartz (white, gray, black) and 0.5-to-1-cm hydraulically equivalent pyrite pebbles (white). Gold particles (not visible) and sand grains are the matrix. Sand, now quartzite, immediately overlies the conglomerate.

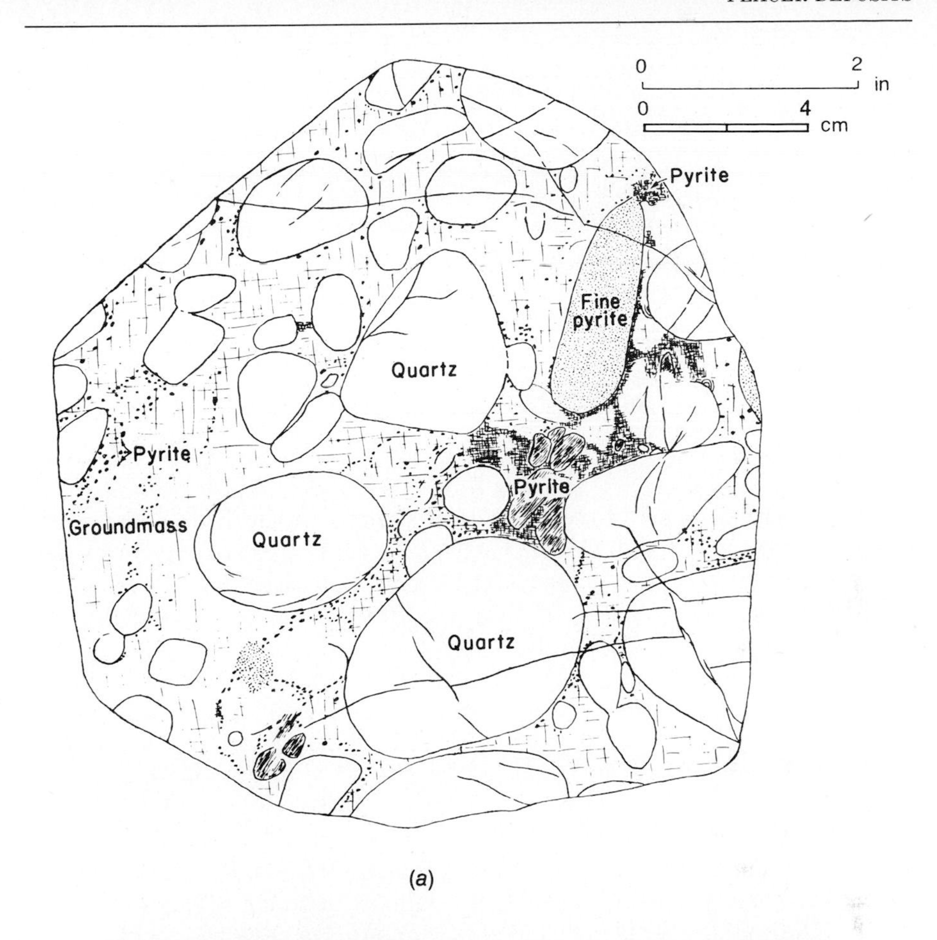
0
2
in
0
4
cm
Pyrite
Fine pyrite
Quartz
Pyrite
Pyrite
Groundmass
Quartz
Quartz

(a)

(b)

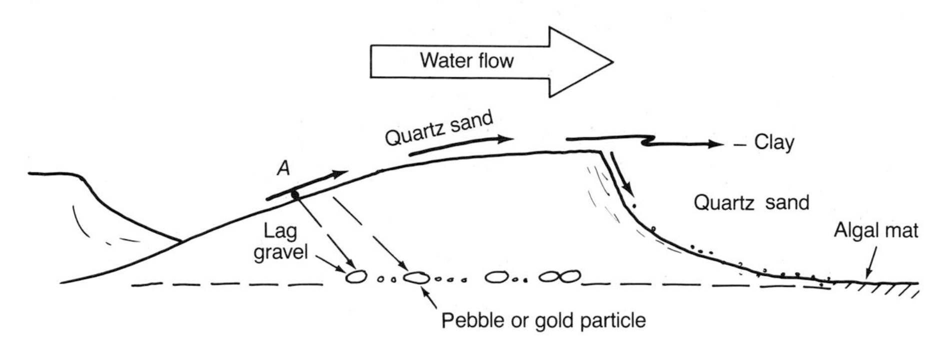

Figure 16-13. Ripple advancing downstream to right with the behavior of different particles indicated. Compare with Figures 16-14 and 16-15.

ries most of the values in the Main Reef Group. In the Klerksdorp area, about 140 km southwest of Johannesburg, the Main Reef Group consists of two minor conglomerates, known as the Commonage Reef and the Ada May Reef, and the major Vaal Reef, which by itself provides 15% of modern world gold production. Each of the conglomerates named in this paragraph was probably deposited at nearly the same time, but at different levels on different fans. Locally minor ore-bearing conglomerates resulted from the erosion and redeposition of previously deposited Witwatersrand rocks. For example, the Black Reef in the West Rand area is a conglomerate at the base of the Transvaal Supergroup. It overlies Witwatersrand units and the Ventersdorp lavas and was derived from Wits strata (Pegg, 1950).

The origin of the Witwatersrand conglomerates and quartzites that occupy the basin has been the subject of almost as much discussion as the origin of the gold. Several environments of deposition have been considered, including marine or tidal lacustrine shorelines, large deltas, enclosed or yoked basins, piedmont or flood plains, regions of alluvial fans, and broad, braided alluvial megachannels (Mellor, 1916; Reinecke, 1930; Gevers, 1961; Brock and Pretorius, 1964a,b; Pretorius, 1976, 1981a; Minter, 1976, 1978, 1982; Smith and Minter, 1980). The Witwatersrand rocks have been mildly metamorphosed to greenschist facies, and except for the presence of quartz pebbles and overall sedimentation dynamics, the conglomerate beds have little in common with modern conglomerates. They are typically dense and tightly cemented. Many of the pebbles were slightly stretched during metamorphism, although in general a high degree of sphericity and roundness has been retained (Figure 16-12). The spaces between pebbles are occupied by locally abundant pyrite, itself detritally rounded to pellet or jelly-bean-like grains, sand-sized grains of quartz, sericite, rutile, zircon, chromite, chlorite, sporadic tourmaline, and many other heavy minerals, including sulfides, sulfosalts, rare diamonds, and of course substantial amounts of native gold and uraninite. Pretorius (1975) includes a list of 78 mineral species found on the Rand in his Table 18. Locally many of the quartz pebbles are a bluish opalescent variety, but most of them are white "bull" quartz typical of greenstone quartz-gold veins or the cherts also common in greenstone belts (Chapter 13). Coarse-grained gold is rare in these deposits; most of the values come from fine-grained nearly microscopic flecks and granules. The gold typically is in hydraulic equilibrium with other minerals with which it occurs, many of them sand-sized. Therefore because its specific gravity is 19.3 g/cm^3, it is measured in micrometers, not millimeters or centimeters. Free gold is fine-grained, but commonly visible.

Attributes of "pay zone" conglomerates, typically in the midfan area (Figure 16-10), are as follows.

1. They are nearly "pebble-supported," that is, pebbles generally touch one another.
2. Pebbles are well rounded, elliptical to spherical, and up to 3 to 4 cm in diameter.
3. Pebbles are generally composed of 2 or 3 rocks or minerals (oligomictic) rather than of many materials (polymictic).
4. The conglomerate is "mature," relatively free of clays and silts.
5. Pyrite is medium- to coarse-grained, generally greater than 5 mm in upstream sites, finer downstream.
6. Pyrite is common, typically 2 to 20% of the matrix, and typically as rounded, abraded granules or pellets.
7. The unit is thick, since greater thickness generally means more reworking, winnowing, and concentration; thicker conglomerates are generally of higher gold grade.

The ore-bearing reefs vary in thickness, but they are generally less than 1 meter thick. Productive beds more than 3 meters thick are uncommon, and are in "upstream" reaches; some of the richest reefs are only centimeters thick and are generally "downstream," more toward the midfan reaches.

Placer minerals in sandbodies, either trough cross-bedded in channels or as overbank, more sheetlike cross-bedded sands farther down on the apron of the fan base, are also important ore units, and here the concept of *lag deposits* or *lag gravels* becomes integral. Adams, Zimpfer, and McLane (1978) studied placer formation and heavy mineral behavior in channel processes in general. Minter (1976) studied the Vaal Reef deposit at Klerksdorp, Buck (1981) examined the Steyn and Saaiplaas reefs in the Welkom goldfield, and Tucker (1980) worked at Randfontein in the West Rand. They showed that most gold platelets and "micronuggets" are nearly hydraulically equivalent to the quartz sand that dominates the midfan and fan base sections, as are pyrite, zircon, uraninite, and other heavy minerals. So the key to this part of the Rand economic geology is the behavior of these grains

Figure 16-14. Foreset beds with visible pyrite defining cross-bedding. Placer gold accumulates with pyrite either in the lag gravels at the base of the set or in the "spoons," where algal mats would also develop. See Figure 16-15 for an inside, X-ray view. The scale is in centimeters. (*Photo by* W. K. Bilodeau.)

in channels in fluvial plains or fans or as they might be reworked and winnowed on a lake floor. If we consider a grain at point A in a small duneform in a channel (Figure 16-13), the process of heavy-mineral concentration in sand or sand-pebble bodies becomes apparent. Flow of water from left to right will sweep sand grains up and over the crest of the ripple, where they roll down to form a foreset bed, commonly with a spoon-shaped relatively low energy toe. If grain A is a clay mineral, its relative hydraulic "lightness" will cause it to be swept out of the system upon exposure to the current. If grain A is a quartz sand grain, it will roll up and be swept over to become part of a new foreset. Hydraulically equivalent pyrite, gold, and uraninite may also be swept over, but if those grains are slightly heavier, they are inclined to be concentrated on the foresets by slight winnowing and by lagging during temporary periods of faster flow or turbulence. Gold will

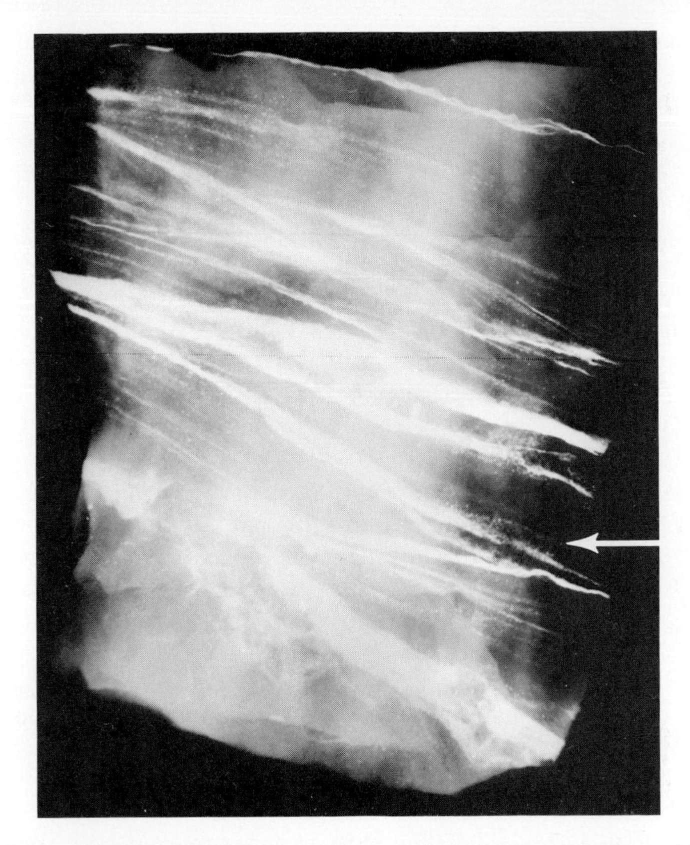

Figure 16-15. X-ray shadowgraph of the specimen of foresets shown in Figure 16-14. Such photos, made by placing 1-to-3-cm-thick slabs of sand body quartzites on X-ray film and exposing them, reveal many sedimentologic details of the cross beds. Note the "spoons" at the lower ends of the foresets (arrow) where coarse pyrite and detrital gold accumulate. (*Photo by* W. K. Bilodeau.)

also be retained in the bottomsets by the mechanical "riffles" of the fibrous plant forms of algal materials. Note that the "spoons" of the foresets represent troughs of least disturbance that algal mats would occupy in lower energy portions of the channels. These foreset–spoon–toe accumulations are shown in the photograph of Figure 16-14, the X-ray shadowgraph of Figure 16-15, and the diagram of Minter (1978) and Smith and Minter (1980), here shown as Figure 16-4*a*. Much more complicated involvements are currently being studied by sedimentologists. Finally if grain *A* were a pebble or a larger, heavier grain of gold or another heavy mineral, it would not be swept over, but rather would simply drop as quartz was washed away around it to be buried by the next advancing ripple as the duneform surfaces move downstream. These basal accumulations, also shown in the accompanying graphics, are called *lag gravels*, and they are vital to understanding placer dynamics. So it is that the Witwatersrand *conglomerates* became so important. Gold contained within the conglomerates is not hydraulically equivalent to the coarse gravels, but rather is slightly hydraulically heavier than sand. It is thus trapped by and concentrated in the conglomerates, generally basally, with quartz grains being washed away. Overall flow-accumulation relationships in a channel are suggested in Figure 16-4*a*. It becomes immediately apparent that understanding paleocurrent directions, evaluating winnowing factors, and knowing whether a mine advance is in low-energy or high-energy channel sediments or even overbank material contributes enormously to mine cost-yield control and ultimate profit. Prior to 1975, entire dip slope panels of reef were stoped, whether or not they were profitable everywhere. Geologic mapping now reveals on a local basis what areas should be mined, what left as pillars, what ignored, and what taken completely.

It has also been discovered, as mentioned earlier, that prominent down-gradient variation results from physical attrition-comminution, progressive sorting, and environmental changes consistent with fan sedimentology. In general, over a 20-to-30-km distance down gradient, the percentage of conglomerate in the section diminishes, the percentage of sand-sized and smaller "fines" increases, total pyrite and pyrite grain size decreases, gold diminishes and becomes finer grained, net uranium and uranium-to-gold ratios increase, and the abundance of algal mat material increases in the quieter, deeper waters. The downstream or distal fan-toe environments, as exemplified in the Saaiplaas mine near Welkom, consist of broad sheets of cross-bedded sand with extensive carbonaceous-uraniferous layers. Such layers only a few millimeters thick, but rich in uranium and gold, may be the target of a 15,000-ft-deep (5-km) drill hole!

Another aspect of gold and uranium occurrence in Rand sediments concerns local postdepositional redistribution of those elements, especially uranium. Near the end of World War II, when interest in fissionable materials was gaining momentum, it was discovered that the frequently mentioned "carbon" of the Rand ores was actually a uraninite-carbon aggregate earlier, but improperly, called *thucholite* (Davidson and Bowie, 1951), a mineraloid whose very name describes its thorium, uranium, carbon, hydrogen, and oxygen composition. Changing economics regarding uranium caused geologists to take a second look at the Rand as a source of supply. It was found to contain much more uranium than was previously thought. The average ore produced from 1953 through 1972 contained about 0.028 wt % U (280 ppm) versus 0.001 wt % Au (10 ppm), a factor of 28 (Pretorius, 1976). Relatively low-grade materials are treated for their uranium content where uranium can be recovered as a by-product of the gold ores, but some uranium is mined from nonauriferous reefs. In these deposits the uranium content pays the mining costs. Thus various grades of uranium ore are recovered.

The carbon, correctly referred to as *kerogen* or simply as "fossil plant matter," has long been recognized as a close companion of gold. Throughout much of the development of the district, this material was used as a guide to the better ores; where the carbon was richest, the gold values were likely to be high. In the so-called "buckshot pyrite" ore—conglomerates containing abundant pyrite in rounded pebblelike forms—pyrite, kerogen, and gold appear to have grown in the order listed, forming concentrically layered granules (Macadam, 1931). In the early 1970s Hallbauer discovered startling morphologies of gold grains when kerogen and silicates were dissolved away from these layered granules or from algal mat samples. Gold faithfully mimics algal forms with beaded, spongelike, strandform, and cellular shapes (Hallbauer, 1975; Hallbauer and Utter, 1977; and Minter, 1976), which require that gold, and uraninite as well, be locally dissolved by organic acids and reprecipitated in organically molded forms. Thus both uranium and gold appear to have been diagenetically "reworked" to locally varying degrees. This complication, and the fact that metamorphism has recrystallized significant amounts of original gold particles, impeded proof of the fundamentally sedimentological nature of the ores for many years.

Serious mining in the Witwatersrand district was begun in 1886, when a gold rush started this activity. Since that time many mines have been developed along the 400 km of strike of the Witwatersrand conglomerate beds on the northerly and southwesterly sides of the Witwatersrand basin. Several mines are operating at depths of more than 3 km. The East Rand Property mine operates at a depth of about 3.5 km. The Western Deeps mine near Carletonville, the deepest in the world, has reached nearly 4 km! Even deeper mines are planned to reach farther down along ore-bearing strata, a distance now locally more than 30 km downdip.

As developed above, appreciation of local sedimentologic features is potentially useful both to exploitation of the ores and to prospecting. The principles being clarified in South Africa can be transferred to the study of placer deposits in general, but in particular to the study of Huronian age rocks (1.9 to 2.5 billion years) elsewhere in the world. Units similar in appearance and containing generally similar gold occurrences with or without uranium have been described from the Jacobina area, Brazil, from the Tarkwa goldfield in Ghana (Sestini, 1973), from Wyoming, United States, and from every continent (Pretorius, 1981a), including several districts in Canada, notably the Blind River district. Here uraninite and thucholite, with little gold, lie in basal conglomerates near Archean basement rocks. Many descriptive and distributional aspects of these rocks and ores (Robertson, 1976) suggest striking similarities with Wits geology. It has also been suggested, however, that the Blind River conglomerates were not extensively reworked, that they have not undergone the long milling and winnowing that upgraded the Witwatersrand equivalents, and that the Blind River units indicate regressive offlap single-cycle sedimentation rather than the transgressive onlap multicycle Wits phenomenon. It could be concluded, therefore, that the more propitious gold exploration targets lie on the *other* side of the Blind River basin, perhaps beneath the Paleozoic cover of Indiana and Ohio. Whether or not this is true, prospecting involving the development of mechanical sedimentary accumulation of heavy minerals should be pressed only after thorough consideration of the geologic literature being generated by students of the Witwatersrand and its ores.

Bibliography

Adams, J., G. L. Zimpfer, and C. F. McLane, 1978. Basin dynamics, channel processes, and placer formation: a model study. *Econ. Geol.* 73:416–426.

Allsop, H. L., 1961. Rb-Sr age measurements on total rock and separated mineral fractions from the Old Granite of the central Transvaal. *J. Geophys. Res.* 66:1499–1508.

Anonymous, 1982. *Basic Placer Mining,* Calif. Div. Mines Geol., Spec. Pub. 41, 16 pp.

Armstrong, F. C., Ed., 1981. Genesis of uranium and gold-bearing Precambrian quartz-pebble conglomerates. USGS Prof. Pap. 1161.

Averill, V., 1946. *Placer Mining for Gold in California.* Calif. Div. Mines Geol. Bull. 135, 477 pp.

Bilibin, Y. A., 1938. *Principles of Placer Geology.* Moscow: G.O.N.T.I., 505 pp.

Boyle, R. W., 1979. *The Geochemistry of Gold and Its Deposits.* Geol Surv. Can. Bull. 280, 584 pp.

Brock, B. B., and D. A. Pretorius, 1964a. Rand basin sedimentation and tectonics, in S. H. Haughton, Ed., *The Geology of Some Ore Deposits of Southern Africa,* vol. 1. Johannesburg: Geol. Soc. So. Afr.

—— and ——, 1964b. An introduction to the stratigraphy and structure of the Rand goldfield, in S. H. Haughton, Ed., *The Geology of Some Ore Deposits of Southern Africa,* vol. 1. Johannesburg: Geol. Soc. So. Afr.

Buck, B., 1981. Sedimentology and heavy mineral behavior in the Steyn and Saaiplaas reefs, Welkom goldfield, Orange Free State, Republic of South Africa (approx. title). Unpub. M.S. Thesis, Univ. Witwatersrand, Johannesburg, So. Afr.

Carpenter, J. H., J. C. Detweiler, J. L. Gillson, E. C. Weichel, Jr., and J. P. Wood, 1953. Mining and concentration of ilmenite and associated minerals at Trail Ridge, Florida. *AIME Trans,* 196:789–795.

Cloud, P. E., Jr., 1968. Atmospheric and hydrospheric evolution on the primitive earth. *Science* 160:729–736.

Cobb, E. H., 1973. Placer deposits of Alaska. USGS Bull. 1374.

Cochran, A., 1982. Fluid inclusion populations in quartz-rich gold ores from the Barberton Greenstone Belt, Eastern Transvaal, South Africa. Unpub. M.S. Thesis, Univ. Ariz., Tucson, 208 pp.

Creitz, E. E., and T. N. McVay, 1949. A study of opaque minerals in Trail Ridge, Florida, dune sands. *AIME Trans.* 181:417–423.

Desborough, G. A., 1970. Silver depletion indicated by microanalysis of gold from placer occurrences, western United States. *Econ. Geol.* 65:304–311.

Davidson, C. F., and S. H. U. Bowie, 1951. On thucolite and related hydrocarbon-uraninite complexes with a note on the origin of the Witwatersrand gold ores. *Geol. Surv. Gr. Brit. Bull.* 3:1–19.

Force, E. R., 1976. Metamorphic source rocks of titanium placer deposits, a geochemical cycle. USGS Prof. Pap. 959–B, pp. B1–B16.

Foster, R. L., E. E. Foord, and P. E. Long, 1978. Mineralogy and composition of Jamison Creek particulate gold, Johnsville mining district, Plumas County, California. *Econ. Geol.* 73:1175–1183.

Fripp, R. E. P., 1976. Stratabound gold deposits in Archean banded iron formation, Rhodesia. *Econ. Geol.* 71:58–75.

Gevers, T. W., 1961. Outline of the geology of southern Africa. *7th Commonw. Min. Metall. Cong Trans.*, Disc., vol. 1, pp. 25–37.

Hails, J. R., 1976. Placer deposits, pp. 213–244 in K. H. Wolf, Ed., *Handbook of Stratabound and Stratiform Ore Deposits*, vol. 3. New York: Elsevier.

Hallbauer, D. K., 1975. The plant origin of the Witwatersrand "carbon." *Miner. Sci. Eng.* 7:111–131.

—— and T. Utter, 1977. Geochemical and mineralogical characteristics of gold particles from recent river deposits and the fossil placers of the Witwatersrand. *Mineral. Deposita* 12:293–306.

Johnson, M. G., 1972. Placer gold deposits of Arizona. USGS Bull. 1355.

Köppel, V. H., and R. Saager, 1974. Lead isotope evidence on the detrital origin of Witwatersrand pyrites and its bearing on the provenance of the Witwatersrand gold. *Econ. Geol.* 69:318–331.

Lynd, L. E., 1960. Titanium, in J. L. Gillson, Ed., *Industrial Minerals and Rocks*, 3rd Ed. New York: AIME.

Macadam, P., 1931. The distribution of gold and carbon in the Witwatersrand bankets. *Geol. Soc. So. Afr. Trans.* 34 (annex):81–88.

Macdonald, E. H., 1983. *Alluvial Mining—the Geology, Technology, and Economics of Placers*. London: Chapman and Hall, 580 pp.

Martens, J. H. C., 1928. Beach deposits of ilmenite, zircon, and rutile in Florida. *Geol. Surv. Fla. Ann. Rept.* 19:124–154.

——, 1935. Beach sands between Charleston, South Carolina, and Miami, Florida. *Geol. Soc. Amer. Bull.* 46:1563–1596.

McKelvey, V. E., and J. R. Balsley, Jr., 1948. Distribution of coastal blacksands in North Carolina, South Carolina, and Georgia, as mapped from an airplane. *Econ. Geol.* 43:518–524.

Mellor, E. T., 1916. The conglomerates of the Witwatersrand. *Inst. Min. Metall. Trans.* 25:226–348.

Minter, W. E. L., 1970. Gold distribution related to the sedimentology of a Precambrian Witwatersrand conglomerate, South Africa, as outlined by moving-average analysis. *Econ. Geol.* 65:963–969.

——, 1976. Detrital gold, uranium, and pyrite concentrations related to sedimentology in the Precambrian Vaal Reef placer, Witwatersrand, South Africa, *Econ. Geol.* 71:157–176.

——, 1978. A sedimentological synthesis of placer gold, uranium, and pyrite concentrations in Proterozoic Witwatersrand sediments, pp. 801–829 in A. D. Miall, Ed., *Fluvial Sedimentation.* Can. Soc. Petrol. Geol. Mem. 5.

——, 1982. The Golden Proterozoic, pp. 115–150 in A. J. Tankard, K. A. Eriksson, D. R. Hunter, M. P. A. Jackson, D. K. Hobday, and W. E. L. Minter, *Crustal Evolution of Southern Africa.* New York: Springer, 523 pp.

Mosley, M. P., and S. A. Schumm, 1977. Stream junctions—a probable location for bedrock placers. *Econ. Geol.* 72:691–694.

Nicolaysen, L. O., A. J. Burger, and W. R. Liebenberg, 1962. Evidence for the extreme age of certain minerals from the Dominion Reef conglomerates and the underlying granite in the Western Transvaal. *Geochim. Cosmochim. Acta* 26:15–23.

Overholt, J. L., G. Vaux, and J. L. Rodda, 1950. The nature of "arizonite." *Am. Mineral.* 35:117–119.

Park, C. F., Jr., and R. A. MacDiarmid, 1975. *Ore Deposits.* San Francisco: Freeman, 530 pp.

Pegg, W. C., 1950. A contribution to the geology of the West Rand area. *Geol. Soc. So. Afr. Trans.* 53:209–227.

Pirkle, E. C., W. A. Pirkle, and W. H. Yoho, 1974. The Green Cove Springs and Boulounge heavy-mineral sand deposits of Florida. *Econ. Geol.* 69:1129–1137.

—— and W. H. Yoho, 1970. The heavy mineral ore body of Trail Ridge, Florida. *Econ. Geol.* 65:17–30.

Pretorius, D. A., 1975. The depositional environment of the Witwatersrand goldfields: a chronological review of speculations and observations. *Miner. Sci. Eng.* 7:18–47.

——, 1976. Gold in the Proterozoic sediments of South Africa: systems, paradigms, and models, pp. 1–27 in K. H. Wolf, Ed., *Handbook of Stratabound and Stratiform Ore Deposits,* vol. 7. New York: Elsevier.

——, 1981a. Gold and uranium in quartz-pebble conglomerates. *Econ. Geol. 75th Anniv. Vol.,* pp. 117–138.

——, 1981b. Pers. Commun.

Puffer, J. H., and H. L. Cousminer, 1982. Factors controlling the accumulation of titanium-iron oxide-rich sands in the Cohansey Formation, Lakehurst area, New Jersey. *Econ. Geol.* 77:379–391.

Raicevic, D., and L. J. Cabri, 1976. Mineralogy and concentration of the Au- and Pt-bearing placers from the Tulameen River area in British Columbia. *Can. Inst. Min. Metall. Bull.*, June, 69:111–119.

Reinecke, L., 1930. Origin of the Witwatersrand system. *Geol. Soc. So. Afr. Trans.* 33:111–133.

Robertson, D. S., 1976. The Blind River uranium deposits: the ores and their settings. Ont. Div. Mines Misc. Pap. 65, 45 pp.

Sestini, G., 1973. Sedimentology of a paleoplacer: the gold-bearing Tarkwaian of Ghana, pp. 275–305 in G. C. Amstutz and A. J. Bernard, Eds., *Ores in Sediments*. New York: Springer, 350 pp.

Slingerland, R. L., 1977. The effects of entrainment on the hydraulic equivalence relationships of light and heavy minerals in sands. *J. Sed. Petrol.* 47:753–770.

Smith, N. D., and N. J. Beukes, 1983. Bar to bank flow convergence zones: a contribution to the origin of alluvial placers. *Econ. Geol.* 78:1342–1349.

—— and W. E. L. Minter, 1980. Sedimentological controls of gold and uranium in two Witwatersrand paleoplacers. *Econ. Geol.* 75:1–14.

Sutherland, D. C., 1982. The transport and sorting of diamonds by fluvial and marine processes. *Econ. Geol.* 77:1613–1620.

Tankard, A. J., K. A. Eriksson, D. R. Hunter, M. P. A. Jackson, D. K. Hobday, and W. E. L. Minter, 1982. *Crustal Evolution of Southern Africa*. New York: Springer, 523 pp.

Teufer, G., and A. K. Temple, 1966. Pseudorutile—a new mineral intermediate between ilmenite and rutile in the natural alteration of ilmenite. *Nature* 211:179–181.

Tucker, R., 1980. Sedimentology of the gold-uranium deposits of the West Rand Cooke section, Randfontein Estates, South Africa. Unpub. M. S. Thesis, Univ. Witwatersrand, Johannesburg, So. Afr.

Vermaak, C.F., 1979. The global status of the South African minerals economy and data summaries of its key commodities. Geol. Soc. So. Afr. Rev. Pap. 1, 57 pp.

Viljoen, R. P., R. Saager, and M. J. Viljoen, 1970. Some thoughts on the origin and processes responsible for the concentration of gold in the early Precambrian of southern Africa. *Mineral. Deposita* 5:164–180.

SEVENTEEN

Deposits Related to Weathering

Many of the ore minerals, especially the sulfides and sulfosalts, are formed in reducing environments and at temperatures and pressures higher than atmospheric. When these minerals are exposed at surface conditions to weathering and erosion, they are inclined to break down chemically, forming new compounds or going in whole or in part into solution. Similarly, neither are many of the common rock-forming minerals stable at the Earth's surface; they too undergo chemical changes as equilibrium is established with the new environment. Most geologists regard weathering as an inorganic process related principally to hydrolysis, hydration, and oxidation reactions between atmospheric and lithospheric components; nearly all studies of weathering have been conducted under this premise. However, Ehrlich (1964) has called attention to the activity of bacteria in weathering processes. He reported that oxidation of arsenopyrite and enargite increases markedly in the presence of thiobacillus-ferrobacillus group bacteria. Many researchers since have included bacterial activity in their considerations (Bohn, McNeal, and O'Connor, 1979), but Eh-pH variations and the composition of soil gases and solutions are still thought to be of paramount importance. Weathering is an absolutely vital process to life as we know it if for no other reason than that it produces the soil that is both the basis of our grain, vegetable, and red meat food chains (Keller, 1956) and a reservoir of gases, moisture, and life forms from microbes to insects that are part of our symbiotic environment. But the purpose of this chapter is to consider how weathering is involved more specifically in ore deposits.

If we think about the "life history" of a single deposit that is exposed by erosion to weathering processes, it becomes apparent that depending upon the materials, environments, dynamics, and products involved, weathering may (1) destroy the deposit by planing it off; (2) leach it of one or more of its valuable constituents, thus rendering it subeconomic; (3) redistribute one or more of its components toward enrichment and enhancement of value; or (4) alter the mineralogy to generate either greater or less value by creating new valuable or deleterious products. If we consider the process overall—and this will summarize the organization of this chapter—(1) weathering can dissolve, transport, and concentrate by redepositing one or more elements, (2) it may convert a worthless material into a useful or valuable one, normally with a change in composition and mineralogy; or (3) it may simply "free" a resistant accessory mineral by disintegrating the rock-forming minerals around it. Examples of redistribution and enrichment to be covered include the generation of nickel laterites, of supergene enrichment in porphyry copper deposits, and of manganese enrichment from bedrock. An example of the second function of weathering is in the conversion of feldspar and feldspathoids into the more amenable bauxite ores of aluminum; the process includes both concentration and amelioration. An example of the third or "freeing" mode is that of releasing cassiterite from tin granites into soil profiles and placers or freeing gold, gem minerals, and even magnetite or ilmenite for natural or engineered hydraulic concentration.

Lelong et al. (1976) provide an excellent summary of weathering, soil genesis, and supergene enrichment. They point out that the chemical effects of meteoric waters on rocks vary according to rock type and thus to the minerals being attacked. Even pure water will dissolve and alter most rock-forming minerals to a limited degree, but meteoric waters contain carbon dioxide and are slightly acidic. Furthermore, soil water generally contains appreciable amounts of humic acids as well as carbonic acid, and its pH can readily be as low as 4.0 or 5.0. These acidic, oxidized waters percolate slowly through the zone of aeration on their way to the groundwater table. In transit, or after reaching the zone of saturation, they may react with carbonate rocks or feldspars and other silicate minerals to become neutral or slightly alkaline, typically with hydrolysis reactions which consume H^+ ions and therefore raise pH in the fluid phase. The rocks thus act as a buffer, or H^+ ion sink. All minerals are soluble under favorable conditions, and they eventually break down or are mechanically removed at the Earth's surface. Mineral solubilities vary widely at standard temperature and pressure, and since the rates of mineral chemical reactions vary independently, there is a wide range of apparent mineral stabilities and metastabilities. Meteoric waters charged with carbon dioxide and oxygen from the atmosphere can oxidize, hydrate, and carbonatize rock-forming silicates. Sulfides are converted to sulfates, most of which are soluble, or they are converted to the more stable oxides, native metals, and carbonates. Iron, manganese, and aluminum form oxides and hydroxides that are relatively insoluble at the surface; lead forms a stable sulfate; lead, zinc, and copper are preserved as carbonates in some environments; copper, zinc, nickel, and chromium can be retained as silicates and oxides; and native gold, copper, and silver may remain stable in the weathered zone. These elements have been recovered as residual concentrates and as simple oxidation products near the weathered rock outcrops that contained them (Emmons, 1917).

The depth to which weathering takes place varies greatly and depends upon climate, porosity and permeability of the rocks, structure, and tectonic and geomorphologic history. Where the water table stands close to the surface, and where relief is minor or circulation of groundwaters is low, oxidation and hydrolysis may be shallow. Many areas of the north, where permafrost, tundra, and swamps prevail and reactions are slow, are weathered to just a few feet below the surface. On the other hand, oxidation is deep in arid regions where temperatures are moderate to high and water tables are low, such as at the Chuquicamata mine in northern Chile and at the Tsumeb mine in South-West Africa (Namibia). At Tsumeb, copper oxidation products have been encountered along faults at depths of nearly 1000 meters. At places in the Southern Piedmont of the United States and in the iron formations of Gabon, adits and shafts to depths of over 100 meters remain in thoroughly weathered, surface-influenced materials. Other instances of great weathering depths will be described below.

The general direction of weathering, then, is in the direction of hydrolysis, hydration, and oxidation. Olivine, feldspars, and feldspathoids yield by reactions of the form

$$\underset{\text{olivine}}{2Mg_2SiO_4} + 4H^+ + 0.5O_2 \rightleftarrows \underset{\text{serpentine}}{Mg_3Si_2O_5(OH)_4} + Mg^{+2}(aq) \qquad (17\text{-}1)$$

and

$$\underset{\text{feldspar}}{3KAlSi_3O_8} + 2H^+ \rightleftarrows \underset{\text{mica}}{KAl_3Si_3O_{10}(OH)_2} + \underset{\text{silica}}{6SiO_2} + \underset{\text{potash}}{2K^+} \qquad (17\text{-}2)$$

or with the same amount of feldspar with four more H^+ and some oxygen,

$$\underset{\text{feldspar}}{3KAlSi_3O_8} + 6H^+ + 0.75O_2 \rightleftarrows \underset{\text{kaolinite}}{1.5[Al_2Si_2O_5(OH)_4]} + \underset{\text{silica}}{6SiO_2} + \underset{\text{potash}}{3K^+} \qquad (17\text{-}3)$$

or with less H^+ and no oxygen,

$$\underset{\text{feldspar}}{3KAlSi_3O_8} + 3H^+ \rightleftarrows \underset{\text{montmorillonite}}{1.5[Al_2Si_4O_{10}(OH)_2]} + \underset{\text{silica}}{3SiO_2} + \underset{\text{potash}}{3K^+} \qquad (17\text{-}4)$$

It is apparent that the type of weathering product can be determined by the pH and Eh of the environment. Amphiboles and biotites are weathered to chlorite or montmorillonite, commonly with some carbonates and leucoxene. Quartz solubility increases with pH from nearly insoluble in strongly acid cool environments through slightly soluble in humid tropical conditions of pH 6 to 8 to readily soluble in alkaline systems of pH above 8 (see Figure 17-10). Magnetite may be either oxidized to hematite or goethite, or remain metastably intact. Olivine (Eq. 17-1) and pyroxene in mafic rocks typically weather to serpentinelike hydrous magnesium silicates. Virtually all ferrous iron ions exposed to soil waters are converted to ferric. As three related diagrams by Garrels and MacKenzie (1971) indicate, the composition of soil gases and liquids is well within the hematite stability field (Figure 17-1). Sulfides in the host rocks behave complicatedly since the mobility of metal ions in the zone of weathering is largely determined by the composition of both the ground waters and the country rocks (Rose, Hawkes, and Webb, 1979). Sulfide-free meteoric waters leach such elements as zinc, molybdenum, and uranium from igneous rocks, leaving behind stable oxidized products of iron, aluminum, titanium, chromium, and in places manganese, nickel, cobalt, copper, lead, and antimony. The host-rock environment is especially important in the oxidation of sulfides; for example, some metals that would be leached from a siliceous host are retained in calcareous rocks. Molybdenum, zinc, and silver are especially soluble in sulfate solutions, but under favorable conditions they form stable oxidation products in limestones. Similarly, copper, which is relatively mobile in sulfate waters that circulate through siliceous igneous rocks, forms practically insoluble carbonate minerals in calcareous environments. Iron and lead oxidize to stable compounds in both siliceous and calcareous rocks and are retained in the zone of weathering. Certain elements are dependent upon the presence or absence of a second element for their stability during weathering. For example, molybdenite oxidizes to the relatively insoluble compound ferrimolybdite [$Fe_2(MoO_4)_3 \cdot 7.5\ H_2O$] in siliceous, iron-rich environments, but molybdenum is readily leached where iron is not plentiful.

Lelong et al. (1976) described the mobility of specific elements and oxides in normal weathering environments (Figure 17-2 and Table 17-1). The elements that form stable hydrolysates (Fe, Ti, Mn) are typically immobile, and those that form soluble oxides or complexes (Na, K, Ca, Mg) are two orders of magnitude more redistributable. The alkali and alkali earth elements have low ionic potentials and form soluble hydrated or simple cations regardless of pH, and are therefore almost universally leached. The divalent transition elements, including the base metals, are soluble especially at low pH, and the sesquioxides (M_2O_3) are almost totally insoluble at normal soil pH.

The fact that certain elements are concentrated in soils during weathering is useful in geochemical analytical prospecting (Rose, Hawks, and Webb, 1979). Even where a metallic element such as copper is comparatively soluble and appears to have been removed, enough remains in the soil that it may be detected by careful geochemical analytical methods. Usually the soil overlying a body of copper mineralization will contain more

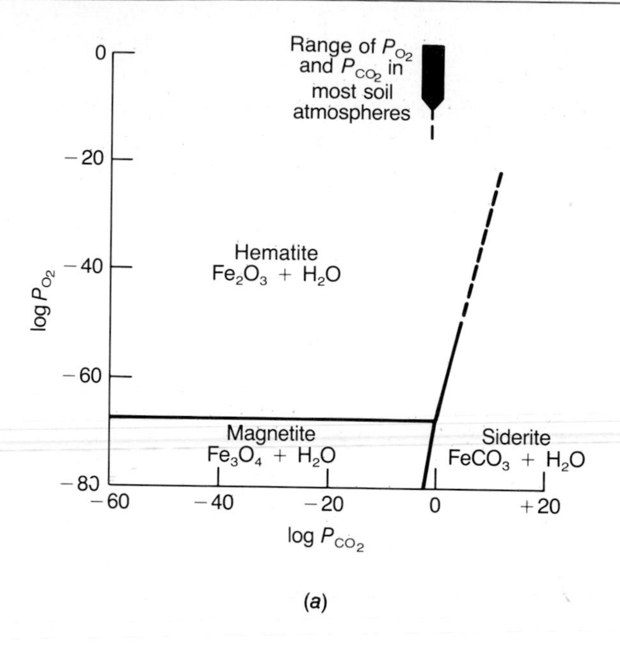

(a)

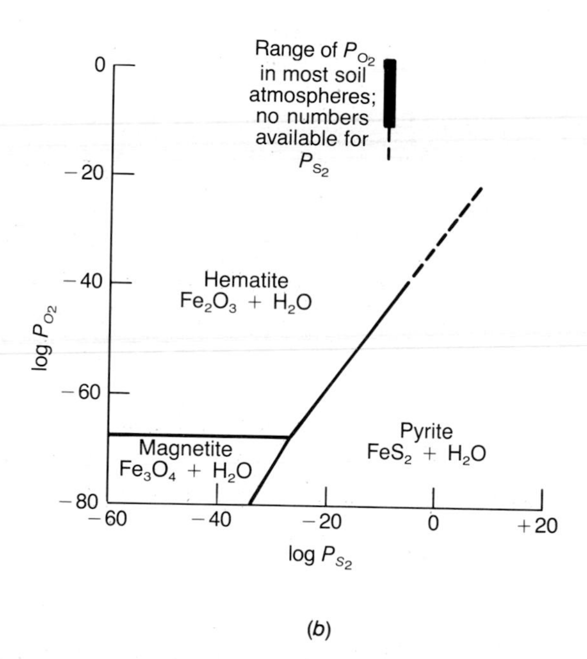

(b)

Figure 17-1. (*a*), (*b*) Partial pressure diagrams showing the stability of some iron minerals as functions of the partial pressure of O_2, CO_2, and S_2 in soils. Magnetite, siderite, pyrite, and pyrrhotite would all be oxidized to hematite if equilibrium were reached. (*c*) (*facing page*) Field of various environments, including soil formation in terms of Eh-Ph. See also Figures 15-9 and 17-10. (*From* Garrels and MacKenzie, 1971.)

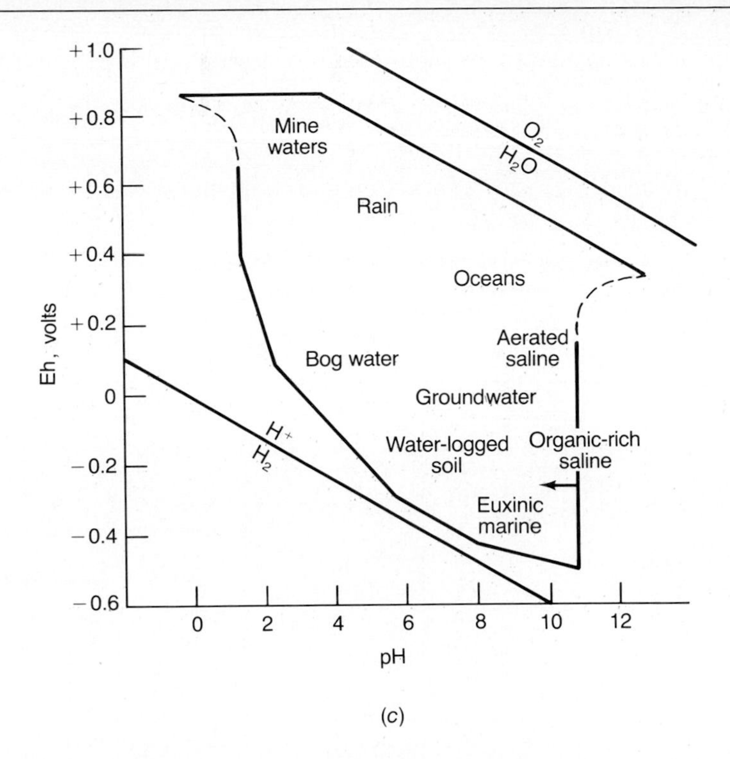

(c)

residual copper than the average soils of the surrounding area. Anomalies can thus be detected. In addition to determinations in soils, four other methods of geochemical study are widely used.

1. Metals may be adsorbed or metabolized in the tissues of plants and show variations and patterns similar to those in the soil; molybdenum uptake in mesquite leaves constitutes a useful geobotanical indicator.

TABLE 17-1. Mobilities of Main Oxides in Weathering Regimens

Group	Oxides	Relative Mobility	
Sesquioxides	Cr_2O_3	60	1–100 = Low
	Al_2O_3	2	
	Fe_2O_3	30	
Dioxides	TiO_2	10–100	
	SiO_2	300	100–500 = Moderate
Alkalies, alkali earths	Na_2O	1000	500–10,000 = High
	K_2O	100–1000	
	CaO	500–2000	
	MgO	300–2000	

2. Mobile trace elements are detectable in stream and lake waters.
3. Metal content in places is increased in streambed sediments downstream from mineralized bodies.
4. The presence of above average amounts of metal can be detected in chip samples of bedrock.

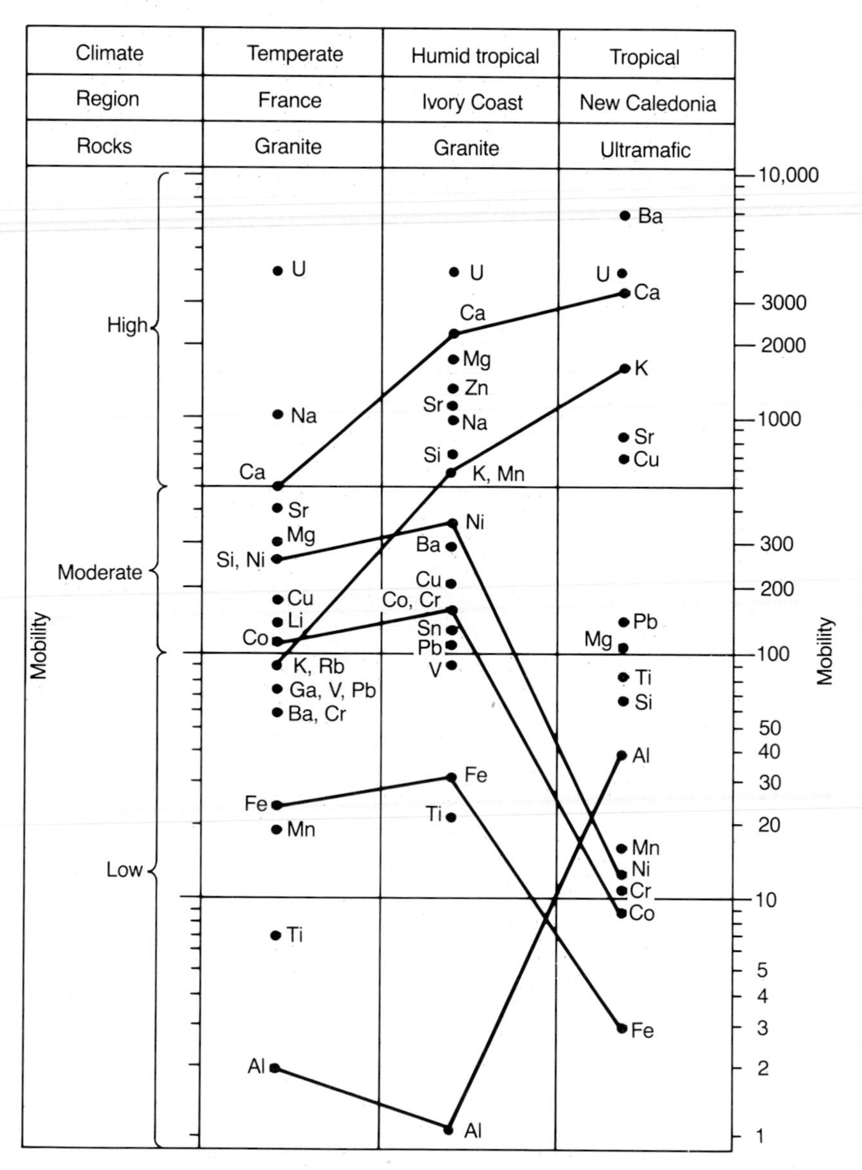

Figure 17-2. "Average" mobility of oxides and ions in selected weathering environments, with several tie lines provided and uranium added. See also Table 17-1. (*After* Lelong et al., 1976.)

In tropical climates, where humic acids are abundant and leaching is especially effective, only the most insoluble oxides remain at the surface. Iron and aluminum form such stable compounds and are so abundant that they are commonly left as residual concentrates. These iron- or aluminum-rich soils are known as *laterites,* the Latin word for "brick earth." Laterite soils are mostly oxides, and thus contrast markedly with the silicate clay soils of the temperate climates (Figure 17-2). Geologists working in temperate climates tend to consider silica as a stable mineral, an "insoluble residue" in weathering, but mapping in the tropics and in the thick red, clayey lateritic soils commonly encountered there illustrates its solubility in warm, alkaline, wet environments (see Figure 17-10). Mafic minerals and feldspars break down in the tropics and release hydrous iron and aluminum compounds that are relatively insoluble under oxidizing conditions and consequently remain at the outcrop. During oxidation, the soluble components are dispersed, and iron and aluminum are oxidized in place. The oxidation products may be transported in solution under low Eh and pH conditions. For example, iron is soluble as the simple ferrous ion or as organic, sulfate, and hydroxide complexes in acid reducing waters (Hem and Cropper, 1959; Hem, 1960; Osborn and Hem, 1961). However, the iron and aluminum oxide-hydroxide oxidation products are generally redeposited as they form because of high Eh and pH. Apparently, framework and layer lattice mineral structures break down in tropical soils, and the oxidation products go into the colloidal state, as evidenced by the presence of orbicular, concretionary structures known as *pisolites*. However, since pisolites also form in ways other than through colloidal solutions, their presence alone is not considered definite proof that the soil had colloidal origin. Most, but not all, laterites are pisolitic (Figure 17-3) or have pisolitic crusts at the grass-roots.

Iron-rich laterites form over ferromagnesian rocks where the rainfall is heavy and the topography subdued. Flat areas and broad swales are ideal because water is retained long enough to dissolve the siliceous components of the soil without eroding the residual materials. Apparently laterites form most readily between the annual high and low positions of the water table; alternating wet and dry seasons are therefore ideal for laterization (Harder, 1952). Laterites have been mined in small amounts as iron ore where limonitic pellets are abundant and fine-grained soils or clays are nearly absent (Percival, 1966). Very large areas covered with this type of laterite are known in Brazil, Cuba, India, central Africa, the Philippine Islands, and elsewhere in the tropics. Many of the iron-rich laterites overlie serpentine bodies or other rocks that originally were rich in iron and deficient in silica. In order to compete successfully with the world's large sedimentary iron-ore deposits and with high-grade magnetite deposits, the iron laterites would have to be nearly pure and strategically located. A significant factor in the cost of mining laterites is the presence of 10 to 30% combined water, which must be transported and then removed during smelting.

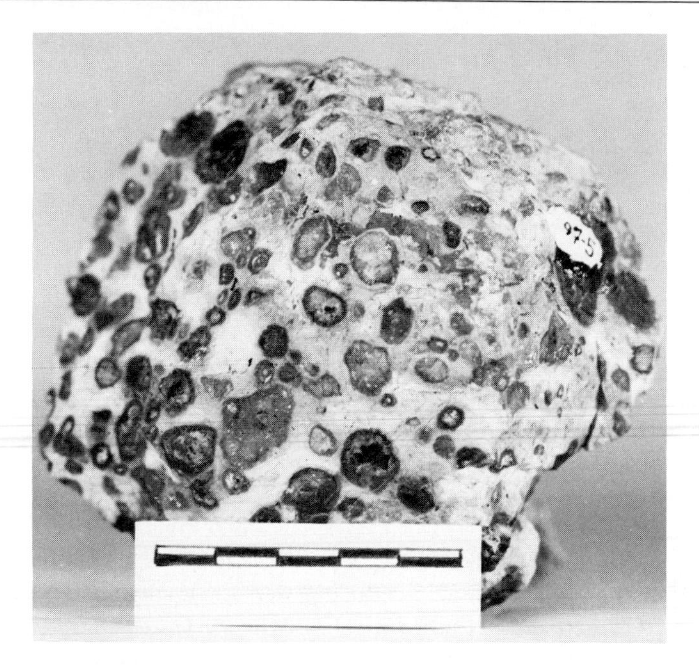

Figure 17-3. Pisolitic aspect of near-surface laterites. See also Figures 17-5 and 17-6, which show the pisolitic zone yielding downward to massive clayey portions. The bar scale is in centimeters. (*Photo by* W. K. Bilodeau.)

In tropical or subtropical environments where the underlying rocks are rich in aluminum and low in iron and silica, concentrations of bauxite are likely to form, especially over syenites and nepheline syenites. Bauxites consist of boehmite [$AlO(OH)$], gibbsite [$Al(OH)_3$], diaspore [$AlO(OH)$], and other hydrous aluminum oxides. The formation of clay minerals may be an intermediate step in the breakdown of some feldspars and feldspathoids to bauxite. Bauxites also form over argillaceous carbonate rocks, in an association with a residual ferruginous clay known as *terra rossa*. Although such concentrations of alumina are generally high in iron, they are mined in places, for example, in the Mediterranean region and in Jamaica. In Jamaica, as described later in this section, bauxites occur both as mantle soil layers on an extensive limestone horizon and as much thicker masses measured in tens of meters in sinkholes and solution cavities in a well-developed karst topography.

Large reserves of aluminous clays are known in many areas, such as those of the kaolins of the states of Georgia, the Carolinas, and Alabama. But it must be remembered that the clay minerals kaolinite, dickite, and others are alumino-silicates, not oxides or hydroxides. Owing to the energy cost of separating alumina from silica, such clays are only a potential ore. Prior to the escalated energy costs of the middle 1970s, several companies hoped to be able to separate kaolin formed in supergene-enriched copper ores as at Butte and Globe-Miami as part of the same hauling-crushing-milling-extraction used to separate copper minerals, but the process of desilication *and* reduction to aluminum metal has proven not feasible at high energy costs.

Manganese oxides are commonly greatly enriched in oxide-hydroxide soil-sediment blankets over bodies of rock that are initially rich in manganese and are deeply weathered. Many metamorphic rocks contain spessartite garnets and other manganese-bearing silicates and carbonates; in places these minerals are present in large amounts. Manganese ores can form where silica and other valueless materials are removed and the relatively insoluble manganiferous oxides remain at the surface, a process of enrichment and conversion to usable mineralogies.

Lateritic pellets of manganese oxides are not as common as similar pellets of limonite, but they are found in many places. In Cuba, the name *granzon* has been applied to these pellets; this term is now widely used. Economic concentrations of granzon are commonly found on the surface near manganese orebodies.

Besides iron, aluminum, and manganese, several other metals form stable oxidation products in weathering regimes. Where they reach economic proportions, these metals are mined as ore, but more often they are merely complicating impurities in iron laterites. Many serpentines contain small amounts of nickel, cobalt, and chromium that are concentrated in the iron-rich laterites, or in the upper parts of the serpentines; in some districts—for example, at Nicaro, Cuba—nickel, cobalt, and chromium have been recovered. The chromium content of laterites is as much as 1.5 to 2.5%, the cobalt content 0.2%. Nickel is concentrated in serpentine just below the laterite, forming ores commonly containing 1.5% nickel. Exceptional deposits, such as those of New Caledonia, contain as much as 6 to 10% nickel.

▸ NICKEL LATERITE DEPOSITS, NEW CALEDONIA

Surface weathering affects mineral deposits in several ways. It may leach away an older ore deposit; it may oxidize the materials in place without changing the grade of the ore; or it may create an ore deposit by the concentration of materials that originally were dispersed throughout the fresh rock. The nickel deposits of New Caledonia are a product of the last-mentioned process; supergene enrichment of copper deposits is considered in the next section.

The island of New Caledonia, South Pacific, is about 400 km long and averages only about 40 km wide (Figure 17-4). Nickel deposits were discovered there by Garnier in 1865, and sporadic production has continued since about 1875. Prior to the discovery of the Sudbury district in Ontario, New Caledonia dominated the world's nickel production.

About one third of New Caledonia's 20,000-km^2 area consists of ultramafic rocks composed of variable amounts of olivine, orthopyroxene, and clinopyroxene. The most common rocks (Troly et al., 1979) are harzburgites (ol-opx) and dunites (ol) with local lherzolites (ol-opx-cpx) and wehrlites (ol-cpx). The ophiolitic ultramafic masses (Figure 17-4) were emplaced in Oligocene time on top of Mesozoic graywackes, shales, and sandstones and Eocene basalts by southwestward-directed thrusting. The veneer of ultramafic rocks may originally have been a single, more extensive slab of oceanic crust thrust over the older units. The original olivine in these rocks was about $Fo_{90}-Fa_{10}$ and contained about 0.3% Ni, and the orthopyroxene was $En_{90}-Fs_{10}$ with about 0.03 to 0.06% Ni (Troly et al., 1979). Most of these rocks are now serpentinized to the hydrous magnesium silicates chrysotile and lizardite, both fibrous serpentines. This hydration event was complex—some of the serpentinization occurred before emplacement of the rocks, on the seafloor; most of it is related to the flat faults and is tectonic in origin;

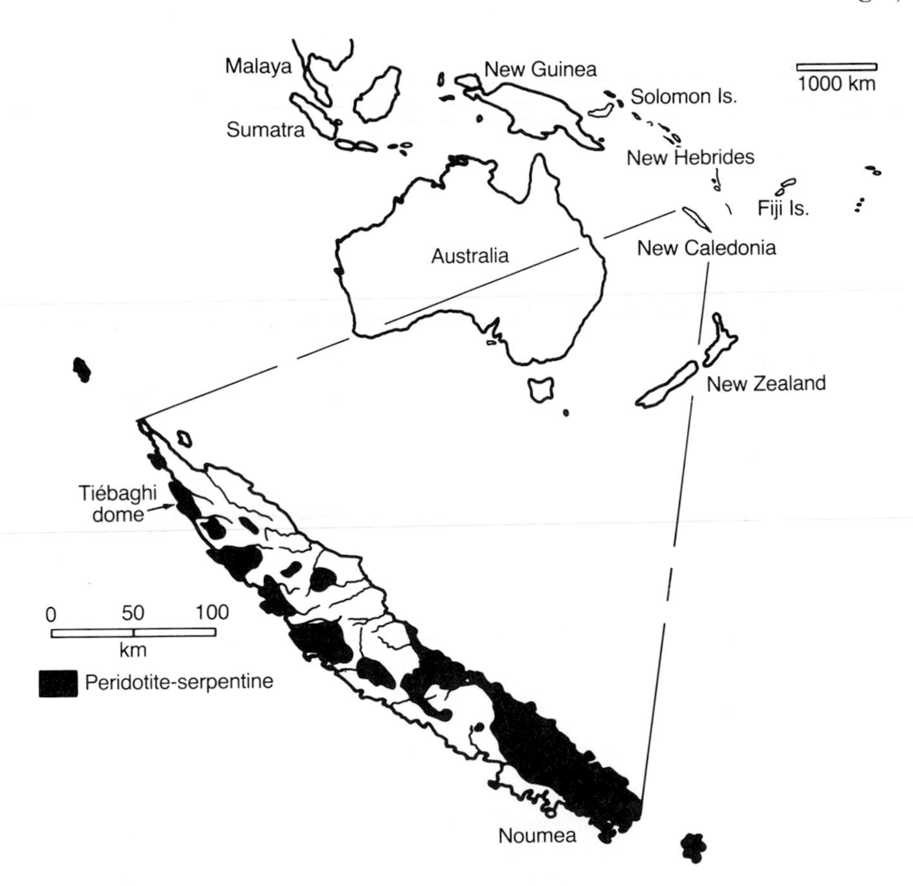

Figure 17-4. Map of New Caledonia showing peridotite-serpentine distribution. Mines from which nickel laterite has been mined are scattered throughout the serpentine areas. Tiébaghi dome is discussed in Chapter 10.

minor amounts are veinlet-controlled and may be postemergence in age. The nickel content of the serpentines is approximately the same as that of the pseudomorphed anhydrous silicates, so serpentinization essentially involves hydration of the original ultramafic minerals accompanied by rearrangement and only minor loss of constituents.

The many descriptions of the mineral deposits of New Caledonia all agree that the nickel was concentrated during laterization of the serpentine or peridotite (Glasser, 1904; Berthelot, 1933; Chételat, 1947; Routhier, 1952, 1963; Trescases, 1973; Troly, Esterle, Pelletier, and Reibell, 1979). The serpentine weathers to a dark-reddish laterite, producing soil profiles like those in other tropical areas (Figure 17-5). Where laterization is complete, the original magnesium silicates of the ultramafics have been destroyed, and silica, calcium, and magnesium removed. Notice in Figure 17-5 that MgO and SiO_2 are almost totally removed from the pisolitic iron crust found at the surface, and that they increase markedly where nickel also increases, the locus of garnierite deposition. The best garnierites are typically devoid of iron at less than 0.3%, but they contain from 10 to 40% MgO, which varies approximately inversely with NiO at 0 to 50%. It is also noteworthy in Figure 17-5 that the nickel ores range from 2 to 3% Ni, and that the best grade is near the bottom of the blanket where the high-nickel serpentine garnierite cements blocks of peridotite, or the serpentinite replacement of it. Cr_2O_3 in chromite, Al_2O_3 from clinopyroxene and plagioclase, and Fe_2O_3 are concentrated in the laterite surface.

Garnierite, a nickeliferous serpentine, is the principal ore mineral in New Caledonia; its composition is $(Mg >> Fe, Ni)_3Si_2O_5(OH)_4$. Garnierite in which nickel replaces magnesium is brown and is called *nouméite*. Another nickeliferous serpentine (népouite) is also an ore mineral; nickel replaces iron in its structure. Nickel-bearing montmorillonite elsewhere called saponite is known locally as *pimelite*, and is a third nickel ore mineral. The colors of the minerals range from bright apple green through light green to white. Where the iron of the silicates is replaced by nickel, the ore is green, and accordingly is known as *green ore*; where the magnesium is replaced, the ore is brown because it retains the color of the iron oxides. Brown ore, known locally as *chocolate ore*, is now the common material being mined. Up to 10% cobalt is present in an amorphous oxide mineraloid called *asbolane*, or cobaltiferous wad. Black manganese oxides identified as pyrolusite and psilomelane have been mined locally in small amounts. Talc and sepiolite $[Mg_4(Si_6O_{15})(OH)_2 \cdot 6H_2O]$ are the most common gangue minerals, and small amounts of chalcedony—representing colloidal silica leached from the laterite zone—are found in fractures cutting the serpentine series.

Studies of the unaltered serpentine and ultramafics show that nickel and cobalt contents in the basement rocks are relatively uniform throughout the island. There seems to be no direct relationship between the grade of the original serpentine or ultramafic rocks and the metal content of the lateritic ores. Thus the presence of ore is strictly a function of favorable

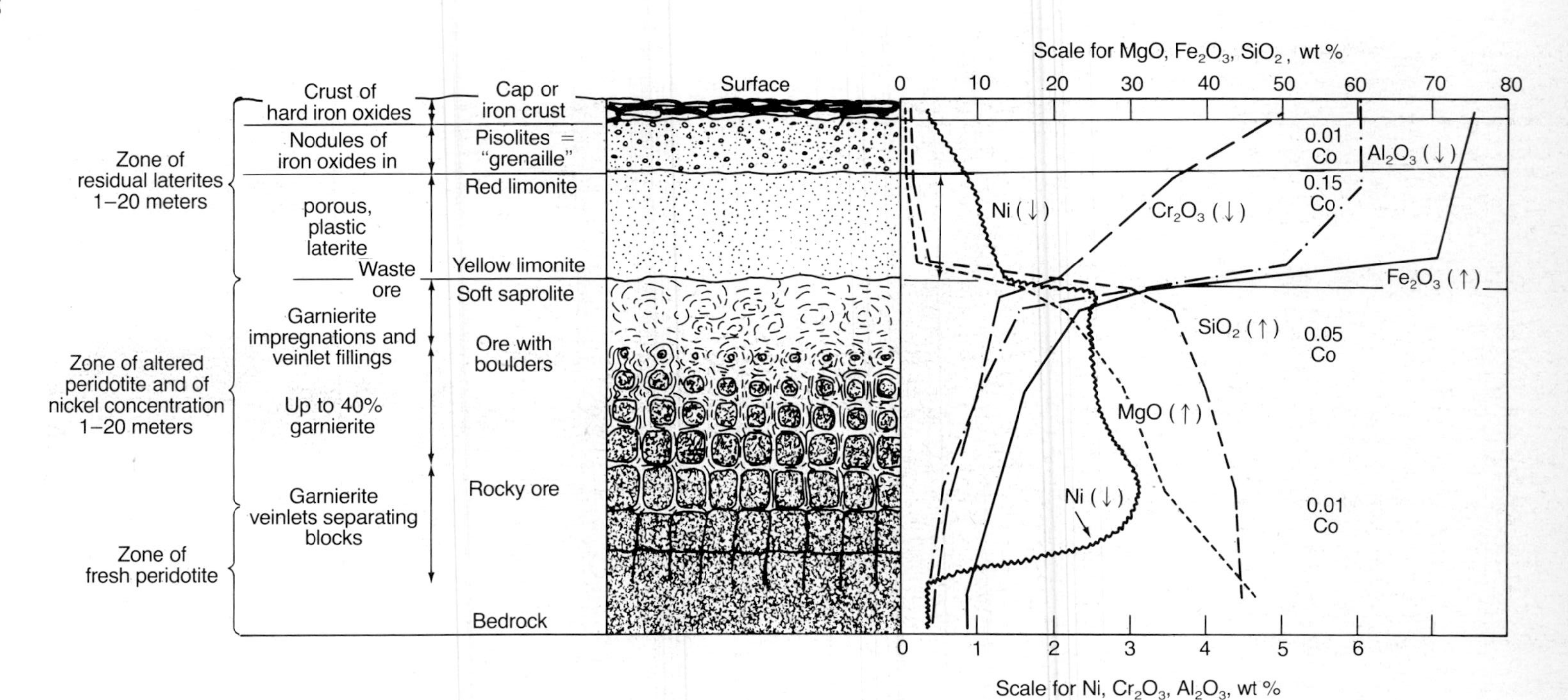

Figure 17-5. Descriptive, occurrence, and chemical variation profile of a typical New Caledonian laterite. Note that there are two chemical scales, and that the highest nickel contents are near the base of the immediate weathering effects. (*Chemical data from* Troly et al., 1979.)

topography and the effect of weathering conditions. The effectiveness of the weathering process is more fully appreciated when we realize that it has concentrated the nickel ten- to thirtyfold from materials that contain only 0.2 to 0.3% Ni (Chételat, 1947; Blanchard, 1944). The efficiency of the nickel-concentration process was controlled largely by topography; the best ores are on gentle slopes and on saddles of spurs extending from the main ridges. Avias (1968) has shown that topography and geomorphology are significant prospecting and evaluation tools. He asserted that the greatest contributors to the development of a nickel laterite body are a stable tropical climate, broad outcrops of ultramafic rocks, a 20° slope and plateau topography that generates stable, slow subsurface water movement, and time. Figure 17-6 gives an ideal topographic profile and an oblique view which show that the best development of nickel enrichment is on slopes with stable, slow hydrologic flowthrough. Webber (1972) showed (Figure 17-7) that there can be various stages of laterization and nickel enrichment that reflect the interplay between erosion rates, hydrologic factors, and permeabilities. Notice that column 5 depicting a laterite in Colombia is almost identical to the depiction of New Caledonian laterite in Figure 17-5. Clearly these variations are considerations in prospecting and evaluation.

In general the zone of nickel concentration lies 0.5 to 6 meters below the surface, but sometimes the cover is 10 to 15 meters thick. The ores are never thick or deep; they "stretch over the Caledonian relief like a thin rug. The mines look more like mountain rice paddies than deep open pits for porphyry copper mining" (page 86, Troly et al., 1979). Yet the island has produced over 75 million tons of ore the grade of which has steadily diminished from nearly 10% Ni in 1900 to 2.5% Ni in recent years. Most of the individual orebodies are small at less than 100,000 tons; the largest deposits contain only about 600,000 tons. Four mine sites are currently active. Foose et al. (1980) show that the average tonnage-grade figures for 64 nickel laterite districts around the world is 40 million tons of 1.41% Ni, so that the New Caledonian district has proven both large and high-grade.

The preceding considers primarily nickel silicate laterite ores, the term referring to the siting of nickel ions in layer lattice silicates. Nickel may also be adsorbed onto and included in the goethites and other hydrous iron oxides so common in laterites, the B type in Figure 17-6. In some places the nickeliferous iron laterite, or nickel oxide ore, reaches grades of 0.8 to 1.5% Ni. Some ores mined in the Dominican Republic and the Philippines are of the oxide type. Nickel silicate ores are of higher grade, as at Riddle, Oregon (Cumberlidge and Chace, 1968), and New Caledonia. But Figures 17-5 and 17-6 show that up to 1% Ni occurs even in the red and yellow iron oxide layers. In short, both ore types are generally present in a single district and the assignment of all parts of a given district as "oxide" or "silicate" is arbitrary. Both oxide and silicate occurrences must be searched out in any exploration program.

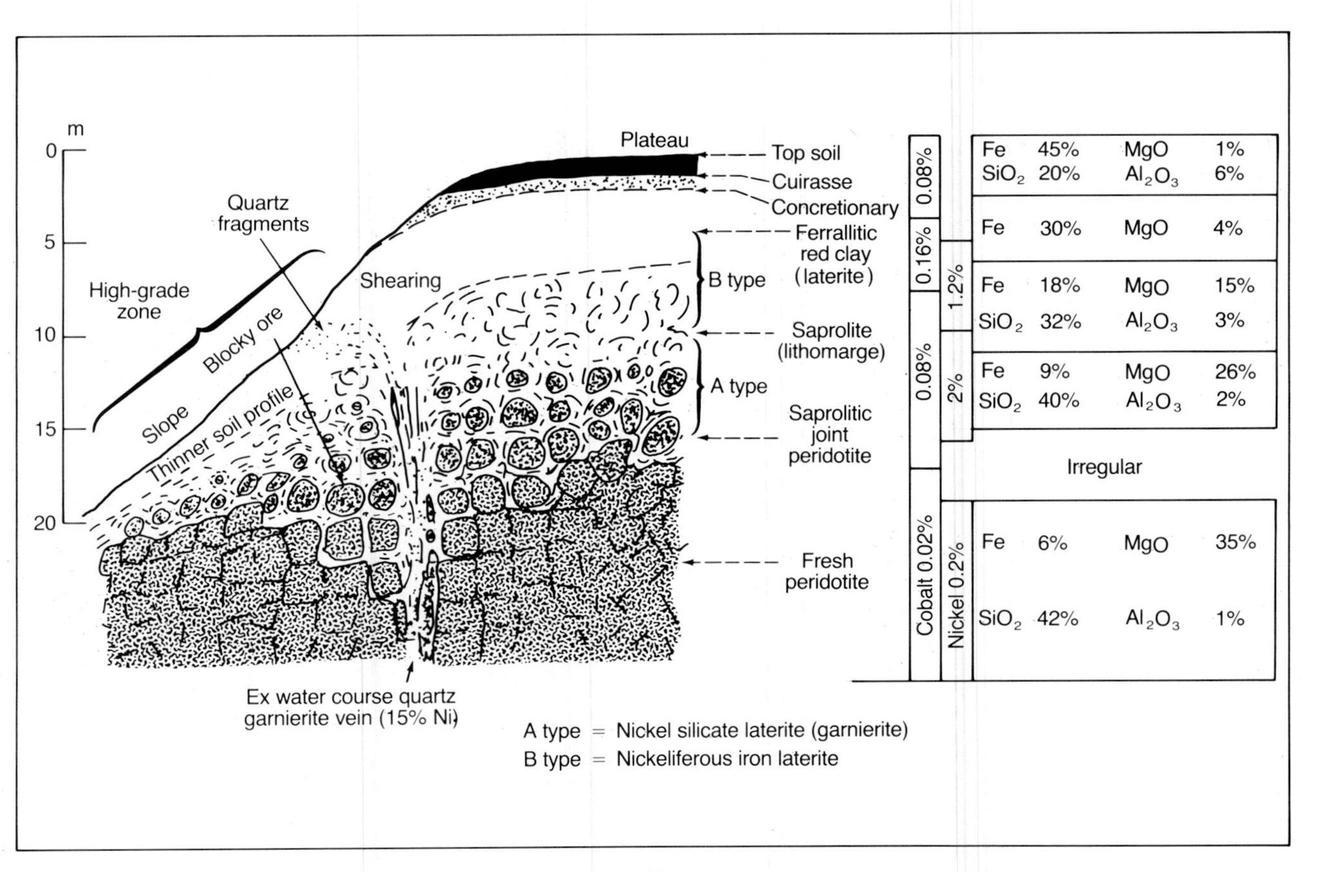

m
0
5
10
15
20
High-grade zone
Quartz fragments
Blocky ore
Slope
Thinner soil profile
Shearing
Plateau
Top soil
Cuirasse
Concretionary
Ferrallitic red clay (laterite)
B type
Saprolite (lithomarge)
A type
Saprolitic joint peridotite
Fresh peridotite
Ex water course quartz garnierite vein (15% Ni)
A type = Nickel silicate laterite (garnierite)
B type = Nickeliferous iron laterite
0.08%
0.16%
0.08%
Cobalt 0.02%
1.2%
2%
Nickel 0.2%
Fe 45% MgO 1%
SiO2 20% Al2O3 6%
Fe 30% MgO 4%
Fe 18% MgO 15%
SiO2 32% Al2O3 3%
Fe 9% MgO 26%
SiO2 40% Al2O3 2%
Irregular
Fe 6% MgO 35%
SiO2 42% Al2O3 1%

(*a*)

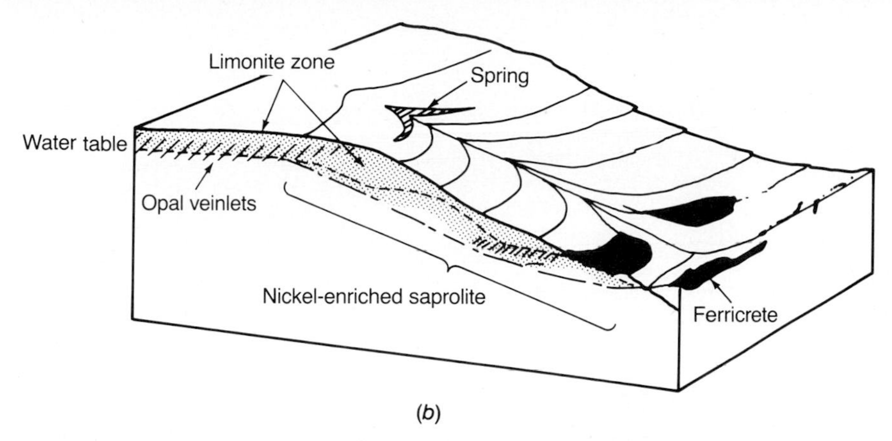

(*b*)

Figure 17-6. (*a*) (*facing page*) Composite profile through a nickel laterite at a plateau margin. (*From* Meillon, 1978.) (*b*) Block diagram of laterite topography and dynamics. (*From* Golightly, 1979.) Both demonstrate topographic-geomorphic controls. Note too that fault water courses can be exceedingly high-grade in nickel.

▸ RESIDUAL MANGANESE DEPOSITS, MORRO DA MINA, BRAZIL

Morro da Mina, in the Lafaiete district, Minas Gerais, Brazil, has been one of the most productive manganese deposits in the western hemisphere. The ore is residual and consists of manganese oxides formed by the weathering of silicate and carbonate basement lithologies. This protore has been called *gondite* (Park, 1956) and *manganese silicate-carbonate protore* (Dorr, Coelho, and Horen, 1956). Dorr and his coworkers state that the protores are metamorphosed sediments that formed a gradational sequence of cherts, mudstones, and manganiferous carbonates. They are in schists of Precambrian age which may be volcanic in origin. The characteristic manganiferous beds have been followed along strike for more than 60 km, and probably continue farther. The irregular distribution of most constituents in the metasediments is compatible with the present erratic distribution of manganoan carbonates in the protores. The sediments have undergone intense regional metamorphism, and locally are invaded by granitic masses and mafic and pegmatitic dikes (Guimaraes, 1935). Igneous metamorphism has been superimposed upon regional metamorphism, particularly near the northwestern part of the mine, a factor that has led observers in the past to relate the mineralization to igneous metamorphic processes. Igneous metamorphism is now thought to have developed patches of the manganoan olivine tephroite (Mn_2SiO_4) and other silicates, mainly at the expense of manganese carbonates.

The mineralogy of the protores was studied intensively by Horen (1955), who found spessartite garnets, rhodochrosite and manganoan calcite, rho-

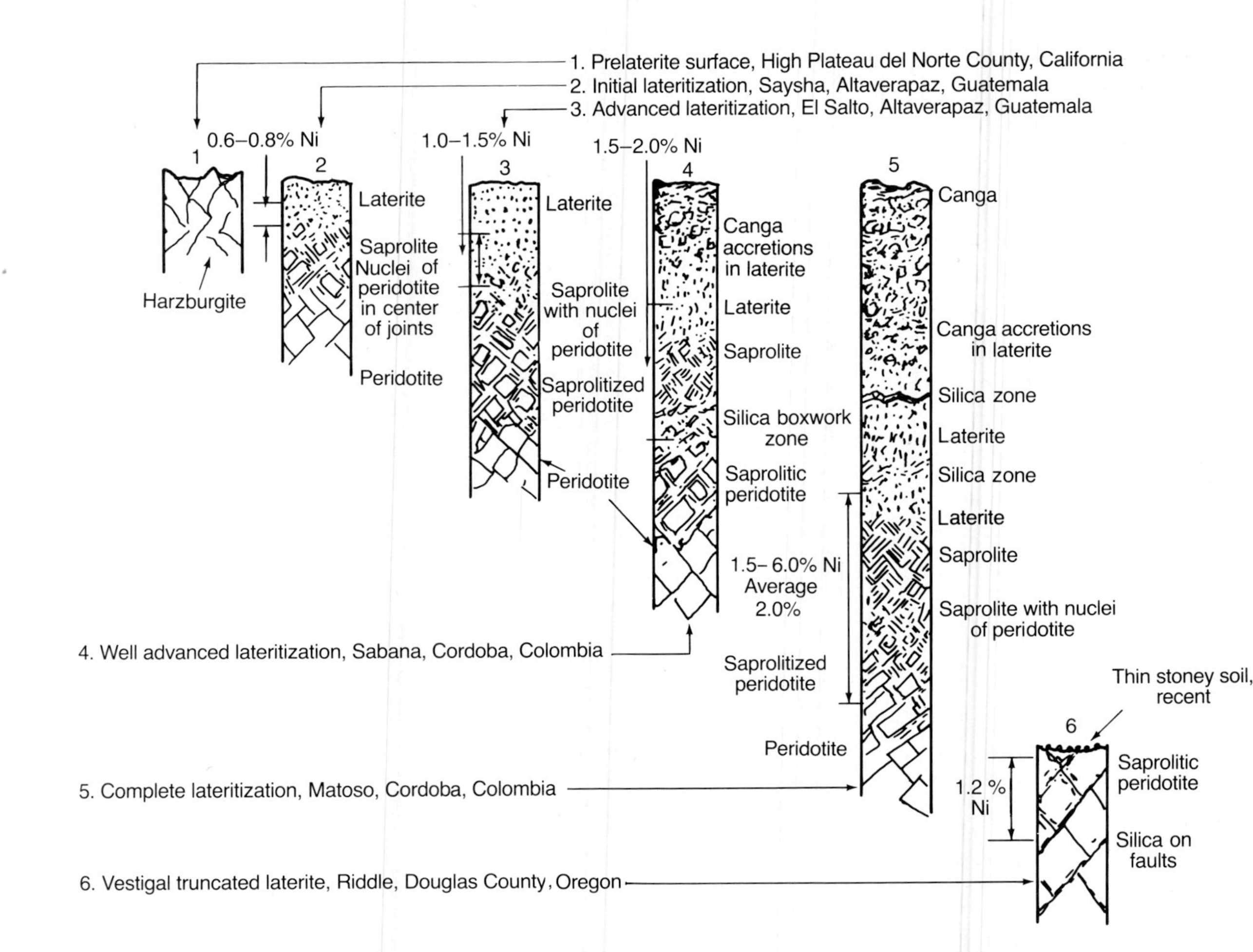

Figure 17-7. Representative profiles of nickel laterites at various stages of development. (*After* Webber, 1972.)

donite ($MnSiO_3$), tephroite, and a long list of minor minerals, particularly manganese silicates. Pyrite and small amounts of other sulfides, notably alabandite (MnS), have been recognized (Park et al., 1951; Odman, 1955; Horen, 1955). Most of the protore is a nondescript, sandy looking, gray-brown rock that contains mainly small crystals of spessartite interspersed in a cement of manganoan carbonate and minor amounts of fine-grained sulfides. In the deeper parts of the mine workings the protore is cut by veins and irregular masses of bright red rhodochrosite, rhodonite, and smaller amounts of other minerals. The protore averages about 30% Mn. In a few places carbonates make up as much as 70% of the rock (Dorr, Coelho, and Horen, 1956), but in many places carbonates are sparse or absent and the rock is massive, dense, and impermeable.

The ore at Morro da Mina consists of a mantle of manganese oxides, products of the removal of silica and carbonate from the protores by weathering. Only in very few places does the ore grade downward into the protore; rather, the contact is usually a sharp, clearly defined, undulating irregular surface that locally extends downward into the protore along fractures, dikes, and water courses. The presence of carbonates and sulfides that are more easily decomposed than the silicates is thought to aid greatly in the weathering of the protores. Where these materials are present, their relatively rapid weathering renders the protore permeable, and decomposition proceeds rapidly and to considerable depths. Ore has been mined to a depth of about 200 meters. Where the carbonates and sulfides are absent in the protores—for example, in those parts of the district where igneous metamorphism has been active—oxidation is limited, and generally the manganese oxides form only a thin film or at most a meter or so of ore. The district has produced a good percentage of the free world's manganese, and deposits like it will continue to be productive into the twenty-first century.

▸ BAUXITE DEPOSITS, JAMAICA

The presence of aluminum in the red soil of Jamaica was recognized as early as 1869 (Sawkins, et al., 1869), but no attention was paid to this until 1942 when R. F. Innes, an agricultural chemist in Jamaica, said that the aluminum content of the samples he analyzed was high enough to warrant exploitation (Hose, 1950). Today, Jamaica is a principal source of aluminum ore (Bramlette, 1947; Schmedemann, 1948; Zans, 1952, 1957, 1959; Hill, 1955; Hose, 1959; Salas, 1959; Kelly, 1961; Blume, 1962; Vincenz, 1964; Comer, 1974), with Surinam and the Little Rock, Arkansas, area on standby. Bauxites are laterites rich enough in Al_2O_3 and low enough in SiO_2 and Fe_2O_3 to be mined as aluminum ore. Jamaica boasts 1.1 billion metric tons of high-grade 55% Al_2O_3, 0 to 5% $SiO_2 + Fe_2O_3$ bauxite.

The oldest rocks of Jamaica are serpentines thought to be of Cretaceous age overlain by amphibole schists and marbles of the Upper Cretaceous Metamorphic Series. Above the Cretaceous layer is a sequence of sandstones, tuffaceous shales that weather deep red, and conglomerates with intercalated layers of limestone. A conglomerate marks the base of the Tertiary layer and is overlain by carbonaceous shales containing sandstone and dark limestone, which in turn are followed by the Tuff Series. The Tuff Series (Figure 17-8) was intruded by andesite porphyry. Both were then eroded, and both were covered by the Yellow Limestone Formation. The White Limestone Formation of Middle Eocene–Lower Miocene age is next in the sequence and is the host rock for the bauxite, which is followed by coastal plain beds of limestone and sand.

The central part of Jamaica is marked by the presence of an anticline known as the Central Inlier. From this central, topographically high anticline the beds dip gently toward the ocean on both sides. Many sink holes have developed in the White Limestone on the flanks of the anticline, and it is in this karst topography that the thickest bauxitic ores are found. Apparently the ore was formed after uplift and deformation, since it fills the karst depressions and blankets the undulating surface of the White Limestone (Zans, 1952).

The Jamaican aluminum ores are soft and earthy, and sometimes shaley and highly porous. They are commonly dark red or yellow brown to creamy and white. At places dark reddish pisolites have been recognized. The bauxite ores contain mainly gibbsite, boehmite, and diaspore; gibbsite is the dominant ore mineral. The ores can be high in iron; generally they carry 18 to 20% iron oxides and are known in the industry as terra rossa. They contain 48 to 50% alumina, 2% titania, and less than 3% silica; combined water makes up about 27%. The iron oxide is hematite or goethite; silica is in clay minerals, largely of the kaolin group. Lower-iron ores typically contain higher alumina. The White Limestone is a chalky, nodular, and well-bedded rock that is dolomitic near its base. It is a pure limestone-dolomite low in silica and iron, but one that many geologists have in the past believed to contain enough impurities—particularly small amounts of volcanic ash—to permit accumulation of the bauxites by weathering since Miocene times (Hartman, 1955; Hose, 1963). The limestone is said to contain on the average no more than 0.2% acid-insoluble material, of which only 0.04% is alumina (Chubb, 1963).

The base of the ore deposits is a sharp irregular contact with the underlying limestones. The orebodies are irregular and range in size from small pockets to basins comprising hundreds of hectares. The average minable thickness is 3 to 10 meters, but thicknesses range from a few centimeters to 40 meters or more.

Ore reserves of Jamaican bauxite are estimated to be several hundred million tons, and the presence of such large deposits on relatively pure carbonate rock has led many geologists to question whether or not the ore could have formed by simple weathering of the limestone. Burns (1961),

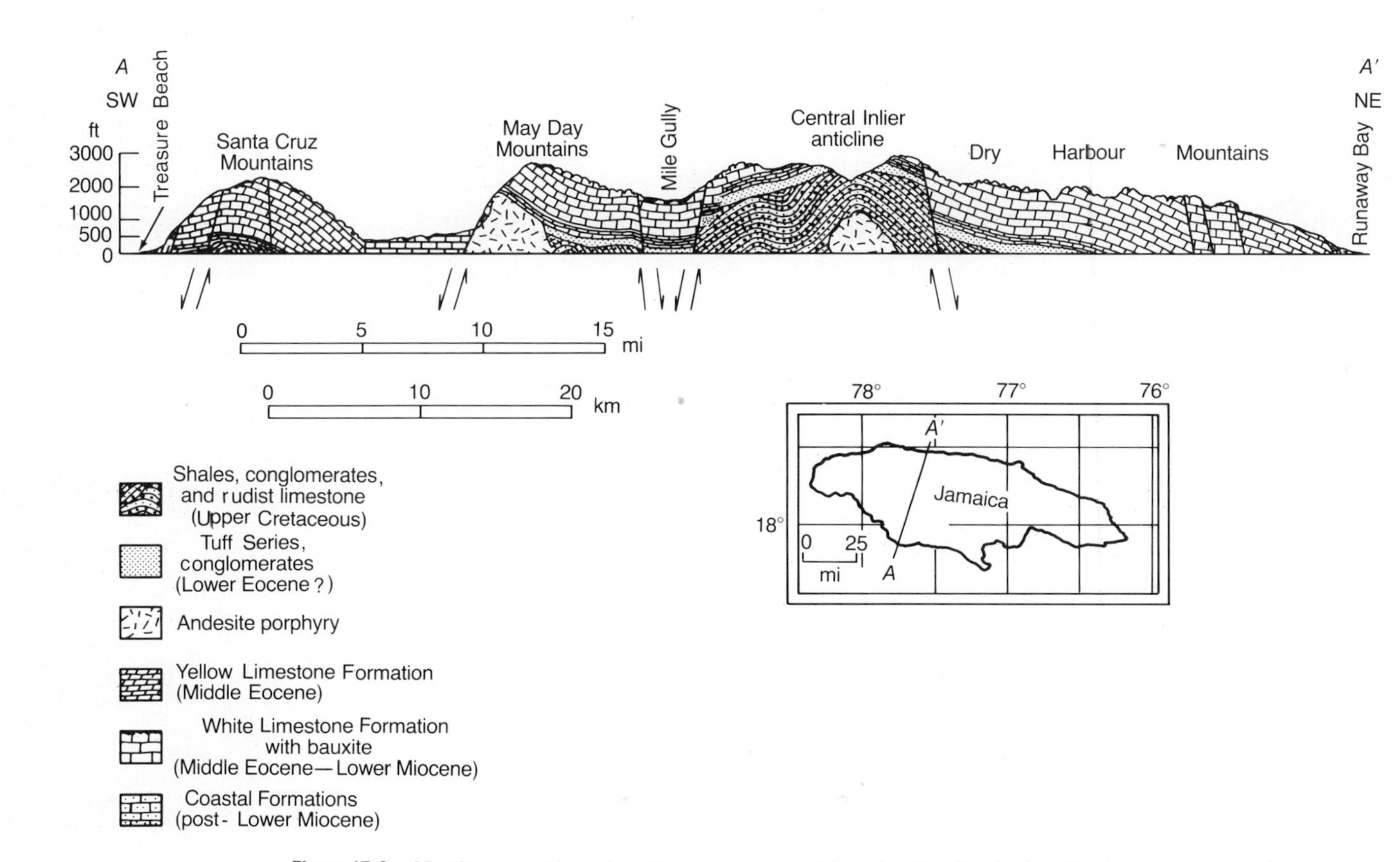

Figure 17-8. Northeast-southwest section across central Jamaica (inset) showing the relations of the bauxite-draped White Limestone Formation with the older strata. Bauxite is exaggerated in black. Note that it occurs only on the White Limestone, in karst solution features and as a soil cover mantle. Vertical exaggeration. 7×. (*From* Zans, 1952.)

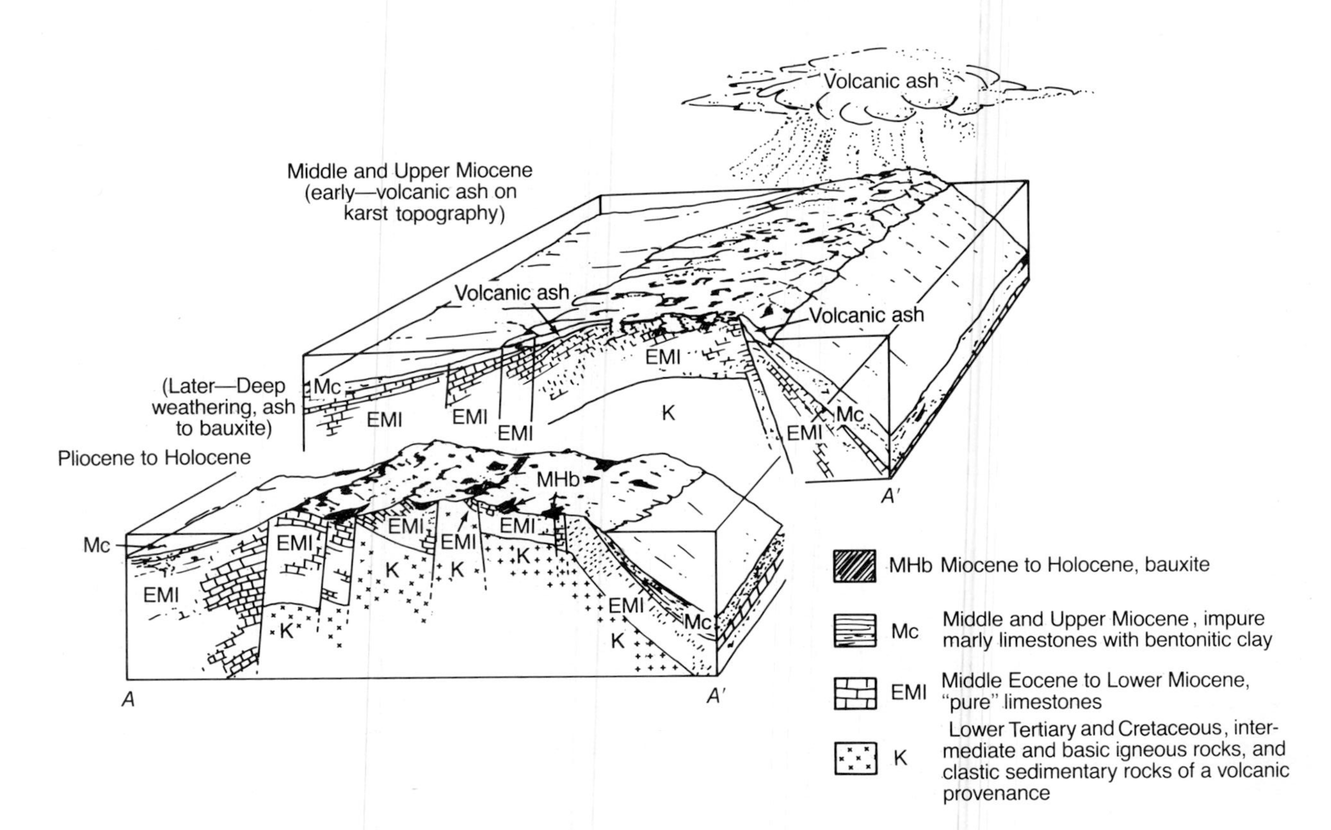

Figure 17-9. Illustration of concepts of Comer and others of the origin of Jamaica aluminum ores. The plane A–A' is nearly the same as that in Figure 17-8. (*After* Comer, 1974.)

who assumed that no loss of ions occurred during weathering, said that an enrichment of 11,600 would be required to produce bauxite of economic grade from the White Limestone, that this factor was unreasonable, and that the White Limestone was not the source of the ore. Waterman (1962) said that a greater thickness than that of the entire White Limestone must have been eroded to make possible the bauxite deposits. He said that the Eocene Tuff Series may have been the source of the ore. Chubb (1963) adopted the idea advanced by Zans (1956) that the source of the bauxites is not the limestone with which they are associated, but the older aluminous silicate rocks of andesitic composition, the Upper Cretaceous, and, in part, the early Eocene andesites and andesitic pyroclastics. Drainage from the central anticline would spread outward through the limestones, and erosion and stream flow would deposit silt and aluminous debris in the pools formed in the karst depressions. On the other hand, Sinclair reaffirmed from trace element analyses of both the White Limestone and the bauxite that the limestone residue was the source of most of the bauxite (Sinclair, 1967).

Comer (1974) stated that the Jamaican bauxites formed from the weathering of Miocene bentonitic volcanic ash (Figure 17-9). A thickness of about 5 meters of ash would be required to give the known amounts of bauxite, and thicknesses of this magnitude are present in the White Limestone along the northern coast of Jamaica. According to Comer, the volcanic ash altered to bauxite by the desilication of glass, plagioclase, biotite, and ferromagnesian minerals. The insoluble residue of Middle Eocene to Lower Miocene host rocks and alluvium from Cretaceous and Lower Tertiary aluminous rocks contributed very little to the bauxite deposits and their associated soils. The most important control on the genesis and distribution of bauxite was the rapid and efficient drainage through the karst topography that leached and removed the silica. His ideas appear correct.

Norton (1973) and Meillon (1978) provide excellent accounts of the chemistry of bauxite formation. As indicated, the alumina and aluminum hydroxides gibbsite [$Al(OH)_3$] and boehmite and diaspore ($AlO{\cdot}OH$ or $Al_2O_3{\cdot}H_2O$) can be derived by complete silica removal from aluminosilicates. Typically feldspars and feldspathoids in syenites yield to kaolin, which is in turn desilicated to bauxite. Iron is also commonly stripped, the more the better. Clearly the ultimate composition of a deeply weathered temperate or tropical soil will depend on original substrate rock type and composition—you cannot make a nickel laterite from a syenite or a bauxite from serpentine—and on the soil-water chemistry. Bauxite requires that both iron and silica be leached, but not alumina (Figure 17-10); iron laterites imply solubility of aluminum, if present, and silica, but not iron; and conventional laterites may involve only desilication. Norton (1973) showed that alumina is most soluble from pH 5 to 7. Iron is leached only at low pH, low Eh, or both, and silica is increasingly soluble at higher pH. The relationships are summarized in Figure 17-10.

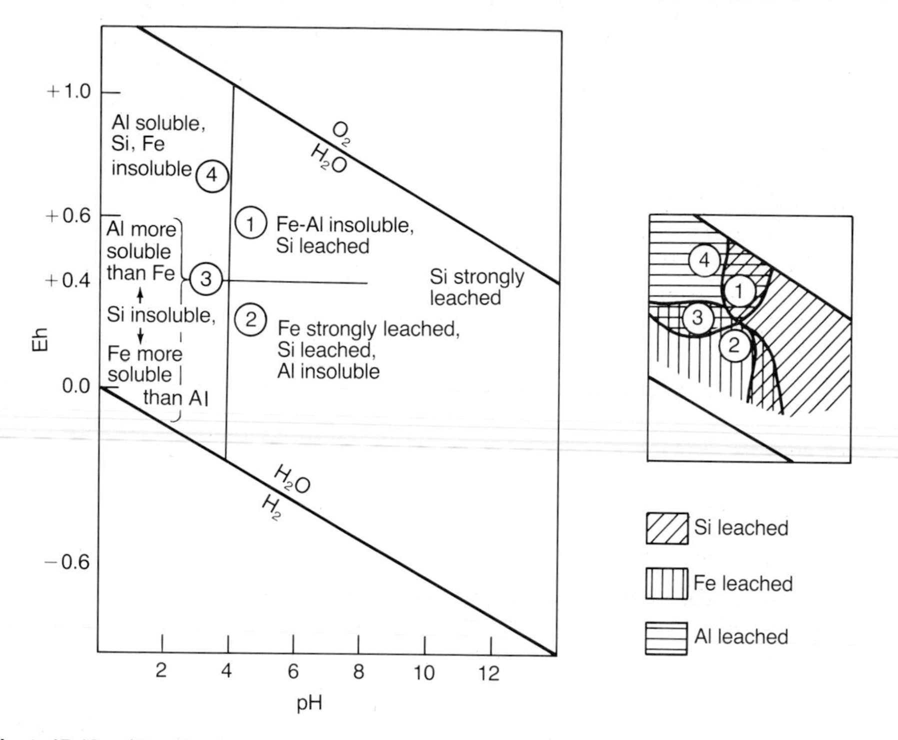

Figure 17-10. Eh-pH relations with respect to bauxite and laterite formation. 1 = environment of laterite oxide-hydroxide formation; 2 = bauxite field; 3 = podzol soil field; 4 = high-iron laterites. (*After* Norton, 1973, Fig. 3.)

▸ SUPERGENE SULFIDE ENRICHMENT

The phenomenon of supergene sulfide enrichment is a remarkable special case of weathering. It depends on a starting material that is porous and permeable to meteoric waters, that contains abundant pyrite to produce oxidizing acids, that contains acid-soluble ore-metal-bearing minerals, and that is underlain by a precipitative environment. Although several elements can be shown to be redistributed, copper is the principal one, and the following discussion pertains primarily to that metal. Low-grade materials that could not originally have been mined can have had their copper content leached and reprecipitated at lower levels in a layer enriched by several factors, and thus constitute orebodies. Many of the mined porphyry copper deposits of the world, such as Morenci and Ray, Arizona, and El Salvador, Chile, owe their economic viability totally to this phenomenon. In many, a profound color contrast between leached, oxidized red-brown material above and gray to white ore below is found (Figures 17-11 and 17-12).

Many metals and ions that are soluble in acidic oxidizing waters above the water table are precipitated where the solutions become more basic and

Figure 17-11. View across the pit at Chuquicamata, Chile (Chapter 11), showing the darker iron-hydroxide-stained leached and oxidized zone (medium gray) above the lighter supergene-enriched zone (light gray to white). The leached zone is waste, containing only a few parts per million copper, while the ore zone averages over 1.2% (12,000 ppm); both contained about 0.7% (7000 ppm) before the supergene enrichment process began. The benches are 50 meters high, so the leached zone is 300 to 400 meters thick. The ore train, for scale, is 59 cars long and is carrying 9000 tons of ore. See also Figure 11-13.

reducing at and below the water table. Accordingly some of the metals dissolved near the surface are precipitated below the groundwater table (Emmons, 1917; Garrels, 1954). This process is important to economic geology and the mining industry because metals leached from the oxidized upper parts of mineral deposits can be redeposited and concentrated at depth. Even more important, this process provides a mechanism by which a small percentage of metal can be leached from a large volume of rock and—provided the conditions below are favorable—can be redeposited as a higher grade deposit in a smaller volume of rock. This process is known as *supergene enrichment.* The primary, subeconomic material underlying the enriched zone—and presumably at one time occupying the leached and the enriched volumes above—is known as *protore.* So these are three fundamental zones to be considered: the *oxidized zone*, the *supergene-enriched zone*, and the *hypogene zone* or *protore*. It must be noted that the coincidence of the threshold between oxidizing conditions above and reducing conditions below has traditionally been taken as the water table. That assumption is probably correct in most instances, and is used here for that reason. But there is considerable evidence in many deposits that the transition can be merely an overall buffering of downward percolating waters

Figure 17-12. View across the Bingham Canyon porphyry copper-molybdenum-gold open-pit mine showing how the form of leached capping—the dark rind just below the skyline—mimics the surface topography. The pit is more than 3 km across; the three horizontal black lines near the center are ore trains. See also Figures I-2, 11-26, and 11-27.

by the chemical mass of the host rocks, not necessarily coincident with the water table.

Figures 17-13, 17-14, and 17-15 show these zones. The oxidized zone from which copper has been stripped and in which all remaining components are fully oxidized is the outcropping surface; the exploration techniques considered hereafter deal only with this. The upper few meters are usually more ferruginous, more massive, and more richly red brown than the oxidized material below, and constitute the *gossan*, or "red thumbprint," or "iron hat." As shown, the leached capping and gossan may extend over several square kilometers.

In this section we first set the stage by examining some deposit cross sections and diagrams to become familiar with the scale, vocabulary, and general dynamics of supergene enrichment, generally with reference to porphyry copper deposits. We then look at some of the chemical reactions that prevail, and the minerals that are produced in porphyries and other deposit types. We then consider briefly the exploration significance and techniques of evaluating supergene effects, and we close with geologic descriptions of copper enrichment at Globe, Arizona, and silver enrichment at Chañarcillo, Chile. The first-time reader is urged to look ahead to Figures 17-13 through 17-17 and Tables 17-2 and 17-3.

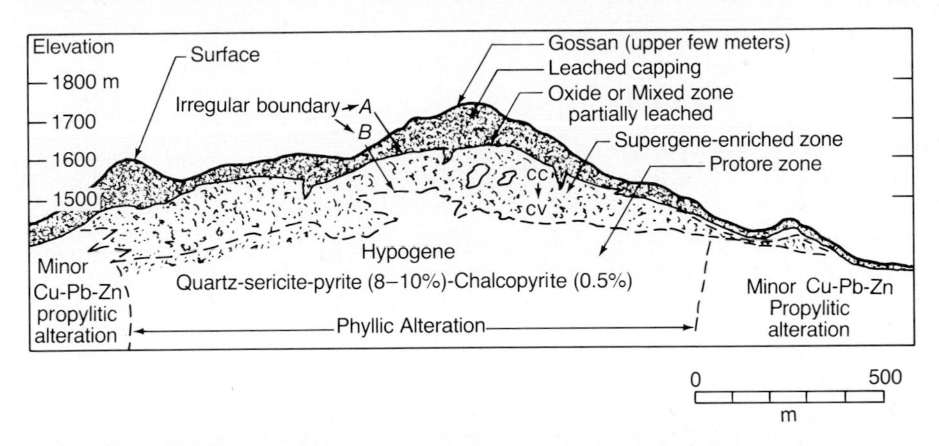

Figure 17-13. Sketch of a supergene enrichment blanket based on a profile at La Caridad, Sonora, Mexico, by Saegart, Sell, and Kilpatrick (1974). Notice that the blanket—an inverted saucer—is a subdued expression of the topography along the upper surface *A*, also presumed to have been the water table. Both *A* and *B* are irregular as functions of porosity, mineral distribution, and local faulting. Hypogene alteration may diminish abruptly at depth as it does at La Caridad or continue downward to a potassic zone as it does at Morenci, Arizona, and most other deposits.

General Aspects

As noted earlier, the lower boundary of the oxidized zone is thought to represent the water table level when enrichment occurred. The base of that zone, then, is commonly a subdued expression of local topography, as it is at La Caridad, Sonora, Mexico (Figure 17-13). It might be expected that a mixed zone existed above that contact as the descending oxidizing rain waters began to "feel" the buffering effect of the reducing zone below. Such mixed, partially leached, partially redeposited zones are common both for the reason suggested and because of tectonic or hydraulic fluctuations of the water table. Precipitation of oxidized mineral species here may substantially enrich a zone of mixed oxide or mixed oxide–relict sulfide assemblages (Bateman, 1950). But at depth or beneath the water table, stagnation and reduction chemically reduce the groundwaters and shift them into the stability fields of the sulfides. Actual pH shift at that reduction boundary in a modern system has been directly measured as shown in Figure 17-14 (Titley, 1978). The sulfides interact, and copper leached above is reprecipitated, commonly producing manifold enrichment. Chalcocite contains 79.8% Cu and covellite 66.4%, so a little of either can be significant. Chalcocite generally is at the top of the blanket where Cu^{+2}/HS^- ratios are higher, covellite at depth where the blanket becomes less enriched and where Cu^{+2}/HS^- is lower. The two formulae (Cu_2S and CuS) reflect the difference. The blanket normally peters out with depths below a few tens of meters of thickness, but traces of secondary chalcocite may be encountered sporad-

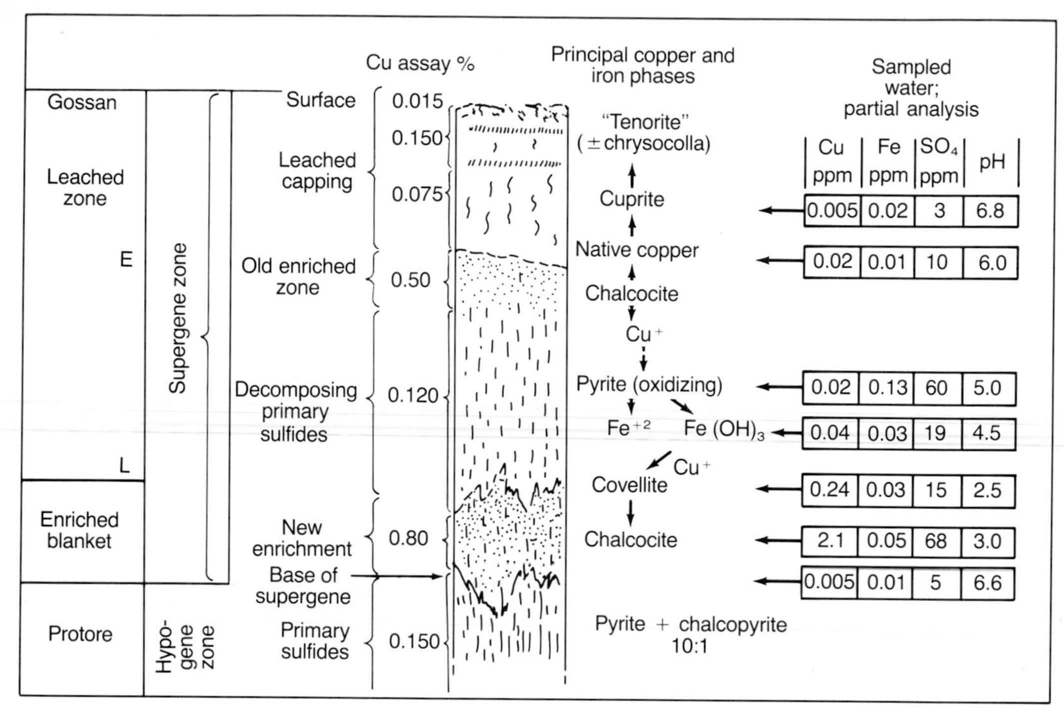

Figure 17-14. Profile of modern, ongoing supergene enrichment near the Plesyumi porphyry copper prospect, New Guinea. (*After* Titley, 1978.) Titley collected waters flowing from the different levels in a ravine and measured pH on the spot. Compare with Figure 17-13. This profile suggests that two water tables have prevailed. An early one at *E* controlled the position of the old enriched zone, and a later one at *L* localized the present blanket. The drop of the water table permits current oxidation-destruction of the old blanket.

ically and sparsely hundreds of meters down. Kaolin and chlorite minerals generally replace feldspars and micas in both the oxidized and the enriched zones, but exceptions are known. Laterally, supergene enrichment stops at the outer edge of phyllic alteration. Beneath the blanket, a return to protore textures and compositions is noted. At some deposits, like Bingham Canyon, Utah, and El Teniente, Chile, the protore is rich enough to be mined. At most others—like Morenci, Arizona, and La Caridad, Sonora—it carries only 0.10 to 0.15% Cu and cannot now be mined.

Figure 17-13 shows supergene "overprint" on only a hypogene phyllic

Figure 17-15. (*Facing page*) Idealized results of the interaction of hypogene and supergene alteration at a porphyry copper deposit. The enriched blanket is deepest, thickest, and richest where pyrite content and permeability are highest, and would be annular in plan view around an ideal deposit like Bajo la Alumbrera, Argentina. At Ray, Arizona, the low-pyrite impermeable center of symmetry called *silicate ore* with reference to its chrysocolla content has been shown to be a potassically altered core as sketched here. A Cordilleran Vein would weather or enrich in a qualitatively similar fashion. All aspects of this diagram are repeated symmetrically around the core.

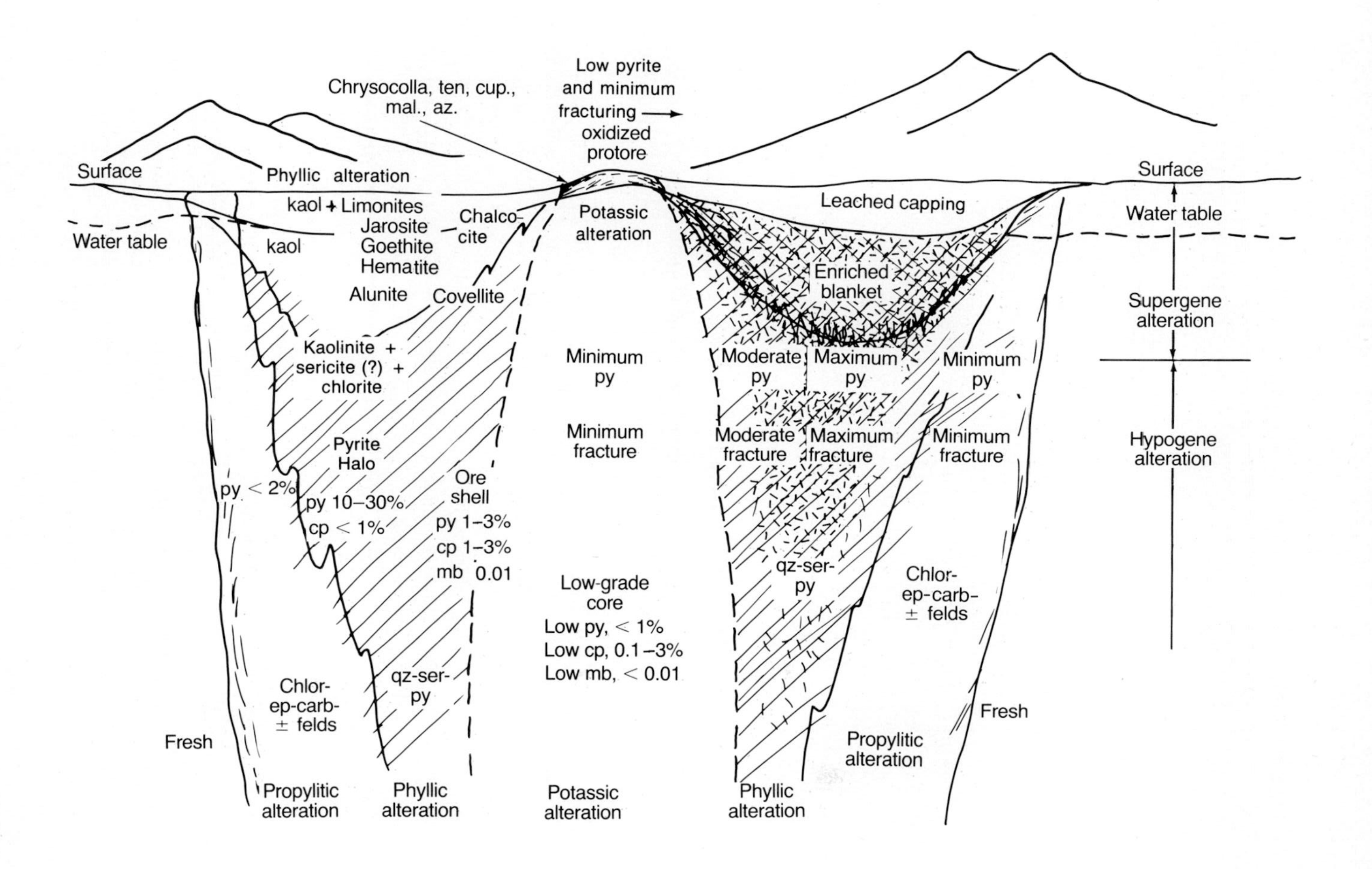
Chrysocolla, ten, cup., mal., az.
Low pyrite and minimum fracturing → oxidized protore
Surface
Phyllic alteration
kaol + Limonites
Jarosite
Goethite
Hematite
Water table
kaol
Chalco- cite
Alunite
Covellite
Kaolinite + sericite (?) + chlorite
Pyrite Halo
py < 2%
py 10–30%
cp < 1%
Ore shell
py 1–3%
cp 1–3%
mb 0.01
Chlor- ep-carb- ± felds
qz-ser- py
Fresh
Propylitic alteration
Phyllic alteration
Potassic alteration
Minimum py
Minimum fracture
Low-grade core
Low py, < 1%
Low cp, 0.1–3%
Low mb, < 0.01
Potassic alteration
Leached capping
Enriched blanket
Moderate py
Maximum py
Minimum py
Moderate fracture
Maximum fracture
Minimum fracture
qz-ser- py
Chlor- ep-carb- ± felds
Fresh
Propylitic alteration
Phyllic alteration
Surface
Water table
Supergene alteration
Hypogene alteration

zone. What if a different level through the schematic PCD cross section of Figure 11-4 is considered? Figure 17-15 is a composite view at a level "lower" than La Caridad so that potassic alteration crops out. It was said earlier that the requirements of supergene processes include high pyrite content, maximum permeability, and surficial, oxygenated waters. A review of Figure 11-4 shows these conditions to be optimal in the "pyrite halo" of the phyllic zone, and it is here that supergene processes are best developed (Figure 17-15). Notice that almost no leaching or dissolution of copper occurs over potassic alteration; here the so-called "copper oxide" minerals like chrysocolla, cuprite, malachite, and others form.

Table 17-2 provides further insight. It implies that the oxidized-leached zone is one of hydrolysis, hydration, and feldspar and mica decomposition to predominant kaolinite as a new phase; relict sericite can still be abundant. Most sulfide minerals are less stable in the zone of weathering than are the rock-forming silicate, oxide, and carbonate minerals. Consequently the sulfides are oxidized, dissolved, or otherwise altered more readily than the surrounding host rocks, leaving only surface exposures of the weathered products. The limonite minerals hematite (Fe_2O_3), goethite [$HFeO_2$ or $Fe(OH)_3$], and jarosite [$K(Fe^{+3},Al)_3(SO_4)_2(OH)_6$] all imply slightly different conditions (Anderson, 1981). Other entries follow from these relationships.

Geochemistry and Mineralogy

We now turn to the geochemistry of supergene processes. Geochemical relations in supergene processes are illustrated by diagrams such as those of Garrels (Figure 17-16; Garrels and Christ, 1965). The fundamental process is oxidation, but since the oxidation of sulfides produces hydrogen ion and sulfate, both pH and Eh are controls. The actual mechanisms of oxidation and dissolution of metal sulfides are not well understood. The sulfide minerals no doubt dissolve ionically under the attack of surface waters, and the constituents then react with dissolved atmospheric oxygen. The oxidation reaction would be much more sluggish if the materials were not first broken down into the ionic state. Sato (1960) suggested that the oxidation agent is aqueous H_2O_2, which can form as an intermediate product during the reduction of oxygen. If the oxidation potential of the environment exceeds that of the $H_2O_2-O_2$ couple, any metal ions present react with the peroxide to form the metal oxide and water. Both theoretical and experimental evidence substantiate this. Moreover, field measurements of the Eh and pH values in the zone of oxidation fall into a narrow zone just above the standard potential for the $H_2O_2-O_2$ couple (Figure 17-17).

Pyrite is the most common hypogene sulfide; few sulfide ore deposits lack it, and supergene processes start with it. The oxidation of pyrite and other iron sulfides generally leaves limonite or hematite and generates sulfuric acid. The normal reaction involved is thought to be

$$\underset{\text{pyrite}}{2FeS_2} + 7O_2 + 2H_2O \rightarrow 2FeSO_4(aq) + 2H_2SO_4(aq) \qquad (17\text{-}5)$$

TABLE 17-2. Elemental, Mineralogic, and Textural Changes in Supergene Enrichment Zones

	Elements or Compounds		Minerals		
	Gained	Lost	Gained	Lost	Textures
Oxidized-leached zone	H_2O 75% Fe* 50%	SiO_2 10% Al_2O_3 15% K_2O 15% S 30% Cu 95% Zn 100%	Hematite Goethite Jarosite Alunite Kaolinite	Pyrite Chalcopyrite Bornite Sphalerite Tecto and phyllosilicates Sericite Carbonates	Boxworks and impregnations of limonite produce rust-ochre-red colors. Cubic pyrite cavities with limonite walls and fences produce egg-crate relicts. Botryoidal hematite after chalcocite.
Water table (?)					
Enrichment zone	H_2O 50% Al_2O_3 20% S 100% Cu 20%	Zn Fe	Kaolinite Chlorite? Sericite Chalcocite Covellite Bornite? Chalcopyrite? Acanthite	Pyrite Chalcopyrite Sphalerite Carbonates	Chalcocite >> covellite replacing pyrite as rims, films, and along fractures. Massive chalcocite rare; chalcocite-covellite restricted to pyrite kaolinite veinlets and relict sericite, generally bright white.
Protore zone	$Zn^{+2}(aq)$ $Fe^{+2}(aq)$				Stockwork of veins, veinlets, and disseminations of pyrite-chalcopyrite in quartz-sericite.

A tabulation of gains, losses, and textures accompanying normal supergene enrichment of a phyllic zone. Percentages are only for relative importance and are approximate. CaO and MgO are generally absent from phyllic protore.

*A relative gain by virtue of loss of other elements; none introduced.

Source: Based in part on Koenig (1978).

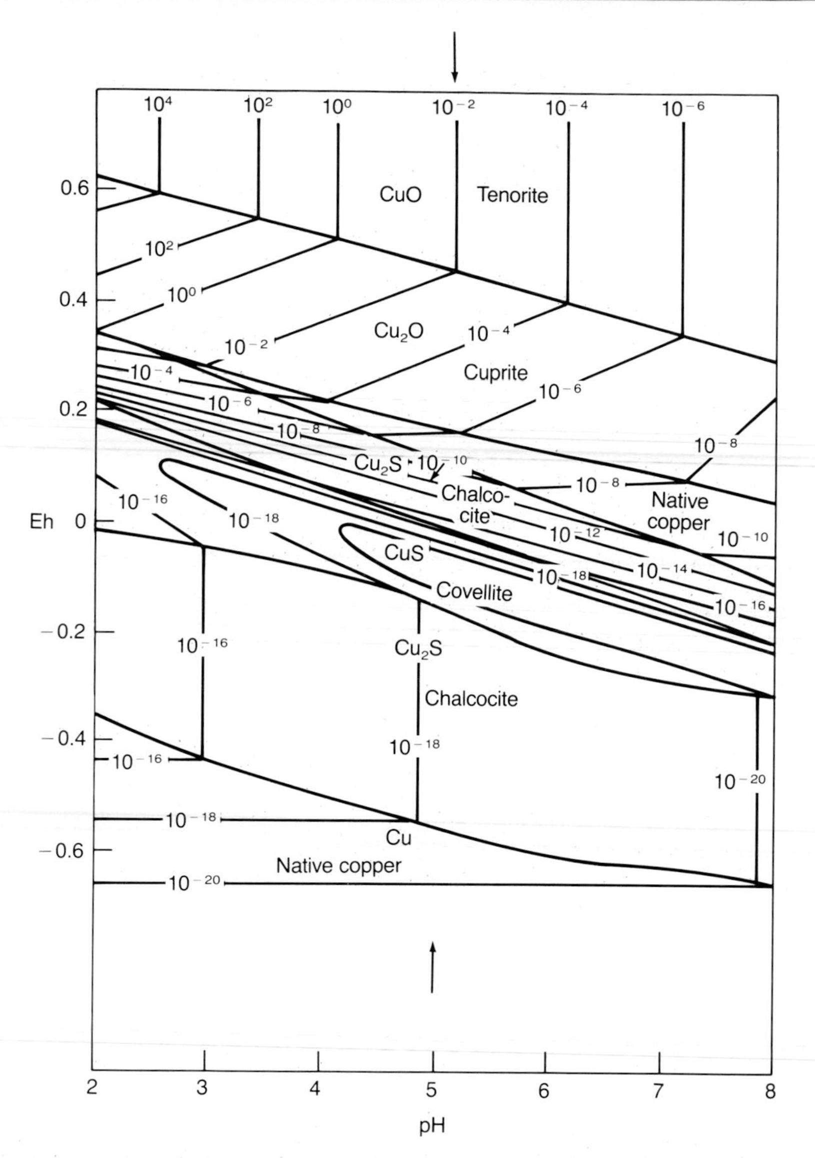

Figure 17-16. Stability relations of some copper minerals at surface conditions and at low $f(CO_2)$ as a function of Eh and pH. Contours describe ($aCu^+ + aCu^{+2}$). Compare a vertical section through this diagram at pH = 5 (arrows) with the mineralogy at Plesyumi in Figure 17-14. See also Figure 17-17 for applications to environment and geology.

Iron is dissolved as the ferrous ion or as ferrous sulfate, and it is then oxidized to ferric sulfate:

$$2FeSO_4(aq) + H_2SO_4(aq) + 0.5O_2 \rightarrow Fe_2(SO_4)_3(aq) + H_2O \qquad (17\text{-}6)$$

Pyrite can also be converted directly to hematite by

$$\underset{\text{pyrite}}{2FeS_2} + 7.5O_2 + 4H_2O \rightarrow \underset{\text{hematite}}{Fe_2O_3} + 4H_2SO_4(aq) \quad (17\text{-}7)$$

or

$$2Fe^{+2}(aq) + 0.5O_2 + 2H_2O \rightarrow \underset{\text{hematite}}{Fe_2O_3} + 4H^+ \quad (17\text{-}8)$$

The iron in massive pyrite bodies (Chapter 13) is likely to be leached and flushed out without leaving much hematite or limonite because the presence of sulfuric acid keeps the pH low and may form a reducing environment that retains the iron in the soluble ferrous state. Conversely, waters rich in oxygen may alter pyrite directly to ferric sulfate, without going through a ferrous sulfate stage:

$$\underset{\text{pyrite}}{2FeS_2} + 7.5O_2 + H_2O \rightarrow Fe_2(SO_4)_3 + H_2SO_4 \quad (17\text{-}9)$$

The Fe^{+2}/Fe^{+3} ratio may vary widely, depending upon both Eh and pH. The kinds of materials that form under set conditions of Eh and pH depend upon relative concentrations and the extent to which various solubility products are satisfied. Compare Equation (17-7) with Equation (17-5)—more water changes the reaction products.

Some iron persists at the outcrop as ferric or ferrous sulfate in arid climates, but generally these sulfates are merely temporary transitional states. Iron migrates in the dissolved state where the solutions remain strongly acidic or deficient in oxygen; a low redox potential favors the stability of ferrous ions in solution, whereas a pH below 3 permits ferric ions to remain dissolved. However, such low redox potentials and extremely acidic conditions are unlikely near the surface and above the water table.

The weathering of iron sulfides involves the oxidation of both iron and sulfur. Sulfur in sulfide minerals generally has a valence of -2. Although the sulfur in FeS_2 would seem to be an exception, it is not, because it occurs as $S_2^{=}$ "molecules" bound to divalent iron. Oxidation to the sulfate radical increases the valence to $+6$. As an intermediate step, oxidation can form native sulfur (S^0). For example, ferric sulfate solutions react with chalcocite to release copper ions in solution and produce native sulfur (Sullivan, 1930; Sato, 1960):

$$\underset{\text{chalcocite}}{Cu_2S} + 4Fe^{+3} + 6SO_4^{=} \rightarrow 2Cu^{+2} + 4Fe^{+2} + 6SO_4^{=} + \underset{\text{native sulfur}}{S^0} \quad (17\text{-}10)$$

Small amounts of native sulfur are found in many leached outcrops.

It is crucial that however the reactions are written, the components of the strong oxidizers sulfuric acid (H_2SO_4) and ferric sulfate [$Fe_2(SO_4)_3$] are abundantly produced by the attack on pyrite by oxygenated rainwater. Ferric sulfate and sulfuric acid generated by the oxidation of iron sulfides act as potent solvents for other metallic sulfides. Ferric sulfate oxidizes

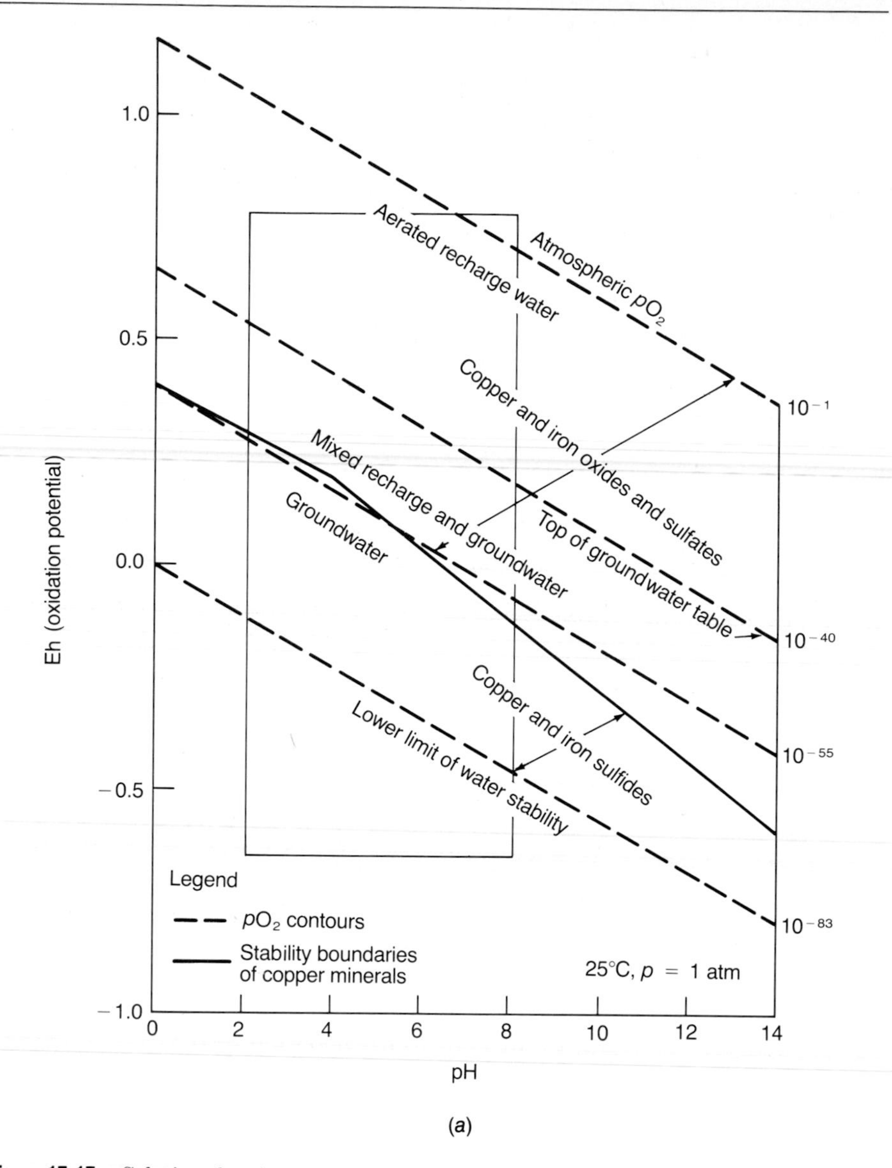

Figure 17-17. Solution chemistry of Figure 17-16 as reflected in (*a*) environments and (*b*) (*facing page*) mineral stabilities. The outline of the area covered by Figure 17-16 is shown by the rectangle. CO_2 is absent from the system of Figure 17-16, and malachite preempts tenorite in this figure. (*From* Anderson, 1981.)

sulfide minerals to soluble sulfates and is itself reduced to ferrous sulfate. At the expense of hydrogen ions and oxygen, ferrous sulfate probably is oxidized quickly back to ferric sulfate:

$$2Fe^{+2} + 2H^{+} + 0.5O_2 \rightarrow 2Fe^{+3} + H_2O \tag{17-11}$$

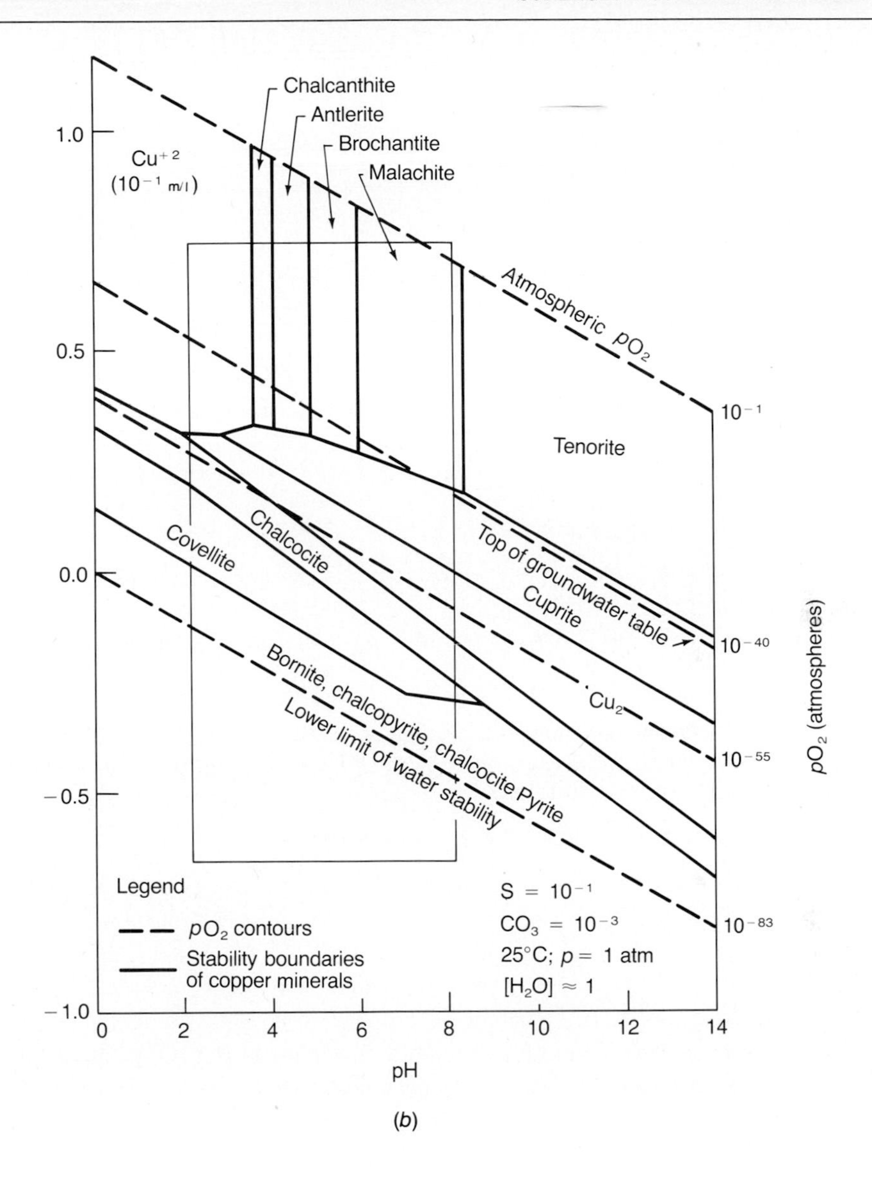

(*b*)

Wherever the acidity of the waters percolating through an ore deposit is maintained, such elements as copper, silver, and zinc are leached. In the absence of iron sulfides, ferric sulfate and sulfuric acid do not form readily, and the oxidation products of other sulfides tend to remain in place. Similarly the neutralization of acidic waters by reactive wall rocks such as carbonates causes dissolved metals to precipitate as stable minerals. Equations (17-12) to (17-34) demonstrate the oxidation of a few common sulfide minerals. Solid reactants and relatively insoluble products are written as stoichiometric compounds; dissolved substances are expressed as ions.

$$\underset{\text{chalcopyrite}}{2CuFeS_2} + 8.5O_2 + 2H_2O \rightarrow Fe_2O_3 + 2Cu^{+2} + 4SO_4^{-2} + 4H^+ \qquad (17\text{-}12)$$

$$\underset{\text{chalcopyrite}}{CuFeS_2} + 8Fe_2(SO_4)_3 + 8H_2O \rightarrow CuSO_4 + 17\,FeSO_4 + 8H_2SO_4 \qquad (17\text{-}13)$$

Equations (17-12) and (17-13) show two reactions involving chalcopyrite, which may break down directly to iron oxide and an acidic solution of cupric sulfate, or which may be taken into solution by ferric sulfate. The sulfuric acid generated in these reactions or generated by oxidizing iron sulfides helps take the copper into solution as cupric sulfate. Bornite behaves similarly:

$$\underset{\text{bornite}}{2Cu_5FeS_4} + 13.5O_2 + 8H_2O \rightarrow Fe_2O_3 + 10Cu^{+2} + 8SO_4^{=} + 16H^+ \qquad (17\text{-}14)$$

$$\underset{\text{bornite}}{Cu_5FeS_4} + 4Fe^{+3} + 4H_2O + 6O_2 \rightarrow 5Cu^{+2} + 5Fe^{+2} + 4SO_4^{=} + 8H^+ \qquad (17\text{-}15)$$

$$\underset{\text{bornite}}{Cu_5FeS_4} + 4H^+ + 9O_2 \rightarrow 5Cu^{+2} + Fe^{+2} + 4SO_4^{=} + 2H_2O \qquad (17\text{-}16)$$

The simple hypogene sulfides that do not contain iron can be oxidized directly or dissolved by ferric sulfate and sulfuric acid produced by the oxidation of associated iron sulfides:

$$\underset{\text{chalcocite}}{Cu_2S} + 2H^+ + SO_4^{-2} + 2.5O_2 \rightarrow 2Cu^{+2} + 2SO_4^{=} + H_2O \qquad (17\text{-}17)$$

$$\underset{\text{chalcocite}}{Cu_2S} + 2Fe^{+3} + 3SO_4^{=} + 1.5O_2 + H_2O \rightarrow 2Cu^+ + 2Fe^{+2} + 4SO_4^{=} + 2H^+ \qquad (17\text{-}18)$$

Native copper is commonly associated with cuprite (Cu_2O) in the upper parts of oxidized copper deposits. In fact, the association is so universal that cuprite without disseminated native copper is rare. The reaction producing native copper and cuprite from chalcocite may be written

$$\underset{\text{chalcocite}}{2Cu_2S} + 8Fe^{+3} + 12SO_4^{=} + 6H_2O + 1.5O_2 \rightarrow \underset{\text{native copper}}{2Cu^0} + \underset{\text{cuprite}}{Cu_2O} + 8Fe^{+2} + 12H^+ + 14SO_4^{=} \qquad (17\text{-}19)$$

The further oxidation of native copper and cuprite can produce tenorite (CuO):

$$\underset{\text{native copper}}{Cu^0} + \underset{\text{cuprite}}{Cu_2O} + O_2 \rightarrow \underset{\text{tenorite}}{3CuO} \qquad (17\text{-}20)$$

or the cupric ions can react with water, also forming tenorite:

$$Cu^{+2} + H_2O \rightarrow \underset{\text{tenorite}}{CuO} + 2H^+ \qquad (17\text{-}21)$$

In the presence of limestone (Figure 17-18*d*), cupriferous solutions react with CO_2 derived from carbonates to form the hydrous carbonates of copper, malachite and azurite, by the reactions

$$\underset{\text{tenorite}}{2CuO} + CO_2 + H_2O \,[\text{or } 2Cu^{+2} + CO_3^{=} + 2(OH)^-] \rightarrow \underset{\text{malachite}}{Cu_2(OH)_2CO_3} \qquad (17\text{-}22)$$

$$\underset{\text{tenorite}}{3CuO} + 2CO_2 + H_2O \,[\text{or } 3Cu^{+2} + 2CO_3^{=} + 2(OH)^-] \rightarrow \underset{\text{azurite}}{Cu_3(OH)_2(CO_3)_2} \qquad (17\text{-}23)$$

in which it is seen that malachite is favored in more dilute, less copper-rich waters or by lower pCO_2. In dehydrating systems, azurite might be expected to be later, and generally is (Beane, 1968). In extremely arid terrain such as the Atacama Desert of Chile, water-soluble minerals—sulfates and chlorides of copper and iron—can remain in the zone of oxidation.

In oxidizing and acid systems, molybdenum is ordinarily removed almost entirely from soils (Hansuld, 1966); only minor amounts can be detected chemically. At places near the base of the weathered zone, or elsewhere if conditions are favorable, molybdic ochre or ferrimolybdite may form. This sulfur-yellow mineral may form according to

$$\underset{\text{molybdenite}}{6MoS_2} + 4Fe^{+3} + 15H_2O + 36O_2 \rightarrow$$
$$\underset{\text{ferrimolybdite}}{2Fe_2(MoO_4)_3 \cdot 15H_2O} + 12SO_4^{=} + 36e \qquad (17\text{-}24)$$

In many districts in which oxidation is limited by low pyrite, lead and molybdenum as an impurity in sphalerite, or, rarely, as molybdenite, combine to form spectacular aggregates of wulfenite crystals:

$$\underset{\text{galena}}{PbS} + \underset{\text{molybdenite}}{MoS_2} + 2H_2O + 7O_2 \rightarrow \underset{\text{wulfenite}}{PbMoO_4} + 3SO_4^{=} + 4H^+ + 2e^- \qquad (17\text{-}25)$$

And molybdenite in a limestone-skarn zone near Ubehebe Peak, California, was oxidized to powellite ($CaMoO_4$), which could be discerned in partly weathered rock with an ultraviolet lamp.

Many other minerals are involved either in the peripheral portions of porphyry base-metal deposits or as principal components of Cordilleran veins also considered in Chapter 11 and in Figure 17-18. Lead in galena can be oxidized directly to anglesite:

$$\underset{\text{galena}}{PbS} + 2O_2 \rightarrow \underset{\text{anglesite}}{PbSO_4} \qquad (17\text{-}26)$$

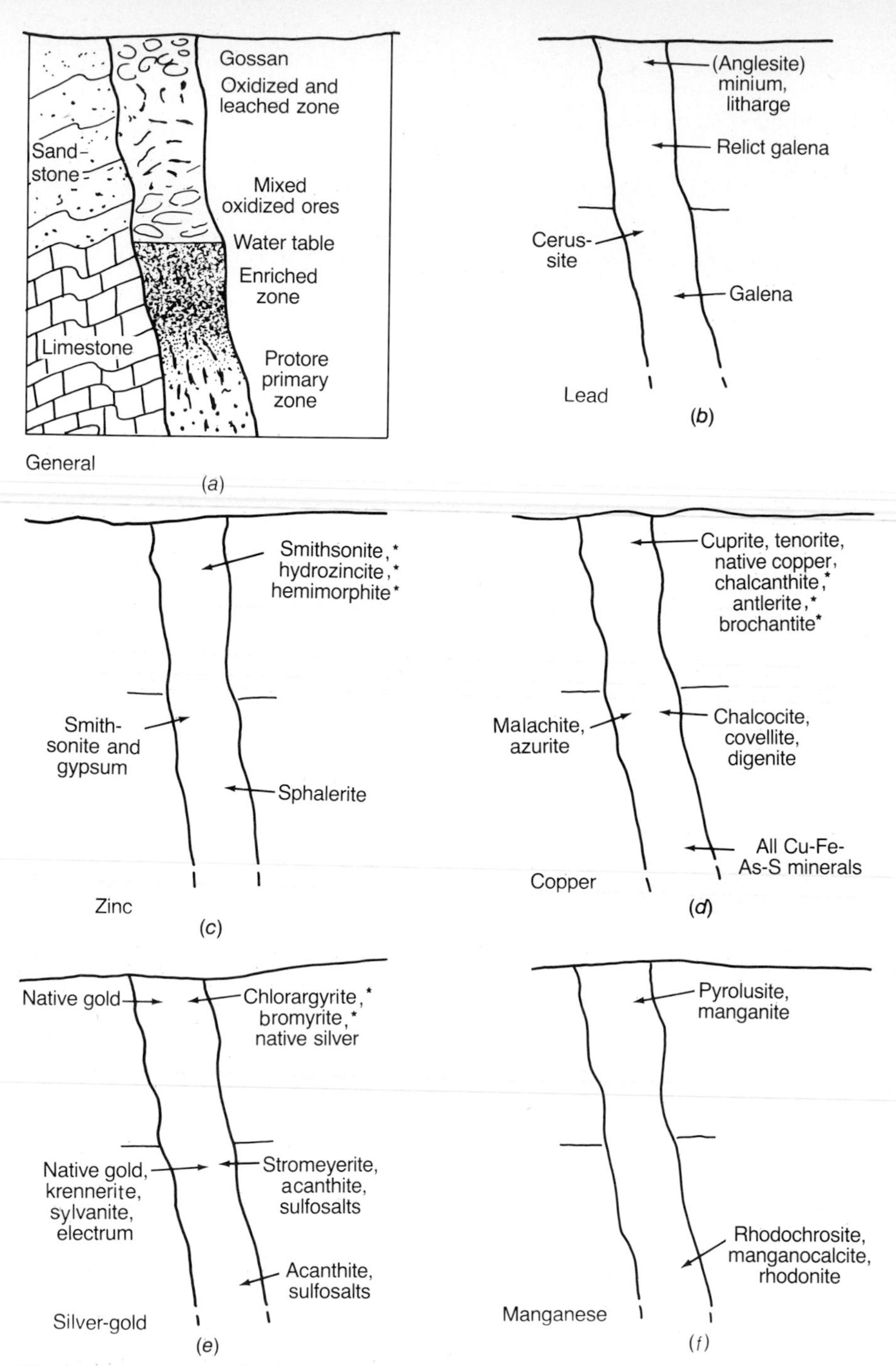

Figure 17-18. Leaching and enrichment characteristics in Cordilleran Veins of various mineralogies. Protore, hypogene minerals are shown at depth. Pyrite to generate sulfuric acid is considered ubiquitous. The footwall is influenced by the carbonate wall rock below the water table. (*a after* Bateman, 1950.)

or galena may be oxidized to anglesite by ferric sulfate:

$$PbS + 2Fe^{+3} + 3SO_4^{=} + 1.5O_2 + H_2O \rightarrow PbSO_4 + 2\,Fe^{+2} + 2H^+ + 3SO_4^{=} \quad (17\text{-}27)$$

In the presence of calcareous rocks, galena oxidizes to the relatively insoluble carbonate cerussite:

$$\underset{\text{galena}}{PbS} + H_2O + CO_2 + 2O_2 \rightarrow \underset{\text{cerussite}}{PbCO_3} + SO_4^{=} + 2H^+ \quad (17\text{-}28)$$

Iron oxide and lead sulfate are relatively insoluble and tend to remain at the outcrops (Fig. 17-18*b*). Galena is soluble in ferric sulfate solutions, but the reaction is so sluggish that unoxidized masses of galena are common in veins where other sulfides have been altered or leached. Leaching of galena is retarded because the common oxidation products anglesite and cerussite are also stable in the zone of weathering, and may encrust relict galena fragments. Where beds of limestone or other buffering rocks are encountered, acidic solutions are neutralized and the stable lead carbonate is formed. Zinc and silver sulfides are also dissolved in the presence of ferric sulfate:

$$\underset{\text{sphalerite}}{ZnS} + 2Fe^{+3} + 3SO_4^{=} + H_2O + 1.5O_2 \rightarrow Zn^{+2} + 2Fe^{+2} + 2H^+ + 4SO_4^{=} \quad (17\text{-}29)$$

$$\underset{\text{acanthite}}{Ag_2S} + 2Fe^{+3} + 3SO_4^{=} + H_2O + 1.5O_2 \rightarrow 2Ag^+ + 2Fe^{+2} + 2H^+ + 4SO_4^{=} \quad (17\text{-}30)$$

Zinc sulfates are very soluble, and zinc is low on Schürmann's series (Table 17-3). As a result, the zinc content of most oxidized orebodies is dispersed in the groundwater system. However, in arid or semiarid climates zinc may be retained in the oxidized zone as smithsonite, hydrozincite [$Zn_5(CO_3)_2(OH)_6$], hemimorphite [$Zn_4Si_2O_7(OH)_2 \cdot H_2O$], or other carbonate and silicate minerals. In a limestone environment, zinc sulfate solutions react to form smithsonite and gypsum:

$$Zn^{+2} + SO_4^{=} + \underset{\text{calcite}}{CaCO_3} + 2H_2O \rightarrow \underset{\text{gypsum}}{CaSO_4 \cdot 2H_2O} + \underset{\text{smithsonite}}{ZnCO_3} \quad (17\text{-}31)$$

Silver behaves much the same as copper in the zone of oxidation but has fewer stable oxidation products. Acanthite, the common silver sulfide, is oxidized by and soluble in ferric sulfate solutions [Equation (17-30)], and silver sulfosalts are broken down similarly. Consequently the silver content of sulfide veins is usually carried downward by meteoric waters. Where the climate is somewhat arid, however, the weakly soluble halogen salts such as chlorargyrite (AgCl), bromyrite (AgBr), iodyrite (AgI), and embolite [Ag(Br,Cl)] are left in the oxidized zone. The silver halides in arid or semi-

TABLE 17-3. Schürmann's Series of Solubilities of Important Sulfides

Affinity of Ion for Sulfur	Metal	Solubility of Sulfide in Water
High	Palladium	Low
	Mercury	
	Silver	
	Copper	
	Bismuth	
	Cadmium	
	Antimony	
	Tin	
	Lead	
	Zinc	
	Nickel	
	Cobalt	
	Iron	
	Arsenic	
	Thallium	
Low	Manganese	High

Schürmann's series depicting relative solubilities and sulfur affinities of the heavy metals. It is also known as the electromotive series. A metal in solution that is higher on the list can galvanically appropriate sulfur from the sulfide of a lower one by replacement. Copper ions, for example, will replace the cations in galena, sphalerite, and especially pyrite, leaving Pb, Zn, and Fe ions in solution and chalcocite or covellite replacement rims on the sulfides (Figure 17-19). Silver enrichment is also clearly possible. The list is in effect a "pecking order" from top to bottom of the sulfides of the metals in supergene environments.

arid regions are due to their insolubility combined with lack of water sufficient to dissolve them, and to a windborne supply of salts available for reaction. Silver carbonates and silver oxides are generally not present, but native silver is common. The native silver forms as a result of the reduction of the silver ion, probably by ferrous iron (Stokes, 1907; Cooke, 1913). Equation (17-32) shows the reaction as it would proceed in the presence of water and sulfate ions:

$$2Ag^{+} + 2Fe^{+2} \rightarrow \underset{\text{native silver}}{2Ag^{0}} + 2Fe^{+3} \tag{17-32}$$

Native gold is inert in most moderately oxidizing environments, but field evidence indicates that solution and short-distance transport of gold do take place under favorable conditions (Emmons, 1912). Gold is slightly soluble as a chloride complex in strongly oxidizing conditions. In acidic solutions and in the presence of a strong oxidizing agent such as MnO_2, gold is oxidized to Au^{+3}, which combines with chlorine ions to form the stable

$(AuCl_4)^-$ complex. Manganese, an abundant and widespread element, is present in many gold ores. Enrichments of gold beneath the surface have been found, suggesting that gold was dissolved and reprecipitated ahead of mechanical erosion. The solution and reprecipitation of gold results in the presence of coarse gold along cracks and small openings throughout the upper parts of veins and adjacent wall rock and may be involved in nugget formation, as mentioned in Chapter 16.

Antimony forms the relatively stable compounds valentinite (Sb_2O_3), bindheimite [$Pb_{1\text{-}2}Sb_{2\text{-}1}(O,OH,H_2O)_{6\text{-}7}$], and stibiconite [$Sb_3O_6(OH)$], also known as cervantite. These minerals are not conspicuous and, owing to iron oxide staining, are easily mistaken for limonite. Nevertheless they have been identified in the oxidized zones of many ore deposits. Where antimony was present in complex sulfantimonide minerals, the oxidation products are likely to be complex also. The oxidation of stibnite probably takes place according to the following equations:

$$\underset{\text{stibnite}}{Sb_2S_3} + 3H_2O + 6O_2 \rightarrow \underset{\text{valentinite}}{Sb_2O_3} + 6H^+ + 3SO_4^= \qquad (17\text{-}33)$$

$$\underset{\text{stibnite}}{3Sb_2S_3} + 10H_2O + 20O_2 \rightarrow \underset{\text{stibiconite}}{2Sb_3O_6(OH)} + 18H^+ + 9SO_4^= \qquad (17\text{-}34)$$

Stibiconite commonly occurs as pseudomorphs after stibnite in jasperoid, especially in Carlin-type deposits (Chapter 12). The stibiconite can be washed out of the cavities during weathering, resulting in bird-foot textures in the jasperoid surface called *turkey tracks*.

Most arsenic compounds, in contrast with those of antimony, are relatively soluble and consequently are leached from the zone of weathering. Traces of rare arsenates, such as conichalcite [$CaCuAsO_4(OH)$] in oxidized copper ores, reflect the former presence of sulfarsenides, but in humid climates these minerals are dissolved with the sulfates. Epithermal and mesothermal veins that carry complex sulfosalt minerals of lead, copper, silver, antimony, arsenic, bismuth, and sulfur contain commensurately complex mixed oxidized supergene minerals.

Hypogene oxide minerals may also be susceptible to weathering, but usually at a slower rate than the sulfides. Chromite and ilmenite commonly persist at the outcrop until they are mechanically removed to enter local placers and stream sands. But given time, they may eventually oxidize or dissolve. Martite (Fe_2O_3 after magnetite), hematite, and goethitic limonite commonly replace magnetite in outcrop, and maghemite (Fe_2O_3) is stable in many lateritic soils. Little has been said in these paragraphs about silicate involvement in sulfide oxidation because the host rocks can be considered essentially inert. In fact, though, supergene processes generally strip alkalies and alkali earths from host rocks if they contain those elements. The reactions typically involve water or hydrogen ion and are of the form of

Equation (17-3) or, with water,

$$\underset{\text{orthoclase}}{2KAlSi_3O_8} + 2H^+ + H_2O \rightarrow \underset{\text{kaolinite}}{Al_2Si_2O_5(OH)_4} + 4SiO_2 + 2K^+ \qquad (17\text{-}35)$$

or

$$\underset{\text{muscovite}}{2KAl_2(AlSi_3)O_{10}(OH)_2} + 2H^+ + 3H_2O \rightarrow \underset{\text{kaolinite}}{3Al_2Si_2O_5(OH)_4} + 2K^+ \qquad (17\text{-}36)$$

Notice that oxidation-reduction is not necessarily involved, although any iron liberated would certainly become oxidized and contribute to limonite in the outcrop. The K^+, or Na^+ in equivalent albite to kaolinite reactions, is lost to groundwaters. The reaction product silica may crystallize as fine-grained quartz or chalcedony, or combine with alumina or copper to form kaolin-chrysocolla-type minerals:

$$Cu^{+2} + SiO_2 + 2H_2O \rightarrow \underset{\text{chrysocolla}}{CuSiO_3 \cdot H_2O} + 2H^+ \qquad (17\text{-}37)$$

$$Al_2O_3 + 2SiO_2 + 2H_2O \rightarrow \underset{\text{kaolinite}}{Al_2Si_2O_5(OH)_4} \qquad (17\text{-}38)$$

Many more complex reactions involving biotite, amphiboles, chlorites, and other minerals can of course be written, many involving only H^+ and therefore not necessarily related to weathering processes.

Oxidation rates vary over a wide range. Cinnabar oxidizes so slowly that for practical purposes it is considered stable in surface outcrops, even though it is thermodynamically unstable there. Galena is commonly protected from oxidation by a jacket of anglesite. By contrast, a few other sulfides oxidize so rapidly that they ignite when exposed to the atmosphere. In deposits of massive sulfides, especially pyrrhotite and marcasite, it is not uncommon for freshly exposed rock to burn by spontaneous combustion. Experience taught the miners of Iron River, Michigan, to leave a skin of iron ore on the walls of stopes in order to prevent the highly pyritiferous black slate wall rocks from igniting. High rock and air temperatures can be produced underground by oxidizing sulfides, making it necessary in many mines to pump refrigerated air underground so that miners can work comfortably. The processes and end products are the same in these situations as they are on surface outcrops. Acidic waters are produced by the oxidation of sulfides; in some districts the mine waters are so acidic that they dissolve mine rails and clothing, creating a major problem for workers and equipment. Similarly, large amounts of noxious and strongly corrosive sulfur dioxide fumes are generated where sulfides burn, and in extreme cases the mines must be abandoned or parts sealed off until the fire exhausts the oxygen supply or burns all of the combustible materials. The burned country rock is generally bleached and altered as if it had been exposed at the surface for a long time.

Precipitation and Enrichment

So far we have said little about the mechanisms of reprecipitation of the minerals that constitute secondary enrichment. It proves true that pyrite is a crucial mineral both above and below the groundwater table. As discussed, it weathers to ferric sulfate and sulfuric acid in the zone of oxidation, enabling meteoric waters to dissolve ore metals. But below the water table, pyrite acts as a host for deposition of the ores because it is relatively soluble and readily relinquishes its sulfur to descending ions of copper, silver, and other metals. Stokes (1907) showed experimentally how the reaction between copper ions and pyrite may take place in supergene copper enrichment. The reaction, known as Stokes' equation, is

$$\underset{\text{pyrite}}{5FeS_2} + 14Cu^{+2} + 14SO_4^{=} + 12H_2O \rightarrow \underset{\text{chalcocite}}{7Cu_2S} + 5Fe^{+2} + 24H^{+} + 17SO_4^{=} \quad (17\text{-}39)$$

The excess acid is neutralized by reaction with silicates, and the chalcocite remains, coating and replacing pyrite in the enriched zone (Figure 17-19). Stokes considered this reaction a generalization of the actual process, and the field evidence supports his interpretation.

In general, metallic elements have certain definite affinities for sulfur, affinities that are related to the solubilities of their sulfide compounds. A metal in solution that has a stronger affinity for sulfur than another metal in a sulfide mineral will preempt the solid phase sulfur and precipitate as a sulfide, expelling the less strongly bonded, more soluble metal into the solution. The sequence of stabilities of the heavy-metal sulfides was first established by Schürmann (1888), and accordingly is known as Schürmann's series (Table 17-3). Ions of any metal in the series will replace another metal sulfide that is lower in the series. Thus copper in solution will replace the iron in pyrite or the zinc in sphalerite. The copper, having appropriated sulfur, remains as chalcocite or covellite, and the displaced iron or zinc ions are removed in solution. In general the farther apart the elements are in the series, the more complete will the replacement be and the greater will be the rate and energetics of reaction. Furthermore, different metals are selectively replaced according to their relative positions in Schürmann's series. For example, silver-bearing solutions react with sphalerite more readily than with covellite or chalcocite. Copper ions in leach fluids pumped and flushed through mine dumps replace scrap iron when the cupriferous waters are recovered and passed through launders containing old cans and car parts; metallic copper is thereby inexpensively gathered, and iron goes into solution.

Schürmann's series has been applied to all ore deposits, both hypogene and supergene. Since Schürmann's experiments were performed under conditions of standard temperature and pressure, his results would be expected to fit better at low temperatures and pressures than at high ones. Hypogene fluids are complicated by so many factors that strict adherence to Schür-

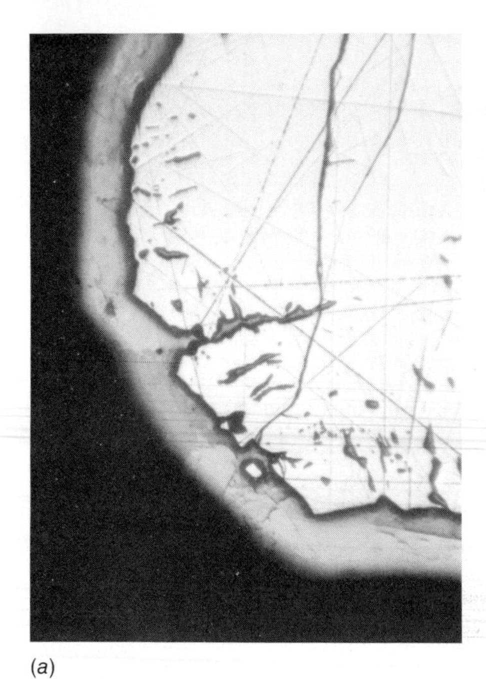

(*a*)

(*b*)

Figure 17-19. (*a*) Photomicrograph of pyrite (white) rimmed and replaced by supergene chalcocite (medium gray). Butte, Montana. 30×. (*Photo by* C. J. Eastoe *and* J. Lang.) (*b*) Spotty dendrites of cupriferous manganese oxides (see text) from a volcanogenic massive sulfide district (Jerome, Arizona) and from a porphyry copper prospect (Agua Tapada, Argentina).

mann's series should not be expected. The simple relations of the series are especially unlikely in high-temperature high-pressure systems, even above 200°C and 1 kbar. Under conditions of supergene deposition, however, where the expected relationships are usually found, Schürmann's series is extremely useful.

Textures

Texturally the results of Schürmann's series are manifest in supergene enriched zones by chalcocite rimming and replacing pyrite, or by chalcocite replacing pyrite along fractures (Figure 17-19). Commonly the chalcocite coatings are mere films perhaps only 1 or 2 molecules thick. When all the pyrite in a specimen is coated with a black monomolecularly thin chalcocite layer, it is easy to overestimate the amount of copper present, but sawing or scratching reveals the thinness of the coating.

Rates and efficacies of supergene reactions are controlled by many factors. In the presence of calcite or other carbonates, oxidizing solutions may be neutralized before the groundwater level is reached, and an enriched oxidized zone will result. The character of the sulfides also influences the reactions. Pyrrhotite, for example, reacts much faster than pyrite; where pyrrhotite is the primary sulfide below the water table, a supergene chalcocite zone is likely to be thin but high-grade. Such a condition was encountered at Ducktown, Tennessee, where the primary ore contained pyrrhotite with subordinate amounts of chalcopyrite. The enriched zone formed just below the groundwater table, in a thickness of 1 to 3 meters; chalcocite replaced pyrrhotite almost entirely, and the ore was unusually rich (Emmons and Laney, 1926). By contrast, where the protore contains pyrite and chalcopyrite, as it does in many disseminated copper deposits, it is not uncommon to find enriched zones extending 200 to 300 meters or more beneath the groundwater level. Normally in a given polished surface one will find chalcopyrite and sphalerite deeply corroded, even if pyrite is only incipiently filmed; secondary copper sulfides replace the primary sulfides of zinc and lead in accordance with Schürmann's series. Many deposits of sphalerite and galena contain small amounts of copper minerals, reworked during weathering and concentrated as supergene products below the water table. Such zones are generally discontinuous and thin.

Supergene chalcocite is commonly soft and powdery. It is referred to as "sooty" chalcocite, to distinguish it from the massive gray crystalline "steely" chalcocite of hypogene ores and because it easily makes a black smudge on the fingers. Supergene chalcocite is not necessarily sooty; it may be massive and megascopically indistinguishable from hypogene chalcocite, although the two may be identified with the mineragraphic microscope by coarser grain size and exsolved bornite or digenite shreds in polished surfaces of the latter. Other copper sulfides also form by supergene processes. Covellite and bornite are not uncommon; chalcopyrite and several of the more complex copper and silver salts of arsenic and antimony also form under special conditions of supergene enrichment. In general, however, supergene sulfides are mineralogically simple compared with hypogene sulfides.

The geochemistry of silver is similar to that of copper, although there are no insoluble silver carbonate analogues of malachite or azurite to retain silver in the oxidized zone of limestone host rocks. Where halogens are available, silver may form insoluble halides like chlorargyrite (AgCl) above the groundwater table, but silver is ordinarily taken into solution as a sulfate. Supergene minerals of silver, as well as the textures and structures of the deposits, are similar to those formed under epithermal conditions. Consequently it is sometimes difficult to distinguish between hypogene and supergene deposits of silver. The need for such a distinction is obvious: supergene deposits are likely to be shallow, blanket- or rodlike occurrences, whereas hypogene deposits may persist in depth. The most common supergene silver sulfide compound is Ag_2S. Above 180°C, Ag_2S forms the isometric mineral argentite; at lower temperatures, the monoclinic acanthite is more stable (Roy et al., 1959). Thus the presence of isometric crystals of Ag_2S indicates a hypogene origin, but the presence of monoclinic Ag_2S shows only that deposition took place below 180°C; hence acanthite cannot be used to distinguish between supergene and low-temperature hypogene ores. Nor can the conversion temperature of acanthite to argentite be applied to most earlier descriptions of silver deposits, because few geologists in the past attempted to distinguish between the polymorphs of Ag_2S. Unless the mineral occurred in elongated crystals of obvious non-isometric habit, it was generally called argentite. The ruby silvers pyrargyrite and proustite are more likely to be hypogene than supergene, but the fact that they can form as supergene minerals means that they cannot be considered positive indicators of hypogene deposition. The general observation that supergene assemblages are typically less complex than hypogene assemblages may be of value in the interpretation of silver deposits. But it should always be remembered that oxidation effects can as well be produced by ascending waters mixing with and adapting to surface environments as by descending supergene ones. Quite possibly some deposits of chrysocolla, malachite, and acanthite, which have been assumed to be supergene, are in fact "hybrid" hypogene plus supergene.

Other sulfides, such as those of lead and zinc, have been reported to form in enriched sulfide zones, but there are no commercial deposits of these supergene ores. Lead sulfate and lead carbonate are relatively insoluble, and they are left in the oxide zone. Zinc also forms a stable carbonate, but the zinc content of sulfide ores is characteristically leached from the zone of oxidation. Many examples of supergene zinc sulfides have been described; perhaps the most noteworthy are at the Horn Silver mine in Utah (Butler et al., 1920) and the Balmat-Edwards district of New York (Brown, 1936). Zinc deposition below the water table forms either wurtzite or a light-colored sphalerite. No deposits of supergene zinc sulfides comparable in size to those of supergene copper and silver are known; evidently most zinc remains in solution to be dispersed in the groundwater system. Masses of supergene smithsonite ($ZnCO_3$) were mined in the early days at Hanover and Magdalena, New Mexico.

Classification of minerals as uniquely supergene or hypogene should be avoided. Certain minerals are characteristic of high-temperature environments and others of low-temperature environments, but there are many exceptions. The availability of the chemical components and the Eh and the pH of the environment are also contributing factors. For example, a certain mineral may be rare or unknown as a product of supergene processes because the proper redox conditions are uncommon in the phreatic zone. But since at least locally almost any Eh is possible below the groundwater table, the possibility that a species may or may not form there should not be decided unconditionally. Although specular hematite is commonly considered a high-temperature mineral, plates of specular hematite as much as 1 cm across are found in limestone above the water table at the Mountain Home property west of Hanover, New Mexico. They are not found at depth and are thought to be of supergene origin. Brown (1936) showed that at Balmat, New York, magnetite forms by the weathering of pyrite. Because magnetite had long been considered a high-temperature mineral, this observation was at first difficult to accept. In the gold deposits of the Southern Piedmont there are doubly terminated quartz crystals in the subsoil above the groundwater table. The quartz crystals contain "phantom' growth zone outlines marked by limonite; similar quartz is not found at depth. Sulfide minerals characteristic of igneous metamorphic zones may indicate hypogene deposition, but again, the possible exception must be anticipated. For example, thermodynamic conditions indicate that pyrrhotite, a typical high-temperature mineral, can form under very restricted supergene conditions (Garrels, 1960). Most geologists would agree that chalcopyrite seldom forms in supergene settings, but the "golden covellite" crystals of chalcopyrite replacing covellite rosettes from Butte, Montana, are almost certainly of supergene origin.

Supergene enrichment is effective in concentrating dispersed metals. As stated, the economic value of many disseminated copper deposits is due to this process. Several factors are involved in the development of a supergene-enrichment zone. Where groundwater is actively circulating in highly permeable rocks, the dissolved metals may be widely dispersed. Where the rocks are not permeable, meteoric waters may not be able to leach the oxidation products and carry them down to the groundwater table or reducing zone. Where erosion is rapid, the reduction interface may be lowered so fast that oxidation and chemical dissolution of the sulfides above cannot keep pace. Where the water table is not lowered through the disseminated mineralization, that is, where the topography is stable, the process comes to a standstill, and enrichment ceases. Thus the development of appreciable concentration requires a balance between the rates of oxidation and erosion as well as a modulated hydraulic groundwater system.

Exploration Significance

The leached, oxidized surface exposures of weathered sulfide deposits known as *gossans* may retain distinctive characteristics of the underlying sulfide materials (Blanchard, 1968). It is desirable to know whether sulfides exist at depth and, further, whether the mineralization is simply pyrite or includes appreciable amounts of copper, zinc, silver, and other valuable sulfides. Consequently many serious efforts have been made to study gossans quantitatively. The criteria by which gossans are evaluated are still not clearly defined, and gossan interpretation remains both an art—or rather an acquired skill—and a science. For every success story there is at least one failure story, but the success stories are numerous enough that no explorationist could ignore them. Kenyon Richard and Harold Courtright of ASARCO successfully plotted subjacent copper ore grades from gossan interpretation before drills arrived at the porphyry copper deposit at Toquepala, Peru. Joseph Durek of Phelps-Dodge selected what proved to be *the best* drill site over the King Peak orebody at Morenci, Arizona, by capping interpretation, and Harrison Schmitt identified orebodies at Esperanza, Mineral Park, and many other localities by such methods. Lacy (1949) studied the oxidation products at Yauricocha, Peru, and was able to relate individual oxide minerals to their source materials. He divided the oxidation products into residual and transported materials and subdivided these according to the textures developed and the minerals from which they were derived. On the basis of this study, Lacy distinguished among gossans that overlie several kinds of sulfide deposits, including lead-zinc ores, lead-zinc-copper ores, copper-pyrite ores, and massive pyrite bodies. Much more careful study is necessary than can be explained here; the reader is referred to Locke (1926), Blanchard (1968), and Anderson (1981). A few key considerations are developed here.

The first impressions transmitted by a gossan are those of the color of the relict limonites and the texture of those limonites in crusts or molds of preexisting sulfides. Resident geologists, or others familiar through long experience in certain areas, use the color and texture of limonite to indicate whether the underlying rocks are favorable, though a quantitative approach is seldom considered (Locke, 1926; Weiss, 1965). Other features include the shapes of the cavities formerly occupied by sulfides; the quantity of kaolinite and other supergene clay minerals; and the presence of metallic oxidation products. Emphasis is usually placed upon the color of the limonite and the configuration of *boxworks,* the meshwork of porous gossan left after much of the original sulfide is leached away.

Anderson (1981) and others have shown that the "limonites" of gossans are composed of the three minerals goethite, hematite, and jarosite. Each has a distinctive color—goethite is orange red, hematite red brown, and jarosite mustard gold. When a piece of capping has been powdered with a knife point or pick, the ratios of the three can be estimated from memory or by comparison with known preprepared mixtures. If jarosite is present, sulfur and potassium had to have been in the preoxidation outcrop, because jarosite is $K(Fe^{+3},Al)_3(SO_4)_2(OH)_6$. Goethite can be the oxidation product of almost any iron oxide or silicate, so mixtures of hematite and jarosite are more meaningful. They typically form distinctive colors such as seal brown and golden Gertie, colors that have special meaning to those who have seen hematite after secondary chalcocite at many localities and jarosite-goethite after pyrite-chalcopyrite. So color is a first means of treatment of a given gossan.

Textures can be equally useful. Cubic cavities bespeak earlier pyrite or galena, and the shapes and morphologies of limonite incrustations can be used to determine the earlier presence of chalcopyrite, sphalerite, pyrrhotite, and several other minerals (Blanchard, 1968). And of course the volume percentages of earlier phases can be estimated once the qualitative inheritance is determined. Blanchard (1968) distinguished between the gossan formed by leaching of massive sulfides and the iron-stained material left by weathering of disseminated mineralization. He pointed out that chalcopyrite oxidizing in an inert gangue will theoretically export all of its copper and two-thirds of its iron in solution, leaving one-third of its iron as limonite in the form of cellular boxwork or sponge. One-to-one mixtures of chalcopyrite and pyrite oxidizing together in an inert gangue will dissolve and export all of the copper and all of the iron in the two minerals. Pyrrhotite yields an excess of acid, but sphalerite yields just enough acid to permit solution of the zinc as zinc sulfate. Certain sulfides, such as chalcocite, bornite, and tetrahedrite, are deficient in sulfur and will not dissolve completely unless external sulfur is introduced, and "green copper oxide" traces will remain in the outcrop. Covellite, sphalerite, and molybdenite contain just enough sulfur to permit solution, while pyrite, chalcopyrite, and pyrrhotite yield free acid.

If we reexamine Equations (17-5) and (17-6), we see that it takes 2 mol of pyrite to produce each mole of ferric sulfate. By Equation (17-13) it then takes 8 mol of ferric sulfate to oxidize 1 mol of chalcopyrite, freeing its copper as soluble $CuSO_4$. So if an original outcrop of otherwise nonreactive rocks has a pyrite-to-chalcopyrite ratio greater than 16:1, all copper will be removed as soluble copper sulfate. If the ratio is less than 16, some copper should remain as cuprite, malachite, or chrysocolla. If one can make that 16:1 assessment, and then estimate the original volume percent pyrite on the basis of the abundance of cavities, then an estimate of original copper grade is possible. So a combination of limonite mineralogy, texture, copper oxidation-product mineralogy, and geochemistry can be highly informative. There are many other "tricks." One, for example, is that of the presence of *spotty dendrites*. Every field geologist is familiar with the delicate lacy, fernlike, arborescent pyrolusite "manganese dendrites" that form in cracks and on joint surfaces in weathered rocks. That lacy development is poisoned by the presence of copper such that only stipples of black spots—spotty dendrites—form in the presence of a few hundred parts per million of copper (Figure 17-19*b*). At 500 to 1000 ppm visible "green copper" minerals are usually present. In their absence, the presence of spotty dendrites is a valuable, instantly discernible, infallible indicator of anomalous copper in a very useful 100-to-500-ppm range.

Many fine-grained oxides, silicates, and carbonates in the zone of oxidation are considered to have traveled at least locally as colloids. Typical products of this kind include opal and chalcedony, smithsonite, reniform hematite and limonite, aluminum hydroxides, and manganese oxides. Wad, the amorphous or cryptocrystalline manganese oxide commonly found in residual deposits and described above, is a good example. Some wad gives no X-ray diffraction pattern and is still colloidal. The composition of wad is indefinite because the colloidal gel tends to incorporate many elements. Unlike most oxide colloids, colloidal manganese oxides carry a negative charge, so that they attract cations of other metals from solution. Small amounts of nickel are common in many deposits of wad, and copper, cobalt, and barium have also been reported. The manganese oxides of the Tocantins district, Brazil, contain as much as 4% Co (Pecora, 1944). The presence of tungsten, strontium, thallium, and a few other metals was considered by Hewett to indicate a hydrothermal, and most probably igneous, association (Hewett and Fleischer, 1960).

The presence of gossan does not necessarily mean that unaltered sulfides exist at depth. Under relatively low Eh and pH conditions, iron goes into solution in the ferrous state and travels appreciable distances from the oxidation zone. Where these ferrous sulfate solutions encounter limestone or a similar basifying medium, the acid is neutralized, and the iron precipitates as ferric oxide or ferric hydroxide. Laterally displaced iron oxide zones of this type are known as "false gossans." Clearly, then, it is of critical importance to distinguish between false gossans and indigenous gossans. And depending on the ratios of original sulfides, the total amount of sulfides, and the inertness of the wall rocks, limonite may remain in the outcrop as residual limonite crusts, it may occur as outward-fading halos of transported limonite around cavities and fractures, or it may be removed altogether.

The depth of oxidation, and to supergene enrichment zones, is another significant problem for mining and exploration geologists. As mentioned, in tectonically stable regions the oxidation zone generally extends down to the water table if the country rocks are permeable. But tectonic activity, a fluctuating water table, or variably impervious wall rocks may modify the pattern of oxidation. The most critical apparent factor is the position and permanence of the water table, because sulfides are generally stable in the slightly alkaline, moderately reducing environment below it. In humid climates the sulfide zone may be present 1 meter or so below the surface; in arid environments, where the water table is likely to be deep, the lower limit of oxidation may extend 200 or 300 meters below the outcrop. At the other extreme, a rapidly lowered water table may leave sulfides isolated well up in the zone of oxidation. This phenomenon is especially common in arid environments where there is not enough water to oxidize and dissolve the sulfides. Complex patterns of "stranded" or "perched" oxidized and enriched blankets are possible, and are encountered at places like Chuquicamata, Chile, and Morenci, Arizona.

The role of permeability in oxidation is demonstrated at the Tsumeb mine, South-West Africa (Namibia), where a nearly vertical ore pipe that cuts through steeply dipping sediments is oxidized in the upper and lower parts but is unweathered at intermediate levels. Here deep oxidation was brought about by groundwaters circulating along a permeable fractured zone that crops out some distance from the mine, but intersects the ore pipe at depth (see Figures 3-32 and 3-34). The unoxidized middle part of the pipe was protected from groundwater by relatively impermeable strata. Although sulfides remain in the oxidized zones, there are relatively sharp changes in the ratio of oxidized to unoxidized minerals. Between the 1200- and 2400-ft levels, the ores are largely sulfides, whereas above and below this interval the ores are oxidized. Oxidation along the deep brecciated stratum is so efficient that the lower part of the pipe is more thoroughly oxidized than much of the shallow weathered zone (Söhnge, 1964).

Descending waters charged with dissolved substances tend to seek density levels. This tendency results in a crude stratification of water (Brown, 1942; McKnight, 1942) and explains how concentrations of oxidation products have formed in certain places where deep circulation of water is doubtful; for example, denser, mineral-laden, and perhaps oxygenated waters may settle below the water table as at OK Tedi, New Guinea, where supergene enriched ores extend well below reasonable water table levels (Titley, 1981).

Inspiration Porphyry Copper Deposit, Arizona Supergene enrichment has been especially important in the history of many disseminated or porphyry copper deposits. Several deposits of disseminated copper in the Inspiration district near Globe-Miami, Arizona, have been affected by supergene enrichment (Ransome, 1919; Tenney, 1935; Peterson et al., 1951; Peterson, 1954, 1962; Olmstead and Johnson, 1966). The area is in southeastern Arizona, near the center of the southwestern United States copper province (Figures 11-1 and 17-20). The principal orebodies at Inspiration form an irregular, elongated body of disseminated copper, which extends for 3 km along a schist-granite contact (Figure 17-21). Evidently, mineralizing fluids that produced normal hypogene alteration-mineralization assemblages ascended along the granite contact where late-magmatic or intrusion tectonics created repeatedly opened veinlets in the contact zone. Supergene enrichment later formed zones locally as much as 225 meters thick, but with an average of 100 meters. About half of the ore is in the Pinal Schist of Precambrian age, and about half is mined from a porphyritic facies of quartz monzonite called the Schultze Granite of Laramide age (62 million years). On the western end of the mineralized zone is the Live Oak orebody; near the center, the Inspiration orebody; and on the eastern end, the Miami orebody. The Live Oak and Inspiration ores are largely overlain by a sheet of granite porphyry. Another orebody under the Miami townsite was found in 1975.

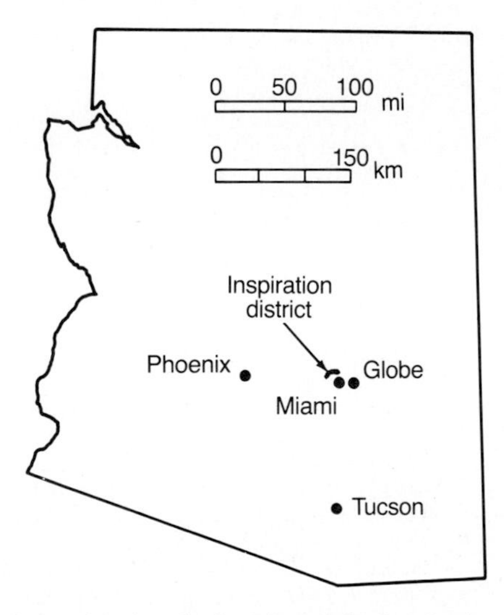

Figure 17-20. Map of Arizona showing the location of the Inspiration district.

Hypogene mineralization formed an assemblage of ore sulfides and alteration products that typify the disseminated copper deposit (see Chapter 11). The unenriched protore at Inspiration averages about 1% Cu (Ransome, 1919) and consists of pyrite, chalcopyrite, and molybdenite distributed along minute veinlets throughout the schist and granite porphyry contact volume. Other orebodies in the district contain much less copper in the

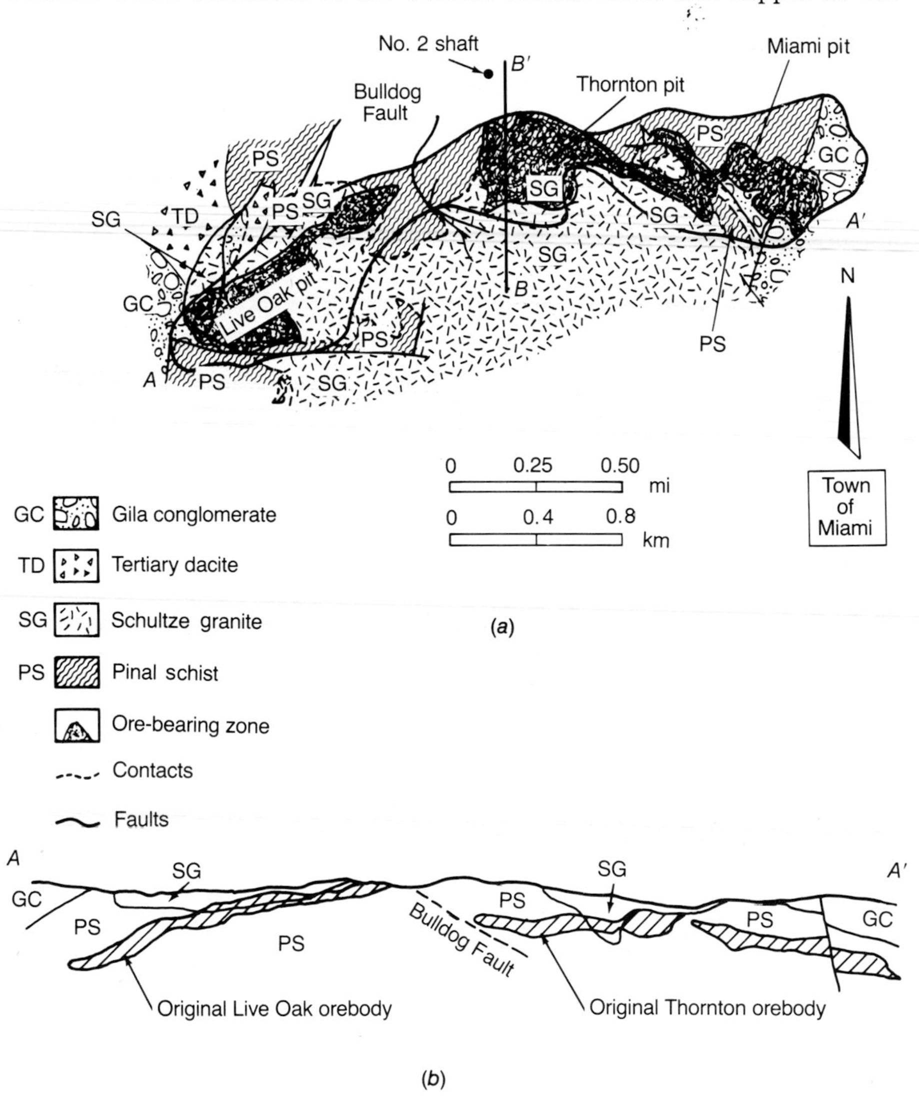

Figure 17-21. (*a*) Outline of the disseminated copper orebodies at Inspiration, Arizona, and their relation to the contact between Pinal Schist and Schultze Granite. The inner, dark-patterned ore distribution is from Ransom (1919); the dashed line is the orebody outline from Olmstead and Johnson (1966), from whom the curved cross section *A–A′* (*b*) is adapted. Section *B–B′* in (*a*) is Figure 17-22, and Figure 17-23 shows assays down No. 2 shaft.

protore than that directly under the Inspiration orebody. The alteration products include pyrite, quartz, sericite, orthoclase, and kaolin (Schwartz, 1947).

Supergene enrichment increased the grade of ore from 1% or less to as much as 5% in localized zones. In the early days of mining, when the upper, high-grade part of the supergene-enriched ore was recovered, the ore averaged more than 2% Cu; today materials that have been only weakly enriched are being mined. The Pinal Schist is more permeable than the Schultze Granite, and protores in the schist were therefore more amenable to supergene enrichment. Leaching was relatively complete in the oxidized zone of the pyritic schist, but the oxidized zone in low-pyrite granite typically retains its copper in the form of malachite and chrysocolla. Accordingly, supergene sulfide enrichment was more thorough in the schist. The supergene-enriched zone starts abruptly beneath the gossan, anywhere from 30 to 200 meters below the surface; it may indicate the position of the water table at the time of supergene enrichment (Figure 17-22). The copper content increases abruptly at the chalcocite zone and tapers off gradually down to the protore (Figure 17-23), reflecting the fact that supergene sulfide enrichment started at the water table and continued downward to the limit of groundwater circulation or until the supply of copper ions in solution was depleted. Incipient enrichment took place along grain boundaries and tiny cracks in pyrite and chalcopyrite. Further replacement of the primary sulfides by chalcocite increased the copper tenor until the upper zone, the

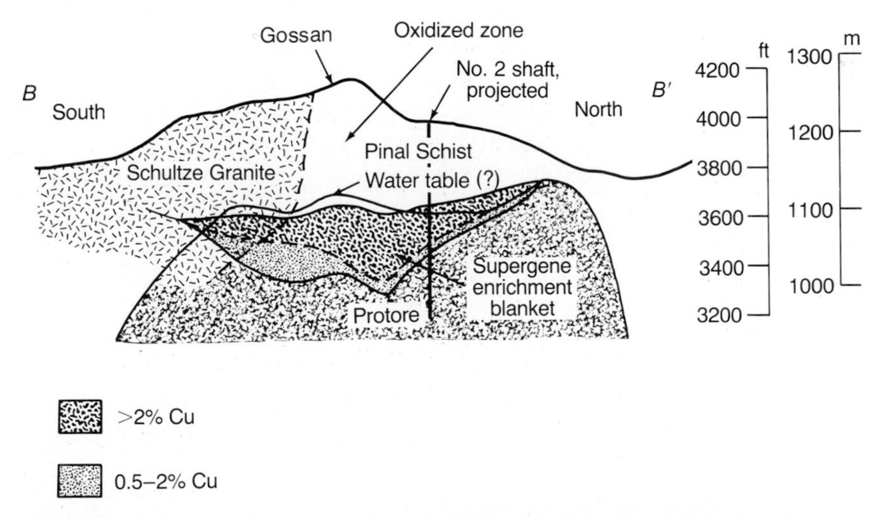

Figure 17-22. West-facing north-south cross section *B–B′* (Figure 17-21) through the Inspiration orebody, showing the shape of the ore zone, the relationship of high-grade ore to lower grade ore, the thickness of the ore-bearing zone in Pinal Schist, and the relationship between the top of the supergene-enriched zone and the presumed groundwater table. (*From* Ransome, 1919.)

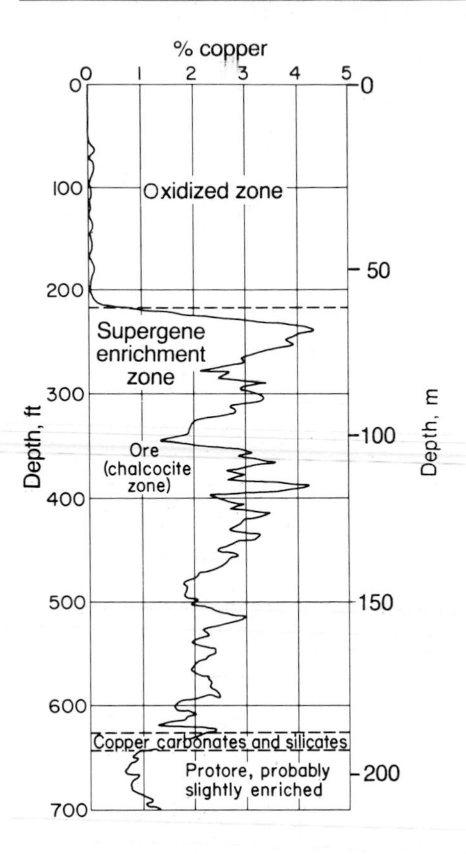

Figure 17-23. Assay graph of No. 2 shaft, Miami mine, Arizona, showing the copper content in the oxidized zone, supergene-enriched zone, and protore. The thin zone of oxidized ore beneath the chalcocite zone is anomalous. (*From* Ransome, 1919.)

richest, contained about 3% Cu. The effects of supergene enrichment are found as far as 400 meters below the surface (Lindgren, 1933).

Chalcocite is practically the only supergene sulfide mineral in the Miami district. It replaced both pyrite and chalcopyrite, but chalcopyrite was more susceptible to replacement. In the upper parts of the supergene zone both the chalcopyrite and the pyrite were completely replaced by chalcocite, and intermediate stages of replacement are represented by chalcocite envelopes around remnant cores of pyrite. Small amounts of covellite are found in places; covellite represents the first step in the enrichment of chalcopyrite. In the Copper Cities deposit, 6 km north of Miami, supergene enrichment did not progress far because the protore was not exposed to weathering until recently (Peterson, 1954). Here pyrite is preserved in the enrichment zone; even the chalcopyrite is only partly replaced by chalcocite.

Almost all the ore mined in the Miami district has been taken from supergene-enriched deposits. In the early days of mining, when the cutoff grade was 1.5% Cu, even the exceptionally rich protore beneath the Inspiration orebody could not be mined economically. With the development of modern mining and metallurgical techniques, the limits of ore grade have gradually been dropped, so that rock containing only 0.5% Cu can be mined profitably under favorable conditions, and some hypogene sulfides have been taken. The Copper Cities deposit is now being mined though it has been enriched only slightly above the tenor of the protore, which averages 0.4% Cu. In each deposit, then, copper has been mined almost exclusively where it has been enriched by supergene activity. Where leaching and supergene enrichment were thorough, a thick high-grade chalcocite zone was produced; these deposits were mined first. Deposits that were not favorably weathered and enriched had to await economic changes before becoming classified as ore. Today some of the marginal materials are heap-leached artificially with ferric sulfate-sulfuric acid and bacteria solutions, and the acid-soluble copper is recovered by replacement of scrap iron in collecting basins in the lower levels of the workings.

Chañarcillo Silver Vein Deposit, Chile Many examples of supergene-enriched silver deposits could be described, in particular some of the famous bonanza deposits of the western hemisphere. Today, however, these deposits are nearly exhausted; most of them have been abandoned, but some are being revived (Chapter 11). Epithermal deposits of this type extend southward from the United States through Mexico, Central America, and along the western slopes of the Andes in South America. An outstanding district was Chañarcillo, where more than 100 million oz of silver were produced (Moesta, 1928; Whitehead, 1919, 1942; Flores-Williams, 1959; Segerstrom, 1962).

Chañarcillo is in the Atacama Desert of Chile, about 50 km south of Copiapó in the arid foothills along the western side of the Interior Valley (Figure 17-24). The district was discovered in 1832 (Miller and Singewald, 1919). During the boom years from 1860 to 1885, it produced about 2.5 million kg of silver. Much of the silver was recovered from high-grade masses; one piece of nearly pure native silver weighed over 90 kg, and another mass of embolite [Ag(Cl,Br)] with native silver weighed 20,000 kg (10 metric tons) and contained 75% Ag!

The rocks in the immediate area of the mine consist of an alternating sequence of Cretaceous limestones and volcanic tuffs intruded by a swarm of diorite dikes and a small granodiorite stock (Segerstrom, 1962). Both the dikes and the stock are highly altered, and parts of the stock consist of endoskarn diopside, wollastonite, epidote, and similar metamorphic minerals. Narrow copper veins lie close to and within the stock; the silver is farther away.

Although Chañarcillo lies on the axis of a broad regional fold, the dips of the rocks are slight. The rocks are cut by many fractures which were probably formed during folding; these fractures contain the ore. Most fractures are parallel to the north-south fold axis, but others cut it at angles of

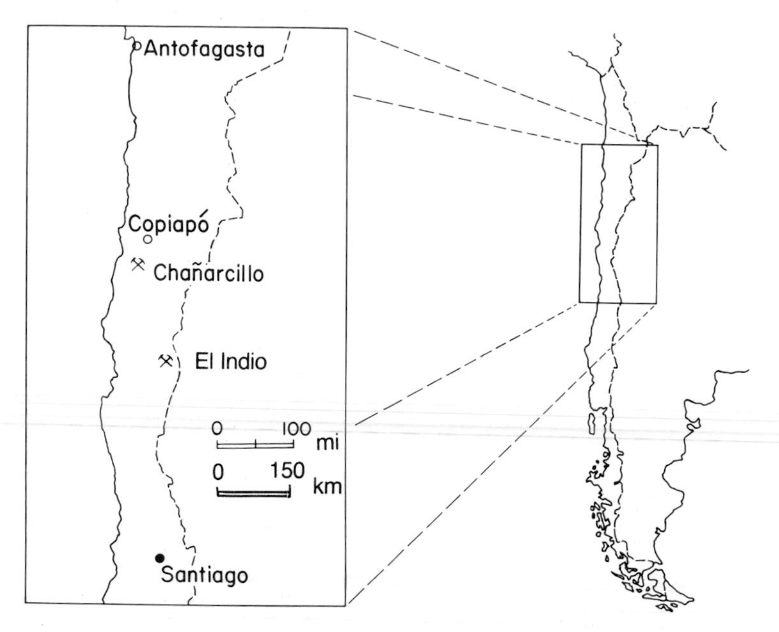

Figure 17-24. Map of Chile showing the location of the Chañarcillo silver and the recently developed El Indio gold districts. See also Figure 11-10 to locate porphyry copper deposits in Chile.

about 45°. The fractures are steeply dipping and of great continuity both in depth and along strike; at least the upper portions of these fissures were open during metallization (Whitehead, 1919, 1942).

The primary veins vary from about 2 cm to 1 meter in width. Hypogene minerals include pyrite, sphalerite, chalcopyrite, galena, arsenopyrite, cobalt arsenides, pearceite ($Ag_{16}As_2S_{11}$), freibergite [$(Cu,Ag)_{12}Sb_4S_{13}$], proustite (Ag_3AsS_3), polybasite ($Ag_{16}Sb_2S_{11}$), and pyrargyrite (Ag_3SbS_3) in a gangue of calcite, barite, quartz, and siderite. The veins extend to the greatest depths explored at about 1000 meters, but the best ores were concentrated in the beds of limestone (Figure 17-25). Small ore zones were also formed where veins intersect fissures and dikes. The major veins are aligned along the axis of the anticline; there seems to be a direct correlation between proximity to the crest of this fold and continuity and richness of the parallel veins.

After the hypogene minerals were deposited, the rocks were faulted. The major structure—an east-west normal fault with about 50 meters of displacement—divided the Chañarcillo district into northern and southern parts. During this period of faulting, the ore-bearing veins were refractured such that subsequent erosion and weathering were able to redistribute the silver minerals in the near-surface parts of the veins.

The shapes and sizes of the primary orebodies were not significantly changed by supergene sulfide enrichment; existing ores were enriched in

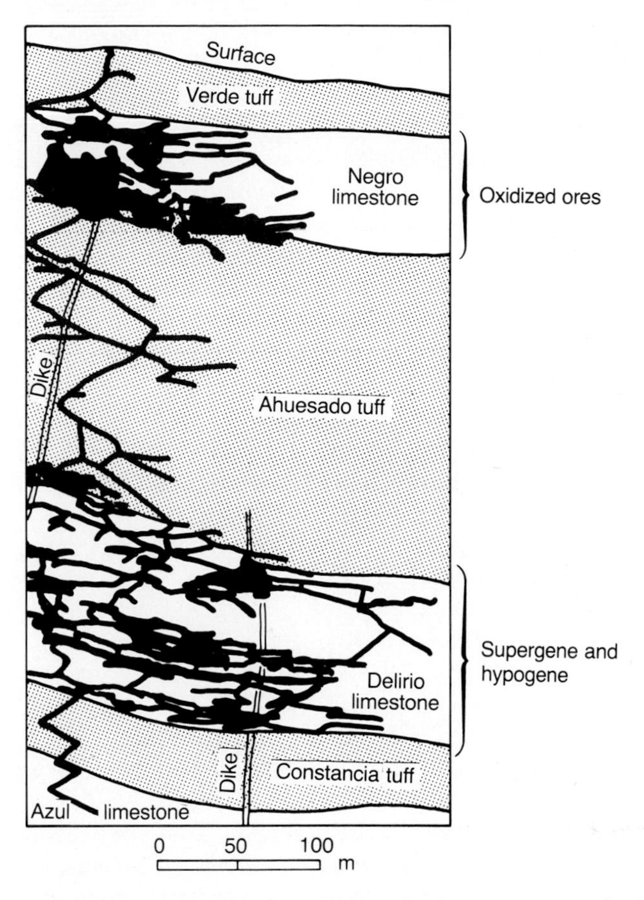

Figure 17-25. Vertical cross section through the Constancia mine, Chañarcillo, Chile, showing by the distribution of stopes (black), the concentration of ore in limestones, and the extensions of ore parallel to the bedding. (*From* Whitehead, 1919.)

silver at the expense of iron, antimony, arsenic, and sulfur. The zone of supergene enrichment ranges in thickness from a minimum of 40 meters in the northern part of the district to a maximum of 200 meters in the south (Whitehead, 1942). Descending meteoric waters leached silver from near-surface veins and precipitated their leached silver content when they reached the primary ore beneath the water table. Accordingly, the upper parts of the hypogene sulfide zones were selectively enriched, and the ore grade diminished from the top to the bottom of the bed. Enrichment was also extensive along faults and water courses. Enrichment obliterated the primary minerals in the richest zones, but pseudomorphous relationships between hypogene and supergene minerals are characteristic in all the enriched ores.

Supergene minerals include stephanite (Ag_5SbS_4), acanthite, dyscrasite (Ag_3Sb), native silver, stromeyerite ($AgCuS$), and minor amounts of pearceite and polybasite (Whitehead, 1919). Enrichment was by replacement of early sulfides by secondary sulfides, antimonides, and native silver; little replacement of the gangue minerals was noted. As a rule, the supergene minerals were richer in silver than the hypogene minerals.

Studies of the supergene sulfide paragenesis have clarified the processes of enrichment. In the early stages of enrichment, the ruby silver minerals pyrargyrite and proustite were placed by acanthite, stromeyerite, stephanite, and small amounts of polybasite and pearceite. Because pyrargyrite was more susceptible to replacement than proustite, it was the first hypogene mineral attacked. Replacement was clearly recorded in concentric bands of supergene minerals formed around unaltered cores of the ruby silvers. Dyscrasite and native silver were restricted to zones of intense enrichment, forming irregular dendritic masses in pyrargyrite, but also in proustite, pearceite, and polybasite as enrichment progressed. Further supergene enrichment eventually resulted in the complete replacement of primary minerals, until native silver and dyscrasite replaced even the earlier supergene minerals in the upper part of the enriched zone. The most intense enrichment was represented by veins of massive dyscrasite and native silver, which usually replaced sulfides but locally filled open spaces or replaced calcite along cleavage planes.

Above the supergene sulfides in the Delirio limestone, and separated from them by 200 meters of Ahuesado tuff (Figure 17-25), is a second bed of limestone which contains oxidized ores. In contrast to the supergene-enriched lodes, the oxidized orebodies in the Negro limestone were changed from their original configurations. The veins were thickened to as much as 10 meters, and became irregular rather than elongated. Oxidation developed silver halides which were zoned according to relative solubilities. Chlorargyrite (AgCl), the least soluble, formed the upper zone, embolite [Ag(Br,Cl)] defined the intermediate zone, and iodyrite (AgI) formed the lower zone (Miller and Singewald, 1919). Bromyrite (AgBr) and iodembolite [Ag(Cl,Br,I)] were also common oxidation products. In places the oxidized ores cropped out, but elsewhere they were overlain by tuff. The oxidized ore minerals were deposited both as replacements of calcite and sulfides and as open-cavity fillings, replacement being the dominant process. As the groundwater level was slowly lowered through the silver-bearing rocks, the oxidized zone gradually encroached on the supergene-enriched sulfides. Thus parts of the oxidized native silver and dyscrasite ores were replaced by silver halides.

Whitehead (1919) studied the chemistry of the enrichment processes and concluded that sulfuric acid and ferric sulfate, mixed with halides that were probably windblown from the Pacific, were the active solvents and reagents. The abundant calcite quickly neutralized the acid, but enough ferric sulfate was present to facilitate the leaching process.

Very little hypogene ore was mined in the Chañarcillo district, even though the primary veins contained 60 to 150 oz (2000 to 5000 ppm) of silver per metric ton. Oxidation and supergene enrichment increased the silver content 25 to 80%, forming deposits with 100 to 240 oz (3100 to 7500 ppm) of silver per ton. The remoteness of Chañarcillo prohibited mining of anything but high-grade ores, and the primary veins were further excluded because of their depth and the presence of considerable water in the lower levels. Thus because of both their nearness to the surface and their higher content of silver, the supergene-enriched zones contained the most favorable mineralization for mining. Since only one metal was involved, Schürmann's series did not control secondary deposition of one metal with respect to another, but rather replacement was in the direction of the highest possible silver-to-sulfur or semimetal ratios.

Bibliography

Anderson, J. A., 1981. Characteristics of leached capping and techniques of appraisal, pp. 275–296 in S. R. Titley, Ed., *Advances in Geology of the Porphyry Copper Deposits, Southwestern North America.* Tucson: Univ. Ariz. Press, 560 pp.

Avias, J., 1968. Pers. Commun.

Bateman, A. M., 1950. *Economic Mineral Deposits.* New York: Wiley.

Beane, R. E., 1968. An investigation of the manner and time of formation of malachite, Unpub. M. S. Thesis, Univ. Ariz., 82 pp.

Berthelot, C., 1933. Le nickel et le chrome dans les colonies françaises. *Chimie & Indus.* 29:718–723.

Blanchard, R., 1944. Derivatives of chromite. *Econ. Geol.* 39:448.

——, 1968. Interpretation of leached outcrops. Nev. Bur. Mines Bull. 66.

Blume, H., 1962. Der Bauxitbergbau auf Jamaika. *Geogr. Rundsch.* 14(6):227–235.

Bohn, H. L., B. L. McNeal, and G. O'Connor, 1979. *Soil Geochemistry.* New York: Wiley, 329 pp.

Bramlette, M. N., 1947. Lateritic ore of aluminum in the Greater Antilles. *Geol. Soc. Amer. Bull.* 58:1248–1249.

Brown, J. S., 1936. Supergene sphalerite, galena, and willemite at Balmat, N.Y. *Econ. Geol.* 31:331–354.

——, 1942. Differential density of ground water as a factor in circulation, oxidation, and ore deposition. *Econ. Geol.* 37:310–317.

Burns, D. J., 1961. Some chemical aspects of bauxite genesis in Jamaica. *Econ. Geol.* 56:1297–1303.

Butler, B. S., G. F. Loughlin, V. C. Heikes, et al., 1920. The ore deposits of Utah. U.S.G.S. Prof. Pap. 111, pp. 521–527.

Chételat, E. de, 1947. La genèse et l'évolution des gisements de nickel de la Nouvelle-Calédonie. *Soc. Geol. France Bull.*, ser. 5, 17:105–160.

Chubb, L. J., 1963. Bauxite genesis in Jamaica. *Econ. Geol.* 58:286–289.

Comer, J. B., 1974. Genesis of Jamaican bauxite. *Econ. Geol.* 69:1251–1264.

Cooke, H. C., 1913. The secondary enrichment of silver ores. *J. Geol.* 21:1–28.

Cumberlidge, J. T., and F. M. Chace, 1968. Geology of the Nickel Mountain mine, Riddle, Oregon, pp. 1650–1662 in J. D. Ridge, Ed., *Ore Deposits of the United States 1933/1967*, Graton-Sales Vols., New York: AIME, 1880 pp.

Dorr, J. V. N., II, I. S. Coelho, and A. Horen, 1956. The manganese deposits of Minas Gerais, Brazil. Symp. sobre Yacimientos de Manganeso, 20th Int. Geol. Congr., Vol. 3, pp. 279–346.

Ehrlich, H. L., 1964. Bacterial oxidation of arsenopyrite and enargite. *Econ. Geol.* 59:1306–1312.

Emmons, W. H., 1912. The agency of manganese in the superficial alteration and secondary enrichment of gold-deposits in the United States. *AIME Trans.* 42:3–73.

——, 1917. The enrichment of ore deposits. USGS Bull. 625.

—— and F. B. Laney, 1926. Geology and ore deposits of the Ducktown mining district, Tennessee. USGS Prof. Pap. 139.

Flores-Williams, H., 1959. *Apuntes de Geología Económica de Yacimientos Minerales.* Santiago: Univ.Chile, Escuela de Geologia.

Foose, M. P., W. D. Menzie, D. A. Singer, and J. T. Hanley, 1980. The distribution and relationships of grade and tonnage among some nickel deposits. USGS Prof. Pap. 1160.

Garrels, R. M., 1954. Mineral species as functions of pH and oxidation-reduction potentials, with special reference to the zone of oxidation and secondary enrichment of sulfide ore deposits. *Geochim. Cosmochim. Acta* 5:153–168.

——, 1960. *Mineral Equilibria.* New York: Harper.

—— and C. L. Christ, 1965. *Solutions, Minerals, and Equilibria.* New York: Harper, 450 pp.

—— and F. T. MacKenzie, 1971. *Evolution of Sedimentary Rocks.* New York: Norton, 397 pp.

Glasser, E., 1904. Rapport à M. le Ministre des Colonies sur les richesses minérales de la Nouvelle-Calédonie. *Ann. Mines* 5:29–154, 503–701.

Golightly, J. P., 1979. Nickeliferous laterites: a general description, pp. 3–23 in D. J. I. Evans, R. S. Shoemaker, and H. Veltman, Eds., *International Laterite Symposium*, New York: AIME, 688 pp.

——, 1981. Nickeliferous laterite deposits. *Econ. Geol. 75th Anniv. Vol.*, pp. 710–735.

Guimaraes, D., 1935. Contribução aoestudo da origem dos depósitos de minério de ferro e manganes do Centro de Minas Gerais. Serv. Fom. Prod. Min. Bull. 8.

Hansuld, J., 1966. Behavior of molybdenum in secondary dispersion media—a new look at an old geochemical puzzle. *Min. Eng.*, Dec, pp. 73–77.

Harder, E. C., 1952. Examples of bauxite deposits illustrating variations in origin, in *Problems of Clay and Laterite Genesis*. New York: AIME.

Hartman, J. A., 1955. Origin of heavy minerals in Jamaican bauxite. *Econ. Geol.* 50:738–747.

Hem, J. D., 1960. Some chemical relationships among sulfur species and dissolved ferrous iron. USGS Water Sup. Pap. 1459-C, pp. 57–73.

—— and W. H. Cropper, 1959. Survey of ferrous-ferric chemical equilibria and redox potentials. USGS Water Sup. Pap. 1459-A, pp. 1–31.

Hewett, D. F., and M. Fleischer, 1960. Deposits of the manganese oxides. *Econ. Geol.* 55:1–55.

Hill, V. G., 1955. The mineralogy and genesis of the bauxite deposits of Jamaica, B.W.I. *Amer. Mineral.* 40:676–688.

Horen, A., 1955. Mineralogy and petrology of the manganese protore at the Merid mine, Minas Gerais, Brazil. *Geol. Soc. Amer. Bull.* 66:1575.

Hose, H. R., 1950. The geology and mineral resources of Jamaica. *Colon. Geol. Min. Res.* 1(1):11–36.

——, 1959. The origin of bauxites in British Guiana and Jamaica. 5th Inter-Guiana Geol. Conf.

——, 1963. Jamaica-type bauxites developed on limestone. *Econ. Geol.* 58:62–69.

Keller, W. D., 1956. Clay minerals as influenced by environments of their formation. *Amer. Assoc. Petrol. Geol. Bull.* 40:2689–2710.

Kelly, W. C., 1961. Some data bearing on the origin of Jamaican bauxite. *Amer. J. Sci.* 259:288–294.

Koenig, B. A., 1978. Oxidation-leaching and enrichment zones of a porphyry deposit—a mineralogic and quantitative chemical study. Unpub. M. S. Thesis, Univ. Ariz., 57 pp.

Lacy, W. C., 1949. Oxidation processes and formation of oxide ore at Yauricocha. Soc. Geol. Peru (Vol. Jubilar XXV Aniv.), pt. II, No. 12.

Lelong, F., Y. Tardy, G. Grandin, J. J. Trescases, and B. Boulange, 1976. Pedogenesis, chemical weathering, and processes of formation of some supergene ore deposits, pp. 93–173 in K. H. Wolf, Ed., *Handbook of Stratabound and Stratiform Deposits*, vol. 6, New York: Elsevier.

Lindgren, W., 1933. *Mineral Deposits*, 4th ed. New York: McGraw-Hill, pp. 849–850.

Locke, A., 1926. *Leached Outcrops as Guides to Copper Ore.* Baltimore: Williams and Wilkins.

McKnight, E. T., 1942. Differential density of ground water in ore deposition. *Econ. Geol.* 37:424–426.

Meillon, J. J., 1978. Economic geology and tropical weathering. *Can. Inst. Min. Metall. Bull.*, July, 71:61–70.

Miller, B. L., and J. T. Singewald, Jr., 1919. *The Mineral Deposits of South America.* New York: McGraw-Hill, 598 pp.

Moesta, F. A., 1928. El mineral de Chañarcillo. *Bol. Minero Santiago Chile* 40:167–182 (transl. of 1870 article).

Norton, S. A., 1973. Laterite and bauxite formation. *Econ. Geol.* 68:353–361.

Osborn, E. T., and J. D. Hem, 1961. Microbiologic factors in the solution and transport of iron. USGS Water Sup. Pap. 1459–H, pp. 213–235.

Odman, O. H., 1955. Morro da Mina manganese deposit and its protore. *Eng. Mineral. Metal.*, Feb., p. 57.

Olmstead, H. W., and D. W. Johnson, 1966. Inspiration geology, pp. 143–150 in S. R. Titley and C. L. Hicks, Eds., *Geology of the Porphyry Copper Deposits.* Tucson: Univ. Ariz. Press, 287 pp.

Park, C. F., Jr., 1956. On the origin of manganese. Symp. sobre Yacimientos de Manganeso., 20th Int. Geol Congr. vol. 1, pp. 75–98.

——, J. V. N. Dorr, P. W. Guild, and A. L. M. Barbosa, 1951. Notes on the manganese ores of Brazil. *Econ. Geol.* 46:1–22.

Pecora, W. T., 1944. Nickel-silicate and associated nickel-cobalt-manganese-oxide deposits near São José do Tocantins, Goiaz, Brazil. USGS Bull. 935–E, pp. 247–305.

Percival, F. G., 1966. The lateritic iron deposits of Conakry, with discussion. *Inst. Min. Metal. Trans.* 75:B85–B91.

Peterson, N. P., 1954. Copper Cities copper deposit, Globe-Miami district, Arizona. *Econ. Geol.* 49:362–377.

——, 1962. Geology and ore deposits of the Globe-Miami district, Arizona. USGS Prof. Pap. 342.

——, C. M. Gilbert, and G. L. Quick, 1951. Geology and ore deposits of the Castle Dome area, Gila County, Arizona. USGS Bull. 971.

Ransome, F. L., 1919. The copper deposits of Ray and Miami, Arizona. USGS Prof. Pap. 115.

Rose, A. W., H. E. Hawkes, and J. S. Webb, 1979. *Geochemistry in Mineral Exploration.* New York: Academic Press, 657 pp.

Routhier, P., 1952. Les gisements de fer de la Nouvelle-Calédonie, *Symp. sur les Gisements de Fer du Monde, 19th Int. Geol. Congr.*, vol. 2.

——, 1963. Etude géologique du versant occidental de la Nouvelle-Calédonie entre le Col de Boghen et la Pointe d'Arama. *Soc. Geol. France Memo.* (n.s.) 32(67):271.

Roy, R., A. J. Majumdarmand, and C. W. Hulbe, 1959. The Ag_2S and Ag_2Se transitions as geologic thermometers. *Econ. Geol.* 54:1278–1280.

Saegart, W. E., J. D. Sell, and B. E. Kilpatrick, 1974. Geology and mineralization of La Caridad porphyry copper deposit, Sonora, Mexico. *Econ. Geol.* 69:1060–1077.

Salas, G. P. 1959. Los depositos de bauxita en Haiti y Jamaica y possibilidades de que exista bauxita en Mexico. Mex. Univ. Nac. Inst. Geol. Bol. 59, pp. 9–42.

Sato, M., 1960. Oxidation of sulfide ore bodies. *Econ. Geol.* 55:928–961, 1202–1231.

Sawkins, J. G., 1869. Reports on the geology of Jamaica, pt. II of the West Indian Survey. Geol. Surv. Great Britain Memo.

Schmedemann, O. C., 1948. Carribean aluminum ores. *Eng. Min. J.* 149(6):78–82.

——, 1950. First Caribbean bauxite development, Reynolds Jamaica Mines, Ltd. *Eng. Min. J.* 151(11):98–100.

Schürmann, E., 1888. Über die Verwandtschaft der Schwermetalle zum Schwefel. *Justus Liebig's Ann. Chem.* 249:326–350.

Schwartz, G. M., 1947. Hydrothermal alteration in the "porphyry copper" deposits. *Econ. Geol.* 42:319–352.

Segerstrom, K., 1962. Regional geology of the Chañarcillo silver mining district and adjacent areas, Chile. *Econ. Geol.* 57:1247–1261.

Sinclair, I. G. L., 1967. Bauxite genesis in Jamaica: new evidence from trace element distribution. *Econ. Geol.* 67:482–486.

Söhnge, P. G., 1964. The geology of the Tsumeb mine, in S. H. Haughton, Ed., *The Geology of Some Ore Deposits of Southern Africa*, vol. 2. Johannesburg: Geol. Soc. So. Afr., 739 pp.

Stokes, N. H., 1907. Experiments on the action of various solutions on pyrite and marcasite. *Econ. Geol.* 2:14–23.

Sullivan, J. D., 1930. Chemistry of leached chalcocite. U.S. Bur. Mines Tech. Pap. 473.

Tenney, J. B., 1935. Globe-Miami district Arizona, in *Copper Resources of the World, 16th Int. Geol. Congr.*, vol. 1.

Titley, S. R., 1978. Geologic history, hypogene features, and processes of secondary sulfide enrichment at the Plesyumi copper prospect, New Britain, Papua New Guinea. *Econ. Geol.* 73:768–784.

——, 1981. Pers. Commun.

Trescases, J. J., 1973. Weathering and geochemical behaviour of the elements of ultramafic rocks in New Caledonia, pp. 149–161 in *Metallogenic Provinces and Mineral Deposits in the Southwestern Pacific*. Austr. Bur. Miner. Res., Geol. Geophys. Bull. 1.

Troly, G., M. Esterle, B. Pelletier, and W. Reibell, 1979. Nickel deposits in New Caledonia—some factors influencing their formation, in D. J. I. Evans, R. S.

Shoemaker, and H. Veltman, Eds., *International Laterite Symposium*. New York: AIME, 688 pp.

Vincenz, S. A., 1964. A note on the radioactivity of Jamaican bauxite and terra rosa. *Overseas Geol. Miner. Res.* 9:295–301.

Waterman, G. C., 1962. Some chemical aspects of bauxite genesis in Jamaica. *Econ. Geol.* 57:829–830.

Webber, B. N., 1972. Supergene nickel deposits. *AIME Trans.* 252:333–347.

Weiss, R. L., 1965. Outcrops as guides to copper ore—four examples. Unpub. Ph.D. Dissert., Columbia Univ., 184 pp.

Whitehead, W. L., 1919. The veins of Chañarcillo, Chile. *Econ. Geol.* 14:1–45.

——, 1942. The Chañarcillo silver district, Chile, pp. 216–220 in W. H. Newhouse, Ed., *Ore Deposits as Related to Structural Features*. Princeton, NJ: Princeton Univ. Press, 280 pp.

Zans, V. A., 1952. Bauxite resources of Jamaica and their development. *Colonia Geol. Miner. Res.* 3(4):307–333; Geol. Surv. Jamaica Pub. 12.

——, 1956. The origin of the bauxite deposits of Jamaica, in *Resumenes de los Trabajos Presentados, 20th Int. Geol. Congr.*

——, 1957. Geology and mineral deposits of Jamaica. Geol. Surv. Dept. Jamaica Pub. 33.

——, 1959. Classification and genetic types of the bauxite deposits. *5th Inter-Guiana Geol. Conf.*, Brit. Guiana Geol. Surv. Proc. 1962, pp. 205–211. Reprinted 1962 as Jamaica Geol. Surv. Pub. 85.

EIGHTEEN

Deposits Related to Regional Metamorphism

The word *metamorphism* can be defined both narrowly and broadly. In its narrow traditional sense, the word connotes the mineralogical and structural adjustment of rocks, minerals, and their textural arrays to temperatures and pressures higher than those under which they formed, commonly in environments including shear stress. The regimes most widely suggested by the term metamorphism are those of igneous, or contact, metamorphism and regional, or dynamothermal, metamorphism. In Chapter 11 we have already considered skarn and related systems at intrusive contacts of igneous rocks with cooler wall rocks. They are an important subset of ore deposits, and are certainly related to metamorphism, but we need not consider them again here. Ore deposits related to regional metamorphism include those of many industrial minerals or rocks such as graphite, garnet, emery, kyanite-sillimanite, pyrophyllite, wollastonite, asbestos, talc, mica, and gems—varieties of corundum (ruby, sapphire), emerald, and garnet. A recently defined type of uranium deposit associated with district-scale migmatitic metamorphism includes Rössing, South-West Africa, Baie Johan Beetz, Quebec, Canada, and others. So it is worthwhile to consider the role of broadscale regional metamorphism, and it is here that the broader definition comes into play. Most metamorphic petrologists would perceive of *regional metamorphism* as involving a series of isograds, the so-called Barrovian zones, extending over tens of kilometers from fresh lithified sedimentary, volcanic, or even igneous rocks through several increasing pressure-temperature regimens, including chlorite (greenschist), biotite, amphibole, and garnet-pyroxene (granulite) isograds. These zones are shown along with best estimates of temperature, pressure, and depth in Figure 18-1. The economic geologist must then consider how these environments and processes interact with ore deposition. Several pertinent questions come to mind:

1. What effect might increasing metamorphic rank have on preexisting sulfide, oxide, or carbonate deposits and on their alteration assemblages?
2. Can we distinguish between metamorphosed, preexisting ore deposits in, for example, a garnet-granulite host and an ore deposit that might have been *formed by* that metamorphism?
3. Do metamorphic processes serve to concentrate sulfides, oxides, and other minerals into economic concentrations, and how would those deposits appear?
4. What kinds of ore deposits *are* created by metamorphism?

Answers to these questions are somewhat blurred when we consider the broader definition of metamorphism recently deployed by some economic geologists, a definition that includes as metamorphism even very-low-temperature phenomena like the dewatering-dehydration of a prism of sediments in a geosyncline, for example. These last environments are discussed in Chapter 20, so in this chapter we use the traditional definition and do not include as "metamorphism" normal late-stage rock-forming environments and events. The student should realize that if metamorphism is defined to include any prograde, progressive heating-compressing-deforming event, then metamorphic hydrothermalism has to be included in the low-temperature end of that process. Again, this part is considered in both Chapters 19 and 20.

Let us briefly consider the four questions posed above. The first involves the effect of metamorphism on older ore deposits, which requires us to consider whether regional metamorphic systems are open or closed to district-scale migration of components. It would be expected that volatile

Figure 18-1. As temperatures and pressures rise, rocks formed at surface conditions are metamorphosed to assemblages characterized by new minerals. Those underscored are useful industrial minerals. Rössing, described in this chapter, formed in the shaded portion of the anatexis-syntexis range. *Anatexis* specifies involvement of one rock type, *syntexis* the involvement of several. Assemblage boundaries are gradational. (*After* Winkler, 1965; Turner, 1968.)

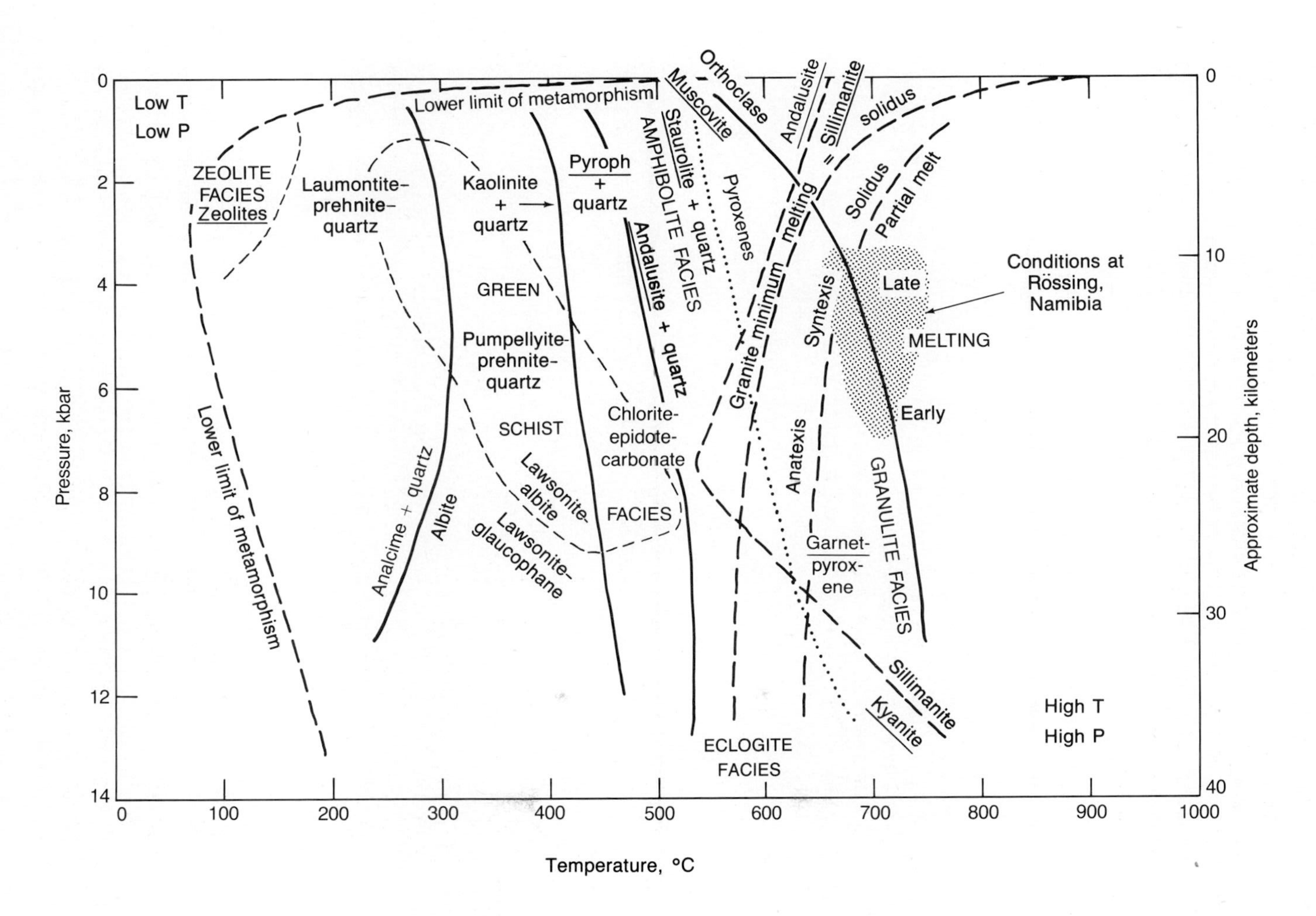

Low T
Low P
High T
High P
Pressure, kbar
0
2
4
6
8
10
12
14
Temperature, °C
0
100
200
300
400
500
600
700
800
900
1000
Approximate depth, kilometers
0
10
20
30
40
ZEOLITE
FACIES
Zeolites
Lower limit of metamorphism
Lower limit of metamorphism
Laumontite-
prehnite-
quartz
Analcime + quartz
Albite
Kaolinite
+ ⟶
quartz
GREEN
Pumpellyite-
prehnite-
quartz
SCHIST
Lawsonite-
albite
Lawsonite-
glaucophane
Pyroph
+
quartz
Andalusite + quartz
Chlorite-
epidote-
carbonate
FACIES
Muscovite
Orthoclase
Staurolite + quartz
AMPHIBOLITE FACIES
Pyroxenes
Andalusite
Sillimanite
Granite minimum melting
solidus
Solidus
Partial melt
Syntexis
Anatexis
Late
MELTING
Early
Conditions at
Rössing,
Namibia
GRANULITE FACIES
Garnet-
pyrox-
ene
ECLOGITE
FACIES
Sillimanite
Kyanite

components like sulfur, carbon dioxide, and water might be fugitive, but the record is mixed even here except for water, which is driven out with increasing rank. If any elements are to be relocated by metamorphism, those of relatively high fugacity or vapor pressure should be most vulnerable. But limestone is metamorphically recrystallized to marble, not to calcium oxide; CO_2 is not necessarily dispatched, and sulfides commonly appear to have persisted into high-rank metamorphic rocks. In the Balmat-Edwards district of New York, sulfides, carbonates, and sulfates have all survived upper amphibolite facies metamorphism without disintegration or mobilization (Engel and Engel, 1962). Metamorphic petrologists used to think in terms of regional metamorphically driven potash metasomatism and "basic fronts" as part of the process called *granitization*, but recent studies have tended to indicate that reequilibration and redistribution of components occur over centimeters or meters rather than kilometers until temperatures close to melting are achieved.

An aspect of these discussions is embodied in observations by Stanton, (1976a–d, 1978) at Broken Hill, New South Wales, Australia (Chapter 13) and by Davis (1972) at the New Brunswick, Canada, massive sulfide deposits. They both showed that stratabound sulfide ores can be intensely metamorphosed and structurally deformed with no more than negligible selective migration of their constituents. Stanton showed at Broken Hill that one can "look through" even granulite metamorphism to determine that original metal ratios are maintained even in stratigraphically close, thin, widespread units measured in centimeters of thickness and square kilometers of area. Copper-lead-zinc ratios in adjacent deformed layers retain their lateral identities such that the Broken Hill geologist or miner knows exactly what stratigraphic unit he is in wherever in the district he might be. In other words, original Cu-Pb-Zn values remained stable and fixed in place while silicates, oxides, and sulfides were metamorphically recrystallized around them. An almost identical circumstance has been discerned at Aggeneys-Gamsberg in the Republic of South Africa, where similarly persistent metal ratios in thin beds are distinctive over wide areas, even with extensive penetrative and flexural deformation and garnet-grade metamorphism. One of the authors visited those deposits with Stanton, and watched as one phenomenon after another reemphasized to Stanton the fact that metamorphism by itself does not necessarily transport sulfides. More (1980) convincingly documented suggestions by Gilmour, Still, and others that the Antler deposit in the Hualpai Mountains of northwest Arizona is a Proterozoic age, complexly folded, deformed, garnet-amphibolite-rank metamorphosed exhalative copper-zinc massive sulfide in which the original sulfides did not migrate, although they were recrystallized, smeared, sheared, and penetratively disconnected. In a review article on regional metamorphic mobilization of sulfides, Vokes (1971) noted that it appears "that there is rather general agreement, among the most recent writers at least, that this metamorphic mobilization involves limited transport distances of the order of millimeters to meters rather than kilometers." He also stated that "there seems to be little disagreement that the ore-forming minerals as a class are no more mobile than rock-forming minerals as a class." It thus appears that metamorphism can reach extreme conditions short of melting without obliteration, dispersal, or kilometer-scale homogenization of preexisting sulfides or sulfide layers. Studies in Scandinavia, Africa, and Canada reinforce this conclusion. If mobilization occurs, it is only close to the point of melting, as at Åmmeberg, Sweden.

Sales and Meyer (1951) showed that a rhyolite dike cutting a Butte, Montana, Cu-Fe-As-S vein served only locally to desulfurize the vein contents from bornite-enargite-pyrite to chalcopyrite-tennantite-hematite without redistribution of metal values, and Mookherjee and Dutta (1970) mapped a diabase dike–massive sulfide ore intersection at the Geco mine in Ontario, Canada, and described similarly local effects with centimeter-scale migration of melted minerals. They concluded that such sulfide melting is "inadequate for regenerating entirely new orebodies." Davis (1972) showed by structural analysis that sulfide layers in New Brunswick showed exactly the same deformation history as the moderately metamorphosed Ordovician metasediments in which they occur, and that the layers thus appeared to be of volcanic-synsediment origin rather than of epigenetic, postdeformation, structurally controlled hydrothermal replacement origin, as they had been thought to be. In so doing, he also demonstrated that sulfides can survive metamorphism without redistribution. Through studies such as his and Stanton's, question 1 is answered. So also is question 2: although extensive chemical, structural, and petrologic study is required, it is normally possible to discern the timing of the concentration of elements into an orebody relative to the age of regional metamorphism.

A final part of the answer to question 2 concerns the metamorphism of alteration effects that might have been associated with preexisting ore deposits. The trained explorationist should be able to predict these effects on the basis of chemical mineralogy and the principles of metamorphic petrology. If a porphyry copper deposit is regionally metamorphosed, certainly the kaolinite- and montmorillonite-family clay minerals in its alteration halos will be dehydrated to aluminosilicates like andalusite or kyanite (Figure 18-1). But the potassic assemblage minerals—quartz, K-feldspar, albite, and biotite—should be expected to survive to high isograds by virtue of their near identity with the low-temperature granitoid silicates. Phyllic alteration, with essential quartz and sericite, can become a phyllite or muscovite schist stable at least to amphibolite facies metamorphism (Figure 18-1), commonly with kink or chevron folds (Figure 18-2). And propylitic assemblage minerals both merge into potassic phases—and thus can be stable at relatively high metamorphic rank—and are identical with greenschist metamorphic assemblages. So except for dehydration and perhaps some

Figure 18-2. Chevron or kink folding—the "ripples" in these pieces of core—developed by regional metamorphism of a quartz-sericite alteration zone around the Haib porphyry copper deposit in South-West Africa (Namibia). Such a rock might as well be called a sericite schist.

minor tectonic deformation, these alteration assemblages should be recognizable at higher metamorphic rank.

One would predict that the metamorphism of an Mg-Fe:Al-Si hydrous silicate like chlorite might produce a mineral like cordierite [$(Mg,Fe)_2Al_4Si_5O_{18}$] or the amphibole anthophyllite [$(Mg, Fe)_7 Si_8O_{22}(OH)_2$] at lower temperatures, and perhaps a garnet like pyrope [$(Mg, Fe)_3Al_2(SiO_4)_3$] or almandite [$Fe_3Al_2(SiO_4)_3$] at higher ones. Cordierite-anthophyllite schists called *dalmatianite* are found in metamorphic rocks near Noranda, Quebec, and at the Antler mine in northwest Arizona, exactly where alteration pipes, which were the chloritized conduits to suprajacent massive sulfide deposits, would be expected to have occurred; and garnet-quartz gneiss called *gar-q* (or *garcue*) is found in the same position in footwall rocks at Broken Hill, Australia, and Gamsberg, South Africa.

Now for questions 3 and 4: if metamorphism does not generally disperse and destroy already formed ore deposits, what effects does it have? Some—like desulfurization—have already been noted. The remarkable willemite-zincite-franklinite iron-zinc silicate-oxide ores in sillimanite-rank metasediments of the Franklin Furnace mine near Sterling, New Jersey, are thought by some (Frondel and Baum, 1974) to have resulted from desulfidation of a preexisting, probably volcanogenic layered pyrite-sphalerite deposit (Callahan, 1966). A large body of information on recrystallization effects in the sulfide minerals of sulfide ores (Stanton, 1972; Clark and Kelly, 1973; Salmon, Clark, and Kelly, 1974) shows general coarsening of grain sizes, development of smooth straight-line mutual 120° grain boundaries, and the purging of some solid solutions by exsolution. And there is evidence that some of the more malleable sulfides can be physically relocated—tongues of nearly pure, deformed, coarse-grained galena jut from a Cu-Zn-Pb massive sulfide body for 1 meter or so outward into dilation crevices in the silicate wall rocks at the Bruce mine near Bagdad, Arizona, and are almost certainly of this metamorphic redistribution-injection origin. Sangster (1971) identified three ways in which metamorphism can create an orebody, none of which need involve more than local movement of components. First, deformation in the form of isoclinal folding can increase ore-layer widths in hinge zones, as was later described by Sawkins and Rye (1974) at the Homestake gold deposits (Chapter 13). Second, selective mobilization—a solid-state plastic flow phenomenon—can substantially increase metal values by mineral flowage in fold hinge zones, generally chalcopyrite and galena more than pyrite or sphalerite. McDonald (1970) reported a 40% increase from 50 to 70% Pb from limbs to crests of folds in sulfide layers at Mt. Isa (Chapter 14). And third, recrystallization can coarsen grain size and promote exsolution of solid solutions to make fine-grained or complex ores more amenable to metal extraction. This aspect has been covered by Gross (1961) in iron ores, by McDonald (1970) in Mt. Isa sulfides, and by Stanton (1976a–d) and Clark and Kelly (1973).

Question 3 asks: Do regional metamorphic processes by themselves concentrate sulfides or oxides into ore deposit configurations? The answer here has to be "only exceptionally"; except for granulite-facies high-rank metamorphic environments that produce migmatites, partial melting, and some ore types described below, there are no environments shown in Figure 18-1 that typically contain sulfide or oxide ores; neither have particular types of sulfide ore deposits been identified that occur only in specific metamorphic rocks. Regional metamorphism, as already stated, is normally a closed-system phenomenon, except for the alkali elements sodium and potassium; otherwise it is more dispersive than concentrative. Lincoln (1981) did find that prehnite-pumpellyite facies burial metamorphism—not actually regional metamorphism—substantially redistributed copper in the Karmutsen tholeiitic basalt flows on Vancouver Island, British Columbia, Canada. Copper dispersed in the flows as chalcopyrite was dissolved and migrated to amygdular cavities, where it reprecipitated as native copper with prehnite. No values approaching economic grades were achieved, but the process may be significant. However, an entire book, *Mineralization in Metamorphic Terranes* (Verwoerd, 1978), describes scores of ore deposits that occur in metamorphic rocks without ascribing an origin involving metamorphic processes to any of them. Intrusive activity into metamorphic terranes can produce "lateral secretion" deposits of the type discussed in Chapters 19 and 20, but these are not the products of regional metamorphic processes themselves.

Question 4 asks what kinds of ore deposits *are* formed by regional metamorphism other than the salutary local effects presented earlier and discussed by Sangster (1971). They are of two types. The first involves local centimeter-to-meter-scale redistribution of mineral components by the chemical mobility induced by higher temperatures and pressures, probably in the presence of pore water; new metamorphic rocks and minerals such as garnets are formed. The second is generated in environments that produce granulite gneisses and *migmatites*, complex interleavings of igneous-looking metamorphic quartzofeldspathic rocks, minerals, and textures. High-temperature, high-pressure metamorphism commonly has approached or reached stages of partial melting in migmatites—low-melting quartzofeldspathic silicates, such as quartz, K-feldspar, albite, and biotite, and oxides, such as magnetite, hematite, uraninite, and scheelite-wolframite, are mobilized and reemplaced, often as the lit-par-lit injections common in migmatites.

We consider some occurrences of the first type in the following paragraphs, and then describe those related to migmatite-granulite complexes.

▸ GORE MOUNTAIN GARNET DEPOSITS, NEW YORK

Upstate New York (Figure 18-3) is a classic area for the study of metamorphic deposits of industrial minerals. In addition to scattered minor occurrences of wollastonite, pyrophyllite, and graphite, major deposits of talc and garnet occur there, generally in 1100-million-year-old Grenville age banded gneiss complexes. Here are the talc deposits of Gouverneur, New York, in a belt of northeast-elongate folded interlayered impure marbles, which contained requisite amounts of magnesia in dolomite, silica in silt and sand, and water, such that metamorphism produced the magnesian mica talc [$Mg_3Si_4O_{10}(OH)_2$] as pure fine-grained masses called *steatite* or impure ones called *soapstone*. Calcium also present in the marbles produced tremolite schist, which is fibrous, unctuous, and "talcose," and along with anthophyllite and serpentine [$Mg_3Si_2O_5(OH)_4$] is one of the marketable "impurities." The Gouverneur talc belt (Figure 18-3) is 10 km long, 100 to 150 meters thick stratigraphically, and extends several hundred meters downdip. The talc is mined by open-pit methods. Other deposits in California, Montana, North Carolina, the Alps of France and Italy, Norway, and India involve hydration or metamorphism of dolomite; deposits in Vermont, the Urals of the U.S.S.R., and South Africa are in hydrated ultramafic rocks like Alpine peridotites and komatiites. These last environments also lead

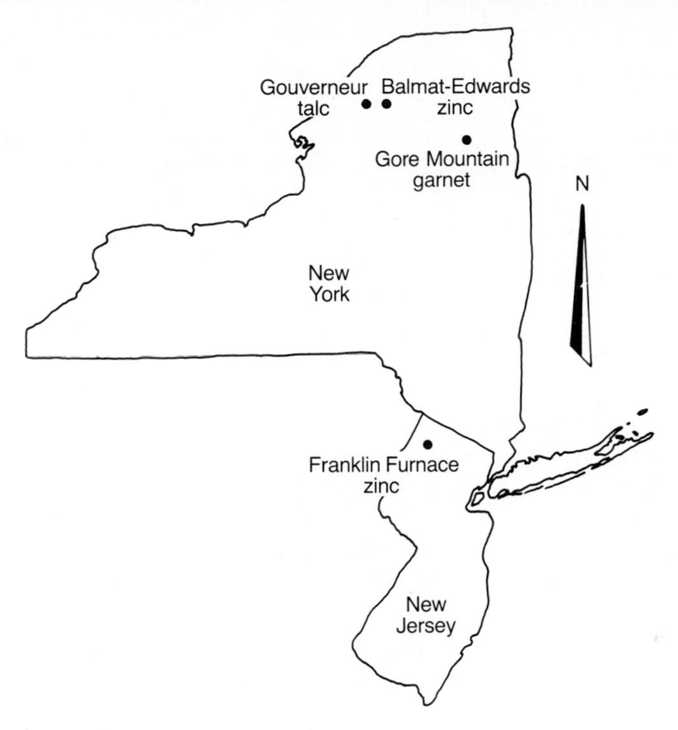

Figure 18-3. Location map of New York and New Jersey showing the locations of the Gouverneur talc, Gore Mountain garnet, Balmat-Edwards zinc, and Franklin Furnace zinc deposits.

to the generation of the fibrous serpentine chrysotile, the cylindrically curled 1 : 1 layer mineral $Mg_3Si_2O_5 \cdot (OH)_4$; it is called *asbestos* and is sought for its weavability and its chemical, thermal, and electrical resistance. Pyrophyllite mined from regionally metamorphosed hydrated rhyolite tuffs in Quebec, North Carolina, and Japan is also "talcose" enough for use in cosmetics.

The garnet deposits of New York are also in Precambrian Grenville Series gneisses, marbles, and schists. Near North Creek, New York, 110 km north of Albany, these metasediments were invaded at the peak of metamorphism by an anorthosite-gabbro complex and a quartz-syenite body which affected both. The garnets (Barton, 1983) are 1-to-30-cm-diameter, cinnamon colored almandite crystals that stud the light-colored plagioclase-hornblende-biotite syenite and, with hornblende-plagioclase rims (Figures 18-4 and 18-5), the dark gray gabbro. Garnets in the gabbro are commonly 5 to 12 cm across, but they may reach half a meter in diameter. They commonly constitute 20 to 60 vol % of the rock, but overall mine grade is 7% garnet and 93% gangue. They appear to owe their origin to metamorphic

Figure 18-4. Outcrop of oval amphibole-rimmed garnets averaging about 10 cm across in a mine face at the Gore Mountain abrasive garnet deposits near North Creek, New York. The host rock is metagabbro. Some components of the almandite garnet have migrated as much as several centimeters to garnet growth nuclei.

Figure 18-5. Single crystal of garnet 15 cm across in the metagabbro of the Gore Mountain garnet mine in upstate New York near Albany. The dark rim is a reaction rim corona of hornblende. The bar scale is in centimeters.

hybridization and mobilization of proper proportions of magnesium, iron, aluminum, and silica at high pressures of 5 to 6 kbar and temperatures of 450 to 700°C (Figure 18-1). The Gore Mountain–Ruby Mountain garnets are $(Mg=Fe)_3Al_2(SiO_4)_3$, have a hardness between 8 and 9 on Mohs' scale, and are particularly valuable for polishing, sandblasting, and sand paper abrasive because of their coarseness, purity, uniformity, and hardness, and because when crushed they splinter to myriad chisel-shaped flakes that are ideal cutting edges. The open pit is 2 km long east-west, up to 200 meters wide across the hybrid zone, and 30 to 40 meters deep. Present production of about 1000 tons of garnet a month is from the Ruby Mountain pit. Gem quality garnet is common, but it is not routinely recovered. The mines have been operated continuously by members of the Barton family, descendants of the deposit's discoverer, since colonial times.

Other abrasive products from regional metamorphic terranes include corundum and *emery*, a fine-grained mixture of corundum and magnetite commonly with minor spinel and plagioclase used for whetstones and grindstones. The "starting materials" may have been either laterites or bauxites on limestones or desilicated mafic rocks, both enriched in iron and alumina. Finally graphite occurs as the ultimate metamorphic product of coal and other organic materials in many high-grade metamorphic terranes, and in veins. These veins have long puzzled geochemists because the transport and the deposition of "native carbon" must require unattainably low $f(O_2)$, even lower than that required for carbon monoxide stability. It has recently been shown by Mancuso and Seavoy (1981) that the graphite veins of Malagasy, Sri Lanka, and elsewhere, which are important sources of high-purity flake graphite, probably resulted from granulite-grade metamorphism of preexisting segregations of organic waxes and oils called *anthraxolite*. These veins have long been described in less metamorphosed areas as stemming from the algal content of banded iron formations, massive sulfide deposit environments, and other biomass environments in Precambrian rocks, but somehow no one ever made the connection between veinform graphite and the much more abundant anthraxolite until Mancuso and Seavoy perceived it. Other important deposits of lump, or pulverulent, graphite occur in contact metamorphic configurations in Sonora, Mexico, where Laramide stocks cut thick Triassic coal seams and charred them.

▸ RÖSSING URANIUM DEPOSITS, SOUTH-WEST AFRICA (NAMIBIA)

The other type of ore deposit formed by metamorphism and metamorphic processes is one involving high-temperature, high-level metamorphism of the kind that produces migmatite terranes, metamorphic core complexes, and possibly S-type partially melted felsic granites. As noted, migmatites are composite bodies composed of high-rank metamorphic rocks like amphibolites or garnet amphibolites, which are intruded by felsic igneous-appearing material, chiefly along subparallel planes. The whole banded body may be folded, and local zones of lit-par-lit injection and assimilation are common. Such bodies are thought to represent metamorphic volumes taken to near the melting point, with local partial melting and autoinjection. If sulfur were present, sulfides might be formed, but only oxides are associated with them. *Metamorphic core complexes* have only recently been defined (Crittenden, Coney, and Davis, 1980) as a lithotectonic environment. These complexes appear to be only late Cretaceous and younger in age, to represent vertical domical uplift with lateral sliding off of higher-level sedimentary and metamorphic rocks along a carapace-shaped discontinuity, and to be related to S-type intrusive igneous rocks, possibly of partial fusion syntexis origin. Minor anomalies of uraninite, cassiterite, and wolframite—that is, U-Sn-W—have been reported, but no economic concentrations are known. S-type granites and porphyries have already been discussed in Chapter 11 in their connection with tin-molybdenum-tungsten-beryllium deposits.

Several major deposits of uranium associated with migmatites and migmatitic metamorphism-magmatism are known. One is at Rössing, South-West Africa (Namibia), and another is at Johan Beetz Bay, 800 km northeast of Quebec City, Canada, on the north shores of the St. Lawrence River. Hauseux (1977) described that occurrence so clearly that her abstract is paraphrased here. The rocks are Precambrian granites, granulites, quartzites, biotite schists and gneisses, and amphibolites in a regionally metamorphosed belt of metasediments and gabbro. They are migmatitically invaded by the Turgeon Lake granite, itself the product of partial melting of sediments. Its uraniferous character is manifested by the presence of uraninite, uraniferous magnetite, phosphuranylite, and metamict radioactive zircons. Uraninite occurs in irregular pods in the tectonic foliation and in bands roughly paralleling original bedding. The granite is like other uraniferous granites with local smoky quartz, brick-red orthoclase, magnetite, and coarse to pegmatitic grain sizes. Metasediment-granite boundaries are diffuse to sharp, and usually conformable to bedding and regional foliation. Uranium was mobilized from original sediments now "granitized" by at least two metamorphic events into this migmatitized foliation and into veins and pods in fold axes. Quite clearly, the multiple injections of uraniferous granitic near-melt material were metamorphically driven, probably when the volume was deeply buried and regionally metamorphosed during the Grenville orogeny.

The most valuable deposit of this type is at Rössing, 20 km northeast of Walvis Bay on the coastline of central South-West Africa (Figure 18-6). The Rössing deposit (Berning, Cooke, Heimstra, and Hoffman, 1976) is in the Damaran metamorphic belt, which is composed of 1000-million-year-old

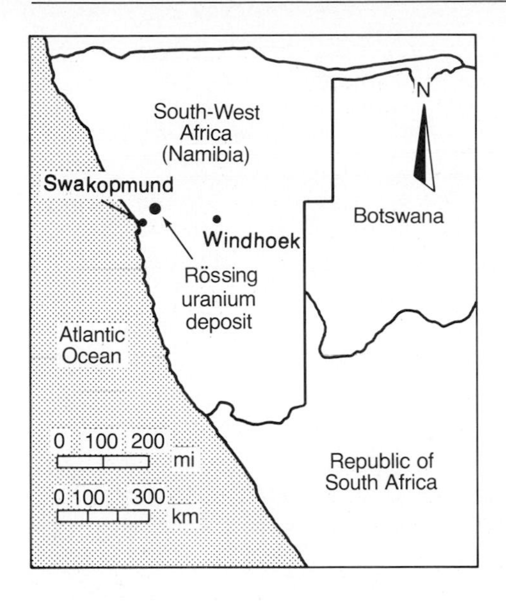

Figure 18-6. Locality plan of the Rössing uranium deposit. The port city of Walvis Bay is immediately southwest of Swakopmund.

late Precambrian sedimentary and volcanic rocks that were intensely deformed and metamorphosed during the Damaran orogenic event 510 million years ago. They are now quartzites, gneisses, and marbles involving garnets, pyroxenes, amphiboles, cordierite, and other high-rank metamorphic rocks and minerals. They were pervasively migmatized and intruded by innumerable dikes, sills, and lit-par-lit impregnations of uraniferous granite, pegmatite, and alaskite, as shown in Figures 18-7 to 18-9. Most of the economic uranium concentration is in alaskite; although neither grade nor tonnage are cited by Berning et al. (1976), grade is probably about 0.05% U_3O_8 with tonnage in the hundreds of millions of tons. The minable deposit is 1 km across by 2.5 km long (Figure 18-7) and is to be mined to an average depth of about 300 meters (Figure 18-8). At district scale, the orebodies are laminar and lie on the flank of a dome in a complex system of gneiss domes and basins. At outcrop scale, intense northeast-southwest isoclinal folding, northwesterly jointing, and recrystallization are evident—early dynamothermal metamorphism to temperatures of 700°C and pressures to 7 kbar and a later pulse at 675 to 750°C and 4 kbar extensively modified original mineralogies and textures to produce a high-rank, folded, laminated gneiss migmatite terrane. Figure 18-9 shows the laminar nature of the migmatites; they were produced by metamorphism and metamorphic melting at conditions described in Figure 18-1. At the peak of metamorphism, the terrane was intruded by quartzofeldspathic "igneous" rocks. Berning et al. (1976) indicate that they were

> *. . . derived by the syntexis of underlying formations, exposures of which are present in the core of the domal structure. The ultimate source of the uranium could thus be the early Precambrian basement that was subjected to syntexis with resultant emplacement of alaskite into stratigraphically higher strata.*

Alaskite is widespread beyond the pit limits, but is not uniformly uraniferous. Uraninite is the dominant primary radioactive mineral as grains averaging only 0.1 mm. It occurs as inclusions in quartz, orthoclase, and biotite, and in fractures, cleavages, and grain boundaries. It is preferentially associated with biotite and zircon. The secondary oxidation-hydration minerals uranophane, torbernite, carnotite, and gummite are also found near the surface. As Figures 18-7 to 18-9 all suggest, the alaskite with its

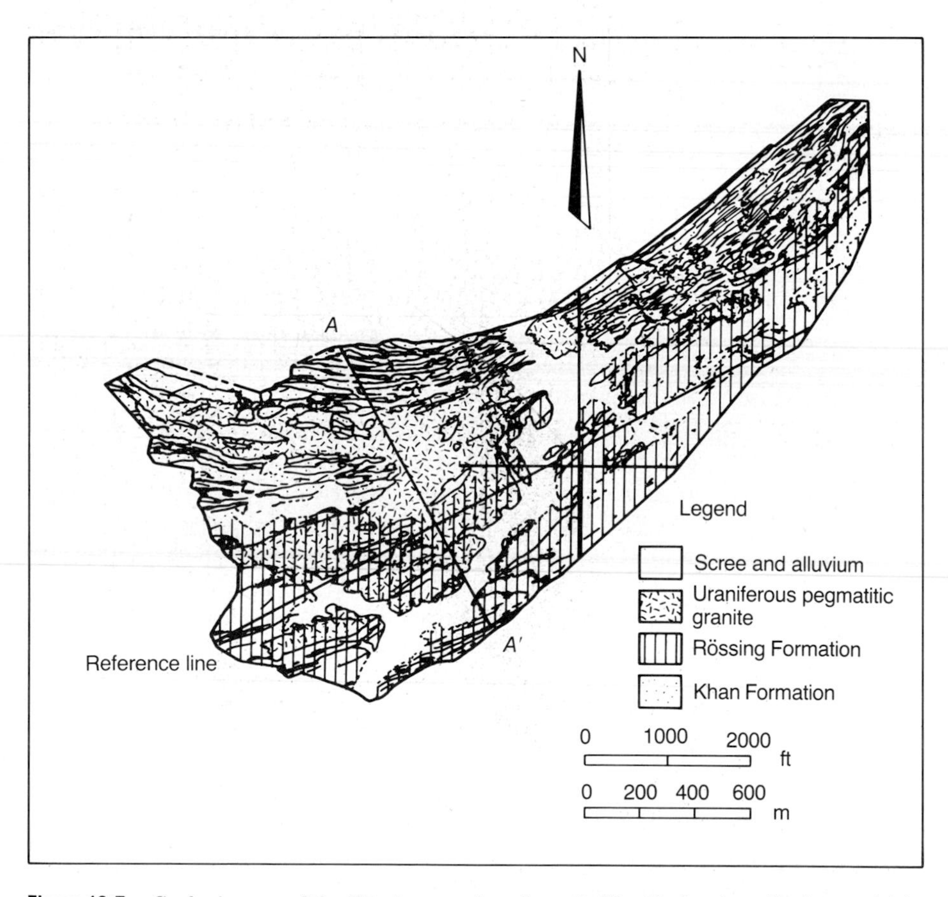

Figure 18-7. Geologic map of the Rössing uranium deposit. The black mineralized material is finely interlayered in the older Khan and Rössing formations and looks like the migmatitic material shown in Figure 18-9. Cross section *A–A′* is Figure 18-9. (*From* Berning et al., 1976.)

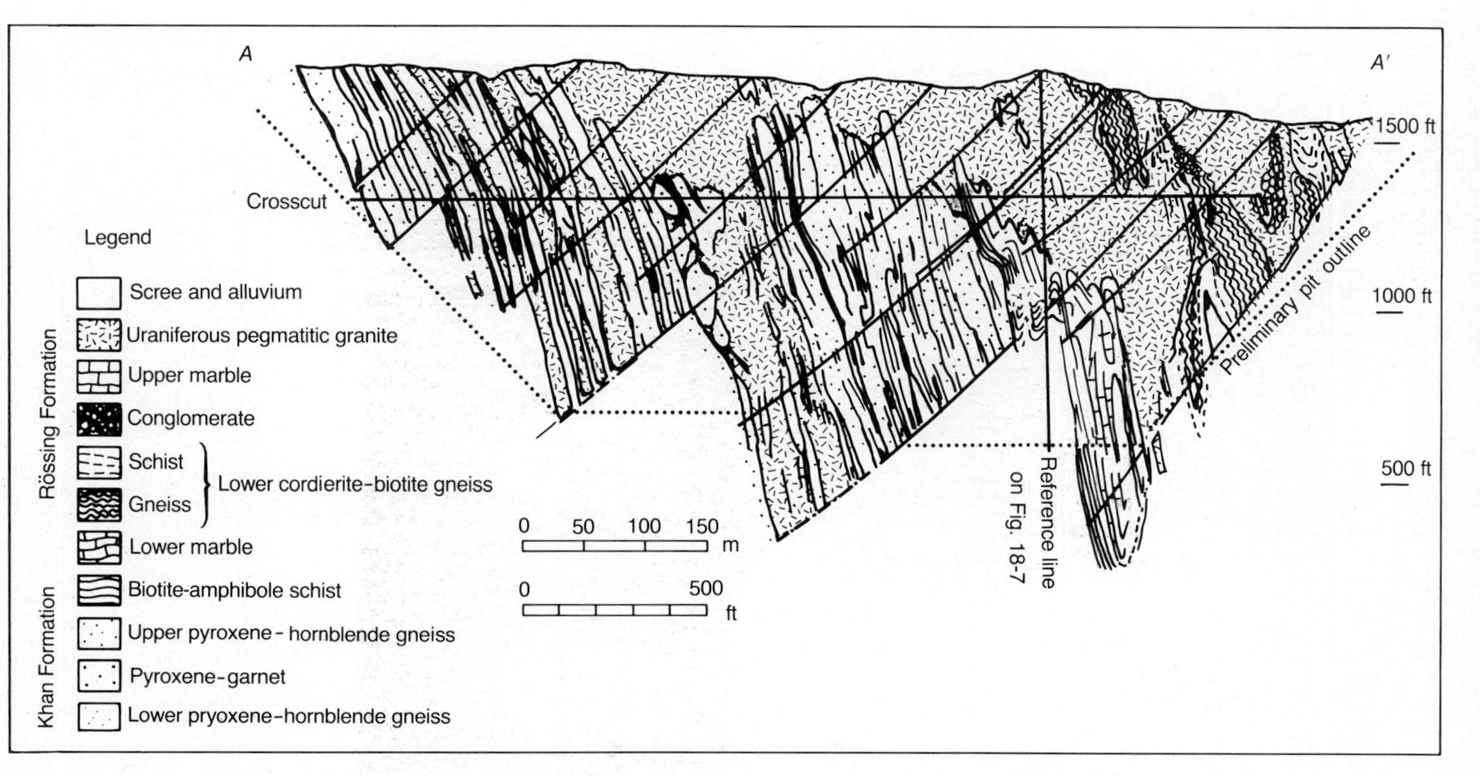

Figure 18-8. Cross section of the Rössing migmatites, facing northeast (see Figure 18-7). Note the intimate laminar intrusive habit. The scale is twice that of Figure 18-7, so more detail can be seen. The black areas contain disseminated uraninite, and a band of oxidized uranium minerals occurs within a few tens of meters of the surface. (*From* Berning et al., 1976.)

Figure 18-9. Dikes of pegmatitic alaskite emplaced into upper biotite gneiss, Dome Gorge, east of Rössing uranium deposit. Such intimate intrusion and interlaying is called *migmatite*. (*From* Berning et al., 1976.)

accessory uraninite has intimately invaded the sedimentary and metamorphic structures of the older host rocks. The final and locally contorted structures evident in the cross section of Figure 18-8 and the finely ribbed to irregular intrusion patterns in the larger scale detail of Figure 18-9 are quite consistent with what would be expected in a terrane metamorphosed at conditions at or near the melting points of its constituent minerals. The anatectic uraniferous melt probably rose about 1 km. The alaskite is not petrographically described by Berning et al. (1976), but it is almost certainly an S-type granitic rock. Exploration for tin and tungsten is indicated, and exploration for uranium, tin, and tungsten is indicated in such migmatitic terranes at other locales in Africa, South America, and elsewhere.

Stumpfl et al. (1976) described unusual late Proterozoic deposits at O'Okiep in Namaqualand, in northwest South Africa, 75 km west of the Aggeneys-Gamsberg mines. They are unusual in that they involve orthopyroxene-plagioclase "noritoid" intrusions into granulite-facies gneisses, carrying disseminations and injected masses of chalcopyrite (1.65% Cu average grade) with accessory cobaltian pyrite and pentlandite. The writers consider emplacement of the noritoids to have occurred by mechanisms of mafic plutonism, concluding that melting was generated by partial fusion of copper-nickel-bearing, originally sedimentary strata at the peak of granulite-grade metamorphism at 6 to 8 kbar pressure (10-to-12-km depth) and at temperatures of 800 to 1000°C.

Bibliography

Barton, P., 1983. Pers. Commun.

Berning, J., R. Cooke, S. A. Heimstra, and U. Hoffman, 1976. The Rössing uranium deposit, South-West Africa. *Econ. Geol.* 71:351–368.

Callahan, W. H., 1966. Genesis of the Franklin-Sterling, N.J., orebodies. *Econ. Geol.* 61:1140–1141.

Clark, B. R., and W. C. Kelly, 1973. Sulfide deformation studies. I. Experimental deformation of pyrrhotite and sphalerite to 2000 bars and 500°C. *Econ. Geol.* 68:332–352.

Crittenden, M. D., Jr., P. J. Coney, and G. H. Davis, Eds., 1980. Cordilleran metamorphic core complexes. Geol. Soc. Amer. Mem. 153, p. 490.

Davis, G. H., 1972. Deformational history of the Caribou stratabound sulfide deposit, Bathurst, New Brunswick, Canada. *Econ. Geol.* 67:634–655.

Engel, A. E. J., and C. G. Engel, 1962. Progressive metamorphism of amphibolite, Northwestern Adirondack Mountains, New York, pp. 37–82 in *Petrologic Studies, Buddington Vol.* Geol. Soc. Amer., 660 pp.

Frondel, C., and J. L. Baum, 1974. Structure and mineralogy of the Franklin Zn-Fe-Mn deposit, N.J. *Econ. Geol.* 69:157–180.

Gilmour, P., 1980. Pers. Commun.

Gross, G. A., 1961. Metamorphism of iron formations and its bearing on their beneficiation. *Can. Inst. Min. Metall. Trans.* 44:24–31.

Hauseux, M. A., 1977. Mode of uranium occurrence in a migmatitic granite terrain, Baie Johan Beetz, Quebec. *Can. Inst. Min. Bull.* 70:110–116.

Lincoln, T. N., 1981. The redistribution of copper during low-grade metamorphism of the Karmutsen volcanics, Vancouver Island, British Columbia. *Econ. Geol.* 76:2147–2161.

Mancuso, J. T., and R. E. Seavoy, 1981. Precambrian coal or anthraxolite: a source for graphite in high-grade schists and gneisses. *Econ. Geol.* 76:951–954.

McDonald, J. A., 1970. Some effects of deformation on sulfide-rich layers in lead-zinc ore bodies, Mount Isa, Queensland. *Econ. Geol.* 65:273–298.

Mookherjee, A., and N. K. Dutta, 1970. Evidence of incipient melting of sulfides along a dike contact, Geco mine, Manitouwadge, Ontario. *Econ. Geol.* 65:706–713.

More, S. W., 1980. The geology and mineralization of the Antler mine and vicinity, Mohave County, Arizona. Unpub. M.S. thesis, Univ. Ariz., 149 pp.

Sales, R. H., and C. Meyer, 1951. Effect of post-ore dike intrusion on Butte ore minerals. *Econ. Geol.* 46:813–820.

Salmon, B. C., B. R. Clark, and W. C. Kelly, 1974. Sulfide deformation studies, II. Experimental deformation of galena to 2000 bars and 400°C. *Econ. Geol.* 69:1–16.

Sangster, D. F., 1971. Metamorphism as an ore-forming process. *Econ. Geol.* 66:499–500.

Sawkins, F. J., and D. M. Rye, 1974. Relationship of Homestake-type gold deposits to iron-rich Precambrian sedimentary rocks. *Inst. Min. Metall. Bull.* 83:856–860.

Stanton, R. L., 1972. *Ore Petrology*. New York: McGraw-Hill, 600 pp.

——, 1978. Tectonic deformations at Broken Hill, New South Wales, Australia, and their significance for interpretation of ore environments: discussions and contributions. *Inst. Min. Metall. Trans.*, 87:B176–180.

Still, A. R., 1980. Pers. Commun.

Stumpfl, E. F., T. N. Clifford, A. J. Burger, and D. vanZyl, 1976. The copper deposits of the O'Okiep district, South Africa: new data and concepts. *Mineral. Deposita* 2:46–70.

Turner, F. J., 1968. *Metamorphic Petrology:Mineralogical and Field Aspects*. New York: McGraw-Hill, 403 pp.

Verwoerd, W. J., Ed., 1978. *Mineralization in Metamorphic Terranes*. Geocongress 75, Stellenbosch. Pretoria, RSA: J. L. Van Schaik, 552 pp.

Vokes, F. M., 1971. Some aspects of the regional metamorphic mobilization of preexisting sulfide deposits. *Mineral. Deposita* 6:122–129.

Winkler, H. G. F., 1965. *Petrogenesis of Metamorphic Rocks*. New York: Springer, 220 pp.

NINETEEN

▼

Deposits Related to Solution-Remobilization

This chapter is controversial because different geologists assign very different levels of importance to solution-remobilization as an ore-forming process, and little truly compelling pertinent data have been marshaled. As has been stated earlier, many explorationists believe that the majority of epigenetic metallic ore deposits stem from igneous activity, and many dismiss the concept that orebody components might have been chemically scavenged from adjacent lithologies, transported a few tens or hundreds of meters, and redeposited and concentrated in fractures or fissures as vein deposits. However, the concept is not a new one, and its importance has been debated since the birth of the science. Ore-deposit genesis explanations in the category of what has been called *lateral secretion* have been advanced since the sixteenth century, peaking with the writings of Sandberger and Van Hise in the early twentieth century and carried into recent literature by Sullivan (1948), Knight (1957), and Boyle (1968, 1979), to name but three. Before the development of sensitive analytical procedures, and in the period before the stable isotope studies of the 1960s and 1970s, there really were no reliable tests available, and virtually all of the early controversies were more rhetorical statements of opinion than scientifically derived conclusions. In recent decades, however, improved understanding of geochemical and physicochemical phenomena has greatly sharpened our perception of what can and cannot transpire in matters relating to local derivation of orebody components. But discussion is by no means over and there are still almost opposite positions regarding the sources of metals in many ore-deposit types. For example, proponents of convective circulation of groundwaters around a shallow pluton as an ore-forming system suggest that all that might be required to produce a porphyry base-metal deposit is the emplacement of a hot epizonal pluton into shallow, normally metalliferous host rocks, the resultant circulation of groundwaters being sufficient to dissolve, transport, and redeposit metals and sulfur to produce orebodies.

It is a matter of record that much of the hydrolytic silicate alteration around porphyry base-metal and epithermal precious-metal deposits involved copious amounts of surface-derived meteoric waters. It has been demonstrated (Norton, 1982) that indeed it would be impossible to exclude meteoric fluids from epizonal ore-forming environments. But is that the same as requiring that metals and sulfur be extracted from wall rocks and effectively redistributed to form orebodies? We just do not know. Does multiple intrusion of a terrane imply successive events of concentration and enrichment of metals in the host rocks, such that there is a multiplier effect of successive enrichment? Does intrusion itself promote the abstraction from deeply adjacent Precambrian basement, for example, of dispersed ions of iron, tungsten, copper, or gold, and the formation of ores? Again, we do not know. We *do* know that fluids circulate, that solution of loosely bonded metallic ions can occur, and that reprecipitation of those ions is reasonable, but we are not yet confident of how quantitatively important such modes of formation might be. The actual physical chemistries of the processes are complex and by and large have not yet been assessed. It is noteworthy that the same general processes are called upon to extract gold from ultramafic rocks to produce the gold ores of Porcupine-McIntyre, copper and zinc from basalts to produce massive sulfides of the Cyprus and Abitibi type, cobalt-nickel-silver from basaltic andesites and associated interflow sediments to produce the Cobalt, Ontario, ores, and copper-molybdenum-lead-zinc-silver-gold-manganese to produce porphyry base-metal deposits from a variety of host-rock types. No compelling correlation between host-rock type and orebody composition in porphyry or Cordilleran Vein type base-metal deposits has yet been discerned. In fact, the conformability and consistency of ore composition in vein deposits in a variety of host rocks tends to deny the influence of the compositions of laterally adjacent host rocks. An unidentified spokesperson for the geological staff of the Cerro de Pasco Corporation (1950) wrote that ore deposition in "the whole Andean Belt reveals a grand, overall unity, notwithstanding the striking diversity of its local components." Although the vein ores "occur in a wide variety of rocks, structures, and associations, the controlling factor in their distribution and hydrothermal genesis is their relation to local hypabyssal intrusives of intermediate composition." Nonetheless, several recent articles have stressed hydrothermal leaching of both major and minor elements from large volumes of wall rocks. For example, Glasson and Keays (1978) showed that gold can be selectively released during cleavage development in sedimentary rocks, as it appears to have been in the auriferous slate belt of central Victoria, Australia. Boyle (1968) presented several calculations to show that diffusion migration of silver is kinetically feasible, although no real chemical arguments bearing on ore formation were given. Clearly we have much to learn about the possible processes of lateral and vertical secretion and their significance in ore-deposit generation.

It should be pointed out that in recent decades we have come to a much clearer understanding of magmatic processes through the study of deuteric phenomena, the partitioning of fluids from melts (Burham and Ohmoto, 1980), and the selective partitioning of metals as chloride complexes into those fluids (Holland, 1972), as they were described in Chapters 3 and 4. So we know reasonably well what melt-related processes can accomplish (Chapter 11), and the retinue of metallization-alteration effects encountered in porphyry and Cordilleran Vein systems, for example, is replicated so consistently regardless of wall-rock type that an overall magmatic-hydrothermal influence seems required. But by the same token, there are other systems that are not part of an obvious orthomagmatic stem, including most of the mesothermal quartz-carbonate-gold vein systems of the world. Many of these occurrences are not directly relatable to contemporaneous epigenetic igneous activity, and a growing body of information and process models suggests that lateral and vertical secretion bordering perhaps on metamorphic remobilization is involved. The particular association of quartz-ankerite-gold veins with talc-carbonate alteration in mafic volcanic rocks has been convincingly related to lateral or vertical secretion. In this chapter we consider the Yellowknife gold deposits, Northwest Territories, Canada, the Porcupine-McIntyre-Kirkland Lake gold orebodies in Ontario, and the cobalt-nickel-silver-arsenic-sulfur ores of the Cobalt district, also in Ontario, Canada, to illustrate some of the principles of orebodies apparently formed by solution-remobilization.

▸ YELLOWKNIFE GOLD DEPOSITS, NORTHWEST TERRITORIES, CANADA

When the Yellowknife gold deposits were brought into production in the years before World War II, Great Slave Lake, Northwest Territories (Figure 19-1), was a remote, near arctic outpost. Although it is certainly still remote to most of us, technology and mineral and energy needs have opened the area. Uranium City and Cluff Lake in the Athabasca uranium area to the southeast (Chapter 20), the Pine Point Mississippi Valley type deposit to the south, and the Echo Bay uranium and silver mines and the gold mines at Contwoyto Lake to the north and northeast now "keep it company." The Yellowknife mines have been productive both of thousands of kilograms of gold and of geologic insight that has been relevant to advances

Figure 19-1. Location map of lakes and several major deposit areas in the Northwest Territories, Alberta, and Saskatchewan, Canada.

in the economic geology of gold in the 1970s and 1980s. The principal contributor has been R. W. Boyle (1955, 1959, 1961, 1979) who developed evidence, in the face of considerable opposition at the time, that components of the ores were extracted from the subjacent and laterally adjacent wall rocks in response to district-scale thermal-metamorphic gradients rather than "from an unknown intrusion at depth." Boyle still encounters vigorous criticism, but his evidence and arguments are forceful, and the student and scholar must consider their applicability there and elsewhere.

The Yellowknife ores consist of a series of northerly to northeasterly striking quartz-gold veins and lenses in shear zones in an Archean amphibolite greenstone belt (Figure 19-2) originally composed largely of andesite, basalt, and minor dacite. The principal shear zones, emphasized in Figure 19-2, are the northeasterly to northerly Con, Giant-Campbell, and Crestaurum shears with the northwesterly Negus-Rycon connecting shear. The gold ores, in lenses and discontinuous veins in these shear zones, are com-

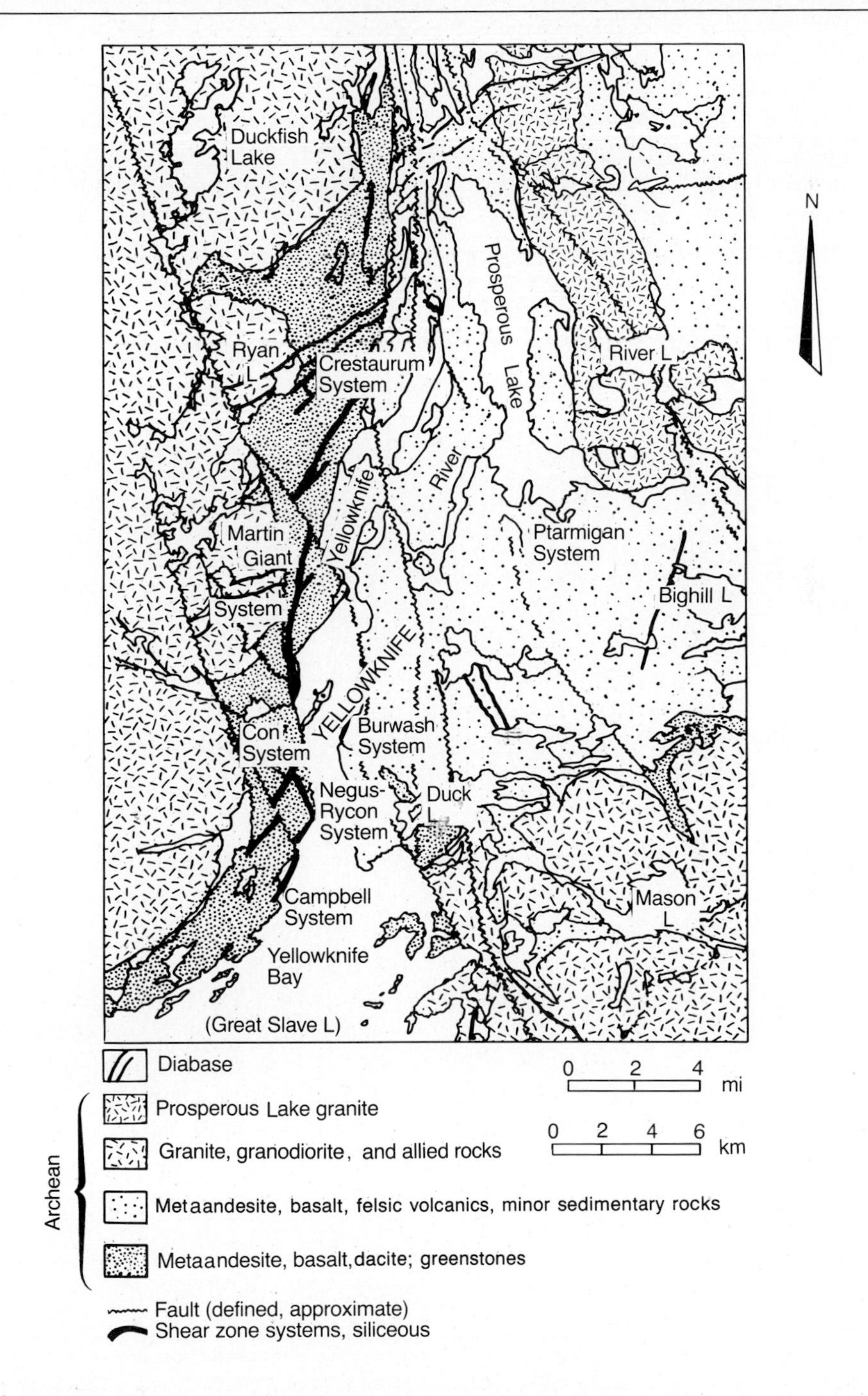

Figure 19-2. Map of the general geology of the Yellowknife district, Northwest Territories. (*After* Boyle, 1959.)

posed of quartz (75 to 85%) with carbonates, pyrite, arsenopyrite, pyrrhotite, stibnite, sphalerite, native gold, aurostibite (AuSb), and minor sulfosalts. The grade of the ore ranges from 15 to 30 ppm (0.5 to 1.0 oz per ton) Au with an Au-Ag ratio of 1 : 6, generally closer to 6. Individual quartz-carbonate-gold veins are predominantly white in color, steeply dipping, and clearly crosscut their altered wall rocks. The veins range from a few cen-

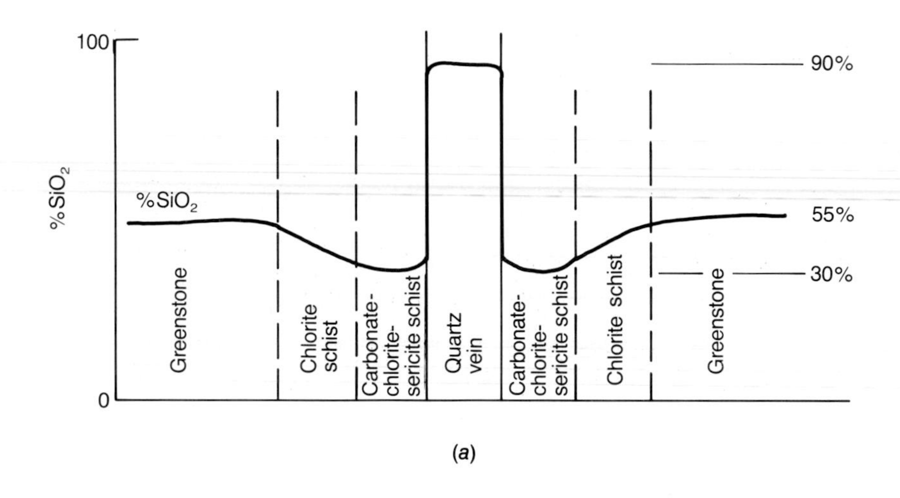

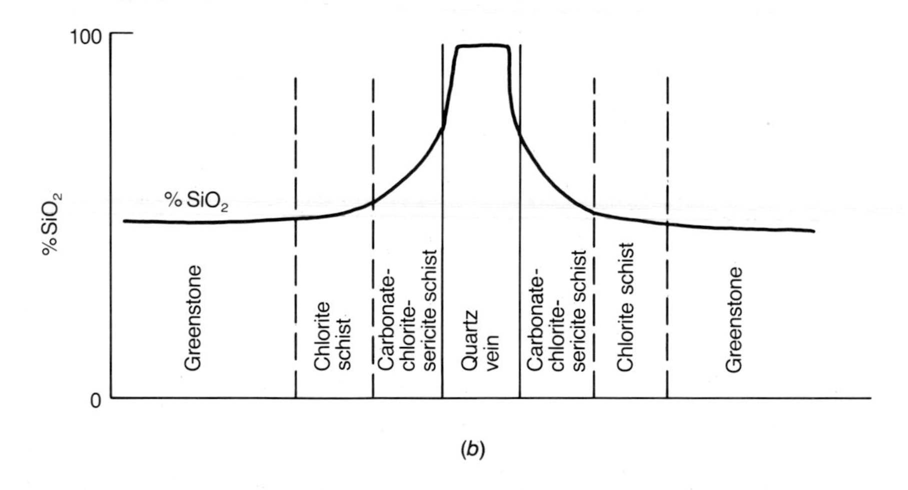

Figure 19-3. (*a*) Actual percent SiO_2 profile across alteration zones at Yellowknife, Northwest Territories. The profile suggests migration of SiO_2 from the altered walls toward and into present sites of quartz veins. (*b*) Expected percent SiO_2 profile across alteration zones if SiO_2 had been introduced and diffused outward from the quartz veins. ((*a*) *after* Boyle, 1955.)

timeters to 2 meters or so wide. Underground mapping has shown that these quartz pods, lenses, and veins are everywhere symmetrically encased in alteration zones. These shear zones, rarely more than 1 or 2 meters wide, are composed of multiple sets of alteration zones a few centimeters wide on either side of small veinlets. Composite shear zones perhaps 50 to 60 meters across are common. Where the individual shear planes are centimeters apart, the shear zones develop an overall symmetry of alteration that can extend for tens of meters. These zones progress from external metamorphosed but unaltered greenstone lithologies—metabasalt, meta-andesite, and metadacite—through outer alteration envelopes of chlorite schist, then carbonate-chlorite schist to the carbonate-sericite schist that encases the quartz-gold veins themselves. Boyle (1955, 1959, 1961) noted that the altered zones are universally impoverished in quartz from background values near 50% in the host rocks through lower values in the chlorite schist to as low as 30% in the chlorite-carbonate schist, a reduction that occurs even in portions of the shear zone without centerline quartz veining. His plot of silica content across a typical composite alteration zone (Figure 19-3*a*) shows a gull-wing shape reflecting the presence of lowest silica values in the altered rocks on either side of the shear zone.

An elongate northerly trending body of granodiorite lies immediately to the west of the greenstone belt and the shear zones. The granodiorite, shown in Figure 19-2 as granite, granodiorite, and allied rocks, is younger than and intrudes metavolcanics dominated by metamorphosed andesite, basalt, and dacite, which host the mineralized veins and shears. Metamorphic effects diminish eastward into the metavolcanic and metasedimentary rocks of the Yellowknife group. The granodiorite appears to be responsible for district-scale contact metamorphism of the greenstone belt that grades away from the contact from an amphibolite facies through epidote-amphibolite to a greenschist chlorite-epidote facies within a band a few kilometers wide and generally parallel to the contact.

Before Boyle's studies, most geologists attributed the gold mineralization and its associated effects directly to the presence of the granodiorite. The mineralization occurs in a belt closest to the pluton, but not in it, and the alteration and metamorphic effects in the greenstone rocks appeared appropriate. But Boyle (1955 et seq.) argued that the gull-wing curve of Figure 19-3*a* requires *abstraction* of silica from the walls, while the bat-wing curve of Figure 19-3*b* describes *addition* of silica. Having compiled hundreds of sets of analyses which conform to the gull-wing shape, Boyle concluded that silica at Yellowknife migrated from the altered walls toward the centerlines of the shear zones where it precipitated. He also shows volatile-containing compounds in the altered rocks to be reduced in amount nearer the granodiorite-greenstone contact, suggesting that such volatiles were purged from the greenstones by the heat of intrusion of the granodiorite and driven, or drawn, into and through the shear zones, as shown diagrammatically in Figure 19-4. Although in his earlier publications Boyle implied that the gold values were also drawn from the walls, he notes in his latest one (Boyle, 1979) that gold values increase upon approach to some of the veins, along with boron, arsenic, antimony, and silver. Perhaps, then, precious-metal values themselves were not actually laterally secreted, although similarities in the geochemical behavior of gold and silica would suggest to recent scholars that they were. At any rate, quartz was certainly laterally secreted. As Boyle (1955) points out, many authors have independently noted distribution of quartz and gold similar to that of Figure 19-3*a* in other gold districts, for example in the Mother Lode system in California (Knopf, 1929). It appears compelling that quartz and carbonate were extracted from the wall rocks at Yellowknife in a manner best depicted in Figure 19-3*a* and best described as lateral secretion, and probable that gold also was so extracted and concentrated.

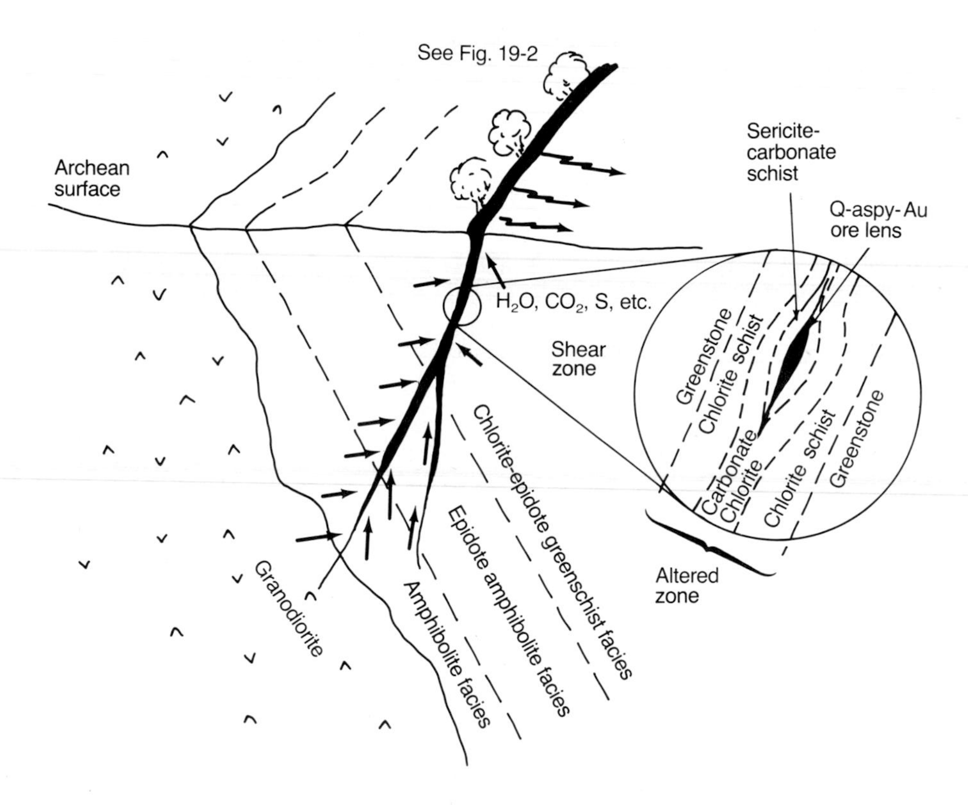

Figure 19-4. Northeast facing schematic of the general relation between metavolcanic rocks (greenstones), the granodiorite, the shear zones, wall-rock alteration in the shear zone, and the apparent movement directions of fluid. (*After* Boyle, 1955, 1959, 1961.)

▸ PORCUPINE-TIMMINS GOLD DEPOSITS, ONTARIO, CANADA

One of the great gold camps of the world trends east-west for about 20 km in the vicinity of Timmins, Ontario (Figure 19-5). These mines have produced more than 2 million kg (55 million oz, 2500 tons, 3 m^3) of gold since the early 1900s, and many of the 32 large mines and scores of minor ones still list appreciable reserves. The Timmins district lies near the western end of the Abitibi Greenstone Belt of Archean rocks, and the Timmins area itself has a complex Archean metallogeny. The gold ores are not in the oldest Archean volcanic units, the Del Oro Group, but occur in the next youngest, still Archean, predominantly mafic Tisdale Group volcanic series (Figure 19-6). The phenomenal Kidd Creek Cu-Zn-Ag massive sulfide deposit, the largest volcanogenic massive sulfide orebody known in the world (Chapter 13), is 25 km north of Timmins in these same younger volcanic

Figure 19-5. Map of Ontario locating the Porcupine-McIntyre district which spawned the town of Timmins. Also shown are the Kirkland Lake town and gold deposits, and the Cobalt, Ontario, Ni-Co-Fe-Ag-As-S deposits.

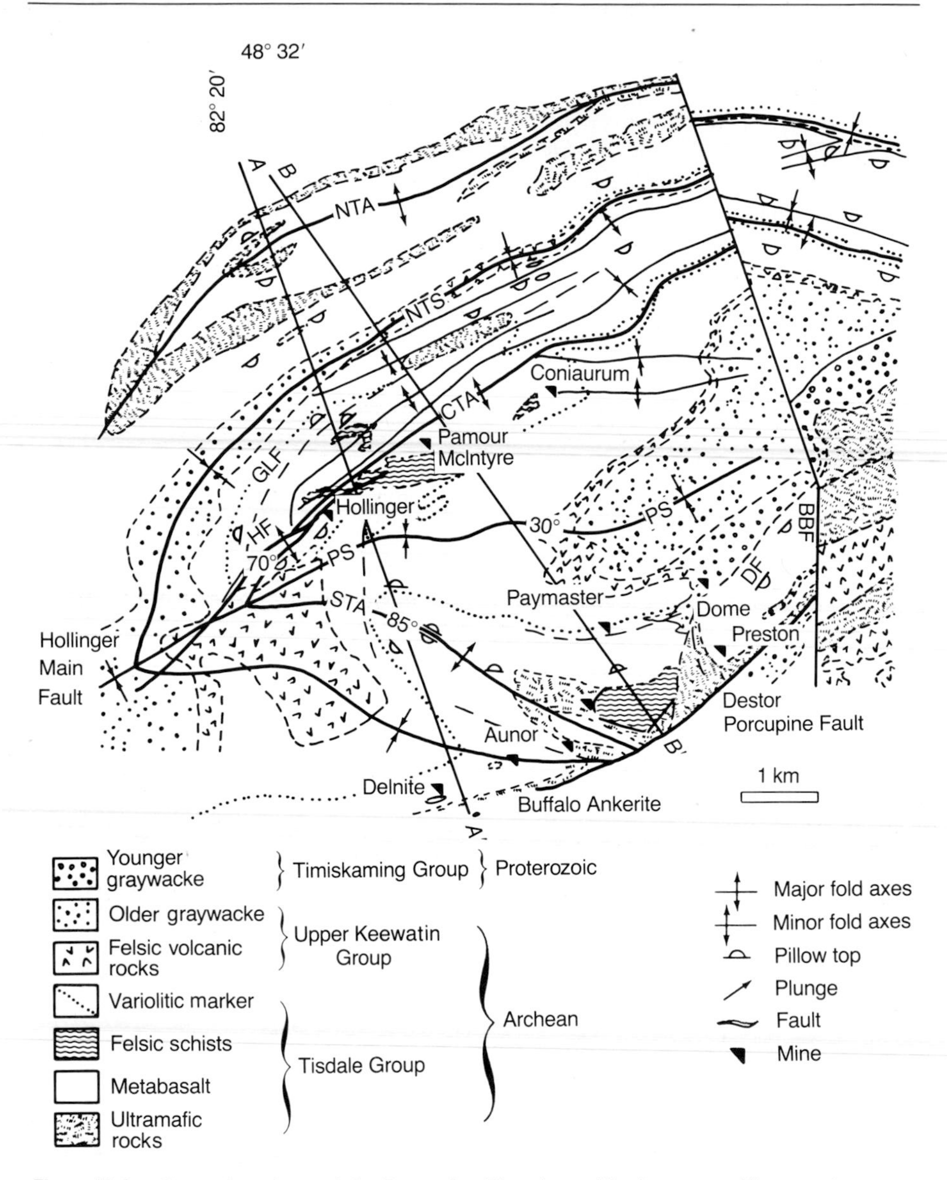

Figure 19-6. General geology of the Porcupine-Timmins gold mines area. The area is dominated lithologically by basalt with ultramafic flows near the base of the section and a variolitic pillow basalt marker layer in the upper portions. The rocks are complexly folded (Figures 19-7 and 19-8), with important anticline-syncline axes shown with letters referring to Figure 19-7. (*From* Davies, 1977.)

units of the Tisdale Group. The area also has recorded gold production from banded iron formations, copper and nickel sulfides from the Langmuir mine 28 km southeast of Timmins (Chapter 9), and copper-gold-silver from the so-called Pearl Lake Porphyry, an unusual Cu-Mo-hematite-anhydrite (exhalite? porphyry?) deposit in a crystal latite tuff or felsic schist in the volcanic succession to be described below; it has been mined from the McIntyre shaft (Davies and Luhta, 1978). Farther to the east, and still in the Abitibi Greenstone Belt, is the Noranda district. We will focus only on the gold deposits that located the town of Timmins—the McIntyre, Hollinger, Dome, Pamour, Coniaurum, and other mines.

Almost all of the gold production of the area has come from quartz-ankerite-carbonate veins that cut through a complexly folded and faulted pile of ultramafic, mafic, and calc-alkaline volcanic greenschist rocks. The lower, generally calc-alkaline Del Oro outpourings are 4000 to 5000 meters thick; the overlying Tisdale Group consists of volcanic shields of komatiites, basaltic komatiites, tholeiitic basalts, and calc-alkaline volcaniclastic greenstone rocks totaling 7000 meters. The geological setting is one of widespread fissure extrusion of primitive, ultramafic to mafic lavas building a thick volcanic platform on the floor of an Archean ocean, with progressive centralization of volcanism at domical areas within subsiding basins. Upon these shields or cones, extrusion of generally more differentiated calc-alkaline andesites, dacites, and rhyolites continued, with exhalation and precipitation of interflow chemical sediments during periods of volcanic quiescence. These rocks were deeply folded and refolded, presumably as part of the Archean greenstone belt lithotectonic event, as shown in Figures 19-6 to 19-8. The folding and refolding produced both throughgoing faulting and extensive strong, penetrative east-northeast shearing, with textbook examples of refraction and expansion of shears and fractures across contacts between more and less brittle lithologies. The actual host rocks to the veins include the volcanic lithologies listed earlier along with diabases and quartz and feldspar porphyritic rocks which have intruded them, and with inter-

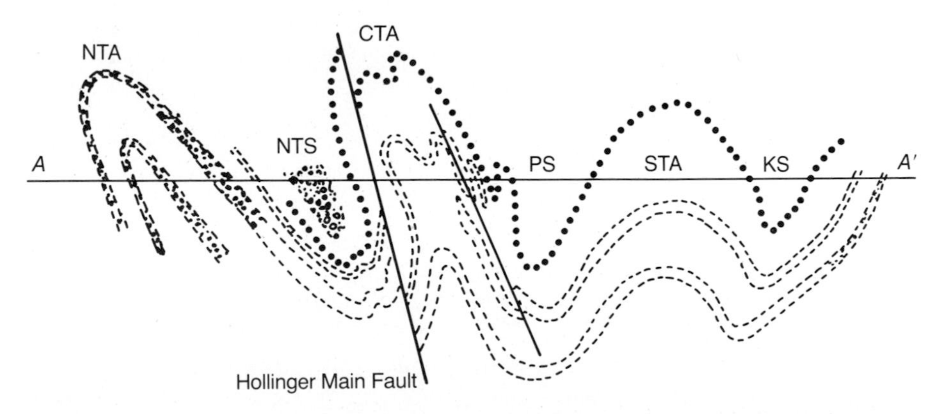

Figure 19-7. Structure section facing east across the Timmins area at *A–A′* on Figure 19-6. The anticlines and synclines are labeled on both figures, and the symbols are the same on both. NTA, CTA, STA = North, Central, South Tisdale Anticline; NTS, PS, KS = North Tisdale, Porcupine, Kayorum Syncline. The rocks underlying these are the Del Oro Group, composed dominantly of calc-alkaline volcanic rocks. (*After* Davies, 1977.)

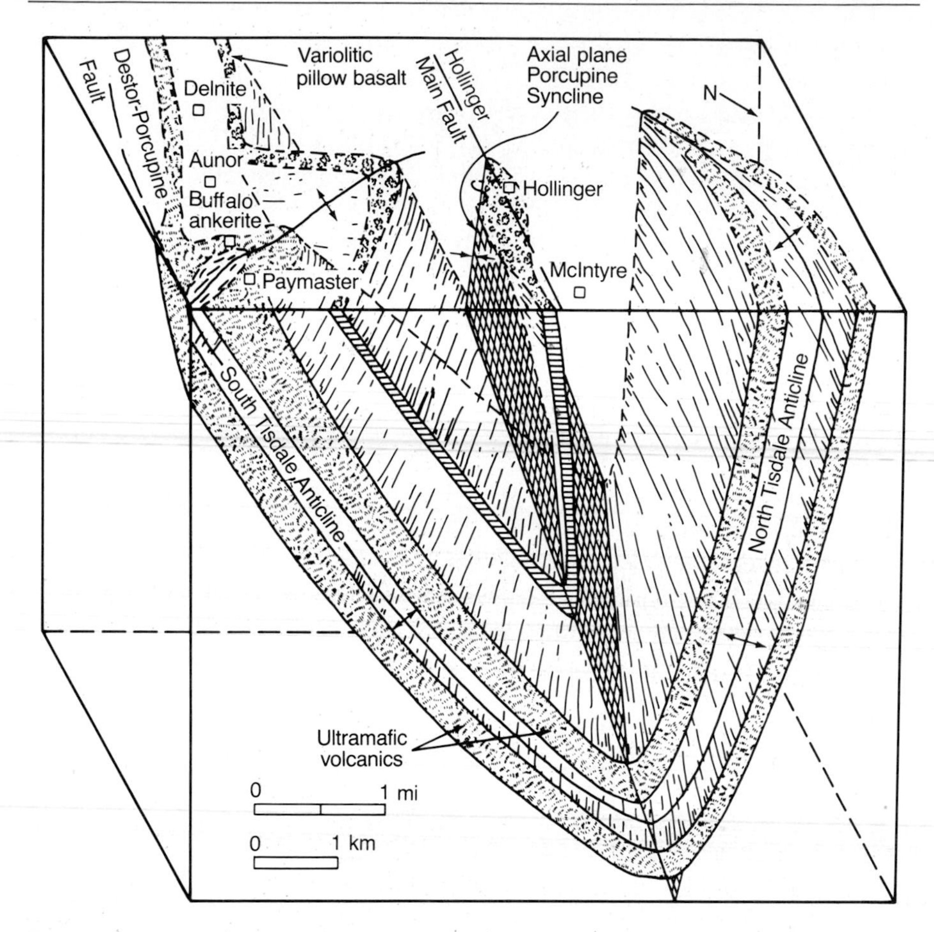

Figure 19-8. Diagrammatic sketch of the main part of the Timmins gold camp, facing southwest. It shows how complex the structural geology is—the North and South Tisdale anticlines are a single anticline refolded around the Porcupine synclinal axis. See also Figures 19-6 and 19-7. (*After* Pyke, 1975.)

layered dacite flows, agglomeratic and tuffaceous dacites and minor rhyolites, and sparse interbeds of argillite, graywacke, and slate. The basalts and ultramafic flows have been described for decades as massive, pillowed, and variolitic (Figures 19-6 and 19-7), with flow-top breccias, local interflow argillites, cherts, banded iron formations, and ferroan carbonate layers included in the section.

Veins in the district have been structurally classified (Jones, 1948; Ferguson, 1968) as well-defined continuous quartz veins that locally pinch, swell, or branch; sinuous veins that appear to have followed or replaced small drag folds in the basalts; tabular veins or shear zones made up of subparallel stringers; and zones of parallel tabular or S-form en-echelon stringers. A typical swarm of veins on the 425-ft level of the Hollinger mine is shown in Figure 19-9, and Figure 19-10 shows typical veins underground at the Hollinger mine (Dunbar, 1948). The principal loci are along anticlinal axial faults and shears and at contacts between rocks with brittleness contrasts. Regarding gold occurrence specifically at the Dome mine, Bell and Rogers (1979) describe "carbonate-rich flow-top . . . light to medium grey . . . crystalline" layers that are generally impersistent laterally, occur in flow-top scoria and vesicular lavas, and are concordant with pillowed basalts above and below. Where folding and shearing have occurred, these interflows have recrystallized and are associated with several modes of mineralization. One type is long, narrow ankerite-gold or quartz-fuchsite-gold veins in these layers and parallel to the general trends of contacts; another type comprises lenticular or irregular quartz veins, en echelon in sheared competent rocks and as stockworks within these layers, in basalts, in felsic schists, and in their altered equivalents.

The mines shown in the figures penetrate down along major faults and folds into the shear zones, and have produced millions of tons of 0.1 to 0.4 oz per ton (3 to 12 ppm) gold ore from steep veins averaging about 1 meter across, which contain quartz and ankeritic carbonates. Silver contents of 1 to 2 ppm give Au-Ag ratios of 5 : 1 to 10 : 1. The ores involve major whitish quartz and carbonates, minor tourmaline, albite, scheelite, pyrite, arsenopyrite, pyrite, pyrrhotite, sericite, and native gold, with traces of gold-silver tellurides, chalcopyrite, galena, sphalerite, and tetrahedrite. Fluid inclusion studies indicate mineralization temperatures of 340 to 370°C (Kerrich and Fryer, 1979). Alteration halos of chlorite, epidote, calcite, pyrite, and sericite, closest to the vein, are present; silicification, carbonatization, and pyritization are also described.

With this geologic complexity, which includes intrusive units, and with the prominent structural control of the orebodies in generally steeply dipping shears and fault fissures, it is not surprising that most early workers ascribed the gold-quartz ores to deep intrusive hydrothermal fluid sources and processes (Dunbar, 1948; Ferguson, 1968). Nonetheless, Hurst (1935) noted that silica and sodium were lower in the altered basalt flanking veins, as Boyle would later find to be true at Yellowknife (Figure 19-3*a*). Hurst suggested lateral secretion as an ore-forming mechanism at Porcupine. But it remained until during the 1970s, and the realization of the association of gold deposits with ultramafic and mafic volcanic rocks and exhalites (Chapter 13), that gold vein deposits in such rocks were genetically reevaluated. Pyke (1975) studied the relationship of gold mineralization and ultramafic volcanic rocks in the Timmins area. Fryer, Kerrich, Hutchinson, Peirce, and Rogers (1979) and Kerrich and Fryer (1979) further consider the problem. Kerrich and Fryer summarize genetic models that have been offered as including (1) derivation of gold from stocks of subvolcanic Archean-age quartz-feldspar porphyry (Ferguson, 1968), (2) derivation of gold from Kenoran age granitic batholiths emplaced 2.4 to 2.6 billion years ago at the end of the Archean (Jones, 1948, George, 1967), (3) lateral secretion of vein-

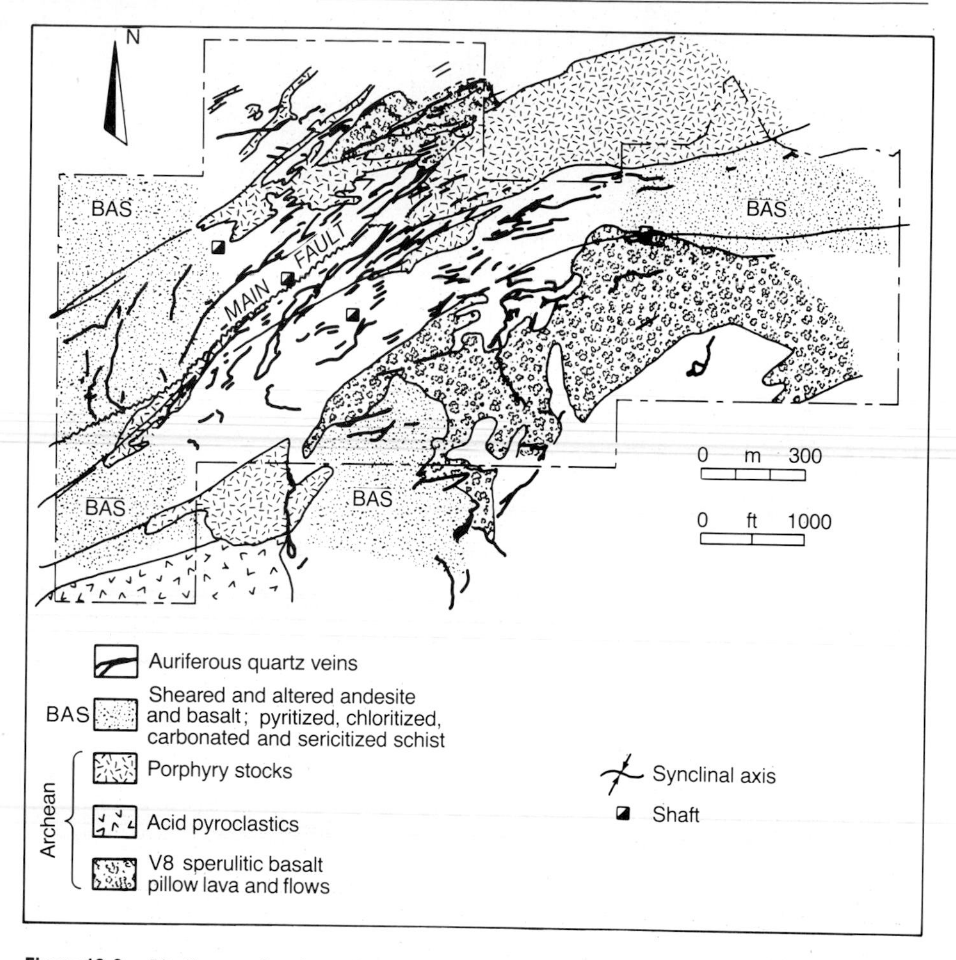

Figure 19-9. Medium-scale view of geologic relationships on the 425 level of the Hollinger mine, Timmins. Notice the obvious structural control of subparallel vein sections. The veins are not individually continuous, and their pattern is consistent with structural-dilation zone geometry and with sedimentation on flow tops. (*After* Boyle, 1979; *from* Jones, 1948.)

filling material from adjacent volcanic rocks (Hurst, 1935), (4) mobilization of gold from subjacent and adjacent ultramafic rocks (Pyke, 1975), and (5) mobilization of gold from exhalites accumulated during periods of volcanic quiescence (Hutchinson, 1976). Recent authors discredit items 1 and 2; the isotopic age of the mineralization, like the porphyries, is reported as 2700 to 3250 million years, or early Archean, but space relationships of the "porphyry association" are too frail. Pyke suggested that ultramafic rocks were especially "rich" in gold at the parts per million level, and therefore might be good source rocks, but later detailed work by Anhaeusser (1976), Kwong and Crocket (1978) and others has shown that ultramafic rocks (distinct

Figure 19-10. Typical McIntyre-Porcupine district veins in the Hollinger mine, Ontario. Such semicontinuous vein zones can be followed for kilometers along strike and downdip. They appear to be both structural fissures and dilation openings. (*From* Dunbar, 1948.)

from basalts) have less than 2 ppb Au and are unlikely first-order sources of gold by lateral secretion mechanisms. Kerrich and Fryer combine various portions of items 3, 4, and 5 in an elaborate model (Figure 19-11) that calls for liberation of water, silica, carbon dioxide, boron, gold, and other ore constituents as basalts and other "greenschist" rocks are heated and metamorphosed at moderate depth under normal high-heat-flux Archean geothermal gradients. Clearly, the mechanisms described border on metamorphism and might have been described in Chapter 18. But since the "metamorphism" is continuous with and a product of accumulation of the pile and is not related to a later dynamothermal event, the mechanism might be expected to be common to many Archean greenstone belts. Hutchinson, Fyfe, and Kerrich (1980), in a paper on deep fluid circulation, scavenging, and ore deposition, suggest that the base metals Cu, Ni, Fe, and Zn and other elements such as Sn, Cr, Pb, W, Pt, and Pd may be leached from volcanic strata and transported through fracture systems to form veins or exhalites. Parts of Figure 19-11 were also reproduced in the 1980 article. A simpler model would involve more local migration of quartz, ankerite, and gold from the interflow sediment layers and auriferous flows into gash-vein dilatant zones and shears in adjacent units. It is too early to say whether Kerrich and Fryer have more correct answers than do Pyke or Boyle, but

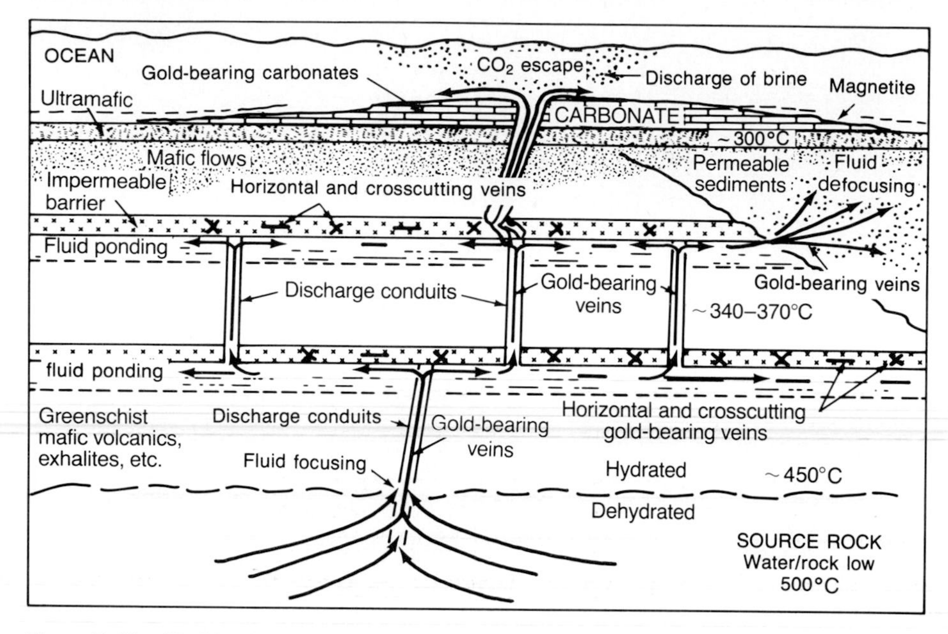

Figure 19-11. Modification of Kerrich and Fryer's (1979) schematic for derivation of McIntyre-Porcupine veins. Gold, silica, and volatiles are freed and carried by water, also freed by deep metamorphic recrystallization of hydrated greenschist lithologies under normal Archean thermal gradients. Fluids may deposit quartz, gold, and carbonate in veins as shown. They may also be ponded and diverted by impermeable interbedded units, and issue either into the sea to form auriferous carbonate exhalites or into permeable sediments.

such processes involving lateral secretion–vertical secretion in the classic sense are now being evaluated quantitatively for the first time.

▸ KIRKLAND LAKE–LARDER LAKE–MALARTIC GOLD DEPOSITS, ONTARIO-QUEBEC, CANADA

Another district contributing to arguments concerning remobilization of gold, quartz, and other ore components is the district that stretches east-west from Kirkland Lake to the area east of Noranda (Figure 19-5). It is geologically complex, as was mentioned in Chapter 13. Ridler (1976) drew an east-west stratigraphic cross section of the Abitibi Supergroup (Figure 13-22) that runs 115 km from the Quebec-Ontario border toward the Timmins area just described. Ultramafic and mafic volcanic Archean rocks are abundantly present as flows draped over the older Round Lake batholith. Ridler's diagram shows basal platform mafic-ultramafic flows of the Skead Group overlain by the Blake River Group, which is more felsic, thickens eastward, and holds the Noranda area massive sulfides, and by clastic sediments and local volcanic piles of trachytic-syenitic flows and stocks in Timiskaming sediments.

In its heydey between 1913 and 1960, the Kirkland Lake area boasted 10 major mines that contributed to a colorful and dynamic district history. The most productive—from west to east the Macassa, Kirkland Lake, Teck-Hughes, Lake Shore, Wright-Hargreaves, Sylvanite, and Toburn—were contiguous producing mines along the 6-km length of a single structural system, the Kirkland Lake fault vein. It was mined to a depth of 2 km and produced nearly 30 million oz (0.9 million kg, 900 metric tons) of gold. It

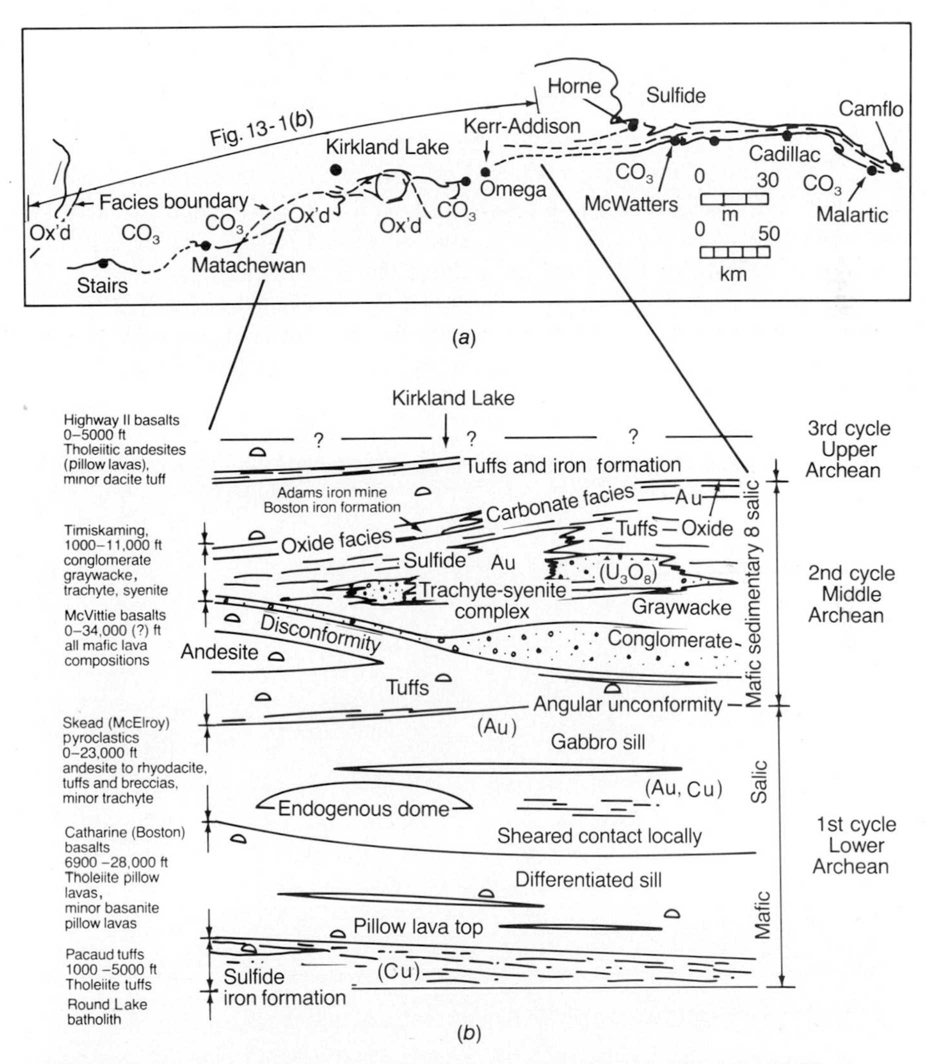

Figure 19-12. (*a*) Facies and (*b*) stratigraphic relationships in part of the Abitibi Greenstone Belt and pertinent to the Kirkland Lake district gold-quartz veins. The location and span of (*a*) is shown as line *Z–Z′* on Figure 13-1*b*. The span of (*b*) is indicated. No intrusive rocks older than the Round Lake batholith are known in the area. (*a from* Ridler, 1976; *b from* Ridler, 1970.)

consisted of a N70°E, steeply southeast dipping quartz vein system which showed classic bifurcation, branching, anastomosing, and pinch and swell characteristics. Snow-white "bull" quartz was the principal gangue mineral, with abundant calcite and minor albite, orthoclase, ankerite, barite, tourmaline, and apatite. The vein fillings are banded, crustified, and dominantly open-space fillings. Accessory opaque minerals include 2% pyrite and minor chalcopyrite, hematite, galena, sphalerite, molybdenite, graphite, and tellurides including calaverite ($AuTe_2$) and petzite (Ag_3AuTe_2). Native gold is in flecks, leaves, and veinlets cutting quartz vein material, wall rocks, and earlier formed vein minerals. Vein widths range from 0.3 to 3 meters, with an average of about 2 m. Grade was about 17 ppm (0.5 oz/ton); native gold ran about 940 fine (6% Ag). The wall rocks are principally syenite and augite syenite porphyry associated with the alkalic volcanics shown in Figure 19-12, but volcanic and sedimentary lithologies are never more than a few tens of meters away in footwall, hanging wall, or down vein. Relatively restricted alteration includes silicification, sericitization, and chloritization, typical products of hydrolysis of anhydrous primary felsic rock silicates. In a few places, the deep-green chromian mica fuchsite-mariposite was found. The veins have been described as mesothermal to "hypothermal," but no obvious temporally satisfactory igneous source rock has been identified.

Figure 19-13. Outcrops along Highway 11, 20 km east of Kirkland Lake. Ridler mapped them as "carbonate exhalite," which contains up to hundreds of parts per million of gold and locally constitutes ore grade. Gold may also be secreted laterally into quartz-gold veins, which have been mined in scores of small nearby workings. Other interpretations are given in the text.

The gold ores extracted from the immediate area of the Kirkland Lake–Larder Lake Break are varied in occurrence, but they appear to be interrelated. In addition to the veins just described, some of the ores in the Kerr-Addison and the Chesterville mines near Virginiatown are simple banded carbonate layered rocks with cherty, fine-grained silica that appear texturally and descriptively to represent essentially undisturbed rock that has been described as exhalite material like that of Figure 19-13. The carbonate is ankeritic, tan to brown, with intermittent silica-rich bands and distinctive local layers containing the chromian mica fuchsite. The gold is invisibly fine-grained native gold with trace amounts of base-metal sulfides, especially pyrite and arsenopyrite. Ridler (1979) stated that

> *. . . individual gold lodes at the Kerr-Addison mine . . . are larger plate-like bodies lying in the plane of original bedding, whether planar or folded. The abundant development of quartz veins* [*in the beds*] *has not significantly altered this fundamental pattern. Analyzed samples of prominently banded chert and pyritic carbonate rock have most of their gold in the pyritic, carbonate-rich beds. Make no mistake, the gold is in the sediment as it is laid down; otherwise the gold-rich pyritic carbonate fragments (which abound) would not exist. It does appear to be essential that some sulfide phase be present.*

A second occurrence in the Kerr-Addison suggests that the gold has been redistributed. Boyle (1979) describes parts of the Kerr-Addison ores as

> *. . . great quartz-carbonate stockworks and disseminated pyritic bodies in intensely sheared and altered zones. Most of the gold is free. Other minerals include pyrite, arsenopyrite, chalcopyrite, millerite, sphalerite, tetrahedrite, galena, and scheelite.*

The average grades are 7 ppm Au and 0.3 ppm Ag, with an Au-Ag ratio of 20 : 1. Yet another occurrence (Hodder, 1981) is that some of the gold in deposits like the Thompson-Bousquet occurrence east of Noranda, Quebec, is found in what appear to be hot-spring-siliceous-sinter deposits called *flow ore*. Gold occurs with quartz and chalcedony in crosscutting textures and structures that strongly resemble the conduit-and-pool morphologies of, for example, Mammoth Hot Springs at Yellowstone National Park, Wyoming, even with the ropy cascade forms of the lips of pools. These flow ores occur along the same horizon as the Kirkland Lake–Kerr-Addison–Malartic deposits. Some flow ore is also found near the Kerr-Addison mine.

The preceding paragraphs have described four types of gold ore in the same general area, namely, (1) bedded-banded carbonate exhalite material, (2) quartz-carbonate stockworks and disseminated pyritic bodies, (3) hot-spring flow ores, and (4) quartz-carbonate vein systems. They all occur along a belt or zone called the Kirkland Lake–Larder Lake Break, which is mapped along a zone 200 km long across eastern Ontario and western Quebec. Its significance is not yet fully understood. It has been identified as a system of faults, as a sedimentary-volcanic stratigraphic discontinuity, as a profound alteration zone, and as a packet of altered ultramafic flows. Ridler (1970, 1976) considers it an exhalite horizon (Figure 19-12), as described in Chapter 13. He mapped it as a package of exhalative chemical sediments extending from west of Matachewan eastward through Kirkland Lake and on to the east through Noranda and into Quebec. He shows it as lying on ultramafic flows at the west (Figure 13-22) and in sediments and volcanics to the east. It overlies a thick pile of trachytes and trachyte tuffs at Kirkland Lake. East of Kirkland Lake, the exhalite as mapped by Ridler is a thick carbonate-exhalite, as shown in outcrop in Figure 19-13. It is as much as 100 meters thick and may contain up to hundreds, even thousands of parts per million of gold (Figure 13-28). It is this material that he describes in the quoted section on the Kerr-Addison ores. Ridler's assignment of these rocks to the exhalite category is not without controversy. Pyke (1975) thinks that they may be at least in part carbonate-altered ultramafic flows, and Strickler (1978) ascribes them to a submarine carbonatite extrusion. Older workers thought the belt to be a great carbonated shear zone, the Kirkland Lake–Larder Lake–Bouzan-Cadillac Break.

Whatever the "Break" proves to be, it constitutes a horizon that extends for hundreds of kilometers, and which is anomalous in its carbonate and gold content. Since the price of gold spurted in 1978, headframes have sprung up like cornstalks from Kirkland Lake past Noranda to Val D'Or, Quebec, a distance of 200 km. And there are many mines of long standing such as the Kerr-Addison near Virginiatown, Ontario, and the mines at Kirkland Lake that have been producing from ores in or near this horizon for decades. If it is an exhalite zone, then the quartz-carbonate stockwork vein ores at the Kerr-Addison and the flow ores can be considered to be local remobilizations of exhalite material. Boyle's stockwork description indicates that quartz, gold, and the other base metals, if they were originally sited by an exhalative process, have been at least locally remobilized and reprecipitated. The nature of the base metals—copper, iron, zinc, nickel, antimony, arsenic, and sulfur—is certainly consistent with an original exhalative mode of origin. Swan, Hausen, and Newell (1982) describe an identical geochemical retinue from Proterozoic exhalative ores from the Yavapai greenschist belt in Arizona. Finally it has been suggested by several district geologists that the quartz-gold ores of Kirkland Lake itself represent solution-remobilization of nearby exhalite gold and silica into fractures and fissures which were propagated in, and held open in, the relatively brittle alkalic igneous rocks. No specific mechanisms have been described in print, and some aspects of the mobilization-precipitation—for instance, the diversity of metals and minerals listed above—are difficult to explain, but the mere fact that the Kirkland Lake ores may be associated with exhalites and are precisely "on strike" with the ten score mines scattered along the Matachewan–Kirkland Lake–Larder Lake–Noranda–Malartic–Val d'Or belt argues that they may well have been remobilized from exhalites into quartz-gold veins, and are therefore products of lateral secretion.

▸ COBALT, SILVER-COBALT DEPOSITS, ONTARIO, CANADA

A much stronger case—but still not an "airtight" one—can be made for remobilization and lateral-vertical secretion of ore-forming constituents at the Cobalt district, Ontario. It is presumably a coincidence that Cobalt is not far from the ore districts just described (Figure 19-5). Ore-forming processes are called upon at Cobalt to have concentrated iron, nickel, cobalt, silver, arsenic, and sulfur and lesser amounts of copper, zinc, and lead from subjacent Archean volcanic rocks into younger veins which cut them and post-Archean units. Vein fillings are demonstrably post-Archean in age—veins carrying the metals listed cut parts of a thick sheet of the Nipissing Diabase, which was emplaced 2150 million years ago. The "cobalt association" of Co-Ni-Ag-Fe-As-S occurs in several dozen deposits in Europe and North America, including Erzberg in Austria, Annaberg-Schneeberg in East Germany, Jáchymov, Czechoslovakia (Figure 1-1), several mines in the Cobalt area itself and elsewhere in eastern and northern Ontario, the Echo Bay mines at Great Bear Lake, Northwest Territories, Canada (Figure 19-1), and two or three occurrences in New Mexico and Arizona, United States. Uranium as pitchblende is part of the paragenesis at some of these deposits (Echo Bay, Port Radium, Jáchymov), but others (Cobalt and Erzberg) have little or no uranium. Silver was discovered in the Cobalt district in 1903 by Fred LaRose, a blacksmith working on the railroad that was being built into farmlands to the north. Nearly vertical veins outcropped across hills in glaciated, forested terrain; a Mr. Trethewey staked an 8-inch (20-cm) wide vein which he found in outcrop and which yielded 340,000 oz (10,500 kg) of silver from an open cut only 17 meters long by 8 meters deep, locally with 1000 oz (31 kg) of silver per ton! Area production peaked in 1911 and was hard hit by depleted reserves and depressed conditions in the 1930s. The 1940s and 1950s saw mostly intermittent leasing and small-scale mining. Several highly profitable mines were open in the 1960s and 1970s, and renewed interest in cobalt and precious metals caused a spurt of exploration activity in the early 1980s. The area has produced 26,000 metric tons of silver, 20,000 of cobalt, 7300 of nickel, and 2300 of copper (Berry, 1971).

There are three ages of rocks at Cobalt (Figure 19-14), the Keewatin volcanic rocks of Archean age, the Cobalt Series sediments which were laid down in the early Proterozoic era, and the above-mentioned Nipissing Diabase of early middle Proterozoic age (2150 million years). The Keewatin units constitute the local basement. They consist of great thicknesses of basalts, basaltic andesites, and andesites as flows, agglomerates, and pyroclastic units. They were described by Jambor (in Berry, 1971) as "mainly

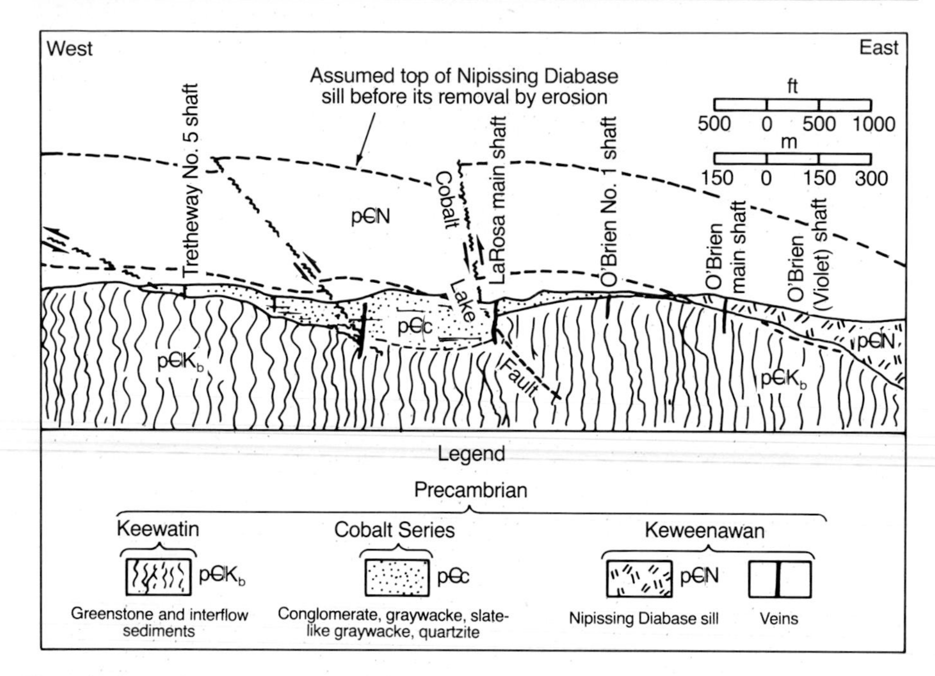

Figure 19-14. Geological cross section showing the major rock types and their relationships at Cobalt, Ontario. Note the sill-like form of the Nipissing Diabase, and the fact that the veins at the O'Brien shafts cut all three rock types. The veins are generally in the plain of the page, and do not show well here. (*After* Boyle, 1968.)

intermediate to mafic flows; some pyroclastic and acid volcanics; minor interflow sediments with chert, sulfides; iron formation; and schist." The minor *interflow sediments* are accumulations of ashy or siliceous sediments enriched in iron, copper, zinc, silver, arsenic, and sulfur. As in the Porcupine-Timmins area, they occur at Cobalt as silty-cherty units of moderate lateral extent, commonly filling irregularities in flow tops and representing accumulation of clastic and chemical sediments during volcanic pauses apparently like the tuffisites described in Chapter 13. They are best considered cherty, metal-rich exhalites interleaved between submarine volcanic flows. The Keewatin volcanic pile at Cobalt contains abundant interflows (Figures 19-15 and 19-16). The volcano-sedimentary package was steeply tilted and eroded to a hilly topography between early and late Archean, and the then nearly vertical northwest-striking flows and interflows were covered—and hills and valleys buried—by the Cobalt Series sediments. The deepest, broadest valleys appear to have been preferentially eroded into sections with abundant interflow units before the gently rolling topography was glaciated and buried by the early Proterozoic, Huronian (Early Aphebian) Cobalt Series sediments. The lower Coleman member of this Gowganda Formation of the Cobalt Series consists of major conglomerates with lesser

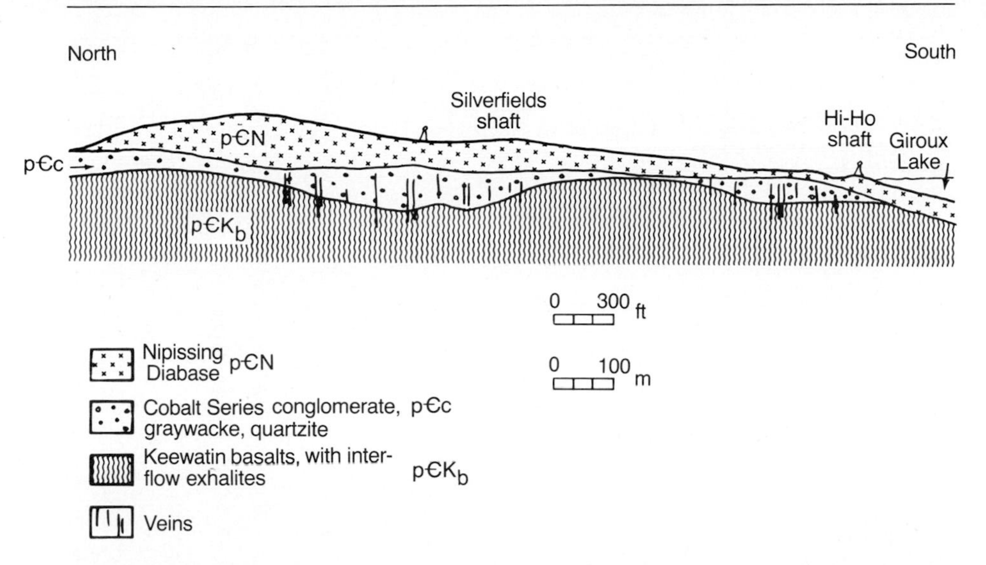

Figure 19-15. Larger scale cross section showing Cobalt, Ontario, district relationships, this one through the Silverfields and Hi-Ho mines. Northeast striking veins are essentially vertical, average 300 meters long by 100 meters high and 10 to 20 cm wide. They are predominantly in the Coleman Formation, extending downward 10 to 20 meters into the interflow-rich Keewatin basalts and andesites. See also Figure 19-16. They locally contain up to hundreds of kilograms of silver per metric ton of ore. (*Composited from field notes and* Boyle, 1968.)

graywackes, quartzites, and arkoses. The basal units were glacial conglomerates and tillites; they yield upward to well-bedded graywackes and quartzites of the Firstbrook member and then to shallow marine quartzites and arkoses of the Lorrain Formation. These nearly flat-lying units were intruded by the third rock type, a 400-meter-thick diabase sill. Post-Keweenawan erosion has cut into this Nipissing Diabase so that its upper surface is mostly gone, and outcrop patterns consist of "windows" through the lower sill contact with Lorrain and Gowganda sediments, locally to "hilltops" of the Keewatin volcanics (Figures 19-14 and 19-15). There are some major regional faults, as shown in Figure 19-14, but they have no known relation to the ores.

Mineralization occurs in a series of nearly vertical veins (Figures 19-15 and 19-17). The sharp-walled veins (Figure 19-18) average less than 1 meter in width and are commonly planar, more so in the Huronian sediments than in the volcanics below or in the diabase above. The vein fillings are composed of iron and nickel arsenides, sulfides, and sulfarsenides and native silver in a dolomite-rhodochrosite-calcite gangue. The assemblages (Petruk, 1968, 1971) are classified as Ni-As, Ni-Co-As, Co-As, Co-Fe-As, and Fe-As, with the Co-As assemblage being most abundant. Prominent minerals include niccolite (NiAs), rammelsbergite ($NiAs_2$), safflorite

(*a*)

(*b*)

Figure 19-16. (*a*), (*b*) Keewatin basement rocks of the Cobalt district in outcrop in nearly vertical layers on a hill overlooking the town of Cobalt. They are composed of basalt and basaltic andesite with interflow layers of anomalously metalliferous sediment as seen here and described in the text. (*c*) (*facing page*) Similar interflow distal exhalite composed of interbedded chert, siderite, and magnetite with minor pyrite from the Ferguson River area near Hudson Bay, District of Keewatin. The upper surface of the bed (nearest the hammer) is regular, and the lower surface drapes over the pillowed surface of the subjacent flow. (*c photo by* Ridler, 1976.)

(c)

[(Co,Fe,Ni)As_2], loellingite ($FeAs_2$) skutterudite [(Fe,Ni)As_3], cobaltite (CoAsS), arsenopyrite (FeAsS), native silver, silver alloys with Sb and Hg, and silver sulfosalts. Both mineralogy and textures are unusual (Figure 19-18); silver occurs as the cores of rosette, flowerlike, and fernlike masses of the arsenides and as veinlets in arsenides and carbonates. The veins are zoned (Petruk, 1968) from Ni-As and Ni-Co-As assemblages below the diabase Co-As and Co-Fe-As assemblages in the midparts through to the Fe-As minerals at the bottom. Scott and O'Connor (1971) report daughter-mineral-rich fluid inclusions indicating saline brines of over 30 wt % NaCl equivalent with two quartz populations giving depositional temperatures of 195 to 260°C and 285 to 360°C after correction for estimated pressures of 0.5 kbar. The ores are inclined to be high-grade in silver near the tops of orebodies, with grade decreasing markedly at the contact between Coleman sedimentary and Keewatin volcanic rocks, and becoming very low a few meters into the underlying Keewatin units, although the veins generally continue downward into those units as Co-Fe-As or carbonate-minor sulfide veins. Silver occurs as specks, leaves, and hackly, spongy masses, of which some weigh up to 800 kg. Miners can tell whether or not a vein is high-grade by the "toque test": a tossed woolen cap will catch and hang only on

(a)

(b)

Figure 19-17. (*a*) Typical, nearly vertical silver-rich, dominantly calcite-dolomite vein in Cobalt Series Coleman Formation graywacke. The vein, seen here underground, is 20 cm wide. (*b*) View of a slot stope, a mined-out vein where it came to the surface. The photo shows the attitude and width of many of the Cobalt veins. See also Figure 19-18.

Figure 19-18. Rough side (above) and a polished face (below) of a hand specimen of a vein showing some of the fernlike rosettes of arsenides and silver (white) in dolomite. The flat sides of the 8.5-cm-wide vein are its walls against Cobalt conglomerate. This specimen from the Silverfields mine assayed 220 kg of silver per ton. The bar scale is in centimeters. (*Photo by* W. K. Bilodeau.)

the rough, hackly face of a native-silver-rich vertical vein wall. Silver and nickel-cobalt arsenides are intimately associated, and silver without Ni-Co-As minerals is unheard of, although the reverse is not true. Minor amounts of native antimony and bismuth are also characteristically associated with silver, and some of the minerals, especially the native silver, are enriched in mercury, some containing as much as 8% Hg. Wall-rock alteration is trivial, consisting of bilaterally symmetrical zones of conspicuous albite-chlorite-epidote-zoisite-sericite in diabase and inconspicuous chlorite-zoisite-carbonate in sediments, neither rarely more than 2 to 3 cm wide (Boyle, 1968; Jambor, 1971). Vein outcrops are colorfully encrusted with pink erythrite [$Co_3(AsO_4)_2 \cdot 8H_2O$] and annabergite [$Ni_3(AsO_4)_2 \cdot 8H_2O$], greenish scorodite ($FeAsO_4 \cdot 2H_2O$), and wad and limonite.

The veins are commonly richest in silver above sections of the basement rocks that contain abundant interflow sediments, and many veins continue structurally downward *into* interflow units. Geologists familiar with the district point out that veins that strike at an angle to subjacent interflows prove to contain substantially richer ore shoots immediately above the intersections. Also, both vein systems and values in them are best developed in the thickest sections of the Coleman Cobalt Series sediments. As seen in Figure 19-15, those thick sections represent valleys in the Archean basement, valleys cut into portions of the section that are interflow-sediment-rich. Thomson (1957) found that fissures, fractures, and veins are largely controlled by old structural features in the Keewatin basement.

The origins of the Cobalt deposits have long been hotly debated. It is tempting to ascribe them to thermal and chemical perturbations generated by intrusion of the Nipissing Diabase sill, especially with respect to a source of nickel and cobalt. But (1) the ores are distinctly post-diabase crystallization in age; (2) they are as abundant at the lowest contact, and hundreds of meters down at the Keewatin-Huronian contact, as at the upper contact of the sill, and less common in the most differentiated internal portions of the sill; (3) the sill shows no evidence of having produced a fluid by crystallization-differentiation; (4) the sill is everywhere splintery fresh, showing no signs of particpation in a hydrothermal system; (5) the diabase might be thought of as a likely source of Ni and Co, but less reasonably of the many other elements involved; and finally, (6) diabase is not integrally associated with deposits of the same type elsewhere in the world. The influence of the diabase thus remains undemonstrated. Most early students of the district have been puzzled, and having noted the presence of the diabase and the strict structural control of the veins, have alluded deferentially to a “deep unknown source.”

Boyle (1968) collected samples of the Keewatin greenstones and their interbeds, the Cobalt Series conglomerate and graywacke, and the Nipissing Diabase, and analyzed them for Ni, Co, Ag, As, Sb, and S as well as for Cu, Pb, Zn, and Bi. The data, in parts per million, are given in Table 19-1.

TABLE 19-1. Rock Geochemistry of the Cobalt District, Ontario

	Ni	Co	Ag	As	S	Sb	Bi	Cu	Pb	Zn
Nipissing Diabase	128	40	0.11	1	600	0	0	83	0	70
Cobalt conglomerate and graywacke	60	25	0	0	180	0	0	20	<5	20
Keewatin basalts	120	68	0.25	10	1250	1	0	210	<5	155
Keewatin interflows	160	74	0.83	140	6%	13	1	420	75	2400
Sulfide concentrate from interflows	1000	1600	7	1100	—	50	—	750	200	4400

Source: After Boyle (1968).

Based on these results, neither the diabase nor the Cobalt Series sediments appear adequate to have contributed metals in any reasonable amount or proportion—even Ni, Co, Ag, and Cu in the diabase are normal background values. But the interflow sediments, and to some extent the basement greenstone volcanics as well, contain at least enough Ni, Co, Ag, As, and S to be interesting, and the pyrite-pyrrhotite-chalcopyrite-enriched concentrate from the interflow pyritiferous slate-schist is abundantly anomalous. This geochemical evidence, along with the distributional evidence cited in earlier paragraphs—especially the net abundance of veins over interflow-rich basement and the presence of ore shoots in veins preferentially above vein-interflow intersections—has been taken by many as evidence of vertical secretion of the ores. Further, there is no suggestion in the district or regional geology of the presence of a conventional source of hydrothermal fluids, a post-Nipissing igneous event, for example; neither is there encouragement in igneous geochemistry in known systems that a "Cobalt-type" fluid would be evolved. In a paper that was not familiar to the present authors when these pages were written, Halls and Stumpfl (1969) were perhaps first to conclude that the ore components at Cobalt were derived from the interflows.

Boyle (1968) concluded with thoughts paraphrased as follows. Because the Cobalt veins are so restricted both laterally and vertically, it seems probable that their constituents were derived from nearby sources. As fractures opened, it seems likely that mobile constituents from the Keewatin greenstones and sediments entered, namely, Ni, Co, As, Ag, Bi, S, and Ca, Mg, and CO_3. The diabase may have contributed some Ni and Co, and its heat may have stimulated redistribution of the elements. "In other words, the veins of Cobalt are essentially secretion veins, and their constituents are late "distillates" from the country rocks." (Boyle, 1968).

Some tough questions still remain. Why do the veins diminish in values a few meters below the Keewatin surface into the interflows? What accounts for rootless veins at the *upper* Nipissing Diabase–Cobalt Series sediment contact? Why were quartz and gold the principal elements mobilized from interflows at Kirkland Lake, and Ni-Co-Ag-Fe-As-S and carbonates the mobile components at Cobalt? Why is there not more Cu-Pb-Zn in the Cobalt veins? How are the high-temperature fluid phases involved? Regardless of such difficulties, the districts described in this chapter are among the best documented instances of lateral and vertical secretion processes known to us, and the principles and products should be considered in evaluating potentially related systems—especially in greenstone belts, but by no means restricted to them—elsewhere in the world. Much more compelling data will doubtlessly be developed in the late 1980s.

Bibliography

Anhaeusser, C. R., 1976. The nature and distribution of Archean gold mineralization in southern Africa. *Miner. Sci. Eng.* 8:46–84.

Bell, D. R., and D. S. Rogers, 1979. A problem related to outlining gold deposits along the greenstone nose at the Dome mine, South Porcupine, Ontario. Presented at Can. Inst. Min. Metall. Meet. Montreal, 13 pp.

Berry, L. G., Ed., 1971. The silver arsenide deposits of the Cobalt-Gowganda region, Ontario. *Can. Miner.* vol. 11, 430 pp.

Boyle, R. W., 1955. The geochemistry and origin of the gold-bearing quartz veins and lenses of the Yellowknife greenstone belt. *Econ. Geol.* 50:51–66.

——, 1959. The geochemistry, origin, and role of carbon dioxide, water, sulfur, and boron in the Yellowknife gold deposits, N.W.T., Canada. *Econ. Geol.* 54:1506–1524.

——, 1961. The geology, geochemistry, and origin of the gold deposits of the Yellowknife district. Geol. Surv. Can. Mem. 310, 193 pp.

——, 1968. *The Geochemistry of Silver and Its Deposits.* Geol. Surv. Can. Bull. 160, 264 pp.

——, 1979. *The Geochemistry of Gold and Its Deposits.* Geol. Surv. Can. Bull. 280, 584 pp.

Burnham, C. W., and H. Ohmoto, 1980. Late-stage processes of felsic magmatism, pp. 1-11 in S. Ishihara and S. Takenouchi, Eds., *Granitic Magmatism and Related Mineralization,* Soc. Min. Geol. Jap., Spec. Issue 8, 247 pp.

Cerro de Pasco Corp. Staff, 1950. Lead and zinc deposits of the Cerro de Pasco Copper Corporation in central Peru, *Proc. 18th Int. Geol. Congr. 1948,* pt. 7, pp. 154–186.

Davies, J. F., 1977. Structural interpretation of the Timmins mining area, Ontario. *Can. J. Earth Sci.,* 14:1046–1053.

—— and L. E. Luhta, 1978. An Archean "porphyry-type" disseminated copper deposit, Timmins, Ontario. *Econ. Geol.* 73:383–396.

Dunbar, W. R, 1948. Structural relations of the Porcupine ore deposits, pp. 442–456 in *Structural Geology of Canadian Ore Deposits.* Montreal: Can. Inst. Min. Metall., 948 pp.

Ferguson, S. A., 1968. Geology and ore deposits of Tisdale Township, Ont. Dept. Mines, Geol. Rept. 58, 177 pp.

Fryer, B. J., R. Kerrich, R. W. Hutchinson, M. G. Peirce, and D. S. Rogers, 1979. Archean precious-metal hydrothermal systems, Dome mine, Abitibi Greenstone Belt, I. Patterns of alteration and metal distribution. *Can. J. Earth Sci.* 16:421–439.

George, P. T., 1967. The Timmins district, in *Centennial Field Excursion N.W. Quebec and N.E. Ontario, Can. Inst. Min. Metall.*, pp. 103–107.

Glasson, M. J., and R. R. Keays, 1978. Gold mobilization during cleavage development in sedimentary rocks from the auriferous belt of central Victoria, Australia: some important boundary conditions. *Econ. Geol.* 73:496–511.

Halls, C., and E. F. Stumpfl, 1969. Geology and ore deposition, western Kerr Lake arch, Cobalt, Ontario. Ninth Commonw. Min. and Metall. Congr., Inst. Min. Metall., Pap. 18, 44 pp.

Hodder, R. W., 1981. Pers. Commun.

Holland, H. D., 1972. Granites, solutions, and base metal deposits. *Econ. Geol.* 67:281–301.

Hurst, M. E., 1935. Vein formation at Porcupine, Ontario. *Econ. Geol.* 30:103–127.

Hutchinson, R. W., 1976. Lode gold deposits: the case for volcanogenic derivation, in *Proc. Pac. N.W. Min. Metall. Conf.*, Oregon Dept. Geol. Min. Indus., pp. 64–105.

——, W. S. Fyfe, and R. Kerrich, 1980. Deep fluid penetration and ore deposition. *Min. Sci. Eng.* 12(3)107–120.

Jambor, J. L., 1971. General geology, pp. 12-33, and Origin of the silver veins, pp. 402–413 in L. G. Berry and J. L. Jambor, Eds., The silver-arsenide deposits of the Cobalt-Gowganda region, Ontario. *Can. Mineral.* 11(pt. 1):1–429.

Jones, W. A., 1948. Hollinger mine, pp. 464–481 in *Structural Geology of Canadian Ore Deposits*, vol. 1. Montreal; Can. Inst. Min. Metall., 948 pp.

Kerrich, R., and B. J. Fryer, 1979. Archean precious-metal hydrothermal systems, Dome mine, Abitibi Greenstone Belt, II. REE and oxygen isotope relations. *Can J. Earth Sci.* 16:440–458.

Knight, C. L., 1957. Ore genesis—the source bed concept. *Econ. Geol.*, 52:808–817.

Knopf, A., 1929. The Mother Lode system of California. USGS Prof. Pap. 157.

Kwong, Y. T. J., and J. H. Crocket, 1978. Background and anomalous gold in rocks of an Archean greenstone assemblage, Kakagi Lake area, N.W. Ontario. *Econ. Geol.* 73:50–63.

Norton, D. L., 1982. Fluid and heat transport phenomena typical of copper-bearing pluton environments: southeastern Arizona, pp. 59–72 in S. R. Titley, Ed.,

Advances in Geology of the Porphyry Copper Deposits, Southwestern North America. Tucson: Univ. Ariz. Press, 560 pp.

Petruk, W., 1968. Mineralogy and origin of the Silverfields silver deposit in the Cobalt area, Ontario. *Econ. Geol.* 63:523–531.

——, 1971. General characteristics of the deposits, pp. 76–107, and Mineralogical characteristics of the deposits and textures of the minerals, pp. 108–139 in L. G. Berry and J. L. Jambor, Eds., The silver-arsenide deposits of the Cobalt-Gowganda region, Ontario. *Can. Mineral.* 11 (pt. 1):1–429.

Pyke, D. R., 1975. On the relationship of gold mineralization and ultramafic volcanic rocks in the Timmins area. Ont. Minist. of Nat. Res. Misc. Pap. 62, 23 pp.

Ridler, R. H., 1970. Relationship of mineralization to volcanic stratigraphy in the Kirkland–Larder Lakes area, Ontario. *Geol. Assoc. Can. Proc.* 21:33–42.

——, 1976. Stratigraphic keys to the gold metallogeny of the Abitibi belt. *Can. Min. J.* 97, June:81–88.

——, 1979. The metallogeny of gold. Keynote address, 85th Adams Club Spec. Symp., McGill Univ., Montreal, Quebec, Feb. 22.

Scott, S. D., and T. P. O'Connor, 1971. Fluid inclusions in vein quartz, Silverfields mine, pp. 263–271 in L. G. Berry and J. L. Jambor, Ed., The silver-arsenide deposits of the Cobalt-Gowganda Region, Ontario. *Can. Mineral.* 11(pt. 1):1–429.

Strickler, S. J., 1978. The Kirkland–Larder Lake stratiform carbonatite. *Mineral. Deposita* 13:355–367.

Sullivan, C. J., 1948. Ore and granitization. *Econ. Geol.* 43:471–498.

Swan, M. M., D. M. Hausen, and R. A. Newell, 1982. Lithological, structural, chemical, and mineralogical patterns in a Precambrian stratiform gold occurrence, Yavapai County, Arizona, pp. 143–157 in D. M. Hausen, and W. C. Park, Ed., *Process Metallurgy*. New York: AIME.

Thomson, R., 1957. Cobalt Camp, pp. 377–388 in *Structural Geology of Canadian Ore Deposits*, vol. 2, Montreal: Can. Inst. Min. Metall., 524 pp.

TWENTY

Epigenetic Deposits of Doubtful Igneous Connection

This chapter deals—as did Chapters 18 and 19—with some deposit types that have proven difficult to classify in the past in part because their genesis has been poorly understood. Members of this group are perhaps more diverse in nature than those described in earlier chapters because what they have in common is the absence of something rather than the presence of a unifying feature. *Epigenetic* in the chapter title refers to the fact that the orebody components to be treated have been lodged in their final configurations after—and commonly much later than—the rocks that contain them were formed. So "epigenetic" eliminates sedimentary, metamorphic, and many igneous deposits. Deposits related to weathering could be included here. As described in Chapter 17, their components have by and large been modified in place or transported such short distances, under such clear-cut circumstances, that they constitute their own valid category. There are some real uncertainties about boundaries between sources and processes of formation of some of the deposits discussed in Chapter 19 and those to be discussed here. In general, the source rock in ore deposits ascribed to lateral secretion is considered to be nearby and identifiable as such, but in those to be covered in this chapter the source rock identity is more remote and less specifically identifiable. An important phrase is 'of doubtful igneous connection.' Lindgren's classification modified by Graton included a category named 'telethermal.' It was most recently described (Park and

MacDiarmid, 1975) as "deposition from nearly spent solutions; temperature and pressure low; upper terminus of the hydrothermal range." *Tele* is the Greek root for distant, so *upper terminus* refers to the outermost, most remote effects of presumed magmatogene fluids. As Park and MacDiarmid stated,

> *Some ore deposits are formed by hydrothermal fluids that have migrated so far from their source that they have lost most of their potential to react chemically with the surrounding rocks. These terminal phases of the hydrothermal plumbing system are called telethermal fluids. The telethermal zone is a shallow environment where temperatures and pressures are low and where the general characteristics of minerals are similar, whether they were precipitated from descending meteoric waters or from ascending hydrothermal fluids diluted by cooler ground waters.*

Clearly, it has been recognized that the environment of formation of these deposits was so distant from the igneous hearth that even its recognition as igneous-related might be difficult. In fact, the only deposits described under 'telethermal' in Park and MacDiarmid are those known as the *Mississippi Valley type*, a group that has been seen since the late 1970s to be divorceable from igneous processes. Other ore types that were listed earlier by Park and MacDiarmid as telethermal are Colorado Plateau type uranium ores and the White Pine, Michigan, copper deposits. They would have included the then undiscovered Athabasca uranium deposits. All of these deposits may now be considered epigenetic deposits involving heated aqueous fluids of various origins unrelated to igneous activity, or at least not demonstrably related to it. Mississippi Valley type deposits are now thought to result from the migration of warm, metalliferous connate-water brines from depressed basins updip into carbonate sediments at basin margins. Sangster (1976) proposed the term *sedimentogenic* for such processes, to emphasize sedimentologic rather than igneous aspects. Colorado Plateau uranium ores were created when groundwaters delivered oxidized, dissolved U^{+6} ions to local near-surface reducing environments, where the ions were reduced and UO_2 precipitated. Athabasca unconformity-type uranium deposits involved warm solutions circulating first downward, then upward along localizing fault planes to produce complex uranium–trace-transition-metal–sulfide ores. Each of these deposit types is described below. The name telethermal can still be applied to some of these deposits, but it must be recognized that use of the term now denotes low-temperature, low-pressure environments so remote from igneous activity that no connection can be established. Heat related to igneous activity may be involved, but even that cannot be ascertained.

MISSISSIPPI VALLEY TYPE DEPOSITS

This deposit group has been the principal source of lead and zinc in both Europe and North America, and deposits of this type probably occur on all continents. Historically (Figure 20-1), the most important American districts include the Tri-State district of Oklahoma-Kansas-Missouri centered on Joplin, Missouri, the Lead Belt of southeast Missouri near the town of Bonne Terre, and the mines of southwest Wisconsin near Shullsburg. Current North American activity involves central and eastern Tennessee, the so-called Viburnum Trend in south central Missouri (Figure 20-3), and the Pine Point area, Northwest Territories, in Canada (Figure 19-1). Great Britain and Europe have scores of deposits, including Cumberland, England; Trepça, Yugoslavia; and Cracow, Poland.

Before we consider recent advances in understanding the genesis of these deposits, let us review their geologic characteristics. E. L. Ohle, in the first of two landmark papers (Ohle, 1959, 1980), enumerated many of these attributes, and the serious student should consult those two papers. The Wolf volumes include a major article by Sangster (1976), and *Economic Geology Monograph* 3 (1967) and a special issue of *Economic Geology* (1977, no. 3) highlight this deposit group. As Ohle (1959) stated,

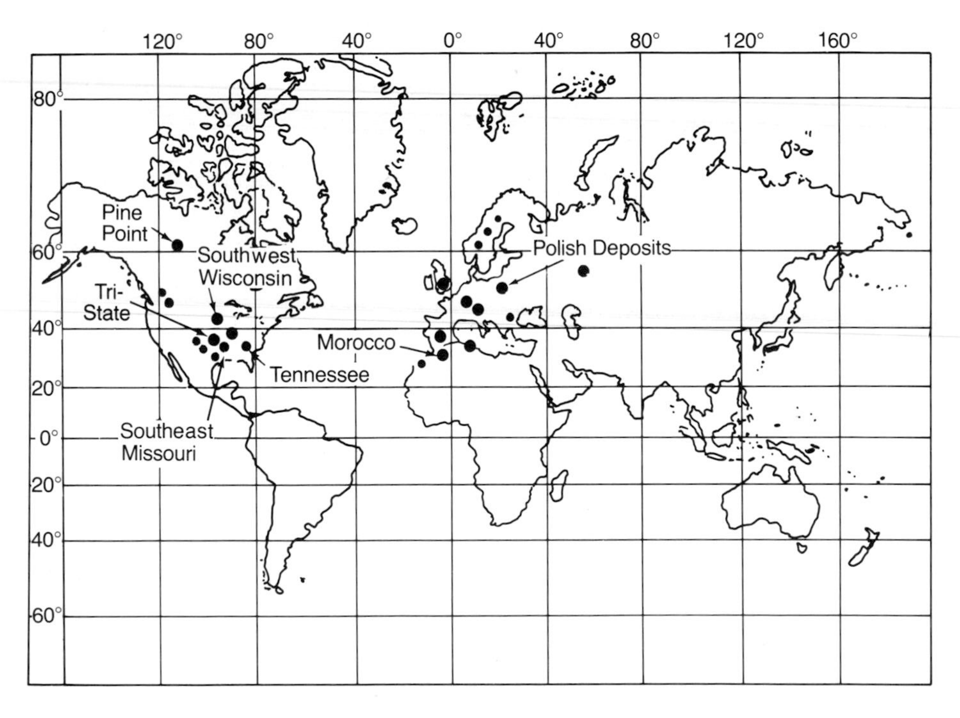

Figure 20-1. Location map of major Mississippi Valley type lead-zinc deposits of the world.

> *. . . a principal thesis . . . is that there is so much similarity in all these deposits that it seems necessary for all of them to have had a similar mode of origin, whatever that might be. Variations in the shape, size, and mineralogy of the orebodies give each district some individuality, but there are enough of the 'typical' characteristics present in all cases to make classification in the* Mississippi Valley type *agreeable to many geologists.*

The Viburnum Trend and Polish deposits are described in detail below.

There are many characteristics common to Mississippi Valley type deposits (MVD). One is a lack of nearby bodies of igneous rock. The districts are commonly laterally extensive, but thin—most districts are less than 100 meters thick vertically, but they extend over hundreds of square kilometers (Figure 20-3, for example). MVD's simple major mineralogy includes only low-silver galena, low-iron sphalerite, barite, and fluorite as ore minerals. Gangue includes only dolomite, calcite, jasperoid, and minor silica; barite and fluorite may be gangue or ore minerals. Pyrite and marcasite are present but normally sparse. Trace elements provide subordinate production of copper from minor but widespread chalcopyrite, nickel and cobalt from rare siegenite [$(Ni,Co)_3S_4$] and millerite (NiS), and cadmium, indium, germanium, and gallium from solid solution in sphalerite. Luzonite [$Cu_3(As,Sb)S_4$] has been reported. The ores contain negligible silver and gold, and are thus of low precious-metal content. They occur preponderantly as bedded strata-bound replacement sheets in dolomite or dolomitized limestone sedimentary host rocks, although the veinform fluorite-calcite ores of Cave-in-Rock, Illinois, the fold-related joint fillings of southwestern Wisconsin, and the solution-collapse breccia fillings of eastern and central Tennessee constitute important modes of occurrence. Galena ores have also been found in sandstone units in or adjacent to carbonate sections. MVDs occur toward the margins of many major sedimentary basins of the world, where sediments onlap cratonic shelves. They are most commonly at shallow depth in structurally passive, anorogenic areas, such as the midcontinent region of the United States, in which flat to gentle dips still prevail. Gentle regional-scale doming is also common. Within the carbonate host rocks, mineralization is related to positive structures of many types, including Precambrian monadnocks not unlike those in the Zambian Copperbelt, which controlled sedimentation in Paleozoic seas by enabling the development of great banks of down-current calcareous sand, atoll structures, algal reefs, and depositional arches and basins. Evidence of solution activity on a grand scale is common, especially in Tennessee. Although some ore is unquestionably of replacement origin, much of it occurs as open-space fillings in solution breccias in which fragments are always stratigraphically dropped, in enlarged joints and fractures in carbonate hosts, and rarely in vugs and cavities along crosscutting fault-vein structures.

Another characteristic of telethermal deposits is the simplicity and often the subtleness of wall-rock alteration effects. Silicification, pyritization, and especially dolomitization and recrystallization are the common types, but how much of this observed association is ore-related alteration and how much is preore, even diagenetic, is often unclear. Indeed if, as noted later, MVDs are formed as a result of connate-water circulation, the separation of the two processes may be impossible, or will be until we are able to date the metal concentrations more precisely.

Dolomitization of host-rock limestone is observed in virtually every district, and whether diagenetic or ore-related, it is helpful in guiding exploration. Dolomite halos may extend anywhere from a few centimeters to several kilometers beyond economic mineral concentrations. Virtually no ore occurs in undolomitized limestone. In certain of the MVD districts, of which the Tri-State area of Kansas-Missouri-Oklahoma is a superb example, intense silicification with jasperoid replacing both limestone and dolomite occurs along brecciated ore "runs." Usually silicification favors dolomite that has replaced original limestone. In many districts these alteration effects have produced distinctively different appearances in originally quite similar beds; recognition of "marker horizons" thus created has been a great aid in structural mapping.

Fluid inclusion studies have revealed salinities of 15 to 30 wt % NaCl and temperatures of 50 to 225°C, with 120°C being about average. Both the moderately high temperatures and the high salinities are surprising, considering the shallow depths of the deposits. In southeast Missouri, for example, the ore-bearing areas were never buried more than 2 km; actual depths were probably considerably less, so abnormally high temperature gradients had to be present at the time of ore formation.

Although the ores are usually referred to as being of simple mineralogy, there are enough minor or trace metals present to show that the ore solutions had the capability of transporting a considerable variety of elements. In southeast Missouri, along with the lead, zinc, copper, sulfur, and silver that are recovered, there are significant amounts of iron, cobalt, nickel, cadmium, indium, germanium, gallium, sodium, chlorine, silicon, and magnesium that were obviously moved at the same time. These same solutions had a remarkable ability to dissolve carbonate rocks and, to a great degree, accomplished the 'ground preparation' that went on simultaneously with ore deposition.

Textures and structures are varied because the ores were deposited by replacement as well as by open-space filling, and the minerals vary from fine- to very coarse-grained. Most mineral museums display spectacular galena, sphalerite, and fluorite specimens with banding and dramatically euhedral crystal growth in cavities that come from these deposits. The evidence for open-cavity filling is an abundance of crystal-lined vugs, comb structures, and rhythmically banded ores. Replacement is dominant in carbonate rocks, although many permeable limestones and dolomites, as well as sandstones, contain ores that primarily fill pore spaces. MVDs are likely to be structurally simple. Because they were formed a long distance from possible magmatic centers and a long way from areas of tectonic activity, circulation of fluids and deposition of ores were controlled by all types of subtle permeability. As stated, most of the ores are in flat-lying beds and show little or no evidence of deposition from fluids ascending from basement rocks directly beneath them. The lateral extent of the ore areas compared to their height indicates that the horizontal component of solution movement probably far exceeded the vertical. Their general appearance and character are such that they have been interpreted as products of meteoric or sedimentary processes, although fluid inclusion data have essentially eliminated this option.

The resolution of genetic problems has proven difficult. Many of the textures and structures of the ores are geometrically similar to those of the rocks that contain them, so questions concerning the relative ages of rocks and ores are difficult. Snyder (1967) marshaled criteria for determining the timing of ore deposition that would help select which genetic mode might apply in a given case. Snyder's criteria for and against syngenetic, diagenetic, and epigenetic timing of ore deposition, with reference to MVDs, are given in Table 20-1.

Geologists familiar with MVDs now almost universally agree that even though the ores are preponderantly stratabound, the ore minerals are epigenetic on the strength of applications of Snyder's criteria.

Assuming that the components of the ores have been introduced into their host rocks, the next question is "how?" Ohle (1970) summarized the modes of formation of Mississippi Valley type lead and zinc deposits. He pointed out six options of genesis for which substantial arguments have been advanced, and has since added a seventh held mostly by European geologists.

1. Original syngenetic deposition of PbS-ZnS with the carbonate host rocks.
2. Original dispersed, low-grade syngenetic deposition in carbonate sections with later concentration by regional metamorphism.
3. Original dispersed syngenetic deposition in carbonates with later concentration by groundwater moving upward in artesian flow.
4. Original dispersed syngenetic deposition with later concentration by downward-moving groundwater.
5. Deposition from fluids of igneous derivation by hydrothermal aqueous or even gas-phase transport.
6. Deposition from connate basinal water that moved updip in response to compaction or other loading pressures.
7. Deposition-concentration associated with weathering and karst development.

TABLE 20-1. Genetic Criteria for Mississippi Valley Lead-Zinc Deposits

Evidence For	Evidence Against
Syngenetic Origin—Ore deposition at the time of host-rock sedimentation	
*Mineralization at a given stratigraphic position or within a restricted range.	*Mineralization unaffected by facies changes or transgressing them.
Textural-structural relationship between mineralization and sedimentary lithologies, facies, and features; "sedimentary ore facies" commonly recognized.	*Mineralization in postdepositional, prelithification structures.
Ore units exclusively both stratiform and stratabound.	*Stable isotopes of ore minerals unlike those of established sedimentary minerals.
Ore units laterally generally uniform in thickness, grade, and metal ratios, with variation governed by sedimentary features.	*Ore structures of marked variation in height, width, or grade.
Diagenetic Origin—Ore deposition in the time between sedimentation and lithification	
*Mineralization within or near a given stratigraphic position.	*Transgressive mineralization unrelated to diagenetic structures.
Textural-structural relationship between mineralization and sedimentary features close but not identical.	*Mineralization of postlithification structures.
*Mineralization texturally late in enlarged pore spaces, veinlets, and so on.	*Similar mineralization in varied lithologies.
Mineralization related to diagenetic textures, structures, and minerals.	*Extensive open-space filling.
*Ore units stratabound, not necessarily stratiform.	*Stable isotopes of ore minerals unlike those of established diagenetic minerals.
Ore units laterally generally uniform with variation governed by sedimentary and diagenetic features.	
Epigenetic Origin—Ore deposition after sedimentation and lithification	
*Mineralization in tectonic, postsedimentation structures.	
*Ore minerals neither stratabound nor stratiform.	
Stable isotopes of ore minerals unlike those of sedimentary and diagenetic minerals, but like those of established epigenetic ones.	
*Wall-rock alteration or trace element dispersion halos into walls.	

*Criteria that apply to the southeast Missouri MVD district.

Source: After Snyder (1967).

We have already seen that options 3, 4, and 7 seem obviated by fluid inclusion data—both salinities and temperatures were too high. Option 1 is precluded by the variety of textural, structural, and temporal details described earlier. Option 2 is not precluded, and is perceivable as merging with option 6 if "regional metamorphism" is supplanted by "compaction" processes; there is no manifestation in MVD settings of regional metamorphism as discussed in Chapter 18. Weiss and Amstutz (1966) proposed that some tuffs, which they identified in the midcontinent Paleozoic section, may have carried adsorbed lead ions which were stripped and concentrated, but their explanation is quantitatively inadequate and has not been adopted.

Options 5 and 6 have proven to be most viable. Lead isotope characteristics—the J-type leads mentioned in Chapter 7—belie "simple" igneous derivation, but the many measurements of lead isotope ratios that indicate ore formation hundreds of millions of years in the future are not easily explained under any of the hypotheses. Some geochemical characteristics (Erickson et al., 1981) suggest that deep-seated alkalic igneous activity may be related, but the link is still tenuous at best. In general, option 5 has been set aside by most MVD geologists in favor of the growing body of information and evidence that supports option 6 and the inclusion of MVDs under this chapter heading. It is necessary to note, however, that several essential elements of option 6—especially the high salinities and high solution temperatures—still need adequate explanation. Igneous heat sources in basement rocks may still prove integral to MVD formation.

Donald E. White, in a 1968 paper, was the first author to exchange magmatogene ideas for "sedimentogene" ones when he assigned the characteristics of fluids included in Tri-State sphalerite to a connate source. The

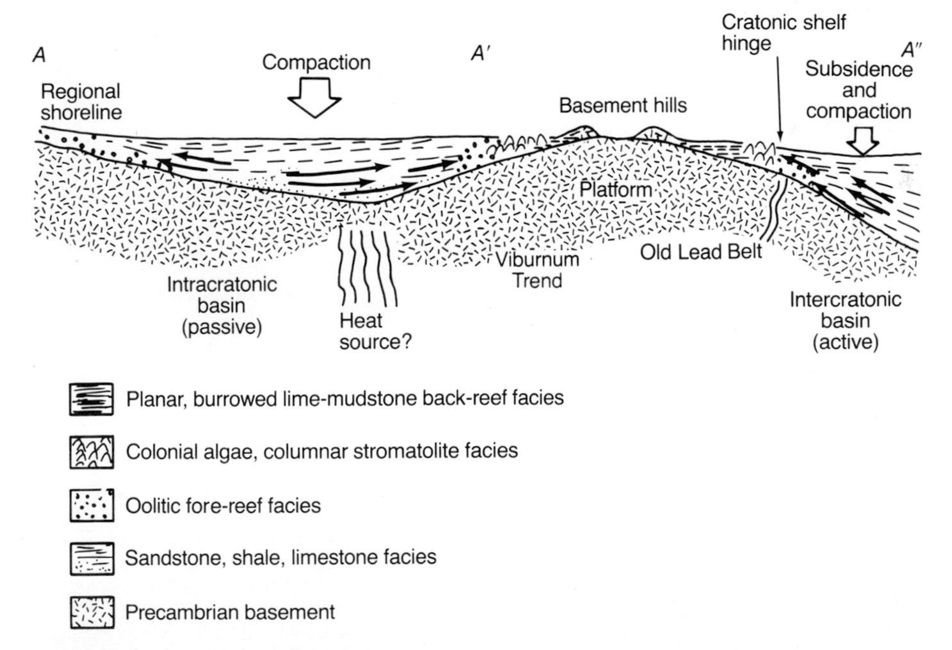

Figure 20-2. Schematic showing dewatering of continental and off-shelf basins and the flow (arrows) of migrating, metal-bearing, locally petroliferous connate brines into carbonate shelf facies. The compaction and fluid movement must have occurred somewhat after the sedimentation. See also Figures 20-3 and 20-4. (*Suggested by* Larsen, 1977.)

general mechanism is sketched in Figure 20-2, a cross section at once schematic and thought to be real. The concept is that original seawater, trapped as connate water while marine sediments accumulated, evolved until ultimate migration. As burial progressed, temperatures increased moderately, and the trapped seawaters reacted with detrital minerals. Water itself was partially consumed by hydration, enriching the remainder in sodium and chloride ions. Partitioning reactions concentrated base metals as chloride complexes as long as reduced sulfur species were kept at low activities of less than 30 ppm below 150°C (Anderson, 1979; Carpenter and Grethen, 1979). Oil field brines have been known for decades to contain locally anomalously high chlorine, zinc, lead, and barium levels, and sphalerite has commonly been reported as a minor mineral in petroleum drill-hole logs. This association with petroleum has led some workers to suggest that the metals may have been transported in organic rather than inorganic complexes. Not surprisingly, the sulfide minerals have been reported where $S^{=}$ was also high. Even in old papers, Ohle (1967) reported, sphalerite was noted in oil-field drill cores from holes in the Arbuckle Formation of central Kansas at 2-km depths, and where $S^{=}$ is low and Zn^{+2}, Pb^{+2}, Ba^{+2}, and Fe^{+2} can be leached and made available for transportation. The compositions of several oil field brines are given in Table 20-2. Not incidentally, it now appears, many MVD ores also contain recognizable amounts of hydrocarbons; specimens of sphalerite crystals in vuggy limestone from the Tri-State district commonly are coated with oil or asphaltic, kerogen-rich compounds (Fowler and Lyden, 1935).

With progressive load or tectonic compaction, formation waters are fugitive (Sharp, 1978). They migrate updip, guided by relatively impervious

TABLE 20-2. Composition of Selected U.S. Oil Field Brines

Locale	Source Rock Age	Dissolved Solids, ppm	Dissolved Metal, ppm					
			Pb	Zn	Cu	Fe	Ba	Hg
Michigan	Silurian	400,000	10	2	—	10	11	—
Mississippi	Cretaceous	320,000	111	357	—	420	59	—
Mississippi	Cretaceous	255,000	80	300	—	298	61	—
Alabama	Jurassic	486,000	215	39	—	467	504	—
Arkansas	Jurassic	351,000	<1	<1	—	3	34	—
Texas	Cretaceous	344,000	226	706	—	1060	1090	—
Texas	Oligocene	75,000	8	190	0–6	140	—	8

Note especially the high Zn-Pb-Ba contents and the fact that, for example, the Arkansas brines might precipitate only barite while others might yield sphalerite and galena. "Dissolved solids" are the sum of Na, Ca, K, and Cl. It is regrettable that heavy-metal analyses of the vast majority of oil field waters are as lacking as hydrocarbon analyses are of hydrothermal fluids.

Source: From Carpenter et al. (1974); Carpenter and Grethen (1979); bottom row from Kharaka, Callender, and Chemerys (1979).

argillaceous intrastrata and passing through permeable sandy layers, probably equilibrating during passage. Doe and Delavaux (1972) reported that ore lead isotopes in southeast Missouri are like those in the basal LaMotte sandstone aquifer (see below) and the Bonneterre formation. Precipitation of sulfides occurs where favorable environments are encountered, presumably reducing sulfurous ones. Banaszak (1979) showed that the fluid requirements outlined by Ohle are probably met. The brines react with silicates upon warming and become acid (pH $\cong$ 4 to 5), sodium-enriched, and as dense as fresh water. Upon rising updip into carbonate rocks, the brines tend to dissolve them. Cooling results in host-rock solution with little pH change, and in deposition of ore sulfides; replacement proceeds early, and open-space filling later. Temperature, pH change, dilution effects, and bacterial activity in the hydrocarbon-rich near-surface environments may all be important. But it is an observable fact that sulfides are deposited in reducing environments in paleoreef, back-reef, and fore-reef environments, and in calcarenites. Recognition of these controls was first systematized by John S. Brown, Frank Snyder, and Ernest Ohle in the late 1950s in the Old Lead Belt of southeast Missouri. Davis (1977) noted that many oil field brines are 'sour' with high H_2S and $S^=$, and that perhaps a two-fluid model prevailed, one where brines with reduced sulfur migrated updip from deep basins to mingle in the reef environment with metal-chloride brines derived from back-reef evaporite sections (Figure 20-2). The scenario is hydraulically difficult but chemically possible.

Clearly, not all the chemical, physical, and hydraulic data are known, but the basic tenet—the involvement of basin-derived, connate, oil field brines as contrasted with igneous-generated fluids however remote from the magma—is better demonstrated every year. Research involving collaboration between petroleum geologists, hydrologists, reservoir engineers, and ore deposit geologists is proceeding, as indicated by the staging of a Penrose Conference on the subject in May 1982. A major continuing problem is that of the actual timeframe of mineralization, MVD ores in general being inscrutable with respect to their absolute age.

As Ohle (1980) concluded, "this review . . . is ended with a plea that all geologists remain inquisitive and critical. There is before us a promising concept. . . ," one that is bound to evolve further.

Viburnum Trend Lead-Zinc Deposits, Missouri

Any of several districts might have been chosen for description as exemplary of MVDs. Pine Point, Northwest Territories, for example, is increasingly well known, and has been described as having involved 50 to 100°C fluids moving updip from a deep off-shelf basin into a reducing reef environment formed at a cratonic hinge to produce ores containing up to 50 wt % combined lead-zinc. It is the type area for the "basin brine" hypothesis. But probably the best known MVD in the world is the southeast

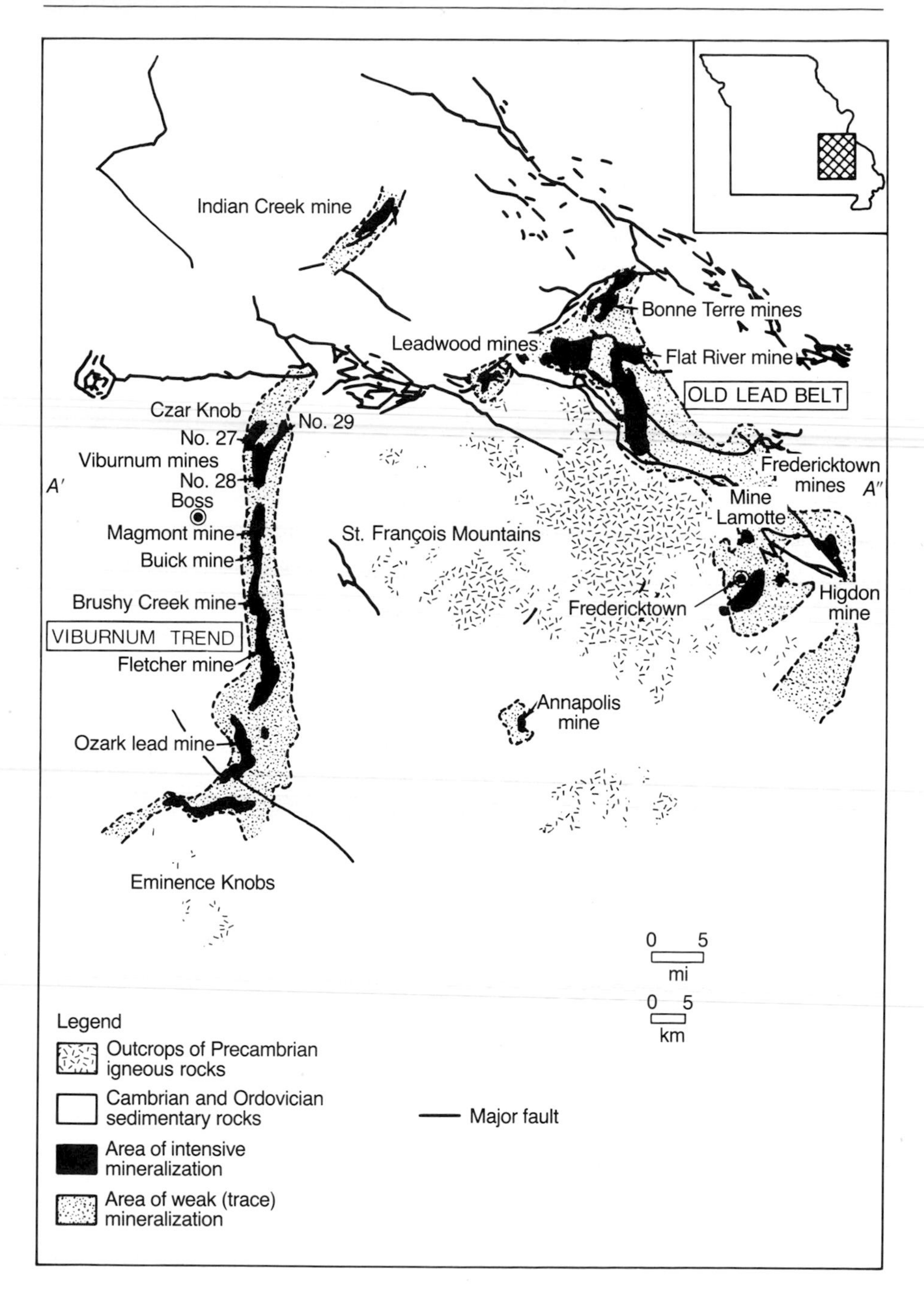

Figure 20-3. Geography and geology of southeast Missouri. Section A'-A'' is schematically that of part of Figure 20-2. The Old Lead Belt and the Viburnum Trend represent Upper Cambrian algal reef-atoll-calcarenite bar complexes on either side of a Precambrian island about the size of today's Jamaica. (*From* Kisvarsanyi, 1977.)

Missouri district. The Old Lead Belt (Figure 20-3) was discovered in the 1700s and was mined as early as 1720 for lead for musket balls by the French Mississippi River explorers and for later use in the French and Indian War; French place names are still common in the Bonne Terre area. The mines produced millions of tons of lead and zinc through 1972, but are now largely exhausted. Geologic-geophysical studies in the Old Lead Belt in the 1950s so clearly established the habitat of the lead-zinc ores that nearby exploration was predictive and successful. The Indian Creek mine (Figure 20-3) was discovered in 1948 and the Viburnum Trend—unknown hitherto—was drilled and developed a few years later. The Viburnum Trend, or New Lead Belt as it is also known, is a north-south belt some 75 km long and 5 to 6 km wide (Figure 20-3) of Cambrian Bonneterre formation sediments. That formation is limestone everywhere else, but near the lead belt it is dolomitized for several kilometers on both sides of reef zones. The average grade of the ores mined from it, from the several major mines designated in Figure 20-3, is 5% Pb and 2% Zn, with 0.2% recoverable Cu, minor cadmium, and almost negligible silver. Production in 1974 provided 85% of the United States newly mined lead consumption and 15% of the world total.

Overall, the Old Lead Belt and the Viburnum Trend constitute a horseshoe-shaped fringe around a cluster of "core" outcrops of middle Precambrian granites, rhyolites, and diabases (Figure 20-3). Thacker and Anderson (1977) noted that these Precambrian rocks—the St. François Mountains of the Ozark Uplift—were a positive area in the Central Stable region of North America, even in Upper Cambrian (St. Croixian) times. As seas transgressed, they deposited the basal La Motte sandstone and a thick overlying section of carbonates. Around the central area of islands surrounded by sunny seas, many of which supported their own stromatolite-fringe atolls (Figure 20-4), there formed a huge reef area of platform-margin *digitate stromatolites,* known in the mining districts as *finger rock* (Figure 20-5). These structures were interpreted much earlier as some sort of hydrothermal alteration effect. The reefs probably looked much like modern algal reefs in which upward-protruding algal 'fingers' are separated by cups of bioclastic debris, oolites, and carbonate sand grains. Laminated, fragment-rich, fore-reef deposits dipped off into deeper water west of the Viburnum Trend and basinward from reefs in the Old Lead Belt (Figure 20-2). Where currents flowed through and past the basement granite islands, generally northeastwardly to northwestwardly, long calcite sandbars formed on the lee sides of islands, and many of these calcarenite ridges later became capped with reefs. Other environments are indicated in Figure 20-4. Burrowed limey mudstones—now dolomitic—occupied the back-reef areas, and the high-energy slope-to-basin fore-reef zones included oolitic facies and primary taluslike sedimentary breccias that yielded basinward to fine-grained, finely bedded, micritic limestone and shale lithologies. After formation of the reef, atoll, and calcarenite bar facies, deepening seas snuffed out the reefs and led to the essentially conformable deposition of the Davis shale and younger units. Postdepositional solution produced abundant cavities and collapse breccias all along the trend, generally in the carbonates immediately above the reef structures and on the basin side.

At some time—no one is really sure when—the lead, zinc, and trace

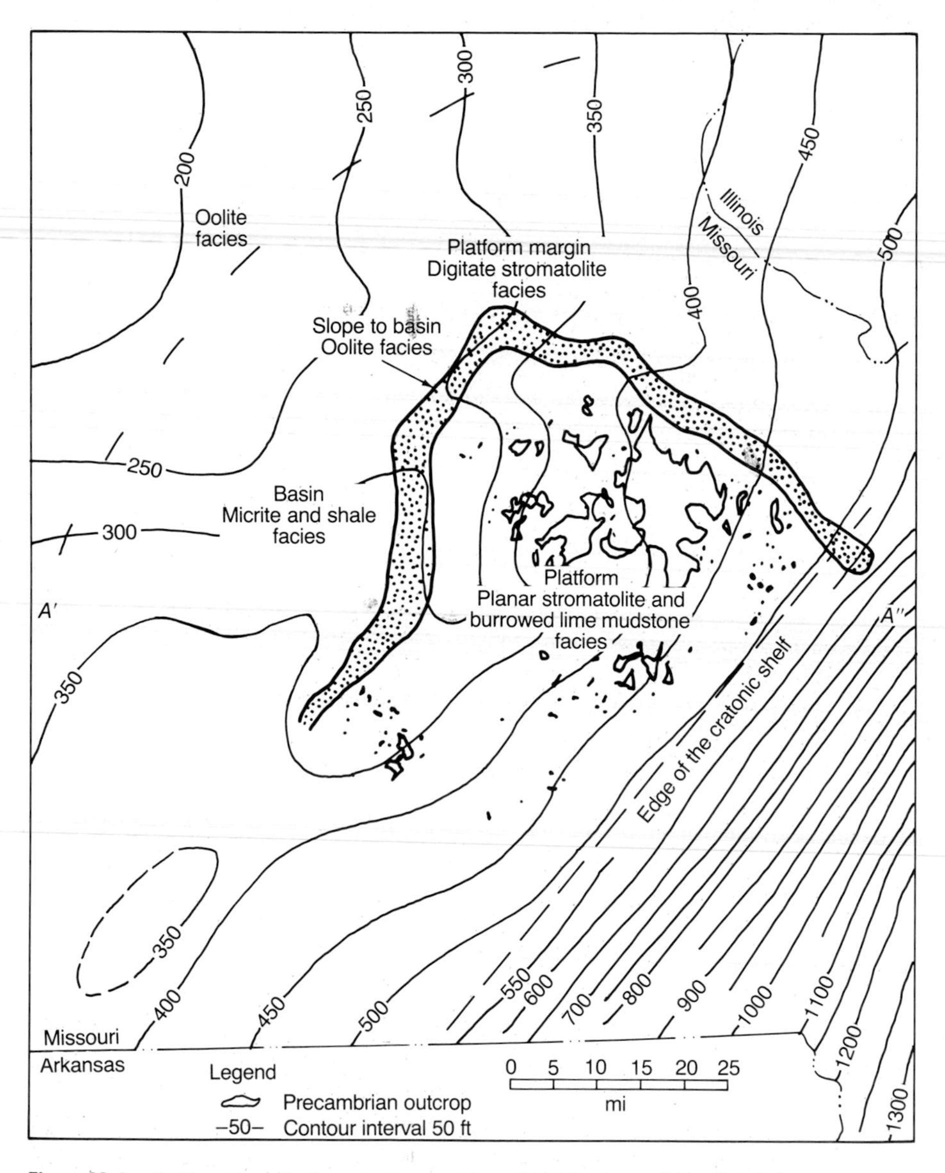

Figure 20-4. Sedimentary facies, environments, and thicknesses of the Cambrian Bonneterre Formation. A'-A'' is from Figure 20-2, as described in Figure 20-3. Note the similarity of the form of the "finger rock" digitate reef facies and the area of intensive mineralization in Figure 20-3. (*From* Larsen, 1977.)

Figure 20-5. Polished slab of "finger rock"—digitate stromatolite facies—in its natural position. The central prong was a colony of algae reaching for the light and warmth of the Cambrian sea surface. Millions of these colonies constituted a bioherm. Bioclastic debris and oolites are draped in the pockets between the fingers. The scale is in inches. (*From* Lyle, 1977.)

element ores were emplaced. Individual mines are well described in articles in the special issue of *Economic Geology* referred to above (1977, pp. 337–490). The principal sites of mineralization at the Viburnum No. 27 mine (Grundmann, 1977) are in the digitate stromatolitic reef horizon, either in permeable calcarenite-filled channels between linear algal mounds or in troughs behind subtidal off-shore algal bars. At the Magmont mine (Sweeney, Harrison, and Bradley, 1977), economic mineralization occurs in solution-induced collapse breccias, primary gravity slump breccias, a favorably permeable silty member, and interreef calcarenite as at the Viburnum No. 27. Rogers and Davis (1977) stress the importance of solution-induced collapse breccias in dolomitized shelf-facies calcarenites that overlie reefs and underlie Davis shales, probably as intertidal drainage channels. Individual breccia bodies are up to 100 meters wide east-west and 30 meters thick, and the ore zone runs for many kilometers north-south. Farther south at the Ozark mine (Mouat and Clendenin, 1977), galena and sphalerite occur both in the sedimentary structures already described and as bedded or disseminated sulfides along the flanks of "highs," in solution collapse breccia matrix, and as "marginal break" ores in fractures and faults bounding breccia bodies (Figure 20-6).

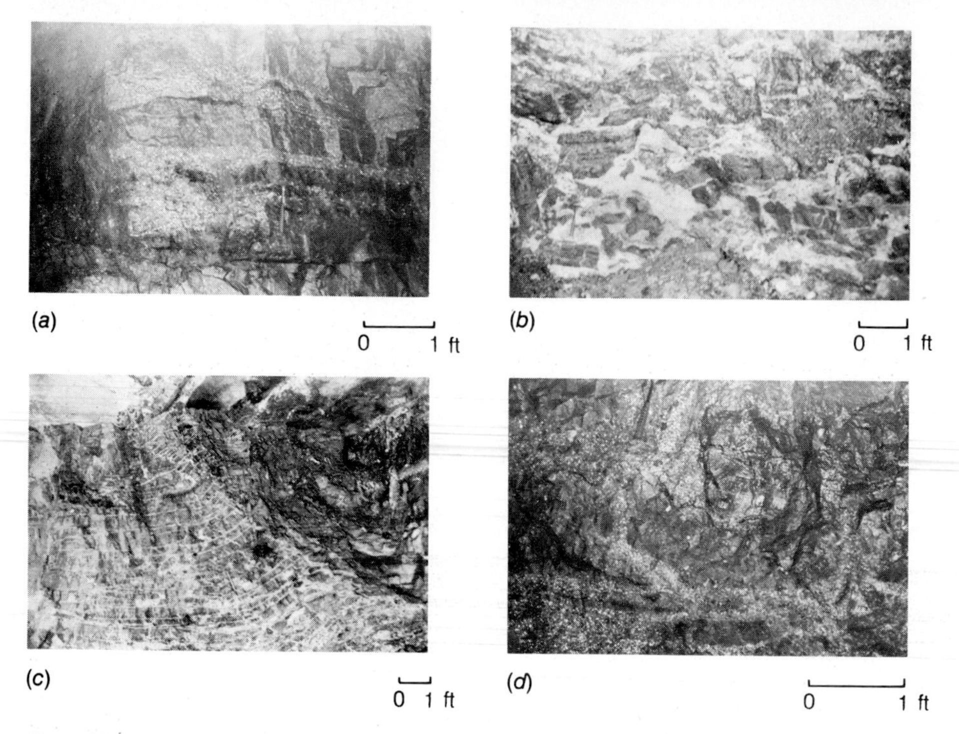

Figure 20-6. Four categories of ore at the Ozark Lead Company mine, Viburnum Trend. (*a*) Bedded galena (white flecks). (*b*) Collapse breccia, with white dolomite, galena interstitial to brown dolomite fragments. (*c*) "Marginal break" ore, with breccia to right and galena and dolomite in bedding planes and joints in incipient solution collapse zones. (*d*) Horizontal bedded galena below disseminated, fracture-controlled ore in dolomite. (*From* Mouat and Clendenin, 1977.)

In all of these occurrences, both open-space filling and replacement are important. It appears that enhancement of primary permeability by dissolution locally guided fluids. Galena was preferentially precipitated in local reducing environments encountered by migrating fluids, for example in the organic-material-rich "fingers" in Figure 20-5 and in organic-detritus-rich off-reef breccias. In many of these, oolites were replaced by galena, as shown in Figure 20-9. Sphalerite was preferred in collapse breccia matrix replacement and void fillings. Zoning of Pb, Zn, and Cu is described in the Viburnum Trend mines, but reasons for it are poorly understood. It seems likely that the ore solutions arrived in pulses of slightly different composition. In general, however, the physical-sedimentologic aspects of the ores are so well understood that the trend has been completely drilled out and evaluated, with little opportunity remaining for significant new discoveries.

Although it is still risky to place full confidence in any one explanation, the dynamics described earlier appear to account well for the Viburnum Trend ores. Solution channels both guided and were enlarged by the movement of solutions that fluid inclusions reveal to be similar but saltier than most oil field brines. The arrows of Figure 20-2 and the conclusions of Banazak (1979) and others now become conceptually operational. The mode of origin of MVDs around the world—or, more correctly, around the sedimentary basins of the world—appears more and more convincingly to be related to the expulsion of connate waters carrying moderate quantities of lead, zinc, copper, cobalt, and nickel ions scavenged from the sedimentary rocks in which they were originally interred. Chlorine ions probably traveled with them, and sulfur may either have been met in the depositional sites or delivered from back-reef sources (Davis, 1977) at the same time.

Some interesting questions remain. What is the age of MVD formation in any given district? How do MVDs fit into the maturation history of the basins? Why are nickel and cobalt characterizing trace elements? Where does the sulfur come from? What is the significance of the zoning patterns? Are hydrocarbons and bacterial activity importantly involved? How did the solutions get so hot and saline? Such questions remain only partially answered, but progress in understanding Mississippi Valley type mineralization is encouraging.

Cracow-Silesian Zinc-Lead Deposits, Poland

The zinc-lead deposits of the Upper Silesia–Cracow area, Poland, appear to be classic examples of Mississippi Valley ores in carbonate rocks. The principal mines are 10 to 25 km northwest of Katowice in southwestern Poland (Figure 20-7). These ores, exploited for more than six centuries, account for a major part of total European production of zinc (36 million

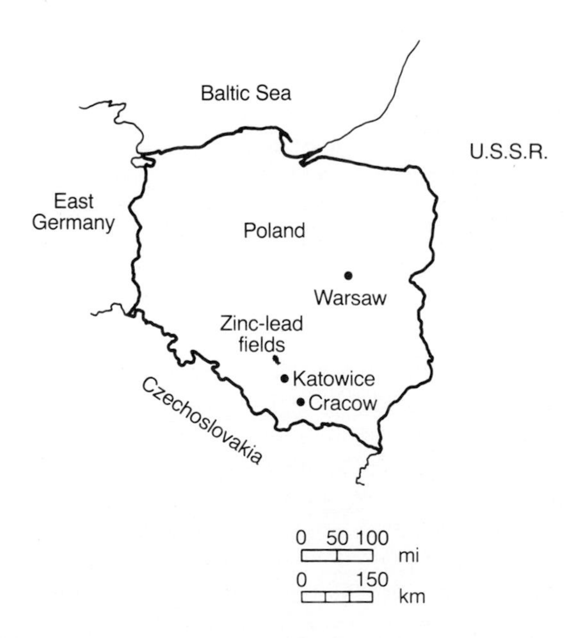

Figure 20-7. Map of Poland showing the location of the Upper Silesian zinc-lead deposits.

tons) and lead (10 million tons). Papers on the district are those of Stappenbeck (1928), Schneiderhöhn (1941), Zwierzycki (1950), Schneider (1964), Galkiewicz (1967), Gruszczyk (1967), Sangster (1976), and Sass-Gustkiewicz, Dzulynski, and Ridge (1982).

The Upper Silesian, or Cracow-Silesian as is more geographically apt, ores are in the Muschelkalk limestone, a Triassic shallow-water marine formation that includes dolomites and shales as well as limestones and is only 200 meters thick. It was deposited in shallow, Bahama Islands–like transgressive Triassic seas dotted with now buried islands of Devonian basement carbonate rocks (Ager, 1980). No evidence of contemporaneous tectonism or igneous activity is known in the region, the ores having been emplaced while the host sediments were horizontal. Concentrations of ore are in the lower 110 meters of the Muschelkalk strata, which were faulted and folded into broad gentle warps during an Early Jurassic orogeny. Lower Muschelkalk beds in the area are preserved only in synclines. Four of these synclines contain most of the ore; they can be discerned (Figure 20-8) from northwest to southeast as the Tarnowskie-Góry, the Bytom, the Czeladz-Bedzin, and the Wilkoszyn-Trzebinia downfolds. Near Cracow, to the southeast, a small amount of ore is present in the Krzeszowice-Siewierz monocline, an extensive structure that causes the Muschelkalk limestone to dip off to the northeast under younger Upper Triassic and Jurassic sediments. Mines near Olkusz are on this same monocline. These structures can all readily be perceived by tracking the Lower Triassic basal Bunter sandstone (TrB) through Figure 20-8—it is a unit not unlike the LaMotte sandstone in Missouri, a time- and distance-transgressive strand deposit laid down by an advancing sea. Schneider (1964) described the Muschelkalk seas as characterized by extensive plateau-type reefs, with marginal belts of barrier reefs with oolitic fore-reef sediments, shaley-marly fore-reef basin facies, and with back reefs of calcarenites, debris, and colonies of algae, and patch reefs. The setting was thus much like that in Cambrian Missouri. To the south was a thick prism of Paleozoic and Triassic sediments in a deep intracratonic basin. Schneider stated that the maximal development of thick reefal limestone-dolomite sections coincides significantly with areas of maximum occurrence of ore. Sass-Gustkiewicz, Dzulynski, and Ridge (1982) made no mention of reefs, saying that the sulfides do not reveal a direct relation to any primary sedimentary structures or specific primary carbonate lithology.

The timing of emplacement of these ores has long been debated. In these Polish deposits, and in similar ones in Italy and Austria, recent reporting stresses open-space filling and replacement textures in karstlike masses, in solution-enhanced primary pore spaces, joints and fractures, and solution collapse breccias, as well as in primary sedimentary breccias and other structures. In these descriptive aspects they sound similar to the MVDs of east and central Tennessee. European geologists agree that much of the ore is in primary sedimentary arrays that require prefolding (pre-

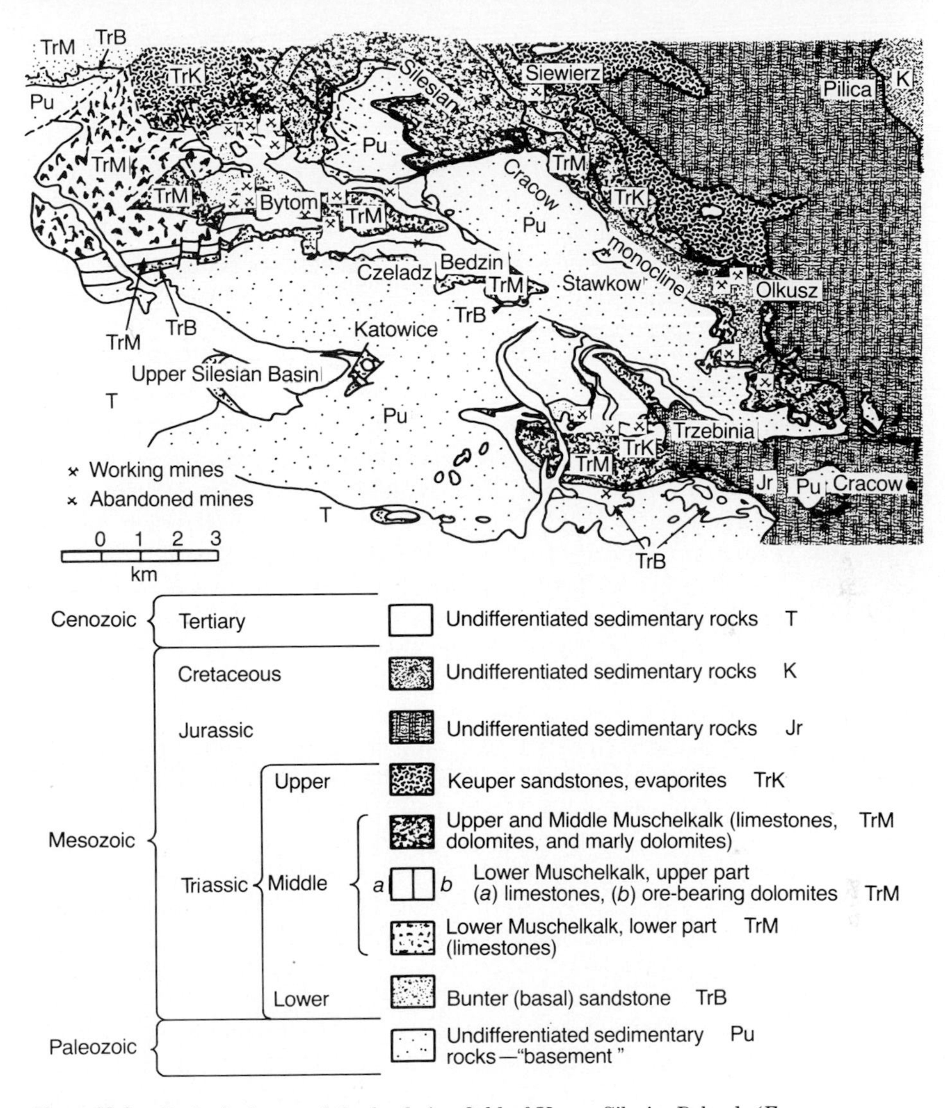

Figure 20-8. Geological map of the lead-zinc field of Upper Silesia, Poland. (*From* Zwierzycki, 1950, Figure 82.)

Jurassic) emplacement of sulfides. There is also agreement that significant amounts of the sulfides are epigenetic in texture, and were introduced into the rocks. The ore minerals actually occur, then, in several modes, all of which involved dolomitization.

Deposition of the ore was preceded by extensive dolomitization, which is most thoroughly developed along a single limestone horizon. The dolomitized zones are widely fractured and brecciated, particularly where they are folded. The ore deposits closely follow the base of the dolomite, where they form layers only 2 to 5 meters thick. Two similar layers are stratigraphically above and within 30 meters of the principal bed in the Bytom syncline, and in several other parts of the region a third layer is recognized. Almost all of the ore lies within 200 meters of the present surface. None of the ore is continuous, even in the principal bed. There are crosscutting faults and fractures which cut both Paleozoic and Mesozoic units, with displacements in the Muschelkalk of generally less than 30 meters. Orebodies are larger and richer near faults and fissures, which were clearly avenues of ingress of fluids. Shattering of the dolomite and mineralization were more extensive along troughs of synclines than along crests of anticlines, so it is doubtful that extensive ore deposits were removed by erosion of the higher level anticlinal folds.

Primary ore minerals include sphalerite, wurtzite, galena, pyrite, marcasite, and small amounts of the arsenic and antimony sulfides jordanite ($Pb_4As_2S_7$), gratonite ($Pb_9As_4S_{15}$), and meneghinite ($Pb_4Sb_2S_7$). An enigmatic regional zoning is shown by the lead-to-zinc ratios; in the rich central areas zinc greatly predominates over lead, but marginal ores contain more lead than zinc. The mines have produced an estimated five to ten times more zinc than lead. A typical analysis of ore from the rich Bytom syncline showed 1.7% Pb and 10.7% Zn. The galena contains very small amounts of silver, some of the sphalerite contains cadmium, and some of the pyrite contains thallium (Zwierzycki, 1950).

Ore textures are varied but uniformly similar in the district. Sass-Gustkiewicz, Dzulynski, and Ridge (1982) note that all of the ore is in original carbonate layers, that dolomitization and sulfide emplacement go hand in hand (Figure 20-9), and that the ores are stratabound in distribution but not stratiform in textural detail. Replacement of limestone by intergrowths of dolomite and sulfides (Figure 20-9) is widespread, as are growths of nets of euhedral sphalerite-galena-marcasite-pyrite microcrystals into variable-sized, solution-enhanced pore spaces in the dolomitized carbonates. Pore-

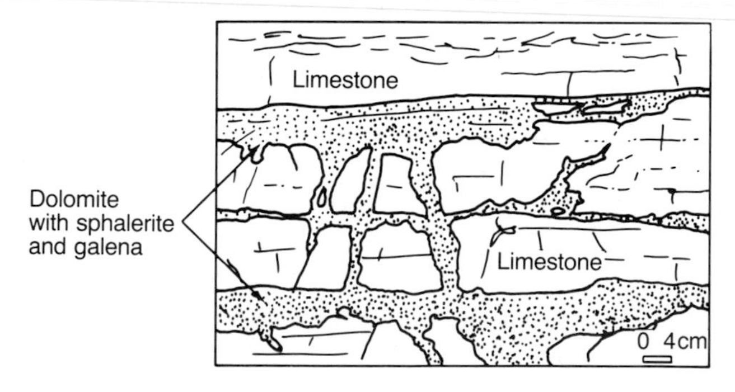

Figure 20-9. Replacement of Muschelkalk limestone by an intergrowth of dolomite and sulfides. Note the suspended blocks of unrotated limestone. (*After* Sass-Gustkiewicz, Dzulynski, and Ridge, 1982.)

space enhancement can be carried in scale from solution-enhanced fractures, joints, and bedding planes (Figure 20-10) to the development of karstlike cavities measured in tens of meters (Figure 20-11). Sass-Gustkiewicz, Dzulynski, and Ridge (1982) describe the ore occurrences (1) as in sandy, disaggregated dolomite (minor), (2) as massive ores, with replacement gen-

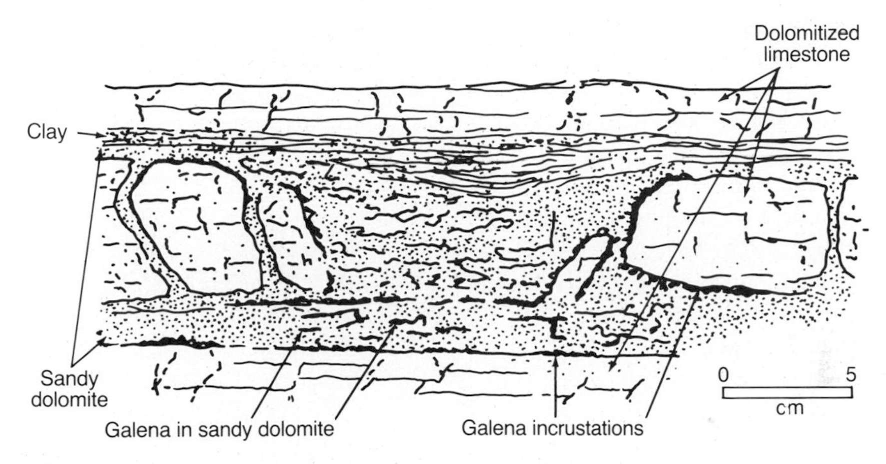

Figure 20-10. Development of euhedral galena (black) lining cavities in dolomitized limestone and infilling irregular planes in disaggregated "sandy" dolomite and clay. (*After* Sass-Gustkiewicz, Dzulynski, and Ridge, 1982.)

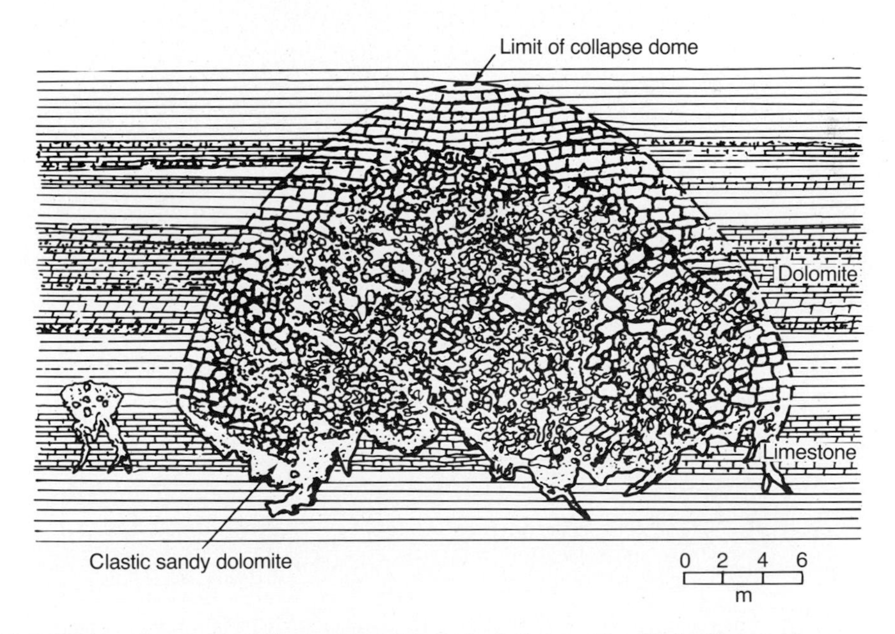

Figure 20-11. Formation of "hydrothermal karst" solution cavities filled with breccia fragments and typically banded, colloform sulfides. Compare with Tennessee occurrence in Figures 3-25 and 3-26. (*After* Sass-Gustkiewicz, Dzulynski, and Ridge, 1982.)

erally in the sequence limestone by dolomite by sphalerite by galena, and (3) as cavity fillings.

Mineral textures in the open cavities and karstic openings are typically colloform, both superficially and in detail (Kutina, 1952; Krusch, 1929). Sphalerite was deposited in rhythmically banded, botryoidal masses that show numerous shrinkage cracks that trend both normal and parallel to orbicular surfaces. Liesegang rings are also present, and spectrographic analyses show slight differences in iron content between light and dark bands. The excellent structural and mineralogical papers published by Haranczyk (1958, 1959, 1963a, b) emphasize the colloform nature of most of the ore. Schneider (1964) presented four photographs of polished slabs of Polish ores that strongly resemble materials shown in the Viburnum Trend papers cited earlier. Schneider stressed the importance of bedded ores, and described special facies of calcarenites, oolites, algal pellets and crusts, bituminous clays, and silts, all associated with sphalerite and galena. Figure 20-12 shows the oolitic form of fine-grained sphalerite. Some of these ooform grains were probably truly oolites, but according to Krusch (1929), at least some of the zinc sulfide was originally deposited in many layered masses as

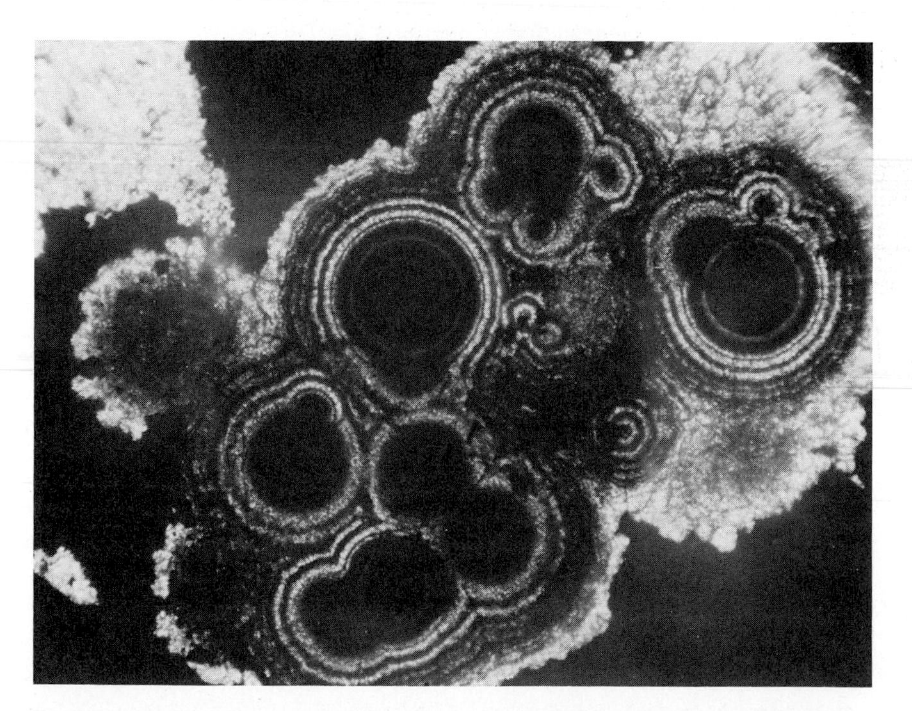

Figure 20-12. Ooform sphalerite or "brunckite" ore. Although true oolites are common, these textures are probably just part of a knobby aggregate of finely banded ZnS deposited in open cavities. Boleslaw mine, Poland. Thin section, crossed nicols. 70×. (*From* Haranczyk, 1959.)

a colloidal gel that gradually crystallized to wurtzite and finally to sphalerite. Marcasite was deposited contemporaneously with some of the zinc sulfide; it too exhibits colloform textures, such as spherical shapes and desiccation cracks. Galena was probably deposited from a true solution; it forms alternating layers with sphalerite and marcasite. Younger generations of galena were deposited after the zinc-iron sulfide phase was completed. Haranczyk's description of the ore in the Boleslaw mine near Olkusz and other nearby properties points out that ZnS is in both isometric and hexagonal forms. ZnS may either be in oolites or form crusts of cryptocrystalline ZnS. Inclusions of galena, some measured in hundredths of a millimeter, form droplets and streaks in the ooform zinc sulfide. Sometimes the galena droplets "run together" and form partly developed skeleton crystals (Haranczyk, 1958, Figures 1 and 2). Massive galena is present at many places; when etched in polished surface, it too may reveal concentric structures signifying replacement of oolites in appropriate environments, or rhythmic, finely controlled precipitation (Figure 20-13).

A somewhat unusual feature of the district is the presence of a clay layer at the base of the dolomitized stratum. Known locally as *vitriolic clay*,

Figure 20-13. Massive galena etched with HNO_3. The laminar ooform grains may be either galena-replaced oolites or colloform banded stalagmite-stalactite-like bodies. Complexly deformed, finely banded galena is younger than the ooform grains, and both were followed by calcite (lower corners). Boleslaw mine, Poland. 70×. (*From* Haranczyk, 1959.)

it is a slightly permeable bed 20 to 40 cm thick that marks the lower limit of mineralization. The clay bed was apparently formed by concentration of argillaceous impurities from the limestone during dolomitization and concomitant recrystallization. Occurring as it does at the base of the dolomite, the clay bed channeled the circulation of metal-bearing fluids. All dolomite and all ore are above the clay layer. The deposition of ores above an impermeable clay and near the base of a permeable dolomite suggested to early workers that descending meteoric groundwaters were more important, but recent papers point out that this vitriolic clay would also serve to direct laterally migrating brines. Ore-bearing fluids also joined the groundwater system via faults, as evidenced by traces of mineralization along faults at depth (Zwierzycki, 1950).

Along the outcropping edges of the synclines and of the Krzeszowice-Siewierz monocline, the ore beds are deeply oxidized to a product known locally as *galman*. In places the zone of galman is as much as 20 to 30 meters thick. The higher solubility of zinc caused the zinc and iron to separate during weathering. Relatively pure iron oxides are concentrated along the outcrops, and zinc enrichment zones lie beneath. Enrichment of cadmium also accompanied oxidation. The most abundant oxidation products are smithsonite, hemimorphite, cerussite, and limonite, but minor amounts of other minerals such as anglesite, goslarite ($ZnSO_4 \cdot 7H_2O$), tarnowitzite [$(Ca,Pb)CO_3$], and phosgenite ($PbCl_2 \cdot PbCO_3$) are also found (Zwierzycki, 1950).

The origin of the Upper Silesian ores has been widely discussed. Certainly the ores were deposited at low to moderate temperatures, and the presence of mineralized fissures continuing from the Muschelkalk limestone downward into the Paleozoic rocks and laterally to the south toward the Triassic basin is strong evidence that ascending solutions deposited the ore. It was argued early that the ores were deposited from descending solutions that were dammed against the vitriolic clay with zinc and lead sulfides precipitated by organic matter in the clay. Stappenbeck (1928) envisioned a process of lateral secretion whereby artesian waters leached syngenetic metals from the limestones and reconcentrated them as ore deposits. Krusch (1929) and Kutina (1952) emphasized the colloform structures and textures that were then, but are no longer, thought to require colloidal transportation.

Haranczyk decided, as a result of careful study, that the ores are of hydrothermal, essentially nonmeteoric origin. Galkiewicz (1967) summarized evidence concerning the origin of the deposits and concluded that they are hydrothermal, although he did not specify their source. Arguments in support of this conclusion are (1) the vertical interval of ore mineralization that extends into the Devonian, Carboniferous, Permian, and Triassic beds, (2) the position of the larger deposits along fractures, (3) the great variability in form and content of the orebodies, (4) the presence of replacement ore textures, (5) the zonality of mineralization, (6) the extensive dolomitization halos around the orebodies, and (7) the temperature of formation of sphalerite at near 120°C. Ridge and Smolarska (1972) point out that the isotope ratios in the Upper Silesian ores are remarkably uniform. It was pointed out in Chapter 7 that lead model ages in other deposits like these are chronometrically unreliable Joplin-anomalous leads.

On the other hand, Gruszczyk (1967) favored a sedimentary origin for the deposits and cited the following evidence: (1) a lack of intrusive or extrusive igneous rocks, (2) the regional extent of the mineralization not only in Poland but throughout much of central Europe, (3) the interdependence of the deposits and facies of the Triassic system, (4) the presence of regional zoning, (5) the subordination of deposits to stratigraphic factors, with deposits in the form of beds and with dislocations being postore, (6) simple mineral composition throughout, and the absence of minerals typical of hydrothermal deposits, and (7) the uniform textures and the explanation of veined and brecciated ores as having been formed by diagenetic processes. Kautzsch also concluded that the Upper Silesian-Cracovian ores, as well as those of many other European districts, are of primary sedimentary origin (Kautzsch, 1967). But Dzulynski and Sass-Gustkiewicz (1977), discussing the Austrian equivalents, refuted their primary sedimentary origin, reaffirming the importance of epigenetic migrating hydrothermal fluids. Sass-Gustkiewicz, Dzulynski, and Ridge (1982) present compelling evidence that the ores were deposited epigenetically and hydrothermally in an ongoing, coupled process involving dissolution and dolomitization controlled by primary and some secondary permeability features and sulfide replacement of carbonates and open-space filling in newly formed cavities, collapse breccias, and karstic openings.

Summarizing the many conflicting opinions and statements concerning the origin of the deposits is difficult. Certainly the ores were formed near the surface at comparatively low temperatures and pressures. Comparison with ores of the Mississippi Valley, especially those of Tennessee, shows a striking similarity. Ascending solutions entered the dolomite in part along faults, and ore deposition took place at low to moderate temperatures (50 to 150°C) after the fluids had mixed and migrated with groundwaters. The ore minerals were deposited in crevices, solution cavities, and joints in the dolomite, just above the clay stratum, or in overlying zones where solution-enhanced and solution-generated brecciation was especially prevalent. The presence of encrustations and stalactites of ore shows that much of the deposition took place in open spaces, but sulfides also replaced the dolomite, as is described from other MVDs. Recent developments in the Mississippi Valley favor a hydrothermal brine origin for the ores, and this, combined with the facts listed by Haranczyk (1963) and Galkiewicz (1967), appears to favor this origin for the Upper Silesian Cracovian deposits. If one rereads Galkiewicz's and Gruszcyk's lists of conclusions just cited, almost every one of the 14 points fits well with the oil field brine concept. Rozkowski (1979) presented a paper on thermal brines issued from the Triassic-Jurassic basin to the south as a potential source of the ores of the Silesian area. Sass-Gustkiewicz, Dzulynski, and Ridge (1982) favor a deeper source below the district, but leave the question of ultimate provenance open.

▸ WESTERN STATES URANIUM DEPOSITS

Another area of progress and success in the 1960s and 1970s has been in uranium geology, that is, in the understanding of uranium geology and geochemistry, in the incorporation of that understanding into exploration concepts, and in the discovery of new deposits. Many new occurrence modes of uranium have been delineated. In part because of them, enormous discoveries have been made in the Athabasca district of Canada described below, in the Jabiluka district of north-central Australia, and elsewhere. This progress has required that new geologic classifications of uranium deposits be developed; those by Cornelius (1976), Ruzicka (1977), McMillan (1977), Kimberley (1978), and Nash, Granger, and Adams (1981) all group radioelement deposits into many diverse categories. S. S. Adams and his associates published reports on geologic and exploration aspects of six types of uranium deposits as part of a project for the U.S. Department of Energy (Adams and Associates, 1981a–f). As they show, uranium ore deposits prove to be of many types and origins, ranging from magmatic, metamorphic, and hydrothermal through sedimentary. Some of the largest and most productive are of sedimentary origin; these include the well-known conglomerates of Witwatersrand, South Africa, and Blind River, Canada (Chapter 16). Others are stratabound in arkosic continental sedimentary rocks. In general the stratabound types have been formed or enriched by the deposition of uranium minerals from circulating groundwaters. There are three important regions in the United States, two of which have provided more than 90% of U.S. domestic production of uranium and vanadium. They remain preeminent in the world uranium reserves picture, also. Almost 700,000 metric tons of a total of 1,700,000 tons of uranium metal in free-world reserves—thus 40%—occur in these stratabound deposits. The most important region (Figure 20-14) has been the Grants, New Mexico, area of uraniferous humate deposits, mined from the 1950s into the 1980s and accounting for 50% of U.S. reserves and production, and the Salt Wash type uranium-vanadium deposits in sandstone, which were most active after World War II in 1950–1954. They were joined in the 1960s by discoveries and geologic delineation of the roll-front deposits of Wyoming, which still contain large reserves. Several deposits geologically similar to both have since been discovered in Tertiary sediments in Texas, and more discoveries can be expected there.

The Salt Wash and roll-front uranium orebodies could be considered the products of lateral secretion (Chapter 19), except that the actual source

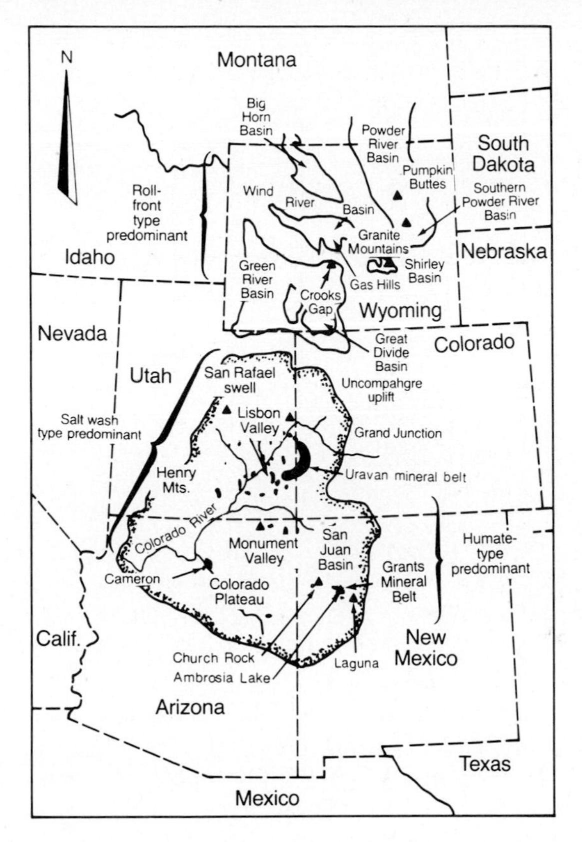

Figure 20-14. Location map of western states uranium deposits, including the Salt Wash, humate, and Wyoming basin roll-front ores, but not showing the Texas roll-front deposits along the coastal plain of that state. (*After* Rackley, 1976.)

volumes of the metals involved are not yet known with certainty. We first consider the roll-front type ores of Wyoming and then the Salt Wash and Grants, New Mexico, types.

As we shall see, all of these deposit types in the western states depend on the same part of the geochemical cycle of uranium. At low temperatures and pressures, uranium in rocks and minerals undergoing weathering and leaching is oxidized from U^{+4} to U^{+6} and becomes soluble in groundwater as $(UO_2)^{+2}$ ion, as one of the uranyl carbonate complex ions (Figure 20-15), or as any of a number of other complexes. As long as these groundwaters remain oxidizing, uranium ions remain mobile; but when they encounter and percolate through reducing environments, the uranyl ions are reduced and uranium is reprecipitated as crystalline uraninite, as colloform bands of pitchblende, as the silicate coffinite $[(U(SiO_4)_{1-x}(OH)_{4x}]$, and as several uranium-vanadium minerals.

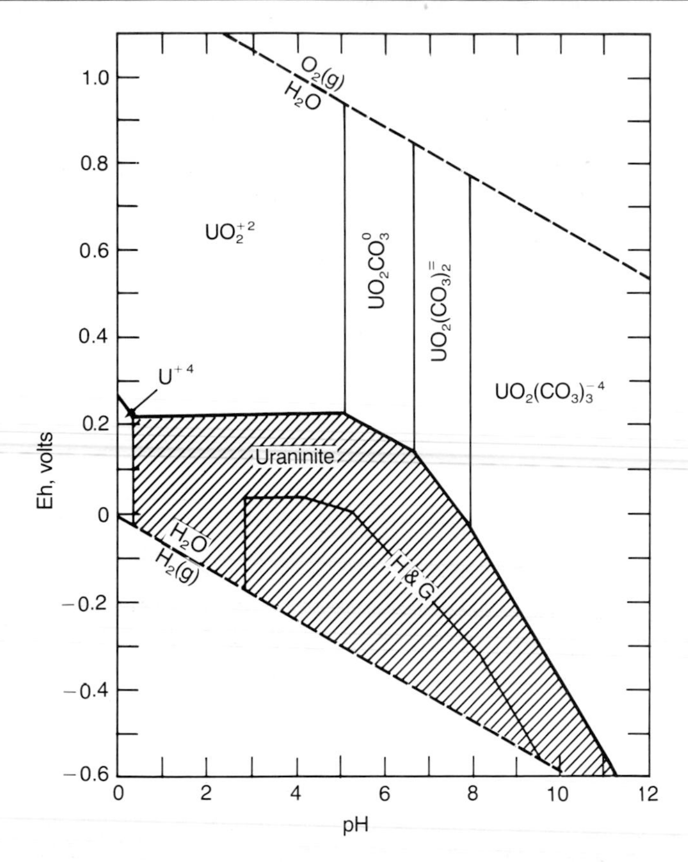

Figure 20-15. Eh-pH diagram of the U-O_2-CO_2-H_2O system at 25°C and $pCO_2 = 10^{-2}$ atm. Uraninite solution boundaries are drawn at $10^{-6}M$ (0.24 ppm) dissolved uranium species. "H & G" denotes the boundary of the uraninite stability field according to Hostetler and Garrels (1962). The shading separates U^{+6} ion dominance and solubility (above) from U^{+4} dominance and uraninite (UO_2) precipitation (below). The carbonate complex ions are important transport media in roll-front deposits. See also Figures 20-16 to 20-18 and consider with Figure 17-2.

Roll-Front Uranium Deposits, Wyoming

As indicated in Figure 20-14, the Wyoming ores are in a region of pronounced local basining and uplift, uplift that has raised flanks peripheral to relatively lowered basin floors. Even before erosion, but almost surely when the gently upturned sediments were breached by erosion, many of them became aquifers (Figure 20-16). Uranium could have been leached from younger volcanic ash, as shown in Figure 20-16, from up-drainage granitic domes or ridge cores, or from the aquifer sands themselves. Migration of the oxidizing fluids down the hydraulic gradient of the aquifer, with aquicludes of silty claystones or shales above and below, carried dissolved U, V, Mo, Se, and S downward along the "oxidation tongue" until

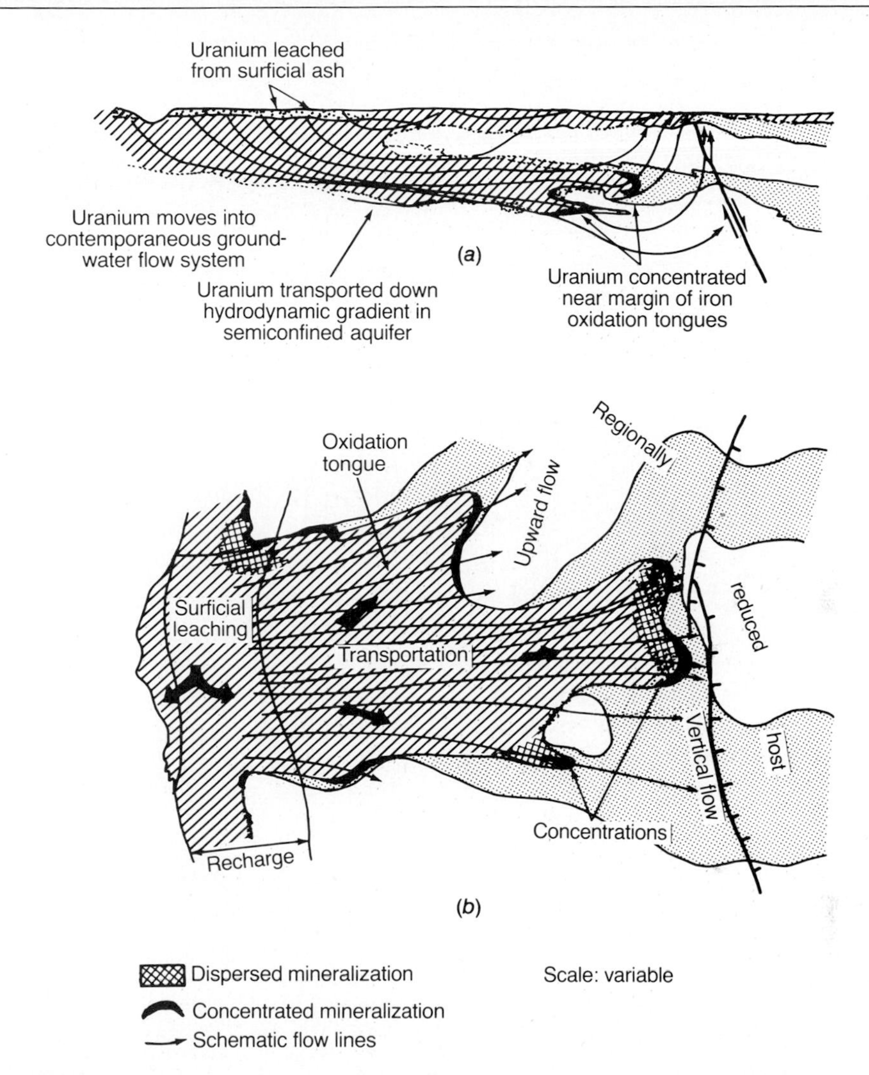

Figure 20-16. (*a*) Cross section and (*b*) plan view of a roll-front type system in the Catahoula Formation of Texas but applicable, in concept, with or without the fault, to roll-front occurrences in Colorado, Wyoming, and elsewhere. (*From* Galloway, 1978.)

the "regionally reduced host" buffered the groundwater at the "redox interface." Roll-front deposition of UO_2 occurred there (Figure 20-17), with zoning of deposition products including leading edge pyrite and calcite (fine dark stipple), ore-zone uraninite and pyrite, and trailing-edge siderite, partially reduced sulfur ($S^{+4} \rightarrow S^0$), goethite, and hematite. It must be recalled that water flows slowly through the roll front during its formation, and that the roll front is a kinetically stable chemical interface, a steady-state reaction front very slowly migrating downdip, not itself a dam or barrier to mechanical flow. The roll-front deposits are typically crescent- or S-shaped,

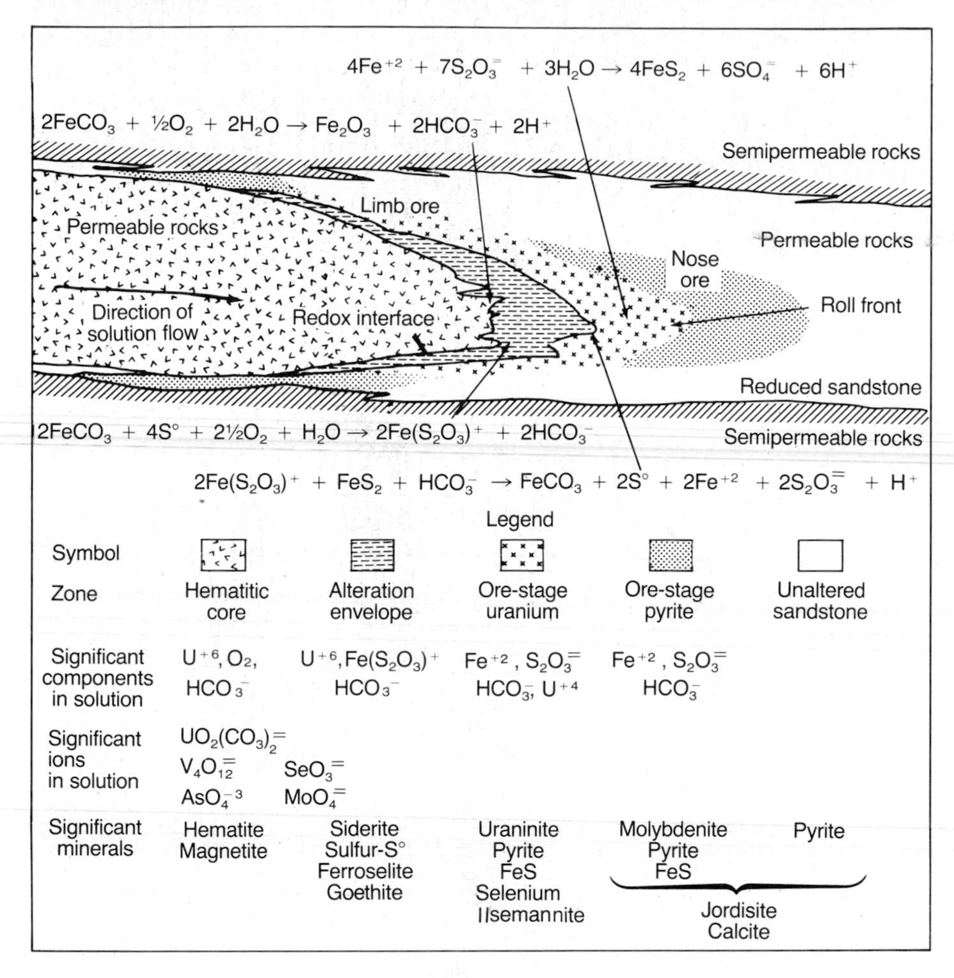

Figure 20-17. "Stop-action" diagram of a roll-front uranium deposit advancing to the right. Groundwater moves through the roll front, which advances slowly downdip at a rate determined by the oxidative capacity of the solution versus the reductive capacity of the permeable host rock. Reactions shown are for iron species. See also Figures 20-16 and 20-18. (*After* Granger and Warren, 1974; DeVoto, 1978.)

but are seldom truly regular. Figure 20-18 is a detailed cross-sectional drawing of a deposit in the Shirley Basin, Wyoming, and Figure 20-16 gives a feeling for its elongation in the third (lateral) dimension. Seldom are the deposits more than 20 meters high, but they may stretch back tens of meters along the upper and lower "limb ore" and be hundreds of meters long overall.

Clearly, there exist several requirements for their formation:

1. A source of uranium
2. Oxidation, mobilization, and transportation of that uranium

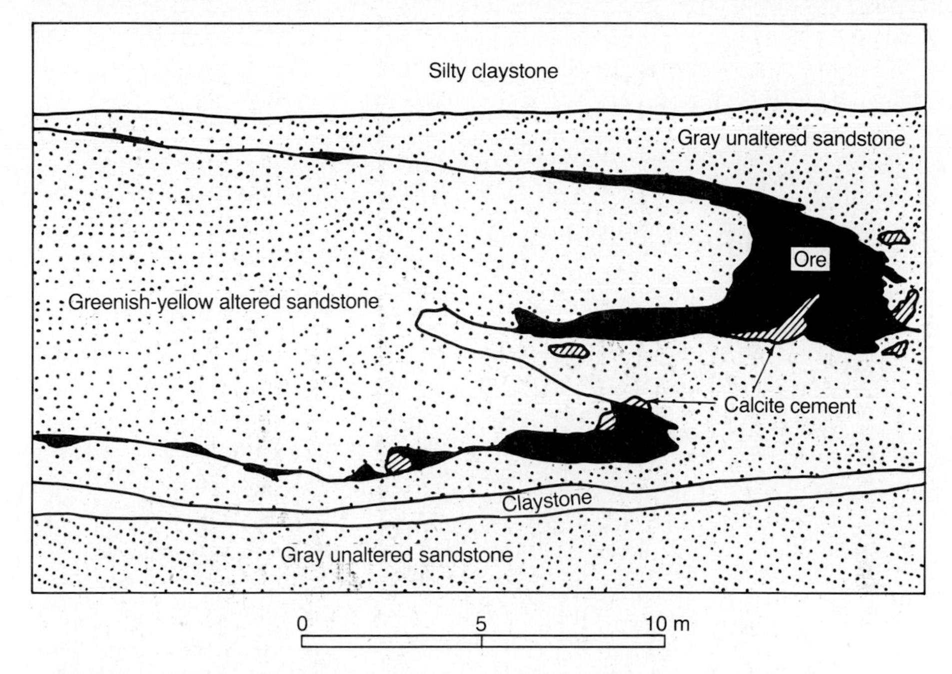

Figure 20-18. Sketch of a roll-front deposit in the Shirley Basin, Wyoming, showing typically irregular fine structure. The relatively impermeable claystones channel groundwater flow from left to right. Uraninite fills interstices and replaces carbonaceous vegetal debris on the cross-bed surfaces. (*From* Harshman, 1972.)

3. Sufficient permeability in a confined aquifer
4. A regionally reduced host aquifer lithology
5. Stable, sustained groundwater flow

If groundwater flow is variable and erratic, the rollfront as a discrete redox interface would doubtlessly be smeared out, breached, and much less regularly crescentic. The deposits can commonly be mined from easily restored open pits in Wyoming and Texas and constitute an important national energy resource. Variations on the roll-front theme occur in arkosic rocks in several western states, as indicated, and should be sought in many areas of the world.

Salt Wash Type Uranium-Vanadium Deposits, Colorado

The numerous uranium-vanadium deposits in the Salt Wash member of the Jurassic Morrison Formation in the Colorado Plateau region of the southwestern United States (Figure 20-14) are good examples of strata-bound uranium deposits formed by flowing groundwaters. Although the deposits have been known since 1899, only after 1945 and because of the expanded development of fissionable materials was there much interest in the region. In recent years the deposits have been of sporadic economic value. The deposits are widely distributed in western Colorado, eastern Utah, northeastern Arizona, and northwestern New Mexico (Figure 20-14). As a result of the intense search for uranium during the past few decades, especially in the 1950s, many articles have been published concerning the deposits. Most of the studies have been detailed descriptions of individual deposits; only a few treat the region as a whole (Hess, 1914; Coffin, 1921; Fischer; 1942, 1950; 1968; 1970; 1974; Kerr, 1958; Kelley, 1963; Rackley, 1976; Nash, Granger, and Adams, 1981). The grade of ore mined ranges from 0.16 to 0.25%, or 1 to 2 kg, of U_3O_8 per metric ton and about 1% V_2O_5. They contain significant U, V, Cu, Ag, Se, and Mo, and erratic Cr, Pb, Zn, As, Co, and Ni (Rackley, 1976). Nash, Granger, and Adams (1981) and Adams and Associates (1981a–f) distinguished the *Salt Wash type* and the *uraniferous humate* deposits of the Grants, New Mexico, mineral belt. Both are described here. In most earlier reports they were combined and called *Colorado Plateau type* or tabular type deposits.

On the Colorado Plateau, uranium and uranium-vanadium deposits of economic value are most abundant in the Shinarump Conglomerate, Chinle Formation, Entrada Sandstone, Todilto Limestone, and Morrison Formation of the Triassic and Jurassic periods, but many other Mesozoic formations, as well as some Paleozoic and Tertiary rocks, also contain ore deposits. Most of these dominantly redbed formations are similar; they consist of continental sandstones, siltstones, conglomerates, and impure limestones (Isachsen et al., 1955; Fischer, 1956), commonly with intercalated bentonitic volcanic tuff layers. The sedimentary units of the Colorado Plateau are generally flatlying to gently tilted. Recent erosion has dissected the "layer cake" upper surface, leaving the canyon lands, mesas, and buttes so familiar to western moviegoers. Many uranium deposits were discovered where they were exposed in vertical canyon walls; many others were drilled out in the rocks underlying mesa tops by following ancient streambeds back from the walls.

It had been realized that most of the orebodies share characteristics, regardless of their mineralogy and geographic or stratigraphic position. Some were small irregular pods sporadically distributed within a favorable rock unit, but the larger deposits formed sinuous ribbons a few thousand meters long and several tens of meters wide but only a few meters thick. Elongated orebodies in these units tended to follow buried stream courses or lenses of conglomeratic material, and in fact were found to be best explored for by searching out not only paleostream channels but also zones within those channels of high original permeability and abundant vegetal "trash"—grasses, tree trunks and branches, and other organic debris. These buff to gray reduced portions of channels were found to occur between upstream and downstream oxidized red sands. Red sandstones also occur as flood-plain and overbank units. Uranium and vanadium minerals commonly fill pore spaces in the reduced sandstones and conglomerates; fragments of wood, lignite, or bone material that acted as centers of reduction were replaced by rich concentrations of uranium minerals—whole pseudomorphed logs of uraninite and brilliantly colored uranium-vanadium oxides were rich prizes encountered in many mines. The principal ore minerals after pitchblende and coffinite are canary yellow carnotite [$K(UO_2)_2(VO_4)_2 \cdot 3H_2O$] and tyuyamunite [$Ca(UO_2)(VO_4)_2 \cdot 5\text{-}8H_2O$], bright yellow-orange-red gummite (a mineraloid mixture of hydrous oxides of U, Th, and Pb), and a variety of yellow, orange, blue, and black vanadium oxides like montroseite [$VO(OH)$]. Minor amounts of pyrite, marcasite, galena, sphalerite, chalcopyrite, bornite, chalcocite, covellite, and many other sulfide minerals also are found. Minor amounts of native selenium are characteristically associated with some uranium and uranium-vanadium ores, and the molybdenum oxide ilsemannite ($Mo_3O_8 \cdot nH_2O$?) is also common (Kerr, 1958).

The ore-bearing formations are nearly horizontal, but they are disturbed in places by moderately strong monoclinal and anticlinal folds and by high-angle faults. Major structural basins, such as the San Juan Basin, the Uinta Basin, and the Piceance Basin, delimit relatively barren areas between which there are clusters of ore deposits. A regional northwesterly structural grain apparently reflects deep-seated igneous and tectonic movements, yet there seems to be no direct relationship between the distribution of uranium and regional structure, except insofar as tectonic history has controlled sedimentation, geomorphology, and igneous activity, and insofar as the ore deposits lie within the uplifted tectonically positive regions between structural basins (Kelley, 1955; Kerr, 1958). Salt domes and various types of igneous intrusives are widely distributed in the Colorado Plateau, and lava flows and pyroclastics are also present. Except for some Precambrian batholiths, the intrusives are all Miocene or younger. No general spatial distribution between igneous activity and ore deposits can be demonstrated, except where uranium mineralization is associated with diatremes and volcanic collapse structures (Shoemaker, 1956).

The origin and source locales of the ore components have been the subjects of much discussion and disagreement. Four principal ideas have been advanced. It has been argued that (1) the ores are syngenetic and were deposited with the sediments or as an early diagenetic cement, (2) the uranium and vanadium were leached from overlying or interlayered tuffs and volcanic rocks, (3) groundwaters leached the metals from the enclosing sands or from nearby granitic sources, forming deposits where the proper Eh-pH conditions prevailed or where groundwater circulation was channeled, and (4) the ores are products of ascending hydrothermal fluids derived from underlying magmas (McKelvey, Everhart, and Garrels, 1955). The syngenetic hypothesis was one of the first proposed (Coffin, 1921; Fischer, 1950), and was widely accepted until the 1950s (Wright, 1955; Fischer, 1956); it was based in large part upon the obvious association between lithologic features and orebodies and upon absence of structural control, of potential igneous sources, and of typical hydrothermal alteration-mineralization minerals and textures. The syngenetic concept was abandoned as the dynamics represented by Figure 20-19 and roll-front processes already described were perceived. Also, lead-lead and potassium-argon isotopic methods have shown the uranium and "alteration" to be variably but universally younger than the host rocks.

The presence of devitrified volcanic ash near most of these uranium deposits, in fact interlayered with many of the redbed units, has supported the hypothesis that the uranium is a product of ash-leaching by meteoric waters (Waters and Granger, 1953). Not only on the Colorado Plateau, but also in other areas of similar uranium deposits, a striking correlation has been noted between the presence of volcanic ash and ore. Uranium deposits in the Dakotas are in lignites that underlie bentonites, the lignites acting as precipitants for the uranium leached by descending meteoric water (Denson and Gill, 1956). But sandstone hosts allow greater dispersion by circulating groundwaters, so the further mechanisms of concentration, such as paleostream channels, are required for Colorado Plateau ores. Recent summary articles (Rackley, 1976) have emphasized the presence of "volcanic material . . . in or overlying the host rock" as a characteristic. Normal traces of uranium in volcanic glass or adsorbed on ash in tuffs might be "stored" in montmorillonite devitrification-hydration phases until release to circulating groundwaters. Hostetler and Garrels (1962) proposed that the ore-bearing fluids were squeezed out of tuff beds by load compaction; that is, connate waters carried the soluble uranium and vanadium salts out of the tuffs and into more permeable sandstone strata. Deposition would have occurred when the migrating solutions encountered a reducing environment in the permeable sands.

It appeared early in the discovery era of these ores that fluids—presumably groundwaters—had flowed through the redbed strata at some time after their deposition. The flow paths were dominantly lateral, and zones of higher permeability—the paleostream channels—were naturally preferred (Figure 20-19). These groundwaters, probably oxidizing early in their flow histories and carrying soluble hexavalent uranium ions, became progressively reduced as they seeped past and through the reducing environments of the stream channels. Uranium ions were reduced to U^{+4} and precipitated as uraninite or pitchblende, coffinite $[U(SiO_4)_{1-x}(OH)_{4x}]$, and other phases. As stated, uranium and vanadium are relatively soluble as hexavalent uranyl dicarbonate and tricarbonate complexes and as tetravalent vanadium complexes (Figure 20-15). Geochemical studies indicate that the metals are most stable in solution in a mildly reducing, neutral to alkaline environment that contains abundant CO_2 (Hostetler and Garrels, 1962). Precipitation took place under more strongly reducing conditions, such as those produced by the action of carbonaceous material or H_2S. The reduced fluids passing "downstream" induced greenish halos of ferrous-iron-dominated clay minerals around the paleochannels for short distances (Figure 20-19). Continued flow of oxidizing waters hydrated and oxidized the uranous oxides and silicates to the bright-colored complex minerals noted above. Thus sinuous, meandering, complex patterns of reduction and oxidation were pathfinders to paleochannels and bonanza ores, both in outcrop and by shallow drilling. Local controls of ore deposition were commonly related to permeability and groundwater movements, and lithologic controls were the most obvious. Thus the ores may be found along valleys in old erosion surfaces, within channel sands, or in sandstones interfingered with shales and mudstones (Wright, 1955; Miller, 1955). Sands containing abundant carbonaceous matter on crossbed and bedding plane surfaces were commonly found to have been selectively enriched in uranium and vanadium.

Another type of uranium deposit occurs on the Colorado Plateau that

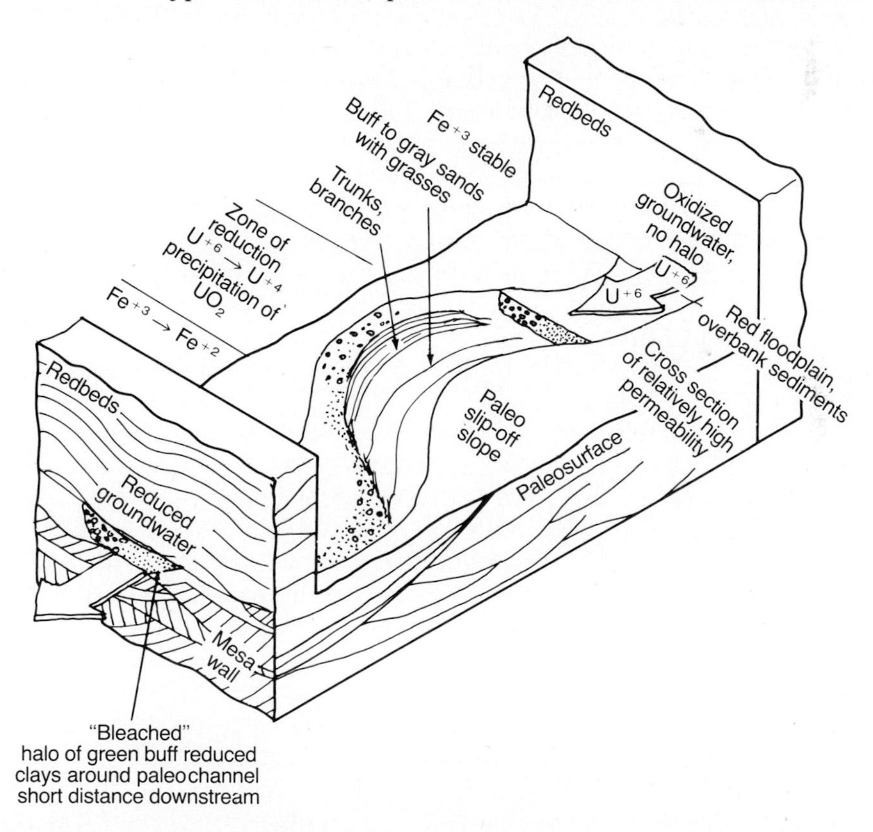

Figure 20-19. Oxidation-reduction dynamics in Salt Wash type paleostream channel mineralization. The buffering capacity of the "zone of reduction" is gradually diminished with the accumulation of U^{+4} and other reduced species. Oxidation may ultimately overwhelm the zone of reduction.

is mentioned here only for completeness—it is one of clearly epigenetic, volcanism-related, structurally controlled deeper level hydrothermal deposits such as those at Marysvale, Utah. It has no known relationship to the sandstone ores that are "rootless" with respect to obvious volcanism and hydrothermalism. Deep exploration in veins and cross cutting structures at Marysvale has disclosed hydrothermal suites of minerals, modified near the surface by groundwaters. The minerals at depth include galena, alunite, hard pitchblende, gersdorffite (NiAsS), sphalerite, chalcopyrite, and several others (Benson et al., 1952). The mineral associations and the presence of cracked hydrocarbons indicate moderate temperatures, in the general range of 100 to 350°C. Wall-rock alteration effects also indicate hydrothermal activity. In places, dolomite, kaolinite, and silica—in inward zonal sequence—have been recognized as alteration products associated with these ore deposits. Additional hydrothermal alteration involves a chrome-mica, recrystallized sedimentary clays, and alunitization, and the widespread presence of molybdenite is suggestive of hydrothermal activity. The ores are spatially and temporally related to igneous features such as diatremes, volcanic vents, and breccia pipes. Tentative radiometric age determinations require an epigenetic origin and imply that the ore deposits and igneous activity are contemporaneous (Kerr, 1958). Although this same hydrothermal fluid involvement in sandstone uranium deposits has been proposed, there is currently no support for the idea and it has been rejected.

Humate Uranium Deposits, Grants Mineral Belt, New Mexico

Figure 20-14 shows a group of deposits in northwestern New Mexico that includes the Church Rock, Ambrosia Lake, and Laguna districts. They have provided more than 50% of U.S. production of uranium, and still contain significant reserves. They have been described as a separate group by Rackley (1976), Nash, Granger, and Adams (1981), and Adams and Associates (Adams and Saucier, 1981b). The following account is based primarily on Nash, Granger, and Adams (1981).

The humate deposits occur in sandstones, arkoses, and siltstones of the Jurassic Morrison Formation. The strata are principally braided stream deposits in a broad alluvial plain in the southern part of the San Juan basin. They are fluvial-lacustrine materials that included volcanic ash (now montmorillonite) washed into shallow lakes, perhaps with some air-fall volcanic ash layers. One wide continuous stream-channel unit called the Jackpile sandstone at the top of the Brushy Basin member of the Morrison has been especially productive. These sediments were not redbeds. Coalified wood, disseminated pyrite, and organic matter from grasses, pollen, and leaves have generated widespread buff, gray, and greenish colors in rocks that apparently were quickly buried so that the organic matter dispersed through them was preserved beneath the water table. Uranium minerals are coextensive with undulant tabular layers of epigenetic organic matter that has

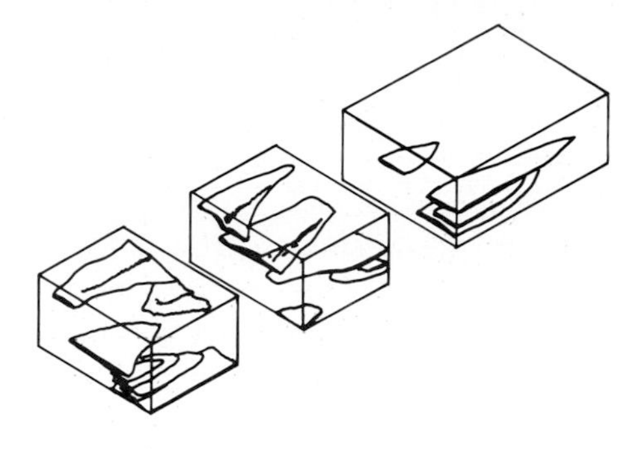

Figure 20-20. Typical relationships of uraniferous humate orebodies (idealized). The orebodies are elongate, lens-shaped masses arranged singly or en echelon vertically and horizontally. The scale is tens to hundreds of meters across, 0 to 20 meters thick. Ambrosia Lake district, New Mexico. (*From* Granger et al., 1961.)

been introduced into its present position in the sands. The carbonaceous matter probably represents digestions of grass mats or vegetation in swampy, boglike ponds or lakes that precipitated humates, the organic compounds of humic acids that give the deposit group its name. The layers average 0.5 to 2 meters thick, locally more, can be tens to hundreds of meters across, and are commonly stacked (Figure 20-20). The uraniferous organic matter in these layers coats sand grains, fills pore spaces, invades fractures and mineral cleavages, and even replaces some quartz grains. Coffinite and pitchblende are present, but most of the uranium appears to be in an organo-uraniferous mineraloid. The humate is thought to have been produced from decaying plant matter along streams, from coalified plant matter deposited in the sands, or from organic ooze from lake or pond mudstones expelled into adjacent sandstones by compaction. Some uranium could have been deposited with the humates, but most of it appears to have been introduced in mildly oxidizing groundwater, adsorbed, and reduced, with partial crystallization to coffinite. In the Grants region, oxidizing roll fronts of much younger age locally advanced into primary uraniferous humate occurrences, resulting in the redistribution of some uranium into typical roll-front deposits.

The deposits constitute a separate type primarily because of the origin, distribution, and abundance of the organic components. The three sources of the uranium that have been proposed are leaching from the host sandstones and ashy components themselves, from the granitic rocks from which the clastics were largely derived, and from overlying tuffaceous units. The second and third options are favored, but little compelling evidence has been gathered.

▸ ATHABASCA TYPE UNCONFORMITY-RELATED URANIUM DEPOSITS

Yet another variety of uranium deposit has only been known since the 1970s. It is represented by at least two large districts, the Ranger-Jabiluka mines and Alligator River district of Northern Territory, Australia, and the Athabasca district (Cluff Lake, Rabbit Lake, Key Lake, and others) in Saskatchewan, Canada (Figure 19-1). Several of the Athabasca deposits were discovered during the period of 1975–1979; information on them is still relatively scant, but discussion of their genesis is vigorous. The general features of the Athabasca deposits translate well to similar deposits elsewhere in the world, so they do indeed seem to represent an ore-deposit type. Once again, they are discussed here because they are clearly epigenetic, are not known to be related to igneous activity, and are not obviously of identified-source lateral-secretion derivation. Athabasca geology has been effectively described by Hoeve and Sibbald (1978), Jones (1980), Clark and Burrill (1981), and Clark, Homeniuk, and Bonner (1982); the Australian occurrences by Battey (1978), Hegge and Roundtree (1978), and the same Clark, Homeniuk, and Bonner paper referred to above. Nash, Granger, and Adams (1981) devote several pages of their uranium geology review article to these unconformity-type deposits, and Adams and Associates (1981e) also describe them.

The key to the formation of these deposits, and thus to their geologic characteristics, is another configuration of the interplay of dissolved U^{+6} ions and reducing environments. According to Hoeve and Sibbald (1978), uranium appears to have been leached from permeable sandstones above an unconformity, with groundwater translation along it; reducing capacity comes from methane-charged fluids moving upward along faults in graphitic zones in the basement rocks and through the unconformable contact boundary. Some geologic description will clarify the dynamics of that setting. Other geologists are not convinced that the uranium is extracted from the arkoses, but suggest that it might come from depth. The following discussion should be read with both interpretations in mind.

Athabasca Uranium Deposits, Saskatchewan, Canada

A glance at the geological map of Canada discloses an elliptical blanket of 1500-million-year-old Proterozoic sandstones called the Athabasca Group (Figure 20-21). They consist of a thick succession of basal conglomerates and fluvial-deltaic sandstones, siltstones, and mudstones of redbed affinity. They were deposited in a shallow basin floored by a weathered paleosol mantle on an Archean granite and a metasedimentary basement. This basement contains local uraninite-bearing veins that are not involved in the deposits described here, important as they may be. The basement is predominantly of 1.8-to-2.2-billion-year middle Proterozoic granitic rocks, but if one were to strip away the Athabascan quartzose sediments, one would find, especially in the southeast portion (Figure 20-21), both granitoid mantled gneiss domes with metasediments dipping off their flanks and wide bands of steeply dipping metapelites and metaarkoses. These supracrustal metasedimentary rocks contain graphitic units at many locales, as crescents

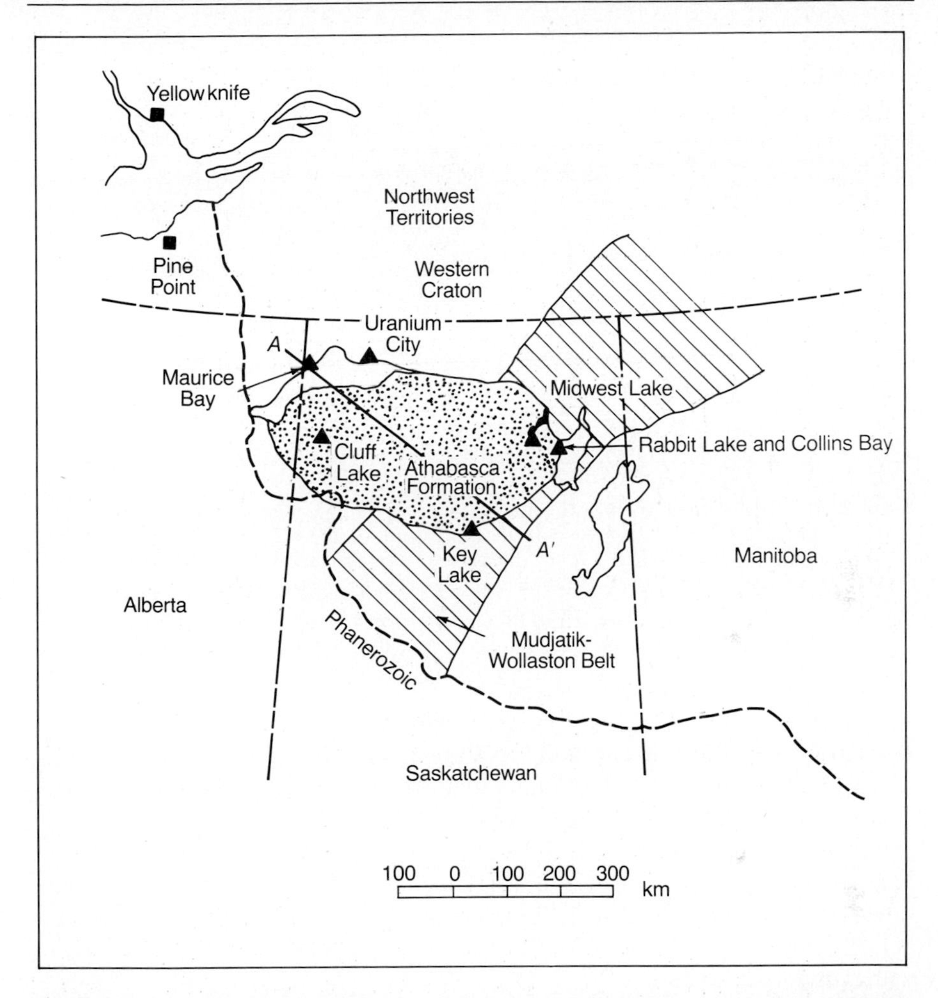

Figure 20-21. Location of the Athabasca Basin and its contained fluvial sediments (stippled) in northern Saskatchewan, Canada. The cross section *A-A'* is shown in Figure 20-22. Major uranium deposits are shown as black triangles. The Mudjatik-Wollaston Belt is a band of deformed metasediments underneath the 200-by-400-km basin. (*From* Clark and Burrill, 1981).

and rings around eroded domes and as jagged bands across gneissic terranes. These granite terranes and strongly deformed amphibole- to granulite-facies metamorphosed rocks, then, are unconformably overlain by the nearly flat-lying, essentially unmetamorphosed permeable Athabasca formation conglomerate and sandstones. This package has been traversed by regional-scale steep faults shown diagrammatically in Figure 20-22. They are dominantly northeast-southwest, but they are east-west to nearly random as well. Particularly where these faults cut downward into graphitic units in the basement rocks, elongate, tabular to ribbonlike uraninite ore-

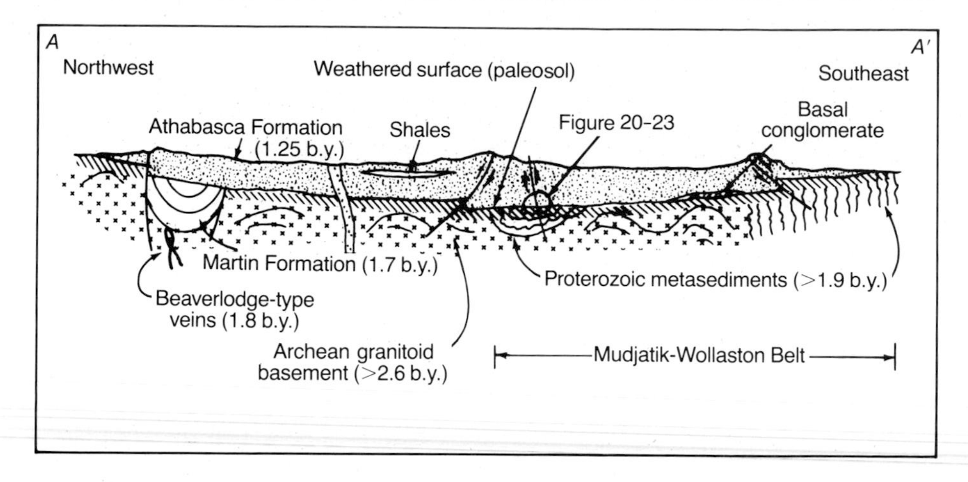

Figure 20-22. Diagrammatic cross section of Athabasca Basin geology. The conglomerates and shales are shown schematically as occurring low and high in the sequence. A late dike (1.25 billion years) and major faults are also shown. The Athabasca Formation is from 0 to 1500 meters thick, the length of the section about 300 km. (*After* Clark and Burrill, 1981.)

bodies have been found at the unconformity surface and up to a few tens of meters above it. At Key Lake the original deposit was 15 to 100 meters wide and high and 3.6 km long; at Midwest Lake, 50 to 200 meters wide and high and 2 km long; at Rabbit Lake 200 meters wide, 215 meters high, and 540 meters long (Clark and Burrill, 1981); and at Collins Bay, two zones are 16 by 40 by 50 meters and 30 by 100 by 1000 meters high, wide, and long, respectively (Jones, 1980). Grades commonly run to 2% U, ten times those of the average western states type ores. Tonnages are also high; the deposit type accounts for 20% of known world reserves, and the proportion will probably increase.

Deposits known in the Athabasca area for many years have been interpreted until recent years as conventionally magmatogene-hydrothermal, although the absence of appropriately aged igneous activity has been perplexing. The ores were known—and are increasingly verified—to contain significant amounts of nickel and arsenic, minor amounts of cobalt and iron, and variable trace amounts of copper, lead, zinc, molybdenum, sulfur, and selenium, an element suite similar to that at Cobalt, Ontario (Chapter 19), and in the western states deposits described earlier. The mineralogy is pitchblende with subordinate coffinite, with alteration envelopes of chloritization, silicification, minor tourmalinization, and calcite-dolomite veinlet impregnation. The alteration includes a tendency for reduction of Fe^{+3} to Fe^{+2} and a shift from red colors to greens and grays. Hydrocarbon "buttons"—small blebs of thucolite—are only found in the orebodies. Fluid inclusion studies indicate temperatures of 120 to 220°C, low pressures at 0.75 kbar, and salinities of 30 wt % NaCl equivalent.

A strictly igneous-related hydrothermal origin became untenable to geologists working there. The trace element profile, the ribbonlike orebody form and shape, the absence of source environments, the localization of the ores and alteration only at the unconformity, the presence of sooty black pyrite rather than the crystalline variety, the restriction of the ores to positions above graphitic basement rocks, and the similarity of deposit styles in many locales in the district all suggested a different mode of origin. Research, continued study, and application of new thinking have indeed framed a whole new concept that seems to apply. Figure 20-23 depicts that model, and Figure 20-24 presents a cross section of an orebody. The process presumably begins with a thermal anomaly in the basement, one most probably created by radioactive decay in the older pegmatites and high-uranium granites. Heated waters traveling up fault planes become strongly reducing where the faults coincide with and cut through graphitic rocks, and probably carry methane, ethane, and other simple hydrocarbons upward. As they pass the level of the unconformity, they carry a "prong" of reductive ca-

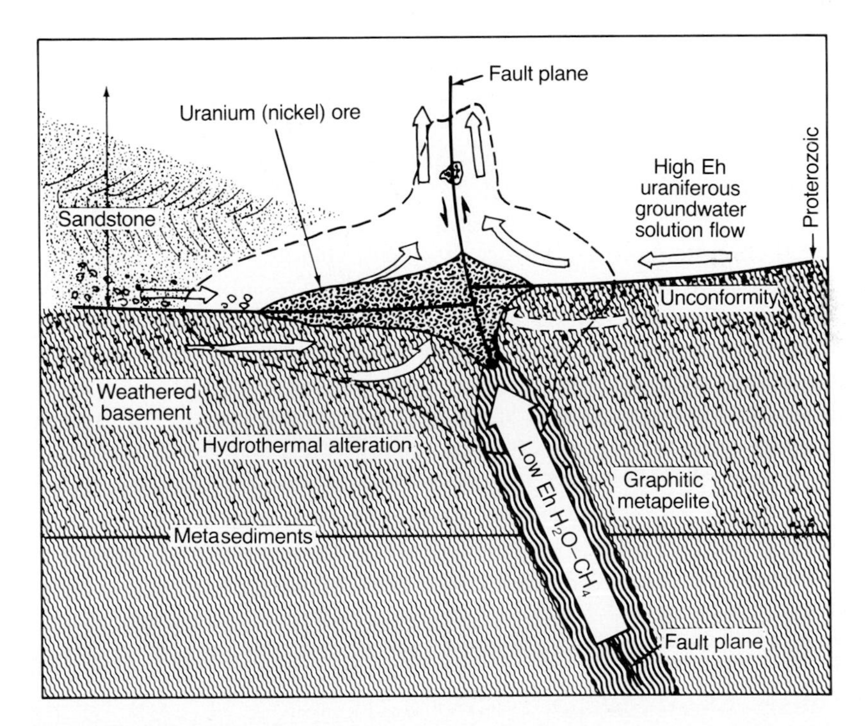

Figure 20-23. Schematic cross section of the formative processes of an unconformity-type uranium deposit. The diagram is based on an Athabasca deposit, but is applicable elsewhere. The ore zone is 100 meters wide and 50 meters high, the alteration zone 1500 by 300 meters. Both extend, ribbonlike, for up to hundreds of meters into and out of the plane of the page along the fault-unconformity-graphite schist intersection. (*After* Clark and Burrill, 1981.)

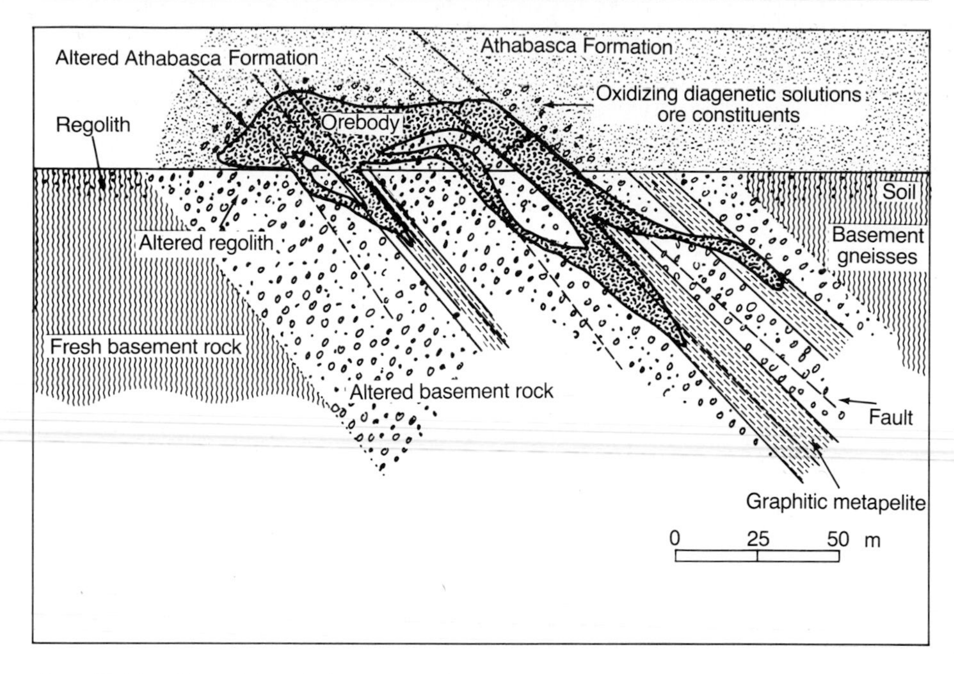

Figure 20-24. Essential geologic aspects of a uranium orebody at the unconformable surface at Rabbit Lake in the Athabasca district. Note the alteration overprint on basement, paleosol, and sandstones alike. (*From* Hoeve and Sibbald, 1978.)

pacity environment with them, roughly the volume within the dashed line of Figure 20-23. A plume of heated water rising along such a fault plane would draw groundwaters from the sandstones above the unconformity into it, and a slow convective cell—a linear one in the third dimension—would be established. If the waters flowing laterally inward and then upward are oxidizing and percolated through sediments like those of the Athabasca Formation, they will have dissolved U^{+6} ions. Then the key environment becomes that of the commingling of reducing, graphite-buffered, methane-ethane-bearing fluids from below and oxidizing, hematite-buffered, U^{+6}-bearing fluids from along the unconformity surface. Where the result is reduction of U^{+6} to U^{+4}, concentrations of uranium as uraninite, coffinite, and thucolite are reasonable, with alteration envelopes of reduced-iron silicates such as chlorite, siderite, and even tourmaline. The presence in these ribbonlike concentrations, the dashed-line volume of Figure 20-23, of veinlets of siderite, dolomite, or calcite suggest that the U^{+6} may have been transported as complex uranyl carbonate ions (Figure 20-15) not unlike those involved in western states ores. After a few million years, the uranium concentration would itself generate enough heat to perpetuate convection until all local permeability is plugged with new minerals. The process would recommence almost spontaneously upon repeated faulting or brecciation. Note that uranium deposition would occur only above unconformities where graphitic units occur beneath them, an observed relationship.

Not all geologists agree with the above-described process model. It accounts for many of the chemical and distributional features of the Athabascan deposits, but many details remain to be worked out and clarified before the ideas of Hoeve, Sibbald, Clark, Burrill, and others can be accepted. If the model is extended beyond the Athabascan circumstances to a more general set of criteria for unconformity-type deposits, the salient features are as follows (Welty, 1980):

1. The basement rocks (1.8 to 2.2 billion years) are older than Middle Proterozoic and are metamorphosed marine or volcanic sediments, typically pelitic or limey shale in original composition, locally with graphite. None are only clean quartzites or redbeds. They commonly are topped with a weathered soil zone, a regolith, although the presence of such a paleosol is not essential.
2. The host rocks overlying the unconformity are Middle Proterozoic or younger, fluvio-deltaic coarse red sandstones and siltstones. They may reach several kilometers in thickness.
3. The deposits invariably occur at or within a few tens to hundreds of meters above the unconformity or a few meters below it, and are horizontally elongate—ribbonlike—along faults through the unconformity.
4. The deposits are epigenetic, with predominant open-space-filling textures in faults, breccia zones, and fractures. Low pressures and low to moderate temperatures are indicated. Where reported, lead-lead isotopic dates are those of the cover rocks above the unconformity, or younger.
5. The uranium mineralogy is simple, generally only pitchblende with coffinite or thucolite. Minor Ni, As, Se, and S and trace V, Fe, Cu, Pb, Co, Mo, and Bi occur.
6. The altered wall rocks show fine-grained chloritization, argillization, albitization, rare hematitization, and late veinlets of carbonates.

The unconformity deposits of Athabasca are generally similar to those in north central Australia, but differences in detail exist. Uranium deposits in other rocks of identical age have been found, as for example those in sandstones and shales in Gabon and those in mixed sediment-volcanic sequences in Labrador. As Adams and Associates (1981e) point out, perhaps the key factor is that 1.8-to-2.2-billion-year-old rocks in general seem to be anomalously rich in uranium and many of them are sulfide-bearing graphitic carbonate-chert-rich sediments. Quite possibly, uranium has been oxidized and dissolved there, carried across the unconformity, mingled with solutions from the basement rocks that are enriched in sulfide reductants, and precipitated. As Adams (1983) points out, such mechanisms make the basement rocks potential "protore" sources of deposits above the unconformities, and confusion about genesis results from the fact that a variety of processes and sources have been involved.

The explorationist's mind cannot help but be stimulated by the varieties of geologic settings that such a rubric connotes, especially when taken with the schema of the western states type occurrences. Wherever uranium can be leached from sedimentary, volcanic, or plutonic rocks and carried to or across a chemical boundary from oxidizing to reducing conditions, there exists a target. The so-called *calcrete* deposits of Australia, South Africa, and elsewhere (Mann and Deutscher, 1978) represent uranium leached from basement rocks and precipitated as carnotite and tyuyamunite in caliche-cemented recent drainages. The enormous volcanic piles of the Cordillera of North, Central, and South America and the terrigenous sediments of all continents become potential sources, and sulfide bodies, petroliferous sands, tar sands, oil shales, regular shales, and euxinic sediments of all kinds become targets. And the migration of fluids by simple gravity, by basin dewatering, and by thermal gradients of many origins permits the broad diversity of genetic types of uranium deposits perceived in the 1970s and still sought in the 1980s.

Exploration techniques for uranium ores are varied because of the special properties of uranium minerals and because radioactivity can be detected instrumentally from a distance. The most widely used techniques employ radioactivity measurement by geophysical devices, such as scintillometers and geiger counters, but prospecting has also been based upon the fluorescence of radioelement minerals, groundwater and stream analyses, electrical geophysical methods, stratigraphic characteristics, biogeochemistry, and geobotany. Regarding the last, a few species of *Astragalus*, also known as "poisonvetch" or "locoweed," grow only where selenium is available in the soil; they are therefore "pathfinder plants" on the Colorado Plateau and in Wyoming where selenium is characteristically associated with uranium-vanadium occurrences (Cannon, 1957; Hawkes and Webb, 1962). Well-water analysis in Wyoming reveals the presence of uranium in micro amounts by alpha-particle or gamma-ray radiometry, and well-water chemistry appropriate to environments behind or in front of roll fronts is determined by Eh-pH measurement. Shallowly buried uranium occurrences can also be sought by a technique called *Track-Etch*, wherein inverted cups with strips of an alpha-particle-sensitive plastic ester are buried a few centimeters below soil surface for a month or two. Soil gas with variable trace amounts of alpha-emitting radon generated by radioactive decay of uranium at greater depth is trapped in the cups, and the sensitive ester strip is "exposed." Alpha-particle tracks in the ester can then be chemically enhanced for optical densitometer measurement much as photosensitive film is developed, and the quantitative distribution of subsurface uranium can be plotted.

Bibliography

Adams, S. S., 1983. Pers. commun.

—— and Associates (E. N. Harshman and S. S. Adams), 1981a. Geology and recognition criteria for roll-type uranium deposits in continental sandstones. U.S. Dept. Energy Rept. GJBX-1, Bendix Corp., Grand Junction, Colo., 185 pp.

—— and —— (S. S. Adams and A. E. Saucier), 1981b. Geology and recognition criteria for uraniferous humate deposits, Grants uranium region, New Mexico. U.S. Dept. Energy Rept. GJBX-2, Bendix Corp., Grand Junction, Colo., 225 pp.

—— and —— (A. Button and S. S. Adams), 1981c. Geology and recognition criteria for uranium deposits of the quartz-pebble conglomerate type. U.S. Dept. Energy Rept. GJBX-3, Bendix Corp., Grand Junction, Colo., 390 pp.

—— and —— (S. S. Adams and R. B. Smith), 1981d. Geology and recognition criteria for sandstone uranium deposits in mixed fluvial-shallow marine sedimentary sequences, South Texas. U.S. Dept. Energy Rept. GJBX-4, Bendix Corp., Grand Junction, Colo., 146 pp.

—— and —— (F. J. Dahlkamp and S. S. Adams), 1981e. Geology and recognition criteria for veinlike uranium deposits of Lower to Middle Proterozoic unconformity and strata-related types. U. S. Dept. Energy Rept. GJBX-5, Bendix Corp., Grand Junction, Colo., 254 pp.

—— and —— (J. T. Thamm, A. A. Kovschak, Jr., and S. S. Adams), 1981f. Geology and recognition criteria for sandstone uranium deposits of the Salt Wash type, Colorado Plateau province. U.S. Dept. Energy Rept. GJBX-6, Bendix Corp., Grand Junction, Colo., 135 pp.

Ager, D. V., 1980. *The Geology of Europe*. New York: Wiley, 535 pp.

Anderson, G. M., 1979. Experimental data on the solubility of sphalerite and galena in brines. Ann. Mtg. AIME Prog., p. 50.

Banaszak, K. J., 1979. A coherent basinal-brine model of the genesis of Mississippi Valley Pb-Zn ores based in part upon absent phases. Ann. Mtg. AIME Prog., p. 43.

Battey, G. C., 1978. Australia's uranium resources, pp. 19–126 in G. M. Phillip, Ed., *Australia's Mineral Energy Resources, Assessment and Potential*. Occas. Publ. Earth Resour. Found., Sydney Univ., no. 1.

Benson, W. E., A. F. Trites, Jr., E. P. Beroni, and J. A. Feeger, 1952. Preliminary report on the White Canyon area, San Juan County, Utah. USGS Circ. 217.

Cannon, H. L., 1957. Description of indicator plants and methods of botanical prospecting for uranium deposits on the Colorado Plateau. USGS Bull. 1030-M.

Carpenter, A. B., and B. L. Grethen, 1979. The origin of metal-rich brines in sedimentary basins. Ann. Mtg. AIME Prog., p. 43.

——, M. L. Trout, and E. E. Pickett, 1974. Preliminary report on the origin and chemical evolution of lead- and zinc-rich oil field brines in central Mississippi. *Econ. Geol.* 69:1191–1206.

Clark, L. A., and G. H. R. Burrill, 1981. Unconformity-related uranium deposits, Athabasca area, Saskatchewan, and East Alligator Rivers area, Northern Territory, Australia. *Can. Inst. Min. Metall. Bull.* 74:63–72.

Clark, R. J. McH., L. A. Homeniuk, and R. Bonner, 1982. Uranium geology in the Athabasca and a comparison with other Canadian Proterozoic basins. *Can. Inst. Min. Metall. Bull.* 75:91–98.

Coffin, R. C., 1921. Radium, uranium and vanadium deposits of southwestern Colorado. Colo. Geol. Surv. Bull. 16.

Cornelius, K. D., 1976. Preliminary rock type and genetic classification of uranium deposits. *Econ. Geol.* 71:941–942.

Davis, J. H., 1977. Genesis of southeast Missouri lead deposits. *Econ. Geol.* 72:443–450.

Denson, N. M., and J. R. Gill, 1956. Uranium-bearing lignite and its relation to volcanic tuffs in eastern Montana and North and South Dakota. USGS Prof. Pap. 300, pp. 413–418.

DeVoto, R. H., 1978. Uranium geology and exploration, lecture notes and references. Colo. School Mines, Golden, Colo., 396 pp.

Doe, B. R., and J. S. Stacey, 1974. The application of lead isotopes to the problems of ore genesis and ore prospect evaluation: a review. *Econ. Geol.* 69:757–776.

—— and M. H. Delevaux, 1972. Source of lead in the S.E. Missouri galena ores. *Econ. Geol.* 67:409–425.

Dzulynski, S., and M. Sass-Gustkiewicz, 1977. Comments on the genesis of eastern alpine Zn-Pb deposits. *Mineral. Deposita* 12:219–234.

Erickson, R. L., E. L. Mosier, S. K. Odland, and M. S. Erickson, 1981. A favorable belt for possible mineral discovery in subsurface Cambrian rocks in southern Missouri. *Econ. Geol.* 76:921–973.

Fischer, R. P., 1942. Vanadium deposits of Colorado and Utah. USGS Bull. 936-P.

——, 1950. Uranium-bearing sandstone deposits of the Colorado Plateau. *Econ. Geol.* 45:1–11.

——, 1956. Uranium-vanadium-copper deposits on the Colorado Plateau. USGS Prof. Pap. 300, pp. 143–154.

——, 1968. The uranium and vanadium deposits of the Colorado Plateau region, in *Ore Deposits of the United States 1933/1967*, Graton-Sales Vols. New York: AIME, pp. 735–746.

——, 1970. Similarities, differences, and some genetic problems of the Wyoming and Colorado Plateau types of uranium deposits in sandstone. *Econ. Geol.* 65:778–784.

——, 1974. Exploration guides to new uranium districts and belts. *Econ. Geol.* 69:362–376.

Fowler, G. M., and L. P. Lyden, 1935. The ore deposits of the Tri-State district—discussion. *Econ. Geol.* 30:565–575.

Galkiewicz, T., 1967. Genesis of the Silesian-Cracow deposits of lead-zinc ores. Econ. Geol. Mon. 3, pp. 156–168.

Galloway, W. E., 1978. Uranium mineralization in a coastal-plain fluvial aquifer system: Catahoula Formation, Texas. *Econ. Geol.* 73:1655–1676.

Granger, H. C., E. S. Santos, B. G. Dean, and F. G. Moore, 1961. Sandstone-type uranium deposits at Ambrosia Lake, New Mexico—an interim report. *Econ. Geol.* 56:1179–1209.

—— and C. G. Warren, 1974. Zoning in the altered tongue associated with roll-type uranium deposits, pp. 185–200 in *Formation of Uranium Ore Deposits, Sedimentary Basins and Sandstone-Type Deposits; North American Deposits*. IAEA Proc. Ser. STI/PUB/374.

Grundmann, W. H., Jr., 1977. Geology of the Viburnum No. 27 mine, Viburnum Trend, southeast Missouri. *Econ. Geol.* 72:349–364.

Gruszcyzk, H., 1967. The genesis of the Silesian-Cracovian zinc-lead ores. Econ. Geol. Mon. 3, pp. 169–177.

Haranczyk, C., 1958. Skeletal and colloform textures of galena from Silesian-Cracovian lead-zinc deposits. *Acad. Sci. Poland Ser. Chem. Geol. Geogr.* 7:55–58.

——, 1959. Brunckite from the Silesian-Cracow zinc and lead deposits. *Acad. Sci. Poland Ser. Chem. Geol. Geogr.* 7:359–366.

——, 1963a. Vertical ore-zoning in the zone of faulting observed in Klucze near Olkusz (Silesian-Cracovian zinc and lead deposits), in *Problems of Postmagmatic Ore Deposition, A Symposium*, vol. 1. Prague: Geol. Surv. Czech., 588 pp.

——, 1963b. Silesian-Cracovian type of Zn-Pb ore deposits and their comagmatic relation to igneous rocks (in Polish). *Rudy Metale Niezelazne* 10:132–139, 187–193.

Harshman, E. N., 1972. Geology and uranium deposits, Shirley basin area, Wyoming. USGS Prof. Pap. 745.

Hawkes, H. E., and J. S. Webb, 1962. *Geochemistry in Mineral Exploration*. New York: Harper, 415 pp.

Hegge, M. R., and J. C. Roundtree, 1978. Geologic setting and concepts on the origin of uranium deposits in the East Alligator river region, N.T., Australia. *Econ. Geol.* 73:1420–1429.

Hess, F. L., 1914. A hypothesis for the origin of the carnotites of Colorado and Utah. *Econ. Geol.* 9:675–688.

Hoeve, J., and T. I. I. Sibbald, 1978. On the genesis of Rabbit Lake and other unconformity-type uranium deposits in northern Saskatchewan, Canada. *Econ. Geol.* 73:1450–1473.

Hostetler, P. B., and R. M. Garrels, 1962. Transportation and precipitation of uranium and vanadium at low temperatures, with special reference to sandstone type uranium deposits. *Econ. Geol.* 57:226–237.

Isachsen, Y. W., T. W. Mitcham, and H. B. Wood, 1955. Age and sedimentary environments of uranium host rocks, Colorado Plateau. *Econ. Geol.* 50:127–134.

Jones, B. E., 1980. The geology of the Collins Bay deposit, Saskatchewan, Canada. *Can. Inst. Min. Metall. Bull.* 73:84–90.

Kautzsch, E., 1967. Genesis of lead-zinc deposits in central Europe. Econ. Geol. Mon. 3, pp. 133–137.

Kelley, V. C., 1955. Regional tectonics of the Colorado Plateau and their relationship to the origin and distribution of uranium. Univ. New Mex. Pub. Geol. 5.

——, Ed., 1963. Geology and technology of the Grants uranium region. New Mex. Bur. Mines Miner. Res. Memo. 15.

Kerr, P. F., 1958. Uranium emplacement in the Colorado Plateau. *Geol. Soc. Amer. Bull.* 69:1075–1111.

Kharaka, Y. K., E. Callender, and J. C. Chemerys, 1979. Heavy metals in migrating oil field waters from the northern Gulf of Mexico basin. Ann. Mtg. AIME Prog., p. 43.

Kimberley, M. M., 1978. Short course in uranium deposits, their mineralogy and origin. Min. Assoc. Can. Short Course Handbook 3, 521 pp.

Kisvarsanyi, G., 1977. The role of the Precambrian igneous basement in the formation of stratabound lead-zinc-copper deposits in southeast Missouri. *Econ. Geol.* 72:435–442.

Krusch, P., 1929. Über kolloidale Vorgänge bei der Entstehung der oberschlesischen Zink-Bleierzlagerstätten. *Z. Oberschles. Berg.*-Hüttenmänn. Vereins Katowice, no. 6–7. Abstr. in *Z. Deut. Geol. Ges.* 81:169–170.

Kutina, J., 1952. Mikroskopischer und spektrographischer Beitrag zur Frage der Entstehung einiger Kolloidalstrukturen von Zinkblende und Wurtzit. *Geologie* 1:436–452.

Larsen, K. G., 1977. Sedimentology of the Bonneterre Formation, southeast Missouri. *Econ. Geol.* 72:408–419.

Lyle, J. R., 1977. Petrography and carbonate diagenesis of the Bonneterre Formation in the Viburnum Trend area, southeast Missouri. *Econ. Geol.* 72:420–434.

Mann, A. W., and R. L. Deutscher, 1978. Genesis principles for the precipitation of carnotite in calcrete drainages in Western Australia. *Econ. Geol.* 73:1724–1737.

McKelvey, V. E., D. L. Everhart, and R. M. Garrels, 1955. Origin of uranium deposits. *Econ. Geol. 50th Anniv. Vol.*, pp. 464–533.

McMillan, R. H., 1977. Uranium in Canada. *Can. Pet. Geol. Bull.* 25:1222–1249.

Miller, L. J., 1955. Uranium ore controls of the Happy Jack deposit, White Canyon, San Juan County, Utah. *Econ. Geol.* 50:156–169.

Mouat, M. M., and C. W. Clendenin, 1977. Geology of the Ozark Lead Company mine, Viburnum Trend area, southeast Missouri. *Econ. Geol.* 72:398–407.

Nash, J. T., H. C. Granger, and S. S. Adams, 1981. Geology and concepts of genesis of important types of uranium deposits. *Econ. Geol. 75th Anniv. Vol.*, pp. 63–116.

Ohle, E. L., 1959. Some considerations in determining the origin of ore deposits of the Mississippi Valley type. *Econ. Geol.* 54:769–789.

——, 1967. The origin of ore deposits of the Mississippi Valley type. Econ. Geol. Mon. 3, pp. 33–39.

——, 1970. Mississippi Valley-type ore deposits, a general review. *Washington Div. Mines Bull.* 61:5–15.

——, 1980. Some considerations in determining the origin of ore deposits of the Mississippi Valley type, pt. II. *Econ. Geol.* 75:161–172.

Park, C. F., Jr., and R. A. MacDiarmid, 1975. *Ore Deposits*, 3rd ed. San Francisco: Freeman, 530 pp.

Rackley, R. I., 1976. Origin of western-states-type uranium mineralization, pp. 89–156 in K. H. Wolf, Ed., *Handbook of Stratabound and Stratiform Ore Deposits*, Vol. 7. New York: Elsevier.

Ridge, J. D., and I. Smolarska, 1972. Factors bearing on the genesis of the Silesian-Cracovian lead-zinc deposits in southern Poland. *24th Int. Geol. Congr. Rept.*, Sec. 6, pp. 216–229.

Rogers, R. K., and J. H. Davis, 1977. Geology of the Buick mine, Viburnum Trend, southeast Missouri. *Econ. Geol.* 72:372–380.

Rozkowski, A., 1979. Thermal brines as a potential source of the ore mineralization of the Cracow-Silesian area. Ann. Mtg. AIME Prog., p. 56.

Ruzicka, V., 1977. Conceptual models for uranium deposits and areas favourable for uranium mineralization. Rept. of Activs., Geol. Surv. Can. Pap. 77-1A.

Sangster, D. F., 1976. Carbonate hosted lead-zinc deposits, pp. 442–456 in K. H. Wolf, Ed., *Handbook of Stratabound and Stratiform Ore Deposits*, Vol. 7. New York: Elsevier.

Sass-Gustkiewicz, M., S. Dzulynski, and J. D. Ridge, 1982. The emplacement of zinc-lead sulfide ores in the Upper Silesian district—a contribution to the understanding of Mississippi Valley-type deposits. *Econ. Geol.* 77:392–412.

Schneider, H. J., 1964. Facies differentiation and controlling factors for the depositional lead-zinc concentrations in the Ladinian geosyncline of the eastern

Alps, pp. 29–46 in G. C. Amstutz, Ed., *Developments in Sedimentology*, Vol. 2. New York: Springer, 184 pp.

Schneiderhöhn, H., 1941. *Lehrbuch der Erzlagerstättenkunde*. Jena: Gustav Fischer, pp. 573–579.

Sharp, J. M., 1978. Energy and momentum transport model of the Ouachita Basin and its possible impact on formation of economic mineral deposits. *Econ. Geol.* 73:1057–1068.

Shoemaker, E. M., 1956. Structural features of the central Colorado Plateau and their relation to uranium deposits. USGS Prof. Pap. 300, pp. 155–170.

Snyder, F. G., 1967. Criteria for origin of stratiform orebodies. Econ. Geol. Mon. 3, pp. 1–13.

Stappenbeck, R., 1928. Ausbildung und Ursprung der oberschlesischen Bleizink-erzlagerstätten. *Arch. Lagerstättenforsch.* 41.

Sweeny, P. H., E. D. Harrison, and M. Bradley, 1977. Geology of the Magmont mine, Viburnum Trend, southeast Missouri. *Econ. Geol.* 72:365–371.

Thacker, J. L., and K. H. Anderson, 1977. The geologic setting of the southeast Missouri lead district—regional geologic history, structure, and stratigraphy. *Econ. Geol.* 72:339–348.

Waters, A. C., and H. C. Granger, 1953. Volcanic debris in uraniferous sandstones and its possible bearing on the origin and precipitation of uranium. USGS Circ. 224.

Weiss, A., and G. C. Amstutz, 1966. Ion-exchange reactions in clay minerals and cation selective membrane properties as possible mechanisms of economic metal concentrations. *Mineral. Deposita* 1:60–66.

Welty, J. W., 1980. Red Creek Canyon, Utah—a Precambrian unconformity-type pitchblende deposit. Unpub. B.A. honors thesis, Williams College, 78pp.

White, D. E., 1968. Environments of generation of some base metal ore deposits. *Econ. Geol.* 63:301–335.

Wright, R. J., 1955. Ore controls in sandstone uranium deposits of the Colorado Plateau. *Econ. Geol.* 50:135–155.

Zwierzycki, J., 1950. Lead and zinc ores in Poland. *18th Int Geol. Cong. Rept.*, pt. 7, pp. 314–324.

TWENTY-ONE

Metallogenic Provinces, Epochs, and Plate Tectonics

We have seen in the last eleven chapters that there is a consanguinity within the realm of ore deposits just as there is, for example, within rock types. We have also seen that many ore-deposit types have preferred, even exclusive tectonic settings or geologic ages. It should therefore be no surprise that ore deposits commonly occur preferentially in swaths of the Earth's crust that are dictated by the dynamics of Earth history and that are commonly constrained in both the geographic-geologic-tectonic distribution sense and the geologic time sense. Given belts of terrane have been favorable loci for the formation of ore-deposit types at different geologic ages. Clearly, understanding the reasons for this clustering has vast exploration and scientific significance.

A *metallogenic province* is defined as a terrane containing mineral deposits that are characterized, at least broadly, by related mineral composition, form, style, and geologic age of mineralization (Petrascheck, 1965). Bateman (1950) defined a metallogenic province as a region characterized by relatively abundant mineralization, one type of which predominates. The term can be used in both senses today. The concept of metallogenic province was used years ago by de Launay (1913), Spurr (1923), and Lindgren (1909, 1933). Considerable impetus has been given to the study of metallogeny by the establishment of an international commission to prepare metallogenic maps of the world (Petrascheck, 1963; Stoll, 1965; Guild, 1974). Russian geologists in particular have worked hard in this field (Bilibin, 1955; Radkevich, 1956; Tvalchrelidze, 1957, 1964, 1967; Smirnov, 1959, 1968). Metallogenic maps of the U.S.S.R. are advanced, and those of other countries are generally available. An example of a metallogenic map of a metallogenic province is given in Figure 21-1. It shows the distribution of various types of ore deposits related to the various island areas of the Caribbean region (Kesler, 1978). Also as examples, three broad-scale metallogenic provinces can be distinguished on a broad scale and with respect to tectonic environment: the Precambrian Shield cratonic areas, the Cordilleran orogenic belts, and the stable interiors, or platform areas. Geologists have now related many aspects of these three types of provinces to plate tectonics, and many of their contained ore deposits have been linked to regional-scale phenomena, as we have seen.

A *metallogenic epoch* is a time interval that was favorable for the deposition of particular useful substances (Lindgren, 1933). Turneaure (1955) used the term to distinguish a geologic period during which mineral deposition in general was most pronounced, as did Kesler (1978); see Figure 21-2. Several authors have emphasized the close relationship between metallogenic provinces and metallogenic epochs. Petrascheck (1965) restricted his use of the concept of provinces to tectonic time intervals within major tectonic events. In some districts, the age of deposits can be fixed only within broad limits, especially if age is based primarily on incomplete and indirect geologic evidence. Although "the Canadian Precambrian Shield" is frequently described by outsiders as a single metallogenic province formed essentially in one metallogenic epoch, geologists who work with mineral deposits in those areas urge that they were constructed during several metallogenic epochs and that their ore deposits were formed at several successive times (Hutchinson, Ridler, and Suffel, 1971).

The size of a metallogenic province can range from that of a single mining district to regions that extend hundreds to thousands of kilometers. The vast Canadian Shield has been defined as a province, as has the Climax–Urad–Henderson–Crested Butte area of central Colorado, a small province characterized by the deposition of Climax-type molybdenite during one brief epoch.

The noun *metallotect* was introduced by Lafitte, Parmingeat, and Routhier (1965) to emphasize a geologic, tectonic, lithologic, or geochemical feature that is believed to have played a role in the concentration of one or more elements and hence is thought to have contributed to the formation of ore deposits. The term is useful in the discussion of both metallogenic provinces and epochs for singling out individual causative aspects.

A few geologists have tended to equate metallogenic provinces with ore-deposit zoning. For example, Rastall (1923) described the zoning at Cornwall (Chapter 6) as 'metallogenic zoning.' Clearly the processes that give rise to regional metallogenic provinces and to district zoning can be closely related, but they are not the same in scale. Deposits within a province may include both zoned and unzoned occurrences; an example is the

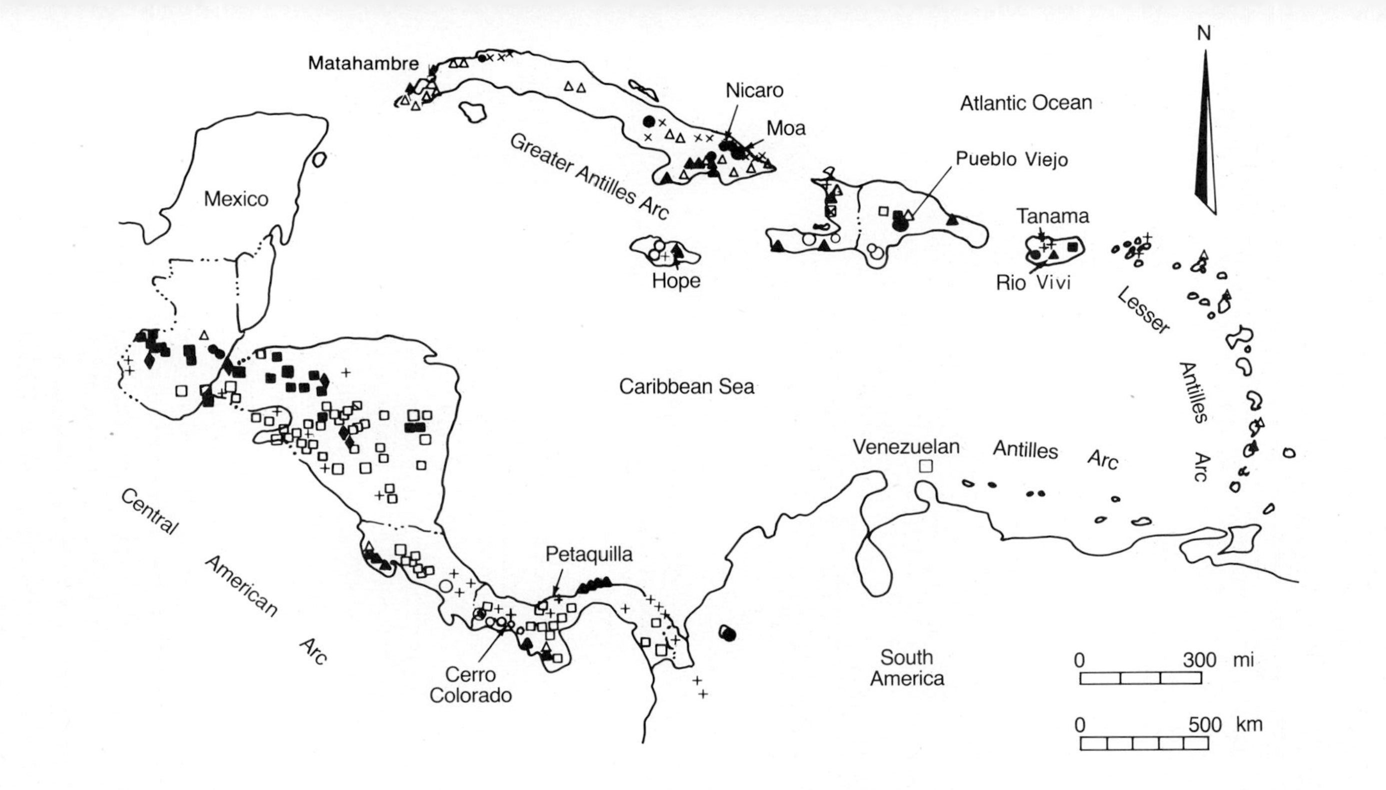

Figure 21-1. Example of a metallogenic map of a metallogenic province. It shows the distribution of various classes or types of ore occurrences in a particular geologic terrane, here the Caribbean region. See also Fig. 21-2. (*After* Kesler, 1978.)

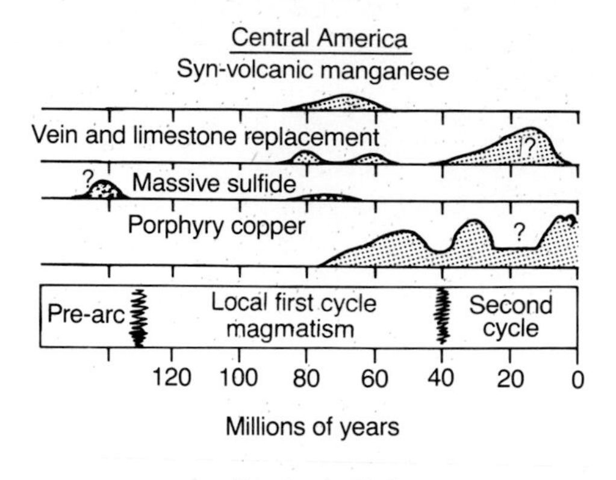

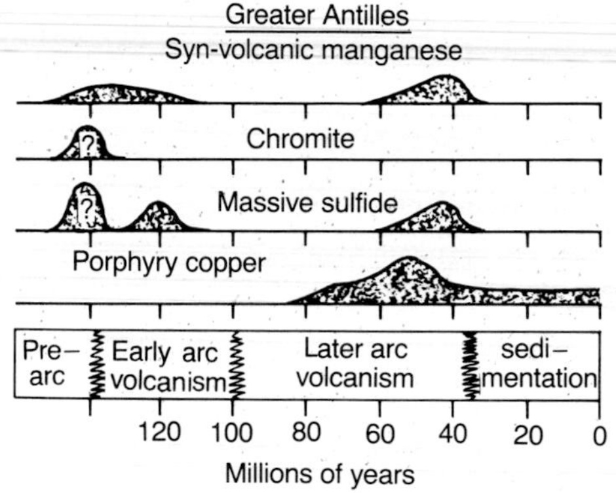

Figure 21-2. Schematic representation of metallogenic epochs. Each blip—for example of porphyry copper deposition at 50 ± 10 million years—represents a pulse of emplacement, a time-constrained metallogenic event or epoch. This one applies to two of the volcanic arcs of the Caribbean region shown in Figure 21-1. (*From* Kesler, 1978.)

Appalachian province of North America where some, but not all, of the deposits show district-scale zoning (Gabelman, 1968; Pardee and Park, 1948).

As early as 1966, White stated that metallogeny during tectonic events can be better understood by considering the relation of metalliferous deposits to (1) major depositional events, (2) regional metamorphic events, (3) plutonic events, and (4) deformational events expressed by regional folds and fracture patterns, and that regional dissection and synthesis can reveal metallogenic attributes. Sullivan (1979), for example, discussed the relationship between intracratonic basins and ore deposits in the context of revealing unifying characteristics of otherwise seemingly unlike ore occurrences that are found in a single broad tectonic setting. Smirnov (1968) described the metallogenic provinces or subdivisions of Russia in an attempt to clarify the relationships of the broad features of the metallogeny of that country. He attributes a polycyclic character to the metallogeny of most mineralized areas, and stated that many metallogenic epochs can and do exist within the limits of a single metallogenic province, as did Kesler (Figure 21-1). In some places, such as south-central Siberia, metallogenic epochs have been repeated from Precambrian times to the Tertiary.

Figure 21-3, after Guild (1974), shows the principal post-Eocene metallogenic provinces of the world in relation to the major lithospheric plates.

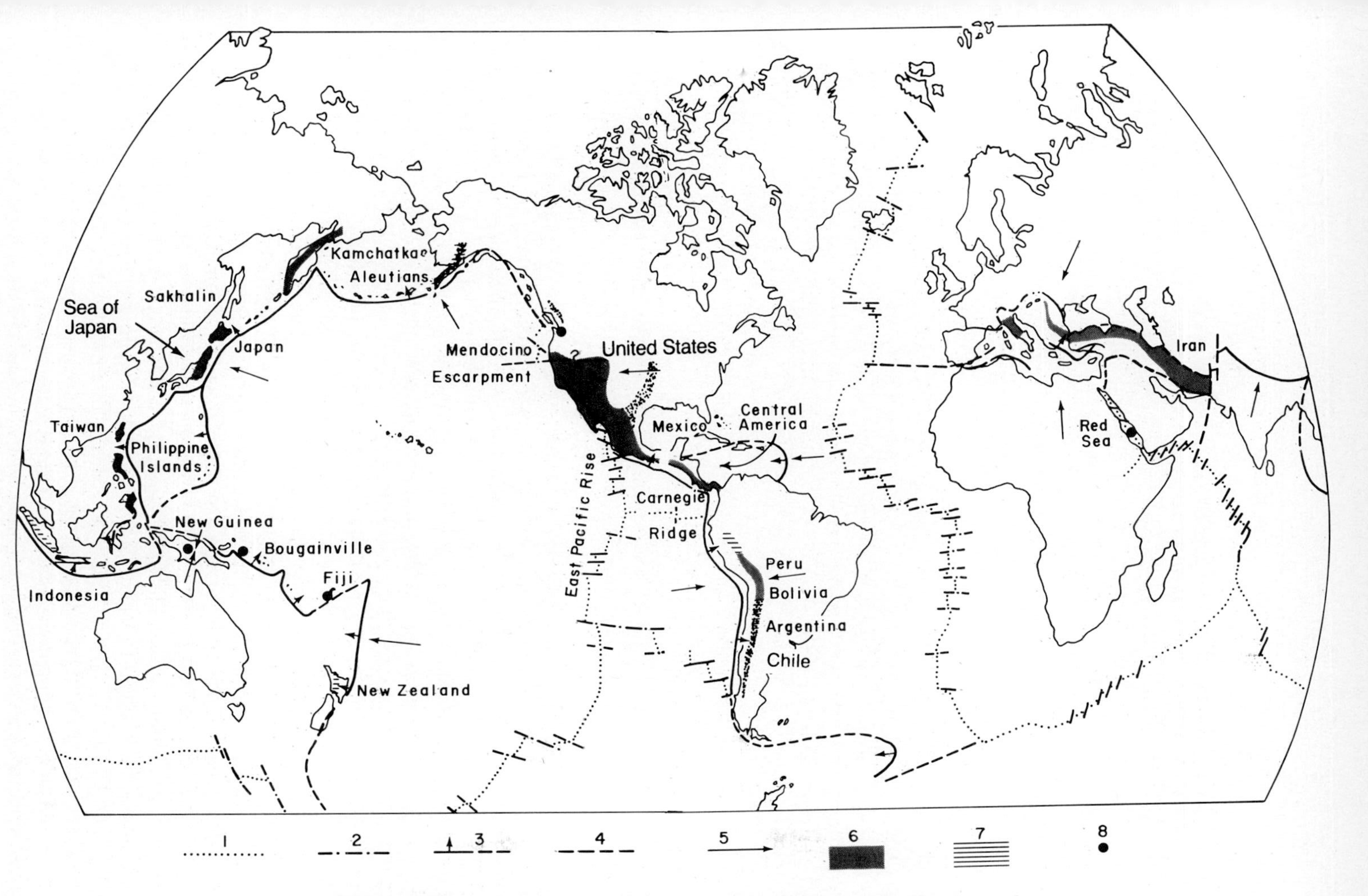

Figure 21-3. Principal post-Eocene metallogenic provinces of the world in relation to the major lithospheric plates. Key to ornament: 1-accreting plate margin, a midoceanic rise; 2-transform fault; 3-consuming plate margin with dip direction of down-going oceanic crustal plate; 4-margin of uncertain nature or location; 5-relative plate motion; 6-area of mineralization of post-Eocene age; 7-minor or suspected post-Eocene mineralization; 8-major or noteworthy isolated ore deposit of post-Eocene age. (*From* Guild, 1974.)

Guild has also made constructive efforts to classify ore deposits relative to major Earth tectonic features. It emphasizes the 'time-lapse' aspect of metallogeny, that at any given time—indeed, at *all* given times—different ore-deposit types have formed in different tectonic settings around the globe. We can thus focus upon the types of deposits generated along compressive, consuming, subducting margins, or we can attempt to predict where—in Japan, Peru, or Mexico—we might expect to find various deposit types. Nishiwaki (1973) published a provocative paper on the metallogenic provinces of Japan, a typical island arc environment. He divided the ore deposits of Japan into four types: (1) the Kuroko deposits, the result of intense submarine volcanic activity of Middle to Late Miocene age (Chapter 13), (2) gold-silver veins of Late Miocene to Pliocene age, (3) Quaternary volcanogenic sulfur deposits with or without pyrite (Chapter 3), and (4) bedded cupriferous iron sulfide deposits, mostly Late Paleozoic in age and metamorphosed by Late Mesozoic orogeny. The last type includes a few minor deposits of Mesozoic age. The distribution and time elements of the first three types are compatible with activities of the present arc-trench system around Japan. The older deposits are explained by inferring a similar arc-trench system roughly parallel to that of the present one. Lessons learned in Japan are being applied, for instance, in southern Arizona, where the lithologic-tectonic elements of a Jurassic island arc terrane are being unraveled from the effects of the younger Laramide arc sweep (Figure 6-7) in order to seek out porphyry base-metal and possible epithermal ore-forming environments that may have occurred in those older units.

Attempts are being made, then, both to relate ore-deposit types to tectonic settings and to determine the distribution of those tectonic settings in space and time, in continental masses and in suspect terranes, in oceanic crust and in continental crust. For example, ore deposits appear to be relatable in many ways to subduction zones beneath continental plates (Mitchell and Garson, 1972; Sillitoe, 1981). Where subduction zones dip under continental masses, the rocks at depth are melted and can thus create magma, which includes various ore constituents (Figure 3-38). The magma, commonly calcalkaline in character, can rise in the continental plate overlying the subduction zone, and at certain places ore deposits can form (Figure 6-7). The Andes mountains of South America contain many active volcanoes that supposedly are related to magma created along the subduction zone at depth beneath the border of the continental mass (Sillitoe, 1973). The numerous intrusive bodies and ore deposits of the Andes mountains and along the coastal areas of South America are thought to be related to this process. Belts or coastline-parallel linear distributions of various related types of ore deposits are thus explained. Further, the volcanic iron ore deposits of Chile and Peru are mostly aligned in a metallogenic province close to the coast, while the copper deposits are mostly in a separate belt farther inland and at considerably higher elevations. The position of lead and zinc deposits is not entirely clear, for in places they lie farther inland than does copper, while elsewhere they are between the iron ores and the copper ores. Westra and Keith (1981) reinforce the former options and underscore the overall zoning relationships.

The potential of discerning metallogenic belts, epochs, and individual deposit-type distributions through plate tectonic–based analysis has been indicated throughout this text. Both to stimulate further thought and to reinforce the potential role of metallogenic concepts in exploration and global understanding, the text will close with a plate tectonic classification of ore deposits (Table 21-1, after Guilbert, 1981) that incorporates several aspects of the classification used in the text and serves as a "logic binder" to interweave tectonics, ore-deposit geology, exploration, and geoscience.

TABLE 21-1. Plate Tectonic–Lithotectonic Classification of Ore Deposits

Classification	Examples Cited in Text
I. Ore deposits associated with midoceanic ridges and ocean floor/oceanic crust formation	
A. Plutonic—Oceanic crust Layer 3	
1. Layered mafic intrusions, chromite	
2. Alpine peridotite, chromite	Moa; Cyprus
3. Chromite-platinoid placers	Goodnews Bay
B. Volcanic—Oceanic crust Layers 1 and 2, hydrothermal-proximal	
1. Cyprus-type massive sulfides	Cyprus
C. Volcanic—Oceanic crust Layer 1, hydrothermal-distal	
1. Manganese-copper-nickel-cobalt nodules	Pacific seafloor
D. Sea breach—Oceanic crust Layers 2 and 3 with supergene activity	
1. Nickel laterites	New Caledonia
II. Ore deposits associated with consuming, subducting margins	
A. Obduction	
1. Alpine peridotites (see IA)	Moa; Cyprus
2. Alpine peridotites with lateritization, nickel (see ID)	New Caledonia
3. Franciscan mélange, mercury-serpentine-gold	New Almaden
4. Ophiolites, massive sulfides (see IB)	
a. Copper-zinc, Cyprus type	Cyprus, Betts Cove
B. Ocean/ocean, island arcs, eugeosynclines	
1. Proximal massive sulfides	
a. Copper-zinc-silver	Abitibi; Kuroko
b. Copper-nickel	Kambalda
c. Antimony	Consolidated Murchison
d. Mercury	Almadén
2. Proximal oxides	
a. Iron formation, Algoma type	Adams-Sherman
3. Distal oxides and sulfides	
a. Gold	Homestake
b. Banded iron formation with gold	Zimbabwe; Kirkland–Larder Lake
c. Iron formation, Algoma type	Minnesota
d. Lead-zinc	Iron King; Mt. Isa

Classification	Examples Cited in Text
4. Porphyry copper-moly-gold	
a. Copper-moly	Atlas
b. Copper-gold	OK Tedi; Panguna
5. Plutonic—ultramafic	
a. Asbestos	Msauli
C. Ocean/continent, trench/arc, Cordilleran orogenics	
1. Magmatic iron (magnetite series, I-type, steep dip)	
a. Plutonic magnetite	El Romeral; Pea Ridge
b. Volcanic magnetite, hematite	Cerro de Mercado; El Laco
c. Igneous metamorphic, magnetite	Santa Rita; Iron Springs
2. Porphyry copper-moly (magnetite series, I-type, steep dip)	
a. Copper	El Salvador; Inspiration
b. Copper-moly	San Manuel–Kalamazoo; Chuqui
c. Moly	Quartz Hill
d. Copper-gold	Bingham Canyon; Bajo La Alumbrera
3. Skarn	
a. Porphyry copper contact, Cu, Zn-Pb, Mo	Bingham Canyon; Santa Rita
b. Hydrothermal skarn	Mission; Linchburg
c. Tungsten	Bold Head
d. Iron skarn and hornfels	El Romeral
4. Cordilleran Vein	
a. Cu-Fe-As-S (magnetite series, I-type, steep dip)	Magma; Casapalca
b. Pb-Zn-Ag (I-S-type, shallow dip)	Coeur d'Alene
c. Gold (?)	
5. Porphyry tin-tungsten (ilmenite series, I-type, shallow dip)	
a. Tin-tungsten porphyries	Llallagua
b. Tin rhyolites	Black Mountain; Mexico
6. Tin-tungsten granites (ilmenite series, S-type, shallow dip [?])	
a. Tin granites	Blue Tier
b. W-Mo-Sn-Be-U granites	Bushveld; Malaysia
7. Metamorphic core complexes (S-type granites)	
a. Tungsten, uranium (?)	Kettle Dome
8. Complex zoned granitic pegmatites (S-type granites?)	Petaca
9. Rare earth (lanthanide) pegmatites	Barringer Hill
10. Uranium granites	
a. Granitic	France; Ross-Adams
b. Pegmatitic	Spruce Pine
c. Migmatitic	Rössing
11. Industrial rocks	
a. Granites	Crotch Island
b. Syenites	

Classification	Examples Cited in Text
12. Lacustrine evaporites	
a. High-level tectonic basins	Chilean nitrates
b. Chain lakes	Searles Lake
D. Ocean/continent—Extension	
1. Porphyry copper, soda-alkalic affinity	Stikine
2. Climax-type moly (A-type, steepening slab [?], rifting)	Climax; Henderson
3. Epithermal-ignimbrite association	
a. Silver-gold	Creede; Pachuca; Oatman
b. Mercury-uranium	McDermitt
c. Antimony	Banxi; Senator
4. "Bulk silver," caldera lake sediments	Waterloo
III. Ore deposits associated with ensialic-ensimatic back-arc basins	
A. Volcanic tendencies—outboard (arc side)	
1. Lead-zinc-copper	Mt. Isa
2. Lead-zinc	Bleïda
B. Sedimentary tendencies—inboard (continent side)	
1. Lead-zinc "black shale hosted"	Sullivan; Rammelsberg
IV. Ore deposits associated with cratons	
A. Geosyncline, miogeosyncline, continental shelf	
1. Industrial rocks	
a. Limestone	
b. Sandstone	
c. Shale, slate	
2. Superior-type iron formations	Lake Superior
B. Littoral margin	
1. Base metal, copper-cobalt-uranium sediments	Zambia; Kupferschiefer
2. Hydrocarbons, petroleum, coal	
3. Fluvial placers	
a. Gold	Calaveras River
b. Uranium	
c. Diamonds	Vaal, Orange rivers
d. Titanium, rare earth elements	
4. Marine placers	
a. Titanium, rare earth	Trail Ridge
b. Diamonds	Namibia
5. Phosphates	Florida
C. Epicontinental	
1. Adjacent to sedimentary basins	
a. Mississippi Valley deposits	Lead Belt–Tri State; Cracow
b. Irish stratabound (?)	Tara; Tynaugh
2. Unconformity related	
a. Uranium	Athabasca
b. Copper-uranium	Colorado Plateau
3. Surface related—Western states uranium	
a. Salt Wash	Salt Wash, Colorado Plateau
b. Roll-front	Shirley Basin
c. Humate	Grants

Classification	Examples Cited in Text
4. Solution-remobilization	
a. Gold	Yellowknife
b. Cobalt-silver-nickel-arsenic	Cobalt
5. Residual or transported clays, soils, organics	
a. Dirt, soil	
b. Laterites, aluminum-iron	Jamaica; Little Rock
c. Clays, fire, ball, and flint	Missouri
d. Kaolins	Georgia; Cornwall
e. Underclays	
f. Coal	
6. Placers, fluvial, lacustrine, residual	
a. Gold-chromite-platinum	Goodnews Bay
b. Chromite-platinum	
c. Gold-uranium	Witwatersrand
V. Ore deposits associated with cratonic openings	
A. Prerift arching, early rifting	
1. Kimberlites	
a. Diamonds	Kimberley
2. Carbonatites	
a. Rare earth elements	Mountain Pass
b. Phosphate-titanium-copper	Palabora
c. Phosphate-titanium-niobium	Magnet Cove
3. Alkalic intrusives	
a. Syenites	
4. Anorthosites	
a. Magnetite	Sanford Lake
b. Titanium	Allard Lake
c. Vanadium	Sanford Lake
5. Layered mafic intrusions	
a. Magnetite-vanadium	Bushveld; Stillwater
b. Chromite-platinoid	Bushveld
c. Copper-nickel	Sudbury; Duluth
6. Moly-silica	
a. Molybdenum	Climax (?); Questa
B. Rifting	
1. Alkali volcanism and lakes	
a. Trona-dawsonite	Green River Basin
b. Carbonatites	Oldoinyo Lengai
C. Proto-oceanic gulf—aulacogens	
1. Copper-zinc-manganese-iron-lead-barium	Red Sea
2. Copper-zinc-silver	Salton Sea
3. Evaporites, potassium-sodium-magnesium-chlorine	Boulby; Castile
D. Narrow shallow ocean (see III)	
E. Open ocean	
1. Manganese-copper-cobalt-nickel marine nodules (see IC)	Pacific seafloor

Plate tectonic–lithotectonic classification of ore deposits, with metallogenic overtones. Deposits in the table are listed as examples only; the lists are by no means intended to be complete catalogs.

Source: From Guilbert (1981).

Bibliography

Bateman, A. M., 1950. *Economic Mineral Deposits.* New York: Wiley, pp. 316-325.

Bilibin, J. A., 1955. *Les Provinces Métallogéniques et les Epoques Métallogéniques.* Moscow: Gosgeolizdat.

de Launay, L., 1913. *Traité de Métallogénie;* vol 1, *Gîtes Minéraux et Métallifères.* Paris: Béranger, pp. 241-288.

Gabelman, J. W., 1968. Metallotectonic zoning in the North American Appalachian region. *23rd Int. Geol. Congr. Rept.*, sec. 7, pp. 17-33.

Guilbert J. M., 1981. A plate tectonic-lithotectonic classification of ore deposits, pp. 1-10 in W. R. Dickinson and W. D. Payne, Eds., *Relations of Tectonics to Ore Deposits in the Southern Cordillera.* Tucson: Ariz. Geol. Soc. Dig. XIV, 288 pp.

Guild, P. W., 1972. Metallogeny and the new global tectonics. *24th Int. Geol. Congr. Rept.*, sec. 4, Mineral Deposits, pp. 17-24.

——, 1974. Distribution of metallogenic provinces in relation to major earth features. Osterr. Akad. Wiss., Vienna, Schriftenreihe Erdwiss. Komm., vol. 1, pp. 10-28.

Hutchinson, R. W., R. H. Ridler, and G. G. Suffel, 1971. Metallogenic relationships in the Abitibi belt, Canada: a model for Archean metallogeny. *Can. Inst. Min. Metall. Trans.* 74:106-115.

Kesler, S. E., 1978. Metallogenesis of the Caribbean region. *J. Geol. Soc.* 135:429-441.

Lafitte, P., F. Permingeat, and P. Routhier, 1965. Cartographie métallogénique, metallotect et geochimie régionale. *Soc. Franç. Miner. Bull.* 88:3-6.

Lindgren, W., 1909. Metallogenic epochs. *Econ. Geol.* 4:409-420.

——, 1933. *Mineral Deposits.* New York: McGraw-Hill, 930 pp.

Mitchell, A. H. G., and M. S. Garson, 1972. Relationship of porphyry copper and circum-Pacific tin deposits to paleo-Benioff zones. *Inst. Min. Metall. Trans.* 81:B10-B25.

Nishiwaki, C., 1973. Metallogenic provinces of Japan. *Bur. Min. Res. Geol. Geophys. Austr. Bull.* 141:81-94.

Pardee, J. T., and C. F. Park, Jr., 1948. Gold deposits of the Southern Piedmont. USGS Prof. Pap. 213.

Petrascheck, W. E., 1963. Die alpin-mediterrane Metallogenese. *Geol. Rundsch.* 53:376-389.

——, 1965. Typical features of metallogenic provinces. *Econ. Geol.* 60:1620-1634.

Radkevich, E. A., 1956. Les zones métallogéniques du primorie et les particularités de leur development:géologie de la partie méridionale de l'Extrême-Orient et de la Transbaikalie. *Geol. Rud. Mestorozhd. Petrog. Geokhim. Trans.*, vyp. 3.

Rastall, R. H., 1923. Metallogenic zones. *Econ. Geol.* 18:104-121.

Sillitoe, R. H., 1973. Environments of formation of volcanogenic massive sulfide deposits. *Econ. Geol.* 68:1321-1325.

——, 1981. Ore deposits in Cordilleran and island arc settings, pp. 49-69 in W. R. Dickinson and W. D. Payne, Eds., *Relations of Tectonics to Ore Deposits in the Southern Cordillera.* Tucson: Ariz. Geol. Soc. Dig. XIV, 288 pp.

Smirnov, V. I., 1959. Essai de subdivision métallogénique de territoire de l'U.R.S.S. *Soc. Geol. France Bull.*, ser. 7, 1:511-526.

——, 1968. The sources of the ore-forming fluid. *Econ. Geol.* 63:380-389.

Spurr, J. E., 1923. *The Ore Magmas.* New York: McGraw-Hill, 915 pp.

Stoll, W. C., 1965. Metallogenic provinces of magmatic parentage. *Min. Mag.* 112:312-323, 394-405.

Sullivan, C. J., 1979. Intracratonic basins and ore deposits. *Can. Inst. Min. Metall. Bull.*, Dec., 72:75-80.

Turneaure, F. S., 1955. Metallogenic provinces and epochs. *Econ. Geol. 50th Anniv. Vol.*, pp. 39-91.

Tvalchrelidze, G. A., 1957. Les époques métallogéniques du Caucase. *Sov. Geol.*, sb. 59.

——, 1964. On genetic types of deposits in composite parts of geosynclines (Caucasus taken as an example), pp. 322-333 in *Problems of Genesis of Ores, Reports of Soviet Geologists, 22nd Int. Geol. Congr. Rept.*

——, 1967. Main metallogenic features of basaltic and granitoid type geosynclines. *Geol. Rud. Mestorozhd. Acad Sci. USSR*, no. 5.

Westra, G., and S. B. Keith, 1981. Classification and genesis of stockwork molybdenum deposits. *Econ Geol.* 76:844-873.

INDEX

- Underlined numbers designate illustrations.
- Common minerals are not completely indexed.
- For minor and trace elements lists in various deposits and deposit types, see Trace elements.
- For geochemical environments of formation of various deposits and deposit types, see Environments of deposition.